Relativity in Modern Physics

Relativity in Modern Physics

Nathalie Deruelle

Jean-Philippe Uzan

Translated by
Patricia de Forcrand-Millard

OXFORD

UNIVERSITY PRESS

OXFORD
UNIVERSITY PRESS

Great Clarendon Street, Oxford, OX2 6DP,
United Kingdom

Oxford University Press is a department of the University of Oxford.
It furthers the University's objective of excellence in research, scholarship,
and education by publishing worldwide. Oxford is a registered trade mark of
Oxford University Press in the UK and in certain other countries

English Edition © Oxford University Press 2018
An earlier version of this book has appeared as *Théories de la Relativité* by
Nathalie Deruelle and Jean-Philippe Uzan © Editions Belin, 2014

The moral rights of the authors have been asserted

First Edition published in 2018

Impression: 1

Published in the United States of America by Oxford University Press
198 Madison Avenue, New York, NY 10016, United States of America

British Library Cataloguing in Publication Data

Data available

Library of Congress Control Number: 2018933841

ISBN 978–0–19–878639–9

Printed and bound by
CPI Group (UK) Ltd, Croydon, CR0 4YY

This book is dedicated to those who have developed relativity in France, in particular, Henri Poincaré, Paul Langevin, Élie Cartan, André Lichnerowicz, Yvonne Choquet-Bruhat, and Thibault Damour

Foreword

The physics community has recently commemorated the centenary of the birth of general relativity, developed by Albert Einstein between 1907 and 1915. This magnificent theory, which unites relativity and gravity within the framework of non-Euclidean geometry, has long been viewed as a unique feature in the physics landscape. Its lofty height has intimidated the majority of physicists, and few have ventured to climb its mathematically steep slopes, especially while the connection of this theory to observable reality remained tenuous.

That era has vanished, because general relativity has now come to play an essential role in the remarkable recent advances in astrophysics and cosmology. Astrophysics has revealed the existence of 'compact' stars like neutron stars, whose matter is so concentrated that the deformation of time and space predicted by general relativity becomes tangible. And then there are the celebrated black holes, which have outgrown the bounds of mathematical abstraction and science fiction to become part of the daily life of astrophysicists. A definitive proof of their existence has recently been provided by the detection of the gravitational waves emitted when they merge, thus opening a new window in astronomy.

On the cosmological side, general relativity lies at the heart of our current understanding of the expansion of the universe. Cosmological observations have recently revealed the presence of a mysterious 'dark energy' which may correspond to the cosmological constant initially introduced by Einstein to obtain a static universe, and then quickly abandoned, before its current revival to explain the observational data. Owing to the enormous progress which has been made in measurement accuracy, relativity has even become a part of the daily life of an ordinary person: the accurate functioning of the GPS actually requires the inclusion of effects due to special and general relativity.

All these fascinating aspects of relativity, and others, are presented and explained rigorously and accurately in this book by Nathalie Deruelle and Jean-Philippe Uzan. The theory of general relativity constitutes the climax of this impressive work, which assembles and summarizes, in a coherent and extremely deep manner, the various relativistic aspects of 'classical' (that is, non-quantum) theoretical physics, starting from the description of Newtonian mechanics and gravitation.

One of the strengths of this book is its revelation of the continued relevance of a number of principles and mathematical tools beyond the conceptual rupture which relativity represents compared to the theories of classical physics. For example, the authors study Newtonian physics in an accelerated reference frame and arbitrary coordinates using the same formalism as that used to describe spacetime in general relativity. The echoes back and forth between the chapters on electromagnetic waves and gravitational waves emphasize both the conceptual differences and the similarities in the treatment of these two concepts. The simultaneous embrace of Newtonian gravity, special relativity, and general relativity allows the reader to be led gradually from familiar, gentle terrain to the highest peaks.

Many of the sections in this book are enriched by research topics, and present viewpoints and results which the reader will find only rarely, if at all, in other textbooks on these subjects. This book therefore constitutes a remarkable reference work, from which the reader can extract a great variety of information as well as many calculation techniques, and it can serve as a point of entry to research articles. It is written equally for the student passionate about theoretical physics and for the researcher wishing to acquire a deeper understanding of relativity, as well as for the curious reader with a scientific background who wishes to independently explore various facets of relativity under the tutelage of experienced guides.

I have had the privilege of sharing my professional life with the two authors of this book, and of being a witness to their insatiable curiosity and passion for transmitting their knowledge to others. I have also had the great fortune to share with Nathalie Deruelle many years of teaching general relativity, and it has been a pleasure to rediscover in this book various paths which she explored during our pedagogical adventure. This book attests admirably to a passionate encounter between a demanding but fascinating scientific domain and two scientists who are simultaneously participants in the continuing development of the field and transmitters of their knowledge to future generations.

David Langlois
June 2018

Contents

Written in collaboration with Félix-Louis Julié

BOOK 1

Space, time, and gravity in Newton's theory

Contents

Book 1

Part I

Kinematics

I do not define *Time*, *Space*, *Place* and *Motion*, as being well known to all. Only I must observe, that the vulgar conceive those quantities under no other notions but from the relation they bear to sensible objects. And thence arise certain prejudices, for the removing of which, it will be convenient to distinguish them into *Absolute* and *Relative*, *True* and *Apparent*, *Mathematical* and *Common*.

Absolute Space, in its own nature, without regard to any thing external, remains always similar and immovable.

Absolute, True, and Mathematical Time, of itself, and from its own nature flows equably without regard to any thing external, and by another name is called Duration...

Sir Isaac Newton, *Philosophiæ Naturalis Principia Mathematica*, London, 1687; English translation by Andrew Motte, *The Mathematical Principles of Natural Philosophy*, London, 1729

1
Cartesian coordinates

In this first chapter we give an elementary and brief presentation of Euclidean geometry, which provides the mathematical framework in which the laws of Newtonian physics are formulated.

1.1 Absolute space and time

In Newtonian physics, 'space' and 'relative, apparent, and common' place are represented by a mathematical ensemble of points, the 'absolute' space $\mathcal{E}_3$, which is postulated to be Euclidean.

Each point of this space is thus characterized by three real numbers, its coordinates, which define its position. In addition, there exist systems of coordinates called Cartesian coordinates such that the *distance* r_{12} between two points with coordinates (X_1, Y_1, Z_1) and (X_2, Y_2, Z_2) is given by the Pythagorean theorem

$$r_{12} = \sqrt{(X_2 - X_1)^2 + (Y_2 - Y_1)^2 + (Z_2 - Z_1)^2} = \sqrt{\sum_{i=1}^{i=3}(X_2^i - X_1^i)^2}. \qquad (1.1)$$

It vanishes only if the two points coincide.

The *length element*, that is, the square of the distance dl (≥ 0) between two infinitesimally separated points with Cartesian coordinates (see Fig. 1.1) X^i and $X^i + dX^i$, which is used to measure lengths of curves and to display the metric properties of figures, is then written as

$$dl^2 = dX^2 + dY^2 + dZ^2 = \sum_{i,j} \delta_{ij}\, dX^i dX^j \equiv \delta_{ij}\, dX^i dX^j. \qquad (1.2)$$

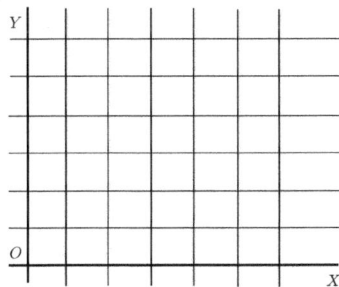

Fig. 1.1 Cartesian coordinates and frame in two dimensions.

Relativity in Modern Physics. Nathalie Deruelle and Jean-Philippe Uzan.
© Oxford University Press 2018. Published in 2018 by Oxford University Press.
DOI: 10.1093/oso/9780198786399.001.0001

The second equality defines the *Kronecker delta* δ_{ij}; in this geometrical context its six components ($\delta_{ij} = 1$ if $i = j$, $\delta_{ij} = 0$ otherwise) are referred to as the coefficients of the Euclidean metric in Cartesian coordinates. The third equality defines the *Einstein summation convention*, according to which repeated indices are summed over (they are then referred to as dummy indices). The origin with coordinates $(0, 0, 0)$ and the three axes (X, Y, Z) constitute a *Cartesian frame $\mathcal{S}$*.

As for the 'apparent' time, it is represented by a real number, the 'absolute' or universal time $t \in R$.

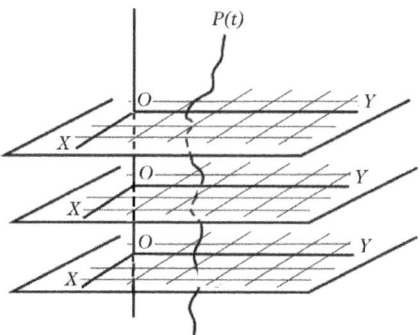

Fig. 1.2 Absolute frame and motion.

It is therefore possible to represent Newtonian spacetime N_4 as a 'foliation', that is, a stack of copies of Euclidean space $\mathcal{E}_3$ ordered by increasing time t: $N_4 = \mathcal{E}_3 \times R$. The ensemble of copies of a point in $\mathcal{E}_3$ then becomes a 'fiber' of N_4 representing a point at absolute rest. In this kinematical context the ensemble of Cartesian frames indexed by t is referred to as an absolute frame.[1] (See Fig. 1.2.)

The motion of a physical object without extent or internal structure, called a point particle, is represented by a curve in spacetime called a *world line* or, equivalently, a *trajectory*, that is, a curve in $\mathcal{E}_3$: $t \mapsto P(t) \in \mathcal{E}_3$, where the parameter t is the absolute time.

1.2 The absolute reference frame

The absolute time t is made concrete by clocks and watches, that is, by repetitive phenomena numbered in increasing order. A good clock is ultimately a device which measures, no matter what sort of motion it undergoes, time intervals, expressed for example in seconds, in accordance with the predictions of the laws of dynamics written as a function of absolute time.

A Cartesian frame of absolute space is materialized in 'relative, apparent, and common' space by a *reference frame*. Specifically, this reference frame is a solid trihedron, that is, an ensemble of physical objects whose relative distances are constant in time and for which an orientation of the axes has been chosen (by using the right-hand rule, for example). It is constructed by means of instruments which qualify as rigid (*i.e.*, they are also solids)

---

[1]We note that t belongs to the real line R and not to the circle S^1: in Newtonian mechanics time has no beginning and no end.

throughout repeated use such as rulers, compasses, and so on, using the Pythagorean theorem and its consequences. Finally, a length standard, for example, the meter, is chosen. This reference frame, which establishes a grid on physical space, is 'good' if all the Euclidean properties of figures are verified to within the measurement accuracy.[2]

The *absolute reference frame* which materializes the absolute frame of Newtonian space-time is a reference frame which must be at rest in order for it to be possible to identify the fibers of N_4 with physical objects at rest. For Newton, the absolute reference frame was formed by the solar system and the stars far enough away to appear fixed, which he postulated to be at absolute rest. We note that since the universe appears to be essentially empty, the points of $\mathcal{E}_3$ are for the most part only 'virtually materialized', a contradiction in terms which shocked Descartes and Kant. Beginning in the seventeenth century, this difficulty was circumvented by introducing the concept of the *aether*, a nebulous medium responsible for materializing $\mathcal{E}_3$.

If space and time do actually embody the structure attributed to them by Newton, we can predict an elementary but important result. Let us imagine two travelers A and B who simultaneously depart from a certain place and return to the same place after their separate peregrinations. The durations of the trips measured by A and B must be the same, *i.e.*, their watches, which are synchronized when they start out, must indicate the same time at arrival.[3]

1.3   Change of Cartesian coordinates

We postulate that if the labeling of the points of $\mathcal{E}_3$ is changed, the distance between two points remains unchanged. Here, we restrict ourselves to transformations $X^i \mapsto X'^i$ which preserve the *form* of the length element, that is, transformations such that $dl^2 \equiv \delta_{ij}\, dX^i dX^j = \delta_{ij}\, dX'^i dX'^j$. Then, by definition, the new coordinates X'^i are also Cartesian, and the transformations are given by

$$X'^i = \mathcal{R}_j{}^i(X^j - d^j) \text{ with } \mathcal{R}_k{}^i \mathcal{R}_l{}^j \delta_{ij} = \delta_{kl}\,. \tag{1.3}$$

We impose the condition $\det \mathcal{R} = +1$, where $\det \mathcal{R}$ is the determinant of the rotation matrix $\mathcal{R}_j{}^i$ (i numbers the lines and j the columns); the transformation therefore preserves the orientation of the axes. Such transformations form the (proper) group of transformations of Cartesian frames, a group with $n(n+1)/2$ parameters, n of them translational and $n(n-1)/2$ rotational, where n is the dimension of the space.[4]

---

[2]If measurements give results which systematically contradict the Euclidean predictions (*e.g.*, if the sum of the angles of a triangle is not equal to π), it can be deduced that the representation of the actual space (the surface of the Earth, for example) by a Euclidean plane is inadequate. For a masterful and concise exposition of the interplay between phenomena and their mathematical representation see the letter of A. Einstein to M. Solovine in, *e.g.*, J. Eisenstaedt (2002).

[3]If in a single experiment this were not the case, it would most likely indicate that their watches are not accurate. However, if a large number of carefully performed experiments gave a result systematically different from the prediction, one would conclude, in agreement with Einstein, that the absolute spacetime of Newton does not adequately represent the actual universe.

[4]Note that we always assume that the *topology* of the absolute space is trivial and that a global orientation exists, thereby excluding spaces of the Möbius-strip type in two dimensions or the Klein-bottle type in three dimensions.

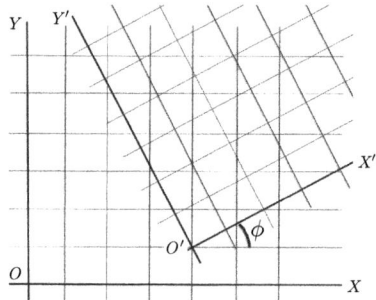

Fig. 1.3 Change of Cartesian coordinates.

In two dimensions the rotation matrix $\mathcal{R}_j{}^i$ is parametrized by an angle ϕ (see Fig. 1.3) and we have

$$\mathcal{R}_X^X = \cos\phi, \quad \mathcal{R}_Y^X = \sin\phi, \quad \mathcal{R}_X^Y = -\sin\phi, \quad \text{and} \quad \mathcal{R}_Y^Y = \cos\phi. \tag{1.4}$$

The Euler angles

In three dimensions the three parameters characterizing a rotation can be taken to be the *Euler angles*. If we use ON to denote the intersection of the $X'OY'$ and XOY planes, these three angles are: the *precession angle* Φ, which is the angle ON makes with the X axis; the *proper rotation angle* Ψ, which is the angle ON makes with the X' axis; and the *nutation angle* Θ, which is the angle between the Z' axis and the Z axis (see Fig. 1.4). The elements of the rotation matrix $\mathcal{R}_j{}^i$ are then given by

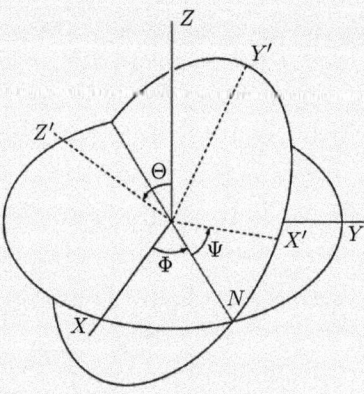

Fig. 1.4 The Euler angles.

$$\begin{cases} \mathcal{R}_1{}^1 = \cos\Phi\cos\Psi - \sin\Phi\sin\Psi\cos\Theta, \\ \mathcal{R}_2{}^1 = \sin\Phi\cos\Psi + \cos\Phi\sin\Psi\cos\Theta, \\ \mathcal{R}_3{}^1 = \sin\Psi\sin\Theta \\[6pt] \mathcal{R}_1{}^2 = -\cos\Phi\sin\Psi - \sin\Phi\cos\Psi\cos\Theta, \\ \mathcal{R}_2{}^2 = \cos\Phi\cos\Psi\cos\Theta - \sin\Phi\sin\Psi, \\ \mathcal{R}_3{}^2 = \cos\Psi\sin\Theta \\[6pt] \mathcal{R}_1{}^3 = \sin\Phi\sin\Theta, \quad \mathcal{R}_2{}^3 = -\cos\Phi\sin\Theta, \\ \mathcal{R}_3{}^3 = \cos\Theta. \end{cases} \tag{1.5}$$

The element $\mathcal{R}^i{}_j$ of the inverse matrix (such that $\mathcal{R}^i{}_k\mathcal{R}^k{}_j = \mathcal{R}^k{}_j\mathcal{R}^i{}_k = \delta^i_j$) is equal to[5] $\mathcal{R}^j{}_i$. We therefore have $\mathcal{R}^1{}_1 = \cos\Phi\cos\Psi - \sin\Phi\sin\Psi\cos\Theta$, $\mathcal{R}^1{}_2 = -\cos\Phi\sin\Psi - \sin\Phi\cos\Psi\cos\Theta$, and so on.

Since the form of the length element is invariant under translations and rotations, the choice of the origin and the direction of the axes of the absolute reference frame cannot be of any importance. This is one way of phrasing the Copernican Principle, the (active) version of which can be stated as follows. Euclidean geometry is universal; a triangular solid, for example, must possess the same geometrical properties no matter where it is located. The universe does not deform it; it is a neutral receptacle of matter that we call homogeneous and isotropic.

We recall that this reference frame must, on the other hand, be at rest. We therefore determine it by trial and error, using the fact that, to within the measurement accuracy, the motions of physical objects must obey the laws of dynamics as they are written in the absolute frame. For low-precision experiments, the 'laboratory walls' can prove to be good enough for materializing the absolute Cartesian frame. However, in the case of more sophisticated experiments it is necessary to use a frame attached to the center of the Earth, and so on, see Section 2.5.

1.4 The group of rigid displacements

In order to simplify the description of certain phenomena or to study them in a moving reference frame, it is sometimes useful to go from the absolute Cartesian frame $\mathcal{S}$ of the spacetime $N_4 = \mathcal{E}_3 \times R$ to a *moving* frame $\mathcal{S}'$, that is, to a family of frames of $\mathcal{E}_3$ indexed by t, the origins and directions of the axes of which vary from leaf to leaf, *i.e.*, in each section $\mathcal{E}_3$ of N_4. This operation differs from the passage from one Cartesian frame to another because the labeling of the points (which are at rest in $\mathcal{S}$) depends on the time in $\mathcal{S}'$. However, we shall continue to postulate that the distance between two points is the same in $\mathcal{S}$ and in $\mathcal{S}'$, that is, that a rigid body remains rigid, no matter what motion it undergoes.

We shall consider only the set of transformations $X^i \to X'^i$ which preserve the form of the length element, with the three axes remaining orthonormal (and having the same orientation) throughout the motion of the frame. This is defined by

---

[5]In fact, $\mathcal{R}^i{}_j\mathcal{R}_k{}^j = \delta^i_k$ implies that $\mathcal{R}^i{}_j = \delta^{ik}\delta_{jl}\mathcal{R}_k{}^l$ (which is a rather heavy-handed way of stating that the element $\mathcal{R}^i{}_j$ is equal to $\mathcal{R}_i{}^j$), because $\mathcal{R}^i{}_j$ is a rotation matrix (*i.e.*, a matrix satisfying $\mathcal{R}_k{}^i\mathcal{R}_l{}^j\delta_{ij} = \delta_{kl}$).

$$X'^i = \mathcal{R}_j{}^i(t)\left(X^j - d^j(t)\right), \quad \text{with} \quad \mathcal{R}_k{}^i(t)\,\mathcal{R}_l{}^j(t)\,\delta_{ij} = \delta_{kl} \quad \text{and} \quad \det\mathcal{R} = +1, \qquad (1.6)$$

where the components in $\mathcal{S}'$ of the rotation matrix $\mathcal{R}_j{}^i(t)$ of the three axes and of the translation of the origin $d^i(t)$ depend on the time (the three Euler angles defined in Section 1.3 are therefore now functions of the time).

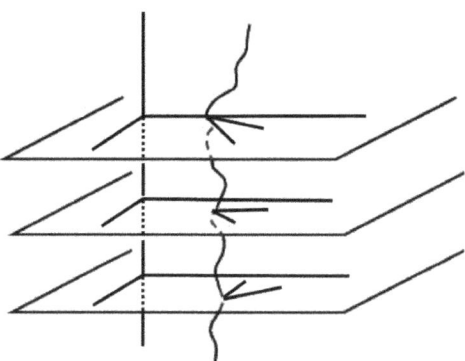

Fig. 1.5 Time-dependent change of frame.

This set of transformations forms the group of *rigid displacements*. They describe the time evolution of the position and orientation of a reference solid trihedron undergoing arbitrary motion relative to the absolute reference frame.

The subgroup of general translations

$$X'^i = \mathcal{R}_j{}^i\left(X^j - d^j(t)\right), \qquad (1.7)$$

where $\mathcal{R}_j{}^i$ is independent of time and $d^j(t)$ is arbitrary, is called the *Milne group*.

The subgroup of translations which are linear in time, called the *Galilean group*, will play a central role in what follows. The passage from the absolute Cartesian frame $\mathcal{S}$ to a frame undergoing uniform rectilinear motion relative to $\mathcal{S}$, called a *Galilean* or *inertial* frame, is carried out by means of a *Galilean transformation*

$$X'^i = \mathcal{R}_j{}^i(X^j - X_0^j - V_0^j t), \qquad (1.8)$$

where $\mathcal{R}_j{}^i$, X_0^i, and V_0^i are constants.

1.5   Kinematics of a point particle (I)

The motion of a point particle P is represented by a curved trajectory of equation $t \in R \mapsto P(t) \in \mathcal{E}_3$, where t is the absolute time. If the Cartesian coordinates of the points $P(t)$ in the absolute frame $\mathcal{S}$ are given by the three functions $X^i(t)$, then their three time derivatives

$$V^i(t) \equiv \frac{dX^i}{dt} \equiv \dot{X}^i, \qquad (1.9)$$

where the dot denotes differentiation with respect to time, are the components of the *velocity* of P in $\mathcal{S}$ at the instant t. Similarly, $a^i(t) \equiv dV^i/dt = d^2 X^i/dt^2 \equiv \ddot{X}^i$ are the components of its *acceleration*.

Under a time-*independent* change of Cartesian frame (see Fig. 1.5), the trajectory equations become $X'^i = X'^i(t) = \mathcal{R}_j^{\ i}\left(X^j(t) - d^j\right)$, so that the components of the velocity and acceleration in $\mathcal{S}'$ then become

$$V'^i \equiv \frac{dX'^i}{dt} = \mathcal{R}_j^{\ i} V^j, \quad a'^i \equiv \frac{dV'^i}{dt} = \mathcal{R}_j^{\ i} a^j. \tag{1.10}$$

It is just as simple to find the law governing the transformation of the velocity and acceleration in going to a *moving* frame. The Cartesian coordinates of P in $\mathcal{S}'$ are given by the three functions $X'^i(t) = \mathcal{R}_j^{\ i}(t)\left(X^j(t) - d^j(t)\right)$, so that, by the Leibniz rule,

$$\begin{cases} V'^i \equiv \dfrac{dX'^i}{dt} = \mathcal{R}_j^{\ i} V^j + \dot{\mathcal{R}}_j^{\ i} X^j - (\mathcal{R}_j^{\ i} d^j)^{\cdot} \\[2mm] a'^i \equiv \dfrac{dV'^i}{dt} = \mathcal{R}_j^{\ i} a^j + 2\dot{\mathcal{R}}_j^{\ i} V^j + \ddot{\mathcal{R}}_j^{\ i} X^j - (\mathcal{R}_j^{\ i} d^j)^{\cdot\cdot}\,. \end{cases} \tag{1.11}$$

We see that the $V'^i(t)$ are not equal to the $\mathcal{R}_j^{\ i}(t)V^j(t)$; the additional terms arise from the motion of the frame and not from the motion of the point particle. The same occurs for the acceleration, *unless* the transformation is Galilean.

Instantaneous rotation of a frame

The rotation matrix $\mathcal{R}_k^{\ j}$ and the inverse matrix $\mathcal{R}^i_{\ l}$ satisfy $\mathcal{R}^i_{\ k}\mathcal{R}_j^{\ k} = \mathcal{R}^k_{\ j}\mathcal{R}_k^{\ i} = \delta^i_j$. Let us introduce functions of time ω'_{ij} such that $\dot{\mathcal{R}}^k_{\ i}\,\mathcal{R}_k^{\ j} = -\mathcal{R}^k_{\ i}\,\dot{\mathcal{R}}_k^{\ j} \equiv \omega'_i{}^j \equiv \delta^{jk}\omega'_{ik}$. Using $\mathcal{R}_k^{\ i}\,\mathcal{R}_l^{\ j}\,\delta_{ij} = \delta_{kl}$, it is easily shown that $\omega'_{ij} = -\omega'_{ji}$. Since the space is three-dimensional, the three independent components of ω'_{ij} can be replaced by three other functions of time ω'^k using $\omega'_{ij} \equiv e_{ijk}\omega'^k$, where e_{ijk} is the Levi-Civita symbol (see its definition and properties below). We therefore have $\dot{\mathcal{R}}_l^{\ k} = -e^k_{\ mi}\,\omega'^m\,\mathcal{R}_l^{\ i}$ and $\dot{\mathcal{R}}^l_{\ i} = e^k_{\ mi}\,\omega'^m\,\mathcal{R}^l_{\ k}$.

In (1.11) for the V'^i let us replace the old coordinates X^j by the new ones $X^j = \mathcal{R}^j_{\ k}\,X'^k + d^j$. Using the above results, we can rewrite the V'^i in the form

$$V'^i = \mathcal{R}_j^{\ i}(V^j - \dot{d}^j) - e^i_{\ jk}\,\omega'^j X'^k. \tag{1.12}$$

Taking the derivative with respect to time, we likewise obtain

$$a'^i = \mathcal{R}_j^{\ i}(a^j - \ddot{d}^j) - 2\,e^i_{\ jk}\,\omega'^j V'^k - e^i_{\ jk}\,\dot{\omega}'^j X'^k + e^i_{\ jk}\,\omega'^j\, e^k_{\ lm}\,X'^l\,\omega'^m, \tag{1.13}$$

where the last term can be written as $X'^i\Omega^2 - \omega'^i(\Omega . X)$ using the notation $\Omega^2 \equiv \delta_{ij}\omega'^i\omega'^j$ and $(\Omega . X) \equiv \delta_{ij}\omega'^i X'^j$. We note that it can also be written as $-\partial U_{\text{centr}}/\partial X'^i$ with $U_{\text{centr}} = -\frac{1}{2}e_{ijk}e^k_{\ lm}\omega'^j\omega'^m X'^l X'^i$.

If the components of the matrix $\mathcal{R}_k^{\ j}$ are written as functions of the Euler angles (see Section 1.3), the three functions ω'^k become

$$\omega'^1 = \dot{\Phi}\sin\Psi\sin\Theta + \dot{\Theta}\cos\Psi, \quad \omega'^2 = \dot{\Phi}\cos\Psi\sin\Theta - \dot{\Theta}\sin\Psi, \quad \omega'^3 = \dot{\Phi}\cos\Theta + \dot{\Psi}. \tag{1.14}$$

This describes the *instantaneous rotation* of the moving frame.

The Levi-Civita symbol

The *Levi-Civita symbol* (in n dimensions) $e_{i_1...i_n}$ is completely antisymmetric (that is, $e_{i_1...i_j...i_k...i_n} = -e_{i_1...i_k...i_j...i_n} \; \forall \; j, \, k$) and is such that $e_{12...n} = +1$. In three dimensions it can be shown that

$$e_{ijk}e_{lmn} = \delta_{il}\delta_{jm}\delta_{kn} + \delta_{im}\delta_{jn}\delta_{kl} + \delta_{in}\delta_{jl}\delta_{km} - \delta_{il}\delta_{jn}\delta_{km} - \delta_{im}\delta_{jl}\delta_{kn} - \delta_{in}\delta_{jm}\delta_{kl}, \qquad (1.15)$$

which implies that $e_{ijk}e^{k}{}_{mn} = \delta_{im}\delta_{jn} - \delta_{in}\delta_{jm}$, $e_{ijk}e^{jk}{}_{l} = 2\delta_{il}$, and $e_{ijk}e^{ijk} = 6$.

If A_l^i is a 3×3 matrix, then $e_{ijk}A_l^i A_m^j A_p^k = \det A \, e_{lmp}$, where $\det A$ is the determinant of A_l^i. Therefore, $\det A = e_{ijk}A_1^i A_2^j A_3^k$.

Rotation about an axis

Show that in going from an absolute frame to a frame rotating with angular velocity $\Omega(t) \equiv d\phi/dt$ (according to the right-hand rule) about the Z axis, the components $(d^2X/dt^2 \equiv \ddot{X}$, $d^2Y/dt^2 \equiv \ddot{Y})$ of the acceleration of a point particle become

$$\begin{cases} \ddot{X}' = +2\Omega\dot{Y}' + \Omega^2 X' + \dot{\Omega}Y' + \cos\phi\ddot{X} + \sin\phi\ddot{Y} \\ \ddot{Y}' = -2\Omega\dot{X}' + \Omega^2 Y' - \dot{\Omega}X' - \sin\phi\ddot{X} + \cos\phi\ddot{Y} \, . \end{cases} \qquad (1.16)$$

1.6 Cartesian vectors and vector fields

The choice of a particular Cartesian frame should be of no kinematical importance as long as the frame is at rest. It is therefore natural to consider the components of the velocity of a point particle $P(t)$ in any frame at rest, that is, the set of $V'^i = \mathcal{R}_j{}^i V^j$ parametrized by time-independent rotation matrices $\mathcal{R}_j{}^i$, as an equivalence class, that is, a unique object corresponding to the *velocity* of $P(t)$, denoted v, the functions $V^i(t)$ and $V'^i(t)$ being just their avatars in the frames $\mathcal{S}$ and $\mathcal{S}'$. Likewise, a^i and $a'^i = \mathcal{R}_j{}^i a^j$ are the avatars of the *acceleration* a of $P(t)$.

In addition to the velocity and acceleration of point particles, Newtonian mechanics will introduce a number of quantities (for example, forces) of the same type, that is, quantities such that if T^i are their components in a Cartesian coordinate system $\mathcal{S}$, their components in another Cartesian system at rest $\mathcal{S}'$ are given by

$$T'^i = \mathcal{R}_j{}^i T^j \, . \qquad (1.17)$$

The equivalence class of the T'^i is the (Cartesian) *vector* T.

Since the new coordinates X'^i are related to the old ones as $X'^i = \mathcal{R}_j{}^i(X^j - d^j)$, an example of a vector is the *separation vector* l_a of two points P_a and P, which has the components $l_a^i = X^i - X_a^i$ in $\mathcal{S}$ and $l_a'^i \equiv X'^i - X_a'^i = \mathcal{R}_j{}^i l_a^j$ in $\mathcal{S}'$.

Therefore, the Euclidean length element gives the distance between two points P_a and P (see Section 1.1), as well as the *norm* l^2 (or the square of the *length* $|l| \equiv \sqrt{l^2}$) of the separation vector l_a: $l^2 \equiv \delta_{ij} l_a^i l_a^j = \delta_{ij} (X^i - X_a^i)(X^j - X_a^j)$. By extension, the *norm* T^2 of any vector T (and its *length* or *modulus* $|T| \equiv \sqrt{T^2}$) can be defined, as well as the *scalar product* $(T.U)$ of two vectors, in the following way: if T^i and U^j are their components in $\mathcal{S}$, then

$$T^2 \equiv |T|^2 \equiv \delta_{ij} T^j T^i = T_i T^i, \quad (T.U) \equiv \delta_{ij} T^j U^i = T_i U^i, \text{ with } T_i \equiv \delta_{ij} T^j \qquad (1.18)$$

(the three T_i are numerically equal to the three T^i). The norm of a vector or the scalar product of two vectors has the same numerical value in $\mathcal{S}$ and $\mathcal{S}'$ because $\delta_{ij} T^j U^i = \delta_{ij} T'^j U'^i$.

Up to now we have defined a vector as a set of three numbers (e.g., l_a or T with components l_a^i or T^i) or three functions of time [e.g., the velocity v with components $V^i(t)$]. It will turn out to be particularly useful to introduce, *à la Faraday*, *vector fields*, that is, sets of three functions of the *coordinates* which transform as (1.17) under a change of Cartesian frame at rest. From this perspective the velocity (or acceleration) becomes a vector field evaluated on the trajectory: $V^i(t) = V^i(X^j(t))$. Likewise, a *scalar field* $\Phi(X^i)$ is a function of coordinates which is invariant under a change of frame: $\Phi'(X'^i) = \Phi(X^i)$. Finally, the separation vector $l_a^i = X^i - X_a^i$ can be viewed as a vector field, a function of the X^i.

We recall that, as we have seen in Section 1.5, the velocity and acceleration do *not* transform as vectors in going to a *moving* frame. On the other hand, the separation vector itself transforms according to (1.17) in the larger group of rigid displacements: $X'^i = \mathcal{R}_j{}^i(t)(X^j - d^j(t))$.

In the next chapter we shall define the mathematical spaces to which the geometrical quantities we have introduced—scalars, vectors, and the metric—belong.

2
Vector geometry

The goal of this chapter is to go from the concept of a vector as an object whose components transform as $T^i \rightarrow \mathcal{R}_j{}^i T^j$ under a change of frame to the 'intrinsic' concept of a vector, T. These concepts are also generalized to 'tensors'.

2.1 Tensor spaces

We recall that a *vector space* E is an ensemble of objects v or w called *vectors* such that $(\alpha v + \beta w)$, where α and β are real numbers, is also a vector, and where vector addition and multiplication by a real number possess the usual properties of commutativity, associativity, and the existence of an identity element and an inverse. By defining a *basis* $\{e_i\}$ and the *dimension* n of E, any vector v of E can be uniquely decomposed as $v = v^i e_i$, where the n real numbers v^i, $i = 1, 2, ...n$, are the *components* of v in the basis $\{e_i\}$ (we use the Einstein summation convention).[1]

We also recall that the *dual space* of E, denoted E^*, is the ensemble of linear maps, or *forms*, which associate a real number with a vector. The dual space E^* can be constructed as follows. To each basis $\{e_i\}$ of E we associate the basis $\{\epsilon^j\}$ of E^*, called the *dual basis* (or *conjugate basis*), using the formula $\epsilon^j(e_i) = \delta_i^j$, where δ_i^j is the Kronecker delta. Then, any form $\lambda \in E^*$ can be decomposed uniquely as $\lambda = \lambda_j \epsilon^j$, where the λ_j are its *components* or *coefficients* in the basis $\{\epsilon^j\}$.

Finally, we recall that the dual of the dual is isomorphic to E, because the action of the form $\lambda = \lambda_i \epsilon^i$ on the vector $v = v^i e_j$, namely, $\lambda(v) = \lambda_i v^i$, can also be viewed as the action of v on λ. We can then write $\lambda(v) \equiv v(\lambda)$. Therefore, vectors are also operators acting on forms to give numbers.

Bilinear forms a associate a real number with a pair of vectors. Their ensemble is denoted as $L_2(E \times E, R)$. They can be defined as follows. For $v = v^i e_i \in E$ and $w = w^j e_j \in E$, by linearity we have $a(v, w) = v^i w^j a(e_i, e_j)$; therefore, knowledge of a is equivalent to knowledge of the n^2 real numbers $a_{ij} \equiv a(e_i, e_j)$, the components (or coefficients) of a in the basis $\{e_i\}$.

The *tensor product* notation $\otimes$ gives rise to an 'automatic' definition of multilinear forms. If we define $\epsilon^i \otimes \epsilon^j$ as a bilinear form such that $(\epsilon^i \otimes \epsilon^j)(v, w) = \epsilon^i(v)\,\epsilon^j(w) = v^i w^j$ [with $(\epsilon^i \otimes \epsilon^j)(e_k, e_l) = \delta_k^i \delta_l^j$], we can then write $a = a_{ij}\, \epsilon^i \otimes \epsilon^j$. The $\epsilon^i \otimes \epsilon^j$ form a basis of $L_2(E \times E, R)$ which can be denoted as $E^* \otimes E^*$.

Analogously, an element b of $L_2(E \times E^*, R) \equiv E^* \otimes E$ associates a real number with a pair consisting of a vector and a form and is written as $b = b_i{}^j \epsilon^i \otimes e_j$, and so on.

---

[1]Occasionally, we will use the common notation of writing vectors as a symbol with an arrow above, in which case the basis vectors will be denoted as $(\vec{i}, \vec{j}, \vec{k})$. A more detailed exposition of tensor calculus can be found, for example, in A. Lichnerowicz (1950).

Relativity in Modern Physics. Nathalie Deruelle and Jean-Philippe Uzan.
© Oxford University Press 2018. Published in 2018 by Oxford University Press.
DOI: 10.1093/oso/9780198786399.001.0001

Therefore, rather than dealing with bi- or multilinear forms, it will be more economical to use *tensors*. For example, $T = T_{ij}{}^k \epsilon^i \otimes \epsilon^j \otimes e_k$ is, by definition, a tensor which is 2-fold *covariant* and singly *contravariant*. A basis of the ensemble of such tensors is $\epsilon^i \otimes \epsilon^j \otimes e_k$, constructed as the tensor product of the basis e_i of E and the conjugate basis ϵ_j of E^*. Therefore, vectors become singly contravariant tensors and forms become singly covariant tensors. In general, we can write

$$T = T^{i_1 \ldots i_p}_{j_1 \ldots j_q} \, e_{i_1} \otimes e_{i_2} \ldots \otimes \, e_{i_p} \otimes \epsilon^{j_1} \otimes \ldots \otimes \, \epsilon^{j_q} \, . \tag{2.1}$$

A tensor which is p-fold contravariant and q-fold covariant is said to be of the type, or *valence*,[2] $\binom{p}{q}$. Tensors of the type $\binom{p}{q}$ form a vector space. Products of the components of two tensors T and T' of the types $\binom{p}{q}$ and $\binom{p'}{q'}$ are the components of the tensor $T \otimes T'$ of type $\binom{p+p'}{q+q'}$, the tensor product of T and T'. The spaces of multilinear forms are thus unified and become the elements of a *tensor algebra* of infinite dimension.

Contraction transforms a tensor of type $\binom{p}{q}$ into a tensor of type $\binom{p-1}{q-1}$ by summation over one covariant index and one contravariant index. For example, the quantities T^{ia}_{kam} obtained by summing the components of a tensor of type $\binom{2}{3}$ over the dummy index a are the components of a tensor of type $\binom{1}{2}$ (and $e_a \otimes \epsilon^a = 1$). The *trace* of a tensor of type $\binom{p}{p}$ is contraction on all its indices: $T = T^{ijk\ldots}_{ijk\ldots}$.

A tensor T^{ijk}_l, for example, is *symmetric in its contravariant indices i and j* if $T^{ijk}_l = T^{jik}_l$ and *antisymmetric* in them if $T^{ijk}_l = -T^{jik}_l$. An *(anti)symmetric* tensor is (anti)symmetric in all its indices. An antisymmetric tensor of type $\binom{0}{p}$ is also referred to as a *p-form*.

2.2 Affine and Euclidean planes

An *affine space* $\mathcal{E}$ of dimension n is an ensemble of points where each *bipoint*, *i.e.*, each pair of points ordered as p_a and p and denoted as (p_a, p), is identified with a vector of a vector space E of the same dimension, denoted $p_a p$ and called the *separation vector*. The bipoints must satisfy the *Chasles relation*: for any point O, $p_a p = p_a O + Op$, and if p is a point of $\mathcal{E}$ and F is a vector of E, then there exists one and only one point q such that $pq = F$. This set of a point and a vector is called a *bound vector*. ('Unbound' vectors, *i.e.*, the elements of E, are therefore referred to as *free vectors*.)

An *affine frame* of $\mathcal{E}$ is the ensemble of a point O as the origin and a basis e_i of E, that is, a set of bound vectors $\mathcal{S} = (O, \{e_i\})$. Therefore, the vector Op associated with the bipoint (O, p) can be decomposed as $Op = X^i e_i$, where the X^i are simultaneously the components of Op in the basis $\{e_i\}$ and the *coordinates* of p in the frame $\mathcal{S}$. The vector $Op \equiv R$ is called the *radius vector* or the *position vector* of the point p. The separation vector $p_a p = Op - Op_a = R - R_a \equiv l_a$ can be decomposed as $l_a = (X^i - X^i_a) e_i$.

Each bipoint is identified with a single vector but, conversely, a 'free' vector F corresponds to an entire equivalence class of bipoints called *equipollents* (p, q), where p is any point and $pq = F$. Two equipollent bipoints (p_1, q_1) and (p_2, q_2) form a *parallelogram*. The operation which associates (p_1, q_1) with (p_2, q_2) is called *parallel transport*.

---

[2] A way of making the valence $\binom{p}{q}$ of a tensor manifest is to write $T^{J_1 \ldots J_p}_{I_1 \ldots I_q}$ (taking care not to confuse this with the components $T^{j_1 \ldots j_p}_{i_1 \ldots i_q}$ of the tensor in some basis).

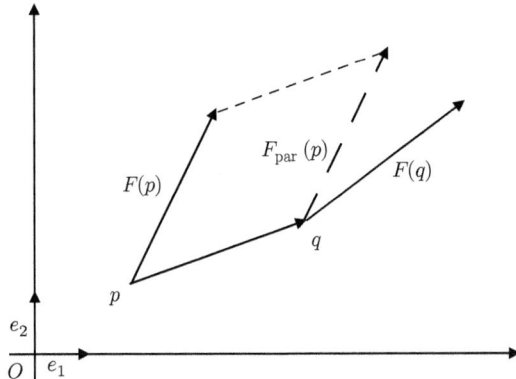

Fig. 2.1 Parallel transport.

It is this identification of bound vectors as bipoints which permits the well known graphical representation of a vector field $F(p)$ [also denoted as $F(Op)$, $F(R)$, or $F(X^i)$] as an oriented segment attached to the point where it is evaluated. And it is the equipollence relation which allows the parallel transport of the vector $F(p)$ to another point q where it will have the same components as at p; see Fig. 2.1.

A *Euclidean space* is an affine space equipped with a *Euclidean metric*, that is, with a bilinear symmetric form (or 2-fold covariant tensor)

$$e = \delta_{ij}\epsilon^i \otimes \epsilon^j, \tag{2.2}$$

where δ_{ij} is the Kronecker delta. The action of e on the basis vectors e_i of E then is $e(e_k, e_l) = e(e_l, e_k) = \delta_{kl}$. Therefore, the basis $\{e_i\}$ is orthonormal, the frame $\mathcal{S} = (O, \{e_i\})$ is *Cartesian*, and the coordinates X^i of p in the frame $\mathcal{S}$ are its Cartesian coordinates. The *signature* of the Euclidean metric, namely, the diagonal elements of the matrix δ_{ij}, is $(+1, +1, ...)$, and its *sign* is the product of these elements, that is, $+1$. The metric defines the *norm* of a vector $v^2 \equiv e(v, v)$, its *length* or *modulus* $|v| = \sqrt{v^2}$, as well as the *distance* between two points, which is equal to the length of their separation vector. Finally, it defines a bijective correspondence between vectors and forms: if a vector v has components v^i, the associated form has the components $\delta_{ij}v^j$ (numerically equal to the v^i).

Let us relate the length element dl^2 defined in Section 1.1 to the metric e. Let p and q be two infinitesimally close points of the absolute space $\mathcal{E}_3$ with coordinates X^i and $(X^i + dX^i)$ in the absolute Cartesian frame $\mathcal{S}$. We denote the vector of the associated vector space as $dp \equiv pq = dX^i e_i$, and so dl^2 is the result of the action of the metric e on the vector dp:

$$e(dp, dp) = dX^i\, dX^j e(e_i, e_j) = dX^i\, dX^j \delta_{ij} \equiv dl^2. \tag{2.3}$$

More generally, the action of e on two vectors v and w defines their *scalar product*: $(v.w) \equiv e(v, w)$.

2.3 Change of basis and frame

A vector v is defined independently of the selected basis e_i or e'_j: $v = v^i e_i = v'^j e'_j$. Therefore, if $e'_j = \mathcal{R}^i_{\ j} e_i$, we will have $v^i = \mathcal{R}^i_{\ j} v'^j$ and $v'^j = \mathcal{R}_i^{\ j} v^i$, where $\mathcal{R}_i^{\ j}$ is the inverse of the matrix $\mathcal{R}^i_{\ j}$ (i.e., $\mathcal{R}^k_{\ i} \mathcal{R}_k^{\ j} = \mathcal{R}^j_{\ k} \mathcal{R}_i^{\ k} = \delta^j_i$).

A change of basis in E induces a change of basis in the dual space E^*: $\epsilon^j \to \epsilon'^j = \mathcal{R}_i^{\ j} \epsilon^i$. The basis $\{\epsilon'^j\}$ remains the conjugate of the basis $\{e'_j\}$: $\epsilon'^j(e'_i) = \delta^j_i$. In the basis $\{\epsilon'^j\}$ a form λ can be decomposed as $\lambda = \lambda_i \epsilon^i = \lambda'_j \epsilon'^j$ with $\lambda_i = \mathcal{R}_i^{\ j} \lambda'_j$ and $\lambda'_j = \mathcal{R}^i_{\ j} \lambda_i$.

A bilinear form transforms in the same way. Writing $a = a_{ij} \epsilon^i \otimes \epsilon^j$ with $a_{ij} \equiv a(e_i, e_j)$ in the basis $\{e_i\}$ and its associated basis $\{\epsilon^i\}$, then in the bases $\{e'_j\}$, $\{\epsilon'^j\}$ we will have $a'_{kl} \equiv a(e'_k, e'_k) = \mathcal{R}^i_{\ k} \mathcal{R}^j_{\ l} a_{ij}$. By extension we obtain the transformation law for tensor components under a change of basis:

$$T'^{i_1...i_p}_{\ \ j_1...j_q} = \mathcal{R}^{i_1}_{\ k_1} ... \mathcal{R}^{i_p}_{\ k_p} \mathcal{R}^{l_1}_{\ j_1} ... \mathcal{R}^{l_q}_{\ j_q} T^{k_1...k_p}_{\ \ l_1...l_q} . \tag{2.4}$$

Therefore, the components of a vector, a form, and more generally a tensor are just their avatars in a given basis, and their superior status is revealed in the transformation laws obeyed by their components. In fact, the entire 'component' approach of geometry is based on this, and the transformation law (2.4) can be viewed as a *definition* of a tensor; see Section 1.6.

Let $\mathcal{S} = (O, \{e_i\})$ and $\mathcal{S}' = (O', \{e'_j\})$ be two frames of an affine space $\mathcal{E}$. The radius vectors $O'p \equiv R'$ and $Op \equiv R$ of the point p in $\mathcal{S}'$ and $\mathcal{S}$ are related as, setting $OO' \equiv d$,

$$O'p = Op - OO', \quad \text{that is,} \quad R' = R - d, \text{ or}$$
$$X'^i e'_i = X^i e_i - d^i e_i, \quad \text{which implies that} \quad X'^i = \mathcal{R}_j^{\ i}(X^j - d^j), \tag{2.5}$$

and we again obtain (1.3). We note that the radius vector of a point p is the same in the two frames only if their origins coincide.

Until now we have not imposed any conditions on the matrix $\mathcal{R}_i^{\ j}$ except that it be invertible. Henceforth, we shall require that it be a proper rotation matrix, that is, that it satisfy the relations

$$\mathcal{R}_i^{\ k} \mathcal{R}_j^{\ l} \delta_{kl} = \delta_{ij} \quad \Longleftrightarrow \quad \mathcal{R}^k_{\ i} \mathcal{R}^l_{\ j} \delta_{kl} = \delta_{ij} , \quad \det \mathcal{R} = 1 . \tag{2.6}$$

[See (1.5) for the explicit expression of this rotation matrix and its inverse as a function of the Euler angles in three-dimensional space.] This subset of changes of affine frame defines the changes of *Cartesian* frame which preserve the value of the components of the Euclidean metric. Indeed, the general expression (2.4) states that if the δ_{ij} are the components of the metric in $\mathcal{S}$, then the components δ'_{ij} in $\mathcal{S}'$ are

$$\delta'_{ij} = \mathcal{R}^k_{\ i} \mathcal{R}^l_{\ j} \delta_{kl} = \delta_{ij} \tag{2.7}$$

owing to (2.6). The components of the Levi-Civita tensor (Section 1.5) are also invariant under rotations in three dimensions:

$$e'_{ijk} = \mathcal{R}_i^{\ l} \mathcal{R}_j^{\ m} \mathcal{R}_k^{\ p} e_{lmp} = \det \mathcal{R} \, e_{ijk} = e_{ijk} . \tag{2.8}$$

Book 1

2.4   Kinematics of a point particle (II)

The motion of a point particle is represented by the curve $P(t)$ in $\mathcal{E}_3$ or, equivalently, by its radius vector $OP(t)$, where t is the absolute time. The *velocity vector* of P in $\mathcal{S}$ is

$$v(t) = \frac{d\,OP}{dt} \equiv \frac{dR}{dt} \equiv \dot{R}. \tag{2.9}$$

If OP is decomposed as $OP = X^i(t)\,e_i$, then $v = V^i(t)\,e_i$, where the three functions of time $V^i(t) \equiv dX^i/dt \equiv \dot{X}^i(t)$ are its components introduced in Section 1.5. Similarly, the *acceleration vector* of P in $\mathcal{S}$ is $a = a^i(t)e_i$, where $a^i(t) \equiv \ddot{X}^i(t)$.

Basis vectors and derivative operators

Here we shall make a remark which will prove to be useful when we deal with non-Cartesian coordinates. The velocity vector v that we have just defined is a 'free' vector belonging to the vector space E_3 which subtends $\mathcal{E}_3$. As such, it is not bound to the point P at which it is evaluated. It is however possible to attach it to that point and to interpret it as the *tangent* to the trajectory at P, in which case it becomes an element of a copy of E_3 associated with P, the *tangent space* of $\mathcal{E}_3$ at P, denoted $E(P)$, with the basis vectors $e_i(P)$ (isomorphic to the e_i, i.e., having the same components).

If we now identify the trajectory with the coordinate line $X^1 = t$, $X^2 = $ const, $X^3 = $ const, we will have $V^1 = dX^1/dt = 1$, $V^2 = V^3 = 0$, that is, $v = e_1(P)$. Therefore, the vector $e_1(P)$ is a velocity, that is, a derivative operator. Let us be more precise and introduce the derivative operator $\frac{\partial}{\partial X^1}$. It acts at P on the coordinates X^i as $\frac{\partial}{\partial X^1}(X^i) = \delta_1^i$, which are nothing but the components of $e_1^i(P)$. More generally, since the operators $\frac{\partial}{\partial X^j}$ are in one-to-one correspondence with the basis vectors $e_j(P)$ of $E(P)$, we see the emergence of an identification of vectors with derivative operators. This will be developed further in Chapter 4.

As we have seen in Section 1.6, the velocity and acceleration are vectors (or singly contravariant tensors) in *time-independent* changes of Cartesian frames $X'^i = \mathcal{R}_j{}^i\,X^j - d^i$. Indeed, it is only if $\mathcal{R}_j{}^i$ and d^i are constants that the components of the velocity and acceleration transform as the components of a vector according to the formula already seen in (1.10):

$$V'^i(t) = \frac{dX'^i}{dt} = \mathcal{R}_j{}^i V^j(t)\,, \quad a'^i(t) = \frac{dV'^i}{dt} = \mathcal{R}_j{}^i a^j(t)\,. \tag{2.10}$$

We therefore can speak of a *velocity vector* v without specifying the affine frame in which it is evaluated only if the basis vectors are at rest. The same is true for the *acceleration vector* a (and its derivatives).

On the other hand, if the frame $\mathcal{S}'$ is *moving* relative to the absolute frame $\mathcal{S}$, then the velocity and acceleration of a trajectory $P(t)$ are *not* represented by the same vectors in $\mathcal{S}$ and $\mathcal{S}'$. Indeed, taking the time derivative of the vector equation $X'^i e_i' = X^i e_i - d^i e_i$, we obtain [see (2.5)]

$$\begin{cases} v' = v - \dot{d} - \Omega \wedge R' \\ a' = a - \ddot{d} - 2\Omega \wedge v' + \Omega \wedge (R' \wedge \Omega) - \dot{\Omega} \wedge R', \end{cases} \tag{2.11}$$

where $\wedge$ denotes the vector product (see Section 2.6 below). Here $R' \equiv O'P = X'^i(t)\,e_i'$ is the radius vector of $P(t)$ in $\mathcal{S}'$; $v' \equiv \dot{X}'^i e_i'$ ($\neq \dot{R}'$ because e_i' depends on time) and $a' \equiv \ddot{X}'^i e_i'$

are its velocity and acceleration with respect to $\mathcal{S}'$; $v \equiv \dot{X}^i \, e_i \, (= \dot{R})$ and $a \equiv \ddot{X}^i \, e_i$ are its velocity and acceleration with respect to $\mathcal{S}$; and $d \equiv OO' = d^i e_i$. Finally, Ω, defined as

$$\dot{e}'_i = \Omega \wedge e'_i, \tag{2.12}$$

is the *rotation vector* of $\mathcal{S}'$ relative to $\mathcal{S}$. The term $-2\Omega \wedge v'$ is the *Coriolis acceleration* and $\Omega \wedge (R' \wedge \Omega)$ is the *centrifugal acceleration*. The latter can also be written as $-\nabla_{R'} U_{\text{centr}}$ with $U_{\text{centr}} = -\frac{1}{2} R'.(\Omega \wedge (R' \wedge \Omega))$; here ∇ is the gradient operator, see Section 2.6.

These equations are the vector versions of (1.11)–(1.13), from which we can also find the components of Ω in $\mathcal{S}'$ ($\Omega = \omega'^i e'_i$) as a function of the Euler angles characterizing the rotation of $\mathcal{S}'$ with respect to $\mathcal{S}$.

We therefore cannot speak of absolute velocity or acceleration, that is, velocity and acceleration independent of the frames in which they are evaluated, whenever these frames are derived one from the other by time-dependent transformations. The only exceptions are Galilean transformations, for which the rotation matrix is a constant and the translation vector is linear in time, and which leave the acceleration unchanged:

$$a' = a \qquad \text{or, in component form,} \qquad a'^i(t) = \mathcal{R}_j{}^i a^j(t) \,. \tag{2.13}$$

As far as the velocity is concerned, it is never represented by the same vector in two frames, one derived from the other by a time-dependent transformation. Under Galilean transformations

$$v' = v - V_0 \qquad \text{or, in component form,} \qquad V'^i(t) = \mathcal{R}_j{}^i \left(V^j(t) - V_0^j \right). \tag{2.14}$$

This *Galilean law of velocity addition* shows that the frames $\mathcal{S}$ and $\mathcal{S}'$ are not kinematically equivalent.

2.5 Examples of moving frames

• *Reference frame corresponding to the 'laboratory walls'.* Let us choose as our (approximately) absolute reference frame the frame attached to the center of the Earth with the e_3 axis pointing toward the North Pole and the other two axes pointing toward two fixed stars. The origin of the laboratory frame is on the surface of the Earth and the directions of its axes are defined by choosing the Euler nutation angle $\Theta = \pi/2$ (see Section 1.3). Then $e'_1 \equiv e'_z$ points along the radius directed from the center of the Earth and $e'_2 \equiv e'_x$ points along the meridian toward the pole (and $e'_3 \equiv e'_y$ points to the west). The angle Ψ is the fixed latitude of the location and $\Phi = \omega \, t$, where ω is the (constant) angular velocity of the Earth's rotation. Then the components of the rotation vector (see Section 1.5) are

$$\omega'^x = \omega \cos \Psi, \quad \omega'^y = 0, \quad \omega'^z = \omega \sin \Psi. \tag{2.15}$$

• *The Kepler frame.* We consider a frame $\mathcal{S}'$ whose origin is attached to a moving point $P = P(t)$, and we decompose the radius vector $R \equiv OP(t)$ in the absolute frame $\mathcal{S}$ as $R = X^i e_i$ with

$$X = r \sin\theta \cos\phi \,, \qquad Y = r \sin\theta \sin\phi \,, \qquad Z = r \cos\theta, \tag{2.16}$$

where the angle $\phi(t) \in [0, 2\pi]$, the angle $\theta(t) \in [0, \pi]$, and $r(t) = \sqrt{(R.R)}$ is the distance from the origin of $\mathcal{S}'$ to the origin of $\mathcal{S}$. We choose e_1' to point along the radius vector R and e_2' and e_3' such that[3]

$$\begin{cases} e_1' = \cos\phi \sin\theta \, e_1 + \sin\phi \sin\theta \, e_2 + \cos\theta \, e_3 \\ e_2' = \cos\phi \cos\theta \, e_1 + \sin\phi \cos\theta \, e_2 - \sin\theta \, e_3 \\ e_3' = -\sin\phi \, e_1 + \cos\phi \, e_2 \,. \end{cases} \qquad (2.17)$$

The components of the rotation vector Ω in $\mathcal{S}'$ are, from (1.14),

$$\omega'^1 = \dot\phi \cos\theta \,, \qquad \omega'^2 = -\dot\phi \sin\theta \,, \qquad \omega'^3 = \dot\theta \,, \qquad (2.18)$$

so that

$$\dot{e}_1' = \dot\theta e_2' + \dot\phi \sin\theta e_3' \,, \quad \dot{e}_2' = -\dot\theta e_1' + \dot\phi \cos\theta e_3' \,, \quad \dot{e}_3' = -\dot\phi(\sin\theta e_1' + \cos\theta e_2') . \qquad (2.19)$$

The conjugate forms ϵ'^i (which permit the metric in $\mathcal{S}'$ to be written as $e = \delta_{ij}\,\epsilon'^i \otimes \epsilon'^j$) are given by similar expressions:

$$\begin{cases} \epsilon'^1 = \cos\phi \sin\theta \, \epsilon^1 + \sin\phi \sin\theta \, \epsilon^2 + \cos\theta \, \epsilon^3 \\ \epsilon'^2 = \cos\phi \cos\theta \, \epsilon^1 + \sin\phi \cos\theta \, \epsilon^2 - \sin\theta \, \epsilon^3 \\ \epsilon'^3 = -\sin\phi \, \epsilon^1 + \cos\phi \, \epsilon^2 \,. \end{cases} \qquad (2.20)$$

This frame will prove useful for solving the Kepler problem in Sections 12.1–12.4.

The Frenet trihedron

The Frenet trihedron (see Fig. 2.2) is a frame attached to a trajectory such that e_1' is tangent to the trajectory, i.e., $e_1' \propto \dot{d}$ where $d(t)$ is the position vector of the point in the absolute frame $\mathcal{S}$ and e_2' is proportional to $\dot{e}_1'$. It is a simple exercise to show that

$$e_1' = \frac{\dot{d}}{\sqrt{\dot{d}.\dot{d}}}, \qquad e_2' = \frac{(\dot{d}.\dot{d})\ddot{d} - (\dot{d}.\ddot{d})\dot{d}}{\sqrt{\dot{d}.\dot{d}}\sqrt{\dot{d} \wedge \ddot{d}}}, \qquad e_3' = \frac{\dot{d} \wedge \ddot{d}}{\sqrt{(\dot{d} \wedge \ddot{d})^2}} \,. \qquad (2.21)$$

If we introduce the *curvature* κ and the *torsion* τ of the trajectory as

$$\kappa \equiv \frac{\sqrt{(\dot{d} \wedge \ddot{d})^2}}{(\dot{d}.\dot{d})^{3/2}}, \qquad \tau \equiv \frac{\dot{d}.(\ddot{d} \wedge \dddot{d})}{(\dot{d} \wedge \ddot{d})^2} \,, \qquad (2.22)$$

we obtain the *Frenet–Serret equations* giving the time derivatives of the basis vectors as well as the rotation vector of the trihedron:

$$\dot{e}_1' = \kappa e_2', \quad \dot{e}_2' = -\kappa e_1' + \tau e_3', \quad \dot{e}_3' = -\tau e_2', \quad \text{and} \quad \Omega = \tau e_1' + \kappa e_3' . \qquad (2.23)$$

---

[3]The e_2' axis points toward the equator along the meridian, which amounts to choosing (see Section 1.3) the Euler nutation angle to be $\Theta = -\pi/2$. We have also set $\Phi(t) \equiv \phi(t)$ and $\Psi(t) \equiv \theta(t) - \pi/2$.

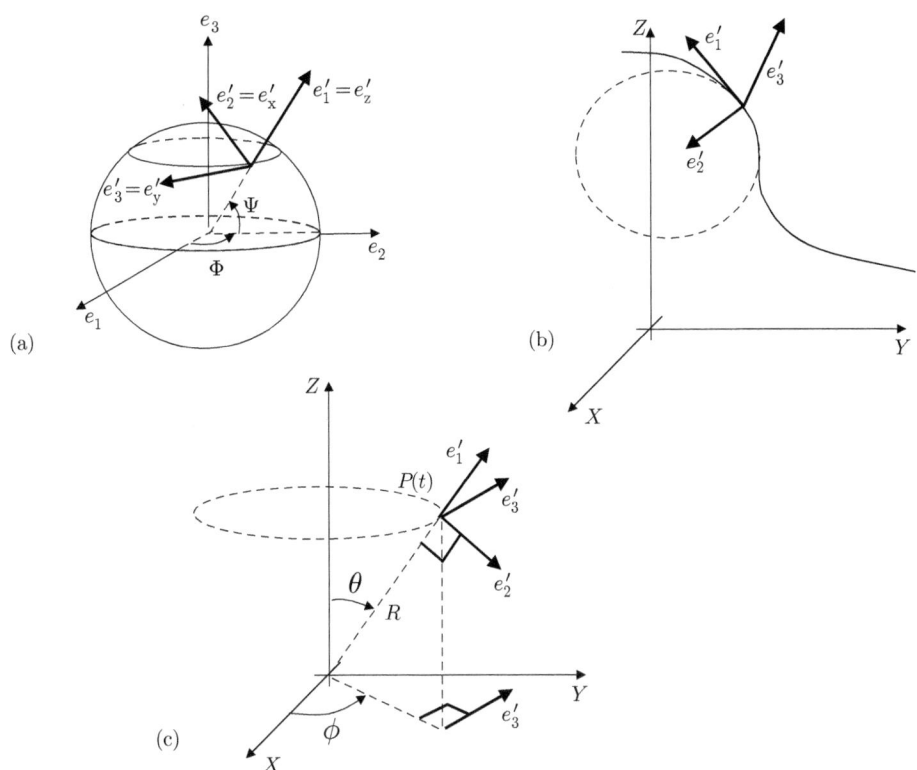

Fig. 2.2 The (a) laboratory, (b) Frenet, and (c) Kepler frames.

2.6   The ABCs of vector calculus

• *Algebraic operations.*   If A and B are two vectors, we can define the following.

The *vector product* $A \wedge B$ is a *form* with the components $(A \wedge B)_i = e_{ijk}A^j B^k$, where e_{ijk} is the Levi-Civita symbol defined in Section 1.5. The components can also be written as $(A^Y B^Z - A^Z B^Y, A^Z B^X - A^X B^Z, A^X B^Y - A^Y B^X)$. The associated vector $(A \wedge B)^i$ has the same components because the metric components are δ_{ij} (see the remark at the end of Section 2.2). It is orthogonal to A and B.

The scalar product of the two vectors is given by

$$(A.B) = \delta_{ij}A^i B^j = A^X B^X + A^Y B^Y + A^Z B^Z.$$

Using the properties of the Levi-Civita symbol given in Section 1.5, it is easily shown that

$$A.(B \wedge C) = C.(A \wedge B), \quad A \wedge (B \wedge C) = (A.C)B - (A.B)C,$$
$$(A \wedge B)^2 = A^2 B^2 - (A.B)^2. \tag{2.24}$$

• *Differential operators.*   Let a function of the point p be variously denoted as $f(p)$, $f(Op)$, $f(R)$, or $f(X^i)$. Its *exterior derivative*, denoted by df, is the form with components

$(\partial_X f, \partial_Y f, \partial_Z f)$, and its *gradient*, denoted by ∇f, is the associated vector (with the same components).[4]

The *curl* of a field of forms $v(R)$ with components $V_i(X, Y, Z)$ is the vector denoted by $\nabla \wedge v$ with components

$$(\nabla \wedge v)^i = e^{ijk} \, \partial V_j / \partial X^k = (\partial_Y V_Z - \partial_Z V_Y, \; -\partial_Z V_X + \partial_X V_Z, \; \partial_X V_Y - \partial_Y V_X).$$

(Since forms and vectors have the same components, the indices can be raised and lowered freely.)

The *divergence* of a vector field $v(R)$ with components $V^i(X^j)$ is the scalar

$$\nabla.v \equiv \partial V^i / \partial X^i = (\partial_X V^X + \partial_Y V^Y + \partial_Z V^Z).$$

The *Laplacian* of a function (or 'scalar field') $f(R)$ is defined as

$$\triangle f \equiv \nabla.\nabla f = \delta^{ij} \frac{\partial^2 f}{\partial X^i \partial X^j} = \partial^2_{XX} f + \partial^2_{YY} f + \partial^2_{ZZ} f.$$

Using the definitions, it is easily shown that

$$
\begin{aligned}
&\nabla.(\nabla \wedge V) \equiv 0, \quad \nabla \wedge (\nabla f) \equiv 0, \\
&\nabla \wedge \nabla \wedge V = -\triangle V + \nabla \nabla.V, \quad \nabla.(V \wedge W) = -V.\nabla \wedge W + W.\nabla \wedge V, \\
&\nabla \wedge (V \wedge W) = (W.\nabla)V - (V.\nabla)W - W(\nabla.V) + V(\nabla.W), \\
&\nabla(V.W) = V \wedge (\nabla \wedge W) + (V.\nabla)W + W \wedge (\nabla \wedge V) + (W.\nabla)V.
\end{aligned}
\tag{2.25}
$$

- *Integration.* Let a curve L be defined parametrically by $X^i = X^i(\lambda)$. The *line integral* of a function $f(X^i)$ along L is $\int_{\lambda_1}^{\lambda_2} f(X^i(\lambda))d\lambda$. Since the Euclidean length element is $dl^2 = \delta_{ij}dX^i dX^j$, the length of the curve is $\int_L dl \equiv \int_{\lambda_1}^{\lambda_2} \sqrt{\delta_{ij}(dX^i/d\lambda)(dX^j/d\lambda)}\, d\lambda$. We note that it is invariant under reparametrization.

We introduce the vector e tangent to the curve with components $dX^i/d\lambda$ and the infinitesimal vector $\vec{dl} \equiv e d\lambda$ (whose norm is dl). The *line integral* of a vector $\vec{v}$ with components $V^i(X^k)$ along L is $\int_L \vec{v}.\vec{dl} \equiv \int \delta_{ij} V^i(X^k(\lambda))(dX^j/d\lambda)d\lambda$.

Let us now consider surface integrals.

The surface area of an infinitesimal rectangle in the XOY plane is $dS = dX\, dY$. Since it is orthogonal to the axis OZ, it is natural to introduce the *surface element* as the form (or vector) $\vec{dS} = (dX\,\vec{i} \wedge dY\,\vec{j}) = (0, 0, dX\, dY)$. The quantity dS then is the modulus of $\vec{dS}$, that is, the scalar product of $\vec{dS}$ and the unit vector orthogonal to the surface element, *i.e.*, $\vec{k} = \vec{i} \wedge \vec{j}$. The equality is easily proved using (2.24).

The generalization is straightforward. Let a surface Σ be defined parametrically as $X^i = X^i(\lambda, \mu)$. We use e_1 and e_2 to denote the two vectors tangent to Σ and having components $\partial X^i / \partial \lambda$ and $\partial X^i / \partial \mu$, while $\vec{dl}_1 \equiv e_1 d\lambda$ and $\vec{dl}_2 \equiv e_2 d\mu$ are the corresponding infinitesimal vectors. The surface element vector is then defined as $\vec{dS} = \vec{dl}_1 \wedge \vec{dl}_2$ and its components are $dS_i = e_{ijk}(\partial X^j / \partial \lambda)(\partial X^k / \partial \mu)d\lambda\, d\mu$.

---

[4] We shall use various notations for the partial derivative: $\frac{\partial f(X,Y,Z)}{\partial X}\Big|_{Y,Z} \equiv \frac{\partial f}{\partial X} \equiv \partial_X f$, and sometimes will denote it simply by a comma: $\partial_X f \equiv f_{,X}$.

The *flux* through Σ of a field of vectors $\vec{v}$ is $\int_\Sigma (\vec{v}.\vec{dS})$.

Let $n = (e_1 \wedge e_2)/|e_1 \wedge e_2|$ be the unit vector normal to Σ, *i.e.*, the vector satisfying $n_i\, \partial X^i/\partial \lambda = n_i\, \partial X^i/\partial \mu = 0$ and $n_i n^i = 1$. The area of the surface Σ is then given by

$$\int_\Sigma dS \equiv \int n.(dl_1 \wedge dl_2) = \int_{\lambda_1}^{\lambda_2} \int_{\mu_1}^{\mu_2} n^i dS_i = \int_{\lambda_1}^{\lambda_2} \int_{\mu_1}^{\mu_2} e_{ijk} n^i \frac{\partial X^j}{\partial \lambda} \frac{\partial X^k}{\partial \mu} d\lambda\, d\mu\,. \qquad (2.26)$$

Finally, let us consider a closed surface ∂V. The volume enclosed by this surface is $\int dX\, dY\, dZ \equiv \int d^3 X$, where the ranges of variation of X, Y, Z are determined by the equation of the surface $\Phi(X, Y, Z) = 0$.

The Stokes and Gauss gradient theorems

The gradient theorem. The integral of the gradient of a function along a path is independent of the path:

$$\int_a^b \vec{\nabla} f.\vec{dl} = f(b) - f(a)\,. \qquad (2.27)$$

The Stokes–Green–Ostrogradsky theorem (actually discovered by Ampère in 1825). If v is a vector, Σ a two-dimensional surface, and $\partial \Sigma$ the boundary of this surface, namely, a closed path, then

$$\int_\Sigma (\nabla \wedge v).\vec{dS} = \int_{\partial \Sigma} \vec{v}.\vec{dl}\,. \qquad (2.28)$$

This is easily demonstrated when Σ is a rectangle in the XOY plane.

The Gauss or divergence theorem (actually discovered by Lagrange in 1764). If v is a vector, V is a three-dimensional volume, and ∂V is its boundary, then

$$\int_V \nabla.v\, d^3 X = \int_{\partial V} \vec{v}.\vec{dS}\,. \qquad (2.29)$$

This is easily demonstrated when V is a parallelepiped.

Choosing $\vec{v} = (X/3, Y/3, Z/3)$ so that $\vec{\nabla}.\vec{v} = 1$, Gauss's theorem can be used, for example, to easily calculate the volume of V when the equation for its boundary ∂V is given in parametric form.

3

Curvilinear coordinates

In this chapter we give a presentation à la *Descartes* of curvilinear coordinates which parallels our discussion of Cartesian coordinates in Chapter 1, and we introduce the concept of the covariant derivative.

3.1 Curvilinear coordinates and tensors

Once we have defined a Cartesian frame and its coordinates X^i, we may wish to label the points of Euclidean space $\mathcal{E}_3$ using a new system $\mathcal{C}$ of *curvilinear* coordinates $x^i = x^i(X^j)$ (also sometimes referred to as *Gaussian* coordinates), which are nonlinearly related to Cartesian coordinates (for example, spherical coordinates). The partial derivatives $\partial x^i/\partial X^j$ and, inversely, $\partial X^j/\partial x^k$, form the Jacobian transformation matrices and are related as $(\partial x^i/\partial X^j)(\partial X^j/\partial x^k) = \delta^i_k$ [and inversely as $(\partial X^i/\partial x^j)(\partial x^j/\partial X^k) = \delta^i_k$]. If the transformation is linear, they reduce to numbers: $\partial x^i/\partial X^j = \mathcal{R}_j{}^i$ and $\partial X^j/\partial x^k = \mathcal{R}^j{}_k$ (the matrix $\mathcal{R}_j{}^i$ is not necessarily a rotation matrix).

The length element dl^2 between two neighboring points (whose value is postulated to be invariant) is no longer manifestly given by the Pythagorean theorem because use of the chain rule for the derivative of a composite function gives

$$dl^2 \equiv \delta_{ij}\, dX^i dX^j = \delta_{ij} \frac{\partial X^i}{\partial x^k} \frac{\partial X^j}{\partial x^l}\, dx^k\, dx^l \equiv e_{kl}\, dx^k\, dx^l\,, \qquad (3.1)$$

where the components e_{kl} of the metric e and its inverse e^{km} (defined as $e_{kl}e^{km} = \delta^m_l$) are now given not by the Kronecker delta, but by the following functions of the coordinates:

$$e_{kl} = \frac{\partial X^i}{\partial x^k} \frac{\partial X^j}{\partial x^l} \delta_{ij}\,, \qquad e^{km} = \frac{\partial x^l}{\partial X^p} \frac{\partial x^m}{\partial X^q} \delta^{pq}\,. \qquad (3.2)$$

The metric in polar coordinates

When passing from the Cartesian coordinates of the plane $X^i = (X, Y)$ to the polar coordinates $x^i = (r, \phi)$ defined as

$$X = r\cos\phi,\quad Y = r\sin\phi, \qquad (3.3)$$

we have

$$e_{rr} = \left(\frac{\partial X}{\partial r}\right)^2 + \left(\frac{\partial Y}{\partial r}\right)^2 = 1,\quad e_{r\phi} = \frac{\partial X}{\partial r}\frac{\partial X}{\partial \phi} + \frac{\partial Y}{\partial r}\frac{\partial Y}{\partial \phi} = 0,$$

$$e_{\phi\phi} = \left(\frac{\partial X}{\partial \phi}\right)^2 + \left(\frac{\partial Y}{\partial \phi}\right)^2 = r^2, \qquad (3.4)$$

Relativity in Modern Physics. Nathalie Deruelle and Jean-Philippe Uzan.
© Oxford University Press 2018. Published in 2018 by Oxford University Press.
DOI: 10.1093/oso/9780198786399.001.0001

which leads to the familiar expression

$$dl^2 = dX^2 + dY^2 = dr^2 + r^2\,d\phi^2. \tag{3.5}$$

In Section 1.6 we defined vector fields as sets of three functions $T^i(X^j)$ transforming as $T^i \mapsto T'^i = \mathcal{R}_j{}^i T^j$ under a change of Cartesian frame. Since the generalization of the rotation matrix $\mathcal{R}_j{}^i$ is the Jacobian matrix $\partial x^i/\partial X^j$, in this broader context we shall define a vector field as three functions which transform under a change of coordinates $X^i \mapsto x^i$ as $T^i \mapsto t^i$ with

$$t^i = \frac{\partial x^i}{\partial X^j} T^j \quad \Longleftrightarrow \quad T^i = \frac{\partial X^i}{\partial x^j} t^j, \tag{3.6}$$

where T^j and $\partial x^i/\partial X^j$ are evaluated at $X^k = X^k(x^l)$.

The polar components of a vector field

In going from Cartesian coordinates (X, Y) to polar coordinates (r, ϕ): $X = r\cos\phi$, $Y = r\sin\phi$ and, inversely, $r = \sqrt{X^2 + Y^2}$, $\tan\phi = Y/X$, we have $t^r = (\partial r/\partial X)T^X + (\partial r/\partial Y)T^Y$ and $t^\phi = (\partial\phi/\partial X)T^X + (\partial\phi/\partial Y)T^Y$, that is,

$$t^r = T^X \cos\phi + T^Y \sin\phi, \quad r\,t^\phi = -T^X \sin\phi + T^Y \cos\phi, \tag{3.7}$$

where T^X and T^Y are evaluated at $X = r\cos\phi$, $Y = r\sin\phi$.

If $l_a^i(X^j) \equiv X^i - X_a^i$ is the vector field defining the separation between some point P and a given point P_a with Cartesian components $l_a^X = X - X_a = r\cos\phi - r_a\cos\phi_a$ and $l_a^Y = r\sin\phi - r_a\sin\phi_a$, its polar components given by (3.7) and, importantly, evaluated at P and *not* at P_a, are written as

$$l_a^r = r - r_a\cos(\phi - \phi_a), \quad l_a^\phi = \frac{r_a}{r}\sin(\phi - \phi_a). \tag{3.8}$$

Therefore, the components of the radius vector and the coordinates of a point can be identified only in a Cartesian system.

More generally, by extension of (2.4), a field of p-fold contravariant and q-fold covariant tensors is defined as a set of functions for which the transformation law under a change of coordinate system is

$$t^{i_1\ldots i_p}_{j_1\ldots j_q} = \frac{\partial x^{i_1}}{\partial X^{k_1}} \cdots \frac{\partial x^{i_p}}{\partial X^{k_p}} \frac{\partial X^{l_1}}{\partial x^{j_1}} \cdots \frac{\partial X^{l_q}}{\partial x^{j_q}} T^{k_1\ldots k_p}_{l_1\ldots l_q}. \tag{3.9}$$

Therefore, the metric and its inverse, whose components transform as in (3.2), are respectively tensors which are 2-fold covariant and 2-fold contravariant.

It is useful to recall that if a tensor is zero in one coordinate system, it is zero in any other.

The scalar product of two vectors is invariant under a change of coordinate system. Indeed, we have

$$(T.U) \equiv \delta_{ij} T^j U^i = e_{ij} t^j u^i = t_i u^i \quad \text{with} \quad t_i \equiv e_{ij} t^j. \tag{3.10}$$

Here the t_i are the components of the form derived from the vector t^i by *lowering an index* and $t_i u^i$ can be viewed as the contraction or trace of a singly contravariant and a singly covariant tensor $t_i u^j$. We note that the functions t_i are *no longer* numerically equal to the t^i: the location of the indices, which in Cartesian coordinates is a formality because $e_{ij} = \delta_{ij}$, here becomes crucial. For example, the components of the form associated with the vector (3.7) of polar components (t^r, t^ϕ) are $t_r = t^r$ and $t_\phi = r^2 t^\phi$, since the metric is $dl^2 = dr^2 + r^2 d\phi^2$.

3.2 The covariant derivative

Under a change of Cartesian frame the components of a vector field $T^i(X^k)$ transform, by definition, as $T^i \mapsto T'^i = \mathcal{R}_j{}^i T^j$. Since the matrix $\mathcal{R}_j{}^i$ and its inverse are constants, the derivatives $\partial T^i / \partial X^k$ transform like those of a Cartesian tensor which is singly contravariant and singly covariant because $\partial T'^i / \partial X'^k = \mathcal{R}_j{}^i \mathcal{R}^l{}_k \left(\partial T^j / \partial X^l \right)$. This is no longer true for a nonlinear coordinate transformation where $T^i \mapsto t^i = \left(\partial x^i / \partial X^j \right) T^j$, because now according to the Leibniz rule and the chain rule for differentiation we have

$$
\begin{aligned}
\frac{\partial t^i}{\partial x^k} &= \frac{\partial}{\partial x^k} \left(\frac{\partial x^i}{\partial X^j} T^j \right) = \frac{\partial x^i}{\partial X^j} \frac{\partial T^j}{\partial x^k} + \frac{\partial^2 x^i}{\partial x^k \partial X^j} T^j \\
&= \frac{\partial x^i}{\partial X^j} \frac{\partial X^l}{\partial x^k} \frac{\partial T^j}{\partial X^l} + \frac{\partial^2 x^i}{\partial X^l \partial X^j} \frac{\partial X^l}{\partial x^k} T^j,
\end{aligned}
\tag{3.11}
$$

where all quantities are evaluated at $X^m(x^p)$. The first term on the right-hand side represents, by definition, the components in the system x^i of the singly contravariant, singly covariant tensor of components $\partial T^i / \partial X^k$ in $\mathcal{S}$; we shall denote them as $\tilde{D}_k t^i$:

$$
\tilde{D}_k t^i \equiv \frac{\partial x^i}{\partial X^j} \frac{\partial X^l}{\partial x^k} \frac{\partial T^j}{\partial X^l}.
\tag{3.12}
$$

These are the components of the *covariant derivative* with respect to x^k of the field of vectors with components t^i. They can be expressed as functions of the ordinary derivatives of the t^i using (3.11), that is,

$$
\tilde{D}_k t^i = \partial_k t^i - \frac{\partial^2 x^i}{\partial X^l \partial X^j} \frac{\partial X^l}{\partial x^k} T^j \quad \text{(where we have used the notation} \quad \partial_j \equiv \frac{\partial}{\partial x^j} \text{)}
\tag{3.13}
$$

or

$$
\tilde{D}_k t^i = \partial_k t^i + \tilde{\Gamma}^i_{km} t^m \qquad \text{with} \qquad \tilde{\Gamma}^i_{km} \equiv \frac{\partial x^l}{\partial X^j} \frac{\partial^2 X^j}{\partial x^k \partial x^m},
\tag{3.14}
$$

where we have transformed the last term in (3.13) using the fact that

$$
\frac{\partial X^j}{\partial x^m} \frac{\partial x^i}{\partial X^j} = \delta^i_m \implies \frac{\partial X^j}{\partial x^m} \frac{\partial^2 x^i}{\partial X^l \partial X^j} + \frac{\partial^2 X^j}{\partial x^m \partial X^l} \frac{\partial x^i}{\partial X^j} = 0.
\tag{3.15}
$$

The functions $\tilde{\Gamma}^i_{km}$ defined in (3.14) are the *connection coefficients*. They are symmetric in their lower indices and there are $n^2(n+1)/2$ of them that is, 18 in three dimensions.[1]

---

[1] Some remarks about notation:
In a more precise fashion (but more pedantic, when there is no possibility of confusion about the coordinate with respect to which the derivative is being taken), the covariant derivative can be written as $\tilde{D}_i = \frac{\tilde{D}}{\partial x^i}$. We shall also sometimes use the semicolon notation $\tilde{D}_i v^j \equiv v^j_{;i}$.

The connection coefficients in polar coordinates

Let us consider the transformation from Cartesian to polar coordinates $X = r\cos\phi$, $Y = r\sin\phi$ and, inversely, $r = \sqrt{X^2 + Y^2}$, $\tan\phi = Y/X$. The connection coefficients defined in (3.14) all vanish except for

$$\tilde{\Gamma}^r_{\phi\phi} = \frac{\partial r}{\partial X}\frac{\partial^2 X}{\partial\phi^2} + \frac{\partial r}{\partial Y}\frac{\partial^2 Y}{\partial\phi^2} = -r \quad \text{and} \quad \tilde{\Gamma}^\phi_{r\phi} = \frac{\partial\phi}{\partial X}\frac{\partial^2 X}{\partial r\partial\phi} + \frac{\partial\phi}{\partial Y}\frac{\partial^2 Y}{\partial r\partial\phi} = \frac{1}{r}. \tag{3.16}$$

Having defined the covariant derivative of a vector, we can now define that of a form of components λ_i in $\mathcal{C}$. Indeed, if the w^i are the components of a vector, then, contracting the indices, $\lambda_i w^i$ is just a function, and if we write $\tilde{D}_j(\lambda_i w^i) = \partial_j(\lambda_i w^i)$ and use the Leibniz rule ($\tilde{D}$ being a derivative) we find

$$\tilde{D}_j\lambda_i = \partial_j\lambda_i - \tilde{\Gamma}^k_{ji}\lambda_k \tag{3.17}$$

and, more generally,

$$\tilde{D}_j t^i_{lm} = \partial_j t^i_{lm} + \tilde{\Gamma}^i_{jk} t^k_{lm} - \tilde{\Gamma}^k_{jl} t^i_{km} - \tilde{\Gamma}^k_{jm} t^i_{lk}. \tag{3.18}$$

We note the important fact that the covariant derivatives $\tilde{D}$ commute: $\tilde{D}_k\tilde{D}_j t^i_{lm} = \tilde{D}_j\tilde{D}_k t^i_{lm}$, because in a Cartesian coordinate system we have[2] $\partial_{kj}T^i_{lm} = \partial_{jk}T^i_{lm}$.

Transformation of the connection coefficients

The components t^i of a vector are expressed as a function of its components t'^i in a different system of curvilinear coordinates x'^i as $t^i = (\partial x^i/\partial x'^l)\, t'^l$. Therefore, the components of its covariant derivative with respect to x^j become

$$\tilde{D}_j t^i \equiv \frac{\partial t^i}{\partial x^j} + \tilde{\Gamma}^i_{jk} t^k = \frac{\partial}{\partial x^j}\left(\frac{\partial x^i}{\partial x'^l}t'^l\right) + \tilde{\Gamma}^i_{jk}\frac{\partial x^k}{\partial x'^l}t'^l$$
$$= \frac{\partial x^i}{\partial x'^l}\frac{\partial t'^l}{\partial x^j} + \frac{\partial^2 x^i}{\partial x^j\partial x'^m}t'^m + \tilde{\Gamma}^i_{jk}\frac{\partial x^k}{\partial x'^m}t'^m. \tag{3.19}$$

The tilde is used to remind us that there exist coordinate systems (the Cartesian coordinates X^i) where the covariant derivative and the ordinary derivative are the same. Thus the 18 connection coefficients can all be expressed in terms of only *three* functions, namely, the functions defining the transformation from Cartesian to curvilinear coordinates $X^i = X^i(x^j)$.

A space in which global systems of Cartesian coordinates do not exist is *curved*. In this case the connection coefficients can be arbitrary functions of the coordinates, and we will use the notation (D_k, Γ^i_{jk}) instead of $(\tilde{D}_k, \tilde{\Gamma}^i_{jk})$.

[2]This will no longer be true in *curved* space where we can no longer define global systems of Cartesian coordinates.

It is an instructive exercise to calculate $D_i D_j t^k - D_j D_i t^k$ as a function of the Γ^i_{jk} and their derivatives. We find $D_i D_j t^k - D_j D_i t^k = R^k_{mij}t^m$, where $R^i_{jkl} \equiv \partial_k\Gamma^i_{lj} - \partial_l\Gamma^i_{kj} + \Gamma^i_{km}\Gamma^m_{lj} - \Gamma^i_{lm}\Gamma^m_{kj}$. The quantity R^i_{jkl} is the *Riemann–Christoffel curvature tensor*. It vanishes identically when the quantities involved carry tildes, that is, in 'flat' Euclidean space (see Book 3, Section 2.2).

Since the $\tilde{D}_j t^i$ are the components of a singly contravariant, singly covariant tensor, using $\tilde{D}_j t^i = (\partial x'^k / \partial x^j)(\partial x^i / \partial x'^l)(\tilde{D}_k t^l)'$ they can be expressed as a function of its components $(\tilde{D}_k t^l)'$ in $\mathcal{C}'$ as $(\tilde{D}_k t^l)' = \partial t'^l / \partial x'^k + \tilde{\Gamma}'^l_{km} t'^m$, where the $\tilde{\Gamma}'^l_{km}$ are the connection coefficients in $\mathcal{C}'$. Therefore,

$$\tilde{D}_j t^i = \frac{\partial x'^k}{\partial x^j} \frac{\partial x^i}{\partial x'^l} \left(\frac{\partial t'^l}{\partial x'^k} + \tilde{\Gamma}'^l_{km} t'^m \right) = \frac{\partial x^i}{\partial x'^l} \frac{\partial t'^l}{\partial x^j} + \frac{\partial x'^k}{\partial x^j} \frac{\partial x^i}{\partial x'^l} \tilde{\Gamma}'^l_{km} t'^m . \tag{3.20}$$

Equating (3.19) and (3.20) and then multiplying by $(\partial x^j / \partial x'^p)(\partial x'^q / \partial x^i)$, we find

$$\tilde{\Gamma}'^q_{pm} = \frac{\partial x^k}{\partial x'^m} \frac{\partial x^j}{\partial x'^p} \frac{\partial x'^q}{\partial x^i} \tilde{\Gamma}^i_{jk} + \frac{\partial x'^q}{\partial x^i} \frac{\partial^2 x^i}{\partial x'^p \partial x'^m}, \tag{3.21}$$

which gives the transformation law of the connection coefficients in going from one curvilinear coordinate system to another. We see that the connection coefficients are not the components of a tensor because they do not transform like tensor components. This is fortunate because, since they all vanish in a system of Cartesian coordinates, they would then all vanish in any system of curvilinear coordinates!

3.3   Parallel transport

As we have seen in Section 2.2, the parallel transport of a vector v from one point to another is a trivial operation in Cartesian coordinates. If $V^i(X_1^j)$ are the components of the vector at point p_1 with coordinates X_1^i, the components of the vector transported to p_2 with coordinates X_2^i will also be $V^i(X_1^j)$: $V^i_{\text{par}}(X_2^j) = V^i(X_1^j)$.

When performing this operation in a system of curvilinear coordinates, it is first necessary to choose the path between the points p_1 and p_2 (with coordinates x_1^i and x_2^i), that is, a trajectory given by $x^i = x^i(\lambda)$, where λ is a parameter.

The *directional* derivative of the field v along the curve, that is, the tangent vector $dx^i/d\lambda \equiv t^i(\lambda)$, is

$$\frac{\tilde{D}v^i}{d\lambda} \equiv \frac{dx^j}{d\lambda} \frac{\tilde{D}v^i}{\partial x^j} = t^j \tilde{D}_j v^i, \tag{3.22}$$

which generalizes the definition in Cartesian coordinates. [It is understood that $\tilde{D}_j v^i$ is evaluated at $x^i = x^i(\lambda)$.]

The parallel transport equation will then be

$$t^j \tilde{D}_j v^i_{\text{par}} = 0 \quad \Longleftrightarrow \quad \frac{dv^i_{\text{par}}}{d\lambda} + \tilde{\Gamma}^i_{jk} t^j v^k_{\text{par}} = 0 \tag{3.23}$$

with the initial condition $v^i_{\text{par}}(p_1) = v^i(p_1)$. This is an ordinary differential equation of first order in λ whose integration gives $v^i_{\text{par}}(\lambda)$. In Cartesian coordinates $x^i \equiv X^i$ where all the connection coefficients vanish, we have $v^i_{\text{par}} \equiv V^i_{\text{par}} = V^i(\lambda_1)$, where the $X_1^i = X^i(\lambda_1)$ are the coordinates of p_1, and we find that the components of the field v parallel-transported to p_2 are the same as those at p_1.

Since in the case of Cartesian coordinates the parallel transport is independent of the path from p_1 to p_2, the same must be true when using curvilinear coordinates. This is demonstrated below using a concrete example.

A constant vector field in polar coordinates

In Cartesian coordinates X^j the equation for parallel transport reduces to $\partial V^i_{\mathrm{par}}/\partial X^j = 0$, which implies that $V^i_{\mathrm{par}} = V^i(p_1)$ is a constant field (see Fig. 3.1). In a system of polar coordinates, the vector field v^i_{par} will be constant if $\tilde{D}_j v^i_{\mathrm{par}} = 0$, that is, if, using the expressions for the connection coefficients given in (3.16),

$$\partial_r v^r_{\mathrm{par}} = 0, \quad \partial_r v^\phi_{\mathrm{par}} + v^\phi_{\mathrm{par}}/r = 0, \quad \partial_\phi v^r_{\mathrm{par}} - r v^\phi_{\mathrm{par}} = 0, \quad \partial_\phi v^\phi_{\mathrm{par}} + v^r_{\mathrm{par}}/r = 0. \tag{3.24}$$

This system of differential equations is *integrable* (as can be checked by verifying that the second derivatives commute: $\partial_{r\phi} v^r_{\mathrm{par}} = \partial_{\phi r} v^r_{\mathrm{par}}$, $\partial_{r\phi} v^\phi_{\mathrm{par}} = \partial_{\phi r} v^\phi_{\mathrm{par}}$) and its solution is

$$v^r_{\mathrm{par}} = a\cos(\phi + \omega), \quad v^\phi_{\mathrm{par}} = -\frac{a}{r}\sin(\phi + \omega), \tag{3.25}$$

where a and ω are two integration constants. One can check that in the Cartesian system (X, Y) the components of this field, given by $V^i_{\mathrm{par}} = (\partial X^i/\partial x^j) v^j_{\mathrm{par}}$, are indeed constant since

$$V^X_{\mathrm{par}} = a\cos\omega, \qquad V^Y_{\mathrm{par}} = -a\sin\omega. \tag{3.26}$$

The fact that the system (3.25) is integrable is crucial for obtaining the result. In *curved* space, where the Christoffel symbols can be any functions of the coordinates, the system is not integrable. The idea of a 'constant field' is no longer meaningful: parallel transport *must* be defined by (3.23) and the result of the parallel transport from p_1 to p_2 depends on the path.

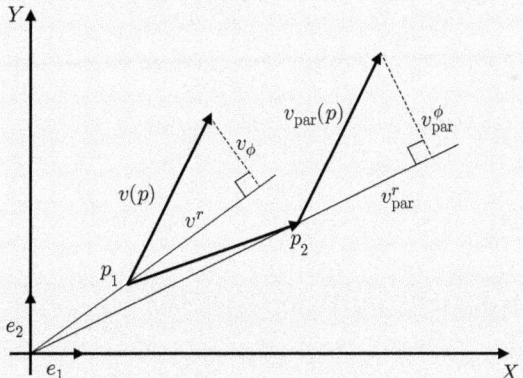

Fig. 3.1 Parallel transport in polar coordinates.

3.4 The covariant derivative and the metric tensor

The connection coefficients of the covariant derivative have been introduced independently of the Euclidean metric, which has components e_{ij} in $\mathcal{C}$. However, the two objects can be related to each other. This is easily seen by differentiating the metric components in (3.2) and forming the appropriate sums and differences to obtain the connection coefficient given in (3.14):

$$\tilde{\Gamma}^i_{jk} = \frac{1}{2}e^{il}\left(\frac{\partial e_{kl}}{\partial x^j} + \frac{\partial e_{jl}}{\partial x^k} - \frac{\partial e_{jk}}{\partial x^l}\right). \tag{3.27}$$

The connection coefficients $\tilde{\Gamma}^i_{jk}$ written in this form are called *Christoffel symbols* and the covariant derivative itself is called the *Levi-Civita covariant derivative*.

The Christoffel symbols in spherical coordinates

In polar coordinates the length element is written as $dl^2 = dr^2 + r^2 d\phi^2$, and so $e_{rr} = 1$, $e_{\phi\phi} = r^2$, $e_{r\phi} = 0$ and, inversely, $e^{rr} = 1$, $e^{\phi\phi} = 1/r^2$, $e^{r\phi} = 0$. From (3.27) it is simple to obtain the connection coefficients $\tilde{\Gamma}^i_{jk}$ already found in (3.16):

$$\tilde{\Gamma}^r_{\phi\phi} = \frac{1}{2}e^{ri}(\partial_\phi e_{\phi i} + \partial_\phi e_{i\phi} - \partial_i e_{\phi\phi}) = -\frac{1}{2}e^{rr}\partial_r e_{\phi\phi} = -r, \quad \tilde{\Gamma}^\phi_{r\phi} = \frac{1}{2}e^{\phi\phi}\partial_r e_{\phi\phi} = \frac{1}{r}, \tag{3.28}$$

all the others being equal to zero.

The length element in spherical coordinates is $dl^2 = dr^2 + r^2(d\theta^2 + \sin^2\theta d\phi^2)$ and from (3.27) we find the Christoffel symbols more readily than from (3.14):

$$\tilde{\Gamma}^\theta_{\theta r} = \tilde{\Gamma}^\phi_{\phi r} = \frac{1}{r}, \quad \tilde{\Gamma}^r_{\theta\theta} = -r, \quad \tilde{\Gamma}^r_{\phi\phi} = -r\sin^2\theta,$$
$$\tilde{\Gamma}^\theta_{\phi\phi} = -\sin\theta\cos\theta, \quad \tilde{\Gamma}^\phi_{\phi\theta} = \frac{\cos\theta}{\sin\theta}. \tag{3.29}$$

It can also be verified explicitly using the definition (3.18) of the covariant derivative of a tensor and (3.27) for the $\tilde{\Gamma}^i_{jk}$ that

$$\tilde{D}_j e_{ik} \equiv 0. \tag{3.30}$$

This result can actually be obtained without calculation by noting that in a Cartesian coordinate system where the covariant and ordinary derivatives are the same, the metric components are constants. Therefore, their (covariant) derivatives are zero, and if they are zero in this system, they are zero in any other.

3.5   Kinematics of a point particle (III)

If the trajectory of a point particle is specified in $\mathcal{S}$ by $X^i = X^i(t)$ where the parameter t is the absolute time, then it is given in $\mathcal{C}$ by $x^i = x^i(t) \equiv x^i(X^j(t))$ so that

$$\begin{cases} v^i \equiv \dfrac{dx^i}{dt} = \dfrac{\partial x^i}{\partial X^j}V^j(t) \\[2mm] \dfrac{dv^i}{dt} = \dfrac{\partial x^i}{\partial X^j}a^j(t) + \dfrac{\partial^2 x^i}{\partial X^j\partial X^k}V^j(t)V^k(t), \end{cases} \tag{3.31}$$

where $V^j(t) \equiv \dot{X}^j$ and $a^j(t) \equiv \ddot{X}^j$ are the components of its velocity and acceleration in $\mathcal{S}$ and where the $\partial x^i/\partial X^j$, etc. are evaluated at $X^k = X^k(t)$.

The three functions $(\partial x^i/\partial X^j)\,a^j(t)$ are, by definition, the components of the acceleration in $\mathcal{C}$. We see from (3.31) that they are equal not to the ordinary time derivatives of the v^i, but to their covariant derivatives: $\tilde{D}v^i/dt \equiv (\partial x^i/\partial X^j)\, a^j(t)$. Using (3.14) and (3.27), we can rewrite this as

$$\frac{\tilde{D}v^i}{dt} \equiv \frac{dv^i}{dt} + \tilde{\Gamma}^i_{jk}v^j v^k$$

$$\text{with} \quad \tilde{\Gamma}^i_{jk} \equiv \frac{\partial x^i}{\partial X^l}\frac{\partial^2 X^l}{\partial x^j \partial x^k} = \frac{1}{2}e^{il}\left(\frac{\partial e_{kl}}{\partial x^j} + \frac{\partial e_{jl}}{\partial x^k} - \frac{\partial e_{jk}}{\partial x^l}\right),$$

(3.32)

where it is understood that all quantities are evaluated at $x^i = x^i(t)$.

If the velocity v^i is viewed as a vector field $v^i(x^j)$ evaluated on the trajectory $x^i = x^i(t)$, we can rewrite the covariant derivative of the velocity as $\tilde{D}v^i/dt = v^j\tilde{D}_j v^i$ and treat it as a directional derivative, the contraction of the vector v^j and the singly contravariant, singly covariant tensor $\tilde{D}_j v^i$, as we have seen in Section 3.3.

The equation

$$\frac{\tilde{D}v^i}{dt} \equiv v^j \tilde{D}_j v^i = 0$$

(3.33)

is the equation describing the parallel transport of the vector field v^i along a curve generated by the field itself, called the *auto-parallel*. It is a straight line in $\mathcal{E}_3$.

Trajectories in polar coordinates and the auto-parallel

Let us consider the transformation from Cartesian coordinates (X,Y) to polar coordinates $X = r\cos\phi$, $Y = r\sin\phi$. Since a trajectory is defined by $X = X(t)$ and $Y = Y(t)$, (3.31) gives

$$\dot{r} = \frac{X\dot{X} + Y\dot{Y}}{\sqrt{X^2 + Y^2}}, \quad \dot{\phi} = \frac{-Y\dot{X} + X\dot{Y}}{X^2 + Y^2}$$

$$r\ddot{r} = \frac{X\ddot{X} + Y\ddot{Y}}{\sqrt{X^2 + Y^2}} + \frac{(Y\dot{X} - X\dot{Y})^2}{(X^2 + Y^2)^{3/2}},$$

$$\ddot{\phi} = \frac{-Y\ddot{X} + X\ddot{Y}}{X^2 + Y^2} - \frac{2(X\dot{X} + Y\dot{Y})(-Y\dot{X} + X\dot{Y})}{(X^2 + Y^2)^2},$$

from which we obtain

$$\ddot{r} - r\dot{\phi}^2 = \frac{X\ddot{X} + Y\ddot{Y}}{\sqrt{X^2 + Y^2}}, \quad \ddot{\phi} + 2\frac{\dot{r}}{r}\dot{\phi} = \frac{-Y\ddot{X} + X\ddot{Y}}{X^2 + Y^2}.$$

If the acceleration $(\ddot{X}, \ddot{Y})$ is zero, the trajectory is rectilinear and uniform and given by the equations $X = X_0 + V_0^X t$, $Y = Y_0 + V_0^Y t$. In polar coordinates this trajectory is written as $r = \sqrt{(X_0 + V_0^X t)^2 + (Y_0 + V_0^Y t)^2}$, $\tan\phi = (Y_0 + V_0^Y t)/(X_0 + V_0^X t)$.

It can be obtained directly using (3.32). Since the trajectory is defined by $r = r(t)$ and $\phi = \phi(t)$, the velocity components are $v^i = (\dot{r}, \dot{\phi})$. The connection coefficients $\tilde{\Gamma}^i_{jk}$ all vanish with the exception of $\tilde{\Gamma}^r_{\phi\phi} = -r$ and $\tilde{\Gamma}^\phi_{r\phi} = 1/r$ (see Sections 3.2 and 3.4). The components of the covariant derivative of the velocity then are

$$\frac{\tilde{D}\dot{r}}{dt} = \ddot{r} - r\dot{\phi}^2, \quad \frac{\tilde{D}\dot{\phi}}{dt} = \ddot{\phi} + 2\frac{\dot{r}}{r}\dot{\phi}.$$

(3.34)

The trajectory will be an auto-parallel if $\tilde{D}\dot{r}/dt = \tilde{D}\dot{\phi}/dt = 0$, that is, if $\ddot{r} - r\dot{\phi}^2 = 0$ and $\ddot{\phi} + 2\dot{r}\dot{\phi}/r = 0$.

The first integral of these equations is

$$r^2\dot{\phi} = L, \quad \dot{r}^2 = 2E - \frac{L^2}{r^2}, \tag{3.35}$$

where L and $E > 0$ are integration constants. A second integration gives the equations found above for a straight line in polar coordinates after appropriate identification of the constants.

3.6   Differential operators and integration

• *The divergence and Laplacian of a vector.*   By extension of its definition in Cartesian coordinates (see Section 2.6), the divergence of a vector field v with components v^i in a system of curvilinear coordinates x^i is defined as $\tilde{D}_i v^i$, where $\tilde{D}_i$ is the associated covariant derivative.

Using (det$e$) to denote the determinant of the coefficients e_{ij} of the Euclidean metric in the coordinates x^i, it is easily seen that

$$\partial_i(\text{det}e) = -(\text{det}e)e_{jk}\partial_i e^{jk} \quad \text{and} \quad \partial_i(\text{det}e) = (\text{det}e)e^{jk}\partial_i e_{jk} , \tag{3.36}$$

so that the traces of the Christoffel symbols (3.27) can be written as $\tilde{\Gamma}^i_{ik} = \frac{1}{2}\partial_k \ln(\text{det}e)$. Thus the divergence of v can also be written as

$$\tilde{D}_i v^i = \frac{1}{\sqrt{\text{det}e}}\partial_i\left(\sqrt{\text{det}e}\, v^i\right). \tag{3.37}$$

Similarly, by extension of its definition in Cartesian coordinates, the Laplacian of a function f is defined as $\tilde{D}_i\tilde{D}^i f$. In a coordinate system x^i we then have (using the fact that $\tilde{D}_i e_{jk} = 0$)

$$\tilde{D}_i\tilde{D}^i f = \tilde{D}_i(e^{ij}\tilde{D}_j f) = \tilde{D}_i(e^{ij}\partial_j f) = \partial_i(e^{ij}\partial_j f) + \tilde{\Gamma}^i_{ik}e^{kj}\partial_j f$$

$$= \frac{1}{\sqrt{\text{det}e}}\partial_i\left(\sqrt{\text{det}e}\, e^{ij}\partial_j f\right). \tag{3.38}$$

• *The Levi-Civita symbol and the volume element.*   In going from Cartesian coordinates X^i to curvilinear coordinates x^i, the components of the Levi-Civita symbol e_{ijk} with $e_{123} = 1$ (see Section 1.5) become

$$\epsilon_{lmn} = \frac{\partial X^i}{\partial x^l}\frac{\partial X^j}{\partial x^m}\frac{\partial X^k}{\partial x^n}\, e_{ijk} = Je_{lmn} , \tag{3.39}$$

where it is easily seen that J is the Jacobian of the transformation $X^i \mapsto x^i$ (i.e., the determinant of the Jacobian matrix $\partial X^i/\partial x^l$). We can relate J to the metric determinant as follows. Under a change of coordinates the metric becomes $e_{ij} = (\partial X^k/\partial x^i)(\partial X^l/\partial x^j)\delta_{kl}$.

The determinant of the left-hand side is $\det e$, while that of the right is J^2 (this is simple to verify in two dimensions). We therefore have

$$\epsilon_{ijk} = \sqrt{\det e}\, e_{ijk}\,, \qquad \epsilon^{ijk} = \frac{1}{\sqrt{\det e}}\, e^{ijk}\,. \tag{3.40}$$

In Euclidean space and Cartesian coordinates the volume element is defined as $dV = dX\,dY\,dZ = \frac{1}{6}e_{ijk}\,dX^i dX^j dX^k$, where e_{ijk} is the Levi-Civita symbol with $e_{123} = 1$.

In going to curvilinear coordinates $X^i \mapsto X^i(x^j)$, using (3.39) and (3.40) and the notation $d^3x = dx\,dy\,dz$ we obtain

$$dV = \frac{1}{6}e_{ijk}\frac{\partial X^i}{\partial x^l}\frac{\partial X^j}{\partial x^m}\frac{\partial X^k}{\partial x^n}dx^l dx^m dx^n = \frac{1}{6}\epsilon_{lmn}dx^l dx^m dx^n$$

$$= \frac{1}{6}\sqrt{\det e}\,e_{lmn}dx^l dx^m dx^n = \sqrt{\det e}\,d^3x\,. \tag{3.41}$$

- *The surface element and Gauss's theorem.* Let us consider a 2-surface Σ defined by $x^i = x^i(y^a)$ where $y^a = \{\lambda, \mu\}$ are two parameters [or, equivalently, defined by a constraint of the form $\Phi(x^i) = 0$]. An element of the surface Σ generalizes (2.26) and is defined as

$$dS_i = \epsilon_{ijk}\frac{\partial x^j}{\partial \lambda}\frac{\partial x^k}{\partial \mu}d\lambda\,d\mu = \sqrt{\det e}\,e_{ijk}\frac{\partial x^j}{\partial \lambda}\frac{\partial x^k}{\partial \mu}d\lambda\,d\mu \tag{3.42}$$

using (3.40) (here it is understood that all quantities are evaluated on the surface).

In the curvilinear coordinates x^i that we are using, the Euclidean length element is written as $dl^2 = e_{ij}dx^i dx^j$. The length element on the surface Σ then is

$$dl^2|_\Sigma = e_{ij}\frac{\partial x^i}{\partial y^a}\frac{\partial x^j}{\partial y^b}dy^a\,dy^b = h_{ab}\,dy^a\,dy^b, \tag{3.43}$$

where $h_{ab} \equiv e_{ij}(\partial x^i/\partial y^a)(\partial x^j/\partial y^b)$ are the components of the *induced metric* on Σ in the coordinates y^a. We shall use $\det h$ to denote its determinant (again with all quantities evaluated on Σ).

Now, the (covariant) unit vector n_i orthogonal to Σ is defined as in Section 2.6 by

$$n_i\frac{\partial x^i}{\partial y^a} = 0 \quad \text{and} \quad e^{ij}n_i n_j = 1 \quad \text{or by} \quad n_i = \frac{\partial_i\Phi}{\sqrt{e^{ij}\partial_i\Phi\partial_j\Phi}}, \tag{3.44}$$

depending on whether the equation for Σ is given by $x^i = x^i(y^a)$ or by $\Phi(x^i) = 0$.

It is an instructive exercise to show that $\epsilon_{ijk}(\partial x^i/\partial\lambda)(\partial x^k/\partial\mu) = n_i\sqrt{\det h}$. Therefore, the surface element can be written in two equivalent ways (see also Section 4.6):

$$dS_i = \sqrt{\det e}\,e_{ijk}\frac{\partial x^j}{\partial \lambda}\frac{\partial x^k}{\partial \mu}d\lambda\,d\mu = \sqrt{\det h}\,n_i\,d\lambda\,d\mu. \tag{3.45}$$

Now that we have the expression (3.37) for the divergence of a vector and the equations for the volume and surface elements (3.41) and (3.45), Gauss's theorem follows. If V is a volume bounded by a surface ∂V, we have

$$\int_V \tilde{D}_i v^i\,dV = \int_{\partial V} v^i dS_i\,. \tag{3.46}$$

The area of a sphere in spheroidal coordinates

In *spheroidal coordinates* (r, θ, ϕ) related to Cartesian coordinates as

$$X = \sqrt{r^2 + a^2}\,\sin\theta\cos\phi, \quad Y = \sqrt{r^2 + a^2}\,\sin\theta\sin\phi, \quad Z = r\cos\theta$$

the surface of a sphere Σ of radius r_0 is given by $r^2 + a^2\sin^2\theta = r_0^2$.
Show that the Euclidean length element is written as

$$dl^2 = \frac{r^2 + a^2\cos^2\theta}{r^2 + a^2}dr^2 + (r^2 + a^2\cos^2\theta)d\theta^2 + (r^2 + a^2)\sin^2\theta\,d\phi^2. \qquad (3.47)$$

Verify (3.45) for this particular case.
Verify that the area of Σ, $A = \int dS$ where dS is the modulus of dS_i, is $A = 4\pi r_0^2$, as it should be.

4
Differential geometry

Here we present some elements of differential geometry, the 'vector' version of Euclidean geometry in curvilinear coordinates, in order to give an intrinsic definition of the covariant derivative and to establish a relation between the moving frames attached to a trajectory introduced in Section 2.5 and the moving frames of Cartan associated with curvilinear coordinates.

4.1 Tangent spaces, vectors, and tangents

As we have seen in Chapter 2, a Euclidean space $\mathcal{E}$ is an affine space, that is, an ensemble of points where any pair of points, or bipoint pq, is identified with a vector of a vector space E with basis $\{e_i\}$, $i = 1, 2, ...n$, and where a point, the origin O, is distinguished. Therefore, $Op = X^i e_i$, where the X^i are simultaneously the Cartesian coordinates of p and the components of the associated vector. In addition, E is equipped with a nondegenerate bilinear form $e = \delta_{ij}\epsilon^i \otimes \epsilon^j$, where $\{\epsilon^i\}$ is the basis canonically associated with the $\{e_i\}$ of its dual space E^*. This metric e defines the distance between the points of $\mathcal{E}$: $e(dp, dp) = \delta_{ij} dX^i dX^j$ with $dp = dX^i e_i$.

The basis vectors of a Cartesian frame, which are bipoints identified with the e_i and 'bound' to the origin O, are tangent to the coordinate axes X^i. Now we can strip the origin O of its special status by associating with *each* point p of $\mathcal{E}$ a 'moving' frame whose basis vectors are tangent to the coordinate lines X^i at p. Such frames can be derived from each other by simple translations of the origin, and their basis vectors $e_i(p)$ obtained by parallel transport from the origin O are identified, for any p, with the vectors e_i whose components are δ_i^j in the coordinate system X^j.

As briefly sketched in Section 2.4, we now consider the derivative operator $\partial/\partial X^{\underline{i}}$ for a given i acting on functions of the coordinates $f(X^j)$ at a point p with coordinates X^j. Its action on the coordinates X^j themselves gives the components of $e_{\underline{i}}$ because $\partial X^j/\partial X^{\underline{i}} = \delta_{\underline{i}}^j$. The operator $\partial/\partial X^{\underline{i}}$ therefore contains as much information as the vector $e_{\underline{i}}$, but since in addition it operates at a given point p, it must be identified with the vector $e_{\underline{i}}(p)$:

$$e_{\underline{i}}(p) \equiv \frac{\partial}{\partial X^{\underline{i}}},$$

which is a basis vector of a vector space attached to the point p and called the *tangent space* at p, $E(p)$. Since the coordinates X^i are Cartesian, this construction is certainly redundant, because all the $e_i(p)$ for a given i are equal for any p.

On the other hand, in the case where the location of p is specified by curvilinear coordinates x^i, the basis vectors of the moving frame at p, which are bipoints tangent to the coordinate lines at p and therefore identifiable as $\partial/\partial x^i$, vary from point to point (that is, they are no longer derived from each other by parallel transport) and they must be distinguished. The vectors $\partial/\partial x^i$ form what is called a *natural basis* of $E(p)$.

Relativity in Modern Physics. Nathalie Deruelle and Jean-Philippe Uzan.
© Oxford University Press 2018. Published in 2018 by Oxford University Press.
DOI: 10.1093/oso/9780198786399.001.0001

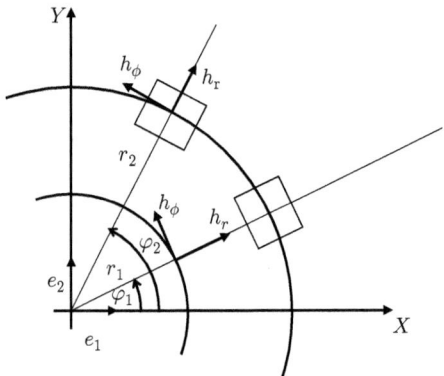

Fig. 4.1 Euclidean space, moving frames, and tangent spaces.

In two dimensions we can visualize the tangent spaces of the Euclidean space in Cartesian coordinates as superimposed sheets of paper whose grids coincide. In polar coordinates (see Fig. 4.1) they can be visualized as small squares of paper (on which the geometry becomes *local* or *differential*) whose grids (the 'moving frames') have an orientation which varies from point to point. We can reorient the small squares of paper and 'glue' them back together, thereby returning to the original Cartesian coordinates.

Given a system of coordinates x^i, a *tangent* (or vector) t at p (for example, a 'force') is an element of the tangent space $E(p)$. It can be decomposed as $t = t^i \partial/\partial x^i$ where the t^i are its natural components. It acts on differentiable functions $f(x^i)$ as

$$t(f) = t^i \frac{\partial f}{\partial x^i}, \tag{4.1}$$

where it is understood that $\partial f/\partial x^i$ is evaluated at[1] p.

In going from coordinates x^i to other curvilinear coordinates x'^i, a tangent at p is written as $t = t^i \partial/\partial x^i = t'^i \partial/\partial x'^i$ and according to (3.9) we have

$$\frac{\partial}{\partial x^i} = \frac{\partial x'^j}{\partial x^i} \frac{\partial}{\partial x'^j} \qquad \text{and} \qquad t'^j = \frac{\partial x'^j}{\partial x^i} t^i,$$

$$\text{and also} \qquad \frac{\partial}{\partial x'^i} = \frac{\partial x^j}{\partial x'^i} \frac{\partial}{\partial x^j} \qquad \text{and} \qquad t^j = \frac{\partial x^j}{\partial x'^i} t'^i, \tag{4.2}$$

where it is understood that the Jacobian matrix of the transformation $\partial x'^k/\partial x^j$ and its inverse $\partial x^i/\partial x'^k$ such that $(\partial x^i/\partial x'^k)(\partial x'^k/\partial x^j) = \delta^i_j$ are evaluated at the point p.

There is one more point in this differential framework that needs clarification, namely, the status of the radius vector $R \equiv Op$ or, more generally, of a free vector $l_a = p_a p$. Following the example in (3.8), we attach the vector $p_a p$ to the point p and view it as a tangent belonging to $E(p)$. Under a general coordinate transformation it then transforms as in (4.2),

---

[1]Therefore, a vector bound to p is simultaneously a point paired with an element of the vector space, a bipoint of an affine space, and a derivative operator belonging to the tangent space of p acting on one function to give another. It is also a singly contravariant tensor acting on forms to give numbers.

and its components, which are $(X^i - X_a^i)$ in the Cartesian frame $\mathcal{S}$, become $l_a^i(X^k) = (\partial x^i/\partial X^j)(X^j - X_a^j)$ in $\mathcal{C}$, where the Jacobian matrix is evaluated at[2] X^k.

4.2 Cotangent spaces and differential forms

Any tangent t at p acting on a function f can by 'duality' be associated with a *form*, denoted $\mathrm{d}f$, an element of the *cotangent* space $E^*(p)$ dual to $E(p)$, which acts on a vector t to give the same result as the action of t on f:

$$\mathrm{d}f(t) = t(f). \tag{4.3}$$

If we take as the function f the kth coordinate of the point p, $f = x^k$, then $t(x^k) = t^k$ according to (4.1) and from the definition of t and linearity we have $\mathrm{d}x^k(t) = \mathrm{d}x^k\left(t^i\partial/\partial x^i\right) = t^i\mathrm{d}x^k\left(\partial/\partial x^i\right) = t^k$, which implies that $\mathrm{d}x^k\left(\partial/\partial x^i\right) = \delta_i^k$. Therefore, the forms $\mathrm{d}x^k$ constitute a basis, the *natural basis*, of the cotangent space[3] $E^*(p)$.

Any form λ of $E^*(p)$ can then be decomposed as $\lambda = \lambda_i\,\mathrm{d}x^i$. In this context where the coordinates x^i are arbitrary, the forms defined in the cotangent space are referred to as *differential forms*. If there exists a function $f(x^i)$ such that $\lambda_i = \partial f/\partial x^i$, that is, if λ_i is the value at p of a function $\lambda_i(x^j)$ such that $\partial\lambda_i/\partial x^j = \partial\lambda_j/\partial x^i$, then we denote it as $\mathrm{d}f$ and we have $\lambda \equiv \mathrm{d}f = (\partial f/\partial x^i)\,\mathrm{d}x^i$ (which justifies the notation used). Such differential forms, of which the $\mathrm{d}x^i$ are examples, are called *exact*.

Knowing the transformation law of basis vectors of the tangent space [see (4.2)], we can immediately derive those of the conjugate exact differential forms $\mathrm{d}x^i$ as well as of the components of any form $\omega = \omega_i\mathrm{d}x^i = \omega_j'\mathrm{d}x'^j$:

$$\mathrm{d}x'^j = \frac{\partial x'^j}{\partial x^i}\,\mathrm{d}x^i \quad \text{and} \quad \omega_i = \frac{\partial x'^j}{\partial x^i}\,\omega_j',$$

$$\text{and also} \quad \mathrm{d}x^j = \frac{\partial x^j}{\partial x'^i}\,\mathrm{d}x'^i \quad \text{and} \quad \omega_i' = \frac{\partial x^j}{\partial x'^i}\,\omega_j. \tag{4.4}$$

4.3 The metric tensor, triads, and frame fields

The introduction of the tangent space at p and its dual, both equipped with their natural conjugate bases, makes it possible to rewrite the Euclidean metric as $e = \delta_{ij}\,\epsilon^i\otimes\epsilon^j = \delta_{ij}\,\mathrm{d}X^i\otimes\mathrm{d}X^j$, where the $\mathrm{d}X^i \equiv \epsilon^i$ are the basis forms associated with the *Cartesian* coordinates X^i. Since we know the transformation law of the forms $\mathrm{d}X^i$ under a nonlinear change of coordinates, we see that the metric transforms as

$$e = \delta_{ij}\,\mathrm{d}X^i \otimes \mathrm{d}X^j = \delta_{ij}\,\frac{\partial X^i}{\partial x^k}\frac{\partial X^j}{\partial x^l}\,\mathrm{d}x^k \otimes \mathrm{d}x^l \tag{4.5}$$

in accordance with (3.1).

---

[2] In this approach the origin O of the affine space loses its privileged status. We also note that in spaces where global Cartesian coordinates do *not* exist (as is the case in the *curved* spaces of general relativity), the position vector can no longer be defined.

[3] Care should be taken not to confuse the form $\mathrm{d}x^i$, a basis vector of the cotangent space at p, with $\mathrm{d}x^i$, the coordinate increment. The form $\mathrm{d}x^i$ acts on the vector $\mathrm{d}p = \mathrm{d}x^j\partial/\partial x^j$ to give the number $\mathrm{d}x^i\colon \mathrm{d}x^i(\mathrm{d}p) = \mathrm{d}x^j$. If the coordinates $x^i \equiv X^i$ are Cartesian, then $\mathrm{d}X^i \equiv \epsilon^i$ is the conjugate basis of $\partial/\partial X^i \equiv e_i$.

Natural bases and spherical coordinates

In the transformation from Cartesian coordinates $X^i = (X, Y, Z)$ to spherical coordinates $x^i = (r, \theta, \phi)$ ($X = r\sin\theta\cos\phi$, $Y = r\sin\theta\sin\phi$, and $Z = r\cos\theta$), the bases of the tangent space and its dual transform as

$$
\begin{cases}
\dfrac{\partial}{\partial r} = \cos\phi\sin\theta\,\dfrac{\partial}{\partial X} + \sin\phi\sin\theta\,\dfrac{\partial}{\partial Y} + \cos\theta\,\dfrac{\partial}{\partial Z} \\[2mm]
\dfrac{\partial}{\partial \theta} = r\left(\cos\phi\cos\theta\,\dfrac{\partial}{\partial X} + \sin\phi\cos\theta\,\dfrac{\partial}{\partial Y} - \sin\theta\,\dfrac{\partial}{\partial Z}\right) \\[2mm]
\dfrac{\partial}{\partial \phi} = r\sin\theta\left(-\sin\phi\,\dfrac{\partial}{\partial X} + \cos\phi\,\dfrac{\partial}{\partial Y}\right)
\end{cases}
\tag{4.6}
$$

$$
\begin{cases}
\mathrm{d}r = \cos\phi\sin\theta\,\mathrm{d}X + \sin\phi\sin\theta\,\mathrm{d}Y + \cos\theta\,\mathrm{d}Z \\[2mm]
\mathrm{d}\theta = \dfrac{1}{r}\left(\cos\phi\cos\theta\,\mathrm{d}X + \sin\phi\cos\theta\,\mathrm{d}Y - \sin\theta\,\mathrm{d}Z\right) \\[2mm]
\mathrm{d}\phi = \dfrac{1}{r\sin\theta}\left(-\sin\phi\,\mathrm{d}X + \cos\phi\,\mathrm{d}Y\right),
\end{cases}
\tag{4.7}
$$

where $\partial/\partial X \equiv e_1, \ldots$ and $\mathrm{d}X \equiv \epsilon^1, \ldots$, because (X, Y, Z) are Cartesian coordinates. It can be verified that these bases are indeed conjugates of each other $[\mathrm{d}r(\partial/\partial r) = 1,\ \mathrm{d}r(\partial/\partial\theta) = 0,\ \text{etc.}]$ and according to (4.5) lead to

$$
e = \mathrm{d}r^2 + r^2\mathrm{d}\theta^2 + r^2\sin^2\theta\,\mathrm{d}\phi^2 \,.
\tag{4.8}
$$

A *triad*, with its three elements denoted as h_i, is a basis of $E(p)$ related to the natural basis as $h_i = L^j{}_i(x^j)\partial/\partial x^j$, where the $L^j{}_i(x^k)$ are a *priori* arbitrary functions of the coordinates [which must nevertheless form a matrix which is (nearly) everywhere invertible].

To this triad of tangent vectors there corresponds a triad of conjugate forms $\theta^i = L_j{}^i(x^k)\mathrm{d}x^j$ with $L^k{}_j L_k{}^i = \delta^i_j$, also called a *frame field*.[4]

A triad of basis vectors $h_i = L^j{}_i\,\partial/\partial x^j$ can be orthonormal if the functions $L^j{}_i(x^k)$ are chosen such that $e(h_i, h_j) \equiv L^k{}_i L^l{}_j e_{kl} = \delta_{ij}$. The conjugate forms $\theta^i = L_j{}^i(x^k)\mathrm{d}x^j$ are not in general exact differential forms, but they allow the metric to be written in the 'quasi-Cartesian' form

$$
e = \delta_{ij}\,\theta^i \otimes \theta^j \,.
\tag{4.9}
$$

Any transformation $\theta'^i = \mathcal{R}_j{}^i\,\theta^j$, where $\mathcal{R}_j{}^i(x^k)$ is a point-dependent rotation matrix, preserves this quasi-Cartesian form of the metric.

Moving frames and spherical coordinates

The natural basis (4.6) of spherical coordinates in which the metric is written $e = \mathrm{d}r^2 + r^2\mathrm{d}\theta^2 + r^2\sin^2\theta\,\mathrm{d}\phi^2$ is not orthonormal because, e.g., $e(\partial/\partial\theta, \partial/\partial\theta) = r^2 \neq 1$, but we can introduce the triad

---

[4] If there exist n functions $x'^i(x^k)$ such that $\theta^i = \mathrm{d}x'^i$, i.e., if we can write $L_j{}^i(x^k)$ in the form $L_j{}^i(x^k) = \partial x'^i/\partial x^j$, which is only possible if $\partial L_j{}^i/\partial x^k = \partial L_k{}^i/\partial x^j$, then the triad h_i forms a basis which is termed *holonomic* and is derived from the natural basis $\partial/\partial x^i$ by the simple change of coordinates $x^i \to x'^i$.

$$h_r = \frac{\partial}{\partial r}, \quad h_\theta = \frac{1}{r}\frac{\partial}{\partial \theta}, \quad h_\phi = \frac{1}{r\sin\theta}\frac{\partial}{\partial \phi}, \tag{4.10}$$

which is orthonormal. The forms of the associated dual basis are

$$\theta^r = \mathrm{d}r, \quad \theta^\theta = r\,\mathrm{d}\theta, \quad \theta^\phi = r\sin\theta\,\mathrm{d}\phi \tag{4.11}$$

and the metric is then written in the form (4.9).

We note the strong similarity between the triad (h_r, h_θ, h_ϕ) and the basis (e'_1, e'_2, e'_3) of the moving frame introduced in (2.17). This is why triads are sometimes called *moving frames*.

Let us make this more precise. The moving frame introduced in (2.17), attached to a curve of $\mathcal{E}_3$, is Cartesian and at a given time t its coordinates globally establish a grid on the leaf labeled by t in Newtonian spacetime, $N_4 = \mathcal{E}_3 \times R$. Therefore, the ϵ'^i of (2.20) are exact differential forms: $\epsilon'^i \equiv \mathrm{d}X'^i$ (where the X'^i are Cartesian coordinates in $\mathcal{S}'$) because the transformation $\mathrm{d}X'^i = \mathcal{R}_j{}^i(t)\,\mathrm{d}X^j$ depends on the parameter t and not on the coordinates X^k. On the other hand, the triad of time-invariant forms θ^i in (4.11) is defined at *any* point, but in the tangent space of the point. These θ^i are *not* exact differential forms because the transformation $\theta^i = \mathcal{R}_j{}^i(X^k)\,\mathrm{d}X^j$ depends on the coordinates X^k. They become exact only on the trajectory where $X^k = X^k(t)$, the tangent space of the point $p(t)$ having been identified with the leaf of N_4 labeled by[5] t.

4.4 Vector fields, form fields, and tensor fields

A vector of the natural basis $\partial/\partial x^i$ (with i fixed) is associated with a particular point p, but it can also be viewed as a *function* of the point p. More precisely, the mapping which associates with any point p the quantity $\partial/\partial x^i$ is called the *vector field* $\partial/\partial x^i$. In general, a field of tangent vectors is the mapping which associates with each point p a tangent t at p. A typical example is the force field which appears in the law of dynamics. It is a mapping of the affine space $\mathcal{E}$ to the ensemble of tangent spaces called the *tangent fiber*. The mapping is continuous if the components $t^i(x^j)$ of this field are continuous functions of the coordinates x^j of the points p. We can analogously define form fields and, more generally, tensor fields. The Euclidean metric viewed as a function of the point is itself a field of bilinear forms or 2-fold covariant tensors. In this context a simple function of the points p is then referred to as a *scalar field*.

All the properties of multilinear forms or tensors derived within the framework of vector geometry (see Chapter 2) are therefore easily carried over to the more general framework we are developing here. Let p vectors $\partial/\partial x^i$ and q 1-forms $\mathrm{d}x^j$ of the natural bases of $E(p)$ and $E^*(p)$ be associated with the coordinates x^i. Their tensor products define a basis of the space of tensors of the type $\binom{p}{q}$ which are p times contravariant and q times covariant. Then any tensor of this type can be written as

$$T = t^{i_1\ldots i_p}_{j_1\ldots j_q}\,\partial_{i_1} \otimes \ldots \otimes \partial_{i_p} \otimes \mathrm{d}x^{j_1} \otimes \ldots \otimes \mathrm{d}x^{j_q}. \tag{4.12}$$

The functions $T^{i_1\ldots i_p}_{j_1\ldots j_q}(x^i)$ are the components of the tensor field $T(p)$ in the natural basis associated with the coordinates x^i.

---

[5]The situation is different in special and general relativity, where time becomes a coordinate and where rigid displacements are absorbed into changes of coordinates.

The transformation laws of basis vectors and 1-forms under a change of coordinate system are known. We can thus derive the transformation law for the components of T already given in (3.9):

$$t'^{i_1...i_p}_{j_1...j_q} = t^{k_1...k_p}_{l_1...l_q} \frac{\partial x'^{i_1}}{\partial x^{k_1}} \cdots \frac{\partial x'^{i_p}}{\partial x^{k_p}} \frac{\partial x^{l_1}}{\partial x'^{j_1}} \cdots \frac{\partial x^{l_q}}{\partial x'^{j_q}}, \qquad (4.13)$$

where all quantities are expressed as functions of $x^m(x'^n)$.

Let us conclude this section with a useful construction, which expands upon the remark at the end of Section 3.1. The existence of a metric e allows us to define a linear mapping $\tilde{e}$ of $E(p)$ in $E^*(p)$ such that, by definition, $\tilde{e}(\partial/\partial x^i) = e_{ij}\,\mathrm{d}x^j$, where e_{ij} are the components of e in the (curvilinear) coordinates x^i. The mapping $\tilde{e}$ therefore establishes a one-to-one correspondence between basis vectors and forms and is isomorphic to e. Therefore, $\tilde{e}$ (or e) associates with a tangent t of components t^i the form $\tilde{e}(t)$ of components $e_{ij}t^j$. Reciprocally, we can define a mapping $\tilde{e}^*$ of $E^*(p)$ in $E(p)$ isomorphic to e which associates with a form λ of components λ_i the tangent $\tilde{e}^*(\lambda)$ of components $e^{ij}\lambda_i$. The metric therefore *raises indices*.

4.5   The covariant derivative (II)

Let us consider an ensemble of points p distinguished by their coordinates x^i and their associated tangent spaces $E(p)$ with natural basis $\partial/\partial x^i$. The mathematical object which allows tangent spaces to be 'connected' to each other and tensors to be 'transported' from one point to another (independently of the existence of a metric) is a 'connection'.

An *affine connection* D associates with any vector v of the tangent space $E(p)$ at p an operator denoted D_v acting on the tensor fields T and possessing the following properties[6]:

- linearity in v, that is, for all $v, w \in E(p)$ and for all $a, b \in R$ we have $D_{av+bw}T = aD_vT + bD_wT$;
- linearity in T, that is, for any tensors T and S of the same type we have $D_v(aT + bS) = aD_vT + bD_vS$;
- satisfaction of the Leibniz rule, namely, any tensors T and S obey $D_v(T \otimes S) = D_vT \otimes S + T \otimes D_vS$;
- if C is the contraction operator, then $C(D_vT) = D_v(CT)$;
- and, finally, compatibility with the notion of a directional derivative associated with tangent vectors, that is, $D_vf = v(f)$ if f is a function of $\mathcal{E}$ in R.

Here D_vT is the *covariant derivative* of T with respect to v. It is a tensor field of the same type as T.

When v and T are the vectors $\partial_i \equiv \partial/\partial x^i$ and ∂_j of the natural basis associated with the coordinates x^i, the covariant derivative D is operationally defined by specifying its *connection coefficients* (also referred to as *affinities*), which are n^3 functions of the x^i denoted by Γ^k_{ij} such that

$$D_{\frac{\partial}{\partial x^i}} \frac{\partial}{\partial x^j} \equiv D_{\partial_i}(\partial_j) \equiv D_i\partial_j = \Gamma^k_{ij}\,\partial_k. \qquad (4.14)$$

---

[6]For a more complete discussion see, for example, Bishop and Goldberg (1980).

We remark that the notation D_i is obviously convenient when there is no possibility of confusion about the coordinate system used, and it is rather inaccurately referred to as the *covariant derivative with respect to*[7] x^i. More generally, we have

$$D_v w = v^i D_i (w^j \partial_j) \equiv v^i (D_i w^j) \partial_j \qquad \text{with} \qquad D_i w^j \equiv \partial_i w^j + \Gamma^j_{ik} w^k, \qquad (4.15)$$

from which we recover (3.13).

The *parallel transport* of a vector w along a line integral of the field v described by the equation $x^i = x^i(\lambda)$ with $v^i = dx^i/d\lambda$ is given by

$$D_v w = 0 , \quad \text{that is,} \quad v^i D_i w^j = 0,$$

$$\text{or also, using (4.15),} \quad \frac{dw^j}{d\lambda} + \Gamma^j_{ik} w^k v^i = 0, \qquad (4.16)$$

which is an ordinary differential equation for the $w^i(\lambda)$ already obtained in (3.23). Therefore, the covariant derivative performs its duty: it transports tensor objects of the same type from one point to another along a given path.

Since the contraction operation implies that $C(\partial_i \otimes dx^j) \equiv \partial_i \otimes dx^i = 1$, we have

$$0 = D_j(\partial_i \otimes dx^i) = (D_j \partial_i) \otimes dx^i + \partial_i \otimes D_j dx^i$$

$$= \Gamma^k_{ji} \partial_k \otimes dx^i + \partial_k \otimes D_j dx^k = \partial_k \otimes (\Gamma^k_{ji} dx^i + D_j dx^k) .$$

We therefore obtain the expression for the covariant derivative of a 1-form as a function of the connection coefficients:

$$D_j dx^k = -\Gamma^k_{ji} dx^i. \qquad (4.17)$$

Moreover, we recover (3.17), namely, $D_j(\lambda_k dx^k) \equiv (D_j \lambda_k) dx^k$ with $D_j \lambda_k = \partial_j \lambda_k - \Gamma^l_{jk} \lambda_l$.

Knowing the derivatives of basis vectors and forms as a function of the Γ^i_{jk}, we can derive the expression for the covariant derivative of a tensor of any type. For example, let us consider a tensor which is singly contravariant and singly covariant, that is, a $\binom{1}{1}$ tensor: $T = T^i_j \partial_i \otimes dx^j$. Its covariant derivative with respect to a vector v is

$$D_v T = v^i D_i(T^j_k \partial_j \otimes dx^k) \equiv v^i (D_i T^j_k) \partial_j \otimes dx^k,$$

$$\text{where} \qquad D_i T^j_k = \partial_i T^j_k + \Gamma^j_{il} T^l_k - \Gamma^l_{ik} T^j_l \qquad (4.18)$$

are the *natural components* of $D_i T$ already given in (3.18).

Now in order to find how the connection coefficients transform under a change of coordinates, we note that the covariant derivative of a vector w along another vector v which is the vector given in (4.15), namely, $D_v w = v^i (D_i w^j) \partial_j$, can just as well be viewed as the

---

[7] We shall also sometimes use a comma to denote partial derivatives as in $\partial T/\partial x^i \equiv \partial_i T \equiv T_{,i}$, and a semicolon to denote covariant derivatives as in $D_{\partial_i} T \equiv D_i T \equiv T_{;i}$.

We note that the connection D and the connection coefficients Γ^i_{jk} are not decorated with a tilde as in Section 3.2 *et seq.*, because here we do not assume *a priori* that there exists a coordinate system in which all the Γ^i_{jk} vanish.

contraction of the vector v with a $\binom{1}{1}$ tensor having components $D_i w^j$, which can be denoted as $Dw \equiv D_i w^j \partial_j \otimes dx^i$. Indeed, since by the definition of the contraction operation $C_{(k,i)} \partial_k \otimes dx^i = 1$, we have

$$C(v, Dw) \equiv C_{(k,i)}(v^k \partial_k \otimes D_i w^j \partial_j \otimes dx^i) = v^k D_k w^j \partial_j \equiv D_v w. \tag{4.19}$$

It is then a simple exercise to find the transformation law for the Γ^i_{jk} already given in (3.21):

$$\Gamma'^j_{ki} = \frac{\partial x^r}{\partial x'^i} \frac{\partial x^l}{\partial x'^k} \frac{\partial x'^j}{\partial x^p} \Gamma^p_{lr} + \frac{\partial^2 x^l}{\partial x'^k \partial x'^i} \frac{\partial x'^j}{\partial x^l}. \tag{4.20}$$

There are as many connections as there are possible choices of the functions Γ^i_{jk}, that is, n^3. If they are symmetric in i and k the connection is said to be *torsion*-free.

Now let us suppose that in addition to a torsion-free connection [that is, $n^2(n+1)/2$ functions Γ^i_{jk}], we are also provided with a *metric*, that is, a field of 2-fold covariant tensors which is symmetric and nondegenerate[8]: $g = g_{ij} dx^i dx^j$.

We can impose the condition that the connection be *compatible* with the metric, that is, that

$$D_v g = 0 \ \forall v \quad \Longleftrightarrow \quad D_i g_{jk} = 0. \tag{4.21}$$

The (symmetric) connection is then called the *Levi-Civita connection* and the connection coefficients, now referred to as *Christoffel symbols*, are related to the metric components by (3.27):

$$\Gamma^i_{jk} = \frac{1}{2} g^{il} (\partial_j g_{kl} + \partial_k g_{lj} - \partial_l g_{jk}). \tag{4.22}$$

Finally, if there exist systems of coordinates X^i where all the Γ^i_{jk} vanish, then the connection is *flat*, the coordinates X^i are *Cartesian*, the metric coefficients are constants, and the Christoffel symbols in the system $x^i = x^i(X^j)$, denoted by $\tilde{\Gamma}^i_{jk}$ in this case, are expressed in terms of only n functions according to (4.20), which was used to define them in (3.25):

$$\tilde{\Gamma}^i_{jk} \equiv \frac{\partial x^i}{\partial X^l} \frac{\partial^2 X^l}{\partial x^j \partial x^k}. \tag{4.23}$$

The Riemann spaces of general relativity will be equipped with Levi-Civita connections which are not flat.[9]

4.6 Vector calculus and differential operators

• *Vector calculus.* By extending the definitions given in Section 2.6, we can define the vector product of two vectors v and w as the 1-form $v \wedge w = e_{ijk} V^j W^k \theta^i$ in the moving frame where the metric is written as $e = \delta_{ij} \theta^i \theta^j$ (see Section 4.3), or else using (3.40)

$$v \wedge w = \epsilon_{ijk} v^j w^k dx^i, \tag{4.24}$$

where v^i and w^j are their components in the natural basis ∂_i and where $\epsilon_{ijk} = \sqrt{\det e} \, e_{ijk}$ with $e_{123} = 1$, the quantity $\det e$ being the determinant of the components e_{ij} of the metric in the coordinates x^i; see Section 3.6.

---

[8]Here we call the metric g rather than e because we do not assume a *priori* that there exist systems of Cartesian coordinates in which its coefficients are given by δ_{ij}.

[9]The tools of differential geometry and tensor calculus which we have introduced in the present and preceding chapters are useful in Newtonian mechanics, but become indispensable in both special and general relativity.

The scalar product of two vectors is $(v.w) = e_{ij}v^i w^j$, and so $\partial_i.\partial_j = e_{ij}$.

Similarly, the scalar product of two forms is $(\lambda.\mu) = e^{ij}\lambda_i\mu_j$ and so $dx^i.dx^j = e^{ij}$.

We therefore find, for example, that the squared norm of the vector product of two vectors can be written as follows using the properties of the Levi-Civita symbol (see Section 1.5):

$$
\begin{aligned}
(v \wedge w)^2 &= e^{il}\epsilon_{ijk}\epsilon_{lmn}v^j w^k v^m w^n = \epsilon^l_{\;jk}\epsilon_{lmn}v^j w^k v^m w^n \\
&= (e_{jm}e_{kn} - e_{jn}e_{km})v^j w^k v^m w^n \\
&= v^2 w^2 - (v.w)^2 .
\end{aligned}
\tag{4.25}
$$

- *Differential operators.* If df is a differential form, the *gradient* of f, denoted ∇f, is the vector associated with it by using the metric e to raise an index. In the coordinates x^i we have

$$
\nabla f \equiv \left(e^{ij}\frac{\partial f}{\partial x^j}\right)\partial_i .
\tag{4.26}
$$

We can similarly define the divergence and the Laplacian according to the discussion in Sections 2.6 and 3.6 as

$$
\nabla.v = D_i v^i = \frac{1}{\sqrt{\det e}}\partial_i(\sqrt{\det e}\, v^i), \quad \triangle f = D_i D^i f = \frac{1}{\sqrt{\det e}}\partial_i(\sqrt{\det e}\, e^{ij}\partial_j f) .
\tag{4.27}
$$

Differential operators in spherical coordinates

As a function of the triad (h_r, h_θ, h_ϕ) introduced in (4.10), we have

$$
\begin{aligned}
\nabla f &= \frac{\partial f}{\partial r}h_r + \frac{1}{r}\frac{\partial f}{\partial \theta}h_\theta + \frac{1}{r\sin\theta}\frac{\partial f}{\partial \phi}h_\phi \\
\nabla.v &= \frac{1}{r^2}\partial_r(r^2 v^r) + \frac{1}{\sin\theta}\partial_\theta(\sin\theta v^\theta) + \partial_\phi v^\phi \\
\triangle f &= \frac{1}{r^2}\partial_r(r^2\partial_r f) + \frac{1}{r^2\sin\theta}\partial_\theta(\sin\theta\partial_r f) + \frac{1}{r^2\sin^2\theta}\partial^2_{\phi\phi}f .
\end{aligned}
\tag{4.28}
$$

The surface element

Since a surface Σ is defined by $x^i = x^i(\lambda, \mu)$ with n_i the unit vector normal to the surface and $\det h$ the determinant of the induced metric, we have the relation given in Section 3.6: $\epsilon_{ijk}(\partial x^j/\partial\lambda)(\partial x^k/\partial\mu) = n_i\sqrt{\det h}$.

This can be demonstrated by generalizing the results of Section 2.6. We introduce two vectors tangent to Σ with components $e^i_\lambda = \partial x^i/\partial\lambda$ and $e^i_\mu = \partial x^i/\partial\mu$, i.e., $e_\lambda = \partial_\lambda$ and $e_\mu = \partial_\mu$, tangent to the coordinate lines λ and μ on Σ. Writing the Euclidean metric as $e = e_{ij}dx^i \otimes dx^j$, the metric induced on Σ will be

$$
e|_\Sigma = e_{ij}\frac{\partial x^i}{\partial y^a}\frac{\partial x^j}{\partial y^b}dy^a \otimes dy^b = e^2_\lambda d\lambda^2 + 2(e_\lambda.e_\mu)d\lambda d\mu + e^2_\mu d\mu^2,
$$

the determinant of which is $\det h = (e_\lambda \wedge e_\mu)^2$ according to (4.25). Moreover, the normal vector is defined as (see Section 2.6) $n = (e_\lambda \wedge e_\mu)/|e_\lambda \wedge e_\mu|$ with, from (4.24), $(e_\lambda \wedge e_\mu)_i = \epsilon_{ijk}e^j_\lambda e^k_\mu = \epsilon_{ijk}(\partial x^j/\partial\lambda)(\partial x^k/\partial\mu)$.

We therefore find $dS_i = \epsilon_{ijk}(\partial x^j/\partial\lambda)(\partial x^k/\partial\mu)d\lambda\, d\mu = n_i\sqrt{\det h}\, d\lambda\, d\mu$.

Part II

Dynamics

Furthermore we may remark that any velocity once imparted to a moving body will be rigidly maintained as long as the external causes of acceleration or retardation are removed, a condition which is found only on horizontal planes...

Galileo Galilei, *Discorsi e Dimostrazioni Matematiche Intorno a Due Nuove Scienze*, Leiden, Elzevir, 1638; English translation by Henry Crew and Alfonso de Salvio, *Discourses and Mathematical Demonstrations Relating to Two New Sciences*, New York, Macmillan, 1914

Every body perseveres in its state of rest, or of uniform motion in a right line, unless it is compelled to change that state by forces impressed thereon.

The alteration of motion is ever proportional to the motive force impressed; and is made in the direction of the right line in which that force is impressed.

To every action there is always opposed an equal reaction: or the mutual actions of two bodies upon each other are always equal, and directed to contrary parts.

Sir Isaac Newton, *Philosophiæ Naturalis Principia Mathematica*, London, 1687; English translation by Andrew Motte, *The Mathematical Principles of Natural Philosophy*, London, 1729

5
Equations of motion

Now that we have set up the Euclidean framework of Newtonian physics and discussed the basic ideas of kinematics, in this chapter we turn to the more prominent aspects of Newtonian dynamics and the principle of Galilean relativity.

5.1 The law of Newtonian dynamics

The fundamental law describing the motion of a point particle P interacting with other particles P_a, called *Newton's second law*, is a differential equation written in the absolute Cartesian frame $\mathcal{S}$ as

$$ma = F \qquad \text{or, in component form,} \qquad m\frac{d^2 X^i}{dt^2} = F^i, \tag{5.1}$$

where a (with Cartesian components $a^i \equiv d^2 X^i/dt^2 \equiv \ddot{X}^i$) is the acceleration of the trajectory of P with equation $X^i = X^i(t)$. The three functions of time F^i are the Cartesian components of a 'Absolute, True, and Mathematical' vector F representing the interaction of P with the points P_a at time t, that is, the 'Relative, Apparent, and Common' *force* exerted on P, the nature and effect of which we need to specify. Finally, the parameter m, a constant, is an attribute of the point P, its *inertial mass*, which can be expressed, for example, in *kilograms* (kg). It characterizes the 'resistance' of P to the action of the other points[1] P_a.

The law of motion in curvilinear coordinates

In Section 3.5 we saw how the acceleration transforms in a change from a system of Cartesian coordinates X^i to a system of curvilinear coordinates $x^i = x^i(X^j)$:

$$a^i \mapsto \quad \frac{\partial x^i}{\partial X^j}a^j \equiv \frac{\tilde{D}v^i}{dt} = \frac{dv^i}{dt} + \tilde{\Gamma}^i_{jk}v^j v^k, \qquad \text{where} \qquad v^i = \frac{\partial x^i}{\partial X^j}V^j$$

$$\text{and} \quad \tilde{\Gamma}^i_{jk} \equiv -\frac{\partial^2 x^i}{\partial X^l \partial X^m}\frac{\partial X^l}{\partial x^j}\frac{\partial X^m}{\partial x^k} = \frac{\partial x^i}{\partial X^l}\frac{\partial^2 X^l}{\partial x^j \partial x^k} = \frac{1}{2}e^{il}\left(\frac{\partial e_{kl}}{\partial x^j} + \frac{\partial e_{jl}}{\partial x^k} - \frac{\partial e_{jk}}{\partial x^l}\right), \tag{5.2}$$

$$\text{where} \quad e_{kl} = \delta_{ij}\frac{\partial X^i}{\partial x^k}\frac{\partial X^j}{\partial x^l}, \quad e^{km} = \delta^{pq}\frac{\partial x^k}{\partial X^p}\frac{\partial x^m}{\partial X^q}.$$

---

[1] We assume, since it appears to agree with experiment, that all the masses are of the same sign, which we take to be positive.

Actually, the origin of the inertia of a body remains one of the mysteries of physics; see, for example, Barbour and Pfister (1995).

Relativity in Modern Physics. Nathalie Deruelle and Jean-Philippe Uzan.
© Oxford University Press 2018. Published in 2018 by Oxford University Press.
DOI: 10.1093/oso/9780198786399.001.0001

Newton's law (5.1) then becomes (see Sections 3.5 and 4.5)

$$m \frac{\tilde{D} v^i}{dt} = f^i \quad \text{or, in intrinsic form,} \quad m \tilde{D}_v v = F, \tag{5.3}$$

where $v^i = (\partial x^i / \partial X^j) V^j$ and $f^i = (\partial x^i / \partial X^j) F^j$ are the components of the velocity and the force in the system $\mathcal{C}$ and $\partial x^i / \partial X^j$ is evaluated on the trajectory of P, i.e., on $X^i = X^i(t)$. Finally, $v = V^i e_i = v^i \partial / \partial x^i$ and $F = F^i e_i = f^i \partial / \partial x^i$. This equation has the same form (it is *invariant*) in any coordinate system. It is the specification of the Christoffel symbols $\tilde{\Gamma}^i_{jk}$ that distinguishes the coordinate system.

For example, in polar coordinates $(X = r \cos \phi,\ Y = r \sin \phi)$, (5.3) is written as (see Section 3.5)

$$m \frac{\tilde{D} \dot{r}}{dt} \equiv m(\ddot{r} - r\dot{\phi}^2) = f^r, \quad m \frac{\tilde{D} \dot{\phi}}{dt} \equiv m \left(\ddot{\phi} + 2 \frac{\dot{r}\dot{\phi}}{r} \right) = f^\phi, \tag{5.4}$$

where $f^r = F^X \cos \phi + F^Y \sin \phi$ and $f^\phi = -F^X \sin \phi + F^Y \cos \phi$.

5.2 Properties of forces

There are some general arguments which allow us to state certain properties of the interactions which can be described by Newtonian physics.

The very fact that we decide to represent an interaction by a vector means that we are limiting ourselves (however, this has never been found to be a restriction!) to phenomena which do not depend on the position or orientation of the reference frame in which they are studied.[2]

Since the algebra of the vector space to which the vectors representing the forces belong is linear, we are *de facto* limiting ourselves to interactions which satisfy the *superposition principle*, which can be stated as follows. If $f_{a'a}$ represents the action of the body $P_{a'}$ on P_a in the absence of any other body, then the vector F_a representing the action of the set of bodies $P_{a'}$ is given by the sum[3] $F_a = \sum_{a'} f_{a'a}$.

Finally, the *law of action and reaction* or *Newton's third law* (the importance of which will become evident in the following chapter) states that the action of a body P_2 on another body P_1, described by f_{21}, must be equal and opposite to the action f_{12} of P_1 on P_2, which translates into the vector condition $f_{21} = -f_{12}$.

The force vector appearing in the law of motion (5.1) is a function of time. More precisely, it is a *functional* of the trajectory of P, which means that at a given time it is a function of the position vector R of the point P as well as, a priori, of the velocity v, the acceleration a, the derivative of the acceleration $\dot{a}$, and so on, of P (*cf.* the example of the magnetic force).

---

[2] This is one aspect of the Copernican principle (see also Section 1.3), namely, the law of motion of a body cannot depend on either the choice or the orientation of the axes of the Cartesian reference frame which is chosen as the absolute reference frame. The 'active' version of this principle states that the law of motion of a isolated system of bodies cannot depend on either the position or the orientation of the system in the absolute frame.

[3] If the $f_{a'a}$ are identified as free vectors, this is truly a vector sum, while if, as in the usual approach, $f_{a'a}$ is identified as a bound vector, that is, as a bipoint visualized as an arrow issuing from P_a, then F_a is generally called the *resultant* force. We note that the bound vector $f_{a'a}$ is evaluated at the point P_a and not at $P_{a'}$.

However, it turns out that a dependence only on R is sufficient for describing gravitation within the Newtonian framework, and in this case $F = F(P)$.

5.3 The principle of Galilean relativity

A *free* point particle is, by definition, not subject to any force. Then according to Newton's law (5.1) we see that it undergoes uniform rectilinear motion in the absolute Cartesian frame.

It should be possible to use this property of free particles to find the absolute Cartesian reference frame, independently of the existence of distant stars assumed to be at absolute rest. In fact, the Cartesian reference frame (for example, the solar system or the laboratory walls) in which the coordinates of free particles vary linearly with time should be the incarnation of the absolute Cartesian frame.[4]

However, if we accept the idea that a free particle is free in any reference frame, then Newton's law is not modified under Galilean transformations because the acceleration is represented by the same vector in any inertial frame (see Section 2.4). In other words, the law of motion of point particles has the same form (it is *invariant*) in all frames undergoing uniform rectilinear motion relative to the absolute frame: $ma = 0$, no matter whether $\mathcal{S}$ is the absolute frame or an inertial frame. Therefore, if in an inertial Cartesian frame $\mathcal{S}_g$ the particle trajectory is uniform and rectilinear with velocity v, then in any other Cartesian frame $\mathcal{S}'_g$ moving with constant velocity V_0 relative to $\mathcal{S}_g$, the trajectory will also be uniform and rectilinear with velocity $v' = v - V_0$, which can vanish if $v = V_0$. This is *Newton's first law*, which was actually stated by Galileo and is also called the *principle of inertia*.

A consequence of this law is that if a free object is at rest in one reference frame, it is not possible to conclude that this frame is the absolute frame and that the object is at absolute rest (and therefore identifiable with a given point of $\mathcal{E}_3$, or, more precisely, a fiber of $N_4 = \mathcal{E}_3 \times R$). It can only be concluded that the frame in question is inertial. The object can be undergoing any uniform translation relative to the absolute reference frame and it is impossible to measure this velocity, at least by means of free particles.

We thus conclude that this *Galilean invariance* of the law of dynamics makes the idea of absolute space meaningless. We can nevertheless imagine that particles subject to interactions might allow absolute space to be determined. However, if an interaction is represented by the same vector F in all Galilean frames, then Newton's law is written in the same way $ma = F$ in *any* inertial frame. But if the law is the same in the absolute Cartesian frame $\mathcal{S}$ and in any inertial Cartesian frame $\mathcal{S}_g$, it cannot be used to distinguish $\mathcal{S}$ from[5] $\mathcal{S}_g$.

The question then becomes the following. Is an interaction (a fundamental one like gravitation or electromagnetism) always represented by the same vector in any inertial frame? For two centuries the answer, based on experiment, was 'yes'. (Of course, in practice it is sometimes necessary to introduce simplifying effective forces which depend on the reference frame, such as a frictional force which is a function of the velocity of the object under study.)

This property in fact seemed to be so general that it came to be stated as a principle, the *principle of Galilean relativity*: *all the laws of Newtonian mechanics must be invariant under*

---

[4]If the particles turn out not to be undergoing uniform translation, this would mean *a priori* that the reference frame or the clocks are not good enough for making precise measurements, or that the particles are not in fact actually free.

[5]This would be possible if a law fixed the initial conditions of motion in the absolute frame. However, Newton's theory is mute on this point.

Galilean transformations. (However, it has now been known for over a century, see Book 2, that this is true only when the relative velocities of the reference frames and the particles are small compared to the speed of light.[6])

Therefore, in Newtonian mechanics absolute space is a phantom on two accounts[7]: its geometrical structure is independent of its matter content (the Pythagorean theorem is valid whether or not the universe is empty), while at the same time it is impossible to 'anchor' it anywhere. In fact, there is no absolute space or absolute reference frame in Newtonian mechanics. What is absolute is its dynamical equivalence class, that is, the ensemble of inertial frames undergoing uniform rectilinear motion with respect to it, which we shall henceforth denote as $\mathcal{S}$ or $\mathcal{S}_g$ without distinguishing between them.

5.4   Moving frames and inertial forces

On the other hand, any frame undergoing acceleration relative to the ensemble of inertial frames can be distinguished from the latter: the motions of free particles will not be represented by uniform rectilinear trajectories, and the dynamical law will not be written as $F = ma$. Since a Cartesian frame $\mathcal{S}'$ is related to an inertial frame $\mathcal{S}$ by a rigid displacement, we have, in component form or the vector equivalent (see Sections 1.5 and 2.4),

$$
\begin{cases}
m\,a' = F - m\,\ddot{d} + m\left(-2\Omega \wedge v' + \Omega \wedge (R' \wedge \Omega) - \dot{\Omega} \wedge R'\right) \\
m\,a'^i = \mathcal{R}_j{}^i F^j + m\left(2\,\dot{\mathcal{R}}_j{}^i V^j + \ddot{\mathcal{R}}_j{}^i X^j - (\mathcal{R}_j{}^i d^j)\right),
\end{cases}
\tag{5.5}
$$

where $\mathcal{R}_j{}^i(t)\,F^j$ are the components in $\mathcal{S}'$ of the vector F representing the interaction of the particles in $\mathcal{S}$. This vector F also represents the interaction in $\mathcal{S}'$ if the interaction does not depend on the acceleration of the reference frame in which it is studied. This is the case for the forces dealt with in Newtonian mechanics, and from now on we shall use the term *vector force* without specifying the frame.[8] The other terms in these expressions are referred to as *inertial forces* and are sometimes termed *fictitious* because they vanish in an inertial frame.

Equation (5.5) gives, in particular, the acceleration in $\mathcal{S}'$ of a free particle $(F = 0)$ whose trajectory is rectilinear and uniform in the inertial frame $\mathcal{S}$ $(X^i = X_0^i + V_0^i\,t)$ This acceleration a' is independent of the mass m, which is natural since the particle motion is a purely kinematical 'effect of perspective'[9] due to the motion of the reference frame and not to the action of a force.

---

[6]Galilean invariance is guaranteed if for example the force representing the action of P_a on P depends only on the separation and, possibly, its derivatives: $f = f(l_a, \dot{l}_a, \ddot{l}_a ...)$, where $l_a \equiv R - R_a$.

[7]According to a clever image of Gilles Châtelet in Les enjeux du mobile [Châtelet (1993)].

[8]This is another facet of the Copernican principle: the force exerted on an object cannot depend on either the choice of reference frame or its motion. This occurs if the vector representing the action of P_a on P depends *only* on the separation: $f = f(l_a)$.

[9]This image is due to J.-M. Lévy-Leblond.

Using this mathematical property of non-invariance of the law of dynamics under the group of rigid displacements, we can deduce the existence of inertial forces in non-inertial reference frames and quantify their effects. For example, the rotation of the plane of oscillation of the Foucault pendulum relative to the walls of the Pantheon measures the *absolute* rotation of the Pantheon, which turns out to coincide with that of the Earth relative to the distant stars. This remarkable consequence of Newtonian physics was amazing to Newton himself as well as to Leibniz, Kant, Mach, Poincaré, Einstein, ...[10]

The Foucault pendulum

In the frame $\mathcal{S}'$ attached (for example!) to the Pantheon walls, the equation of motion of the mass m at the end of a pendulum is $a' = F/m - 2\Omega \wedge v'$ to first order in Ω, i.e., if we neglect the centrifugal force quadratic in Ω and take Ω to be constant. The Earth's rotation vector Ω, of modulus ω, is directed along the polar axis, from which we find (see Section 2.5 and Fig. 5.1) $\Omega = \omega(\cos\Psi e'_x + \sin\Psi e'_z)$, where e'_x is the basis vector of $\mathcal{S}'$ pointing from south to north, e'_z points along the vertical upward from the point, and Ψ is the latitude at the point ($\Psi = 0$ at the equator). The force F applied to the mass is the sum of the tension of the cable (which is fixed at $X' = Y' = 0$, $Z' = l$) and the local constant gravitational force. For small oscillations it reduces to $F/m = -(g/l)X' e'_x - (g/l)Y' e'_y$, where $X'(t)$ and $Y'(t)$ specify the position of the mass in the horizontal plane and g is the acceleration of gravity. The equations of motion then become

$$\ddot{X}' + \frac{g}{l}X' - 2(\omega\sin\Psi)\dot{Y}' = 0, \quad \ddot{Y}' + \frac{g}{l}Y' + 2(\omega\sin\Psi)\dot{X}' = 0, \tag{5.6}$$

the solution of which to first order in ω is

$$X'(t) = X'_0[\cos(\omega t\sin\Psi)\cos(t\sqrt{g/l}) + (\sqrt{l/g}\,\omega\sin\Psi)\sin(\omega t\sin\Psi)\sin(t\sqrt{g/l})],$$

$$Y'(t) = X'_0[(\sqrt{l/g}\,\omega\sin\Psi)\cos(\omega t\sin\Psi)\sin(t\sqrt{g/l}) - \sin(\omega t\sin\Psi)\cos(t\sqrt{g/l})]. \tag{5.7}$$

Let us consider the times $t_n = 2\pi n\sqrt{l/g}$, where $2\pi\sqrt{l/g}$ is the proper period of oscillation of the pendulum. Then we have $X'_n = X'_0\cos(t_n\omega\sin\Psi)$, $Y'_n = -X'_0\sin(t_n\omega\sin\Psi)$, which shows that the plane of the oscillations of the pendulum turns relative to $\mathcal{S}'$ from the north toward the east with period

$$P = \frac{2\pi}{\omega\sin\Psi}. \tag{5.8}$$

At the poles $P = 1$ day, at the equator the plane of the pendulum remains fixed (within the approximations we have made), and in Paris ($\Psi = 48°51'$) $P \sim 32$ hours.

---

[10]We also note that the existence of inertial forces makes it difficult to construct a rigid, rotating reference frame of large size. In fact, it is necessary to make the objects of mass m of which the system is composed subject to forces $F = m[-\Omega \wedge (R' \wedge \Omega) + \dot{\Omega} \wedge R' + \ddot{d}]$ which grow with R'.

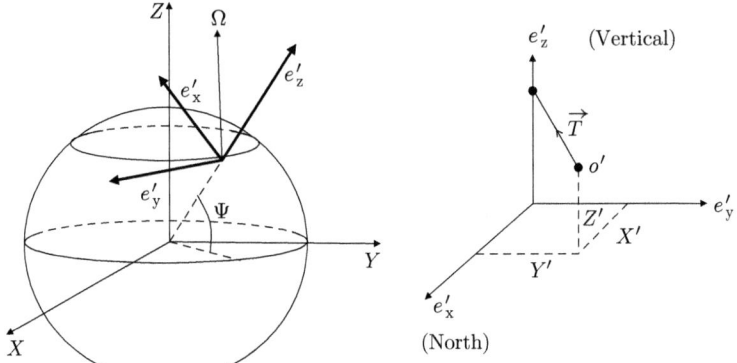

Fig. 5.1 Deviation toward the east and the Foucault pendulum.

The deviation toward the east

The acceleration of a particle in the frame attached to the laboratory walls reduces to $a' = a - 2\Omega \wedge v'$ if we neglect the centrifugal acceleration and take Ω to be constant. If a is constant, integration at zeroth order in Ω gives $v' = at$ (assuming that the initial velocity is zero). Substituting this expression for v' into the equation for a', we obtain the latter to first order in Ω: $a' = a - 2\Omega \wedge at$, the first integral of which gives $v' = at - \Omega \wedge at^2$ and the second gives $O'P = \frac{1}{2}at^2 + H - \frac{1}{3}\Omega \wedge at^3$, where H is the initial position vector of the particle in the laboratory.

If we take $a = -ge'_z$ and $H = he'_z$, using $\Omega = \omega(\cos \Psi e'_x + \sin \Psi e'_z)$ (see Section 2.5) we find $O'P = X'^i e'_i$ with $X' = 0$, $Y' = -\frac{1}{3}g\omega t^3 \cos \Psi$, and $Z' = -\frac{1}{2}gt^2 + h$. Since the time to fall is $\sqrt{2h/g}$, the particle will have deviated from the vertical (we recall that we are neglecting the centrifugal force) by an amount

$$Y' = -\omega \left(\frac{8h^3}{9g}\right)^{1/2} \cos \Psi. \tag{5.9}$$

The negative sign indicates deviation toward the east. Taking the gravitational acceleration at the Earth's surface to be $g = 9.81$ m/s^2, $\omega = 2\pi/24$ hours, $h = 68$ m (the height of the Pantheon), and $\Psi = 48°51'$, we find $|Y'| \approx 8$ mm, the value actually measured by Camille Flammarion in 1903.

6
Dynamics of massive systems

In this chapter we present the laws of motion of an *ensemble* of point masses forming a solid body whose shape is invariant, or a fluid whose shape can vary with time.

6.1 The equations of motion of a solid

The law describing the motion of a point mass is $ma = f$, where a is the acceleration of its trajectory $P(t)$ in a Cartesian inertial frame $\mathcal{S}$, m is its inertial mass, and f is a vector representing the force on the particle. An ensemble of point masses constitutes a *solid* if the distances between the points can be assumed constant. Then there exists a Cartesian frame $\mathcal{S}'$ attached to the solid which is obtained from $\mathcal{S}$ by a rigid displacement in which all the points P are at rest.

Let $R = OP$ be the radius vector of the point P in $\mathcal{S}$, $d = OO'$ be the vector to the origin of the frame $\mathcal{S}'$ in which P is at rest, and $R' = O'P$ be the position vector of P in $\mathcal{S}'$. Differentiating the identity $R = R' + d$ with respect to time twice, we find the following, using the fact that v' and a' vanish (see Section 2.4):

$$v = \dot{d} + \Omega \wedge R', \quad a = \ddot{d} - \Omega \wedge (R' \wedge \Omega) + \dot{\Omega} \wedge R', \quad \text{with } \dot{R}' \equiv \Omega \wedge R', \qquad (6.1)$$

where Ω is the *rotation vector* of the solid, that is, of $\mathcal{S}'$. We choose the *center of mass*, defined as

$$\sum mR' = 0, \qquad (6.2)$$

where the sum runs over all the points P, to be the origin O' of $\mathcal{S}'$. We note that the center of mass does not necessarily correspond to a mass point of the solid.

Now from the dynamical law $ma = f$ with a given in (6.1) and using (6.2) (Ω is independent of the points P) we find

$$M\ddot{d} = F, \qquad (6.3)$$

where $M = \sum m$ and $F \equiv \sum f$ is the vector sum of the forces applied at the various points of the solid. This equation governs the motion of the center of mass in the inertial frame $\mathcal{S}$. We note that the vector F represents the *external* force applied to the solid because the sum of the internal forces is zero (which lies at the origin of the rigidity of the solid) owing to the equality of action and reaction.[1]

---

[1] This demonstrates the importance of this law.

Relativity in Modern Physics. Nathalie Deruelle and Jean-Philippe Uzan.
© Oxford University Press 2018. Published in 2018 by Oxford University Press.
DOI: 10.1093/oso/9780198786399.001.0001

We now introduce

$$K \equiv \sum R' \wedge f \qquad \text{and} \qquad J \equiv \sum mR' \wedge v. \tag{6.4}$$

The vector K represents the *moment* of the (external) forces applied to the solid, and J is its *angular momentum*. According to the law of dynamics $K = \sum mR' \wedge a$, so that the equation of motion, which in $\mathcal{S}$ governs the rotational motion of the solid about its center of mass, is

$$\dot{J} = K. \tag{6.5}$$

Indeed,

$$\dot{J} = \sum m\dot{R}' \wedge v + \sum mR' \wedge a$$

or

$$\sum m\dot{R}' \wedge v = \sum m(\Omega \wedge R') \wedge (\dot{d} + \Omega \wedge R') = 0$$

if $\sum mR' = 0$. Q.E.D.

We still need to find the relation between J and Ω. We have

$$J \equiv \sum mR' \wedge v = \sum mR' \wedge (\dot{d} + \Omega \wedge R') = \sum m[R'^2\Omega - R'(R'.\Omega)] \tag{6.6}$$

or, in component form (see Section 2.4), $J = J'^i e'_i$, where e'_i are the basis vectors of the frame attached to the solid and

$$J'_i = I_{ij}\,\omega'^j \qquad \text{with} \qquad I_{ij} \equiv \sum m(\delta_{ij}\,X'^k X'_k - X'_i X'_j). \tag{6.7}$$

The axes X'^i can be chosen such that the *inertia tensor* I_{ij} is diagonal. The three elements on the diagonal I_1, I_2, and I_3 are the *principal moments of inertia*. If they are all the same, the solid in question is a *spherical top*; if $I_1 = I_2$ and $I_3 = 0$ the solid is a *rotor*, and if $I_1 = I_2 \neq I_3$ it is a *symmetric top*.

The moments of inertia of a homogeneous ellipsoid

Let us consider an ellipsoid bounded by the surface $(X_1^2/a_1^2) + (X_2^2/a_2^2) + (X_3^2/a_3^2) = 1$. Making the change of variable $X_i = a_i x_i$, the principal moments of inertia become those of a unit sphere (for which the volume element in spherical coordinates is given in Section 3.6) of density ϱ. Using $3\int x_i^2 dx_1 dx_2 dx_3 = \int r^4 dr \sin\theta d\phi = 4\pi r^5/5$ and the fact that the volume of the ellipsoid is $V = 4\pi a_1 a_2 a_3/3$, we have, if the ellipsoid is homogeneous (that is, ϱ is constant),

$$I_1 = \varrho \int (X_2^2 + X_3^2)dX_1 dX_2 dX_3 = \varrho a_1 a_2 a_3 \int (a_2^2 x_2^2 + a_3^2 x_3^2)dx_1 dx_2 dx_3 = \frac{M}{5}(a_2^2 + a_3^2) \tag{6.8}$$

and similar expressions for I_2 and I_3, where $M = \varrho V$ is the total mass of the ellipsoid.

If the Earth is described as a homogeneous oblate spheroid of equatorial radius $a_1 = a_2 \equiv r_e = 6378$ km and polar radius $a_3 \equiv r_p = 6357$ km, we find $(I_3 - I_1)/I_3 = (r_e^2 - r_p^2)/2r_e^2 \simeq 3.3 \times 10^{-3}$. Measurements (by observation of satellite trajectories and the precession of the equinoxes) give $I_1 = 0.3302\,M_\oplus r_e^2$, $I_3 = 0.3312\,M_\oplus r_e^2$, or $(I_3 - I_1)/I_3 = 3.0 \times 10^{-3}$ (the difference can be attributed to the fact that the Earth is not homogeneous).

The Euler equations of motion of a solid

The equation of motion of a solid $\dot{J} = K$ in the basis e_i' of a frame attached to the solid can be written as (see Section 2.4)

$$\dot{J} = \frac{d}{dt}(J'^i e_i') = \dot{J}'^i e_i' + J'^i \dot{e}_i' = \dot{J}'^i e_i' + J'^i \Omega \wedge e_i' = \dot{J}'^i e_i' + J'^i \omega'^j e_j' \wedge e_i' = (\dot{J}'^i - e^i{}_{jk} J'^j \omega'^k) e_i'. \quad (6.9)$$

Since $J'^1 = I_1 \omega'^1$ and so on, where the I_i are the principal moments of inertia of the solid, the equations of motion can in the end be written as

$$I_1 \frac{d\omega'^1}{dt} - \omega'^2 \omega'^3 (I_2 - I_3) = K'^1, \quad (6.10)$$

with the other two obtained by cyclic permutation.

6.2 Examples of motions of a solid

Let us consider the case $K = 0$ (no torque). We see from (6.5)–(6.7) that the rotation vector Ω of a spherical top or a rotor (parallel to the angular momentum J) remains at a fixed direction in $\mathcal{S}$ no matter what sort of motion its center of mass undergoes. Such an object is called a *gyroscope*.

Now let us assume that the solid is sufficiently compact so that the spatial variations of the external forces f applied at the points making it up are negligible. We then can write $f = m_g g$, where g is a constant vector (a gravitational field, for example) and m_g is the coupling of a given point of the solid to this field (for example, the gravitational mass). We then have $K \equiv \sum R' \wedge f = (\sum m_g R') \wedge g$. If it turns out that $m_g = m$, then the 'center of gravity' and the center of mass coincide, and $K = 0$, so that although it is acted on by an external force, the rotation vector of a gyroscope remains at a fixed direction in $\mathcal{S}$.

Symmetric tops and the polhode

The angular momentum of a free solid is constant and it is always possible to choose an inertial frame such that $J = j e_3$. Moreover, J is related to the angular velocity $\Omega = \omega'^i e_i'$ of the symmetric top as $J = I_1(\omega'^1 e_1' + \omega'^2 e_2') + I_3 \omega'^3 e_3'$, where e_i' are the basis vectors of the frame attached to the solid and I_1 and I_3 are its moments of inertia. Writing $e_3 = \mathcal{R}_3{}^j e_j'$ and ω'^i as functions of the Euler angles (see Sections 1.3 and 1.5), we find

$$j \sin \Psi \sin \Theta = I_1(\dot{\Phi} \sin \Psi \sin \Theta + \dot{\Theta} \cos \Psi)$$

$$j \cos \Psi \sin \Theta = I_1(\dot{\Phi} \cos \Psi \sin \Theta - \dot{\Theta} \sin \Psi) \quad (6.11)$$

$$j \cos \Theta = I_3(\dot{\Phi} \cos \Theta + \dot{\Psi}).$$

For $\Theta \neq 0$ the first two equations can be written as $I_1 \dot{\Theta} = 0$, $j = I_1 \dot{\Phi}$. Therefore, the symmetry axis of the top maintains a constant angle Θ relative to the angular momentum J and the line of nodes turns at the angular velocity of *regular precession* $\dot{\Phi} = j/I_1$. The last equation gives the velocity of the rotation of the top about its symmetry axis, $\omega'^3 \equiv \dot{\Phi} \cos \Theta + \dot{\Psi} = j \cos \Theta / I_3$, as well as the proper rotational velocity

$$-\dot{\Psi} = \omega'^3 \frac{I_3 - I_1}{I_1}. \quad (6.12)$$

These results can also be derived from the Euler equations (6.10). The equation for ω'_3 again gives ω'^3 equal to a constant, and those for ω'_2 and ω'_1 are equivalent to $d\omega'^1/dt + \omega_0\,\omega'^2 = 0$ and $d\omega'^2/dt - \omega_0\,\omega'^1 = 0$ with $\omega_0 = \omega'^3(I_3 - I_1)/I_1$, which have the solution $\omega'^1 = A\cos(\omega_0 t + \phi)$ and $\omega'^2 = A\sin(\omega_0 t + \phi)$, where A and ϕ are two integration constants. Likewise, the vector Ω, and therefore also the vector J, turns about the axis of the top e'_3 at the angular speed ω_0, which is equal to $-\dot{\Psi}$ (see Fig. 6.1).

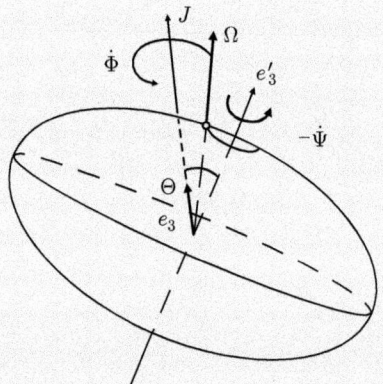

Fig. 6.1 Precession of a symmetric top.

The circle described on the surface of the Earth by the tip of the vector corresponding to rotation about its axis of inertia (that is, by the point where the centrifugal acceleration vanishes), or the *polhode* (path of the pole), which was predicted by Euler in 1765, was first measured in 1891. Its radius is about ten meters. If the Earth is treated as a homogeneous ellipsoid of revolution, then from (6.8) and the fact that $2\pi/\omega'^3 = 1$ day, we find $P \equiv (2\pi/\omega'^3)(I_3/(I_3 - I_1)) \approx 304$ days. Using the measured values $I_1 = 0.330\,2\,M_{\oplus}r_e^2$ and $I_3 = 0.331\,2\,M_{\oplus}r_e^2$, we obtain $P \approx 331$ days. The period which is actually measured (the 'Chandler wobble') is about 432 days, with the difference from the predicted value being largely attributable to the Earth's elasticity (Newcomb).

6.3   The Euler equations of fluid motion

A *perfect fluid* is characterized by its (inertial) mass density $\varrho(t, x^i)$, its pressure $p(t, x^i)$ which phenomenologically describes its internal collisions, and a velocity field $v(t, x^i)$ giving its velocity at x^i at time t. The *streamlines* are integral curves of the velocity field; the streamlines and trajectories of the fluid elements only coincide if the flow is stationary, that is, if $\partial v^i/\partial t = 0$.

Suppose that we are working in a Cartesian inertial frame (see Section 2.6 for the definitions of the differential operators used).

The *continuity equation* or *mass conservation equation*

$$\frac{\partial\varrho}{\partial t} + \nabla.(\varrho v) = 0 \tag{6.13}$$

expresses the fact that the variation of the mass inside a given volume is equal to the quantity of matter which enters the volume minus the quantity which leaves.

Specifically, let us consider a parallelepiped V *at rest* in this frame. The mass it contains is $m = \int_V \varrho(t, X^i)d^3X$, where $d^3X \equiv dX^1dX^2dX^3$. Its variation during a time interval

dt is $dm = dt \int_V (\partial \varrho / \partial t) \, dX^1 dX^2 dX^3$ and is equal to the amount which leaves the volume, that is, $-dt \int_S (\varrho V^1) dX^2 dX^3$ minus two similar terms. Using the divergence theorem (see Section 2.6) this is $-dt \int_V \partial_1(\varrho V^1) dX^1 dX^2 dX^3 - \dots = -dt \int_V \partial_i(\varrho V^i) d^3X$. Therefore, $\int_V (\partial \varrho / \partial t + \nabla.(\varrho v)) \, d^3X = 0$ for any size of the parallelepiped V, so that the integrand must be equal to zero. Q.E.D.

The *Euler equation* is Newton's equation $F = m \, dv/dt$ for a fluid element of mass m:

$$\frac{dv}{dt} \equiv \frac{\partial v}{\partial t} + (v.\nabla)v = \frac{1}{\varrho}(f - \nabla p). \tag{6.14}$$

Indeed, the force exerted on the small parallelepiped τ bounded by S and containing mass m (sufficiently small that its velocity can be defined unambiguously) is, by the definition of the pressure, $-\int_S p \, dS = -\int_\tau \nabla p \, d^3X = -(\nabla p)\tau$ according to the divergence theorem. The quantity ϱ is the mass density $(m = \varrho\tau)$, and f is the density of the force field acting on the fluid. This force may be external, or internal, that is, due to forces of the rest of the fluid on the element in question which are not included in the pressure (a gravitational force, for example).

Equations (6.13) and (6.14) for the motion of a fluid have the two following simple consequences.[2]

Using the continuity equation (6.13) and the divergence theorem, it is easy to prove the following useful relation valid for any function $g(t, X^i)$:

$$\frac{d}{dt} \int \varrho g \, d^3X = \int \varrho \frac{dg}{dt} d^3X. \tag{6.15}$$

For $g = 1$ we find the statement that the total mass is constant: $M \equiv \int \varrho \, d^3X$.

For $g = v = \dot{R}$ we find, using the Euler equation (6.14) and applying the divergence theorem a second time,

$$\frac{d^2}{dt^2} \int \varrho R \, d^3X = \int f \, d^3X \equiv F_{\text{ext}}, \tag{6.16}$$

that is, the equation of motion of the fluid as a whole.

The Eulerian and Lagrangian derivatives

Let us consider, in some Cartesian frame, a tensor field $G(t, X^i)$ (which may be a scalar, a vector, and so on). Its *Eulerian derivative* $\partial G/\partial t$ is its rate of variation at a given point in space. The differential $dG = \partial_i G \, dX^i + (\partial G/\partial t)dt$ represents its variation under an infinitesimal change of t and X^i. If we require that dX^i arise from the displacement of the fluid, we then have $dX^i = V^i dt$, where the V^i are the Cartesian components of the velocity field $v(t, X^i)$ at the instant and at the point in question. Therefore,

$$\frac{dG}{dt} = \frac{\partial G}{\partial t} + V^i \frac{\partial G}{\partial X^i} = \frac{\partial G}{\partial t} + v.\nabla G. \tag{6.17}$$

This *Lagrangian derivative* dG/dt (also called the *particle* or *convective* derivative) represents the variation of G along the trajectory of a fluid element.

If $G = \varrho$, the mass density of the fluid, then $d\varrho/dt = \partial \varrho/\partial t + v.\nabla \varrho = -\varrho \nabla.v$ owing to the continuity equation. A flow is *incompressible* if $d\varrho/dt = 0$, and its velocity field then has zero divergence: $\nabla.v = 0$.

---

[2]We shall postpone the study of wave propagation to Section 17.3.

Finally, an *equation of state* relates the pressure and the density. For a *barotropic fluid* it is simply

$$p = p(\varrho). \tag{6.18}$$

As an example, for a *polytrope* we have $p \propto \varrho^\gamma$, where γ is its *adiabatic index*.

The continuity equation and the Euler equation are written in tensor notation as

$$\frac{\partial \varrho}{\partial t} + \tilde{D}_i(\varrho v^i) = 0, \quad \frac{\partial v^i}{\partial t} + v^j \tilde{D}_j v^i = \frac{1}{\varrho}(f^i - e^{ij}\partial_j p), \tag{6.19}$$

where the e^{ij} are the coefficients of the inverse Euclidean metric in the coordinates x^i and $\tilde{D}$ is the associated covariant derivative (see Section 3.2).

Newton's bucket

In an accelerated frame $\mathcal{S}'$ the Euler equation can be written as (see Section 5.4)

$$\frac{\partial v'}{\partial t} + v'.\nabla v' = \frac{1}{\varrho}(f - \nabla p) - 2\Omega \wedge v' + \Omega \wedge (R' \wedge \Omega) - \dot{\Omega} \wedge R' - \ddot{d}. \tag{6.20}$$

Let us consider a fluid (water, for example) at rest in a reference frame undergoing uniform rotation (for example, a bucket turning about the vertical). The external force f is that of terrestrial gravity $f = \varrho g$, where g is a constant vector. The Euler equation reduces to $0 = (f - \nabla p)/\varrho + \Omega^2 X'$, where X' is the distance to the rotation axis. Let $[X' = X'(\lambda), Z' = Z'(\lambda)]$ be the equations of the water surface Σ. Its tangent vector is $(\dot{X}', \dot{Z}')$, where here the dot denotes differentiation with respect to the parameter λ. The components of the gravitational and centrifugal accelerations are $(0, -g)$ and $(\Omega^2 X', 0)$. Their sum must be orthogonal to Σ, and so $\Omega^2 X' \dot{X}' - g\dot{Z}' = 0$, from which we derive the equation for Σ: $Z' = X'^2 \Omega^2/(2g)$. It is a parabola (see Fig. 6.2).

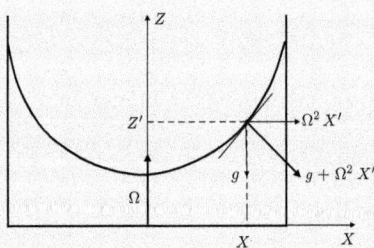

Fig. 6.2 Newton's bucket.

Therefore, even though it is at rest relative to the bucket walls, the surface of the water is not horizontal. Its curvature measures the absolute rotation of the walls, which amazed Newton himself and was exhaustively discussed by Mach (who wondered if Σ would remain concave if the bucket were several leagues thick...).

7

Conservation laws

In this chapter we define the conserved quantities associated with an isolated dynamical system, that is, the quantities which remain constant during the motion of the system. The law of *momentum* conservation follows directly from Newton's third law. The law of *angular momentum* conservation holds if the forces acting on the elements of the system depend only on the separation of the elements. Finally, the conservation of *total energy* requires in addition that the forces be derivable from a *potential*.

7.1   Momentum and the center of mass

The superposition principle for forces allows Newton's law of motion for a body P_a acted on by other bodies $P_{a'}$ in an inertial Cartesian frame $\mathcal{S}$ to be written as

$$m_a a_a = F_a \quad \text{with} \quad F_a = \sum_{a' \neq a} f_{a'a} \, . \tag{7.1}$$

For example, in the case of three bodies $m_1 a_1 = f_{21} + f_{31}$, $m_2 a_2 = f_{32} + f_{12}$, and $m_3 a_3 = f_{13} + f_{23}$, and so $\sum_a m_a a_a \equiv \sum_a m_a dv_a/dt = 0$ owing to Newton's third law $f_{aa'} = -f_{a'a}$. The vector corresponding to the *total momentum* $\mathcal{P}$ defined as

$$\mathcal{P} = \sum_a m_a v_a \tag{7.2}$$

is therefore constant.[1]

Integrating again with respect to time, we find that the motion of the *center of mass* C (or the *center of inertia*, or also the *barycenter*) of a set of point masses P_a whose radius vector $OC \equiv \rho$ is defined as

$$\rho = \frac{\sum_a m_a R_a}{M} \quad \text{with} \quad M = \sum_a m_a \, , \tag{7.3}$$

where $R_a \equiv OP_a$ are the radius vectors of the points P_a, is uniform and rectilinear in $\mathcal{S}$:

$$\rho = \frac{\mathcal{P}}{M} t + \rho_0 \tag{7.4}$$

with ρ_0 a constant.

---

[1] It is understood that the sum runs over all the points representing the matter system in question. If only a subset of these points is summed over, it describes a subsystem which is not *isolated*, but subject to an *external* force (due to the other bodies) and $d\mathcal{P}/dt = \sum F_{\text{ext}}$. The equations of motion of such a subsystem when it is a solid have been given in Section 6.1.

We note that if the vectors $m_a v_a$ are attached to the trajectories of the particles a, then $\mathcal{P}$ is their *resultant*, obtained after they are parallel-transported to a reference point.

Relativity in Modern Physics. Nathalie Deruelle and Jean-Philippe Uzan.
© Oxford University Press 2018. Published in 2018 by Oxford University Press.
DOI: 10.1093/oso/9780198786399.001.0001

The *center-of-mass frame* is the frame $\mathcal{S}_{\text{cm}}$ undergoing uniform translation with respect to $\mathcal{S}$ with its origin coinciding with the center of mass C. In this frame $CP_a \equiv R_{a\,\text{cm}} = R_a - \rho$, and the total momentum is zero:

$$\sum_a m_a R_{a\,\text{cm}} = 0 \quad \Longrightarrow \quad \mathcal{P}_{\text{cm}} \equiv \sum_a m_a v_{a\,\text{cm}} = 0. \tag{7.5}$$

We note that the fact that the total momentum is constant in any Galilean frame can be proved using only the equality of action and reaction in any frame ($f_{aa'} = -f_{a'a}$).

Furthermore, under a change of frame by a transformation from the group of rigid displacements ($R = R' + d$, $v = v' + \dot{d} + \Omega \wedge R'$, see (1.7)), the total momentum transforms as follows if we set $\mathcal{P}' \equiv \sum mv'$ and $M\rho' = \sum mR'$:

$$\mathcal{P} = \mathcal{P}' + M(\dot{d} + \Omega \wedge \rho'). \tag{7.6}$$

We therefore see that the total momentum is zero in any rotating frame if its origin coincides with the center of mass.

A final remark: it is of course possible to express the total momentum in a system of curvilinear coordinates x^i, where the values of its components $(\partial x^i / \partial X^j)\mathcal{P}^j$ are no longer constant but depend on the point. If we start from the law of dynamics written in a system of curvilinear coordinates, that is, $m_a \tilde{D}_{v_a} v_a = F_a$, then we can obtain the total momentum conservation law only *after* the parallel transport of all the vectors to the same reference point. The simplest way to proceed is of course to return to a system of Cartesian coordinates.[2]

Particle collisions and inertial mass measurement

The numerical value to be assigned to the inertial mass of a given body can in principle be obtained by performing experiments involving collisions of the body with a test mass serving as a standard. One measures the change of velocity of the two bodies in an inertial frame and uses the momentum conservation law:

$$(m\,v + m_{\text{ref}}v_{\text{ref}})_1 = (m\,v + m_{\text{ref}}v_{\text{ref}})_2 \quad \Longrightarrow \quad m(v_2 - v_1) = m_{\text{ref}}(v_{\text{ref2}} - v_{\text{ref1}}). \tag{7.7}$$

This method is not readily applicable to elementary particles or stars, whose masses are estimated by comparing the predicted trajectories to the observed ones (for example, in a cyclotron or in the solar system).

7.2   Angular momentum

Let us again start from Newton's second law written in a Cartesian inertial frame $\mathcal{S}$, $m_a a_a = \sum_{a' \neq a} f_{a'a}$, and form the vector product with the position vector R_a of the point P_a. Newton's third law then implies that

$$\frac{d}{dt}\left(\sum_a R_a \wedge m_a v_a \right) = \sum_{a,a' \text{ with } a < a'} l_{a'a} \wedge f_{a'a}, \tag{7.8}$$

---

[2]In the curved spaces of general relativity where Cartesian coordinate systems do not exist, the parallel transport of vectors from one point to another will depend on the path followed, and the definition of the total momentum of a matter system will have to be formulated within a different framework.

where $l_{a'a} \equiv R_a - R_{a'}$ (this is very easy to see in the case of three bodies). The right-hand side of (7.8) represents the *torque* of the forces applied to the system. We note that since the resultant of the force vectors is zero, the torque is independent of the choice of origin O of the inertial frame which is used. If it is zero (as is the case if $f_{a'a}$ is proportional to the separation vector $l_{a'a}$), then the *angular momentum* $\mathcal{M}_O$ of the system relative to O, defined as

$$\mathcal{M}_O = \sum_a R_a \wedge m_a v_a, \qquad (7.9)$$

is constant.

The Cartesian components of $\mathcal{M}_O$ are therefore constants and can be written (omitting the indices a and O) as $\mathcal{M}_i = \sum m e_{ijk} X^j V^k$ or $\mathcal{M}_X = \sum m(Y\dot{Z} - \dot{Y}Z)$, and so on. They can of course be expressed in a system of curvilinear coordinates; for example, in spherical coordinates

$$\begin{cases} \mathcal{M}_X = -\sum_a m_a\, r_a^2 (\dot{\theta}_a \sin \phi_a + \dot{\phi}_a \sin \theta_a \cos \theta_a \cos \phi_a) \\[2mm] \mathcal{M}_Y = \sum_a m_a\, r_a^2 (\dot{\theta}_a \cos \phi_a - \dot{\phi}_a \sin \theta_a \cos \theta_a \sin \phi_a) \\[2mm] \mathcal{M}_Z = \sum_a m_a\, r_a^2 \dot{\phi}_a \sin^2 \theta_a\, . \end{cases} \qquad (7.10)$$

If we view the angular momentum $\mathcal{M}$ as a covariant vector (*i.e.*, a form), then its components in a system of curvilinear coordinates x^i depend on the coordinates of the reference point at which it is evaluated and are given by $\mathcal{M}'_i = (\partial X^j / \partial x^i)\mathcal{M}_j$, which are no longer constants.

Under a change of frame by a rigid displacement $(R = R' + d,\ v = v' + \dot{d} + \Omega \wedge R')$ the angular momentum transforms as follows, setting $\mathcal{M}'_{O'} = \sum R' \wedge mv'$, $\mathcal{P}' \equiv \sum mv'$, and $M\rho' = \sum mR'$:

$$\mathcal{M}_O = \mathcal{M}'_{O'} + d \wedge \mathcal{P}' + M(d \wedge \dot{d} + \rho' \wedge \dot{d} + d \wedge (\Omega \wedge \rho')) + \sum mR' \wedge (\Omega \wedge R'). \qquad (7.11)$$

Therefore, for example, in the center-of-mass frame where $\dot{d} = 0$, $\Omega = 0$, $\mathcal{P}' = 0$, and $\sum mR' = 0$, it is independent of the point relative to which it is calculated (it was denoted as J in Section 6.1).

7.3 Energy

Finally, let us take the scalar product of Newton's equation $m_a a_a = \sum_{a' \neq a} f_{a'a}$ and the velocity v_a of the point P_a. Newton's third law then implies that

$$\frac{d}{dt} T = \sum_{a<a'} f_{a'a} \cdot (v_a - v_{a'}), \quad \text{where} \quad T \equiv \sum_a \frac{1}{2} m_a v_a^2 \qquad (7.12)$$

is the *total kinetic energy* of the system. In the general case the integral over time of the right-hand side depends on the trajectories followed by the system points. It will be independent of them if the integrand is a total derivative. This is the case if $f_{a'a}$ is derived from a *potential* energy $w_{a'a}$ which is a function of only the separation $l_{a'a} = R_a - R_{a'}$ of the points P_a and $P_{a'}$ (and possibly the time), that is, if

$$f_{a'a} = -\nabla_a w_{a'a} \quad \text{or, in component form,} \quad f_{i\,a'a} = -\frac{\partial w_{a'a}}{\partial X_a^i}. \tag{7.13}$$

(We note that such a class of force vectors automatically satisfies Newton's third law.) The *total potential energy* of the system is defined as

$$W = \sum_{a<a'} w_{a'a}, \quad \text{which implies} \quad F_a = -\nabla_a W. \tag{7.14}$$

Since $w_{a'a}$ is a function of $(R_a - R_{a'})$ (and possibly the time), we have

$$\frac{dW}{dt} = \sum_{a<a'} \frac{dw_{a'a}}{dt} = \sum_{a<a'} \left(\nabla_a w_{a'a}.(v_a - v_{a'}) + \frac{\partial w_{a'a}}{\partial t} \right)$$

$$= -\sum_{a<a'} f_{a'a}.(v_a - v_{a'}) + \frac{\partial W}{\partial t} = -\frac{dT}{dt} + \frac{\partial W}{\partial t}. \tag{7.15}$$

If in addition the potential energy does not depend explicitly on the time (in which case the forces are said to be *conservative*), we have $\partial W/\partial t = 0$ and the *total energy* of the system $\mathcal{E}$, the sum of the kinetic and potential energies

$$\mathcal{E} = T + W, \tag{7.16}$$

is constant in the selected inertial frame.

The potential energy W is invariant under a change of frame by a rigid displacement $(R = R' + d, \; v = v' + \dot{d} + \Omega \wedge R')$ if it depends only on the relative separations of the particles. As far as the kinetic energy is concerned, setting $T' = \sum \frac{1}{2}mv'^2$, $\mathcal{M}'_{O'} = \sum R' \wedge mv'$, $\mathcal{P}' \equiv \sum mv'$, and $M\rho' = \sum mR'$, it transforms as

$$T = T' + \frac{1}{2}M\dot{d}^2 + \dot{d}.\mathcal{P}' + \Omega.\mathcal{M}'_{O'} + M\dot{d}.(\Omega \wedge \rho') + \frac{1}{2}\sum m(\Omega \wedge R')^2. \tag{7.17}$$

We therefore see that the kinetic energies differ only by a constant in going from one inertial frame to another. In the case of a pure rotation $(d = 0)$ the expression can again be simplified, using (7.11) and the fact that $\Omega.(R \wedge (\Omega \wedge R)) = (\Omega \wedge R)^2$:

$$T = T' - \frac{1}{2}\sum m(\Omega \wedge R')^2 + \Omega.\mathcal{M}_O. \tag{7.18}$$

The kinetic and potential energies are scalars and therefore have the same value in any system of curvilinear coordinates, which indicates that (in contrast to the momentum and angular momentum conservation laws) it is easy to find the law of total energy conservation starting from Newton's equation written in any system of curvilinear coordinates. In fact, forming the scalar product with $v_i \equiv e_{ij}v^j$ and using f^i to denote the components of the force in the coordinates x^i (see, for example, Section 5.1), we have

$$v_i\left(\frac{dv^i}{dt} + \tilde{\Gamma}^i_{jk}v^jv^k\right) = \frac{v_i f^i}{m} \quad \text{with} \quad \begin{cases} v_iv^jv^k\,\tilde{\Gamma}^i_{jk} = \dfrac{1}{2}v^iv^jv^k\,\dfrac{\partial e_{ij}}{\partial x^k} \\[2mm] \dfrac{1}{2}\dfrac{d}{dt}(v_iv^i) = v_i\dfrac{dv^i}{dt} + \dfrac{1}{2}v^iv^jv^k\dfrac{\partial e_{ij}}{\partial x^k} \end{cases}, \tag{7.19}$$

and so $\left(\frac{1}{2}mv_iv^i\right)^{\cdot} = v_i f^i$. We then proceed as above to obtain $\mathcal{E} \equiv T + W = \text{const}$ if W depends only on the relative distances.

Velocity-dependent potential energy

If the potential energy depends not only on the positions but also on the relative velocities of the system elements, $w_{a'a} = w(R_a - R_{a'}, v_a - v_{a'})$ (here we do not consider a possible t dependence), we will have

$$\frac{d}{dt}\sum_{a<a'} w_{a'a} = \sum_{a<a'}\left(\frac{\partial w_{a'a}}{\partial R_a}.(v_a - v_{a'}) + \frac{\partial w_{a'a}}{\partial v_a}.(\dot{v}_a - \dot{v}_{a'})\right), \tag{7.20}$$

where we have written $\nabla_a w_{a'a} \equiv \partial w_{a'a}/\partial R_a$ and the vector with components $\partial w_{a'a}/\partial V^i$ as $\partial w_{a'a}/\partial v_a$. From the Leibniz rule we then find

$$\frac{d}{dt}\sum_{a<a'}\left(w_{a'a} - \frac{\partial w_{a'a}}{\partial v_a}.(v_a - v_{a'})\right) = \sum_{a<a'}\left(\frac{\partial w_{a'a}}{\partial R_a} - \frac{d}{dt}\frac{\partial w_{a'a}}{\partial v_a}\right).(v_a - v_{a'}). \tag{7.21}$$

Therefore, if the force vector $f_{a'a}$ is derived from a potential energy as

$$f_{a'a} = -\left(\frac{\partial w_{a'a}}{\partial R_a} - \frac{d}{dt}\frac{\partial w_{a'a}}{\partial v_a}\right), \tag{7.22}$$

from the definition of T in (7.20) we have (since the third law is satisfied)

$$T + \sum_{a<a'}\left(w_{a'a} - \frac{\partial w_{a'a}}{\partial v_a}.(v_a - v_{a'})\right) = \text{const}. \tag{7.23}$$

It should be noted that the force defined in (7.22) depends on the relative positions, velocities, *and* accelerations of the points of the system (unless $w_{aa'}$ is linear in the velocities). Similarly, if the potential energy depends on the relative positions, velocities, *and* accelerations, the force is obtained from the expression

$$f_{a'a} = -\left(\frac{\partial w_{a'a}}{\partial R_a} - \frac{d}{dt}\frac{\partial w_{a'a}}{\partial v_a} + \frac{d^2}{dt^2}\frac{\partial w_{a'a}}{\partial a_a}\right) \tag{7.24}$$

and the dynamical law is then in general a fourth-order differential equation.

Forces of the type (7.22) or (7.24), which depend only on the *relative* positions, velocities, accelerations, and so on, are invariant under Galilean transformations, and, more generally, under Milne transformations (any acceleration of the origin without rotation of the axes), but are not invariant under rotation. Only forces depending only on the relative positions, and derived from potential energies which also depend only on the relative positions, are invariant under the group of rigid displacements.

Bernoulli's equation

Let us multiply the Euler equation for the motion of a fluid, $\partial v/\partial t + (v.\nabla)v = (f - \nabla p)/\varrho$, by the velocity field. Using the relation $(v.\nabla)v = \frac{1}{2}\nabla v^2 + (\nabla \wedge v)\wedge v$, we find

$$v.\left(\frac{\partial v}{\partial t} + \frac{1}{2}\nabla v^2\right) = \frac{v}{\varrho}.(f - \nabla p). \tag{7.25}$$

If part of the flow is stationary, that is, if $\partial v/\partial t = 0$, and if $f = -\varrho\nabla U$ where the *potential* U depends only on R (as will be the case in gravitation), then $v.\nabla = d/dt$ and, along the trajectory of a fluid element, that is, along a streamline, since the fluid is stationary we have

$$\frac{1}{2}v^2 + U + \int \frac{dp}{\varrho} = \text{const}. \tag{7.26}$$

This expression, called *Bernoulli's equation*, is the law of energy conservation along a streamline, which extends throughout the fluid if the fluid in addition is irrotational (that is, if $\nabla \wedge v = 0$).

7.4   Virial theorems

Virial theorems relate the time-averaged values of kinetic and potential energies.

By definition, the time average of a function $f(t)$ is

$$\bar{f} = \frac{1}{2\tau}\int_{-\tau}^{\tau} dt\, f(t), \tag{7.27}$$

where τ is one period or tends to infinity depending on the situation. If $f(t) = dF(t)/dt$, then $\bar{f} = 0$ if $F(t)$ is bounded or periodic.

In an inertial frame the kinetic energy of a system is, by definition, $T = \frac{1}{2}\sum_a m_a v_a^2$, where m_a and v_a are the masses and velocities of its elements. This can be rewritten as $T = \sum_a \left(\frac{1}{2}m_a R_a v_a\right)^{\cdot} - \frac{1}{2}\sum_a m_a R_a a_a$, where the R_a are the radius vectors of the elements (and $v_a = \dot{R}_a$, $a_a = \dot{v}_a$). Consequently, if the motion of the system is bounded, from Newton's second and third laws we have

$$\overline{T} = -\frac{1}{2}\sum_a m_a \overline{R_a.a_a}$$

$$= -\frac{1}{2}\sum_a \sum_{a'\neq a} \overline{R_a.f_{a'a}} = \frac{1}{2}\sum_{a<a'} \overline{(R_a - R_{a'}).f_{a'a}}. \tag{7.28}$$

If now $f_{a'a}$ is derivable from a potential energy $w_{a'a}$ as $f_{a'a} = -\nabla_a w_{a'a}$ and if $w_{a'a}$ is a homogeneous function of degree k of the distance $r_{a'a} = |R_a - R_{a'}|$ between the particles, i.e., if $w_{a'a} = (\text{const})\, r_{a'a}^k$, then $(R_a - R_{a'}).f_{a'a} = k\, w_{a'a}$ and so

$$2\overline{T} = k\overline{W}, \tag{7.29}$$

where $W = \sum_{a<a'} w_{a'a}$ is the total potential energy of the system. If $k = -1$ (the case of the gravitational force) we have $2\overline{T} + \overline{W} = 0$. Equation (7.29) is called the *scalar virial theorem* (Clausius, 1870).

Tensorial virial theorem

Let us consider the moment of inertia tensor of a system[3]: $\tilde{I}^{ij} = \sum m X^i X^j$, where the X^i are the Cartesian coordinates of the points making up the system in an inertial frame. We differentiate twice with respect to time. From the fundamental law of motion $F = ma$ we have (where the V^i are the components of the velocity)

$$\ddot{\tilde{I}}^{ij} = 2 \sum_a m_a V_a^i V_a^j + \sum_a (F_a^i X_a^j + F_a^j X_a^i). \tag{7.30}$$

Using the superposition principle ($F_a = \sum_{a' \neq a} f_{a'a}$) and Newton's third law ($f_{aa'} = -f_{a'a}$), and denoting the components of the separation vector as $l_{a'a}^i = X_a^i - X_{a'}^i$, we have

$$\ddot{\tilde{I}}^{ij} = 2 \sum_a m_a V_a^i V_a^j - \sum_{a<a'} (l_{a'a}^i f_{a'a}^j + l_{a'a}^j f_{a'a}^i). \tag{7.31}$$

If now $f_{a'a}$ is derivable from a potential energy $w_{a'a}$ as $f_{a'a} = -\nabla_a w_{a'a}$ and if $w_{a'a}$ is a homogeneous function of degree k of the distance $r_{a'a} = |l_{a'a}|$ between the particles, that is, if $w_{a'a} = \text{const } r_{a'a}^k$, then

$$\ddot{\tilde{I}}^{ij} = 2 \sum_a m_a V_a^i V_a^j - 2k \sum_{a<a'} \frac{w_{a'a}}{r_{a'a}^2} l_{a'a}^i l_{a'a}^j. \tag{7.32}$$

By analogy with the definitions of the kinetic and potential energies, we can define the *kinetic energy tensor* T^{ij} and the *potential energy tensor* W^{ij} as

$$T^{ij} \equiv \sum_a \frac{1}{2} m_a V_a^i V_a^j, \quad W^{ij} \equiv \sum_{a<a'} \frac{w_{a'a}}{r_{a'a}^2} l_{a'a}^i l_{a'a}^j, \tag{7.33}$$

so that (7.32) becomes

$$\frac{1}{2} \ddot{\tilde{I}}^{ij} = 2T^{ij} - k W^{ij}. \tag{7.34}$$

This is the *tensorial virial theorem*. If the motion is stationary (that is, if $\ddot{\tilde{I}}^{ij} = 0$), or taking the average for bounded motion, we have

$$2T^{ij} = k W^{ij}, \tag{7.35}$$

where the trace is the scalar virial theorem given in (7.29). For gravity $k = -1$.

Let us conclude with the remark that in our treatment of the forces we have neglected those due to internal collisions, that is, we have ignored the pressure term $-\nabla p$ present in the Euler equation (see Section 6.3). The pressure can be included by adding to (7.31) the term $-\int_V \partial_i p\, X_j d^3 X$, where the X^i are the coordinates of a fluid element and the integral runs over the entire volume occupied by the fluid. Integrating by parts and using the divergence theorem, we find $-\int_V \partial_i p\, X_j d^3 X = \delta_{ij} \int_V p\, d^3 X$ because the pressure vanishes at the surface of the fluid. Therefore, the virial theorem (7.34) is generalized to fluids as

---

[3]This tensor is related to the one introduced in Section 6.1 as $I_{ij} = \tilde{I} \delta_{ij} - \tilde{I}_{ij}$.

$$\frac{1}{2}\ddot{I}^{ij} = 2T^{ij} - k\,W^{ij} + \delta^{ij}\,\Pi\,, \qquad \text{where} \qquad \Pi \equiv \int_V p\,d^3X \tag{7.36}$$

is the *internal energy* and

$$I^{ij} \equiv \int_V \varrho X^i X^j\,, \quad T^{ij} \equiv \frac{1}{2}\int_V \varrho V^i V^j\,d^3X\,, \quad W^{ij} \equiv \frac{1}{2}\int_V w^{ij}\,d^3X$$

with

$$w^{ij} = \int_V \frac{(X^i - X'^i)(X^j - X'^j)}{|X - X'|^2}\,w\,d^3X'\,,$$

where $w = \text{const}\,|X - X'|^k$.

8

Lagrangian mechanics

According to Newton's first law, the trajectory representing the motion of a free particle between two points p_1 and p_2 is a straight line, that is, the shortest path joining p_1 and p_2. In other words, out of all the possible paths between p_1 and p_2, the trajectory effectively followed by a free particle is the one that minimizes the length.

However, even though the use of the principle of extremal length of the paths between two points gives the straight line joining the points, this does not mean that the straight-line path is traveled with constant velocity in an inertial frame. Moreover, the trajectory describing the motion of a particle subject to a force is not uniform and rectilinear and therefore does not minimize the length of the path joining the points.

A great accomplishment of the mechanicians of the seventeenth (Fermat) and eighteenth (Jean Bernoulli, D'Alembert, Maupertuis, Euler, Lagrange, and others) centuries was to show that the Newtonian law of motion of a particle subject to a gradient force derived from a 'potential energy' can in fact always be obtained from an extremal principle, or 'principle of least action'.

8.1 The Euler–Lagrange equations

The *Lagrangian formalism* is constructed as follows.[1]

Let us first place ourselves in an inertial frame.

The *action* $S[P_s(t)] = \int_{t_1}^{t_2} L[P_s(t)] \, dt$ associated with a particle which departs from the point $p_1 = P(t_1)$ at the instant t_1 and arrives at $p_2 = P(t_2)$ at t_2 is a functional of the paths $P_s(t)$, parametrized by s, joining p_1 and p_2 (we shall take $P_{s=0}(t) \equiv P(t)$ to be the trajectory effectively followed by the particle). Here $L[P_s(t)]$ is the *Lagrangian* associated with the particle. If $x^i = x^i(t, s)$ are the equations of these paths in some system of curvilinear coordinates, the Lagrangian at a given time t_0 will depend on three functions of the path $x^i(t_0, s)$ and, also, *a priori*, on all their time derivatives at t_0: $\dot{x}^i(t, s)|_0$, $\ddot{x}^i(t, s)|_0$, and so on, where the dot denotes differentiation with respect to the time t. These are the data required for constructing the entire path $P_s(t)$ starting from the point $P_s(t_0)$. The action therefore depends on the entire history of the path s. However, to obtain in the end Newton's equation of motion which is of second order in the time, it is sufficient, as will be seen *a posteriori*, to require that the Lagrangian be a function of the three $x^i(t, s)$ and their first time derivatives only. If the system is composed of N particles (a) it possesses $3N$ *degrees of freedom* which constitute its *configuration space*, and its Lagrangian depends on the $3N$ fonctions $x_a^i(t, s)$ and their $3N$ time derivatives. It is usual to denote them collectively as $q(t, s)$ and $\dot{q}(t, s)$, and to refer to them as the *generalized coordinates* and *generalized velocities* of the system. The action is then written as $S = \int_{t_1}^{t_2} L(q, \dot{q}) \, dt$. It is a function of the parameter s.

---

[1] For a more complete discussion see, for example, Basdevant (2005), Landau and Lifshitz (1976), or Lanczos (1970). For the mathematical aspects see Bourguignon (1995) and Arnold (1997).

Relativity in Modern Physics. Nathalie Deruelle and Jean-Philippe Uzan.
© Oxford University Press 2018. Published in 2018 by Oxford University Press.
DOI: 10.1093/oso/9780198786399.001.0001

The *principle of least action* stipulates that the action be an *extremum* for the actual motion:

$$\delta S \equiv \frac{d}{ds} S \Big|_0 \, ds = 0,$$

$$\text{with} \quad S = \int_{t_1}^{t_2} L(q, \dot{q}) \, dt, \quad \text{where} \quad q = q(t, s) \quad \text{and} \quad \dot{q}(t, s) \equiv \frac{\partial q}{\partial t}, \tag{8.1}$$

and where δS is the variation of the action in going from the path with $s = 0$ to a path an infinitesimal distance away $x^i(t, 0) \mapsto x^i(t, ds)$. Let us calculate δS by introducing the notation $\delta q \equiv \frac{\partial x_a^i}{\partial s}\big|_0 \, ds$ and $\delta \dot{q} \equiv \frac{\partial^2 x_a^i}{\partial s \partial t}\big|_0 \, ds$:

$$\delta S = \int_{t_1}^{t_2} \delta L \, dt = \int_{t_1}^{t_2} \left(\frac{\partial L}{\partial q} \delta q + \frac{\partial L}{\partial \dot{q}} \delta \dot{q} \right) dt$$

$$= \left(\frac{\partial L}{\partial \dot{q}} \delta q \right) \Big|_{t_1}^{t_2} + \int_{t_1}^{t_2} \left(\frac{\partial L}{\partial q} - \frac{d}{dt} \frac{\partial L}{\partial \dot{q}} \right) \delta q \, dt, \tag{8.2}$$

where we have used the commutativity of the second derivatives $[\delta \dot{q} = (\delta q)\dot{}]$ and performed an integration by parts, and where it is understood that all quantities are evaluated at $s = 0$. Since the variations of the paths vanish at t_1 and t_2 but otherwise are arbitrary, δS is zero if the integrand is zero. Therefore, the trajectory effectively followed by the system must satisfy the $3N$ *Euler–Lagrange equations*:

$$\frac{\delta L}{\delta q} \equiv \frac{d}{dt} \frac{\partial L}{\partial \dot{q}} - \frac{\partial L}{\partial q} = 0. \tag{8.3}$$

These are ordinary second-order differential equations for the $3N$ functions $q(t, s = 0) \equiv q(t)$ only if L is linear in $\dot{q}$. The function $\delta L / \delta q$ is called the *variational derivative*[2] with respect to q of $L(q, \dot{q})$.

We note that these equations remain valid if the Lagrangian depends explicitly on time $[L = L(q, \dot{q}, t)]$, and that the Lagrangians are defined up to a total derivative with respect to time. If

$$L'(q, \dot{q}, t) = L(q, \dot{q}, t) + df(q, t)/dt,$$

the actions S and S' differ by only a constant which vanishes when they are varied.

8.2 The laws of motion (II)

We still need to specify the generalized coordinates as well as the expression for the Lagrangian for each physical problem studied.

First let us consider a free particle in an inertial Cartesian frame $\mathcal{S}$ (see Fig. 8.1). Then $q = X^i(t, s)$ and $\dot{q} \equiv V^i(t, s)$. If the action is chosen to be the path length, $S[P_s(t)] = \int_{t_1}^{t_2} \sqrt{\delta_{ij} V^i V^j} \, dt$, the Euler–Lagrange equations (8.3) again tell us that the trajectory effectively followed is a straight line, but they do not tell us the velocity along the trajectory.

---

[2]If L is a functional of q, $\dot{q}$, and $\ddot{q}$, its variational derivative is $\frac{\delta L}{\delta q} = -\frac{d^2}{dt^2} \frac{\partial L}{\partial \ddot{q}} + \frac{d}{dt} \frac{\partial L}{\partial \dot{q}} - \frac{\partial L}{\partial q}$, and is in general fourth-order. Minimizing the action then gives $\frac{\delta L}{\delta q} = 0$ with the condition that not only the generalized coordinates q but also the generalized velocities $\dot{q}$ be fixed at t_1 and t_2.

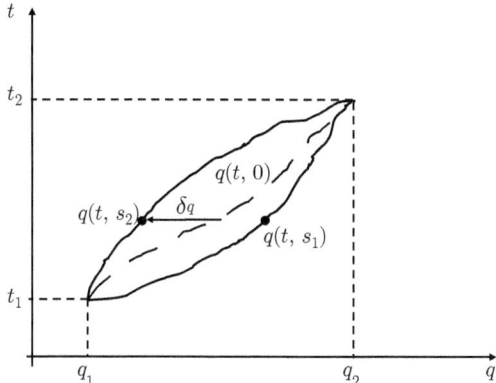

Fig. 8.1 Paths in spacetime.

The reason for this indeterminacy is that the length of a path is the same whether it is parametrized by the time t or by some function of t. To obtain the *uniform* rectilinear motion of a free particle it is necessary to get rid of this *reparametrization invariance* and extremize a different functional of the paths, for example, $S[P_s(t)] = \frac{1}{2} \int_{t_1}^{t_2} (\delta_{ij} V^i V^j) \, dt$. This is not a random choice. Indeed, since the corresponding Lagrangian $(L = \frac{1}{2} \delta_{ij} V^i V^j)$ is quadratic in the velocity, its variational derivative is just the acceleration a of the particle, which is invariant under Galilean transformations, so that the equation of motion satisfies the principle of relativity.[3]

Now, while remaining in an inertial frame $\mathcal{S}$ but using a system of curvilinear coordinates $\mathcal{C}$, let us consider N interacting particles (a). We have $q = x_a^i(t, s)$ and $\dot{q} = v_a^i(t, s)$. Inspired by the preceding arguments, we choose as our Lagrangian

$$L(x^i, v^i) = \frac{1}{2} \sum m \, e_{ij} \, v^i v^j - W(x^i) \tag{8.4}$$

[with the indices (a) understood], the difference of the particle *kinetic* and *potential energies* in $\mathcal{S}$, where m_a is the inertial mass of the particle (a) and e_{ij} are the components of the Euclidean metric in $\mathcal{C}$.

We then have $\partial L/\partial x^i = \frac{1}{2} v^j v^k \partial_i e_{jk} - \partial_i W$ and $\partial L/\partial v^i = e_{ij} v^j$, which implies that $(\partial L/\partial v^i)^{\cdot} = e_{ij} \dot{v}^j + v^j v^k \partial_k e_{ij}$. Consequently,

$$\frac{d}{dt} \frac{\partial L}{\partial v^i} - \frac{\partial L}{\partial x^i} = e_{ij} \dot{v}^j + \frac{1}{2} v^j v^k (\partial_k e_{ij} + \partial_j e_{ik} - \partial_i e_{jk}) + \partial_i W. \tag{8.5}$$

Therefore, the Euler–Lagrange equations (8.3), after raising an index and using the definition of the Christoffel symbols, give the equations of motion of a particle (a) of the system in the form

---

[3]If, for example, we take the Lagrangian to be $L(v) = v^4$, the Euler–Lagrange equation always implies $v = $ const, but the variational derivative $(4v^2 v)^{\cdot}$ will *not* be invariant under a Galilean transformation. Another aspect of the same property is that if the Lagrangian is quadratic in the velocity, then $L(v) = v^2$ and $L'(v) = (v - V_0)^2$, where V_0 is a constant, differ only by a total derivative with respect to time $[L' = L + (V_0^2 t - 2V_0.R)^{\cdot}]$. They therefore have the same variational derivative (that is, $\dot{v}$), which incorporates the requirement that the velocity of a free particle can be defined only up to a constant relative to the absolute frame.

$$\frac{dv^i}{dt} + \Gamma^i_{jk} v^j v^k = \frac{1}{m} F^i \quad \text{with} \quad \begin{cases} \Gamma^i_{jk} \equiv \frac{1}{2} e^{il} \left(\frac{\partial e_{kl}}{\partial x^j} + \frac{\partial e_{jl}}{\partial x^k} - \frac{\partial e_{jk}}{\partial x^l} \right) \\ \\ F^i \equiv -e^{ij} \frac{\partial W}{\partial x^j} \end{cases}, \quad (8.6)$$

where it is understood that all quantities are evaluated at $x^i = x^i_{(a)}(t, s=0) \equiv x^i(t)$ and where the e^{ij} are the components of the inverse metric in the coordinates x^i. We therefore again find the equations of motion given in Section 5.1, and on the left-hand side we recognize the covariant derivative of the velocity.

The formulation of a problem in mechanics therefore amounts to specifying the potential energy $W(x^i_a)$. In order for this potential energy not to spoil the Galilean invariance of the Euler–Lagrange equations, it is sufficient that it depend only on the *relative* positions, velocities, accelerations, *etc.* of the particles making up the system. However, in order for the forces derived from it to be invariant under the group of rigid displacements, it is necessary that it depend only on the relative positions; see Section 7.3.

Acceleration and the covariant derivative

Let us consider the Lagrangian $L[P(t)] = L(x^i(t), v^i(t))$. The quantity

$$\vartheta = \left(\frac{d}{dt} \frac{\partial L}{\partial v^i} - \frac{\partial L}{\partial x^i} \right) dx^i \quad (8.7)$$

(where the coefficients are the variational derivatives of L with respect to the x^i) is a differential form. This can be proved by using the transformation law for the basis forms under the change of coordinates $x^i \to x'^i$ (see Section 4.2) as well as the fact that since $v'^j = \frac{\partial x'^j}{\partial x^i} v^i$, then $\frac{\partial L}{\partial v^i} = \frac{\partial L}{\partial v'^j} \frac{\partial x'^j}{\partial x^i}$ and $\frac{\partial L}{\partial x^i} = \frac{\partial L}{\partial x'^j} \frac{\partial x'^j}{\partial x^i} + \frac{\partial L}{\partial v'^j} \frac{\partial^2 x'^j}{\partial x^i \partial x^k} v^k$.

If we choose $L_{\text{kin}} = \frac{1}{2} e_{ij} v^i v^j$, the form ϑ is called the *quantity of acceleration*, and the vector associated with it by raising an index is just the covariant derivative of the velocity $\tilde{D}_v v$:

$$\tilde{D}_v v = e^{ij} \left(\frac{d}{dt} \frac{\partial L_{\text{kin}}}{\partial v^i} - \frac{\partial L_{\text{kin}}}{\partial x^i} \right) \frac{\partial}{\partial x^j} = \left(\frac{dv^i}{dt} + \Gamma^i_{jk} v^j v^k \right) \frac{\partial}{\partial x^i}. \quad (8.8)$$

Therefore, the principle of least action also gives a precise geometric status to $\tilde{D}_v v$.

The principle of least action can also be used to easily find the equations of motion of a particle in a moving reference frame. The passage from one inertial Cartesian frame to another moving frame is effected by a rigid displacement such that the new generalized coordinates are related to the old ones as $X'^i(t, s) = \mathcal{R}_j{}^i(t)(X^j(t, s) - d^j(t))$ and the radius vectors as $R' = R - d$. A gradient is an invariant vector under such a transformation and, after elementary rearrangements (see below), the Euler–Lagrange equations (8.3), where the generalized coordinates are the X'^i, lead to the equations of motion already encountered in Section 5.3, but now with the force being a gradient $F = -\nabla W$:

$$\begin{cases} a'^i = \mathcal{R}_j{}^i \left(-\dfrac{1}{m} \dfrac{\partial W}{\partial X'^k} \delta^{jk} - \ddot{d}^j \right) - 2e^i{}_{jk}\omega'^j V'^k - e^i{}_{jk}\dot{\omega}'^j X'^k \\[2mm] \qquad + e^i{}_{jk}\omega'^j e^k{}_{lm} X'^l \omega'^m \\[2mm] a' = -\dfrac{1}{m}\nabla W - \ddot{d} - 2\Omega \wedge v' + \Omega \wedge (R' \wedge \Omega) - \dot{\Omega} \wedge R' . \end{cases} \qquad (8.9)$$

The Euler–Lagrange equations in an accelerated frame

Let us show how the equations of motion in an accelerated frame are obtained in the Lagrangian formalism.

Since the rotation matrix satisfies $\delta_{ij}\mathcal{R}_k{}^i(t)\mathcal{R}_l{}^j(t) = \delta_{kl}$, the kinetic part of the Lagrangian in $\mathcal{S}$ is written as (see Section 1.5) $L_{\text{kin}} = \frac{1}{2}\delta_{ij}V^i V^j = \frac{1}{2}\delta_{ij}f^i f^j$ with $f^i = \mathcal{R}_k{}^i V^k = V'^i - \dot{\mathcal{R}}_k{}^i X^k + \dot{d}'^i$ and $X^k = \delta^{kl}\delta_{mn}\mathcal{R}_l{}^n(X'^m + d'^m)$, so that $\partial L_{\text{kin}}/\partial V'^m = \delta_{jm}(V'^j - \dot{\mathcal{R}}_k{}^j X^k + \dot{d}'^j)$ and $\partial L_{\text{kin}}/\partial X'^m = \delta_{jm}\dot{\mathcal{R}}_k{}^j V^k$. Therefore,

$$\frac{d}{dt}\frac{\partial L_{\text{kin}}}{\partial V'^m} - \frac{\partial L_{\text{kin}}}{\partial X'^m} = \delta_{jm}(\dot{V}'^j - 2\dot{\mathcal{R}}_k{}^j V^k - \ddot{\mathcal{R}}_k{}^j X^k + \ddot{d}'^j), \qquad (8.10)$$

which, after introducing the instantaneous rotation of the frame, leads to (8.9) (see Section 1.5).

To obtain the vector form of the Euler–Lagrange equation, we write the kinetic part of the Lagrangian in $\mathcal{S}$ as $2L_{\text{kin}} = v^2 = (v' + \dot{d} + \Omega \wedge R')^2$, where the squared quantities are interpreted not as scalar products $[v^2 \equiv (v.v)]$ but as *duality brackets* $v^2 \equiv \bar{v}(v)$, where $\bar{v}$ is the form associated with the vector v by lowering an index (a trivial operation in this case because $\bar{v} = \bar{V}_i\epsilon^i$ with $\bar{V}_i = \delta_{ij}V^j$ numerically equal to V^i). Therefore, $\partial L_{\text{kin}}/\partial R' \equiv (\partial L_{\text{kin}}/\partial X'^i)\,\epsilon'^i = -\bar{\Omega} \wedge \bar{v}$ and $\partial L_{\text{kin}}/\partial v' \equiv (\partial L_{\text{kin}}/\partial V'^i)\,\epsilon'^i = \bar{v}$, which (using the fact that $\dot{\epsilon}'^i = \bar{\Omega} \wedge \epsilon'^i$ by the definition of the vector Ω) implies that $(\partial L_{\text{kin}}/\partial v')\dot{} = (\partial L_{\text{kin}}/\partial V'^i)\,\epsilon'^i + \bar{\Omega} \wedge \bar{v} = \dot{\bar{v}}$. Then the variational derivative with respect to R' becomes

$$\left(\frac{d}{dt}\frac{\partial L_{\text{kin}}}{\partial V'^i} - \frac{\partial L_{\text{kin}}}{\partial X'^i} \right)\epsilon'^i = \dot{\bar{v}} - \bar{\Omega} \wedge \bar{v} + \bar{\Omega} \wedge \bar{v} = \dot{\bar{v}} . \qquad (8.11)$$

Now we need only to go from forms to vectors and write $\dot{v} \equiv a$ as a function of R', v', and a' using $a = a' + \ddot{d} + 2\Omega \wedge v' - \Omega \wedge (R' \wedge \Omega) + \dot{\Omega} \wedge R'$ to recover (8.9).

The principle of least action can therefore be used to obtain the equations of motion not only in any coordinate system, but also in any frame. However, except in inertial frames (for which $\Omega = 0$ and $\ddot{d} = 0$), it does not allow them to be written in a unique form which is valid in any moving frame. On the other hand, the theory of special relativity, where the time is no longer a parameter but becomes a fourth coordinate, yields a genuine unification of changes of coordinate system and changes of frame.

Constraints and Lagrange multipliers

Let us consider a particle subject to a constraint, that is, to a force which we shall not make explicit and whose effect on the particle is to fix the evolution of one of its degrees of freedom. Such a constraint is written as $f(x^i, t) = 0$. (For example, $X^2 + Y^2 = 1$ means that the particle must move on the surface of a cylinder.) Now to determine the motion of the particle subject to this constraint in another force field we can of course solve the equation $f(x^i, t) = 0$ and extract, for example, $x^1 = x^1(x^2, x^3)$, so that the Lagrangian of the particle now depends on only two degrees of freedom (it is a useful exercise to solve the problem of free motion on a cylinder in this manner). A more elegant method is to introduce the Lagrangian

$$L'(x^i, v^i, \lambda, t) = L(x^i, v^i, t) + \lambda f(x^i, t), \tag{8.12}$$

where the parameter λ is called a *Lagrange multiplier*, and set to zero the variational derivatives with respect to the three x^i and λ:

$$\frac{d}{dt} \frac{\partial L}{\partial \dot{x}^i} - \frac{\partial L}{\partial x^i} - \lambda \frac{\partial f}{\partial x^i} = 0, \quad f(x^i, t) = 0. \tag{8.13}$$

(The example of motion on a cylinder will convince the reader of the superiority of the Lagrange method in this case.)

Incompressible fluids

The kinetic energy of a fluid element of mass m and small volume τ is $\frac{1}{2}\varrho v^2 \tau$, where ϱ is the mass density and v is the velocity of the element (these are unambiguously defined if τ is sufficiently small).

A flow is *incompressible* if ϱ is constant along the trajectory of the fluid element. Since $m = \varrho \tau$ is constant, τ is thereby constrained to be constant also. The Lagrangian describing the fluid element is then $\frac{1}{2}\varrho v^2 \tau - p(\tau - \tau_0)$, where p is a Lagrange multiplier and τ_0 is the small initial volume to which the element is confined. The variation of this Lagrangian when the path of the element is varied is $\delta(\frac{1}{2}\varrho v^2 \tau - p(\tau - \tau_0)) = \varrho\tau v.\delta v + \frac{1}{2}v^2\delta(\varrho\tau) - p\,\delta\tau = \varrho\tau v.\delta v - p\,\delta\tau$ because $\varrho\tau = m$ is constant.

Now the calculation of the variation of τ when all the particles making up the element undergo an average displacement from X^i to $X^i + \delta X^i$ is mathematically identical to the calculation of this variation when passing from the Cartesian coordinates X^i to the curvilinear coordinates $x^i = X^i + \xi^i$, where $\xi^i(x^j, t) \equiv \delta X^i$ is a field of infinitesimal vectors. In the coordinates x^i the volume element is $\sqrt{\det e}\, d^3x$, where $\det e$ is the determinant of the components of the Euclidean metric in the coordinates x^i (see Section 3.6). To calculate $\det e$ as a function of ξ^i, we use the fact that the metric is a 2-fold covariant tensor and therefore transforms as (see Section 3.6)

$$e_{ij} = \frac{\partial X^k}{\partial x^i} \frac{\partial X^l}{\partial x^j} \delta_{kl} = (\delta_i^k - \partial_i \xi^k)(\delta_j^\delta - \partial_j \xi^\delta)\delta_{k\delta} = \delta_{ij} - \partial_i \xi_j - \partial_j \xi_i \tag{8.14}$$

to first order in ξ^i. From this we deduce that $\det e = 1 - 2\partial_i \xi^i$ (always to first order) and so $\delta\tau = -\partial_i \xi^i \tau$.

The variation of the Lagrangian describing a fluid element then becomes $(\varrho v.\delta v + p\,\partial_i \xi^i)\tau$, so that the variation of the Lagrangian describing the entire fluid—the sum of these elementary Lagrangians—is

$$\delta L = \int_V \left(\varrho v.\delta v + p\,\partial_i(\delta X^i) \right) dV = \int_V \left(\varrho V_i \delta V^i - \partial_i p\,\delta X^i \right) dV, \tag{8.15}$$

where the integral runs over the entire volume occupied by the fluid and we have replaced a sum over τ by an integral over V and transformed the last term using the divergence theorem and the fact that δX^i vanishes on the surface of the fluid. Since $V^i = dX^i/dt$, a final integration by parts now gives the equation of motion of the incompressible fluid in the form

$$\varrho\frac{dv}{dt} = -\nabla p, \tag{8.16}$$

which is the Euler equation of a fluid if p is interpreted as the pressure.[4]

8.3 Conservation laws (II)

The 'principle of Galilean relativity' stipulates that a mechanical interaction must be represented by the same vector in all the frames of the group of rigid displacements (see Section 5.3). This is an aspect of the 'Copernican principle', according to which interactions cannot depend on the reference frame in which they are described. The 'active' version of this principle is that the interactions governing the dynamics of an isolated system must be independent of the system location and spatial orientation (see Section 5.2).

Moreover, we have seen in Chapter 7 that the total momentum of an isolated system is conserved if the forces acting on it satisfy Newton's third law (equality of action and reaction). We have also seen that its total angular momentum is conserved if in addition the inter-particle forces are proportional to the particle position vectors, and, finally, that the total energy is conserved if these forces are derivable from a potential which depends on the particle separations and is independent of time. Now, a quantity which depends only on relative separations is invariant under translation and rotation. It is therefore clear that the Copernican principle and the existence of first integrals of the equations of motion are closely related. Noether succeeded in explaining this link within the framework of the Lagrangian formalism.

Let $S = \int dt\, L(x_a^i, \dot{x}_a^i, t)$ be the action, in an inertial frame, of a system of N particles a, where $x^i = x_a^i(t,s)$, $i = 1,2,3$, is the equation of a possible path of the particle a parametrized by s and where $\dot{x}_a^i \equiv \partial x_a^i/\partial t$. In Section 8.1 we have seen how this action varies when the paths are varied, that is, when $s \mapsto s + ds$ [cf. (8.2)]. The change of S when the system is displaced globally is obtained in the same way. Indeed, such an operation is described by the transformation $x_a^i(t,s) \mapsto x_a'^i(t,s) = x_a^i(t,s) + \xi^i$, where the ξ^i are the components, in the coordinates x^i, of a field of infinitesimal vectors so that

$$\delta_\xi S = \int_{t_1}^{t_2} dt \sum_a \left(\frac{\partial L}{\partial x_a^i} - \frac{d}{dt}\frac{\partial L}{\partial \dot{x}_a^i}\right)\xi^i + \left(\sum_a \frac{\partial L}{\partial \dot{x}_a^i}\xi^i\right)\Bigg|_{t_1}^{t_2}. \tag{8.17}$$

If the Euler–Lagrange equations of motion are satisfied, the first term vanishes and the variation of the action reduces to

$$\delta_\xi S = \left(\sum_a \frac{\partial L}{\partial \dot{x}_a^i}\xi^i\right)\Bigg|_{t_1}^{t_2}. \tag{8.18}$$

In Cartesian coordinates $x^i = \{X,Y,Z\}$ a translation along the X axis is defined by $X_a' = X_a + \xi$, $Y_a' = Y_a$, $Z_a' = Z_a$, where ξ is constant. If the action of the system, and therefore

---

[4]The derivation of the Euler equation using variational principles has a long history. The approach we have followed here is apparently that of Lagrange; *cf.* Lanczos (1986).

its Lagrangian and its dynamics, is invariant under this translation, *i.e.*, $\delta_\xi S = 0$, then $\sum_a \partial L / \partial \dot{X}_a \equiv \mathcal{P}_X$ is a constant. Translations along the Y and Z axes are treated the same way. Then from the requirement of invariance of the action under general translations $X'^i_a = X^i_a + \xi^i$ we deduce the existence of three conserved quantities

$$\sum_a \frac{\partial L}{\partial V^i_a} = \mathcal{P}_i . \tag{8.19}$$

Since the Lagrangian of a system is given in Cartesian coordinates by (see Section 8.2)

$$L = \frac{1}{2} \sum_a m_a \delta_{ij} V^i_a V^j_a - W(X^i_a) , \tag{8.20}$$

we see that the quantities in (8.19) are just the components of the total momentum (7.2):

$$\mathcal{P} = \sum_a m_a v_a . \tag{8.21}$$

The latter is therefore constant if the Lagrangian (that is, the potential energy W) is invariant under translations. In Section 7.1 we saw that $\mathcal{P}$ is constant if the forces satisfy Newton's third law. Requiring the equality of action and reaction is therefore equivalent to requiring translational invariance of the potential energy from which the forces are derived.

In spherical coordinates $x^i = \{r, \theta, \phi\}$ a rotation about the OZ axis is defined by $r' = r$, $\theta' = \theta$, $\phi' = \phi + \xi$, where ξ is constant. The action is invariant under such a displacement if [*cf.* (8.18)] $\sum_a \partial L / \partial \dot{\phi}_a = \mathcal{M}_Z$, where $\mathcal{M}_Z$ is a constant. The Lagrangian is written in spherical coordinates as (*cf.* (8.6) and Section 3.1)

$$L = \frac{1}{2} \sum_a m_a (\dot{r}^2_a + r^2_a \dot{\theta}^2_a + r^2_a \sin^2 \theta_a \, \dot{\phi}^2_a) - W(r_a, \theta_a, \phi_a, t) , \tag{8.22}$$

and so

$$\mathcal{M}_Z = \sum_a m_a r^2_a \sin^2 \theta_a \, \dot{\phi}_a \tag{8.23}$$

is just the Z component of the angular momentum of the system defined in (7.10). We note that (8.23) can also be derived from (8.20) for the Lagrangian in Cartesian coordinates. Indeed, the spherical components of the vector field ξ^i_{sph} are $(0, 0, \xi)$, and so their Cartesian components are $\xi^i_{\text{Cart}} = (\partial X^i / \partial x^j) \xi^j_{\text{sph}}$ or $\xi^i_{\text{Cart}} = \xi(-Y, X, 0)$. Requiring that (8.18) vanish then gives $\sum m(-V^X Y + V^Y X) = \text{const}$, which is just $\mathcal{M}_Z$ in Cartesian coordinates.

We proceed in the same way for rotations about the X axis [for which $\xi^i_{\text{Cart}} = \xi(0, Z, -Y)$] and the Y axis [$\xi^i_{\text{Cart}} = \xi(Z, 0, -X)$], and we find that requiring invariance of the action under general rotations leads to conservation of the total angular momentum $\mathcal{M}_O$ relative to the origin of the frame, with

$$\mathcal{M}_O = \sum_a m_a R_a \wedge v_a . \tag{8.24}$$

Isometries of the Euclidean metric

Let $e_{ij}(p)$ be the values of the metric coefficients at a point p with coordinates x^k in some system of curvilinear coordinates. At a neighboring point q with coordinates $x^k + \xi^k$ these coefficients are, to first order in ξ, $e_{ij}(p) + \xi^k \partial_k e_{ij}$. Now let us perform a change of coordinate $x^k = x'^k + \xi^k$ such that in the new system q has the same coordinates as p in the old one. In the new coordinates the metric coefficients at q will be (to first order in ξ)

$$e'_{kl}(q) = \frac{\partial x^i}{\partial x'^k} \frac{\partial x^j}{\partial x'^l} e_{ij}(q) = (\delta^i_k + \partial_k \xi^i)(\delta^j_l + \partial_l \xi^j)(e_{ij}(p) + \xi^k \partial_k e_{ij})$$
$$= e_{kl}(p) + e_{kp}\partial_l \xi^p + e_{lp}\partial_k \xi^p + \xi^p \partial_p e_{kl}.$$

The difference $e'_{kl}(q) - e_{kl}(p)$ is called the *Lie derivative* of the metric.

The displacement is an *isometry* and the vector ξ^i is a *Killing vector* if the Lie derivative of the metric vanishes: $e'_{lm}(q) = e_{lm}(p)$ (this means that the change $\xi^k \partial_k e_{ij}$ due to the displacement can be compensated for by a change of coordinates). In order for this to be possible, ξ^i must satisfy the *Killing equations*:

$$e_{kp}\partial_l \xi^p + e_{lp}\partial_k \xi^p + \xi^p \partial_p e_{kl} = 0 \qquad \Longleftrightarrow \qquad \tilde{D}_i \xi_j + \tilde{D}_j \xi_i = 0. \qquad (8.25)$$

(The equivalence is easily shown by explicit calculation of the covariant derivative; see Section 3.2.)

It is of course simplest to find the Killing vectors of the Euclidean metric by using Cartesian coordinates, so that the Killing equations reduce to $\partial_i \xi_j + \partial_j \xi_i = 0$, the general solution of which is

$$\xi^i = d^i + \omega^i{}_j X^j \qquad \text{with} \qquad \omega_{ij} = -\omega_{ji}, \qquad (8.26)$$

where d^i is a constant vector and $\omega_{ij} \equiv \delta_{ik}\omega^k{}_j$ is an antisymmetric matrix with constant coefficients. The Euclidean metric in three dimensions therefore possesses six Killing vectors, three translational ones proportional to $\xi^i = (1,0,0)$, $\xi^i = (0,1,0)$, and $\xi^i = (0,0,1)$, and three rotational ones proportional to $\xi^i = (-Y, X, 0)$, $\xi^i = (0, Z, -Y)$, and $\xi^i = (Z, 0, -X)$.

The fact that the Euclidean metric possesses three translational Killing vectors expresses the *homogeneity* of Euclidean space, while the fact that it possesses three rotational ones expresses the *isotropy* of Euclidean space.

From the Killing equations it follows that $\tilde{D}_i \tilde{D}_j \xi_k = 0$ (because the Euclidean covariant derivatives commute). To determine a Killing field everywhere, it is therefore sufficient to know the value of the field as well as its first derivatives at a point, which corresponds to six initial conditions (since the nine first derivatives obey the six Killing equations). Consequently, the maximum number of Killing vectors that a metric can possess is six (and, more generally, $n(n+1)/2$ in n dimensions): Euclidean space is *maximally symmetric*.

Finally, let us calculate the time derivative of a Lagrangian $L(q, \dot{q}, t)$:

$$\frac{dL}{dt} = \frac{\partial L}{\partial t} + \frac{\partial L}{\partial q}\dot{q} + \frac{\partial L}{\partial \dot{q}}\ddot{q} = \frac{\partial L}{\partial t} + \frac{\partial L}{\partial q}\dot{q} + \frac{d}{dt}\left(\frac{\partial L}{\partial \dot{q}}\dot{q}\right) - \frac{d}{dt}\left(\frac{\partial L}{\partial \dot{q}}\right)\dot{q}$$

$$= \frac{\partial L}{\partial t} + \frac{d}{dt}\left(\frac{\partial L}{\partial \dot{q}}\dot{q}\right)$$

if the equations of motion are satisfied. Therefore, if we require that the Lagrangian of a system have no explicit time dependence, that is, that $\partial L/\partial t = 0$ (which amounts to requiring that a dynamical experiment on a closed system be reproducible), we then obtain a final conservation law, the law of energy conservation:

$$\mathcal{E} = \sum_a \frac{\partial L}{\partial v^i_{(a)}} v^i_{(a)} - L. \tag{8.27}$$

Since the Lagrangian is given by (8.4), we again find that $\mathcal{E}$ is the sum of the kinetic and potential energies, *cf.* (7.16):

$$\mathcal{E} = \sum_a \frac{1}{2} m_a v_a^2 + W. \tag{8.28}$$

Let us make a concluding remark. The Euler–Lagrange equations can be written as

$$\frac{dp_i^{(a)}}{dt} = \frac{\partial L}{\partial x^i_{(a)}}, \qquad \text{where} \qquad p_i^{(a)} = \frac{\partial L}{\partial v^i_{(a)}} \tag{8.29}$$

is the *conjugate momentum* of the $x^i_{(a)}$. Therefore, if in a system of well chosen coordinates the Lagrangian does not depend on a generalized coordinate $x^i_{(a)}$ (which is then referred to as a *cyclic coordinate*), the corresponding conjugate momentum is constant, and supplies a first integral of the motion.

9
Hamiltonian mechanics

The complexity of the Newtonian equations of motion for N interacting bodies, in celestial mechanics in particular, led to the development in the late 18th and early 19th centuries (by Poisson, Hamilton, Jacobi, ...) of a formalism which reduces these equations to first-order differential equations. In this chapter we give a brief overview of this formalism, which is known as *Hamiltonian mechanics*.[1]

9.1 Hamilton's equations

Lagrangian mechanics describes the evolution of a system of N particles subject to forces which are gradients of a potential energy. It does this using a Lagrangian L which is a function of the $3N$ generalized coordinates $q^i(t)$ and their time derivatives $\dot{q}^i$ specifying the particle positions and velocities. The equations of motion are the $3N$ second-order Euler–Lagrange equations (see Section 8.1)

$$\frac{d}{dt}\frac{\partial L}{\partial \dot{q}^i} - \frac{\partial L}{\partial q^i} = 0. \tag{9.1}$$

Let us introduce the quantity p_i, called the *conjugate momentum* of q^i and defined as

$$p_i \equiv \frac{\partial L}{\partial \dot{q}^i}. \tag{9.2}$$

If $L = \frac{1}{2}m\dot{q}^2 - W(q,t)$ and $q \equiv X$ is a Cartesian coordinate, then p reduces to the particle momentum. If the expressions in (9.2) are invertible, that is, if we can extract $\dot{q}^i = \dot{q}^i(q^j, p_j)$ from them, the equations of motion (9.1) can be rewritten as a system of $6N$ differential equations of first order in time:

$$\dot{p}_i = f_i(q^j, p_j, t), \quad \dot{q}^i = g^i(q^j, p_j, t), \tag{9.3}$$

where the g^i are obtained by inverting (9.2) and $f_i \equiv \partial L / \partial q^i$, the $\dot{q}^j$ being replaced by the g^j.

In 1835 Hamilton found more explicit expressions for f_i and g^i. To show how he did this, let us perform some operations on the differential of the Lagrangian $L(q, \dot{q}, t)$ (omitting the indices i):

$$dL = \frac{\partial L}{\partial \dot{q}}d\dot{q} + \frac{\partial L}{\partial q}dq + \frac{\partial L}{\partial t}dt = p\,d\dot{q} + \frac{\partial L}{\partial q}dq + \frac{\partial L}{\partial t}dt$$

$$\Longleftrightarrow \quad d(p\,\dot{q} - L) = \dot{q}\,dp - \frac{\partial L}{\partial q}dq - \frac{\partial L}{\partial t}dt, \tag{9.4}$$

---

[1] A more complete discussion of Hamiltonian mechanics can be found in the references cited at the beginning of the preceding chapter.

Relativity in Modern Physics. Nathalie Deruelle and Jean-Philippe Uzan.
© Oxford University Press 2018. Published in 2018 by Oxford University Press.
DOI: 10.1093/oso/9780198786399.001.0001

where we have used the definition of p and the Leibniz rule. When the equations of motion (9.1)–(9.2) are satisfied, we have $\partial L/\partial q = \dot{p}$ and (9.4) becomes (restoring the indices i with summation over repeated indices understood)

$$dH = \dot{q}^i \, dp_i - \dot{p}_i \, dq^i - \frac{\partial L}{\partial t} dt \,, \quad \text{where} \quad H \equiv p_i \dot{q}^i - L \tag{9.5}$$

is the *Hamiltonian* of the system. Since the $\dot{q}^i$ are expressed as a function of the q^j and p_j by inverting (9.2), the Hamiltonian is a function of the $6N$ 'coordinates' $\{q^i, p_j\}$ forming the *phase space*.

The passage from L to H is a *Legendre transformation*.

Legendre transformations

Let us consider a function $F(x_i, \lambda_i)$ of the variables x_i and parameters λ_i. We introduce new variables y_i defined as

$$y_i = \frac{\partial F}{\partial x_i} \,. \tag{9.6}$$

After inversion, i.e., finding $x_i = x_i(y_j, \lambda_j)$ (which we assume to be possible), we define a new function $G = G(y_i, \lambda_i)$ as

$$G = \sum_i x_i y_i - F \,. \tag{9.7}$$

Next let us consider the change of G produced by infinitesimal variations of the y_i and λ_i. Using the definition of G, it can be written as

$$\begin{aligned}
dG &= \sum_i \frac{\partial G}{\partial y_i} dy_i + \sum_i \frac{\partial G}{\partial \lambda_i} d\lambda_i = \sum_i (x_i dy_i + y_i dx_i) - dF \\
&= \sum_i x_i dy_i - \sum_i \frac{\partial F}{\partial \lambda_i} d\lambda_i + \sum_i \left(y_i - \frac{\partial F}{\partial x_i} \right) dx_i \,.
\end{aligned} \tag{9.8}$$

The last term of (9.8) vanishes owing to the definition (9.6) of the new variables y_i. And so (9.8) states that

$$\frac{\partial G}{\partial y_i} = x_i \,, \quad \frac{\partial G}{\partial \lambda_i} = -\frac{\partial F}{\partial \lambda_i} \,. \tag{9.9}$$

Therefore, *Hamilton's equations* of motion, a rewriting of the Euler–Lagrange equations, can be read off from (9.5):

$$\dot{q}^i = \frac{\partial H}{\partial p_i} \,, \quad \dot{p}_i = -\frac{\partial H}{\partial q^i} \,, \quad \frac{\partial H}{\partial t} = -\frac{\partial L}{\partial t} \,. \tag{9.10}$$

The functions $-\partial H/\partial q^i$ and $\partial H/\partial p_i$ are then the functions f_i and g^i of (9.3). We note that

$$\frac{dH}{dt} = \frac{\partial H}{\partial t} + \sum_{i=1}^{3N} \left(\frac{\partial H}{\partial q^i} \dot{q}^i + \frac{\partial H}{\partial p_i} \dot{p}_i \right) = \frac{\partial H}{\partial t} \tag{9.11}$$

if the equations of motion (9.10) are satisfied.

The Hamiltonian of a system of particles

In an inertial frame the Lagrangian of a system of N particles is given by the difference of the kinetic and potential energies (see Section 8.2):

$$L(x_a^i, \dot{x}_a^i) = \sum_{a=1}^{N} \frac{1}{2} m_a \, e_{ij} \, \dot{x}_a^i \dot{x}_a^j - W(x_a^i), \tag{9.12}$$

where the e_{ij} are the metric components in the coordinates x^i. The conjugate momenta and the Hamiltonian are therefore given by

$$p_i^a = m_a e_{ij} \dot{x}_a^j, \quad H = \sum_{a=1}^{N} \frac{p_a^2}{2m_a} + W(x_a^i), \tag{9.13}$$

where $p^2 = e^{ij} p_i p_j$. Hamilton's equations then read

$$\dot{x}_a^i = \frac{1}{m_a} e^{ij} p_j, \quad \dot{p}_i^a = -\frac{\partial W}{\partial x_a^i}, \tag{9.14}$$

which is of course nothing but a rewriting of the Euler–Lagrange equations (8.5). When the equations of motion are satisfied, the Hamiltonian is just the energy of the system and (9.11) allows us to recover the fact that the energy is conserved throughout the motion when the Hamiltonian does not depend explicitly on time.

9.2 Canonical transformations

Given a Lagrangian and having constructed the corresponding Hamiltonian, Hamilton's equations (9.10) amount to simply a rewriting of the Euler–Lagrange equations (9.1).

The feature that makes the Hamiltonian formulation superior is that the dimension of the (q^i, p_i) phase space is double that of the q^i configuration space, so that in addition to *point transformations* $q^i \to q^i(Q^j, t)$, it is possible to perform more general transformations $q^i \to q^i(Q^j, P_j, t)$, $p_i \to p_i(Q^i, P_j, t)$ in order to simplify solving the equations of motion.

Among all the possible transformations, those which preserve the *form* of Hamilton's equations naturally have a special status: they are the *canonical transformations*. In order to find the conditions they must satisfy, let us consider the actions S and S' of the system written respectively in the variables (q, p) and (Q, P). Since (again omitting the indices i)

$$S \equiv \int L \, dt = \int (p \, \dot{q} - H) dt = \int (p \, dq - H \, dt) \tag{9.15}$$

by definition of the Hamiltonian [cf. (9.5)], it is first of all necessary that S' be written as $S' \equiv \int (P \, dQ - H' \, dt)$ so that Hamilton's equations preserve their form, and, in addition, that S' differs from S by only a constant, since S describes the same dynamical system. We must therefore have

$$p \, dq - H \, dt = P \, dQ - H' \, dt + dF \quad \text{or} \quad dF = p \, dq - P \, dQ + (H' - H)dt, \tag{9.16}$$

where $F = F(q, Q, t)$ is an arbitrary function [indeed, $\int dF = F(q(t_2), Q(t_2), t_2) - F(q(t_1), Q(t_1), t_1)$ is just a constant]. Then, having chosen a *generating function* $F(q, Q, t)$, we see from (9.16) that the transformation will be canonical if

$$p = \frac{\partial F}{\partial q}, \quad P = -\frac{\partial F}{\partial Q}, \quad \text{and} \quad H' = H + \frac{\partial F}{\partial t}. \tag{9.17}$$

The explicit form of the transformation is then obtained by inverting $P = -\partial F(q, Q, t)/\partial Q$ to obtain $q = q(Q, P, t)$. Inserting the latter into $p = \partial F(q, Q, t)/\partial q$ gives $p = p(Q, P, t)$. Therefore, the Hamiltonian H' can be written explicitly in terms of the variables (Q, P, t). Since the invariance of Hamilton's equations is guaranteed, the equations of motion become

$$\dot{Q}^i = \frac{\partial H'}{\partial P_i}, \quad \dot{P}_i = -\frac{\partial H'}{\partial Q^i}. \tag{9.18}$$

Example of a canonical transformation

Let us consider the Lagrangian of the *harmonic oscillator* $L(q, \dot{q}) = \frac{1}{2}\dot{q}^2 - \frac{1}{2}q^2$, where $q(t)$ is the position of the oscillator on, for example, the X axis of an inertial frame. The Euler–Lagrange equations $(\partial L/\partial \dot{q})\dot{} = \partial L/\partial q$ are $\ddot{q} = q$, the solution of which is sinusoidal motion $q = q_0 \sin(t - t_0)$.

The conjugate momentum associated with q is $p = \partial L/\partial \dot{q} = \dot{q}$ and the Hamiltonian is $H \equiv p\dot{q} - L = \frac{1}{2}p^2 + \frac{1}{2}q^2$.

We consider the generating function $F(q, Q) = \frac{1}{2}q^2 \cos Q/\sin Q$. Equations (9.17) then tell us that $p = q \cos Q/\sin Q$ and $P = \frac{1}{2}q^2/\sin^2 Q$, from which we find that the associated canonical transformation is

$$q = \sqrt{2P}\sin Q \quad \text{and} \quad p = \sqrt{2P}\cos Q. \tag{9.19}$$

Since the function F does not depend explicitly on time, we find that the Hamiltonian of the harmonic oscillator reduces to

$$H' = H = \frac{1}{2}p^2 + \frac{1}{2}q^2 = P. \tag{9.20}$$

Hamilton's equations $\dot{Q} = \partial H'/\partial P$ and $\dot{P} = -\partial H'/\partial Q$ reduce to $\dot{Q} = 1$ and $\dot{P} = 0$, which have the simple solution $Q = t - Q_0$ and $P = P_0$. Returning to the initial variables (q, p) given in (9.19), we again find that the oscillator undergoes sinusoidal motion: $q = \sqrt{2P_0}\sin(t - Q_0)$ and $p = \sqrt{2P_0}\cos(Q - Q_0)$.

Finally, we can check the calculation by verifying that p is indeed equal to $\dot{q}$.

This example gives a hint of how effective canonical transformations can be for simplifying the equations of motion.

Let us return to (9.16). From the Leibniz rule we have $P\,dQ = d(PQ) - Q\,dP$, and so we can rewrite this equation as

$$df = p\,dq + Q\,dP + (H' - H)dt, \tag{9.21}$$

where $f \equiv F + PQ$ is a new generating function which now depends on the q and P. Therefore, the canonical transformation is defined as

$$p = \frac{\partial f}{\partial q}, \quad Q = \frac{\partial f}{\partial P}, \quad \text{and} \quad H' = H + \frac{\partial f}{\partial t}. \tag{9.22}$$

9.3   The Hamilton–Jacobi equation

As we have seen in Section 9.2, the canonical transformations $q^i = q^i(Q^j, P_j, t)$, $P_i = p_i(Q^j, P_j, t)$, determined, for example, from a given generating function $f(q^i, P_i, t)$, can simplify the expression for a Hamiltonian $H = H(q^j, p_j, t)$ as well as make it easier to solve the equations of motion.

One might therefore ask if it is possible to choose f such that the new Hamiltonian $H'(Q^i, P_i, t)$ vanishes, so that the equations of motion in the new coordinates will be trivial:

$$H'(Q^i, P_i, t) = 0 \qquad \Longrightarrow \qquad \dot{Q}^i = 0, \ \dot{P}_i = 0. \tag{9.23}$$

The answer is 'yes' as seen from (9.22), which can be written in the more explicit form

$$H' = H(q^i, p_i, t) + \frac{\partial f(q^i, P_i, t)}{\partial t},$$

$$\text{with} \quad p_i = \frac{\partial f(q^i, P_i, t)}{\partial q^i} \quad \text{and} \quad Q^i = \frac{\partial f(q^i, P_i, t)}{\partial P_i}. \tag{9.24}$$

Given the expression for p_i as a function of f, we see that in order for H' to vanish it is necessary and sufficient that the function $f(q^i, P_i, t)$ satisfy the so-called *Hamilton–Jacobi equation*:

$$H\left(q^i, \frac{\partial f}{\partial q^i}, t\right) + \frac{\partial f}{\partial t} = 0. \tag{9.25}$$

This is a partial differential equation for f of first order in the $3N$ variables q^i and the time t. Its *complete solution* contains as many integration constants as independent variables, that is, $3N + 1$. One of them is simply additive because only derivatives of f appear. The other $3N$ can be identified as the P_i which, according to the equations of motion (9.23), are indeed constants.

Once the function $f(q^i, P_i, t)$ is known, the last expression in (9.24), where the Q^i are also constants according to the equations of motion (9.23), gives by inversion the general solution of the initial equations of motion, that is, the $3N$ positions $q^i(t)$ as a function of the $6N$ constants Q^i and P_j.

The Hamilton–Jacobi method: an example

Let us again consider the example of the harmonic oscillator with the Lagrangian $L = \frac{1}{2}\dot{q}^2 - \frac{1}{2}q^2$ and the associated Hamiltonian $H = \frac{1}{2}p^2 + \frac{1}{2}q^2$ (see Section 9.2). The Hamilton–Jacobi equation (9.25) in this case becomes

$$\frac{1}{2}q^2 + \frac{1}{2}\left(\frac{\partial f}{\partial q}\right)^2 + \frac{\partial f}{\partial t} = 0, \tag{9.26}$$

which is easy to solve. We proceed by *separation of variables* and seek a solution of the form $f = -Pt + S_0(q)$, where P is a constant. Then $S_0(q)$ satisfies an ordinary differential equation

$$\left(\frac{dS_0}{dq}\right)^2 = 2P - q^2, \tag{9.27}$$

which can be solved by quadrature. We obtain, for $P > 0$ and ignoring an additive constant,

$$f = -Pt + P\left(u + \frac{1}{2}\sin 2u\right) \quad \text{with} \quad \sin u = \frac{q}{\sqrt{2P}}. \tag{9.28}$$

The last expression in (9.24), $Q = \partial f/\partial P$ where Q is another constant, then again gives harmonic motion [noting that $\cos u(\partial u/\partial P) = -q/(2P)^{3/2}$]:

$$q = \sqrt{2P}\sin(t + Q). \tag{9.29}$$

We see from this example that the canonically conjugate variables (Q, P) are identified with the pair $(-t_0, E)$, where E is the oscillator energy and t_0 is the initial time.

The generating function f, the solution of the Hamilton–Jacobi equation (9.25), can be related to the action of the system $S = \int L(q, \dot{q}, t)dt$. In order to do this we return to the variational principle (see Section 8.1) and write

$$\delta S = \int_{t_0}^t \delta L \, dt = \left(\frac{\partial L}{\partial \dot{q}}\delta q\right)\bigg|_{t_0}^t + \int_{t_0}^t \left(\frac{\partial L}{\partial q} - \frac{d}{dt}\frac{\partial L}{\partial \dot{q}}\right)\delta q, \tag{9.30}$$

again omitting the $3N$ indices i. To obtain the Euler–Lagrange equations of motion we consider all possible paths starting from q_0 at t_0 and ending at q at t. The actual (*i.e.,* observable) particle trajectories are those which extremize S: $\delta S = 0$.

Let us now consider particles leaving from q_0 at t_0 with arbitrary initial velocities (but which are nearly equal to each other). The integral then vanishes because the particles obey the equations of motion. We are left with the first term, whose lower limit is zero but whose upper limit is nonzero because at time t the particles are spread out and their positions differ by δq. The action thus becomes a function of the trajectories q and the time t, $S = S(q, t)$. From (9.30) we then have

$$\frac{\partial S}{\partial q} = \frac{\partial L}{\partial \dot{q}} \equiv p. \tag{9.31}$$

Moreover, $S = \int L \, dt$. We therefore obtain on the one hand $dS = L \, dt$, while on the other $dS = (\partial S/\partial q)dq + (\partial S/\partial t)dt$. Equating these gives $(\partial S/\partial t)dt = -(\partial S/\partial q)dq + L \, dt$, or, using (9.31) and the definition of the Hamiltonian $dH = p \, dq - L \, dt$,

$$\frac{\partial S}{\partial t} + H\left(q, \frac{\partial S}{\partial q}, t\right) = 0, \tag{9.32}$$

which is just the Hamilton–Jacobi equation (9.25). Therefore, the action S of the system, evaluated on the actual trajectories all starting from a given point of the configuration space, satisfies the same equation as the generating function f which associates with it a zero Hamiltonian.

9.4 Poisson brackets

Let us consider two arbitrary functions f and g of $6N$ variables (q^i, p_i). Their *Poisson bracket* is defined as

$$\{f, g\} = \frac{\partial f}{\partial p_i} \frac{\partial g}{\partial q^i} - \frac{\partial f}{\partial q^i} \frac{\partial g}{\partial p_i} \tag{9.33}$$

(summation over the indices i is understood). This implies

$$\{f, q^i\} = \frac{\partial f}{\partial p_i}, \qquad \{f, p_i\} = -\frac{\partial f}{\partial q^i}, \tag{9.34}$$

and

$$\{q^i, q^j\} = 0, \quad \{p_i, p_j\} = 0, \quad \{p_i, q^j\} = \delta_i^j. \tag{9.35}$$

It can also be verified that the Poisson brackets possess the following properties:

$$\begin{aligned}
\{f, g\} &= -\{g, f\} \\
\{f_1 f_2, g\} &= f_1\{f_2, g\} + f_2\{f_1, g\} \\
\{f, \{g, h\}\} &+ \{g, \{h, f\}\} + \{h, \{f, g\}\} = 0,
\end{aligned} \tag{9.36}$$

where the last expression (the only one whose proof is somewhat tedious) is called the *Jacobi identity*.

Moreover, if the variables q^i and p_i as well as f and g have an explicit time dependence $f = f(q^i(t), p_i(t), t)$, $g = g(q^i(t), p_i(t), t)$, it can easily be shown that

$$\frac{d\{f, g\}}{dt} = \left\{ \frac{df}{dt}, g \right\} + \left\{ f, \frac{dg}{dt} \right\}. \tag{9.37}$$

Now let us identify the variables (q^i, p_i) as the positions and conjugate momenta of a system of N interacting particles. Hamilton's equations (9.10) governing their motion can be rewritten in a very symmetric form using (9.34):

$$\dot{q}^i = \{H, q^i\}, \quad \dot{p}_i = \{H, p_i\}. \tag{9.38}$$

The total time derivative of a function $f(q^i(t), p_i(t), t)$ then reduces to

$$\frac{df}{dt} = \frac{\partial f}{\partial t} + \left(\frac{\partial f}{\partial q^i} \dot{q}^i + \frac{\partial f}{\partial p_i} \dot{p}_i \right) = \frac{\partial f}{\partial t} + \{H, f\}. \tag{9.39}$$

Therefore, given a Hamiltonian $H(q^i, p_i)$, if we find a function $f(q^i, p_i)$ whose Poisson bracket with H vanishes, then f is a constant of the motion, that is, a first integral of Hamilton's equations. If now f and g are two constants of the motion, then from (9.37) their bracket $\{f, g\}$ is also a constant of the motion. This is the *Poisson theorem*. It can serve as a powerful tool for integrating the equations of motion (though the new constant of the motion is not always independent of the preceding ones).

An important property of the Poisson brackets is that they are invariant under a canonical transformation. More precisely, if $q^i \mapsto q^i(Q^j, P_j, t)$ and $p_i \mapsto p_i(Q^j, P_j, t)$, then, as we shall show below,

$$\{Q^i, Q^j\}_{(p,q)} \equiv \frac{\partial Q^i}{\partial p_k}\frac{\partial Q^j}{\partial q^k} - \frac{\partial Q^i}{\partial q^k}\frac{\partial Q^j}{\partial p_k} = 0 = \{Q^i, Q^j\}_{(P,Q)}$$

$$\{P_i, P_j\}_{(p,q)} \equiv \frac{\partial P_i}{\partial p_k}\frac{\partial P_j}{\partial q^k} - \frac{\partial P_i}{\partial q^k}\frac{\partial P_j}{\partial p_k} = 0 = \{P_i, P_j\}_{(P,Q)} \tag{9.40}$$

$$\{P_i, Q^j\}_{(p,q)} \equiv \frac{\partial P_i}{\partial p_k}\frac{\partial Q^j}{\partial q^k} - \frac{\partial P_i}{\partial q^k}\frac{\partial Q^j}{\partial p_k} = \delta_{ij} = \{P_i, Q^j\}_{(P,Q)}.$$

Invariance of the Poisson brackets

Let us introduce the column vector η formed from the $3N$ variables q^i and the $3N$ conjugate momenta p_i, $\eta = \binom{q^i}{p_i}$. Hamilton's equations can then be written in the *symplectic form*

$$\dot{\eta} = J\frac{\partial H}{\partial \eta}, \quad \text{where} \quad \frac{\partial H}{\partial \eta} \quad \text{is the column vector} \quad \begin{pmatrix} \frac{\partial H}{\partial q^i} \\ \frac{\partial H}{\partial p_i} \end{pmatrix} \quad \text{and} \quad J \equiv \begin{pmatrix} 0 & I \\ -I & 0 \end{pmatrix} \tag{9.41}$$

(J is a $6N \times 6N$ block matrix).

We now make a change of coordinates in the phase space $q^i \mapsto q^i(Q^j, P_j)$ and $p_i \mapsto p_i(Q^j, P_j)$, that is, in matrix form, $\zeta = \zeta(\eta)$ with $\zeta = \binom{Q^i}{P_i}$. (Here for simplicity we limit ourselves to the case of a transformation which does not depend explicitly on time.) Differentiating with respect to time, we find

$$\dot{\zeta} = M\dot{\eta}, \quad \text{where} \quad M \equiv \begin{pmatrix} \frac{\partial Q^i}{\partial q^j} & \frac{\partial Q^i}{\partial p_j} \\ \frac{\partial P_i}{\partial q^j} & \frac{\partial P_i}{\partial p_j} \end{pmatrix} \tag{9.42}$$

is the $6N \times 6N$ Jacobian matrix of the transformation.

If Hamilton's equations are satisfied, $\dot{\eta}$ is given by (9.41), and (9.42) becomes $\dot{\zeta} = MJ(\partial H/\partial \eta)$. However, because

$$\frac{\partial H}{\partial q^i} = \frac{\partial H}{\partial Q^j}\frac{\partial Q^j}{\partial q^i} + \frac{\partial H}{\partial P_j}\frac{\partial P_j}{\partial p_i}$$

and so on, we have $(\partial H/\partial \eta) = \tilde{M}(\partial H/\partial \zeta)$, where $\tilde{M}$ is the transpose of M. The equations of motion therefore become

$$\dot{\zeta} = MJ\tilde{M}\frac{\partial H}{\partial \zeta}. \tag{9.43}$$

The transformation $\eta \mapsto \zeta = \zeta(\eta)$ will be canonical if (9.43) has the same form as (9.41), that is, if the Jacobian matrix M is such that

$$MJ\tilde{M} = J. \tag{9.44}$$

Given the expressions for M and J, (9.44) is the matrix version of the invariance condition for the Poisson brackets (9.40). Q.E.D.

The Poisson bracket and canonical transformation

Let us return to the example of the harmonic oscillator studied in Sections 9.2 and 9.3. We consider the Hamiltonian $H(Q, P) = P$ and ask whether or not it is possible to write it in the form $H(q, p) = \frac{1}{2}p^2 + \frac{1}{2}q^2$. We see that it is possible if we set $q = \sqrt{2P} \sin Q$ and $p = \sqrt{2P} \cos Q$. Now we need to check that this is indeed a canonical transformation.

We can seek the corresponding generating function $F(q, Q)$ using [see (9.17)] $p = \partial F/\partial q$ and $P = -\partial F/\partial Q$, so that here we have $q \cos Q/\sin Q = \partial F/\partial q$ and $q^2/(2 \sin^2 Q) = -\partial F/\partial Q$, which is easily integrated to give $F = q^2 \cos Q/(2 \sin Q)$, in accordance with the expression found in Section 9.2.

However, we can avoid resorting to integration by using the invariance of the Poisson brackets. In this approach it is necessary to calculate only the *derivatives* $\partial p/\partial Q$, $\partial p/\partial P$, and so on, and it is a routine exercise to show that we do indeed have $\{q, q\}_{P,Q} = 0$, $\{p, p\}_{P,Q} = 0$, and $\{p, q\}_{P,Q} = 1$.

10
Kinetic theory

The motion of N particles is governed by $6N$ equations of motion of first order in time, written either in Hamiltonian form or in terms of Poisson brackets. Clearly, as the number of particles grows it becomes necessary to resort to a statistical description. Here we present the equations governing the evolution of the particle *distribution* and relate the macroscopic thermodynamical quantities to the *distribution function*.

10.1 Liouville's theorem and equation

The *phase space* is the $6N$-dimensional space in which each point Q distinguished by $6N$ coordinates $\{q^i, p_i\}$ ($3N$ positions and $3N$ momenta) represents a *state*. A trajectory $Q(t)$ in this space represents the motion of a system of N particles. This trajectory is determined by specifying an initial state $Q(t_0)$ and a Hamiltonian $H(q^i, p_i)$ which governs the dynamics of the system. A different trajectory corresponds to different initial conditions. Since Hamilton's equations are first-order equations, the trajectories never intersect. An ensemble of trajectories constitutes a *flow*, the features of whose evolution we wish to determine.

Let us consider the evolution of an elementary volume of this flow, defined simply as the product (summation over the $3N$ indices i is understood)

$$d\Gamma = dq^i dp_i \tag{10.1}$$

at time t. At time $t + dt$ the point $Q(t)$ of the flow (q^i, p_i) has evolved to $(q'^i, p'_i) = (q^i, p_i) + (\dot{q}^i, \dot{p}_i)dt = (q^i, p_i) + (\partial H/\partial p_i, \partial H/\partial q^i)dt$, and the elementary volume at time $t + dt$ is $d\Gamma' = dq'^i dp'_i$. Liouville showed that this volume is conserved during the time evolution:

$$d\Gamma = d\Gamma'. \tag{10.2}$$

The proof of this statement is based on the fact that the calculation of $d\Gamma'$ starting from $d\Gamma$ amounts to calculating the Jacobian of the change of variable $(q^i, p_i) \to (q'^i, p'_i)$. Using Hamilton's equations, one can then show that this Jacobian is 1.[1] This conclusion arises solely from the structure of Hamilton's equations and does not depend on the explicit form of the Hamiltonian. However, it should be noted that this property does not hold in the configuration space $(q^i, \dot{q}^i)$, except of course if the q^i are the Cartesian coordinates of the particles (because then the canonical velocities and momenta are proportional: $p = mv$).

Now let us introduce a *distribution* function $f^{(N)}(q^i, p_i, t)$ such that $f^{(N)}(q^i, p_i, t)d\Gamma$ is the probability that at time t the state of the system will be in a volume $d\Gamma$ about (q^i, p_i). This positive function is normalized:

---

[1] Because the Jacobian is expressed in terms of the Poisson brackets of the p and q, which is equal to 1. For a detailed discussion see, for example, Landau and Lifshitz (1976).

Relativity in Modern Physics. Nathalie Deruelle and Jean-Philippe Uzan.
© Oxford University Press 2018. Published in 2018 by Oxford University Press.
DOI: 10.1093/oso/9780198786399.001.0001

$$\int f^{(N)}(q^i, p_i, t)d\Gamma = 1.$$ (10.3)

To see how $f^{(N)}$ evolves in time, we use the fact that if the system is at (q^i, p_i) at time t, it will be at (q'^i, p'_i) at time $t + dt$, that is,

$$f^{(N)}(q^i, p_i, t)\, d\Gamma = f^{(N)}(q^i + \dot{q}^i dt, p_i + \dot{p}_i dt, t + dt)\, d\Gamma'.$$ (10.4)

Using Hamilton's equations and the Liouville theorem (10.2) following from them, we can write this condition as

$$\frac{\partial f^{(N)}}{\partial t} + \sum_{i=1}^{3N}\left(\frac{\partial H}{\partial p_i}\frac{\partial f^{(N)}}{\partial q^i} - \frac{\partial H}{\partial q^i}\frac{\partial f^{(N)}}{\partial p_i}\right) = 0,$$ (10.5)

or, by the definition of the Poisson brackets,

$$\frac{df^{(N)}}{dt} = \frac{\partial f^{(N)}}{\partial t} + \left\{H, f^{(N)}\right\} = 0.$$ (10.6)

This equation stating the conservation of the probability density is the *Liouville equation*.

10.2 The Boltzmann–Vlasov equation

The Hamiltonian of a system of N particles is [see (9.13)]

$$H = \sum_a \frac{p_a^2}{2m_a} + W(x^i).$$

Let us take the case where the potential energy is written as $W = \sum w_{aa'}$ (with implicit summation over a and a' from 1 to N, $a < a'$) with $f_{aa'} = -\nabla_a w_{aa'}$ being the interaction force between particles a and a'. We also assume that this force depends only on the relative positions $l_{aa'} = R_{a'} - R_a$, i.e., that $w_{aa'}$ depends only on the relative separation $|R_{a'} - R_a| = r_{aa'}$.

To simplify our arguments we shall from now on use an inertial frame and Cartesian coordinates, so that $p_a^i \equiv m_a V_a^i = m_a dX_a^i/dt$ (here $i = \{1, 2, 3\}$).

The Liouville equation (10.5) then takes the following form in terms of the vectors p_a and R_a with components p_a^i and X_a^i:

$$\frac{\partial f^{(N)}}{\partial t} + \sum_{a=1}^{N}\left(\frac{p_a}{m_a}\cdot\frac{\partial f^{(N)}}{\partial R_a} - \frac{\partial W_a}{\partial R_a}\cdot\frac{\partial f^{(N)}}{\partial p_a}\right) = 0 \quad \text{with} \quad W_a \equiv \sum_{a'\neq a} w_{aa'}.$$ (10.7)

Here the term $(\partial W_a/\partial R_a).(\partial f^{(N)}/\partial p_a)$ is understood as $(\partial W_a/\partial X_a^i)(\partial f^{(N)}/\partial p_i^a)$. We now introduce the function for the *one-particle probability density*, defined as

$$f^{(1)}(u_1, t) = \int f^{(N)}(u_1, \dots, u_N, t)d^6 u_2 \dots d^6 u_N,$$ (10.8)

$$\text{where} \quad u_a \equiv \{R_a, p_a\} \quad \text{and} \quad d^6 u_a \equiv d^3 X_a d^3 p_a.$$

The quantity $f^{(1)}(u_1, t)d^6 u_1$ represents the probability of finding the particle $a = 1$ in a volume $d^6 u_1$ about u_1 for any positions and momenta (or velocities) of the other $N - 1$

particles, *i.e.*, for any trajectories in the phase space. The evolution equation for $f^{(1)}$ can be obtained from (10.7). If we assume that $f^{(N)}$ tends to 0 when p_a and R_a go to infinity, then $\int (\partial f^{(N)}/\partial R_a) d^3 X_a = 0$ and $\int (\partial f^{(N)}/\partial p_a) d^3 p_a = 0$ for $a = 2 \ldots N$, so that

$$\frac{\partial f^{(1)}}{\partial t} + \frac{p_1}{m_1} \cdot \frac{df^{(1)}}{\partial R_1} = \int \frac{\partial W_1}{\partial R_1} \cdot \frac{\partial f^{(N)}}{\partial p_1} d^6 u_2 \ldots d^6 u_N \,. \tag{10.9}$$

Let us now assume that the particles are indistinguishable, which means they have the same mass $m_a \equiv m$. This implies that $f^{(N)}$ is invariant under permutation of the u_a. Consequently, the right-hand side of (10.9) reduces to

$$\int \frac{\partial W_1}{\partial R_1} \cdot \frac{\partial f^{(N)}}{\partial p_1} d^6 u_2 \ldots d^6 u_N = (N-1) \int \frac{\partial w_{12}}{\partial R_1} \cdot \frac{\partial f^{(N)}}{\partial p_1} d^6 u_2 \ldots d^6 u_N$$
$$\equiv (N-1) \int \frac{\partial w_{12}}{\partial R_1} \cdot \frac{\partial f^{(2)}}{\partial p_1} d^6 u_2 \,, \tag{10.10}$$

where the function $f^{(2)}(u_1, u_2, t)$, obtained by integrating $f^{(N)}$ over $u_3 \ldots u_N$, is the *two-particle distribution function*, that is, $f^{(2)} d^6 u_1 d^6 u_2$ represents the probability of finding one particle in the volume $d^6 u_1$ about u_1 *and* another particle in the volume $d^6 u_2$ about u_2 for any positions and momenta of the other $N-2$ particles. Therefore,

$$\frac{\partial f^{(1)}}{\partial t} + \frac{p_1}{m} \cdot \frac{\partial f^{(1)}}{\partial R_1} = (N-1) \int \frac{\partial w_{12}}{\partial R_1} \cdot \frac{\partial f^{(2)}}{\partial p_1} d^6 u_2 \,. \tag{10.11}$$

Since we are assuming that the particles are indistinguishable, $f^{(1)}$ is related to the so-called *distribution function* $f(R, p, t)$ as

$$f(R, p, t) = N f^{(1)}(R_1, p_1, t) \quad \text{such that} \quad \int f(R, p, t) d^3 X \, d^3 p = N \,. \tag{10.12}$$

Here $f(R, p, t) d^6 u$ is the number of particles located at X^i and $X^i + dX^i$ whose velocity components are located between V^i and $V^i + dV^i$.

Equation (10.11) involves $f^{(2)}$. We can find the evolution equation of $f^{(2)}$ in a similar manner, and it clearly will involve the three-particle distribution function $f^{(3)}$. In this way we obtain a hierarchy of equations called the BBGKY hierarchy,[2] which is no easier to solve than the original Liouville equation. However, it is possible to obtain approximate solutions of (10.11) by truncating the hierarchy on the basis of physical arguments.

Let us begin by decomposing $f^{(2)}$ as (this is always possible)

$$f^{(2)}(u_1, u_2, t) = f^{(1)}(u_1, t) f^{(1)}(u_2, t) + g(u_1, u_2, t), \tag{10.13}$$

where g is the *two-particle correlation function* measuring the excess of the probability of finding one particle at u_2 knowing that there is one at u_1, relative to the probability that

---

[2] For Bogoliubov, Born, Green, Kirkwood, and Yvon. More details can be found in, for example, Binney and Tremaine (2008).

would be expected for a system of independent particles. Equation (10.11) is then written in terms of f as

$$
\frac{\partial f(u_1, t)}{\partial t} + \frac{p_1}{m} \cdot \frac{\partial f(u_1, t)}{\partial R_1}
$$

$$
= \frac{(N-1)}{N} \frac{\partial f(u_1, t)}{\partial p_1} \cdot \frac{\partial}{\partial R_1} \int w_{12} f(u_2, t) d^6 u_2 + N(N-1) \int \frac{\partial w_{12}}{\partial R_1} \cdot \frac{\partial g}{\partial p_1} d^6 u_2 .
\tag{10.14}
$$

Explicitly, we have

$$
\int w_{12} f(u_2, t) d^6 u_2 = \int w_{12} \left(\int f(R_2, p_2, t) d^3 p_2 \right) d^3 X_2 = \int w_{12} \, n(R_2, t) d^3 X_2 ,
\tag{10.15}
$$

where, by the definition of the distribution function f,

$$
n(R, t) \equiv \int f(R, p, t) \, d^3 p
$$

is the particle density about the point R. Therefore, assuming that $N \gg 1$ and dropping the index 1, we have

$$
\frac{\partial f}{\partial t} + \frac{p}{m} \cdot \frac{\partial f}{\partial R} - \frac{\partial f}{\partial p} \cdot \frac{\partial W}{\partial R} = N^2 C(u, t) ,
\tag{10.16}
$$

where

$$
W(R_1) \equiv \int w_{12} \, n(R_2, t) d^3 X_2 \quad \text{and} \quad C(u_1, t) \equiv \int \frac{\partial w_{12}}{\partial R_1} \cdot \frac{\partial g}{\partial p_1} d^6 u_2 .
$$

The quantity $C(u, t)$ characterizes the interdependence of the particles. We have therefore gone from the Liouville equation (10.5) for a function of $6N$ variables to an equation governing the evolution of the distribution function f (in six variables) assuming only that the interaction between the particles depends on their separations alone, that the particles are indistinguishable, and that there is a large number of them.

In order to solve (10.16) we must model the 'collision' term $C(u, t)$. If the time between two collisions is large compared to the dynamical time, which can be assumed, for example, when dealing with the gravitational dynamics of a galaxy or a globular cluster, we can take $C = 0$ and the equation becomes

$$
\frac{\partial f}{\partial t} + \frac{p}{m} \cdot \frac{\partial f}{\partial R} - \frac{\partial f}{\partial p} \cdot \frac{\partial W}{\partial R} = 0 \quad \text{or} \quad \frac{\partial f}{\partial t} = \left\{ f, \frac{p^2}{2m} + W \right\} .
\tag{10.17}
$$

This is called the collisionless *Boltzmann equation* (or the *Vlasov equation*). It is considerably easier to solve than the Liouville equation.

Equations (10.16) and (10.17) involve the potential energy of all the particles. We are therefore using a *mean-field approximation*.

10.3 The Jeans equations

Let us consider an ensemble of identical particles described, in an inertial frame and Cartesian coordinates (where $p^i = mV^i$), by the distribution function $f(R, p, t)$, whose evolution is governed by the Boltzmann–Vlasov equation (10.17). The *Jeans equations* are the equations obtained by taking various averages over velocities. The *kth-order moment* of the equation is

$$\frac{\partial}{\partial t}\int f\, V^{j_1}\ldots V^{j_k} d^3V + \frac{\partial}{\partial X^i}\int fV^i\, V^{j_1}\ldots V^{j_k} d^3V$$

$$-\frac{1}{m}\frac{\partial W}{\partial X^i}\delta^{ik}\int \frac{\partial f}{\partial V^k}V^{j_1}\ldots V^{j_k}d^3V = 0, \tag{10.18}$$

where the integral runs over the velocities $d^3V = dV^1 dV^2 dV^3$.

At order $k = 0$ the first and second terms respectively involve

$$n(r,t) = m^3 \int f(r,p,t)d^3V, \qquad \bar{V}^i(r,t) \equiv \frac{m^3}{n}\int f(r,p,t)V^i d^3V, \tag{10.19}$$

which represent the numerical density and the average velocity of the flow. If $f(r,p,t)$ tends to zero for large V^i, the divergence theorem implies that the last term vanishes. We then find

$$\frac{\partial n(r,t)}{\partial t} + \nabla.\,[n(r,t)\,\bar{v}(r,t)] = 0\,, \tag{10.20}$$

where $\bar{v}$ is the field of vectors with Cartesian components $\bar{V}^i$. This continuity equation corresponds to conservation of the number of particles in the flow.

To calculate the moment of order $k = 1$, we note that the third term in (10.18) involves

$$m^3 \int \frac{\partial f}{\partial V^k}V^j d^3V = -m^3 \int \frac{\partial V^j}{\partial V^k}fd^3V = -\delta^j_k\, n(r,t), \tag{10.21}$$

after an integration by parts and assuming again that $f(r,p,t)$ tends to zero for large V^i. We then find

$$\frac{\partial}{\partial t}\left[n(r,t)\bar{V}^j(r,t)\right] + \frac{\partial}{\partial X^i}\left[n(r,t)\overline{V^jV^i}(r,t)\right] + n(r,t)\delta^{ji}\frac{1}{m}\frac{\partial W}{\partial X^i} = 0, \tag{10.22}$$

with

$$\overline{V^jV^i}(r,t) = \frac{m^3}{n}\int f(r,p,t)V^iV^j d^3V. \tag{10.23}$$

This term can be decomposed as $\overline{V^jV^i} = \bar{V}^i\bar{V}^j + \sigma^{ij}$, with σ^{ij} characterizing the velocity spread about the average $\bar{v}$. Inserting this decomposition into (10.22) and subtracting the continuity equation (10.20) multiplied by $\bar{V}^j$, we find

$$\frac{\partial \bar{V}^j(r,t)}{\partial t} + \bar{V}^i\frac{\partial \bar{V}^j}{\partial X^i} = -\delta^{ji}\frac{1}{m}\frac{\partial W}{\partial X^i} - \frac{1}{n}\frac{\partial}{\partial X^i}\left[n(r,t)\sigma^{ij}(r,t)\right]. \tag{10.24}$$

The term $n\sigma^{ij}$ can be interpreted as the anisotropic pressure tensor, and so this equation is equivalent to the Euler equation of fluid mechanics (see Section 6.3).

10.4 The Maxwell distribution and the thermodynamical limit

The expressions derived in the preceding sections are independent of the form of the distribution function. In the case of a gas of free particles, that is, a gas described by the Boltzmann equation (10.17) without any external potential, the distribution function can be derived from first principles.

In a stationary situation the Boltzmann equation (10.17) implies that $\nabla f = 0$, so that f is a function of only the velocity $f = f(v)$. The particle density in a volume d^3v of phase space is

$$f(v)d^3v = f_x(v_x)f_y(v_y)f_z(v_z)\, dv_x dv_y dv_z \tag{10.25}$$

if the variables v_x, v_y, and v_z are independent. The isotropy of space implies that $f(v)$ is a function only of $|v|$. Then (10.25) implies that

$$f_x(v_x)f_y(v_y)f_z(v_z) = \varphi(|v|^2).$$

This equation can be solved by differentiating with respect to one of the components, v_x for example, which gives $f'_x(v_x)f_y(v_y)f_z(v_z) = 2v_x d\varphi/d|v|^2$ so that $(df_x/dv_x)/(2v_x f_x) = (d\varphi/d|v|^2)/\varphi(|v|^2)$. Each side is therefore equal to a constant A and by convention we take $A = -\frac{1}{2}m\beta$, where β has the dimension of an inverse energy. This equation can be integrated to give $\ln f_x = Av_x^2 + B$. Since f_x and φ are normalized, we deduce that

$$f_x(v_x) = \sqrt{\frac{m\beta}{2\pi}}e^{-\frac{1}{2}\beta m v_x^2} \quad \Rightarrow \quad \varphi(v^2) = \left(\frac{m\beta}{2\pi}\right)^{3/2} e^{-\frac{1}{2}\beta m v^2}. \tag{10.26}$$

This is the *Maxwell distribution*.

The Maxwell distribution can be used to calculate the average kinetic energy of a particle: $e_k = \frac{1}{2}m\langle v^2\rangle = \frac{3}{2}\beta^{-1}$, where we have used $\langle v_x^2\rangle = \int dv_x\, v_x^2 f_x(v_x) = 1/(m\beta)$. For a gas containing N particles, the internal energy U is then given by

$$U \equiv Ne_k = \frac{3}{2}N\beta^{-1} \quad \Longrightarrow \quad \rho \equiv \frac{U}{V} = \frac{3N}{2V}\beta^{-1}. \tag{10.27}$$

The pressure of this gas is calculated from the force exerted on the container walls by particle collisions. A particle bouncing off a wall exerts a force $fdt = 2m(v.n)n$, where n is the normal to the wall. Summing over all the particles colliding with a surface S during time dt gives $F\,dt = 2m(v.n)[S(v.n)dt](N/V)f(v)d^3v$. Integrating over the (positive) velocities, we find that the pressure $p \equiv \langle F\rangle/S$ is given by

$$p = \frac{2mN}{V}\frac{1}{2}\int (v.n)^2 f(v)d^3v = \frac{mN}{V}\frac{1}{3}\langle v^2\rangle = \frac{N}{V}\beta^{-1}. \tag{10.28}$$

Combining this with (10.27), we obtain the equation of state of the fluid:

$$pV = \frac{2}{3}U, \quad p = \frac{2}{3}\rho. \tag{10.29}$$

The constant β is interpreted by relating (10.28) to the thermodynamical equation of state of a perfect monatomic gas. Indeed, the gas we have discussed here is identical to a perfect gas, which has the equation of state $pV = Nk_{\mathrm{B}}T$, where T is the *thermodynamical temperature* and k_{B} is the *Boltzmann constant* (we can also write $pV = nRT$, where n is the number of moles and $R = k_{\mathrm{B}}N_{\mathrm{A}}$, N_{A} being Avogadro's number). We then see that β is related to T as

$$\beta^{-1} = k_{\mathrm{B}}T. \tag{10.30}$$

In thermodynamics, (10.27) is sometimes generalized to $U = 2Ne_{\mathrm{k}}/(3w) = N/(w\beta)$, where the coefficient w, not necessarily equal to $2/3$, is assumed to take into account any additional degrees of freedom (like rotation, *etc.*). The equation of state (10.29) for a perfect gas is then generalized to

$$pV = wU, \quad p = w\rho. \tag{10.31}$$

The equation of state $p = w\rho$ will prove useful for modeling the structure of stars and large objects as well as cosmological fluids.

Part III

Gravitation

Having observed this I came to the conclusion that in a medium totally devoid of resistance all bodies would fall with the same speed.

Galileo Galilei, *Discorsi e Dimostrazioni Matematiche Intorno a Due Nuove Scienze*, Leiden, Elzevir, 1638; English translation by Henry Crew and Alfonso de Salvio, *Discourses and Mathematical Demonstrations Relating to Two New Sciences*, New York, Macmillan, 1914

There is a power of gravity tending to all bodies, proportional to the several quantities of matter which they contain.

This force is ever proportional to the body whose force it is; and differs nothing from the inactivity of the mass, but in our manner of conceiving it.

The force of gravity towards the several equal particles of any body is reciprocally as the square of the distance of places from the particles...

Sir Isaac Newton, *Philosophiæ Naturalis Principia Mathematica*, London, 1687; English translation by Andrew Motte, *The Mathematical Principles of Natural Philosophy*, London, 1729

11
The law of gravitation

We have described the geometrical framework representing Newtonian space and time in Part I and the laws of Newtonian dynamics in Part II of this book. Now we embark on the study of Newton's law of gravitation.

11.1 Gravitational mass and inertial mass

The *inertial mass*[1] m_I of a point particle is the parameter characterizing the particle which appears in Newton's law of motion:

$$F = m_I a \,, \tag{11.1}$$

where a is the acceleration of the particle trajectory and F is a vector representing the force on the particle; see Section 5.1. The inertial mass is a measure of the 'resistance' of the point particle to an applied force. The numerical value of the inertial mass of a body can in principle be obtained from collision experiments (see Section 7.1) by assigning to a reference body a unit inertial mass of one *kilogram* or, more rigorously, one 'inertial kilogram'.

Since the 'power of gravity' is an experimental fact, the *gravitational mass* m_G is another parameter which is naturally associated with massive objects subject to or producing a 'gravitational force'.

This concept therefore has a dual nature.

• The *passive* gravitational mass characterizes the response of a body to the gravitational action of external objects. More precisely, it defines the *weight* of a body, that is, the gravitational force F acting on the body as

$$F = m_G g \,, \tag{11.2}$$

where g is the *acceleration of gravity* characterizing the external gravitational action independently of the body subject to it: $g = (F/m_G)_a$ for any body a. Therefore, if the weight F of the body is known, we can in principle deduce both g and the gravitational mass m_G by assigning a unit mass of one 'gravitational kilogram' to a reference body.

---

[1] The terms *inertial mass* and *gravitational mass* were first used by H. Bondi.

Relativity in Modern Physics. Nathalie Deruelle and Jean-Philippe Uzan.
© Oxford University Press 2018. Published in 2018 by Oxford University Press.
DOI: 10.1093/oso/9780198786399.001.0001

The principle of weight measurement

It is possible in principle to determine the numerical value of an object's weight (without knowing the law of gravitation) by means of Newton's second law $F = m_I a$ by measuring in an inertial reference frame the acceleration acquired by the body in an external gravitational field, where it is assumed that the inertial mass of the body is known by some other means. It is also possible to determine the weight by *weighing* the body in statics experiments using, for example, the tension on a spring. (In an inertial frame it is not necessary to know the inertial mass.)

• The *active* gravitational mass characterizes the object which creates a gravitational field. It can therefore be called the 'gravitational charge'.

The gravitational force of a body A on another body B, F_{AB}, is proportional to $m_{G,A}^{active} m_{G,B}^{passive}$. However, Newton's third law of equal action and reaction $F_{AB} = -F_{BA}$ implies that $m_{G,A}^{active} m_{G,B}^{passive} = m_{G,B}^{active} m_{G,A}^{passive}$, so that for any body $m_G^{active}/m_G^{passive}$ is a constant which can be set equal to 1. Therefore, passive and active gravitational masses are identical and we shall no longer distinguish between them. (Similarly, in electrostatics we do not distinguish the passive and active electric charges appearing in Coulomb's law.)

On the other hand, there is nothing in the structure of Newtonian physics that requires electric or gravitational 'charges' to be equal to inertial masses. For example, the ratio of the electric charge and the inertial mass varies from body to body and can even be zero (for a 'neutral' body).

11.2 Equality of gravitational and inertial mass

It is an experimental fact, established by Galileo (in his experiments on inclined planes rather than the probably apocryphal experiment in the Leaning Tower of Pisa), that in the absence of friction, all objects, no matter what their inertial mass, or the nature of their constituents, or the internal energy or cohesive forces of their constituents, fall in the same way in an external gravitational field (in contrast to, for example, the behavior in an electric field of two individual charges of opposite sign and of their neutral ensemble). The ratio of the gravitational and inertial masses is therefore the same for any object, and we can take it to be unity:

$$m_I = m_G \equiv m \,. \tag{11.3}$$

The parameter characterizing any possible deviation from this *universality of free fall* is defined as

$$\eta \equiv 2\frac{|a_1 - a_2|}{|a_1 + a_2|} \,, \tag{11.4}$$

where a_1 and a_2 are the accelerations of two bodies in free fall in an external gravitational field. Given that $F = m_I a$ by Newton's second law and $F = m_G g$ by the definition of weight and gravity, we have $a = (m_G/m_I)g$ and (11.4) can be rewritten as

$$\eta = 2\frac{|m_G^1/m_I^1 - m_G^2/m_I^2|}{m_G^1/m_I^1 + m_G^2/m_I^2} \,. \tag{11.5}$$

Present-day experiments, which are among the most precise in physics, give $\eta < 10^{-12}$. Therefore, to this precision we can take the gravitational and inertial masses to be identical and express them both using a common unit, the *kilogram*.

It is important to note that this equality of gravitational and inertial mass holds *a priori* only for bodies in an external gravitational field so that their proper gravity can be ignored.

Experiments involving pendulums

The equation of motion of a body suspended at the end of a cable of length L and oscillating in a gravitational field g is, in the limit of small angular displacements θ,

$$\ddot{\theta} + \omega^2 \theta = 0, \quad \text{where} \quad \omega \equiv \omega_0 \sqrt{\frac{m_G}{m_I}} \quad \text{and} \quad \omega_0 \equiv \sqrt{\frac{g}{L}}. \tag{11.6}$$

Comparing the oscillations of two different bodies A and B suspended by cables of the same length, we find

$$\eta \approx 2 \frac{|\omega_B - \omega_A|}{\omega_0}. \tag{11.7}$$

This method was invented by Galileo, who estimated that $\eta = 0$ with an accuracy of 1%. Newton repeated the experiment and by taking into account air resistance obtained an accuracy of 10^{-3}. In 1827 Bessel reached an accuracy of 10^{-5}.

Experiments involving torsion balances

In 1922 Eötvös used a torsion balance essentially consisting of a rod connecting two different bodies A and B and suspended by a cable. In a rotating reference frame in which the Earth is at rest, the suspended bodies are subject to two forces, namely, the Earth's gravitational attraction $m_G g$ pointing toward the center of the Earth and the centrifugal force $m_I c$ tending to make the pendulum move away from the Earth's rotational axis. If η is nonzero, the cable experiences a couple of amplitude

$$C \simeq \left(\frac{m_A m_B}{m_A + m_B} \right)_I \frac{g.(l \wedge c)}{|g + c|} \eta, \tag{11.8}$$

where $l \equiv AB$ is the separation vector. When the balance is turned by $180°$, l changes sign, thus allowing measurement of η. Eötvös and his collaborators Pekár and Fekete obtained $\eta = 0$ with an accuracy of 10^{-8}.

The experiment was repeated by Dicke and his collaborators Krotkov and Roll in 1966 at Princeton with g the acceleration due to the solar gravitational field and c that due to the orbital rotation of the Earth (so that it was c which changed sign every 12 hours). They obtained an accuracy of 10^{-11}. In 1972 Braginski and Panov in Moscow reached 10^{-12}.

Finally, the Eöt-Wash experiment performed at the University of Washington in 2000 obtained an accuracy[2] of 13.2×10^{-13}.

---

[2]For more details, see, for example, Uzan and Lehoucq (2005).
The outer space experiment MICROSCOPE, launched in 2016, reached an accuracy of 10^{-15}, see P. Touboul *et al.* (2017). The planned mission STEP (Satellite Test of the Equivalence Principle) is expected to reach an accuracy of 10^{-18}.

11.3   Newton's gravitational force and field

Newton[3] postulated that the gravitational force exerted on a point particle P by an ensemble of particles P_a is represented by a vector $F(t)$ written in the absolute Cartesian frame $\mathcal{S}$ as

$$F = -m_{\mathrm{G}} \sum_a \frac{Gm_{Ga}}{r_a^3} l_a, \tag{11.9}$$

where $l_a = P_a P$ is the separation vector between P_a and P (the force is attractive) with components $l_a^i = X^i(t) - X_a^i(t)$, where $X_a^i(t)$ and $X^i(t)$ are the coordinates of P_a and P; r_a is the distance $r_a^2 = \delta_{ij}\, l_a^i l_a^j = e(l_a, l_a)$; G is *Newton's constant*; and, finally, m_{G} and m_{Ga} are the (gravitational) masses of P and P_a.

In any other frame or coordinate system the distance r_a is given by the same function of time. In going from $\mathcal{S}$ to an accelerated frame $\mathcal{S}'$ or from a system of Cartesian coordinates X^i to curvilinear coordinates x^i, the components of the separation vector l_a transform as (omitting the subscript a)

$$l'^i = \mathcal{R}_j{}^i l^j \,, \qquad l'^i = \frac{\partial x^i}{\partial X^j} l^j \,, \tag{11.10}$$

where $\partial x^i/\partial X^j$ is evaluated on the trajectory of P (in this case the vector l_a is *bound* to P). Moreover, it turns out that the gravitational force can be represented by the same vector in any frame. It is therefore written in the form (11.9) in any frame, inertial or noninertial, Cartesian or non-Cartesian.

The gravitational force can be defined as a *field*, in which case the vector l_a, bound to P, is viewed as a function of the point P of coordinates X^i, and not as a functional of the trajectory of a point mass $P(t)$. It can therefore be derived from a *gravitational potential U* (defined up to an additive constant) as

$$F = -m_{\mathrm{G}}\, \nabla U \ \text{ with } \ U \equiv - \sum_a \frac{Gm_{Ga}}{r_a}, \tag{11.11}$$

where ∇U is the gradient of the function U with components $(\nabla U)^i = e^{ij}(\partial U/\partial x^j)$ in any system of curvilinear coordinates; see Section 4.6.

To summarize, the equations of motion of a particle of inertial mass m_{I} and gravitational mass m_{G} in the gravitational potential (11.11) are written in vector form as (see Sections 5.1 and 5.4)

---

[3]The contribution of Robert Hooke to establishing the law of gravitation has been under debate since the 17th century, but, as stated by Émilie de Breteuil, marquise du Châtelet, in her *Explication abrégée du système du monde, et explication des principaux phénomènes astronomiques tirée des Principes de M. Newton* [*Abbreviated explanation of the System of the World, and explanation of the principal astronomical phenomena taken from the Principia of Mr. Newton*] (edited by Alexis Clairaut in 1759): "It should not be thought that this idea casually mentioned in Hook's (sic) book diminishes the glory of Newton (...). The example of Hook and that of Kepler serve to show how great a distance there is between a glimpse of the truth and a demonstrated truth, and how little the grandest illuminations of the spirit serve the sciences when they cease to be guided by Geometry."

$$
\begin{cases}
m_\mathrm{I}a \ \ = -m_\mathrm{G}\nabla U & \text{in any inertial frame} \\[2mm]
m_\mathrm{I}\tilde{D}_v v = -m_\mathrm{G}\nabla U & \text{in any system of curvilinear coordinates} \\[2mm]
m_\mathrm{I}a' \ \ = -m_\mathrm{G}\nabla U + m_\mathrm{I}\left(-2\Omega\wedge v' + \Omega\wedge(R'\wedge\Omega) - \dot{\Omega}\wedge R' - \ddot{d}\right) & \\[2mm]
& \text{in any frame}\,,
\end{cases}
\tag{11.12}
$$

where the equality $m_\mathrm{I} = m_\mathrm{G} \equiv m$ is imposed by experiment; see Section 11.2.

The principal consequence of the equality of the gravitational and inertial masses is that the equation of motion of a point mass P in a gravitational field is independent of its mass. Let us write it in an accelerated frame belonging to the *Milne group* (see Section 1.4), that is, a frame without rotation $\Omega = 0$, $X' = X$, $Y' = Y$, $Z' = Z - d(t)$, and set $\ddot{d} = g$. Then (11.12) becomes

$$
a' = -\nabla U - g = -\nabla(U + g\,Z) = -\nabla(U + g\,Z')\,.
\tag{11.13}
$$

We can then 'efface' the gravitational field acting on P by attributing its motion to a rotationless accelerated reference frame in free fall such that $g = -\nabla U$, where ∇U is evaluated on the trajectory of P. In this frame $a' = 0$ and the motion of P is that of a free particle (the effacement of the field occurs only for P and the particles close enough to P where ∇U can be considered constant). Conversely, a gravitational field can be created artificially (*i.e.*, in the absence of any gravitational mass) by going to a frame undergoing an acceleration g identified with the gradient of a gravitational potential. Finally, we note that the rotational motion of a solid in a constant gravitational field is also free motion because the torque K exerted on the solid is zero; see Section 6.2.

This (local) equivalence between the gravitational force and inertial forces, which is accidental in Newtonian physics, will be elevated to a *postulate* in general relativity called the *equivalence principle* (more precisely, the postulated equality of gravitational and inertial mass along with the universality of free fall that it implies will be referred to as the *weak equivalence principle*).

Measurement of mass and of Newton's constant

In the laboratory frame (see Section 2.5), let us consider an object of gravitational mass m suspended at the end of a pendulum and subjected to its own weight $-mge_Z$ and to the attractive force $(GmM/d^2)e_X$ of another gravitational mass M located a distance de_X away. The pendulum is displaced by an angle α given by

$$
\tan\alpha \approx \alpha = \frac{GM}{gd^2}\,.
\tag{11.14}
$$

In a first series of experiments the angles α and α_ref are compared for two masses M and M_ref, which makes it possible in principle to find the value of the gravitational mass M from[4] $M = M_\mathrm{ref}\alpha/\alpha_\mathrm{ref}$.

---

[4]In contrast to the methods mentioned in Section 11.1, this method of measuring gravitational masses requires knowing that the gravitational force depends only on the separation of the bodies.

The acceleration of terrestrial gravity g is given by (11.9) as $g = -\sum Gm_a l_a / r_a^3$, with the sum running over all the mass points making up the Earth. However, Newton's theorem (see Section 11.4) provides the simple result

$$g = -\frac{GM_\oplus}{r_\oplus^2},$$ (11.15)

where $M_\oplus$ and $r_\oplus$ are the gravitational mass and the radius of the Earth. Therefore,

$$\alpha \approx \frac{M}{M_\oplus} \frac{r_\oplus^2}{d^2}.$$ (11.16)

By measuring α and knowing d and $r_\oplus$, we can deduce the gravitational mass of the Earth as a function of M. This method was proposed by Maskelyne in 1772 to measure the Earth's mass.

This idea was taken up by Cavendish in 1798 for an experiment using a torsion balance to compare the gravitational attraction due to the Earth to that due to reference masses. He found the average density of the Earth to be 5.31 g/cm³, close to the value 5.52 accepted today.

These results can in principle be used to find G in the following way. Once we know the gravitational mass (from the first series of experiments) and the inertial mass (from collision experiments) of a body, g can be obtained by measuring the acceleration a acquired in the Earth's field: $g = m_I a / m_G = 9.81$ m/s². Then, knowing the Earth's radius and mass from the Cavendish experiment, we arrive at $G = -g r_\oplus^2 / M_\oplus$.

Cavendish obtained $G = 6.71 \times 10^{-11}$ m³/(kg s²), a value close to that accepted today. In 1889 Boys reached an accuracy of 0.1 % by miniaturizing the experiment.

Measurements of G using torsion balances continue to be carried out today. The measurement accuracy is limited in particular by the inelasticity of the cable. The value accepted since 2002 is[5]

$$G = 6.674\,08 \times 10^{-11}\,\text{m}^3/(\text{kg s}^2)$$ (11.17)

with a relative uncertainty of 4.7×10^{-5}.

11.4   The Poisson equation and the gravitational Lagrangian

When matter is described as a continuous medium of (gravitational) mass density ϱ, the potential U in (11.11) becomes

$$U(P) = -G \int \frac{\varrho(t, P')}{r_{P'P}} dV',$$ (11.18)

where $r_{P'P} = |P'P| = |R - R'|$ is the distance between the point where U is evaluated and the point P' of the extended body creating the field. The volume element is $dV = dX^1 dX^2 dX^3$ in Cartesian coordinates and $dV = \sqrt{\det e}\, dx^1 dx^2 dx^3$ in curvilinear coordinates x^i, and $\det e$ is the determinant of the metric coefficients (so that in spherical coordinates $e = dr^2 + r^2(d\theta^2 + \sin^2\theta d\phi^2)$ and therefore $\sqrt{\det e} = r^2 \sin\theta$; see Section 3.6).

---

[5]The values of constants can be found at the site http://physics.nist.gov/cuu/Constants/.

Spherical symmetry: Newton's theorem

It is a simple exercise[6] to show that if the P' mass distribution is spherically symmetric, then the Newtonian potential U at a point P is equal to that created by a point of the same total mass located at the center of the distribution.

Indeed, if we place the body at the origin of the frame, by symmetry we can place the point P at $(0, 0, r)$. The Cartesian components of the position vector of a point P' in spherical coordinates are $(r' \sin\theta \cos\phi, r' \sin\theta \sin\phi, r' \cos\theta)$. Then the potential at P is

$$U(r) = -G \int \frac{\varrho(r', t)}{\sqrt{r'^2 + r^2 - 2rr'\cos\theta}}\, r'^2 \sin\theta\, dr'\, d\theta\, d\phi, \tag{11.19}$$

which is easily integrated to give

$$U(r) = -\frac{GM}{r} \quad \text{if } r > r_0, \tag{11.20}$$

where $M = 4\pi \int_0^{r_0} \varrho r'^2 dr'$ is the total mass of the body and r_0 is its radius.

This is a remarkable theorem.

It shows that the gravitational force $F = -m\nabla U$ at the surface of a spherical body, the Earth for example, is the *same* as it would be if all the mass were concentrated at the center.

Moreover, it shows that the gravitational mass of a composite body [the coefficient M which appears in the expression for the potential (11.20)] is equal to the inertial mass $\int \varrho r^2 \sin\theta\, dr d\theta d\phi$, because $\varrho_{\text{inertial}} = \varrho_{\text{grav}}$ owing to the 'weak' equivalence principle. The equality of the gravitational and inertial masses therefore holds not only for a test body placed in an external field, but also for bodies whose proper gravity cannot be ignored.[7]

Finally, it shows that the internal motion of the constituents of a spherically symmetric mass distribution does not affect the external gravitational field.

Since $U = -\frac{GM_{\text{in}}(r)}{r} - 4\pi G \int_r^{r_0} \varrho(r') r' dr'$ if $r < r_0$, we similarly find that the potential inside a spherical cavity where $\varrho = 0$ is constant, so that the field inside the cavity vanishes.

Laplace and Poisson derived from (11.18) a differential relation between U and ϱ:

$$\triangle U = 4\pi G \varrho, \tag{11.21}$$

where $\triangle$ is the Laplacian operator (see Section 4.6; in tensor notation[8] $\triangle U \equiv e^{ij} \tilde{D}_i \partial_j U$).

We note that the local form (11.21) of the equation for U admits solutions other than (11.18) which incorporates the choice of a boundary condition. If, for example, ϱ depends

---

[6] ...which, however, delayed the publication of the *Principia* by 25 years because in order to prove it, Newton first had to invent (in competition with Leibniz) integral calculus. We note that it is impossible, except in special cases, to uniformly distribute a finite number of points on a sphere, and so the idea of a point distribution having *spherical symmetry* actually makes sense only in the continuum limit. See, for example, Saff and Kuijlaars (1997).

[7] The fact that the equality of gravitational and inertial mass holds not only for a test body but also for bodies whose contribution to the gravitational field cannot be neglected is specific to the $1/r$ behavior of the Newtonian potential (11.18). In general relativity this is referred to as the *strong equivalence principle*. Present-day theories of gravitation which satisfy this principle are Einstein's general relativity, the Nordström theory constructed within the framework of special relativity, and Newton's theory (see also Section 12.4).

[8] Those familiar with the *Dirac delta distribution* $\delta_3(x^i) \equiv \delta(x)\delta(y)\delta(z)$, where $\int \delta(x)dx = 1$ and $\triangle(1/r) = -4\pi\delta_3(x^i)$ with $\triangle$ denoting the Laplacian, will notice that the Poisson equation (11.21) is readily derived from (11.11) because $\sum m_a \delta_3(x^i - x_a^i) = \varrho$.

only on time (the case of a uniform matter distribution throughout space), then (11.18) [and also (11.19) in this case] is not defined, whereas (11.21) has the solution $U = \frac{2}{3}\pi G \varrho r^2$. As we shall see in Chapter 16, this will play a role in Newtonian cosmology.

The gravitational force between two particles can be written as $f_{a'a} = -\nabla_a w_{a'a}$ where $w_{a'a} = -G m_a m_{a'}/r_{a'a}$ is (see Section 7.3) the *potential energy* (or the *gravitational energy*) of the pair P_a, $P_{a'}$ (it is also m_a times the potential $U = -G m_{a'}/r_{a'a}$ created at P_a by $P_{a'}$). Therefore, the gravitational potential energy of an ensemble of particles is

$$W = -\sum_{a,a' \text{ with } a<a'} \frac{G m_a m_{a'}}{r_{a'a}} = \frac{1}{2}\sum_a m_a U_a \quad \text{with} \quad U_a = -\sum_{a\neq a'} \frac{G m_{a'}}{r_{a'a}}, \quad (11.22)$$

which when generalized to a continuous distribution becomes

$$W = \frac{1}{2}\int \varrho U \, dV \quad \text{with} \quad \triangle U = 4\pi G \varrho. \quad (11.23)$$

We can use the Poisson equation to eliminate ϱ from the expression for W; indeed, we have

$$4\pi G \int \varrho U \, dV = \int U \triangle U \, dV = \int \nabla(U.\nabla U) dV - \int (\nabla U)^2 dV$$
$$= -\int (\nabla U)^2 dV \,,$$

where we have used the divergence theorem and the fact that, since $U \propto 1/r$ at infinity (as long as there is no matter there), the surface integral vanishes. Therefore,

$$W = \frac{1}{2}\int \varrho U \, dV = -\frac{1}{8\pi G}\int (\nabla U)^2 dV. \quad (11.24)$$

Now that we have the expression for the potential energy of a gravitational system, we can define its Lagrangian as $L = T - W$, where T is its kinetic energy, that is, in an inertial frame,

$$L = \frac{1}{2}\sum_a m_a v_a^2 + \sum_{a<a'} \frac{G m_a m_{a'}}{r_{a'a}} \quad (11.25)$$

for a system of point particles. We can check that the Euler–Lagrange equations $(\partial L/\partial v_a)^{\cdot} = \nabla_a L$ derived from this are indeed equivalent to the equations of motion (11.12) (with $m_I = m_G$).

Owing to (11.24), the Lagrangian of a continuous matter distribution can a *priori* be written in various ways, because W can be written as

$$W = \int dV \left(\frac{(1+c)}{2}\varrho U + \frac{c}{8\pi G}(\nabla U)^2 \right),$$

where c is an arbitrary constant. However, the value to choose is $c = 1$, which gives

$$L = \int dV \left(\frac{1}{2}\varrho v^2 - \varrho U - \frac{1}{8\pi G}(\nabla U)^2 \right). \quad (11.26)$$

It is indeed this Lagrangian which gives the Euler equation for a fluid subject to an external gravitational field (see Sections 6.3 and 8.2) when the path followed by the fluid elements is varied while holding the potential U fixed.

This Lagrangian has the additional virtue that the Poisson equation (11.21) can be derived from it by varying the *configuration* of the gravitational potential U. If $U \mapsto U + \delta U$, then

$$(\nabla U)^2 \mapsto (\nabla U)^2 + 2\nabla U . \nabla \delta U = (\nabla U)^2 + 2\nabla(\nabla U.\delta U) - 2\triangle U \delta U .$$

After using the divergence theorem we find

$$\delta L = - \int dV \left(\varrho - \frac{1}{4\pi G}\triangle U \right) \delta U , \tag{11.27}$$

so that the principle of least action ($\delta L = 0$ for any variation δU of the configuration) again gives the Poisson equation.

Newtonian gravity: unanswered questions

The main faults of the Newtonian equation of motion for massive bodies in a gravitational field were pointed out very early, by Newton himself and his contemporaries. For example:

1. The gravitational force acting on a body at time t depends on the positions of the other bodies at the *same* time. Therefore, the interaction between far-separated objects is instantaneous. This worried Newton, who could not find an explanation for it (hence his well known statement '*hypotheses non fingo*').

2. The gravitational force is long-range and universal (all massive bodies experience it). It is therefore impossible to isolate a body from the attraction of other gravitational bodies, except by moving them off to infinity. A body can never be considered to be completely free, but the construction of inertial reference frames is *based* on the notion of a free particle.

3. The acceleration of a body is defined relative to the origin of the reference frame used. Leibniz refused to treat an immaterial point as special and believed that the laws of motion should involve only the relative separations of bodies. By subtracting two by two the N equations governing the motion of N gravitational objects, one indeed ends up with $(N - 1)$ independent equations involving only the separation vectors $l_{aa'}$ and their second derivatives $\ddot{l}_{aa'} = \ddot{l}^i_{aa'}e'_i$ in any frame belonging to the Milne group of *arbitrary* (i.e., not necessarily uniform) translations. The observation of the relative motions of gravitational bodies therefore does not permit the determination of the translational motion of the reference frame relative to the absolute frame, and it is always possible to find one frame in which one of the gravitational bodies is at rest. Then gravity effectively *disappears* for this body.

General relativity will address these objections. The speed of propagation of the gravitational interaction will be finite (equal to the speed of light), and all bodies will be 'free' in a gravitational field, their trajectories extremizing the distance between the points of a space which is not Euclidean but 'curved' by gravitation.

12
The Kepler problem

In 1618 Kepler established three laws governing the motion of a planet around the Sun:

- the motion is planar and the trajectories are ellipses;

- the area swept out by the radius vector per unit time is constant;

- the cube of the semi-major axis a is proportional to the square of the period P, $a^3 = (\text{const})P^2$.

Newton explained the underlying dynamics in 1665.

12.1 The reduced equations of motion

In accordance with Newton's theorem and the equality of gravitational and inertial mass, the equations of motion of two spherically symmetric gravitational bodies A and B are, in an inertial Cartesian frame,

$$a_A = -\frac{Gm_B}{r_{BA}^3} l_{BA} \quad \text{and} \quad a_B = -\frac{Gm_A}{r_{AB}^3} l_{AB} . \qquad (12.1)$$

If R_A and R_B are the radius vectors of the centers of the bodies A and B, the separation vector is defined as $l_{AB} \equiv AB \equiv R_B - R_A = -l_{BA}$ and its length is $r_{AB} = r_{BA}$. Decomposing the radius vectors as $R_A = X_A^i(t)e_i$ and $R_B = X_B^i(t)e_i$, the components of the accelerations $a_A \equiv d^2R_A/dt^2$ and $a_B \equiv d^2R_B/dt^2$ become $a_A^i = \ddot{X}_A^i$ and $a_B^i = \ddot{X}_B^i$, where the dot denotes differentiation with respect to the universal time t.

These equations imply that the *center of mass* having position vector ρ with coordinates ρ^i,

$$\rho^i \equiv \frac{m_A X_A^i + m_B X_B^i}{M}, \quad \text{where} \quad M \equiv m_A + m_B , \qquad (12.2)$$

undergoes uniform translational motion (indeed, the entire system is not subject to any force, and so the barycentric frame is Galilean), $\ddot{\rho} = \ddot{\rho}^i e_i = 0$. Setting $l_{AB} \equiv l$ and $r_{AB} \equiv r$, (12.1) also implies that

$$\frac{d^2 l}{dt^2} = -\frac{GM}{r^3} l . \qquad (12.3)$$

Therefore, the two-body problem reduces to the motion of an effective point Q of arbitrary mass in the gravitational field of a mass M located at the origin of the frame. Once $l(t)$ is known, the radius vectors of the two bodies are given by[1]

---

[1] Solar system observations are sufficiently accurate that it is necessary to take into account the motion of the Sun about the center of mass of the system. Newton postulated that this center of mass is at absolute rest and embodies the origin of the absolute reference frame. Leibniz considered only (12.3), valid in any frame of the Milne group undergoing arbitrary translation relative to the absolute frame, to be admissible and ignored the equation giving the motion of the center of mass, which is immaterial and unobservable.

Relativity in Modern Physics. Nathalie Deruelle and Jean-Philippe Uzan.
© Oxford University Press 2018. Published in 2018 by Oxford University Press.
DOI: 10.1093/oso/9780198786399.001.0001

$$R_A - \rho = -\frac{m_B}{m_A + m_B} l , \quad R_B - \rho = \frac{m_A}{m_A + m_B} l . \tag{12.4}$$

A brute-force method of solving (12.3) is to expand it in the Cartesian components of l: $l^i(t) = \{X(t), Y(t), Z(t)\}$. Then by setting $X(t) = r(t) \sin \theta(t) \cos \phi(t)$, $Y(t) = r(t) \sin \theta(t) \sin \phi(t)$, and $Z(t) = r(t) \cos \theta(t)$, we obtain

$$
\begin{cases}
\ddot{r} - r(\dot{\theta}^2 + \sin^2 \theta \, \dot{\phi}^2) = -\dfrac{GM}{r^2} \\[2mm]
\ddot{\theta} + 2 \dfrac{\dot{r}}{r} \dot{\theta} - \dot{\phi}^2 \cos \theta \sin \theta = 0 \\[2mm]
\ddot{\phi} + 2 \dot{\theta} \, \dot{\phi} \, \dfrac{\cos \theta}{\sin \theta} + 2 \dfrac{\dot{r}}{r} \dot{\phi} = 0 .
\end{cases}
\tag{12.5}
$$

Another, faster, method is to introduce a moving frame attached to the trajectory of Q, defined in Section 2.5, such that $l = r e_1'$. We find that $\dot{l} = \dot{r} e_1' + r \dot{e}_1'$, that is, $\dot{l} = \dot{r} e_1' + r \dot{\theta} e_2' + r \dot{\phi} \sin \theta \, e_3'$. A second differentiation then leads to (12.5).

A third method is to write the vector equation (12.3) in spherical coordinates and is a good exercise on the use of the covariant derivative. Setting $X^i = \{X, Y, Z\}$ and $x^i = \{r, \theta, \phi\}$ with $X = r \sin \theta \cos \phi$, $Y = r \sin \theta \sin \phi$, and $Z = r \cos \theta$, the calculation of the covariant derivative ($\tilde{D}_v v$ with $v = \dot{l}$) again gives (12.5). The calculation is rather tedious if (3.14) for the connection coefficients is used, and much more rapid if, instead, (3.27) is used, once the metric in spherical coordinates ($e = dr^2 + r^2 d\theta^2 + r^2 \sin^2 \theta d\phi^2$) has been obtained.

12.2 The ellipses of Kepler

Inspection of the system (12.5) shows that the motion is planar. We can always choose the frame such that initially $\theta = \pi/2$ and $\dot{\theta} = 0$; then the second equation in (12.5) states that $\theta = \pi/2$ at all times. The plane is specified by two angles, i and Ω (the angles of nutation and precession, denoted as Θ and Φ in Section 1.3). In the polar coordinates r, ϕ of the orbital plane, where ϕ is the angle between l and the X axis, (12.5) then becomes

$$\ddot{\phi} + 2 \frac{\dot{r}}{r} \dot{\phi} = 0 , \quad \ddot{r} - r \dot{\phi}^2 = -\frac{GM}{r^2} , \tag{12.6}$$

the first integrals of which are easily found to be

$$\frac{d\phi}{dt} = \frac{L}{r^2} , \quad \left(\frac{dr}{dt} \right)^2 = 2E + \frac{2GM}{r} - \frac{L^2}{r^2} , \tag{12.7}$$

where the two integration constants L and E are called the *angular momentum* and *energy* per unit mass of Q. Since $r^2 \dot{\phi}$ is the areal velocity, Kepler's second law reads as the law of angular momentum conservation.

To obtain the trajectory, we take the ratio of $(dr/dt)^2$ and $(d\phi/dt)^2$, set $u = 1/r$, and differentiate (the Binet method), which gives

$$\frac{d^2 u}{d\phi^2} + u = \frac{1}{p} \quad \text{with} \quad p \equiv \frac{L^2}{GM} , \tag{12.8}$$

whose solution are the conic sections with equation

$$r = \frac{p}{1 + e\cos(\phi - \omega)} \quad \text{with} \quad e^2 \equiv 1 + \frac{2EL^2}{G^2 M^2}, \tag{12.9}$$

where the relation between e and E is obtained by requiring that the trajectory satisfy (12.7). For $E < 0$ this is the equation of an ellipse of *eccentricity* e and *focal parameter* p, whose *major axis*, of length $2a$ $[p = a(1 - e^2)]$, makes an angle ω with the X axis (ω is the proper rotation angle denoted Ψ in Section 1.3). This is Kepler's first law.

For $E < 0$ and $L \neq 0$ the time dependence of the trajectory is obtained in parametric form as[2]

$$\begin{cases} r = a(1 - e\cos\eta) \\ \tan\dfrac{(\phi - \omega)}{2} = \sqrt{\dfrac{1 + e}{1 - e}}\,\tan\dfrac{\eta}{2} \end{cases} \quad \text{with} \quad \sqrt{\frac{GM}{a^3}}(t - T) = \eta - e\sin\eta. \tag{12.10}$$

Therefore, the orbital period is $P = 2\pi\sqrt{a^3/GM}$ and its square is proportional to the cube of the major axis. This is Kepler's third law.[3]

The motion depends on the six integration constants $c_a = (i, \Omega, a, e, \omega, T)$, as required.[4]

The geometry of an ellipse

Figure 12.1 illustrates the principal characteristics of an ellipse:

- P, the point closest to the focus F, corresponds to $\phi - \omega \equiv \nu = 0$;
- A, the point farthest from F, is reached for $\nu = \pi$;
- the semi-latus rectum p characterizes the size of the ellipse;
- the eccentricity e characterizes how flat the ellipse is;
- the semi-major axis a and the semi-minor axis b are related to p and e as

$$a = \frac{p}{(1 - e^2)}, \qquad b = \frac{p}{\sqrt{1 - e^2}}; \tag{12.11}$$

---

[2]This is done by rewriting the second equation in (12.7) as $\dot{r}^2 = -\frac{2E}{(r/a)^2}\left[\frac{r}{a} - (1 - e)\right]\left[(1 + e) - \frac{r}{a}\right]$ and then making the change of function $r = a(1 - e\cos\eta)$ [which immediately gives $t(\eta)$]. The equation for ϕ is then derived from (12.9): $\cos(\phi - \omega) = \frac{\cos\eta - e}{1 - e\cos\eta}$, which is equivalent to the second equation in (12.10).

[3]This magnificent dynamical derivation of Kepler's laws by Newton in 1665 was not in fact fully satisfactory, because it involves the assumption that the gravitational bodies are point-like objects. However, the ratio of the Sun's radius to the Earth's orbital radius is not negligible (it is of order 5×10^{-3}). Newton was thus endeavored to prove that the attraction of a spherical body is identical to that of a point of equal mass located at the center of the sphere, which, as mentioned above in Section 11.4, delayed the publication of the *Principia* until 1687.

[4]Of course, it is necessary to add to these the six constants characterizing the uniform motion of the center of mass. In the Cartesian center-of-mass frame the time dependence of the components of the radius vector l is (see Section 1.3, where $\Phi \to \Omega$, $\Psi \to \omega$, and $\Theta \to i$)

$$X = \mathcal{R}^1{}_1 X' + \mathcal{R}^1{}_2 Y', \quad Y = \mathcal{R}^2{}_1 X' + \mathcal{R}^2{}_2 Y', \quad Z = \mathcal{R}^3{}_1 X' + \mathcal{R}^3{}_2 Y',$$

with $X' = r\cos(\phi - \omega) = a(\cos\eta - e)$, $Y' \equiv a\sqrt{1 - e^2}\sin\eta$, and $\sqrt{\frac{GM}{a^3}}(t - T) = \eta - e\sin\eta$.

- the area enclosed by the ellipse is $S = \pi ab$;
- the distances from A and P to the focus F, r_A and r_P, are related to p and e as

$$p = \frac{2 r_P r_A}{r_A + r_P}, \quad e = \frac{r_A - r_P}{r_A + r_P};$$

- the circle of radius a and center O is called the *apsidal circle* and $OF = ae$;
- the definitions of the points B, B', and H can be read from the figure; with a bit of trigonometry we find that $HB/HB' = \sqrt{1 - e^2}$. These relations allow ν and η to be related. Using $FH = r\cos\nu$, $HB = r\sin\nu$, $OH = a\cos\eta$, and $HB' = a\sin\eta$, we find (recalling that $\nu \equiv \phi - \omega$)

$$r\cos\nu = a(\cos\eta - e), \quad r\sin\nu = a\sqrt{1 - e^2}\sin\eta, \quad \text{and} \quad r = a(1 - e\cos\eta). \tag{12.12}$$

The relation between the angles ν and η then is

$$\tan\frac{\nu}{2} = \sqrt{\frac{1 + e}{1 - e}} \tan\frac{\eta}{2}. \tag{12.13}$$

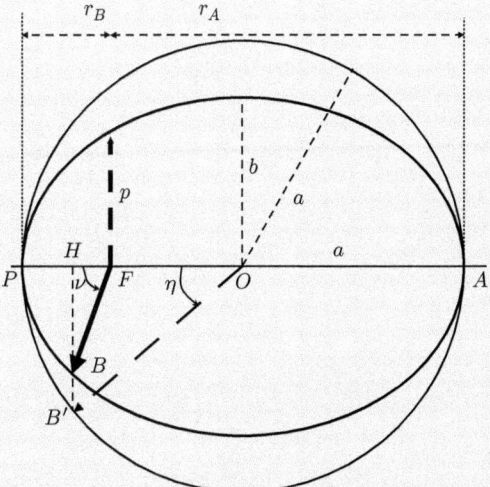

Fig. 12.1 The geometry of an ellipse.

Orbital parameters

Equations (12.9) and (12.10) can be used to describe the motion of a planet in the solar system (neglecting the attraction of other bodies) (see Fig. 12.2). Then m_B is the mass of the Sun and m_A is the planet mass. In the terminology of astronomy, P and A, respectively the points closest and farthest from the focus, are the *perihelion* and the *aphelion* (or, in less 'heliocentric' terms, the *periastron* and the *apastron*), and

- the angle i is the inclination of the orbital plane relative to the *ecliptic* (the plane of the Earth's orbit);

- the angle Ω is the angle between the intersection of these two planes and the *vernal equinox* (the Sun–Earth axis at the spring equinox);

- the angle ω is the *longitude of the periastron*;

- the angle $\nu \equiv (\phi - \omega)$ is the *true anomaly*;

- η is the *eccentric anomaly*;

- T is the *time the body is at periastron*;

- $n = \sqrt{GM/a^3}$ is the *mean angular motion*;

- finally, $\ell \equiv n(t - T)$ is the *mean anomaly*.

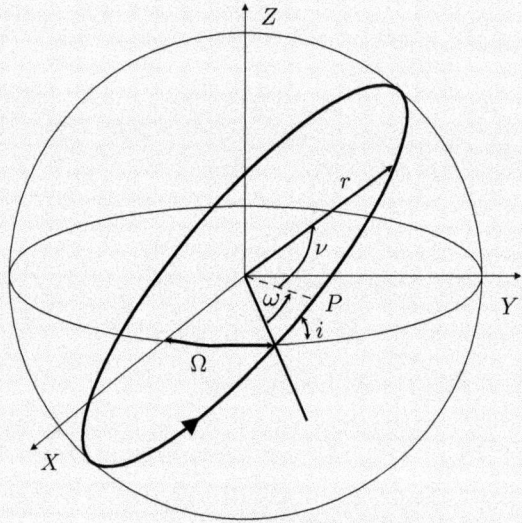

Fig. 12.2 Orbital parameters.

12.3   The Kepler problem in the Lagrangian formalism

If X_A^i, X_B^i, $V_A^i = \dot{X}_A^i$, and $V_B^i = \dot{X}_B^i$ are the Cartesian coordinates and velocities of two bodies of mass m_A and m_B and r is their separation, the gravitational Lagrangian of the system in an inertial frame is [see (11.25)]

$$L(X_A^i, X_B^i, V_A^i, V_B^i) = \frac{1}{2} m_A \, \delta_{ij} V_A^i V_A^j + \frac{1}{2} m_B \, \delta_{ij} V_B^i V_B^j + \frac{G m_A m_B}{r} , \qquad (12.14)$$

and its variation gives (12.1). It can also be written in terms of the center-of-mass coordinates with components $l^i = X_B^i - X_A^i$ and the velocities as

$$L(\rho^i, l^i, \dot{\rho}^i, \dot{l}^i) = L_{cm} + L_r \quad \text{with} \quad \begin{cases} L_{cm}(\rho^i, \dot{\rho}^i) = \dfrac{1}{2}M\,\delta_{ij}\dot{\rho}^i\dot{\rho}^j \\[2mm] L_r(l^i, \dot{l}^i) = \dfrac{1}{2}\mu\,\delta_{ij}\dot{l}^i\dot{l}^j + \dfrac{G\mu M}{r}, \end{cases} \tag{12.15}$$

where $M = m_A + m_B$ and where $\mu = m_A m_B/M$ is the *reduced mass* of the system and L_r is the *reduced Lagrangian*. The Euler–Lagrange equations then show that the two-body problem reduces to the composition of uniform translation of the center of mass and the motion $\ddot{l}^i = -GMl^i/r^3$, governed by (12.3), of a point Q of mass μ in the field of a mass M located at the origin of the reference frame.

In Section 12.1 we discussed three methods of obtaining the equation of motion (12.5).

A fourth method is to write the reduced Lagrangian $L_r \equiv T - W$ in spherical coordinates [*cf.* (8.22)]:

$$L_r(r, \theta, \phi, \dot{r}, \dot{\theta}, \dot{\phi}) = \frac{1}{2}\mu(\dot{r}^2 + r^2\dot{\theta}^2 + r^2\sin^2\theta\,\dot{\phi}^2) + \frac{G\mu M}{r}. \tag{12.16}$$

The Euler–Lagrange equations then lead to (12.5), but with the advantage that the third of these equations now becomes $(r^2\sin^2\theta\,\dot{\phi})' = 0$, which immediately provides a first integral, since the momentum conjugate to ϕ is conserved because ϕ is cyclic.

A (final) method of integrating the equation of motion makes better use of the conservation laws of the system. Since the potential depends only on the distance to the center of the frame, the angular momentum of the point Q, $\mathcal{M}/\mu = l \wedge \dot{l}$, is conserved (see Section 7.2). The vector l then is always perpendicular to the constant vector $\mathcal{M}$, and so the motion occurs in the plane perpendicular to $\mathcal{M}$ specified by the two angles i and Ω (i is the angle between the vector $\mathcal{M}$ and the OZ axis, and Ω is the angle of intersection of the orbital plane and the OX axis). The frame can then be changed so that this plane has the new equation $\theta' = \pi/2$. We thus recover the result that the motion is planar. Since the component $\mathcal{M}_z \equiv \mu L$ written in spherical coordinates is constant [*cf.* (7.10)], we have $r^2\dot{\phi} = L$. Finally, since the potential does not depend explicitly on time the energy $\mathcal{E} \equiv \mu E = T + W = \frac{1}{2}\mu(\dot{r}^2 + r^2\dot{\phi}^2) - G\mu M/r$ of the point Q is conserved (see Section 8.3). To sum up, the laws of angular momentum and energy conservation show without any integration that the motion is planar and also give the two first integrals (12.7).

The remarkable fact that the orbits $E < 0$ are closed curves is due to the $1/r^{n-1}$ dependence with $n = 2$ of the Newtonian potential (see Section 12.4). This feature can be interpreted as a consequence of an additional conservation law specific to the Newtonian potential. In fact, it is easily verified that if $\ddot{l} = -GMl/r^{n+1}$ with $n = 2$, the *Laplace vector* h is conserved[5]:

$$(l \wedge \dot{l}) \wedge \dot{l} + \frac{GM}{r}\,l \equiv -GM\,h = \text{const}. \tag{12.17}$$

The explicit expressions for the components of this vector are easily obtained using the 'Kepler frame' (see Section 2.5, with $\theta = \pi/2$). Setting $l = re_1'$, we have $\dot{l} = \dot{r}e_1' + r\dot{\phi}e_3'$, and so, using the first integrals in (12.7), $GMh = L\dot{r}\,e_3' + e_1'\left(GM - L^2/r\right)$. Returning to the e_i

---

[5]This is also called the *Runge–Lenz* vector (Runge, 1919) in electrostatics, where the Coulomb potential has the same $1/r$ behavior.

frame $(e_1' = \cos\phi\, e_1 + \sin\phi\, e_2,\ e_3' = -\sin\phi\, e_1 + \cos\phi\, e_2)$ and decomposing the vector h as $h = e(\cos\omega\, e_1 + \sin\omega\, e_2)$, we find

$$
\begin{aligned}
GM\, e \cos\omega &= -L\dot{r}\sin\phi + \left(GM - \frac{L^2}{r}\right)\cos\phi \\
GM\, e \sin\omega &= L\dot{r}\cos\phi + \left(GM - \frac{L^2}{r}\right)\sin\phi
\end{aligned}
\tag{12.18}
$$

or

$$
GMe\cos(\phi - \omega) = GM - \frac{L^2}{r}, \quad e^2 = 1 + \frac{2EL^2}{G^2M^2}.
\tag{12.19}
$$

The first equation is that of the Kepler ellipse, obtained without any integration, and the second defines the modulus e of the vector h as the eccentricity.

The Laplace vector

Poincaré showed that a *generic* dynamical system eventually fills its entire phase space, that is, it eventually passes through all the spatial points and takes on all the velocities allowed by the total momentum, energy, and angular momentum conservation laws. In the case of a particle bound by a central potential, this means that ultimately the particle passes through all the points on the annulus in the plane of its trajectory contained between the apastron and the periastron. If the motion is periodic, this annulus reduces to a curve (an ellipse in the present case), which means that the dimension of the hypersurface of the phase space in which the motion occurs decreases from 2 to 1. This implies that there is an additional constraint on the motion, that is, an additional conservation law, which in the case at hand is the conservation of the Laplace (Runge–Lenz) vector.

The existence of this additional symmetry has remarkable consequences. It causes the Hamilton–Jacobi equation of the Newtonian potential to be separable not only in spherical coordinates but also in parabolic coordinates.[6] Moreover, as shown by Fock in 1935, the wave functions of the hydrogen atom are related to the spherical harmonics in four dimensions, that is, they are related to the group of rotations in four dimensions whose generators have been shown by Bargmann to be the angular momentum $\mathcal{M}$ and the Laplace vector.[7]

12.4   Central forces

Let us consider a body of mass m subject to a central force, that is, a force depending only on the distance r from the origin of an inertial frame and derivable from a potential energy $W(r)$. The equation of motion of the body then is

$$
m\ddot{l} = F \quad \text{with} \quad F = -\nabla W = -\frac{dW}{dr}\frac{l}{r}.
\tag{12.20}
$$

The specific angular momentum $L \equiv l \wedge \dot{l}$ is conserved, $\dot{L} = 0$, so that the motion is restricted to the plane perpendicular to L. In the polar coordinates r, ϕ of this plane, the reduced Lagrangian takes the form

---

[6]See, for example, Landau and Lifshitz (1976).

[7]For more on this see, for example, McIntosh (1971).

$$L_r = \frac{m}{2}(\dot{r}^2 + r^2\dot{\phi}^2) - W. \tag{12.21}$$

(Since the coordinate ϕ is cyclic, it does not appear explicitly here.) The Lagrange equations $\partial L_r/\partial q = (\partial L_r/\partial \dot{q})\dot{}$ are

$$(r^2\dot{\phi})\dot{} = 0, \quad m(\ddot{r} - r\dot{\phi}^2) = -\frac{dW}{dr}. \tag{12.22}$$

The first integral for ϕ is obtained immediately and corresponds to the conservation of the specific angular momentum L, and the first integral for r corresponds to conservation of the specific energy E:

$$L = r^2\dot{\phi}, \quad E = \frac{1}{2}(\dot{r}^2 + r^2\dot{\phi}^2) + V = \frac{1}{2}\dot{r}^2 + \frac{L^2}{2r^2} + V, \quad \text{with} \quad V = \frac{W}{m}. \tag{12.23}$$

Integration by quadrature gives

$$t = \int \frac{dr}{\sqrt{2(E-V) - L^2/r^2}}, \quad \phi = \int \frac{(L/r^2)dr}{\sqrt{2(E-V) - L^2/r^2}}. \tag{12.24}$$

The motion is that of a particle evolving in one dimension in the effective potential $V_{\text{eff}} = V + L^2/2r^2$, where the second term represents the centrifugal energy. The turning points $\dot{r} = 0$ satisfy $E = L^2/2r^2 + V$. They delimit the domain of the allowed motion. If this domain is of the form $r_{\text{max}} \geq r \geq r_{\text{min}}$, the trajectory is restricted to an annulus of inner radius r_{min} and outer radius r_{max}. For the trajectory to be periodic it is necessary that

$$\Delta\phi = 2 \int_{r_{\text{min}}}^{r_{\text{max}}} \frac{(L/r^2)dr}{\sqrt{2(E-V) - L^2/r^2}} = 2\pi \frac{n}{p}, \tag{12.25}$$

where n and p are both integers. In 1873 J. Bertrand showed that only two families of potential lead to periodic trajectories: the Newtonian potential $V \propto 1/r$, and the Hooke potential of the harmonic oscillator $V \propto r^2$.

Central forces in the Hamiltonian formalism

The problem of motion in a central force field can be studied within the framework of the Hamiltonian formalism (see Section 9.1). We extract the conjugate momenta from the Lagrangian (12.21) $p_\phi = mr^2\dot{\phi}$ and $p_r = m\dot{r}$, so that the Hamiltonian $H(q, p) = \sum p\dot{q} - L$ becomes

$$H = \frac{1}{2m}\left(p_r^2 + \frac{p_\phi^2}{r^2}\right) + W(r). \tag{12.26}$$

The Hamilton equations ($\dot{q} = \partial H/\partial p$, $\dot{p} = -\partial H/\partial q$) then are

$$\dot{r} = \frac{p_r}{m}, \quad \dot{p}_r = \frac{p_\phi^2}{mr^3} - \frac{dW}{dr}, \quad \text{and} \quad \dot{\phi} = \frac{p_\phi}{mr^2}, \quad \dot{p}_\phi = 0, \tag{12.27}$$

which is equivalent to (12.22). Turning now to the Hamilton–Jacobi equation (9.32), it can be written as

$$\frac{1}{2m}\left(\left(\frac{\partial S}{\partial r}\right)^2 + \frac{1}{r^2}\left(\frac{\partial S}{\partial \phi}\right)^2\right) + W(r) + \frac{\partial S}{\partial t} = 0 \qquad (12.28)$$

and solved by separation of variables and quadrature: we set $S/m = -Et + L\phi + S_0(r)$, the substitution of which into (12.28) gives $S_0 = \int dr \sqrt{2(E - V(r)) - L^2/r^2}$ (we have set $W = mV$). The resulting trajectory equations $\partial S/\partial E = -mt_0$ and $\partial S/\partial L = m\phi_0$ are indeed equivalent to (12.24).

Central forces and Newton's theorem

As we have seen in Section 11.4, Newton showed that the gravitational potential (and the force derived from it) created by a body possessing spherical symmetry is the same as if the body were a point mass. In addition, he showed that the gravitational mass of the body is equal to its inertial mass, thereby generalizing, to the case of bodies whose proper gravity cannot be ignored, the experimentally verified equality of the gravitational and inertial masses of their elementary constituents.

This property does *not* hold for central forces different from the Newton force.

Let us consider the potential $U(P) = -\sum_a G\, m_a f(|P_a P|)$, where f is some function of the distance $|P_a P|$ between the point P and the point P_a at which the 'elementary' particle (a) of gravitational mass m_a equal to its inertial mass is located. Passing to the continuum limit, we have $U(R) = -G \int \varrho(R', t) f(|R - R'|) dV'$, where R is the radius vector of the point P, R' is that of a point of the body, and ϱ is the gravitational/inertial mass density.

If the mass distribution is spherically symmetric, U can be calculated by placing P on the Oz axis such that $R = (0, 0, r)$. In spherical coordinates where $R' = (r' \sin\theta \cos\phi, r' \sin\theta \sin\phi, r' \cos\theta)$, we will have $|R - R'| = \sqrt{r^2 + r'^2 - 2rr' \cos\theta}$, so that the potential becomes (r_0 being the radius of the body)

$$U(r) = -2\pi G \int_0^{r_0} dr'\, r'^2 \varrho(r', t)\, I$$

$$\text{with} \quad I = \int_0^\pi d\theta \sin\theta f(\sqrt{r^2 + r'^2 - 2rr' \cos\theta}) = \frac{F(r + r') - F(|r - r'|)}{rr'}.$$

Let us set $y = \sqrt{r^2 + r'^2 - 2rr' \cos\theta}$ and $F(y) = \int dy\, y f(y)$.

We see first of all that the potential for $r > r_0$ will be proportional to $f(r)$ and therefore the same as if the body were a point if $F(r + r') - F(r - r')$ does *not* depend on r. Since $F'(y) = y f(y)$, we must have $(r + r')f(r + r') = (r - r')f(r - r')$ for all r'. Therefore, it is necessary that $f(y) = 1/y$, that is, the potential must be Newton's potential.

If now $f(y) = 1/y$, then $F(y) = y$ so that $I = 2/r$, which implies that $U(r) = -GM/r$, where the gravitational mass M, given by $M = 4\pi \int_0^{r_0} dr'\, r'^2 \varrho(r', t)$, is equal to the inertial mass of the body. We then again find Newton's theorem.

Therefore the 'strong equivalence principle' (that is, the equality of the gravitational and inertial masses of *all* bodies) and the *effacement* property (that is, in this context, the possibility of treating a spherically symmetric massive body as a point mass) are closely related and specific to Newton's $1/r^2$ law of gravity.

13
The N-body problem

In 1886 Karl Weierstrass submitted the following question to the scientific community on the occasion of a mathematical competition to mark the sixtieth birthday of King Oscar II of Sweden:

   Given a system of arbitrarily many mass points that attract each other according to Newton's laws, try to find, under the assumption that no two points ever collide, a representation of the coordinates of each point as a series in a variable which is some known function of time and for all of whose values the series converges uniformly.

   The jury (which included Weierstrass, Cayley, Hermite, and Chebyshev) awarded the prize to Henri Poincaré not for solving the problem—he showed that the equations of motion for more than two gravitational bodies are not in general integrable—but for coming up with the first ideas of what later became known as chaos theory.

13.1   The Laplace effect

The equality of the gravitational mass m and the inertial mass m^{I} is very well tested experimentally, as we saw in Section 11.2. Newton's law of gravitation implies that it also holds for celestial bodies of considerably larger mass, whose proper gravity cannot be neglected (see Sections 11.4 and 12.4). In 1825 Laplace proposed to test the law by studying the coupled Earth–Moon system in the gravitational field of the Sun.

   The equations of motion of the Sun (S), the Earth (E), and the Moon (M), considered as point objects, can be written in an inertial frame in obvious notation as

$$
\begin{cases}
m_{\mathrm{S}}^{\mathrm{I}} \ddot{R}_{\mathrm{S}} = -G m_{\mathrm{S}} m_{\mathrm{E}} \dfrac{l_{\mathrm{ES}}}{r_{\mathrm{ES}}^{3}} - G m_{\mathrm{S}} m_{\mathrm{M}} \dfrac{l_{\mathrm{MS}}}{r_{\mathrm{MS}}^{3}} , \\[2ex]
m_{\mathrm{E}}^{\mathrm{I}} \ddot{R}_{\mathrm{E}} = -G m_{\mathrm{E}} m_{\mathrm{S}} \dfrac{l_{\mathrm{SE}}}{r_{\mathrm{SE}}^{3}} - G m_{\mathrm{E}} m_{\mathrm{M}} \dfrac{l_{\mathrm{ME}}}{r_{\mathrm{ME}}^{3}} , \\[2ex]
m_{\mathrm{M}}^{\mathrm{I}} \ddot{R}_{\mathrm{M}} = -G m_{\mathrm{M}} m_{\mathrm{S}} \dfrac{l_{\mathrm{SM}}}{r_{\mathrm{SM}}^{3}} - G m_{\mathrm{M}} m_{\mathrm{E}} \dfrac{l_{\mathrm{EM}}}{r_{\mathrm{EM}}^{3}} .
\end{cases}
\tag{13.1}
$$

Setting $l \equiv l_{\mathrm{EM}} \equiv R_{\mathrm{M}} - R_{\mathrm{E}}$ and $r \equiv r_{\mathrm{EM}}$, the acceleration of the Moon relative to the Earth then is

$$
\ddot{l} = -G \left(\frac{m_{\mathrm{M}}}{m_{\mathrm{M}}^{\mathrm{I}}} m_{\mathrm{E}} + \frac{m_{\mathrm{E}}}{m_{\mathrm{E}}^{\mathrm{I}}} m_{\mathrm{M}} \right) \frac{l}{r^{3}} + G m_{\mathrm{S}} \left(\frac{m_{\mathrm{E}}}{m_{\mathrm{E}}^{\mathrm{I}}} \frac{l_{\mathrm{SE}}}{r_{\mathrm{SE}}^{3}} - \frac{m_{\mathrm{M}}}{m_{\mathrm{M}}^{\mathrm{I}}} \frac{l_{\mathrm{SM}}}{r_{\mathrm{SM}}^{3}} \right).
\tag{13.2}
$$

If we write $m/m^{\mathrm{I}} = 1 + \epsilon$, this can be rewritten as

$$
\ddot{l} = -G m_{*} \frac{l}{r^{3}} + G m_{\mathrm{S}} \left(\frac{l_{\mathrm{SE}}}{r_{\mathrm{SE}}^{3}} - \frac{l_{\mathrm{SM}}}{r_{\mathrm{SM}}^{3}} \right) + G m_{\mathrm{S}} \left(\epsilon_{\mathrm{E}} \frac{l_{\mathrm{SE}}}{r_{\mathrm{SE}}^{3}} - \epsilon_{\mathrm{M}} \frac{l_{\mathrm{SM}}}{r_{\mathrm{SM}}^{3}} \right),
\tag{13.3}
$$

Relativity in Modern Physics. Nathalie Deruelle and Jean-Philippe Uzan.
© Oxford University Press 2018. Published in 2018 by Oxford University Press.
DOI: 10.1093/oso/9780198786399.001.0001

where we have introduced the effective mass of the Earth–Moon pair:

$$m_* = m_E(1 + \epsilon_M) + m_M(1 + \epsilon_E).$$

Equation (13.3) is exact. The first, principal, term describes the Earth–Moon gravitational attraction. The second gives rise to tidal effects, which add linearly to the other effects; we shall ignore them. Finally, in the third term the difference between l_{SE} and l_{SM} can be neglected, so that the equation of motion to be solved is reduced to the form

$$\ddot{l} = -Gm_* \frac{l}{r^3} + \delta a \frac{l_{SE}}{r_{SE}} \quad \text{with} \quad \delta a \equiv \frac{Gm_S}{r_{SE}^2}(\epsilon_E - \epsilon_M). \tag{13.4}$$

To solve this equation we proceed as for the two-body problem. Vector multiplication by l and the introduction of the specific angular momentum $L \equiv l \wedge \dot{l}$ gives

$$\dot{L} = \delta a \, l \wedge \frac{l_{SE}}{r_{SE}}. \tag{13.5}$$

After scalar multiplication by l we have (using $l.\ddot{l} = r\ddot{r} - L^2/r^2$)

$$\ddot{r} = -\frac{Gm_*}{r^2} + \frac{L^2}{r^3} + \delta a \frac{l}{r} . \frac{l_{SE}}{r_{SE}} . \tag{13.6}$$

In order to make rapid progress, let us make the crude approximation that the orbits of the Earth about the Sun and the Moon about the Earth are circular *and* lie in the same plane. Then in the frame of the plane centered on the Earth (e_X, e_Y) we have the following in lowest order:

$$l = r(e_X \cos\omega_M t + e_Y \sin\omega_M t), \quad l_{ES} = r_{ES}(e_X \cos\omega_E t + e_Y \sin\omega_E t), \tag{13.7}$$

where ω_M and ω_E are the angular frequencies of the Moon's orbit about the Earth and the Sun's orbit about the Earth. Equations (13.5) and (13.6) can then be simplified to give

$$\dot{L} = -\delta a \, r \sin\Omega t, \quad \ddot{r} = -\frac{Gm_*}{r^2} + \frac{L^2}{r^3} + \delta a \cos\Omega t, \tag{13.8}$$

where we have set $\Omega = \omega_M - \omega_E$.

We now proceed by iteration: $r = r_0 + \delta r$ and $L = L_0 + \delta L$, and we recall that the laws of the two-body problem give $\omega_M^2 = Gm_*/r_0^3 = L_0^2/r_0^4$. The first equation in (13.8) then can be integrated directly to give

$$\delta L = \frac{r_0}{\Omega} \delta a \cos\Omega t . \tag{13.9}$$

Substituting this into the second equation expanded to first order in δr, we find

$$\delta\ddot{r} + \omega_M^2 \delta r = \left(1 + 2\frac{\omega_M}{\Omega}\right) \delta a \cos\Omega t , \tag{13.10}$$

the solution of which is

$$\delta r = \frac{1 + 2\frac{\omega_M}{\Omega}}{\omega_M^2 - \Omega^2} \delta a \cos\Omega t . \tag{13.11}$$

Therefore, the effect of a nonzero δa is to polarize the Earth–Moon system toward the Sun.

Numerically, $\omega_E = 2\pi/1$ year and $\omega_M = 2\pi/27\,\text{days} \sim 13.4\,\omega_E \sim 2.7 \times 10^{-6}\,\text{s}^{-1}$. Therefore, $\delta r_{\max} \sim 39\delta a/2\omega_M^2$. Moreover, $Gm_S/r_{SE}^2 \sim 5.9 \times 10^{-6}\,\text{km/s}^2$, and so

$$\delta r \approx 1.8 \times 10^{10}(\epsilon_E - \epsilon_M)\cos(\omega_M - \omega_E)t\,, \tag{13.12}$$

with δr expressed in meters.

The time for a laser beam to complete a round-trip between a network of terrestrial observatories and four lunar mirrors has been measured experimentally with an accuracy of 1 ps, corresponding to 1 cm in the Earth–Moon separation. The constraint $\delta r < 1$ cm then becomes

$$|\epsilon_E - \epsilon_M| \equiv \left| \left(\frac{m}{m_I} \right)_E - \left(\frac{m}{m_I} \right)_M \right| < 5.5 \times 10^{-13}\,. \tag{13.13}$$

Of course, in a more careful analysis one must verify that there is no other perturbing effect with the same signature, for example, the same period. The main perturbation turns out to be the tidal effect which we have neglected in (13.4).[1]

13.2 The restricted three-body problem

Let us consider the motion of three gravitationally interacting bodies m, m_1, and m_2 in the limit $m \ll m_1$, $m \ll m_2$. This amounts to taking m to be a test particle which does not affect the motion of the other two bodies. In addition, we assume that the three bodies move in the same plane. Finally, we assume that the motion of m_1 and m_2 is circular.

Then in their center-of-mass frame the trajectories of m_1 and m_2 are (see Section 12.1)

$$R_1(t) = \frac{m_1}{M}d\,(\cos\omega t\,e_X + \sin\omega t\,e_Y)\,, \quad R_2(t) = -\frac{m_2}{M}d\,(\cos\omega t\,e_X + \sin\omega t\,e_Y)\,, \tag{13.14}$$

where $M = m_1 + m_2$, d is the separation of the two masses, and $d^3\omega^2 = GM$ (Kepler's third law). The motion of the mass m, with radius vector $R = Xe_X + Ye_Y$, is then governed by the Lagrangian

$$L(X, Y, \dot{X}, \dot{Y}) = \frac{m}{2}(\dot{X}^2 + \dot{Y}^2) + Gm\left(\frac{m_1}{|R - R_1|} + \frac{m_2}{|R - R_2|} \right)\,. \tag{13.15}$$

Changing to dimensionless quantities *via* the transformations $R_{1,2} \to R_{1,2}d$ and $t \to t/\omega$, setting $\mu = m_2/M$, and choosing the system of units such that $G = 1$, the Lagrangian can be rewritten as

$$L = md^2\omega^2 L_0\,, \quad \text{with} \quad L_0 = \frac{1}{2}(\dot{X}^2 + \dot{Y}^2) + \left(\frac{1-\mu}{|R - R_1(t)|} + \frac{\mu}{|R - R_2(t)|} \right)\,. \tag{13.16}$$

Here L_0 depends explicitly on the time, and so the energy is not conserved. This is due to the fact that we have constrained m_1 and m_2 to remain in circular orbits without taking into account the gravitational field created by m.

---

[1]We note that in the absence of the Laplace effect ($m_I = m$) and when the tidal term can be neglected, that is, when the size of the Earth–Moon 'laboratory' can be considered sufficiently small that the Sun's field is constant, then the motion of the Earth–Moon system [*cf.* (13.3)] will *not* depend on the presence of the Sun. This is another aspect of the effacement property shared by the gravitational theories of Newton, Nordström, and Einstein.

The simplest way to determine the dynamics is to go to the rotating frame which lies in the plane of the two masses and whose OX' axis is the line joining m_1 and m_2: $e'_X = \cos \omega t\, e_X + \sin \omega t\, e_Y$. Then, using the same system of units, the positions of m_1 and m_2 will respectively be $(1 - \mu, 0)$ and $(-\mu, 0)$, and the equation of motion is (omitting the primes)

$$\frac{d^2 R}{dt^2} = \nabla \left(\frac{1 - \mu}{|R - R_1|} + \frac{\mu}{|R - R_2|} \right) + 2 \frac{dR}{dt} \wedge \Omega + \Omega \wedge (R \wedge \Omega), \tag{13.17}$$

where the vector $\Omega = e_Z$ in the chosen units. The centrifugal term can then be written as $\Omega \wedge (R \wedge \Omega) = R = \nabla (x^2 + y^2)/2$, where (x, y) are the components of R. Then (13.17) becomes

$$\ddot{x} - 2\dot{y} = \frac{\partial \Phi}{\partial x} \;, \quad \ddot{y} + 2\dot{x} = \frac{\partial \Phi}{\partial y} \;, \tag{13.18}$$

where the function Φ is defined as

$$\Phi(x, y) = \frac{1}{2}(x^2 + y^2) + \frac{\mu}{\sqrt{(x + \mu)^2 + y^2}} + \frac{1 - \mu}{\sqrt{(x - 1 + \mu)^2 + y^2}} \;. \tag{13.19}$$

Introducing $\rho_1 = \sqrt{(x + \mu)^2 + y^2}$ and $\rho_2 = \sqrt{(x + \mu - 1)^2 + y^2}$, the function Φ can also be written as

$$\Phi = \frac{3 - \mu(1 - \mu)}{2} + \mu \left(\frac{1}{2} + \frac{1}{\rho_1} \right) (\rho_1 - 1)^2 + (1 - \mu) \left(\frac{1}{2} + \frac{1}{\rho_2} \right) (\rho_2 - 1)^2. \tag{13.20}$$

Multiplying the first expression in (13.18) by $\dot{x}$ and the second by $\dot{y}$, we obtain

$$2\Phi - (\dot{x}^2 + \dot{y}^2) = J, \tag{13.21}$$

where J is the *Jacobi constant*. The other first integral of this system is not known, and so we have to resort to numerical integration. For a given value of J we must have $\dot{x}^2 + \dot{y}^2 = 2\Phi - J > 0$, so that the plane is divided into two regions, one allowed and one forbidden, separated by the curve $\Phi = J/2$, which is called the *Hill curve*.

The Lagrange points

In 1772 Lagrange proved that there exist five equilibrium points for a test particle in the gravitational field of two other bodies (the restricted three-body problem).These points are called *libration points* or *stationary Lagrange points*. They are defined by $\ddot{x} = \ddot{y} = \dot{x} = \dot{y} = 0$, which amounts to solving the equation (since (13.21) implies that $\Phi = $ const) $\partial \Phi / \partial x = \partial \Phi / \partial y = 0$; cf. (13.18).

If $y = 0$, then $\rho_1 = |x + \mu|$ and $\rho_2 = |x + \mu - 1|$, and the equation $\partial \Phi / \partial x = 0$ reduces to an equation of third degree whose three solutions $x_1 < -\mu < x_3 < 1 - \mu < x_2$ define the Lagrange points L_1, L_3, and L_2. These three points are unstable, but it turns out to be useful to position solar observation satellites at the point L_1 of the Sun–Earth system and satellites for observations pointing away from the Sun at L_2.

If $y \neq 0$, we see from (13.20) that Φ possesses two minima at $\rho_1 = \rho_2 = 1$. These Lagrange points L_4 and L_5 have the coordinates $x = 1/2 - \mu$, $y = \pm\sqrt{3}/2$ (in the rotating frame in which Φ is expressed), and together with the centers of the two masses m_1 and m_2 form two equilateral triangles. It can be shown that L_4 and L_5 are stable if $\mu < \mu_c = (1 - \sqrt{1 - 4/27})/2 \sim 0.03852$ and unstable otherwise. A large number of minor planets, called *trojans*, have accumulated at the points L_4 and L_5 of the Sun–Jupiter system.

13.3 Gauss equations of perturbations

The motion of the planets of the solar system is only Keplerian to a first approximation. It is necessary to take into account the presence of other planets and asteroids, as well as the fact that the bodies are non-spherical. Euler (1748), Lagrange (1782), Gauss, and many others have contributed to the solution of this problem. Modern trajectory calculations include hundreds of objects and are performed by computer programs which have become standard tools in space engineering.

Here we shall limit ourselves to the study of the motion of two gravitational bodies with spherical symmetry under the action of a perturbing force, for which the equations are

$$a_A = -\frac{Gm_B}{r_{BA}^3}l_{BA} + \frac{\mathcal{F}_A}{m_A} \quad \text{and} \quad a_B = -\frac{Gm_A}{r_{AB}^3}l_{AB} + \frac{\mathcal{F}_B}{m_B}, \tag{13.22}$$

where $l_{AB} = R_B - R_A$, $a_A \equiv \ddot{R}_A$, and $a_B \equiv \ddot{R}_B$, with R_A and R_B the radius vectors of A and B in an inertial frame. If $\mathcal{F}$ is a force internal to the system (due, for example, to the deformation of one of the bodies), then the principle of action and reaction (obeyed by Newton's law of gravity) imposes the constraint $\mathcal{F}_A = -\mathcal{F}_B$, so that the motion of the center of mass is not affected by the perturbation and the problem reduces to solving the equation of motion of a fictitious point Q:

$$\ddot{l} = -\frac{GM}{r^3}l + \mathcal{A}, \tag{13.23}$$

where $l \equiv l_{AB}$ and $\mathcal{A} = \mathcal{F}/\mu$ with μ the reduced mass of the system and M the total mass. The problem also reduces to solving (13.23) in the case of an external force (due to the presence of a third body, for example) if $m_A \gg m_B$ so that the center of mass coincides with the center of the body A.

Equation (13.23) can be solved by iteration. To zeroth order the motion is Keplerian and the trajectory of Q is an ellipse $l = l_K(t, c_a)$ depending on six parameters (see Section 12.2):

$$c_a = \{a, e, i, \omega, \Omega, T\}. \tag{13.24}$$

To first order the perturbation $\mathcal{A}$ is evaluated on the Keplerian trajectory of Q and its effect is modeled by changing the ellipse parameters: $c_i \mapsto c_i(t)$. The ellipse whose elements are $c_a(t)$ is called the *osculating ellipse* and describes Keplerian motion close to the actual motion in a neighborhood of t. For example, a variation of the sole longitude of the periastron translates into precession of the ellipse in its plane.

To obtain the evolution equations of the orbital parameters we write $l = l_K(t, c_a)$, where now the six orbital parameters c_a are functions of time. Since we started with three unknown functions $l^i(t)$ which we have replaced by six new functions $c_a(t)$, we can add three independent constraints on the c_a without spoiling the generality of the approach. It is common to require that the velocity also has a Keplerian form, that is, that $\dot{l} = \partial l_K/\partial t$. Since $\dot{l} = \partial l_K/\partial t + \dot{c}_a \partial l_K/\partial c_a$, we then have $\dot{c}_a \partial l_K/\partial c_a = 0$. Therefore, $\ddot{l} = \partial^2 l_K/\partial t^2 + \dot{c}_a \partial^2 l_K/(\partial c_a \partial t)$ with $\partial^2 l_K/\partial t^2 = -(GM/r^3)l_K$, and so the equation of motion (13.23) becomes

$$\dot{c}_a \frac{\partial l_K}{\partial c_a} = 0, \quad \dot{c}_a \frac{\partial \dot{l}_K}{\partial c_a} = \mathcal{A}. \tag{13.25}$$

Forming the scalar product of the first equation with $\partial \dot{l}_K/\partial c_b$ and of the second with $\partial l_K/\partial c_b$ and then subtracting them from each other, we arrive at (summation over a is everywhere understood)

$$[c^b, c^a]\dot{c}_a = \mathcal{A}.\frac{\partial l_K}{\partial c_b}, \quad \text{where} \;\; [c^b, c^a] \equiv \frac{\partial \dot{l}_K}{\partial c_b}.\frac{\partial l_K}{\partial c_a} - \frac{\partial \dot{l}_K}{\partial c_a}.\frac{\partial l_K}{\partial c_b} \tag{13.26}$$

is the *Lagrange bracket* of c_b and c_a, which is easily shown to be independent of time (so that it can be evaluated at $t = 0$). We note that if the perturbing force can be derived from a potential $\mathcal{F} = -\mu \nabla U_{\text{pert}}$, (13.26) can be simplified to

$$[c^b, c^a]\dot{c}_a = -\frac{\partial U_{\text{pert}}}{\partial c_b}. \tag{13.27}$$

An easy calculation using the properties of Keplerian ellipses (see Section 12.2) gives

$$[e, \omega] = -\sqrt{\frac{GMa}{1 - e^2}}\, e, \quad [e, \Omega] = [e, \omega]\cos i, \quad [i, \Omega] = -\sqrt{GMa(1 - e^2)}\, \sin i,$$

$$[a, \omega] = \frac{1}{2}\sqrt{\frac{GM(1 - e^2)}{a}}, \quad [a, \Omega] = [a, \omega]\cos i, \quad [a, T] = -\frac{1}{2}\frac{GM}{a^2} \tag{13.28}$$

(all the other independent brackets vanish). The $\partial l_K^i/\partial c_j$ are obtained in the same way, for example, $\partial X_K/\partial a = X_K/a$. We also introduce $\mathcal{R}$, $\mathcal{S}$, and $\mathcal{W}$, the components of the perturbation $\mathcal{A}$ along the radius vector and the directions perpendicular to it (one in the orbital plane, and the other orthogonal to it).[2] Assembling all these results, inversion of (13.28) finally gives the *Gauss equations*:

$$\begin{cases}
\dfrac{da}{dt} = 2\sqrt{\dfrac{a^3}{GM(1 - e^2)}}\left[\mathcal{R}e\sin\nu + (1 + e\cos\nu)\mathcal{S}\right] \\[2ex]
\dfrac{de}{dt} = \sqrt{\dfrac{a(1 - e^2)}{GM}}\left[\mathcal{R}\sin\nu + \dfrac{(e + 2\cos\nu + e\cos^2\nu)}{1 + e\cos\nu}\mathcal{S}\right] \\[2ex]
\dfrac{di}{dt} = \sqrt{\dfrac{a(1 - e^2)}{GM}}\dfrac{\cos(\omega + \nu)}{1 + e\cos\nu}\mathcal{W} \\[2ex]
\dfrac{d\Omega}{dt} = \sqrt{\dfrac{a(1 - e^2)}{GM}}\dfrac{\sin(\omega + \nu)}{1 + e\cos\nu}\dfrac{\mathcal{W}}{\sin i} \\[2ex]
\dfrac{d\omega}{dt} = \sqrt{\dfrac{a(1 - e^2)}{GM}}\left[-\dfrac{\mathcal{R}}{e}\cos\nu + \dfrac{(2 + e\cos\nu)\sin\nu}{e(1 + e\cos\nu)}\mathcal{S} - \dfrac{\sin(\omega + \nu)}{1 + e\cos\nu}\operatorname{ctg} i\,\mathcal{W}\right] \\[2ex]
\dfrac{d\ell}{dt} = n + \sqrt{\dfrac{a}{GM}}\dfrac{1 - e^2}{e(1 + e\cos\nu)}\left[\mathcal{R}(-2e + \cos\nu + 2\cos^2\nu) - \sin\nu(2 + e\cos\nu)\mathcal{S}\right],
\end{cases} \tag{13.29}$$

---

[2]Therefore, $\mathcal{F}^X/\mu = \mathcal{R}\frac{X_K}{r} - \mathcal{S}(\mathcal{R}_1{}^1 \sin\nu - \mathcal{R}_1{}^2 \cos\nu) + \mathcal{R}_1{}^3\,\mathcal{W}$, with similar expressions for $\mathcal{F}^Y$ and $\mathcal{F}^Z$ (see footnote 4 of Section 12.1 and Section 1.3).

where $\ell \equiv n(t-T)$, $\nu \equiv \phi-\omega$, and $n \equiv \sqrt{\frac{GM}{a^3}}$. These equations form a closed system governing the time evolution of the six osculating elements (13.24). Here the time dependence of the true anomaly ν is Keplerian at the order in which we are working[3]: $\dot{\nu} = \sqrt{\frac{GM}{a^3(1-e^2)^3}}(1 + e \cos \nu)^2$. We remark that a can be replaced by p, which satisfies the equation

$$\frac{dp}{dt} = 2r\sqrt{\frac{p}{GM}}\,\mathcal{S}\,. \tag{13.30}$$

The Delaunay elements

A remarkable accomplishment of Delaunay in his study of lunar motion (1850–1870) was finding combinations of osculating elements $c_a = \{q_l, p_l\}$ with $1 \leq l \leq 3$ called *canonial combinations*, whose Lagrange brackets reduce to

$$[q^l, q^m] = [p^l, p^m] = 0\,, \quad [q^l, p^m] = \delta^{lm}\,. \tag{13.31}$$

These Delaunay combinations are

$$q^1 = \sqrt{GMa}\,, \quad q^2 = \sqrt{GMp}\,, \quad q^3 = \cos i \sqrt{GMp}$$
$$p^1 = \ell \equiv n(t-T)\,, \quad p^2 = \omega\,, \quad p^3 = \Omega\,. \tag{13.32}$$

(They are not unique; for example, the pair $\{q^1, p^1\}$ can be replaced by $\{GM/2a, T\}$.) Therefore, the Delaunay elements bear the same relation to the Lagrange brackets as the generalized coordinates and their conjugate momenta bear to the Poisson brackets in Hamiltonian mechanics; see Section 9.1. Their equations of motion (13.27) indeed have a Hamiltonian form:

$$\frac{\partial H}{\partial p_l} = \dot{q}_l\,, \quad \frac{\partial H}{\partial q_l} = -\dot{p}_l\,, \tag{13.33}$$

where H can be written as

$$H = \frac{1}{2}\frac{G^2 M^2}{q_1^2} + \delta H(q_l, p_l)\,, \tag{13.34}$$

the first term being the Hamiltonian of the system of two point masses and δH the perturbation.

The example of a radial perturbation

Let us consider the gravitational potential

$$U(r) = -\frac{GM}{r}\left(1 + \frac{\epsilon}{n+1}\frac{R^n}{r^n}\right)\,, \tag{13.35}$$

---

[3] Now that we have the first-order approximation, we can of course continue and construct the second-order one by iteration. As shown by Poincaré, the series are not convergent. However, for each type of perturbation the divergence is more or less rapid depending on which combinations of osculating elements are chosen. See, for example, Brouwer and Clemence (1961) or Hagihara (1970). See also Duriez (2003).

where ϵ is a small parameter and R is a constant. The associated force $F = -m\nabla U$ is a central force with modulus

$$F = -\frac{GMm}{r^2}\left(1 + \epsilon\frac{R^n}{r^n}\right).$$ (13.36)

The components of the perturbation $\mathcal{A}$ then are

$$\mathcal{R} = -\epsilon\frac{GM}{r^2}\frac{R^n}{r^n}\,, \quad \mathcal{S} = 0\,, \quad \mathcal{W} = 0\,.$$ (13.37)

The Gauss equations (13.29) and (13.30) then imply that i, Ω, and p remain constant. The equations for ω and e reduce to

$$\dot{\omega} = \epsilon\frac{\sqrt{GMp}}{e}\frac{R^n}{r^{n+2}}\cos\nu\,, \quad \dot{e} = -\epsilon\sqrt{GMp}\frac{R^n}{r^{n+2}}\sin\nu$$ (13.38)

with, to first order, $r = p/(1 + e_0\cos\nu)$ and $\dot{\nu} = \sqrt{GMp}/r^2$, which immediately gives

$$\omega(\nu) = \omega_0 + \frac{\epsilon}{e_0}\frac{R^n}{p^n}\int_0^\nu d\nu\,\cos\nu(1 + e_0\cos\nu)^n\,, \quad e(\nu) = e_0 - \epsilon\frac{R^n}{p^n}\int_0^\nu d\nu\,\sin\nu(1 + e_0\cos\nu)^n\,.$$ (13.39)

This shows, in particular, that e does not exhibit a secular drift, $\Delta e \equiv e(2\pi) - e_0 = 0$ for any n.

Let us consider the special case $n = 2$ (the perturbing force could be a tidal effect in the symmetry plane of an ellipsoid). We obtain

$$\Delta\omega = 2\pi\epsilon\frac{R^2}{p^2}\,, \quad e = e_0 + \epsilon\frac{R^2}{3e_0p^2}[(1 + e_0\cos\nu)^3 - (1 + e_0)^3]\,.$$ (13.40)

Therefore, at the periastron $e = e_0$ and at the apastron $e = e_0 - 2\epsilon R^2(3 + e_0^2)/3p^2$. The orbit is an oval with constant p, flatter at the apastron than at the periastron and turning by an angle $\Delta\omega$ at each revolution.

By attributing the perturbing forces to the Newtonian attraction of the planets, in 1846 Le Verrier and Adams independently succeeded in obtaining agreement between calculation and observation of the trajectory of Uranus by postulating the existence of another planet, Neptune, which was discovered shortly afterwards by Galle at the predicted position, a spectacular success. Gauss had in fact shown in 1801 that one could have predicted the existence of Uranus, discovered accidentally by Herschel, using the same method. On the other hand, the anomalous advance of the perihelion of Mercury completely baffles the Newtonian theory.

Advance of the perihelion of Mercury

The observed advance of the perihelion of Mercury is of the order of 5600 arcseconds per century. Most of it, 5026 arcseconds, is explained by the effect of precession of the equinoxes; see Section 14.1 below. The perturbations due to the gravitational field of the other planets account for another 531 arcseconds. There remain about 43 arcseconds per century which are unexplained in the Newtonian theory.

Causes of the advance of the perihelion of Mercury (in arcseconds per century)

Equinoxes	Venus	Earth	Mars	Jupiter	Saturn	Others	Total	Observed	Anomaly
5025.6	277.8	90.0	2.5	153.6	7.3	0.2	5557.0	5599.7	42.7

Numerous hypotheses were advanced (in particular, Le Verrier postulated the existence of a new planet, Vulcan, near the Sun). However, while they did explain the anomaly of Mercury, either they involved objects which were not optically detected, or they gave rise to unacceptable perturbations in the orbits of the other planets. The problem was solved only in 1915, by Einstein in his theory of general relativity. He obtained

$$\Delta\omega = \frac{6\pi GM}{pc^2},$$
(13.41)

where c is the speed of light. For Mercury ($a = 58 \times 10^6$ km, $e = 0.21$) we find $\Delta\omega = 42.7$ arcseconds per century.

Book 1

14

Deformations of celestial bodies

In the preceding chapter we assumed that celestial bodies are spherically symmetric. Here we lift this condition in order to study various gravitational effects arising from their non-sphericity.

14.1 Quadrupole expansion of the potential

The gravitational potential created at a point P with radius vector R (of length r) by a body of density ϱ is given in any reference frame by (see Section 11.4)

$$U(R) = -G \int \frac{\varrho(R')}{|R - R'|} dV' = -\frac{G}{r} \int \frac{\varrho(R')}{\sqrt{1 - 2\frac{R'.R}{r^2} + \frac{r'^2}{r^2}}} dV', \tag{14.1}$$

where R' is the radius vector (of length r') of a point P' of the distribution and $dV = \sqrt{\det e}\, d^3x$ with $\det e$ the determinant of the metric coefficients in the coordinates x^i. If $R \gg R'$, that is, if P is far outside the body, then $U(R)$ can be expanded as

$$
\begin{aligned}
U(R) = &-\frac{G}{r} \int \varrho(R') dV' - \frac{GR}{r^3} \cdot \int \varrho(R') R' dV' \\
&- \frac{G}{2r^5} \int \varrho(R')[3(R.R')^2 - r^2 r'^2] dV' + \mathcal{O}\left(\frac{1}{r^4}\right).
\end{aligned}
\tag{14.2}
$$

The first term involves the total mass, $M = \int \varrho(R') dV'$, and the second involves $\int \varrho(R') R' dV'$ (the continuous version of $\sum m_a R_a$), which is zero if the origin of the frame coincides with the center of mass of the body. Finally, introducing the *quadrupole moment*

$$Q_{ij} = \int \varrho(R')(3X'_i X'_j - r'^2 \delta_{ij}) dV', \tag{14.3}$$

where X'^i are the Cartesian components of the position vector R', the expansion of the potential can be written as follows in the center-of-mass frame:

$$U(R) = -\frac{GM}{r} - \frac{G}{2r^5} Q_{ij} X^i X^j + \mathcal{O}\left(\frac{1}{r^4}\right), \tag{14.4}$$

where X^i are the Cartesian coordinates of P. The quadrupole moment is related to the *inertia tensor* of the distribution (see Section 6.1) $I_{ij} = \int \varrho(R')(\delta_{ij} r'^2 - X'_i X'_j) dV'$ as $Q_{ij} = -3I_{ij} + I\delta_{ij}$, where I is the trace of I_{ij}. Therefore, to lowest order the quadrupole term characterizes the effect of the non-sphericity of a body on the gravitational potential it creates.

Relativity in Modern Physics. Nathalie Deruelle and Jean-Philippe Uzan.
DOI: 10.1093/oso/9780198786399.001.0001

Now let us suppose that the body possesses a symmetry of revolution about the OZ axis. Then (see Section 6.1) I_{ij} is diagonal and $I_{XX} = I_{YY} \equiv I_1$, so that, setting $I_{ZZ} = I_3$, we have $Q_1 = Q_2 = I_3 - I_1$ and $Q_3 = -2Q_1$. Therefore, $Q_{ij}X^iX^j = (I_3 - I_1)(X^2 + Y^2 - 2Z^2)$ becomes $Q_{ij}X^iX^j = (I_3 - I_1)r^2(1 - 3\cos^2\theta)$ in spherical coordinates, where θ is the angle the radius vector R makes with the OZ axis. The quadrupole expansion of the potential is then written as

$$U(r,\theta) = -\frac{GM}{r}\left(1 - J_2\left(\frac{R_0}{r}\right)^2 P_2(\cos\theta)\right), \quad \text{where} \quad J_2 \equiv \frac{I_3 - I_1}{MR_0^2}, \tag{14.5}$$

R_0 is the equatorial radius of the distribution, and $P_2(\cos\theta) = \frac{1}{2}(3\cos^2\theta - 1)$ is the second-order Legendre polynomial. For the Earth the measured value of J_2 is $J_2 \simeq 1.08 \times 10^{-3}$; see Section 6.1.

The oblateness of the Sun and the motion of Mercury

Let us describe the Sun as a spheroid of revolution about the OZ axis so that the gravitational potential it creates is approximately given by (14.5). The gravitational force per unit mass that it exerts on a planet (whose own field we neglect) is $F/m = -\nabla U$ and is the sum of the Keplerian term and a perturbation whose components in spherical coordinates are (*cf.* the expression for ∇ in terms of the triad $\{h_r, h_\theta, h_\phi\}$ in Section 4.6)

$$\mathcal{A} = \frac{3GMJ_2}{r^2}\left(\frac{R}{r}\right)^2\left(\frac{3\cos^2\theta - 1}{2}h_r - \cos\theta\sin\theta\, h_\theta\right), \tag{14.6}$$

where r is the distance from the Sun to the planet and M and R are the solar mass and radius. We therefore see that the planet remains in the equatorial plane $\theta = \pi/2$ if it was located there initially. In this particular case where the perturbation is radial and has the form in (13.37), we have $\mathcal{R} = -\epsilon GMR^2/r^4$ with $\epsilon = 3J_2/2$. The advance of the perihelion that follows from this was obtained in (13.40):

$$\Delta\omega = \frac{2\pi\epsilon R^2}{p^2} \quad \text{or} \quad \Delta\omega = \frac{3\pi J_2 R^2}{p^2}, \tag{14.7}$$

where $p = a(1-e^2)$ is the parameter of the osculating ellipse. In (13.41) we gave the advance of the perihelion of Mercury predicted by general relativity: $\Delta\omega = 6\pi GM/(pc^2)$, which is in complete agreement with observation. It is difficult to measure the coefficient J_2, but the currently accepted value ($J_2 \approx 2 \times 10^{-7}$) gives an advance too small to be actually measurable. Therefore, the oblateness of the Sun does not spoil the agreement between general relativity and the observation of Mercury.

The precession of the equinoxes

Let us consider a reference frame $\mathcal{S}$ centered on the Earth and attached to the plane of the *ecliptic* in which the Sun undergoes its annual motion (which we assume to be circular).[1] We attach a second reference frame $\mathcal{S}'$ to the Earth, the axis EZ' coinciding with the Earth's axis of proper rotation. The orthogonal plane $X'EY'$ is the *equatorial plane*. In the approximation

---

[1]The reference frame $\mathcal{S}$ is not inertial—it is a frame of the Milne group; see Section 1.4. However, in this 'freely falling' frame Newton's equations apply without the introduction of inertial forces; see Section 11.3.

where both bodies are spherical, the two Euler angles Φ and Θ characterizing the relative position of these two planes, called the *precession* and *nutation* angles (Section 1.3), remain constant. The *proper rotation* angle Ψ varies as $\Psi = \omega_0 t$, where $\omega_0 = 2\pi/24$ h is the angular velocity of the Earth. The *precession of the equinoxes* is a slow variation of the precession angle Φ due to the reaction of the Earth's non-sphericity on the equation of motion of its angular momentum (see Fig. 14.1).

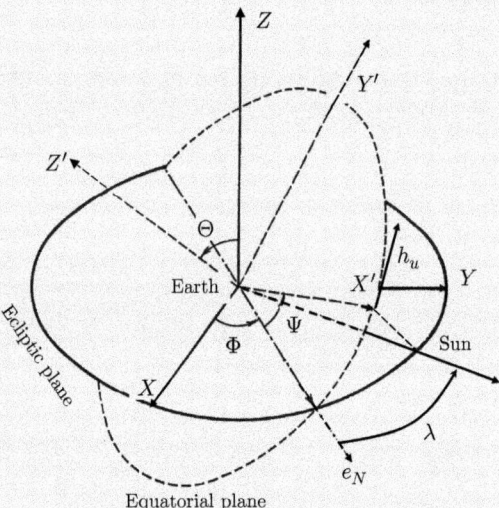

Fig. 14.1 The precession of the equinoxes.

We shall describe the Earth as a spheroid of revolution of mass M and moments of inertia $I_1 = I_2$ and I_3. The gravitational potential it creates is given to quadrupole order by (14.5): $U(r,\theta) = -GM/r + G(I_3 - I_1)(3\cos^2\theta - 1)/(2r^3)$, where r is the distance from the center of the Earth to the point S and θ is the angle the radius vector ES makes with the symmetry axis EZ' (here we neglect the *polhode motion*, see Section 6.2, so that the Earth's rotation and symmetry axes are the same). The force $F = -M_\odot \nabla U = -M_\odot(h_r(\partial U/\partial r) + (h_\theta/r)(\partial U/\partial\theta))$ (see Section 4.6) exerted by the Earth on the Sun splits into a radial component originating in the motion of the Sun about the Earth and an orthoradial component which exerts a torque $ES \wedge F$ on the Sun:

$$ES \wedge F = r\, h_r \wedge F = -M_\odot \frac{\partial U}{\partial\theta} h_r \wedge h_\theta = \frac{3GM_\odot(I_3 - I_1)}{r^3}\cos\theta\sin\theta\, h_r \wedge h_\theta\,. \tag{14.8}$$

The vector $h_u \equiv h_r \wedge h_\theta$ and therefore the vector $ES \wedge F$ lie in the equatorial plane. According to the law of action and reaction, the Sun exerts an opposing torque on the Earth $K \equiv -ES \wedge F$. Finally, it is an easy exercise in geometry to show that the components of K in $\mathcal{S}'$ are

$$K = -\frac{3GM_\odot(I_3 - I_1)\sin\Theta}{2r^3}\left[\cos\Theta(1 - \cos 2\lambda)e_{\mathrm{N}} - \sin 2\lambda e_2'\right], \tag{14.9}$$

where $e_{\mathrm{N}} = e_1'\cos\Psi - e_2'\sin\Psi$ is the direction of the *line of nodes* and λ is the angle between this line and the radius vector of the Sun.

Since the Earth is a rotating body (here modeled by a symmetric top; see Section 6.1), the Euler equations of motion in $\mathcal{S}'$ of its angular momentum ($\dot{J} = K$, where $J'^1 = I_1 \omega'^1$ and so on) are written as

$$I_1 \frac{d\omega'^1}{dt} + \omega'^2 \omega'^3 (I_3 - I_1) = K'^1, \quad I_1 \frac{d\omega'^2}{dt} - \omega'^3 \omega'^1 (I_3 - I_1) = K'^2, \quad I_3 \frac{d\omega'^3}{dt} = K'^3, \quad (14.10)$$

$$\omega'^1 = \dot{\Phi} \sin \Psi \sin \Theta + \dot{\Theta} \cos \Psi, \quad \omega'^2 = \dot{\Phi} \cos \Psi \sin \Theta - \dot{\Theta} \sin \Psi, \quad \omega'^3 = \dot{\Phi} \cos \Theta + \dot{\Psi}. \quad (14.11)$$

The vector K is periodic in λ. Here we are interested in the *secular effect* produced by its average over λ (that is, its average over the year):

$$\overline{K'^1} = -\frac{3GM_\odot (I_3 - I_1) \sin \Theta \cos \Theta}{2r^3} \cos \Psi, \quad \overline{K'^2} = \frac{3GM_\odot (I_3 - I_1) \sin \Theta \cos \Theta}{2r^3} \sin \Psi. \quad (14.12)$$

Moreover, using the fact that $\dot{\Phi}, \dot{\Theta}$, and $\dot{\Psi}$ are nearly constant and that $\dot{\Psi} \gg (\dot{\Phi}, \dot{\Theta})$, we can write [cf. (14.11)] $\omega'^3 \approx \dot{\Psi}$, $\dot{\omega}'^1 \approx \dot{\Psi}(\dot{\Phi} \cos \Psi \sin \Theta - \dot{\Theta} \sin \Psi)$, and $\omega'^2 \approx -\dot{\Psi}(\dot{\Phi} \sin \Psi \sin \Theta + \dot{\Theta} \cos \Psi)$.

The third expression in (14.10) then gives $\omega'^3 \approx \dot{\Psi} \approx$ const, which means that the daily rotation of the Earth is not changed. The linear combination of the first two equations which eliminates $\overline{K'^1}$ and $\overline{K'^2}$ gives $\dot{\Theta} = 0$, showing that the angle between the two planes (equatorial and ecliptic) remains constant. Finally, the remaining equation gives the variation of the precession angle:

$$\dot{\Phi} = -\frac{3G(I_3 - I_1)}{2\dot{\Psi} I_3} \cos \Theta \, \frac{M_\odot}{r^3}. \quad (14.13)$$

If we add to the contribution of the Sun that of the Moon (assuming that the Moon's orbit lies in the plane of the ecliptic, which is a crude approximation), then $M_\odot / r^3 \mapsto M_\odot / r^3 + M_{\text{Moon}} / r_{\text{EM}}^3$.

The numerical values are $(M_\odot / r^3) + (M_{\text{Moon}} / r_{\text{EM}}^3) = 3.2 M_\odot / r^3$, $(I_3 - I_1)/I_3 = 3.0 \times 10^{-3}$, $\dot{\Psi} = 2\pi/24$ h, $\Theta = 23$ degrees, and $GM_\odot / r^3 = (2\pi/\text{year})^2$ (Kepler's third law). This leads to $2\pi/\dot{\Phi} \approx 26,000$ years, which agrees (within the accuracy of the approximations we have made) with observation. (The first observation was made by Hipparchus in the second century B.C.)

14.2 Causes of non-sphericity of bodies

The gravitational attraction of the Moon (or the Sun) differs from one point of the Earth to another simply because the distance to the attracting body is not the same. Assuming that the attractor of mass m_* located at A is spherical, the gravitational potential that it induces at a point P on the Earth, $U = -Gm_*/|AP|$, can be expanded as

$$U(R) = -\frac{Gm_*}{d} \left(1 + \frac{r}{d} \cos \varphi + \frac{r^2}{d^2} \frac{3 \cos^2 \varphi - 1}{2} + \cdots \right), \quad (14.14)$$

where $OA = d \, e_X$ is the Earth–attractor vector, $OP = R = r \, h_r$ is the radius vector of a point P of the Earth, and the angle φ is defined as $\cos \varphi = e_X . h_r$ (see Fig. 14.2).

The acceleration induced by the attractor at P in spherical coordinates is

$$g(R) = -\nabla U = -(\partial_r U) \, h_r - \frac{1}{r} (\partial_\varphi U) \, h_\varphi. \quad (14.15)$$

Therefore, to zeroth and first order $g_0 = 0$ and $g_1 = (Gm_*/d^2) e_X$, which is just the field created by the attractor at O. This is a uniform field which leads to a global displacement of the matter (and therefore of its center of mass) without deforming the planet. The first term which has a differential effect is

$$g_2 = \frac{Gm_* r}{2d^3} \left[(3\cos\varphi - 1))h_r - 3\sin\varphi\cos\varphi h_\varphi \right] = \frac{Gm_* r}{d^3} \left[2\cos\varphi e_X - \sin\varphi e_Y \right]. \tag{14.16}$$

We must add to this acceleration that due to the Earth's own gravitational field $-g_E h_r = -(GM_E/r^3) R$, which is derived from the potential $U_{\mathrm{proper}}(R) = -GM_E/r$ in the approximation where the Earth is assumed to be spherical. Introducing the coefficient ζ, called the *equilibrium tidal amplitude*, the first term of the perturbing potential of the attractor can be rewritten as

$$U_{2\mathrm{tidal}} = -\zeta g_E P_2(\cos\varphi), \quad \zeta = \frac{m_*}{M_E}\left(\frac{r_E}{d}\right)^3 r_E, \tag{14.17}$$

where r_E is the average radius of the Earth and $P_2(\cos\varphi) \equiv (3\cos^2\varphi - 1)/2$.

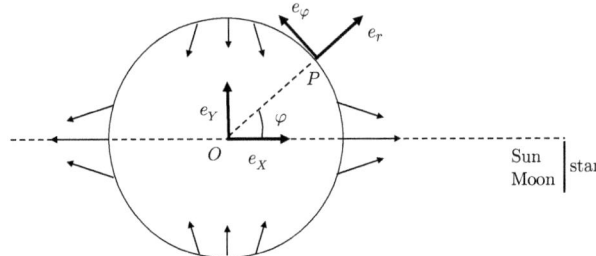

Fig. 14.2 Gravitational field induced by an attracting body at the surface of the Earth.

Tides in the fluid model of the Earth

The response of the Earth to a perturbing potential can be obtained by assuming that the planet surface is an equipotential (in which case the situation is analogous to that of Newton's bucket; see Section 6.3). This approximation of a 'fluid Earth' amounts to considering the deformation of an 'ocean' of negligible mass covering a spherical, rigid Earth. To the order in which we are working, the total potential at the surface of the Earth is [cf. (14.17)]

$$U_{\mathrm{total}}(R) = U_{\mathrm{proper}}(R) + U_{2\mathrm{tidal}}(R) = -\frac{GM_E}{r} - \zeta g_E P_2(\cos\varphi). \tag{14.18}$$

Let us expand the 'ocean' surface as $r(\varphi) = r_E[1 + \varepsilon(\varphi)]$. In lowest order, $U_{\mathrm{total}} = U_0$ gives $GM_E/r_E = U_0$. In linear order, $\varepsilon(\varphi) = (\zeta/r_E)P_2(\cos\varphi)$, and the equation of the ocean surface is

$$r = r_E + \zeta P_2(\cos\varphi). \tag{14.19}$$

This surface is an ellipsoid which in the limit $\zeta \ll r_E$ is close to an oblate spheroid of semi-major axis $a = r_E + \zeta$ in the direction of the attractor, and whose perpendicular cross section is circular of radius $b = c = r_E - \zeta/2$ (which explains why ζ is called the equilibrium tidal amplitude). For the Moon and the Sun, numerical calculation gives $\zeta_{\mathrm{Moon}} = 0.36$ m and $\zeta_{\mathrm{Sun}} = 0.16$ m.

This simple model neglects the fact that the Sun (or the Moon) orbits the Earth. However, the approximation is quite good because the amplitude of the inertial force $d^2 n^2 \sim GM_E/d \ll GM_E/r_E$. We have solved the problem assuming a static configuration, which is also a good approximation (the slowest proper vibrational modes of the Earth have a period of about an hour, which is small compared to the 12 hours of the excitation). We have calculated the Earth's proper potential assuming that the Earth is a sphere, but actually its deformation should be included and it gives a correction of order 10%. We have assumed that the Earth is a fluid, and a more accurate study of its response should involve modeling using the mechanics of continuous media. Our ignorance of the mechanical properties of the planet can be dealt with by introducing the Love numbers (see Section 14.3).

The Earth is also deformed by its proper rotation. The centrifugal force derived from the potential $U = -\frac{1}{2}R.(\Omega \wedge (R \wedge \Omega))$ (see Section 2.4) is

$$U_{\rm rot}(R) = -\frac{1}{2}\omega_E^2 r^2 \sin^2\theta = \frac{1}{3}\omega_E^2 r^2 [P_2(\cos\theta) - 1], \qquad (14.20)$$

where $\omega_E = 2\pi/24$ h and θ is the angle between the radius vector and the axis of rotation: $e_Z.h_r = \cos\theta$.

Therefore, the perturbing potentials due to rotation and tidal effects, (14.17) and (14.20), become

$$U_{\rm rot}(R) = -\frac{1}{2}\omega_E^2 r^2 \sin^2\theta \quad \text{and} \quad U_{\rm 2tidal}(R) = -\frac{m_*}{M_E + m_*}r^2 n^2 P_2(\cos\varphi), \qquad (14.21)$$

where we have used Kepler's law to transform the second term. Here $n = 2\pi/27$ days if the perturbing body is the Moon. The potential due to tides is 10^5 times weaker than that due to rotation. Moreover, the forces associated with these two sources act in different planes (see Fig. 14.3).

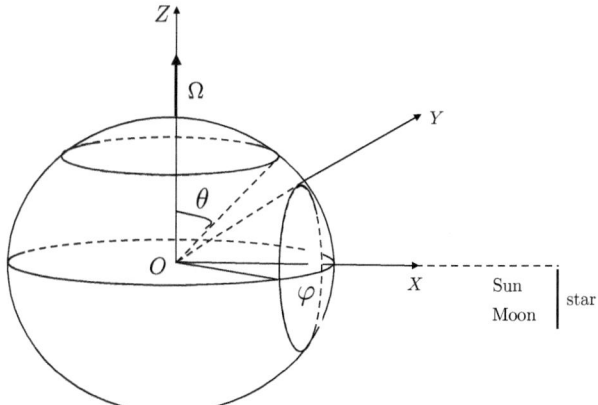

Fig. 14.3 Comparison of rotational and tidal effects.

Deformation of the Earth due to rotation

The 'ocean' of negligible mass covering the Earth, assumed to be spherical, is deformed by the proper rotation of the Earth. In spherical coordinates the total potential at the Earth's surface can then be written as

$$U_{\text{total}}(R) = U_{\text{proper}} + U_{\text{rot}} = -\frac{GM_E}{r} + \frac{1}{3}\omega_E^2 r^2 \left[P_2(\cos\theta) - 1\right], \qquad (14.22)$$

where $\omega_E = 2\pi/24$ h and the angle θ is the angle between the radius vector and the rotation axis, $e_Z.h_r = \cos\theta$. Expanding the ocean surface as $r = a[1 + \varepsilon(\theta)]$, the requirement that it be an equipotential implies that $(GM_E/a)\varepsilon = (\omega_E^2 a^2/2)\sin^2\theta + \omega_E^2 a^2\varepsilon\sin^2\theta$ to linear order. Since numerically $\omega_E^2 a^2 \ll GM_E/a$, we have

$$\varepsilon = \frac{1}{2}\frac{\omega_E^2 a^3}{GM_E}\sin^2\theta \equiv \frac{q}{2}\sin^2\theta. \qquad (14.23)$$

The Earth's surface is therefore a spheroid whose axis of revolution coincides with the rotation axis. Using $\sqrt{GM_E/a^3} \sim 2\pi/r24$ h, we obtain $q \approx 3.4 \times 10^{-3}$.

If now we describe the Earth itself as an ellipsoid of revolution about the OZ axis, the proper potential becomes

$$U_{\text{proper}}(R) = -\frac{GM_E}{r}\left[1 - J_2\left(\frac{a}{r}\right)^2 P_2(\cos\theta)\right] \qquad (14.24)$$

[cf. (14.5) for the definition of J_2]. Expanding the equation of the surface as $r(\theta) = a[1 + \varepsilon(\theta)]$, we obtain *Clairaut's formula*:

$$\varepsilon = \left(\frac{3}{2}J_2 + \frac{1}{2}q\right)P_2(\cos\theta) \equiv f P_2(\cos\theta). \qquad (14.25)$$

Setting J_2 equal to the measured value $J_2 = 1.08 \times 10^{-3}$, we find $f \approx 3.3 \times 10^{-3}$. Therefore, the Earth is flattened at the poles: $f = (r_e - r_p)/r_e$, which, using $r_e = 6378$ km, corresponds to a deformation of 21.2 km, in good agreement with the measured value.

The Roche limit

Let us consider a body A subject to a tidal force created by another body B as well as by the centrifugal force due to its proper rotation. One can state that A will not break up as long as the perturbing body B is located at a distance larger than the *Roche limit* r_R at which the sum of the tidal and centrifugal forces balances the gravitational force at the surface of A:

$$\frac{2GM_B r_A}{r_R^3} + \omega_A^2 r_A = \frac{GM_A}{r_A^2}. \qquad (14.26)$$

If the proper rotation of A is synchronous, that is, if $\omega_A^2 = GM_B/r_R^3$, we have

$$r_R = \left(\frac{3M_B}{M_A}\right)^{1/3} r_A. \qquad (14.27)$$

The approximation is rather crude and the numerical factor of 3 is very approximate. To determine it more accurately it is necessary to take into account the non-sphericity of the bodies along with their response to the perturbing potentials within the framework of the mechanics of continuous media.[2]

---

[2]See, for example, Pascoli (2000).

14.3 The figure of the Earth

To calculate the proper potential of the Earth, we need to be able to determine its deformation due to a perturbing potential, as discussed in the preceding section. Doing this accurately requires knowledge of the internal structure of the Earth and use of the techniques of the mechanics of continuous media. In this approach the internal stress–strain relationships of a body are described by various phenomenological parameters and it becomes possible to study the response of the body (deformation, oscillations, *etc.*) to the field of an external force.

Let us consider a planet located in an external perturbing potential U_{pert} such that the force exerted on each mass element of the planet located at $R = r\,h_r$ is $dF(R) = -dm\nabla U_{\mathrm{pert}}$. We expand the potential in spherical harmonics as $U_{\mathrm{pert}}(R) = \sum_{\ell m} w_{\ell m}(r)Y_{\ell m}(\theta, \phi)$. In the case of a planet of center O located in the perturbing field of a spherical star A, this expansion reduces to $U_{\mathrm{pert}}(R) = \sum_{\ell} u_\ell(r)P_\ell(OA.R)$, where P_ℓ is the Legendre polynomial of order ℓ. This simplification corresponds to symmetry of revolution about the line connecting the planet and the perturbing star (OA).

The radial and orthoradial deformations of the planet due to the external field can be parametrized as $\delta_{\mathrm{def}\,r}(R) = \sum \varepsilon_\ell(r)P_\ell$ and $\delta_{\mathrm{def}\,\theta}(R) = \sum \kappa_\ell(r)P_\ell$. The proper gravitational potential of the planet is parametrized as $U_{\mathrm{proper}}(R) = \sum v_\ell(r)P_\ell$. We therefore have

$$U_{\mathrm{pert}}(R) = \sum_\ell u_\ell(r)P_\ell \quad \text{induces} \quad \begin{cases} \delta_{\mathrm{def}\,r}(R) = \sum \varepsilon_\ell(r)P_\ell \\[2mm] \delta_{\mathrm{def}\,\theta}(R) = \sum \kappa_\ell(r)P_\ell \end{cases} \tag{14.28}$$

which induces $U_{\mathrm{proper}}(R) = \sum v_\ell(r)P_\ell$.

Following a phenomenological approach, Love (1909) and Shida (1912) proposed that the coefficients $(\varepsilon_\ell, \kappa_\ell, v_\ell)$ be expanded as

$$\varepsilon_\ell(r) = H_\ell(r)\frac{u_\ell(r)}{g(r)}, \quad \kappa_\ell(r) = L_\ell(r)\frac{u_\ell(r)}{g(r)}, \quad v_\ell(r) = K_\ell(r)u_\ell(r) \tag{14.29}$$

and to define $\quad h_\ell = H_\ell(r_{\mathrm{surf}}), \quad k_\ell = K_\ell(r_{\mathrm{surf}}), \quad \ell_\ell = L_\ell(r_{\mathrm{surf}}),$

where $g(r)$ is the norm of the local gravitational acceleration a distance r from the center, and h_ℓ, k_ℓ, and ℓ_ℓ are the *Love numbers*. In general, a Love number refers to any dimensionless parameter which relates the coefficients of the multipole expansion of a physical quantity (pressure, *etc.*) to the coefficients of the expansion of the external potential. These numbers are just a way of encoding our ignorance of the internal structure of the body in question.

In the case of a planet which is spherical, homogeneous (that is, the density and the parameters describing the deformations are independent of r), and incompressible (so that the divergence of the displacement field vanishes), the Love numbers can be calculated explicitly (this is called the 'Kelvin approximation'). Since the planet is assumed to be incompressible, it remains homogeneous throughout the deformation, and calculation of the gravitational potential at its surface gives[3]

$$U_{\mathrm{proper}}(R) = \sum \frac{3g}{2\ell+1}\varepsilon_\ell(r)P_\ell \quad \Longrightarrow \quad k_\ell = \frac{3}{2\ell+1}h_\ell. \tag{14.30}$$

---

[3]See, for example, Melchior (1973) and Murray and Dermott (1999).

To obtain the other Love numbers it is necessary to solve the equilibrium equation of the solid (the analog of the Euler equation with the stresses). It can be shown that[3]

$$h_\ell = \frac{2\ell+1}{2(\ell-1)} \frac{1}{1 + \frac{(2\ell^2+4\ell+3)\mu}{2g\varrho R}} \;, \quad \ell_\ell = \frac{3}{2\ell(\ell-1)} \frac{1}{1 + \frac{(2\ell^2+4\ell+3)\mu}{2\ell g\varrho R}} \;. \tag{14.31}$$

Here μ is one of the two 'Lamé coefficients' (λ, μ) which phenomenologically describe the stress–strain relationships in a solid (for an incompressible solid $\lambda = \infty$ and μ is the shear modulus; for a fluid $\mu = 0$).

In a more realistic model the Love numbers cannot be calculated because they depend on the entire internal structure of the planet. However, they can be measured[4]: $k_2 = 0.299$ (the Earth perturbed by the Sun or the Moon), $k_2 = 0.14$ (Mars perturbed by the Sun), and $k_2 = 0.030$ (the Moon perturbed by the Earth). Equation (14.30) and the value of k_2 are useful for studying the secular lengthening of the day; see the discussion that follows.

The Earth is deformed by tidal effects. It turns on its axis with a frequency of $\omega_E = 2\pi/24$ h, while the Moon revolves around the Earth with a frequency of $n = 2\pi/27$ d. The tidal bulge, carried along by the Earth's rotation, appears to be ahead by a constant angle Δ relative to the Moon (see Fig. 14.4). In fact, the bulge is delayed on its retrograde motion tending to restore it to the Earth–Moon axis. This delay is due to relaxation effects in the interior of the Earth (friction, viscosity, *etc.*). As Fig. 14.4 illustrates, the fact that the symmetry axis of the bulge is not aligned with the Earth–Moon axis means that the Moon exerts a torque on the bulge, which slows down the Earth's rotation and thus lengthens the terrestrial day.

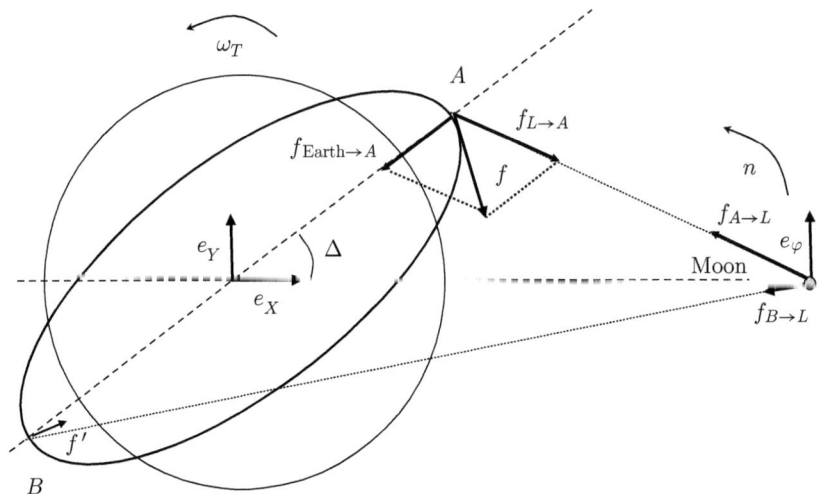

Fig. 14.4 Deformation of the Earth under the action of the Moon.

---

[4]For example, the amplitude of the ocean tides and the deviation of the vertical relative to the normal to the surface of the planet can be used to measure $1 + k_\ell - h_\ell$. The spatial variations of the strength of terrestrial gravity depend on $1 + (2h_\ell/\ell) - \ell k_\ell/(\ell+1)$. Finally, measurement of the vertical relative to a set of fixed stars gives $1 + k_\ell - \ell_\ell$.

More precisely, the torque K_E induced by the perturbing potential of the Moon $U_2 = -\zeta g_E P_2(\cos\varphi)$ [(14.17)] is equal and opposite to the torque K_M exerted by the Earth on the Moon (assumed to be spherical and located at $R_M = d\,e_X$):

$$-K_E = K_M = R_M \wedge F(R_M), \quad F(R_M) = -m_* \nabla U_{\text{proper}}(R)|_{R=R_M}. \tag{14.32}$$

Only the non-radial part of the Earth's potential, given in lowest order by its quadrupole moment,[5] contributes:

$$U_{\text{proper}}(R) = -\frac{3}{5}h_2\zeta g_E \left(\frac{r_E}{r}\right)^3 P_2(\cos\varphi) = -k_2\zeta g_E \left(\frac{r_E}{r}\right)^3 P_2(\cos\varphi). \tag{14.33}$$

Here k_2 is the second Love number (14.31) and ζ is defined in (14.17). We then find that

$$K_M(R_M) = \frac{3}{2}k_2 G m_*^2 \frac{r_E^5}{d^6} \sin 2\Delta\, e_z \equiv \mathcal{K}\, e_z. \tag{14.34}$$

When $\Delta > 0$, K_E slows down the Earth's rotation thereby decreasing the rotational energy, $\dot{E}_E = K_E.(\omega_E e_z) = -\mathcal{K}\omega_E < 0$. In return, the Moon is accelerated and its orbital energy increases, $\dot{E}_M = \mathcal{K}n > 0$. The energy balance is maintained by loss of the total energy $\dot{E} = -\mathcal{K}(\omega_E - n) < 0$ to dissipation as heat owing to friction. The same result would be obtained for $\Delta < 0$ with the difference that the energy of the Earth's rotation would increase while the orbital energy of the Moon would decrease.

The bulge of the Earth due to the Moon

The delay of the bulge can be understood by analogy with a damped oscillator excited by an external periodic force. The evolution equation is $\ddot{x} + \omega_0^2 x + \dot{x}/\tau = a_0 \cos\omega t$ with $\omega_0^2 = k/m$, and the response to the excitation is $x(t) = A\cos(\omega t + \Delta)$ with $A = a_0/\sqrt{(\omega_0^2 - \omega^2)^2 + (\omega/\tau)^2}$ and $\sin\Delta = -(\omega/\tau)/\sqrt{(\omega_0^2 - \omega^2)^2 + (\omega/\tau)^2}$.

Since the frictional force opposes the motion, Δ is negative and the response is always delayed relative to the excitation.

The angle Δ can be related to the energy dissipation $Q \equiv 2\pi E_0/\delta E$, where E_0 is the maximum potential energy stored during a cycle and δE is the energy dissipated in a cycle. In this case $E_0 = \int_0^A kx\,dx = m\omega_0^2 A^2/2$. The work performed by the frictional force during time δt is $-m\dot{x}\delta x/\tau$, so that the rate of energy dissipation is $\dot{E} = -\dot{x}^2/\tau$. Therefore, during a cycle $\delta E = 2\pi\langle\dot{E}\rangle/\omega = \pi m A^2\omega/\tau$ and so $Q = \omega_0^2\tau/\omega$. In the limit $\omega_0^2 \gg \omega^2 \gg \omega/\tau$, we obtain $\sin\Delta = -1/Q$.

The Earth's rotation and the recession of the Moon

The total energy of the system is the sum of the Earth's rotational energy $I_E\omega_E^2/2$, where I_E is the Earth's moment of inertia, and the Moon's orbital energy $-GM_E m_*/2d$. Using Kepler's third law $G(M_E + m_*) = n^2 d^3$, we find

---

[5]The quadrupole part of the proper potential external to the Earth has the form $Ar^{-3}P_2(\cos\varphi)$. The constant A is obtained by matching this potential to that obtained by solving the interior problem at the Earth's surface (14.30).

$$\dot{E} = I_{\rm E}\omega_{\rm E}\dot{\omega}_{\rm E} + \frac{1}{2}\mu n^2 d\dot{d}, \tag{14.35}$$

where μ is the reduced mass. The total angular momentum $J = (I_{\rm E}\omega_{\rm E} + \mu n d^2)e_z$ is conserved because the system is isolated: $\dot{J} = 0 = I_{\rm E}\dot{\omega}_{\rm E} + \mu n d\dot{d}/2$ (the derivative is calculated bearing in mind that $n^2 d^3$ is a constant), and so (14.35) becomes

$$\dot{E} = -\frac{1}{2}\mu n d\dot{d}(\omega_{\rm E} - n). \tag{14.36}$$

Since $\dot{E} < 0$ we have $\operatorname{sgn}\Delta = \operatorname{sgn}\dot{d} = -\operatorname{sgn}\dot{\omega}_{\rm E}$. Therefore, when $\Delta > 0$ the semi-major axis of the satellite orbit increases and the planet recedes. This is what happens in the Earth–Moon system. The opposite case occurs in, for example, the Mars–Phobos system.

Equating (14.36) to $\dot{E} = -\operatorname{sgn}\Delta\,\mathcal{K}(\omega_{\rm E} - n)$ and using Kepler's laws again, we obtain

$$\dot{d} = \frac{3}{2}(\operatorname{sgn}\Delta)\,k_2\left(\frac{m_*}{M_{\rm E}}\right)\left(\frac{r_{\rm E}}{d}\right)^5 n\,d\sin 2\Delta. \tag{14.37}$$

The conservation of total angular momentum then leads to the conclusion that

$$\dot{\omega}_{\rm E} = -\frac{3}{4}(\operatorname{sgn}\Delta)\frac{k_2}{\alpha_{\rm E}}\mu\frac{m_*}{M_{\rm E}^2}\left(\frac{r_{\rm E}}{d}\right)^3 n^2\sin 2\Delta, \tag{14.38}$$

where the moment of inertia has been written as $I_{\rm E} = \alpha_{\rm E}M_{\rm E}R_{\rm E}^2$, with $\alpha_{\rm E}$ a dimensionless number.

Numerically, $r_{\rm E} = 6.37 \times 10^6$ m, $d = 3.8 \times 10^8$ m, $M_{\rm E} = 6 \times 10^{24}$ kg, and $m_* = 7.3 \times 10^{22}$ kg. The Love number is $k_2 = 0.299$ and measurement gives $\sin\Delta = 1/12$. We therefore obtain $\dot{d} = 3.5$ cm/yr, which corresponds to about 10^{-9} m/s, in agreement with the value derived from lunar laser telemetry. The period of rotation of the Earth $T = 2\pi/\omega_{\rm E}$ then increases by $\dot{T} = 3 \times 10^{-3}$ s/century. This is consistent with the value for the lengthening of the day measured on the geological time scale.

Synchronous rotation of the Moon

The variation of the Moon's angular velocity $\omega_{\rm M}$ due to tidal effects produced by the Earth can be obtained from (14.38) by a simple symmetry argument. In the limit $m_* \ll M_{\rm E}$,

$$\dot{\omega}_{\rm M} = -\frac{3}{2}(\operatorname{sgn}(\omega_{\rm M} - n))\frac{k_{2\rm M}}{\alpha_{\rm M}}\frac{M_{\rm E}}{m_*}\left(\frac{r_{\rm M}}{d}\right)^3 n^2\sin 2\Delta_{\rm M}$$

$$\sim \frac{k_{2\rm M}}{k_{2\rm E}}\frac{\alpha_{\rm E}}{\alpha_{\rm M}}\left(\frac{r_{\rm M}}{r_{\rm E}}\right)^3\left(\frac{M_{\rm E}}{m_*}\right)^3\frac{\sin 2\Delta_{\rm M}}{\sin 2\Delta_{\rm E}}\dot{\omega}_{\rm M}. \tag{14.39}$$

Assuming that $\alpha_{\rm M} \sim \alpha_{\rm E}$ and using $k_{2\rm M} = 0.030$, $\sin\Delta_{\rm M} \sim 1/27$, and $R_{\rm M} = 1.74 \times 10^6$ m, we find $\dot{\omega}_{\rm M} \sim 500\,\dot{\omega}_{\rm E}$, which shows that the Moon has become synchronized before the Earth will.

15
Self-gravitating fluids

In this chapter we present a brief description of 'perfect fluids', which are characterized by their mass density $\varrho(t, x^i)$, pressure $p(t, x^i)$, and velocity field $v(t, x^i)$. The motion and equilibrium configurations of the fluid are determined by the equation of state, for example, $p = p(\varrho)$ for a barotropic fluid, and by the gravitational potential $U(t, x^i)$ created at a point x^i by the other fluid elements.

15.1 The Euler and Poisson equations

Given an equation of state, the equations of the problem to be solved are the continuity equation, the Euler equation, and the Poisson equation. In an inertial frame and using coordinates x^i, these equations can be written as (see Sections 6.3 and 11.4)

$$\frac{\partial \varrho}{\partial t} = -\nabla . (\varrho v), \quad \frac{dv}{dt} \equiv \frac{\partial v}{\partial t} + (v . \nabla) v = -\nabla U - \frac{1}{\varrho} \nabla p, \quad \triangle U = 4 \pi G \varrho, \quad p = p(\varrho). \quad (15.1)$$

The fluid is stationary if none of its variables depends explicitly on time, and the fluid is static if in addition $v = 0$.

The *matching conditions* at the surface Σ of the fluid $x^r = $ const can be read off from (15.1). If ϱ undergoes a finite discontinuity, the Poisson equation implies that U and $\partial_r U$ are continuous there (because the integral of a discontinuous function is continuous). The continuity equation then imposes the condition that v^r vanish on Σ (to eliminate the term involving $\partial_r \varrho$, which is a distribution). Finally, the Euler equation states that $\partial_r p$ has at most one finite discontinuity on Σ, and consequently p is continuous there, so that $p = 0$ on Σ because $p = 0$ outside the fluid.

The gravitational energy of the fluid is given by [*cf.* (11.24)]

$$W_{\text{grav}} = \frac{1}{2} \int \varrho U \, dV = -\frac{1}{8 \pi G} \int (\nabla U)^2 dV, \quad (15.2)$$

where the first integral runs over the entire volume of the fluid, the second runs over the entire space (one is derived from the other using the Poisson equation with the condition that U falls off at least as $1/r$ at infinity; see Section 11.4), and the volume element is $dV = \sqrt{\det e} \, dx^1 dx^2 dx^3$, with $\det e$ the determinant of the metric coefficients in the coordinates x^i.

15.2 Static models with spherical symmetry

Let us consider a spherically symmetric matter distribution confined to $r < r_0$, where r is the radial coordinate of a spherical coordinate system, creating a gravitational field which is also spherically symmetric such that $U = U(t, r)$. Outside the matter the Laplace equation

Relativity in Modern Physics. Nathalie Deruelle and Jean-Philippe Uzan.
© Oxford University Press 2018. Published in 2018 by Oxford University Press.
DOI: 10.1093/oso/9780198786399.001.0001

$\triangle U = 0$ reduces to $d^2U/dr^2 + (2/r)(dU/dr) = (r^2U')'/r^2 = 0$ (the expressions for the various differential operators can be found in Section 4.6), the solution of which is

$$r > r_0: \qquad U = -\frac{GM}{r}, \tag{15.3}$$

where GM is an integration 'constant' depending only on time (the additive constant here can be set equal to zero because the expressions in (15.1) involve only spatial derivatives of U). We remark that according to Newton's theorem (see Section 11.4), the solution is valid for any motion of the matter creating this field as long as it remains spherically symmetric.

Now let us assume in addition that the matter distribution is static: $\varrho = \varrho(r)$, $v = 0$. Inside the matter distribution the Euler and Poisson equations (15.1) then reduce to

$$r < r_0: \qquad \frac{dp}{dr} = -\varrho\frac{dU}{dr}, \quad \frac{d}{dr}\left(r^2\frac{dU}{dr}\right) = 4\pi G\varrho\, r^2. \tag{15.4}$$

These equations can be rewritten as a system of first-order differential equations for $U(r)$ and the *mass function* $m(r)$:

$$\frac{dm}{dr} \equiv 4\pi\varrho r^2, \quad \frac{dU}{dr} = \frac{Gm}{r^2}, \quad \frac{d\varrho}{dr} = -\frac{\varrho}{dp/d\varrho}\frac{dU}{dr}, \tag{15.5}$$

where we have imposed the condition that $U(r)$ remain finite at the origin. Given an equation of state $p = p(\varrho)$, the system (15.4) or (15.5) can be integrated (numerically if necessary) with the initial conditions

$$m(0) = 0, \quad U(0) = U_0, \quad \varrho(0) = \varrho_0. \tag{15.6}$$

The value chosen for U_0 is not important because only derivatives of U appear in (15.4); we can *a posteriori* choose it such that U vanishes at infinity. We then see that for a given equation of state there exists an entire family of models parametrized by the central density ϱ_0.

The integration ends at the point where the pressure vanishes, which defines the radius r_0 of the distribution and its mass $M = m(r_0)$.

A 'star' of constant density

In the case of constant density, (15.4) can be integrated by inspection:

$$U(r) = D + \frac{2\pi G}{3}\varrho r^2, \quad p(r) = p_0 - \frac{2\pi G}{3}\varrho^2 r^2, \tag{15.7}$$

where we have excluded a $1/r$ term from the potential so that the latter remains finite at the origin. The exterior solution is given by (15.3). The continuity equations for U and dU/dr and the vanishing of p at the fluid surface determine D, GM, and p_0:

$$M = \frac{4\pi}{3}\varrho r_0^3, \quad D = -\frac{3GM}{2r_0}, \quad p_0 = \frac{GM\varrho}{2r_0}, \tag{15.8}$$

so that

$$U(r) = -\frac{GM}{2r_0}\left(3 - \frac{r^2}{r_0^2}\right), \quad p(r) = \frac{GM\varrho}{2r_0}\left(1 - \frac{r^2}{r_0^2}\right).$$ (15.9)

Therefore, in contrast to the general case where the solution depends only on the central density, this model depends on two parameters, the density ρ and the radius r_0 of the object (or its mass).

The gravitational lifetime of the Sun

The gravitational potential inside a static body for spherical symmetry and constant density ϱ is [cf. (15.9)] $U = -(GM/2r_0)\left(3 - r^2/r_0^2\right)$, where r_0 is the radius of the object and $M = 4\pi\varrho r_0^3/3$ is its mass. Its gravitational energy then is

$$W_{\text{grav}} = \frac{1}{2}\int \varrho U dV = -\frac{GM\varrho}{4r_0}\int\left(3 - \frac{r^2}{r_0^2}\right)r^2\sin\theta\,d\phi\,d\theta\,dr \quad \text{or} \quad W_{\text{grav}} = -\frac{3}{5}\frac{GM^2}{r_0}.$$ (15.10)

For the Sun we find $W_{\text{grav}} = -2.3 \times 10^{41}$ J.

Now, the flux of energy radiated by the Sun can be measured and is found to be $\phi \approx 1.5$ kW per m^2 at the surface of the Earth. Therefore, the *luminosity*, defined as $L_\odot = 4\pi D^2\phi$ where D is the Earth–Sun distance, is $L_\odot \approx 3.8 \times 10^{26}$ W.

If the energy radiated by the Sun were due to a loss of gravitational energy, we would have $dW_{\text{grav}}/dt = -L_\odot$, corresponding to a (short!) solar lifetime of $\tau \approx -W_{\text{grav}}/L_\odot \approx 2 \times 10^7$ yr. On the other hand, the solar lifetime is estimated to be 5×10^9 yr if its radiation is attributed to nuclear reactions.

15.3 Polytropes and the Lane–Emden equation

The equation of state of a *polytrope* is $p = K\varrho^\gamma$, where K and $\gamma \equiv 1 + \frac{1}{n}$ are constants.[1] The Euler equation (15.4) can then be integrated to find the density as a function of the potential:

$$\varrho = \varrho_0\theta^n \quad \text{with} \quad \theta \equiv \frac{C_1 - U}{K(n+1)}\varrho_0^{-\frac{1}{n}},$$ (15.11)

where C_1 is an integration constant and ϱ_0 is an arbitrary constant introduced for convenience. The Poisson equation (15.4) then becomes the *Lane–Emden equation*:

$$\frac{1}{\xi^2}\frac{d}{d\xi}\left(\xi^2\frac{d\theta}{d\xi}\right) = -\theta^n,$$ (15.12)

with ξ related to r as $r = \alpha\xi$ with $\alpha \equiv \sqrt{K(n+1)/(4\pi G)}\,\varrho_0^{\frac{1-n}{2n}}$. The initial conditions at $\xi = 0$ are $\theta(0) = 1$ (which defines the constant ϱ_0 as the central density of the star) and $(d\theta/d\xi)|_0 = 0$ (so that the potential U is smooth at the origin). Analytic solutions of the Lane–Emden equation are known only for $n = 0$ ($\theta = 1 - \xi^2/6$, the case of constant density), $n = 1$ ($\theta = \sin\xi/\xi$), and $n = 5$ ($\theta = (1+\xi^2/3)^{-1/2}$). In other cases it is necessary to resort to

---

[1]The constant n is called the *polytropic index* and γ is called the *adiabatic index*. The higher γ is, the 'stiffer' the equation of state. The values $\gamma = 5/3$ and $\gamma = 4/3$ describe *white dwarfs*. For a detailed discussion of the physics of compact stars, see, for example, Grandclément (2008) and also Collins (2003).

Taylor-series expansions $\theta(\xi) = 1 - \xi^2/6 + n\xi^4/120 - (8n^2 - 5n)\xi^6/15120 + \dots$ or numerical integration. For $n > 5$ ($\gamma < 6/5$), it turns out that $\theta(\xi)$ never vanishes and therefore describes configurations of infinite extent.

Once $\theta(\xi)$ (that is, ϱ or p) is known, the stellar radius is known as a function of K, n, and ϱ_0, namely, $r_0 \equiv \alpha\bar{\xi}$ with $\theta(\bar{\xi}) = 0$. The mass is also known: $M = 4\pi \int_0^{r_0} \varrho r^2 dr = 4\pi \int_0^{r_0} \varrho_0\theta^n r^2 dr = 4\pi\varrho_0\alpha^3 \int_0^{\bar{\xi}} \theta^n \xi^2 d\xi$, which can be rewritten using the Lane–Emden equation (15.12) as

$$M = -4\pi\varrho_0\alpha^3\bar{\xi}^2 \left.\frac{d\theta}{d\xi}\right|_{\xi=\bar{\xi}} \qquad \text{with} \qquad \alpha \equiv \sqrt{\frac{K(n+1)}{4\pi G}}\,\varrho_0^{\frac{1-n}{2n}}. \tag{15.13}$$

We note that if $n = 3$ ($\gamma = 4/3$), M does not depend on the central density; this is the *Chandrasekhar mass*.[2]

The gravitational potential U is given by $U = -K(n+1)\varrho_0^{1/n}\theta + C_1$ in the interior of the star and is proportional to $1/r$ outside it. The continuity conditions on U and its derivative give $U = -GM/r$ with M from (15.13) (in agreement with Newton's theorem, see Section 11.4) and $C_1 = -GM/r_0$.

The gravitational energy $W_{\text{grav}} = 2\pi \int \varrho U r^2 dr$ is also a function of the parameters of the problem, that is, of K, n, and ϱ_0. In fact, there is a simple relation between M, r_0, and W_{grav}, namely,[3]

$$W_{\text{grav}} = -\frac{3}{5-n}\frac{GM^2}{r_0} \qquad \text{or also} \qquad W_{\text{grav}} = -\frac{3(\gamma-1)}{5\gamma-6}\frac{GM^2}{r_0}. \tag{15.14}$$

For $n = 0$ we recover (15.10) for a homogeneous sphere.

The *internal energy* density ϵ is defined by[4] $p = (\gamma-1)\epsilon$. A similar calculation gives $W_{\text{int}} = \int \epsilon\, dV = (5\gamma-6)^{-1}GM/r_0^2$. Therefore, the total energy of the star or its *binding energy* is

---

[2]After numerical integration of the Lane–Emden equation (15.12) in this particular case ($n = 3$), one finds $M = 2.02(4/\sqrt{\pi})(K/G)^{3/2}$. The constant K can be estimated only using quantum mechanics; see the references in the preceding footnote. One then finds $M = 1.46\,M_\odot$.

[3]Using the Lane–Emden equation (15.11) we find

$$\frac{W_{\text{grav}}}{2\pi\alpha^3\varrho_0} = \int_0^{\bar{\xi}} \xi^2\theta^n\, U\, d\xi = \frac{GM}{r_0}\bar{\xi}^2\bar{\theta}' - K(n+1)\varrho_0^{1/n}\, I \quad \text{with} \quad I = \int_0^{\bar{\xi}} \zeta^2\theta^{n+1} d\zeta,$$

where we have set $\theta' = d\theta/d\xi$ and $\bar{\theta}' = (d\theta/d\xi)|_{\xi=\bar{\xi}}$. The integral I is calculated using (15.11) and integration by parts: first one shows that $I = \int_0^{\bar{\xi}} \xi^2\theta'^2 d\xi$ and also that $I = -\frac{n+1}{3}\int_0^{\bar{\xi}} \xi^3\theta'\theta^n d\xi$, and then

$$\int_0^{\bar{\xi}} \xi^3\theta'\theta^n d\xi = \int_0^{\bar{\xi}} \xi^2\theta'(\xi^2\theta')' d\xi/\xi = \tfrac{1}{2}(\bar{\xi}^3\bar{\theta}'^2 + \int_0^{\bar{\xi}} \xi^2\theta'^2 d\xi).$$

We therefore have $I = \frac{n+1}{6}(\bar{\xi}^3\bar{\theta}'^2 + I)$, from which we extract $I = \frac{n+1}{5-n}\bar{\xi}^3\bar{\theta}'^2$.

Finally, we use (15.13) to express $\bar{\theta}'$ as a function of M, we introduce $r_0 = \alpha\bar{\xi}$, and we replace α by its value as a function of ϱ_0.

[4]Indeed, the force exerted on a fluid element of mass $m = \varrho\tau$ contained in a small volume τ bounded by the surface S is $-\int_S p\, dS$ by the definition of the pressure (see Section 6.3). The energy density associated with the small volume τ then is $-p\, dV$. Since $\tau = m/\varrho = \int dV$, we have $dV = -(m/\varrho^2)d\varrho$. Therefore, $-p\, dV = mp\,d\varrho/\varrho^2$. For a polytrope $p = K\varrho^\gamma$, and so $-\int_\tau p\, dV = \tau p/(\gamma-1)$. Therefore, the internal energy density is $\epsilon = p/(\gamma-1)$.

We note that the internal energy of a star of constant density ($n = 0$) is zero.

$$W = W_{\text{grav}} + W_{\text{int}} = -\frac{3-n}{5-n}\frac{GM^2}{r_0} \qquad \text{or also} \qquad W = -\frac{3\gamma-4}{5\gamma-6}\frac{GM^2}{r_0}. \tag{15.15}$$

The system is bound if $\gamma > 4/3$.

Finally, we note that the Lane–Emden equation possesses a property called *homology*, that is, it is invariant under rescalings $\xi \mapsto A\xi$ and $\theta \mapsto B\theta$ if $A^2 B^{n-1} = 1$ with A and B being constants. This is an indication that it can be transformed into a first-order equation. It is easy to see that if we introduce (following Chandrasekhar) the scale-invariant functions

$$u = -\frac{\xi\theta^n}{\theta'}, \quad v = -\frac{\xi\theta'}{\theta} \quad \text{where} \quad \theta' \equiv \frac{d\theta}{d\xi}, \tag{15.16}$$

then the Lane–Emden equation can be rewritten as

$$\frac{u}{v}\frac{dv}{du} = -\frac{u+v-1}{u+nv-3} \tag{15.17}$$

with the initial conditions $v = 0$ and $u = 3$ [obtained by a finite series expansion of the solution of the Lane–Emden equation (15.12)]. Equation (15.17) can be used to study models of stellar *families*.

15.4 The isothermal sphere

The equation of state of a perfect gas at constant temperature is $p = w\varrho$ with w a constant; see Section 10.4. The Euler equation for spherical symmetry (15.4) describing this *isothermal sphere* can be integrated to give $\varrho = \varrho_0\, e^{(-U/w)}$, and the Poisson equation becomes $(r^2 U')' = 4\pi G\varrho_0 r^2\, e^{(-U/w)}$. This can be cast in the form of the Lane–Emden equation

$$\frac{1}{\xi^2}\frac{d}{d\xi}\left(\xi^2\frac{d\psi}{d\xi}\right) = e^{-\psi} \tag{15.18}$$

after setting $U \equiv w\psi$ and $r = \alpha\xi$ with $\alpha = \sqrt{w/(4\pi G\varrho_0)}$. The initial conditions are $\psi(0) = 0$ (which defines ϱ_0 as the central density) and $(d\psi/d\xi)|_0 = 0$. Equation (15.18) can be integrated numerically.[5] It turns out that the density decreases and has fallen by about one half at $\xi_0 = 3$, the *King radius*.

We note that the ansatz $\psi = 2\log\xi - \ln 2$, referred to as the 'singular isothermal sphere' because $\varrho = 2\varrho_0/\xi^2$ diverges at the origin, solves this equation and gives a good approximation of the regular solution for large ξ.

Equation (15.18) possesses the property of *homology* since it is invariant under the transformations $\xi \mapsto e^{B/2}\xi$, $\psi \mapsto \psi + B$, where B is a constant. Introducing the invariant scaling functions (called *Milne functions*)

$$u = \frac{\xi e^{-\psi}}{d\psi/d\xi}, \quad v = \xi\frac{d\psi}{d\xi}, \tag{15.19}$$

we can rewrite it (since the singular solution has been excluded) as a first-order equation:

---

[5]See, for example, Binney and Tremaine (2008).

$$\frac{u}{v}\frac{dv}{du} = -\frac{u-1}{u+v-3}\,, \tag{15.20}$$

with the initial conditions $v = 0$, $u = 3$ [obtained by a finite series expansion of the solution of (15.18)].

Kinetic theory and the isothermal sphere

Let us consider a system of N identical particles of mass m interacting only gravitationally. We assume that the system is isolated, so that its total energy $\mathcal{E}$ as well as N are constant (the *microcanonical* description). In a stationary state such a system is described statistically by a distribution function $f(R, v)$ such that $f d^3 X d^3 V$ represents the number of particles in the volume $d^3 X$ centered at X^i whose components of the velocity v lie between V^i and $V^i + dV^i$ (X^i are the Cartesian components of R in an inertial frame).

If the two-point correlation function of the distribution can be neglected, then (see Section 10.2) f satisfies the stationary Boltzmann equation, which can be written as

$$v.\nabla f - \frac{\partial f}{\partial v}.\nabla U = 0\,, \tag{15.21}$$

where U is the average gravitational potential created by the distribution.

The total energy of the system is the sum of its kinetic and gravitational energies (see Section 11.4; here we denote the kinetic energy by K rather than T):

$$\mathcal{E} = K + W \quad \text{with} \quad \begin{cases} K = \dfrac{m}{2} \displaystyle\int v^2 f(R, v)\, d^3 X d^3 V \\[2mm] W = \dfrac{1}{2} \displaystyle\int \varrho U d^3 X = -\dfrac{1}{8\pi G} \int (\nabla U)^2 d^3 X. \end{cases} \tag{15.22}$$

The distribution function (normalized using $\int dx\, e^{-ax^2/2} = \sqrt{2\pi/a}$)

$$f(R, v) = \frac{\varrho_0}{m} \left(\frac{\beta m}{2\pi} \right)^{3/2} e^{-\beta m(\frac{1}{2}v^2 + U)}\,, \tag{15.23}$$

where ϱ_0 and β are constants, solves the Boltzmann equation and extremizes the entropy, defined as $S = -\int f \ln f\, d^3 X d^3 V$, for $\mathcal{E}$ and N constant.[6] [If $U = 0$ we recover the Maxwell distribution (10.31).] We then deduce that the mass density ϱ and the Poisson equation which must be satisfied by the average potential U, given by $\varrho = m \int f(R, v) d^3 V$ and $\triangle U = 4\pi G\varrho$, are

$$\varrho = \varrho_0 e^{-\beta m U}\,, \quad \triangle U = 4\pi G \varrho_0 e^{-\beta m U}. \tag{15.24}$$

The quantities ϱ and U satisfy the Euler and Poisson equations (15.4) if the equation of state is $p = \varrho/(\beta m)$. Therefore, the distribution function (15.23) describes an isothermal sphere having the equation of state $p = w\varrho$ with $w = 1/(\beta m)$.

---

[6]Introducing the Lagrange multipliers α and β, the extremization equation for S can be written as (see Section 8.2) $\delta S - \beta \delta \mathcal{E} - \alpha \delta N = 0$. Only the calculation of δW requires a bit of care. We have $\delta W = \frac{1}{2} \int (U \delta \varrho + \varrho\, \delta U) d^3 X$ and also $\delta W = -\frac{1}{8\pi G} \int \delta(\nabla U)^2 d^3 X = \int \varrho\, \delta U d^3 X$, after integration by parts and

By elementary integration, we also deduce from (15.23) and (15.22) that the kinetic energy of this perfect gas is $K = 3N/(2\beta)$, so that we can identify β as the inverse temperature: $\beta = 1/(kT)$, where k is the Boltzmann constant; see also Section 10.4. Finally, the scalar virial theorem $2K + W = 0$ (see Section 7.4) provides the easiest method of obtaining the gravitational energy W of the distribution. To sum up,

$$K = \frac{3}{2}NkT, \quad W = -2K, \quad \mathcal{E} = -K,\tag{15.25}$$

so that the *specific heat* of the system $C_V = d\mathcal{E}/dT|_{V,N}$, which characterizes the increase of the energy when the temperature is raised, is given by

$$C_V = -\frac{3}{2}Nk < 0.\tag{15.26}$$

Therefore, as it loses energy ($\mathcal{E} < 0$) the system becomes warmer and contracts. This implies that a self-gravitating system cannot be in thermodynamical equilibrium given the hypotheses we have made.[7]

15.5 Maclaurin spheroids

The gravitational potential inside a homogeneous ellipsoid of revolution bounded by the surface

$$X^2 + Y^2 + \frac{Z^2}{1 - e^2} = a^2\tag{15.27}$$

was found by Maclaurin[8] by solving the Poisson equation $\triangle U = 4\pi G\varrho$ with $\varrho = \mathrm{const}$:

$$U(X,Y,Z) = -\pi G\varrho\sqrt{1 - e^2}\left[a^2 I - (X^2 + Y^2)A_1 - Z^2 A_3\right]\tag{15.28}$$

with

$$\begin{cases} I = 2\dfrac{\mathrm{Arcsin}\, e}{e} \\[2mm] A_1 = \dfrac{\mathrm{Arcsin}\, e - e\sqrt{1 - e^2}}{e^3}, \quad A_3 = 2\dfrac{e - \sqrt{1 - e^2}\,\mathrm{Arcsin}\, e}{e^3\sqrt{1 - e^2}}. \end{cases}$$

(The constant I has been chosen to make the exterior potential go to zero at infinity.) The equipotentials $U = \mathrm{const}$ are also ellipsoids of revolution, but the surface (15.27) is not an equipotential.

---

using the Poisson equation. Therefore, $\int \varrho\,\delta U d^3 X = \int U\delta\varrho\,d^3 X$ and $\delta W = \int U\delta\varrho\,d^3 X = m\int U\delta f\,d^3 X d^3 V$. The extremization equation then becomes

$$\int\left[\ln f + \beta m\left(\frac{1}{2}v^2 + U\right) + (\alpha + 1)\right]\delta f\,d^3 X d^3 V = 0,$$

the solution of which is (15.23) after redefinition of the constant α.

[7]An instability of this type can lead to a *gravothermal catastrophe*; see Lynden-Bell and Wood (1968). A more recent discussion can be found in, for example, Chavanis (2003).

[8]See Chandrasekhar (1969) for a detailed discussion.

In the limit where the eccentricity tends to zero, (15.28) can be expanded as

$$U(r,\theta) = -2\pi G\varrho a^2 \left[1 - \frac{r^2}{3a^2} - \frac{e^2}{3}\left(1 + \frac{2}{5}\frac{r^2}{a^2}P_2(\cos\theta)\right)\right] + \mathcal{O}(e^4),\tag{15.29}$$

where $r^2 = X^2 + Y^2 + Z^2$, $\cos\theta = \frac{Z}{r}$, and $P_2(x) = (3x^2 - 1)/2$ is the second-order Legendre polynomial. On the surface of the spheroid this quadrupole series coincides with that derived from (14.18).

Now that we have the solution of the Poisson equation, we can solve the continuity equation and the Euler equation (15.1) by seeking a stationary solution describing a fluid undergoing rigid rotation about its symmetry axis OZ. To determine its velocity in the inertial center-of-mass frame $\mathcal{S}$, we go to a frame rotating with the fluid, in which the velocity $v' = v - \Omega \wedge R = 0$, where $R = Xe_X + Ye_Y + Ze_Z$ corresponds to a point of the fluid and $\Omega = \omega e_Z$ is its rotation vector. Then $v = \Omega \wedge R = \omega(-Ye_X + Xe_Y)$ such that $(v.\nabla v) = V^i\partial_i V^j e_j = -\omega^2(Xe_X + Ye_Y) = -\frac{1}{2}\omega^2\nabla(X^2 + Y^2)$. Therefore, the Euler equation (15.1) becomes

$$\nabla\left(U + \frac{p}{\varrho} - \frac{\omega^2}{2}(X^2 + Y^2)\right) = 0,\tag{15.30}$$

the solution of which is, using (15.28),

$$\frac{p}{\varrho} = -\pi G\varrho\sqrt{1 - e^2}\left[(X^2 + Y^2)\left(A_1 - \frac{\omega^2}{2\pi G\varrho\sqrt{1 - e^2}}\right) + Z^2 A_3 - \text{const}\right].\tag{15.31}$$

The surface of the star is given by (15.27) and also by $p = 0$. These two surfaces coincide if

$$A_1 - \frac{\omega^2}{2\pi G\varrho\sqrt{1 - e^2}} = (1 - e^2)A_3\tag{15.32}$$

[and $(1 - e^2)a^2 A_3 = \text{const}$], or equivalently if

$$\frac{\omega^2}{2\pi G\varrho} = -\frac{3(1 - e^2)}{e^2} + \frac{\sqrt{1 - e^2}}{e^3}(3 - 2e^2)\text{Arcsin } e.\tag{15.33}$$

In the limit $e \to 0$ we have $\omega^2 \approx 8\pi G\varrho e^2/15$ (a result obtained by Newton). The curve $\omega(e)$ reaches its maximum $\omega_{\max}^2/(\pi G\varrho) = 0.449$ at $e = 0.930$ and falls to 0 again at $e = 1$. Therefore, for a given ϱ and rotational velocity ω ($< \omega_{\max}$), (15.33) has two solutions $e_{1,2}$ corresponding to two different equilibrium solutions.

The mass of the spheroid is $M = (4\pi/3)a^3\sqrt{1 - e^2}$ and its moments of inertia are $I_1 = I_2 = Ma^2(2 - e^2)/5$ and $I_3 = 2Ma^2/5$ (see Section 6.1). The component of its angular momentum along the OZ axis is $\mathcal{M}_Z = (2M/5)\omega a^2$ (see Section 7.2). Its kinetic energy is $T = \frac{1}{2}\int \varrho v^2 dV$, or, since $v = \Omega \wedge R = \omega(-Ye_X + Xe_Y)$,

$$T = \frac{1}{2}\int \varrho v^2 dV = \frac{\omega^2}{2}\int \varrho(X^2 + Y^2)dV = \frac{1}{2}\omega^2 I_3$$

$$= \frac{16\pi^2}{15}G\varrho^2 a^5\frac{(1 - e^2)}{e}\text{Arcsin } e\left(\frac{3}{2e^2} - 1 - \frac{3\sqrt{1 - e^2}}{2e\text{Arcsin } e}\right).\tag{15.34}$$

Its gravitational energy is

$$
W_{\text{grav}} = \frac{1}{2}\int \varrho\, U dV = -\frac{1}{2}\pi G\varrho\sqrt{1-e^2}\left[a^2 IV - A_1 I_3 - \frac{A_3}{2}(I_1 + I_1 - I_3)\right]
$$

$$
= -\frac{16\pi^2}{15}G\varrho^2 a^5\frac{(1-e^2)}{e}\operatorname{Arcsin} e .
$$
(15.35)

The ratio $T/|W_{\text{grav}}|$ is an increasing function of the eccentricity. For $e = 0$ (a sphere) $T/|W_{\text{grav}}| = 0$, and for $e = 1$ (a disk) $T/|W_{\text{grav}}| = 1/2$.

Jacobi ellipsoids

Given the eccentricity ($0 \le e \le 1$) of a homogeneous ellipsoid of revolution of mass density ϱ, we then know the speed of rigid rotation ω about its symmetry axis which maintains it in gravitational equilibrium (Maclaurin, 1742); cf. (15.33). This speed is independent of the length of the major axis a.

In 1834 Jacobi showed that if e exceeds its *bifurcation* value $e = 0.813$ [corresponding to $\omega^2/(\pi G\varrho) \ge 0.374$ and $T/|W_{\text{grav}}| \ge 0.1375$], there exists, along with the Maclaurin spheroid, another equilibrium shape in rigid rotation. It is ellipsoidal, that is, it is described by the equation $X^2 + \frac{Y^2}{1-e_2^2} + \frac{Z^2}{1-e^2} = a^2$, where the value of e_2 as well as the rotational velocity are determined[9] by e. Dedekind (1860) and Riemann (1892) generalized these results to the case where the rotation of the fluid is not constrained to be rigid.

For a given mass and angular momentum, a Maclaurin spheroid turns out to have a total energy $\mathcal{E} = T + W$ [cf. (15.34) and (15.35)] greater than that of the corresponding Jacobi ellipsoid. It is therefore unstable (this instability is termed *secular* because it requires energy dissipation). Direct analysis (Riemann, 1860) of perturbations of the equilibrium configuration shows that a *dynamical* (i.e., without energy dissipation) instability develops for $e \ge 0.953$ [or $\omega^2/(\pi G\varrho) = 0.4402$].

In 1885 Poincaré performed a similar analysis of perturbations of the equilibrium configurations of Jacobi ellipsoids and demonstrated the existence of a point of bifurcation to pear-shaped configurations for $e = 0.881$ or $\omega^2/(\pi G\varrho) = 0.284$. It has been found that this point also corresponds to the development of the dynamical instability (Cartan, 1924).

---

[9]The equations can be found in Chandrasekhar (1969). For more recent developments, see, for example, Christodoulou *et al.* (1995).

16
Newtonian cosmology

As we have seen in Chapter 11, there exist two formulations of Newton's law of gravitation. The first, due to Newton himself, gives the $1/r^2$ *force* between two masses, and the second, due to Laplace and Poisson, gives the differential equation which must be satisfied by the *potential* from which the force is derived. The two formulations are not equivalent when the matter density does not vanish at infinity but instead becomes, for example, constant. The construction of models of the universe is therefore ambiguous in Newtonian theory.

However, some results presented here will be recovered within the framework of general relativity, and they will in addition make it possible to lay the foundation of the theory of the formation of large-scale structures in the universe such as galaxies and galactic clusters.

16.1 The model of an expanding sphere

Let us imagine a matter distribution spherically symmetric about the origin O of an inertial frame which interacts only gravitationally. If gravity is described by Newton's law, the spherical layer of particles labeled by the index (a) located a distance $r_a(t)$ from the center is subject to the force due to a mass M_a equal to the mass lying inside the sphere of radius r_a and located at the origin: $\ddot{r}_a = -GM_a(t)/r_a^2$ (according to Newton's theorem; see Section 11.4). Let us now assume that no particle overtakes or is overtaken by another. Then M_a is independent of time and the first integral of the motion is

$$\dot{r}_a^2 = -v_a^2 + \frac{2GM_a}{r_a}, \tag{16.1}$$

where v_a^2 is an integration constant with the dimensions of velocity squared characterizing the layer in question labeled (a).

For $v_a^2 > 0$ the solution of (16.1) is written in parametric form as[1]

$$r_a = \frac{r_{ma}}{2}(1 - \cos \eta), \quad t - t_{0a} = \frac{r_{ma}}{2v_a}(\eta - \sin \eta), \tag{16.2}$$

where t_{0a} is a second integration constant and $r_{ma} \equiv 2GM_a/v_a^2$.

If t_{0a} and r_{ma}/v_a are independent of (a) the motion will be self-similar, that is, different layers of particles will reach their maximum displacement at the same time and fall back to the origin at the same time. Setting $t_{0a} \equiv t_0$ and $r_{ma}/(2v_a) \equiv a_0/c$, where c is a constant with dimensions of velocity introduced for convenience, (16.2) can be rewritten as

$$r_a = \frac{v_a}{c}a(t) \quad \text{with} \quad a(\eta) = a_0(1 - \cos \eta) \quad \text{and} \quad t - t_0 = \frac{a_0}{c}(\eta - \sin \eta). \tag{16.3}$$

---

[1]When $v_a = 0$, the solution of (16.1) is $r_a = C_a(t - t_{0a})^{\frac{2}{3}}$ with $C_a^3 = \frac{9}{2}GM_a$. The matter cloud expands indefinitely, as for $v_a^2 < 0$.

Relativity in Modern Physics. Nathalie Deruelle and Jean-Philippe Uzan.
© Oxford University Press 2018. Published in 2018 by Oxford University Press.
DOI: 10.1093/oso/9780198786399.001.0001

Equation (16.1) becomes $\dot{a}^2/a^2 + c^2/a^2 = 2GM_a/r_a^3$ or

$$\frac{\dot{a}^2}{a^2} + \frac{c^2}{a^2} = \frac{8\pi G}{3}\varrho \qquad \text{if we set} \qquad \varrho = \frac{3c^2 a_0}{4\pi G a^3(t)} \,. \tag{16.4}$$

If we treat the galaxies as the particles of a uniform cloud which is spherically symmetric about the origin of an inertial frame, our model (16.3) and (16.4) will describe a universe which expands and eventually collapses on itself. Newton, who believed in eternal Creation (despite the contradiction in terms) concluded that such a cloud, since it is unstable, cannot represent the universe in its totality.

Newton's objection can be assuaged by assuming that the cloud is undergoing a global rotation. The problem then becomes considerably more complicated,[2] but one can imagine that the centrifugal force could counterbalance gravitation, resulting in a stable configuration. This idea was developed by Kant and Laplace in building their models of the solar system (which in the eighteenth century *was* the universe).

If we accept the idea that the universe may not be static, Newton's objection also falls. The universe then has a history, a beginning and eventually an end. As pointed out by Milne and McCrea in 1934, this Newtonian model is mathematically equivalent to some relativistic models derived from the Einstein theory of gravitation, with the function $a(t)$ then referred to as the *scale factor* of the universe.

We note that the velocity distribution is isotropic with regard to *all* the particles of the cloud. Indeed, since the motion is self-similar, the position vector of particle P_i of layer (i) is $OP_i = (v_i/c)a(t)$, where v_i is a constant vector. Then with P_j a particle of the layer (j), we have $P_iP_j = (v_j - v_i)a(t)/c$, so that the velocity vector $v_{ij} = d(P_iP_j)/dt$ of particle j relative to that of particle i is

$$v_{ij} = H(t)\, P_iP_j \,, \tag{16.5}$$

where $H(t)$ is the *Hubble parameter*:

$$H(t) \equiv \frac{1}{a}\frac{da}{dt} \,. \tag{16.6}$$

For a time t short compared to the characteristic evolution time of the universe, that is, $t \ll a_0/c$, the Hubble parameter $H(t)$ is nearly constant and the recession speed of 'galaxy' j relative to that of galaxy i is linear in the distance. This is *Hubble's law*.

However, isotropy of the velocity field does not imply that all particles are equivalent. Indeed, observation of a point on the outer edge of the cloud would allow an observer to be located relative to the center. Moreover, the reference system of this observer, which is *co-moving* with the particle to which it is attached, is not inertial—it is in free fall.

16.2 The pitfalls of the infinite Newtonian universe

The external radius of the cloud discussed above is not specified. We can therefore let it tend to infinity without changing any of our conclusions.[3] It would, however, be erroneous

---

[2]See, for example, Binney and Tremaine (2008).

[3]The first person to have exploded the spherical world of the Greeks to pass from *"the closed world to the infinite universe"* (in the words of Koyré) seems to have been Thomas Digges in *A Perfect Description of the Celestial Orbs according to the most ancient doctrine of the Pythagoreans, lately revived by Copernicus and by geometrical demonstrations approved*, London (1576).

to conclude that such a configuration of infinite extent no longer has a center, at least if gravitation is described by a force.

Let us consider, as did Newton, a cloud uniformly filling the entire universe, that is, a cloud of constant density, at least on average, throughout space. *Each* galaxy, or group of galaxies, then becomes a center. Since the $1/r^2$ attractive forces acting on it due to galaxies diametrically opposite cancel, we may be tempted to conclude that the galaxy in question is undergoing free motion and, paradoxically, that gravity disappears for it. However, the integral of the forces does not in fact converge (the summations over the angles and the distance do not commute; equivalently, Gauss's theorem does not hold).

The same problem arises if we describe gravitation by a force which is the gradient of a potential $F = -m\nabla U$, $U = -\int (\varrho(R',t)/|R-R'|)dV$. The potential U must remain finite in the passage to the limit, which requires that the matter density decrease sufficiently rapidly at infinity (faster than $1/r^2$). Therefore, the density must depend on r, and the special nature of the origin then remains.

Therefore, strictly speaking, a universe which is infinite and has no center cannot be modeled using Newtonian theory. The universe must have a center, either because it has infinite extent, or because its density is a function of the distance.

These models of island-universes whose history may not be eternal while time runs on forever, and which are organized about a center for which it is impossible to know if it is at rest or in uniform translation in the absolute reference frame, are a rather disappointing aspect of the superb edifice built by Newton. ... The Newtonian theory of gravity does not allow realization of the dream of Giordano Bruno[4]: the universe of Newton, whether finite or not, must have a center. Strictly speaking, it is the only free point of the universe, in uniform translation relative to the absolute frame. This is absurd![5]

16.3   The 'Friedmann' equation

In order to build satisfactory Newtonian cosmological models, it is necessary to broaden the theory, first of all by describing gravity as a local field theory in the form of the Poisson law, which involves only the acceleration of gravity $g \equiv f/\varrho$, so that the problem of convergence of the potential (like $\nabla U = -g$) does not arise. Then the Galilean group must be enlarged to the Milne group.

Let us assemble the equations governing the Newtonian 'cosmological fluid' in an inertial frame (see Section 15.1):

---

[4] *"There is a single universal space, a single vast immensity which we may freely call Void: in it are innumerable globes like this on which we live and grow... there are in this space those countless bodies such as our earth and other earths, our sun and other suns, which all revolve within this infinite space, through finite and determined spaces or around their own centres."* This perspective did not appeal to Kepler, who wrote to Galileo in 1610: *"This very cogitation carries with it I don't know what secret, hidden horror; indeed one finds oneself wandering in this immensity, to which are denied limits and center and therefore also all determinate places."* [Citations of G. Bruno and J. Kepler, in Koyré (1957)].

[5] As Leibniz said: *"These gentlemen maintain therefore, that Space is a real absolute Being. But this involves them in great difficulties... the Fiction of a material finite Universe, moving forward in an infinite empty Space cannot be admitted. It is altogether unreasonable and impracticable. For, besides that there is no real Space out of the material Universe, such an Action would be without any Design in it: It would be working without doing any thing, agendo nihil agere. There would happen no Change, which could be observed by Any Person whatsoever. These are Imaginations of Philosophers who have incomplete notions, who make Space an absolute Reality."* [Cited in Koyré (1957)].

$$\Delta U = 4\pi G\varrho, \quad \frac{\partial v}{\partial t} + (v.\nabla)v = -\frac{1}{\varrho}\nabla p - \nabla U, \quad \frac{\partial \varrho}{\partial t} + \nabla.(\varrho v) = 0, \qquad (16.7)$$

along with the equation of state $p = p(\varrho)$.

We seek a solution of the form $\varrho = \varrho(t)$, $p = p(t)$ (corresponding to the assumption that the matter is homogeneous throughout the universe), and $v = H(t)R$, where R is the radius vector (of modulus r) from the origin to the 'galaxy' under consideration, which corresponds to isotropy (with regard to *all* the points) of the distribution. For $H > 0$ the matter distribution is expanding, and the galaxies are moving away from each other with a speed proportional to their distance.

We already know the solution of the Poisson equation (see, for example, Section 15.2): $U = \frac{2}{3}\pi G\varrho(t)\, r^2$. Setting $H \equiv \dot{a}/a$, where $a(t)$ is the scale factor, the continuity equation gives

$$\varrho(t) = \frac{\varrho_0}{a^3}, \qquad (16.8)$$

with ϱ_0 a constant of integration. Finally, from the Euler equation we find the 'Friedmann' equation (we use the quotes because this equation was discovered by Friedmann in 1922 in the very different context of general relativity, where it is assumed that space itself is expanding):

$$H^2 + \frac{Kc^2}{a^2} = \frac{8\pi G}{3}\varrho, \qquad (16.9)$$

where Kc^2 is an integration constant. These equations are the same as those obtained in Section 16.1, and it should be noted that the solution does not involve the pressure. If $K = 0$, (16.8) and (16.9) are easy to integrate and we find

$$a = (6\pi G\varrho_0)^{\frac{1}{3}} t^{\frac{2}{3}}, \quad H = \frac{2}{3t}, \quad \varrho = \frac{1}{6\pi G\, t^2}. \qquad (16.10)$$

In general relativity this is referred to as the Einstein–de Sitter solution.

In this cosmological model all the particles (galaxies) are accelerated relative to each other. So then how is it possible to materialize the inertial frame in which the equations are supposed to be valid?

We can follow Leibniz and say that only relative motion is important. Then each point-like galaxy defines a local frame in free fall in which the equations are assumed to hold. This dynamical equivalence between frames in free fall and inertial frames means that gravitation can be locally effaced. The fact that this is possible stems from the equality of gravitational and inertial mass and the invariance of the equations of motion ($\Delta U = 4\pi G\rho$, $a = -\nabla U$) under the Milne group: $x \to x' = x - d(t)$ with the condition that the potential is transformed at the same time as $U \to U' = U - x.\ddot{d}$; see Section 11.3. It should be noted that under these transformations of the Milne group the gravitational potential is no longer a frame-independent, scalar quantity.

16.4 The evolution of perturbations

With the reservations discussed above, Newtonian mechanics can nevertheless be used to study the evolution of *perturbations* of a homogeneous, isotropic cloud.

Starting from the continuity, Euler, and Poisson equations, we assume that the density, pressure, velocity field, and gravitational field are the sum of the uniform, isotropic background solution described in Section 16.3 and a perturbation $(\varrho_1, p_1, v_1, g_1)$. It is easy to linearize the equations of motion about the background solution and find the equations for the perturbations[6]:

$$\begin{cases} \dfrac{\partial \varrho_1}{\partial t} + 3H\varrho_1 + HR.\nabla\varrho_1 + \varrho\nabla.v_1 = 0 \\[2mm] \dfrac{\partial v_1}{\partial t} + Hv_1 + H(R.\nabla)v_1 = g_1 - \dfrac{1}{\varrho}\nabla p_1 \\[2mm] \nabla \wedge g_1 = 0, \ \ \nabla.g_1 = -4\pi G\varrho_1 \\[2mm] p_1 \ \ = v_s^2\varrho_1, \end{cases} \tag{16.11}$$

where in the last equation $v_s \equiv \sqrt{dp/d\varrho}$ is the *speed of sound*. We solve these equations by decomposing the perturbations into a sum of Fourier modes written generically in the form

$$f_1(R,t) = f_q(t)\exp\left(i\frac{R.q}{a(t)}\right), \tag{16.12}$$

where the 'co-moving wave vector' q (independent of time) characterizes the mode whose 'co-moving' wavelength (that is, relative to the scale factor) is $\lambda(t) \equiv 2\pi a(t)/q$ (in Section 17.3 we shall give a more detailed description of waves). Then after rearrangement (16.11) becomes

$$\begin{cases} \dot{\delta} = \dfrac{q^2\epsilon}{a} \\[2mm] \dot{v}_q^\perp + Hv_q^\perp = 0, \ \ \dot{\epsilon} + H\epsilon = \left(-\dfrac{v_s^2}{a} + \dfrac{4\pi G\varrho a}{q^2}\right)\delta \\[2mm] q \wedge g_q = 0, \ \ iq.g_q = -4\pi Ga\varrho_q, \end{cases} \tag{16.13}$$

where we have introduced the *density contrast* $\delta \equiv \varrho_q/\varrho$ and have decomposed v_q as $v_q = v_q^\perp + iq\epsilon$ with $v_q^\perp.q = 0$. Eliminating ϵ, we obtain the evolution equation of the density contrast:

$$\ddot{\delta} + 2H\dot{\delta} + \left(\dfrac{v_s^2 q^2}{a^2} - 4\pi G\varrho\right)\delta = 0. \tag{16.14}$$

Given the solution of (16.14), we can find the pressure perturbation $p_q = v_s^2\varrho\,\delta$ and the expression for the *compression modes* $\epsilon = a\,\dot{\delta}/q^2$. The *rotational modes* $\vec{v}_q^\perp$ decrease as $1/a$ [*cf.* (16.13)]. Finally, the perturbation of the gravitational acceleration is obtained by solving (16.13) directly: $g_q = 4i\pi G\varrho a\delta q/q^2$.

The *Jeans approximation* (1902) consists of neglecting the expansion of the matter cloud in which the perturbations propagate, that is setting $H = 0$ as well as $a = 1$ and $\varrho = \text{const}$ in

---

[6]See, for example, Weinberg (1972), Peter and Uzan (2013), or Mukhanov (2005).

(16.14). We then see that perturbations whose wavelength is smaller than the *Jeans length* λ_J, defined as

$$\lambda_J = v_s \sqrt{\frac{\pi}{G\varrho}}, \quad k_J = \frac{\sqrt{4\pi G\varrho}}{v_s}, \quad M_J \equiv \frac{4}{3}\pi\varrho\lambda_J^3 \tag{16.15}$$

(where we have also introduced the corresponding wave vector and mass), oscillate with constant frequency $\omega \equiv v_s q \sqrt{1 - (k_J/q)^2}$ and propagate with 'phase' velocity $\omega/q \equiv v_s \sqrt{1 - (k_J/q)^2} \to v_s$ as long as gravity can be neglected.

Inversely, perturbations of wavelength longer than the Jeans length grow exponentially in a characteristic time $1/\omega$ which for very long wavelengths or $v_s \to 0$ is $t_{\mathrm{grav}} = 1/\sqrt{4\pi G\varrho}$. Therefore, in the Jeans approximation the static background cosmological solution is highly unstable.

Let us now take into account the expansion of the cloud in which the perturbations propagate.

Again, if the size of the perturbations $\lambda = 2\pi a(t)/q$ is larger than the Jeans length (16.15), where $v_s \approx 0$ if the pressure term is negligible, then the solution of (16.14), in the simple case of $K = 0$ so that $a(t)$, $H(t)$, and $\varrho(t)$ are given by (16.10), is the sum of a decaying mode and a growing mode:

$$\delta = C_1 t^{-1} + C_2 t^{2/3} \qquad \Longleftrightarrow \qquad \delta = D_1 a^{-2/3} + D_2 a. \tag{16.16}$$

Therefore, in an expanding cloud the *zero modes*, that is, the perturbations of size larger than the Jeans length, begin to grow (but only as the scale factor, that is, as a power of the time). Then when their wavelength $\lambda = 2\pi a(t)/q \propto t^{2/3}$ becomes smaller than the Jeans length (which, since $\varrho \propto 1/t^2$, now grows linearly in time and therefore faster than λ), they begin to oscillate with decreasing amplitude, since, as is easily shown, the adiabatic solution of (16.14) then is

$$\delta \propto \frac{1}{\sqrt{a}} \exp\left(\pm i\, q \int \frac{v_s}{a} dt\right). \tag{16.17}$$

These results serve as a starting point for studying the growth of the large-scale structure of the universe.

16.5 Olbers's paradox

Let us consider a static, homogeneous, and infinite universe in which the density of luminous objects, all alike and having the same absolute luminosity L_0, is n. In a thin spherical layer of radius r and thickness dr we then have $4\pi n r^2 dr$ 'stars' whose apparent luminosity at the center is $4\pi n L_0 dr$, given that the luminosity falls off as $1/r^2$. Therefore, the total luminosity at the center is $4\pi n L_0 \int dr$ and tends to infinity. However, the night sky is dark (or nearly so, because the Earth is in fact immersed in the microwave radiation of a black body at 3 kelvin).

This paradox was popularized by Olbers in 1826, but Thomas Digges seems to have been the first to conceive of it in 1576 [the version presented here is due to Chéseaux (1744)].

Resolution of Olbers's paradox

In order to solve Olbers's paradox we need to abandon one of the hypotheses leading to it.

It can be argued that the universe is not infinite (Kepler, 1610); that the density is 'fractal' (C. Charlier, 1908 and E. Fournier d'Albe, 1907); that stars have a finite lifetime and their light takes time to reach us (Kelvin, 1901); that the universe has an infinite lifetime but is not static, so that a Doppler effect occurs (the 'stationary state' model; Bondi, 1957); or that the universe is expanding slowly and has finite age (the Big Bang model). Finally, we note that the loophole suggested by Chéseaux, namely, that the light is absorbed by the intervening stars, in fact solves nothing owing to considerations of thermodynamical equilibrium (Herschel, 1831).[7]

---

[7]To learn more, see, for example, Harrison (1987).

17
Light in Newtonian theory

Astronomy (and physics in general) is certainly not a complete science without a theory of light. However, the nature of light and its kinematical properties were not completely understood until the advent of Maxwell's theory, special and general relativity, and quantum field theory. The answers to these questions provided by the Newtonian theory were only partial and sometimes even contradictory. Here we shall present a few aspects of this topic.[1]

17.1 Light and gravity

If asked the question, "Does gravity influence the propagation of light?", the proponents of a corpuscular theory (like Newton himself, and also Michell, Laplace, Blair, Soldner...) would have answered, "Yes, of course."

Let us recall the first integral of the equation of motion of a particle in the gravitational field of a spherically symmetric body of mass M (see Section 12.2):

$$r^2 \frac{d\phi}{dt} = L, \quad \left(\frac{dr}{dt}\right)^2 = 2E + \frac{2GM}{r} - \frac{L^2}{r^2}, \tag{17.1}$$

where r and ϕ are the polar coordinates of the particle in an inertial reference frame whose origin is at the center of the body, and L and E are its specific angular momentum and energy. We shall study radial motion, i.e., $L = 0$. The particle will escape the attraction of the central body if E is positive or zero. Its initial velocity at r_0, the radius of the central body, must therefore be greater than the *escape velocity*: $dr/dt|_0 = \sqrt{2GM/r_0}$, or ~ 11 km/s for the Earth and ~ 630 km/s for the Sun. Assuming that light particles are emitted with velocity c, we deduce that if they are emitted from a body of radius smaller than the *gravitational radius*

$$r_S = \frac{2GM}{c^2}, \tag{17.2}$$

they will not reach infinity. For a star of one solar mass $r_S \sim 3$ km, and for a star with the mass of the Earth $r_S \sim 1$ cm.

Römer and the speed of light

In 1676 Ole Römer invoked the finiteness of the speed of light to explain the fluctuations of the dates of emersion of the satellite Io, which depend on the variation of the distance between Jupiter and the Earth from one eclipse to another. If the Sun, Earth, and Jupiter lie along a straight line, there are two possible configurations. For the first, S-E-J, the travel time is

---

[1]See Eisenstaedt (2005) and Uzan and Lehoucq (2005).

Relativity in Modern Physics. Nathalie Deruelle and Jean-Philippe Uzan.
© Oxford University Press 2018. Published in 2018 by Oxford University Press.
DOI: 10.1093/oso/9780198786399.001.0001

$(D_{\text{SJ}} - D_{\text{SE}})/c$, and then, six months later, for the second configuration E-S-J the travel time is $(D_{\text{SJ}} + D_{\text{SE}})/c$. The difference of these is $2D_{\text{SE}}/c$. Römer obtained 22 minutes, which implies that $c \sim 215,000$ km/s using today's data for the size of the solar system.

Now let us consider a light particle deflected by a star, the Sun for example, grazing its surface $r = r_0$ with speed c (see Fig. 17.1). Its trajectory is a hyperbola given by (see Section 12.2)

$$r = \frac{p}{1 + e \cos \phi} \quad \text{with} \quad p \equiv \frac{L^2}{GM}, \tag{17.3}$$

where $L = r_0 c$ [from the first equation in (17.1)] and $1/r_0 = (GM/r_0^2 c^2)(1 + e)$ [or $e \simeq r_0 c^2/(GM) \gg 1$]. This hyperbola is very close to the straight line $X = r \cos \phi = r_0$. Its asymptotes are determined by $1 + e \cos \phi_a = 0$, or $\phi_a \simeq \pm\pi/2 + 1/e$. The light particle is therefore deflected by an angle $\Delta\phi$ given by

$$\Delta\phi \simeq \frac{2GM}{r_0 c^2}. \tag{17.4}$$

For the solar data we find $\Delta\phi \sim 0.9''$.

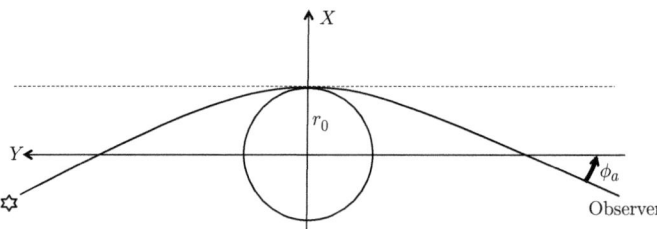

Fig. 17.1 The bending of light rays.

These curious possibilities of the existence of 'dark' stars and light bending were respectively discovered by Michell in 1784 and Soldner in 1801, within the framework of Newton's corpuscular theory of light then in fashion.

In the nineteenth century the wave theory of light triumphed over the corpuscular theory owing to the work of Young and Fresnel (following Huygens and Euler). In the wave theory the question of an interaction between light and gravitation was not addressed. Einstein's invention of the photon within the framework of quantum mechanics in the early twentieth century did not lead to the rebirth of these eighteenth-century ideas, because it turned out that a photon *always* travels at the same velocity c, and that only the *probability* of its presence can be defined. General relativity proposes a coherent theory of the interaction of (non-quantum) light and gravitation. It gives a value for the bending of light which is *twice* the Newtonian prediction. It also defines r_{S} as the *Schwarzschild radius* or the 'event horizon' of a 'black hole'.

17.2   Stellar aberration

Let us place ourselves in the 'absolute' reference frame of the solar system and choose the axes of the plane of the ecliptic such that the components of the velocity of light coming from a fixed star are $(0, c\cos\alpha, c\sin\alpha)$ (c is negative), and consider the time when the Earth's velocity has components $(0, v, 0)$ (see Fig. 17.2). In the frame attached to the Earth defined by $(X' = X, Y' = Y - vt, Z' = Z)$, the velocity of light has the components $(0, c\cos\alpha - v, c\sin\alpha) = (0, c'\cos\alpha', c'\sin\alpha')$ with

$$\tan\alpha' = \frac{\tan\alpha}{1 - v/(c\cos\alpha)} \qquad \text{or} \qquad \alpha' \approx \alpha + \frac{v\sin\alpha}{c}. \tag{17.5}$$

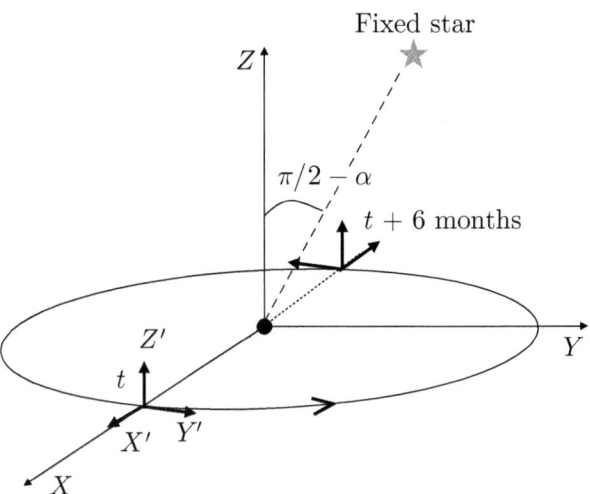

Fig. 17.2 Stellar aberration.

Six months later the Earth's velocity is $(0, -v, 0)$ and α' will have changed ('wandered') by $\Delta\alpha' \approx 2v\sin\alpha/|c|$, where $v/c \approx 10^{-4} \approx 20$ arcseconds. This is the phenomenon of stellar aberration discovered and explained by Bradley in 1728.

In a corpuscular theory of light, we have, strictly speaking, $c = c_\infty + v_*$, where v_* is the speed of the star in the absolute reference frame and $c_\infty^2 = c_0^2 - 2GM/r_0$, with c_0 the speed of emission of the 'photon' at the surface of the star of mass M and radius r_0. Therefore, $\Delta\alpha'$ for a given α' should vary from one star to another, which is not the case. This can be explained in the wave theory: the speed of a wave is independent of the speed of the source (see Section 17.3). Therefore, reasoned Fresnel, v should be replaced by $|v - V_0|$, where V_0 is the speed of the wave-supporting aether in the solar system. Since aberration is observed, $|v - V_0| \neq 0$: the Earth does not drag the aether, rather, there must exist an 'aether wind' blowing at a speed of about 30 km/s if the aether is at rest in the reference frame of the solar system.

17.3 Wave propagation

Let us recall the Euler and continuity equations which in the absence of an external field govern the motion of a perfect fluid of mass density ϱ, pressure p, and velocity v (see Section 6.3):

$$\frac{\partial \varrho}{\partial t} + \nabla \cdot (\varrho v) = 0, \quad \frac{\partial v}{\partial t} + (v.\nabla)v = -\frac{1}{\varrho}\nabla p, \quad p = p(\varrho). \tag{17.6}$$

The ansatz $\varrho = \varrho_0$, $p = p_0 = p(\varrho_0)$, $v = V_0$, where ϱ_0, p_0, and V_0 are constants, solves these equations and describes a homogeneous fluid (for example, air) propagating at velocity V_0 in an inertial frame $\mathcal{S}$. Setting $\varrho = \varrho_0 + \varrho_1(t, X^i)$, $p = p_0 + p_1(t, X^i)$, and $v = V_0 + v_1(t, X^i)$ and linearizing the equations, we find (after taking the time derivative of the continuity equation and eliminating $\partial v_1/\partial t$ using the Euler equation) the equation describing the propagation of the density perturbation:

$$-\frac{1}{c_s^2}\frac{\partial^2 \varrho_1}{\partial t^2} + \triangle\varrho_1 - \frac{2V_0}{c_s^2}.\nabla\frac{\partial \varrho_1}{\partial t} - \frac{1}{c_s^2}V_0.\nabla(V_0.\nabla \varrho_1) = 0, \tag{17.7}$$

where $c_s^2 = dp/d\varrho|_0$ is *the speed of sound*. The terms involving V_0 are eliminated by going to the frame $\mathcal{S}'$ moving at velocity V_0 with respect to $\mathcal{S}$, defined as $X'^i = X^i - V_0^i t$, so that $\frac{\partial}{\partial t} \to \frac{\partial}{\partial t} - V_0\frac{\partial}{\partial X'}$ and $\frac{\partial}{\partial X} \to \frac{\partial}{\partial X'}$. Therefore, the sound propagation equation is not invariant under Galilean transformations (which is not surprising, since it was obtained by perturbing a particular solution).

A particular solution of (17.7) is

$$\varrho_1 = a_k \cos\Phi \quad \text{with} \quad \Phi \equiv k_i X^i - (|k|c_s + k_i V_0^i)t - \phi_k, \tag{17.8}$$

where k is a vector of modulus $|k| = \sqrt{k_i k^i}$, and a_k and ϕ_k are constants. A perturbation of this type is a completely *delocalized* wave.

Let us imagine a 'snapshot' of the perturbation at a given time t. It varies sinusoidally in space along the lines determined by k, called the *wave vector*. The *wavelength* of the perturbation is $\lambda \equiv 2\pi/|k|$. The *wave fronts*, that is, the set of points where the phase is constant at a given instant, are planes orthogonal to the wave vector k.

In each plane of $\mathcal{E}_3$ orthogonal to k, that is, on each wave front, the density perturbation (17.8) has the same amplitude and oscillates in time with *angular frequency*

$$\omega \equiv (|k|c_s + k_i V_0^i) \tag{17.9}$$

($P \equiv 2\pi/\omega$ is the associated *period* and $\nu \equiv 1/P = \omega/2\pi$ is the *frequency*).

Let us consider the case where the wave vector lies along the X axis, $k^i = (k, 0, 0)$. Then the perturbation (17.8) is constant if its *phase* $\Phi = kX - \omega t - \phi_k$ is constant. Therefore, if it has a certain value in the plane $X - X_0$ at time t_0, it will have the same value in another plane located a distance ΔX_0 away after a time interval such that $\omega\Delta t_0 = k\Delta X_0$. The wave surfaces, that is, the perturbation itself, *propagate* in the positive direction of the X axis at the phase velocity

$$c_{\text{ph}} = \frac{\omega}{|k|}\frac{k}{|k|}. \tag{17.10}$$

The solution (17.8) is called a *monochromatic plane wave*. The adjective 'chromatic' is not really suitable for sound waves, but it is commonly used.

Any linear superposition of monochromatic plane waves (17.8)

$$\varrho_1 = \int d^3k\, a(k) \cos\Phi \quad \text{with} \quad \Phi \equiv k_i X^i - (|k|c_s + k_i V_0^i)t - \phi(k) \tag{17.11}$$

is also a solution of (17.7) and constitutes a *wave packet*. Inversely, any square-integrable perturbation can be decomposed into *Fourier modes* as in (17.11). The more strongly peaked the function $a(k)$ is at some value of k, the more the wave is spread out spatially. Inversely, the broader $a(k)$ is, the more the packet is localized on the trajectory given by $\partial\Phi/\partial k_i = 0$, which the packet travels along at the *group velocity*

$$c_{\mathrm{g}} = c_{\mathrm{s}}\frac{k}{|k|} + V_0 \tag{17.12}$$

(see below for a more precise definition). The phase and group velocities (17.10) and (17.12) coincide only in the frame where the air is at rest or if V_0 and k are collinear; see (17.9). The functions $a(k)$ and $\phi(k)$ encode the initial conditions of the perturbation, that is, they represent the action of the *source* of the perturbation.

Phase velocity and group velocity

Let us consider a function $F(X,t) = \int dk\, a(k)\, e^{i(kX-\omega(k)t)}$, a sum of monochromatic waves individually propagating at the *phase velocity* $v_{\mathrm{p}} = \omega/k$. We assume that $a(k)$ is a function with maximum at $k = k_0$ such that only values close to k_0 contribute to the integral. We can then expand the integrand about k_0 and write $F(X,t,k_0) \approx e^{i(k_0 X-\omega(k_0)t)} \int dk\, a(k)\, e^{i(k-k_0)(X-v_{\mathrm{g}}t)}$ with $v_{\mathrm{g}} \equiv d\omega/dk|_0$, which can be rewritten as

$$F(X,t,k_0) \approx A(k_0, X - v_{\mathrm{g}}t)e^{i(k_0 X-\omega(k_0)t)}, \tag{17.13}$$

where $A(k_0, X - v_{\mathrm{g}}t) \equiv \int dk\, a(k)\, e^{i(k-k_0)(X-v_{\mathrm{g}}t)}$. The function $F(X,t,k_0)$ then is a plane wave of amplitude A and describes a wave packet whose envelope moves at the *group velocity*

$$v_{\mathrm{g}} \equiv \left.\frac{d\omega}{dk}\right|_0 . \tag{17.14}$$

The Doppler effect

Let $\mathcal{S}_{\mathrm{r}}$ be an inertial frame moving at speed v_{r} relative to $\mathcal{S}$: $X_{\mathrm{r}}^i = X^i - v_{\mathrm{r}}^i t$. Since the density perturbation ϱ_1 is a scalar, it is the same in $\mathcal{S}$ and $\mathcal{S}_{\mathrm{r}}$: $\varrho_1 = \int d^3k\, a(k) \cos\Phi$, where $\Phi \equiv k_i X_{\mathrm{r}}^i - \omega_{\mathrm{r}}t - \phi(k)$ with $\omega_{\mathrm{r}} = |k|c_{\mathrm{s}} + k_i(V_0^i - v_{\mathrm{r}}^i)$. We then deduce that in the reference frame $\mathcal{S}_{\mathrm{r}}$ of a stationary receiver, the wave packet moves at group velocity $c_{\mathrm{r}} \equiv d\omega_{\mathrm{r}}/dk$, which is *independent* of the emitter velocity and given by

$$c_{\mathrm{r}} = c_{\mathrm{s}}\frac{k}{|k|} + V_0 - v_{\mathrm{r}}, \quad \tan\alpha_{\mathrm{r}} = \frac{\tan\alpha}{1 - |v_{\mathrm{r}} - V_0|/(c_{\mathrm{s}}\cos\alpha)}. \tag{17.15}$$

In the second equation for the *aberration*, α and α_{r} are the angles the vectors k and c_{r} make with $(v_{\mathrm{r}} - V_0)$. This should be compared to (17.5) giving the stellar aberration in the corpuscular theory. We recall that V_0 is the global speed of the fluid in the frame $\mathcal{S}$.

Finally, the fact that the phase Φ has the same value in any inertial frame allows the frequency of a wave measured by a stationary receiver in $\mathcal{S}_{\mathrm{r}}$ to be related to that measured in the frame $\mathcal{S}_{\mathrm{e}}$ where the emitter of the perturbation is stationary:

$$\nu_{\rm r} = \nu_{\rm e} \frac{1 + k_i(V_0^i - v_{\rm r}^i)/(|k|c_{\rm s})}{1 + k_i(V_0^i - v_{\rm e}^i)/(|k|c_{\rm s})}.$$

(17.16)

This expression, discovered by Doppler, depends (except in first order in $1/c_{\rm s}$) not on the relative velocity of the receiver and emitter, but on their velocities relative to the fluid, so that it is possible, for example, to measure changes of the velocity of the receiver with respect to the fluid.

In Newtonian wave theory, light was thought to propagate in the aether just like sound propagates in air.

17.4   The Fizeau experiment

Let us imagine a refractive medium, water, for example, at rest in the laboratory frame. The speed of light in the medium is measured to be c/n, where n is the *index of refraction* of the medium. If we now give the medium a speed u, what will the speed of light c' be with respect to the laboratory? Observations of stellar aberration imply that the Earth does not drag the aether in its wake, and so one might be tempted to conclude that refractive media also do not drag it, and so $c' = c/n$. However, in 1818 Fresnel showed (by 'ingenious' reasoning, in the words of Sommerfeld) that refractive media must partially drag the aether according to[2]

$$c' = \frac{c}{n} \left[1 + \frac{nu}{c} \left(1 - \frac{1}{n^2} \right) + \mathcal{O}(u^2/c^2) \right].$$

(17.17)

Fizeau verified this formula (which does *not* depend on the Earth's velocity) as follows. A U-shaped tube of length $2D$ has water running through it at a speed $u \approx 7$ m/s. Two light beams are sent through the tube, one in the direction along the current and the other opposite to it, and they are made to interfere at the exit from the tube. The travel times of the light beams are $t_\pm = 2D/c'_\pm \approx (2Dn/c) \left[1 \mp (nu/c) \left(1 - 1/n^2 \right) \right]$, and so the ratio of the shift Δi of the interference pattern and the spacing between fringes i is

$$\frac{\Delta i}{i} = \frac{c(t_- - t_+)}{\lambda} \approx \frac{4Dn^2u}{\lambda c} \left(1 - \frac{1}{n^2} \right),$$

(17.18)

which was measured by Fizeau in 1850.

Optical interference

Let us consider two monochromatic plane waves (sound or light waves, for example) with angular frequencies ω_a and wave vectors k_a:

$$E_a = E_{0a} \cos(\Phi_a + \psi_a) \quad \text{with} \quad \Phi_a \equiv \vec{k}_a.\vec{X} - \omega_a t + \phi_a.$$

(17.19)

The sources of these waves emit packets of quasi-monochromatic waves separated by irregular time intervals, a fact which can be encoded in the phases ψ_a, which then become random variables. If the random nature of the emission can be ignored, the two sources are said to be *coherent* and the phases ψ_a can be set equal to zero (in practice, this implies that the sources both come from a single primary source split into two beams). The phases ϕ_a specify the initial conditions.

---

[2]See Darrigol (2003). This formula explains why the refraction angle in a prism for light coming from the stars, the speed of which is in principle varying owing to the orbital motion of the Earth, is actually constant.

At a given point $\vec{X}$ the two coherent waves *interfere*, and the resultant wave is the *sum of* E_1 and E_2 if the equations governing their propagation are linear. This is the case with optical waves, which ultimately are governed by the linear Maxwell equations, and also sound waves, which are defined as *perturbations* of the Euler equations; see Section 17.3.

A measuring device placed at $\vec{X}$ (a screen, for example) will be sensitive to the *intensity I* of the wave, that is, to the time-averaged value of the square of its amplitude. This is because in Maxwell's theory the wave energy is proportional to its squared amplitude and the characteristic observation times are very long compared to its period or even the time duration of the wave packets. Therefore, for coherent waves we have

$$I \equiv \langle E_{01} \cos \Phi_1 + E_{02} \cos \Phi_2 \rangle^2$$

$$= \left\langle E_{01}^2 \frac{1 + \cos 2\Phi_1}{2} + E_{02}^2 \frac{1 + \cos 2\Phi_2}{2} + E_{01} E_{02} \left(\cos(\Phi_1 + \Phi_2) + \cos(\Phi_1 - \Phi_2) \right) \right\rangle.$$

Given the phases Φ_a above, we find that on average all the cosine terms vanish, except for the last one when the two waves have the same angular frequency $\omega_1 = \omega_2$ (this occurs if the beams come from a single primary source). The measured intensity then is

$$I = I_1 + I_2 + 2\sqrt{I_1 I_2} \cos(\Delta\Phi), \quad \text{where} \quad \Delta\Phi \equiv (\vec{k}_2 - \vec{k}_1).\vec{X} + (\phi_2 - \phi_1). \tag{17.20}$$

Let us consider the example of the *Young slits*, where the observation point is at $\vec{X} = (D, y)$ and the sources are at $\vec{S}_{1,2} = (0, \pm d/2)$. The wave vectors $\vec{k}_{1,2}$ have the same modulus $2\pi/\lambda$, because $\omega_2 = \omega_1$ and the two beams propagate in the same medium. We then have (see Fig. 17.3)

$$\vec{k}_{1,2} = \frac{2\pi}{\lambda} \frac{(D, y \mp d/2)}{\sqrt{D^2 + (y \mp d/2)^2}}.$$

If y and d are of the same order of magnitude and much smaller than D, we see that $(\vec{k}_2 - \vec{k}_1).\vec{X}$ is of order 2 and does not contribute to the phase difference. Regarding the second term in (17.20), at $\vec{S}_{1,2}$ the phases Φ_a of the waves are the same, so that [cf. (17.19)] $\vec{k}_1.\vec{S}_1 + \phi_1 = \vec{k}_2.\vec{S}_2 + \phi_2$ because $\omega_2 = \omega_1$. To linear order we therefore have

$$\Delta\phi \sim \phi_2 - \phi_1 \quad \text{with} \quad \phi_2 - \phi_1 = 2\pi \frac{dy}{\lambda D}. \tag{17.21}$$

The bright fringes correspond to $\Delta\Phi = 2\pi n$, where n is an integer, and the fringe spacing is $i = \lambda D/d$.

In the Fizeau experiment described above the initial phases of the waves are such that $\vec{k}_1.\vec{S}_1 + \phi_1 = \vec{k}_2.\vec{S}_2 + \phi_2 - \omega(t_- - t_+)$ with $\omega = 2\pi c/\lambda$ and where the extra term increases with the water velocity. The fringe spacing therefore remains constant but the interference pattern [cf. (17.21)] is shifted by $c(t_- - t_+)D/d$.

In the *Newton's rings* experiment a light beam falls perpendicularly on a spherical lens of radius R resting on a mirror (see Fig. 17.3). Part of the beam is reflected by the lens and the rest passes through the lens and is reflected by the mirror. In this case the point $\vec{X}$ is on the lens and we have $\vec{X} = (r, d)$, where r is the distance to the axis and $r^2 + (R - d)^2 = R^2$, so that $d \approx r^2/2D$ to lowest order in $1/R$. The phases ϕ_1 and ϕ_2 differ by π owing to the reflection on the mirror. Finally, the wave vectors have opposite directions: $\vec{k}_2 = -\vec{k}_1$ with $\vec{k}_2 = (0, 2\pi/\lambda)$. Then (17.20) gives $\Delta\Phi = 4\pi d/\lambda + \pi$. The dark fringes correspond to $\Delta\Phi = (2n + 1)\pi$ with integer n and their radii are $r = \sqrt{n\lambda R}$.

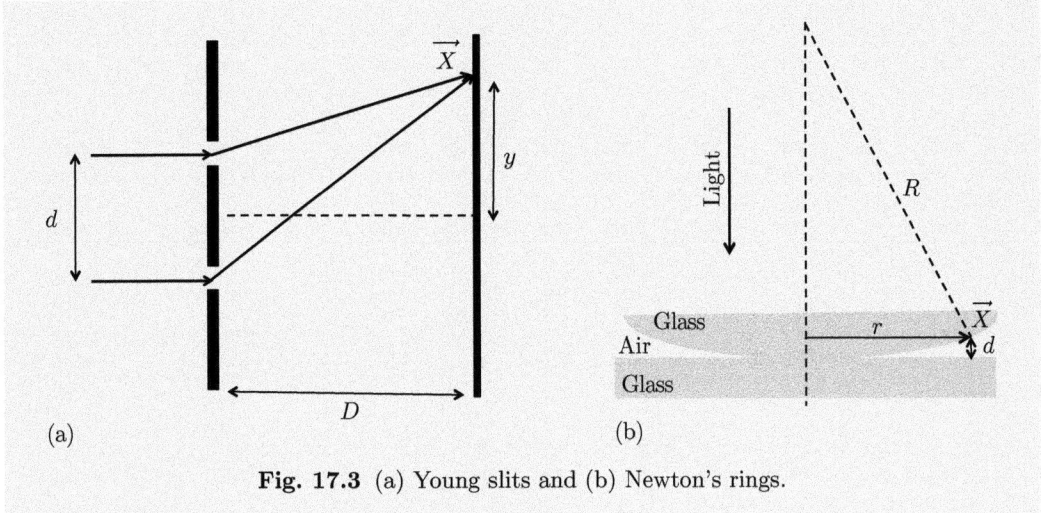

Fig. 17.3 (a) Young slits and (b) Newton's rings.

17.5   The Michelson–Morley experiment

In the laboratory frame the speed of light is $c_l = c - v$, where c is the speed of light relative to the aether (which is assumed to be at rest in the reference frame attached to the Sun) and v is the orbital velocity of the Earth. We consider an interferometer one of whose arms, of length l, is parallel to v. On this arm the time required for the light to make a round trip is $T_1 = l/(c-v) + l/(c+v) = 2lc/(c^2 - v^2)$. On the other arm, perpendicular to the first one, we have $c_l = c - v = (c^1 - v, c^2)$ with $c^1 = v$ and $(c^1)^2 + (c^2)^2 = c^2$, from which we find $|c_l| = \sqrt{c^2 - v^2}$ and the round-trip time is[3] $T_2 = 2l/\sqrt{c^2 - v^2}$. Therefore, $T_1 - T_2 \approx (l/c)(v^2/c^2)$. When the interferometer is turned by 90 degrees, the arms are interchanged and the path difference is doubled, so that the expected shift Δi of the interference pattern is

$$\frac{\Delta i}{i} \approx \frac{2l}{\lambda}\frac{v^2}{c^2}, \tag{17.22}$$

where i is the fringe spacing. Therefore, $\Delta i/i \approx 0.4$ for $l \approx 10$ m, $\lambda \approx 0.5$ micron, and $v/c \approx 10^{-4}$.

No shift was observed.[4]

FitzGerald noted that *if* the motion shortened the interferometer arm parallel to v by a factor of $\sqrt{1 - v^2/c^2}$, there would be no shift of the fringes. This *length contraction* was explained by Lorentz in a detailed study of electrostatic cohesive forces but at the price of making numerous hypotheses about the structure of matter.[5] As we all know, in the end all these brilliant fixes could not measure up to the new view of space and time proposed by Einstein in 1905.

---

[3] And not $2l/c$, an error dating from 1881 and corrected by Alfred Potier, professor at the École Polytechnique. This persuaded Michelson to redo the experiment with Morley in 1886.

[4] As commented upon by Poincaré (1906), "*It seems that this impossibility of experimentally demonstrating the absolute motion of the Earth is a general law of Nature.*"

[5] "*An explanation was necessary, and was forthcoming; they always are; hypotheses are what we lack the least.*" Poincaré (1902).

List of books and articles cited in the text

Arnold, V. I. (1997). *Mathematical methods of classical mechanics* (2nd edn). Springer, New York.

Barbour, J. and Pfister, H. eds. (1995). *Mach's principle: from Newton's bucket to quantum gravity (Einstein Studies)*. Birkhäuser, Boston.

Basdevant, J.-L. (2005). *Variational principles and dynamics* [in French]. Éditions Vuibert, Paris.

Binney, J. and Tremaine, S. (2008). *Galactic dynamics* (2nd edn). Princeton University Press, Princeton, NJ.

Bishop, R. L. and Goldberg, S. I. (1980). *Tensor analysis on manifolds*. Dover, New York.

Bourguignon, J.-P. (1995). *Variational principles* [in French]. Éditions de l'École Polytechnique, Paris.

Brouwer, D. and Clemence, G. M. (1961). *Methods of celestial mechanics*. Academic Press, New York.

Chavanis, P. H. (2003). Gravitational instability of isothermal and polytropic spheres. *Astron. Astrophys.*, **401**, 15–42; also, http://arxiv.org/abs/astro-ph/0207080.

Collins, G. W. (2003). *The fundamentals of stellar astrophysics*. http://ads.harvard.edu/books/1989fsa..book/.

Chandrasekhar, S. (1969). *Ellipsoidal figures of equilibrium*. Yale University Press, New Haven.

Châtelet, G. (1993). English translation: *Figuring space: philosophy, mathematics, and physics*. Springer, Dordrecht (1999).

Christodoulou, D. *et al.* (1995). Phase-transition theory of instabilities. III. The third-harmonic bifurcation on the Jacobi sequence and the fission problem. *Ap. J.*, **446**, 500–509; also, http://arxiv.org/abs/astro-ph/9505008.

Darrigol, O. (2003). *Electrodynamics from Ampère to Einstein*. Oxford University Press, Oxford.

Duriez, L. (2003). *Course in classical celestial mechanics* [in French]. http://lal.univ-lille1.fr/Documents_pedagogiques/index.html.

Eisenstaedt, J. (2002). *Einstein and general relativity* [in French]. CNRS Éditions, Paris.

Eisenstaedt, J. (2005). *Before Einstein: relativity, light, and gravitation* [in French]. Éditions du Seuil, Paris.

Galileo, G. (1638). *Discorsi e dimostrazioni matematiche intorno a due nuove scienze.* Elzevir, Leiden. [English translation by Crew, H. and de Salvio, A. (1914). *Dialogs concerning two new sciences.* Macmillan, New York.]

Grandclément, P. (2008). *Compact objects* [in French]. http://www.luth.obspm.fr/article400.html?lang=fr.

Hagihara, Y. (1970). *Celestial mechanics.* MIT Press, Cambridge.

Harrison, E. (1987). *Darkness at night.* Harvard University Press, Cambridge.

Koyré, A. (1957). *From the closed world to the infinite universe.* Johns Hopkins Press, Baltimore.

Lanczos, C. (1986). *The variational principles of mechanics* (4th edn). Dover, New York.

Landau, L. and Lifshitz, E. (1976). *Mechanics* (3rd edn). Elsevier Butterworth–Heinemann, Oxford.

Lichnerowicz, A. (2016). *Elements of tensor calculus.* Dover, New York. [French original: *Éléments de calcul tensoriel,* Armand Colin, 1950].

Lynden-Bell, D. and Wood, R. (1968). The gravo-thermal catastrophe in isothermal spheres and the onset of red-giant structure for stellar systems. *Mon. Not. R. Astr. Soc.,* **138**, 495–525.

McIntosh, H. V. (1971). *Group theory and its applications,* E.M. Loebl ed. Vol. II. pp. 75–144. Academic Press, New York.

Melchior, P. (1973). *Planetary physics and dynamics* [in French], Vols. III and IV. Éditions Vander, Brussels.

Mukhanov, V. (2005). *Physical foundations of cosmology.* Cambridge University Press, Cambridge.

Murray, C. D. and Dermott, S. F. (1999). *Solar system dynamics.* Cambridge University Press, Cambridge.

Newton, I. (1687). *Philosophiænaturalis principia mathematica.* London. [English translation by Motte, A. (1729). *The mathematical principles of natural philosophy.* Benjamin Motte, London.]

Pascoli, G. (2000). *Fundamental astronomy* [in French]. Éditions Dunod, Paris.

Peter, P. and Uzan, J.-P. (2013). *Primordial cosmology.* Oxford University Press, Oxford.

Poincaré, H. (1902). *La science et l'hypothèse* [English translation in *The value of science: essential writings of H. Poincaré*. Modern Library, Random House, New York (2001)].

Poincaré, H. (1906). *Sur la dynamique de l'électron*, Palermo Rendiconti **21**, 129 [English translation: *On the dynamics of the electron*, https://en.wikisource.org/wiki/Translation:On_the_Dynamics_of_the_Electron_(July)].

Saff, E. and Kuijlaars, A. B. J. (1997). Distributing many points on a sphere. *Math. Intelligencer*, **19**, 5–11.

Touboul, P. *et al.* (2017). *MICROSCOPE* mission: First results of a space test of the equivalence principle. *Phys. Rev. Lett.*, **119**, 231101.

Uzan, J.-P. and Lehoucq, R. (2005). *The fundamental constants* [in French]. Éditions Belin, Paris.

Weinberg, S. (1972). *Gravitation and cosmology*. John Wiley, New York.

Book 1

BOOK 2

Special relativity and Maxwell's theory

Contents

Book 2

Book 2

A note on the units

In special relativity the limiting velocity c, identified as the speed of light, is a universal constant. We shall use the system of units in which its numerical value is 1. We then end up with only two fundamental dimensional quantities: the time expressed in seconds (s) and the mass expressed in kilograms (kg). The meter (m) and all other quantities are derived from these. Here we list some useful conversion factors.

SI base units:

1 meter $= \frac{1}{2.99792458 \times 10^8}$ s $= 3.33564 \times 10^{-9}$ s

1 ampere (A) $= 1.82637 \times 10^{-8}$ kg$^{1/2}$·s$^{-1/2}$

1 kelvin (K) $= 1.53616 \times 10^{-40}$ kg
(the Boltzmann constant is set equal to 1)

SI derived units:

1 newton (N) $= 1$ kg·m·s^{-2} $= 3.33564 \times 10^{-9}$ kg·s^{-1}

1 joule (J) $= 1$ N·m $= 1.11265 \times 10^{-17}$ kg

1 coulomb (C) $= 1$ A·s $= 1.82637 \times 10^{-8}$ kg$^{1/2}$·s$^{1/2}$

1 volt (V) $= 1$ J/C $= 6.09214 \times 10^{-10}$ kg$^{1/2}$·s$^{-1/2}$

1 V/m $= 0.182638$ kg$^{1/2}$·s$^{-3/2}$

1 tesla (T) ($= 10^4$ gauss) $= 1$ N/(A·m) $= 5.47534 \times 10^7$ kg$^{1/2}$·s$^{-3/2}$

In addition:

the charge of the electron:
$e = -1.602176485 \times 10^{-19}$ C $= -2.92617 \times 10^{-27}$ kg$^{1/2}$·s$^{1/2}$

1 eV $= 1.78266 \times 10^{-36}$ kg

the mass of the electron: $m_e = 9.10956 \times 10^{-31}$ kg

the mass of the proton $\approx 1838\, m_e$

the vacuum permeability: $\mu_0 = 4\pi \times 10^{-7}$ N/A^2 $= 4\pi = 376.7$ ohm (Ω)
(1 $\Omega = 1$ V/A; the value of the ampere given above is derived from this)

the vacuum permittivity: $\epsilon_0 = \frac{1}{\mu_0(2.99792458 \times 10^8)^2}$ C^2/(N·m^2) $= 1/4\pi$

And finally:

$\alpha = \frac{1}{137.0359997}$ is the fine structure constant
(its value is independent of the choice of units)

$\hbar \equiv e^2/\alpha$, or $\hbar = 1.2 \times 10^{-51}$ kg·s, is Planck's constant

$a_{\text{Bohr}} = e^2/m_e\alpha^2$, or $a_{\text{Bohr}} = 1.77 \times 10^{-19}$ s $= 5.29 \times 10^{-11}$ m, is the Bohr radius

Part I

Kinematics

Gentlemen! The concepts about time and space, which I would like to develop before you today, have grown on experimental physical grounds. Herein lies their strength. Their tendency is radical. Henceforth, space for itself, and time for itself shall completely reduce to a mere shadow, and only some sort of union of the two shall preserve independence.

Hermann Minkowski, *Raum und Zeit*. Lecture delivered before the Congress of Natural Philosophers, Cologne, 21 September 1908; English translation by Meghnad Saha, in: *The Principle of Relativity*, Calcutta, 1920

Book 2

1
Minkowski spacetime

Minkowski spacetime, the geometrical framework in which the laws of relativistic dynamics are formulated, is a very simple mathematical extension of three-dimensional Euclidean space. In this chapter we shall give a concise presentation of its main features. We shall stress the interpretation of the fourth dimension, which in special relativity is the time. Time now loses the 'universal' and 'absolute' nature that it had in the Newtonian theory.

1.1 An absolute spacetime

The two founding principles

In 1905 Albert Einstein created a new mechanics based on two postulates[1]: the *principle of relativity*, which states that the laws of physics must have the same form in any inertial reference frame (that is, in any frame where free particles undergo uniform rectilinear motion), and therefore they cannot be used to distinguish a particular frame; and the *principle of invariant light speed in a vacuum*, which states that light propagates at a velocity of constant modulus $|c|$ in *all* reference frames. (This postulate arose from the relation between the speed of light and the vacuum permeability and permittivity which follows from Maxwell's equations: $c = 1/\sqrt{\epsilon\mu}$.)

Therefore, the law governing the transformation from one reference frame to another cannot be the Galilean law, because the latter implies that the speed of light is *not* the same in two frames moving relative to each other. It states that if the speed of light is $|c|$ in $\mathcal{S}$, it must be $|c'| = |c - V_0|$ in a frame $\mathcal{S}'$ moving at speed V_0 relative to $\mathcal{S}$.

Now, the Galilean transformations follow directly from the structure of space and time assumed in Newtonian physics, and so rejecting Galilean transformations implies abandoning the idea of representing space by a Euclidean space $\mathcal{E}_3$, time by a universal parameter t, and spacetime by a $\mathcal{E}_3 \times R$.

In special relativity, 'relative, apparent, and common' (in the words of Newton) space *and* time are represented by a mathematical set of points p called *events*, which constitute the *Minkowski spacetime*: $\{p\} = \mathcal{M}_4$. $\mathcal{M}_4$ is postulated to be a *pseudo*-Euclidean four-dimensional space. This means that each event p is associated with a set of four real numbers, its *spacetime coordinates*, and that there exist, among the various possible labelings, *Minkowski* (or pseudo-Cartesian) coordinate systems (T, X, Y, Z), denoted more compactly as X^μ ($\mu = 0, 1, 2, 3$), such that the length element or *interval*, denoted ds^2, between two points with coordinates X^μ and $X^\mu + dX^\mu$ is defined by a generalized Pythagorean theorem:

---

[1]The English translation of Einstein's seminal paper, *Zur Elektrodynamik bewegter Körper* can be found in Einstein (1905a).

Relativity in Modern Physics. Nathalie Deruelle and Jean-Philippe Uzan.
© Oxford University Press 2018. Published in 2018 by Oxford University Press.
DOI: 10.1093/oso/9780198786399.001.0001

$$ds^2 = -dT^2 + dX^2 + dY^2 + dZ^2$$

$$= \sum_{\mu,\nu} \eta_{\mu\nu} \, dX^\mu dX^\nu = \eta_{\mu\nu} \, dX^\mu dX^\nu = dX_\nu \, dX^\nu. \tag{1.1}$$

The second equation defines the 10 coefficients $\eta_{\mu\nu}$ of the *Minkowski metric* in Minkowski coordinates: $\eta_{00} = -1$, $\eta_{0i} = 0$ $(i = 1, 2, 3)$, and $\eta_{ij} = \delta_{ij}$, where δ_{ij} is the Kronecker delta. In the third expression of (1.1) we have used the Einstein summation convention of summing over repeated indices, and in the fourth we define the operation of lowering an index: $dX_\nu \equiv \eta_{\mu\nu} \, dX^\nu$ (and so $dX_1 = dX$ but $dX_0 = -dT$). The origin O with coordinates $(0, 0, 0, 0)$ and the four axes $\{T, X, Y, Z\}$ constitute an orthonormal Minkowski frame[2] $\mathcal{S}$.

The interval can be positive, in which case the distance between p and $p + dp$ is $ds \equiv \sqrt{ds^2}$. If it is negative the distance is denoted as $d\tau \equiv \sqrt{-ds^2}$. Finally, it may vanish, in which case p and $p + dp$ have zero separation, but are nevertheless distinct because they have different coordinates.

Owing to the fact that its metric is specified *a priori*, the spacetime $\mathcal{M}_4$ of special relativity (as opposed to that of general relativity, where the metric is a dynamical quantity) can be regarded as an *absolute spacetime*.

Review of vector geometry

Minkowski spacetime is an affine space. The coordinates X^μ of the point p can therefore also be viewed as the components of the position vector $Op \equiv X^\mu e_\mu$, where O is the origin $(0, 0, 0, 0)$ and the four vectors e_μ form the basis, associated with the coordinates X^μ, of the vector space M_4 spanning $\mathcal{M}_4$. The ensemble of the origin O and the basis vectors e_μ constitutes a frame $\mathcal{S}$ of $\mathcal{M}_4$. Similarly, dX^μ represents a coordinate increment (not necessarily infinitesimally small) or, equivalently, the μth component of the vector $dp = dX^\mu e_\mu$. The quantity dX_μ is the μth component of the form $dX_\mu \epsilon^\mu$, where the four forms ϵ^μ form the basis associated with the e_μ of M_4^*, the dual space of M_4: $\epsilon^\mu(e_\nu) = \delta_\nu^\mu$, where δ_ν^μ is the *Kronecker delta* (equal to 1 if $\mu = \nu$ and 0 otherwise). Finally, (1.1) means that M_4 is equipped with a pseudo-Euclidean metric, that is, the bilinear symmetric form (or 2-fold covariant tensor)

$$\ell = \eta_{\mu\nu} \epsilon^\mu \otimes \epsilon^\nu. \tag{1.2}$$

We have discussed various concepts of vector geometry in Chapter 2 of Book 1. For example, we recall that the metric ℓ acts on the vector $dp = dX^\mu e_\mu$ to give ds^2 as

$$\ell(dp, dp) = dX^\mu dX^\nu \, \ell(e_\mu, e_\nu) = dX^\mu \, dX^\nu \, [\eta_{\rho\sigma} \, (\epsilon^\rho \otimes \epsilon^\sigma)](e_\mu, e_\nu)$$
$$= dX^\mu dX^\nu \, \eta_{\rho\sigma} \, [\epsilon^\rho(e_\mu)] \, [\epsilon^\sigma(e_\nu)] = dX^\mu dX^\nu \, \eta_{\rho\sigma} \, \delta_\mu^\rho \, \delta_\nu^\sigma = \eta_{\mu\nu} \, dX^\mu dX^\nu = ds^2.$$

The first equality expresses the linearity of ℓ, the second defines it, the third defines the tensor product, the fourth defines a dual basis, and the fifth uses the properties of the Kronecker delta and the definition of the Minkowski metric.

The metric also defines the scalar product of two different vectors $u = u^\mu e_\mu$ and $v = v^\nu e_\nu$ as

$$(u \cdot v) \equiv \ell(u, v) = \eta_{\mu\nu} u^\mu v^\nu \equiv u_\mu v^\mu, \tag{1.3}$$

where $u_\mu \equiv \eta_{\mu\nu} u^\nu$ are the components of the form $u_\mu \epsilon^\mu$ dual to the vector u.

---

[2]The signature chosen for the Minkowski metric is therefore $(-1, +1, +1, +1)$.

From now on, Greek indices run from 0 to 3 and Latin indices from 1 to 3 (or to n if the spatial dimension is not specified).

We also recall that for any spatial dimension n, a tensor T which is p-fold contravariant and q-fold covariant is a multilinear form which acts on p forms and q vectors to give a number:

$$T = T^{i_1 \cdots i_p}_{j_i \cdots j_q} e_{i_1} \otimes e_{i_2} \cdots \otimes e_{i_p} \otimes \epsilon^{j_1} \otimes \cdots \epsilon^{j_q}, \tag{1.4}$$

where $T^{i_1 \cdots i_p}_{j_i \cdots j_q}$ are the components of this tensor in the basis $e_{i_1} \otimes \cdots e_{i_p} \otimes \epsilon^{j_1} \cdots \otimes \epsilon^{j_q}$, that is, the result of the action of T on p forms of the basis ϵ^i and q vectors of the basis e_j. (For a more detailed discussion see Sections 2.1 and 2.3 of Book 1.)

Let us conclude with a few definitions, the physical interpretation of which will be discussed in the following section.

The sections $T = \text{const}$ of Minkowski spacetime are Euclidean spaces $\mathcal{E}_3$ and the Cartesian coordinates $X^i = (X, Y, Z)$ are referred to as 'spatial' coordinates.[3]

The quantity $T \equiv X^0$ is called the *time coordinate*.

The ensemble of events p at null distance from, for example, the origin O, that is, such that $\ell(Op, Op) = 0$, forms a cone of equation $-T^2 + X^2 + Y^2 + Z^2 = 0$, called the *light cone* (see Fig. 1.1).

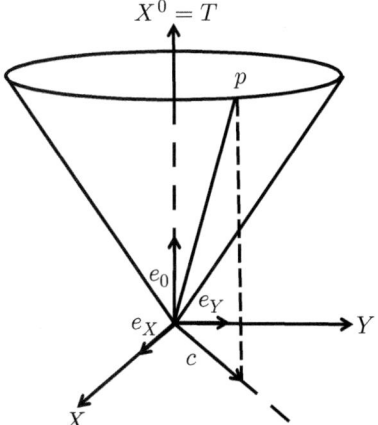

Fig. 1.1 Minkowski spacetime.

If $\ell(Op, Op) > 0$, the interval between O and p is *space-like*; if $\ell(Op, Op) < 0$, it is *time-like*. If $\ell(Op, Op) = 0$, it is *light-like* and Op is an isotropic vector or *null vector*, not to be confused with a vector all of whose components are zero; we also use c to denote the unit 3-vector of its projection on the plane $T = 0$. The ensemble of events p (said to lie 'inside

---

[3]When there is no possibility of confusion, $\mathcal{S}$ will designate either a Minkowski frame of $\mathcal{M}_4$ or a three-dimensional Cartesian frame of $\mathcal{E}_3$, the ensemble consisting of the spatial origin and the three basis vectors e_i.

We can of course perform any change of coordinate in these spatial sections; for example, we can transform to spherical coordinates, where the length element is written as

$$ds^2 = -dT^2 + dr^2 + r^2(d\theta^2 + \sin^2 \theta \, d\phi^2).$$

(See the discussion of differential geometry in Chapters 3 and 4 of Book 1.)

the light cone') for which the distance to O is time-like and $T > 0$, represents the *future* of O (or the *past* if $T < 0$).

1.2   Inertial reference frames

Since a section $X^0 = \text{const}$ of Minkowski spacetime is a Euclidean space $\mathcal{E}_3$, a Cartesian frame of $\mathcal{E}_3$ can be realized as in Newtonian physics by choosing an oriented physical trihedron (for example, the laboratory walls), constructed using the ordinary Pythagoras theorem and its consequences. The three coordinates X^i, the 'spatial' coordinates, will thus specify the location of the event p. The 'time' coordinate $X^0 \equiv T$ will be realized by the time given by accurate clocks located at $X^i = \text{const}$, that is, clocks which are *at rest* relative to the grid of spatial coordinates (this restriction is essential!).

In addition, for the realization of a Minkowski frame we require that the physical reference frame $\mathcal{S}$ thus constructed, the ensemble of a Cartesian trihedron and clocks at rest, be *inertial*, that is, that the motion of free particles measured using the time T of its clocks be uniform and rectilinear. If this were not the case, it would indicate, absent proof to the contrary, that the particles are in fact subject to forces, or that the clocks are of poor quality or are moving relative to each other.

Finally, the (future) light cone issuing from an event p, that is, the 2-sphere issuing from the site X^i at time T and moving away at speed c (of modulus 1), is realized by a light flash (or a thin spherical shell of particles moving radially at speed c).

"c=1"

At first sight it might appear strange that the four basis vectors e_μ are realized in 'relative, apparent, and common' space by the ticking of a clock plus three orthogonal unit rulers. However, in the theory of relativity the concepts of length, of ruler, and, in general, of rigid body are secondary. For example, spatial localization is effected in practice not by using 'rulers', but by triangulation using laser telemeters which are at rest with respect to each other (if they are at rest, the spatial geometry obtained (when gravity is ignored!) must be Euclidean). Therefore, only the time unit needs to be defined, since the length unit is derived from it by multiplication by a universal constant c.

Numerically, this fundamental constant is equal to the speed of light in the vacuum. For this reason, light plays a special role in special relativity, which is why there is an entire terminology based on it. One speaks of, for example, 'light signals' instead of 'particles traveling at speed c' and so on. However, it should be emphasized that the theory is *not* based on these identifications: the fundamental speed c could just as well be realized by objects other than light, or by nothing at all.[4] It can always be set equal to 1, in which case lengths are expressed in, for example, seconds.

It is useful to note that such an inertial frame is not *a priori* unique, and that we never need to choose the direction of the reference trihedron axes, nor are we ever led to single out any particular one of them.[5]

In addition, since it must not depend on the coordinates chosen, the value of the interval ds^2 will be zero in any Minkowski frame if it is zero in one. Therefore, a light signal must

---

[4]See, for example, Ellis and Uzan (2005).

[5]The idea of 'absolute' rest, inherent in the representation of space and time in Newtonian physics, is therefore unnecessary in relativity theory. It is equally unnecessary to introduce an *aether* which would give physical reality to all the points of $\mathcal{M}_4$.

propagate with speed c of modulus 1 in *all* inertial frames, in contrast to the Newtonian prediction.

Minkowski spacetime is therefore a suitable framework for incorporating the two Einstein postulates stated above.

The Michelson–Morley experiment

In a reference frame attached to the Earth and regarded as inertial, the speed of light will have modulus 1, $|c| = 1$, as in any other frame. The times required for two light signals to make a round-trip along the arms of length l of an interferometer, one parallel and one perpendicular to the orbital velocity of the Earth, will be the same: $2l/|c|$. They will be the same after the apparatus is rotated. The path difference between the two signals will then be the same in the two configurations, and so no shift of the interference pattern is expected, in agreement with experiment (1881), and in contrast to the predictions of A. A. Michelson and his contemporaries. No FitzGerald–Lorentz-type shortening of the arms (see Section 17.5 of Book 1) will be required to explain this result, which instead is interpreted as a direct consequence of the Einstein postulates.

Another difference between Einstein's physics and Newton's physics is that in the latter, *all* (good) clocks, no matter what their motion, are assumed to measure the same time T. In special relativity nothing *a priori* requires that the time measured by a moving clock coincides with the time T measured by the clocks of $\mathcal{S}$. There is no longer any reason for the concepts of "now" and of simultaneity to be universal; for example, it can no longer be asserted that the duration of a trip measured by the traveler will be the same as that measured by the person who stays at home, which was a straightforward statement in Newtonian physics.

1.3 Lorentz transformations

A change of the labeling $X^\mu \to X'^\mu$ of the points p of $\mathcal{M}_4$ which preserves the form of the interval $(ds^2 = \eta_{\mu\nu}\, dX^\mu dX^\nu = \eta_{\mu\nu}\, dX'^\mu dX'^\nu)$ and therefore describes the passage from one system of Minkowski coordinates $\mathcal{S}$ to another $\mathcal{S}'$ is called a *Poincaré transformation*. Such a transformation is written as

$$X'^\nu = \Lambda_\mu{}^\nu (X^\mu - d^\mu) \qquad \text{with} \qquad \Lambda_\rho{}^\mu \Lambda_\sigma{}^\nu \eta_{\mu\nu} = \eta_{\rho\sigma}, \tag{1.5}$$

where d^μ and $\Lambda_\mu{}^\nu$ are independent of X^μ. This change of coordinates is accompanied by a change of the origin of the affine space, $O \mapsto O'$ with $OO' = d^\mu e_\mu$, as well as a change of basis of the vector space M_4, $e_\mu = \Lambda_\mu{}^\nu e'_\nu$, and its dual $\epsilon^\nu = \Lambda^\nu{}_\mu \epsilon'^\mu$, where $\Lambda^\nu{}_\mu$ is the inverse matrix of $\Lambda_\mu{}^\nu$, such that[6] $\Lambda_\mu{}^\nu \Lambda^\mu{}_\rho = \delta^\nu_\rho$.

These transformations are the Minkowski generalization of changes of Cartesian frame in Euclidean geometry (see Book 1, Sections 1.3 and 2.4). They depend on 10 parameters:

---

[6] We recall (see Book 1, Section 2.3) that in a linear transformation the components of a p-fold contravariant and q-fold covariant tensor become

$$T'^{i_1\cdots i_p}_{j_i\cdots j_q} = T^{k_1\cdots k_p}_{l_i\cdots l_q} \Lambda_{k_1}{}^{i_1} \cdots \Lambda_{k_p}{}^{i_p} \Lambda^{l_1}{}_{j_1} \cdots \Lambda^{l_q}{}_{j_q}.$$

The components $\eta_{\mu\nu}$ of the Minkowski metric, the Kronecker delta δ^μ_ν, and the Levi-Civita symbol $e_{\mu\nu\rho\sigma}$ (the properties of which are given below in Section 7.2) are three quantities which are invariant under restricted Lorentz transformations.

the four components of the translation vector d^μ and the six independent components of the 'pseudo'-rotation matrix $\Lambda_\mu{}^\nu$, which actually has $4 \times 4 = 16$ components constrained by the 10 relations in (1.5). Three parameters, for example, the Euler angles, describe spatial rotations, while the other three describe 'pseudo'-rotations or *boosts*.

If we restrict ourselves to rotations and pseudo-rotations ($d^\mu = 0$), the transformations (1.5) are referred to as *Lorentz transformations*. If $\det \Lambda = +1$ the transformation is termed *proper*, and if $\Lambda_0{}^0 > 1$ it is termed *orthochronous*. Transformations satisfying both these conditions are called *restricted* Lorentz transformations. It is easy to show (see also Section 1.5) that they form a non-commutative group, the *Lorentz group*.

Let us consider the particular Lorentz transformation called a *Lorentz boost*:

$$T' = T \cosh \psi - X \sinh \psi, \quad X' = -T \sinh \psi + X \cosh \psi, \quad Y' = Y, \quad Z' = Z, \tag{1.6}$$

where ψ is a constant, sometimes called the *rapidity*. We then have

$$ds^2 = -dT^2 + dX^2 + dY^2 + dZ^2 = -dT'^2 + dX'^2 + dY'^2 + dZ'^2,$$

with T' a time coordinate and X'^i three spatial coordinates.

In order for the two frames $\mathcal{S}$ and $\mathcal{S}'$ to remain equivalent, the three Cartesian coordinates X'^i must specify the location of the event p. As far as the coordinate T' is concerned, it must represent the time as measured by clocks at rest in the new reference frame. Therefore, $X' = 0 \iff X = T \tanh \psi$ describes the motion in $\mathcal{S}$ of the spatial origin of $\mathcal{S}'$, and the constant $\tanh \psi \equiv V_0 \in \,]-1, +1[$ is interpreted as the speed of the frame $\mathcal{S}'$ measured at the time T of $\mathcal{S}$: $\mathcal{S}'$ is undergoing uniform translation relative to $\mathcal{S}$ along the X axis.

As a function of V_0, the transformation (1.6) takes the form

$$T' = \frac{T - V_0 X}{\sqrt{1 - V_0^2}}, \quad X' = \frac{X - V_0 T}{\sqrt{1 - V_0^2}} \iff T = \frac{T' + V_0 X'}{\sqrt{1 - V_0^2}}, \quad X = \frac{X' + V_0 T'}{\sqrt{1 - V_0^2}}. \tag{1.7}$$

(To simplify the notation, it is common to introduce the coefficient $\Gamma \equiv 1/\sqrt{1 - V_0^2}$, called the *Lorentz factor*.)

In Newtonian physics it was necessary to distinguish between changes of Cartesian frame of $\mathcal{E}_3$ and the law for passing from one frame to another undergoing uniform translation relative to the first. In special relativity these ideas merge together, because time is now a coordinate. We therefore use the term *inertial frame* to describe the ensemble of Minkowski frames related to each other by Poincaré transformations; it is in these frames that the motion of a free particle must be represented as a straight line.

We note that the law (1.7) reduces to the Galilean transformation ($T' = T$, $X' = X - V_0 T$) when $V_0 \ll 1$ and if we restrict ourselves to events deep inside the light cone issuing from the origin O such that $X = \mathcal{O}(V_0 T)$.

We also note (see Fig. 1.2) that the new basis vectors ($e'_0 = \cosh \psi\, e_0 + \sinh \psi\, e_1$ and $e'_1 = \sinh \psi\, e_0 + \cosh \psi\, e_1$) are orthogonal in the sense of the Minkowski metric $[\ell(e'_0, e'_1) = 0]$, but on (Euclidean) paper they make an acute angle and merge with the bisectrix when[7] $V_0 \to 1$.

---

[7]The apparent absence of equivalence between the (T, X) and (T', X') axes is due to a Euclidean 'cultural bias'. In some studies (mostly older ones), an *imaginary time* $\mathcal{T} = iT$ is introduced in order to cast the Minkowski length element in an apparent Euclidean form: $ds^2 = d\mathcal{T}^2 + dX^2 + dY^2 + dZ^2$.

Here we also recall that we can always use curvilinear coordinates to describe the spatial sections of inertial frames; *cf.* footnote 3 of this chapter. We shall reserve the term 'Minkowski frame' for frames whose spatial coordinates are Cartesian.

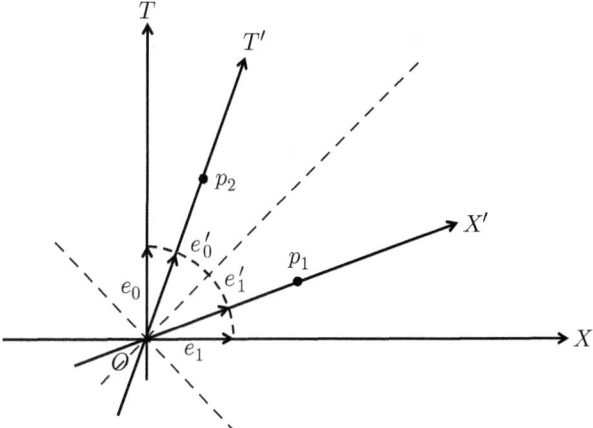

Fig. 1.2 A Lorentz boost.

Lastly, we note that light cones are invariant under Lorentz transformations, as they must be, owing to the fact that a null interval in one frame is a null interval in any other.

We can realize a Lorentz boost by introducing a new physical trihedron undergoing uniform translation with respect to the first one. The location of the event p is then specified by the three spatial coordinates X'^i of the new reference frame. As far as the new time coordinate T' is concerned, it represents the time measured by clocks which are *at rest* in the new frame, that is, which are undergoing uniform rectilinear motion relative to the first frame. Then, according to (1.7) the time indicated by a clock at rest will not be the same as the time indicated by an identical clock which is moving.

The fact that T' measures the time in $\mathcal{S}'$ and T the time in $\mathcal{S}$ also shows that the idea of simultaneity is not universal. For example, as seen from Fig. 1.2, two events O and p_1 which are simultaneous in the frame $\mathcal{S}'$ are not so when measured by the time T of $\mathcal{S}$. This is surprising when one is used to the idea of an 'absolute' time. If two events are separated by a space-like interval, there exists a reference frame in which they are simultaneous, while if they are separated by a time-like interval, there exists a frame in which they occur at the same place.

Therefore, physical space and time are viewed as a single 'block' which does not involve any 'flow' of time and which each inertial frame 'slices' according to its proper time axis and its associated space. Singling out a particular time axis is no more justified than singling out a particular spatial axis, for example, the X axis.[8]

We conclude with an important remark. While they do unify the spatial rotation of axes and passage from one inertial frame to another, Lorentz transformations do *not* encompass the passage from an inertial frame to an accelerated frame, that is, they do not provide a relativistic generalization of the Newtonian group of rigid displacements. We shall return to the discussion of accelerated frames in special relativity in Chapter 5 below.

---

[8]A heuristic introduction to Lorentz transformations can be found in, for example, Lévy-Leblond (1976) and (1979), as well as Langlois (2011).

1.4   Time dilation

Let us consider two events O and A occurring at the same place but separated by a time interval $\Delta \tau$ in an inertial frame S (see Fig. 1.3a). By direct application of (1.7) we find that in S' these two events are separated by a time interval

$$\Delta T' = \frac{\Delta \tau}{\sqrt{1 - V_0^2}}. \tag{1.8}$$

This is the phenomenon of *time dilation*: the events O and A might, for example, represent the ticking of a clock in S and $\Delta \tau$ might define the second (or the lifetime of a particle); this clock moves at speed $(-V_0)$ in S' and runs slowly relative to the clocks at rest in S' which measure the time T'. Therefore, a phenomenon of duration $\Delta \tau$ occurring at a given site in an inertial frame S will last a time $\Delta T'$ in S' and, according to (1.7), will unfold between two different sites separated by a distance $\Delta X' = -V_0 \Delta \tau / \sqrt{1 - V_0^2}$ (it can be checked that indeed $-\Delta T'^2 + \Delta X'^2 = -\Delta \tau^2$).

It also follows directly from (1.7) that a clock of S' will be slowed down in a symmetric manner relative to a clock of S: $\Delta T = \Delta \tau / \sqrt{1 - V_0^2}$, where now $\Delta \tau$ measures, for example, the second in S' (see Fig. 1.3).

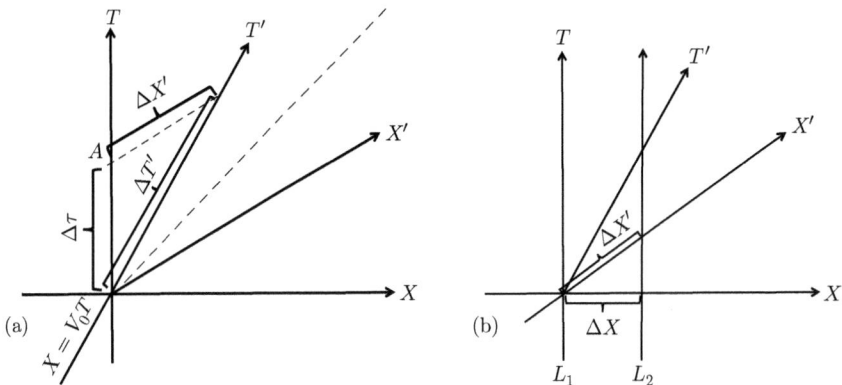

Fig. 1.3 Time dilation and length contraction.

The muon lifetime (I)

Muons are unstable particles which decay on the average in a time $\Delta \tau = 2.1948$ μs in their rest frame S. In Newtonian physics, the clocks of any reference frame must measure the same lifetime and, even in a frame in which the muons move at nearly the speed of light, they should in principle not be able to travel farther than $c \Delta \tau \approx 650$ m.

In 1941 B. Rossi and D. B. Hall, in measuring the cosmic muon flux at various altitudes, showed that the muon path length was much larger than 650 m, so that the muon lifetime $\Delta T'$ measured in the reference frame S' of the Earth had to be much greater than 2.2 μs. The experiment was performed again in a quantitative fashion in 1963 by D. H. Frisch and J. H. Smith, who measured a lifetime $\Delta T' \approx 8.8 \Delta \tau$ for muons moving with Lorentz factor $1/\sqrt{1 - V_0^2} \approx 8.8$, in agreement with the relativistic prediction (to within about 10%); see also Section 2.2 below.

Length contraction

Let us imagine a *rigid ruler* which is at rest in $\mathcal{S}$, that is, an object for which the world lines of its extremities are represented by the lines L_1 and L_2 which in $\mathcal{S}$ are parallel to the T axis; see Fig. 1.3b. The *proper length* is the distance ΔX between L_1 and L_2 at constant T. In $\mathcal{S}$ the lines L_1 and L_2 have the equations $X = 0$ and $X = \Delta X$, $\forall\, T$. In $\mathcal{S}'$, where the ruler is undergoing uniform translation at velocity $-V_0$, the equations of motion of the extremities become [cf. (1.7)]

$$
\begin{cases}
T' = \dfrac{T}{\sqrt{1 - V_0^2}}, & X' = -\dfrac{V_0 T}{\sqrt{1 - V_0^2}} \quad \text{or} \quad X' = -V_0 T' \ \text{ for } L_1 \\[4mm]
T' = \dfrac{T - V_0 \Delta X}{\sqrt{1 - V_0^2}}, & X' = \dfrac{\Delta X - V_0 T}{\sqrt{1 - V_0^2}} \quad \text{or} \quad X' = -V_0 T' + \Delta X \sqrt{1 - V_0^2} \ \text{ for } L_2,
\end{cases}
\tag{1.9}
$$

and so at a given time T' measured by the clocks of $\mathcal{S}'$

$$
\Delta X' = \Delta X \sqrt{1 - V_0^2},
\tag{1.10}
$$

which gives rise to the (imprecise) saying 'a moving ruler is shorter than a ruler at rest'.[9]

1.5 Thomas rotation

As an example of the properties of the Lorentz group,[10] here we present two different methods of obtaining the law for the transformation from one Minkowski frame $\mathcal{S}$ to another $\mathcal{S}'$ when the velocity V_0 of $\mathcal{S}'$ relative to $\mathcal{S}$ lies in the XOY plane and makes an angle ϕ with OX. In the first method we (1) go from $\mathcal{S}$ to a frame $\mathcal{S}^{(1)}$ by rotating the X and Y axes by an angle ϕ such that the $X^{(1)}$ axis becomes collinear with V_0. Then we (2) go from $\mathcal{S}^{(1)}$ to $\mathcal{S}^{(2)}$ by performing a Lorentz boost of velocity V_0 along $X^{(1)}$. Finally, we (3) go from $\mathcal{S}^{(2)}$ to the final frame $\mathcal{S}'$ by rotating the $X^{(2)}$ and $Y^{(2)}$ axes by an angle $(-\phi)$ so that the X' and Y' axes are parallel to the X and Y axes of the initial frame $\mathcal{S}$. The sequence of transformations is (setting $V_0 = \tanh \psi$)

$$
\begin{cases}
T^{(1)} = T, & X^{(1)} = X \cos\phi + Y \sin\phi, \quad Y^{(1)} = -X \sin\phi + Y \cos\phi, \\[2mm]
T^{(2)} = T^{(1)} \cosh\psi - X^{(1)} \sinh\psi, & X^{(2)} = -T^{(1)} \sinh\psi + X^{(1)} \cosh\psi, \quad Y^{(2)} = Y^{(1)}, \\[2mm]
T' = T^{(2)}, & X' = X^{(2)} \cos\phi - Y^{(2)} \sin\phi, \quad Y' = X^{(2)} \sin\phi + Y^{(2)} \cos\phi,
\end{cases}
\tag{1.11}
$$

and their composition gives

---

[9]The phenomenon of length contraction is conceptually less important than that of time dilation because the idea of a rigid body is superfluous in relativity. Moreover, although one can devise many amusing undergraduate exercises, it has in fact never been observed directly.

[10]A more mathematical and detailed presentation of the properties of the Lorentz group can be found in, for example, Gourgoulhon (2013).

$$\begin{cases} T' = T \cosh\psi - X \sinh\psi \cos\phi - Y \sinh\psi \sin\phi \,, \\[4pt] X' = -T \sinh\psi \cos\phi + X[1 + \cos^2\phi(\cosh\psi - 1)] + Y \sin\phi \cos\phi(\cosh\psi - 1) \,, \qquad (1.12) \\[4pt] Y' = -T \sinh\psi \sin\phi + X \sin\phi \cos\phi(\cosh\psi - 1) + Y[1 + \sin^2\phi(\cosh\psi - 1)] \,. \end{cases}$$

In the second method of going from the frame $\mathcal{S}$ to the frame $\mathcal{S}'$ we (1) go from $\mathcal{S}$ to a frame $\mathcal{S}^{(a)}$ by a Lorentz boost of velocity V_a along the X axis (V_a is the speed measured at time T of $\mathcal{S}$). Then we (2) go from $\mathcal{S}^{(a)}$ to a frame $\mathcal{S}^{(b)}$ by a second Lorentz boost, this time of velocity V_b along the $Y^{(a)}$ axis, where V_b is the speed measured at time $T^{(a)}$. Finally, since the spatial axes of $\mathcal{S}^{(b)}$ are *not* parallel to those of $\mathcal{S}$, we (3) go from $\mathcal{S}^{(b)}$ to the final frame $\mathcal{S}'$ by rotating the $X^{(b)}$ and $Y^{(b)}$ axes by an angle θ so that the X' and Y' axes are collinear with the X and Y axes of the initial frame $\mathcal{S}$. The transformations are (setting $V_a = \tanh\psi_a$ and $V_b = \tanh\psi_b$)

$$\begin{cases} T^{(a)} = T \cosh\psi_a - X \sinh\psi_a \,, \quad X^{(a)} = -T \sinh\psi_a + X \cosh\psi_a \,, \quad Y^{(a)} = Y \,, \\[4pt] T^{(b)} = T^{(a)} \cosh\psi_b - Y^{(a)} \sinh\psi_b \,, \quad X^{(b)} = X^{(a)} \,, \\[4pt] \qquad\qquad\qquad\qquad Y^{(b)} = -T^{(a)} \sinh\psi_b + Y^{(a)} \cosh\psi_b \,, \qquad\qquad (1.13) \\[4pt] T' = T^{(b)} \,, \quad X' = X^{(b)} \cos\theta - Y^{(b)} \sin\theta \,, \quad Y' = X^{(b)} \sin\theta + Y^{(b)} \cos\theta \,, \end{cases}$$

and their composition gives

$$\begin{cases} T' = T \cosh\psi_a \cosh\psi_b - X \sinh\psi_a \cosh\psi_b - Y \sinh\psi_b \,, \\[4pt] X' = -T[\sinh\psi_a \cos\theta - \cosh\psi_a \sinh\psi_b \sin\theta] \\[4pt] \qquad + X[\cosh\psi_a \cos\theta - \sinh\psi_a \sinh\psi_b \sin\theta] - Y \cosh\psi_b \sin\theta \,, \\[4pt] Y' = -T[\sinh\psi_a \sin\theta + \cosh\psi_a \sinh\psi_b \cos\theta] \\[4pt] \qquad + X[\cosh\psi_a \sin\theta + \sinh\psi_a \sinh\psi_b \cos\theta] + Y \cosh\psi_b \cos\theta \,. \end{cases}$$

This is identical to (1.12) if

$$\begin{cases} \cosh\psi = \cosh\psi_a \cosh\psi_b \,, \\[6pt] \sin\phi = \dfrac{\sinh\psi_b}{\sqrt{\cosh^2\psi_a \cosh^2\psi_b - 1}} \,, \quad \sin\theta = -\dfrac{\sinh\psi_a \sinh\psi_b}{\cosh\psi_a \cosh\psi_b + 1} \,, \end{cases} \qquad (1.14)$$

which can also be written as

$$\begin{cases} V_0^2 = V_a^2 + V_b^2 - V_a^2 V_b^2 \,, \\[6pt] \sin\phi = \dfrac{V_b\sqrt{1 - V_a^2}}{V_0} \,, \quad \sin\theta = -\dfrac{V_a V_b}{1 + \sqrt{1 - V_0^2}} \,. \end{cases} \qquad (1.15)$$

Equations (1.14) and (1.15) give, as functions of V_a and V_b, the modulus V_0 of the velocity of $\mathcal{S}'$ relative to $\mathcal{S}$, its angle ϕ with the X axis, and also the angle θ of the *Thomas rotation*. The fact that this angle is nonzero (except in the Galilean limit) shows that the composition of two

Lorentz boosts in non-collinear directions is identical to a single Lorentz boost followed by a rotation of the spatial axes. We also see that the expression for the angle ϕ is not symmetric in V_a and V_b, and therefore depends on the order in which the successive translations of the frames are done.

Let us conclude with the simpler case of the composition of two collinear Lorentz boosts, the first of velocity $V_1 \equiv \tanh \psi_1$ measured at time T of $\mathcal{S}$ and the second of velocity $V_2 \equiv \tanh \psi_2$ measured at time $T^{(1)}$ of the intermediate frame. An elementary calculation shows that this composition of transformations will be a Lorentz boost of velocity $V_0 \equiv \tanh \psi$ given by

$$V_0 = \frac{V_1 + V_2}{1 + V_1 V_2} \quad \text{or} \quad \psi = \psi_1 + \psi_2. \tag{1.16}$$

In this case the rapidities, but not the velocities, add.

2

The kinematics of a point particle

In the preceding chapter we considered only objects undergoing uniform translation relative to each other. Here we shall discuss the kinematics of point particles undergoing any type of motion. We shall introduce the concept of *proper time*, the geometric representation of the time measured by an accelerated clock.

2.1 World lines

A *world line* is a curve L in $\mathcal{M}_4$, $\lambda \mapsto p(\lambda)$, whose equations in a Minkowski frame $\mathcal{S}$ with coordinates X^μ are $X^\mu = X^\mu(\lambda)$. Its tangent at p is the vector $dp/d\lambda$ with components $dX^\mu/d\lambda$ in $\mathcal{S}$. A world line represents the motion of a *material point* or *point particle P*, that is, an object whose spatial extent and internal structure can be ignored. More precisely, each event $p(\lambda) \in L$ represents the location X^i of the object at the time T of an inertial frame $\mathcal{S}$.

If this world line is time-like everywhere,[1] that is, if (see Section 1.1) $\ell(dp, dp) = \eta_{\mu\nu}dX^\mu dX^\nu = ds^2 < 0$ at any point p of L, we can choose the parameter λ to be the curvilinear abscissa τ of L, $d\tau = \sqrt{-ds^2}$. The tangent $u \equiv dp/d\tau$ with components $U^\mu = dX^\mu/d\tau$ in $\mathcal{S}$ is then called the *4-velocity*, to distinguish it from the *3-velocity* V with components[2] $V^i \equiv dX^i/dT$.

Since $\ell(u, u) = \eta_{\mu\nu}U^\mu U^\nu = -1$ by definition of the curvilinear abscissa, we have $U^0 = \sqrt{1 + U^i U_i}$ (choosing U^0 to be positive so that τ increases with T), and so

$$V^i \equiv \frac{dX^i}{dT} = \frac{U^i}{U^0} = \frac{U^i}{\sqrt{1 + U^j U_j}} \qquad \Longrightarrow \qquad V^2 = \frac{U^i U_i}{1 + U^j U_j} < 1, \qquad (2.1)$$

where $V^2 = V.V = |V|^2 = \delta_{ij}V^i V^j = V_i V^i$. Therefore, the maximum 3-velocity that a (non-tachyonic) particle can have is equal to the speed of light. Inversely,

$$U^0 \equiv \frac{dX^0}{d\tau} = \frac{1}{\sqrt{1 - V^2}}, \quad U^i \equiv \frac{dX^i}{d\tau} = \frac{V^i}{\sqrt{1 - V^2}}. \qquad (2.2)$$

The *4-acceleration* of P is the vector $\gamma = du/d\tau$ with components $\gamma^\mu \equiv dU^\mu/d\tau$ in $\mathcal{S}$. In terms of the 3-velocity V^i and the 3-acceleration $a^i \equiv dV^i/dT$ we have

---

[1]World lines which are not time-like everywhere describe *tachyons*, hypothetical and problematical particles which we do not consider here.

[2]In what follows, when there is no possibility of confusion, the term *velocity* will be used for both 4- and 3-velocity. It should of course always be borne in mind that $u = U^\mu e_\mu$ is a vector of M_4 while $V = V^i e_i$ is a vector of E_3. Likewise, the Minkowski scalar product defined in (1.3) $(u \cdot \gamma) \equiv \ell(u, \gamma) = \eta_{\mu\nu}U^\mu\gamma^\nu$ should not be confused with the Euclidean scalar product $a.V \equiv e(a, V) = \delta_{ij}\,a^i V^j$. Finally, it will sometimes be necessary to distinguish between world lines $X^\mu = X^\mu(\tau)$ and *trajectories* $X^i = X^i(T)$.

Relativity in Modern Physics. Nathalie Deruelle and Jean-Philippe Uzan.
© Oxford University Press 2018. Published in 2018 by Oxford University Press.
DOI: 10.1093/oso/9780198786399.001.0001

$$\gamma^0 = \frac{a \cdot V}{(1 - V^2)^2}, \quad \gamma^i = \frac{1}{1 - V^2}\left(a^i + V^i \frac{a \cdot V}{1 - V^2}\right), \tag{2.3}$$

where $a \cdot V \equiv \delta_{ij} a^i V^j \equiv a_i V^i$. Since $\eta_{\mu\nu} U^\mu U^\nu = -1$ implies that $\frac{d}{d\tau}(\eta_{\mu\nu} U^\mu U^\nu) = 0$, the 4-velocity and acceleration are orthogonal, that is, in the various notations we are using,

$$(u \cdot \gamma) \equiv \ell(u, \gamma) \equiv \eta_{\mu\nu} U^\mu \gamma^\nu \equiv U_\mu \gamma^\mu = 0,$$
$$V \cdot \gamma \equiv e(V, \gamma) \equiv \delta_{ij} V^i \gamma^j \equiv V_i \gamma^i = \gamma^0. \tag{2.4}$$

Uniformly accelerated rectilinear motion

Uniformly accelerated rectilinear motion is defined by $\gamma_\mu \gamma^\mu = g^2$, where g is a constant and $\gamma^\mu = dU^\mu/d\tau = d^2 X^\mu/d\tau^2$ are the components of the 4-acceleration of the particle in an inertial frame $\mathcal{S}$ and $X^\mu(\tau)$ are its 4 coordinates. Choosing the spatial axis X to be parallel to the motion, we can write the 4-velocity U^μ as $U^0 = \cosh f(\tau)$, $U^1 = \sinh f(\tau)$, so that $U_\mu U^\mu = -1$. The condition $-(dU^0/d\tau)^2 + (dU^1/d\tau)^2 = g^2$ then requires that $f = g\tau$. Then, after choosing the initial conditions and with h a constant, we have

$$\begin{cases} \gamma^\mu \equiv \dfrac{dU^\mu}{d\tau} = g(\sinh g\tau, \ \cosh g\tau); \quad U^\mu \equiv \dfrac{dX^\mu}{d\tau} = (\cosh g\tau, \ \sinh g\tau) \\[2mm] \qquad\qquad T = \dfrac{1}{g}\sinh g\tau, \quad X = \dfrac{1}{g}\cosh g\tau + h. \end{cases} \tag{2.5}$$

The trajectory is a branch of the hyperbola $(X - h)^2 - T^2 = 1/g^2$.

The 3-velocity is $V \equiv dX/dT = \tanh g\tau = gT/\sqrt{1 + g^2 T^2}$ and tends to ± 1, that is, to the speed of light for T and $\tau \to \pm\infty$. The 3-acceleration is

$$a \equiv \frac{d^2 X}{dT^2} = \frac{g}{\cosh^3 g\tau} = \frac{g}{(1 + g^2 T^2)^{3/2}}.$$

In the limit of small velocities ($g\tau \ll 1$) $a \sim g$ and we recover the Newtonian parabola: $T \sim \tau$, $v \sim gT$, $X - h \sim 1/g + \frac{1}{2} gT^2$.

Uniform circular motion

Let us consider a particle P constrained to follow uniform circular motion

$$X^\mu(T) = (T, \ r_0 \cos\Omega T, \ r_0 \sin\Omega T, \ 0)$$

in a Minkowski frame $\mathcal{S}$. Its 3-velocity and 3-acceleration

$$V^i \equiv \frac{dX^i}{dT} = r_0 \Omega(-\sin\Omega T, \cos\Omega T, 0), \quad a^i \equiv \frac{d^2 X^i}{dT^2} = -r_0 \Omega^2(\cos\Omega T, \sin\Omega T, 0) \tag{2.6}$$

are orthogonal, $a \cdot V = 0$, and have constant modulus $|V| = r_0 \Omega$ and $|a| = r_0 \Omega^2$.

The components of its 4-velocity and 4-acceleration are given by (2.2) and (2.3):

$$U^\mu \equiv \frac{dX^\mu}{d\tau} = \frac{(1, -r_0 \Omega \sin\Omega T, r_0 \Omega \cos\Omega T, 0)}{\sqrt{1 - r_0^2 \Omega^2}}, \quad \gamma^\mu \equiv \frac{d^2 X^\mu}{d\tau^2} = -\frac{r_0 \Omega^2(0, \cos\Omega T, \sin\Omega T, 0)}{1 - r_0^2 \Omega^2}. \tag{2.7}$$

Here γ has constant modulus $\sqrt{\gamma_\mu \gamma^\mu} = r_0 \Omega^2 / (1 - r_0^2 \Omega^2)$. The quantity $U^0 = dT/d\tau$ relates the curvilinear abscissa τ to the time T of $\mathcal{S}$ as $\tau = \sqrt{1 - V^2}\, T$. In order for it to be defined, that is, in order for $U^\mu U_\mu = -1$ and for the world line to be time-like, we must have $V^2 < 1$, that is, $r_0 \Omega$ must remain smaller than 1, so that for a given Ω the particle cannot orbit arbitrarily far from the origin. When $|V| \to 1$, that is, when $r_0 \Omega \to 1$, the length of the world line $\tau = \sqrt{1 - V^2}\, T$ tends to zero: $\tau \to 0$ for finite T.

2.2   Proper time

Let us now consider the interpretation of the curvilinear abscissa τ. By definition, τ measures the length of the world line L representing the motion of the point particle P. Therefore, the distance between two events p_1 and p_2 of L is (see Section 1.1)

$$\Delta\tau = \int_1^2 \sqrt{-ds^2} = \int_1^2 \sqrt{-\eta_{\mu\nu} \frac{dX^\mu}{dX^0} \frac{dX^\nu}{dX^0}}\, dX^0 = \int_1^2 \sqrt{1 - V^2}\, dT . \tag{2.8}$$

This is a geometrical invariant: its numerical value (expressed, for example, in seconds) must be the same, no matter what coordinate system is used to define the trajectory of P.

Now let us consider the case where the 3-velocity V is constant in $\mathcal{S}$ so that $\Delta\tau = \sqrt{1 - V^2}\Delta T$, where ΔT is the time interval separating p and $p + \Delta p$ measured by clocks at rest in $\mathcal{S}$. We introduce the inertial frame $\mathcal{S}'$ in which the particle is at rest. We then see that $\Delta\tau$ is just the time interval separating p and $p + \Delta p$ measured using the time T' of the clocks of $\mathcal{S}'$. If $\Delta\tau$ represents the ticking of a clock (or the lifetime of a particle) in $\mathcal{S}'$, then this time interval measured by the clocks of $\mathcal{S}$ will be longer: $\Delta T = \Delta\tau/\sqrt{1 - V^2}$, in accordance with the phenomenon of time dilation discussed in Section 1.4. Therefore, τ is the time measured by the clocks moving along with P.

If the 3-velocity of a point particle is not constant, we can generalize this result by *postulating* that the time measured by an accelerated clock co-moving with P is given by τ. The curvilinear abscissa τ then measures the *proper time* of the clock.

It should, however, be noted that this identification of the curvilinear abscissa with the time measured by an accelerated clock requires that a (good) clock, even though it is an extended object, can be accelerated without its operation being affected.

Extremization of the proper time and the geodesic

The world line of a free point particle is a time-like straight line of $\mathcal{M}_4$. It is also a *geodesic*, that is, the *longest* path between two events. Its equation can be obtained by extremizing the length or the proper time $\tau(s)$, which was denoted as $\Delta\tau$ in (2.8):

$$\tau(s) = \int_{\lambda_1}^{\lambda_2} d\lambda \sqrt{-\eta_{\mu\nu} \dot{X}_s^\mu \dot{X}_s^\nu} \quad \text{with} \quad \dot{X}_s^\mu \equiv \frac{dX_s^\mu}{d\lambda}, \tag{2.9}$$

where λ parametrizes one of the paths $X^\mu = X_s^\mu(\lambda)$ joining p_1 and p_2, and s parametrizes the various paths.

If we use $\delta X_s^\mu \equiv (\partial X_s^\mu / \partial s)|_0\, ds$ and $\delta \dot{X}^\mu \equiv (\partial^2 X_s^\mu / \partial s \partial \lambda)|_0\, ds$ to denote the variations of X_s^μ and $\dot{X}_s^\mu$ in going from the path with $s = 0$ to a neighboring path, the variation of the proper time (2.9), $\delta \tau \equiv (d\tau/ds)|_0\, ds$, will be written as follows omitting the label[3] s:

$$\delta\tau = -\int_{\lambda_1}^{\lambda_2} d\lambda\, \frac{\eta_{\mu\nu}\dot{X}^\nu \delta\dot{X}^\mu}{\sqrt{-\eta_{\rho\sigma}\dot{X}^\rho\dot{X}^\sigma}} = -\left.\frac{\eta_{\mu\nu}\dot{X}^\nu\delta X^\mu}{\sqrt{-\eta_{\rho\sigma}\dot{X}^\rho\dot{X}^\sigma}}\right|_{\lambda_1}^{\lambda_2} + \int_{\lambda_1}^{\lambda_2} d\lambda \left[\frac{d}{d\lambda}\left(\frac{\eta_{\mu\nu}\dot{X}^\nu}{\sqrt{-\eta_{\rho\sigma}\dot{X}^\rho\dot{X}^\sigma}}\right)\right]\delta X^\mu. \quad (2.10)$$

Since the variations δX^μ vanish at λ_1 and λ_2, the first term of the second equality is zero. The proper time will then be an extremum if the integrand of the second term vanishes. Since the integral (2.9) is reparametrization-invariant, that is, is unchanged for any transformation $\lambda \mapsto f = f(\lambda)$, we can set $\eta_{\rho\sigma}\dot{X}^\rho\dot{X}^\sigma = -1$, which returns us to parametrizing the geodesic by its proper time (after variation!), and its equation of motion then becomes trivially that of a straight line of $\mathcal{M}_4$:

$$\frac{dU^\mu}{d\tau} = 0, \quad \text{where} \quad U^\mu = \frac{dX^\mu}{d\tau}. \quad (2.11)$$

Tangent inertial frames and 'observers'

Since the 3-velocity $V(\tau)$ of a point particle can always be considered to be constant during a short interval $\Delta\tau$, it is possible to associate with each event p with curvilinear abscissa τ a *tangent inertial frame*, also called an *instantaneous inertial frame*, that is, a Minkowski frame $\mathcal{S}(\tau)$ having (constant) 3-velocity $V = V(\tau)$ relative to $\mathcal{S}$ and spatial axes parallel to those of $\mathcal{S}$. As long as the particle velocity does not change significantly during the time $\Delta\tau$, the curvilinear abscissa τ will, as we have seen, be identified with the times of clocks at rest in $\mathcal{S}(\tau)$. At $(\tau + \Delta\tau)$ it will be identified with the times of other clocks, those of $\mathcal{S}(\tau + \Delta\tau)$, and so on. (In Section 5.1 below we shall give a more sophisticated discussion of such 'local' frames.) The proper time is therefore defined by some authors as the time read from this succession of clocks undergoing uniform translations.

We can further postulate that it is also the time measured by an *accelerated* clock co-moving with the point particle. For example, if the period of circular motion of radius r_0 is $P \equiv 2\pi/\Omega$ according to the clocks of $\mathcal{S}$, it will be $P_{\text{proper}} = \sqrt{1 - r_0^2\Omega^2}\, P$ according to a clock accompanying the moving particle (see Section 2.1). If $r_0\Omega \to 1$, that is, if $|V| \to 1$, then $P_{\text{proper}} \to 0$, meaning that the clock makes a complete turn 'in no time' (according to the time *it* measures). It should also be noted that this postulate implies that if the clock is an extended object, it can spin on itself without its operation being affected.

In many books on this subject, an 'observer' (at rest and supplied with a clock) is associated with an inertial frame, and similarly for the ensemble of a point particle and an accelerated clock. Such an anthropomorphic viewpoint can help to guide intuition. However, we should not lose sight of the fact that an 'accelerated observer' might in fact be an elementary particle, or an atomic clock, and so on.[4]

The muon lifetime (II)

Let us consider a particle P constrained to undergo uniform circular motion $X^\mu(T) = (T, r_0 \cos\Omega T, r_0 \sin\Omega T, 0)$ in a Minkowski frame $\mathcal{S}$. The modulus of its 3-velocity is constant:

---

[3]See Book 1, Section 8.1 for an introduction to variational principles.

[4]For a coherent and mathematical treatment of the concept of observer, see Gourgoulhon (2013).

$|V| = r_0\Omega$, so that its proper time τ is related to the time T of $\mathcal{S}$ as $\tau = \sqrt{1 - V^2}\,T$; see Section 2.1.

Let us assume that the particle is created at $p(\tau)$ and decays at $p(\tau + \Delta\tau)$. We set $\Delta T = \Delta\tau/\sqrt{1 - V^2}$.

Owing to the meaning assigned to it, $\Delta\tau$ represents the proper lifetime of the particle, that is, the lifetime measured by a clock moving along with the particle (or the lifetime measured by the family of clocks of inertial frames tangent to the particle's world line). Then ΔT will correspond to the value of this same lifetime as measured by the clocks of $\mathcal{S}$.

The muon lifetime in a reference frame where the muon is at rest is 2.1948 μs (see also Section 1.4). We must then have $\Delta\tau = 2.1948$ μs and $\Delta T = 2.1948\,\mu s/\sqrt{1 - V^2}$.

In an experiment performed at CERN in 1976, J. Bailey *et al.* measured the lifetime ΔT of muons using the clocks of the reference frame in which they were moving, that is, the storage ring. Their speed was $0.999419\,c$ (corresponding to a Lorentz factor of $1/\sqrt{1 - V^2} = 29.33$). It was found that $\Delta T = 64.37$ μs, in perfect agreement (up to two parts in a thousand) with the prediction.

Since $\Omega\,\Delta T/(2\pi) = V\Delta T/(2\pi r_0) \approx \Delta T/(2\pi r_0) \approx 460$ turns for radius of the storage ring $r_0 = 7$ m, the muon motion cannot be considered to be uniform and rectilinear. The muon 3-acceleration is $a = V^2/r_0 \approx 1.3 \times 10^{16}$ m/s^2 and the modulus of the spatial part of the 4-acceleration [*cf.* (2.7)] is $|\gamma| = a/(1 - V^2) \approx 1.1 \times 10^{19}$ m/s^2. Since the experimental result is in agreement with the prediction, this indicates that the functioning of the 'muon clock', that is, the properties of the weak interaction responsible for the muon decay, is not affected by this acceleration.[5]

The absolute time of Newton now exits the stage.

2.3   The Langevin twins

Let us imagine a point particle which, after leaving the spatial origin of an inertial frame $\mathcal{S}$ at $T = 0$ (event p_1), returns after ΔT (event p_2) (see Fig. 2.1a). The duration of the trip according to a clock at rest in $\mathcal{S}$ is ΔT. On the other hand, the proper time $\Delta\tau$ that passes between p_1 and p_2 according to a clock co-moving with the particle is given by (2.8) and, no matter what motion the particle undergoes, $\Delta\tau$ will always be less than ΔT. In particular, if the particle moves away from the spatial origin at constant velocity V and then makes an instantaneous U-turn to return with velocity $-V$, (2.8) can be integrated immediately to give

$$\Delta\tau = \Delta T\sqrt{1 - V^2}. \qquad (2.12)$$

Mathematically, the result (2.12) just corresponds to application of the Pythagorean theorem in Minkowski geometry, and the fact that $\Delta\tau < \Delta T$ is the Minkowski version of the 'triangle inequality': the line connecting the events p_1 and p_2 is the *longest* possible path.

This result was popularized in France by Paul Langevin in the 1920s. If the clocks P and P' are 'twins' (for example, identical particles), the traveling twin P' will at the end of the trip be younger than the sedentary twin P.[6] This result was considered paradoxical. It was

---

[5]Muons are charged particles, and so they emit radiation during their circular motion (the so-called *synchrotron* radiation; see Section 18.3 below). Therefore, the 'muon clock' is also not affected by this emission of electromagnetic waves.

[6]Returning to the experiment of Bailey *et al.* described in Section 2.2: after one trip around the storage ring, the traveling muon has aged an amount equal to $1/460$ of its lifetime, while the muon that stays at rest in the laboratory has aged $29.33/460$ of its lifetime.

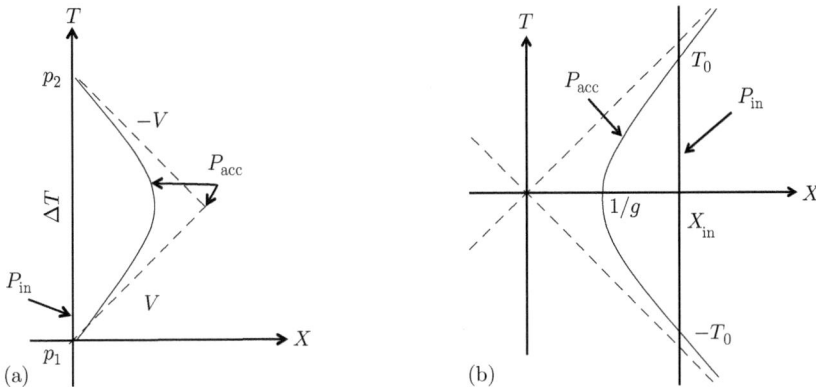

Fig. 2.1 The Langevin twins.

argued that if the lengths of the world lines of P and P' are calculated in the frame $\mathcal{S}'$ where P' is at rest, then it is P which becomes the traveler, which is correct. Then, (2.8) is saying that when P rejoins P', it will be P and not P' who will have aged less! However, this part of the argument is false because (2.8) only applies if $\mathcal{S}'$ is an inertial frame. But the motion of P' cannot be inertial all the time. In contrast to P, it must at some instant or another be accelerated relative to the ensemble of inertial frames, at least to do the U-turn.

In order to calculate the lengths of the world lines *in* $\mathcal{S}'$, it is necessary to know how to pass from an inertial frame to an accelerated frame. Below in Chapter 5 (see, for example, Section 5.2) we shall study some examples of such transformations which turn Minkowski coordinates into curvilinear coordinates.

An example of the twin 'paradox'

We consider a point particle P_{in} at $X = X_{\text{in}} > 1/g$ in an inertial frame $\mathcal{S}$ (the sedentary twin) and another undergoing uniform acceleration P_{acc} (the traveler), whose world line is the hyperbola given by (2.5), namely, $gT = \sinh g\tau$, $gX = \cosh g\tau$ (see Fig. 2.1b). The proper time of the sedentary twin is the time T of $\mathcal{S}$, and τ is that of the traveler.

The world lines intersect at $T = \pm T_0$, the instant the traveler leaves the sedentary twin and the instant they are reunited. The interval separating these two events is $\Delta\tau_{\text{in}} = 2T_0$ for P_{in} (the sedentary twin). The duration measured by P_{acc} (the traveler) is $\Delta\tau_{\text{acc}} = 2\tau_0$ with $gT_0 = \sinh g\tau_0$, and so

$$\Delta\tau_{\text{acc}} = \frac{2}{g}\text{ArgSinh}\frac{g\Delta\tau_{\text{in}}}{2} \quad (< \Delta\tau_{\text{in}}). \tag{2.13}$$

The Langevin twins in a closed space

A length element defines the *local* geometry of a space but says nothing about its global structure or *topology*. For example, the Euclidean plane $E_2 = R \times R$ (whose topology is termed *trivial*) has the same local geometry as the cylinder $R \times S^1$ or the torus $S^1 \times S^1$.

The trivial topology of Minkowski spacetime is $R \times E_3$, but other topologies are possible. For example, if (T, X, Y, Z) is identified with $(T+L, X, Y, Z)$ by means of a translation of length

L along the T axis, the topology of $\mathcal{M}_4$ becomes $S^1 \times E_3$, and there then exist closed time-like curves which contradict the causality hypothesis, according to which an effect cannot precede its cause.

We therefore limit ourselves to topologies $\mathcal{M}_4 = R \times \Sigma$ where the spatial sections Σ are obtained by identification of certain points of E_3. In the example of Fig. 2.2, the space (here two-dimensional) has the topology of a torus $S^1 \times S^1$. It can be shown[7] that there exist 18 different topologies having the same local structure as E_3.

If we decide to represent space and time by Minkowski spacetimes possessing such topologies, the twin paradox seems to resurface (see Fig. 2.2b): twin 2 is not inertial but twins 3 and 4 are and, owing to the non-trivial topology of the space, they return to their point of departure where they meet up again with twin 1 without ever having changed their direction or their speed.

However, there is no paradox because the twins 1, 3, and 4 are in fact not equivalent. It is not possible to reverse their trajectories by a change of frame because they are topologically different: trajectory 1 can be shrunk to a point by a continuous deformation, whereas trajectories 3 and 4 loop around the space.

The non-equivalence of the inertial frames of 1, 3, and 4 arises from the fact that the identification of the points (T, X) and $(T, g(X))$ of the spatial sections defines a particular foliation of spacetime and singles out directions. The events (T, X) and $(T, g(X))$ are two representations of the same event in the inertial frame of twin 1 only because the hypersurfaces Σ are hypersurfaces of constant time only for this twin.

Therefore, the choice of a spatial topology introduces the idea of an absolute space and therefore the idea of absolute rest. The oldest twin will always be the inertial twin whose motion is compatible with the choice of the spatial topology because it is this twin who will have the longest world line between the two events.[7]

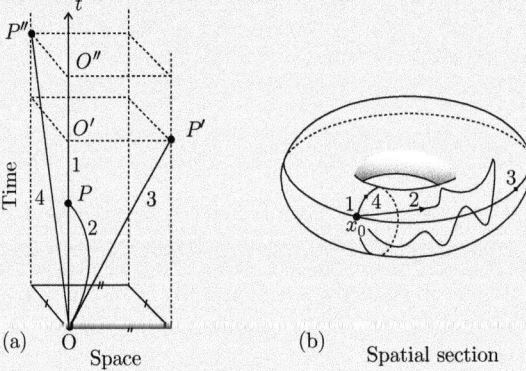

Fig. 2.2 The Langevin twins in a closed space.

2.4   Transformation of velocities and accelerations

Let us consider the time-like world line L of a point particle P. In an inertial frame $\mathcal{S}$ with Minkowski coordinates X^μ the equations of L are $X^\mu = X^\mu(\tau)$, where τ is the proper time of P. The components of the 4-velocity and 4-acceleration in $\mathcal{S}$ are $U^\mu \equiv dX^\mu/d\tau$ and $\gamma^\mu \equiv d^2 X^\mu/d\tau^2$, and from the definition of the proper time we have $\eta_{\mu\nu} U^\mu U^\nu = -1$.

---

[7]For more detail see Uzan (2002).

Under a Lorentz transformation the coordinates of P in $\mathcal{S}'$ become $X'^\nu(\tau) = \Lambda_\mu{}^\nu X^\mu(\tau)$, where $\Lambda_\mu{}^\nu$ is a Lorentz rotation matrix, so that

$$U'^\nu \equiv \frac{dX'^\nu}{d\tau} = \Lambda_\mu{}^\nu U^\mu, \quad \gamma'^\mu \equiv \frac{dU'^\nu}{d\tau} = \Lambda_\mu{}^\nu \gamma^\mu, \tag{2.14}$$

because the $\Lambda_\mu{}^\nu$ are constants. (It can be verified that indeed $\eta_{\mu\nu} U'^\mu U'^\nu = -1$ because $\Lambda_\rho{}^\mu \Lambda_\sigma{}^\nu \eta_{\mu\nu} = \eta_{\rho\sigma}$.) Therefore, the components U'^μ and U^μ are the avatars of the same vector in two different bases: $u = (dX^\mu/d\tau)e_\mu = (dX'^\mu/d\tau)e'_\mu$, and the same holds for the acceleration γ. We can then speak of the 4-velocity and 4-acceleration in the absolute sense, without specifying the inertial frame in which they are evaluated. This was a property of the 3-acceleration also in Newtonian physics, but it did *not* hold for the 3-velocity, which was not represented by the same vector in two different inertial frames.

In the particular case of a boost of velocity $V_0 \equiv \tanh\psi$ along the X axis, (2.14) becomes [cf. (1.6)]

$$U'^0 = U^0 \cosh\psi - U^1 \sinh\psi, \quad U'^1 = -U^0 \sinh\psi + U^1 \cosh\psi,$$
$$U'^2 = U^2, \quad U'^3 = U^3, \tag{2.15}$$

so that the 3-velocity $V'^i \equiv dX'^i/dT' = U'^i/U'^0$ of a point particle measured at the time of $\mathcal{S}'$ is written as a function of $V^i = dX^i/dT = U^i/U^0$ and V_0 as

$$V'^X = \frac{V^X - V_0}{1 - V^X V_0}, \quad V'^Y = \frac{V^Y \sqrt{1 - V_0^2}}{1 - V^X V_0}, \quad V'^Z = \frac{V^Z \sqrt{1 - V_0^2}}{1 - V^X V_0}. \tag{2.16}$$

The formulas are of course symmetric, because we also have

$$V^X = \frac{V'^X + V_0}{1 + V'^X V_0}, \quad V^Y = \frac{V'^Y \sqrt{1 - V_0^2}}{1 + V'^X V_0}, \quad V^Z = \frac{V'^Z \sqrt{1 - V_0^2}}{1 + V'^X V_0}. \tag{2.17}$$

We recover the Galilean velocity composition law ($T' = T$, $V'^X = V^X - V_0$, $V'^Y = V^Y$, $V'^Z = V^Z$) if $V_0^2 \ll 1$ and $V^X V_0 \ll 1$. Equations (2.16) also imply that

$$V'^2 = 1 - \frac{(1 - V_0^2)(1 - V^2)}{(1 - V_0 . V)^2}. \tag{2.18}$$

We therefore again find that if the modulus of the particle velocity V tends to 1 in $\mathcal{S}$, it will also tend to 1 in $\mathcal{S}'$.

Similarly, we obtain the relation between the components of the 3-accelerations $a^i \equiv d^2 X^i/dT^2$ and $a'^i \equiv d^2 X'^i/dT'^2$:

$$a'^X = \frac{(1 - V_0^2)^{\frac{3}{2}}}{(1 - V^X V_0)^3} a^X, \quad a'^Y = \frac{1 - V_0^2}{(1 - V^X V_0)^3} \left[(1 - V^X V_0)a^Y + V^Y V_0\, a^X \right] \tag{2.19}$$

(and a similar expression for a'^Z). In the Galilean limit, that is, for $V_0^2 \ll 1$ and $V^i V_0 \ll 1$, the 3-acceleration becomes invariant: $a'^i = a^i$.

Composition of velocities and 'rapidity'

The velocity composition law (2.16)–(2.17) is not additive. However, it can be shown that under certain conditions there exists a *functional* of the velocities that is additive. These conditions on the velocity composition law (denoted $\oplus$) are:

(1) $\oplus$ must possess a neutral element O such that $u \oplus O = O \oplus u = u$ for all u;

(2) $\oplus$ must possess an *absorbing element* c (of modulus 1 for suitably chosen units) such that $u \oplus c = c \oplus u = c$ for all u;

(3) $\oplus$ must be associative, $u \oplus (v \oplus w) = (u \oplus v) \oplus w$;

(4) $\frac{d(u \oplus v)}{du}$ and $\frac{d(u \oplus v)}{dv}$ must exist and be continuous in u and v; and

(5) $\frac{d(u \oplus v)}{du} > 0$ and $\frac{d(u \oplus v)}{dv} > 0$ so that $u \neq O$, $v \neq O$, $u \neq c$, and $v \neq c$.

If these five conditions are satisfied, it is always possible to construct a differentiable and strictly monotonic 'rescaling' function such that $f(u \oplus v) = f(u) + f(v)$ which permits redefinition of the velocities so that they become additive quantities. Here $f(u)$ is the *rapidity*.[8] Zero rapidity corresponds to zero velocity, and infinite rapidity corresponds to velocity equal to c. The law of velocity composition is then given by $u \oplus v = f^{-1}[f(u) + f(v)]$. The relativistic law $u \oplus v = (u+v)/(1+uv)$ is a special case of this transformation corresponding to the choice $f = \arctan(u)$ for the rapidity; see also Section 1.5.

E. Whittaker gives an argument justifying this particular choice.[9] Considering the composition law $k = u \oplus v$ as a function $g(u, v, k) = 0$ and supposing that g is symmetric in u, v, and $-k$ (because this relation can be interpreted as a function of the relative velocities of three reference frames: A relative to B, B relative to C, and C relative to A), he shows that the relativistic law is the only law linear in u, v, and k. Thus, starting from the general form $g(u, v, k) = l + m(u + v - k) + n(uv - uk - vk) - puvk = 0$, l and n must vanish (because for $v = 0$ we have $k = u$ for all u, and so $l - nu^2 = 0$ for all u); moreover, for $v = 1$ we have $k = 1$ (by the definition of the absorbing velocity), and so $m = p$.

The Fresnel formula and the Fizeau experiment

Special relativity is constructed in such a way as to incorporate the postulate that the speed of light *in the vacuum* (that is, in the absence of a physical medium) is the same in all inertial frames. This is what an experiment like the Michelson–Morley experiment shows to be true.

On the other hand, the effective 3-velocity of light in a refractive medium depends on the reference frame in which it is measured because it is not equal to the limiting velocity c; its modulus is on average $|c|/n = 1/n$ in the reference frame where the medium, of refractive index n, is at rest.[10] Since n is always greater than 1, we can view light propagating in a refractive medium as physical particles whose world lines are time-like and whose 3-velocity components therefore transform as (2.16) under a change of inertial frame. The speed of light c' (relative to

---

[8] This theorem was proved by Whitrow (1935). The name 'rapidity' was proposed by Lévy-Leblond (1976).

[9] Whittaker (1949).

[10] The index of refraction $n = \sqrt{\epsilon \mu}$ is related to the *permittivity* ϵ and the *permeability* μ, the coefficients characterizing the medium, which appear in the Maxwell equations of electrodynamics; see Chapter 16 below. In the units we have chosen, $\epsilon_0 = 1/4\pi$ and $\mu_0 = 4\pi$ are the permittivity and permeability of the vacuum, a 'medium' whose state of being at rest or not is useless to speculate on because the speed of light is *always* the same.

For a study of light propagation in matter and its various applications which are not treated in the present book see, for example, Jackson (1975) or Raimond (2000).

the laboratory) when the medium is moving with speed u parallel to the direction of propagation is then given by (2.16) with $V_0 = -u$:

$$c' = \frac{1/n + u}{1 + u/n} = \frac{1}{n} + u\left(1 - \frac{1}{n^2}\right) + \mathcal{O}(u^2). \tag{2.20}$$

This expression was obtained by Fresnel (1818) within the framework of the Newtonian wave theory assuming partial dragging of the aether by the refractive medium, and verified by Fizeau in 1850 (see Book 1, Section 17.4). Two light beams travel down a U-shaped tube of length $2D$ in which water flows at speed u, with one beam traveling in the same direction as the water current and the other traveling in the opposite direction. The two beams then interfere at the exit of the tube. The travel times of the two light signals measured by the time of $\mathcal{S}$ are $T_\pm = 2D/c'_\pm$, which leads to a shift Δi of the interference pattern as measured by the fringe spacing i of $\Delta i/i \approx \left(4Dn^2u/\lambda c\right)\left(1 - 1/n^2\right)$.

In special relativity the difference of the travel times is of purely kinematical origin and the experiment tests the velocity composition law.

3
The kinematics of light

In this chapter we embark on the study of light in the vacuum in special relativity. Here we shall represent light by world lines of zero length, that is, by particles which propagate at speed c. Such a description does not include the electromagnetic or quantum wave nature of light, but it does allow the interpretation of experiments which measure light travel times (such as the Sagnac experiment) or aberration effects due to motion of the receiver.

3.1 Light lines

In Section 1.1 we defined the light cone, the ensemble of points at zero distance from a given event. As we saw in Section 1.2, the future light cone of apex p, that is, the 2-sphere issuing from X^i at time T and moving away at the speed of light $c = 1$, can represent a 'light flash'. The generatrices of the cone are light-like lines, that is lines of zero length, leaving p, and one of these lines can represent the world line of a *signal* or *light corpuscle*[1] emitted at X^i at time T.

More generally, a *light line* representing the motion of a light corpuscle is a curve L of $\mathcal{M}_4$, $\lambda \mapsto p(\lambda)$, whose equations in a Minkowski frame $\mathcal{S}$ with coordinates X^μ are $X^\mu = X^\mu(\lambda)$, and whose tangent at p, $k \equiv dp/d\lambda$ of components $k^\mu = dX^\mu/d\lambda$ in $\mathcal{S}$, is a vector which is not necessarily constant but has zero norm at any point p of L:

$$\ell(k, k) = \eta_{\mu\nu} k^\mu k^\nu = 0 \qquad \Longleftrightarrow \qquad k^0 = \sqrt{k_i k^i} \qquad (3.1)$$

(with the convention that k^0 is positive). Therefore, a light line is defined at each point by its direction cosines, that is, k^i/k^0 (of which there are two since k^i/k^0 is normalized to 1), while the component k^0 remains arbitrary.

The 3-velocity and 3-acceleration of a light line

Let us consider a light line. The components $k^\mu = dX^\mu/d\lambda$ of its tangent vector, of zero norm, satisfy $k^0 = \sqrt{k_i k^i}$. The 3-velocity of the light corpuscle represented by this line is $c^i \equiv dX^i/dT = k^i/k^0 = k^i/\sqrt{k^j k_j}$, which has modulus 1. The 3-acceleration is given by

$$a^i \equiv \frac{d^2 X^i}{dT^2} = \frac{dk^j}{d\lambda} \frac{1}{(k_l k^l)} \left[\delta^i_j - \frac{k^i k_j}{(k_l k^l)} \right] \qquad \Longrightarrow \qquad a \cdot c = 0.$$

---

[1] Here we shall use the terms light 'signal' or 'corpuscle' indiscriminately to refer to the representation of light as a particle propagating at speed c. Since the term 'light corpuscle' is outdated, we shall occasionally replace it by the term *photon*, but it should be clearly understood that our description of light here has, with some exceptions, nothing to do with quantum theory, and that a 'light corpuscle' may refer to a packet of 'classical' photons of well-determined world line.

Relativity in Modern Physics. Nathalie Deruelle and Jean-Philippe Uzan.
© Oxford University Press 2018. Published in 2018 by Oxford University Press.
DOI: 10.1093/oso/9780198786399.001.0001

Therefore, the 'photon' trajectory is a straight line if the light propagates 'freely' (because the trajectory of any free particle is a straight line). It is a broken line if the photon interacts with mirrors which reflect it (in agreement with Snell's law; see Section 16.2 below). Finally, in the continuum limit (which implies an infinite series of mirrors) and if the trajectory is planar, it can only be a circle traveled around at speed $|c| = 1$.

In Section 2.1 we introduced time-like lines for which the tangent vector u is normalized to -1, $\ell(u, u) = -1$, so that the component $U^0 = dT/d\tau$ relates the time T of the frame $\mathcal{S}$ to the proper time measured by a clock co-moving with the particle. We also saw that when the particle is *ultrarelativistic*, that is, when its 3-velocity tends to the speed of light, the proper time 'freezes': $d\tau \to 0$ for finite dT. The components of the 4-velocity u, see (2.2), then diverge. Therefore, a light line defined by (3.1) can be viewed as the limit of the world line of a material particle, and the fact that k^0 is arbitrary means that we decide not to associate a proper time with it. In the following chapter we shall see how to interpret k^0.

3.2 The Sagnac effect

Let us consider the following three trajectories in an inertial frame $\mathcal{S}$ with Minkowski coordinates $X^\mu = (T, X, Y, Z)$:

$$\begin{cases} X_P^\mu = (T,\ r_0 \cos \Omega T,\ r_0 \sin \Omega T,\ 0) \\ X_\pm^\mu = (T,\ r_0 \cos(\Omega + \omega_\pm)T,\ r_0 \sin(\Omega + \omega_\pm)T,\ 0). \end{cases} \tag{3.2}$$

The first represents the motion of a material point P_{acc} traveling in a circle and the other two represent the motions of two particles emitted at time $T = 0$ at P_{acc}, also constrained to move in a circle in the prograde ($\omega_+ > 0$) and retrograde ($\omega_- < 0$) directions relative to P_{acc}. The times taken by $P_\pm$ to rejoin P_{acc} measured using the time T of the 'laboratory' $\mathcal{S}$ are $T_\pm = \pm 2\pi/\omega_\pm$, and so

$$\Delta T \equiv T_+ - T_- = 2\pi \frac{\omega_+ + \omega_-}{\omega_+ \omega_-}. \tag{3.3}$$

The world line of P_{acc} is time-like, and so its 4-velocity is (see Section 2.1)

$$U_P^\mu = \frac{1}{\sqrt{1 - r_0^2 \Omega^2}} [1,\ -r_0 \Omega \sin \Omega T,\ r_0 \Omega \cos \Omega T,\ 0]. \tag{3.4}$$

The proper time of P_{acc} is related to T as $U_P^0 = dT/d\tau_P$, or $\tau_P = \sqrt{1 - r_0^2 \Omega^2}\, T$, and so the advance of P_+ compared to P_- measured using the time of the receiver P_{acc} is

$$\Delta \tau_P = \sqrt{1 - r_0^2 \Omega^2}\, \Delta T = 2\pi \sqrt{1 - r_0^2 \Omega^2}\, \frac{\omega_+ + \omega_-}{\omega_+ \omega_-}. \tag{3.5}$$

Now if the particles are light corpuscles constrained to follow a circular trajectory by means of mirrors,[2] the tangents to their light lines will be given by

---

[2] Which must remain tangent to the circumference of the circle; see Section 3.1.

$$k^\mu = k^0[1, \; -r_0(\Omega + \omega_\pm)\sin(\Omega + \omega_\pm)T, \; r_0(\Omega + \omega_\pm)\cos(\Omega + \omega_\pm)T, \; 0]$$

and must be of zero length: $k_\mu k^\mu = 0$, which gives

$$r_0\omega_\pm = \pm 1 - r_0\Omega \qquad \Longrightarrow \qquad \frac{\omega_+ + \omega_-}{\omega_+\omega_-} = \frac{2r_0^2\Omega}{1 - r_0^2\Omega^2}. \tag{3.6}$$

Therefore, the expressions for the advances of P_+ compared to P_- as measured using the time of the receiver $P_{\rm acc}$ or the laboratory time are

$$\begin{cases} \Delta\tau_P = \dfrac{4\pi r_0^2\Omega}{\sqrt{1 - r_0^2\Omega^2}} = \dfrac{4S\Omega}{\sqrt{1 - \beta^2}} \\[2mm] \Delta T = \dfrac{4\pi r_0^2\Omega}{1 - r_0^2\Omega^2} = \dfrac{4S\Omega}{1 - \beta^2} \end{cases} \qquad \text{with} \quad S = \pi r_0^2, \;\; \beta = r_0\Omega. \tag{3.7}$$

This is the *Sagnac effect.*

The Sagnac effect (I)

Measurement of the delay (3.7) in practice requires using the wave properties of light. Let λ be the wavelength (or period) of the beams $P_\pm$ measured *in the laboratory*. If they are allowed to interfere at $P_{\rm acc}$, one observes a shift Δi of the interference pattern (relative to the pattern obtained when $\Omega = 0$)

$$\frac{\Delta i}{i} = \frac{\Delta T}{\lambda} = \frac{4S\Omega}{\lambda(1 - \beta^2)} = \frac{4S\Omega}{\lambda} + \mathcal{O}\left(\beta^2\right), \tag{3.8}$$

where i is the fringe spacing and S is the surface bounded by the beams (which, as is easily shown, can follow a closed broken line rather than a circle).

In the Newtonian corpuscular theory of light the time for a rotating device to make a trip around the origin must be the same for the two directions because the speeds have the same modulus relative to the rotating reference frame, and so the prediction is that there is no shift. However, in the wave theory of light the speed of light is constant *relative to the aether* in which the rotating reference frame moves and a shift is expected.

The experiment was proposed by O. Lodge in 1897 and performed by G. Sagnac in 1913 following calculations made in 1905. Sagnac measured[3] the effect (3.8) with an accuracy of 10^{-2}.

The Michelson–Gale–Pearson experiment

The effect (3.8) should be observed whenever the receiver is rotating relative to an inertial frame, and therefore also in an interferometer fixed on the Earth, whose diurnal rotation (relative to the quasi-inertial reference frame of the solar system) can thereby be measured. This experiment, also proposed by O. Lodge in 1893, was performed in 1925 by A. A. Michelson, H. G. Gale, and F. Pearson using an apparatus of 612 by 339 meters which the light beams

---

[3]The Sagnac effect is now measured using high-precision (10^{-12}) ring laser gyroscopes (the first experiment was that of W. M. Macek and D. T. Davis in 1963), and is commonly used in navigation systems. It has also been measured using massive particles like neutrons (S. A. Werner *et al.*, 1979), calcium atoms (Ch. Bordé *et al.*, 1991), and electrons (F. Hasselbach and M. Nicklaus, 1993). In this case λ is the de Broglie wavelength of the particles.

traveled around in opposite directions before interfering; they did indeed observe a shift of the interference fringes.[4]

These experiments, like that of the Foucault pendulum (see Book 1, Section 5.4), measure *absolute* rotations relative to the ensemble of inertial frames, which have a special status in special relativity just as in Newtonian physics. In Section 5.3 we shall see how they can be described using rotating frames.

3.3 Aberration formulas

In a Lorentz rotation, the components of a null vector $k = k^\mu e_\mu = k'^\mu e'_\mu$ transform like any vector:

$$k'^\nu = \Lambda^\nu_\mu k^\mu. \tag{3.9}$$

[It can be checked that $\eta_{\mu\nu} k^\mu k^\nu = 0$ implies that $\eta_{\mu\nu} k'^\mu k'^\nu = 0$ because $\Lambda^\mu_\rho \Lambda^\nu_\sigma \eta_{\mu\nu} = \eta_{\rho\sigma}$: a vector which is null in one frame is null in any other.]

If the transformation is a boost, (3.9) reduces to [cf. (1.6)]

$$k'^0 = k^0 \cosh\psi - k^1 \sinh\psi, \ k'^1 = -k^0 \sinh\psi + k^1 \cosh\psi, \ k'^2 = k^2, \ k'^3 = k^3, \tag{3.10}$$

where $V_0 \equiv \tanh\psi$ is the speed of the frame $\mathcal{S}'$, which moves along the X axis of $\mathcal{S}$.

Let us consider a light source in the XOY plane of an inertial frame $\mathcal{S}$ whose radius vector makes an angle α with the OX axis (see Fig. 3.1). The null vector tangent to the light line of a photon emitted from this source has the components $k^\mu = k^0(1, \cos\alpha, \sin\alpha, 0)$. In the frame $\mathcal{S}'$ moving with speed V_0 along the OX axis, its components $k'^\mu = k'^0(1, \cos\alpha', \sin\alpha', 0)$ are given by (3.10), and so the angles α and α' are related as $k'^0 \cos\alpha' = -k^0 \sinh\psi + k^0 \cos\alpha \cosh\psi$, $k'^0 \sin\alpha' = k^0 \sin\alpha$ or

$$\tan\alpha' = \frac{\sqrt{1-V_0^2}}{1 - V_0/\cos\alpha} \tan\alpha. \tag{3.11}$$

Using a bit of trigonometry, we can write these expressions in a form which explicitly displays the symmetry between the frames $\mathcal{S}$ and $\mathcal{S}'$, for example,

$$\tan\frac{\alpha'}{2} = \sqrt{\frac{1+V_0}{1-V_0}} \tan\frac{\alpha}{2}, \ \text{ or } \ \cos\alpha' = \frac{\cos\alpha - V_0}{1 - V_0 \cos\alpha} \iff \cos\alpha = \frac{\cos\alpha' + V_0}{1 + V_0 \cos\alpha'}. \tag{3.12}$$

The zeroth components, the interpretation of which will be given in the following chapter, are related as

$$k'^0 = \frac{k^0(1 - V_0\cos\alpha)}{\sqrt{1-V_0^2}}, \ \ k^0 = \frac{k'^0(1 + V_0\cos\alpha')}{\sqrt{1-V_0^2}}. \tag{3.13}$$

We see from (3.12) that for any α, $\alpha' \to \pi$ if $V_0 \to +1$ and $\alpha' \to 0$ if $V_0 \to -1$. Therefore, a light source which is isotropic in the frame where it is at rest emits, in the frame where it is in rapid motion, primarily in the forward direction (*i.e.*, in its direction of propagation). We shall return to this effect in Section 17.1 when we study the electromagnetic field created by a moving charge.

---

[4]In 1979 S. A. Werner and J. L. Staudenmann observed the same effect on the phase of the neutron wave function.

Bradley's formula

Let us place ourselves in the quasi-inertial frame S of the solar system and consider the light (that is, a continuous flux of light corpuscles) coming from a star, represented by the null vector with components $k^\mu = k^0(1, \cos\alpha, \sin\alpha, 0)$ (with $\alpha \in [\pi/2, \pi]$), at the time when the Earth is moving along the X axis with speed $V_\oplus$ relative to S. In the frame S' attached to the Earth, the angle is α' given by (3.12) and (3.13), with V_0 replaced by $V_\oplus$. Six months later, $V_\oplus$ will have changed sign and α' will have changed by $\Delta\alpha'$ with

$$\cos(\alpha' + \Delta\alpha') - \cos\alpha' = \frac{2V_\oplus \sin^2\alpha}{1 - V_\oplus^2 \cos^2\alpha} = \frac{2V_\oplus \sin^2\alpha'}{1 + 2V_\oplus \cos\alpha' + V_\oplus^2}, \text{ or } \Delta\alpha' \approx 2V_\oplus|\sin\alpha'|.$$

We therefore recover, to first order in $V_\oplus$, Bradley's formula for stellar aberration (1728), obtained within the framework of the Newtonian corpuscular theory of light (see Book 1, Section 17.2). In special relativity the aberration formula (3.12)–(3.13) is just an 'effect of perspective' due to Lorentz rotation, and the question of whether or not to take into account the source velocity does not arise.

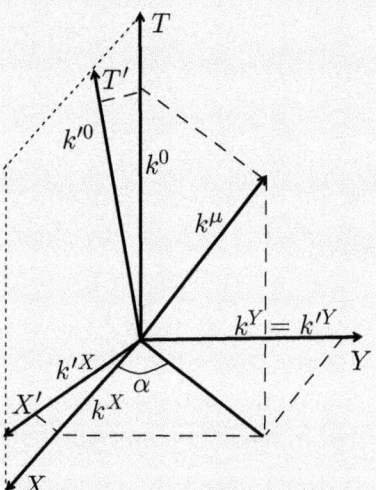

Fig. 3.1 The aberration of light.

'Superluminal' jets

Let us consider a light source S (for example, a 'spot' from a jet emitted by a quasar) which has 3-velocity V in an inertial frame where the 'astronomer' A is at rest. It is always possible to choose the spatial axes such that the world lines are, respectively, $S = (T, 0, VT, 0)$ and $A = (T, R, 0, 0)$. The light corpuscles emitted by S at T_{em} in $X_{\text{em}}^i = (0, VT_{\text{em}}, 0)$ follow lines of zero length: $X^i - X_{\text{em}}^i = T - T_{\text{em}}$, and the time of reception by A is given by $T_{\text{rec}} = T_{\text{em}} + \sqrt{R^2 + V^2 T_{\text{em}}^2}$.

During the interval ΔT_{em} the source moves a distance $d = (0, V\Delta T_{\text{em}}, 0)$, but A measures the projection of this vector on the line of sight, that is (see Fig. 3.2a) $d_a = V\Delta T_{\text{em}}|\sin\theta|$, where θ is the angle between SA and the Y axis. The apparent speed measured by A then is $V_a = d_a/\Delta T_{\text{rec}}$ or, since (to first order) $\Delta T_{\text{rec}} = \Delta T_{\text{em}} \left(1 + V^2 T_{\text{em}}^2/\sqrt{R^2 + V^2 T_{\text{em}}}\right) = \Delta T_{\text{em}}(1 + V\cos\theta)$,

$$V_a = \frac{V|\sin\theta|}{1 + V\cos\theta}, \tag{3.14}$$

the maximum value of which, obtained for $\cos\theta = -V$, is $V_{a\max} = V/\sqrt{1 - V^2}$. The speeds V_a and $V_{a\max}$ can be greater than 1 and the jets can appear to be superluminal.

Now let us consider an entire light surface expanding at radial speed[5] V. The apparent speed of each point of the sphere is given by (3.14). The apparent image of the source is an ovoid, given by the projection on a surface $T = \text{const}$ of the intersection of the past light cone issuing from the world line of A at T_{rec} and the cone representing the evolution of the source (see Figs. 3.2b and c).

An easy calculation then gives the equation of the ovoid and the angle θ_m for which the line of sight is tangent to it at S_m. If $V T_{\text{em}} \ll R$, we then find that the apparent speed of S_m is just the apparent maximum speed obtained above, namely, $V_m = V/\sqrt{1 - V^2}$.

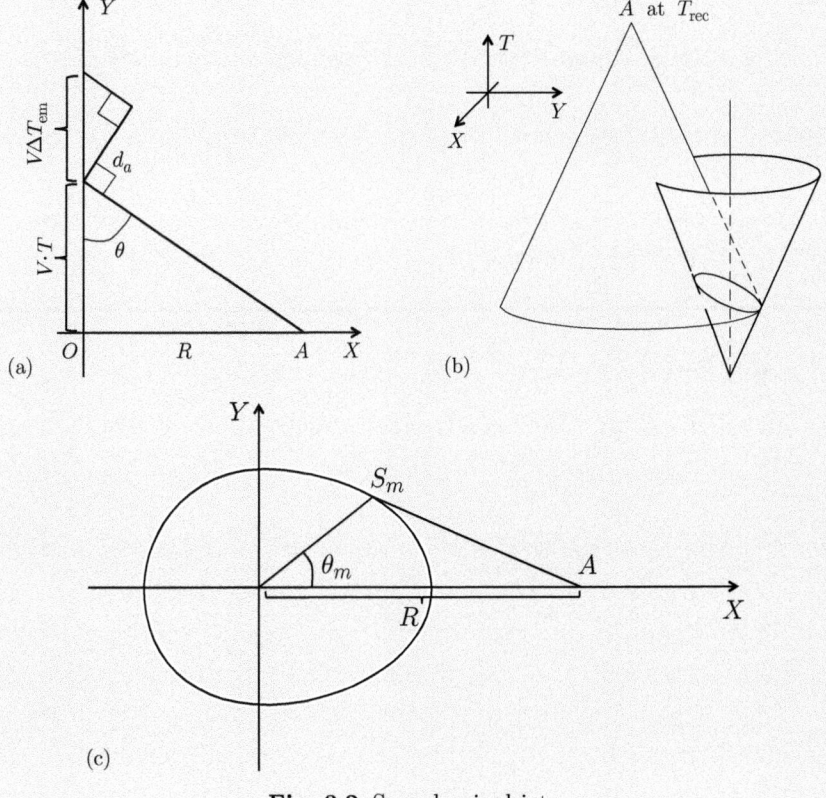

Fig. 3.2 Superluminal jets.

[5] Martin Rees (1966).

Reflection on a moving mirror

In the inertial frame $\mathcal{S}$ where the mirror is at rest, Snell's law applies (see Section 16.2): the angles of incidence and reflection of the light are equal and lie in the same plane.

First let us deal with the case where the frame $\mathcal{S}'$ is undergoing uniform translation at speed V_0 tangentially to the mirror (the mirror speed then is $-V_0$ in $\mathcal{S}'$). In the frame $\mathcal{S}$ the components of the null vectors tangent to the lines of the incident and reflected light are

$$k_{\rm inc}^{\mu} = k^0(1, \cos\alpha, \sin\alpha, 0), \quad k_{\rm ref}^{\mu} = k^0(1, \cos\alpha, -\sin\alpha, 0),$$

where α is the angle of incidence and $-\alpha$ the angle of reflection. Equation (3.11) then gives the angles of incidence and reflection in the frame $\mathcal{S}'$ where the mirror is moving:

$$\tan\alpha_{\rm inc}' = \tan\alpha \frac{\sqrt{1 - V_0^2}}{1 - V_0/\cos\alpha}, \quad \tan\alpha_{\rm ref}' = -\tan\alpha \frac{\sqrt{1 - V_0^2}}{1 - V_0/\cos\alpha} \quad \Longrightarrow \quad \alpha_{\rm inc}' = -\alpha_{\rm ref}'.$$

Therefore, Snell's law remains unchanged when the mirror is undergoing tangential uniform rectilinear motion (and also $k_{\rm inc}'^0 = k_{\rm ref}'^0$). This result also follows simply from the fact that the geometry of the problem is invariant under this tangential motion of the mirror.

Let us now turn to the case where the frame $\mathcal{S}'$ moves perpendicularly to the reflective surface of the mirror. In the frame $\mathcal{S}$ the components of the null vectors tangent to the lines of the incident and reflected light are

$$k_{\rm inc}^{\mu} = k^0(1, -\cos\alpha, \sin\alpha, 0), \quad k_{\rm ref}^{\mu} = k^0(1, \cos\alpha, \sin\alpha, 0),$$

where now α is the angle of reflection and $\pi - \alpha$ is the angle of incidence. Again, (3.11) gives the angles of incidence and reflection in the frame $\mathcal{S}'$ where the mirror is in motion:

$$\tan\alpha_{\rm inc}' = -\tan\alpha \frac{\sqrt{1 - V_0^2}}{1 + V_0/\cos\alpha}, \quad \tan\alpha_{\rm ref}' = \tan\alpha \frac{\sqrt{1 - V_0^2}}{1 - V_0/\cos\alpha} \quad \Longrightarrow \quad \alpha_{\rm inc}' \neq \pi - \alpha_{\rm ref}'.$$

Snell's law stating the equality of the angles of incidence and reflection therefore no longer applies when the mirror moves perpendicularly to its reflecting surface. At lowest order in V_0 we have $\alpha_{\rm inc}' = \pi - \alpha_{\rm ref}' + \epsilon$ with $\epsilon = 2V_0 \sin\alpha_{\rm ref}'$. In addition,

$$k_{\rm ref}'^0 = k_{\rm inc}'^0 \frac{1 - V_0\cos\alpha}{1 + V_0\cos\alpha}, \tag{3.15}$$

the interpretation of which will be discussed in Section 4.1.

4

The wave vector of light

In the preceding chapter we showed that the motion of 'light corpuscles' whose structure is ignored can be represented by simple world lines of zero length. Here we shall show how such lines can also describe an undulatory aspect of light,[1] namely, its frequency.

4.1 The wave vector and spectral shifts

A simple way of taking into account the 'color' attribute of light is to identify the time component k^0 of the vector k^μ tangent to a light line (which we left arbitrary in the preceding chapter) with the constant frequency ω of the wave associated with the light corpuscle:

$$k^0 = \omega. \tag{4.1}$$

Here $\nu \equiv \omega/2\pi$ is the frequency, $P \equiv 2\pi/\omega$ is the period, and $\lambda = P$ (in the vacuum) is the wavelength. The vector k^μ is then called the *4-wave vector*.[2]

Let us consider the case where the light propagates in the XOY plane of a frame $\mathcal{S}$ making an angle α with the X axis, and then change inertial frame. Owing to (4.1), we now interpret (3.13) and (3.15) as spectral shift formulas:

$$\omega' = \frac{\omega(1 - V_0 \cos\alpha)}{\sqrt{1 - V_0^2}}, \quad \omega = \frac{\omega'(1 + V_0 \cos\alpha')}{\sqrt{1 - V_0^2}}. \tag{4.2}$$

This is the relativistic version of the Doppler–Fizeau formula: a wave of frequency ω in the frame $\mathcal{S}$ has frequency ω' as measured using the time of $\mathcal{S}'$. We note that if $\alpha = \pi/2$, then $\omega' = \omega/\sqrt{1 - V_0^2}$, which can also be written as $\lambda' = \sqrt{1 - V_0^2}\,\lambda$. This is the *transverse Doppler effect*, the avatar of the phenomenon of time dilation or length contraction.

We note that (4.2) can be obtained without explicit use of the Lorentz transformation (3.10). We introduce the 4-velocity u of the world line of the spatial origin of the frame $\mathcal{S}'$. Its components in $\mathcal{S}$ are $U^\mu = (1, V_0, 0, 0)/\sqrt{1 - V_0^2}$, and its components in $\mathcal{S}'$ are $U'^\mu = (1, 0, 0, 0)$. Since the scalar product of two vectors is invariant, we have

$$\eta_{\mu\nu} k^\mu U^\nu = \eta_{\mu\nu} k'^\mu U'^\nu, \tag{4.3}$$

where we recall that $k^\mu = \omega(1, \cos\alpha, \sin\alpha, 0)$ and $k'^\mu = \omega'(1, \cos\alpha', \sin\alpha', 0)$. Equation (4.2) for ω' then follows.

---

[1] We have already alluded to this in Sections 1.2 (the Michelson experiment), 2.4 (the Fizeau experiment), and 3.2 (the Sagnac effect).

[2] As will be discussed in Chapter 15 below, a monochromatic plane wave is delocalized and therefore characterized by a 'bundle' of wave vectors of given frequency ω, or a 'free' vector. However, in practice a beam of light is always of finite width, and if we can view it as a 'light ray', the wave vectors become 'bound', that is, tangent to a light trajectory.

Relativity in Modern Physics. Nathalie Deruelle and Jean-Philippe Uzan.
© Oxford University Press 2018. Published in 2018 by Oxford University Press.
DOI: 10.1093/oso/9780198786399.001.0001

The Doppler–Fizeau formula

Equation (4.2) gives the spectral shift of a light wave as a function of the motion of the receiver. To first order in V_0 it reduces to $\omega' = \omega(1 - V_0 \cos \alpha)$. This expression was obtained by Doppler (1842) and Fizeau (1848) by introducing a special reference frame in which an aether carrying the light waves was at rest (see Book 1, Section 16.5). The relativistic calculation does not involve any privileged frame.

The Ives–Stilwell experiment

At second order in V_0 the relativistic Fizeau formula (4.2) becomes $\omega' = \omega(1 - V_0 \cos \alpha + V_0^2/2)$, which differs from the Newtonian prediction (see Book 1, Section 16.5). In 1938 H. E. Ives and G. R. Stilwell produced two beams of hydrogen atoms, one traveling toward a receiver ($\alpha = 0$) and the other traveling away from it ($\alpha = \pi$), and measured the average frequency of one of their emission lines, $\bar{\omega}' = (\omega'_- + \omega'_+)/2$. They obtained a result which agrees with (4.2), namely, $\bar{\omega}' = \omega V_0^2/2$, where V_0 is the beam velocity and ω is the frequency of the emission line of the atoms at rest.

4.2   Light signals and spectral shifts

In the preceding section we encoded the information about the frequency of a monochromatic light wave in the zeroth component of its wave vector. An alternative method of taking into account the wave nature of light is based on the fact that the emission of *successive* light corpuscles by the source also defines the period of a light signal.

As an illustration, let us consider the example of a light source and a receiver moving along the X axis of a frame $\mathcal{S}$. Let L_e and L_r be their world lines and U_e^μ and U_r^μ the components of their 4-velocities (see Fig. 4.1).

At the instant $\tau_{\rm em}$ of its proper time, the source emits a light signal (event $P_{\rm em1}$), at $\tau_{\rm em} + \Delta\tau_{\rm em}$ it emits a second signal (event $P_{\rm em2}$), and so on. By hypothesis, $\Delta\tau_{\rm em} \equiv 2\pi/\omega_{\rm em}$ represents the period of the emitted light and $\omega_{\rm em}$ its frequency; at the time given by the clocks of $\mathcal{S}$ the interval separating the two events is $\Delta T_{\rm em} = \Delta\tau_{\rm em}/\sqrt{1 - V_e^2(\tau_{\rm em})}$ (with the condition that $\Delta\tau_{\rm em}$ is so short that the velocity can be assumed constant).

The two signals then propagate on their respective light cones, and the two generatrices which cut L_r at $P_{\rm rec1}$ and $P_{\rm rec2}$ are the world lines of the two light corpuscles which reach the receiver at the instants $\tau_{\rm rec}$ and $\tau_{\rm rec} + \Delta\tau_{\rm rec}$ of its proper time. The quantity $\Delta\tau_{\rm rec} = 2\pi/\omega_{\rm rec}$ will be the period and $\omega_{\rm rec}$ the frequency measured by the receiver; at the time of $\mathcal{S}$ the interval separating $P_{\rm rec1}$ and $P_{\rm rec2}$ is $\Delta T_{\rm rec} = \Delta\tau_{\rm rec}/\sqrt{1 - V_r^2(\tau_{\rm rec})}$ (again with the condition that the velocity can be assumed constant).

We still need to relate $\Delta T_{\rm rec}$ to $\Delta T_{\rm em}$. If the light corpuscles propagate toward negative X (see Fig. 4.1a), the equation of the world line of the first one is $T = -X + T_{\rm em1} + X_{\rm em1}$, so that $\Delta T_{\rm rec} + \Delta X_{\rm rec} = \Delta T_{\rm em} + \Delta X_{\rm em}$ or

$$\Delta T_{\rm rec} = \Delta T_{\rm em}(1 + V_e(\tau_{\rm em}))/(1 + V_r(\tau_{\rm rec})).$$

If they travel toward positive X (Fig. 4.1b), we find similarly

$$\Delta T_{\rm rec} = \Delta T_{\rm em}(1 - V_e(\tau_{\rm em}))/(1 - V_r(\tau_{\rm rec}))$$

(these are the formulas of the 'Newtonian' Doppler–Fizeau effect).

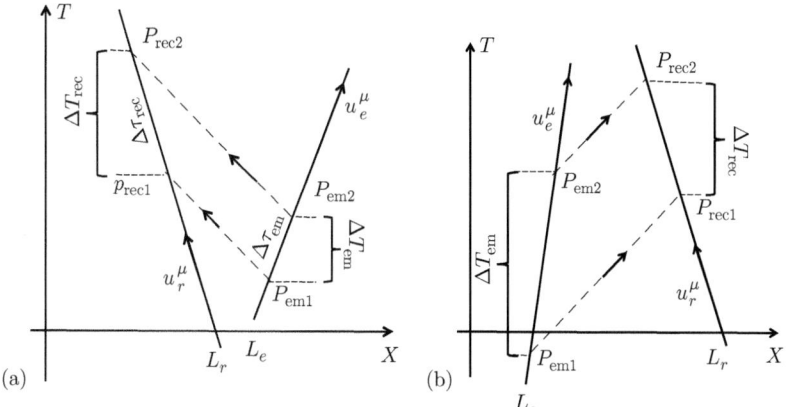

Fig. 4.1 The Doppler–Fizeau effect.

Assembling the results, in the first case (Fig. 4.1a) we have

$$\Delta\tau_{\mathrm{rec}} = \sqrt{\frac{1 - V_r(\tau_{\mathrm{rec}})}{1 + V_r(\tau_{\mathrm{rec}})}} \sqrt{\frac{1 + V_e(\tau_{\mathrm{em}})}{1 - V_e(\tau_{\mathrm{em}})}} \Delta\tau_{\mathrm{em}}$$

$$= \sqrt{\frac{1 + V'_e}{1 - V'_e}} \Delta\tau_{\mathrm{em}} \quad \text{with} \quad V'_e \equiv \frac{V_e(\tau_{\mathrm{em}}) - V_r(\tau_{\mathrm{rec}})}{1 - V_e(\tau_{\mathrm{em}})V_r(\tau_{\mathrm{rec}})}.$$

(4.4)

The formula in the second case (Fig. 4.1b) can be obtained simply by changing the signs of V_e and V_r. When the velocities of the source and the receiver are constant, V'_e is the 3-velocity of the source relative to the receiver given by the velocity composition law (2.16).

If, for example, the source is at rest $V_e = 0$, and if $\Delta\tau_{\mathrm{em}}$ is its period in $\mathcal{S}$, then (4.4) gives the period $\Delta\tau_{\mathrm{rec}}$ of the signals measured by the receiver, that is, in the frame $\mathcal{S}'$ where it is at rest. In terms of the frequencies we then have $\omega' = \sqrt{(1 + V_r)/(1 - V_r)}\,\omega$, in complete agreement with the Fizeau formula (4.2) with $\alpha = \pi$ and $V_0 = V_r$. It is easily seen that the same occurs for the other cases in Fig. 4.1.

4.3 An example of a particle horizon

In this section we illustrate the idea of a particle *horizon* as well as the limits of validity of the spectral shift formulas (4.2) and (4.4) by the example of two objects which exchange light signals. One of the objects is inertial and the other is undergoing uniformly accelerated rectilinear motion.

We consider two particles, one, P_{in}, at rest at $X = 1/g$ in $\mathcal{S}$ and the other, P_{acc}, undergoing uniform acceleration. The world line of the latter is given by

$$T = \frac{1}{g}\sinh g\tau, \quad X = \frac{1}{g}\cosh g\tau,$$

(4.5)

see (2.5) and Fig. 4.2. We shall use L_{in} and L_{acc} to denote the world lines of P_{in} and P_{acc}.

We assume that P_{in} sends a light signal at $T = T_{\mathrm{em}}$. This signal propagates on a light cone, and the equation of the generatrix which cuts L_{acc} is $X = T - T_{\mathrm{em}} + 1/g$. Since L_{acc}

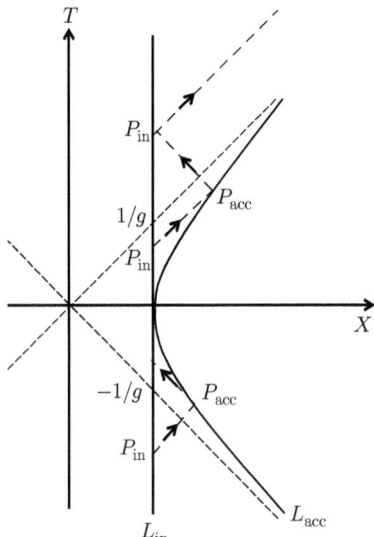

Fig. 4.2 The horizon of a uniformly accelerated particle.

is asymptotic to $X = T$ for $T \to +\infty$, we see that if $T_{\mathrm{em}} > 1/g$, none of the light corpuscles will reach P_{acc} (of course, with the condition that the particle is accelerated forever).

Even though the particle P_{acc} is accelerated, at each instant of time it is possible to associate with it an inertial frame $\mathcal{S}'$ tangent to its world line. The light signals it emits propagate along the common light cones in $\mathcal{S}$ and $\mathcal{S}'$, and the equations of the generatrices which cut L_{in} are $X = -T + \mathrm{const}$. Since L_{acc} is asymptotic at $X = -T$ for $T \to -\infty$, we see that none of these photons can reach P_{in} before $T = -1/g$.

Therefore, P_{acc} enters the *horizon* of P_{in} at $T = -1/g$ (before this time, no signal coming from P_{acc} can reach P_{in}), and at $T = +1/g$ it is P_{in} which leaves the horizon of P_{acc} (after this time, no signal coming from P_{in} can reach P_{acc}).

Qualitatively, the reason for this phenomenon is that the speed of P_{acc} approaches the speed of light asymptotically. When $T \to -\infty$, the signals emitted by the particle P_{acc} cannot get ahead of it and P_{in} cannot receive them. When $T \to +\infty$, P_{acc} moves away at the speed of light and the signals emitted by P_{in} cannot catch up to it.

To discuss this more quantitatively we return to our reasoning of the preceding section. The interval $\Delta\tau_{\mathrm{acc}}^{e}$ of the proper time of P_{acc} which separates the sending of two light signals corresponds to a time interval in $\mathcal{S}$ of $\Delta T_{\mathrm{em}} = \Delta\tau_{\mathrm{acc}}^{e}/\sqrt{1 - V_{e}^{2}}$, where V_{e} is the speed of P_{acc} in $\mathcal{S}$ at the moment of emission ('time dilation': for $\Delta\tau_{\mathrm{acc}}^{e}$ a constant, $\Delta T_{\mathrm{em}} \to \infty$ when $V_{e} \to \pm\infty$). Since in addition the two signals are sent from different places, the interval separating the reception of the two photons by P_{in} is $\Delta T_{\mathrm{rec}} = \Delta T_{\mathrm{em}}(1 + V_{e})$ (the Doppler effect with $+$ sign because the photons travel toward negative X). This interval is also the interval effectively measured by P_{in}, which is at rest in $\mathcal{S}$. We therefore have

$$\Delta\tau_{\mathrm{in}}^{r} = \Delta T_{\mathrm{rec}} = \Delta T_{\mathrm{em}}(1 + V_{e}) = \Delta\tau_{\mathrm{acc}}^{e}\sqrt{\frac{1 + V_{e}}{1 - V_{e}}}. \tag{4.6}$$

If $\Delta\tau^e_{\text{acc}}$ is the period of the light emitted by P_{acc}, we see that when P_{acc} enters the horizon of P_{in} ($V_e \to -1$), the period measured by P_{in} tends to 0 and the frequency of the signal is infinitely shifted towards the blue. Inversely, when P_{in} leaves the horizon of P_{acc} ($V_e \to 1$), the frequency at which P_{in} receives the photons is infinitely shifted towards the red. We recall that this calculation makes sense only if the velocity of P_{acc} does not vary appreciably during the emission. We shall see below that this imposes the condition $g\Delta\tau^e_{\text{acc}} \ll 1$ (for example, for $g = 10^n \times 10$ m/s^2 we would have $\Delta^e_{\text{acc}}\tau \ll 10^{-n}$ yr, which is not very restrictive.).

Now let us suppose that it is P_{in} which sends the signals. It is then P_{in} whose velocity is temporarily constant relative to P_{acc} and equal to $-V_r$, where V_r is the speed of P_{acc} in $\mathcal{S}$ at the instant the photon is received. Inverting the roles, we find

$$\Delta\tau^r_{\text{acc}} = \Delta\tau^e_{\text{in}} \sqrt{\frac{1+V_r}{1-V_r}}, \tag{4.7}$$

where $\Delta\tau^r_{\text{acc}}$ is the interval, measured by P_{acc}, separating the reception of the photons which P_{in} has emitted at its proper time interval $\Delta\tau^e_{\text{in}}$. If $\Delta\tau^e_{\text{in}}$ is the period of the light emitted by P_{in}, we see that when $V_r \to -1$ the frequency of the signal received by P_{acc} is infinitely shifted towards the blue. Inversely, when $V_r \to 1$ we have $\Delta\tau^r_{\text{acc}} \to \infty$, and the photons received by P_{acc} are spaced farther and farther apart. We shall see below that this reasoning is valid as long as $g\Delta\tau^r_{\text{acc}} \ll 1$. Therefore, the spectral shift *cannot* be given by (4.7) when $V_r \to 1$.

Light signals and wave vectors

Equations (4.6) and (4.7) can also be obtained by describing the light signals using wave vectors and, as in (4.3), making use of the invariance of their scalar products with the 4-velocities of P_{in} and P_{acc}.

If k is the wave vector associated with a light corpuscle, then $-(k \cdot u_{\text{in}}) \equiv \omega_{\text{in}}$ is the frequency measured by P_{in} and $-(k \cdot u_{\text{acc}}) \equiv \omega_{\text{acc}}$ is that measured by P_{acc}.

Let us now consider the case where P_{in} sends signals to P_{acc}. The components of u_{acc} in $\mathcal{S}$ are $U^\mu_{\text{acc}} = (1, V_r)/\sqrt{1-V_r^2}$. The components of k are $k^\mu = \omega^e_{\text{in}}(1,1)$, since $U^\mu_{\text{in}} = (1,0)$. Therefore,

$$-(k \cdot u_{\text{acc}}) \equiv \omega^r_{\text{acc}} = -\eta_{\mu\nu}k^\mu U^\nu_{\text{acc}} = \omega^e_{\text{in}}\frac{1-V_r}{\sqrt{1-V_r^2}} = \omega^e_{\text{in}}\sqrt{\frac{1-V_r}{1+V_r}}, \tag{4.8}$$

in agreement with (4.7).

Next we take the case where it is P_{acc} which sends the signals to P_{in}. In the inertial frame $\mathcal{S}'$ tangent to the world line of P_{acc} at the moment the signal is emitted, the components of u_{in} are $U'^\mu = (1, -V_e)/\sqrt{1-V_e^2}$. The components of k are $k'^\mu = \omega^e_{\text{acc}}(1,-1)$. Therefore,

$$-(k \cdot u_{\text{in}}) \equiv \omega^r_{\text{in}} = \eta_{ij}k'^i U'^j_{\text{in}} = \omega^e_{\text{acc}}\frac{1-V_e}{\sqrt{1-V_e^2}} = \omega^e_{\text{acc}}\sqrt{\frac{1-V_e}{1+V_e}}, \tag{4.9}$$

also in agreement with (4.6).

Finally, to test the validity of (4.6)–(4.9), let us do an exact calculation of the time interval $\Delta\tau^r_{\text{in}}$ separating the reception (by P_{in} at rest at $X = 1/g$) of light signals emitted by a uniformly accelerated particle P_{acc} at its proper time interval $\Delta\tau^e_{\text{acc}}$, without assuming that $g\Delta\tau^e_{\text{acc}} \ll 1$.

The world line of P_{acc} is given in (4.5) and $\Delta\tau_{\text{acc}}^e = \tau_{e_2} - \tau_{e_1}$, where τ_{e_1} and τ_{e_2} are the proper times of P_{acc} when the first and second light signals are emitted. The world line of the first photon moving toward P_{in} is $T = -X + X_{e_1} + T_{e_1} = -X + \exp(g\tau_{e_1})/g$. It reaches the world line of P_{in} at $gT_{r_1} = -1 + \exp(g\tau_{e_1})$. We then have $g\Delta T_r = \exp(g\tau_{e_2}) - \exp(g\tau_{e_1}) = \exp(g\tau_{e_1})[\exp(g\Delta\tau_{\text{acc}}^e) - 1]$. Since P_{in} is at rest in $\mathcal{S}$, its proper time τ_{in} is identified with the time T, the time of clocks at rest in $\mathcal{S}$, and we have $\Delta\tau_{\text{in}}^r = \Delta T_r$. Finally, recalling that the 3-velocity of P_{acc} is $V = \tanh g\tau$, we find that, for any time lapse between the emission of the two signals by P_{acc},

$$g\Delta\tau_{\text{in}}^r = \sqrt{\frac{1 + V_{e_1}}{1 - V_{e_1}}}\,[\exp(g\Delta\tau_{\text{acc}}^e) - 1] = \sqrt{\frac{1 + V_{e_2}}{1 - V_{e_2}}}\,[1 - \exp(-g\Delta\tau_{\text{acc}}^e)], \qquad (4.10)$$

which indeed reduces to (4.6) when $g\Delta\tau_{\text{acc}}^e \ll 1$.

A calculation just like the preceding one gives the exact formula for the time interval $\Delta\tau_{\text{acc}}^r$ separating the reception by P_{acc} of light signals emitted by the particle at rest P_{in} as a function of the interval $\Delta\tau_{\text{in}}^e$:

$$g\Delta\tau_{\text{acc}}^r = -\ln\left(1 - g\Delta\tau_{\text{in}}^e\sqrt{\frac{1 + V_{r1}}{1 - V_{r1}}}\right) = \ln\left(1 + g\Delta\tau_{\text{in}}^e\sqrt{\frac{1 + V_{r2}}{1 - V_{r2}}}\right), \qquad (4.11)$$

which reduces to (4.7) if $g\Delta\tau_{\text{acc}}^r \ll 1$.

We see from this expression that for a finite emission period $\Delta\tau_{\text{in}}^e$, the period measured by the accelerated detector P_{acc} does tend to infinity when $V_{r2} \to 1$, that is, when it moves away from the emitter with a speed approaching the speed of light, in agreement with (4.7). But (4.7) gives an incorrect result for the *manner* in which this period tends to infinity, because from the exact expression (4.11) we have

$$g\Delta\tau_{\text{acc}}^r \sim -\frac{1}{2}\ln(1 - V_{r2}). \qquad (4.12)$$

However, it should be borne in mind that these results hold only if the particle P_{acc} is accelerated forever![3]

---

[3]These results are less academic in general relativity, when the horizon is that of a black hole; see, for example, Misner *et al.* (1973) and Book 3, Section 6.4.

5

Accelerated frames

Labeling the points of Minkowski spacetime using curvilinear coordinates rather than Minkowski coordinates is mathematically just as simple as in Euclidean space. However, the *interpretation* of such a change of coordinates as passage from an inertial frame to an accelerated frame is more subtle. Here we study some examples and show how, within the framework of special relativity, Newtonian inertial accelerations turn into mere geometrical quantities.

5.1 Local frames and Fermi coordinates

Review of Newtonian spacetime

In Newtonian mechanics it is natural to consider four types of change of coordinates or reference frame:

(1) A change of Cartesian frame (rotation of the three axes and translation of the origin), which leaves velocity and acceleration vectors unchanged;

(2) Passage from one inertial frame to another by means of the Galilean group (uniform translation of the origin and the three spatial axes), which modifies the velocity but leaves the acceleration and the law of Newtonian dynamics unchanged;

(3) Passage to an accelerated frame by means of the larger group of rigid displacements (any rotation of the three axes and translation of the origin), which introduces inertial accelerations into the law of dynamics;

(4) Passage to curvilinear coordinates in a given Cartesian frame, which leads to writing the law of dynamics in terms of the covariant derivative.

This relatively complicated situation is due to the structure of Newtonian spacetime: the coordinate transformations (1) and (4) operate in the Euclidean space $\mathcal{E}_3$, while the operations (2) and (3) define families of frames, one for each leaf of $N_4 = \mathcal{E}_3 \times R$ (see Book 1, Part I for more details).

In pseudo-Euclidean Minkowski spacetime $\mathcal{M}_4$, the (linear) Poincaré transformations (which leave the 4-acceleration *and* the 4-velocity invariant) unify the change of Cartesian frame (1) and the passage from one inertial frame to another (2). The set of formulas characterizing these transformations is (see Section 1.3)

$$X'^\nu = \Lambda_\mu{}^\nu (X^\mu - d^\mu); \quad e_\mu = \Lambda_\mu{}^\nu e'_\nu, \quad \epsilon'^\nu = \Lambda_\mu{}^\nu \epsilon^\mu, \tag{5.1}$$

where d^μ is a constant vector and the Lorentz matrices $\Lambda_\mu{}^\nu$, with constant coefficients, satisfy $\Lambda_\rho{}^\mu \Lambda_\sigma{}^\nu \eta_{\mu\nu} = \eta_{\rho\sigma}$. The vectors and forms e_μ and ϵ^μ form the bases associated with the Minkowski coordinates X^μ of the vector space underlying $\mathcal{M}_4$ and its dual. These transformations cannot be used to go from an inertial frame to an 'accelerated frame'.

One way of going to an accelerated frame is to use the concept of tangent inertial frame introduced in Section 2.2. Let us consider an accelerated particle P_{acc} (see Fig. 5.1) whose

Relativity in Modern Physics. Nathalie Deruelle and Jean-Philippe Uzan.
© Oxford University Press 2018. Published in 2018 by Oxford University Press.
DOI: 10.1093/oso/9780198786399.001.0001

position vector in $\mathcal{S}$ is $OO' \equiv d(\tau) = d^\mu(\tau)e_\mu$. If τ is its proper time, its 4-velocity $u \equiv \dot{d} = U^\mu e_\mu$ is constrained by $(u \cdot u) \equiv \eta_{\mu\nu} U^\mu U^\nu = -1$. At the point O' defined by τ we consider an *inertial frame which is tangent* to the world line at this point. It is obtained by a Poincaré transformation

$$X'^\nu = \Lambda^\nu_{\ \mu}(\tau)\,(X^\mu - d^\mu(\tau)); \quad e_\mu = \Lambda^\nu_{\ \mu}(\tau)\,e'_\nu, \quad \epsilon'^\nu = \Lambda^\nu_{\ \mu}(\tau)\,\epsilon^\mu \tag{5.2}$$

such that $e'_0 = u(\tau)$, which fixes four components of the inverse matrix as a function of the three independent components of the 4-velocity as $\Lambda^\mu_{\ 0} = U^\mu(\tau)$. The three vectors of the spatial basis e'_i are determined by choosing the three parameters defining spatial rotations, for example, the Euler angles (so that the Lorentz matrix $\Lambda^\nu_{\ \mu}$ is determined by a set of six parameters, as required). Since the 4-acceleration $\gamma = \dot{u}$ is orthogonal to u we can choose $e'_1 = \gamma/\sqrt{(\gamma \cdot \gamma)}$.

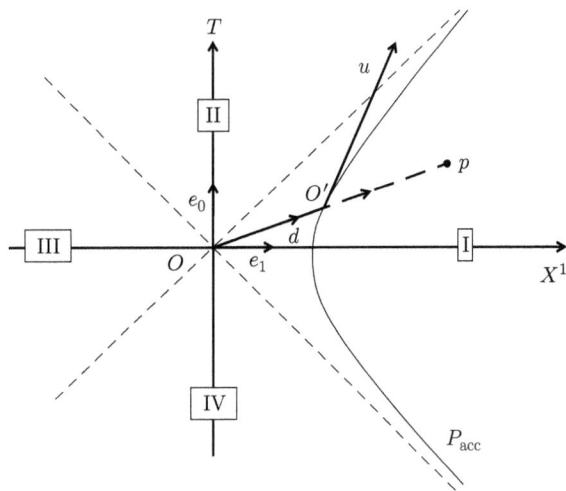

Fig. 5.1 Tangent inertial frame and Fermi coordinates.

Now we consider an arbitrary point p. Its radius vector can be decomposed as $Op = OO' + O'p$ or $X^\mu e_\mu = d^\mu e_\mu + X'^\mu e'_\mu$. If p is not too far from the world line of $P_{\rm acc}$, there will exist one and only one instant τ for which $O'p$ is orthogonal to u and for which we have $X^\mu e_\mu = d^\mu e_\mu + X'^i e'_i$, $i = (1, 2, 3)$. The event p can thus be defined either by its Minkowski coordinates X^μ, or by specifying τ and X'^i. We shall denote these four coordinates, called *Fermi coordinates*, as x^μ: $x^0 \equiv \tau$, $x^i \equiv X'^i$. They are related to the X^μ as

$$X^\mu = d^\mu(x^0) + \Lambda^\mu_{\ i}(x^0)\,x^i. \tag{5.3}$$

The coordinates x^μ therefore represent the position of a point of Minkowski spacetime in a *local frame* whose origin is attached to a particular world line and whose basis vectors vary from point to point. This frame is the relativistic analog of a frame from the group of rigid displacements of Newtonian physics. An important point is that the relations $X^\mu = X^\mu(x^\nu)$ are *not* linear and, as we shall see explicitly from some examples, the components of the metric in the coordinates x^μ are $\ell_{\mu\nu} \neq \eta_{\mu\nu}$.

The Minkowski metric in Fermi coordinates

We assume for simplicity that the accelerated motion of the point O' in the inertial frame (T, Z) is vertical, $d^\mu(\tau) = (d^0(\tau), d^Z(\tau))$, where τ is the proper time. It is easy to show that the basis vectors of the Fermi frame attached to O' are $e'_0 = \dot{d}$ and $e'_z = -\ddot{d}/\sqrt{\ddot{d} \cdot \ddot{d}}$ (for $\ddot{d}^Z < 0$).

Let us consider a point p. Using the relation $Op = OO' + O'p$ (that is, $Te_0 + Ze_Z = d^0 e_0 + d^Z e_Z + T'e'_0 + Z'e'_z$) and choosing O' such that $T' = 0$, it can then be shown that the transformation from Minkowski coordinates (T, Z) to Fermi coordinates $\tau \equiv t$ and $Z' \equiv z$ is given by [using the fact that $(\dot{d} \cdot \dot{d}) = -1$]

$$T = d^0(t) + z\,\dot{d}^Z, \quad Z = d^Z(t) + z\sqrt{1 + (\dot{d}^Z)^2}. \tag{5.4}$$

Finally, it can then be shown that the length element of the Minkowski metric is given by

$$ds^2 = -dT^2 + dZ^2 = -\left(1 + \frac{z\,\ddot{d}^Z}{\sqrt{1 + (\dot{d}^Z)^2}}\right)^2 dt^2 + dz^2. \tag{5.5}$$

5.2 The example of Rindler coordinates

Let us consider a uniformly accelerated particle P_{acc} in an inertial frame $\mathcal{S}$, that is, in a system of Minkowski coordinates $X^\mu = (T, Z)$ where the length element is given by $ds^2 = -dT^2 + dZ^2$.

The world line of this particle $d^\mu(\tau) = (d^0(\tau), d^Z(\tau))$ is the $Z > 0$ branch of the hyperbola $Z^2 - T^2 = 1/g^2$, where g is its acceleration. We can write this in parametric form with τ the proper time of P_{acc} (see Sections 2.1 and 4.3):

$$d^0(\tau) = \frac{1}{g}\sinh g\tau, \quad d^Z(\tau) = \frac{1}{g}\cosh g\tau \tag{5.6}$$

$$\implies U^\mu \equiv \dot{d}^\mu = (\cosh g\tau,\ \sinh g\tau), \quad \gamma^\mu \equiv \ddot{d}^\mu = g(\sinh g\tau,\ \cosh g\tau).$$

Let us construct the inertial frames $\mathcal{S}_g$ tangent to the world line of P_{acc}; see Fig. 5.1. The basis vectors (e'_0, e'_Z) of these frames $\mathcal{S}_g$ are derived from the basis vectors (e_0, e_Z) of $\mathcal{S}$ by Lorentz transformations such that the e'_0 are at each instant τ parallel to the 4-velocity u of P_{acc}. The e'_Z are orthogonal to the e'_0, and so [cf. (5.6)]

$$e'_0 = e_0 \cosh g\tau + e_Z \sinh g\tau = u, \quad e'_Z = e_0 \sinh g\tau + e_Z \cosh g\tau = \frac{\gamma}{g}. \tag{5.7}$$

The coordinate transformations $(T, Z) \mapsto (T', Z')$ are obtained using the identity $T'e'_0 + Z'e'_Z = d + Te_0 + Ze_Z$ and written as

$$T' = T \cosh g\tau - Z \sinh g\tau, \quad Z' = -T \sinh g\tau + Z \cosh g\tau - \frac{1}{g}. \tag{5.8}$$

Now let us consider a point p constrained to lie in quadrant I of Minkowski spacetime ($Z > 0$, $-Z < T < Z$; see Fig. 5.1). The Fermi coordinates ($t \equiv \tau$, $z \equiv Z$) of p associated with the world line of P_{acc} satisfy, by definition, $T' = T \cosh gt - Z \sinh gt = 0$, $z = -T \sinh gt + Z \cosh gt - 1/g$, or

$$gT = (1 + gz) \sinh gt, \quad gZ = (1 + gz) \cosh gt$$

$$gz = -1 + g\sqrt{Z^2 - T^2}, \quad gt = \text{argth}\,(T/Z). \tag{5.9}$$

The transformation $(T, Z) \to (t, z)$, a special case of (5.3) and (5.4), is not linear.

Now that we have completed this construction, we can forget about the scaffolding and treat (5.9) as defining the passage (in quadrant I) from Minkowski coordinates (T, X) to the curvilinear (or *Gaussian*) coordinates (t, x), known as *Rindler coordinates*.[1]

Review of curvilinear coordinates

Going beyond the framework of linear transformations involves introducing local bases of the vector space tangent to each point p of $\mathcal{M}_4$: $\frac{\partial}{\partial x^\mu} = \left(\frac{\partial X^\nu}{\partial x^\mu}\right)\frac{\partial}{\partial X^\nu}$ and the associated bases of the cotangent dual space: $dx^\mu = (\partial x^\mu/\partial X^\nu)dX^\nu$. These expressions are the exact analog in the spacetime $\mathcal{M}_4$ of the expressions which define coordinate transformations in $\mathcal{E}_3$; see Book 1, Chapters 3 and 4. The transformation laws for the components of vectors, forms, and tensors follow. For example (see Book 1, Sections 3.1 and 3.2), a vector v will be written as $v = V^\mu e_\mu \equiv V^\mu \frac{\partial}{\partial X^\mu}$ in Minkowski coordinates X^μ and $v = v^\mu \frac{\partial}{\partial x^\mu}$ in the coordinates x^μ with

$$v^\mu = \frac{\partial x^\mu}{\partial X^\nu} V^\nu. \tag{5.10}$$

Similarly, the length element and the Minkowski metric are given in the new coordinate system by

$$ds^2 = \ell_{\mu\nu}\,dx^\mu dx^\nu, \quad \ell = \ell_{\mu\nu}\,dx^\mu \otimes dx^\nu, \tag{5.11}$$

where the components of the metric tensor $\ell_{\mu\nu}$ and its inverse $\ell^{\mu\rho}$ satisfying $\ell_{\mu\nu}\ell^{\mu\rho} = \delta_\nu^\rho$ are related to the components $\eta_{\mu\nu}$ of the Minkowski metric in Minkowski coordinates as

$$\ell_{\mu\nu} = \eta_{\rho\sigma}\frac{\partial X^\rho}{\partial x^\mu}\frac{\partial X^\sigma}{\partial x^\nu}, \quad \ell^{\mu\rho} = \eta^{\lambda\sigma}\frac{\partial x^\mu}{\partial X^\lambda}\frac{\partial x^\rho}{\partial X^\sigma}. \tag{5.12}$$

In the particular case considered here of the transformation (5.9), the vectors of the natural basis of the tangent space at a point p are

$$\begin{cases} \dfrac{\partial}{\partial t} = \dfrac{\partial T}{\partial t}\dfrac{\partial}{\partial T} + \dfrac{\partial Z}{\partial t}\dfrac{\partial}{\partial Z} = (1 + gz)\left(\cosh gt\dfrac{\partial}{\partial T} + \sinh gt\dfrac{\partial}{\partial Z}\right) \\[2mm] \qquad = (1 + gz)(e_0 \cosh gt + e_Z \sinh gt) \\[3mm] \dfrac{\partial}{\partial z} = \dfrac{\partial Z}{\partial z}\dfrac{\partial}{\partial Z} + \dfrac{\partial T}{\partial z}\dfrac{\partial}{\partial T} = \cosh gt\dfrac{\partial}{\partial Z} + \sinh gt\dfrac{\partial}{\partial T} = e_0 \sinh gt + e_Z \cosh gt. \end{cases} \tag{5.13}$$

We can also introduce the non-holonomic basis (h_z, h_t) (that is, not associated with a coordinate system; see Book 1, Section 4.3) and the associated frame field (θ^z, θ^t):

---

[1]These were introduced by A. Einstein and N. Rosen in 1935 but popularized by W. Rindler in 1956; see Rindler (1991).

$$h_z = \frac{\partial}{\partial z}, \quad h_t = \frac{1}{1+gz}\frac{\partial}{\partial t}, \quad \theta^z = dz, \quad \theta^t = (1+gz)dt. \tag{5.14}$$

In this new system the Minkowski length element is written as [see also (5.5)]

$$ds^2 = -dT^2 + dZ^2 = -\left(\frac{\partial T}{\partial t}dt + \frac{\partial T}{\partial z}dz\right)^2 + \left(\frac{\partial Z}{\partial t}dt + \frac{\partial Z}{\partial z}dz\right)^2 = -(1+gz)^2dt^2 + dz^2, \tag{5.15}$$

and the Minkowski metric (5.11) becomes

$$\ell = -dT^2 + dZ^2 = -(1+gz)^2dt^2 + dz^2 = -(\theta^t)^2 + (\theta^z)^2. \tag{5.16}$$

The local frames (θ^t, θ^z) attached to $z = 0$ (the world line of P_{acc}) are just the tangent inertial frames (e'_0, e'_z) introduced in (5.7) with[2] $t = \tau$.

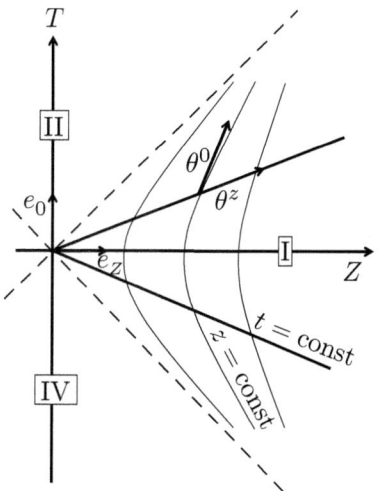

Fig. 5.2 Rindler coordinates.

Each line of coordinate $z = \text{const}$ is a hyperbola of $\mathcal{M}_4$ and represents the world line of a particle of constant 4-acceleration $g/(1+gz)$. Each line $t = \text{const}$ is a space-like straight line stemming from the origin; see Fig. 5.2.

The Rindler coordinates are manifestly *not* derived from the Minkowski coordinates (T, Z) by a Lorentz transformation. Instead we see a resemblance to the passage from Cartesian coordinates (u, v) to polar coordinates $(r, \phi : u = r\cos\phi, v = r\sin\phi)$ in Euclidean geometry:

$$dl^2 = dv^2 + du^2 = r^2d\phi^2 + dr^2. \tag{5.17}$$

The analogy becomes complete if we set $1 + gz = gr$, $gt = i\phi$; $X = u$, $T = iv$ with $i^2 = -1$.

---

[2]By analogy with the identification made in Newtonian physics, see Book 1, Section 4.3, the tetrad is often referred to as a *moving frame*, which gives rise to colorful expressions found in the literature such as 'the tetrad carried by a uniformly accelerated observer' in Misner *et al.* (1973), p. 169.

However, we note that while the polar coordinates (r, ϕ) cover the entire Euclidean plane (u, v), this is not the case with the Rindler coordinates (t, z), which cover only quadrant I of the Minkowski plane (T, Z). Going in the reverse direction from the Rindler coordinates (t, z) to the coordinates (T, Z) by the transformation (5.9) and then extending the domain of variation of (T, Z) to the *entire* Minkowski plane is an example of a *maximal analytic extension* of a spacetime.

The equation for the world line of a free particle undergoing uniform translation in the inertial frame $\mathcal{S}$, $Z = VT + Z_0$, becomes the following in a uniformly accelerated frame [*cf.* (5.9)]:

$$\frac{g}{1 + gz} = \frac{\cosh gt - V \sinh gt}{Z_0}. \tag{5.18}$$

Therefore, the equation for the world line of the particle P_{in} at rest at $Z_0 = 1/g$ in $\mathcal{S}$ becomes $1 + gz = 1/\cosh gt$ in Rindler coordinates. However, since $gT = \tanh gt$, this curve represents the world line of P_{in} in the interval $-1/g < T < 1/g$ only; see Fig. 5.3.

In addition, a light corpuscle emitted from P_{acc} at $Z = 0$ and time t_{e} is represented by a world line of zero length $[(1 + gz)dt = -dz]$ and intersects the world line of P_{in} at $z = z_{\text{r}}$, $t = t_{\text{r}}$ given by

$$\exp(-2gt_{\text{r}}) = 2\exp(-gt_{\text{e}}) - 1, \quad 1 + gz_{\text{r}} = \exp(gt_{\text{e}})\sqrt{2\exp(-gt_{\text{e}}) - 1}. \tag{5.19}$$

Armed with these results, it is a simple exercise to redo the calculations of the spectral shifts and lifetime in Sections 2.3 and 4.3, this time in the Rindler frame.

The Rindler frame and spectral shifts

As an example, let us redo the calculation of Section 2.3 of the duration of the round-trip of the uniformly accelerated 'twin' P_{acc} in the Rindler frame where the twin is at rest. The twin world line is $z = 0$ and the proper time is [*cf.* (5.15)] $\tau_{\text{acc}} \equiv t$. The world line of the inertial twin P_{in} is given by (5.18) with $V = 0$, $B = Z_{\text{in}}$ or $1 + gz = gZ_{\text{in}}/\cosh gt$. The proper time is obtained by evaluating the length element (5.15) on the world line, which gives $d\tau_{\text{in}} = gZ_{\text{in}}dt/\cosh^2 t$ or $\tau_{\text{in}} = Z_{\text{in}} \tanh gt$. The world lines intersect at $\pm t_0$ so that $gZ_{\text{in}} = \cosh gt_0$. The duration of the round-trip of the accelerated twin P_{acc} is $\Delta\tau_{\text{acc}} = 2t_0$. For P_{in} it is $\Delta\tau_{\text{in}} = 2Z_{\text{in}} \tanh gt_0 = 2\sinh gt_0/g$. We then recover the result (2.13): $g\Delta\tau_{\text{acc}} = \text{ArgSinh}(\Delta\tau_{\text{in}}/2)$.

As a second example, let us consider two uniformly accelerated particles P_{e} and P_{r}, at rest at z_{e} and z_{r} in the Rindler frame (see Fig. 5.3). P_{e} sends to P_{r} a light signal of duration $\Delta\tau_{\text{e}}$. The coordinate time interval which corresponds to this proper time interval can be read off from (5.15): $\Delta t = \Delta\tau_{\text{e}}/(1 + gz_{\text{e}})$. Since P_{r} is at rest and the metric coefficients do not depend on t, the coordinate duration of the signal at its arrival at P_{r} will also be Δt, but the proper time interval $\Delta\tau_{\text{r}}$ that P_{r} observes will be given by $\Delta\tau_{\text{r}} = \Delta t(1 + gz_{\text{r}})$ or

$$\Delta\tau_{\text{r}} = \Delta\tau_{\text{e}} \frac{1 + gz_{\text{r}}}{1 + gz_{\text{e}}}. \tag{5.20}$$

Since $(1 + gz)$ is the (00) component of the Minkowski metric in Rindler coordinates, we can rewrite (5.20) in a more general form which gives the frequency shift when the emitter and receiver are at rest in the coordinate system (x^0, x^i) and the components of the Minkowski metric $\ell_{\mu\nu}$ are independent of x^0:

$$\nu_{\text{rec}} = \sqrt{\frac{\ell_{00}(\text{em})}{\ell_{00}(\text{rec})}} \, \nu_{\text{em}}. \tag{5.21}$$

Now if we introduce the accelerations (in $\mathcal{S}$) g_e and g_r of P_e and P_r, such that $z_e = 1/g_e - 1/g$ and $z_r = 1/g_r - 1/g$, (5.20) becomes the following with $z_r = z_e + h$ (or $h = 1/g_r - 1/g_e$):

$$\Delta\tau_r = \Delta\tau_e(1 + g_e h). \tag{5.22}$$

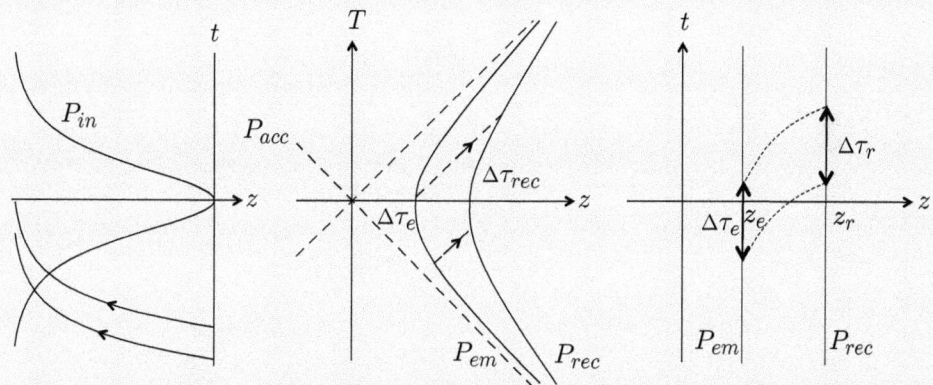

Fig. 5.3 Spectral shifts in inertial and Rindler frames.

Of course, the same problem can be studied using the Minkowski coordinates (T, Z) instead of the Rindler ones. Then the equations for the world lines of P_e and P_r respectively at $Z_e = 1/g_e$ and $Z_r = 1/g_r$ at $T = 0$ are

$$d_e^\mu = \left(\frac{1}{g_e}\sinh g_e\tau, \ \frac{1}{g_e}\cosh g_e\tau\right), \quad d_r^\mu = \left(\frac{1}{g_r}\sinh g_r\tau, \ \frac{1}{g_r}\cosh g_r\tau\right)$$

(with $g_e > g_r$). Here P_e sends light signals at its proper time interval $\Delta\tau_e = \tau_{e_2} - \tau_{e_1}$. The signals are received by P_r at its proper time interval $\Delta\tau_r = \tau_{r_2} - \tau_{r_1}$. The world line of the first signal is $T = Z - Z_{e_1} + T_{e_1}$. Therefore, $g_e\exp(-g_r\tau_{r_1}) = g_r\exp(-g_e\tau_{e_1})$, so that $\Delta\tau_r = (g_e/g_r)\Delta\tau_e$. As a consequence, the proper time interval $\Delta\tau_r$ separating the reception by P_r of the two signals emitted by P_e at its proper time interval $\Delta\tau_e$ is indeed given by (5.20) if we set $h = 1/g_r - 1/g_e$.

5.3 Rotating reference frames and the geometrization of inertia

We consider an inertial frame $\mathcal{S}$ with its Minkowski coordinates $X^\mu = (T, X, Y, Z)$ satisfying $ds^2 = -dT^2 + dX^2 + dY^2 + dZ^2$ and make a change of coordinate $X^\mu = (T, X, Y, Z) \mapsto x^\mu =$

(t, r, ψ, z) to pass to a frame rotating about the Z axis:

$$T = t, \quad X = r\cos(\psi + f(t)), \quad Y = r\sin(\psi + f(t)), \quad Z = z. \tag{5.23}$$

In this new system the Minkowski length element is written as [see (5.11) and (5.12)]

$$ds^2 = -(1 - r^2\Omega^2)dt^2 + 2r^2\Omega\, dt\, d\psi + dr^2 + r^2\, d\psi^2 + dz^2, \tag{5.24}$$

where $\Omega \equiv df/dt$ is the angular speed of rotation of the frame.

The Sagnac effect (II)

Let us redo the calculation of the Sagnac delay in Section 3.2, now using the curvilinear coordinates (5.23).

Let a particle P_{acc} be constrained to follow circular motion of radius r_0 and constant frequency Ω in $\mathcal{S}$. The equation of its world line is $x^\mu = (t, r_0, 0, 0)$. The expression for its proper time as a function of the coordinate time $T = t$ can be read off from (5.24): $\tau_P = \sqrt{1 - r_0^2\Omega^2}\, t$.

Now we consider two light corpuscles $P_\pm$, one prograde and the other retrograde, constrained to follow world lines given by $x^\mu = (t, r_0, \omega_\pm t, 0)$, where $\omega_\pm$ are (respectively positive and negative) constants. The coordinate times $t_\pm$ taken by these two light corpuscles to make a complete circuit (which are also the times measured by clocks at rest in the frame $\mathcal{S}$) are $T_\pm = t_\pm = \pm 2\pi/\omega_\pm$. Their difference measured at the proper time of P_{acc} is $\Delta\tau_P = 2\pi\sqrt{1 - r_0^2\Omega^2}\,(\omega_+ + \omega_-)/(\omega_+\omega_-)$.

The tangent vectors $k^\mu = k^0(1, 0, \omega_\pm, 0)$ must be null vectors: $l_{\mu\nu}k^\mu k^\nu = 0$ or $l_{00} + 2l_{0\psi}\omega_\pm + l_{\psi\psi}\omega_\pm^2 = 0$, from which we can extract [see (5.24)] $\frac{\omega_+ + \omega_-}{\omega_+\omega_-} = \frac{2r_0^2}{1 - \Omega^2 r_0^2}$. We therefore recover the equations for the Sagnac delay (3.7): $\Delta T = 4\pi r_0^2\Omega/(1 - r_0^2\Omega^2)$ and $\Delta\tau_P = 4\pi r_0^2/\sqrt{1 - r_0^2\Omega^2}$.

A uniformly rotating disk

Defining the spatial distance dl between two points P and $P + dP$ respectively located at (r, ψ, z) and $(r + dr, \psi + d\psi, z)$ as half the proper time measured by P for the light to make a round-trip from P to $P + dP$, we have

$$dl^2 = \frac{r^2 d\psi^2}{1 - \Omega^2 r^2} + dr^2. \tag{5.25}$$

Indeed, the light follows world lines of zero length for which $ds^2 = 0$, and so [cf. (5.24)]

$$dt = (\Omega r^2 d\psi \pm \sqrt{r^2 d\psi^2 + (1 - \Omega^2 r^2)dr^2})/(1 - \Omega^2 r^2).$$

The duration of a round-trip then is $2\sqrt{r^2 d\psi^2 + (1 - \Omega^2 r^2)dr^2}/(1 - \Omega^2 r^2)$. Now, the proper time τ of P is related to the coordinate t as $d\tau = \sqrt{1 - \Omega^2 r^2}dt$, from which we find (5.25).

Therefore, the circumference C of a circle centered at the origin as a function of its radius r is given by

$$C = \int_0^{2\pi} \frac{r\, d\psi}{\sqrt{1 - \Omega^2 r^2}} = \frac{2\pi r}{\sqrt{1 - \Omega^2 r^2}}.$$

We also see from the expression for the metric (5.24) that a particle at $r = r_0$ cannot be at rest in the rotating frame if $r_0 > 1/\Omega$ and must necessarily have an angular velocity $d\psi/dt \equiv \omega < 0$. This 'dragging' phenomenon is easily explained. In the inertial frame the linear speed of the particle is $r_0(\Omega + \omega)$, which must remain smaller than the speed of light.

Now that we have introduced the idea of an accelerated reference frame in special relativity and given some examples, we can define the concept of inertial acceleration and show that it is 'encoded' in special relativity in the *covariant derivative*.

Review of covariant differentiation

In general, the equation of the world line of a point particle is given by $X^\mu = X^\mu(\tau)$ in Minkowski coordinates X^μ and by $x^\mu = x^\mu(\tau) \equiv x^\mu(X^\nu(\tau))$ in a coordinate system $x^\mu = x^\mu(X^\nu)$. The components of its 4-velocity and its derivative in the coordinates x^μ then are

$$\begin{cases} u^\mu \equiv \dfrac{dx^\mu}{d\tau} = \dfrac{\partial x^\mu}{\partial X^\nu} U^\nu \\[2mm] \dfrac{du^\mu}{d\tau} = \dfrac{\partial x^\mu}{\partial X^\nu} \gamma^\nu + \dfrac{\partial^2 x^\mu}{\partial X^\nu \partial X^\rho} U^\nu U^\rho, \end{cases}$$

where $U^\nu \equiv dX^\mu/d\tau$ and $\gamma^\mu \equiv dU^\mu/d\tau$ are the components of the 4-velocity and acceleration in the Minkowski coordinates X^μ. These expressions are identical to those giving the transformation of the Newtonian velocity and its derivative under a general coordinate transformation (see Book 1, Chapter 4), and show that the components $(\partial x^\mu/\partial X^\nu)\gamma^\nu$ of the acceleration in the coordinates x^μ are not the ordinary time derivatives of u^μ, but rather their *covariant* derivatives: $\tilde{D}u^\mu/d\tau \equiv du^\mu/d\tau - (\partial^2 x^\mu/\partial X^\nu \partial X^\rho) U^\nu U^\rho$, which can also be written as (Book 1, Section 3.5)

$$\frac{\tilde{D}u^\mu}{d\tau} \equiv u^\nu \tilde{D}_\nu u^\mu = u^\nu \left(\partial_\nu u^\mu + \tilde{\Gamma}^\mu_{\nu\rho} u^\rho \right) = \frac{du^\mu}{d\tau} + \tilde{\Gamma}^\mu_{\nu\rho} u^\nu u^\rho \quad \text{with} \quad \tilde{\Gamma}^\mu_{\nu\rho} \equiv \frac{\partial^2 X^\sigma}{\partial x^\nu \partial x^\rho} \frac{\partial x^\mu}{\partial X^\sigma}. \tag{5.26}$$

It can further be shown that the *connection coefficients* or *Christoffel symbols* $\tilde{\Gamma}^\mu_{\nu\rho}$ can be written as a function of the components $\ell_{\mu\nu} = \eta_{\rho\sigma} \frac{\partial X^\rho}{\partial x^\mu} \frac{\partial X^\sigma}{\partial x^\nu}$ of the Minkowski metric in the coordinates x^μ as

$$\tilde{\Gamma}^\mu_{\nu\rho} = \frac{1}{2} \ell^{\mu\sigma} \left(\frac{\partial \ell_{\rho\sigma}}{\partial x^\nu} + \frac{\partial \ell_{\sigma\nu}}{\partial x^\rho} - \frac{\partial \ell_{\nu\rho}}{\partial x^\sigma} \right). \tag{5.27}$$

We also recall that the covariant derivative of a 1-form of components j_μ is given by (Book 1, Section 3.2 et seq.)

$$\tilde{D}_\rho j_\mu = \partial_\rho j_\mu - \tilde{\Gamma}^\sigma_{\rho\mu} j_\sigma, \tag{5.28}$$

and the covariant derivative of the Minkowski metric is zero:

$$\tilde{D}_\mu \ell_{\nu\rho} \equiv \partial_\mu \ell_{\nu\rho} - \tilde{\Gamma}^\sigma_{\mu\nu} \ell_{\sigma\rho} - \tilde{\Gamma}^\sigma_{\mu\rho} \ell_{\nu\sigma} = 0. \tag{5.29}$$

Let us consider a free particle. Its motion is uniform and rectilinear, that is, the components U^μ of its 4-velocity u in a system of Minkowski coordinates are constant, and those of its 4-acceleration are zero: $\gamma^\mu \equiv dU^\mu/d\tau = 0$. Its world line is a straight line of $\mathcal{M}_4$. In a general coordinate system x^μ the components u^μ of the 4-velocity are no longer constant, but those of its covariant derivative (5.26) are zero because $\gamma^\mu = 0$. Therefore, the equation of motion of a free particle in the coordinates x^μ is

$$\frac{\tilde{D}u^\mu}{d\tau} \equiv \frac{du^\mu}{d\tau} + \tilde{\Gamma}^\mu_{\nu\rho} u^\nu u^\rho = 0. \tag{5.30}$$

This is the equation of a line in $\mathcal{M}_4$ in Gaussian coordinates, and it is also the equation which extremizes the path between two events, and so it is called the *geodesic equation*. It generalizes (2.11).

Let us imagine, for example, that we are in a rotating frame where the Minkowski length element is given by (5.24). Now we make a new coordinate transformation $x^\mu = (t, r, \psi, z) \mapsto x'^\mu = (t, x, y, z)$ so that

$$t = T, \quad x = r \cos \psi, \quad y = r \sin \psi, \quad z = Z. \tag{5.31}$$

In this coordinate system the Minkowski length element is written as

$$ds^2 = -\left(1 - \Omega^2(x^2 + y^2)\right) dt^2 + 2\Omega \, dt (x\, dy - y\, dx) + dx^2 + dy^2 + dz^2. \tag{5.32}$$

An easy calculation gives the Christoffel symbols from the general expressions (5.26) or (5.27) (without assuming that Ω is constant):

$$\tilde{\Gamma}^x_{tt} = -\Omega^2 x - \frac{d\Omega}{dt} y, \quad \tilde{\Gamma}^y_{tx} = -\tilde{\Gamma}^x_{ty} = \Omega, \quad \tilde{\Gamma}^y_{tt} = -\Omega^2 y + \frac{d\Omega}{dt} x \tag{5.33}$$

(the others are obtained by symmetry or else are zero), so that the equation of motion of a free particle (5.30) becomes

$$\frac{d^2 x}{dt^2} = +2\Omega \frac{dy}{dt} + \Omega^2 x + \frac{d\Omega}{dt} y, \quad \frac{d^2 y}{dt^2} = -2\Omega \frac{dx}{dt} + \Omega^2 y - \frac{d\Omega}{dt} x. \tag{5.34}$$

This can be rewritten in three-dimensional form as

$$a = -2\Omega \wedge v + \Omega \wedge (R \wedge \Omega) - \frac{d\Omega}{dt} \wedge R, \tag{5.35}$$

where we have introduced the quantities $R \equiv (x, y, 0)$, $v \equiv dR/dt$, and $a \equiv dv/dt$, and Ω is a 3-vector parallel to e_Z. We recognize (5.35) as the expression for the Newtonian inertial acceleration, where $t = T$ is the time measured by clocks which are at rest in the inertial frame $\mathcal{S}$; cf. Book 1, Section 2.5.

The Christoffel symbols therefore simultaneously encode the chosen spatial coordinate system [here a Cartesian one because at $t = $ const the length element (5.32) reduces to $dx^2 + dy^2 + dz^2$] and the inertial accelerations arising from the fact that the axes (x, y, z) are rotating relative to the axes (X, Y, Z). Minkowski spacetime thus offers the geometrization of inertial forces as a sort of *bonus*. (In Section 10.4 we shall see that this geometrization of inertial forces also makes it possible to imagine a geometrization of the force of gravitation.)

Inversely, we can ask the question: if the components $\ell_{\mu\nu}(x^\rho)$ of the Minkowski metric are given in a system of curvilinear coordinates x^μ, how can we return to an inertial frame and its associated Minkowski coordinates X^μ, thereby getting rid of the inertial accelerations? The answer is simple in principle: the Christoffel symbols $\tilde{\Gamma}^\mu_{\nu\rho}$ are known as a function of $\ell_{\mu\nu}(x^\rho)$ and its inverse matrix from (5.27): $\tilde{\Gamma}^\mu_{\nu\rho} = \frac{1}{2}\ell^{\mu\sigma}(\partial_\nu \ell_{\rho\sigma} + \partial_\rho \ell_{\sigma\nu} - \partial_\sigma \ell_{\rho\nu})$. On the other hand, they are related to the Minkowski coordinates X^μ by (5.26), which can also be written as

$$\frac{\partial^2 X^\sigma}{\partial x^\mu \partial x^\nu} = \tilde{\Gamma}^\rho_{\mu\nu} \frac{\partial X^\sigma}{\partial x^\rho}. \tag{5.36}$$

Once these *linear* equations for $X^\sigma(x^\nu)$ have been solved, we need only to choose the integration constants, which is done by imposing the condition $\eta_{\rho\sigma} \frac{\partial X^\rho}{\partial x^\mu} \frac{\partial X^\sigma}{\partial x^\nu} = \ell_{\mu\nu}$. We will be left with 10 of them, the parameters of the Poincaré group whose choice determines a particular inertial frame.

We conclude by noting that, in order to view the 10 functions $\ell_{\mu\nu}(x^\rho)$ as components of the Minkowski metric, these functions cannot be completely arbitrary—they have to be expressible in terms of four independent functions $X^\mu(x^\nu)$ as $\ell_{\mu\nu} = \eta_{\rho\sigma} \frac{\partial X^\rho}{\partial x^\mu} \frac{\partial X^\sigma}{\partial x^\nu}$. If this is not the case, these 10 functions will still characterize a metric, but it will be a metric of a spacetime which is richer than Minkowski spacetime, that is, a *curved* spacetime.

5.4 The abandonment of accelerated reference solids

We are now left with the problem of understanding what the curvilinear coordinates x^μ actually represent, or, equivalently, the question of how to realize them by a reference frame in actual, 'relative, apparent, and common' physical space.

This does not pose a conceptual problem in Newtonian physics, where accelerated reference frames are obtained by rigid displacement, so that they can be realized by accelerating a reference *solid* in which, by construction, the distance between any two points remains fixed and the three axes remain orthonormal over time.

In special relativity the Minkowski coordinate $T = X^0$ represents the time of a clock which is at rest in the inertial frame S under consideration. On the other hand, the coordinate $t = x^0$ does *not* in general represent the time of a clock at rest at $x^i = $ const, because we have postulated that it is the proper time which plays this role. This proper time is given by $\tau = \int \sqrt{1 - V^2}\, dT$, where V is the 3-velocity of the clock relative to the inertial frame. Measured in terms of the coordinates x^μ, it is obtained from the length element by setting $dx^i = (dx^i/dt)dt$, which leads to

$$\tau = \int \sqrt{-ds^2} = \int \sqrt{-\ell_{\mu\nu} \frac{dx^\mu}{dt} \frac{dx^\nu}{dt}}\, dt = \int \sqrt{-\ell_{00}(t, x^i)}\, dt, \tag{5.37}$$

because the 3-velocity dx^i/dt of the clock vanishes.

What about the spatial coordinates x^i? Can they represent a three-dimensional Cartesian reference solid? In general, the answer is no.[3]

The 'rigidity' of the Rindler frame

Let us consider, in an inertial frame S with Minkowski coordinates (T, X) (we restrict ourselves to a single spatial dimension), two adjacent lines with spatial coordinates $x(T, X) = x_0$ and $x(T, X) = x_0 + \Delta x$. In S they have a velocity $V \equiv dX/dT$ given by $dx/dT = \partial x/\partial T + (\partial x/\partial X)V = 0$. In addition, ΔX is their proper distance measured in S, that is, at constant T. In the inertial frame S_g moving at speed V relative to S where they are momentarily at rest, their proper distance then is $\Delta X_g = \Delta X/\sqrt{1 - V^2}$ (length contraction). A solid can therefore be defined by requiring that ΔX_g be equal to Δx, the distance between the coordinate lines of the accelerated frame.

The Rindler coordinates satisfy this criterion. The world lines of two adjacent lines are given by $g^2(X^2 - T^2) = (1 + gx_0)^2$ and $g^2(X^2 - T^2) = [1 + g(x_0 + \Delta x)]^2$. Their speed in S is $V = dX/dT = T/X$. Their proper distance ΔX measured in S, that is, at constant T, is

---

[3]Pauli (1921) gives a clear discussion of the concept of rigid motion introduced by M. Born, as well as its limitations.

given by $gX\Delta X = (1 + gx_0)\Delta x$. In the inertial frame tangent to their world line it is (length contraction) $\Delta X_g = \Delta X/\sqrt{1 - V^2}$. From this we obtain

$$\Delta X_g = \frac{\Delta X}{\sqrt{1 - V^2}} = \frac{(1 + gx_0)\Delta x}{gX}\frac{1}{\sqrt{1 - T^2/X^2}} = \frac{(1 + gx_0)\Delta x}{g\sqrt{X^2 - T^2}} = \Delta x.$$

In this precise sense the coordinate line $x = $ const can be considered to be a 'rigid axis' and we can imagine a Rindler frame realized by a solid accelerated along the X axis of the inertial frame $\mathcal{S}$. This feature is peculiar to the Rindler frame. In the general case, it is necessary to abandon the idea of a 'rigid' frame and a reference 'solid' unless they are inertial.

Since the concept of accelerated reference *solid* is rather subtle in the theory of relativity, we are led to allow *any* coordinate system x^μ to be associated with a physical reference frame, with the 'non-rigid' coordinates x^i labeling the position of its points at coordinate time $t = x^0$, whose physical interpretation in terms of the proper time measured by a clock has to be specified case by case. Within this larger framework we therefore abandon the concepts of rigid frame and rigid solid introduced in Newtonian physics.

Part II

Dynamics

$$E = mc^2$$

Albert Einstein,
Ist die Trägheit eines Körpers von seinem Energieinhalt abhängig?
[Does the inertia of a body depend upon its energy content?]
Annalen der Physik, **18**, 639–643 (1905)

Book 2

6
Dynamics of a point particle

In this chapter, after attributing an inertial 'mass–energy' to particles and distinguishing between the action of an external field and of long-range and short-range internal forces, we present the 4-momentum conservation law for massive particles and light particles in inertial reference frames and give some examples which illustrate the role played by this law in collisions.

6.1 Free particles

In Section 1.2 we stated that an inertial reference frame is a frame in which a 'free' massive object, that is, an object which appears not to be subject to any force and does not interact with anything, undergoes uniform rectilinear motion. Thus, the laboratory walls and their clocks can in some cases serve as an inertial reference frame. In other cases it is necessary to use a system attached to the center of the Earth, or to the center of the solar system, and so on.

These rules for constructing an inertial reference frame have allowed us to establish a correspondence between *Minkowski frames* (mathematical objects) and *inertial reference frames* (physical objects).

We have thus been able to illustrate certain properties of relativistic kinematics (for example, time dilation, the velocity composition law, the Sagnac effect) by experiments (the muon lifetime and the experiments of Fizeau and Sagnac).

The *first law* of relativistic dynamics is the translation into mathematics of this physical concept of a free particle which we have already used; the equation of motion of a free particle is that of a time-like straight line of $\mathcal{M}_4$. In an inertial reference frame it is trivially

$$\frac{d^2 X^\mu}{d\tau^2} = 0 \qquad \Longleftrightarrow \qquad \frac{dU^\mu}{d\tau} = 0, \tag{6.1}$$

where $U^\mu \equiv dX^\mu/d\tau$ is the particle 4-velocity normalized to unity, $\eta_{\mu\nu} U^\mu U^\nu = -1$, so that τ measures its proper time. In coordinates $x^\mu = x^\mu(X^\nu)$ it is written as (see Section 5.3)

$$\frac{\tilde{D}u^\mu}{d\tau} \equiv \frac{du^\mu}{d\tau} + \tilde{\Gamma}^\mu_{\nu\rho} u^\nu u^\rho = 0 \quad \text{with} \quad \tilde{\Gamma}^\mu_{\nu\rho} \equiv \frac{1}{2} \ell^{\mu\sigma} (\partial_\nu \ell_{\rho\sigma} + \partial_\rho \ell_{\sigma\nu} - \partial_\sigma \ell_{\nu\rho}), \tag{6.2}$$

where $u^\mu \equiv dx^\mu/d\tau$ are the components of the 4-velocity in the new coordinates and $\tilde{D}u^\mu/d\tau$ is its covariant derivative, with the Christoffel symbols $\tilde{\Gamma}^\mu_{\nu\rho}$ simultaneously describing the chosen system of spatial coordinates (for example, spherical coordinates) and the frame acceleration (for example, a frame undergoing uniform rotation).

As we have seen in Sections 2.2 and 5.3, eqns (6.1) and (6.2) for straight lines of $\mathcal{M}_4$ are also the geodesic equations which extremize the proper time taken to travel between two points.

Relativity in Modern Physics. Nathalie Deruelle and Jean-Philippe Uzan.
© Oxford University Press 2018. Published in 2018 by Oxford University Press.
DOI: 10.1093/oso/9780198786399.001.0001

A free particle in the Rindler frame

In the Rindler coordinates (t, x) the Minkowski metric is written as [cf. (5.16)] $ds^2 = -(1 + gx)^2 dt^2 + dx^2$, where g is a constant. The nonzero Christoffel symbols (6.2) are $\tilde{\Gamma}^t_{tx} = g/(1 + gx)$ and $\tilde{\Gamma}^x_{tt} = g(1 + gx)$. The geodesic equation then becomes

$$\frac{d^2 t}{d\tau^2} + \frac{2g}{1 + gx}\frac{dt}{d\tau}\frac{dx}{d\tau} = 0, \quad \frac{d^2 x}{d\tau^2} + g(1 + gx)\left(\frac{dt}{d\tau}\right)^2 = 0.$$

The first equation can be integrated to give $(1 + gx)^2 dt/d\tau = E$, where E is a constant, and the second becomes $d^2 x/d\tau^2 + gE^2/(1 + gx)^3 = 0$, which we integrate to obtain $(dx/d\tau)^2 - E^2/(1 + gx)^2 = C$ with $C = -1$ so that $-(1 + gx)^2 (dt/d\tau)^2 + (dx/d\tau)^2 = -1$. Therefore, $dx/dt = (1 + gx)\sqrt{1 - (1 + gx)^2/E^2}$, which has the solution $g/1 + gx = (\cosh gt - A \sinh gt)/B$ with $E = gB/\sqrt{1 - A^2}$. We thus recover (5.18) for a straight line in Rindler coordinates, that is, the trajectory of a free particle.

6.2 Interactions

To establish the laws of dynamics of an interacting body, that is, the equations determining its world line, it is useful to classify interactions as *external fields* on the one hand, and *long-range* (infinite range) and *short-range forces* on the other.

If the presence of the body under study P, called the test body, only negligibly modifies the motion of the bodies with which it interacts, the interaction is considered as an *external field*. Then we postulate that the motion of P is given by the *second law* of relativistic dynamics, which is written as follows[1] in a Minkowski frame $\mathcal{S}$:

$$m\,\gamma = F, \quad \text{or also} \quad m\frac{d^2 X^\mu}{d\tau^2} = F^\mu. \tag{6.3}$$

The constant m is an attribute of P, its *inertial mass*, expressed, for example, in *kilograms*. We assume that all massive objects have mass of the same sign, which we choose to be positive. The vector $\gamma = du/d\tau$ is the 4-acceleration of P. The quantity F is called the *force 4-vector*. It is *a priori* a functional of the world line of P, but we shall require that it depend only on the position and at most the 4-velocity of P: $F = F(p(\tau), u(\tau))$. Therefore, for given initial conditions, namely, the spatial location and 3-velocity of P in $\mathcal{S}$ at $\tau = \tau_0$, the integration of (6.3) for known F will determine $X^\mu(\tau)$, that is, the world line of P. We note that since γ is orthogonal to the 4-velocity u, $(F \cdot u) \equiv F^\mu U_\mu$ must vanish, and so (6.3) involves only three independent components.[2]

---

[1]The law (6.3), which generalizes Newton's law, was proposed by M. Planck in 1906. Below we shall discuss two examples of an external force: the force exerted by a scalar field on a mass (Section 10.1), and the Lorentz force of an electromagnetic field on a charge (Section 11.3).

[2]The zeroth component of (6.3) is not independent because $\gamma^0 = \gamma^i V_i \equiv \gamma . V$; cf. Section 2.1. It is written as

$$\frac{1}{\sqrt{1 - V^2}}\frac{d}{dT}\frac{m}{\sqrt{1 - V^2}} = F^i V_i \implies \frac{d}{dT}\left(\frac{1}{2}mV^2\right) = F.V \quad \text{for} \quad |V| \ll 1.$$

In the nonrelativistic limit it therefore states that the change of the kinetic energy is equal to the work done by the force.

We note that the law (6.3) implies that we choose to represent the forces by vectors of M_4 and that they therefore must not depend on either the position or the orientation *or* the uniform translational motion of the inertial reference frame in which they are studied. (We recall that in Newtonian physics nothing requires *a priori* that the forces be represented by the same vectors of E_3 in different inertial reference frames. The fact that they are or are not is imposed by experiment and not, as here, by the structure of the theory.)

Also, the law (6.3), which takes the same form in any inertial frame, states that it is impossible to single out any particular frame. It is the specification of the initial conditions of the motion in a particular frame which distinguishes that frame.

On the other hand, the equivalence class of inertial reference frames preserves its special status, as it is only in those frames that the law (6.3) is invariant. In fact, in any coordinate system the law (6.3) is written in terms of the covariant derivative $\tilde{D}$ as

$$ m\,\tilde{D}_u\,u = F \qquad \Longleftrightarrow \qquad m\,\frac{\tilde{D}u^\mu}{d\tau} = f^\mu \qquad \Longleftrightarrow \qquad \frac{du^\mu}{d\tau} + \tilde{\Gamma}^\mu_{\nu\rho}u^\nu u^\rho = \frac{f^\mu}{m}, \qquad (6.4) $$

where $f^\mu = (\partial x^\mu/\partial X^\nu)F^\nu$ and $u^\mu = (\partial x^\mu/\partial X^\nu)U^\nu$ are the components of F and the velocity u in the coordinates x^μ, while F^ν and U^ν are the components in the inertial reference frame $\mathcal{S}$. Choosing (6.4) rather than (6.3) as the law of dynamics means that we are 'betting' that the forces are represented by the same vectors F in *all* frames, inertial or not. In the *covariant* form (6.4) the law of dynamics, in contrast to (6.3), is then the same in *all* frames. As we have seen in Section 5.3, the inertial accelerations due to the fact that the system x^μ is accelerated relative to the inertial reference frame $\mathcal{S}$ are encoded in the Christoffel symbols $\tilde{\Gamma}^\mu_{\nu\rho}$.

Now let us turn to the case where the external field approximation is not valid, and consider the dynamics of a body interacting *via long-range* forces, which curve the world lines of all the bodies involved, even if they stay far apart (electromagnetism and gravitation are two obvious examples of a long-range interaction).

Since we have excluded tachyons, so that any information must travel by signals whose velocity does not exceed the speed of light, the force $F_a(\tau)$ exerted by $P_{a'}$ on P_a cannot depend on events of the world line $L_{a'}$ of $P_{a'}$ occurring after the *retarded position* of $\hat{P}_{a'}$, the intersection of the light cone originating from $p_a(\tau)$ and $L_{a'}$; see Fig. 6.1a. As far as the retarded point $\hat{p}_{a'}$ is concerned, it is itself influenced by the events of L_a before the corresponding retarded position, the intersection of the cone originating from $\hat{p}_{a'}$ and from L_a, and so on. The result of this 'Jacob's ladder' is that the force F_a exerted by $P_{a'}$ on P_a no longer reduces to a vector function of only the positions and velocities, but becomes a full functional of the world lines of the interacting bodies, and it is not possible, except within the framework of an iterative scheme, to reduce the law (6.3) to a second-order differential equation.[3]

Conversely, a force is *short-range* if the bodies interact only when their straight world lines intersect. The event point of the intersection is their *collision* point, and its past cone includes all the world lines of the incoming particles, while its future cone includes those of the outgoing particles. The tangents to the world lines are discontinuous at the collision point (which is what signals that an interaction has taken place). The incoming and outgoing

---

[3] For explicit examples of such an iteration scheme, see Section 21.3 for the case of the electromagnetic interaction and Section 10.4 for the scalar interaction.

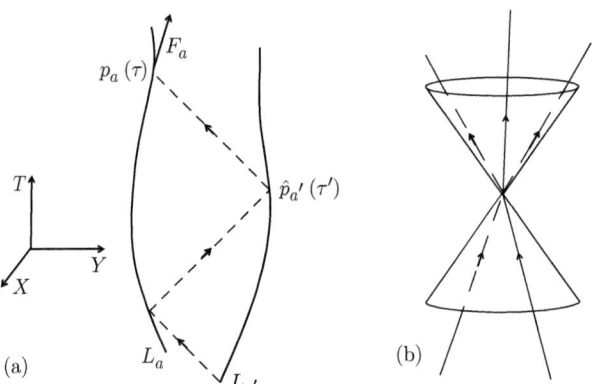

Fig. 6.1 Long- and short-range interactions.

particles are not necessarily of the same type, nor are the numbers of incoming and outgoing world lines necessarily the same; see[4] Fig. 6.1b.

In the rest of this chapter we shall focus on short-range interactions.

6.3   The momentum conservation law

Let $\tau \mapsto q(\tau)$ be the world line L of a massive object Q with equation $X^\mu = X^\mu(\tau)$ in a Minkowski frame $\mathcal{S}$ of coordinates X^μ. The vector $u = dq/d\tau$, of components $U^\mu = dX^\mu/d\tau$, normalized as $(u \cdot u) \equiv \eta_{\mu\nu} U^\mu U^\nu = -1$, is its 4-velocity and τ is its proper time.

The fundamental quantity describing the interaction of this object with other objects is

$$p = mu, \text{ whose components are } P^\mu = mU^\mu \text{ in the inertial reference frame } \mathcal{S}, \quad (6.5)$$

where m is its inertial mass. Since the vector u is normalized to unity, we have

$$p^2 = -m^2. \quad (6.6)$$

The components of p are written as [*cf.* Section 2.1]

$$P^0 \equiv E = \frac{m}{\sqrt{1 - V^2}}, \quad P^i = \frac{mV^i}{\sqrt{1 - V^2}} \quad \Longrightarrow \quad P^i = P^0 V^i, \quad (6.7)$$

where $V^i \equiv dX^i/dT$ is the 3-velocity of the object in $\mathcal{S}$. In the lowest orders in V we have

$$P^i = mV^i + \mathcal{O}(V^3), \quad P^0 \equiv E = m + \tfrac{1}{2}mV^2 + \mathcal{O}(V^4). \quad (6.8)$$

Since mV^i and $\tfrac{1}{2}mV^2$ in Newtonian physics are the particle *momentum* and *kinetic energy*, the vector p is called the *4-momentum* of the particle, with P^i its momentum and $P^0 - m$ its kinetic energy (both relativistic). In the (locally) inertial reference frame where the particle

---

[4]Of course, the interactions cannot be *strictly* pointlike. For example, nuclear interactions have a range L of order 10^{-5} m. In Sections 10.1 and 10.2 we shall see that if their strength decreases as $e^{-r/L}$ in one frame, it decreases exponentially in any other inertial reference frame.

is at rest, the components of the momentum reduce to $P^0 = m$ and $P^i = 0$, so that the mass m is also (in the units we have chosen) the *rest energy*, or *self energy*, of the particle.[5]

Now let us imagine a light line representing the motion of a light corpuscle.[6] By extension of (6.5) we can also associate with it a 4-momentum proportional to its wave 4-vector:

$$p = Ck \quad \text{such that} \quad p^2 = 0, \tag{6.9}$$

whose components are

$$P^0 \equiv E = C\omega, \; P^i = Ck^i \; \text{ with } \; \omega = \sqrt{k_i k^i} \implies P^i = P^0 c^i, \tag{6.10}$$

where ω is the frequency of the associated wave (*cf.* Section 4.1) and c^i has modulus 1. The quantity $P^0 \equiv E$ is the energy and P^i is the momentum of a light corpuscle. The constant C, which is expressed in kg-s and therefore has the dimension of an *action*, is the characteristic of the light corpuscle (just as the mass m is the characteristic of a body moving at a speed less than c).

If we wish to describe the momentum of a light corpuscle as the limit of that of a particle whose speed tends to c, $V^i \to c^i$ [*cf.* (6.7)], we are led to assign a mass $m = 0$ to it in order that the particle energy and momentum P^0 and P^i remain finite in the passage to the limit. Therefore, *'photons', that is, elementary light corpuscles, have zero mass.* This says nothing about the value of C, but it indicates that it must be the same for *all* photons.

Let us consider a collision. The total 4-momentum of the incident particles is the sum of the individual momenta, and the total 4-momentum of the outgoing particles is defined in the same way.

We *postulate* that these are equal:

$$p^{\text{in}} \equiv \sum_a p_a^{\text{in}}, \quad p^{\text{out}} \equiv \sum_a p_a^{\text{out}}, \quad p^{\text{in}} = p^{\text{out}}. \tag{6.11}$$

This is the *law of momentum conservation* (that is, of energy and 3-momentum conservation).

We recall that this law is valid only for short-range interactions. We also note that in general it is not sufficient for determining the individual momenta of the outgoing particles when those of the incoming particles are known. In order to do this it is necessary to specify the nature of the short-range force giving rise to the interaction.[7]

---

[5] The expression $P^0 = m$, more commonly written as $E = mc^2$, quoted at the beginning of Part II of this book, is probably the most famous formula in physics. The mass–energy equivalence appeared explicitly for the first time in the article Einstein (1905b), Ist die Trägheit eines Körpers von seinem Energieinhalt abhängig? (Does the inertia of a body depend upon its energy content?), where Einstein wrote, "*If a body gives off the energy L in the form of radiation, its mass diminishes by L/c^2*" and "*The mass of a body is a measure of its energy-content; if the energy changes by L, the mass changes in the same sense by $L/(9 \times 10^{20})$, the energy being measured in ergs, and the mass in grammes.*"

[6] As already mentioned in Sections 4.1 and 4.2, and as we shall see in detail in Chapters 15 and 16, the representation of an electromagnetic wave by a *single* light line is an idealization in the Maxwell theory of light. However, it allows us to give a very simple explanation of, for example, the Compton effect; see Section 6.5.

[7] The law (6.11) is due to G. N. Lewis and R. C. Tolman (1909).

On the principle of action and reaction

We recall that in Newtonian physics the momentum conservation law of a system of particles follows from the principle of the *equality of action and reaction* (Newton's *third law*) and is valid at any time, *including* during the interaction; see Book 1, Section 7.1. This principle is *not* applicable in special relativity owing to the fact that in the absence of a universal time it is impossible to specify *when* this equality must hold.

For the same reason, to define the total momentum of a system in the case of long-range interactions where the particle 4-velocities are not constant, it is necessary to specify the point on the world lines at which the 4-velocities are evaluated. The simplest approach is to evaluate them at the time T of a given inertial reference frame. Then the 4-momentum of a system of particles depends on the chosen frame (just as the 3-velocity of a particle depends on the inertial reference frame chosen in Newtonian physics). In Sections 10.4 and 12.4 we shall see that it is the sum of the 4-momenta of the particles *and* of the field carrying the interaction which is conserved.

On the other hand, in the case of short-range interactions where the individual momenta before and after the interaction are constant, the total momenta as well as the law (6.11) do not depend on the choice of inertial reference frame.

6.4   Collisions

There are innumerable applications of the conservation law (6.11) in particle physics.[8] Here we shall present a few illustrative examples.

First of all, we note that the relation (6.6) can be used to define the *effective mass* of a system of particles as

$$M_{\text{eff}}^2 \equiv -p_{\text{tot}}^2 = E_{\text{tot}}^2 - \pi_{\text{tot}}^2, \tag{6.12}$$

where $\pi_{\text{tot}} \equiv \sum_a \pi_a$ and $E_{\text{tot}} \equiv \sum_a E_a = \sum_a \sqrt{\pi_a^2 + m_a^2}$. In the present section we shall use the notation $E \equiv P^0$ and π for the momentum 3-vector of components P^i. According to (6.11), M_{eff}^2 is a *relativistic invariant*, that is, a scalar which has the same value *before* and *after* the collision in any inertial reference frame.

The *center-of-mass (cm) frame* is the reference frame in which $\pi_{\text{tot}} = 0$, so that $M_{\text{eff}} = E_{\text{tot}}^{\text{cm}}$ is the total energy of the system in this frame and

$$M_{\text{eff}} = \sum_a \sqrt{\pi_{a(\text{cm})}^2 + m_a^2} \geq \sum_a m_a. \tag{6.13}$$

The equality holds only if $\pi_a^{(\text{cm})} = 0$ for all the particles.

In the special case of a two-particle system the definition (6.12) becomes the following, using the expressions (6.7) for massive particles and (6.10) for photons:

$$M_{\text{eff}}^2 = m_1^2 + m_2^2 + 2E_1 E_2(1 - V_1.V_2), \quad M_{2\gamma} = 2E_1 E_2(1 - \cos\theta). \tag{6.14}$$

We see that $M_{2\gamma} = 0$ if the two photons propagate in the same direction ($\theta = 0$), even if there is no reference frame in which they are at rest.

---

[8]An introduction to the dynamics of particle collisions can be found in, for example, Langlois (2011) or Rougé (2005).

As our first application of the relativistic invariance of p^2_{tot}, let us calculate a reaction threshold. When the sum of the masses of the final state is greater than that of the initial state, it is necessary to supply some minimum kinetic energy in order for the reaction to occur. For example, let us consider antiproton production in the reaction $p+p \longrightarrow p+p+p+\bar{p}$ (O. Chamberlain and E. Segré, 1955). The energy in the center of mass of the final system must be at least $4m_{\text{proton}}$ [*cf.* (6.13)]. If the experiment is performed sending a proton of energy $E_{\text{incident}} = \sqrt{\pi^2 + m^2_{\text{proton}}}$ in the laboratory frame towards a fixed target ($E_{\text{targ}} = m_{\text{proton}}$, $\pi = 0$ in the same frame) the relativistic invariant is easily found to be

$$-p^2_{\text{tot,in}} = \left(\sum p^0_a\right)^2 - \left(\sum p^i_a\right)^2 = \left(\sqrt{\pi^2 + m^2_{\text{proton}}} + m_{\text{proton}}\right)^2 - \pi^2$$

$$= 2m^2_{\text{proton}} + 2m_{\text{proton}}E_{\text{inc}} \,.$$

Therefore, for the reaction to be possible we must have

$$-p^2_{\text{tot,in}} = -p^2_{\text{tot,final}} > 16\,m^2_{\text{proton}} \quad \Longrightarrow \quad E_{\text{inc}} > 7\,m_{\text{proton}}. \tag{6.15}$$

The *threshold energy* is the minimum kinetic energy of the incident particle needed for the reaction to occur, $E_{\text{thr}} \equiv E_{\text{inc}} - m$, or $E_{\text{thr}} = 6\,m_{\text{proton}} \sim 5.6$ GeV in our example. This calculation shows the usefulness of the relativistic invariant, because we have calculated it in the laboratory frame for the initial state and in the cm frame for the final state, thereby avoiding having to perform Lorentz transformations.

Our second example is the conversion of rays γ into electron–positron pairs (e^-, e^+). The reaction $\gamma \longrightarrow e^- + e^+$ cannot occur in a vacuum, because $\pi^2 = E^2$ for the incident photon while in the cm frame the final system has $\pi = 0$, so that p_{tot} cannot be conserved. However, photon conversion is possible in the presence of a nucleus N of mass M_N, and the reaction becomes $\gamma + N \longrightarrow N + e^- + e^+$. We have $-p^2_{\text{tot,in}} = (E_{\text{incident}} + M_N)^2 - E^2_{\text{incident}}$ in the laboratory frame, where N is initially at rest, and $-p^2_{\text{tot,fin}} = (E_{e^+} + E_{e^-} + E_N)^2$ in the cm frame. Given that $E_{e^\pm} \geq m_e$ and $E_N \geq M_N$, we deduce that $-p^2_{\text{tot,in}} = -p^2_{\text{tot,fin}} \geq (M + 2m_e)^2$, so that $E_{\text{inc}} \geq 2m_e(1 + m_e/M_N) \sim 2m_e$, and there is no need to know the nucleus motion after the interaction.[9]

Finally, we should also mention that the relativistic invariant (6.12) can be used to predict the presence in certain reactions of particles which cannot be detected. For example, it is observed that a radioactive nucleus N_1 is transformed into a nucleus N_2 with the emission of a continuous spectrum of electrons (β *decay*). This is possible only if the interaction involves a third decay product. Otherwise, that is, if $N_1 \longrightarrow N_2 + e^-$, we would have

$$-p^2_{\text{tot,in}} = M^2_{N_1} = -p^2_{\text{tot,fin}} = \left(\sqrt{M^2_{N_2} + \pi^2} + \sqrt{m^2_e + \pi^2_e}\right)^2,$$

because in the cm frame of the final state $\pi_{N_2} = -\pi_e$. This relation fixes π_e and also the energy of the emitted electrons $E_e = \sqrt{m^2_e + \pi^2_e}$, and no continuous spectrum would be

---

[9]Photon conversion into (e^+, e^-) pairs is used in high-energy astrophysics. Photons of energy greater than 100 keV cannot be detected by telescopes using focusing mirrors. In the EGRET (Energetic Gamma-Ray Experiment Telescope) instrument installed on the Compton Gamma-Ray Observatory satellite (1991–2000), the photon is converted in the upper part of the telescope into an electron–positron pair whose energy is then measured.

observed. Therefore, there must be at least one more particle in the final state, which in this case would be an electron antineutrino $\bar{\nu}_e$.

6.5 Compton scattering

In this section we shall illustrate the conservation law (6.11) by the Compton experiment, that is, the collision of a light corpuscle with a particle, and the concept of the *quantum of action* that can be derived from it.[10]

Let us imagine a light corpuscle and a particle characterized by their 4-momenta $p_{\gamma,\text{in}}$ with $(p_{\gamma,\text{in}} \cdot p_{\gamma,\text{in}}) = 0$ and $p_{e,\text{in}}$ with $(p_{e,\text{in}} \cdot p_{e,\text{in}}) = -m^2$, where m is the particle mass (see Fig. 6.2). We work in the inertial reference frame where the particle is at rest at the origin before the collision and choose the X axis to be parallel to the incident light corpuscle. Then

$$P^\mu_{\gamma,\text{in}} = E_{\text{in}}(1,1,0,0), \quad P^\mu_{e,\text{in}} = m(1,0,0,0), \tag{6.16}$$

where $E_{\text{in}} \equiv C\omega_{\text{in}}$ is the energy of the incident light corpuscle and ω_{in} is its frequency [*cf.* (6.9) and (6.10)]. After the collision the particle acquires a velocity V which by suitable choice of axes lies in the XOY plane and makes an angle ϕ with the OX axis:

$$P^\mu_{e,\text{out}} = \frac{m}{\sqrt{1-V^2}}(1, V\cos\phi, V\sin\phi, 0). \tag{6.17}$$

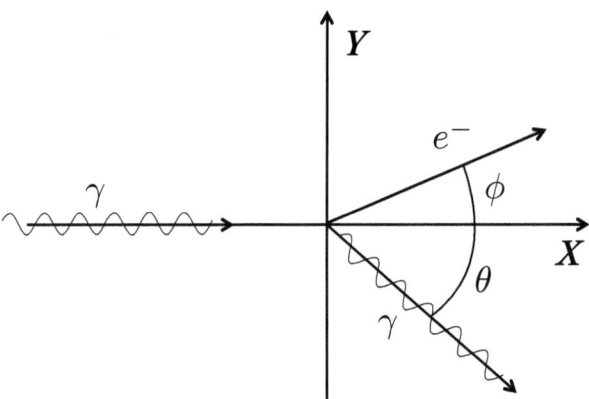

Fig. 6.2 Compton scattering.

At this stage we must make our first hypothesis. The conservation law (6.11) does not forbid the production of a bunch of N light corpuscles in the interaction, but we shall assume that each incoming corpuscle corresponds to a single outgoing one.[11]

The conservation law (6.11) then requires that the Z component of the outgoing light corpuscle momentum be zero. Therefore,

$$P^\mu_{\gamma,\text{out}} = E_{\text{out}}(1, \cos\theta, \sin\theta, 0), \tag{6.18}$$

---

[10]Compton (1923).

[11]For a discussion of this point see Hernandez (2005).

where θ is the angle its 3-velocity makes with the OX axis and $E_{\text{out}} \equiv C\omega_{\text{out}}$ is its energy, with ω_{out} its frequency after the collision. The law for the remaining components reads

$$\begin{cases} E_{\text{in}} + m = E_{\text{out}} + \dfrac{m}{\sqrt{1 - V^2}} \\[2mm] E_{\text{in}} = E_{\text{out}} \cos\theta + \dfrac{mV\cos\phi}{\sqrt{1 - V^2}} \\[2mm] 0 = E_{\text{out}} \sin\theta + \dfrac{mV\sin\phi}{\sqrt{1 - V^2}}\,. \end{cases} \tag{6.19}$$

This is a system of three equations for the four unknowns E_{out}, ϕ, V, and θ characterizing the final system. Choosing θ as a free parameter, after a bit of algebra we find

$$E_{\text{out}} = \frac{E_{\text{in}}}{1 + \frac{E_{\text{in}}}{m}(1 - \cos\theta)}, \tag{6.20}$$

so that

$$\begin{cases} \cot\phi = -\left(1 + \dfrac{E_{\text{in}}}{m}\right)\tan\dfrac{\theta}{2}, \quad V = \dfrac{1}{\sqrt{1 + A^2}} \\[3mm] \text{with } A^2 = \left(\dfrac{E_{\text{in}}}{m}\right)^2 \left[2 + \dfrac{E_{\text{in}}}{m}\left(2 + \dfrac{E_{\text{in}}}{m}\right)(1 - \cos\theta)\right] \dfrac{(1 - \cos\theta)}{\left[1 + \frac{E_{\text{in}}}{m}(1 - \cos\theta)\right]^2}\,. \end{cases} \tag{6.21}$$

Comparison of (6.20) and (6.21) with the experimental data (light scattering by a graphite plate) confirms the collision model that has been used and makes it possible to determine [by measuring $E_{\text{out}}(\theta)$, with E_{in} known] that the particle scattering the light is an electron,[12] $m = m_e$. However, (6.20) and (6.21) *a priori* say nothing about the wavelength of the scattered light.

However, since $E_{\text{in}} = C\omega_{\text{in}}$ and $E_{\text{out}} = C\omega_{\text{out}}$ where C is assumed independent of ω, from (6.20) we deduce that

$$\lambda_{\text{out}} - \lambda_{\text{in}} = \frac{2\pi C}{m_e}(1 - \cos\theta), \tag{6.22}$$

where we have introduced the wavelength $\lambda = 2\pi/\omega$. This expression (6.22) can be tested experimentally by measuring λ_{in} and λ_{out} using spectrographs. The agreement between theory and experiment confirms the hypothesis and also gives the value of C:

$$C = \frac{e^2}{\alpha} \equiv \hbar \quad \text{or} \quad \lambda_{\text{out}} - \lambda_{\text{in}} = \frac{h}{m_e}(1 - \cos\theta), \tag{6.23}$$

where e is the electron charge, $\alpha \approx 1/137$ is the *fine-structure constant*, $\hbar$ is the *Planck constant*, and $h/m_e \equiv 2\hbar\pi/m_e$ is the *Compton wavelength*.

---

[12]The electron can be considered as quasi-free because it is loosely bound to the light carbon atom of graphite. Therefore, as in Newtonian physics, the numerical values of the inertial masses can in principle be determined by means of collision experiments; *cf.* Book 1, Section 7.1.

The Compton experiment is therefore interpreted as a collision between an electron and a *fundamental* 'light corpuscle', the *photon*, and the action C is the *quantum of action* $\hbar$:

$$P^0 = \hbar\omega, \quad P^i = \hbar\omega c^i, \quad P_\mu P^\mu = 0. \tag{6.24}$$

For the effect to be measurable it is necessary [see (6.23)] that λ_{in} be of order h/m_e, or, numerically, $\lambda_{\text{in}} \approx 2.4 \times 10^{-12}$ m, which corresponds to hard x-rays or γ-rays. Therefore, electromagnetic waves of sufficiently short wavelength can be described as light corpuscles whose world line is a light line.

The quantum of energy $E = h\nu$

The idea of a quantum of energy was introduced by Planck in 1900 to explain the spectrum of a black body, and the hypothesis that light energy is composed of a finite number of quanta was made by Einstein in 1905 in his explanation of the photoelectric effect discovered by Lénard in 1902. In his article of 1923, Compton, in contrast to the presentation we have given here, did not derive the value of the quantum of action from his experiment, but instead used the Einstein formula $E = h\nu$ to interpret the results. The term *photon* was coined by G. N. Lewis in 1926. Finally, the value of the fine-structure constant is a fundamental experimental quantity which so far no theory has been able to explain.

In 1924 in his thesis *Recherches sur la théorie des quanta [Research on the theory of the quanta]* de Broglie proposed that since a wave vector and a momentum related as $p^\mu = \hbar k^\mu$ can be associated with a photon, the same should be true for any particle of mass m. With any particle should be associated a wave vector k^μ related to the particle 4-momentum as $k^\mu = p^\mu/\hbar$, with the constant of proportionality $\hbar$ assumed to be the same for *all* particles. Then $k^2 = -m^2/\hbar^2$ or $\omega^2 - k_i k^i = m^2/\hbar^2$. These wave properties of matter were confirmed in 1926 by Davisson and Germer, who observed that electrons scattered by a crystal form a diffraction pattern on a screen with the fringe spacing given by the de Broglie formula.

We conclude this chapter by mentioning that the description of a collision between a photon and an electron in a frame where the electron is initially moving (rather than at rest as we assumed above) can be obtained by Lorentz transformation of the results given above. This results in an *increase* of the photon energy. This is called *inverse Compton scattering* and is important in astrophysics.[13]

The Klein–Nishina and Thomson formulas

The law of total momentum conservation does not provide the direction θ of the outgoing photon in the Compton experiment. To obtain this it is necessary to describe the interaction between the photon and the electron *during* the collision. This falls within the domain of *quantum electrodynamics*.[14] In 1929, Klein and Nishina used quantum electrodynamics to derive the *effective cross section* of the process, that is, the number of photons scattered per unit time in a solid angle *do* of direction θ relative to the number of photons per unit time and unit area of the scattered beam:

$$\frac{d\sigma}{do} = \frac{1}{2}\left(\frac{e^2}{m_e}\right)^2 x^2 \left(x + 1/x - \sin^2\theta\right), \text{ with } x \equiv \frac{1}{1 + \epsilon(1 - \cos\theta)} \quad \text{and} \quad \epsilon \equiv \frac{\hbar\omega}{m_e} = \frac{e^2\omega}{\alpha\, m_e}. \tag{6.25}$$

---

[13] See, for example, Rybicki and Lightman (1985).

[14] See, for example, Weinberg (2005).

Here $x = E_{\text{out}}/E_{\text{in}}$ [*cf.* (6.20)] and $\hbar\omega = h/\lambda$ is the energy of the incoming photon. The total effective cross section is

$$\sigma \equiv 2\pi \int_0^\pi \frac{d\sigma}{do} \sin\theta \, d\theta = \pi \left(\frac{e^2}{m_e}\right)^2 \frac{1}{\epsilon^3} \left[\frac{2\epsilon\left(2 + 8\epsilon + 9\epsilon^2 + \epsilon^3\right)}{(1 + 2\epsilon)^2} - \left(2 + 2\epsilon - \epsilon^2\right)\log(1 + 2\epsilon)\right]. \quad (6.26)$$

In the *classical limit* of long wavelengths, $\epsilon = h/(m_e\lambda) \to 0$ and the Klein–Nishina effective cross section tends to the Thomson cross section obtained using the Maxwell theory [see Section 18.4]:

$$\frac{d\sigma}{do} \to \frac{1}{2}\left(\frac{e^2}{m_e}\right)^2 (1 + \cos^2\theta) \quad \text{and} \quad \sigma \to \frac{8\pi}{3}\left(\frac{e^2}{m_e}\right)^2. \quad (6.27)$$

The limit $\epsilon \to 0$ can also be interpreted as follows. The electron action is $\mathbf{S} = -m\sqrt{1 - V^2}\,T$ [this is the three-dimensional version of (2.8)], and the characteristic collision time is $T \sim 1/\omega$, where ω is the photon frequency. In addition, $V < 1$. We therefore obtain the order of magnitude $\mathbf{S}/\hbar \sim m/(\hbar\omega) = 1/\epsilon$. The *classical limit* where quantum effects can be neglected and the Maxwell theory applies then corresponds to

$$\frac{\mathbf{S}}{\hbar} \to \infty. \quad (6.28)$$

7
Rotating systems

In this chapter we continue our presentation of the laws of relativistic dynamics for systems of point particles, beginning with the law of angular momentum conservation in collisions. We also formulate the equations of motion for particles possessing an internal rotation or 'spin'.

7.1 Angular momentum and center of mass

We consider an ensemble of free particles each characterized by its (constant) momentum p_a. The total momentum $p = \sum_a p_a$ does not depend on the inertial frame which is used, but the angular momentum will depend on the frame, because its definition involves position vectors between a reference event and events q_a on the particle world lines, which are chosen to be *simultaneous* in a given frame.

Let us work in a particular inertial frame $\mathcal{S}$ where the components of the momenta p_a are P_a^μ. We take the reference event to be the origin O of this frame and choose the events q_a to lie at the intersections of the plane $T = $ const with the particle world lines; see Fig. 7.1. The components of the position vectors Oq_a will be X_a^μ. The *angular momentum tensor* of the system in $\mathcal{S}$ is

$$\mathcal{M}_\mathcal{S}^{\mu\nu} \equiv \sum_a (X_a^\mu P_a^\nu - X_a^\nu P_a^\mu) \quad \text{with} \quad X_a^\mu = (T, X_a^i(T)). \tag{7.1}$$

Since $P_a^i = m_a V_a^i / \sqrt{1 - V_a^2}$, where m_a is the mass of the particle a and V_a^i is the 3-velocity, the spatial components $\mathcal{M}_\mathcal{S}^{YZ}$ and so on, reduce to the components of the Newtonian angular momentum 3-vector $\mathcal{M}_O^i$ when $|V_a| \ll 1$ (see Book 1, Section 7.2).

Since the particles are free we have $X_a^i = V_a^i T + X_{a0}^i$, where X_{a0}^i are their positions at $T = 0$. In addition, $P_a^i \propto V_a^i$, and, finally, $\mathcal{M}_\mathcal{S}^{ij}$ is antisymmetric. Therefore, the terms involving $V_a^i V_a^j T$ are eliminated and the spatial components of the angular momentum are constant:

$$\mathcal{M}_\mathcal{S}^{ij} = \sum_a (X_{a0}^i P_a^j - X_{a0}^j P_a^i). \tag{7.2}$$

Since $P_a^i = P^0 V_a^i$, the same occurs for the cross-term components:

$$\mathcal{M}_\mathcal{S}^{0i} \equiv T \sum_a P_a^i - \sum_a X_a^i P_a^0 = T \sum_a V_a^i P_a^0 - \sum_a (V_a^i T + X_{a0}^i) P_a^0 = - \sum_a X_{a0}^i P_a^0. \tag{7.3}$$

The *angular momentum conservation law* states that if the particles collide, their angular momenta $\mathcal{M}_\mathcal{S}^{\mu\nu}$ will be the same before and after the collision.[1]

---

[1] We recall that in Newtonian physics the angular momentum conservation law follows from the principle of action and reaction, and is valid also during an interaction if the force is collinear with the particle separation

Relativity in Modern Physics. Nathalie Deruelle and Jean-Philippe Uzan.
© Oxford University Press 2018. Published in 2018 by Oxford University Press.
DOI: 10.1093/oso/9780198786399.001.0001

Introducing the total energy $P^0 \equiv \sum_a P_a^0$ and the total momentum $P^i \equiv \sum_a P_a^i = \sum_a P_a^0 V_a^i$, we can rewrite (7.3) as

$$R^i = V^i T + C_{\mathcal{S}}^i, \quad \text{where} \quad R^i \equiv \frac{\sum_a X_a^i P_a^0}{P^0}, \quad V^i \equiv \frac{P^i}{P^0}, \quad \text{and} \quad C_{\mathcal{S}}^i \equiv \frac{M_{\mathcal{S}}^{0i}}{P^0}. \tag{7.4}$$

Since $|V| < 1$, this equation can be interpreted as the trajectory of the *center of mass* in $\mathcal{S}$. The world line $L_{\mathcal{S}}$ has 4-velocity $U^\mu = \left(1, V^i\right)/\sqrt{1 - V^2}$ and intersects the $T = 0$ plane at the point $q_{\mathcal{S}}$ with coordinates $(0, C_{\mathcal{S}}^i)$; see Fig. 7.1.

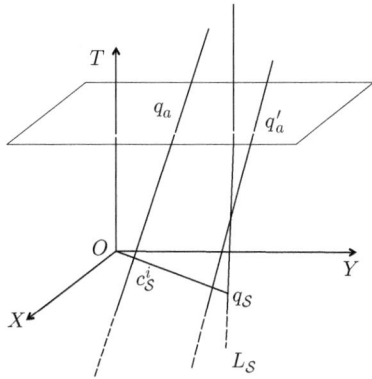

Fig. 7.1 World line of a center of mass.

It is important to note (see the demonstration below) that the 4-velocities of the centers of mass do not depend on the inertial frame chosen, and so their world lines are parallel straight lines of $\mathcal{M}_4$. However, they do *not* coincide: the point $q_{\mathcal{S}'} \notin L_{\mathcal{S}}$.

World lines of centers of mass

It is easy to show that centers of mass have the same velocity in all inertial frames. Using $V_{\mathcal{S}}^i \equiv P^i/P^0$ to denote the components in $\mathcal{S}$ of the velocity of the center of mass relative to $\mathcal{S}$, we have

$$V_{\mathcal{S}}'^X = (V_{\mathcal{S}}^X - V_0)/(1 - V_0 V_{\mathcal{S}}^X) = (P^X - V_0 P^0)/(P^0 - V_0 P^X),$$
$$V_{\mathcal{S}}'^Y = V^Y \sqrt{1 - V_0^2}/(1 - V_0 V_{\mathcal{S}}^X) = P^Y \sqrt{1 - V_0^2}/(P^0 - V_0 P^X)$$

in the frame $\mathcal{S}'$ moving with speed V_0 along the X axis; *cf.* (2.16). The components in $\mathcal{S}'$ of the velocity of the center of mass relative to $\mathcal{S}'$ will be, by definition, $V_{\mathcal{S}'}'^i \equiv P'^i/P'^0$. Since the 4-momentum is a 4-vector we have

$$P'^0 = (P^0 - V_0 P^X)/\sqrt{1 - V_0^2}, \quad P'^X = (P^X - V_0 P^0)/\sqrt{1 - V_0^2}, \quad P'^Y = P^Y.$$

vector; see Book 1, Section 7.2. The conservation law in (7.2) and (7.3) does not add anything to the study of the Compton effect in Section 6.5 because, as is easily shown, in that case it reduces to the conservation of total momentum.

Therefore,

$$V_{\mathcal{S}'}^{\prime X} = (P^X - V_0 P^0)/(P^0 - V_0 P^X) = V_{\mathcal{S}}^{\prime X}, \quad V_{\mathcal{S}'}^{\prime Y} = P^Y \sqrt{1 - V_0^2}/(P^0 - V_0 P^X) = V_{\mathcal{S}}^{\prime Y},$$

or $V_{\mathcal{S}}^{\prime i} = V_{\mathcal{S}'}^{\prime i}$. (Q.E.D.)

Now, to show that the world lines of the centers of mass relative to $\mathcal{S}$ and $\mathcal{S}'$, $L_{\mathcal{S}}$ and $L_{\mathcal{S}'}$, are parallel but never coincide, we write the condition for this to occur, namely, $q_{\mathcal{S}'} \in L_{\mathcal{S}}$. We have

$$C_{\mathcal{S}'}^{\prime X} = C_{\mathcal{S}}^X \sqrt{1 - V_0^2}/(1 - V_0 V^X), \quad C_{\mathcal{S}'}^{\prime Y} = C_{\mathcal{S}}^Y + V_0 V^Y C_{\mathcal{S}}^X/(1 - V_0 V^X),$$

where

$$C_{\mathcal{S}}^X \equiv -\sum_a X_{0a}^{\mathcal{S}} P_a^0/P^0, \quad C_{\mathcal{S}}^Y \equiv -\sum_a Y_{0a}^{\mathcal{S}} P_a^0/P^0$$

and similar expressions with primes on all quantities for $C_{\mathcal{S}'}^{\prime X}$ and $C_{\mathcal{S}'}^{\prime Y}$. We can now express $X_{0a}^{\prime \mathcal{S}'}$ and $Y_{0a}^{\prime \mathcal{S}'}$ as functions of $X_{0a}^{\mathcal{S}}$ and $Y_{0a}^{\mathcal{S}}$, which gives

$$X_{0a}^{\prime \mathcal{S}'} = X_{0a}^{\mathcal{S}} \sqrt{1 - V_0^2}/(1 - V_0 V_a^X), \quad Y_{0a}^{\prime \mathcal{S}'} = Y_{0a}^{\mathcal{S}} + V_0 V_a^Y X_{0a}^{\mathcal{S}}/(1 - V_0 V_a^X).$$

Substituting these expressions into those for $C_{\mathcal{S}'}^{\prime X}$ and $C_{\mathcal{S}'}^{\prime Y}$, we find

$$C_{\mathcal{S}'}^{\prime X} = C_{\mathcal{S}}^X \sqrt{1 - V_0^2}/(1 - V_0 V^X)$$

and

$$C_{\mathcal{S}'}^{\prime Y} = (C_{\mathcal{S}}^Y - V_0 \mathcal{M}_{\mathcal{S}}^{XY}/P^0)/(1 - V_0 V^X) \neq C_{\mathcal{S}}^Y + V_0 V^Y C_{\mathcal{S}}^X/(1 - V_0 V^X).$$

Therefore, $q_{\mathcal{S}'} \notin L_{\mathcal{S}}$.

The *center-of-mass frame* is the inertial frame $\mathcal{S}$ (defined up to a rotation of the spatial axes) in which the total 3-momentum $P^i = \sum_a P_a^i$ vanishes. The center of mass is therefore at rest there ($V^i = P^i/P^0 = 0$), and its 4-velocity has components $U^\mu = (1, 0, 0, 0)$. We can choose the origin such that the world line coincides with the time axis: $C_{\mathcal{S}}^i = 0$.

The cross components of the angular momentum tensor relative to the cm frame are zero [$\mathcal{M}^{0i} = 0$; cf. (7.3) and (7.4)], and for small velocities their spatial components reduce to those of the Newtonian angular momentum vector of the center of mass.

7.2   Intrinsic angular momentum

In a frame $\mathcal{S}'$ related to $\mathcal{S}$ by a Poincaré transformation [$X^{\prime \mu} = \Lambda_\rho{}^\mu(X^\rho - d^\rho)$; cf. Section 1.3], the components of the angular momentum tensor relative to $\mathcal{S}$ are $\mathcal{M}_{\mathcal{S}}^{\prime \mu \nu} = \Lambda_\rho{}^\mu \Lambda_\sigma{}^\nu \mathcal{M}_{\mathcal{S}}^{\rho \sigma}$, where $\mathcal{M}_{\mathcal{S}}^{\rho \sigma}$ are its components[2] in $\mathcal{S}$.

Now let us consider the angular momentum tensor in $\mathcal{S}'$. Its components $\mathcal{M}_{\mathcal{S}'}^{\prime \mu \nu}$ in $\mathcal{S}'$ are [cf. (7.1)] $\mathcal{M}_{\mathcal{S}'}^{\prime \mu \nu} \equiv \sum_a (X_a^{\prime \mu} P_a^{\prime \nu} - X_a^{\prime \nu} P_a^{\prime \mu})$, where the $P_a^{\prime \mu}$ are the momentum components in $\mathcal{S}'$. Since $P_a^{\prime \mu} = \Lambda_\rho{}^\mu P_a^\rho$, we have

$$\mathcal{M}_{\mathcal{S}'}^{\prime \mu \nu} = \sum_a \Lambda_\rho{}^\mu \Lambda_\sigma{}^\nu [(X_a^\rho - d^\rho)P_a^\sigma - (X_a^\sigma - d^\sigma)P_a^\rho] = \mathcal{M}_{\mathcal{S}}^{\prime \mu \nu} - (d^{\prime \mu} P^{\prime \nu} - d^{\prime \nu} P^{\prime \mu}), \quad (7.5)$$

---

[2]A geometric presentation of intrinsic angular momentum can be found in Gourgoulhon (2010).

where P'^μ are the components of the total momentum of the system in $\mathcal{S}'$. Therefore, as stated in the preceding section, the angular momentum in $\mathcal{S}'$ differs from that in $\mathcal{S}$.

Now we introduce the (covariant) vector $J_\mathcal{S}$ with components

$$J_{\mu\,\mathcal{S}} = -\frac{1}{2}e_{\mu\nu\rho\sigma}\mathcal{M}_\mathcal{S}^{\nu\rho}\,U_{\mathrm{cm}}^\sigma, \tag{7.6}$$

where $e_{\mu\nu\rho\sigma}$ is the Levi-Civita symbol (see below for its properties) and $\mathcal{M}_\mathcal{S}^{\nu\rho}$ and U_{cm}^σ are the components in $\mathcal{S}$ of the angular momentum tensor in $\mathcal{S}$ and the 4-velocity of the center of mass. We note that the vector J_μ has only three independent components because $J_\mu U_{\mathrm{cm}}^\mu = 0$.

Owing to (7.5), in another frame $\mathcal{S}'$ the components will be

$$\begin{aligned}
J'_{\mu\,\mathcal{S}} &= -\frac{1}{2}e_{\mu\nu\rho\sigma}\mathcal{M}_\mathcal{S}^{\prime\nu\rho}U_{\mathrm{cm}}^{\prime\sigma} = -\frac{1}{2}e_{\mu\nu\rho\sigma}\left[\mathcal{M}_{\mathcal{S}'}^{\prime\nu\rho} + (d'^\nu P'^\rho - d'^\rho P'^\nu)\right]U_{\mathrm{cm}}^{\prime\sigma}\\
&= J'_{\mu\,\mathcal{S}'} + K'_\mu \quad\text{with}\quad J'_{\mu\,\mathcal{S}'} \equiv -\frac{1}{2}e_{\mu\nu\rho\sigma}\mathcal{M}_{\mathcal{S}'}^{\prime\nu\rho}U_{\mathrm{cm}}^{\prime\sigma}\\
&\text{and}\quad K'_\mu = -e_{\mu\nu\rho\sigma}d'^\nu P'^\rho U_{\mathrm{cm}}^{\prime\sigma}.
\end{aligned} \tag{7.7}$$

In the cm frame where $U_{\mathrm{cm}}^{\prime\sigma} = (1,0,0,0)$, the components of the vector K are all zero. We indeed have $K'_0 = -e_{0ij0}d'^i P'^j = 0$ and $K'_i = -e_{i\mu k0}d'^\mu P'^k = 0$ owing to the properties of the Levi-Civita symbol and because $P'^k = 0$. The vector K then is a null vector because its components $K_\mu = \Lambda^\nu{}_\mu K'_\nu$ will vanish in any frame $\mathcal{S}$. The vector J therefore will not depend on the frame $\mathcal{S}$ in which the angular momentum is defined:

$$J_{\mu\,\mathcal{S}} = J_{\mu\,\mathcal{S}'} \equiv J_\mu. \tag{7.8}$$

The vector J is the *intrinsic angular momentum*. In the center-of-mass frame its components are

$$J_0 = 0, \quad J_i = \frac{1}{2}e_{ijk}\mathcal{M}_{\mathrm{cm}}^{jk}, \tag{7.9}$$

which is the generalization of Newton's definition; see Book 1, Section 6.1 [this justifies the factor of $(-1/2)$ in the definition (7.6)].

The Levi-Civita symbol

The Levi-Civita symbol $e_{\mu\nu\rho\sigma}$ is the completely antisymmetric pseudo-tensor such that $e_{0123} = +1$. Its indices are raised using the inverse Minkowski metric $\eta^{\mu\nu}$. Therefore, $e^0{}_{123} = e^{0123} = -1$. It possesses the following properties:

$$e^{\mu\nu\rho\sigma}e_{\alpha\beta\gamma\delta} = -\det\begin{pmatrix} \delta_\alpha^\mu & \delta_\beta^\mu & \delta_\gamma^\mu & \delta_\delta^\mu \\ \delta_\alpha^\nu & \delta_\beta^\nu & \delta_\gamma^\nu & \delta_\delta^\nu \\ \delta_\alpha^\rho & \delta_\beta^\rho & \delta_\gamma^\rho & \delta_\delta^\rho \\ \delta_\alpha^\sigma & \delta_\beta^\sigma & \delta_\gamma^\sigma & \delta_\delta^\sigma \end{pmatrix}, \quad e^{\mu\nu\rho\sigma}e_{\alpha\beta\gamma\sigma} = -\det\begin{pmatrix} \delta_\alpha^\mu & \delta_\beta^\mu & \delta_\gamma^\mu \\ \delta_\alpha^\nu & \delta_\beta^\nu & \delta_\gamma^\nu \\ \delta_\alpha^\rho & \delta_\beta^\rho & \delta_\gamma^\rho \end{pmatrix}, \tag{7.10}$$

as well as

$$e^{\mu\nu\rho\sigma}e_{\alpha\beta\rho\sigma} = -2(\delta^\mu_\alpha\delta^\nu_\beta - \delta^\mu_\beta\delta^\nu_\alpha), \quad e^{\mu\nu\rho\sigma}e_{\alpha\nu\rho\sigma} = -6\,\delta^\mu_\alpha \,,$$

and, finally,

$$e^{\mu\nu\rho\sigma}e_{\mu\nu\rho\sigma} = -24 \,.$$

We recall that the Levi-Civita symbol with three indices e_{ijk} (which is completely anti-symmetric with $e_{123} = 1$, and whose indices are raised using the Euclidean metric δ^{ij} so that $e^{123} = +1$) possesses similar properties:

$$e_{ijk}e_{lmn} = \delta_{il}\delta_{jm}\delta_{kn} + \delta_{im}\delta_{jn}\delta_{kl} + \delta_{in}\delta_{jl}\delta_{km} - \delta_{il}\delta_{jn}\delta_{km} - \delta_{im}\delta_{jl}\delta_{kn} - \delta_{in}\delta_{jm}\delta_{kl}, \quad (7.11)$$

which implies that $e_{ijk}e^k{}_{mn} = \delta_{im}\delta_{jn} - \delta_{in}\delta_{jm}$, $e_{ijk}e^{jk}{}_l = 2\delta_{il}$, and $e_{ijk}e^{ijk} = 6$ (see Book 1, Section 1.5).

7.3 Spin dynamics

The discussion of the two preceding sections has led us to the definition of the intrinsic angular momentum of a system of free particles, (7.6) and (7.8).

Now that this definition has been elaborated, we can forget about how we arrived at it and introduce the concept of a particle, not necessarily free, which possesses not only an inertial mass, but also an intrinsic angular momentum or *spin*[3] j, a vector orthogonal to its 4-velocity u: $(j \cdot u) = 0$. In an inertial frame $\mathcal{S}$ where its components are J_μ and those of u are U^μ we have $J_\mu U^\mu = 0$.

Relativistic 'tops'

In Newtonian physics it is possible to define the concept of a rigid body, an ensemble of massive points whose relative spatial separations remain constant in time. It is also possible to introduce a reference frame attached to this solid which is derived from inertial frames by a transformation belonging to the group of rigid displacements. A vector Ω characterizes the rotation of this frame, and therefore that of the solid, relative to inertial frames. Concomitantly, we can introduce the angular momentum vector J of the system relative to its center of mass. This J is a quantity which is dynamically related to Ω *via* the inertia tensor of the solid. If Ω and J are parallel, the body can be regarded as a spherical top, while if they are not, the top is asymmetric. Newton's laws of dynamics imply that the angular momentum J is constant if no torque is exerted on the body. Therefore, no matter what the motion of the center of mass is, the vector J will keep its orientation fixed relative to inertial frames. If the top is a sphere, its rotation vector Ω, which can be measured, will also keep its orientation fixed. This corresponds to a gyroscope; see Book 1, Section 6.1.

In classical (non-quantum) physics, a relativistic 'particle' carrying a 'spin' j (the relativistic version of the vector J) is not a fundamental object but rather a composite one, just as in Newtonian physics. Moreover, it must be an extended object. The order of magnitude of the spin is $j \simeq mrv$, where m is the mass of the object, r is its size (the frame this is measured in is not important here), and v is its linear rotation velocity. Since $v < 1$ we have $r > j/m$.

In addition, since the idea of an accelerated rigid body cannot be defined in special relativity (except in special cases; *cf.* Section 5.4), it is not possible to directly relate the angular momentum j to a global angular velocity of the system as in Newtonian physics. The only case where

---

[3]It should be clear that the concept of 'spin' introduced here has nothing to do with quantum mechanics.

this is possible is if the 'particle' is a sphere (in an inertial frame where it does not spin). Indeed, from symmetry we can state that its rotational velocity Ω, which is measurable, must be proportional to j.

In an inertial frame the equation of motion of a particle of mass m carrying a spin and interacting with an external field therefore splits into two parts: one is the equation determining the motion of its center of mass $p(\tau)$ given in Section 6.2, $m\gamma = F$, where $\gamma \equiv \dot{u} \equiv dp/d\tau$ is the 4-acceleration, $u \equiv dp/d\tau$ with $(u \cdot u) = -1$ is the 4-velocity, and F is the force 4-vector. The other is the evolution equation of its spin, which must also be postulated:

$$\frac{dj}{d\tau} - u(j \cdot \gamma) = C \quad \text{with} \quad (j \cdot u) = 0 \quad \text{and} \quad (C \cdot u) = 0, \tag{7.12}$$

or, in component form,

$$\frac{dJ_\mu}{d\tau} - U_\mu(J_\nu \gamma^\nu) = C_\mu \quad \text{with} \quad J_\mu U^\mu = 0 \quad \text{and} \quad C_\mu U^\mu = 0. \tag{7.13}$$

The term proportional to u is needed in order for j to remain orthogonal to u. The quantity C is the *torque* applied to the spin, which like the force must be specified; just like j, it has only three independent components.[4]

In the inertial frame tangent to the trajectory where $U^\mu = (1, 0, 0, 0)$ we have $J_\mu = (0, J)$ and, as in Newtonian physics, (7.12) reduces to $dJ/dt = C$, where t is the time measured by clocks at rest in the tangent inertial frame. If the particle is free ($\gamma = 0$), (7.13) also reduces to $dJ_\mu/d\tau = C_\mu$. However, in contrast to the Newtonian prediction, a free spin ($C = 0$) does not preserve a fixed direction in an inertial frame if it is accelerated. This is the so-called *Thomas precession*, which will be discussed in the following section.

The equations of motion of a spin in any frame are obtained by making the above equations covariant:

$$\frac{\tilde{D}j^\mu}{d\tau} - u^\mu \left(j_\nu \frac{\tilde{D}u^\nu}{d\tau} \right) = c^\mu \quad \text{with} \quad j_\mu u^\mu = 0 \quad \text{and} \quad c_\mu u^\mu = 0, \tag{7.14}$$

where $\tilde{D}$ is the covariant derivative associated with the coordinates x^μ, τ is the proper time of the particle, and u^μ is the particle 4-velocity such that $\ell_{\mu\nu} u^\mu u^\nu = -1$, $\ell_{\mu\nu}$ being the components of the Minkowski metric in the coordinates x^μ. Finally, $j_\mu = (\partial X^\nu / \partial x^\mu) J_\nu$, $u^\mu = (\partial x^\mu / \partial X^\nu) U^\nu$, and $c^\mu = (\partial x^\mu / \partial X^\nu) C^\nu$ are the components of the spin j, the velocity u, and the torque c, J_ν, U^ν, and C^ν being their components in the Minkowski frame $\mathcal{S}$.

7.4 Thomas precession

Let a particle carrying a spin be constrained to travel along a circular trajectory at constant speed. How does the orientation of the spin vary in this motion?[5]

---

[4]The concept of a particle carrying a spin was introduced by J. Frenkel in 1926 and the equation of motion (7.13) is due to M. Mathisson (1937).
  We shall see an example of a torque C in Section 13.4.

[5]The precession of the axes of a solid body moving in a circular orbit was discovered by E. Borel in 1913. He calculated the effect in lowest order in the speeds using the non-commutative composition law for

Given an inertial frame $\mathcal{S}$ and its Minkowski coordinates $X^\mu = (T, X, Y, Z)$, the equations of motion of the particle spin reduce to the following in the absence of a torque [see (7.13)]:

$$\frac{dJ^\mu}{d\tau} = U^\mu \left(J_\nu \frac{dU^\nu}{d\tau} \right) \quad \text{with} \quad J_\mu U^\mu = 0, \tag{7.15}$$

where J^μ and U^μ are the components of the spin and the 4-velocity in $\mathcal{S}$.

The particle world line is constrained to be $X^\mu = (T, r_0 \cos\Omega T, r_0 \sin\Omega T, 0)$, and the particle 4-velocity is (cf. Section 2.1) $U^\mu \equiv dX^\mu/d\tau = \Gamma(1, -r_0\Omega\sin\Omega T, r_0\Omega\cos\Omega T, 0)$ with $\Gamma \equiv (1 - r_0^2\Omega^2)^{-1/2}$ so that $U_\mu U^\mu = -1$. The constraint $J_\mu U^\mu = 0$ gives $J_0 = -J^0 = r_0\Omega(J_X \sin\Omega T - J_Y \cos\Omega T)$. The equations of motion for the spatial components are $dJ^Z/dT = 0$ and

$$\begin{cases} \dfrac{dJ^X}{dT} = +r_0^2\Omega^3\Gamma^2 \sin\Omega T (J^X \cos\Omega T + J^Y \sin\Omega T) \\[2mm] \dfrac{dJ^Y}{dT} = -r_0^2\Omega^3\Gamma^2 \cos\Omega T (J^X \cos\Omega T + J^Y \sin\Omega T), \end{cases} \tag{7.16}$$

the solution of which such that $J^Y = 0$ at $T = 0$ is[6]

$$\begin{cases} J^X = J_1(\cos\Omega T \cos\Gamma\Omega T + \Gamma\sin\Omega T \sin\Gamma\Omega T) \\[2mm] J^Y = J_1(\sin\Omega T \cos\Gamma\Omega T - \Gamma\cos\Omega T \sin\Gamma\Omega T) \\[2mm] J^Z = J_2, \quad J^0 = -J_1 r_0 \Gamma\Omega \sin\Gamma\Omega T. \end{cases} \tag{7.17}$$

After one revolution, at $\Delta T = 2\pi/\Omega$, J^Y has increased by $\Delta J^Y = -J_1\Gamma\sin(2\pi\Gamma)$. In the limit of small speeds we can write $\Gamma \approx 1 + r_0^2\Omega^2/2$ and $\Delta J^Y \approx -2\pi J_1 r_0^2\Omega^2/2$. Therefore, the 3-vector (J^X, J^Y, J^Z) rotates about the Z axis (orthogonal to the trajectory) with angular velocity

$$\omega_{\text{Thomas}} \approx -\frac{r_0^2\Omega^3}{2}, \tag{7.18}$$

where the sign indicates that the rotation is in the direction opposite to the orbital rotation.

Thomas precession and a rotating frame

We consider a particle with a spin constrained to undergo circular motion of frequency Ω in an inertial frame $\mathcal{S}$. The particle world line is $X^\mu = (T, r_0 \cos\Omega T, r_0 \sin\Omega T, 0)$.

Now we pass to the coordinate system where the particle is at rest: $X^\mu = (T, X, Y, Z) \mapsto x^\mu = (t, r, \psi, z)$ so that $T = t$, $X = r\cos(\psi + \Omega t)$, $Y = r\sin(\psi + \Omega t)$, and $Z = z$. Next we set

---

non-collinear Lorentz transformations. The effect was rediscovered in atomic physics by L. Thomas in 1926, thus the name Thomas rotation; cf. Section 1.5. It is also sometimes called Wigner rotation in reference to the work of E. Wigner (1939) on the structure of the Lorentz group. Here we follow the treatment of the subject by Misner *et al.* (1973).

[6] One way of obtaining this is to first rewrite (7.16) in the form $dJ^X/dT = \Gamma^2\Omega^3 r_0^2 f \sin\Omega T$, $dJ^Y/dT = -\Gamma^2\Omega^3 r_0^2 f \cos\Omega T$ with $f \equiv J^X \cos\Omega T + J^Y \sin\Omega T$. We then differentiate $f(T)$ twice, which gives $d^2 f/dT^2 = -\Gamma^2\Omega^2 f$, the solution of which is $f = A\cos\Gamma\Omega T + B\sin\Gamma\Omega T$. Therefore, $J^X \cos\Omega T + J^Y \sin\Omega T = A\cos\Gamma\Omega T + B\sin\Gamma\Omega T$. We differentiate again to obtain $J^X \sin\Omega T - J^Y \cos\Omega T = \Gamma(A\sin\Omega T - B\cos\Omega T)$. The solution (7.17) follows.

$x^\mu \mapsto x'^\mu = (t, x, y, z)$ with $x = r \cos \psi$ and $y = r \sin \psi$. In this system the spatial coordinates are Cartesian, the particle world line is $x'^\mu = (t, r_0, 0, 0)$, and the Minkowski line element is written as (*cf.* Section 5.3)

$$ds^2 = -(1 - \Omega^2(x^2 + y^2))dt^2 + 2\Omega\, dt(xdy - ydx) + dx^2 + dy^2 + dz^2. \tag{7.19}$$

In this new frame the covariant components j'_μ of the spin are given as a function of its components $J_\mu = \eta_{\mu\nu}J^\nu$ in $\mathcal{S}$ obtained in (7.17) by $j'_\mu = (\partial x^\nu/\partial x'^\mu)(\partial X^\rho/\partial x^\nu)J_\rho$ and have the especially simple expression

$$j'_0 = 0, \quad j'_x = J_1 \cos \Gamma \Omega t, \quad j'_y = -\Gamma J_1 \sin \Gamma \Omega t, \quad j'_z = 0, \tag{7.20}$$

where $\Gamma \equiv (1 - r_0^2\Omega^2)^{-1/2}$. In the nonrelativistic limit where $\Gamma \to 1$, the time evolution of the spin components, of fixed direction in $\mathcal{S}$, is nothing but a reflection of the rotation of the system $x^\mu = (t, x, y, z)$.

The result (7.20) can be obtained by direct calculation in the rotating frame $\mathcal{C}' = (t, x, y, z)$. The Christoffel symbols of the metric (7.19) were given in (5.33) and on the particle world line are $\tilde{\Gamma}^x_{tt} = -r_0\Omega^2$ and $\tilde{\Gamma}^y_{tx} = -\tilde{\Gamma}^x_{ty} = +\Omega$. In addition, the contra- and covariant components of the 4-velocity are $u'^\mu = \Gamma(1, 0, 0, 0)$ and $u'_\mu = (-1/\Gamma, 0, \Gamma\Omega r_0, 0)$. The constraint $j'_\mu u'^\mu = 0$ implies that $j'_0 = 0$. Finally, recalling [*cf.* (5.30)] that $\tilde{D}j'_\mu/d\tau = dj'_\mu/d\tau - \tilde{\Gamma}^\nu_{\mu\rho}u'^\rho j'_\nu$ and [*cf.* (5.28)] that $\tilde{D}u'^\mu/d\tau = du'^\mu/d\tau + \tilde{\Gamma}^\mu_{\nu\rho}u'^\nu u'^\rho$, the equations of motion of the spin (7.17) become $dj'_z/dt = 0$ and

$$\frac{dj'_x}{dt} = \Omega\, j'_y, \quad \frac{dj'_y}{dt} = -\Gamma^2\Omega\, j'_x, \tag{7.21}$$

the solution of which is indeed (7.20).

Thomas precession and the Gravity Probe B experiment

In the Gravity Probe B (GPB) experiment (2004), gyroscopes were placed in orbit with the goal of measuring their precession during their motion about the Earth.[7]

If the Earth's gravitational field is described using Newton's theory, by equating the centrifugal and gravitational forces for a circular orbit we find $\Omega^2 r_0 = GM_\oplus/r_0^2$, where G is Newton's constant and $M_\oplus$ is the Earth's mass. Therefore [*cf.* (7.18)],

$$\omega_{\text{Thomas}} = -\frac{1}{2r_0}\left(\frac{GM_\oplus}{R_\oplus}\right)^{3/2}\left(\frac{R_\oplus}{r_0}\right)^{3/2}. \tag{7.22}$$

Since the gravitational radius of the Earth is $GM_\oplus/R_\oplus \approx 7 \times 10^{-10}$ with $R_\oplus \approx 6400$ km and the satellite altitude is 642 km, we obtain $\omega_{\text{Thomas}} \approx -3.4 \times 10^{-13}/\text{s} \approx -2.2$ arcsec/yr.

The precession measured by the GPB experiment, called the *geodetic precession*, is $\omega_{\text{geod}} = +6.6$ arcsec/yr with an accuracy of 1%. The flagrant disagreement between ω_{geod} and ω_{Thomas} is an indication of the impossibility of incorporating gravity into special relativity. Needless to say, the measured result is in complete agreement with the prediction of general relativity.

---

[7]A detailed description of the experiment can be found at http://einstein.stanford.edu/.

8
Fields and matter

Here we shall discuss the laws of relativistic dynamics for continuous media, namely, the 'fields' which mediate interactions in relativistic theories, and also fluids.

8.1 Equations of motion of a free field

The concept of *field* introduced by Faraday and formalized by Maxwell lies at the heart of contemporary physics. The intuitive idea behind this concept is, on the one hand, that massive bodies, owing to their internal constitution (that is, their *charge*), impregnate space with what are called 'fields', that is entities which are revealed by the presence of other bodies possessing the same type of charge. The other aspect of this idea is that the interactions between these bodies, which determine their motion, are conveyed by these fields. This physical concept of a field is represented mathematically by one or several functions of points p in Minkowski spacetime, which we shall call $\Phi(p)$.

The phenomenology of a field may vary from one inertial frame to another, but, to satisfy the principle of relativity, we require—or rather we guess—that the manifestations of a field in various particular frames are avatars of a single geometrical object, a *tensor field*, which are related to each other by Lorentz transformations. (See Section 1.3 for a review of the tensor transformation laws.)

In a given inertial frame and its associated Minkowski coordinates X^μ, a field *configuration* is a 'photograph' of the field at an instant of time T given by the clocks of the frame, that is, its spatial components $\Phi^{i_1 \cdots i_p}_{j_\mu \cdots j_q}(X^i)|_T$. The field dynamics is the time evolution of this spatial configuration.

The dynamics of a free field, that is, its equations of motion, is determined by specifying in an inertial frame $\mathcal{E}$ a *Lagrangian density* $\mathcal{L}(\Phi, \partial_\mu \Phi)$, which is a functional of the field and its derivatives,[1] and we choose, from among all the configurations evolving between two given configurations, those which extremize the action

$$\mathbf{S}_{\mathrm{f}}[\Phi(X^\mu, s)] = \int d^4 X \, \mathcal{L}(\Phi, \partial_\mu \Phi). \tag{8.1}$$

Here $d^4 X \equiv dX^0 \, d^3 X = dX^0 \, dX^1 dX^2 dX^3$ defines the *volume element* of $\mathcal{M}_4$ in Minkowski coordinates and s parametrizes, at a given $X^0 \equiv T$, the various configurations $\Phi(X^\mu, s)$. Using the notation $\delta\Phi \equiv (\partial\Phi/\partial s)|_0 \, ds$ and $\delta\partial_\mu\Phi \equiv (\partial^2\Phi/\partial s\partial X^\mu)|_0 \, ds = \partial_\mu\delta\Phi$ for the

---

[1] For an introduction to the Lagrangian and Hamiltonian mechanics of a point, see, for example, Book 1, Chapters 8 and 9.

We also recall that when there is no possible ambiguity regarding the coordinate system which is used, it will be convenient to denote partial derivatives as $\frac{\partial\Phi}{\partial X^\mu} \equiv \partial_\mu\Phi \equiv \Phi_{,\mu}$.

Relativity in Modern Physics. Nathalie Deruelle and Jean-Philippe Uzan.
© Oxford University Press 2018. Published in 2018 by Oxford University Press.
DOI: 10.1093/oso/9780198786399.001.0001

variations of Φ and its derivatives in passing from the configuration $s = 0$ to a neighboring configuration, the variation of the action $\delta \mathbf{S}_{\mathrm{f}} \equiv (d\mathbf{S}_c/ds)|_0 \, ds$ is written as

$$
\begin{aligned}
\delta \mathbf{S}_{\mathrm{f}} &= \int d^4 X \, \delta \mathcal{L} = \int d^4 X \left[\frac{\partial \mathcal{L}}{\partial \Phi} \delta \Phi + \frac{\partial \mathcal{L}}{\partial \partial_\mu \Phi} \delta \partial_\mu \Phi \right] \\
&= \int d^4 X \left[\frac{\partial \mathcal{L}}{\partial \Phi} - \partial_\mu \left(\frac{\partial \mathcal{L}}{\partial \partial_\mu \Phi} \right) \right] \delta \Phi + \int d^4 X \, \partial_\mu \left(\frac{\partial \mathcal{L}}{\partial \partial_\mu \Phi} \delta \Phi \right) \\
&= \int d^4 X \left[\frac{\partial \mathcal{L}}{\partial \Phi} - \partial_\mu \left(\frac{\partial \mathcal{L}}{\partial \partial_\mu \Phi} \right) \right] \delta \Phi + \int d^3 X \left(\frac{\partial \mathcal{L}}{\partial \partial_0 \Phi} \delta \Phi \right) \Bigg|_{T_1}^{T_2} \\
&\quad + \int_{T_1}^{T_2} dT \int_S \frac{\partial \mathcal{L}}{\partial \partial_i \Phi} \delta \Phi \, dS_i \,,
\end{aligned}
\tag{8.2}
$$

where S is the boundary of the 3-space (for example, a cube) and dS_i is an element of the surface $[dS_i = (dY\,dZ, dX\,dZ, dX\,dY)$ in Cartesian coordinates]. The first boundary term is zero because the configurations are fixed at T_1 and T_2, while the second, obtained by using Gauss's theorem, is also zero if we consider configurations which are fixed at spatial infinity or if restriction is made to configurations which vanish sufficiently rapidly. Therefore, the equations of motion of the field are the *Euler–Lagrange equations*

$$
\frac{\partial \mathcal{L}}{\partial \Phi} - \partial_\mu \left(\frac{\partial \mathcal{L}}{\partial \partial_\mu \Phi} \right) = 0 \,.
\tag{8.3}
$$

If $\mathcal{L}$ is nonlinear in $\partial_\mu \Phi$, these are second-order differential equations.[2]

8.2 The energy–momentum tensor of a free field

A field Φ and its derivatives $\partial_\mu \Phi$ are tensors under Lorentz transformations, that is, are geometrical objects with an intrinsic definition independent of the inertial frame. For the equations of motion (8.3) to also satisfy the relativity principle, that is, to be the same in any inertial frame, it is necessary that the action (8.1), a functional of Φ, be a scalar, that is, its numerical value for a given configuration must be the same no matter which Minkowski coordinate system is used to express the components of Φ and its derivatives.

The volume element $d^4 X$ is invariant under a Lorentz transformation.[3] The action will therefore be a scalar if the integrand, that is, the Lagrangian density $\mathcal{L}(\Phi, \partial_\mu \Phi)$, is itself a

---

[2]Equations (8.3) remain of second order if $\mathcal{L}$ also depends on $\partial_\mu \partial_\nu \Phi$ when the terms containing these second derivatives take the form of a divergence $\partial_\mu V^\mu$, where $V^\mu = V^\mu(\Phi, \partial_\nu \Phi)$, which when transformed into a surface term using Gauss's theorem does not contribute to the equations of motion. We indeed have

$$
\delta \int d^4 X \, \partial_\mu V^\mu = \int d^4 X \, \partial_\mu \delta V^\mu = \int d^4 X \, \partial_\mu \left(\frac{\partial V^\mu}{\partial \Phi} \delta \Phi + \frac{\partial V^\mu}{\partial \partial_\nu \Phi} \delta \partial_\nu \Phi \right) = 0
$$

by Gauss's theorem if the variations of the field configurations *and* their derivatives, $\delta \Phi$ and $\delta \partial_\nu \Phi$, are required to vanish on the boundaries of the domain.

[3]By definition, the components of the metric $\eta_{\mu\nu}$ are not changed by a Lorentz transformation; its determinant is -1 in the coordinates X^μ or X'^μ, and so $d^4 X = dX^0 dX^1 dX^2 dX^3 = dX'^0 dX'^1 dX'^2 dX'^3$ (see Book 1, Section 2.6). It should be noted that throughout this section not only must the reference frame be inertial, but also the spatial coordinates must be Cartesian. In Section 8.5 we shall see how to generalize the definitions below to curvilinear spatial coordinates *and* to accelerated reference frames.

scalar, which means that it must not depend *explicitly* on the coordinates X^μ. Using the equations of motion (8.3), we then have[4]

$$\frac{\partial \mathcal{L}}{\partial X^\mu} = \frac{\partial \mathcal{L}}{\partial \Phi}\partial_\mu\Phi + \frac{\partial \mathcal{L}}{\partial \partial_\nu\Phi}\partial_\mu\partial_\nu\Phi = \partial_\nu\Big(\frac{\partial \mathcal{L}}{\partial \partial_\nu\Phi}\Big)\partial_\mu\Phi + \frac{\partial \mathcal{L}}{\partial \partial_\nu\Phi}\partial_\mu\partial_\nu\Phi$$

$$= \partial_\nu\Big(\partial_\mu\Phi\frac{\partial \mathcal{L}}{\partial \partial_\nu\Phi}\Big) \qquad \Longleftrightarrow \qquad \partial_\nu\Big(-\partial_\mu\Phi\frac{\partial \mathcal{L}}{\partial \partial_\nu\Phi} + \delta_\mu^\nu\mathcal{L}\Big) = 0\,. \tag{8.4}$$

This is an aspect of *Noether's theorem*: any invariance of the action, here invariance under a change of inertial frame, corresponds to a *conservation law*, that is, the vanishing of the divergence of some quantity. The quantity defined here is the *Noether energy–momentum tensor*, also called the *canonical tensor*:

$$\Theta_\mu{}^\nu \equiv -\partial_\mu\Phi\frac{\partial \mathcal{L}}{\partial \partial_\nu\Phi} + \delta_\mu^\nu\mathcal{L} \quad \text{with} \quad \partial_\nu\Theta_\mu{}^\nu = 0\,. \tag{8.5}$$

(One reason for the name given to $\Theta_\mu{}^\nu$ is that if Φ depends only on the time, then $-\Theta_0{}^0 = \dot{\Phi}\frac{\partial \mathcal{L}}{\partial \dot{\Phi}} - \mathcal{L}$ is identified as the energy as defined in the Lagrangian mechanics of a point; *cf.* Book 1, Section 8.3.)[5]

The tensor obtained by raising an index of the Noether tensor,

$$\Theta^{\mu\nu} \equiv \eta^{\mu\rho}\Theta_\rho{}^\nu = -\partial^\mu\Phi(\partial \mathcal{L}/\partial \partial_\nu\Phi) + \eta^{\mu\nu}\mathcal{L},$$

is not *a priori* symmetric: $\Theta^{\mu\nu} \neq \Theta^{\nu\mu}$. It can however be made so. Let us consider the tensor

$$T^{\mu\nu} \equiv \Theta^{\mu\nu} + \partial_\rho\sigma^{\mu\nu\rho}, \qquad \text{where} \quad \begin{cases} \partial_\rho\sigma^{[\mu\nu]\rho} = \partial^{[\mu}\Phi\dfrac{\partial \mathcal{L}}{\partial \partial_{\nu]}\Phi} \\[2mm] \sigma^{\mu\nu\rho} = -\sigma^{\mu\rho\nu}\,. \end{cases} \tag{8.6}$$

The first condition, where the brackets denote antisymmetrization $f_{[\mu\nu]} \equiv \frac{1}{2}(f_{\mu\nu} - f_{\nu\mu})$, guarantees the symmetry of $T^{\mu\nu}$, $T^{\mu\nu} = T^{\nu\mu}$, and the second guarantees that it is conserved, $\partial_\nu T^{\mu\nu} = 0$, because

$$\partial_{\nu\rho}\sigma^{\mu\nu\rho} = -\partial_{\nu\rho}\sigma^{\mu\rho\nu} = -\partial_{\rho\nu}\sigma^{\mu\rho\nu} = -\partial_{\nu\rho}\sigma^{\mu\nu\rho} \qquad \Longrightarrow \qquad \partial_{\nu\rho}\sigma^{\mu\nu\rho} = 0\,. \tag{8.7}$$

(The first equation follows from the antisymmetry of $\sigma^{\mu\nu\rho}$, the second from the symmetry of the second derivatives, and the third is just a redefinition of the dummy indices.) The tensor $T^{\mu\nu}$ which is symmetric and conserved is the object that we shall refer to as the *energy–momentum tensor*.[6]

---

[4]The precise meaning of the notation is the following. The quantity $\partial \mathcal{L}/\partial X^0$, for example, is the derivative of $\mathcal{L}$ with respect to X^0, with the other coordinates held fixed. If $\mathcal{L}$ depends explicitly on X^0 [and not only through $\Phi(X^\mu)$ and its derivatives], it is necessary to add to (8.1) the derivative of $\mathcal{L}$ with respect to X^0 for fixed Φ and $\partial_\mu\Phi$.

[5]Since it is possible to add a total divergence to the Lagrangian without changing the equations of motion (*cf.* footnote 2 of the preceding section), the Noether tensor is not unique. If we add $\partial_\mu V^\mu$ to $\mathcal{L}$, then we must add $\partial_\rho(\delta_\mu^\nu V^\rho - \delta_\mu^\rho V^\nu)$ to $\Theta_\mu{}^\nu$.

[6]Since the tensor $\sigma^{\mu\nu\rho}$ is antisymmetric in its last two indices, it *a priori* possesses 24 independent components, and the six symmetry conditions (8.6) on $T^{\mu\nu}$ do not determine all of them. In addition, these symmetry conditions may be imposed only after the field equations are satisfied. In Section 12.4 we shall give an explicit example of the symmetrization of the Noether tensor.

We have just seen that the energy–momentum tensors (8.5) and (8.6) are conserved if the equations of motion are satisfied. Conversely, imposing the condition of conservation of the energy–momentum tensor implies the equations of motion, with the condition that $\mathcal{L}$ not depend explicitly on the X^μ and that Φ not be constant. We have

$$
\partial_\nu \Theta_\mu{}^\nu = \partial_\nu \left(-\partial_\mu \Phi \frac{\partial \mathcal{L}}{\partial \partial_\nu \Phi} + \delta_\mu^\nu \mathcal{L} \right) = -\partial_{\nu\mu}\Phi \frac{\partial \mathcal{L}}{\partial \partial_\nu \Phi} - \partial_\mu \Phi\, \partial_\nu \left(\frac{\partial \mathcal{L}}{\partial \partial_\nu \Phi} \right) + \partial_\mu \mathcal{L}
$$

$$
= -\partial_{\nu\mu}\Phi \frac{\partial \mathcal{L}}{\partial \partial_\nu \Phi} - \partial_\mu \Phi\, \partial_\nu \left(\frac{\partial \mathcal{L}}{\partial \partial_\nu \Phi} \right) + \frac{\partial \mathcal{L}}{\partial \Phi}\partial_\mu \Phi + \partial_{\mu\nu}\Phi \frac{\partial \mathcal{L}}{\partial \partial_\nu \Phi} \tag{8.8}
$$

$$
= \partial_\mu \Phi \left(\frac{\partial \mathcal{L}}{\partial \Phi} - \partial_\nu \left(\frac{\partial \mathcal{L}}{\partial \partial_\nu \Phi} \right) \right),
$$

where on the last line we recognize the left-hand side of the Euler–Lagrange equations (8.3).

Now let us integrate the equation $\partial_\nu T^{\mu\nu} = 0$ on the hyperplane $T = \text{const}$, that is, over the entire space V. Using Gauss's theorem, we find

$$
0 = \int_V d^3X\, \partial_\nu T^{\mu\nu} = \int_V d^3X (\partial_0 T^{\mu 0} + \partial_i T^{\mu i}) = \frac{d}{dT}\int_V d^3X\, T^{\mu 0} + \int_S T^{\mu i} dS_i,
$$

$$
\text{or} \quad \frac{d}{dT} P^\mu = -\int_S T^{\mu i} dS_i, \quad \text{where} \quad P^\mu \equiv \int_V d^3X\, T^{\mu 0}, \tag{8.9}
$$

with S the boundary of the 3-space and dS_i its surface element. If the solution Φ of the field equations falls off sufficiently rapidly at spatial infinity so that the surface integral vanishes, the field is termed non-radiative and we have

$$
P^\mu = \text{const.} \tag{8.10}
$$

Therefore, a conservation law allows us to construct *conserved quantities*, here the field 4-*momentum*. Equation (8.10) is the exact analog of the 4-momentum of a free particle being constant.

Owing to the antisymmetry of $\sigma^{\mu\nu\rho}$ in the last two indices, we have

$$
P^\mu \equiv \int_V d^3X\, T^{\mu 0} = \int_V d^3X \left(\Theta^{\mu 0} + \partial_\rho \sigma^{\mu 0 \rho} \right) = \int_V d^3X\, \Theta^{\mu 0} + \int_S \sigma^{\mu 0 i} dS_i. \tag{8.11}
$$

Therefore, the 4-momentum of the field is the same when it is defined using the conservation of $T^{\mu\nu}$ or $\Theta^{\mu\nu}$, as long as the tensor $\sigma^{\mu\nu\rho}$ decreases sufficiently rapidly at infinity so that the surface integral vanishes. Thus in this case we can define the 4-momentum P^μ, which is constant when the field is non-radiative, whether or not the energy–momentum tensor is symmetric.

However, only when it is symmetric will we also have

$$
\partial_\rho (X^\mu T^{\nu\rho} - X^\nu T^{\mu\rho}) = 0, \tag{8.12}
$$

where X^μ are the coordinates of a point, because $\partial_\nu T^{\mu\nu} = 0$ and $\partial_\mu X^\nu = \delta_\mu^\nu$. Integrating (8.12) over a 3-volume V of the hyperplane $T = \text{const}$ and using Gauss's theorem, we find

$$\frac{d}{dT}\mathcal{M}^{\mu\nu} = -\int_S (X^\mu T^{\nu i} - X^\nu T^{\mu i})dS_i\,, \quad \text{where } \mathcal{M}^{\mu\nu} \equiv \int_V d^3X\,(X^\mu T^{\nu 0} - X^\nu T^{\mu 0})\,, \quad (8.13)$$

$\mathcal{M}$ being the *angular momentum* tensor of the field. If the field falls off sufficiently quickly at spatial infinity, it is constant.

Transformation of the 4-momentum

The energy–momentum of a field is represented by the same tensor in any inertial frame in Minkowski coordinates, and the transformation law of its components $T^{\mu\nu}$ in going from one Minkowski frame to another is known. On the other hand, the 4-momentum vector P^μ introduced in (8.9), being defined by an integral over a spatial 3-volume, depends *a priori* on the choice of inertial frame (it is the clocks of S which define the $T = \text{const}$ cross-sections).

Now let us return to the conservation law $\partial_\nu T^{\mu\nu} = 0$ and integrate it over a *4-volume* $\mathcal{M}$ sandwiched between two space-like hypersurfaces Σ_- and Σ_+, and bounded laterally by the time-like 'cylinder' $\mathcal{B} = L \times S_2$, where S_2 is the 2-sphere at infinity and L is arbitrarily 'short'; see Fig. 8.1. Applying Gauss's theorem, we find

$$0 = \int_\mathcal{M} d^4X\,\partial_\nu T^{\mu\nu} = \int_{\partial\mathcal{M}} T^{\mu\nu} dS_\nu\,, \tag{8.14}$$

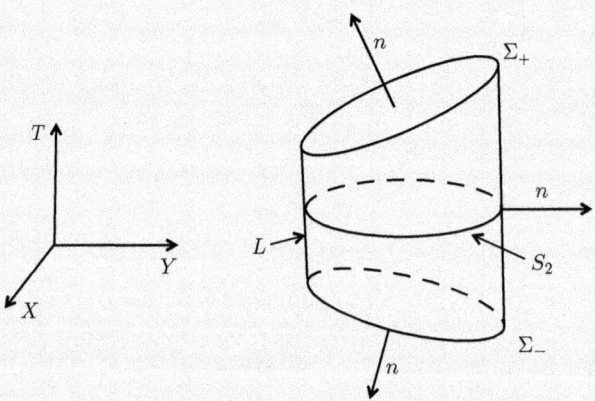

Fig. 8.1 4-momentum: integration region.

where $\partial\mathcal{M} = \Sigma_- + \Sigma_+ + \mathcal{B}$ and dS_ν is the 3-surface element on each of the hypersurfaces. If now

$$\int_\mathcal{B} T^{\mu\nu} dS_\nu = 0 \quad \forall\ \mathcal{B}\,, \quad \text{then} \quad \int_\Sigma T^{\mu\nu} dS_\nu \equiv P^\mu = \text{const}\,. \tag{8.15}$$

The 4-momentum P^μ thus constructed, where Σ is an arbitrary space-like hypersurface, does not depend on the choice of inertial frame. It reduces to (8.9) if $\Sigma \equiv V$ is the $T = \text{const}$ cross-section of a given frame. In Section 6.3 we saw that the total 4-momentum of a system of free particles is represented by the same vector in any inertial frame. It is therefore the same for that of a free field.

On the other hand, the angular momentum depends on the choice of frame just like that of an ensemble of particles; see Section 7.1.

8.3 Fluid equations of motion

The kinematics of a fluid, a continuous ensemble of particles gathered into *fluid elements*, is described by a velocity field u of components $U^\mu(X^\nu)$ in a Minkowski inertial frame $\mathcal{S}$, whose integral curves define the world lines, each line $q = q(\tau)$ representing the motion of a fluid element. We shall require that the world lines always be time-like, and we normalize u as $(u \cdot u) = -1$ so that $u(q)$ can be viewed as the 4-velocity $u = dq/d\tau$ of the fluid element at q.

The fluid dynamics which describes phenomenologically the internal collisions in the fluid is postulated to be encoded in a (symmetric) *energy–momentum* tensor of components $T^{\mu\nu}$ in $\mathcal{S}$, which can always be decomposed as

$$T^{\mu\nu} = \epsilon\, U^\mu U^\nu + (U^\mu Q^\nu + U^\nu Q^\mu) + \Pi^{\mu\nu}, \tag{8.16}$$

where $\epsilon(X^\mu)$ is a scalar function, the vector field $Q^\mu(X^\nu)$ is perpendicular to U^μ ($Q_\mu U^\mu = 0$), and the tensor $\Pi^{\mu\nu}(X^\rho)$ can be further decomposed as

$$\Pi^{\mu\nu} = \pi^{\mu\nu} + p\, h^{\mu\nu}, \tag{8.17}$$

where $h^{\mu\nu}$ is the *projection tensor*: $h^{\mu\nu} \equiv \eta^{\mu\nu} + U^\mu U^\nu$ ($\Rightarrow h^{\mu\nu} U_\nu = 0$). Here $p(X^\mu)$ is a scalar function and $\pi^{\mu\nu}(X^\rho)$ is a traceless, transverse tensor, such that $\pi^\mu{}_\mu = 0$ and $\pi^{\mu\nu} U_\nu = 0$. We have therefore exchanged the ten components of $T^{\mu\nu}$ for the two scalars ϵ and p, the three independent components of the vector Q^μ, and the five independent components of the tensor $\pi^{\mu\nu}$. In the frame where the fluid element at X^μ is momentarily at rest so that $U^\mu = (1, 0, 0, 0)$, we have

$$T^{00} = \epsilon, \quad T^{0i} = Q^i, \quad T^{ij} = p\, \delta^{ij} + \pi^{ij}, \quad T^i_i = 3p. \tag{8.18}$$

A fluid is termed *perfect* if $Q^\mu = 0$ and $\pi^{\mu\nu} = 0$. Its energy–momentum tensor then is

$$T^{\mu\nu} = (\epsilon + p)U^\mu U^\nu + p\, \eta^{\mu\nu}. \tag{8.19}$$

The equation of motion of the 'free' fluid, that is, a fluid not subject to any external long-range force,[7] is obtained by *requiring* that its energy–momentum tensor be conserved:

$$\partial_\nu T^{\mu\nu} = 0. \tag{8.20}$$

In the case of a perfect fluid, to which we restrict ourselves from now on, this equation can be expanded as

$$U^\mu U^\nu \partial_\nu(\epsilon + p) + (\epsilon + p)(U^\nu \partial_\nu U^\mu + U^\mu \partial_\nu U^\nu) + \partial^\mu p = 0. \tag{8.21}$$

Contracting it with U^μ, we first obtain (since $U^\mu U_\mu = -1$)

$$\partial_\nu(\epsilon\, U^\nu) + p\, \partial_\nu U^\nu = 0, \tag{8.22}$$

which allows us to rewrite (8.21) as

$$(\epsilon + p)U^\nu \partial_\nu U^\mu + U^\mu U^\nu \partial_\nu p + \partial^\mu p = 0. \tag{8.23}$$

This equation involves only three independent components because it is orthogonal to U^μ. Therefore, (8.21) is equivalent to the set of equations (8.22) and (8.23).

---

[7]We thereby exclude the gravitational interaction as well as charged fluids in an electromagnetic field.

Equation (8.23) is the *relativistic Euler equation*. When $p \ll \epsilon$ and the fluid 3-velocity $V \ll 1$, it reduces to the Newtonian Euler equation $dV/dt \equiv \partial V/\partial t + (V \cdot \nabla V)V = -\nabla p/\epsilon$, where the time T of the inertial frame is viewed as the universal time t of Newton. Equation (8.22) becomes the continuity (or mass conservation) equation: $\partial \epsilon/\partial t + \nabla \cdot (\epsilon V) = 0$; see Book 1, Section 6.3. Extending the terminology used in Newtonian physics, we shall refer to ϵ as the *energy density* and p as the *pressure*.

The equation of state of a relativistic fluid

The equations (8.22) and (8.23) are four equations in five unknowns, namely, ϵ, p, and the three components of the vector U^μ (constrained by $U^\mu U_\mu = -1$).[8] To completely determine the fluid motion it is therefore necessary to have an additional equation, the *equation of state*, which relates the pressure and the density. A fluid is *barotropic* if its pressure depends only on the density, and the equation of state is usually written in terms of the *adiabatic index* γ as the system

$$\frac{d\epsilon}{dn} = \frac{\epsilon + p}{n}, \quad \frac{dp}{dn} = \gamma \frac{p}{n}, \tag{8.24}$$

where the parameter n is the *particle density*. Then (8.22) can be rewritten as the particle number conservation law:

$$\partial_\mu(n\,U^\mu) = 0. \tag{8.25}$$

If the index γ is constant, the fluid is a *polytrope* and the system (8.24) can be integrated to give

$$p(n) = \kappa\,n^\gamma, \quad \epsilon(n) = \frac{\kappa\,n^\gamma}{\gamma - 1} + m\,n, \tag{8.26}$$

where κ and m are integration constants having the dimensions of mass. In the Newtonian limit where $\epsilon \approx m\,n$ we have $p \propto \epsilon^\gamma$. On the other hand, in the ultra-relativistic limit (for $\gamma \neq 1$)

$$p \to w\epsilon \quad \text{with} \quad w \equiv \gamma - 1. \tag{8.27}$$

Finally, perturbing the Euler continuity equations (8.22) and (8.23) about the static solution $U^\mu = (1,0,0,0)$, $\epsilon = \epsilon_0$, $p = p_0$, we easily find the equation of motion of the perturbation ϵ_1 (by differentiating the continuity equation with respect to time and eliminating $\nabla \dot{V}$ using the gradient in the Euler equation):

$$-\frac{\partial^2 \epsilon_1}{\partial T^2} + c_s^2 \triangle \epsilon_1 = 0, \tag{8.28}$$

where $c_s \equiv \sqrt{dp/d\epsilon}|_0$ is the *speed of sound*. Therefore, the equation for *sound propagation* is the same as in Newtonian physics; see Book 1, Section 17.3. It is not invariant under a change of inertial frame, which allows determination of the frame in which the fluid is at rest.[9]

---

[8] It is usual to decompose the energy density as $\epsilon \equiv \rho + \epsilon_{\text{int}}$, where ρ is the *rest-mass density* (or *proper energy density*) and ϵ_{int} is the *internal energy density*. The enthalpy is defined as $H \equiv \epsilon + p$ and the *specific enthalpy* is H/ρ. The quantity Q^μ is the *momentum density* (or the *heat flux*). Finally, $\pi^{\mu\nu}$ is the traceless anisotropic part of the *constraint tensor* $\Pi^{\mu\nu}$.

For a more thorough presentation of relativistic hydrodynamics see, for example, Gourgoulhon (2013). Its applications in astrophysics are discussed in Martì and E. Müller (2003). Finally, Ollitrault (2008) discusses its applications to heavy-ion physics.

[9] If $c_s = 1$, (8.28) becomes the *Klein–Gordon equation* and is invariant under a Lorentz transformation; see Section 9.1. In this case the fluid, whose equation of state is $p = \epsilon$ and which is termed *rigid*, resembles the *aether* or the *vacuum*. It is pointless to specify whether this 'medium' is at rest or not, because the speed of sound then is *always* the same.

8.4 The particle energy–momentum tensor

The main idea of the present chapter is that the dynamics of a 'continuous medium', a free field or fluid, is encoded in its energy–momentum tensor $T^{\mu\nu}$, from which one we can extract both the equations of motion by requiring conservation ($\partial_\nu T^{\mu\nu} = 0$), and the conserved quantities, for example, the 4-momentum $P^\mu = \int d^3X\, T^{0\mu}$. Here we shall show that it is possible to do the same in the case of an ensemble of point particles.

We shall see that the energy–momentum tensor for describing particles in an inertial Minkowski frame $\mathcal{S}$ is

$$T^{\mu\nu}(X^\rho) = \sum_a m \int_L d\tau\, \delta_4(X^\rho - X^\rho(\tau))\, U^\mu U^\nu, \tag{8.29}$$

where $X^\mu(\tau)$ is the world line L of the particle a of mass m parametrized by its proper time τ, $U^\mu = dX^\mu/d\tau$ is its speed (with $\eta_{\mu\nu}U^\mu U^\nu = -1$), and $\delta_4(X^\mu) \equiv \delta(T)\delta_3(X^i)$ and $\delta_3(X^i) \equiv \delta(X)\delta(Y)\delta(Z)$. Here δ the *Dirac distribution*, a 'function' which is zero everywhere except at the origin and has unit integral, $\int dX\delta(X) = 1$; its main properties are given in Section 9.2.

Since $\int d^3X\delta_3(X^i) = 1$ and $U^0 = dT/d\tau$, we see immediately that

$$\int d^3X\, T^{\mu 0} = \sum_a m \int_L d\tau\, \delta(T - T(\tau))\, U^\mu U^0 = \sum_a m\, U^\mu \equiv P^\mu \tag{8.30}$$

(the velocities are evaluated at the time T). Therefore, the 4-momentum of the system is expressed, as desired, as a volume integral over an energy–momentum density.

In addition, given that $\int d\tau\delta(T - T(\tau)) = 1/U^0(T)$ (see Section 9.2), we see that (8.29) can also be written as

$$T^{\mu\nu} = \sum_a m\frac{U^\mu U^\nu}{U^0}\, \delta_3(X^i - X^i(T)), \quad \text{where } U^0 = \frac{1}{\sqrt{1 - V^2(T)}}, \quad U^i = \frac{V^i(T)}{\sqrt{1 - V^2(T)}}, \tag{8.31}$$

so that

$$\partial_\nu T^{\mu\nu} = \partial_0 T^{\mu 0} + \partial_i T^{\mu i}$$

$$= \sum_a m \left[\frac{\partial}{\partial T}\left(U^\mu \delta_3(X^j - X^j(T))\right) + U^\mu V^i \frac{\partial}{\partial X^i}\delta_3(X^j - X^j(T))\right] \tag{8.32}$$

$$= \sum_a m\frac{dU^\mu}{dT}\delta_3(X^j - X^j(T)) = \sum_a m \int_L d\tau \frac{dU^\mu}{d\tau}\delta_4(X^\mu - X^\mu(\tau))$$

because $\frac{\partial}{\partial T}\left(\delta_3(X^j - X^j(T))\right) + V^i\frac{\partial}{\partial X^i}\delta_3(X^j - X^j(T)) \equiv 0$. We thus see that requiring that the energy–momentum tensor be conserved gives the particle motion, namely, the particle velocities U^μ are constant.

To relate the description of an ensemble of particles to that of a fluid, we now rewrite (8.29) in the form

$$T^{\mu\nu} = \epsilon\, U^\mu U^\nu, \qquad \text{where} \qquad \epsilon(X^\mu) \equiv \sum_a m \int_L d\tau\, \delta_4(X^\mu - X^\mu(\tau))\,. \tag{8.33}$$

[This amounts to replacing the 4-velocities by a vector field $U^\mu(X^\nu)$, which is possible owing to the presence of the Dirac distribution.] We also see [*cf.* (8.19)] that the pressure of a fluid of free particles is zero (which is to be expected, because the particles, being free, do not undergo collisions). The equations of motion (8.22) and (8.23) derived from the conservation of $T^{\mu\nu}$ reduce to

$$\partial_\nu(\epsilon\, U^\nu) = 0 \qquad \text{and} \qquad U^\nu \partial_\nu U^\mu = 0\,. \tag{8.34}$$

Substituting (8.33) for ϵ, the first of these equations expresses mass conservation $\sum_a m = $ const (after integrating and using Gauss's theorem) and the second again states that the particle velocities are constant.

Let us end by considering an ensemble of particles undergoing collisions, that is, interacting *via* short-range forces. Between collisions their 3-velocities V^i are constant. We assume that there exists a reference frame in which these velocities are zero on average $\langle V^i \rangle = 0$ (this average is either a time average of the velocity of a single particle, or the average over the angles of the velocities of a large number of particles; they are the same according to the 'ergodicity theorem'). In this frame the averages of the non-zero components of the energy–momentum tensor (8.31) are

$$\begin{cases} \epsilon \equiv \langle T^{00} \rangle = \left\langle \displaystyle\sum_a \frac{m}{\sqrt{1-V^2}} \delta_3(X^i - X^i(T)) \right\rangle \\[3mm] p \equiv \dfrac{1}{3}\langle T^i_i \rangle = \dfrac{1}{3} \left\langle \displaystyle\sum_a \frac{mV^2}{\sqrt{1-V^2}} \delta_3(X^i - X^i(T)) \right\rangle\,. \end{cases} \tag{8.35}$$

Whatever the precise mathematical interpretation of these averages is, we have that in the nonrelativistic limit $(V \to 0)$ and the ultra-relativistic limit $(V \to 1)$ the respective equations of state are

$$\epsilon \sim \epsilon_0 + \frac{3}{2}p\,, \quad \epsilon \simeq 3p\,, \tag{8.36}$$

where $\epsilon_0 = \sum_a m \langle \delta_3(X^i - X^i(T)) \rangle$ is the energy density of the fluid at rest.

A fluid without pressure is called *dust*. The Euler equation (8.23) or (8.34) implies that its flow is uniform and rectilinear. The absence of pressure corresponds to the absence of collisions between the particles of the fluid, and in this case the congruence of the world lines of individual particles appearing in (8.29) or (8.33) is identified with the flow itself.

A fluid with the equation of state $\epsilon = 3p$ is called a *radiation fluid*. As we have just seen, it is composed of an ensemble of ultrarelativistic particles which could also be photons; see Section 12.4.

8.5   Conservation laws in an accelerated frame

In a system of curvilinear coordinates the equations of motion of a field or a fluid are derived, as in a Minkowski frame, from the conservation law for the energy–momentum tensor, but using the 'covariantized' version, that is, the version with the ordinary derivatives ∂_μ replaced by the covariant derivatives $\tilde{D}_\mu$:

$$\tilde{D}_\nu T^{\mu\nu} = 0 \quad \Longleftrightarrow \quad \partial_\nu T^{\mu\nu} + \tilde{\Gamma}^\mu_{\rho\nu} T^{\rho\nu} + \tilde{\Gamma}^\nu_{\rho\nu} T^{\mu\rho} = 0 , \tag{8.37}$$

where $\tilde{\Gamma}^\mu_{\nu\rho}$ are the Christoffel symbols in the chosen coordinates x^μ. In order to derive directly (that is, without going to Minkowski coordinates as in Section 8.2) from (8.37) the fact that the energy or angular momentum of the system is constant, we need to specify their geometrical origin, namely, the homogeneity and isotropy of Minkowski spacetime.

Isometries and Killing vectors of $\mathcal{M}_4$

Under an infinitesimal change of coordinate $x^\rho \to \tilde{x}^\rho = x^\rho + \xi^\rho$ the components $\ell_{\mu\nu}(x^\rho)$ of the metric in the system $\{x^\rho\}$ at the point P of coordinates x^ρ are related to its components $\tilde{\ell}_{\mu\nu}(\tilde{x}^\rho)$ at P in the system $\{\tilde{x}^\rho\}$ by the transformation law for 2-fold covariant tensors, namely,

$$\ell_{\mu\nu}(x^\rho) = \frac{\partial \tilde{x}^\rho}{\partial x^\mu} \frac{\partial \tilde{x}^\sigma}{\partial x^\nu} \tilde{\ell}_{\rho\sigma}(\tilde{x}^\lambda) = (\delta^\rho_\mu + \partial_\mu \xi^\rho)(\delta^\sigma_\nu + \partial_\nu \xi^\sigma) \tilde{\ell}_{\rho\sigma}(\tilde{x}^\lambda). \tag{8.38}$$

Now writing in lowest order $\tilde{\ell}_{\rho\sigma}(\tilde{x}^\lambda) = \tilde{\ell}_{\rho\sigma}(x^\lambda) + \xi^\lambda \partial_\lambda \ell_{\rho\sigma}$, that is, introducing the metric at the point $\tilde{P}$ which in the system $\{\tilde{x}^\rho\}$ has the same coordinates as the point P in the system $\{x^\rho\}$, we find (see Book 1, Section 8.3)

$$\ell_{\mu\nu}(x^\rho) - \tilde{\ell}_{\mu\nu}(x^\rho) = \xi^\lambda \partial_\lambda \ell_{\mu\nu} + \ell_{\lambda\mu} \partial_\nu \xi^\lambda + \ell_{\nu\lambda} \partial_\mu \xi^\lambda = \tilde{D}_\nu \xi_\mu + \tilde{D}_\mu \xi_\nu , \tag{8.39}$$

which defines the *Lie derivative* of the metric. The displacement $P \mapsto \tilde{P}$ is an *isometry* and ξ^ρ is a *Killing vector* if the Lie derivative of the metric vanishes, that is, if the *Killing equations* $\tilde{D}_\mu \xi_\nu + \tilde{D}_\nu \xi_\mu = 0$ are satisfied.

The Killing vectors of $\mathcal{M}_4$ are easily obtained using Minkowski coordinates, in which the Killing equations reduce to $\partial_\mu \xi_\nu + \partial_\nu \xi_\mu = 0$. The general solution is $\xi^\mu = d^\mu + \omega^\mu{}_\nu X^\nu$, where d^μ and $\omega_{\mu\nu} = -\omega_{\nu\mu}$ are a constant vector and a constant antisymmetric matrix. Therefore, $\mathcal{M}_4$ possesses ten Killing vectors, four corresponding to translations ($\xi^\mu_{(t)} = \delta^\mu_t$), three to spatial rotations, and three to boosts. This is the maximum number of Killing vectors that a space of dimension four can possess (*cf.* Book 1, Section 8.3), and so $\mathcal{M}_4$ is a *maximally symmetric* space.

The ten Killing vectors of $\mathcal{M}_4$ can be used to derive conserved quantities directly from the conservation law (8.37). Forming the scalar product with a Killing vector ξ^μ, we have

$$0 = \xi^\mu \tilde{D}_\nu T^\nu{}_\mu = \tilde{D}_\nu(\xi^\mu T^\nu{}_\mu) - T^{\nu\mu} \tilde{D}_\nu \xi_\mu = \tilde{D}_\nu(\xi^\mu T^\nu{}_\mu) \tag{8.40}$$

because $T^{\nu\mu}$ is symmetric and ξ^μ is a Killing vector.

Now we integrate this equation, as in (8.14), over a 4-volume $\mathcal{M}$ enclosed between two space-like hypersurfaces Σ_- and Σ_+ and bordered laterally by the time-like cylinder $\mathcal{B} = L \times S_2$, where S_2 is the 'sphere' at infinity and L is arbitrarily 'short' (see Fig. 8.1). Since for any vector v^μ we have $\sqrt{-\ell}\, \tilde{D}_\mu v^\mu = \partial_\mu(\sqrt{-\ell}\, v^\mu)$, where ℓ is the determinant of $\ell_{\mu\nu}$, we find, using Gauss's theorem and recalling that the volume element is $\sqrt{-\ell}\, d^4x$ (see, for example, Book 1, Sections 2.6, 3.6, and 4.6),

$$0 = \int_{\mathcal{M}} d^4x \, \partial_\nu (\sqrt{-\ell}\, \xi^\mu T^\nu{}_\mu) = \int_{\partial\mathcal{M}} \xi^\mu T^\nu_\mu dS_\nu \,,$$

$$\text{with} \quad dS_\mu = \sqrt{-\ell}\, e_{\mu\nu\rho\sigma} \frac{\partial x^\nu}{\partial y^1} \frac{\partial x^\rho}{\partial y^2} \frac{\partial x^\sigma}{\partial y^3} dy^1 dy^2 dy^3 = \sqrt{|h|}\, n_\mu d^3 y \,,$$

(8.41)

where $\partial\mathcal{M} = \Sigma_- + \Sigma_+ + \mathcal{B}$ and where $e_{\mu\nu\rho\sigma}$ is the totally antisymmetric Levi-Civita symbol with $e_{0123} = 1$. The boundary $\partial\mathcal{M}$ is defined by the equations $x^\mu = x^\mu(y^a)$, where h is the determinant of the metric induced on $\partial\mathcal{M}$ and n_μ is the unit 4-vector (*i.e.*, $\ell_{\mu\nu} n^\mu n^\nu = \pm 1$) normal to $\partial\mathcal{M}$ pointing outward from the domain.

If the integral over $\mathcal{B}$ vanishes, that is, if there is no flux at spatial infinity, then

$$\int_{\Sigma} \xi^\mu T^\nu_\mu dS_\nu = \text{const} \,.$$

(8.42)

If ξ^μ is the Killing vector corresponding to time translations [with components $\xi^\mu = (1,0,0,0)$ in an inertial frame where $\sqrt{-\ell} = 1$ and where (8.42) reduces to $-\int_\Sigma d^3X \, T^0_0$], this constant is the energy of the system. The other Killing vectors lead to the definitions of the 3-momentum, the angular momentum, and the center of mass.

9
The classical scalar field

Scalar fields are paradigmatic in relativistic field theory. Here we shall present some of their properties as part of an introduction to the study of the electromagnetic field to be taken up in Part III of this book. We shall also see how a *complex* scalar field can confer an effective mass to a 'gauge' field.

9.1 The Klein–Gordon equation

In Section 8.1 we gave the Euler–Lagrange equations of motion of a free field without specifying the Lagrangian density $\mathcal{L}(\Phi, \partial_\mu \Phi)$. Here we shall consider the special case of a *massive* scalar field $\varphi(X^\mu)$ which is real and whose action in an inertial Minkowski frame is given by

$$\mathbf{S}_{\mathrm{f}} = \int d^4 X \, \mathcal{L} \quad \text{with} \quad \mathcal{L} = -\frac{1}{2}\left(\partial_\mu \varphi \, \partial^\mu \varphi + \mu^2 \varphi^2\right). \tag{9.1}$$

To have the action expressed in kg-s, the field φ must be in $\sqrt{\mathrm{kg/s}}$; μ is in s^{-1}, but $\mu\hbar$ is in kg, which is why the scalar field is termed 'massive'.

The Euler–Lagrange equation of motion (8.3) governing the dynamics of this free field then becomes the *Klein–Gordon equation*:

$$\Box \varphi - \mu^2 \varphi = 0 \quad \text{with} \quad \Box \varphi \equiv \partial^\mu \partial_\mu \varphi \equiv \partial_\mu^\mu \varphi = -\frac{\partial^2 \varphi}{\partial T^2} + \triangle \varphi. \tag{9.2}$$

Here $\Box$ is the *d'Alembertian* and $\triangle$ is the Laplacian (in Cartesian coordinates $\triangle = \delta^{ij}\partial_i\partial_j \equiv \partial_i^i$; see Book 1, Section 4.6 for its expression in other coordinate systems, for example, spherical coordinates).[1] It can be checked that the Klein–Gordon equation, obtained by varying a scalar quantity, is indeed invariant under Lorentz transformations. Indeed, under the transformation $X^\mu \mapsto X'^\mu = \Lambda_\nu{}^\mu X^\nu$ we have $\frac{\partial}{\partial X^\mu} = \Lambda_\mu{}^\nu \frac{\partial}{\partial X'^\nu}$, and so

$$\Box \equiv \eta^{\mu\nu} \frac{\partial}{\partial X^\mu}\frac{\partial}{\partial X^\nu} = \eta^{\mu\nu}\Lambda_\mu{}^\rho \Lambda_\nu{}^\sigma \frac{\partial}{\partial X'^\rho}\frac{\partial}{\partial X'^\sigma} = \eta^{\rho\sigma}\frac{\partial}{\partial X'^\rho}\frac{\partial}{\partial X'^\sigma}$$

because $\eta^{\mu\nu}\Lambda_\mu{}^\rho \Lambda_\nu{}^\sigma = \eta^{\rho\sigma}$ by the definition of Lorentz rotations.

The Noether energy–momentum tensor is derived from (8.5) and the form of the action (9.1):

$$T_{\mu\nu} \equiv \eta_{\mu\rho}\Theta_\nu{}^\rho = \partial_\mu\varphi\partial_\nu\varphi - \frac{1}{2}\eta_{\mu\nu}\left(\partial_\rho\varphi\,\partial^\rho\varphi + \mu^2\varphi^2\right). \tag{9.3}$$

We see immediately that it is symmetric and easily verify that it is conserved ($\partial_\nu T^{\mu\nu} = 0$) if the Klein–Gordon equation (9.2) is satisfied. Reciprocally, the conservation of $T_{\mu\nu}$ implies the Klein–Gordon equation (with the condition that the field φ is not constant).

---

[1] Instead of the mass term $\frac{1}{2}\mu^2\varphi^2$, it is possible to introduce into the Lagrangian a more general 'potential' $V(\varphi)$. The Klein–Gordon equation (9.2) then becomes $\Box\varphi - dV/d\varphi = 0$ and is no longer linear in φ.

Relativity in Modern Physics. Nathalie Deruelle and Jean-Philippe Uzan.
© Oxford University Press 2018. Published in 2018 by Oxford University Press.
DOI: 10.1093/oso/9780198786399.001.0001

The variational derivative and Hamilton's equations

The action (9.1) of the scalar field is written as $\mathbf{S}_f = \int dT\, L$, where the Lagrangian L is

$$L = \int d^3X\, \mathcal{L} \quad \text{with} \quad \mathcal{L} = \frac{1}{2}\left(\dot\varphi^2 - (\nabla\varphi)^2 - \mu^2\varphi^2\right).$$

L is a function of a triple infinity of generalized coordinates, namely, the values of φ at each point X^i of space along with their velocity extensions $\dot\varphi$.

The *variational derivative* of L with respect to $\dot\varphi$ at a given spatial point X^i reduces to the ordinary derivative of the Lagrangian density $\mathcal{L}$:

$$\frac{\delta L}{\delta\dot\varphi} = \frac{\partial\mathcal{L}}{\partial\dot\varphi} = \dot\varphi.$$

The variational derivative of L with respect to φ is then given by

$$\delta L = \delta \int d^3X\, \mathcal{L} = -\int d^3X \left(\partial^i\varphi\, \delta\partial_i\varphi + \mu^2\varphi\,\delta\varphi\right) = -\int d^3X \left(\partial^i\varphi\, \partial_i\delta\varphi + \mu^2\varphi\,\delta\varphi\right)$$

$$= -\int_S \partial^i\varphi\, dS_i\,\delta\varphi + \int d^3X(\triangle\varphi - \mu^2\varphi)\delta\varphi$$

after integration by parts and use of Gauss's theorem. Then since $\delta\varphi$ is taken to be zero on the boundary of the domain,

$$\frac{\delta L}{\delta\varphi} = \triangle\varphi - \mu^2\varphi.$$

The Klein–Gordon equation (9.2) thus also arises from the Euler–Lagrange equations written in the form

$$\frac{d}{dT}\frac{\delta L}{\delta\dot\varphi} = \frac{\delta L}{\delta\varphi}. \tag{9.4}$$

Let us now pass to the Hamiltonian formulation. The *conjugate momentum* of φ and the *Hamiltonian* are

$$\pi \equiv \frac{\delta L}{\delta\dot\varphi} = \dot\varphi; \quad H = \int d^3X\,\mathcal{H} \quad \text{with} \quad \mathcal{H} \equiv \pi\dot\varphi - \mathcal{L} = \frac{1}{2}\left(\pi^2 + (\nabla\varphi)^2 + \mu^2\varphi^2\right).$$

The Hamiltonian is a functional of $\{\varphi, \pi\}$ which can be varied in two different ways. First,

$$\delta\int dT\, H = \frac{1}{2}\delta\int dT d^3X\,\left(\pi^2 + (\nabla\varphi)^2 + \mu^2\varphi^2\right) = \int dT d^3X\,\left[\pi\delta\pi - (\triangle\varphi - \mu^2\varphi)\delta\varphi\right], \tag{9.5}$$

after integrating by parts and ignoring the boundary term. Second (again integrating by parts and ignoring the boundary term),

$$\delta\int dT\, H = \delta\int dT d^3X\,(\pi\dot\varphi - \mathcal{L}) = \int dT d^3X\,(\dot\varphi\delta\pi - \dot\pi\delta\varphi) - \delta\mathbf{S}_f. \tag{9.6}$$

Equating (9.5) and (9.6), we again find that for the variation of the action to be an extremum, $\delta\mathbf{S}_f = 0$, the Klein–Gordon equation (9.2) must be satisfied: $\pi = \dot\varphi$, $\dot\pi = (\triangle\varphi - \mu^2\varphi)$.

Therefore, the equation of motion of the field also follows from the *Hamilton equations*:

$$\pi = \frac{\delta L}{\delta \dot\varphi}, \quad \dot\pi = -\frac{\delta L}{\delta \varphi}. \tag{9.7}$$

It also follows from the Poisson equations[2]

$$\dot\varphi = \{H, \varphi\}, \quad \dot\pi = \{H, \pi\},$$

where the Poisson bracket of A and B is defined by $\{A, B\} = \dfrac{\delta A}{\delta \pi}\dfrac{\delta B}{\delta \varphi} - \dfrac{\delta A}{\delta \varphi}\dfrac{\delta B}{\delta \pi}$.

9.2 Fourier expansion of the free field

In general, a scalar field $\varphi(X^\mu)$, like any real, *square-integrable* function, can be expanded as

$$\varphi(T, X^i) = \int \frac{d^3 k}{(2\pi)^{\frac{3}{2}}}\, \varphi_{\mathrm k}(T)\, e^{ik_i X^i} \quad \text{with} \quad \varphi^*_{-\mathrm k}(T) = \varphi_{\mathrm k}(T), \tag{9.8}$$

where k is a 3-vector with components k^i and the star indicates complex conjugation. The field $\varphi_{\mathrm k}(T)$ is the *Fourier transform* of $\varphi(T, X^i)$ and $\varphi_{\mathrm k}(T)\, e^{ik_i X^i}$ is a *Fourier mode*.

The Fourier transform

Let f be a square-integrable function of a single variable (*i.e.* $\int dx |f(x)|^2$ is finite). Its *Fourier transform* $f_k \equiv \hat f(k)$ is defined as

$$\hat f(k) = \int_{-\infty}^{+\infty} \frac{dx}{\sqrt{2\pi}} e^{-ikx} f(x). \tag{9.9}$$

In three dimensions $\hat f(k^i) = (2\pi)^{-3/2} \int d^3 x\, e^{-ik_i x^i} f(x^i)$. If f is a real function ($f = f^*$), we have $\hat f^*(-k) = \hat f(k)$.

The transform of a Gaussian $g(x) = \alpha\, e^{-\beta x^2}$ is the Gaussian $\hat g(k) = (\alpha/\sqrt{2\beta})e^{-k^2/4\beta}$.

If $\int dx\, |g(x)|^2 = 1$ (*i.e.* if $\alpha^2 \sqrt\pi = \sqrt{2\beta}$), then, on the one hand, $\int dk\, |\hat g(k)|^2 = 1$ also, while on the other[3] $\Delta x\, \Delta k = \frac{1}{2}$ with $(\Delta x)^2 \equiv \int dx\, |g(x)|^2 x^2$ and $(\Delta k)^2 \equiv \int dk\, |\hat g(k)|^2 k^2$.

The Fourier transform of $\delta_\sigma(x) = \dfrac{e^{-x^2/2\sigma^2}}{\sigma\sqrt{2\pi}}$ is $\hat\delta_\sigma(k) = \dfrac{e^{-k^2\sigma^2/2}}{\sqrt{2\pi}}$, and, inversely, the transform of $C(x) = \dfrac{e^{-x^2\sigma^2/2}}{\sqrt{2\pi}}$ is $\hat C(k) = \delta_\sigma(k)$. Since the Dirac distribution (see below) can be approximated by $\delta_\sigma(x)$ with $\sigma \to 0$, we see that its Fourier transform is a constant, and so we can write, at least formally,

---

[2] The basics of Lagrangian and Hamiltonian mechanics are presented in, for example, Book 1, Chapters 8 and 9.

[3] To demonstrate these properties of Gaussians we use the fact that $I \equiv \int du\, e^{-au^2} = \sqrt{\pi/a}$, which can be proved by calculating $I^2 = \int du\, dv\, e^{-a(u^2+v^2)}$ in polar coordinates.

When the function $f(x)$ is not a Gaussian, its width Δx and the width Δk of its Fourier transform satisfy the Heisenberg–Kennard inequality: $\Delta x\, \Delta k > \frac{1}{2}$.

$$1 = \int dx\, \delta(x)\, e^{-ikx} \quad \text{and} \quad \delta(x) = \int \frac{dk}{2\pi} e^{ikx}. \tag{9.10}$$

Now it is easy to show that if $\hat{f}(k)$ is the Fourier transform of $f(x)$ then, inversely,

$$f(x) = \int \frac{dk}{\sqrt{2\pi}} \hat{f}(k) e^{+ikx}.$$

We can similarly prove the *Parseval–Plancherel theorem*: $\int dx\, f_1^*(x) f_2(x) = \int dk\, \hat{f}_1^*(k) \hat{f}_2(k)$.

The Dirac distribution

The *Dirac distribution* $\delta(x - x_0)$ operates on a 'well-behaved' function $f(x)$ (*i.e.*, one that falls off sufficiently quickly at infinity) to give the value of $f(x)$ at $x = x_0$. We write

$$\int_{-\infty}^{+\infty} dx\, \delta(x - x_0) f(x) = f(x_0)$$

and interpret $\delta(x - x_0)$ as a 'function' which is zero everywhere except at x_0 and whose integral is 1. It can be approximated by ordinary functions, for example, by the Gaussian $\delta_\sigma(x) = \frac{1}{\sigma\sqrt{2\pi}} e^{-x^2/2\sigma^2}$ with $\sigma \to 0$.

In two dimensions $\delta_2(x^i - x_0^i) \equiv \delta(x - x_0)\delta(y - y_0)$, and so on, and we have

$$\int \delta_n(x^i)\, d^n x = \left(\int \delta(x)\, dx \right)^n = 1.$$

Some properties of the Dirac distribution:

a. $\delta(ax) = \frac{1}{|a|}\delta(x)$ and, more generally, $\delta(f(x)) = \sum_i \frac{1}{|f'(a_i)|}\delta(x - a_i)$, where the a_i are the zeros of f;

b. $\int dx\, \delta'(x) f(x) = -f'(0)$ (the prime denotes the derivative with respect to x);

c. $\Theta'(x) = \delta(x)$, where the function $\Theta(x)$ is the *Heaviside distribution*: $\Theta(x) = 0$ for $x < 0$ and $\Theta(x) = 1$ for $x \geq 0$;

d. In the Fourier space we can formally write $\delta(x) = \frac{1}{2\pi} \int dp\, e^{-ipx}$; see (9.10);

e. $\triangle(1/r) = -4\pi\delta_3(x^i)$, where $\triangle$ denotes the Laplacian.

In the Fourier space where $D(x^i) \equiv \frac{1}{(2\pi)^3} \int d^3 p\, e^{-ip_i x^i} \bar{D}(p^i)$, the equation $\triangle D(x^i) = -4\pi\delta_3(x^i)$ gives $\bar{D}(p^i) = \frac{4\pi}{p^2}$. We then have[4]

$$D(x^i) = \int \frac{d^3 p\, e^{-ip_i x^i}}{2\pi^2 p^2} = \frac{1}{2\pi^2} \int dp\, \sin\theta\, d\theta\, d\phi\, e^{-ipr\cos\theta} = \frac{2}{\pi} \int dp\, \frac{\sin pr}{pr} = \frac{1}{r}.$$

It can be shown in the same way that the solution of $\triangle\varphi - \mu^2\varphi = -4\pi\delta_3(X^i)$ is $\varphi = \frac{e^{-\mu r}}{r}$.

---

[4]From contour integration or the residue theorem.

With the field φ expanded as in (9.8), the Klein–Gordon equation (9.2) becomes an ordinary differential equation for each mode φ_k:

$$\ddot{\varphi}_k + \omega_k^2 \varphi_k = 0 \quad \text{with} \quad \omega_k = \sqrt{|k|^2 + \mu^2} \quad \text{and} \quad |k|^2 \equiv k_i k^i \,. \tag{9.11}$$

This equation of motion for the Fourier modes can also be obtained directly starting from the action (9.1) by substituting into it the field expansion[5] (9.8):

$$\mathbf{S}_f = \int dT\, L \quad \text{with} \quad L = \int d^3k\, \mathcal{L}_k \quad \text{and}$$

$$\mathcal{L}_k = \frac{1}{2}\left[\dot{\varphi}_k \dot{\varphi}_k^* - (|k|^2 + \mu^2)\varphi_k \varphi_k^*\right] = \frac{1}{2}\left[(\dot{\varphi}_{\mathcal{R}}^2 + \dot{\varphi}_{\mathcal{I}}^2) - (|k|^2 + \mu^2)(\varphi_{\mathcal{R}}^2 + \varphi_{\mathcal{I}}^2)\right]. \tag{9.12}$$

Therefore, $\mathbf{S}_f$ becomes a functional of $\varphi_{\mathcal{R}} \equiv \mathcal{R}e[\varphi_k(T)]$ and $\varphi_{\mathcal{I}} \equiv \mathcal{I}m[\varphi_k(T)]$, and the Euler–Lagrange equations $\frac{d}{dT}(\partial\mathcal{L}_k/\partial\dot{\varphi}_{\mathcal{R}}) = \partial\mathcal{L}_k/\partial\varphi_{\mathcal{R}}$ again give (9.11).

The momenta conjugate to $\varphi_{\mathcal{R}}$ and $\varphi_{\mathcal{I}}$ are $\pi_{\mathcal{R}} \equiv \partial\mathcal{L}_k/\partial\dot{\varphi}_{\mathcal{R}} = \dot{\varphi}_{\mathcal{R}}$ and $\pi_{\mathcal{I}} \equiv \partial\mathcal{L}_k/\partial\dot{\varphi}_{\mathcal{I}} = \dot{\varphi}_{\mathcal{I}}$, and so the Hamiltonian $H = \int d^3k\,(\pi_{\mathcal{R}}\dot{\varphi}_{\mathcal{R}} + \pi_{\mathcal{I}}\dot{\varphi}_{\mathcal{I}}) - L$ becomes (writing $\pi_k = \pi_{\mathcal{R}} + i\pi_{\mathcal{I}}$)

$$H = \int d^3k\, \mathcal{H}_k \quad \text{with} \quad \begin{cases} \mathcal{H}_k = \dfrac{1}{2}\left[\pi_{\mathcal{R}}^2 + \pi_{\mathcal{I}}^2 + (|k|^2 + \mu^2)(\varphi_{\mathcal{R}}^2 + \varphi_{\mathcal{I}}^2)\right] \\[2mm] \qquad = \dfrac{1}{2}\left[\pi_k \pi_k^* + (|k|^2 + \mu^2)\varphi_k \varphi_k^*\right], \end{cases} \tag{9.13}$$

and the Hamilton equations $\dot{\varphi}_{\mathcal{R},\mathcal{I}} = \partial\mathcal{H}_k/\partial\pi_{\mathcal{R},\mathcal{I}}$, $\dot{\pi}_{\mathcal{R},\mathcal{I}} = -\partial\mathcal{H}_k/\partial\varphi_{\mathcal{R},\mathcal{I}}$ again give the equation of motion (9.11).

The equation of motion (9.11) is just that of the harmonic oscillator, for which the general solution guaranteeing that the field is real can be written as

$$\varphi_k(T) = a_k \frac{e^{-i\omega_k T}}{\sqrt{2\omega_k}} + a_{-k}^* \frac{e^{i\omega_k T}}{\sqrt{2\omega_k}} \quad \text{with} \quad \omega_k = \sqrt{|k|^2 + \mu^2}\,, \tag{9.14}$$

where a_k is an integration constant. (The factor $1/\sqrt{2\omega_k}$, which appears a bit incongruous here, proves useful when one embarks on quantizing the field.) Each mode $\varphi_k(T)e^{ik_i X^i}$ is

---

[5] Using the properties of the Fourier transform and the Dirac distribution given above, we have

$$\int d^3X\, \varphi^2 = \int d^3X d^3k\, d^3k'\, \varphi_k \varphi_{k'} e^{i(k+k').X}/(2\pi)^3$$

$$= \int d^3k\, d^3k'\, \varphi_k \varphi_{k'} \delta_3(k+k') = \int d^3k\, \varphi_k \varphi_{-k} = \int d^3k\, \varphi_k \varphi_k^*,$$

where $(k.X) \equiv k_i X^i$. Similarly, $\int d^3X\, \dot{\varphi}^2 = \int d^3k\, \dot{\varphi}_k \dot{\varphi}_k^*$. Finally,

$$\int d^3X\, (\nabla\varphi)^2 = -\int d^3X\, d^3k\, d^3k'\, (k.k')\varphi_k \varphi_{k'} e^{i(k+k').X}/(2\pi)^3 = -\int d^3k\, d^3k'\, (k.k')\varphi_k \varphi_{k'} \delta_3(k+k')$$

$$= +\int d^3k\, k^2 \varphi_k \varphi_{-k} = +\int d^3k\, k^2 \varphi_k \varphi_k^*.$$

It is important to note that this Fourier expansion is possible only because the Klein–Gordon equation is linear and the action is quadratic in φ. If the mass term is replaced by a more general potential $V(\varphi)$, cf. footnote 1 above, the Fourier expansion will make sense only after linearization of the equation of motion $\Box\varphi - dV/d\varphi = 0$ about a *minimum* of $V(\varphi)$ (so that ω is real).

thus a 'monochromatic' plane wave, that is, a wave with *wave* vector k. The relation between the *frequency* ω_k and the 3-vector k is the *dispersion relation*.

All in all, the square-integrable general solution of the Klein–Gordon equation (9.2) is, after rewriting the integral in the second term in (9.14) with $k^i \rightarrow -k^i$,

$$\varphi(T, X^i) = \int \frac{d^3k}{(2\pi)^{\frac{3}{2}}} \left(a_k \frac{e^{-i\omega_k T}}{\sqrt{2\omega_k}} e^{ik_i X^i} + a_k^* \frac{e^{i\omega_k T}}{\sqrt{2\omega_k}} e^{-ik_i X^i} \right) \quad \text{with} \quad \omega_k = \sqrt{|k|^2 + \mu^2}. \quad (9.15)$$

Therefore any linear superposition of monochromatic plane waves such as (9.15) solves the Klein–Gordon equation and represents a *wave packet* which is more or less compact depending on how spread out the function a_k is about the value k_0. This wave packet propagates at the group velocity $c_g^i = d\omega_k/dk_i = k^i/\omega_k$, which is equal to the speed of light if $\mu = 0$, and less than the speed of light otherwise; see Book 1, Section 17.3.

The energy–momentum tensor (9.3) can also be expanded in Fourier modes, and, using footnote 5, we find that the Hamiltonian and the energy of the field coincide when the equations of motion are satisfied:

$$P^0 \equiv \int d^3X \, T^{00} = H \quad \text{when} \quad H = \frac{1}{2} \int d^3k \, \omega_k (a_k a_k^* + a_k^* a_k) = \int d^3k \, \omega_k \, |a_k|^2, \quad (9.16)$$

where $\omega_k = \sqrt{|k|^2 + \mu^2}$ is the frequency of each oscillator, $|a_k|^2 \equiv a_k^* a_k$, and the last equality ceases to be valid when the coefficients a_k become non-commuting quantum operators.

9.3   Complex fields, charge, and symmetry breaking

Now let us consider the following action, a functional of the *complex* scalar field $\phi(X^\mu)$:

$$\mathbf{S}_\phi = \int d^4X \mathcal{L}_\phi, \quad \text{where} \quad \mathcal{L}_\phi = -\partial_\mu \phi^* \partial^\mu \phi - V(\phi^* \phi). \quad (9.17)$$

The Euler–Lagrange equation of motion $\partial_\mu(\partial\mathcal{L}_\phi/\partial\partial_\mu\phi^*) = \partial\mathcal{L}/\partial\phi^*$ can be written as

$$\Box\phi - \frac{dV}{d\phi^*} = 0. \quad (9.18)$$

After multiplying by ϕ^* and subtracting the complex conjugate of the resulting expression, we find that the *Wronskian* or *current* Θ^μ is conserved:

$$\partial_\mu\Theta^\mu = 0 \quad \text{with} \quad \Theta^\mu \equiv -i(\phi \, \partial^\mu\phi^* - \phi^* \partial^\mu\phi). \quad (9.19)$$

Integrating over spacetime and following the same arguments as in deriving (8.9), we find that if the field falls off sufficiently rapidly at spatial infinity the *charge*

$$Q \equiv -i \int d^3X (\phi \, \dot{\phi}^* - \phi^* \dot{\phi}) \quad (9.20)$$

will be a constant of the motion (Q is expressed in kg-s and so has the same dimensions as $\hbar$).

The same result can be obtained using the Noether theorem. We see that the action and the Lagrangian density (9.17) as well as the equation of motion (9.18) are invariant under the transformation [of the U(1) type] $\phi \mapsto \phi\, e^{i\alpha} = \phi(1+i\alpha+\cdots)$, where α is a real constant. We then have

$$0 = \delta \mathbf{S}_\phi = \int d^4X \delta\mathcal{L}_\phi = -\int d^4X\, \partial_\mu(\delta\phi^*\partial^\mu\phi + \delta\phi\,\partial^\mu\phi^*)$$

$$+ \int d^4X \left[\delta\phi^*\left(\Box\phi - \frac{dV}{d\phi^*} \right) + \delta\phi\left(\Box\phi^* - \frac{dV}{d\phi} \right) \right] = -\alpha \int d^4X\, \partial_\mu \Theta^\mu,$$

if the equation of motion is satisfied and because $\delta\phi = i\alpha\,\phi$, which again leads to the result that the charge Q is a constant.

Now let us write $\phi = (\phi_1 + i\phi_2)/\sqrt{2}$, where ϕ_1 and ϕ_2 are real.

If the potential is $V = \mu^2 \phi^*\phi = \frac{1}{2}\mu^2\phi_1^2 + \frac{1}{2}\mu^2\phi_2^2$, the equation of motion becomes a set of two linear Klein–Gordon equations ($\Box\phi_1 - \mu^2\phi_1 = 0$, $\Box\phi_2 - \mu^2\phi_2 = 0$), the general solution of which in Fourier space was given in (9.15) (where $\varphi \to \phi_1$ or ϕ_2 and $a_k \to a_k^1$ or a_k^2). Performing the calculations of Section 9.2, footnote 5, we can write the Hamiltonian (9.16) and the charge (9.20) as

$$H = \int d^3k\, \omega_k(|a_k^1|^2 + |a_k^2|^2), \quad Q = -i \int d^3k\, (a_k^{1*}a_k^2 - a_k^1 a_k^{2*}), \tag{9.21}$$

or, after the *Bogoliubov transformation* $a_k^1 = (a_k + b_k)/\sqrt{2}$, $a_k^2 = -i(a_k - b_k)/\sqrt{2}$,

$$H = \int d^3k\, \omega_k(|a_k|^2 + |b_k|^2), \quad Q = -i \int d^3k\, (|a_k|^2 - |b_k|^2). \tag{9.22}$$

Within the framework of quantum field theory the coefficients a_k and b_k respectively become the 'annihilation operators' of a positively charged 'particle' and a negatively charged 'antiparticle'.

Let us now consider the case where the potential takes the form of a 'Mexican hat':

$$V(\phi^*\phi) = V_0 - \mu^2\phi^*\phi + \lambda(\phi^*\phi)^2 \tag{9.23}$$

with $\lambda > 0$; see Fig. 9.1. Its minimum is located on the circle $\phi_1^2 + \phi_2^2 = \mu^2/\lambda$; in the language of quantum field theory the 'vacuum' is 'degenerate'.[6] We expand it about a particular point on this circle, for example, $\phi_2 = 0$, setting $\phi_2 = \varphi_2$ and $\phi_1 = \mu/\sqrt{\lambda} + \varphi_1$. The Lagrangian density (9.17) is then written as $\mathcal{L}_\phi = -\frac{1}{2}(\partial_\mu\varphi_1)^2 - \frac{1}{2}(\partial_\mu\varphi_2)^2 - V(\varphi_1, \varphi_2)$ with

$$V(\varphi_1, \varphi_2) = V_0 - \frac{\mu^4}{4\lambda} + \mu^2\varphi_1^2 + \mu\sqrt{\lambda}\,\varphi_1(\varphi_1^2 + \varphi_2^2) + \frac{\lambda}{4}(\varphi_1^2 + \varphi_2^2)^2. \tag{9.24}$$

Since this is just a rewriting, the symmetry of the action (under the transformation $\phi \mapsto \phi = \phi\, e^{i\alpha}$) is of course preserved, as well as the constancy of the charge Q defined in (9.20). However, if we truncate the series at the quadratic term,

---

[6]The minimum value of the potential is $V_{\min} = V_0 - \mu^4/(4\lambda)$. The equations of motion and their solutions do not depend on it, but this potential energy density has a 'weight' owing to the mass–energy equivalence ($E = mc^2$), and 'gravitates' owing to the equivalence principle; see Section 10.4. This is the *cosmological constant problem*, which is still unresolved.

$$\mathcal{L}_\phi = -\frac{1}{2}(\partial_\mu \varphi_1)^2 - \frac{1}{2}(\partial_\mu \varphi_2)^2 - V_{\text{quad}} \quad \text{with} \quad V_{\text{quad}}(\varphi_1, \varphi_2) = V_0 - \frac{\mu^4}{4\lambda} + \mu^2 \varphi_1^2 , \quad (9.25)$$

this will no longer be true: in the language of quantum field theory, the symmetry (which is termed 'global' because α does not depend on the point) is 'spontaneously broken'. The equations of motion then reduce to linear Klein–Gordon equations:

$$\Box \varphi_1 - 2\mu^2 \varphi_1 = 0 , \quad \Box \varphi_2 = 0 . \quad (9.26)$$

The first of these describes a scalar field of mass $\sqrt{2}\mu$, and the second a massless scalar field, the *Nambu–Goldstone boson*.[7]

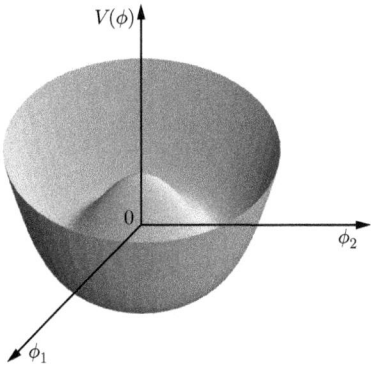

Fig. 9.1 The Goldstone potential.

9.4 The BEH mechanism

The fact that global symmetry breaking leads to the appearance of a massless, and therefore long-range, scalar field (see Section 10.1) is problematic because such a field is not observed experimentally. However, as we shall see, the BEH mechanism[8] will make it 'disappear'.

To see this we consider the coupling of the complex scalar field with action given in (9.17) and (9.23) to a massless vector field A_μ with a Maxwellian action (see Section 12.3)

$$\mathbf{S}_A = \int d^4 X \mathcal{L}_A, \quad \text{where} \quad \mathcal{L}_A = -\frac{1}{4} F_{\mu\nu} F^{\mu\nu} \quad \text{with} \quad F_{\mu\nu} \equiv \partial_\mu A_\nu - \partial_\nu A_\mu . \quad (9.27)$$

The action $\mathbf{S}_A$ is invariant under the transformation $A_\mu \mapsto A_\mu + \partial_\mu \alpha/q$, where α is an arbitrary dimensionless function of the coordinates and the constant $q\hbar$ has the dimensions of electric charge. The action (9.17) and (9.23) of the field ϕ is, as we have seen, invariant under the transformation $\phi \mapsto \phi\, e^{\mathrm{i}\alpha}$, where α is a constant.

We still need to construct the action of the interaction.

---

[7]The paradigmatic model of the 'spontaneous breaking of a global symmetry' presented here is due to J. Goldstone (1961).

[8]Named for its inventors, Englert and Brout (1964) and Higgs (1964).

If we *require* that the total action be invariant when the constant α becomes a *function* of the coordinates, it is easily seen that it is necessary to replace the ordinary derivative of ϕ, $\partial_\mu \phi$, in the action (9.17) and (9.23) by the 'covariant' derivative $\partial_\mu \phi - iqA_\mu \phi$, so that the total action is written as

$$\mathbf{S} = \int d^4 X \, \mathcal{L} \quad \text{with}$$

$$\mathcal{L} = -(\partial_\mu \phi^* + iqA_\mu \phi^*)(\partial^\mu \phi - iqA^\mu \phi) - \left(V_0 - \mu^2 \phi^* \phi + \lambda(\phi^* \phi)^2\right) - \frac{1}{4}F_{\mu\nu}F^{\mu\nu}.$$

(9.28)

Therefore, requiring that the 'global symmetry' becomes *local* necessitates the introduction of the field A_μ, which for this reason is called a *gauge field*, and also dictates the form of the interaction term.

Let us write ϕ as $\phi = (\phi_1 + i\phi_2)/\sqrt{2}$ and, as in Section 9.3, expand the action to quadratic order about a particular point of the potential minimum setting $\phi_1 = \mu/\sqrt{\lambda} + \varphi_1$ and $\phi_2 = \varphi_2$. We then obtain

$$\begin{cases} \mathcal{L}_{\text{quad}} = -V_0 + \dfrac{\mu^4}{4\lambda} - \dfrac{1}{2}(\partial_\mu \varphi_1)^2 - \mu^2 \varphi_1^2 - \dfrac{1}{2}M^2 W_\mu^2 - \dfrac{1}{4}G_{\mu\nu}G^{\mu\nu} \\[2mm] \text{with} \quad M \equiv \dfrac{q\mu}{\sqrt{\lambda}}, \quad W_\mu \equiv A_\mu - \dfrac{\partial_\mu \varphi_2}{M}, \quad \text{and} \quad G_{\mu\nu} \equiv \partial_\mu W_\nu - \partial_\nu W_\mu. \end{cases}$$

(9.29)

In this Lagrangian we again find the 'cosmological constant' $V_0 - \mu^4/(4\lambda)$ and the massive field φ_1 in the absence of coupling [*cf.* (9.25)], but we also see, on the one hand, that the massless Nambu–Goldstone boson φ_2 has disappeared, while, on the other, the effective gauge field W_μ has acquired a mass $M\hbar = q\mu\hbar/\sqrt{\lambda}$ and is not coupled to φ_1.

The BEH mechanism which we have sketched here within the framework of scalar electrodynamics makes it possible, when generalized to the electroweak theory, to give a mass to the gauge bosons $W^\pm$ and Z discovered at CERN in 1983. The BEH boson φ_1 (sometimes called the Higgs particle) was discovered in 2012, also at CERN. It has a mass of 125 GeV.

On the 'renormalizability' of the theory

The redefinition of the field made in (9.29) ($A_\mu \mapsto W_\mu$) is due to P. Higgs. It amounts to subjecting the field A_μ to a particular type of gauge transformation ($\alpha = -q\,\varphi_2/M$), termed 'unitary'. It has the advantage of explicitly eliminating the Nambu–Goldstone boson, but the resulting theory of a massive scalar field and a massive gauge field (called a Proca field when it is a vector field; see Section 14.3), which are decoupled (to quadratic order), is not an a priori 'renormalizable' *quantum* field theory. The choice of the 'Landau gauge' made by F. Englert and R. Brout indicates that the theory is 'renormalizable', but then it is the unitarity (that is, the gauge invariance of the gauge boson) which is no longer manifest. The *tour de force* of G. t'Hooft and M. Veltmann in 1972 was to show that the theory is in fact both unitary and renormalizable to all orders.[9]

---

[9]To learn more, see Englert (2012) and, for example, de Wit (2008).

10

The Nordström theory

In the preceding chapter we described a free scalar field and a scalar field interacting with other fields. Now we turn to the description of the interaction of a scalar field with *particles* which 'feel' it, that is, 'charged' particles. If the field is massless, and therefore long-range, and if the particle charge corresponds to its inertial mass, we have what is known as the Nordström theory, a coherent theory of gravity which, however, disagrees with experiment.

10.1 The coupling of a scalar field to a particle

As in Section 9.1, we consider a real scalar field of 'mass' μ whose action in an inertial Minkowski frame can be written as

$$\mathbf{S}_{\mathrm{f}} = \frac{1}{4\pi G} \int d^4 X \, \mathcal{L} \quad \text{with} \quad \mathcal{L} = -\frac{1}{2} \left(\partial_\mu \varphi \, \partial^\mu \varphi + \mu^2 \varphi^2 \right). \tag{10.1}$$

For $\mathbf{S}_{\mathrm{f}}$ to be expressed in kg-s, and if φ is dimensionless, the constant G must have the same dimensions as the Newton constant (that is, s/kg), while the constant μ is expressed in s^{-1}.

Just as for a free particle of inertial mass m, the equation of motion of a particle interacting with the external field φ can be obtained from a variational principle if an interaction term is added to the free-particle action.

Since the world line of a free particle is a geodesic of $\mathcal{M}_4$ (see Sections 2.2 and 5.3) which extremizes its length, its action is

$$\mathbf{S}_1[p_s(\lambda)] = -m \int d\lambda \, (-\eta_{\mu\nu} \dot{X}_s^\mu \dot{X}_s^\nu)^{1/2}, \tag{10.2}$$

where m is its inertial mass and $X_s^\mu(\lambda)$ is the world line of the path s in a Minkowski frame of coordinates X^μ. As for the interaction term of the action it can be written, for example, as

$$\mathbf{S}_{\mathrm{i}}[p_s(\lambda)] = -m_g \int_{\lambda_1}^{\lambda_2} d\lambda \, \mathcal{U}[\varphi(X_s^\mu)] \sqrt{-\eta_{\mu\nu} \dot{X}_s^\mu \dot{X}_s^\nu}, \tag{10.3}$$

where $\mathcal{U}[\varphi]$ is an *a priori* arbitrary function of $\varphi(X^\mu)$ and m_g is a constant characterizing the coupling of the particle to the field and has the dimensions of a mass if $\mathcal{U}$ is dimensionless. (We note that just as for the action of a free particle, $\mathbf{S}_{\mathrm{i}}$ is invariant under reparametrization.) The calculation of the variation of $\mathbf{S}_{\mathrm{i}}[p_s(\lambda)]$ when the path between the two points is varied is basically the same as that of Section 2.2:

Relativity in Modern Physics. Nathalie Deruelle and Jean-Philippe Uzan.
© Oxford University Press 2018. Published in 2018 by Oxford University Press.
DOI: 10.1093/oso/9780198786399.001.0001

$$\delta \mathbf{S}_i = m_g \left. \frac{\mathcal{U}\,\eta_{\mu\nu}\dot{X}^\nu \delta X^\mu}{\sqrt{-\eta_{\rho\sigma}\dot{X}^\rho \dot{X}^\sigma}}\right|_{\lambda_1}^{\lambda_2}$$

$$-m_g \int_{\lambda_1}^{\lambda_2} d\lambda \left[\frac{d}{d\lambda}\left(\frac{\mathcal{U}\,\eta_{\mu\nu}\dot{X}^\nu}{\sqrt{-\eta_{\rho\sigma}\dot{X}^\rho\dot{X}^\sigma}}\right) + \sqrt{-\eta_{\mu\nu}\dot{X}^\mu\dot{X}^\nu}\,\partial_\mu \mathcal{U}\right]\delta X^\mu. \tag{10.4}$$

The equation of motion of the particle in the field φ extremizes the sum of the actions: $\delta(\mathbf{S}_l + \mathbf{S}_i) = 0$. Choosing the parameter λ to be the proper time τ such that $\eta_{\mu\nu}\dot{X}^\mu\dot{X}^\nu = -1$, the equation of motion in the Minkowski inertial frame $\mathcal{S}$ is then written as

$$m\frac{dU^\mu}{d\tau} = F^\mu \quad \text{with} \quad F^\mu = -\frac{m_g}{1 + \frac{m_g}{m}\mathcal{U}}(\partial^\mu \mathcal{U} + U^\mu U^\nu \partial_\nu \mathcal{U}). \tag{10.5}$$

(We check that $F_\mu U^\mu = 0$.)[1]

Now to obtain the equation of motion of the field φ *created* by the particle, we again use a variational principle, this time extremizing with respect to configurations φ the action of the field coupled to the particle, namely, $\mathbf{S}_f[\varphi] + \mathbf{S}_i[\varphi]$, where $\mathbf{S}_f$ is the Klein–Gordon action (10.1) and the interaction term (10.3) becomes $\mathbf{S}_i[\varphi] = -m_g \int d\lambda \mathcal{U}(\varphi)(-\eta_{\mu\nu}\dot{X}^\mu\dot{X}^\nu)^{1/2}$.

Since now we must vary $\mathbf{S}_i$ with respect to the field φ [and not the path $X^\mu(\lambda)$], we can fix the parametrization of the source world line in the action. We take it equal to the proper time $\lambda = \tau$, $\eta_{\mu\nu}\dot{X}^\mu\dot{X}^\nu = -1$. Therefore, $\mathbf{S}_i[\varphi] = -m_g \int d\tau \mathcal{U}(\varphi)$. We still need to transform this integral into an integral over spacetime. For this we write m_g as $m_g = m_g \int d^4X\,\delta_4(X^\mu - X^\mu(\tau))$, where $\delta_4(X^\mu) \equiv \delta(T)\delta_3(X^i)$, each δ being a Dirac delta distribution; see Section 9.2. The interaction term is then rewritten as

$$\mathbf{S}_i[\varphi] = -m_g \int d^4X\,d\tau\,\mathcal{U}(\varphi)\delta_4(X^\mu - X^\mu(\tau)). \tag{10.6}$$

The extremization of $\mathbf{S}_f[\varphi] + \mathbf{S}_i[\varphi]$ then follows immediately and leads to the Klein–Gordon equation coupled to a point source:

$$\Box\varphi - \mu^2\varphi = 4\pi G m_g \int d\tau\,\frac{d\mathcal{U}}{d\varphi}\delta_4(X^\mu - X^\mu(\tau)). \tag{10.7}$$

Now let us work in a frame where the source is at rest at the origin. Its proper time then is that of the frame, $\tau = T$, and (10.7) reduces to

$$\Box\varphi - \mu^2\varphi = 4\pi G m_g\frac{d\mathcal{U}}{d\varphi}\delta_3(X^i). \tag{10.8}$$

The static, spherically symmetric solution of (10.8) is obtained using the fact that the d'Alembertian in this case reduces to the Laplacian, and in spherical coordinates the Laplacian reduces to $\triangle f = (r^2 f')'/r^2$, where the prime denotes differentiation with respect to

---

[1]The Lorentz force obtained by coupling the particles to a vector field (the 'electromagnetic potential' A_μ) which determines the motion of an electric charge is another example of an external force (*cf.* Section 11.3 below).

the radial coordinate r. If the interaction is linear, that is, if $\mathcal{U} = \varphi$, the solution which in the sense of a distribution converges at infinity (but which is *not* square-integrable) is (see Section 9.2)

$$\varphi = -\frac{Gm_g}{r}e^{-\mu r}. \tag{10.9}$$

If $\mu = 0$, the interaction is *long-range* (that is, of infinite range), otherwise it is *short-range* with range $L \equiv \mu^{-1}$.

If the interaction is *not* linear, $\mathcal{U}$ can be given in the form of a Taylor series expansion

$$\mathcal{U} = \varphi + \tfrac{1}{2}a_2\varphi^2 \cdots, \tag{10.10}$$

where the coefficient a_2 is *a priori* arbitrary. Then (10.8) is solved by iteration. The first iteration again gives the solution (10.9). In the next order $d\mathcal{U}/d\varphi = 1 + a_2\varphi$, where $\varphi = -Gm_g e^{-\mu r}/r$ diverges at the origin, and so the right-hand side of (10.8) is no longer defined. This problem, which is inherent in nonlinear theories, is 'solved' by *renormalizing* $d\mathcal{U}/d\varphi$ to 1, so that for any function $\mathcal{U}$ we can take (10.9) as the *renormalized* solution of (10.8).

Short-range interactions

The solution (10.9) was obtained in a particular frame in which the source of the field is at rest at the origin. In a different inertial frame the source will be undergoing uniform rectilinear motion. As we shall see in detail in Section 17.1, the solution (10.9) can be written in any inertial frame as

$$\varphi(X^\mu) = -\frac{Gm_g}{\varrho_\mathcal{R}}e^{-\mu\varrho_\mathcal{R}}, \quad \text{where } \varrho_\mathcal{R} \equiv -U_\mu l^\mu_\mathcal{R} \text{ with } l^\mu_\mathcal{R} \equiv X^\mu - X^\mu_\mathcal{R} \text{ and } \eta_{\mu\nu}l^\mu_\mathcal{R}l^\nu_\mathcal{R} = 0. \tag{10.11}$$

Here U^μ is the 4-velocity of the field source in the frame $\mathcal{S}$ and $l^\mu_\mathcal{R}$ is the null vector pointing from X^μ toward the past and intersecting the world line of the source at $X^\mu_\mathcal{R}$ (see Fig. 10.1). Using the fact that $\partial_\mu l^\nu_\mathcal{R} = \delta^\nu_\mu + (U^\nu l_\mu/\varrho)_\mathcal{R}$ (see Section 17.1), we can verify that (10.11) is indeed the solution of the Klein–Gordon equation.

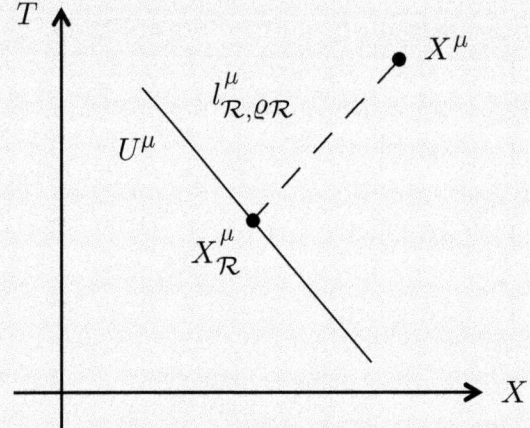

Fig. 10.1 Retarded quantities.

We also see that the exponential falloff of the field is preserved: a short-range interaction in an inertial frame is short-range in *any* other frame.

We note that the value at the point X^μ of the field φ (also called the *Liénard–Wiechert potential*; Section 17.1) depends on the *retarded* position $X^\mu_\mathcal{R}$ of the source, whether or not μ is zero, that is, whether or not the interaction propagates at a speed equal to or less than the speed of light. This is not contradictory because the solution (10.9) is *static*: it was 'established' throughout space in the infinitely distant past and the retardation effects have been 'erased'.

10.2 The field–matter system

The action which completely describes the dynamics of a system of particles as well as of the scalar field that they create and to which they are coupled is the sum of the action of the free particles, the particle–field interaction term, and the free-field action:

$$
\begin{aligned}
\mathbf{S}_{\text{tot}} = &-\sum_a m \int d\lambda \sqrt{-\eta_{\mu\nu}\dot{X}^\mu \dot{X}^\nu} - \sum_a m_g \int d\lambda\, \mathcal{U}(\varphi)\sqrt{-\eta_{\mu\nu}\dot{X}^\mu \dot{X}^\nu} \\
&-\frac{1}{8\pi G}\int d^4 X \left(\partial_\mu \varphi \partial^\mu \varphi + \mu^2 \varphi^2\right),
\end{aligned}
\tag{10.12}
$$

where the interaction term is also written as $\mathbf{S}_i[\varphi] = -\sum_a m_g \int d^4 X d\tau \mathcal{U}(\varphi)\delta_4(X^\mu - X^\mu(\tau))$, and where $\mathcal{U}$ is a function of φ to be specified. The extremization of (10.12) relative to variations of the path $X^\mu_s(\lambda)$ gives the equations of motion for each particle a [*cf.* (10.5)], and its extremization with respect to variations of the field configuration gives the Klein–Gordon equation coupled to sources [*cf.* (10.8)]. Let us rewrite these:

$$
\begin{cases}
m\dfrac{dU^\mu}{d\tau} = F^\mu \quad \text{with} \quad F^\mu = -\dfrac{m_g}{1 + \frac{m_g}{m}\mathcal{U}}(\partial^\mu \mathcal{U} + U^\mu U^\nu \partial_\nu \mathcal{U}) \\[2ex]
\Box \varphi - \mu^2 \varphi = 4\pi \sum_a G m_g \int d\tau\, \dfrac{d\mathcal{U}}{d\varphi}\delta_4(X^\mu - X^\mu(\tau)).
\end{cases}
\tag{10.13}
$$

The equations (10.13) are the analog of the Lorentz and Maxwell equations in electromagnetism; see Sections 11.3 and 12.3. They form a complete system determining both the motion of the sources and the field configuration once the function $\mathcal{U}(\varphi)$ is specified along with the initial conditions and at infinity.

The energy–momentum tensor of the field is, *cf.* (9.3),

$$
T^{\text{field}}_{\mu\nu} = \frac{1}{4\pi G}\left[\partial_\mu \varphi\, \partial_\nu \varphi - \frac{1}{2}\eta_{\mu\nu}\left(\partial_\rho \varphi\, \partial^\rho \varphi + \mu^2 \varphi^2\right)\right].
$$

Let us calculate its divergence taking into account the presence of sources. First of all we have

$$
\partial_\nu T^{\mu\nu}_{\text{field}} = \frac{\partial^\mu \varphi}{4\pi G}\left(\Box \varphi - \mu^2 \varphi\right),
\tag{10.14}
$$

which, taking into account the field equation in (10.13), becomes

$$
\partial_\nu T^{\mu\nu}_{\text{field}} = \sum_a m_g \int d\tau\, \partial^\mu \mathcal{U}\, \delta_4(X^\mu - X^\mu(\tau)).
\tag{10.15}
$$

Now using the equation of motion of the particles in (10.13) to replace $\partial^\mu \mathcal{U}$ and noticing that $U^\nu \partial_\nu \mathcal{U} = \dot{\mathcal{U}}$, (10.15) can be rewritten as

$$\partial_\nu T^{\mu\nu}_{\text{field}} = -\sum_a m \int d\tau\, \delta_4(X^\mu - X^\mu(\tau)) \frac{dU^\mu}{d\tau}$$
$$-\sum_a m_g \int d\tau\, \delta_4(X^\mu - X^\mu(\tau)) \frac{d(U^\mu \mathcal{U})}{d\tau} \,. \tag{10.16}$$

In the first term on the right-hand side we recognize the divergence of the energy–momentum tensor of point particles; *cf.* Section 8.4, (8.29) and (8.32). The second term is also written as the divergence of an energy–momentum tensor of the *interaction*. In the end we have

$$\partial_\mu (T^{\mu\nu}_{\text{field}} + T^{\mu\nu}_{\text{mat}}) = 0,$$
$$\text{where} \quad T^{\mu\nu}_{\text{mat}} = \sum_a m \int d\tau\, \delta_4(X^\mu - X^\mu(\tau)) U^\mu U^\nu \left(1 + \frac{m_g}{m} \mathcal{U}\right). \tag{10.17}$$

Integrating over the 3-volume V of the hyperplane $T = \text{const}$, we find, following (8.9),

$$\frac{d}{dT} P^\mu = -\int_S T^{\mu i}_{\text{field}}\, dS_i, \quad \text{where} \quad P^\mu \equiv P^\mu_{\text{field}} + P^\mu_{\text{mat}}$$
$$\text{with} \quad P^\mu_{\text{field}} \equiv \int_V d^3X\, T^{\mu 0}_{\text{field}} \quad \text{and} \quad P^\mu_{\text{mat}} \equiv \sum_a \left(1 + \frac{m_g}{m} \mathcal{U}\right) m U^\mu. \tag{10.18}$$

(It is understood that the 4-velocities U^μ as well as the function $\mathcal{U}$ are evaluated at the points where the world lines and the hypersurface $X^0 = T$ intersect.) Therefore, in the absence of *radiation*, that is, flux through the boundary S of V, the total momentum P^μ, the sum of the field momentum, the particle momenta, *and* the interaction momentum, is constant owing to the field and particle equations (10.13).

Let us take the case where $P^i = 0$. Then $P^0 \equiv M_{\text{in}}$ is the rest energy, that is, the *inertial mass* of the field–source system.

Since $T^{00}_{\text{field}} = \frac{1}{8\pi G}(\dot\varphi^2 + (\nabla\varphi)^2 + \mu^2\varphi^2)$, it is written as

$$M_{\text{in}} = \frac{1}{8\pi G} \int_V d^3X \left(\dot\varphi^2 + (\nabla\varphi)^2 + \mu^2\varphi^2\right) + \sum_a \left(1 + \frac{m_g}{m} \mathcal{U}(\varphi)\right) \frac{m}{\sqrt{1 - V^2}}. \tag{10.19}$$

Therefore, the inertial mass of an interacting system is the sum of the inertial masses $\sum_a m$ of its 'fundamental' constituents (or the constituents treated as being fundamental), their kinetic energy $\sum_a m(U^0 - 1)$, the interaction energy[2] $\sum_a m_g \mathcal{U} U^0$, and the energy of the field they create P^0_{field}. If the system radiates, this mass will not be constant. Moreover, it is not manifestly positive. Finally, we note that it must be 'renormalized' because the field φ and the function $\mathcal{U}$ are evaluated everywhere, *including* at the particle positions, where they diverge. The definition of the inertial mass of an interacting system is therefore much more involved in special relativity than in Newtonian physics, where it reduces to the sum of the inertial masses of the constituents.

---

[2]This term is absent in electromagnetism; see Section 12.4.

Short-range interactions and collision laws

If the scalar field φ is massive ($\mu \neq 0$), its range is limited to the vicinity of the particles; *cf.* (10.9). The same will generally be true of $\mathcal{U}(\varphi)$. Therefore, the force in (10.13) of a particle a' on a particle a will operate only if the two particles are separated by a distance $L < \mu^{-1}$. They are free before and after this *collision*.

For the same reason, the energy–momentum tensor of the field is not significantly different from zero except in a small region near the particles where the conservation law (10.18) reduces to the conservation of the total momentum of the particles studied in Section 6.3: $P^\mu = \sum m\, U^\mu =$ const (ignoring the divergent terms in $P_{\text{field}}^\mu \equiv \int d^3 X T_{\text{field}}^{\mu 0}$ and in the momentum of the interaction $\sum m_g \mathcal{U} U^\mu$).

Equation (10.19) then reduces to the equation $M_{\text{in}} = \sum m/\sqrt{1 - V^2}$ (it is understood that $\sum m V_a/\sqrt{1 - V^2} = 0$). The definition of the inertial mass of a system therefore poses no difficulty when the constituents interact *via* a short-range force.

10.3 The Nordström force

Let us return to (10.13) giving the equation of motion of a particle of 'charge' m_g coupled to a scalar field φ by means of a function $\mathcal{U}$. If we try to describe gravity by such a 'potential', then m_g will be the *passive gravitational mass* of the body P (and m will be its inertial mass). The experimental fact that all bodies 'fall' in the same way in a gravitational field forces us to choose $m_g = m$. The equation of motion in (10.13) then becomes (G. Nordström, 1912)

$$\frac{dU^\mu}{d\tau} = -\frac{1}{1 + \mathcal{U}} (\partial^\mu \mathcal{U} + U^\mu U^\nu \partial_\nu \mathcal{U}). \tag{10.20}$$

When the velocities are small and $\mathcal{U} \ll 1$, (10.20) reduces to Newton's second law $dV/dT = -\nabla \mathcal{U}$ and, to recover Newtonian gravity, we must have $\mathcal{U} = -GM/r$ in lowest order, where r is the distance to the source (which is at rest in the reference frame in question), M is its *gravitational mass* (or *active gravitational mass*), and G is Newton's constant. The field φ then is massless, $\mu = 0$; *cf.* (10.9).

The fact that (10.20) is a *kinematical* rather than dynamical equation (because the mass of the bodies subject to the field does not appear in it) is called the *weak equivalence principle*. We may ask whether or not it also applies to light corpuscles. One or the other must be true: either light is insensitive to gravity and propagates on the cones of $\mathcal{M}_4$, or it obeys (10.20). However, the result in both cases is the same, because the world line of an ultra-relativistic particle always approaches a light-like line (see the explicit calculation below). This means that the light coming from a distant source will not be deflected by a source at rest in an inertial frame. Since observations confirm the bending of light predicted by general relativity, we conclude that gravitation cannot be described by a scalar field theory formulated within the framework of special relativity.

Light deflection

When the function $\mathcal{U}$ depends only on the radial coordinate r, the solution of (10.20) follows step by step that of the Kepler problem in the Newtonian theory (see Book 1, Sections 12.2 and 17.1). The motion is planar and we use polar coordinates: $X = r \cos \phi$, $Y = r \sin \phi$. We first find that $\partial_0 \mathcal{U} = 0$ and $U^j \partial_j \mathcal{U} \equiv \dot{X} \partial_X \mathcal{U} + \dot{Y} \partial_Y \mathcal{U} = \dot{\mathcal{U}}$ (where the dot denotes differentiation with respect to the proper time τ).

The zeroth component of (10.20) is then written as $(1+\mathcal{U})\ddot{T} = -\dot{\mathcal{U}}\dot{T}$, the solution of which is $(1+\mathcal{U})\dot{T} = E$, where E is a constant. The combination $(\ddot{Y}\cos\phi - \ddot{X}\sin\phi)$ gives $(1+\mathcal{U})(2\dot{r}\dot{\phi} + r\ddot{\phi}) = -r\dot{\phi}\dot{\mathcal{U}}$, which can be integrated by inspection to give $(1+\mathcal{U})\dot{\phi} = L/r^2$, where L is a second constant of integration. Finally, the normalization of the 4-velocity $(-\dot{T}^2 + \dot{r}^2 + r^2\dot{\phi}^2) = -1)$ gives $\dot{r}^2$. We note that the particle 3-velocity is given by $(\dot{r}^2 + r^2\dot{\phi}^2)/\dot{T}^2$ and tends to $V_\infty^2 = (E^2-1)/E^2$ at infinity.

Next, following Binet, we set $r = 1/u$, form the ratio $\dot{r}^2/\dot{\phi}^2$, and then take the derivative with respect to the angle ϕ. This gives

$$\left(\frac{du}{d\phi}\right)^2 + u^2 = \frac{1}{L^2}\left[E^2 - (1+\mathcal{U})^2\right] \implies \frac{d^2u}{d\phi^2} + u = \frac{1}{p}(1+\mathcal{U})\frac{d\mathcal{U}}{d\varphi}, \qquad (10.21)$$

where we have introduced the variable $\varphi \equiv -GM/r$ and set $p \equiv L^2/GM$.

At this point we must specify the φ dependence of the function $\mathcal{U}$. At lowest order it must correspond to the Newtonian potential $-GM/r \equiv \varphi$, and so the second equation in (10.21) reduces to the Newtonian equation $d^2u/d\phi^2 + u = 1/p$, the solution of which is the hyperbola $r = p/(1 + e\cos\phi)$ with $1 + e = L^2/(GMr_0)$, where r_0 is the minimum distance of approach (for example, the solar radius).

We still need to express L^2 as a function of the speed of the particle at infinity. Since the initial conditions are $dr/d\phi = 0$ at $\phi = 0$, the first expression in (10.21) gives $L^2/r_0^2 = E^2 - (1+\mathcal{U}|_0)^2 \approx \frac{1}{1-V_\infty^2}$ if $\mathcal{U}|_0 \ll 1$ and $V_\infty \to 1$. Therefore, $e \approx (r_0/GM)(1 - V_\infty^2)^{-1}$ (instead of the Newtonian result $e \approx r_0/GM$; cf. Book 1, Section 17.1). The hyperbola is therefore very close to the straight line $r\cos\phi = r_0$. Its asymptotes are determined by $1 + e\cos\phi_a = 0$ or $\phi_a \simeq \pm\pi/2 + 1/e$ and the particle is deflected by an angle $\Delta\phi = 2/e$ or

$$\Delta\phi = \frac{2GM}{r_0}(1 - V_\infty^2) \to 0. \qquad (10.22)$$

Therefore, as announced, an ultra-relativistic particle is not deflected in the scalar theory of gravity.

Advance of the perihelion

In the Nordström theory of the gravitational field, the equation of motion of a particle reduces to (10.21), where $u = 1/r$ and p is a constant of integration. In addition, if the source remains at the origin of an inertial frame, then $\mathcal{U} = \varphi + \frac{1}{2}a_2\varphi^2 \cdots$ with $\varphi = -GM/r$. When the mass M of the source (the Sun, for example) is much larger than that of the particle (Mercury, for example), this external-field model is suitable for describing the dynamics of the system.

In lowest order the solution of (10.21) reduces to the Keplerian ellipse $u = (1 + e\cos\phi)/p$, where $p \equiv a(1 - e^2)$, a being the major axis and e the eccentricity. In the next order (10.21) is written as

$$\frac{d^2u}{d\phi^2} + u(1 + C) = \frac{1}{p}, \quad \text{where} \quad C = \frac{GM(1 + a_2)}{p}. \qquad (10.23)$$

The solution is $u = \frac{1 + e\cos(\sqrt{1+C}\phi)}{(1+C)p}$, the equation of an ellipse which precesses by an angle $\Delta\omega = 2\pi/\sqrt{1+C} - 2\pi$ in each revolution, where C is small so that:

$$\Delta\omega = -\Delta\omega_{RG}\frac{(1+a_2)}{6}, \quad \text{where} \quad \Delta\omega_{RG} \equiv \frac{6\pi GM}{a(1-e^2)}. \tag{10.24}$$

$\Delta\omega_{RG}$ is the value confirmed by observation and predicted by general relativity. We therefore see that we can choose a_2 $(= -7)$ such that the Nordström theory correctly predicts the advance of the perihelion (see Fig. 10.2). (However, we recall that it predicts an incorrect result for the deflection of light rays, as we have seen.)

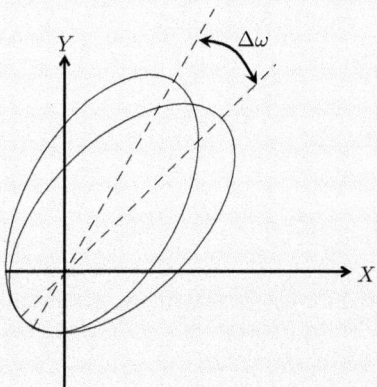

Fig. 10.2 Precession of the perihelion.

Book 2

10.4 Inertial and gravitational masses of a two-body system

Let us consider two point particles orbiting each other using the Nordström theory which describes gravity by means of a massless scalar field φ. According to the 'weak equivalence principle', their gravitational masses are equal to their inertial masses, $m = m_g$. When their velocities are small the gravitational field they create is also weak and is the solution of the Klein–Gordon equation (10.13). The latter then reduces to the Poisson equation with $\mathcal{U} \approx \varphi$, or $\triangle\varphi - \mu^2\varphi = 4\pi G \sum_a m\delta_3(X^i - X^i(T))$, the solution of which is the quasi-static Newton potential $\varphi = -\sum_a Gm/r$ if $\mu = 0$, where r is the distance between the frame origin and the mass m. Therefore, the inertial mass of the system (10.19),

$$M_{\text{in}} = \frac{1}{8\pi G}\int d^3X \left(\dot{\varphi}^2 + (\nabla\varphi)^2 + \mu^2\varphi^2\right) + \sum_a \left(1 + \frac{m_g}{m}\mathcal{U}(\varphi)\right)\frac{m}{\sqrt{1-V^2}},$$

reduces in this particular case to

$$M_{\text{in}} = \frac{1}{8\pi G}\int d^3X \, (\nabla\varphi)^2 + \sum_a m\left(1 + \frac{1}{2}V^2 + \varphi\right) + \mathcal{O}(G^2). \tag{10.25}$$

[Since the motion is Newtonian, in lowest order we have $V^2 = \mathcal{O}(G)$.]

The first term in (10.25) is the *opposite* of the Newtonian potential energy (see Book 1, Section 11.4). We calculate it by integrating by parts and using the Poisson equation. Up to a surface term we have

$$\frac{1}{8\pi G} \int d^3X \, (\nabla\varphi)^2 = -\frac{1}{8\pi G} \int d^3X \, \varphi \triangle \varphi$$

$$= \frac{1}{2} \sum_a m_a \int d^3X \delta_3(X^i - X_a^i(T)) \sum_b \frac{Gm_b}{r_b} = \frac{Gm_a m_b}{R}, \tag{10.26}$$

where R is the distance between the two bodies and we have ignored the divergent terms. The second term in (10.25), again ignoring divergent terms, is

$$\sum_a m \left(1 + \frac{1}{2}V^2 + \varphi\right) = m_a + m_b + \frac{1}{2}m_a V_a^2 + \frac{1}{2}m_b V_b^2 - 2\frac{Gm_a m_b}{R}$$

$$= m_a + m_b + \frac{1}{2}\frac{m_a m_b}{m_a + m_b}V^2 - 2\frac{Gm_a m_b}{R} \tag{10.27}$$

$$= m_a + m_b - \frac{3}{2}\frac{Gm_a m_b}{R},$$

where the last line is evaluated in the center-of-mass system and for circular motion (see below and Book 1, Chapter 11). Adding (10.26) and (10.27), we then find that in the lowest, Newtonian, order the inertial mass of the system is the sum of the masses of the constituents: $M = m_a + m_b$. However, in the next order it is necessary to add the contributions of the field energy (10.26) as well as the kinetic and interaction energies (10.27) of the particles, which gives

$$M_{in} = m_a + m_b - \frac{1}{2}\frac{Gm_a m_b}{R} + \mathcal{O}(G^2). \tag{10.28}$$

The extra term turns out to be equal to the sum of the *Newtonian* potential and kinetic energies. Equation (10.28) is an illustration of the mass–energy equivalence.

Now let us calculate the *gravitational* mass of this system of two particles orbiting each other. For this we need to find the gravitational field created by the system far away in the post-Newtonian order, which means solving the Klein–Gordon equation (10.13) (with $\mu = 0$) iteratively:

$$\Box\varphi = 4\pi \sum_a Gm \int d\tau \frac{d\mathcal{U}}{d\varphi} \delta_4(X^\mu - X^\mu(\tau)) \tag{10.29}$$

with $\mathcal{U} = \varphi + \frac{1}{2}a_2\varphi^2 + \cdots$.

To lowest order in G, where $\mathcal{U} = \varphi$, the solution of (10.29) is the retarded Liénard–Wiechert potential given in (10.11): $\varphi(X^\mu) = -\sum_a Gm/\varrho_\mathcal{R}$, where $\varrho_\mathcal{R} \equiv -(X^\mu - X_\mathcal{R}^\mu)U_{\mu\mathcal{R}}$, with $X_\mathcal{R}^\mu$ the intersection of the past cone with apex at the point X^μ and the world line of the particle m, and $U_\mathcal{R}^\mu$ the velocity at this point; see Fig. 10.3.

At the next order in G, $\frac{d\mathcal{U}}{d\varphi}(X^\mu) = 1 - \sum_a Gma_2/\varrho_\mathcal{R}$, which we 'renormalize' to $\frac{d\mathcal{U}}{d\varphi}(z^\mu(\tau))$ $= 1 - Gm'a_2/\rho$ when evaluating it on the world line of m at the point $z^\mu(\tau)$. Here ρ is defined by $\rho \equiv -(z^\mu - \hat{z}'^\mu)\hat{U}'_\mu$, with $\hat{z}'^\mu$ the intersection of the past cone with apex z^μ and the world line of m', and $\hat{U}'^\mu$ the velocity at this point. Therefore, to this order[3] the solution of (10.29) is

---

[3]In Section 17.2 we shall see how to obtain the solution using the retarded propagator.

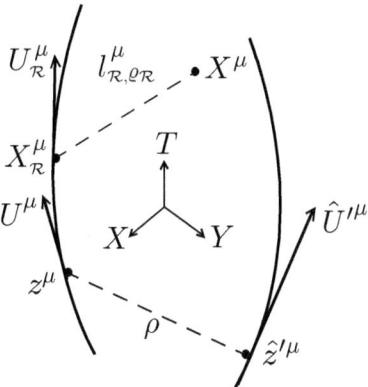

Fig. 10.3 Retarded quantities.

$$\varphi(X^\mu) = -\sum_a \frac{Gm}{\varrho_\mathcal{R}} + \sum_a \frac{G^2 mm' a_2}{\varrho_\mathcal{R}\, \rho_\mathcal{R}} + \mathcal{O}(G^3)\,. \tag{10.30}$$

Later on in Chapter 20 we shall give a detailed derivation of φ far from the system and for small velocities in the similar case of electromagnetism. The main steps in the (elementary) calculation are the following.

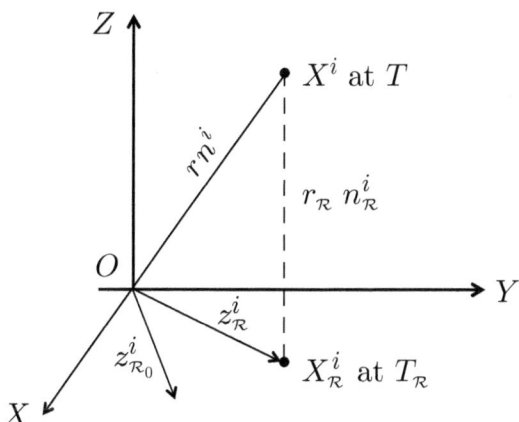

Fig. 10.4 Three-dimensional decomposition of retarded quantities.

(1) We change over to three-dimensional notation where, for example (see Section 17.1 and Fig. 10.4),

$$\varrho_\mathcal{R} = r_\mathcal{R}(1 - n_\mathcal{R}\,.V_\mathcal{R})/\sqrt{1 - V_\mathcal{R}^2}$$

with $r_\mathcal{R} = T - T_\mathcal{R}$ and $r_\mathcal{R} n_\mathcal{R}^i = X^i - X_\mathcal{R}^i$.

(2) We introduce an origin inside the system and write $r_{\mathcal{R}} n_{\mathcal{R}} = r\, n - z_{\mathcal{R}}$, where $r\, n$ is the radius vector of the point with component X^i of length r and $z_{\mathcal{R}}$ is the radius vector of the mass m at $T_{\mathcal{R}}$; see Fig. 10.4. At lowest order in z/r we have $r_{\mathcal{R}} n_{\mathcal{R}} = r\, n$. Then we expand the potential to order $1/r$ and order $V_{\mathcal{R}}^2$ (see Section 20.2).

(3) Finally, we take all the quantities at the time $T_{\mathcal{R}_0}$ such that $T - T_{\mathcal{R}_0} = r$ is the time taken by the gravitational field (which propagates at the speed of light) to go from the origin to the point in question [at the required order $V_{\mathcal{R}} = V_{\mathcal{R}_0} + \dot{V}_{\mathcal{R}_0}(n\,.z_{\mathcal{R}_0})$, where $z_{\mathcal{R}_0}$ and $V_{\mathcal{R}_0}$ are the radius vector and the 3-velocity of the mass m at the time $T_{\mathcal{R}_0}$; see Section 20.2].

The result, to be compared with (20.9), is (where RN is the separation 3-vector between the two bodies and R is their distance)

$$\varphi = -\sum_a \frac{Gm}{r}\left\{ 1 + (n\,.V_{\mathcal{R}_0}) + \left[(n\,.\dot{V})(n\,.z) + (n\,.V)^2 - \frac{1}{2}V^2 \right]_{\mathcal{R}_0} \right\}$$
$$+2\frac{Gm_a m_b a_2}{rR} + \mathcal{O}(G^3).$$
(10.31)

The final step is to go to the center-of-mass frame where $z_a = -m_b RN/M$ and $z_b = m_a RN/M$ with $M = m_a + m_b$. At the order considered the motion of the two masses is Newtonian: $\dot{V} = -GMN/R^2$ and $V^2 = GM/R$ [where $V = (RN)$], and so, taking the average over an orbital period (*cf.* Section 18.1), we have $\overline{(n\,.V)^2 - (GM/R)(n\,.N)^2} = 0$.

With all the calculations done, the gravitational potential (10.30) evaluated far from a system of two bodies slowly orbiting around each other and taking the average over an orbital period is given by

$$\varphi = -\frac{GM}{r} \quad \text{with} \quad M \equiv m_a + m_b - \frac{1}{2}\frac{Gm_a m_b}{R}\left(1 + 4a_2\right).$$
(10.32)

Here M is the *gravitational mass* (or *active gravitational mass*) of the system. We see that it is equal to the *inertial* mass of the system given in (10.28) if $a_2 = 0$, that is, if the coupling of the field φ to the particles is linear: $\mathcal{U} = \varphi$ (at least in the order studied). Therefore, the linear Nordström theory of gravity satisfies, like general relativity, what is called the *strong equivalence principle.*[4]

Inertia and gravitation

A feature specific to the gravitational interaction (called the *equivalence principle* in general relativity) is that all bodies 'fall' in the same way in a gravitational field. The equations of motion, such as Newton's equation ($a = -\nabla U$) or its relativistic analog in the Nordström theory [(10.20), which we rewrite below] do not involve the mass of the body subject to the field and therefore are kinematical rather than dynamical equations. The same property is found in inertial forces: the equation of motion of a free particle, whether in an inertial frame or not, does not depend on its mass.

---

[4]However, we recall that this is its only virtue, because, in conflict with observation, it does not predict any bending of light rays; *cf.* Section 10.3. In addition, for $a_2 = 0$ it predicts a *lag* of the perihelion instead of the observed advance, *cf.* Section 10.3. Finally, it predicts a spin precession which disagrees with the observed geodetic precession; *cf.* Section 7.4.

However, as we have seen in Section 5.3, in special relativity inertial accelerations are 'geometrized', because they are encoded in the Christoffel symbols of the Minkowski metric.

We can therefore also envision the geometrization of gravitation, that is, its absorption into the Christoffel symbols. However, the Christoffel symbols of the Minkowski metric can always be made to cancel out *everywhere*, by going from an accelerated frame to an inertial Minkowski frame where they all vanish. In order not to efface gravity everywhere, the Christoffel symbols encoding gravity must therefore be associated with metrics which are richer than the Minkowski metric, so that they cannot be made to vanish everywhere by a change of coordinates. The spaces associated with such metrics are *curved* spaces.

Let us take the example of the equation of motion (10.20) of a particle in a gravitational field in the Nordström theory:

$$\frac{dU^\mu}{d\tau} = -\frac{1}{1+\mathcal{U}}(\partial^\mu \mathcal{U} + U^\mu U^\nu \partial_\nu \mathcal{U}), \tag{10.33}$$

where $U^\mu \equiv dX^\mu/d\tau$ are the components of the particle 4-velocity in an inertial frame, τ is its proper time, and $\mathcal{U}$ is the gravitational potential, which is Newtonian at lowest order. As is easily verified (A. Einstein and A. Fokker, 1914), this equation can be rewritten as

$$\frac{Du^\mu}{d\bar{\tau}} \equiv \frac{du^\mu}{d\bar{\tau}} + \Gamma^\mu_{\nu\rho} u^\nu u^\rho = 0 \quad \text{with} \quad \Gamma^\mu_{\nu\rho} \equiv \frac{1}{2}g^{\mu\sigma}(\partial_\nu g_{\rho\sigma} + \partial_\rho g_{\sigma\nu} - \partial_\sigma g_{\nu\rho}), \tag{10.34}$$

where $g_{\mu\nu} \equiv (1+\mathcal{U})^2\eta_{\mu\nu}$, so that $\Gamma^\mu_{\nu\rho} = (\delta^\mu_\nu \partial_\rho \mathcal{U} + \delta^\mu_\rho \partial_\nu \mathcal{U} - \partial^\mu \mathcal{U} \eta_{\nu\rho})/(1+\mathcal{U})$, and the velocity $u^\mu \equiv dX^\mu/d\bar{\tau} = U^\mu/(1+\mathcal{U})$ is such that $g_{\mu\nu}u^\mu u^\nu = -1$. This rewriting is important because it shows that the Nordström equation of motion (10.33) is actually the geodesic equation in a spacetime whose metric, with coefficients $g_{\mu\nu}$ in the coordinates X^μ, is *not* the Minkowski metric (a metric $g = \Omega\ell$ where Ω is a function of the points is said to be *conformal* to the Minkowski metric). Indeed, there is no change of coordinates which can reduce the $\Gamma^\mu_{\nu\rho}$ given in (10.34) to zero [because 'at best' (10.34) reduces to (10.33) and not to $dU^\mu/d\tau = 0$]. The spacetime in which the motion of a particle is a geodesic (that is, 'free') in the Nordström gravitational field is therefore a curved space.

We have seen that the Nordström theory is refuted by observation. To arrive at a geometrical theory which satisfactorily describes gravitation, it is necessary to consider, as is done in general relativity, spacetimes which differ from Minkowski spacetime by more than a simple multiplication by a conformal factor.

Part III

Electromagnetism

The velocity of transverse undulations in our hypothetical medium, calculated from the electro-magnetic experiments of MM. Kohlrausch and Weber, agrees so exactly with the velocity of light calculated from the optical experiments of M. Fizeau, that we can scarcely avoid the inference that light consists in the transverse undulations of the same medium which is the cause of electric and magnetic phenomena.

James Clerk Maxwell, *On physical lines of force. Philosophical Magazine, Volumes 21 & 23 Series 4, 1861, Part III, Prop. XVI*

Book 2

11

The Lorentz force

Here we shall begin our study of the Maxwell theory by defining the electromagnetic potential and field as well as the force exerted by an external field on a particle carrying an electric charge. We will integrate the equation of motion for the cases of uniform rectilinear and circular motion.

We shall follow the four-dimensional approach, which is more efficient and transparent, but will also give the main results in 3-vector notation.

11.1 The electromagnetic potential and field

In the Maxwell theory the fundamental mathematical quantity representing electromagnetic phenomena is a field of covariant vectors (or a field of 1-forms), the *electromagnetic potential*, with components $A_\mu(X^\nu)$ in a given inertial frame with Minkowski coordinates X^ν.

A quantity derived from the potential plays a central role because it is this quantity rather than the potential itself which directly represents observable phenomena: this is the *Faraday tensor* or *electromagnetic field tensor*, defined as

$$F_{\mu\nu} \equiv \frac{\partial A_\nu}{\partial X^\mu} - \frac{\partial A_\mu}{\partial X^\nu} \equiv \partial_\mu A_\nu - \partial_\nu A_\mu \equiv A_{\nu,\mu} - A_{\mu,\nu}\,, \tag{11.1}$$

where in the last equality we have used the notation $f_{,\mu} \equiv \partial_\mu f \equiv \frac{\partial f}{\partial X^\mu}$.

We note that if A_μ is replaced by $A_\mu + \partial_\mu g$, where $g(X^\mu)$ is some function, $F_{\mu\nu}$ remains unchanged. This is a first glimpse of the *gauge invariance* of the Maxwell theory.

A contravariant vector can be associated with the potential by raising the index using the coefficients of the inverse Minkowski metric: $A^\mu \equiv \eta^{\mu\nu} A_\nu$ (or $A^0 = -A_0$, $A^i = A_i$). Similarly, the Faraday tensor, which is antisymmetric and 2-fold covariant (a 2-form), can be associated with tensors of the type $\binom{1}{1}$ which are mixed or 2-fold contravariant: $F^\mu{}_\nu \equiv \eta^{\mu\rho} F_{\rho\nu}$ (so that $F^0{}_\mu = -F_{0\mu}$) and $F^{\mu\nu} \equiv \eta^{\mu\rho}\eta^{\nu\sigma} F_{\rho\sigma} = -F^{\nu\mu}$. Another useful quantity is the *dual* of the Faraday tensor:

$$^*F_{\rho\sigma} \equiv \frac{1}{2} F^{\mu\nu} e_{\mu\nu\rho\sigma}\,, \tag{11.2}$$

where $e_{\mu\nu\rho\sigma}$ is the Levi-Civita symbol.[1]

---

[1]Note that $^{**}F_{\mu\nu} = -F_{\mu\nu}$. The properties of the Levi-Civita symbol were given in Section 7.2. The ones which will be most useful here are

$$e_{ijk}e^k{}_{ml} = \delta_{im}\delta_{jl} - \delta_{il}\delta_{jm}, \quad e_{ijk}e^{jk}{}_l = 2\delta_{il} \text{ and } e_{0123} = e_{123} = +1.$$

Relativity in Modern Physics. Nathalie Deruelle and Jean-Philippe Uzan.
© Oxford University Press 2018. Published in 2018 by Oxford University Press.
DOI: 10.1093/oso/9780198786399.001.0001

Let us introduce the commonly used three-dimensional notation.

Since it is antisymmetric, the Faraday tensor possesses six independent components which we denote as

$$F_{0i} \equiv -E_i , \quad F_{jk} \equiv B^i e_{ijk} , \tag{11.3}$$

or, equivalently,

$$^*F_{0i} = +B_i , \quad ^*F_{jk} = E^i e_{ijk} . \tag{11.4}$$

Here E^i is the *electric field*, B^i is the *magnetic field*, and e_{ijk} is the Levi-Civita symbol. The fields E^i and B^i can also be expressed as a function of the purely spatial components of $F_{\mu\nu}$ and its dual as

$$\begin{cases} B_k = +\dfrac{1}{2}F^{ij}e_{ijk} & \Longleftrightarrow \quad B_1 = +F_{23}, \quad B_2 = -F_{13}, \quad B_3 = +F_{12} \\[2mm] E^k = +\dfrac{1}{2}{}^*F_{ij}e^{ijk} & \Longleftrightarrow \quad E_1 = +{}^*F_{23}, \quad E_2 = -{}^*F_{13}, \quad E_3 = +{}^*F_{12} . \end{cases} \tag{11.5}$$

In addition, it is usual to split up the potential as

$$A^\mu \equiv (\Phi, A^i) , \tag{11.6}$$

where $\Phi \equiv +A^0 = -A_0$ is the *Coulomb potential* and A^i is the *magnetic potential*. The definition (11.1) can then be rewritten as

$$E_i = -\partial_0 A_i - \partial_i \Phi , \quad B^i = e^{ijk}\partial_j A_k . \tag{11.7}$$

Finally, we can introduce the 3-vectors (A, E, B), whose components (A^i, E^i, B^i) transform as $A'^i = \mathcal{R}_j{}^i A^j$, etc., under a rotation of the *spatial* axes of a given inertial Cartesian frame. Then (11.7) can be written in the 3-vector form

$$E = -\frac{\partial A}{\partial T} - \nabla\Phi , \quad B = \nabla \wedge A . \tag{11.8}$$

Review of three-dimensional vector calculus

The vector product of two vectors A and B is a vector denoted by $A \wedge B$ with components $(A \wedge B)^i = e^i{}_{jk}A^j B^k$. The scalar product is $A.B \equiv \delta_{ij}A^i B^j = A^i B_i$. The gradient operator is denoted ∇ and has components ∂_i, and the Laplacian is the operator $\triangle = \nabla.\nabla$.

The following expressions can then be proved as an exercise (see also Book 1, Sections 2.6, 3.6, and 4.6):

$$\nabla \wedge \nabla f \equiv 0, \quad \nabla.\nabla \wedge A \equiv 0, \quad \nabla \wedge (fA) = f\nabla \wedge A + \nabla f \wedge A,$$

$$\nabla.(A \wedge B) = -A.\nabla \wedge B + B.\nabla \wedge A,$$

$$\nabla \wedge (A \wedge B) = (B.\nabla)A - (A.\nabla)B + A\nabla.B - B\nabla.A,$$

$$\nabla \wedge \nabla \wedge A = -\triangle A + \nabla\nabla.A, \quad \nabla(A.B) = A.\nabla B + B.\nabla A + B \wedge \nabla \wedge A + A \wedge \nabla \wedge B,$$

$$A.(B \wedge C) = C.(A \wedge B), \quad A \wedge (B \wedge C) = (A.C)B - (A.B)C, \quad (A \wedge B)^2 = A^2 B^2 - (A.B)^2.$$

11.2 Transformation of the field under a Lorentz rotation

Since the potential is required to be a vector so that electromagnetic phenomena satisfy the principle of relativity, it transforms as a vector under a special Lorentz transformation: $A'_\mu = \Lambda^\nu{}_\mu A_\nu$ and $A'^\mu = \Lambda_\nu{}^\mu A^\nu$, where $\Lambda_\nu{}^\mu$ is the Lorentz matrix and $\Lambda^\nu{}_\mu$ is its inverse; cf. Section 1.3. The Faraday tensor transforms as

$$F'_{\mu\nu} = \Lambda^\rho{}_\mu \Lambda^\sigma{}_\nu F_{\rho\sigma}, \quad F'^\mu{}_\nu = \Lambda_\rho{}^\mu \Lambda^\sigma{}_\nu F^\rho{}_\sigma, \quad F'^{\mu\nu} = \Lambda_\rho{}^\mu \Lambda_\sigma{}^\nu F^{\rho\sigma}. \tag{11.9}$$

The first thing we can deduce is that the scalar $F_{\mu\nu}F^{\mu\nu}$ and the pseudo-scalar $^*F^{\rho\sigma}F_{\rho\sigma}$ are invariants[2]:

$$F_{\mu\nu}F^{\mu\nu} = F'_{\mu\nu}F'^{\mu\nu}, \quad {}^*F^{\rho\sigma}F_{\rho\sigma} = {}^*F'^{\rho\sigma}F'_{\rho\sigma}. \tag{11.10}$$

Let us rewrite this in three-dimensional notation.

Under a special Lorentz transformation representing the change from an inertial reference frame $\mathcal{S}$ to a frame $\mathcal{S}'$ undergoing uniform translation at speed V_0 relative to $\mathcal{S}$ along the X^1 axis, the components of the vector A^μ transform as

$$\Phi' = \frac{\Phi - V_0 A_1}{\sqrt{1 - V_0^2}}, \quad A'_1 = \frac{A_1 - V_0\Phi}{\sqrt{1 - V_0^2}}, \quad A'_2 = A_2, \quad A'_3 = A_3. \tag{11.11}$$

The electric and magnetic fields then become

$$\begin{cases} E'_1 = E_1, \quad E'_2 = \dfrac{E_2 - V_0 B_3}{\sqrt{1 - V_0^2}}, \quad E'_3 = \dfrac{E_3 + V_0 B_2}{\sqrt{1 - V_0^2}} \\[4mm] B'_1 = B_1, \quad B'_2 = \dfrac{B_2 + V_0 E_3}{\sqrt{1 - V_0^2}}, \quad B'_3 = \dfrac{B_3 - V_0 E_2}{\sqrt{1 - V_0^2}}. \end{cases} \tag{11.12}$$

Finally, (11.10) can be written in 3-vector form as[3]

$$E^2 - B^2 = E'^2 - B'^2, \quad E.B = E'.B'. \tag{11.13}$$

Therefore, in the change of inertial frame from $\mathcal{S}$ to $\mathcal{S}'$, a purely electric field E (i.e., with $B = 0$) will transform into a pair $(E', B' = -V_0 \wedge E')$, and a purely magnetic field B will transform into a pair $(B', E' = V_0 \wedge B')$. Reciprocally, if in a frame $\mathcal{S}'$ the fields E' and B' are orthogonal, there will exist another frame $\mathcal{S}$ moving in a direction perpendicular to E' and B' in which either E (if $|E'| < |B'|$) or B (if $|B'| < |E'|$) is zero. On the other hand, if E and B are orthogonal and have the same modulus in one frame, they will remain orthogonal and will have the same modulus in any other inertial frame, according to (11.13). These transmutations of electric field into magnetic field and *vice versa* are clearly just kinematic effects arising from the change of inertial frame.

---

[2] We also see that $^*F^{\rho\sigma}F_{\rho\sigma}$ is a 4-divergence:

$$^*F^{\rho\sigma}F_{\rho\sigma} \equiv \frac{1}{2}e_{\mu\nu\rho\sigma}F^{\mu\nu}F^{\rho\sigma} = 2e^{\mu\nu\rho\sigma}(\partial_\mu A_\nu)(\partial_\rho A_\sigma) = 2\partial_\mu V^\mu \text{ with } V^\mu \equiv e^{\mu\nu\rho\sigma}A_\nu\,\partial_\rho A_\sigma.$$

[3] More precisely, $F_{\mu\nu}F^{\mu\nu} = 2(B^2 - E^2)$ and $^*F^{\rho\sigma}F_{\rho\sigma} = 4\,E.B$. The pseudovector $V^\mu \equiv e^{\mu\nu\rho\sigma}A_\nu\,\partial_\rho A_\sigma$ splits into $V^0 = -A.B$, $V^i = -\Phi B^i + e^{ijk}A_j E_k$. In order to show in the three-dimensional formalism that $^*F^{\rho\sigma}F_{\rho\sigma}$ is a divergence, i.e., that $\partial_0 V^0 + \partial_i V^i = 2E.B$, we need to use the first group of Maxwell equations (see Section 12.1) as well as the definitions of E and B as functions of Φ and A. On the other hand, the proof is immediate in the four-dimensional formalism; see footnote 2.

11.3   The equation of motion of a charge

An *electric charge* is an elementary object of inertial mass m whose motion is affected by the presence of an electromagnetic field created by other charges. This charge is quantified by a number q. If $X^\mu(\tau)$ is its world line in a Minkowski frame $\mathcal{S}$ and $U^\mu \equiv dX^\mu/d\tau$ is its velocity (with $U^\mu U_\mu = -1$), its equation of motion is postulated to be the *Lorentz equation*

$$m\frac{dU^\mu}{d\tau} = F^\mu \quad \text{with} \quad F^\mu = q\,F^\mu{}_\nu\,U^\nu. \tag{11.14}$$

Here $F_{\mu\nu}$ is the Faraday tensor due to the other charges and F^μ is the *Lorentz force*.[4]

   This law of motion can also be obtained from a variational principle. Let us choose the action of a charge to be the sum of the free-particle action (that is, $\mathbf{S}_1[p_s(\lambda)] = -m\int_{\lambda_1}^{\lambda_2} d\lambda\,\sqrt{-\eta_{\mu\nu}\dot{X}_s^\mu \dot{X}_s^\nu}$; cf. Sections 2.2 and 10.1) and the interaction term

$$\mathbf{S}_i[p_s(\lambda)] = q\int_{\lambda_1}^{\lambda_2} d\lambda\,A_\mu \dot{X}_s^\mu\,, \tag{11.15}$$

where $A_\mu = A_\mu(X_s^\nu(\lambda))$ is evaluated on the world line of the path s in an inertial frame $\mathcal{S}$ with Minkowski coordinates X^μ and $\dot{X}_s^\mu \equiv dX_s^\mu/d\lambda$. The calculation of the variation of $\mathbf{S}_i[p_s(\lambda)]$ when the path between the two points is varied is the same as in Sections 2.2 and 10.1:

$$\begin{aligned}
\delta\mathbf{S}_i &= q\int_{\lambda_1}^{\lambda_2} d\lambda(A_\mu \delta\dot{X}^\mu + \dot{X}^\mu \delta A_\mu) \\
&= qA_\mu \delta X^\mu\big|_{\lambda_1}^{\lambda_2} - q\int_{\lambda_1}^{\lambda_2} d\lambda(\dot{X}^\nu \partial_\nu A_\mu - \dot{X}^\mu \partial_\nu A_\mu)\delta X^\mu \\
&= -q\int_{\lambda_1}^{\lambda_2} d\lambda\,(\partial_\nu A_\mu - \partial_\mu A_\nu)\dot{X}^\nu \delta X^\mu = q\int_{\lambda_1}^{\lambda_2} F_{\mu\nu}\dot{X}^\nu\,\delta X^\mu d\lambda\,.
\end{aligned} \tag{11.16}$$

The equation of motion of the charge extremizes the sum of the actions: $\delta(\mathbf{S}_1 + \mathbf{S}_i) = 0$. Choosing the parameter λ to be the proper time τ so that $\dot{X}^\mu \equiv U^\mu$, the variation of $\mathbf{S}_1$ reduces to $\delta\mathbf{S}_1 = -m\int(dU_\mu/d\tau)\,\delta X^\mu d\tau$, and we thus find that the motion is governed by[5] (11.14).

   We have required A_μ to be a covariant vector, that is, that it transform as $A'_\mu = \Lambda^\nu{}_\mu A_\nu$ under a Lorentz rotation. The Faraday tensor $F^\mu{}_\nu$ is mixed and therefore becomes $F'^\mu{}_\nu = \Lambda_\rho{}^\mu \Lambda^\sigma{}_\nu F^\rho{}_\sigma$. The Lorentz force $F^\mu = qF^\mu{}_\nu U^\nu$, the contraction of a mixed tensor and the velocity vector, then is also a vector: $F'^\mu = \Lambda_\rho{}^\mu F^\rho$. Finally, the acceleration vector $\gamma^\mu \equiv dU^\mu/d\tau$ becomes $\gamma^\mu = \Lambda_\rho{}^\mu \gamma^\rho$, so that after multiplication by $\Lambda^\nu{}_\mu$ the Lorentz equation has the same form in $\mathcal{S}'$ as in $\mathcal{S}$: $m\gamma'^\nu = F'^\nu$.

   Let us rewrite this in three-dimensional notation.

---

[4]We have $F^\mu U_\mu = 0$ due to the antisymmetry of $F_{\mu\nu}$. In addition, $qF_{\mu\nu}$, that is, qE or qB, has the dimensions of a force and is therefore expressed in kg·s^{-1} (since we have set $|c| = 1$).

[5]We note that the action (11.15) is invariant under reparametrization. This is essential in order for the choice of parametrization of the world lines to remain free.

In terms of the 3-velocity and 3-acceleration $V^i \equiv dX^i/dT$ and $a^i \equiv dV^i/dT$, the Lorentz force and the acceleration become

$$
\begin{cases}
F^0 = \dfrac{q}{\sqrt{1-V^2}}\, E.V\,, \quad F^i = \dfrac{q}{\sqrt{1-V^2}}\left(E^i + (V \wedge B)^i\right) \\[2ex]
\dfrac{dU^0}{d\tau} = \dfrac{V.a}{(1-V^2)^2}\,, \quad \dfrac{dU^i}{d\tau} = \dfrac{a^i}{1-V^2} + \dfrac{V^i(V.a)}{(1-V^2)^2}\,,
\end{cases}
\tag{11.17}
$$

which makes it possible to write the spatial components of the Lorentz equation (11.14) in various forms, for example,

$$
ma = q\sqrt{1-V^2}\left(E + V \wedge B - V(V.E)\right),
$$

$$
\text{where} \quad m\frac{d}{dT}\left(\frac{V}{\sqrt{1-V^2}}\right) = q\left(E + V \wedge B\right).
\tag{11.18}
$$

The zeroth component is also written as

$$
mV.\frac{d}{dT}\frac{V}{\sqrt{1-V^2}} = q\,E\,.V\,.
\tag{11.19}
$$

In another inertial frame $\mathcal{S}'$ the equation of motion is (11.18), where all the physical quantities and variables are primed, *including* the time T', which is the time measured by a clock at rest in $\mathcal{S}'$.

'Fictitious time' vs. 'pure and simple' time

Within the framework of Newtonian physics, the time T appearing in the three-dimensional versions of the Lorentz equation (11.18) is the absolute time. If we go to another inertial frame $\mathcal{S}'$ moving with velocity V_0 relative to the first, the Galilean transformation law for the 3-velocities requires that $V \mapsto V' = V - V_0$, $a \mapsto a' = a$, and the right-hand side of (11.18), which depends on the 3-velocity of the charge, does *not* satisfy the principle of relativity if E and B are represented by the same 3-vectors in the two frames ($E' = E$, $B' = B$). It will take a different form depending on V_0, and the problem will be to understand in which *particular* inertial frame it will have the form (11.18). For it to preserve the same form in $\mathcal{S}'$ to first order in the 3-velocities V_0 et V, we must take

$$
E \mapsto E' = E + V_0 \wedge B + \mathcal{O}(V_0^2) \quad \text{and} \quad B \mapsto B' = B - V_0 \wedge E + \mathcal{O}(V_0^2).
$$

The electric and magnetic fields (in contrast to, for example, Newton's gravitational field) will then not be represented by the same 3-vectors in two different inertial frames.

To preserve the invariance of the equation of motion (11.18) under Galilean transformations at higher orders in the 3-velocities, we need to be more clever. Following H. A. Lorentz, we introduce a 'fictitious' time $T' = T/\sqrt{1-V_0^2}$ and a contraction of the bodies in the direction of their motion, in short, as shown by H. Poincaré, we 'pretend' to pass from one inertial frame to another as given by the Lorentz formulas, while the 'true' formula is actually the Galilean one: $X' = X - V_0T$, where T is the absolute time of Newton.

The viewpoint is radically different in special relativity. The times T and T' are, according to Einstein's equations, 'purely and simply' the 'true' times, and the time dilation and length contraction are just kinematic effects which no longer need to be explained by the dynamics of charged objects. The Lorentz transformation law follows from the new representation of spacetime, and the invariance of the equations of electromagnetism in going from one inertial frame to another is manifest if we write them in terms of Minkowski vectors and tensors, as we have seen above.[6]

The Lagrangian and Hamiltonian of a charge in a field

The action of a charge in an external field is [*cf.* (11.15)]

$$\mathbf{S} = -m \int d\tau + q \int A_\mu U^\mu d\tau \equiv \int L \, dT \quad \text{with} \quad L = -m\sqrt{1 - V^2} - q\Phi + q\,A.V \,, \quad (11.20)$$

where the *Lagrangian L* is a function of the 3-velocity V of the charge and its position (through the spatial dependence of the potentials Φ and A). The *conjugate 3-momentum* associated with V is defined as

$$\pi \equiv \frac{\partial L}{\partial V} = \frac{mV}{\sqrt{1 - V^2}} + q\,A, \quad (11.21)$$

and it is easy to see that the Euler–Lagrange equation $d\pi/dT = \nabla L$ does again give the Lorentz equation (11.18) [using the fact that $dA/dT = \partial A/\partial T + V.\nabla A$ along with the relation $\nabla(A.V) = V.\nabla A + V \wedge \nabla \wedge A$ and the definitions (11.8) of E and B].

Equation (11.21) can be inverted to give V in terms of π, and we find that the Hamiltonian H of the charge, defined as $H \equiv \pi.V - L$, is expressed as a function of the position and π as

$$H = \sqrt{m^2 + (\pi - qA)^2} + q\Phi \,. \quad (11.22)$$

The Hamilton equations $\partial H/\partial \pi = V$, $\nabla H = -d\pi/dT$ then also give the Lorentz equation (11.18).

See, for example, Book 1, Chapters 8 and 9 for an introduction to the Lagrangian and Hamiltonian formalisms.

11.4 Charge in a uniform and constant field

As examples of the solution of the Lorentz equation, we shall discuss the elementary situations where the electromagnetic field reduces, in a given inertial frame, to either a uniform constant electric field or magnetic field, that is, a field independent of both time and position.

Let us imagine a charge in an inertial frame $\mathcal{S}$ moving along the lines of an electric field parallel to the X axis. The only nonzero component of the Faraday tensor then is $F^{01} \equiv E = \text{const}$, and the modulus of the acceleration $\gamma^\mu \equiv dU^\mu/d\tau$ of the charge will be a constant owing to the Lorentz equation (11.14):

$$\begin{aligned}
\gamma^\mu \gamma_\mu &= \frac{q^2}{m^2} (F^\mu_{\ \nu} U^\nu)(F_{\mu\rho} U^\rho) = \frac{q^2}{m^2} \left(F^0_{\ 1} F_{01} (U^1)^2 + F^1_{\ 0} F_{10} (U^0)^2 \right) \\
&= \frac{q^2 E^2}{m^2} \left(-(U^1)^2 + (U^0)^2 \right) = g^2 \quad \text{with} \quad g \equiv \frac{qE}{m} \,.
\end{aligned} \quad (11.23)$$

---

[6]For a historical approach to electrodynamics see, for example, Darrigol (2000) and Darrigol (2005).

It is easy to understand this result when we realize that in inertial frames tangent to the charge world line, the electric field is not changed by the Lorentz transformation, as can be read off from (11.12): $E_1' = E_1$.

The Lorentz equation reduces to $dU^1/d\tau = (q/m)F^1_{\ 0}U^0 \iff \ddot{X} = g\dot{T}$, and, similarly, $\ddot{T} = g\dot{X}$ with $U_\mu U^\mu = -1$, that is, $\dot{T}^2 = 1 + \dot{X}^2$. We therefore have $\ddot{X} = g\sqrt{1 + \dot{X}^2}$, which can be integrated to give $\dot{X}^2 = g^2(X - X_0)^2 - 1$. The solution of the latter is $X = X_0 + \cosh[g(\tau - \tau_0)]/g$. Therefore, after the constants of integration are chosen the trajectory of the charge will be (see also Section 2.1)

$$\left(T = \frac{1}{g}\sinh g\tau \ , \ X = \frac{1}{g}\cosh g\tau\right) \iff X = \frac{1}{g}\sqrt{1 + g^2T^2}. \qquad (11.24)$$

Charge in a critical electric field

We consider the case where the field E acts over a distance $L_{min} = 2m/(qE)$, the value beyond which quantum mechanics[7] predicts the possibility of pair production of a particle and its antiparticle of mass m and charges q and $-q$ from the vacuum. We then have $gL_{min} \equiv (qE/m)L_{min} = 2$. If the initial 3-velocity of the charge is zero, its value when leaving the field is obtained from (11.24) and is $V = \sqrt{3}/2$, or $V \approx 0.87c$. Its final energy is $mU^0_{final} = m\cosh g\tau_{final} = m\,gL_{min} = 2m$.

This being the case, pair production is only important if, in addition to acting over a distance $L > L_{min}$, the field is of the order of the *Schwinger critical value* $E_{crit} = m^2/(q\hbar)$, where $\hbar$ is the *Planck constant*.[8] For an electron

$$E_{crit} = m^2\alpha/e^3 \approx 2.4 \times 10^{17}\mathrm{kg}^{1/2}\mathrm{s}^{-3/2} \approx 1.3 \times 10^{18}\mathrm{V/m}.$$

[For comparison, the Coulomb field at the Bohr orbit of the hydrogen atom is given by $E_{Bohr} = e/a_{Bohr}^2$, where $a_{Bohr} = e^2/(m\alpha^2)$ is the *Bohr radius*, and so we have $E_{crit} = E_{Bohr}/\alpha^3 \approx 2.5 \times 10^6 E_{Bohr}$.]

The minimum distance over which this critical field must act in order to have pair production is $L_{min} = 2m/(qE_{crit}) = 2\hbar/m$. For an electron

$$L_{min} = 2e^2/(m\alpha) = 2.6 \times 10^{-21}\ \mathrm{s} = 7.7 \times 10^{-13}\ \mathrm{m}\ (\text{or } L_{min} = 2\alpha\,a_{Bohr}).$$

The energy gained per unit length $L \gg L_{min}$ by a charge in a critical field is $m(U^0 - 1)/L = m(g_{crit}L - 1)/L \approx mg_{crit} = m^2/\hbar$. For an electron,

$$m^2/\hbar = m^2\alpha/e^2 = 7.1 \times 10^{-10}\ \mathrm{kg/s} = 1.3 \times 10^9\ \mathrm{GeV/m},$$

to be compared with the hundreds of MeV/m accessible at present-day linear accelerators.[9]

---

[7]See, for example, Berestetskii, Lifshitz, and Pitaevskii (1971).

[8]In the units we are using, the Planck constant is given by $\hbar = e^2/\alpha$, where α is the *fine-structure constant* ($\alpha \approx 1/137$) and e is the electron charge. Numerically, $\hbar = 1.2 \times 10^{-51}$ kg·s $= 1.05 \times 10^{-34}$ J·s. The *Bohr radius* discussed below is $a_{Bohr} = e^2/(m\alpha^2) = 5.29 \times 10^{-11}$ m.

[9]Electron–positron pair production has been observed in relativistic heavy ion collisions, see A. Belkacem *et al.* (1993), as well as in laser beam collisions, see D. L. Burke *et al.* (1997). A detailed description of these processes can of course only be obtained using quantum electrodynamics.

Now let us consider a charge traveling in a constant magnetic field B along the Z axis. The only nonzero component of the Faraday tensor then will be $F_{12} \equiv B$, and the Lorentz equation will reduce to

$$\frac{dU^0}{d\tau} = 0, \quad \frac{dU^1}{d\tau} = \omega\, U^2, \quad \frac{dU^2}{d\tau} = -\omega\, U^1, \quad \frac{dU^3}{d\tau} = 0 \quad \text{with} \quad \omega \equiv \frac{qB}{m}, \tag{11.25}$$

the solution (satisfying $U^\mu U_\mu = -1$) of which is, for appropriate choice of the time origin,

$$T = \frac{\omega}{\Omega}\tau, \quad X = r_0 \sin\omega\tau, \quad Y = r_0 \cos\omega\tau, \quad Z = U_Z\tau$$

$$\text{with} \quad \Omega = \frac{\omega}{\sqrt{1 + r_0^2\omega^2 + U_Z^2}}, \tag{11.26}$$

where r_0 and U_Z are integration constants. The charge therefore travels along a helix of radius r_0 and pitch $2\pi U_Z/\omega$ about the magnetic field, at an angular velocity ω measured using its proper time and Ω measured using the time T of the inertial frame. We note that Ω can also be written in terms of the 3-velocity $V^i \equiv dX^i/dT$ as

$$\Omega = \omega\sqrt{1 - V^2} \tag{11.27}$$

because $V^2 = (r_0^2\omega^2 + U_Z^2)/(1 + r_0^2\omega^2 + U_Z^2)$. For $V \to 1$ we must have $r_0\,\omega \to \infty$ (for finite U_Z), and from (11.26) we see that the frequency Ω measured at the time T of the inertial frame then tends to $1/r_0$. The energy per unit mass is $U^0 \equiv 1/\sqrt{1 - V^2} = \omega/\Omega$ and, if it is large, we have $\omega/\Omega \approx r_0\,\omega = (q/m)r_0 B$.

When the charge is an electron, the quantity $\omega \equiv eB/m$ is referred to as the *cyclotron frequency*, and $\Omega \equiv \omega\sqrt{1 - V^2}$ is the *synchroton frequency*. The distance r_0 in terms of the 3-velocity is obtained from (11.26) and (11.27):

$$r_0 = \frac{V\sin\theta}{\omega\sqrt{1 - V^2}} \quad \text{with} \quad V_Z \equiv V\cos\theta. \tag{11.28}$$

This is the *Larmor radius*.

Synchrotron motion

In large modern accelerators like LEP at CERN, the electron energy P^0 can reach 100 GeV, corresponding to an energy per unit mass (or celerity) $P^0/m \equiv U^0 \equiv 1/\sqrt{1 - V^2} \approx 2 \times 10^5$. Using $\theta = \pi/2$, for a radius $r_0 = 5$ km, (11.28) gives $\omega \approx 1.2 \times 10^{10}$/s and (11.27) gives $\Omega \approx 6 \times 10^4$/s. Finally, from (11.25) we find the value of the (effective) magnetic field: $B \approx 3.7 \times 10^6\ \mathrm{kg}^{1/2}/\mathrm{s}^{3/2} \approx 0.05$ tesla.

Another example: the magnetic field of the muon storage ring at CERN where the time-dilation experiment of J. Bailey *et al.* was performed in 1979 was $B = 1.472$ tesla. Since the muon has the same charge as the electron and a mass 206.7 times heavier, (11.25) gives $\omega \approx 1.25 \times 10^9$/s. For a ring radius of $r_0 = 7$ m, (11.28) gives (with $\theta = \pi/2$) the celerity $1/\sqrt{1 - V^2} \approx 29.26$, and so, from (11.26) and (11.27) we have $T/\tau = \omega/\Omega = 1/\sqrt{1 - V^2} \approx 29.26$. The muon lifetime measured at the laboratory time T is about 30 times longer than its proper lifetime, as was verified with an accuracy of 0.1 % (see Section 2.2).

12
The Maxwell equations

In this chapter we present the Maxwell equations determining the electromagnetic field created by an ensemble of charges, and also derive them from a variational principle. We study their invariances: gauge invariance and invariance under Poincaré transformations, which allows us to derive the conservation laws for the total charge of the system and also for the system energy, momentum, and angular momentum.

12.1 The first group of Maxwell equations

The six components of the (antisymmetric) Faraday tensor $F_{\mu\nu}$ are not all independent because they are expressed as a function of the four components of the potential A_μ: $F_{\mu\nu} = \partial_\mu A_\nu - \partial_\nu A_\mu$. Moreover, since one of the four components of the potential can be chosen arbitrarily without changing $F_{\mu\nu}$ (this is the 'gauge invariance' mentioned in Section 11.1), there are $6 - (4-1) = 3$ constraints on the components of $F_{\mu\nu}$. It can be verified that they satisfy the following identity, referred to as the *first group of Maxwell equations*:

$$\partial_\mu F_{\nu\rho} + \partial_\nu F_{\rho\mu} + \partial_\rho F_{\mu\nu} \equiv 0 \quad \Longleftrightarrow \quad e^{\mu\nu\rho\sigma}\partial_\nu F_{\rho\sigma} \equiv 0 \quad \Longleftrightarrow \quad \partial_\nu {}^*F^{\mu\nu} = 0 \qquad (12.1)$$

(there are three, not four, independent components because the divergence vanishes: $e^{\mu\nu\rho\sigma}\partial_{\mu\nu}F_{\rho\sigma} \equiv 0$).

The identity (12.1) can be written in terms of the electric ($F_{0i} = -E_i$) and magnetic ($F_{jk} = B^i e_{ijk}$) fields as

$$\partial_i B^i = 0\,, \quad e^{ijk}\partial_j E_k = -\partial_0 B^i\,, \qquad (12.2)$$

or, in 3-vector notation,

$$\nabla.B = 0\,, \quad \nabla \wedge E = -\frac{\partial B}{\partial T}\,. \qquad (12.3)$$

The first equation in (12.3) is *Gauss's law of magnetism*, and the second is *Faraday's law of induction*. Since Gauss's law is included in Faraday's law [in the sense that $\partial(\nabla.B)/\partial T = \partial_{i0}B^i \equiv 0$; cf. (12.2)], there are indeed only three constraint equations.

12.2 Current and charge conservation

A *continuous* ensemble of moving charges is characterized by a *current vector*, $j^\mu(X^\nu)$. The action describing, in an inertial frame X^ν, the interaction of this current with an electromagnetic field is chosen to be of the form

$$\mathbf{S}_\mathrm{i} = \int A_\mu j^\mu \, d^4X\,, \qquad (12.4)$$

where $d^4X \equiv dX^0 dX^1 dX^2 dX^3$ is the volume element in Minkowski coordinates X^μ and the integral runs over all of spacetime.

Relativity in Modern Physics. Nathalie Deruelle and Jean-Philippe Uzan.
© Oxford University Press 2018. Published in 2018 by Oxford University Press.
DOI: 10.1093/oso/9780198786399.001.0001

The Lorentz force $F^\mu = qF^\mu{}_\nu U^\nu$ acting on a point charge is invariant under the gauge transformation $A_\mu \to A_\mu + \partial_\mu g$ (where g is arbitrary) which leaves the electromagnetic field tensor $F_{\mu\nu} = \partial_\mu A_\nu - \partial_\nu A_\mu$ unchanged; see Section 11.1. In order for the action (12.4), which gives rise to the force acting on a charge *distribution*, to be invariant also, we must have

$$0 = \mathbf{S_i}(A_\mu + \partial_\mu g) - \mathbf{S_i}(A_\mu) = \int j^\mu \partial_\mu g \, d^4X = \int \partial_\mu(gj^\mu) \, d^4X - \int g \, \partial_\mu j^\mu \, d^4X. \qquad (12.5)$$

The first term vanishes owing to the Gauss theorem for any function g which vanishes on the boundary of the integration region, while the second term must vanish $\forall\, g$, and so

$$\partial_\mu j^\mu = 0, \qquad (12.6)$$

which is the law of *current conservation*. Integration over a 3-volume V of the hyperplane $T = \mathrm{const}$ gives $\int d^3X\, \partial_0 j^0 + \int d^3X\, \partial_i j^i = 0$, or, using Gauss's theorem,

$$\frac{dQ}{dT} = -\int_S j^i \, dS_i, \qquad \text{where} \qquad Q \equiv \int_V j^0 d^3X. \qquad (12.7)$$

The quantity $d^3X = dX^1 dX^2 dX^3$ is the 3-volume element and S is the 2-surface bounding V with surface element dS_i. When there is no flux through S, that is, when the charge distribution remains confined, the *total charge* Q of the system contained in V is constant. This is an application of *Noether's theorem* (see Sections 8.2 and 9.3): any invariance of the action (here, gauge invariance) is associated with a conserved quantity (here, the total charge).[1]

The current which describes point charges is

$$j^\mu = \sum_a q \int_L d\tau \, \delta_4(X^\nu - X^\nu(\tau)) \, U^\mu, \qquad (12.8)$$

where $X^\mu(\tau)$ is the world line L of the particle a of charge q parametrized by its proper time τ, $U^\mu = dX^\mu/d\tau$ is the particle velocity ($\eta_{\mu\nu} U^\mu U^\nu = -1$), and $\delta_4(X^\mu) \equiv \delta(T)\delta(X)\delta(Y)\delta(Z)$, with δ the Dirac delta distribution [a 'function' which vanishes everywhere except at the origin and has unit integral $\int dX \delta(X) = 1$; see Section 9.2]. Since $\int d^4X \, \delta_4(X^\mu) = 1$, the interaction part of the action (12.4) is indeed that of point charges; *cf.* (11.15):

$$\mathbf{S_i} \equiv \int d^4X A_\mu j^\mu = \sum_a q \int d^4X \, d\tau \, \delta_4(X^\nu - X^\nu(\tau)) \, A_\mu U^\mu = \sum_a q \int d\tau \, A_\mu U^\mu. \qquad (12.9)$$

Moreover, since $\int d^3X \, \delta_3(X^\mu) = 1$ and $U^0 = dT/d\tau$, we have

$$\int d^3X \, j^0 = \sum_a q \int d\tau \, \delta(T - T(\tau)) \, U^0 = \sum_a q = Q. \qquad (12.10)$$

The total charge of the system is thus expressed, as desired, as a volume integral of a current. It can also be shown that $\partial_\mu j^\mu = 0$ using the fact that $\partial_0 \delta(X - X(T)) = -(dX/dT)(\partial \delta(u)/\partial u)$

---

[1]To show that Q is a scalar and therefore constant in any inertial frame, we proceed as in Section 8.2, integrating (12.6) over a *four*-volume $\mathcal{M}$ sandwiched between two space-like hypersurfaces. If the flux through the lateral 'cylinder' is zero, we have $\int_\Sigma j^\mu \, dS_\mu \equiv Q = \mathrm{const}$. When Q is written in this form the fact that it is independent of the choice of reference frame is manifest. If $\Sigma \equiv V$ is the hypersurface $T = \mathrm{const}$ of an inertial frame, we recover (12.7).

with $u = X - X(T)$, and so on (the calculation was discussed in detail in Section 8.4 in proving the conservation of the energy–momentum tensor of point particles).

In the three-dimensional version the components of the current j^μ,

$$j^0 \equiv \rho = \sum_a q\, \delta_3(X^j - X^j(T)), \quad j^i = \sum_a q\, V^i \delta_3(X^j - X^j(T)), \tag{12.11}$$

where V^i is the 3-velocity, coincide with the Newtonian definitions of the charge density and current. (However, it should be noted that it is $\rho\, d\tau/dT$ and not ρ which is a 4-scalar.)

12.3   The second group of Maxwell equations

A charge distribution described by the current vector j^μ in the inertial frame with coordinates X^μ creates an electromagnetic field which we here postulate to be governed by the *second group of Maxwell equations*:

$$\partial_\nu F^{\mu\nu} = 4\pi j^\mu, \tag{12.12}$$

where $F_{\mu\nu} = \partial_\mu A_\nu - \partial_\nu A_\mu$ is the Faraday tensor and A_μ is the potential describing the field. They are invariant under the gauge transformations $A_\mu \to A_\mu + \partial_\mu g$ and include the current conservation law $\partial_\mu j^\mu = 0$ because $F_{\mu\nu}$ is antisymmetric.

These equations also follow from a variational principle. Let us choose the action of the field to be the sum of the interaction term $(\mathbf{S_i} = \int A_\mu j^\mu\, d^4X)$ and the following functional of the potential:

$$\mathbf{S_f} = -\frac{1}{16\pi} \int F_{\mu\nu} F^{\mu\nu}\, d^4X, \tag{12.13}$$

where $d^4X \equiv dX^0 dX^1 dX^2 dX^3$ and the integral runs over all of spacetime. If we vary, as in Section 8.1, the potential A_μ, we obtain

$$\delta(\mathbf{S_i} + \mathbf{S_f}) = \int d^4X \left(j^\mu \delta A_\mu - \frac{1}{8\pi} F^{\mu\nu}(\partial_\mu \delta A_\nu - \partial_\nu \delta A_\mu) \right)$$

$$= -\frac{1}{4\pi} \int \partial_\mu (F^{\mu\nu} \delta A_\nu) d^4X - \frac{1}{4\pi} \int d^4X\, (\partial_\nu F^{\mu\nu} - 4\pi j^\mu)\, \delta A_\mu.$$

The first term transforms into a surface integral according to Gauss's theorem. The Maxwell equations (12.12) therefore extremize the action for any variation of the potential configuration which vanishes on the domain boundary. These are the Euler–Lagrange equations (8.3) associated with[2] $\mathbf{S_f} + \mathbf{S_i}$.

The Maxwell equations can also be written as a function of the potential as

$$\Box A_\mu - \partial_\mu(\partial_\nu A^\nu) = -4\pi j_\mu, \tag{12.14}$$

where $\Box$ is the d'Alembertian, $\Box \equiv \eta^{\mu\nu} \partial_\mu \partial_\nu \equiv -\frac{\partial^2}{\partial T^2} + \triangle$, and $\triangle \equiv \delta^{ij} \frac{\partial^2}{\partial X^i \partial X^j}$ is the Laplacian.

---

[2] The choice (12.13) for the action is necessitated by the fact that it must lead to the Maxwell equations, which are verified experimentally, but it is not unique. For example, the addition to the integrand of a (gauge-invariant) term $^*F_{\mu\nu} F^{\mu\nu}$ does not modify the equation of motion because this pseudo-scalar is a divergence (see Section 11.2, footnote 2) whose variation is zero.

A remark about the dimensions: the choice (12.13) for the action, as well as the Maxwell equations (12.12) which follow from it, implies that $F_{\mu\nu} F^{\mu\nu}$, that is, E^2 and B^2 have the dimensions of an energy density and are therefore expressed in kg/s^3 (since we have set $|c| = 1$).

Since (12.14) is linear, its general solution is the sum of the general solution of the homogeneous equation ($j^\mu = 0$) describing *electromagnetic waves* (see Chapters 15 and 16 below), and a particular solution of the complete equation.

The choice of gauge

As we have already seen and will see many times below, the Maxwell equations are not changed by gauge transformations $A_\mu \to \tilde{A}_\mu = A_\mu + \partial_\mu g$, where g is an arbitrary function of the coordinates X^μ. *Gauge fixing* means choosing a particular g.

Once a solution A_μ of (12.14) is known, we can, for example, choose g such that $\partial_0 g = -A_0$ (which defines it up to a function of the spatial coordinates). We then have $\tilde{A}_0 = 0$, which defines the *temporal gauges* or *Hamiltonian gauges*. In other words, we can solve (12.14) by *requiring* that $A_0 = 0$.

Alternatively, we can choose g such that $\triangle g = -\partial_i A^i$ so that $\partial_i \tilde{A}^i = 0$. This is what is called the *Coulomb gauge* or *radiation gauge* (sometimes also the *transverse gauge*). It is unique if we require that g be regular everywhere and vanish at spatial infinity.

We should also mention the existence of the *axial gauge* defined by $A_3 = 0$.

Finally, we can require that $\partial_\mu \tilde{A}^\mu = 0$, which means that g must satisfy $\Box g = -\partial_\mu A^\mu$, the solution of which is $g = g_0 + h$, where g_0 is a particular solution and h is a *harmonic function* solution of $\Box h = 0$. The ensemble of resulting potentials $\tilde{A}_\mu = A_\mu + \partial_\mu g_0 + \partial_\mu h$ form the *Lorenz gauges*.[3] They preserve the manifestly Lorentz-invariant form of the Maxwell equations.

The Maxwell equations (12.12) can be written in three-dimensional form in terms of the electric and magnetic fields ($F^{0i} = E^i$, $F_{ij} = e_{ijk}B^k$) as

$$\partial_i E^i = 4\pi j^0 \,, \quad -\partial_0 E^i + e^{ijk}\partial_j B_k = 4\pi j^i \,, \tag{12.15}$$

or in 3-vector notation as

$$\nabla.E = 4\pi\rho \,, \quad \nabla \wedge B = \frac{\partial E}{\partial T} + 4\pi j \,, \tag{12.16}$$

where $\rho \equiv j^0$ and j are the (Newtonian) charge density and the current 3-vector. The first equation (12.16) is *Gauss's law*, and the second, where the term $\partial E/\partial T$ is called the *displacement current*, is the *Ampère–Maxwell law*.

Lorentz invariance

Using the relation $\nabla \wedge \nabla A = -\triangle A + \nabla\nabla.A$, we easily find that the complete Maxwell equations, $\nabla.B = 0$, $\nabla \wedge E = -\partial B/\partial T$ [*cf.* (12.3) and (12.16)], imply that

$$-\frac{\partial^2 E}{dT^2} + \triangle E = 4\pi\left(\frac{\partial j}{\partial T} + \nabla\rho\right), \quad -\frac{\partial^2 B}{dT^2} + \triangle B = -4\pi\nabla \wedge j \,. \tag{12.17}$$

As we shall see in detail in Chapter 15, the electric and magnetic fields outside the charges (*i.e.*, where $j = \rho = 0$) travel at unit velocity (or velocity c, numerically equal to the speed of light), as verified by H. Hertz. This is why Maxwell concluded that light is an electromagnetic field.

---

[3](Ludvig Valentin) Lorenz, and not, as is sometimes erroneously written, (Hendrik Antoon) Lorentz. For more details about the passage from one gauge to another see, for example, Jackson (2002).

All the experiments on electromagnetism require that the equations (12.17) have the same form in any inertial frame because, in contrast to what happens, for example, with sound (see Book 1, Section 17.3), it turns out to be impossible to find a frame in which an 'aether' carrying light waves is at rest. However, in the Newtonian framework E and B are 3-vectors and T is the absolute time, and it is problematical to require that the equations (12.17) remain invariant in passing from one inertial frame to another if such passage is governed by Galilean transformations (except by requiring that the fields not be represented by the same 3-vectors in two different frames, and introducing a 'fictitious' time, and so on, as mentioned in Section 11.3.)

On the other hand, when expressed in tensor form $e^{\mu\nu\rho\sigma}\partial_\nu F_{\rho\sigma} = 0$ [cf. (12.1) and (12.12)], the Maxwell equations preserve by construction the same form under Lorentz transformations, which in special relativity describe the passage from one inertial frame to another. Hence, the solution of (12.17) in the vacuum is a *wave packet* (see Chapter 15 below) propagating at the speed of light in *any* Minkowski frame, in agreement with experiment. It is therefore useless to specify whether or not the medium in which the waves propagate is at rest, and the *rigid fluid* corresponding to the aether (see footnote 9 of Section 8.3) is relegated to limbo.

Duality and magnetic charges

The complete Maxwell equations are, *cf.* (12.12) and (12.1),

$$\partial_\nu F^{\mu\nu} = 4\pi\, j^\mu, \quad \partial_\nu\, {}^*F^{\mu\nu} = 0\,, \tag{12.18}$$

where ${}^*F_{\rho\sigma} \equiv \frac{1}{2}F^{\mu\nu}e_{\mu\nu\rho\sigma}$ and $F_{\mu\nu} = \partial_\mu A_\nu - \partial_\nu A_\mu$. To write them in a more symmetric form, we can follow Dirac (1931) and make the generalization

$$\partial_\nu F^{\mu\nu} = 4\pi\, j^\mu_{\text{e}}, \quad \partial_\nu\, {}^*F^{\mu\nu} = 4\pi j^\mu_{\text{m}}\,, \tag{12.19}$$

where we assign the particles, aside from their mass, a *magnetic charge* along with the electric charge. (In this case the Faraday tensor can no longer be derived from a potential A_μ, except outside the magnetic charges.) Written in the form (12.19), the field equations possess a symmetry referred to as *duality*. If for a constant θ we write

$$\tilde{F}_{\mu\nu} = F_{\mu\nu}\cos\theta - {}^*F_{\mu\nu}\sin\theta$$

(which implies that ${}^*\tilde{F}_{\mu\nu} = {}^*F_{\mu\nu}\cos\theta + F_{\mu\nu}\sin\theta$), the equations (12.19) will preserve the same form $\partial_\nu \tilde{F}^{\mu\nu} = 4\pi\, \tilde{j}^\mu_{\text{e}}$ and $\partial_\nu {}^*\tilde{F}^{\mu\nu} = 4\pi\tilde{j}^\mu_{\text{m}}$ as long as new currents are defined as

$$\tilde{j}^\mu_{\text{e}} = j^\mu_{\text{e}}\cos\theta - j^\mu_{\text{m}}\sin\theta\,, \quad \tilde{j}^\mu_{\text{m}} = j^\mu_{\text{m}}\cos\theta + j^\mu_{\text{e}}\sin\theta\,.$$

Dyons are particles whose electric and magnetic charges are proportional to each other: $j^\mu_{\text{m}} = j^\mu_{\text{e}}\tan\vartheta$. Such particles are equivalent to particles without a magnetic charge because $\tilde{j}_{\text{m}}$ can be eliminated by the dual transformation $\theta = -\vartheta$, so that (12.19) becomes (12.18). Finally, a particle with zero electric charge is called a *magnetic monopole*.

The existence of dyons or magnetic monopoles has up to now not been confirmed.[4]

---

[4]A complete discussion of electromagnetic duality is given by Figueroa-O'Farrill (1998).

12.4   The field energy–momentum tensor

As we saw in Section 8.2, we can associate with any Lagrangian density $\mathcal{L}$, which is a functional of a field Φ but does not depend explicitly on the coordinates, a *Noether energy–momentum tensor* $\Theta_\mu{}^\nu$ given by $\Theta_\mu{}^\nu \equiv -\partial_\mu \Phi \frac{\partial \mathcal{L}}{\partial \partial_\nu \Phi} + \delta_\mu^\nu \mathcal{L}$. In electromagnetism the role of the field Φ is played by the potential A_μ and the Lagrangian density is [*cf.* (12.13)] $\mathcal{L}(A_\mu) = -(1/16\pi) F_{\mu\nu} F^{\mu\nu}$. We therefore find

$$\Theta_\mu{}^\nu = \frac{1}{4\pi} \left(F_{\mu\rho} F^{\nu\rho} - \frac{1}{4} \delta_\mu^\nu F^{\rho\sigma} F_{\rho\sigma} \right) + \frac{1}{4\pi} F^{\nu\rho} \partial_\rho A_\mu . \tag{12.20}$$

From Section 8.2 we also know that this tensor is conserved ($\partial_\nu \Theta_\mu{}^\nu = 0$) if the field equations, which in the present case are the Maxwell equations in the vacuum ($\partial_\nu F^{\rho\nu} = 0$), are satisfied.

Since the divergence of the last term in (12.20) is zero owing to the field equations and the antisymmetry of $F^{\mu\nu}$, we shall define the *energy–momentum tensor of the electromagnetic field* as

$$T_{\mu\nu} = \frac{1}{4\pi} \left(F_{\mu\rho} F_\nu{}^\rho - \frac{1}{4} \eta_{\mu\nu} F^{\rho\sigma} F_{\rho\sigma} \right), \quad \text{such that} \quad \partial_\nu T^{\mu\nu} = 0 \ \text{if} \ \partial_\nu F^{\rho\nu} = 0, \tag{12.21}$$

which has the advantage of being symmetric and gauge-invariant.[5] The tensor (12.21) has zero trace $T_\mu{}^\mu = 0$, and it can also be written as

$$T_{\mu\nu} = \frac{1}{8\pi} \left(F_{\mu\rho} F_\nu{}^\rho + {}^* F_{\mu\rho} {}^* F_\nu{}^\rho \right) . \tag{12.22}$$

Let us now take into account the presence of the charges giving rise to the field. Using only the antisymmetry of $F_{\mu\nu}$ and the first group of Maxwell equations $\partial_\rho {}^* F^{\mu\rho} = 0$, we have

$$\partial_\rho T^{\mu\rho} = -\frac{1}{4\pi} F^\mu{}_\rho \, \partial_\nu F^{\rho\nu} . \tag{12.23}$$

Reciprocally, if we *require* that the energy–momentum tensor of the field be conserved, *i.e.*, that $\partial_\nu T^{\mu\nu} = 0$, the identity (12.23) then implies that $\partial_\nu F^{\mu\nu} = 0$. In other words, the conservation of the energy–momentum tensor of the electromagnetic field implies the second group of Maxwell equations outside the charges.

If we require that the Maxwell equations be satisfied in the presence of charges $\partial_\nu F^{\mu\nu} = 4\pi j^\rho$, we will have

$$\partial_\rho T^{\mu\rho} = -F^\mu{}_\rho j^\rho . \tag{12.24}$$

Replacing j^ρ by its expression (12.8) and using the Lorentz equation $m \, dU^\mu/d\tau = q F^\mu{}_\nu U^\nu$, we can rewrite (12.24) as

$$\partial_\rho T^{\mu\rho} = -\sum_a m \int d\tau \, \delta_4 (X^\nu - X^\nu(\tau)) \frac{dU^\mu}{d\tau} . \tag{12.25}$$

---

[5] Since, *modulo* the Maxwell equations, $F^{\nu\rho} \partial_\rho A_\mu = \partial_\rho \sigma^{\mu\nu\rho}$ with $\sigma^{\mu\nu\rho} = A^\mu F^{\nu\rho}$, we see that $\Theta^{\mu\nu}$ and $T^{\mu\nu}$ differ by only a total divergence, which does not modify the definition of the field momentum as long as it falls off sufficiently rapidly at infinity; see (8.11).

We also note that the addition to the Lagrangian density of the divergence ${}^* F_{\mu\nu} F^{\mu\nu}$ mentioned above in footnote 3 will lead to an energy–momentum tensor different from (12.20).

We recognize the divergence of the energy–momentum tensor of massive particles on the right-hand side [*cf.* Section 8.4, (8.29) and (8.32)]. Hence, we finally have

$$\partial_\mu (T^{\mu\nu} + T_{\text{part}}^{\mu\nu}) = 0, \quad \text{where} \quad T_{\text{part}}^{\mu\nu} = \sum_a m \int d\tau \, \delta_4 (X^\mu - X^\mu(\tau)) U^\mu U^\nu . \tag{12.26}$$

This conservation law should be compared with that obtained in (10.17) in the case of a scalar interaction.

Integration over a 3-volume V with Cartesian volume element d^3X of the hyperplane $T = \text{const}$ of a given inertial frame gives

$$\frac{d}{dT} P^\mu = - \int_S T^{\mu i} dS_i , \quad \text{where} \quad P^\mu \equiv P_{\text{field}}^\mu + P_{\text{charges}}^\mu$$

$$\text{with} \quad P_{\text{field}}^\mu \equiv \int_V d^3X \, T^{\mu 0} \quad \text{and} \quad P_{\text{charges}}^\mu \equiv \sum_a m U^\mu , \tag{12.27}$$

where the 4-velocities U^μ are evaluated at the points where the world lines intersect the hypersurface $X^0 = T$. Therefore, in the absence of *radiation* (that is, flux through the boundary S of V), the total momentum P^μ, the sum of the momenta of the field and the charges, will be constant owing to the Maxwell and Lorentz equations.[6]

Let us consider the case $P^i = 0$. Then $P^0 \equiv \sum_a m U^0 + P_{\text{field}}^0$ is the rest energy, that is, the *inertial mass* of the system. If the system radiates, this mass is not constant.

We can similarly obtain the loss of angular momentum of the ensemble of the field and the particles which create it. In the center-of-mass frame where the total 3-momentum P^i vanishes, the total intrinsic angular momentum is given by [*cf.* (7.9) and (8.13)]

$$J_i = \int d^3X \, e_{ijk} X^j (T^{k0} + T_{\text{part}}^{k0}), \tag{12.28}$$

where $X^j \equiv r n^j$ are the components of the radius vector of length r of the reference point. Its time derivative is

$$\frac{dJ_i}{dT} = \int d^3X \, e_{ijk} X^j \partial_0 (T^{k0} + T_{\text{part}}^{k0}) = - \int d^3X \, e_{ijk} X^j \partial_l (T^{kl} + T_{\text{part}}^{kl})$$

$$= - \int d^3X \, \partial_l [e_{ijk} X^j (T^{kl} + T_{\text{part}}^{kl})] + \int d^3X \, e_{ilk} (T^{kl} + T_{\text{part}}^{kl}) \tag{12.29}$$

$$= -r \int e_{ijk} n^j T^{kl} \, dS_l,$$

where we have used the conservation law (12.26), integrated by parts, and exploited the symmetry of the energy–momentum tensors. Then we used Gauss's theorem and the fact that there are no charges at infinity.

Since the surface of the sphere at infinity grows as the squared distance r to the charges and the energy–momentum tensor is quadratic in the fields, we see from (12.27) that the

---

[6] In Section 8.2 we saw how this momentum transforms in a change of inertial frame. See also footnote 1 of this chapter.

system will radiate if the Faraday tensor falls off as $1/r$. For the radiated angular momentum to be finite [cf. (12.29)], it is necessary that this $1/r^2$ *radiation* part of the energy–momentum tensor, $T^{\mu\nu}_{\text{rad}}$, be such that $e_{ijk}n^j T^{kl}_{\text{rad}}dS_l$ vanish. To obtain the radiation rate it is necessary to know the Faraday tensor to order $1/r^2$ inclusive [see also (12.34) below].

Let us now go to the three-dimensional notation.

It is usual to write

$$T^{00} \equiv W, \quad T^{0i} \equiv S^i, \quad T^{ij} \equiv \sigma^{ij}, \tag{12.30}$$

where W is known as the *energy density* of the field, S^i as the *Poynting vector*, and σ^{ij} as the *constraint tensor*. In terms of the electric and magnetic fields we have

$$W = \frac{E^2 + B^2}{8\pi}, \quad S = \frac{E \wedge B}{4\pi},$$

$$\text{and} \quad \sigma_{ij} = \frac{1}{4\pi}\left(-E_i E_j - B_i B_j + \frac{1}{2}\delta_{ij}(E^2 + B^2)\right). \tag{12.31}$$

The *isotropic pressure* p of the field is defined as $p = \frac{1}{3}\sigma^i_i$ and is related to the energy density as

$$W = 3p. \tag{12.32}$$

The 'equation of state' of an electromagnetic field is therefore the same as that of a gas of ultrarelativistic particles;[7] *cf.* (8.36).

The energy and momentum of an interacting system given by (12.27) can be written as

$$P^0 = \frac{1}{8\pi}\int d^3X\,(E^2 + B^2) + \sum_a \frac{m}{\sqrt{1 - V^2}},$$

$$P = \frac{1}{4\pi}\int d^3X\,E \wedge B + \sum_a \frac{mV}{\sqrt{1 - V^2}}. \tag{12.33}$$

If $P = 0$, $P^0 \equiv M_{\text{in}}$ is the inertial mass of the system. It is *a priori* positive, but must be 'renormalized' because the field (E, B) must be evaluated everywhere, *including* at the charges where it diverges. We have already seen this in the case of the scalar interaction, *cf.* Section 10.2, and will see it again in Section 13.1 and Chapter 4 below.

Finally, the energy and angular momentum radiated per unit time given by (12.27) and (12.29) are written as follows using (12.31) for T^{ij}:

$$\frac{dP^0}{dT} = -\frac{r^2}{4\pi}\int (E \wedge B).n\,do, \quad \frac{dJ}{dT} = \frac{r^3}{4\pi}\int n \wedge [(E.n)E + (B.n)B]do, \tag{12.34}$$

where r is the distance from the center of mass to the sphere at infinity and do is the element of solid angle [$do = \sin\theta d\theta d\phi$ in spherical coordinates, where the components of the unit vector n entering into the definition of dS_i are $n = (\sin\theta\cos\phi, \sin\theta\sin\phi, \cos\theta)$].

---

[7]Starting from the Maxwell equations ($\nabla \wedge B = \partial_0 E + 4\pi j$ and $\nabla \wedge E = -\partial_0 B$), it is easy to see that

$$\frac{\partial W}{\partial T} + \nabla.S = -j.E,$$

which is just the 3-vector version of the zeroth component of (12.24). It is important, however, to avoid writing the left-hand side in four-dimensional form [i.e., $\partial_\mu W^\mu$ with $W^\mu = (W, S^i)$], because $W^i = T^{0i}$ are the components of a tensor and do not transform as a vector under Lorentz transformations; see (12.35)–(12.37) below. Moreover, $j.E$ is not a scalar of M_4.

Transformation of the energy–momentum tensor

In the special Lorentz transformation (1.7) representing the passage from the inertial frame X^μ to another X'^μ undergoing uniform translation with velocity V_0 relative to the first along the X^1 axis, the components of the 2-fold covariant tensor $T_{\mu\nu}$ transform as $T'_{\mu\nu} = \Lambda^\rho{}_\mu \Lambda^\sigma{}_\nu T_{\rho\sigma}$. We then have

$$W' = \frac{W - 2V_0 S^1 + V_0^2 \sigma^{11}}{1 - V_0^2}, \tag{12.35}$$

$$S'^1 = \frac{S^1(1 + V_0^2) - V_0 W - V_0 \sigma^{11}}{1 - V_0^2}, \quad S'^2 = \frac{S^2 - V_0 \sigma^{12}}{\sqrt{1 - V_0^2}}, \quad S'^3 = \frac{S^3 - V_0 \sigma^{13}}{\sqrt{1 - V_0^2}}, \tag{12.36}$$

$$\sigma'^{11} = \frac{\sigma^{11} - 2V_0 S^1 + V_0^2 W}{1 - V_0^2}, \quad \sigma'^{12} = \frac{\sigma^{12} - V_0 S^2}{\sqrt{1 - V_0^2}}, \quad \sigma'^{13} = \frac{\sigma^{13} - V_0 S^3}{\sqrt{1 - V_0^2}}, \tag{12.37}$$

and $\sigma'^{22} = \sigma^{22}$, $\sigma'^{23} = \sigma^{23}$, $\sigma'^{33} = \sigma^{33}$. Note that $T_{\mu\nu}$ transforms as the product of the components of two vectors, because it is symmetric.

13
Constant fields

In this chapter we review the basic ideas of *electrostatics* (Coulomb's law) and *magnetostatics* (the Biot–Savart law). We study the motion of a charge in a Coulomb field in detail. Then we discuss the Rutherford scattering formula which established the 'planetary' model of the atom, the Bohr–Sommerfeld quantization which displayed the limits of the theory, and, finally, the spin coupling explaining the atomic fine structure.

13.1 Coulomb's law

Let us consider charges in an inertial frame $\mathcal{S}$ which are at rest or at least have velocities small enough to be negligible compared to the speed of light. The associated current is $j^\mu = (\rho(X^i), 0)$, and we shall be interested in static solutions of the Maxwell equation (12.14) for the potential [that is, $\Box A_\mu - \partial_\mu(\partial_\nu A^\nu) = -4\pi j_\mu$]. A particular solution of the spatial components of the equation is $A_i = 0$ (we are therefore working in the Coulomb gauge; see Section 12.3) corresponding to zero magnetic field. The time component of the potential then reduces to the Poisson equation

$$\triangle \Phi = -4\pi\rho \,. \tag{13.1}$$

(This equation should be compared with the equation for the gravitational potential in Book 1, Section 11.3, or with the static Klein–Gordon equation of Section 10.2.) If we have a distribution of point charges q_a, then ρ is given by (12.11), $\rho = \sum_a q_a \delta_3(X^i - X_a^i(T))$, and the solution is [using the fact that $\triangle(1/r) = -4\pi\delta_3(X^i)$; cf. Section 9.2]

$$\Phi = \sum_a \frac{q_a}{r_a} \qquad \Longrightarrow \qquad E \equiv -\nabla\Phi = \sum_a \frac{q_a}{r_a^3} l_a \,, \tag{13.2}$$

where l_a, the length of r_a, is the 3-vector from the charge q_a to the reference point and has components $l_a^i = X^i - X_a^i$. The solution (13.2) for the electric field is *Coulomb's law*.

The energy of the field created by this distribution, $P_{\text{field}}^0 = \int d^3 X\, T^{00}$, reduces to [cf. (12.30) and (12.31)]

$$P_{\text{field}}^0 = \frac{1}{8\pi}\int d^3 X\, E^2 = -\frac{1}{8\pi}\int d^3 X\, E\,.\nabla\Phi$$
$$= -\frac{1}{8\pi}\int d^3 X\, \nabla\,.(E\Phi) - \frac{1}{8\pi}\int d^3 X\, \Phi\triangle\Phi, \tag{13.3}$$

where we have used only the fact that $E = -\nabla\Phi$. The first term of the last equation vanishes owing to Gauss's theorem, because the solution of the field equations (13.2) is such that $(E\Phi)$ falls off as $1/r^3$ for a confined charge distribution. It should be noted that P_{field}^0, the integral of a positive-definite density (E^2), is positive.

Relativity in Modern Physics. Nathalie Deruelle and Jean-Philippe Uzan.
© Oxford University Press 2018. Published in 2018 by Oxford University Press.
DOI: 10.1093/oso/9780198786399.001.0001

Then from (13.1) and (13.2) we find

$$P^0_{\text{field}} = -\frac{1}{8\pi} \int d^3X \, \Phi \triangle \Phi = \frac{1}{2} \int d^3X \, \Phi \rho = \frac{1}{2} \sum_a q_a \Phi_a, \qquad (13.4)$$

where Φ_a is the potential created by all the charges at X^α_a (the location of the charge q_a), *including* the charge q_a itself: $\Phi_a = \sum_{a'} q_{a'}/r_{aa'}$, where $r_{aa'}$ is the spatial separation of the charges a and a'. This expression diverges because it contains a self-energy term $(a' = a)$. It is usual to ignore such divergent terms and work with the so-called *electrostatic energy*

$$P^0_{\text{electrostatic}} = \frac{1}{2} \sum_a q_a \sum_{a' \neq a} \frac{q_{a'}}{r_{aa'}} = \sum_{a,a' \text{ with } a<a'} \frac{q_a q_{a'}}{r_{aa'}}. \qquad (13.5)$$

In contrast to P^0_{field}, this energy is not necessarily positive.

The 'classical' electron radius

The Coulomb potential Φ created by a single charge with spherical symmetry will depend only on the distance r to the center, and so the Poisson equation (13.1) becomes $\frac{d^2\Phi}{dr^2} + \frac{2}{r}\frac{d\Phi}{dr} = -4\pi\rho$. If in addition ρ is constant for $r \leq r_0$ and zero beyond this radius, the integration can be done by inspection and we find (see, for example, Book 1, Section 15.2)

$$\Phi(r) = \frac{Q}{2r_0}\left(3 - \frac{r^2}{r_0^2}\right) \quad \text{for} \quad r \leq r_0 \quad \text{and} \quad \Phi(r) = \frac{Q}{r} \quad \text{for} \quad r \geq r_0. \qquad (13.6)$$

Here we have excluded a $1/r$ term in the expression for the potential in order that the potential remain finite at $r = 0$, and we have imposed the condition that the potential and its derivative be continuous at $r = r_0$. Finally, $Q \equiv (4\pi/3)\rho r_0^3$ is the total charge of the distribution.

The energy of the field created by this charge is $P^0_{\text{field}} = \frac{1}{2}\int d^3X \, \Phi\rho$ or, substituting (13.6) for Φ,

$$P^0_{\text{field}} = \frac{3}{5}\frac{Q^2}{r_0}. \qquad (13.7)$$

If the electron is viewed as such a charged sphere and if we assume that its mass m is entirely electromagnetic in origin, $m = P^0_{\text{field}}$, then from (13.7) we find its radius:

$$r_e = \frac{3e^2}{5m}, \quad \text{where} \quad \frac{e^2}{m} \approx 9.4 \times 10^{-24} \text{ s} \approx 2.8 \times 10^{-15} \text{ m}. \qquad (13.8)$$

The quantity r_e is called the *classical electron radius*. The proton radius is smaller than this by a factor of about 2000, and for the sake of comparison we recall that the Bohr radius is $a_{\text{Bohr}} = e^2/(m\,\alpha^2)$ with $\alpha \approx 1/137$.[1]

---

[1] The discovery of the electron spin (G. Uhlenbeck and S. Goudsmit, 1925) put an end to the idea, espoused by H. A. Lorentz, that the electron mass could be entirely electromagnetic in origin. Indeed, the requirement that the angular momentum of a sphere, of order $mr_e v_{\text{eq}}$, be of order $\hbar = e^2/\alpha$ implies a velocity at the equator v_{eq} of order $1/\alpha$ or 137 times the speed of light.

13.2 Quadrupole expansion of the potential

The Coulomb potential generated by a static charge distribution[2] of density ρ at a point P with radius 3-vector R (of length r when the origin is chosen to lie at the center of the distribution) is given by (see Book 1, Section 14.1)

$$\Phi(R) = \int \frac{\rho(R')}{|R - R'|} dV' = \frac{1}{r} \int \frac{\rho(R')}{\sqrt{1 - 2\frac{R' \cdot R}{r^2} + \frac{r'^2}{r^2}}} dV', \qquad (13.9)$$

where R' is the radius vector of a point P' of the distribution (r' is its length) and $dV \equiv \sqrt{\det e}\, d^3x$, $\det e$ being the determinant of the Euclidean metric coefficients in the coordinates x^i. If $r \gg r'$, i.e., if the point P is far outside the charge distribution, it is possible to expand $\Phi(R)$ as

$$\Phi(R) = \frac{1}{r} \int \rho(R')dV' + \frac{R}{r^3} \cdot \int \rho(R')R'dV'$$
$$+ \frac{1}{2r^5} \int \rho(R') \left[3(R \cdot R')^2 - r^2 r'^2\right] dV' + \mathcal{O}\left(\frac{1}{r^4}\right), \qquad (13.10)$$

which can be rewritten in the following form in terms of the Cartesian coordinates X^i of P:

$$\Phi(X^i) = \frac{Q}{r} + \frac{d_i X^i}{r^3} + \frac{D_{ij} X^i X^j}{2r^5} + \mathcal{O}\left(\frac{1}{r^4}\right). \qquad (13.11)$$

Here $Q \equiv \int \rho(R')dV' = \sum_a q_a$ is the total charge and d^i is the *dipole moment* of the distribution:

$$d \equiv \int \rho(R')R'dV' \quad \text{or also} \quad d = \sum_a q_a R_a. \qquad (13.12)$$

We see here that d will depend on the choice of coordinate origin unless $Q = 0$, and that it coincides with the center of mass of the distribution if the charge-to-mass ratio is the same for all the charges: $q_a/m_a = \text{const}$. Finally, D_{ij} is the *quadrupole moment*:

$$D_{ij} \equiv \int \rho(R')(3X_i'X_j' - r'^2\delta_{ij})dV' \quad \text{or also} \quad D^{ij} = \sum_a q_a(3X_a^i X_a^j - \delta^{ij}r_a^2), \qquad (13.13)$$

where X'^i are the Cartesian components of the position vector R' and X_a^i are the coordinates of the radius vector of the charge q_a of length r_a.

13.3 A charge in a Coulomb field

The motion of a charge q in the Coulomb field of a charge Q held fixed at the origin of an inertial frame is governed by the Lorentz equation, which can be written in 3-vector notation as [see (11.18) and (13.2)]

---

[2]Or a distribution of slowly moving charges for which Maxwell's equations always reduce approximately to $\triangle\Phi = -4\pi\rho$.

$$\frac{d}{dT} \frac{mV}{\sqrt{1 - V^2}} = \frac{qQ}{r^3} R, \tag{13.14}$$

where R is the radius 3-vector of length r of the charge and $V = \dot{R}$ (the dot denotes the derivative with respect to T).

The problem can be solved like the Kepler problem (see Book 1, Section 12.1 *et seq.*). By symmetry, the motion is planar and we take $R = (r \cos \phi, r \sin \phi)$. Taking on the one hand the vector product of (13.14) with R and, on the other, the scalar product with V, we obtain the conservation laws for the specific angular momentum $\mathcal{L}$ and energy $\mathcal{E}$ of the system:

$$\frac{\dot{\phi}}{\sqrt{1 - V^2}} = \frac{\mathcal{L}}{r^2}, \qquad \frac{1}{\sqrt{1 - V^2}} = \mathcal{E} - \frac{qQ}{mr} \tag{13.15}$$

with $V^2 = \dot{r}^2 + r^2 \dot{\phi}^2$. When the motion is circular ($r = r_0$, $\dot{r} = 0$, and $\ddot{r} = 0$), after a bit of algebra we find the frequency $\Omega \equiv \dot{\phi}$:

$$\Omega = \frac{-qQ/mr_0}{\sqrt{2} r_0} \left[-1 + \sqrt{1 + \frac{4}{(-qQ/mr_0)^2}} \right]^{1/2}$$

$$\implies \Omega \sim \frac{\sqrt{-qQ/mr_0}}{r_0} \text{ when } -\frac{qQ}{mr_0} \to 0, \quad \text{and} \quad \Omega \sim \frac{1}{r_0} \text{ when } -\frac{qQ}{mr_0} \to \infty. \tag{13.16}$$

In the case of the hydrogen atom for which $Q = -q = e$ and using $r_0 \equiv a_{\text{Bohr}} = e^2/m\alpha^2$, we find $-qQ/mr_0 = \alpha^2 \ll 1$ and $\Omega \approx m\alpha^3/e^2$.

Bound motion in a Coulomb field

When the motion of the charge q in the Coulomb field (13.14) is not circular, we can extract the first integrals from (13.15):

$$\dot{r}^2 = 1 - \frac{(1 + \mathcal{L}^2/r^2)}{(\mathcal{E} - qQ/mr)^2}, \qquad \dot{\phi} = \frac{\mathcal{L}}{r^2 (\mathcal{E} - qQ/mr)}. \tag{13.17}$$

Then we apply the Binet method: we form the ratio $\dot{r}^2/\dot{\phi}^2$, set $u = 1/r$, and differentiate, which gives

$$\frac{d^2 u}{d\phi^2} + u \left[1 - \left(\frac{qQ}{m\mathcal{L}} \right)^2 \right] = -\left(\frac{qQ}{m\mathcal{L}} \right) \left(\frac{\mathcal{E}}{\mathcal{L}} \right). \tag{13.18}$$

From this for $-qQ > 0$ and $|qQ/(m\mathcal{L})| < 1$ we find the solution satisfying (13.17):

$$r = \frac{p}{1 + e \cos \nu} \qquad \text{with} \qquad \nu \equiv (\phi - \tilde{\omega}) \sqrt{1 - \left(\frac{qQ}{m\mathcal{L}} \right)^2}. \tag{13.19}$$

Here $\tilde{\omega}$ is an integration constant and the parameter p and eccentricity e are given by

$$p = \frac{\mathcal{L}}{(-qQ/m\mathcal{L})} \frac{1}{\mathcal{E}} \left[1 - \left(\frac{qQ}{m\mathcal{L}} \right)^2 \right], \quad e^2 = \frac{1}{\mathcal{E}^2} \left[1 + \frac{\mathcal{E}^2 - 1}{(qQ/m\mathcal{L})^2} \right].$$ (13.20)

If $e < 1$ this trajectory is an ellipse which precesses by an angle of $\Delta\omega$ in each revolution (see Fig. 10.2) with

$$\Delta\omega = \frac{2\pi}{\sqrt{1 - \left(\frac{qQ}{m\mathcal{L}} \right)^2}} - 2\pi.$$ (13.21)

The time dependence is also easily obtained. We find that it can be written in the quasi-Newtonian form (for $e < 1$; see Book 1, Section 12.2)

$$\begin{cases} r = a(1 - e\cos\eta) \\ \operatorname{tg}\dfrac{\nu}{2} = \sqrt{\dfrac{1+e}{1-e}} \operatorname{tg}\dfrac{\eta}{2} \end{cases} \quad \text{with} \quad n(T - T_0) = \eta - e_t \sin\eta,$$ (13.22)

where T_0 is an integration constant, ν is given in (13.19), and the major axis is $a \equiv p/(1 - e^2)$. The 'mean motion' n and the new eccentricity e_t are given by

$$a = \frac{(-qQ/m)\mathcal{E}}{1 - \mathcal{E}^2}, \quad n = \frac{\left(1 - \mathcal{E}^2 \right)^{\frac{3}{2}}}{(-qQ/m)}, \quad e_t = e\mathcal{E}^2,$$ (13.23)

where e is a known function of $\mathcal{E}$ and $\mathcal{L}$; cf. (13.20).

For completeness, we need to relate the proper time τ of the charge to the time T in the frame using $dT/d\tau = 1/\sqrt{1 - V^2}$. From (13.15) and (13.22) we find

$$n(\tau - \tau_0) = \sqrt{\frac{e_t}{e}} (\eta - e\sin\eta).$$ (13.24)

The Rutherford effective cross section

The trajectory of a charge in a Coulomb field (when $qQ < 0$ and $|qQ/(m\mathcal{L})| < 1$, the situation to which we restrict ourselves) is given by (13.19), where the parameter p and the eccentricity e are related to the specific energy $\mathcal{E}$ and angular momentum $\mathcal{L}$ of the charge as in (13.20). If $e > 1$ this trajectory reduces to a shifted hyperbola (see Fig. 13.1). A bit of trigonometry shows that it has outgoing asymptote $Y\cos\phi_{\text{out}} = X\sin\phi_{\text{out}} - \rho$ and incoming one $Y\cos\phi_{\text{in}} = X\sin\phi_{\text{in}} + \rho$ (if $\phi_{\text{out}} < \pi$), with ϕ_{in} and ρ defined as

$$\operatorname{tg} \left[\phi_{\text{out}} \sqrt{1 - \left(\frac{qQ}{m\mathcal{L}} \right)^2} \right] = -\sqrt{e^2 - 1}, \quad \phi_{\text{in}} = \frac{2\pi}{\sqrt{1 - (qQ/m\mathcal{L})^2}} - \phi_{\text{out}},$$

$$\rho = \frac{p}{\sqrt{e^2 - 1}\sqrt{1 - (qQ/m\mathcal{L})^2}}.$$

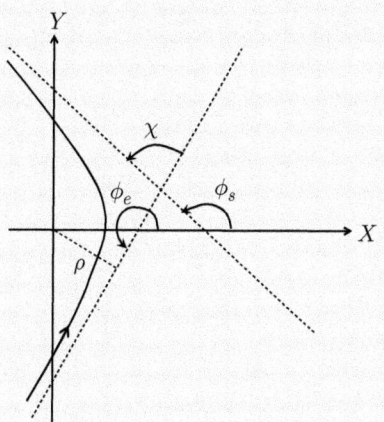

Fig. 13.1 Scattering by a Coulomb field.

Here ρ is the *impact parameter*, that is, the distance to the center at which the particle would have passed had it not been deflected. The modulus of the asymptotic 3-velocity of the charge is related to its energy as $\mathcal{E} = 1/\sqrt{1 - V_\infty^2}$. With a bit of algebra we can write $\mathcal{L}$ and $e^2 - 1$, and therefore the angle ϕ_{out}, as functions of ρ and V_∞ as

$$\text{tg}\left[\phi_{\text{out}}\sqrt{1 - \frac{1}{x^2}}\right] = -V_\infty\sqrt{x^2 - 1}, \qquad \text{where} \qquad x \equiv \frac{V_\infty}{\sqrt{1 - V_\infty^2}}\frac{\rho}{(-qQ/m)}. \tag{13.25}$$

The deflection angle of the charge is (see Fig. 13.1) $\chi = 2\pi - (\phi_{\text{in}} - \phi_{\text{out}})$ or

$$\chi = 2\phi_{\text{out}} - \pi\left(\frac{2x}{\sqrt{x^2 - 1}} - 1\right). \tag{13.26}$$

If n is the density of charges of speed V_∞ which cross, per unit time and unit area, the cross section perpendicular to the incident beam, then there will be $2\pi\rho\, d\rho\, n$ particles in the ring lying between ρ and $\rho + d\rho$. The *effective cross section* σ is defined as

$$d\sigma = 2\pi\rho\, d\rho = 2\pi\rho\frac{d\rho}{d\chi}d\chi = \rho\frac{d\rho}{d\chi}\frac{1}{\sin\chi}do = \frac{\rho}{\sin\chi\frac{d\chi}{d\rho}}do, \tag{13.27}$$

where $do = 2\pi\sin\chi d\chi$ is the element of solid angle of the opening χ. Therefore, using (13.25) to replace ρ by x, we have

$$\frac{d\sigma}{do} = -\left(\frac{qQ}{m}\right)^2\frac{1 - V_\infty^2}{V_\infty^2}\frac{x}{\frac{d}{dx}\cos\chi}, \tag{13.28}$$

where χ is known as a function of x from (13.26).

For large x we have $\text{tg}\phi_{\text{out}} \sim -xV_\infty$ and $\chi \sim 2\phi_{\text{out}} - \pi$, and (13.28) can be simplified to

$$\left|\frac{d\sigma}{do}\right| \sim \left(\frac{qQ}{2m}\right)^2\frac{1 - V_\infty^2}{V_\infty^4}\frac{1}{\sin^4\frac{\chi}{2}}. \tag{13.29}$$

This expression includes the *Rutherford formula* (that is, the nonrelativistic case $V_\infty \ll 1$ with xV_∞ and therefore χ finite):

$$\left|\frac{d\sigma}{do}\right| \sim \left(\frac{qQ}{2mV_\infty^2}\right)^2 \frac{1}{\sin^4\frac{\chi}{2}}, \tag{13.30}$$

and its relativistic generalization for small deflection angles $[V_\infty = \mathcal{O}(1)$ and large x or small $\chi]$:

$$\left|\frac{d\sigma}{do}\right| \sim \left(\frac{2qQ}{m}\right)^2 \frac{1 - V_\infty^2}{V_\infty^4} \frac{1}{\chi^4}. \tag{13.31}$$

The formula obtained by Rutherford in 1911 explained the results of the experiments performed by Geiger and Marsden in 1909 to measure the deflection of α particles by a thin gold foil. The underlying planetary model of the atom then replaced the 'plum-pudding' model of J. J. Thomson, in which the electrons move in a distribution of positive charge of the size of the atom (see Section 18.4).

Sommerfeld quantization

The equations of motion of a charge in a Coulomb field can also be obtained by varying the action of the system; *cf.* (11.20):

$$\mathbf{S} = -m\int d\tau + q\int A_\mu U^\mu d\tau \equiv \int L\, dT \quad \text{with} \quad L = -m\sqrt{1-V^2} - \frac{qQ}{r}. \tag{13.32}$$

Since in spherical coordinates $V^2 = \dot{r}^2 + r^2\dot{\phi}^2$, the conjugate momenta associated with ϕ and r are

$$p_\phi \equiv \frac{\partial L}{\partial \dot{\phi}} = \frac{mr^2\dot{\phi}}{\sqrt{1-V^2}}, \qquad p_r \equiv \frac{\partial L}{\partial \dot{r}} = \frac{m\dot{r}}{\sqrt{1-V^2}}, \tag{13.33}$$

and the Lagrange equations, $\dot{p}_\phi = \partial L/\partial\phi$ and $\dot{p}_r = \partial L/\partial r$, indeed lead to (13.17) and (13.18) [using $V \cdot \left(V/\sqrt{1-V^2}\right)^{\cdot} = r\left(\dot{r}/\sqrt{1-V^2}\right)^{\cdot} - r\dot{r}\dot{\phi}^2/\sqrt{1-V^2}$].

The order of magnitude of the action (13.32) is $-(qQ/r_0)T$, where T is the characteristic time of the system, that is, the period of rotation of the charge q: $T \sim 1/\Omega$ with $\Omega \sim \sqrt{-qQ/mr_0^3}$ in the nonrelativistic limit; *cf.* (13.16). If we take its ratio to the quantum of action $\hbar = e^2/\alpha$, for an electron $(q = -e)$ in the field of a proton $(Q = e)$ we find

$$\frac{\mathbf{S}}{\hbar} \sim \alpha\frac{\sqrt{m\,r_0}}{e} \qquad \Longrightarrow \qquad \frac{\mathbf{S}}{\hbar} \sim 1 \quad \text{for} \quad r_0 = a_{\text{Bohr}} \equiv \frac{e^2}{m\alpha^2}. \tag{13.34}$$

As we have already mentioned at the end of Section 6.5, quantum effects become important when the action becomes comparable to $\hbar$, that is, when the orbital radius is comparable to the Bohr radius, while the orbital velocity is still small compared to the speed of light: $\Omega a_{\text{Bohr}} = \alpha^2 \ll 1$.

Generalizing the orbital quantization prescription of Bohr to take into account the *fine structure* of the hydrogen atom, Sommerfeld in 1916 imposed the requirement that, for an electron $(q = -e)$ in the field of a proton $(Q = e)$,

$$\int_0^{2\pi} p_\phi\, d\phi = 2\pi n_\phi\, \hbar, \qquad 2\int_{r_{\text{min}}}^{r_{\text{max}}} p_r\, dr = 2\pi n_r\, \hbar, \tag{13.35}$$

where n_ϕ and n_r are integers. Since $p_\phi = m\mathcal{L}$ and $p_r = m\sqrt{A + 2B/r + C/r^2}$ with $A = -(1 - \mathcal{E}^2)$, $B = e^2\mathcal{E}/m$, and $C = e^4/m^2 - \mathcal{L}$, the Sommerfeld conditions lead to

$$\mathcal{L} = \frac{n_\phi e^2}{m\alpha}, \quad \mathcal{E} = \left[1 + \left(\frac{\alpha}{n_r + \sqrt{n_\phi^2 - \alpha^2}}\right)^2\right]^{-\frac{1}{2}} = 1 - \frac{\alpha^2}{2n^2} + \frac{\alpha^4}{2n^4}\left(\frac{3}{4} - \frac{n}{n_\phi}\right) + \mathcal{O}(\alpha^6), \quad (13.36)$$

where $n \equiv n_r + n_\phi$.

By a mischievous accident, the *Sommerfeld formulas* are the same (with n_ϕ replaced by $j + 1/2$) as those which solve the Dirac equation, which includes the electron spin.[3] Thus the energy of the transition between the 2s and 2p states corresponding to $n_r = 1$ and $n_\phi = 1$ and 2 (or $j = 1/2$ and $j = 3/2$) is $E_{22} - E_{21} = m_e \alpha^4/32$, in agreement with experiment.

13.4 Spin in a Coulomb field

As we have seen in Section 7.3, a particle can be assigned, in addition to its mass, a *spin* 4-vector J_μ orthogonal to its 4-velocity ($J_\mu U^\mu = 0$), the equation of motion of which in an inertial frame was given in (7.13):

$$\frac{dJ_\mu}{d\tau} - U_\mu(J_\nu \gamma^\nu) = C_\mu,$$

where γ^μ is the 4-acceleration and C_μ is a possible *torque* which is also orthogonal to U^μ.

If no torque is applied but the particle is constrained to undergo circular motion of radius r_0 and frequency Ω, then, as we saw in Section 7.4, the spin precesses at the *Thomas frequency*, which for small velocities is [*cf.* (7.18)] $\omega_{\text{Thomas}} = -r_0^2 \Omega^3/2$. If this particle is an electron in its Bohr orbit, that is, if $r_0 = e^2/(m\alpha^2)$ and [see Section 13.3] $\Omega \approx m\alpha^3/e^2$, then $\omega_{\text{Thomas}} \approx -m\alpha^5/(2e^2)$.

Now let us consider a particle of charge q in an external electromagnetic field. Its acceleration is given by the Lorentz force $\gamma^\mu = (q/m)F^\mu_{\ \nu}U^\nu$. We assume in addition that the particle carries a spin and this spin is coupled to the field. Just like the Lorentz force, the torque C_μ must be postulated; a natural choice (which agrees with experiment) is

$$C_\mu = \frac{gq}{2m}(F_{\mu\nu}J^\nu + U_\mu U^\rho J^\nu F_{\rho\nu}), \quad (13.37)$$

where the constant g is the *Landé factor*. The first term is of the same type as the Lorentz force, and the second guarantees that $C_\mu U^\mu = 0$. Substituting these expressions for C_μ and γ^μ into the equation of motion, we obtain the BMT equation (V. Bargmann, L. Michel, and V. L. Telegdi, 1959):

$$\frac{dJ_\mu}{d\tau} = \frac{q}{m}\left[\frac{g}{2}F_{\mu\nu}J^\nu + \left(\frac{g}{2} - 1\right)U_\mu U^\rho J^\nu F_{\rho\nu}\right] \quad \text{with} \quad J_\mu U^\mu = 0. \quad (13.38)$$

From experiment (for example, the anomalous Zeeman effect) we have $g = 2$ for an electron (excluding quantum electrodynamical corrections).

Now let us consider an electron in a circular orbit in a Coulomb field. Its world line is $X^\mu(T) = (T, \ r_0 \cos \Omega T, \ r_0 \sin \Omega T, \ 0)$ and the condition $J_\mu U^\mu = 0$ implies that $J^0 =$

---

[3]See, for example, Weinberg (2005) and Granovskii (2004).

$-r_0\Omega(J^X \sin \Omega T - J^Y \cos \Omega T)$. The Faraday tensor on the world line reduces to $F_{X0} = (Q\Omega/r_0^2) \cos \Omega T$ and $F_{Y0} = (Q\Omega/r_0^2) \sin \Omega T$. Therefore, (13.38) can be written as

$$
\begin{cases}
\dfrac{dJ^X}{dT} = x\Omega\Gamma f(T) \cos \Omega T, \quad \dfrac{dJ^Y}{dT} = x\Omega\Gamma f(T) \sin \Omega T \\[2mm]
\text{with } f(T) \equiv J^X \sin \Omega T - J^Y \cos \Omega T, \quad \Gamma \equiv \dfrac{1}{\sqrt{1 - r_0^2\Omega^2}}, \quad \text{and} \quad x \equiv \dfrac{-qQ}{mr_0}.
\end{cases}
\tag{13.39}
$$

This system can be solved exactly like (7.17) and we find $J^Z = \text{const}$ and

$$
\begin{cases}
J^X = J_1 \left[\dfrac{1}{\sqrt{1 - x\Gamma}} \sin \Omega T \sin(\Omega T \sqrt{1 - x\Gamma}) + \cos \Omega T \cos(\Omega T \sqrt{1 - x\Gamma}) \right] \\[3mm]
J^Y = J_1 \left[-\dfrac{1}{\sqrt{1 - x\Gamma}} \cos \Omega T \sin(\Omega T \sqrt{1 - x\Gamma}) + \sin \Omega T \cos(\Omega T \sqrt{1 - x\Gamma}) \right].
\end{cases}
\tag{13.40}
$$

At $T = 0$ the spin lies in the XOZ plane, and after one revolution, at $T = 2\pi/\Omega$, J^Y has increased by $-J_1 \frac{\sin(2\pi\sqrt{1-x\Gamma})}{\sqrt{1-x\Gamma}} \approx 2\pi \frac{J_1 x\Gamma}{2}$ if $x\Gamma \ll 1$, so that the angular velocity of the spin precession is

$$
\omega \approx \frac{x\Gamma\Omega}{2} = \frac{1}{2}\left(\frac{-qQ}{mr_0}\right) \frac{\Omega}{\sqrt{1 - (r_0\Omega)^2}}.
\tag{13.41}
$$

In the case of an electron in its Bohr orbit at $r_0 = e^2/m\alpha^2$ we have $(\Omega r_0)^2 = x = \alpha^2 \ll 1$, and so $\omega \approx +m\alpha^5/(2e^2)$, in perfect agreement with the observed *fine structure* of hydrogen.[4]

13.5   The Biot–Savart law

Let us consider a confined charge distribution in a given inertial frame, and study the *average* or *static* magnetic field that it creates. Since the time average of any time derivative vanishes,[5] Maxwell's equation (12.16) for the magnetic field (that is, $\nabla \wedge B = \partial_0 E + 4\pi j$) reduces to *Ampère's law*

$$
\nabla \wedge \overline{B} = 4\pi\overline{j},
\tag{13.42}
$$

where the *steady current* 3-vector $\overline{j}$ describes the charge distribution 'on the average'. This equation implies that (see Section 11.1 for a review of vector calculus)

$$
\nabla.\overline{j} = 0,
\tag{13.43}
$$

which is the averaged version of the current conservation law.

---

[4]See, for example, Itzykson and Zuber (2006).

We note that $\omega = -\omega_{\text{Thomas}} = \frac{1}{2}\omega_{\text{SO}} = \omega_{\text{Larmor}}$, where ω_{Thomas} is the precessional velocity of a spin which is *not coupled* to the field, ω_{SO} is the precession obtained assuming that the spin of modulus $\hbar/2$ is coupled to the magnetic field existing in the inertial frame tangent to the electron, and ω_{Larmor} is the precessional velocity of an orbital angular momentum in the same field. The story of these factors of 2 can be found in Basdevant (2007).

[5]Actually, $\overline{\partial_0 f} \equiv \frac{1}{T}\int_0^T \frac{\partial f}{\partial T} dT = \frac{f(T) - f(0)}{T} \to 0$ when $T \to \infty$ if $f(T)$ remains bounded.

Since $B = \nabla \wedge A$, (13.42) can also be written in the Lorenz gauge ($\partial_\mu A^\mu = 0$; see Section 12.3) as[6]

$$\nabla . \overline{A} = 0, \quad \triangle \overline{A} = -4\pi\overline{j}. \tag{13.44}$$

The second expression in (13.44) is a Poisson equation whose solution at the point P given by the radius 3-vector R is (*cf.* Section 13.1)

$$\overline{A} = \int \frac{\overline{j}(R')}{|R - R'|} dV' \quad \Longrightarrow \quad \overline{B} = \int \frac{\overline{j}(R') \wedge (R - R')}{|R - R'|^3} dV', \tag{13.45}$$

where R' is the radius vector of a point P' of the distribution. The solution obtained above for the magnetic field is called the *Biot–Savart law*.

Let us now consider the situation far from the distribution.

We can write $|R - R'| = r(1 - R.R'/r^2) \cdots$, where r is the length of the radius vector R, and so we have

$$\overline{A}^i = \frac{1}{r} \int dV' \, \overline{j}'^i + \frac{1}{r^3} X^j \int dV' \, \overline{j}'^i X'_j + \mathcal{O}\left(\frac{1}{r^3}\right), \tag{13.46}$$

where the prime on the current indicates that it is evaluated at X'^i.

At this point it is convenient to specify the properties of the 3-vector $\overline{j}$ and to note that for any function $f(X^i)$ we have[7]

$$\int dV \, \overline{j}^i \partial_i f = 0. \tag{13.47}$$

We thus deduce that the 3-vector $\overline{j}$ does not possess any component normal to the 2-surface S, with surface element $dS_i = n_i dS$, bounding the distribution. Indeed, (13.47) can be rewritten as follows, after integration by parts and using Gauss's theorem and the fact that the average current has zero divergence:

$$\int_S f \, (\overline{j}^i n_i) dS = 0 \quad \Longrightarrow \quad \overline{j}^i n_i|_S = 0 \tag{13.48}$$

because the function f is arbitrary. Then, choosing $f = X^j$ and $f = X^j X^k$, we see from (13.47) that

$$\int dV \, \overline{j}^j = 0, \quad \int dV \, (\overline{j}^j X^k + \overline{j}^k X^j) = 0. \tag{13.49}$$

The first term of the magnetic potential (13.46) therefore vanishes, a feature which we can attribute to the absence of a *magnetic charge* (see Section 12.3). Now from the second term of (13.46) we subtract half the second integral in (13.49) so that

---

[6]The calculations of this section are most easily done using three-dimensional notation (where, for example, (13.42) is written as $e^{ijk}\partial_j \overline{B}_k = 4\pi\overline{j}^i$) and using wherever possible the property of the Levi-Civita symbol $e^{ijk}e_{klm} = \delta^i_l \delta^j_m - \delta^i_m \delta^j_l$.

[7]This property can be proved, at least formally, as follows. If the distribution is described as an ensemble of point charges, we have (before the averaging; see Section 12.1) $j^i(T, X^i) = \sum_a q_a \delta_3(X^i - X^i_a(T))V^i_a$. The average of j^i is hard to define, but we can write $\overline{\int dV \, j^i \partial_i f} = \sum_a q_a \partial_i f \, \overline{V_a} = 0$ because $\overline{V_a} = 0$. Equation (13.47) then follows if we argue that the average of the integral is equal to the integral of the average.

$$X^j \int dV' \, \overline{j}'^i \, X'_j = \frac{1}{2} X^j \int dV' (\overline{j}'^i X'_j - \overline{j}'^j X'_i) \,. \tag{13.50}$$

If we define the *magnetic moment* of the distribution as

$$\mathcal{M}^i \equiv \frac{1}{2} e^{ijk} \int dV' \, X'_j \, \overline{j}'_k \qquad \text{or also} \qquad \mathcal{M} \equiv \frac{1}{2} \int dV' R' \wedge \overline{j}', \tag{13.51}$$

it is easily shown using the properties of the Levi-Civita symbol that

$$X^j \int dV' \, \overline{j}'^i \, X'_j = e^{ijk} \mathcal{M}_j X_k \,. \tag{13.52}$$

In the end, the magnetic potential can be written as

$$\overline{A}^i = \frac{1}{r^3} e^{ijk} \mathcal{M}_j X_k + \mathcal{O}\left(\frac{1}{r^3}\right) \qquad \text{or also} \qquad \overline{A} = \frac{\mathcal{M} \wedge R}{r^3} \,. \tag{13.53}$$

From this it is easy to derive the magnetic field by setting[8] $n^i \equiv X^i/r$:

$$\overline{B} = \frac{3n(\mathcal{M}.n) - \mathcal{M}}{r^3} + \mathcal{O}\left(\frac{1}{r^4}\right). \tag{13.54}$$

The magnetic moment

When the charges are point charges, the magnetic moment defined in (13.51) becomes

$$\mathcal{M} = \frac{1}{2} \sum_a qR \wedge V, \tag{13.55}$$

where R_a and V_a are the radius vectors and velocities of the charges. If the charge-to-mass ratio is the same for all the particles, $q_a/m_a = q/m$, then $\mathcal{M} = (q/2m)J$, where $J = \sum_a mR \wedge V$ is the angular momentum of the system. It is therefore constant if the system is isolated.

The situation is the same in the case of only two charges, whatever their ratios q_a/m_a are. In fact, in the system of the center of mass ρ where, to the order in which we are working, $R = m'\rho/M$ and $R' = -m\rho/M$, we have $\mathcal{M} = \frac{1}{2} \frac{qm'^2 + q'm^2}{mm'M} J$ with $J = \mu\rho \wedge \dot{\rho}$ and $M = m + m'$, $\mu = mm'/M$.

---

[8]We can as before make the expansion

$$\overline{B}^i = e^{ijk} \partial_j \overline{A}_k = e^{ijk} e_k{}^{mn} \partial_j \left(\frac{\mathcal{M}_m X_n}{r^3}\right) = (\delta^{im} \delta^{jn} - \delta^{in} \delta^{jm}) \mathcal{M}_m \, \partial_j \left(\frac{X_n}{r^3}\right)$$

$$= -\frac{\mathcal{M}^i}{r^3} + \frac{3X^i \mathcal{M}_m X^m}{r^5} \,.$$

14
The free field

Here we shall study the structure of Maxwell's equations in the vacuum and the action from which they are derived while emphasizing the consequences of their gauge invariance. Gauge invariance, on the one hand, allows one of the components of the magnetic potential to be chosen freely and, on the other, imposes a constraint on the initial conditions such that in the end the general solution has only two 'degrees of freedom'. We then develop the Hamiltonian formalisms in the Maxwell theory and compare them to the formalisms using non-gauge-invariant or massive vector fields.

14.1 Two degrees of freedom

The fundamental quantity describing electromagnetic phenomena in the Maxwell theory is the potential A_μ, whose equations of motion in the vacuum and in a system of Minkowski coordinates X^μ are [*cf.* (12.14)]

$$\Box A_\mu - \partial_\mu(\partial_\nu A^\nu) = 0 \,. \tag{14.1}$$

Equation (14.1) is invariant under the *gauge transformation* $A_\mu \mapsto \tilde{A}_\mu = A_\mu + \partial_\mu g$, where g is an arbitrary function of the coordinates. If A_μ satisfies (14.1), then $\tilde{A}_\mu$ will also satisfy it.

It is therefore possible to impose a condition on $\tilde{A}_\mu$, for example, the *Lorenz condition* $\partial_\mu \tilde{A}^\mu = 0$ which preserves the manifestly Lorentz-invariant form of the Maxwell equations. However, as already mentioned above in Section 12.3, this condition does not completely fix the gauge, *i.e.*, it does not completely determine the function g, because g, which satisfies $\partial_\mu A^\mu + \Box g = 0$, is defined up to a 'harmonic' function h: $g = g_0 + h$ with $\Box h = 0$. Fixing the gauge completely requires breaking the manifest Lorentz invariance and working in a given inertial frame, that is, distinguishing a particular time. It then becomes possible, as we shall see below, to rewrite (14.1) in $\mathcal{S}$ in a form which is manifestly gauge invariant.

To do this we use the potential decomposition $A_\mu \equiv (A_0, A)$, where A is a 3-(co)vector with components A_i introduced by J. Bardeen[1]:

$$A_0 \equiv -\Phi \,, \quad A \equiv \bar{A} + \nabla \mathcal{A} \quad \text{with} \quad \nabla.\bar{A} = 0 \,. \tag{14.2}$$

For a given A, these equations determine the function $\mathcal{A}$ and the two components of $\bar{A}$. Actually, $\mathcal{A}$ is *the* everywhere-regular solution of the elliptic equation $\triangle \mathcal{A} = \nabla.A$, and so $\bar{A}$ follows: $\bar{A} = A - \nabla \mathcal{A}$. In Bardeen's terminology, the four components of the potential A_μ have been split into two *scalar modes* ($A_0 \equiv -\Phi$ and $\mathcal{A}$) and two *vector modes* ($\bar{A}$, which has only two independent components). Here, the terms 'scalar' and 'vector' are used in the

---

[1] Within the similar framework of linearized general relativity; see Bardeen (1980).

Relativity in Modern Physics. Nathalie Deruelle and Jean-Philippe Uzan.
© Oxford University Press 2018. Published in 2018 by Oxford University Press.
DOI: 10.1093/oso/9780198786399.001.0001

sense of $\mathcal{E}_3$. Under a gauge transformation $A_\mu \to A_\mu + \partial_\mu g$ these modes transform as (the dot represents the derivative with respect to time T of $\mathcal{S}$)

$$\Phi \mapsto \Phi - \dot{g}, \quad \mathcal{A} \mapsto \mathcal{A} + g, \quad \bar{A} \mapsto \bar{A}. \tag{14.3}$$

The two vector modes $\bar{A}$ and the scalar mode

$$\Psi \equiv \Phi + \dot{\mathcal{A}} \tag{14.4}$$

then are gauge invariant because $\Psi \mapsto \Psi$, $\bar{A} \mapsto \bar{A}$.

If we now substitute the decomposition $A_\mu = (-\Phi, \bar{A} + \nabla \mathcal{A})$ (which implies that $\partial_\nu A^\nu = \dot{\Phi} + \triangle \mathcal{A}$) into the Maxwell equations (14.1), we see that they can be rewritten in a form which is no longer manifestly Lorentz invariant but which, in return, is manifestly gauge invariant:

$$\triangle \Psi = 0, \quad \Box \bar{A} = 0. \tag{14.5}$$

The only solution to the first equation, the *constraint* equation, which is regular everywhere and vanishes at infinity is

$$\Psi = 0. \tag{14.6}$$

Therefore, the gauge 'hits twice': the first time in allowing two components of the potential to be combined to make a single gauge-invariant quantity (Ψ), and the second time *via* the field equations containing the constraint $\triangle \Psi = 0$ whose only solution is (14.6). The second equation in (14.5) is a propagation equation for the two independent components of $\bar{A}$, which are the two *degrees of freedom* of *electromagnetic waves*:

$$-\frac{\partial^2 \bar{A}}{\partial T^2} + \triangle \bar{A} = 0. \tag{14.7}$$

Fixing the gauge consists of choosing an element of the equivalence class of the modes $\Psi = \Phi + \dot{\mathcal{A}}$ and $\bar{A}$ which we have found. If we choose $\mathcal{A} = 0$, we have $\Phi = 0$ (because $\Psi = 0$), which corresponds to the *Hamiltonian gauge*. Since we have $\partial_i A^i = \partial_i \bar{A}^i = 0$ as well, this gauge is also the *Coulomb gauge*, and since $\partial_\nu A^\nu \equiv \dot{\Phi} + \triangle \mathcal{A} = 0$, it satisfies the Lorenz condition.

On the Cauchy problem

We consider a system of second-order, *linear* and *hyperbolic*, partial differential equations of the type

$$g^{\mu\nu} D_\mu D_\nu \Phi + A^\mu D_\mu \Phi + B\Phi + C = 0. \tag{14.8}$$

The condition 'hyperbolic' means that $g^{\mu\nu}$ is a *Lorentz* metric, that is, a metric of signature $(-, +, +, \cdots)$, like the Minkowski metric. Here Φ is an arbitrary tensor field, A^μ, B, and C are a vector and two arbitrary functions, and D_μ is the covariant derivative associated with $g_{\mu\nu}$ (in our case of Minkowski metric in Cartesian coordinates X^μ, we have $g_{\mu\nu} = \eta_{\mu\nu}$, and the covariant derivative reduces to an ordinary derivative, $D_\mu = \partial_\mu$).

There is an entire series of theorems (the first ones going back to Cauchy and Kovalevskaya)[2] showing that the *Cauchy problem* for this system is *well posed*. This means that (1) the

---

[2]See R. M. Wald (1984), Chapter 10, for a general overview. Mathematically precise definitions of the adjectives 'smooth' and 'close' and a demonstration of the theorem can be found in, for example, Hawking and Ellis (1973), Chapter 7.

specification of Φ and its normal derivatives on a space-like hypersurface Σ (e.g., its time derivatives on the $T = 0$ surface of an inertial frame) is sufficient for uniquely determining its further evolution (for initial data, called *Cauchy data*, which are sufficiently 'smooth'); (2) the solutions generated by 'close' initial conditions are close; and (3) the field propagates in a causal manner, that is, two solutions generated by two ensembles of Cauchy data which differ only in a finite volume S of Σ are the same outside the future cone with apex S.

The Maxwell equations (14.1) in the vacuum for the electromagnetic potential A^μ, that is, $\partial_\mu \partial_\nu (\eta^{\mu\nu} A^\rho - \eta^{\mu\rho} A^\nu) = 0$, are not *a priori* of the type (14.8), but they become so if they are written in their manifestly gauge-invariant form (14.5), namely, $\triangle \Psi = 0$ and $\Box \bar{A}_i = 0$, where $A_i = \partial_i \mathcal{A} + \bar{A}_i$ with $\partial_i \bar{A}^i = 0$ and $\Psi \equiv \Phi + \dot{\mathcal{A}}$. The Cauchy problem for the two degrees of freedom of the electromagnetic field is therefore well posed in the Maxwell theory.

More generally, the Maxwell equations (14.1) written in 3-vector notation where $A_\mu = (-\Phi, A)$ are

$$\nabla \cdot \dot{A} = -\triangle \Phi, \quad \ddot{A} = \triangle A - \nabla \left(\dot{\Phi} + \nabla \cdot A \right). \tag{14.9}$$

We see that Φ can be chosen arbitrarily on the initial hypersurface $T = 0$. However, once Φ is chosen, $\nabla \cdot A$ is then constrained by the first equation. The Cauchy problem therefore splits into two: the initial-condition problem of finding the Cauchy data which satisfy the constraint, and the evolution problem of integrating the second equation with the solution of the preceding problem as the initial conditions. After thus finding the solution, we have that the constraint is automatically satisfied for all T (actually, the divergence of the second equation is the time derivative of the first). This makes it possible, for example, to test the convergence of numerical integration algorithms.

14.2 The gauge-invariant action

In Section 12.3 we saw that the Maxwell equations in the vacuum (14.1) can be derived from a variational principle, that is, by extremizing the action (12.13) (from which in this section we omit the factor 4π)

$$\mathbf{S} = -\frac{1}{4} \int d^4X \, F_{\mu\nu} F^{\mu\nu} \quad \text{with} \quad F_{\mu\nu} = \partial_\mu A_\nu - \partial_\nu A_\mu \tag{14.10}$$

with respect to the potential A_μ. Here we shall show that the gauge-invariant version of the Maxwell equations in the vacuum (14.5) can also be derived directly by extremizing (14.10).

Introducing the Bardeen decomposition of the potential,

$$A_\mu = (-\Phi, \partial_i \mathcal{A} + \bar{A}_i) \quad \text{with} \quad \partial_i \bar{A}^i = 0 \quad \text{and} \quad \Psi \equiv \Phi + \dot{\mathcal{A}}, \tag{14.11}$$

we find $F_{0i} = \dot{\bar{A}}_i + \partial_i \Psi$ and $F_{ij} = \partial_i \bar{A}_j - \partial_j \bar{A}_i$, so that, using the fact that $\partial_i \bar{A}^i = 0$,

$$\begin{aligned}
\mathbf{S} &= \frac{1}{2} \int d^4X \left(\dot{\bar{A}}_i \dot{\bar{A}}^i - \frac{1}{2} F_{ij} F^{ij} + (\partial_i \Psi)(\partial^i \Psi) \right) + \int dT \int d^3X \, \partial_i (\Psi \dot{\bar{A}}^i) \\
&= \frac{1}{2} \int d^4X \left(\dot{\bar{A}}_i \dot{\bar{A}}^i - \frac{1}{2} F_{ij} F^{ij} + (\partial_i \Psi)(\partial^i \Psi) \right),
\end{aligned} \tag{14.12}$$

because the last term on the first line is a divergence which is transformed into a surface term using Gauss's theorem, and so does not contribute to the equations of motion. Moreover, it is zero if we require that the potential fall off sufficiently rapidly on the 2-sphere at infinity. The Lagrangian L and the Lagrangian density $\mathcal{L}$ are then defined as

$$\mathbf{S} = \int dT\, L \quad \text{with} \quad L = \int d^3 X\, \mathcal{L} \quad \text{and} \quad \mathcal{L} = \frac{1}{2}\left(\dot{\bar{A}}_i \dot{\bar{A}}^i - \frac{1}{2}F_{ij}F^{ij} + (\partial_i \Psi)(\partial^i \Psi)\right). \quad (14.13)$$

The density $\mathcal{L}$ is a functional (at each point X^i) of the generalized coordinates $q \equiv \{\bar{A}_i, \Psi\}$, but only the velocity extension of $\bar{A}_i$, $\dot{q} \equiv \dot{\bar{A}}_i$, appears.[3] This being the case, the variation of $\mathbf{S}$ with respect to a change of the configuration of Ψ and $\bar{A}_i$ is obtained as in Sections 8.1 and 12.3 and is given by (recalling that $F_{ij} = \partial_i \bar{A}_j - \partial_j \bar{A}_i$)

$$\delta\mathbf{S} = \int_{T_1}^{T_2} dT \int_V dV \left[\delta\bar{A}_i \,\Box\, \bar{A}^i - \delta\Psi \triangle \Psi\right]$$
$$+ \int dV \left.(\dot{\bar{A}}^i \delta\bar{A}_i)\right|_{T_1}^{T_2} + \int_{T_1}^{T_2} dT \int_S \left(\partial^i \Psi \,\delta\Psi - F^{ij}\,\delta\bar{A}_j\right) dS_i, \quad (14.14)$$

where we have used Gauss's theorem to write the last term as an integral over the 2-sphere at spatial infinity S with surface element dS_i. The surface terms vanish if the configurations are fixed at T_1 and T_2 as well as at spatial infinity. Therefore, the configurations extremizing $\mathbf{S}$ satisfy the equations of motion

$$\Box\,\bar{A}_i = 0 \quad \text{and} \quad \triangle \Psi = 0, \quad (14.15)$$

which are just (14.5). *Q.E.D.* We note that they can be written as Euler–Lagrange equations:

$$\frac{d}{dT}\frac{\delta L}{\delta \dot{\bar{A}}_i} = \frac{\delta L}{\delta \bar{A}_i}, \quad 0 = \frac{\delta L}{\delta \Psi} \quad (14.16)$$

with L given in (14.13), with the understanding that the variational derivatives imply integrations by parts if spatial derivatives of the variables appear in the Lagrangian density; for example, $\delta L/\delta\Psi = -\triangle\Psi$; see Section 9.1.

14.3 Hamiltonian formalisms

• The equations of motion (14.15), as we shall now see, can also be obtained using a gauge-invariant Hamiltonian formalism (see Section 9.1 for the similar case of the scalar field).

The Lagrangian density $\mathcal{L}$ was given in (14.13), and so we start by defining the conjugate momentum of $\bar{A}_i$ as

$$\bar{\pi}^i \equiv \frac{\partial \mathcal{L}}{\partial \dot{\bar{A}}_i} = \dot{\bar{A}}^i \quad (14.17)$$

and note that Ψ does not possess a conjugate momentum. The *Hamiltonian density* and the *Hamiltonian* are defined as

---

[3]See Book 1, Sections 8.1 and 9.1 for introductions to the Lagrangian and Hamiltonian formalisms in the mechanics of a point particle. See also the example of the scalar field in Sections 8.1 and 9.1 of the present book.

$$\mathcal{H} \equiv \bar{\pi}^i \dot{\bar{A}}_i - \mathcal{L} = \frac{1}{2}\left(\bar{\pi}_i \bar{\pi}^i + \frac{1}{2}F_{ij}F^{ij} - (\partial_i \Psi)(\partial^i \Psi)\right), \quad H \equiv \int d^3X\, \mathcal{H}. \tag{14.18}$$

They are functionals of $\{\bar{A}_i, \bar{\pi}^i\}$ and Ψ, and we can perform the variation in two different ways. First,

$$\delta \int dT\, H \equiv \frac{1}{2}\delta \int dT\, d^3X \left(\bar{\pi}_i \bar{\pi}^i + \frac{1}{2}F_{ij}F^{ij} - (\partial_i \Psi)(\partial^i \Psi)\right)$$

$$= \int dT\, d^3X (\bar{\pi}_i \delta\bar{\pi}^i - \triangle\bar{A}^i \delta\bar{A}_i + \triangle\Psi\delta\Psi) + \int dT \int_S (F^{ij}\delta\bar{A}_j - \partial^i\Psi\delta\Psi) dS_i \tag{14.19}$$

$$= \int dT\, d^3X (\bar{\pi}_i \delta\bar{\pi}^i - \triangle\bar{A}^i \delta\bar{A}_i + \triangle\Psi\delta\Psi)$$

(recalling that $F_{ij} = \partial_i \bar{A}_j - \partial_j \bar{A}_i$ and ignoring the surface term). Second,

$$\delta \int dT\, H \equiv \delta \int dT\, d^3X (\bar{\pi}^i \dot{\bar{A}}_i - \mathcal{L})$$

$$= \int dT\, d^3X (\dot{\bar{A}}_i \delta\bar{\pi}^i - \dot{\bar{\pi}}^i \delta\bar{A}_i) - \delta\mathbf{S} + \int d^3X (\bar{\pi}^i \delta\bar{A}_i)\Big|_{T_1}^{T_2} \tag{14.20}$$

$$= \int dT\, d^3X (\dot{\bar{A}}_i \delta\bar{\pi}^i - \dot{\bar{\pi}}^i \delta\bar{A}_i) - \delta\mathbf{S}.$$

Therefore, if we require that the action be extremal, $\delta\mathbf{S} = 0$, by equating (14.19) and (14.20) we find

$$\bar{\pi}_i = \dot{\bar{A}}_i, \quad \triangle\bar{A}^i = \dot{\bar{\pi}}^i, \quad \triangle\Psi = 0, \tag{14.21}$$

which after eliminating $\bar{\pi}^i$ gives $\Box\bar{A}_i = 0$ and $\triangle\Psi = 0$, that is, (14.15). Q.E.D. These can be written in the form of Hamilton equations, with the variational derivatives having the same meaning as in (14.16):

$$\frac{\delta H}{\delta\bar{\pi}^i} = \dot{\bar{A}}_i, \quad \frac{\delta H}{\delta\bar{A}_i} = -\dot{\bar{\pi}}^i, \quad \frac{\delta H}{\delta\Psi} = 0. \tag{14.22}$$

Knowing that the only regular solution of $\triangle\Psi = 0$ is the constraint $\Psi = 0$, we can at this point introduce the *reduced Hamiltonian*

$$H_{\text{reduced}} = \int d^3X\, \mathcal{H}_{\text{reduced}}, \quad \text{where} \quad \mathcal{H}_{\text{reduced}} = \frac{1}{2}\bar{\pi}^i \bar{\pi}_i + \frac{1}{2}(\partial_i \bar{A}_j)(\partial^i \bar{A}^j). \tag{14.23}$$

The density $\mathcal{H}_{\text{reduced}}$ is just (14.18), simplified by integrating by parts the term $(\partial_j \bar{A}_i)(\partial^i \bar{A}^j)$ and using the constraint $\Psi = 0$, which is allowed because the variation of (14.23) does indeed give the equations of motion of the two degrees of freedom of the field, namely, $\Box\bar{A}_i = 0$. The reduced Hamiltonian (14.23) is the starting point for the quantization of the electromagnetic field.

Finally, we note that 'on the *mass shell*' or '*on-shell*', that is, when the equations of motion (14.21) are satisfied, the Hamiltonian density (14.18) is identified as the energy density of the field as defined in (12.30) and (12.31), namely, $T^{00} = (E^2 + B^2)/2$ (up to the factor of 4π

omitted in this section). First, we have $F_{ij}F^{ij} = 2B^2$ (always), and then $E_i \equiv -\dot{A}_i - \partial_i\Phi = -\dot{A}_i - \partial_i\Psi$, and so $E_i = -\dot{A}_i = -\bar{\pi}_i$ by the definition of the momentum $\bar{\pi}^i$ and because on the mass shell $\Psi = 0$:

$$\mathcal{H}_{\text{on-shell}} = \frac{1}{2}(E^2 + B^2). \tag{14.24}$$

- We have just constructed a Hamiltonian formalism by directly writing the action in terms of gauge-invariant variables. This is not necessary.[4]

We again start from the action (14.10) with the potential decomposed as $A_\mu = (-\Phi, A)$, where A is the magnetic potential. The Lagrangian density is then written as

$$\mathcal{L} \equiv -\frac{1}{4}F_{\mu\nu}F^{\mu\nu} = \frac{1}{2}(\dot{A} + \nabla\Phi).(\dot{A} + \nabla\Phi) - \frac{1}{2}(\nabla \wedge A).(\nabla \wedge A). \tag{14.25}$$

The Coulomb potential Φ does not have a conjugate momentum. That of A is

$$\pi \equiv \frac{\partial\mathcal{L}}{\partial\dot{A}} = \dot{A} + \nabla\Phi, \tag{14.26}$$

so that the Hamiltonian density $\mathcal{H} \equiv \pi.\dot{A} - \mathcal{L}$ is

$$\mathcal{H} = \frac{1}{2}\pi.\pi + \frac{1}{2}(\nabla \wedge A).(\nabla \wedge A) - \pi.\nabla\Phi. \tag{14.27}$$

The Hamilton equations are

$$0 = \frac{\delta\mathcal{H}}{\delta\Phi} = \nabla.\pi, \quad \dot{A} = \frac{\delta\mathcal{H}}{\delta\pi} = \pi - \nabla\Phi, \quad \dot{\pi} = -\frac{\delta\mathcal{H}}{\delta A} = \nabla \wedge \nabla \wedge A, \tag{14.28}$$

where $\delta\mathcal{H}/\delta A$ is easily calculated by writing $2(\nabla \wedge A).(\nabla \wedge A) = (\partial_i A_j - \partial_j A_i)(\partial^i A^j - \partial^j A^i)$ and using the relation $\nabla(\nabla.A) - \triangle A = \nabla \wedge B$ with $B \equiv \nabla \wedge A$.

We extract $\pi = \nabla\Phi + \dot{A} \equiv -E$ from the second equation. The two other equations then are just the second group of Maxwell equations $\nabla.E = 0$ and $\dot{E} = \nabla \wedge B$.

Finally, to recover the equations in their manifestly gauge-invariant form, we make the Bardeen decomposition $A = \nabla\mathcal{A} + \bar{A}$ with $\nabla.\bar{A} = 0$, $\Psi \equiv \Phi + \dot{\mathcal{A}}$ and rewrite (14.26) and (14.28) as

$$\pi = \nabla\Psi + \dot{\bar{A}}, \quad 0 = \nabla.\pi = \triangle\Psi \quad \begin{cases} \dot{\pi} = \nabla\dot{\Psi} + \ddot{\bar{A}} \\ \quad = \nabla \wedge \nabla \wedge \bar{A} = \triangle\bar{A}, \end{cases} \tag{14.29}$$

which is indeed equivalent to (14.15). *Q.E.D.*

---

[4]In fact, it is impossible in the case of nonlinear field theories like general relativity; see, for example, Wald (1984), Appendix E, whose approach we follow here.

For a systematic treatment of systems with constraints see Dirac (1964) and Henneaux and Teitelboim (1992).

We note that on the mass shell, that is, when (14.28) is satisfied, the Hamiltonian density (14.27) and the Hamiltonian are written as

$$\mathcal{H}_{\text{on-shell}} = \frac{1}{2}(E^2 + B^2) - \nabla.(\Phi E)$$

$$\implies \quad H_{\text{on-shell}} \equiv \int d^3X \, \mathcal{H}_{\text{on-shell}} = \frac{1}{2}\int d^3X (E^2 + B^2). \tag{14.30}$$

Therefore, the on-shell Hamiltonian density is not necessarily positive-definite. On the other hand, the Hamiltonian is positive *modulo* the surface terms, whose vanishing (or positivity) must be verified for the particular solution chosen.

The non-gauge-invariant kinetic term

The Lagrangian density of the electromagnetic field is

$$\mathcal{L} = -\tfrac{1}{4}F_{\mu\nu}F^{\mu\nu} = -\tfrac{1}{2}(\partial_\mu A_\nu)(\partial^\mu A^\nu) + \tfrac{1}{2}(\partial_\mu A_\nu)(\partial^\nu A^\mu).$$

It is gauge-invariant. Let us generalize this as

$$\mathcal{L} = -\frac{\alpha}{2}(\partial_\mu A_\nu)(\partial^\mu A^\nu) + \frac{\beta}{2}(\partial_\mu A_\nu)(\partial^\nu A^\mu)$$

$$= \frac{1}{2}(\beta-\alpha)\dot\Phi^2 + \frac{\alpha}{2}\dot A_i \dot A^i + \beta \dot A^i \partial_i \Phi + \frac{\alpha}{2}(\partial_i \Phi)(\partial^i \Phi) - \frac{\alpha}{2}(\partial_i A^j)(\partial^i A_j) + \frac{\beta}{2}(\partial_i A_j)(\partial^j A^i), \tag{14.31}$$

where we have decomposed the potential as $A_\mu = (-\Phi, A_i)$. [A possible third term of the form $(\partial_\mu A^\mu)(\partial_\nu A^\nu)$ can be reduced to $(\partial_\mu A_\nu)(\partial^\nu A^\mu)$ by integration by parts: $(\partial_\mu A^\mu)(\partial_\nu A^\nu) = (\partial_\mu A_\nu)(\partial^\nu A^\mu) + \partial_\mu(A^\mu \partial_\nu A^\nu - A^\nu \partial_\nu A^\mu)$.]

The conjugate momenta are given by

$$\pi_\Phi \equiv \frac{\partial \mathcal{L}}{\partial \dot\Phi} = (\beta - \alpha)\dot\Phi, \quad \pi^i \equiv \frac{\partial \mathcal{L}}{\partial \dot A_i} = \alpha \dot A^i + \beta \partial^i \Phi. \tag{14.32}$$

We thus see that the absence of a conjugate momentum for Φ (if $\alpha = \beta$) is related to the gauge-invariance of the Lagrangian.

Let us take as an example the case $\beta = 0$ and $\alpha = 1$. The Hamiltonian density $\mathcal{H} \equiv \pi_\Phi \dot\Phi + \pi^i \dot A_i - \mathcal{L}$ reduces to

$$\mathcal{H} = -\frac{1}{2}(\pi_\Phi^2 + \partial_i \Phi \partial^i \Phi) + \frac{1}{2}(\pi_i \pi^i + \partial_i A_j \partial^i A^j). \tag{14.33}$$

The system has four degrees of freedom, a scalar field and a vector field, both massless. We note that the kinetic term of Φ is *negative*: this is a *ghost*. The energy of the system is therefore not positive-definite, which in general leads to instabilities when the system is coupled to other degrees of freedom.[5]

---

[5] The quantization of such a Hamiltonian is also problematic, as shown by Pais and Uhlenbeck (1950). See also, for example, Woodard (2007).

The Proca Hamiltonian

The Proca theory of a vector field of mass m is based on the Lagrangian density

$$\mathcal{L} = -\frac{1}{4} F_{\mu\nu} F^{\mu\nu} - \frac{1}{2} m^2 A_\mu A^\mu, \quad \text{where} \quad F_{\mu\nu} \equiv \partial_\mu A_\nu - \partial_\nu A_\mu. \tag{14.34}$$

(Such a field appears in the description of the BEH mechanism; see Section 9.4.) The Euler–Lagrange equations lead to the generalization of the second group of Maxwell equations:

$$\partial_\nu F^{\mu\nu} = -m^2 A^\mu \quad \Longrightarrow \quad \partial_\mu A^\mu = 0. \tag{14.35}$$

(Therefore, the Lorenz 'condition' is now imposed by the equations of motion.) If the potential is decomposed as $A_\mu = (-\Phi, A_i)$, the Lagrangian density can be written as

$$\mathcal{L} = \frac{1}{2} \dot{A}_i \dot{A}^i + \dot{A}^i \partial_i \Phi + \frac{1}{2} (\partial_i \Phi)(\partial^i \Phi) - \frac{1}{4} F_{ij} F^{ij} + \frac{1}{2} m^2 \Phi^2 - \frac{1}{2} m^2 A_i A^i, \tag{14.36}$$

and the equations of motion $\frac{\delta \mathcal{L}}{\delta \Phi} = 0$ and $\frac{d}{dT} \frac{\delta L}{\delta \dot{A}_i} = \frac{\delta L}{\delta A_i}$ are equivalent to (14.35):

$$\partial_i (\dot{A}^i + \partial^i \Phi) = m^2 \Phi, \quad \Box A_i = \partial_i (\dot{\Phi} + \partial_j A^j) + m^2 A_i \quad \Longrightarrow \quad \dot{\Phi} + \partial_j A^j = 0. \tag{14.37}$$

As in the Maxwell theory, Φ has no conjugate momentum, and that of A_i is $\pi^i \equiv \partial \mathcal{L}/\partial \dot{A}_i = \dot{A}^i + \partial^i \Phi$. The Hamiltonian density $\mathcal{H} \equiv \pi^i \dot{A}_i - \mathcal{L}$ is given by

$$\mathcal{H} = \frac{1}{2} \pi_i \pi^i - \pi^i \partial_i \Phi + \frac{1}{4} F_{ij} F^{ij} - \frac{1}{2} m^2 \Phi^2 + \frac{1}{2} m^2 A_i A^i. \tag{14.38}$$

Let us introduce the magnetic field B such that $F_{ij} F^{ij} = 2B^2$ and the electric field $E_i \equiv -\dot{A}_i - \partial_i \Phi$. The equations of motion $\pi^i = -E^i$ and $m^2 \Phi = -\partial_i E^i$ are indeed equivalent to (14.37) and (14.35). Therefore, the on-shell Hamiltonian density becomes

$$\mathcal{H} = \frac{1}{2} (E^2 + B^2) + \frac{1}{2} m^2 A^2 + \frac{(\nabla E)^2}{2m^2} + \nabla . (E\Phi), \tag{14.39}$$

and so the Hamiltonian $H \equiv \int d^3 X\, \mathcal{H}$ is positive if the chosen solution falls off sufficiently rapidly on the two-sphere at spatial infinity.

However, the mass term breaks the Maxwell gauge invariance, which opens up an interesting Pandora's box when the theory is quantized.[6]

---

[6]See, for example, Itzykson and Zuber (2006), and also Ruegg and Ruiz-Altaba (2004).

15
Electromagnetic waves

In this chapter we study the solutions of the Maxwell equations in the vacuum: monochromatic plane waves and their polarizations, plane waves, and the motion of a charge in the field of a wave (which is the principle upon which particle detection is based). We conclude by giving the conditions for the *geometrical optics* limit, and then establish the connection between electromagnetic waves and the kinematic description of light discussed in the first part of this book.

15.1   Monochromatic plane waves: propagation

As we saw in the preceding chapter, the Maxwell equations in the vacuum for the potential A_μ reduce, in the Hamiltonian gauge of an inertial frame $\mathcal{S}$, to

$$A_\mu = (0, \bar{A}_i) \quad \text{with} \quad \partial_i \bar{A}^i = 0 \quad \text{and} \quad \Box \bar{A}_i = 0 \,. \tag{15.1}$$

If they are square integrable, which we assume to be the case, the two independent components of the vector $\bar{A}_i$ can be decomposed into Fourier modes as

$$\bar{A}_i(T, X^i) = \int \frac{d^3 k}{(2\pi)^{\frac{3}{2}}} \hat{A}_i(T, k^i) e^{ik_j X^j} \,, \quad \text{with} \quad \hat{A}_i^*(T, -k^i) = \hat{A}_i(T, k^i) \tag{15.2}$$

to guarantee that the variables are real (see the introduction to the Fourier transform in Section 9.2). The fact that $\bar{A}_i$ has zero divergence is expressed by the transversality condition $k^i \hat{A}_i = 0$.

For each 3-vector k of components k^i we introduce two unit vectors called the *polarization vectors* $\epsilon_{k,\lambda}$, $\lambda = \{1,2\}$, such that $\epsilon_{1,2}.\epsilon_{1,2} = 1$, which are mutually orthogonal ($\epsilon_{k,\lambda} \, \epsilon_{k,\lambda'} = \delta_{\lambda\lambda'}$), and orthogonal to k^i ($\epsilon_\lambda^i k_i = 0$). We fix the orientation in the plane orthogonal to k^i by requiring, for example, that $\epsilon_1 = e_Y$ and $\epsilon_2 = e_Z$ if $k^i = (k, 0, 0)$. We can therefore incorporate the transversality condition in the decomposition (15.2) by writing

$$\bar{A}^i(T, X^i) = \sum_\lambda \int \frac{d^3 k}{(2\pi)^{\frac{3}{2}}} q_{k,\lambda}(T) \, \epsilon_{k,\lambda}^i e^{ik_j X^j} \quad \text{with} \quad q_{-k,\lambda}^*(T) = q_{k,\lambda}(T) \,. \tag{15.3}$$

We have thereby traded the two functions of time and spatial coordinates $\bar{A}_i$ for a sextuple infinity of functions of time $q_{k,\lambda}(T)$.

The function (15.3) is divergence-free by construction ($\partial_i \bar{A}^i = 0$ because $k_i \epsilon_{k,\lambda}^i = 0$) and, for its d'Alembertian to be zero, the functions $q_{k,\lambda}(T)$ must satisfy the ordinary differential equation

$$\ddot{q}_{k,\lambda} + |k|^2 q_{k,\lambda} = 0 \,, \quad \text{where} \quad |k|^2 \equiv k_i k^i \,. \tag{15.4}$$

Relativity in Modern Physics. Nathalie Deruelle and Jean-Philippe Uzan.
© Oxford University Press 2018. Published in 2018 by Oxford University Press.
DOI: 10.1093/oso/9780198786399.001.0001

Fourier decomposition of the Hamiltonian

As we have seen in Section 14.3, the equation of motion (15.1) is derived from the reduced Hamiltonian

$$H_{\text{reduced}} = \int d^3X\, \mathcal{H}, \quad \text{where} \quad \mathcal{H} = \frac{1}{2}\bar{\pi}^i\,\bar{\pi}_i + \frac{1}{2}(\partial_i\bar{A}_j)(\partial^i\bar{A}^j)\,.$$

In Fourier space, using the properties of the Dirac distribution (see Section 9.2) and the reality condition for the fields, we have

$$H_{\text{reduced}} = \frac{1}{16\pi^3}\int d^3X\, d^3k\, d^3k'\,[\hat{\pi}^j(T,k'_i)\,\hat{\pi}_j(T,k_i) - (k'_l k^l)\hat{A}_j(T,k'_i)\hat{A}^j(T,k_i)]e^{i(k_i+k'_i)X^i}$$

$$= \frac{1}{16\pi^3}\int d^3X\, d^3k\, d^3k'\,[\hat{\pi}^j(T,-k'_i)\,\hat{\pi}_j(T,k_i) + (k'_l k^l)\hat{A}_j(T,-k'_i)\hat{A}^j(T,k_i)]e^{i(k_i-k'_i)X^i}$$

$$= \frac{1}{2}\int d^3k\, d^3k'\,[\hat{\pi}^j(T,-k'_i)\,\hat{\pi}_j(T,k_i) + (k'_l k^l)\hat{A}_j(T,-k'_i)\hat{A}^j(T,k_i)]\delta(k_i-k'_i)$$

$$= \frac{1}{2}\int d^3k\,[\hat{\pi}^j(T,-k_i)\,\hat{\pi}_j(T,k_i) + (k_l k^l)\hat{A}_j(T,-k_i)\hat{A}^j(T,k_i)]$$

$$= \frac{1}{2}\int d^3k\,[\hat{\pi}^{*j}(T,k_i)\,\hat{\pi}_j(T,k_i) + (k_l k^l)\hat{A}^*_j(T,k_i)\hat{A}^j(T,k_i)]\,.$$

The Fourier transforms $\hat{A}_j(T,k_i)$ and $\hat{\pi}_j(T,k_i)$, being transverse ($k_i\hat{A}^i = k_i\hat{\pi}^i = 0$), can be written as

$$\hat{A}^i = \sum_\lambda q_{k,\lambda}(T)\,\epsilon^i_{k,\lambda}\,, \quad \hat{\pi}^i = \sum_\lambda p_{k,\lambda}(T)\,\epsilon^i_{k,\lambda}\,,$$

where $\epsilon^i_{k,\lambda}$ are the two polarization vectors orthogonal to the wave vector k^i, so that the reduced Hamiltonian becomes

$$H_{\text{reduced}} = \frac{1}{2}\sum_\lambda \int d^3k\,(p_{k,\lambda}p^*_{k,\lambda} + |k|^2\,q_{k,\lambda}q^*_{k,\lambda})\,. \tag{15.5}$$

This Hamiltonian is just the Hamiltonian of a sextuple infinity of identical harmonic oscillators of frequency $\omega = |k|$. The Hamilton equations again give the equations of motion (15.4) (see the similar case of the scalar field in Section 9.2). The Hamiltonian (15.5) is the starting point for the quantization of the electromagnetic field.

The solution of the equation of motion (15.4) guaranteeing the reality of $\bar{A}_i$ is

$$q_{k,\lambda} = a_{k,\lambda}\,e^{-i|k|T} + a^*_{k,\lambda}\,e^{+i|k|T}\,, \tag{15.6}$$

where $a_{k,\lambda}$ are integration constants, that is, two arbitrary functions[1] of k^i. After rewriting the integral of the second term in (15.6) for $k^j \to -k^j$, we find that the square-integrable general solution of the vacuum Maxwell equations for the potential in the Hamiltonian gauge is a superposition of 'monochromatic' plane waves $A^i_{k,\lambda}(X^\mu)$:

---

[1]In quantum electrodynamics it is $\sqrt{2|k|}a_{k,\lambda}$ (rather than $a_{k,\lambda}$), which becomes an 'annihilation operator'; see Sections 9.2 and 9.3.

$$A_\mu = (0, \bar{A}_i) \quad \text{with} \quad \bar{A}^i(X^\mu) = \sum_\lambda \int \frac{d^3k}{(2\pi)^{3/2}} A^i_{k,\lambda}(X^\mu) \quad \text{and}$$

$$A^i_{k,\lambda}(X^\mu) \equiv a_{k,\lambda}\, \epsilon^i_{k,\lambda}\, e^{ik_\mu X^\mu} + \text{c.c.}, \quad \text{where} \quad k^\mu = (|k|, k^i) \quad \text{and} \quad k_i \epsilon^i_{k,\lambda} = 0. \tag{15.7}$$

Therefore, the monochromatic plane waves form a basis (in the sense of distributions, because they are not square-integrable) in which any solution of the vacuum Maxwell equations can be expanded. If desired, we can avoid the introduction of complex quantities by writing $A^i_{k,\lambda}(X^\mu)$ as

$$A^i_{k,\lambda}(X^\mu) = \bar{a}_{k,\lambda}\, \epsilon^i_{k,\lambda} \cos(k_\mu X^\mu - \phi_{k,\lambda}) \quad \text{with} \quad k^\mu = (|k|, k^i), \tag{15.8}$$

where the constants $\bar{a}_{k,\lambda}$ and $\phi_{k,\lambda}$ are real.

The electric and magnetic fields are defined as $E_i = -\partial_0 A_i$ (in the Hamiltonian gauge) and $B^i = \epsilon^{ijk}\partial_j A_k$. When the potential is given by the monochromatic plane wave (15.8) they can be written as

$$E^i = -|k|\, \bar{a}_{k,\lambda}\epsilon^i_{k,\lambda} \sin(k_\mu X^\mu - \phi_{k,\lambda}), \quad B^i = -k^j\, e^i{}_{jk}\bar{a}_{k,\lambda}\epsilon^k_{k,\lambda} \sin(k_\mu X^\mu - \phi_{k,\lambda}). \tag{15.9}$$

In 3-vector notation we have

$$(E\,.n) = 0, \quad B = n \wedge E \quad \text{with} \quad n \equiv \frac{k}{|k|}. \tag{15.10}$$

The components of the energy–momentum tensor are given by (see Section 12.4) $T^{00} \equiv W = \frac{1}{8\pi}(E^2 + B^2)$, $T^{0i} \equiv S^i$ with $S \equiv \frac{1}{4\pi}(E \wedge B)$ and $T_{ij} \equiv \sigma_{ij} = \frac{1}{4\pi}(-E_iE_j - B_iB_j + \frac{1}{2}\delta_{ij}(E^2 + B^2))$. When the potential is given by the monochromatic plane wave (15.8) they become

$$W = \frac{|k|^2}{4\pi}\bar{a}^2_{k,\lambda} \sin^2(k_\mu X^\mu - \phi_{k,\lambda}), \quad S = W\,n, \tag{15.11}$$

$$\sigma_{ij}\, n^i n^j = W, \quad \sigma_{ij}\, n^j = S_i, \quad \sigma_{ij} - n_i n_j(\sigma_{kl}\, n^k n^l) = 0,$$

where the last expression is easily shown if we work in the frame in which $n = (1, 0, 0)$, $E = (0, |E|, 0)$, and $B = (0, 0, |E|)$, because then the only nonzero component of σ_{ij} is $\sigma_{XX} = W$. We note that $\langle W \rangle = \frac{1}{8\pi}|k|^2 \bar{a}^2_{k,\lambda}$ when averaged over $k_\mu X^\mu$. Finally, the various components of $T^{\mu\nu}$ can be grouped together to give[2]

$$T^{\mu\nu} = \frac{W}{|k|^2}k^\mu k^\nu. \tag{15.12}$$

---

[2]Knowing the transformation law for the $T^{00} \equiv W$ component of the energy–momentum tensor (see Section 12.4) as well as that of the 4-vector k^μ under a change of inertial frame $\mathcal{S} \to \mathcal{S}'$, we see that the components of the energy–momentum tensor in $\mathcal{S}'$ are given by $T'^{\mu\nu} = \frac{W'}{|k'|^2}k'^\mu k'^\nu$. Therefore, $\frac{W}{|k|^2}$ is a scalar.

The manifestly Lorentz-invariant form

We have defined a monochromatic plane wave as the particular solution of the vacuum Maxwell equations in the Hamiltonian gauge in a given inertial frame, $A_\mu = (0, A^i_{k,\lambda})$, where $A^i_{k,\lambda}(X^\mu)$ is given in (15.7) or (15.8). As we saw in Section 14.1, this gauge is also the Coulomb gauge (because $\partial_i \bar{A}^i = 0$) and it satisfies the Lorenz condition $\partial_\mu A^\mu = 0$. Therefore, the gauge is completely fixed, but A_μ is not in a manifestly Lorentz-invariant form.

In a different inertial frame $\mathcal{S}'$, which, for example, moves along the X axis at speed V_0, the gauge is no longer the Hamiltonian gauge because $A'_0 = V_0 A_X / \sqrt{1 - V_0^2} \neq 0$. However, it will always belong to the class of Lorenz gauges because the condition is invariant under Poincaré transformations. The manifestly Lorentz-invariant form of the potential describing a monochromatic plane wave then is

$$A_\mu = a_\mu \cos(k_\nu X^\nu - \phi) \quad \text{with} \quad k_\mu k^\mu = 0 \quad \text{and} \quad a_\mu k^\mu = 0, \qquad (15.13)$$

where a_μ is a constant 4-vector and ϕ is a constant. Choosing the Hamiltonian gauge means requiring that $a_0 = 0$ in the selected frame.

The Faraday tensor associated with the potential (15.13) is

$$F_{\mu\nu} \equiv \partial_\mu A_\nu - \partial_\nu A_\mu = -(k_\mu a_\nu - k_\nu a_\mu) \sin(k_\rho X^\rho - \phi).$$

[It can be written as the exterior product of the forms (*i.e.*, of the covariant vectors) k and a as $F = \sin(k_\rho X^\rho - \phi)\, a \wedge k$; see Section 22.1 *et seq.* below.]

The energy–momentum tensor $T_{\mu\nu} \equiv \frac{1}{4\pi}(F_{\mu\rho} F_\nu{}^\rho - \frac{1}{4}\eta_{\mu\nu} F_{\rho\sigma} F^{\rho\sigma})$ reduces to

$$T^{\mu\nu} = \frac{1}{4\pi} \sin^2(k_\rho X^\rho - \phi)(a_\rho a^\rho) k^\mu k^\nu, \qquad (15.14)$$

which is the manifestly Lorentz-invariant form of (15.12).

15.2 Monochromatic plane waves: polarization

A monochromatic plane wave propagating in the k direction in an inertial frame $\mathcal{S}$ is completely described in the Hamiltonian gauge by the magnetic potential [*cf.* (15.8)]

$$A = \epsilon_1 a_1 \cos(|k|T + \psi_1) + \epsilon_2 a_2 \cos(|k|T + \psi_2), \qquad (15.15)$$

where we have omitted the indices k in order to simplify the notation. Here ϵ_1 and ϵ_2 are two unitary vectors which are orthogonal to each other and to the wave vector k (chosen such that $\epsilon_1 = e_Y$, $\epsilon_2 = e_Z$ if $k = |k|e_X$), and $\psi_\lambda \equiv (\phi_\lambda - k_i X^i)$ ($\lambda = 1, 2$) is the phase at $T = 0$ of each component of A at a point in space. Every quantity derived from the potential (for example, the electric and magnetic fields) will have the same structure.

Let us consider, in a Euclidean plane with basis vectors ϵ_1 and ϵ_2 and origin O (which can be identified with the point having coordinates X^i in the space $\mathcal{E}_3$ representing the physical space), the endpoint P of a bound vector $OP = A$.

If $\psi_1 = \psi_2 \equiv \psi$, then

$$A = (a_1 \epsilon_1 + a_2 \epsilon_2) \cos(|k|T + \psi), \qquad (15.16)$$

and the point P will oscillate with a period equal to that of the wave, $2\pi/|k|$, along the line making an angle ϕ with ϵ_1 such that $\tan \phi = a_2/a_1$. In this case the polarization of the wave is said to be *linear*.

If $\psi_1 = \psi_2 \pm \pi/2 \equiv \psi$ and $a_1 = a_2 \equiv a$, then

$$A = a\left[\epsilon_1 \cos(|\mathbf{k}|T + \psi) \mp \epsilon_2 \sin(|\mathbf{k}|T + \psi)\right]. \tag{15.17}$$

In this case, during one period the point P traces a circle of radius a in the prograde direction if $\psi_1 = \psi_2 - \pi/2$, or in the retrograde direction if $\psi_1 = \psi_2 + \pi/2$. The polarization is called *right-* or *left-handed circular polarization* (these are also referred to as *positive* and *negative* *helicities*).

In the general case the point P traces an ellipse (see Fig. 15.1).

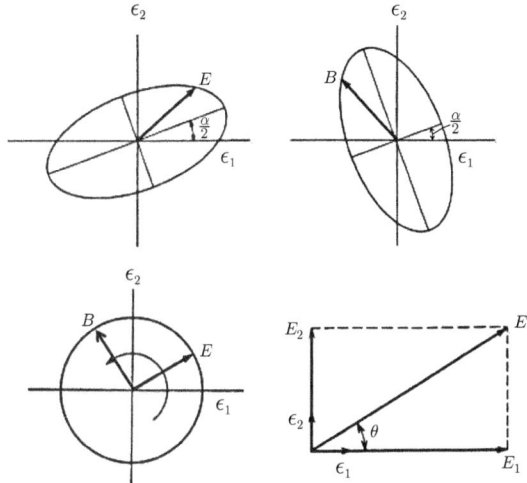

Fig. 15.1 Elliptical, circular, and linear polarizations.

Let $A_1 \equiv a_1 \cos(|\mathbf{k}|T + \psi_1)$ and $A_2 \equiv a_2 \cos(|\mathbf{k}|T + \psi_2)$ be the components of A in the basis (ϵ_1, ϵ_2). The equation of the ellipse is obtained by eliminating the time and becomes

$$\sin^2(\psi_2 - \psi_1) = \left(\frac{A_1}{a_1}\right)^2 + \left(\frac{A_2}{a_2}\right)^2 - 2\left(\frac{A_1}{a_1}\right)\left(\frac{A_2}{a_2}\right)\cos(\psi_2 - \psi_1). \tag{15.18}$$

To find the orientation and the major axis of the ellipse, we rotate the axes by an angle α, $(\epsilon_1, \epsilon_2) \rightarrow (\xi_1, \xi_2)$, such that $\epsilon_1 = \xi_1 \cos\alpha - \xi_2 \sin\alpha$ and $\epsilon_2 = \xi_1 \sin\alpha + \xi_2 \cos\alpha$. We then find

$$A = X_1\xi_1 + X_2\xi_2 \quad \text{and} \quad \begin{cases} A_1 = X_1\cos\alpha - X_2\sin\alpha \\ A_2 = X_1\sin\alpha + X_2\cos\alpha, \end{cases} \tag{15.19}$$

where X_1 and X_2 are the components of A in the basis (ξ_1, ξ_2). We then substitute (15.19) for A_1 and A_2 as functions of X_1 and X_2 into (15.18). The cross term proportional to X_1X_2 vanishes if the angle α is chosen such that

$$\tan 2\alpha = \frac{U}{Q}, \quad \text{where} \quad \begin{cases} U \equiv 2a_1 a_2 \cos(\psi_2 - \psi_1) \\ Q \equiv a_1^2 - a_2^2. \end{cases} \tag{15.20}$$

Now substituting this value of α into the coefficients of X_1^2 and X_2^2, with a bit of trigonometry we find that (15.18) for the ellipse takes the 'canonical' form

$$\left(\frac{X_1}{a_+}\right)^2 + \left(\frac{X_2}{a_-}\right)^2 = 1, \quad \text{where} \quad a_\pm = \frac{V}{\sqrt{2}\,(I \mp \sqrt{A + U^2})^{1/2}} \tag{15.21}$$

$$\text{with} \quad I \equiv a_1^2 + a_2^2, \quad V \equiv 2a_1 a_2 \sin(\psi_2 - \psi_1).$$

The parameters (U, Q, V, I) are the *Stokes parameters*. They are related to each other as

$$I^2 = Q^2 + U^2 + V^2. \tag{15.22}$$

15.3 Plane waves

A *plane wave* is a solution of the vacuum Maxwell equations which depends on only one of the Cartesian spatial coordinates, for example, X. In this case the Maxwell equations in the Hamiltonian gauge, $\Box \bar{A}_i = 0$, reduce to $-\partial^2 \bar{A}_i / \partial T^2 + \partial^2 \bar{A}_i / \partial X^2 = 0$, or also $\partial^2 \bar{A}_i / (\partial u \partial v) = 0$ after making the change of variable $u = T - X$, $v = T + X$. The solution therefore has the form $A_\mu = (0, \bar{A}_i)$ with

$$\bar{A}^i = \bar{A}_+^i(T - X) + \bar{A}_-^i(T + X) \quad \text{and} \quad \partial_i \bar{A}_\pm^i = 0. \tag{15.23}$$

A solution of this type can be decomposed on a basis of monochromatic plane waves as

$$\bar{A}_\pm^i = \sum_\lambda \int \frac{dk}{\sqrt{2\pi}} (a_{k,\lambda} \epsilon_{k,\lambda}^i e^{-ik(T \mp X)} + \text{c.c.}), \tag{15.24}$$

where the two polarization vectors $\epsilon_{k,\lambda}$ reduce in this case to the basis vectors e_Y and e_Z. The quantity $\bar{A}_+^i$ describes a wave propagating in the positive X direction at the speed of light, and $\bar{A}_-^i$ is the same in the negative direction.

Spherical waves

We can analogously define a *spherical wave* as a solution of the vacuum Maxwell equations depending only on the time T and the radial distance r to the origin. Given that $\triangle f = \frac{1}{r} \frac{\partial^2 (rf)}{\partial r^2}$, the wave equation takes the form $-\partial_T^2 (rA_i) + \partial_r^2 (rA_i) = 0$, the general solution of which is the sum of a wave propagating from the origin and a wave converging toward the origin, $A_\mu = (0, \bar{A}_i)$, with

$$\bar{A}_i(r, T) = \frac{\bar{A}_i^+(r - T)}{r} + \frac{\bar{A}_i^-(r + T)}{r}.$$

The electric and magnetic fields are defined in the Hamiltonian gauge as $E_i = -\partial_0 A_i$ and $B^i = e^{ijk} \partial_j A_k$. In the case of a plane wave propagating along the vector e_X and therefore depending only on $u = T - X$ they are given by

$$E = -\frac{d\bar{A}}{du} \quad \text{with} \quad (E \cdot e_X) = 0, \quad B = e_X \wedge E. \tag{15.25}$$

The energy–momentum tensor of the wave possesses the same properties as that of a monochromatic plane wave [*cf.* (15.11)]:

$$W = \frac{E^2}{4\pi}, \quad S = W e_X, \quad \sigma_{XX} = W. \tag{15.26}$$

Finally, just like a monochromatic plane wave, the potential describing a plane wave can be cast in manifestly Lorentz-invariant form [*cf.* (15.13)]:

$$A_\mu = A_\mu(\xi), \quad \text{where} \quad \xi \equiv -k_\mu X^\mu \quad \text{with} \quad k_\mu k^\mu = 0 \quad \text{and} \quad A_\mu k^\mu = 0. \tag{15.27}$$

The null vector k^μ, which is constant, determines the direction of propagation of the plane wave [for example, $k^\mu = (k, k, 0, 0)$ for a wave propagating in the positive X direction], and the condition $A_\mu k^\mu$ guarantees that the gauge is a Lorenz gauge because it implies that $\partial_\mu A^\mu = 0$. Choosing the Hamiltonian gauge amounts to requiring that $A_0 = 0$ in the chosen frame.

The Faraday tensor associated with the potential (15.27) is

$$F_{\mu\nu} \equiv \partial_\mu A_\nu - \partial_\nu A_\mu = -k_\mu \frac{dA_\nu}{d\xi} + k_\nu \frac{dA_\mu}{d\xi}. \tag{15.28}$$

[It can be written as an exterior product; see Section 22.1 *et seq.*: $F = (dA/d\xi) \wedge k$.] It satisfies $F_{\mu\nu} k^\nu = 0$, which means that E and B are orthogonal to the direction of propagation. Since $F_{\rho\sigma} F^{\rho\sigma} = 0$, which means that E and B have the same modulus, the energy–momentum tensor of the wave is simply

$$T_{\mu\nu} \equiv \frac{1}{4\pi}\left(F_{\mu\sigma} F_\nu{}^\sigma - \frac{1}{4}\eta_{\mu\nu} F_{\rho\sigma} F^{\rho\sigma}\right) = \frac{1}{4\pi}\left(\frac{dA}{d\xi} \cdot \frac{dA}{d\xi}\right) k_\mu k_\nu. \tag{15.29}$$

15.4 Motion of a charge in a plane wave

Let us consider a plane wave propagating in a direction determined by the spatial components of a null vector k^μ. It is described by a potential $A_\mu(\xi)$, where $\xi \equiv -k_\mu X^\mu \equiv -(k \cdot X)$, belonging to the class of Lorenz gauges if $(A \cdot k) = 0$ [*cf.* (15.27)]. Its Faraday tensor is given in (15.28). Then the Lorentz equation of motion of a charge q of mass m and 4-velocity U^μ in the field of this wave is

$$\frac{dU^\mu}{d\tau} = \frac{q}{m} F^\mu{}_\nu U^\nu = \frac{q}{m}\left[-k^\mu\left(U \cdot \frac{dA}{d\xi}\right) + (k \cdot U)\frac{dA^\mu}{d\xi}\right]. \tag{15.30}$$

Since the scalar product of the second term with k^μ is zero, we have $(k \cdot U) = (k \cdot U_{\text{in}})$, where U^μ_{in} is the initial 4-velocity, that is, the 4-velocity at $\tau = 0$, where τ is the proper time of the charge. Since $\xi(\tau) = -(k \cdot X(\tau))$, we have $\xi = -(k \cdot U_{\text{in}})\tau$. Therefore, the quantity ξ [equal, for example, to $k(T - X)$ if the wave travels along the X axis] when evaluated on the world line is proportional to the proper time τ of the charge. Equation (15.30) can therefore be rewritten as

$$\frac{dU^\mu}{d\xi} = \frac{q}{m}\left[\frac{k^\mu}{(k \cdot U_{\text{in}})}\left(U \cdot \frac{dA}{d\xi}\right) - \frac{dA^\mu}{d\xi}\right]. \tag{15.31}$$

If we now take the scalar product of (15.31) and the vector A_μ, since $(A \cdot k) = 0$ we obtain

$$\left(A \cdot \frac{dU}{d\xi}\right) = -\frac{q}{m}\left(A \cdot \frac{dA}{d\xi}\right), \quad \text{that is,} \quad \left(U \cdot \frac{dA}{d\xi}\right) = \frac{d}{d\xi}(A \cdot U) + \frac{q}{2m}\frac{d}{d\xi}(A \cdot A). \tag{15.32}$$

Equation (15.31) can then be integrated to give the 4-velocity of the charge:

$$U^\mu = U^\mu_{\text{in}} - \frac{q}{m}(A^\mu - A^\mu_{\text{in}}) - \frac{q^2}{2m^2}\frac{k^\mu}{(k \cdot U_{\text{in}})}(A - A_{\text{in}})^2, \tag{15.33}$$

where A_{in} is the value of the potential at $\tau = 0$, and we have simplified the expression using the fact that the 4-velocity is normalized to -1.

A charge in a linearly polarized wave

Let us choose the frame such that the wave, which we assume monochromatic, propagates along the X axis, $k^\mu = (k, k, 0, 0)$, and is polarized along the Y axis. Then in the Hamiltonian gauge we have $A_\mu = (0, 0, a \cos \xi, 0)$ with $\xi = k(T - X)$, k being the wave frequency and ka the amplitude of the associated field. If we choose the initial condition on the charge motion to be $U^\mu_{\text{in}} = (1, 0, 0, 0)$, then $(k \cdot U_{\text{in}}) = -k$ and $\xi(\tau) = k\tau$.

The components of the 4-velocity $U^\mu = k(dX^\mu/d\xi)$ are given by (15.33), which becomes

$$U^X = \frac{\beta^2}{2}(\cos \xi - 1)^2, \quad U^Y = -\beta(\cos \xi - 1), \quad U^0 = 1 + k\frac{dX}{d\xi}, \quad \text{with} \quad \beta \equiv \frac{qa}{m}. \quad (15.34)$$

These equations can be integrated by inspection. If we assume that the charge is at the origin at $\tau = 0$, we have

$$kX = \frac{\beta^2}{8}(6\xi - 8\sin \xi + \sin 2\xi), \quad kY = \beta(\xi - \sin \xi), \quad kT = \xi + kX. \quad (15.35)$$

The charge traces a figure eight in the XOY plane, about the line starting from the origin along which it is carried with the average velocity

$$\langle V^X \rangle = \frac{3\beta^2}{4(1 + \frac{3\beta^2}{4})}, \quad \langle V^Y \rangle = \frac{\beta}{1 + \frac{3\beta^2}{4}}.$$

In the nonrelativistic limit $(\beta \to 0)$ the charge drifts slowly along the Y axis, that is, along the electric field, and oscillates with the same frequency k along a flat figure eight parallel to the field. In the ultrarelativistic limit where β is large, the charge accompanies the wave along the X axis while describing large figure eights with period $2\pi/k$ measured using its proper time, and period $(2\pi/k)(1 + 3\beta^2/4)$ measured using the time of the inertial frame (see Fig. 15.2).

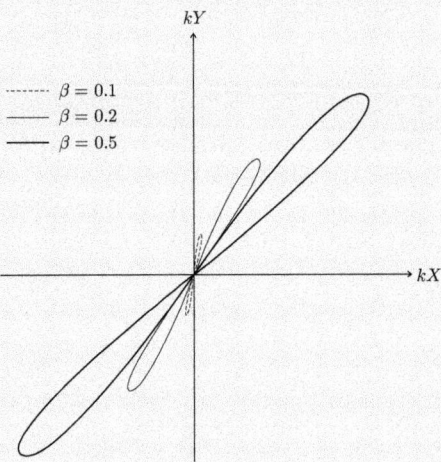

Fig. 15.2 A charge in a linearly polarized monochromatic wave.

To describe the motion of the charge in the inertial frame $\mathcal{S}'$ where it is at rest on the average, we need to combine Lorentz transformations along the X axis and then along the Y axis [see (1.12)], with the angle ϕ given by $\tan\phi \equiv \langle V^Y \rangle / \langle V^X \rangle = 4/(3\beta)$ and the rapidity ψ by $\tanh\psi = \langle V \rangle = \beta\sqrt{1+9\beta^2/16}/(1+3\beta^2/4)$. Then we need to substitute (15.35) for the world line in $\mathcal{S}$. A rather tedious trigonometric calculation gives

$$kX' = -\frac{\beta^2}{4(1+9\beta^2/16)}\left[3\sqrt{1+\beta^2/2}+1+\left(\frac{3\beta^2}{16\sqrt{1+\beta^2/2}}-1\right)\cos\xi\right]\sin\xi,$$

$$kY' = -\frac{\beta}{1+9\beta^2/16}\left[\sqrt{1+\beta^2/2}-\frac{3\beta^2}{16}+\frac{\beta^2}{16}\left(\frac{1}{\sqrt{1+\beta^2/2}}+3\right)\cos\xi\right]\sin\xi,$$

$$kT' = \sqrt{1+\beta^2/2}\,\xi+\frac{\beta^2}{8\sqrt{1+\beta^2/2}}\sin 2\xi\,.$$

The trajectory is again a figure eight, as in the previous case.

A charge in a circularly polarized wave

Let us choose the frame such that the wave, assumed to be monochromatic, propagates along the X axis, $k^\mu = (k,k,0,0)$. If it is circularly polarized (see 15.2), we can take $A_\mu = (0,0,a\cos\xi,a\sin\xi)$ with $\xi = k(T-X)$, k being the wave frequency and ka the amplitude of the associated field. If the initial condition on the charge motion is taken to be $U^\mu_{\text{in}} = (1,0,0,0)$, we have $(k\cdot U_{\text{in}}) = -k$ and $\xi(\tau) = k\tau$.

The components of the 4-velocity $U^\mu = k(dX^\mu/d\xi)$ are given by (15.33):

$$U^Y = -\frac{qa}{m}(\cos\xi-1),\quad U^Z = -\frac{qa}{m}\sin\xi,\quad U^X = -\left(\frac{qa}{m}\right)^2(1-\cos\xi),\quad U^0 = 1+k\frac{dX}{d\xi}. \tag{15.36}$$

Integrating this gives

$$kX = \beta^2(\xi-\sin\xi),\quad kY = \beta(\xi-\sin\xi),\quad kZ = \beta(\cos\xi-1),\quad kT = \xi+X \quad\text{with}\quad \beta \equiv \frac{qa}{m}. \tag{15.37}$$

On the average, the charge travels along the line $O\bar{X}$ of the XOY plane, making an angle ϕ with OX such that $\tan\phi = 1/\beta$, at the velocity

$$\langle V^X \rangle = \frac{\beta^2}{1+\beta^2},\quad \langle V^Y \rangle = \frac{\beta}{1+\beta^2}\,.$$

In addition, in the $\bar{X}OZ$ plane the charge traces ellipses perpendicular to the average motion.

In the nonrelativistic limit $\beta \to 0$ the charge drifts slowly along the Y axis and traces circles with the same frequency as the wave. In the ultrarelativistic limit the charge accompanies the wave along the X axis and oscillates along the X axis with period $2\pi/k$ measured using its proper time, and $(2\pi/k)(1+\beta^2)$ measured using the time of the inertial frame.

To describe the motion of the charge in the inertial frame $\mathcal{S}'$ where it is at rest on the average, we combine Lorentz transformations along the X axis and then along the Y axis (see 15.2), where the angle ϕ is now given by $\tan\phi \equiv \langle V^Y \rangle / \langle V^X \rangle = 1/\beta$, and the rapidity ψ by $\tanh\psi = \langle V \rangle = \beta\sqrt{1+\beta^2}$. We then insert (15.37) for the world line into $\mathcal{S}$. We easily find

$$kX' = -\frac{\beta^2\sin\xi}{\sqrt{1+\beta^2}},\quad kY' = -\frac{\beta\sin\xi}{\sqrt{1+\beta^2}},\quad kZ' = \beta(\cos\xi-1),\quad kT' = \sqrt{1+\beta^2}\,\xi.$$

In this frame the charge describes a circle of radius β/k in the $\bar{X}'OZ'$ plane, where the axis $O\bar{X}'$ makes an angle ϕ with the OX' axis such that $\tan\phi = 1/\beta$.

15.5   The geometrical optics limit

Just like the waves of the scalar field studied in Section 9.2 and the sound waves studied in Book 1, Section 17.3, the *monochromatic plane waves* of (15.7) or (15.8), which are solutions of the vacuum Maxwell equations for the potential, as well as the electric and magnetic fields (15.10) derived from them, are completely delocalized: at any given instant T they vary sinusoidally along lines determined by the wave vector k, and in each plane perpendicular to k they oscillate in the time T of the frame at frequency $\omega = |\mathbf{k}|$ (or *period* $P \equiv 2\pi/\omega$).

Let us consider the case where the direction $k^i = (k, 0, 0)$ is the X axis. Then $A^i_{\mathbf{k},\lambda}(X^\mu)$ is constant if its phase $k_\mu X^\mu = -k(T - X)$ is constant. Therefore, if the potential has a certain value at (X_0, T_0), it will have the same value at another point located a distance ΔX_0 away after a time lapse of $\Delta T_0 = \Delta X_0$. The potential $A^i_{\mathbf{k},\lambda}(X^\mu)$ thus describes a *wave* which propagates in the positive X direction at the speed of light. Its *wavelength* is $\lambda \equiv 2\pi/k$. As shown by (15.11) or (15.14) for the associated energy–momentum tensor, this wave transports energy along its direction of propagation.

The *wave fronts*, sets of points of equal phase at a given time, are planes orthogonal to the direction of propagation, and *light rays* are lines parallel to the direction of propagation.

A *plane wave* (15.23), the superposition of *monochromatic* plane waves propagating, for example, in the X direction [*cf.* (15.24)], is more or less localized on the planes $X = \pm T$ according to how spread out the functions $a_{\mathbf{k},\lambda}$ are about a value k_0 of k.

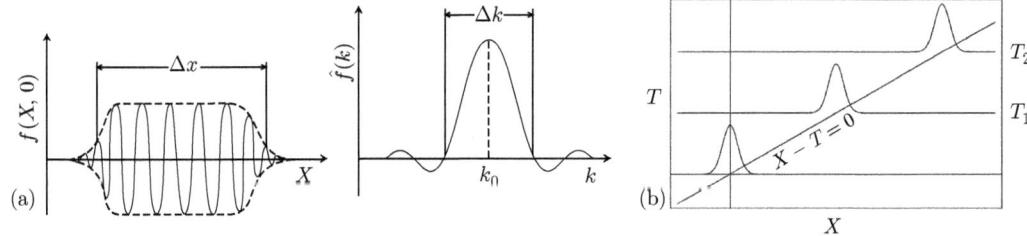

Fig. 15.3 Structure of an electromagnetic plane wave.

A Gaussian wave packet

Let us consider the case of a plane wave propagating along the X axis and polarized in the Y direction:

$$\bar{A}^i_\pm = \sum_\lambda \int \frac{dk}{\sqrt{2\pi}} (a_{k,\lambda}\epsilon^i_{k,\lambda}e^{-ik(T\mp X)} + \text{c.c.}),$$

with the amplitude distribution given by the Gaussian $\Sigma_\lambda a_{k,\lambda}\epsilon_{k,\lambda} = Ce_Y\, e^{-D^2(k-k_0)^2}$. Since the exponential differs significantly from 0 only if $D|k - k_0| < 1$, the superposition will involve a larger number of different wavelengths the smaller D is. The plane wave is also a Gaussian (see the discussion of the properties of Fourier transforms in Section 9.2), $\bar{A}_i = (0, \bar{A}_\pm, 0)$ with

$$\bar{A}_\pm = \frac{\sqrt{2}C}{D} e^{-\frac{(T \mp X)^2}{4D^2}} \cos[k_0(T \mp X)].$$

This wave, of frequency k_0, is the more localized on the line of zero length $T = \pm X$ the smaller D is.

Let us now consider the electromagnetic potential

$$A_i = \Phi_i(T, X^i) e^{i\phi(T, X^i)} \tag{15.38}$$

in a more general manner, assuming that the (real) phase ϕ varies much more rapidly than the amplitude Φ_i, that is, that

$$\partial_\mu \Phi_i \ll \Phi_i \partial_\mu \phi. \tag{15.39}$$

If we think of the wavelength as the typical distance over which the phase varies, this assumption implies that the amplitude and direction of propagation of the potential are nearly constant over several wavelengths.

In the Hamiltonian gauge where $A_0 = 0$, the magnetic potential (15.38) satisfies the vacuum Maxwell equations (namely, $\partial_i A^i = 0$ and $\Box A_i = 0$) if

$$\partial_i \Phi_i + i\Phi^i \partial_i \phi = 0, \quad \Box \Phi_i + 2i\, \partial^\mu \phi \partial_\mu \Phi_i + \phi_i \left[i\, \Box \phi - \partial_\mu \phi \partial^\mu \phi \right] = 0. \tag{15.40}$$

Under the assumption (15.39) these equations simplify to

$$\Phi^i \partial_i \phi = 0, \quad \Box \phi = 0, \quad \partial_\mu \phi\, \partial^\mu \phi = 0, \tag{15.41}$$

where the last equation is referred to as the *eikonal equation*.

Let us introduce the vector

$$k_\mu = \partial_\mu \phi. \tag{15.42}$$

The eikonal equation requires it to be null. Moreover, we have $0 = \partial_\nu(\partial_\mu \phi \partial^\mu \phi) = 2\partial^\mu \phi\, \partial_{\mu\nu} \phi$, so that $k^\mu \partial_\mu k_\nu = 0$, and so the curves with tangent vector k^μ are light lines. These are null lines of $\mathcal{M}_4$, and their projections on the planes $T = $ const are light rays. If the amplitude Φ_i, which is not constrained by (15.41), is localized, the light rays can reduce to a thin beam of straight lines.

Therefore, given the condition (15.39), that is, in the limit where the wavelength[3] $\lambda \to 0$, the laws of light propagation in the Maxwell theory reduce to those of *geometrical optics*, where electromagnetic waves are replaced by light rays, as we have done in[4] Chapters 3 and 4.

---

[3]This is often an excellent approximation for visible light, whose typical frequency is of order 10^{14} Hz and typical wavelength is 10^{-7} m.

[4]We recall, however, that a light ray is an abstract object. It can be represented approximately by a light beam, but it is impossible to make the beam infinitesimally thin because diffraction phenomena arise.

16
Waves in a medium

In the preceding chapters we have discussed Maxwell's equations in their 'fundamental' (or *microscopic*) form, where the electromagnetic field is evaluated in the vacuum, that is, *outside* the current and charge densities creating the field. To obtain a 'mean' or macroscopic description of electromagnetic phenomena *inside* matter, we need to use a phenomenological approach. This is the subject of the present chapter.

16.1   The Maxwell equations in a medium

The electromagnetic field inside a medium induces charge and current distributions called *polarization*, which are the response of the matter to the field. The charge and current densities can be decomposed into the sum of free densities (that is, imposed from outside the medium, and which create the field) and induced densities. We write $\rho = \rho_{\text{free}} + \rho_{\text{pol}}$, with the polarization charge density given by $\rho_{\text{pol}} = -\nabla.P$, where P is the *polarization*. Similarly, we write $j = j_{\text{free}} + j_{\text{pol}} + j_{\text{mag}}$, with the polarization current given by $j_{\text{pol}} = \partial P/\partial T$ (thus the continuity equation $\dot{\rho}_{\text{pol}} + \nabla.j_{\text{pol}} = 0$ is satisfied) and the magnetization current by $j_{\text{mag}} = \nabla \wedge M$, where M is the *magnetization*.

The second group of Maxwell equations (12.16), namely, $\nabla.E = 4\pi\rho$, $\nabla \wedge B = \partial E/\partial T + 4\pi j$, can then be rewritten in terms of the 3-vectors corresponding to the *electric displacement field D* and the *magnetic field strength H* defined as

$$D \equiv \frac{E}{4\pi} + P, \quad H \equiv \frac{B}{4\pi} - M, \tag{16.1}$$

where $\nabla.D = \rho_{\text{free}}$ and $\nabla \wedge H = \partial D/\partial T + j_{\text{free}}$. These equations, like the microscopic Maxwell equations, involve only free sources, but at the price of introducing the effective fields D and H which take into account the properties of the medium.

The determination of the vectors P and M as a function of the polarization currents (for which various microscopic models can be used) is a complicated problem, and D and H are *a priori* nonlocal and nonlinear functionals of the primary fields[1] E and B. However, in the weak-field limit the response of the medium can be assumed to be linear, and if in addition the medium is homogeneous and isotropic the relations $D = D[X^i, T, E, B]$ and $H = H[X^i, T, E, B]$ can be approximated by $D = \varepsilon E$ and $H = B/\mu$, where μ and ε are two constant parameters: the *susceptibility* and the *permittivity* characterizing the medium (in a vacuum $\epsilon = 1/4\pi$ and $\mu = 4\pi$ in our units). We also introduce the *impedance* of the medium $Z \equiv \sqrt{\mu/\varepsilon}$, which is 4π in the vacuum (or 376.7 Ω in S.I. units). With these approximations and definitions, the Maxwell equations (12.3) and (12.16) are replaced by

---

[1] We refer the reader to Jackson (1975) for examples.

Relativity in Modern Physics. Nathalie Deruelle and Jean-Philippe Uzan.
© Oxford University Press 2018. Published in 2018 by Oxford University Press.
DOI: 10.1093/oso/9780198786399.001.0001

$$\nabla.B = 0, \quad \nabla \wedge E = -\frac{\partial B}{\partial T}, \quad \nabla.D = \rho, \quad \nabla \wedge H = \frac{\partial D}{\partial T} + j, D = \varepsilon E, \quad H = \frac{B}{\mu}, \quad (16.2)$$

where ρ and j are the free charge and current densities. Only the second group of Maxwell equations has been modified, because only they involve the properties of the medium. The first group remains unchanged, and so the magnetic potential can be defined as for the vacuum. Equations (16.2) are often called the *macroscopic form of Maxwell's equations*. It is important to emphasize the fact that they have not been derived directly from their fundamental version, because the response of a medium to an electromagnetic excitation is described phenomenologically using the coefficients ε and μ. It should also be noted that they are only valid in the frame where the medium is at rest and they are not Lorentz invariant; they cannot be written in four-dimensional notation or derived from an action using the variational principle.

Combining the equations for $\nabla \wedge H$ and $\nabla \wedge E$, we find that

$$(E.\partial_T D + H.\partial_T B) + \nabla.(E \wedge H) = -j.E$$

(see Section 11.1 for a review of vector calculus). As long as the relations between D and E on the one hand and H and B on the other are linear and independent of time, we have $E.\partial_T D = \partial_T(E.D/2)$ and $H.\partial_T B = \partial_T(B.H/2)$, and so

$$\partial_T W + \nabla.S = -j.E, \quad \text{where} \quad W = \frac{1}{2}(E.D + H.B) \quad \text{and} \quad S = E \wedge H, \quad (16.3)$$

which (outside the free charges) can be interpreted as an energy conservation equation. In the vacuum case $D = E/4\pi$ and $H = B/4\pi$, and so we recover the expression obtained in Section 12.4, footnote 7. However, the energy density W and the Poynting vector S are no longer the components of an energy–momentum tensor.

Waves in a homogeneous medium

In a homogeneous and isotropic medium without free charges or currents, the macroscopic Maxwell equations imply that the electric and magnetic fields propagate as

$$\left(\frac{\partial^2}{\partial T^2} - \frac{1}{\mu\varepsilon}\triangle\right) E = 0. \quad (16.4)$$

This equation is similar to (12.17) obtained from the microscopic Maxwell equations and describes a wave propagating at speed $c_{EM} = 1/\sqrt{\mu\varepsilon} \equiv 1/n$, which defines the *index of refraction* n ($n = 1$ for the vacuum). The big difference from (12.17) is that the operator $\partial_T^2 - n^2\triangle$ does not reduce to the d'Alembertian, which again illustrates the fact that the relativistic invariance is broken. In the case of a monochromatic plane wave of wave vector k, (16.4) is equivalent to the *dispersion relation*

$$|k|^2 - n^2\omega^2 = 0, \quad \text{where} \quad n = \sqrt{\mu\varepsilon}. \quad (16.5)$$

In general, $\mu\varepsilon$ is not real and positive and so k is complex, which corresponds to dissipation of the wave in the medium in which it is propagating (when n depends on the wavelength the medium is termed *dispersive*). When $\mu\varepsilon$ is real and positive, the dispersion relation implies that $|k| = \sqrt{\mu\varepsilon}\,\omega = n\omega$. The phase velocity then is $c_\phi \equiv \omega/|k| = 1/n$, and the group velocity is $c_g \equiv d\omega/d|k| = c_\phi$, so that $c_\phi c_g = 1/n^2$.

Book 2

Waves in a conducting medium

In a *conductor* the susceptibility and permittivity are the same as in the vacuum, but the presence of an electric field E induces a current according to *Ohm's law*: $j = \sigma E$. This law can be viewed as the first term of a Taylor series expansion of a general law of the form $j = j(E)$ which depends on the structure of the conductor and reflects its microscopic physics. The *conductivity* σ is in general a position-dependent tensor. It reduces to a number for a homogeneous and isotropic medium.

Maxwell's equations (16.2) with $D = E/4\pi$ and $H = B/4\pi$ imply that [cf. (12.17), where we have used $\nabla \wedge j = \sigma \nabla \wedge E = -\sigma\, \partial B/\partial T$]

$$\triangle E - \frac{\partial^2 E}{\partial T^2} = 4\pi\sigma \frac{\partial E}{\partial T}, \quad \triangle B - \frac{\partial^2 B}{\partial T^2} = 4\pi\sigma \frac{\partial B}{\partial T}. \tag{16.6}$$

For a monochromatic plane wave propagating in the direction e_X we have $E \propto e^{\mathrm{i}(\omega T - kX)}$, and (16.6) reduces to the dispersion relation $k^2 - \omega^2 + 4\pi\sigma \mathrm{i}\omega = 0$ and k is complex. Setting $k = k_{\mathrm{R}} + \mathrm{i}k_{\mathrm{I}}$, we have $k_{\mathrm{R}}^2 - k_{\mathrm{I}}^2 = \omega^2$ and $k_{\mathrm{R}}k_{\mathrm{I}} = -2\pi\sigma\omega$. The fields therefore evolve as $E \propto e^{k_{\mathrm{I}}X} e^{\mathrm{i}(k_{\mathrm{R}}X - \omega T)}$ with

$$k_{\mathrm{I}} = -\omega \left[\frac{\sqrt{1 + \left(\frac{4\pi\sigma}{\omega}\right)^2} - 1}{2} \right]^{\frac{1}{2}}, \quad k_{\mathrm{R}} = +\omega \left[\frac{\sqrt{1 + \left(\frac{4\pi\sigma}{\omega}\right)^2} + 1}{2} \right]^{\frac{1}{2}}. \tag{16.7}$$

The amplitude falls off exponentially with X, which defines the *skin depth* of the conductor: $1/k_{\mathrm{I}} \sim (2\pi\sigma)^{-1}$ in the limit $\sigma \ll \omega$. A wave of this type is called an *evanescent wave*.

16.2   Matching conditions and the Snell–Descartes laws

The matching conditions on the electromagnetic field at the interface between two different media (for example, a pair of lenses) can be obtained from the macroscopic Maxwell equations (16.2) with the use of the Gauss and Stokes theorems (see, for example, Book 1, Section 2.6).

Using $\nabla.D = \rho$ and considering the elementary volume shown in Fig. 16.1 below, we find

$$\int_V \rho\, d^3X = \int_V \nabla.D\, d^3X = \int_{\partial V} D.dS = [D.n]_\pm |dS_1| + \int_{S_2} D.dS_2,$$

where we have used the notation $[X]_\pm = X_+ - X_-$. The contribution of the second integral tends to 0 when the thickness of V tends to 0. We then have $[D.n]_\pm = \sigma$, where σ is the surface density of free charge at the interface. In the same way, $\nabla.B = 0$ gives $[B.n]_\pm = 0$. Moreover, (16.2) for H implies, for the path shown in Fig. 16.1, that

$$\int_C H.d\ell = \int_\Sigma (\nabla \wedge H).dS = \frac{d}{dT}\int_\Sigma D.dS + \int_\Sigma j.dS.$$

Now letting ϵ go to 0, we see that the left-hand side gives $[H \wedge n]_\pm \delta\ell$, while the right-hand side reduces to $j_{\mathrm{s}}\delta\ell$, where j_{s} is the surface current. We thus obtain the constraint $[H \wedge n]_\pm = j_{\mathrm{s}}$ on the components perpendicular to n. Similarly, we find $[E \wedge n]_\pm = 0$ by using (16.2) for E.

Therefore, at the interface between two media the normal components of B and D on the one hand and the tangential components of E and H on the other are continuous as long as there is no surface current or charge at the interface:

$$[D.n]_\pm = [B.n]_\pm = 0, \quad [E \wedge n]_\pm = [H \wedge n]_\pm = 0. \tag{16.8}$$

These matching conditions apply to static fields as well as to waves (see Fig. 16.1).

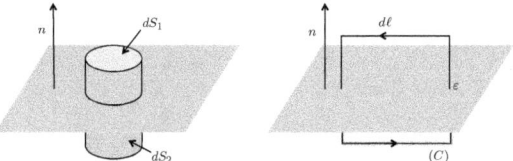

Fig. 16.1 Matching conditions.

Now let us consider the particular case of a monochromatic plane wave propagating in a medium of index of refraction n_1 which is separated from a medium with index n_2 by an infinitesimally thin, planar, and infinite interface (located at $Z = 0$). This incident wave of wave vector k_I, $E_I = E_{I0} e^{i(k_I.R - \omega_I T)}$, where R is the radius vector with components X^i, in general gives rise to a reflected wave $[E_R = E_{R0} e^{i(k_R.R - \omega_R T)}]$ and a transmitted wave $[E_T = E_{T0} e^{i(k_T.R - \omega_T T)}]$; see Fig. 16.2.

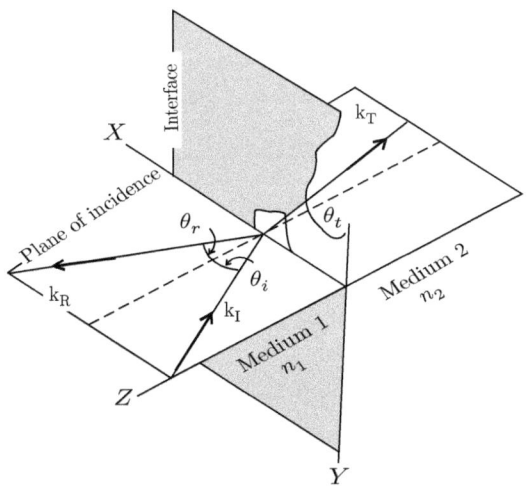

Fig. 16.2 The Snell–Descartes laws.

Since the tangential component of $E_I + E_R$ must be equal to that of E_T for all times, we have $\omega_I = \omega_T = \omega_R \equiv \omega$. The dispersion equation in each medium then gives $|k_R|^2 = \omega^2 n_1^2 = |k_I|^2$ and $|k_T|^2 = \omega^2 n_2^2 = |k_I|^2 (n_2/n_1)^2$. On the other hand, the continuity of the tangential component of E at every point on the interface implies that $k_I.R|_{Z=0} = k_T.R|_{Z=0} = k_R.R|_{Z=0}$. The three wave vectors are therefore coplanar, and the plane containing them is called the *plane of incidence* (it is well defined as long as the incidence is not normal). Using the notation of Fig. 16.2, the relation between the wave vectors can be written as $|k_I| \sin \theta_i = |k_R| \sin \theta_r = |k_T| \sin \theta_t$. We then deduce that

$$\theta_r = \theta_i , \quad \frac{\sin \theta_i}{\sin \theta_t} = \frac{n_2}{n_1}. \tag{16.9}$$

These are respectively called *Snell's law* and *Descartes's law*. They do not depend either on the exact nature of the interface, or on properties of the wave such as its polarization.

The Fresnel laws

The matching conditions (16.8) imply that the tangential components of E and H are continuous, that is, $(E_{I0} - E_{R0}) \wedge e = E_{T0} \wedge e$ and (because $B = k \wedge E / |k|$ and $H = B / \mu$) $(k_I \wedge E_{I0} + k_R \wedge E_{R0}) \wedge e / \mu_1 = (k_T \wedge E_{T0}) \wedge e / \mu_2$, where e is the vector normal to the interface.

When the wave is polarized normal to the plane of incidence, these two relations give (see Fig. 16.3) $E_{I0} + E_{R0} - E_{T0} = 0$ and $\frac{E_{I0} - E_{R0}}{Z_1} \cos \theta_i = \frac{E_{T0}}{Z_2} \cos \theta_t$, where we recall that the impedances are defined as $Z \equiv \sqrt{\mu / \varepsilon}$ and the indices of refraction as $n = \sqrt{\varepsilon \mu}$. From this we derive the two *Fresnel equations* relating the amplitudes of the reflected and transmitted waves to that of the incident wave:

$$\left(\frac{E_{R0}}{E_{I0}} \right)_{\perp} = \frac{Z_2 \cos \theta_i - Z_1 \cos \theta_t}{Z_2 \cos \theta_i + Z_1 \cos \theta_t}, \quad \left(\frac{E_{T0}}{E_{I0}} \right)_{\perp} = \frac{2 Z_2 \cos \theta_i}{Z_2 \cos \theta_i + Z_1 \cos \theta_t}, \tag{16.10}$$

where the subscript $\perp$ indicates that E is perpendicular to the plane of incidence. The transmitted wave is therefore always in phase with the incident wave. This is not the case with the reflected wave. If we limit ourselves to the case where $\mu_1 = \mu_2$, for $n_1 < n_2$ we have $\theta_t < \theta_i$ and $\cos \theta_i < \cos \theta_t$, and the reflected wave is out of phase by π. If $n_1 > n_2$ we have $\theta_t > \theta_i$ and $\cos \theta_i > \cos \theta_t$, so that the reflected wave is in phase with the incident wave.

When the wave is polarized parallel to the plane of incidence, we similarly find (see Fig. 16.3) $(E_{I0} - E_{R0}) \cos \theta_i - E_{T0} \cos \theta_t = 0$ and $\frac{E_{I0} + E_{R0}}{Z_1} = \frac{E_{T0}}{Z_2}$. The Fresnel equations then are

$$\left(\frac{E_{R0}}{E_{I0}} \right)_{\parallel} = \frac{Z_1 \cos \theta_i - Z_2 \cos \theta_t}{Z_2 \cos \theta_t + Z_1 \cos \theta_i}, \quad \left(\frac{E_{T0}}{E_{I0}} \right)_{\parallel} = \frac{2 Z_2 \cos \theta_i}{Z_2 \cos \theta_t + Z_1 \cos \theta_i}. \tag{16.11}$$

Again, the transmitted wave is in phase with the incident wave. The reflected wave is in phase if $n_1 \cos \theta_t < n_2 \cos \theta_i$, which can be rewritten using the Snell–Descartes laws as $\sin(\theta_t - \theta_i) \cos(\theta_t + \theta_i) > 0$. This condition depends not only on the ratio n_1 / n_2, as for a wave polarized perpendicularly to the plane of incidence, but also on θ_i, and so the reflected wave can be out of phase by π for any value of n_2 / n_1. Brewster's angle, the angle of incidence for which a wave passes through the interface between two different media without any reflection, is obtained for $\theta_i + \theta_t = \pi/2$ so that[2]

$$\tan \theta_{i_B} = \frac{n_2}{n_1}. \tag{16.12}$$

The case of an elliptically polarized wave is obtained by the superposition of these two cases. For an unpolarized wave incident at Brewster's angle, the reflected wave will be polarized perpendicularly to the plane of incidence, which can be used to produce a linearly polarized wave.

---

[2] For an air–glass interface ($n_1 \sim 1$, $n_2 \sim 1.5$) Brewster's angle is $\theta_{i_B} = 56.3$ deg. Brewster's angle can be used to determine the index of refraction of a medium by measuring the angle.

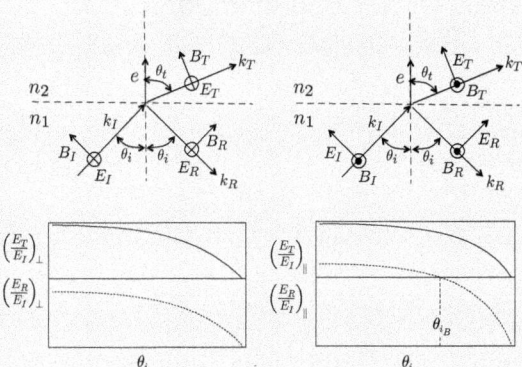

Fig. 16.3 Fresnel's laws.

Reflection and transmission coefficients

The energy flux of an electromagnetic field per unit surface area is given by the time average of the Poynting vector $S = E \wedge H$; cf. (16.3). For a plane wave $H = B/\mu = k/|k| \wedge E/\mu$, and so $\bar{S} = (|E_0|^2/(2\mu))k/|k|$, where E_0 is the field amplitude and $k/|k|$ is the direction of propagation of the wave. The *reflection* and *transmission coefficients* characterize the energy flow (the intensity) across the interface and are defined as

$$R \equiv \left| \frac{\bar{S}_R.e}{\bar{S}_e.e} \right| = \left(\frac{E_{R0}}{E_{I0}} \right)^2, \quad T \equiv \left| \frac{\bar{S}_T.e}{\bar{S}_e.e} \right| = \left(\frac{E_{T0}}{E_{I0}} \right)^2 \frac{Z_1 \cos \theta_t}{Z_2 \cos \theta_i}, \tag{16.13}$$

where e is the normal to the plane. [We recall that the impedances are defined as $Z \equiv \sqrt{\mu/\varepsilon}$ and the indices of refraction as $n = \sqrt{\varepsilon\mu}$.] We have $R + T = 1$ [using (16.10) or (16.11)], which is equivalent to the law of energy conservation (16.3). Fresnel's laws (16.10) and (16.11) give $R_\perp$, $T_\perp$ and $R_\parallel$, $T_\parallel$ explicitly for incident waves whose polarization is perpendicular or parallel to the plane of incidence.

When $n_1 > n_2$ the Snell–Descartes laws give $\theta_r > \theta_i$, and so $\theta_r = \pi/2$ is reached for $\sin \theta_{i_0} = n_2/n_1$. For this angle of incidence the transmitted wave propagates parallel to the interface and no energy crosses the interface. Therefore, θ_{i_0} is the angle of total reflection.

For $\theta_i > \theta_{i_0}$ we have $\sin \theta_r > 1$, and so θ_r is complex and $\cos \theta_r = i\sqrt{(\sin \theta_i/\sin \theta_{i_0}) - 1}$. The transmitted wave then is

$$\exp(ik_T.X) = \exp i|k_T| (X \sin \theta_r + Z \cos \theta_r)$$

$$= \exp[-|k_T|\sqrt{(\sin \theta_i/\sin \theta_{i_0}) - 1}Z] \, \exp[i|k_T|(\sin \theta_i/\sin \theta_{i_0})X].$$

The refracted wave thus is exponentially attenuated perpendicularly to the plane of the surface and propagates parallel to it. Therefore, even if the field is nonzero in region 2, there is no energy transfer across the interface.

16.3   The geometrical optics approximation

As in Section 15.5, we decompose the electric and magnetic fields as

$$E = E_0(X^i, T)e^{i\phi(X^i,T)}, \quad B = B_0(X^i, T)e^{i\phi(X^i,T)} \tag{16.14}$$

and define the wave 3-vector $k \equiv \nabla\phi$ and the frequency[3] $\omega \equiv -\partial_T\phi$.

Now let us assume that the wavelength, defined as $2\pi/|k| = \mathcal{O}(2\pi/\omega)$, is small compared to the characteristic scales on which the amplitudes E_0 and B_0, the properties of the medium (ε and μ), and k^i and ω vary, so that the fields are well approximated over several wavelengths by a plane wave propagating in a homogeneous medium. Substituting the decomposition (16.14) into the Maxwell equations (16.2) and neglecting the spatial derivatives of the amplitudes and of the characteristics of the medium, we find that $k_i E_0^i = k_i B_0^i = 0$, $k \wedge E_0 = -\omega B_0$, and $k \wedge B_0 = n^2 \omega E_0$. Combining the last two expressions [and using the relation $k \wedge (k \wedge E_0) = -k^2 E_0$ because $k.E_0 = 0$], we obtain the dispersion relation $k^2 - n^2\omega^2 = 0$, or, in terms of the phase ϕ,

$$(\nabla\phi)^2 - n^2(\partial_T\phi)^2 = 0. \tag{16.15}$$

This eikonal equation resembles the one derived for the vacuum case in Section 15.5, but now it is valid in a medium with $n \neq 1$, and the Lorentz invariance is broken [that is, it is no longer possible to write this equation in the form $(\partial_\mu\phi)^2 = 0$].

A wave which is initially monochromatic will remain so as long as the properties of the medium do not depend on time; see (16.2). In this case, which corresponds to most applications, the phase can be written as $\phi(X^i, T) = \omega[\mathcal{S}(X^i) - T]$ and the eikonal equation takes the form

$$(\nabla\mathcal{S})^2 = n^2 \quad \Longleftrightarrow \quad \nabla\mathcal{S} = ne_s, \tag{16.16}$$

where e_s is a 3-vector of unit norm. The surfaces $\mathcal{S} = $ const are the wave fronts, and the vector normal to these surfaces is $\nabla\mathcal{S}$. Therefore, in the limit of small wavelengths, the electromagnetic wave propagation law reduces to the geometric law (16.16) describing the evolution of the wave vector $k \equiv \omega\nabla\mathcal{S}$, which is no longer constant on the characteristic scale of spatial variation of the index n. See Fig. 16.4.

The Poynting vector (16.3) is $S = E \wedge H = (E \wedge B)/\mu$, and averaged over several periods it becomes $\bar{S} = (E_0 \wedge B_0)/2\mu$. Writing B_0 using the Maxwell equations (which in our approximations and notation become $\nabla\mathcal{S} \wedge E_0 = -B_0$ and $\nabla\mathcal{S} \wedge B_0 = n^2 E_0$), we find that $\bar{S} = (E_0^2/2\mu)\nabla\mathcal{S}$. The average of the energy (16.3) is given by $\bar{W} = \varepsilon E_0^2/4 + B_0^2/4\mu$. Maxwell's equations imply that the two terms of the sum are equal, and so $\bar{W} = \varepsilon E_0^2/2 = n^2 E_0^2/2\mu$. We then have

$$\bar{S} = \frac{\bar{W}}{n^2}\nabla\mathcal{S} = v_\phi\bar{W}e_s, \tag{16.17}$$

---

[3]For a monochromatic plane wave $\phi = k_\mu X^\mu = k_i X^i - \omega T$, and so $\partial_i\phi = k_i$ and $-\partial_T\phi = \omega$ are constants.

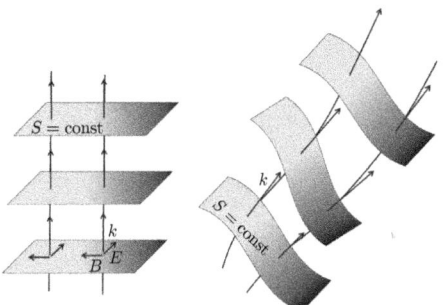

Fig. 16.4 Wave fronts.

using (16.16) and recalling that the phase velocity is $v_\phi = 1/n$. The direction of the energy propagation is collinear with the phase gradient, that is, collinear with the *local* wave 3-vector and therefore normal to the wave fronts.

In the eikonal approximation the amplitudes B_0 and E_0 are not constrained. We can therefore assign to them a profile such that the solution of the field equations in this approximation will be a light ray propagating along a curve tangent at each point to the vector e_s. These curves are the light rays with equation $X^i = X^i_{\text{ray}}(s)$ satisfying

$$\frac{dX_{\text{ray}}}{ds} = e_s[X_{\text{ray}}(s)] \tag{16.18}$$

if the arc length s is chosen to be the curvilinear abscissa along the ray. Since $ne_s = \nabla\mathcal{S}$, the identity $d(ne_s)/ds \equiv e_s.\nabla(ne_s)$ becomes

$$d(ne_s)/ds = e_s.\nabla(\nabla\mathcal{S}) = (n^{-1}\nabla\mathcal{S}.\nabla)(\nabla\mathcal{S}) = [\nabla(\nabla\mathcal{S})^2]/2n = [\nabla n^2]/2n$$

using (16.16). Therefore, (16.18) now is

$$\frac{d}{ds}\left[n(X_{\text{ray}})\frac{dX_{\text{ray}}}{ds}\right] = \nabla n. \tag{16.19}$$

The wave propagation problem thus reduces to a purely geometrical problem of determining integral curves. Expanding (16.19), we obtain $de_s/ds = [\nabla n - (e_s.\nabla n)e_s]/n$. Since $de_s/ds = N/R_c$, where R_c is the radius of curvature and N is the unit vector along the principal normal, we find that $R_c^{-1} = N.\nabla \ln n$, which tells us that light rays bend in the direction of increasing index of refraction.

16.4 Fermat's principle

In the preceding section we saw that if we choose the electric and magnetic fields to have the form $E = E_0(X^i, T)e^{i\omega(\mathcal{S}(X^i)-T)}$ and $B = B_0(X^i, T)e^{i\omega(\mathcal{S}(X^i)-T)}$, then the Maxwell equations (16.2) tell us that $\nabla\mathcal{S} = ne_s$, where e_s is a unit 3-vector, with the condition that the amplitudes E_0 and B_0 and the index $n(X^i)$ of the medium remain practically constant over several wavelengths λ, where $\lambda \equiv 2\pi/\omega$.

From the relation $ne_s = \nabla \mathcal{S}$ we find that $\nabla \wedge (ne_s) = 0$. Integrating this equation over an arbitrary surface and then using the Stokes theorem, we find that the integral of ne_S on any closed contour vanishes. Therefore, the integral

$$\int_{\gamma, P_1}^{P_2} ne_s.dl_\gamma \tag{16.20}$$

along a path γ connecting the points P_1 and P_2 has a value which is independent of the path, where dl_γ is the length element along this path. The integral (16.20) is given by $\mathcal{S}(P_2) - \mathcal{S}(P_1)$ and is therefore proportional to the phase difference between the two points.

The inequality $e_s.dl_\gamma \leq |e_s| \wedge |dl_\gamma| = |dl_\gamma| \equiv d\ell_\gamma$ implies that

$$\int_{\gamma P_1}^{P_2} ne_s.dl_\gamma \leq \int_{\gamma P_1}^{P_2} nd\ell_\gamma, \tag{16.21}$$

where the equality is attained only if the path γ corresponds to a light ray, because then e_s is collinear with dl_γ according to (16.18).

This is *Fermat's principle*, which states that among all the paths γ joining P_1 and P_2, the light will travel along the path which minimizes the *optical path*[4] $\int_{P_1}^{P_2} nds$. This principle can be restated as follows: *among all the paths joining two points, light will travel along the path which minimizes the travel time*. In the vacuum $n = 1$, and we find that light rays are straight lines. For a light ray the phase difference between two points is therefore proportional to the optical trajectory $\phi_2 - \phi_1 = \omega(\mathcal{S}_2 - \mathcal{S}_1) = \omega \int_{P_1}^{P_2} nds$. The ray equation can be obtained independently of the chosen orientation, so that a ray traveling from P_1 to P_2 is identical to one traveling from P_2 to P_1.

It can be verified that the equations of light rays can be obtained using the Euler–Lagrange equations applied to the Lagrangian $\int n(X^i)ds = \int n(X^i)|V|dT \equiv \int \mathcal{L}\, dt$ with $V^i = dX^i/dT$. We have $\partial\mathcal{L}/\partial X = |V|\nabla n$ and $\partial\mathcal{L}/\partial V = nV/|V| = ndX/ds$, and so the Euler–Lagrange equations are $|V|\nabla n = \frac{d}{dT}(ndX/ds)$, or $\nabla n = \frac{d}{ds}(ndX/ds)$, which is just (16.19).

The Snell–Descartes laws revisited

As an application of Fermat's principle, let us consider a light ray arriving at an interface located at $Z = 0$ between two homogeneous media of indices n_1 and n_2 (e_s is therefore constant in each zone) at angle of incidence θ_1. If we take a rectangular path lying in the plane of incidence, the XZ plane, which is infinitesimally thin in the X direction, we obtain $0 = \int_\gamma ne_s.dl_\gamma = n_1e_s.(-e_X) + n_2e_s.e_X = -n_1\sin\theta_1 + n_2\sin\theta_2$. We therefore again arrive at the Snell–Descartes laws (16.9). However, geometrical optics does not tell us that there is a reflected wave, and it cannot predict the relative intensities of the transmitted and incident waves; see the Fresnel laws (16.16) and (16.17).

---

[4]The *optical path* is defined as the distance light travels in the vacuum during the time needed to travel the distance between P_1 and P_2 in a medium of index of refraction n. Given that $s_{P_1 P_2} = \int_{P_1}^{P_2} ds$ and that the propagation speed is $1/n$, we find that the travel time, and therefore the optical path, is given by $\ell_{P_1 P_2} = \int_{P_1}^{P_2} nds$. Here $\ell_{P_1 P_2}$ is the distance between P_1 and P_2 in a space where the distance between two neighboring points is given by $d\ell^2 = n^2\delta_{ij}dX^i dX^j$ instead of by the Euclidean length element $d\ell^2 = \delta_{ij}dX^i dX^j$. The light ray is a curve of minimum length in such a space, that is, a *geodesic*.

16.5 The various descriptions of light

In the preceding chapters we have described light in several ways: (1) as an electromagnetic wave, (2) as an ensemble of particles with unit velocity, and (3) as a light ray. It is useful to summarize how these descriptions are related to each other, as well as their respective domains of validity (see Fig. 16.5).

We have seen that the Maxwell equations of electromagnetism possess solutions corresponding to waves propagating at the speed of light. As Maxwell concluded, light is an electromagnetic wave. This wave description is verified experimentally by interference and diffraction phenomena,[5] which arise at scales of the order of the wavelength $L \sim \lambda$. The geometrical optics limit takes us from these waves to the concept of light rays which are integral curves of the field of vectors normal to the wave fronts. This description is valid in the limit $\lambda \ll L$.

Special relativity introduces a fundamental velocity c. By identifying c as the speed of light, we then describe light by the world lines of zero length of a particle which has zero mass. The projections of these light lines on a surface $T = \text{const}$ are the light rays of the eikonal approach. This particle, which we have referred to as a 'light corpuscle', is no longer hypothetical once it is identified with the *photon* (or an ensemble of photons), as is done in interpreting, for example, the photoelectric effect and Compton scattering. This corpuscular

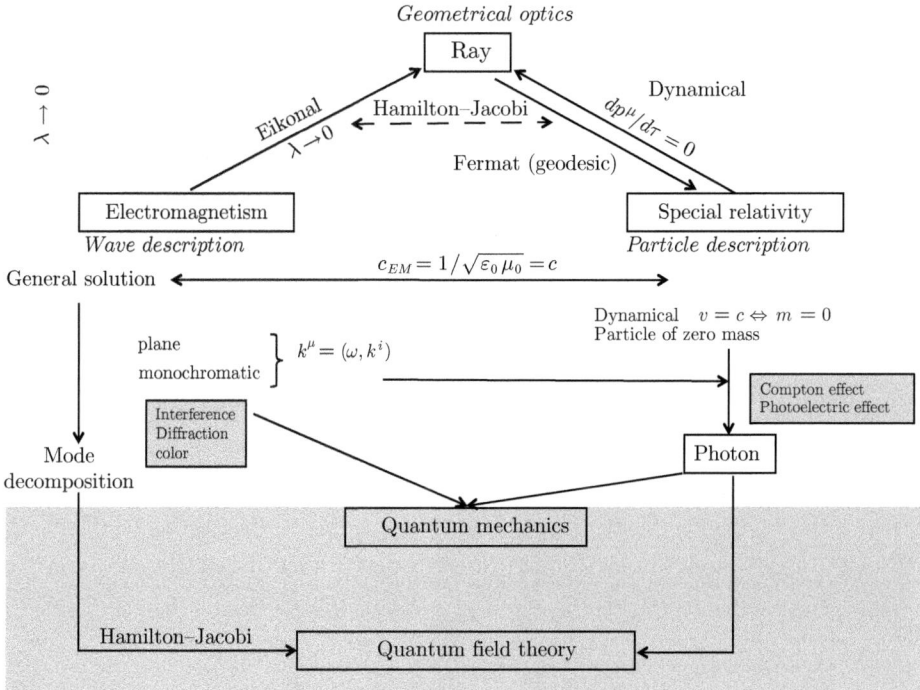

Fig. 16.5 The various descriptions of light.

---

[5]We do not discuss these in this book. See, for example, Landau and Lifshitz (1980); see also Book 1, Section 17.4.

description of light is needed for systems whose characteristic size is of the order of the atomic radius, $L \sim a_{\mathrm{Bohr}}$.

These descriptions are related to each other. A light ray is both the trajectory (projected on a surface $T = \mathrm{const}$) of a photon, and the characteristic lines of the electromagnetic wave obtained using the eikonal equation. Moreover, as we have seen, the wave 4-vector k^{μ} and the photon 4-momentum must be proportional, with the constant of proportionality being Planck's constant. However, we stress the fact that a plane wave is not a photon because it describes a continuous energy flux, and also that a wave packet contains a large number of photons.[6]

At scales such that $L \sim \lambda \sim a_{\mathrm{Bohr}}$, the particle and wave descriptions are both necessary. Light must therefore be described using quantum electrodynamics, which associates a wave function with it. The description of a free field based on the Hamilton equations discussed in Section 14.3 is the starting point for the quantization of the electromagnetic field.

---

[6]For example, for a laser producing a wave of frequency 1.8×10^{14} rad/s such that $E_0 \sim 1.5$ MV/m, the energy flux is $I = c\varepsilon_0^2 E_0^2 \sim 10^{-2}$ W/m^2. Since each photon has an energy $\hbar\omega$, we find that the photon flux is of order 5×10^{16} s^{-1} m^{-2}.

A light signal emitted by such a laser in a given direction can be considered equivalent to a light corpuscle; for example, for a cross sectional area of 10^{-6} m^2 and a duration of 10^{-6} s, this 'corpuscle' will contain 10^4 photons.

Part IV

Electrodynamics

One cannot escape the feeling that these mathematical formulae have an independent existence and an intelligence of their own, that they are wiser than we are, wiser even than their discoverers, that we get more out of them than was originally put into them.

Heinrich Rudolf Hertz, quoted by E. T. Bell in *Men of Mathematics*, 1937

17
The field of a moving charge

We begin our study of electromagnetic radiation by calculating the field created by a single moving charge, first when in uniform motion, and then when it is accelerated. After studying some special cases, we define the asymptotic, *radiation*, part of the field, which will permit us in the following chapters to calculate the energy radiated by a system of charges.

17.1   The field of a charge in uniform motion

In Section 13.1 we saw that the Coulomb potential created by a point charge or a spherically symmetric charge q which is at rest at the origin of an inertial frame $\mathcal{S}$ is written as $A^\mu = (q/r, 0)$, where r is the spatial distance from the charge to a point $X^\mu = (T, X^i)$. This is the Lorenz gauge because $\partial_\mu A^\mu = 0$.

We note that r is also equal to the time taken by light to travel from the charge to the point, and that $r = T - T_\mathcal{R}$, where $X^\mu_\mathcal{R} = (T_\mathcal{R}, 0)$ are the coordinates of the *retarded point*, the intersection of the past light cone with apex X^μ and the world line of the charge; see Fig. 17.1a.

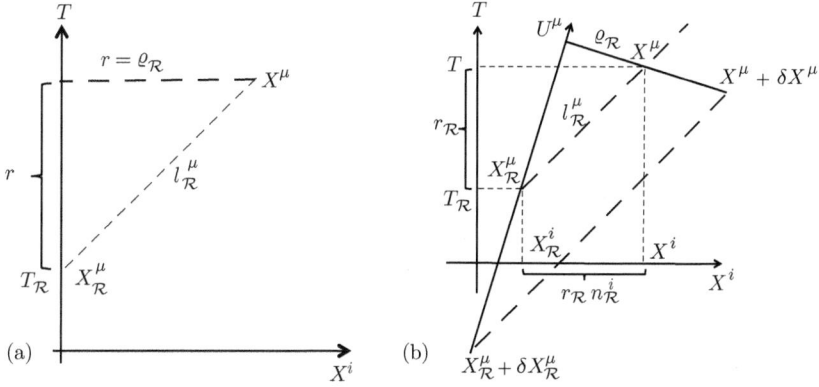

Fig. 17.1 Retarded quantities.

The scalar $\varrho_\mathcal{R} \equiv -U_\mu(X^\mu - X^\mu_\mathcal{R})$ is just r because the charge velocity is $U^\mu = (1,0)$. The potential can then be rewritten as

$$A^\mu = \frac{q\, U^\mu}{\varrho_\mathcal{R}}, \quad \text{where} \quad \varrho_\mathcal{R} \equiv -U_\mu\, l^\mu_\mathcal{R} \quad \text{with} \quad l^\mu_\mathcal{R} \equiv X^\mu - X^\mu_\mathcal{R} \quad \text{and} \quad \eta_{\mu\nu} l^\mu_\mathcal{R} l^\nu_\mathcal{R} = 0\,. \quad (17.1)$$

Relativity in Modern Physics. Nathalie Deruelle and Jean-Philippe Uzan.
© Oxford University Press 2018. Published in 2018 by Oxford University Press.
DOI: 10.1093/oso/9780198786399.001.0001

The advantage of this rewriting is that since it is covariant, the expression holds in *any* inertial frame (where $\varrho_{\mathcal{R}}$ no longer reduces to the spatial distance between the charge and the point; see Fig. 17.1b), and thus gives the potential of a charge undergoing any uniform rectilinear motion. The potential written in the form (17.1) is referred to as the *retarded Liénard–Wiechert potential*.[1]

Now, to find the electromagnetic field it is necessary to calculate the derivative of the Liénard–Wiechert potential (17.1), that is, determine how far the retarded point $X_{\mathcal{R}}^{\mu}$ moves on the world line of a charge when the field point X^{μ} is moved a distance δX^{μ}.

Derivatives of retarded quantities

Since the point $X_{\mathcal{R}}^{\mu} \equiv X^{\mu}(\tau_{\mathcal{R}})$ is the intersection of the past cone with apex X^{μ} and the charge world line (see Fig. 17.1b), we have $\eta_{\mu\nu}(X^{\mu} - X_{\mathcal{R}}^{\mu})(X^{\nu} - X_{\mathcal{R}}^{\nu}) = 0$. From this we see that when the field point is displaced by δX^{μ} we have $\eta_{\mu\nu}(X^{\mu} - X_{\mathcal{R}}^{\mu})(\delta X^{\nu} - \delta X_{\mathcal{R}}^{\nu}) = 0$. Since $\delta X_{\mathcal{R}}^{\nu} = U_{\mathcal{R}}^{\nu}\delta\tau_{\mathcal{R}} = U_{\mathcal{R}}^{\nu}\partial_{\rho}\tau_{\mathcal{R}}\,\delta X^{\rho}$, where $U_{\mathcal{R}}^{\mu}$ is the speed of the charge at the retarded time and $\tau_{\mathcal{R}}$ is its proper time, we have $\eta_{\mu\rho}(X^{\mu} - X_{\mathcal{R}}^{\mu})\delta X^{\rho} - U_{\mu\mathcal{R}}(X^{\mu} - X_{\mathcal{R}}^{\mu})\partial_{\rho}\tau_{\mathcal{R}}\,\delta X^{\rho} = 0$, or

$$\partial_{\rho}\tau_{\mathcal{R}} = -\frac{l_{\rho}}{\varrho}\bigg|_{\mathcal{R}} \quad \text{and} \quad \frac{\partial X_{\mathcal{R}}^{\nu}}{\partial X^{\mu}} = -\frac{U^{\nu}l_{\mu}}{\varrho}\bigg|_{\mathcal{R}} \quad \text{with} \quad \varrho_{\mathcal{R}} \equiv -U_{\mu\mathcal{R}}l_{\mathcal{R}}^{\mu} \quad \text{and} \quad l_{\mathcal{R}}^{\mu} \equiv X^{\mu} - X_{\mathcal{R}}^{\mu}. \tag{17.2}$$

These expressions are valid even when the velocity U^{μ} is not constant.

When the velocity U^{μ} is constant we find $\partial_{\mu}\varrho_{\mathcal{R}} = -U_{\mu} + l_{\mu\mathcal{R}}/\varrho_{\mathcal{R}}$ [see (17.2)], from which we obtain the derivatives of the potential $\partial_{\mu}A_{\nu}$ and the Faraday tensor $F_{\mu\nu} \equiv \partial_{\mu}A_{\nu} - \partial_{\nu}A_{\mu}$:

$$\partial_{\mu}A_{\nu} = \frac{qU_{\nu}}{\varrho_{\mathcal{R}}^{2}}\left(U_{\mu} - \frac{l_{\mu\mathcal{R}}}{\varrho_{\mathcal{R}}}\right), \quad F_{\mu\nu} = \frac{q(U_{\mu}l_{\nu\mathcal{R}} - U_{\nu}l_{\mu\mathcal{R}})}{\varrho_{\mathcal{R}}^{3}}. \tag{17.3}$$

It can be checked that $\partial_{\mu}A^{\mu} = 0$, and that the Faraday tensor can be written as an exterior product, $F = (q/\varrho_{\mathcal{R}}^{3})\,U \wedge l_{\mathcal{R}}$; see Section 22.1 *et seq*.

Now we need to express the retarded quantities as functions of the coordinates of the point X^{μ}. Since the charge is in uniform motion, the equation for its world line is $X^{\mu} = U^{\mu}\tau$ for a suitable choice of the time origin. The retarded point $X_{\mathcal{R}}^{\mu} = U^{\mu}\tau_{\mathcal{R}}$ is the intersection of the charge world line and the retarded cone with apex X^{μ}: $\eta_{\mu\nu}(X^{\mu} - U^{\mu}\tau_{\mathcal{R}})(X^{\nu} - U^{\nu}\tau_{\mathcal{R}}) = 0$, which gives $\tau_{\mathcal{R}} = -(X \cdot U) - \sqrt{(X \cdot U)^{2} + X^{2}}$, so that $\varrho_{\mathcal{R}} = \sqrt{(X \cdot U)^{2} + X^{2}}$. Therefore,

$$A^{\mu} = \frac{qU^{\mu}}{\sqrt{(X \cdot U)^{2} + X^{2}}}, \quad F_{\mu\nu} = \frac{q\,(U_{\mu}X_{\nu} - U_{\nu}X_{\mu})}{[(X \cdot U)^{2} + X^{2}]^{\frac{3}{2}}}, \tag{17.4}$$

---

[1]Of course, we could have equally well introduced the *advanced point* $X_{\mathcal{A}}^{\mu}$, the intersection of the charge world line and the future light cone with apex X^{μ}. In that case the potential would take the form of an *advanced potential* $A^{\mu} = -qU^{\mu}/\varrho_{\mathcal{A}}$, with $\varrho_{\mathcal{A}} \equiv -U_{\mu}(X^{\mu} - X_{\mathcal{A}}^{\mu})$ (we have $\varrho_{\mathcal{A}} = -\varrho_{\mathcal{R}}$ if U^{μ} is a constant). Even more generally, we could have written $A^{\mu} = aqU^{\mu}/\varrho_{\mathcal{R}} - (1 - a)qU^{\mu}/\varrho_{\mathcal{A}}$ with a an arbitrary constant (if $a = 1/2$, A_{μ} is a *symmetric potential*). The choice (17.1) is dictated by the requirement of causality: phenomena at X^{μ} must be expressed in terms of events occurring in the *past* of X^{μ} and not in its future. In what follows we shall construct only retarded quantities.

As we shall clearly see in Chapters 20 and 21, the experimentally well verified prediction that a system of charges radiating electromagnetic waves thereby loses energy and in the end becomes unstable follows directly from this choice of retarded solutions. The motion of a system of charges obtained from symmetric solutions is conservative.

where $(X \cdot U) \equiv X^\mu U_\mu$ and $X^2 \equiv X_\mu X^\mu$. We note that if $U^\mu \to \infty$, *i.e.*, if the charge 3-velocity tends to the speed of light so that the world line becomes tangent to the light cone with apex at the origin, then $F_{\mu\nu}$ tends to zero everywhere except if $(X \cdot U) = 0$, that is, except in the 3-plane tangent to the cone containing the charge world line. In this 3-plane $F_{\mu\nu}$ diverges.

Let us give the three-dimensional versions of the expressions we have obtained. We decompose $l_\mathcal{R}^\mu$ as (see Fig. 17.1b)

$$l_\mathcal{R}^0 = T - T_\mathcal{R} \equiv r_\mathcal{R}, \quad l_\mathcal{R}^i = X^i - X_\mathcal{R}^i \equiv r_\mathcal{R} n_\mathcal{R}^i, \tag{17.5}$$

where $n_\mathcal{R}^i$ is a unit 3-vector (because $\eta_{\mu\nu} l_\mathcal{R}^\mu l_\mathcal{R}^\nu = 0$). Since $U^\mu = \left(1/\sqrt{1 - V^2}, V/\sqrt{1 - V^2}\right)$, where V is the 3-velocity of the charge, $\varrho_\mathcal{R}$ defined in (17.1) can be written as

$$\varrho_\mathcal{R} = r_\mathcal{R} \frac{(1 - n_\mathcal{R} . V)}{\sqrt{1 - V^2}}, \tag{17.6}$$

where $n_\mathcal{R} . V \equiv n_\mathcal{R}^i V_i$ denotes the three-dimensional scalar product. Therefore, the 3-vector version of the Liénard–Wiechert potential (17.1) $A^\mu \equiv (\Phi, A)$ is

$$\Phi = \frac{q}{r_\mathcal{R}(1 - n_\mathcal{R} . V)}, \quad A = V\Phi, \tag{17.7}$$

where we recall that the subscript $\mathcal{R}$ indicates that the location of the charge is evaluated at the retarded time $T_\mathcal{R}$ given by (17.5); see Fig. 17.2a. The expressions for the electric and magnetic fields $E_i = -F_{0i}$ and $B^i = e^{ijk} \partial_j A_k$ can also be obtained from (17.3):

$$E = \frac{q(1 - V^2)}{r_\mathcal{R}^2 (1 - n_\mathcal{R} . V)^3}(n_\mathcal{R} - V), \quad B = n_\mathcal{R} \wedge E = V \wedge E. \tag{17.8}$$

The relations between B and E are an expression of the fact that the Faraday tensor is the exterior product of two 1-forms U and $l_\mathcal{R}$.

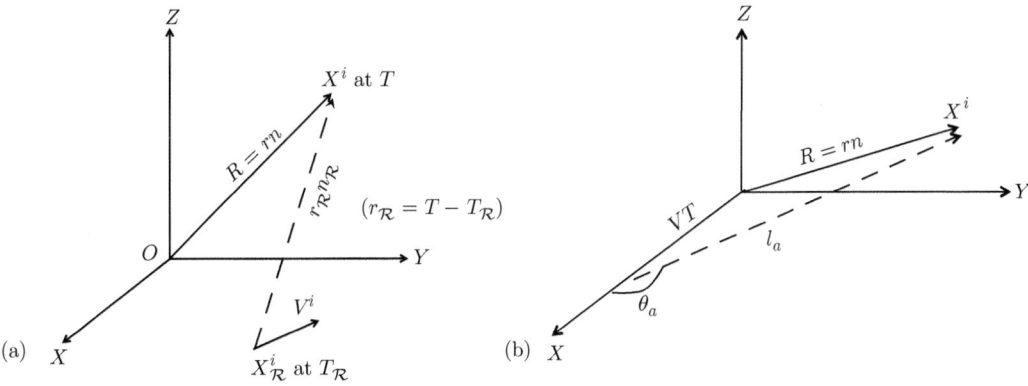

Fig. 17.2 The three-dimensional quantities.

Let us now turn to the expressions (17.4) for the potential and the field as functions of the radius vector R of the point. Since $(X \cdot X) = -T^2 + r^2$ and $(X \cdot U) = (-T + V.R)/\sqrt{1 - V^2}$, they can be written as

$$\Phi = \frac{q}{\sqrt{(R - VT)^2 + (R.V)^2 - r^2 V^2}},$$

$$E = \frac{q(1 - V^2)}{[(R - VT)^2 + (R.V)^2 - r^2 V^2]^{3/2}} (R - VT),$$

(17.9)

where $R.V \equiv X^i V_i$ and $r \equiv \sqrt{X^i X_i}$ is the length of the vector R. The magnetic potential and field can be derived from[2] $A = V\Phi$ and $B = V \wedge E$.

Finally, let $l_a \equiv R - VT$ be the separation 3-vector with components $l_a^i = (X - VT, Y, Z)$ if the axes are chosen such that the particle moves along the OX axis; see Fig. 17.2b. Its length is r_a and θ_a is the angle between the charge trajectory and the direction l_a. Then we have $Y^2 + Z^2 = r_a^2 \sin^2 \theta_a$, and also $r_a \sqrt{1 - V^2 \sin^2 \theta_a} = \sqrt{(X - VT)^2 + (1 - V^2)(Y^2 + Z^2)}$, which is easily shown to be equal to $\sqrt{(R - VT)^2 + (R.V)^2 - r^2 V^2}$. Therefore, the potential and the field (17.9) can also be written as

$$\Phi = \frac{q}{r_a \sqrt{1 - V^2 \sin^2 \theta_a}}, \quad E = \frac{q \, l_a}{r_a^3} \frac{1 - V^2}{(1 - V^2 \sin^2 \theta_a)^{3/2}}.$$

(17.10)

From these expressions we recover the fact that when the charge velocity is close to the speed of light, the field is proportional to $(1 - V^2)$ and therefore tends to zero everywhere except in the vicinity of the plane orthogonal to the trajectory where the particle is located ($\theta_a = \pi/2$), where it diverges as $(1 - V^2)^{-1/2}$.

17.2 The field of an accelerated charge

In the preceding section we found the potential created by a charge in uniform motion starting from the Coulomb expression for the potential in the frame where it is at rest, and then making a Lorentz transformation. We can follow exactly the same reasoning for any type of motion by introducing inertial frames tangent to the charge world line. In this way we obtain the retarded Liénard–Wiechert potential created by a charge in any type of motion:

$$A^\mu = \frac{q U_{\mathcal{R}}^\mu}{\varrho_{\mathcal{R}}}, \quad \text{where} \quad \varrho_{\mathcal{R}} \equiv -U_{\mu \mathcal{R}} \, l_{\mathcal{R}}^\mu \quad \text{with} \quad l_{\mathcal{R}}^\mu \equiv X^\mu - X_{\mathcal{R}}^\mu \quad \text{and} \quad \eta_{\mu\nu} l_{\mathcal{R}}^\mu l_{\mathcal{R}}^\mu \equiv (l_{\mathcal{R}} \cdot l_{\mathcal{R}}) = 0, \quad (17.11)$$

the subscript $\mathcal{R}$ indicating that the quantity is evaluated at the intersection $X_{\mathcal{R}}^\mu$ of the past cone with apex at the point X^μ and the world line of the charge in question; see Fig. 17.1b.

---

[2] Equation (17.9) can also be obtained by a Lorentz transformation. Let us now denote the frame in which the charge is at rest at the origin as $\mathcal{S}'$, so that the potential components are $A'^\mu = (q/r', 0)$. Then we go to the frame $\mathcal{S}$ moving at velocity $-V$ along the X axis, in which the charge has velocity V. The potential transforms as a contravariant vector: $A'^\mu = (\Phi', 0) \rightarrow A^\mu = (\Phi'/\sqrt{1 - V^2}, V/\sqrt{1 - V^2} \Phi')$. Now $r' = \sqrt{X'^2 + Y'^2 + Z'^2}$ must be expressed as a function of the coordinates of $\mathcal{S}$ using the Lorentz transformation $T = (T' + VX')/\sqrt{1 - V^2}$, $X = (X' + VT')/\sqrt{1 - V^2}$, which again leads to (17.9).

The retarded propagator and a continuous distribution

The Liénard–Wiechert potentials (17.11) can also be obtained by direct integration of the Maxwell equations.

If the source of the electromagnetic field is a point charge q, the Maxwell equations can be written in Lorenz gauge as (see Sections 12.2 and 12.3)

$$\Box A^\mu = -4\pi j^\mu = -4\pi q \int_L d\tau\, \delta_4(X^\nu - X^\nu(\tau))\, U^\mu(\tau) \quad \text{with} \quad \partial_\nu A^\nu = 0. \tag{17.12}$$

Here $X^\mu = X^\mu(\tau)$ are the equations of the charge world line L, $U^\mu = dX^\mu/d\tau$ is the charge velocity ($U^2 = -1$), and $\delta_4(X^\nu - X^\nu(\tau))$ is the Dirac delta distribution (its main properties were given in Section 9.2).

In order to solve this equation,[3] we introduce the *retarded propagator* $D(X^\mu)$, which is the solution of $\Box D(X^\mu) = -4\pi \delta_4(X^\mu)$.

In the Fourier space, where $D(X^\mu) = \frac{1}{(2\pi)^4} \int d^4p\, e^{-ip_\mu X^\mu} \bar{D}(p^\mu)$ and writing $\delta_4(X^\mu) = \frac{1}{(2\pi)^4} \int d^4p\, e^{-ip_\mu X^\mu}$, the equation for $D(X^\mu)$ becomes $\bar{D}(p^\mu) = \frac{4\pi}{p_\mu p^\mu}$. We then have

$$D(X^\mu) = \frac{1}{4\pi^3} \int d^4p \frac{e^{-ip_\mu X^\mu}}{p^2 - p_0^2} = \frac{1}{4\pi^3} \int dp^0 \frac{e^{ip_0 T}}{p^2 - p_0^2} e^{-ipr\cos\theta} p^2\, dp \sin\theta\, d\theta\, d\phi,$$

where $r^2 = X^i X_i$. Since $\int dp^0 \frac{e^{ip_0 T}}{-p_0^2 + p^2} = \frac{-i\pi e^{ipT}}{p}$ (after choosing the sign), for $T > 0$ we obtain

$$D(X^\mu) = -\frac{1}{2\pi r} \int dp\, e^{ipT}(e^{ipr} - e^{-ipr}) = \frac{\delta(T-r)}{r},$$

$$\text{so that} \quad \Box \frac{\delta(T-r)}{r} = -4\pi \delta_4(X^\mu). \tag{17.13}$$

Therefore, if we set $g(\tau) \equiv T - T(\tau) - r(\tau)$ with $r(\tau) = \sqrt{(X_i - X_i(\tau))(X^i - X^i(\tau))}$, (17.12) can be written as follows using (17.13):

$$\Box A^\mu = q \int d\tau\, \Box \frac{\delta(g(\tau))}{r(\tau)} U^\mu(\tau), \quad \text{the solution of which is}$$

$$A^\mu = q \int d\tau\, \delta(g(\tau)) \frac{U^\mu(\tau)}{r(\tau)} = q \int d\tau \frac{\delta(\tau - \tau_{\mathcal{R}})}{\frac{dg}{d\tau}\big|_{\mathcal{R}}} \frac{U^\mu(\tau)}{r(\tau)} = \frac{q\, U^\mu(\tau_{\mathcal{R}})}{\frac{dg}{d\tau}\big|_{\mathcal{R}} r(\tau_{\mathcal{R}})},$$

where $\tau_{\mathcal{R}}$ is the zero of $g(\tau)$: $T - T(\tau_{\mathcal{R}}) - r(\tau_{\mathcal{R}}) = 0$. In addition, $\frac{dg}{d\tau} = -U^0 + (X^i - X^i(\tau))\frac{U_i}{r(\tau)}$, so that $\frac{dg}{d\tau}\big|_{\mathcal{R}} = \frac{\varrho_{\mathcal{R}}}{r(\tau_{\mathcal{R}})}$ with $\varrho_{\mathcal{R}} \equiv -U_{\mu\mathcal{R}}(X^\mu - X^\mu(\tau_{\mathcal{R}}))$. We then find that the retarded Liénard–Wiechert potential is given by $A^\mu = \frac{q U^\mu_{\mathcal{R}}}{\varrho_{\mathcal{R}}}$.

If now we assume that the current distribution is continuous, we can write it as $j^\mu(X^\nu) = \int d^4X'\, \delta_4(X^\nu - X'^\nu) j^\mu(X'^\rho)$. We then use (17.13) to find the solution of the Maxwell equation $\Box A^\mu = -4\pi j^\mu$ in the form

$$A^\mu(T, X^i) = \int d^3X' \frac{j^\mu(T - r', X'^i)}{r'}, \quad \text{where} \quad r' = \sqrt{(X^i - X'^i)(X_i - X'_i)}. \tag{17.14}$$

---

[3] For a detailed presentation of *Green functions* see, for example, Raimond (2000).

To calculate the derivatives of the Liénard–Wiechert potential we use (17.2), in particular, $\partial_\mu T_{\mathcal{R}} = l_{\mu\mathcal{R}}/\varrho_{\mathcal{R}}$, from which for arbitrary velocities we easily find

$$\partial_\mu A_\nu = q\left(\frac{U_\nu}{\varrho^2}\right)_{\mathcal{R}}\left[U_\mu - \frac{l_\mu}{\varrho}\right]_{\mathcal{R}} - q\left(\frac{l_\mu}{\varrho^2}\right)_{\mathcal{R}}\left[\gamma_\nu + \frac{U_\nu(l\cdot\gamma)}{\varrho}\right]_{\mathcal{R}}, \tag{17.15}$$

as well as the Faraday tensor

$$F_{\mu\nu} = \frac{q}{\varrho_{\mathcal{R}}^3}[U_\mu l_\nu - U_\nu l_\mu]_{\mathcal{R}} + \frac{q}{\varrho_{\mathcal{R}}^2}\left[\gamma_\mu l_\nu - \gamma_\nu l_\mu + (U_\mu l_\nu - U_\nu l_\mu)\frac{(l\cdot\gamma)}{\varrho}\right]_{\mathcal{R}}, \tag{17.16}$$

where $\gamma^\mu \equiv dU^\mu/d\tau$ is the 4-acceleration of the charge. We can check that indeed $\partial_\mu A^\mu = 0$, and we see that the Faraday tensor can also be written as an exterior product of two 1-forms (see Section 22.1 *et seq.*):

$$F = \frac{q}{\varrho_{\mathcal{R}}^2}\left[\frac{U}{\varrho}\left(1 + (l\cdot\gamma)\right) + \gamma\right]_{\mathcal{R}} \wedge l_{\mathcal{R}}. \tag{17.17}$$

The 3-vector versions of the Liénard–Wiechert potential and field (17.11) and (17.16) generalize (17.7) and (17.8):

$$\begin{cases} \Phi = \dfrac{q}{r_{\mathcal{R}}(1 - n_{\mathcal{R}}.V_{\mathcal{R}})}, \quad A = \dfrac{qV_{\mathcal{R}}}{r_{\mathcal{R}}(1 - n_{\mathcal{R}}.V_{\mathcal{R}})}, \\[2ex] E = \dfrac{q(1-V_{\mathcal{R}}^2)}{r_{\mathcal{R}}^2(1-n_{\mathcal{R}}.V_{\mathcal{R}})^3}(n_{\mathcal{R}}-V_{\mathcal{R}}) + \dfrac{q}{r_{\mathcal{R}}(1-n_{\mathcal{R}}.V_{\mathcal{R}})^3}\, n_{\mathcal{R}} \wedge [(n_{\mathcal{R}}-V_{\mathcal{R}}) \wedge a_{\mathcal{R}}], \\[2ex] B = n_{\mathcal{R}} \wedge E, \end{cases} \tag{17.18}$$

where we again recall that the label $\mathcal{R}$ indicates that the position $X_{\mathcal{R}}^i = X^i(T_{\mathcal{R}})$, the 3-velocity $V_{\mathcal{R}}^i = V^i(T_{\mathcal{R}})$, and the 3-acceleration $a_{\mathcal{R}}^i = a^i(T_{\mathcal{R}})$ of the charge are evaluated at the retarded time $T_{\mathcal{R}}$ given implicitly by $T_{\mathcal{R}} = T - r_{\mathcal{R}}$ with $X^i - X^i(T_{\mathcal{R}}) \equiv r_{\mathcal{R}}n_{\mathcal{R}}^i$, where $n_{\mathcal{R}}$ is a unit 3-vector: $n_{\mathcal{R}}.n_{\mathcal{R}} = 1$; see[4] Fig. 17.2a.

The field of a uniformly accelerated charge

The world line of a particle uniformly accelerated along the X axis of an inertial frame $\mathcal{S}$ is a hyperbola whose equation can be written as (see Section 2.1) $gT = \sinh g\tau$, $gX = \cosh g\tau$, where τ is the proper time and g is a constant. As we saw in Section 11.4, this is the world line of a charge in a constant electric field.

Since the null vector $l_{\mathcal{R}}^\mu \equiv X^\mu - X_{\mathcal{R}}^\mu$ has components $l_{\mathcal{R}}^\mu = (T - \frac{1}{g}\sinh gT_{\mathcal{R}}, X - \frac{1}{g}\cosh gT_{\mathcal{R}}, Y, Z)$, we have $\varrho_{\mathcal{R}} \equiv -(l\cdot U)_{\mathcal{R}} = T\cosh gT_{\mathcal{R}} - X\sinh gT_{\mathcal{R}}$ and $(l\cdot\gamma)_{\mathcal{R}} = -gT\sinh gT_{\mathcal{R}} + gX\cosh gT_{\mathcal{R}} - 1$, so that the Faraday tensor (17.16) becomes

---

[4]We note that $n \wedge (n-V) \wedge a = (n.a)(n-V) - (1-n.V)a$; see the review of vector calculus in Section 11.1.

$$
\begin{cases}
F^{01} = \dfrac{qg^2}{(gT\cosh g\tau_{\mathcal{R}} - gX\sinh g\tau_{\mathcal{R}})^3}\left[gT\sinh g\tau_{\mathcal{R}} - gX\cosh g\tau_{\mathcal{R}} - g^2(T^2 - X^2)\right], \\[12pt]
F^{02} = \dfrac{qg^4 XY}{(gT\cosh g\tau_{\mathcal{R}} - gX\sinh g\tau_{\mathcal{R}})^3}\,, \quad F^{03} = \dfrac{qg^4 XZ}{(gT\cosh g\tau_{\mathcal{R}} - gX\sinh g\tau_{\mathcal{R}})^3}\,, \\[12pt]
F_{23} = 0\,, \quad F_{13} = \dfrac{qg^4 TZ}{(gT\cosh g\tau_{\mathcal{R}} - gX\sinh g\tau_{\mathcal{R}})^3}\,, \quad F_{12} = \dfrac{qg^4 TY}{(gT\cosh g\tau_{\mathcal{R}} - gX\sinh g\tau_{\mathcal{R}})^3}\,.
\end{cases}
\tag{17.19}
$$

Now to express $F_{\mu\nu}$ as a function of only X^μ, it is necessary to solve the equation of the cone $\eta_{\mu\nu}l^\mu_{\mathcal{R}}l^\nu_{\mathcal{R}} = 0$, or

$$
(gT - \sinh g\tau_{\mathcal{R}})^2 = (gX - \cosh g\tau_{\mathcal{R}})^2 + g^2(Y^2 + Z^2)\,.
$$

Rather than finding $\tau_{\mathcal{R}}$ as a function of X^μ, it is preferable to obtain T as a function of the observation point X^i and $\tau_{\mathcal{R}}$, that is, the charge position:

$$
gT = \sinh g\tau_{\mathcal{R}} + \sqrt{(gX - \cosh g\tau_{\mathcal{R}})^2 + g^2(Y^2 + Z^2)}\,,
\tag{17.20}
$$

where the choice of the plus sign in front of the square root guarantees that $\tau_{\mathcal{R}}$ is the retarded solution.

Let us place ourselves, for example, on the axis of the trajectory ($Y = Z = 0$). Equation (17.20) then gives $e^{g\tau_{\mathcal{R}}} = g(T + X)$, and the field (17.19) reduces to

$$
F^{01} \equiv E = -\frac{4qg^2}{[1 + g^2(T + X)(T - X)]^2} \quad \text{for } T+X > 0, \quad \text{and} \quad E = 0 \quad \text{for } T+X < 0.
\tag{17.21}
$$

Therefore, the field at the origin ($X = Y = Z = 0$) is zero until $T = 0$, when it jumps to a *finite* value. The charge then is at $X = 1/g$, its distance of closest approach, but the field it creates at $T = 0$ is determined by its position at the retarded time, when it was arriving from infinity at a speed approaching the speed of light. The field then decreases and tends to zero as the charge moves away and approaches again the speed of light.

The field of a charge in circular motion

The world line of a particle moving uniformly around a circle of radius r_0 at frequency Ω in the $Z = 0$ plane of an inertial frame $\mathcal{S}$ is (see Section 2.1) $X^\mu = (T, r_0\cos\Omega T, r_0\sin\Omega T, 0)$, and so the particle velocity $U^\mu \equiv dX^\mu/d\tau$ (with $U_\mu U^\mu = -1$) and acceleration are

$$
U^\mu = \frac{1}{\sqrt{1 - r_0^2\Omega^2}}(1, -r_0\Omega\sin\Omega T, r_0\Omega\cos\Omega T, 0), \quad \gamma^\mu = -\frac{r_0\Omega^2}{1 - r_0^2\Omega^2}(0, \cos\Omega T, \sin\Omega T, 0).
$$

If this motion is caused by the action of a magnetic field B, we will have (see Section 11.4) $\Omega = \omega/\sqrt{1 + r_0^2\omega^2}$, where $\omega \equiv qB/m$ with m the mass of the charge. If the motion is due to a central Coulomb field, Ω is given by (13.16).

The null vector $l^\mu_{\mathcal{R}} \equiv X^\mu - X^\mu_{\mathcal{R}}$ has the components $l^\mu_{\mathcal{R}} = (T - T_{\mathcal{R}},\ X - r_0\cos\Omega T_{\mathcal{R}},\ Y - r_0\sin\Omega T_{\mathcal{R}},\ Z)$. Therefore,

$$
\begin{cases}
\varrho_{\mathcal{R}} \equiv -(l\cdot U)_{\mathcal{R}} = \dfrac{T - T_{\mathcal{R}} + r_0\Omega(X\sin\Omega T_{\mathcal{R}} - Y\cos\Omega T_{\mathcal{R}})}{\sqrt{1 - r_0^2\Omega^2}}\,, \\[12pt]
(l\cdot\gamma)_{\mathcal{R}} = \dfrac{r_0\Omega^2\left[r_0 - (X\cos\Omega T_{\mathcal{R}} + Y\sin\Omega T_{\mathcal{R}})\right]}{1 - r_0^2\Omega^2}\,,
\end{cases}
$$

from which we easily find the expression for the components of the Faraday tensor (17.16). For example,

$$F_{03} = -\frac{q\,Z[1 - r_0\Omega^2(X\cos\Omega T_{\mathcal{R}} + Y\sin\Omega T_{\mathcal{R}})]}{[T - T_{\mathcal{R}} + r_0\Omega(X\sin\Omega T_{\mathcal{R}} - Y\sin\Omega T_{\mathcal{R}})]^3}\,,\tag{17.22}$$

where $T_{\mathcal{R}}$ is obtained (implicitly) as a function of X^μ by solving the equation of the cone $\eta_{\mu\nu}(X^\mu - X_{\mathcal{R}}^\mu)(X^\nu - X_{\mathcal{R}}^\nu) = 0$, or, setting $r = \sqrt{X^2 + Y^2 + Z^2}$,

$$T - T_{\mathcal{R}} - \sqrt{r^2 - 2r_0(X\cos\Omega T_{\mathcal{R}} + Y\sin\Omega T_{\mathcal{R}}) + r_0^2} = 0\,.\tag{17.23}$$

Let us place ourselves at the origin. Then $T_{\mathcal{R}} = T - r_0$, and the nonzero components of the field are

$$\begin{cases} F_{10} = E_X = \dfrac{q}{r_0^2}\left[r_0\Omega\sin\Omega T_{\mathcal{R}} - (1 - r_0^2\Omega^2)\cos\Omega T_{\mathcal{R}}\right] \\[2mm] F_{20} = E_Y = -\dfrac{q}{r_0^2}\left[r_0\Omega\cos\Omega T_{\mathcal{R}} + (1 - r_0^2\Omega^2)\sin\Omega T_{\mathcal{R}}\right] \\[2mm] F_{12} = B_Z = \dfrac{q\Omega}{r_0}\,. \end{cases}\tag{17.24}$$

The time average of the electric field is zero. The magnetic field is the *Ampère field* created by the charge current.

17.3 The radiation field

Let us return to (17.16) for the Faraday tensor created by a charge in any type of motion. The first term, which is the only one when the charge is in uniform motion, falls off as $1/r^2$ at spatial infinity and reduces to the Coulomb electric field if the charge is at rest; see Section 17.1. The second term, which falls off as $1/r$, is the *radiation* part of the field.

More precisely, let us consider a charge which is *confined* near the spatial origin of a particular inertial frame $\mathcal{S}$. If the observation point is very far away, the null vector $l_{\mathcal{R}}^\mu \equiv X^\mu - X_{\mathcal{R}}^\mu$, where $X_{\mathcal{R}}^\mu$ is the intersection of the past cone with apex X^μ and the charge world line, can be approximated by the vector $l^\mu \equiv X^\mu - X_{\mathcal{R}_0}^\mu$, which is also null. Here $X_{\mathcal{R}_0}^\mu = (T_{\mathcal{R}_0} = T - r, 0)$ is the intersection of the cone with the spatial origin $X^i = 0$; see Fig. 17.3.

The *radiation field* created by such a charge is then defined as the principal part of (17.16) at large distances, or

$$F_{\text{rad}}^{\mu\nu} = \frac{q}{\varrho_{\mathcal{R}}^2}\left[\gamma_{\mathcal{R}}^\mu l^\nu - \gamma_{\mathcal{R}}^\nu l^\mu + (U_{\mathcal{R}}^\mu l^\nu - U_{\mathcal{R}}^\nu l^\mu)\frac{(l\cdot\gamma_{\mathcal{R}})}{\varrho_{\mathcal{R}}}\right]\quad\text{with}\quad l^\mu = (r, X, Y, Z),$$

$$\text{or also}\quad F_{\text{rad}} = \mathcal{F}_{\text{rad}}\wedge l\quad\text{with}\quad \mathcal{F}_{\text{rad}} \equiv \frac{q}{\varrho_{\mathcal{R}}^2}\left[\frac{U}{\varrho}(l\cdot\gamma) + \gamma\right]_{\mathcal{R}},\tag{17.25}$$

where $r = \sqrt{X^2 + Y^2 + Z^2}$ is the (assumed large) distance to the origin, $\varrho_{\mathcal{R}} \simeq -(l\cdot U_{\mathcal{R}})$, and the velocity $U_{\mathcal{R}}^\mu$ and acceleration $\gamma_{\mathcal{R}}^\mu$ of the charge are always evaluated at the retarded point $X_{\mathcal{R}}^\mu$ (except when the motion is slow, and at lowest order where they can be evaluated at time $T - r$; see Chapter 20).

We note that F_{rad} is orthogonal to l: $(F_{\text{rad}}\cdot l) = 0$. Therefore, the radiation part of the Faraday tensor has the structure of that of a plane wave; *cf.* (15.28).

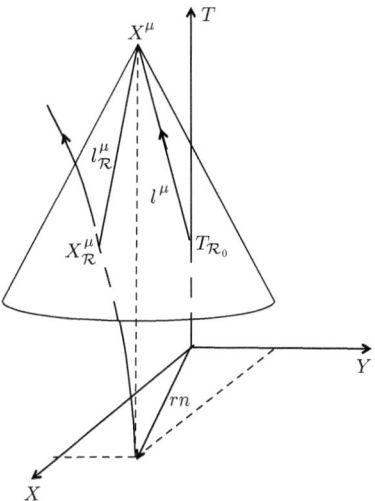

Fig. 17.3 The radiation field.

The 3-vector version of (17.25) is [*cf.* (17.18)]

$$E_{\text{rad}} = \frac{q}{r(1 - n.V_{\mathcal{R}})^3} \, n \wedge [(n - V_{\mathcal{R}}) \wedge a_{\mathcal{R}}], \quad B_{\text{rad}} = n \wedge E_{\text{rad}}, \quad \text{and} \quad E_{\text{rad}}.n = 0. \quad (17.26)$$

Here $R \equiv rn$ with $(n.n) = 1$ is the radius vector of the point, $V_{\mathcal{R}}$ and $a_{\mathcal{R}}$ are the charge 3-velocity and acceleration at the retarded time $T_{\mathcal{R}} = T - r_{\mathcal{R}}$ with $X^i - X^i(T_{\mathcal{R}}) \equiv r_{\mathcal{R}} n_{\mathcal{R}}^i$, and $n_{\mathcal{R}}.n_{\mathcal{R}} = 1$.

The radiation field of a uniformly accelerated charge

We consider a charge which is uniformly accelerated along the OX axis of an inertial frame $\mathcal{S}$. The equation of its world line is $gT = \sinh g\tau$, $gX = \cosh g\tau$, where τ is its proper time and g is a constant. The components of the electromagnetic field tensor that it creates were given in (17.19).

We limit ourselves to times when the charge is close to the spatial origin, that is, to values of τ such that $g\tau$ is not too large compared to 1. The point at which the radiation field is evaluated is located far away in the future cone of this segment of the world line (see Fig. 17.3). By symmetry we can always take $Z = 0$ and set $X = r\cos\phi$, $Y = r\sin\phi$. Finally, we have $T - T_{\mathcal{R}_0} = r$, where $T_{\mathcal{R}_0}$ is the time at which the light must be emitted from the origin in order to reach the field point (see Fig. 17.3). However, since $T_{\mathcal{R}_0} \ll r$ we have $T = r$ in leading order, and so the radiation part of (17.19) reduces to $E_X^{\text{rad}} = \mathcal{F}_{\text{rad}} \sin\phi$, $E_Y^{\text{rad}} = -\mathcal{F}_{\text{rad}} \cos\phi$, $B_Z^{\text{rad}} = -\mathcal{F}_{\text{rad}}$ with

$$\mathcal{F}_{\text{rad}} = -\frac{q}{r} \frac{g\sin\phi}{(\cosh g\tau_{\mathcal{R}} - \cos\phi \sinh g\tau_{\mathcal{R}})^3} = -\frac{q}{r} \frac{a_{\mathcal{R}} \sin\phi}{(1 - V_{\mathcal{R}} \cos\phi)^3}, \quad (17.27)$$

where $V = \tanh g\tau$ and $a = g(1 - V^2)^{3/2} = g/\cosh^3 g\tau$ are the 3-velocity and 3-acceleration of the charge, and the retarded time $\tau_{\mathcal{R}}$ is given by eqn (17.20), that is, $gT = \sinh g\tau_{\mathcal{R}} +$

$\sqrt{(gX - \cosh g\tau_{\mathcal{R}})^2 + g^2(Y^2 + Z^2)}$, which, as long as the charge remains near the origin, reduces to $\sinh g\tau_{\mathcal{R}} = g(T - r)$, so that

$$\mathcal{F}_{\text{rad}} = -\frac{q}{r} \frac{g \sin\phi}{\left(\sqrt{1 + g^2(T-r)^2} - g(T-r)\cos\phi\right)^3}. \tag{17.28}$$

We note [cf. (17.27)] that this field becomes exponentially small ($\mathcal{F}_{\text{rad}} \propto e^{-3g\tau_{\mathcal{R}}}$) when the charge velocity approaches the speed of light. This is a consequence of the fact that the charge acceleration g is *constant*, and so its 3-acceleration a tends to zero.

The radiation field of a charge in circular motion

The equation of the world line of a particle moving uniformly around a circle of radius r_0 at frequency Ω in the $Z = 0$ plane of an inertial frame $\mathcal{S}$ is $X^\mu = (T, r_0 \cos\Omega T, r_0 \sin\Omega T, 0)$. In Section 17.2 we saw how to calculate the electromagnetic field it creates everywhere.

At a point far from the charge the radiation field reduces to (17.25), and, setting $X = r\sin\theta\cos\phi$, $Y = r\sin\theta\sin\phi$, and $Z = r\cos\theta$ with $r \gg r_0$ and using the fact that $T - T_{\mathcal{R}} \approx r$ at leading order, we find

$$\begin{cases} E^X_{\text{rad}} = -\dfrac{qr_0\Omega^2}{r} \dfrac{\sin^2\theta\cos\phi\cos(\Omega T_{\mathcal{R}} - \phi) - \cos\Omega T_{\mathcal{R}} + r_0\Omega\sin\theta\sin\phi}{[1 + r_0\Omega\sin\theta\sin(\Omega T_{\mathcal{R}} - \phi)]^3} \\[4mm] E^Y_{\text{rad}} = -\dfrac{qr_0\Omega^2}{r} \dfrac{\sin^2\theta\sin\phi\cos(\Omega T_{\mathcal{R}} - \phi) - \sin\Omega T_{\mathcal{R}} - r_0\Omega\sin\theta\cos\phi}{[1 + r_0\Omega\sin\theta\sin(\Omega T_{\mathcal{R}} - \phi)]^3} \\[4mm] E^Z_{\text{rad}} = -\dfrac{qr_0\Omega^2}{r} \cos\theta\sin\theta \dfrac{\cos(\Omega T_{\mathcal{R}} - \phi)}{[1 + r_0\Omega\sin\theta\sin(\Omega T_{\mathcal{R}} - \phi)]^3}. \end{cases} \tag{17.29}$$

The magnetic field is given by $B_{\text{rad}} = n \wedge E_{\text{rad}}$, where $n \equiv R/r$, R being the radius vector to the point. It can be checked that indeed $(E_{\text{rad}}.n) = 0$.

In the ultrarelativistic limit, that is, when $r_0\Omega \to 1$, the field diverges when $\sin\theta = 1$ and $\sin(\Omega T_{\mathcal{R}} - \phi) = -1$, that is, in the plane of motion of the charge and when the charge is aligned with the observation direction.

In the *wave zone*, that is, for $r \gg \lambda$ or $r\Omega \gg 1$, we can take the average of the field over $T_{\mathcal{R}}$. Calculating the integrals,[5] we find

$$\langle E^{\text{rad}}_X \rangle = \langle \mathcal{F}_{\text{rad}} \rangle \sin\phi, \quad \langle E^{\text{rad}}_Y \rangle = -\langle \mathcal{F}_{\text{rad}} \rangle \cos\phi, \quad \langle E^{\text{rad}}_Z \rangle = 0$$

$$\langle B^{\text{rad}}_X \rangle = \langle \mathcal{F}_{\text{rad}} \rangle \cos\theta\cos\phi, \quad \langle B^{\text{rad}}_Y \rangle = \langle \mathcal{F}_{\text{rad}} \rangle \cos\theta\sin\phi, \quad \langle B^{\text{rad}}_Z \rangle = -\langle \mathcal{F}_{\text{rad}} \rangle \sin\theta$$

with

$$\langle \mathcal{F}_{\text{rad}} \rangle = \frac{qr_0^2\Omega^3}{2r} \frac{\sin\theta(2 - r_0^2\Omega^2\sin^2\theta)}{(1 - r_0^2\Omega^2\sin^2\theta)^{\frac{5}{2}}}. \tag{17.30}$$

---

[5] We use $\dfrac{1}{2\pi} \displaystyle\int_0^{2\pi} \dfrac{dx}{(1 - A\cos x)^3} = \dfrac{1}{2} \dfrac{1 + A^2}{(1 - A^2)^{\frac{5}{2}}}$ and $\dfrac{1}{2\pi} \displaystyle\int_0^{2\pi} \dfrac{dx \cos x}{(1 - A\cos x)^3} = \dfrac{3}{2} \dfrac{A}{(1 - A^2)^{\frac{5}{2}}}$.

The radiation field of a charge accelerated by a wave

In Section 15.4 we saw that the 4-velocity of a charge q of mass m in a monochromatic plane wave of frequency k, where the field has amplitude ka, propagating along the X axis and linearly polarized in the Y direction, is given by

$$U^X = \frac{\beta^2}{2}(\cos\xi - 1)^2, \quad U^Y = -\beta(\cos\xi - 1), \quad U^0 = 1 + U^X,$$

where $\beta \equiv qa/m$ and $\xi \equiv k\tau$. Here τ is the proper time of the charge. The acceleration is found to be $\gamma^\mu = k\,dU^\mu/d\xi$.

The field radiated by this charge at the point (T, X, Y, Z) is given by (17.25) with (setting $r = \sqrt{X^2 + Y^2 + Z^2}$)

$$\begin{cases} \varrho \equiv -(l \cdot U) = r + Y\beta(\cos\xi - 1) + \dfrac{1}{2}\beta^2(r - X)(\cos\xi - 1)^2, \\[2mm] (l \cdot \gamma) = k\beta \sin\xi\,[Y + \beta(r - X)(\cos\xi - 1)]. \end{cases}$$

In the nonrelativistic limit $\beta \to 0$, the radiation field simplifies to

$$E^X = \frac{qk\beta}{r}\frac{XY}{r^2}\sin\xi, \quad E^Y = -\frac{qk\beta}{r}\frac{r^2 - Y^2}{r^2}\sin\xi, \quad E^Z = \frac{qk\beta}{r}\frac{ZY}{r^2}\sin\xi, \tag{17.31}$$

where $\xi \approx kT_{\mathcal{R}} \approx k(T - r)$. The radiated field therefore has the same frequency as the wave causing the charge to move.

In the ultrarelativistic limit $\beta \to \infty$, we find that the field does not grow as β^2 as might a *priori* be expected. Instead, it converges as $1/\beta^3$ everywhere except on the X axis along which the charge is carried by the incident wave:

$$\begin{cases} E^X = \dfrac{4qk}{r\beta^3}\dfrac{Y}{(r - X)^2}\dfrac{\sin\xi}{(1 - \cos\xi)^4}, \quad E^Y = \dfrac{4qk}{r\beta^3}\dfrac{r^2 - rX - Y^2}{(r - X)^3}\dfrac{\sin\xi}{(1 - \cos\xi)^4}, \\[3mm] E^Z = -\dfrac{4qk}{r\beta^3}\dfrac{YZ}{(r - X)^3}\dfrac{\sin\xi}{(1 - \cos\xi)^4}, \end{cases} \tag{17.32}$$

where the time of the frame is related to ξ as[6] $kT \approx \frac{\beta^2}{8}(6\xi - 8\sin\xi + \sin 2\xi)$.

---

[6]For a more detailed study of the *multipole expansion* of the radiation field see, for example, Raimond (2000).

18

Radiation by a charge

In this chapter we study the energy radiated by a single charge. After deriving the *Larmor formulas*, we study the paradigmatic cases of the radiation of a linearly accelerated charge, the synchrotron radiation of a charge in circular motion, and the radiation of a charge accelerated by an electromagnetic wave (Thomson scattering).

We shall discover in passing that the hydrogen atom as described by the Rutherford model of an electron orbiting a proton is highly unstable in the Maxwell theory.

18.1   The Larmor formulas

Let us recall the expression for the radiation part of the Faraday tensor of an accelerated charge whose motion remains confined near the origin of the reference frame (see Section 17.3):

$$F_{\text{rad}}^{\mu\nu} = \frac{q}{\varrho_{\mathcal{R}}^2} \left[\gamma_{\mathcal{R}}^\mu l^\nu - \gamma_{\mathcal{R}}^\nu l^\mu + (U_{\mathcal{R}}^\mu l^\nu - l^\nu U_{\mathcal{R}}^\mu) \frac{(l \cdot \gamma_{\mathcal{R}})}{\varrho_{\mathcal{R}}} \right] \quad \text{with} \quad l^\mu = (r, X, Y, Z), \qquad (18.1)$$

where $r = \sqrt{X^2 + Y^2 + Z^2}$ is the distance to the origin, $\varrho_{\mathcal{R}} \simeq -(l \cdot U_{\mathcal{R}})$ in the order in which we are working, and $U_{\mathcal{R}}^\mu$ and $\gamma_{\mathcal{R}}^\mu$ are the velocity and acceleration of the charge at the retarded point $X_{\mathcal{R}}^\mu$, the intersection of the world line and the past cone with apex at the field observation point[1] X^μ.

The radiation part of the energy–momentum tensor of the field created by a single charge is therefore given by[2]

$$T_{\mu\nu}^{\text{rad}} = \frac{1}{4\pi} \left(F_{\mu\rho} F_\nu{}^\rho - \frac{1}{4} \eta_{\mu\nu} F^{\rho\sigma} F_{\rho\sigma} \right)_{\text{rad}} = \frac{q^2}{4\pi} \frac{l_\mu l_\nu}{\varrho_{\mathcal{R}}^4} \left[\gamma_{\mathcal{R}}^2 - \frac{(l \cdot \gamma_{\mathcal{R}})^2}{\varrho_{\mathcal{R}}^2} \right]. \qquad (18.2)$$

The conservation law for the energy–momentum tensor of the field and the charges creating it leads to (see Section 12.4)

$$\frac{dP^\mu}{dT} = -\int_S T^{\mu i} dS_i, \qquad (18.3)$$

where P^μ is the momentum of the field and the particles and T is the time in the selected inertial frame $\mathcal{S}$. The surface S is the sphere at infinity with surface element $dS_i = n_i \, dS$, where

---

[1] As examples, we recall that if the charge is uniformly accelerated, (18.1) leads to (17.28) for the field, while if the charge is in circular motion, it leads to (17.29).

[2] The fact that $T_{\mu\nu}^{\text{rad}}$ is proportional to $l_\mu l_\nu$ arises from the fact that $F_{\text{rad}} = \mathcal{F}_{\text{rad}} \wedge l$ with $(l \cdot l) = 0$ and $(\mathcal{F}_{\text{rad}} \cdot l) = 0$ [cf. (17.25)], so that $(F \cdot F)_{\text{rad}} \equiv (F_{\rho\sigma} F^{\rho\sigma})_{\text{rad}} = 0$. We note also that the expression for $T_{\mu\nu}^{\text{rad}}$ is particularly simple because we are dealing with the radiation of a *single* particle. The radiation of a *system* of (slowly moving) charges will be studied in Chapters 20 and 21.

Relativity in Modern Physics. Nathalie Deruelle and Jean-Philippe Uzan.
© Oxford University Press 2018. Published in 2018 by Oxford University Press.
DOI: 10.1093/oso/9780198786399.001.0001

$dS = r^2 do$ with $do = \sin\theta \, d\theta \, d\phi$ the element of solid angle and n^i the unit 3-vector perpendicular to S and pointing toward the exterior of the domain: $n^i = (\sin\theta\cos\phi, \sin\theta\sin\phi, \cos\theta)$, so that[3] $l^\mu = (r, rn^i)$. Substituting (18.2) into (18.3), we find that the momentum of the system evolves as

$$P^\mu(T) = -\frac{q^2}{4\pi} \int \left(\frac{r}{\varrho_\mathcal{R}}\right)^4 \left[\gamma_\mathcal{R}^2 - \frac{(l\cdot\gamma_\mathcal{R})^2}{\varrho_\mathcal{R}^2}\right] \frac{l^\mu}{r} \, do \, dT \quad \text{with} \quad \begin{cases} l^\mu = (r, rn^i) \\ \varrho_\mathcal{R} \simeq -(l\cdot U_\mathcal{R}). \end{cases} \tag{18.4}$$

The integrand is evaluated at the retarded point $X_\mathcal{R}^\mu$, which is not known as a function of the coordinates X^μ until a particular world line is specified explicitly. To treat the general case, we pass from integration over T to integration over $\tau_\mathcal{R}$. For this we use the fact that $l_\mathcal{R}^\mu$ is null, which implies [cf. (17.2)] that $\partial_\rho \tau_\mathcal{R} = -(l_\rho/\varrho)_\mathcal{R}$, or, when r, θ, ϕ are constants, $d\tau_\mathcal{R} = -(l_0/\varrho)_\mathcal{R} \, dT \simeq (r/\varrho_\mathcal{R})dT$. We then have

$$P^\mu(\tau) = -\frac{q^2}{4\pi} \int \left(\frac{r}{\varrho}\right)^3 \left(\gamma^2 - \frac{(l\cdot\gamma)^2}{\varrho^2}\right) \frac{l^\mu}{r} \, do \, d\tau \quad \text{with} \quad \begin{cases} l^\mu = (r, rn^i) \\ \varrho = -(l\cdot U), \end{cases} \tag{18.5}$$

where we have suppressed the subscript $\mathcal{R}$ on the dummy variable $\tau_\mathcal{R}$.

To calculate the integral over the angles now with τ fixed, we work in the inertial frame tangent to the trajectory, in which $U^\mu = (1,0,0,0)$ and $\gamma^0 = 0$, so that $\varrho = r$ and $(l\cdot\gamma) = n_i\gamma^i$. Using $\frac{1}{4\pi}\int n_i n_j \, do = \frac{1}{3}\delta_{ij}$ and $\int n_i n_j n_k \, do = 0$ (see below), we find that in this frame $P^0(\tau) = -(2q^2/3)\int \gamma^2 \, d\tau$ and $P^i = 0$. The expression for P^μ in the original inertial frame where the particle velocity is $U^\mu = dX^\mu/d\tau$ then is

$$P^\mu(\tau) = -\frac{2q^2}{3} \int \gamma^2 \, U^\mu \, d\tau. \tag{18.6}$$

Calculation of spatial averages

To obtain (18.6) we need to calculate $\int n_i n_j \, do$. A brute-force method is to decompose n_i as $n_i = (\sin\theta\cos\phi, \sin\theta\cos\phi, \cos\theta)$. Then $\int n_1 n_2 \, do = \int \sin^2\theta\cos\phi\sin\phi \, \sin\theta d\theta \, d\phi = 0$, and so on. It is much quicker to notice that the result can only be proportional to the Kronecker delta: $\int n_i n_j \, do \propto \delta_{ij}$. The proportionality factor is obtained by taking the trace, and we find (in three dimensions)

$$\frac{1}{4\pi} \int n_i n_j \, do = \frac{\delta_{ij}}{3}. \tag{18.7}$$

The average of the product of four unit vectors is obtained in the same way:

---

[3] As we saw in Section 12.4, the energy–momentum and angular momentum conservation laws follow from the set of Maxwell and Lorentz equations describing a *closed* system of charges and the field they create. In the special case of a single charge, these equations state that this charge is in uniform rectilinear motion and that its (Coulomb) field is not radiative.

Therefore, in order to be able to legitimately apply these laws to our case of a single charge accelerated by an *external* field, we must imagine that this field is created by charges forming a part of the system under study, but whose contribution to $T_{\mu\nu}^{\text{rad}}$ can be ignored, either because they are massive enough to be only negligibly accelerated, or because their contribution to the radiation can be distinguished from that of the charge under study.

$$\frac{1}{4\pi}\int n_i n_j n_k n_l\, do = \frac{1}{15}(\delta_{ij}\delta_{kl}+\delta_{ik}\delta_{jl}+\delta_{il}\delta_{jk}). \tag{18.8}$$

Let us conclude by giving the three-dimensional versions of the above expressions. We have $l^\mu = (r, rn^i)$ and, *cf.* (17.6) and (2.3),

$$\varrho_{\mathcal{R}} \simeq r\,\frac{1-n.V_{\mathcal{R}}}{\sqrt{1-V_{\mathcal{R}}^2}}\;,\qquad \gamma_{\mathcal{R}}^\mu = \left[\frac{a.V}{(1-V^2)^2}\;,\;\frac{1}{1-V^2}\left(a^i+V^i\frac{a.V}{1-V^2}\right)\right]_{\mathcal{R}}\;, \tag{18.9}$$

where $V^i = dX^i/dT$ and $a^i = d^2X^i/dT^2$ are the charge 3-velocity and acceleration, and $n.V \equiv n^i V_i$ denotes the scalar product of 3-vectors. A short calculation using, among other things, the fact that $\gamma^2 = [(a.V)^2 + a^2(1-V^2)]/(1-V^2)^3$, then tells us that the energy density (18.2) radiated by the system of the charge and the field it creates can be written as

$$T_{\mathrm{rad}}^{00} \equiv W_{\mathrm{rad}} = \frac{q^2}{4\pi r^2}\left[\frac{a_{\mathcal{R}}^2}{(1-n.V_{\mathcal{R}})^4}+\frac{2(n.a_{\mathcal{R}})(V_{\mathcal{R}}.a_{\mathcal{R}})}{(1-n.V_{\mathcal{R}})^5}-\frac{(1-V_{\mathcal{R}}^2)(n.a_{\mathcal{R}})^2}{(1-n.V_{\mathcal{R}})^6}\right]. \tag{18.10}$$

The radiated momentum density is given by $T_{\mathrm{rad}}^{0i} \equiv S_{\mathrm{rad}}^i = W_{\mathrm{rad}}n^i$. [Of course, this energy density $W \equiv (E^2 + B^2)/(8\pi)$ and the Poynting vector $S \equiv (E \wedge B)/(4\pi)$ can also be obtained directly using eqn (17.26) for the radiation field.]

Since $d\tau_{\mathcal{R}} = \sqrt{1-V_{\mathcal{R}}^2}\,dT_{\mathcal{R}}$ and [*cf.* (17.2) as well as (18.5) and (18.9)] $d\tau_{\mathcal{R}} = -(l_0/\varrho_{\mathcal{R}})dT \simeq (r/\varrho_{\mathcal{R}})dT \simeq \sqrt{1-V_{\mathcal{R}}^2}\,dT/(1-n_0V_{\mathcal{R}})$, we have $dT_{\mathcal{R}}/dT \simeq 1/(1-n_0V_{\mathcal{R}})$, so that (18.5) becomes

$$P^0(T) = -\frac{q^2}{4\pi}\int\left[\frac{a^2}{(1-n.V)^3}+\frac{2(n.a)(V.a)}{(1-n.V)^4}-\frac{(1-V^2)(n.a)^2}{(1-n.V)^5}\right]do\,dT, \tag{18.11}$$

where we have again suppressed the label $\mathcal{R}$ of the dummy variable[4] $T_{\mathcal{R}}$.

To calculate the integral over the solid angle do, we can choose the axes such that V points along the Z axis. Then $do = \sin\theta\,d\theta\,d\phi$, $(n.V) = V\cos\theta$, and $(n.a) = a_X\sin\theta\cos\phi + a_Y\sin\theta\sin\phi + a_Z\cos\theta$. After elementary integrations[5] we find

$$\begin{cases} P^0(T) = -\dfrac{2q^2}{3}\displaystyle\int dT\left[\dfrac{a^2-(a\wedge V)^2}{(1-V^2)^0}\right] \\[4mm] P^i(T) = -\dfrac{2q^2}{3}\displaystyle\int dT\left[\dfrac{a^2-(a\wedge V)^2}{(1-V^2)^3}\right]V^i, \end{cases} \tag{18.12}$$

which, owing to (18.9), is just the three-dimensional version of (18.6). Equations (18.6) and (18.12) are the *Larmor formulas.*

---

[4] However, this simplification of the notation should not cause us to lose sight of the fact that (18.11) now gives the radiated power as a function of the emission time.

[5] Using $\int_{-1}^{+1}\frac{dx}{(1-Vx)^5} = \frac{2(1+V^2)}{(1-V^2)^4}$, $\int_{-1}^{+1}\frac{x\,dx}{(1-Vx)^5} = \frac{2V(5+V^2)}{3(1-V^2)^4}$, $\int_{-1}^{+1}\frac{x^2\,dx}{(1-Vx)^5} = \frac{2(1+5V^2)}{3(1-V^2)^4}$, and $\int_{-1}^{+1}\frac{x^3\,dx}{(1-Vx)^5} = \frac{2V(1+V^2)}{(1-V^2)^4}$.

The integrand can also be written as $a^2 - (a\wedge V)^2 = a^2(1-V^2) + (V.a)^2$ (*cf.* the review of vector calculus in Section 11.1).

Angular momentum radiated by an accelerated charge

The rate at which angular momentum is radiated by an accelerated charge is, *cf.* (12.34),

$$\frac{dJ}{dT} = \frac{r^3}{4\pi} \int_S n \wedge [(E.n)E + (B.n)B] \, do, \tag{18.13}$$

where S is the sphere at infinity with surface element $dS = r^2 do$, $do = \sin\theta \, d\theta \, d\phi$ is the element of solid angle, and $n^i = (\sin\theta\cos\phi, \sin\theta\sin\phi, \cos\theta)$ is the unit 3-vector perpendicular to S pointing toward the exterior of the domain.

It is clear that knowledge of only the radiation field is not sufficient for calculating this rate. In fact, from (17.26) we have $(E_{\mathrm{rad}}.n) = (B_{\mathrm{rad}}.n) = 0$ (fortunately, as otherwise dJ/dT would diverge, because E_{rad} and B_{rad} fall off only as $1/r$ at infinity).

Therefore, to calculate dJ/dT, we must return to the exact expression for the field created by a moving charge, (17.18), namely, $E = E_{\mathrm{Coul}} + E_{\mathrm{rad}}$ with, for observation point far from the charge,

$$E_{\mathrm{Coul}} \simeq \frac{q(1 - V_{\mathcal{R}}^2)}{r^2(1 - n.V_{\mathcal{R}})^3}(n - V_{\mathcal{R}}),$$

$$E_{\mathrm{rad}} \simeq \frac{q}{r(1 - n.V_{\mathcal{R}})^3}[(n.a_{\mathcal{R}})(n - V_{\mathcal{R}}) - (1 - n.V_{\mathcal{R}})a_{\mathcal{R}}],$$

and $B \simeq n \wedge E$. We then have

$$\frac{dJ}{dT} = \frac{r^3}{4\pi} \int_S (E_{\mathrm{Coul}}.n)\, n \wedge E_{\mathrm{rad}} \, do$$

$$= -\frac{q^2}{4\pi} \int_S \frac{1 - V_{\mathcal{R}}^2}{(1 - n.V_{\mathcal{R}})^5}[(n.a_{\mathcal{R}})n \wedge V_{\mathcal{R}} + (1 - n.V_{\mathcal{R}})n \wedge a_{\mathcal{R}}] \, do.$$

Now if we pass from integration over T to integration over $T_{\mathcal{R}}$ and drop the dummy index $\mathcal{R}$ as in obtaining (18.11), we find the angular momentum radiated as a function of the emission time:

$$J = -\frac{q^2}{4\pi} \int_S \frac{1 - V^2}{(1 - n.V)^4}[(n.a)n \wedge V + (1 - n.V)n \wedge a] \, do \, dT. \tag{18.14}$$

The integral over the angles is done as in going from (18.11) to (18.12) (see footnote 5 above) and leads to the Larmor formula for the radiated angular momentum:

$$J(T) = -\frac{2q^2}{3} \int dT \, \frac{V \wedge a}{1 - V^2}. \tag{18.15}$$

18.2 Radiation by a linearly accelerated charge

When the charge moves along the direction e_X with 4-velocity U^μ and 4-acceleration γ^μ, we easily see that (18.2) for the radiation part of the energy–momentum tensor simplifies to[6]

---

[6]We have $\varrho = r(U^0 - U^X \cos\psi)$, where ψ is the angle between the direction of observation and the X axis; $(l \cdot \gamma) = r(\gamma^0 - \gamma^X \cos\psi)$; and $U^X = \sqrt{(U^0)^2 - 1}$, $\gamma^0 = \gamma\sqrt{(U^0)^2 - 1}$, and $\gamma^X = \gamma U^0$. Then $\gamma^2 \varrho^2 - (l \cdot \gamma)^2 = \gamma^2 r^2 \sin^2\psi$.

Book 2

$$T_{\mu\nu}^{\rm rad} = \frac{q^2}{4\pi r^2} \frac{l_\mu l_\nu}{r^2} \frac{\gamma_{\mathcal{R}}^2 \sin^2 \psi}{(U_{\mathcal{R}}^0 - U_{\mathcal{R}}^X \cos \psi)^6}, \quad \text{with} \quad l^\mu = r(1, n^i) \quad \text{and} \quad \cos \psi = n.e_X. \qquad (18.16)$$

The radiated energy is given by (18.5):

$$P^0 = -\frac{q^2}{2} \int \frac{\gamma^2 \sin^2 \psi}{(U^0 - U^X \cos \psi)^5} \sin \psi \, d\psi \, d\tau. \qquad (18.17)$$

Now let us give the three-dimensional versions of these expressions.

The radiated energy density is obtained from (18.16) or (18.10):

$$W_{\rm rad} = \frac{q^2}{4\pi r^2} \frac{a_{\mathcal{R}}^2 \sin^2 \psi}{(1 - V_{\mathcal{R}} \cos \psi)^6}. \qquad (18.18)$$

Since $W = (E^2 + B^2)/(8\pi)$, in the special case of constant acceleration this formula can also be obtained using the expression for the radiation field (17.27). Equations (18.11), (18.12), and (18.6) giving the power radiated by the charge then become

$$\frac{dP^0}{dT \, do} = -\frac{q^2}{4\pi} \frac{a^2 \sin^2 \psi}{(1 - V \cos \psi)^5}, \quad \frac{dP^0}{dT} = -\frac{2q^2}{3} \frac{a^2}{(1 - V^2)^3} = -\frac{2q^2}{3} \gamma^2. \qquad (18.19)$$

The radiated power is a maximum for the solid angle which extremizes $dP^0/(dT \, do)$, that is, $\cos \psi_{\max} = \frac{\sqrt{1 + 15V^2} - 1}{3V}$ (for $V > 0$). In the ultrarelativistic limit we find the following by truncating the series expansion:

$$\begin{cases} \psi_{\max} \sim \dfrac{\sqrt{1 - V^2}}{2} \\[2mm] \left. \dfrac{dP^0}{dT \, do} \right|_{\max} \sim \dfrac{2048}{3125\pi} \dfrac{q^2 \gamma^2}{(1 - V^2)} = \dfrac{2048}{3125\pi} \dfrac{q^2 a^2}{(1 - V^2)^4} \end{cases} \quad \text{when} \quad V \to 1. \qquad (18.20)$$

Therefore, the power radiated at $\psi_{\max} \approx 0$ diverges as $(1 - V^2)^{-4}$, and the total power [see the second expression in (18.19)] diverges as $(1 - V^2)^{-3}$ if $V \to 1$ with a finite. On the other hand, if γ^2 remains finite (as occurs if the charge is accelerated by a constant electric field; see Section 11.4), then the power radiated at $\psi_{\max}$ diverges as $(1 - V^2)^{-1}$ and the total radiated power remains finite.

Radiation losses in a linear accelerator

Let x be the energy per unit length that can be transferred to a charge q in a linear accelerator. We assume that this energy gain is the result of uniformly accelerated motion, that is, that the charge world line is given by $gT = \sinh g\tau$, $gX = \cosh g\tau$, where g is the modulus of the acceleration. By definition, $x \equiv mU^0/X = (m/X)(dT/d\tau)$, and so $x = mg$, where m is the particle mass.

The total radiated power given by the Larmor formula is then written as [see (18.19)]

$$\frac{dP^0}{dT} = -\frac{2q^2}{3} g^2 = -\frac{2}{3} \left(\frac{qx}{m} \right)^2.$$

Introducing the energy of the charge, $\mathcal{E} \equiv mU^0 = m\sqrt{1 + g^2 T^2}$, the integral gives

$$P^0 = -\frac{2q^2}{3}\frac{x}{m^2}\sqrt{\mathcal{E}^2 - m^2}.$$

For an electron $dP^0/dT \approx 1.6 \times 10^{-6} x^2$ MeV/s and $P^0 \approx 3.6 \times 10^{-5} x\,\mathcal{E}$ GeV, with x in MeV/m and $\mathcal{E}$ in GeV.

Therefore, to transfer an energy of $\mathcal{E} = 100$ GeV to an electron using a linear accelerator capable of producing an acceleration of $x = 100$ MeV/m, the required length of the accelerator is 1 km and the energy lost is $P^0 \approx 0.37$ GeV.

(If the acceleration g is due to a constant electric field E, we have $g = qE/m$ (see Section 11.4), and so $E = x/m$ or $E \approx 3.34 \times 10^{-3} x$ V/m, with x expressed in MeV/m.)

18.3   Radiation of a charge in circular motion

Let us now consider a charge in circular motion ($X = r_0 \cos\Omega T$, $Y = r_0 \sin\Omega T$, $Z = 0$) at constant angular velocity Ω due, for example, to a constant external magnetic field B. In this case (see Section 11.4) $\Omega = \omega/\sqrt{1 + r_0^2\omega^2}$, where $\omega \equiv qB/m$, m being the mass of the charge. Since now the 3-velocity $V = r_0\Omega(-\sin\Omega T, \cos\Omega T, 0)$ and the 3-acceleration $a = -r_0\Omega^2(\cos\Omega T, \sin\Omega T, 0)$ are orthogonal, (18.10) for the energy density becomes

$$T_{\text{rad}}^{00} = \frac{q^2 r_0^2 \Omega^4}{4\pi r^2}\frac{(r_0\Omega - \sin\theta\sin\psi)^2 + \cos^2\theta(1 - r_0^2\Omega^2)}{(1 - r_0\Omega\sin\theta\sin\psi)^6} \quad \text{with} \quad \psi \equiv \phi - \Omega T_{\mathcal{R}}, \quad (18.21)$$

where θ is the angle between the Z axis and the observation direction and $\tan\phi \equiv Y/X$. This expression is of course equal to $(E_{\text{rad}}^2 + B_{\text{rad}}^2)/(8\pi)$ with E_{rad} and B_{rad} given by (17.29). However, it should be noted that the time average of T_{rad}^{00} is *not* equal to $(\langle \mathcal{F}_{\text{rad}}\rangle)^2/4\pi$ with $\langle \mathcal{F}_{\text{rad}}\rangle$ from (17.30), because $\langle \mathcal{F}_{\text{rad}}^2\rangle \neq (\langle \mathcal{F}_{\text{rad}}\rangle)^2$.

Equation (18.11) giving the power radiated by the charge in the direction of observation as a function of the time of emission is written as

$$\frac{dP^0}{dT\,do} = -\frac{q^2 r_0^2 \Omega^4}{4\pi}\frac{(r_0\Omega - \sin\theta\sin\psi)^2 + \cos^2\theta(1 - r_0^2\Omega^2)}{(1 - r_0\Omega\sin\theta\sin\psi)^5} \quad \text{with} \quad \psi \equiv \phi - \Omega T. \quad (18.22)$$

In the nonrelativistic limit $r_0\Omega \ll 1$ this becomes $\frac{dP^0}{dT\,do} \approx -\frac{q^2 r_0^2\Omega^4}{4\pi}(\sin^2\theta\sin^2\psi + \cos^2\theta)$. It varies sinusoidally in time and its Fourier transform is peaked at the frequency 2Ω. In the ultrarelativistic case $r_0\Omega \to 1$ the field again varies periodically, but its Fourier spectrum turns out to be not only wide, but also centered at the frequency[7]

$$\omega_{\text{rad sync}} \approx 0.29 \times \frac{3}{2}\frac{\Omega}{(1 - r_0^2\Omega^2)^{3/2}}.$$

Calculating the integrals,[8] we obtain the average of (18.22) over the time T:

---

[7]The detailed calculation of the spectrum can be found in, for example, Landau and Lifshitz (1980) or Rybicki and Lightman (1985).

[8]Using $\int_0^{2\pi}\frac{d\psi}{(1 - a\sin\psi)^5} = \frac{\pi}{4}\frac{8 + 3a^2(8 + a^2)}{(1 - a^2)^{9/2}}$, $\int_0^{2\pi}\frac{\sin\psi\,d\psi}{(1 - a\sin\psi)^5} = \frac{\pi}{4}\frac{5a(4 + 3a^2)}{(1 - a^2)^{9/2}}$, and $\int_0^{2\pi}\frac{\sin^2\psi\,d\psi}{(1 - a\sin\psi)^5} = \frac{\pi}{4}\frac{4 + 27a^2 + 4a^4}{(1 - a^2)^{9/2}}$.

$$\left\langle \frac{dP^0}{dT do} \right\rangle = -\frac{q^2 r_0^2 \Omega^4}{16} \frac{\left[8 - 4\sin^2\theta - \Omega^2 r_0^2 (1 + 3\Omega^2 r_0^2)\sin^4\theta\right]}{(1 - r_0^2 \Omega^2 \sin^2\theta)^{7/2}}. \tag{18.23}$$

The ratio of the average radiated intensities in the plane of the trajectory $(\theta = \pi/2)$ and perpendicular to this plane $(\theta = 0)$ is

$$\frac{\langle \frac{dP^0}{dT do}\rangle_{\pi/2}}{\langle \frac{dP^0}{dT do}\rangle_0} = \frac{4 + 3\Omega^2 r_0^2}{8(1 - \Omega^2 r_0^2)^{5/2}}. \tag{18.24}$$

Therefore, a charge in rapid circular motion radiates principally in the plane of its orbit.

Finally, the total radiated power is the integral of (18.23) over $do = \sin\theta \, d\theta \, d\phi$, which reduces,[9] as it should, to (18.6) or (18.12). The radiated angular momentum is calculated from (18.15):

$$\frac{dP^0}{dT} = -\frac{2q^2 r_0^2 \Omega^4}{3(1 - r_0^2 \Omega^2)^2} = -\frac{2q^2}{3}\gamma^2, \quad \frac{dJ^Z}{dT} = \frac{2q^2 r_0^2 \Omega^3}{3\sqrt{1 - r_0^2 \Omega^2}}. \tag{18.25}$$

Radiation losses in a synchrotron

The power radiated by a charge traveling in a circular orbit is given by (18.25). Introducing its velocity $|V| = r_0\Omega$ and its energy $\mathcal{E} \equiv mU^0 = m/\sqrt{1 - V^2}$, we find that this power and the energy radiated per revolution are given by

$$\frac{dP^0}{dT} = -\frac{2q^2}{3m^4}\frac{V^4 \mathcal{E}^4}{r_0^2}, \quad P^0 = -\frac{4\pi q^2}{3m^4}\frac{V^3 \mathcal{E}^4}{r_0}. \tag{18.26}$$

Let us take the example of LEP at CERN (see also Section 11.4). For an energy $\mathcal{E} = 100$ GeV and a radius $r_0 = 5$ km, the radiated power is $dP^0/dT \approx 3.5 \times 10^{-23}$ kg/s $\approx 2 \times 10^7$ MeV/s $\approx 3 \times 10^{-6}$ watt per electron. The energy lost in a revolution is $P^0 \approx 1.8$ GeV, so that an electron loses about 2% of its energy in each revolution.

We note that the energy lost in a synchrotron grows as the quartic power of the energy $\mathcal{E}$ of the particle, whereas in a linear accelerator it grows only linearly; cf. Section 18.2.

Radiation of the hydrogen atom

The frequency of an electron of charge $q = -e$ in its Bohr orbit, that is, at a distance $r_0 = e^2/ma^2$ from the proton of charge $Q = e$ which is assumed at rest, was given in (13.16): $\Omega \approx ma^3/e^2$. The radiated power given by the Larmor formula (18.2), that is, $dP^0/dT \approx -(2e^2/3)a^2$ for orbital velocity small compared to 1, then becomes the following (using $a^2 = r_0^2\Omega^4$):

$$\frac{dP^0}{dT} \approx -\frac{2}{3}\frac{m^2\alpha^8}{e^2}. \tag{18.27}$$

For a *single* hydrogen atom this is $dP^0/dT \approx 5.2 \times 10^{-25}$ kg/s $\approx 4.7 \times 10^{-8}$ watt. This prediction, which is in flagrant contradiction with experiment, was initially viewed as an argument against

---

[9]Using $\int_0^\pi \frac{\sin\theta}{(1 - V^2\sin^2\theta)^{7/2}} = \frac{2(15 - 10V^2 + 3V^4)}{15(1 - V^2)^3}$, $\int_0^\pi \frac{\sin^3\theta}{(1 - V^2\sin^2\theta)^{7/2}} = \frac{4(5 - V^2)}{15(1 - V^2)^3}$, and $\int_0^\pi \frac{\sin^5\theta}{(1 - V^2\sin^2\theta)^{7/2}} = \frac{16}{15(1 - V^2)^3}$.

the Rutherford planetary model of the atom (1911; see Section 13.3). In fact, it rather reveals the limits of classical electrodynamics which, as we have mentioned, for example, at the end of Section 11.3, does not apply when the ratio $\mathbf{S}/\hbar$, with $\mathbf{S}$ the electron action, becomes of order unity.

18.4 Thomson scattering

In Section 15.4 we studied the motion of a charge in the field of a linearly polarized monochromatic plane wave, and in Section 17.3 we found the radiation part of the field created by a moving charge. Here we shall study the energy radiated to infinity.

The incident wave is described at the observation point (T, X, Y, Z) by the potential $A_\mu = (0, 0, a \cos k(T - X), 0)$. It propagates along the X axis with frequency k and is linearly polarized in the Y direction. Its energy density is given by (15.11):

$$W_{\text{in}} = \frac{k^2 a^2}{4\pi} \sin^2 k(T - X) \quad \text{or, on average,} \quad \langle W_{\text{in}} \rangle = \frac{k^2 a^2}{8\pi}. \tag{18.28}$$

The 4-velocity of the charge accelerated by this wave is (see Section 17.3)

$$U^X = \frac{\beta^2}{2}(\cos \xi - 1)^2, \quad U^Y = -\beta(\cos \xi - 1), \quad U^0 = 1 + U^X, \tag{18.29}$$

where $\xi \equiv k\tau$ and $\beta \equiv qa/m$, τ is the proper time of the particle, and q and m are its charge and mass. The total radiated power is given by the Larmor formula (18.6) using the time T of the frame rather than the proper time of the charge:

$$\frac{dP^0}{dT} = -\frac{2q^2}{3}\gamma^2, \tag{18.30}$$

where $\gamma^\mu = k dU^\mu/d\xi$ is the acceleration of the charge, so that from (18.29) we have $\gamma^2 = k^2 (qa/m)^2 \sin^2 \xi$.

The *Thomson total cross section* σ_{tot} (chosen to be positive) is the ratio of the average of this radiated energy and the averaged energy density of the incident wave $\langle W_{\text{in}} \rangle$ (18.28). It is therefore given by

$$\sigma_{\text{tot}} = \frac{8\pi}{3}\left(\frac{q^2}{m}\right)^2. \tag{18.31}$$

The energy radiated per unit time T in the direction n is given by (18.5):

$$\frac{dP^0}{do\,dT} = -\frac{q^2}{4\pi}\left(\frac{r}{\varrho}\right)^3\left(\gamma^2 - \frac{(l \cdot \gamma)^2}{\varrho^2}\right)\frac{1}{U^0} \quad \text{with} \quad \begin{cases} l^\mu = (r, rn^i) \\ \varrho = -(l \cdot U), \end{cases} \tag{18.32}$$

where [*cf.* (18.29)] $U^0 = 1 + \frac{\beta^2}{2}(\cos \xi - 1)^2$ and

$$\varrho \equiv -(l \cdot U) = r + Y\beta(\cos \xi - 1) + \frac{1}{2}\beta^2(r - X)(\cos \xi - 1)^2$$

$$(l \cdot \gamma) = k\beta \sin \xi \left[Y + \beta(r - X)(\cos \xi - 1)\right]. \tag{18.33}$$

It is interesting to note that integration of (18.32) and (18.33) over the angles must and indeed does give (18.30) for any β, that is, for any velocity transferred to the charge by the incident wave.

The *differential cross section* is the time average of the radiated energy (18.32) divided by the average energy density of the incident wave:

$$\frac{d\sigma}{do} \equiv -\frac{1}{\langle W_{\text{in}} \rangle} \frac{1}{2\pi} \int_0^{2\pi} \frac{dP^0}{do\, dT}\, d\xi\,. \tag{18.34}$$

In the limit where the average velocity transferred to the charge is small, this expression, in lowest order in β, reduces to the *Thomson cross section*:

$$\frac{d\sigma}{do} = \left(\frac{q^2}{m}\right)^2 (1 - \sin^2\theta \sin^2\phi) + \mathcal{O}(\beta), \tag{18.35}$$

where we have changed over to spherical coordinates: $X = r\sin\theta\cos\phi$, $Y = r\sin\theta\sin\phi$, $Z = r\cos\theta$, where ψ satisfying $\cos\psi \equiv \sin\theta\sin\phi$ is the angle between the observation direction n and the Y axis, which at this order is the direction of the electric field and of the charge motion.

We have considered an incident wave polarized along the Y axis and propagating along the X axis, but we would have obtained the same expression (18.35) for a wave propagating in the Z direction. To describe the radiation of a charge in the field of an *unpolarized* wave, it is therefore sufficient to take the average of (18.35) over the angle ϕ, which gives

$$\left\langle \frac{d\sigma}{do} \right\rangle = \left(\frac{q^2}{m}\right)^2 \frac{1 + \cos^2\theta}{2} + \mathcal{O}(\beta), \tag{18.36}$$

where now θ is the angle between the direction of the incident wave and the scattered wave.

Thomson scattering and the Compton effect

The cross sections (18.31) and (18.34) are in principle valid for any energy density of the incident wave, and therefore for any velocity transferred by the wave to the charge.

However, as we saw in Section 6.5, light can also be described as photons, in which case the scattering cross section is given by the Klein–Nishina formula (6.25) and (6.26). We see that these expressions obtained using quantum electrodynamics do not in general coincide with the classical expressions we have obtained here. However, they both converge to the Thomson limit (18.31) and (18.36) if the charge velocity remains small compared to the speed of light in the case of the classical expression (18.34) and (18.35), or if the photon energy is small compared to the mass of the charge in the case of the Klein–Nishina formula.

Therefore, as already pointed out in Section 6.5, the relevant parameter for judging the validity of the classical formulas is $\mathbf{S}/\hbar$, where $\mathbf{S}$ is the action of the charge. In the present case $\mathbf{S} \sim mV^2T$, where m is the electron mass, $V \sim \beta = ea/m$ is the electron velocity, and $T \sim 1/k$ is the characteristic time of the problem. Therefore, since $\hbar = e^2/\alpha$, we have $\mathbf{S}/\hbar \sim a^2\alpha/(km) \sim (W_{\text{in}}/k^3)(\alpha/m)$, where W_{in} is the energy density of the incident wave and W_{in}/k^3 is the wave energy contained in a volume the size of the wavelength $1/k$. The classical expressions (18.31), (18.34), and (18.35) therefore are only applicable if this wave energy is large compared to the electron mass. If it is not, the Klein–Nishina formula must be used.

19
The radiation reaction force

A charge placed in an external electromagnetic field is subject to the Lorentz force and is therefore accelerated (see the examples of such motion in Sections 11.4 and 15.4). Moreover, as we have seen in the preceding chapter, an accelerated charge radiates electromagnetic waves, and the system of the charge plus the field it creates loses energy according to the Larmor formula (18.6) or (18.12). It is therefore to be expected that the charge slows down. Here we shall study the reaction force acting on a charge due to the radiation it emits, along with the related questions of *renormalization* and the physical interpretation.

In addition, we shall again see that a hydrogen atom, this time described by the Thomson model, is unstable in the Maxwell theory.

19.1 The Abraham–Lorentz–Dirac force

A heuristic method of describing the expected slowing of an accelerated charge in response to the radiation it emits is to modify the Lorentz equation (11.14) by including a radiation *reaction force*. We therefore write the equation of motion of a charge *including* its own field as

$$m\frac{dU^\mu}{d\tau} = q(F_{\text{ext}}^{\mu\nu} + F_{\text{self}}^{\mu\nu})U_\nu\,. \tag{19.1}$$

We see that we need to evaluate the self-retarded potential $A_{\text{self}}^\mu = qU_{\mathcal{R}}^\mu/\varrho_{\mathcal{R}}$ due to the charge itself, as well as the resulting Faraday tensor $F_{\mu\nu}^{\text{self}}$ [given in (17.16)], *on the world line L* of the charge.

Since A_{self}^μ and $F_{\mu\nu}^{\text{self}}$ diverge on L, we start by working near L and introduce the point X_0^μ of L such that (see Fig. 19.1)

$$X^\mu - X_0^\mu \equiv \epsilon\, n^\mu, \quad \text{with } n^\mu U_\mu = 0 \text{ and } n^\mu n_\mu = 1, \tag{19.2}$$

where $U^\mu \equiv U^\mu(\tau_0)$ is the velocity of the charge at X_0^μ and

$$l_{\mathcal{R}}^\mu \equiv X^\mu - X_{\mathcal{R}}^\mu = \epsilon\, n^\mu + (\tau_0 - \tau_{\mathcal{R}})U^\mu - \frac{1}{2}(\tau_0 - \tau_{\mathcal{R}})^2\gamma^\mu + \frac{1}{6}(\tau_0 - \tau_{\mathcal{R}})^3\dot\gamma^\mu + \cdots, \tag{19.3}$$

where the dot denotes differentiation with respect to τ. Requiring that $(l^\mu l_\mu)_{\mathcal{R}} = 0$, we obtain $(\tau_0 - \tau_{\mathcal{R}})$ by successive iterations:

$$\tau_0 - \tau_{\mathcal{R}} = \epsilon - \frac{1}{2}\epsilon^2(n\cdot\gamma) + \frac{\epsilon^3}{2}\left[\frac{3}{4}(n\cdot\gamma)^2 + \frac{1}{3}(n\cdot\dot\gamma) - \frac{1}{12}\gamma^2\right] + \cdots \tag{19.4}$$

Relativity in Modern Physics. Nathalie Deruelle and Jean-Philippe Uzan.
© Oxford University Press 2018. Published in 2018 by Oxford University Press.
DOI: 10.1093/oso/9780198786399.001.0001

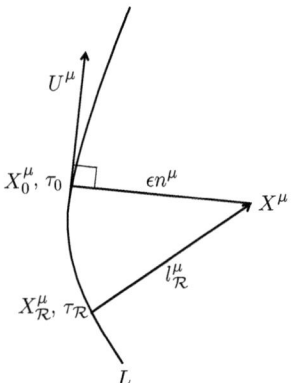

Fig. 19.1 Laurent expansion of the field.

The Laurent series expansions of various retarded quantities then follow (see below for details) and we find

$$\frac{1}{q} A^\mu_{\text{self}} = \frac{U^\mu}{\epsilon} - \gamma^\mu - \frac{U^\mu (n \cdot \gamma)}{2}$$
$$+ \epsilon \left[\frac{3}{8} U^\mu (n \cdot \gamma)^2 + \frac{1}{3} U^\mu (n \cdot \dot\gamma) - \frac{1}{8} U^\mu \gamma^2 + \gamma^\mu (n \cdot \gamma) + \frac{1}{2} \dot\gamma^\mu \right] + \cdots$$

(19.5)

Laurent expansion of retarded quantities

Given the expansion (19.4) of $(\tau_0 - \tau_{\mathcal{R}})$, we easily obtain

$$\begin{cases} l^\mu_{\mathcal{R}} = \epsilon(n^\mu + U^\mu) - \frac{1}{2}\epsilon^2 \left[(n \cdot \gamma)U^\mu + \gamma^\mu \right] \\ \qquad + \frac{1}{2}\epsilon^3 \left[\frac{3}{4}(n \cdot \gamma)^2 U^\mu + \frac{1}{3}(n \cdot \dot\gamma)U^\mu - \frac{1}{12}\gamma^2 U^\mu + \gamma^\mu (n \cdot \gamma) + \frac{1}{3}\dot\gamma^\mu \right] + \cdots \\ U^\mu_{\mathcal{R}} = U^\mu - \epsilon\gamma^\mu + \frac{1}{2}\epsilon^2 \left[(n \cdot \gamma)\gamma^\mu + \dot\gamma^\mu \right] + \cdots \\ \varrho_{\mathcal{R}} = \epsilon + \frac{1}{2}\epsilon^2 (n \cdot \gamma) + \epsilon^3 \left[\frac{1}{8}\gamma^2 - \frac{1}{3}(n \cdot \dot\gamma) - \frac{1}{8}(n \cdot \gamma)^2 \right] + \cdots \end{cases}$$

To obtain the Faraday tensor from (19.5) for the retarded potential, it is necessary to know how far the point X^μ_0 moves along the world line when X^μ is shifted by δX^μ. By varying the expression $(X^\mu - X^\mu_0)U_\mu = 0$ and using $\delta X^\mu_0 = U^\mu \delta\tau_0$ and $\delta U^\mu = \gamma^\mu \delta\tau_0$, we find

$$\partial_\mu \tau_0 = -\frac{U_\mu}{1 + \epsilon(n \cdot \gamma)} = -U_\mu \left[1 - \epsilon(n \cdot \gamma) + \epsilon^2 (n \cdot \gamma)^2 + \ldots \right],$$

and so

$$\partial_\mu U^\nu = -\gamma^\nu U_\mu \left[1 - \epsilon(n \cdot \gamma) + \cdots \right], \quad \partial_\mu \epsilon = n_\mu,$$
$$\partial_\mu n^\nu = \frac{1}{\epsilon} \left(-n_\mu n^\nu + \delta^\nu_\mu + U^\nu U_\mu - U^\nu U_\mu (n \cdot \gamma) + \cdots \right).$$

(19.6)

[The second equation follows from varying $\epsilon = (X^\mu - X^\mu_0)n_\mu$ using $(n \cdot u) = 0$ and $(n \cdot \delta n) = 0$, and the third follows from varying $X^\mu - X^\mu_0 = \epsilon n^\mu$.]

The self electromagnetic field $F_{\mu\nu}^{\text{self}} = \partial_\mu A_\nu^{\text{self}} - \partial_\nu A_\mu$ created by the charge near its world line is derived from (19.5) for the potential using (19.6). We find

$$\frac{1}{q} F_{\mu\nu}^{\text{self}} = \frac{U_\mu n_\nu - U_\nu n_\mu}{\epsilon^2} + \frac{1}{\epsilon}\left[(n\cdot\gamma)(n_\mu U_\nu - n_\nu U_\mu) + \frac{1}{2}(\gamma_\mu U_\nu - \gamma_\nu U_\mu)\right]$$

$$+\frac{1}{8}\left[3(n\cdot\gamma)^2 + \gamma^2\right](U_\mu n_\nu - U_\nu n_\mu) + \frac{1}{2}(n_\mu\dot\gamma_\nu - n_\nu\dot\gamma_\mu) \tag{19.7}$$

$$+\frac{2}{3}(U_\mu\dot\gamma_\nu - U_\nu\dot\gamma_\mu) + \frac{3}{4}(n\cdot\gamma)(U_\mu\gamma_\nu - U_\nu\gamma_\mu) + \cdots,$$

from which we derive the Laurent expansion of the quantity of interest in (19.1), namely,

$$\frac{1}{q} F_{\text{self}}^{\mu\nu} U_\nu = \frac{n^\mu}{\epsilon^2} - \frac{1}{\epsilon}\left[n^\mu(n\cdot\gamma) + \frac{1}{2}\gamma^\mu\right] + n^\mu\left[\frac{3}{8}(n\cdot\gamma)^2 - \frac{3}{8}\gamma^2\right]$$

$$-\frac{2}{3}\gamma^2 U^\mu + \frac{3}{4}(n\cdot\gamma)\gamma^\mu + \frac{2}{3}\dot\gamma^\mu + \mathcal{O}(\epsilon). \tag{19.8}$$

At this stage the question becomes how to interpret (19.8) when $\epsilon \to 0$. One possible answer,[1] which is ultimately justified by studying systems of interacting charges (see the following chapter), is to take the average over n^μ. To do this we work in the inertial frame tangent to the world line where the vector n^μ is a spatial vector ($n^0 = 0$) and $\langle n_i n_j \rangle = \frac{1}{3}\delta_{ij}$, and then return to the original frame. This gives

$$\frac{1}{q}\langle F_{\text{self}}^{\mu\nu} U_\nu \rangle = -\frac{5}{6}\frac{\gamma^\mu}{\epsilon} + \frac{2}{3}\left(\dot\gamma^\mu - \gamma^2 U^\mu\right), \tag{19.9}$$

so that, *modulo* this substitution, the equation of motion (19.1) of a charge in an external field taking into account its own field is

$$m\gamma^\mu = q F_{\text{ext}}^{\mu\nu} U_\nu + \frac{2q^2}{3}\left(\dot\gamma^\mu - \gamma^2 U^\mu\right), \tag{19.10}$$

as proposed by Abraham (1905), Lorentz (1892), and Dirac (1938). Here the mass has been renormalized: $m + 5q^2/(6\epsilon) \to m$, which in practice amounts to ignoring the divergent term in (19.9). The second term in (19.10) is called the *radiation reaction force*.

In the limit where the charge velocity is small compared to the speed of light, (19.10) reduces to

$$ma = F_{\text{ext}} + \frac{2q^2}{3}\dot a, \tag{19.11}$$

where $a \equiv dV/dT$ is the 3-acceleration of the charge, $\dot a$ is its derivative with respect to T, and $F_{\text{ext}} = q(E + V \wedge B)$ is the Lorentz force due to the external field (E, B).

---

[1] For a review of the various regularization schemes used in classical electrodynamics, see, for example, Damour (1975).

The Thomson model of the atom

In the 'plum pudding' model of the atom proposed by J. J. Thomson in 1904, the electron, of mass m and charge $-e$, immersed in a 'pudding' of charge $+e$, undergoes harmonic motion: $d^2X/dT^2 + \omega_0^2 X = 0$, where the frequency ω_0 is taken to be the Bohr frequency $\omega_0 = m\alpha^3/e^2$, and α is the fine-structure constant; see Section 13.3. This oscillating charge radiates, and the counter-reaction to this radiation is given by the Abraham–Lorentz–Dirac force (19.11). Therefore, the electron equation of motion taking into account this force becomes

$$\frac{d^2X}{dT^2} + \omega_0^2 X - \tau_0 \frac{d^3X}{dT^3} = 0 \quad \text{with} \quad \tau_0 \equiv \frac{2e^2}{3m}, \tag{19.12}$$

the solution of which is the real part of $X_0 e^{-i\omega T}$ with $-\omega^2 + \omega_0^2 - i\omega^3 \tau_0 = 0$. If we assume that the reaction force remains weak so that ω remains close to ω_0, this third-degree equation for ω reduces to $-\omega^2 + \omega_0^2 - i\omega_0^3 \tau_0 = 0$. This approximation is equivalent to replacing (19.12) by

$$\frac{d^2X}{dT^2} + \omega_0^2 X + \omega_0^2 \tau_0 \frac{dX}{dT} = 0. \tag{19.13}$$

The electron motion is exponentially damped: $X \approx X_0 e^{-i\omega_0 T} e^{-\frac{\omega_0 \tau_0}{2} T}$ with $\omega_0 \tau_0 / 2 = \alpha^3 / 3 \approx 1.3 \times 10^{-7}$. The Thomson atom is therefore unstable (like the Rutherford atom, as we have seen in Section 18.3 and will study again in Section 21.2).

Now let us assume that, in addition, the electron is immersed in an electromagnetic wave. At lowest order in V the Lorentz force it is subjected to is $-eE$. If the atom is much smaller than the wavelength of the field, the spatial structure of the wave can be neglected and we can set $E = E_0 e^{i\omega T}$. The electron equation of motion then becomes that of a forced oscillator:

$$\frac{d^2X}{dT^2} + \omega_0^2 X + \omega_0^2 \tau_0 \frac{dX}{dT} = -\frac{eE_0}{m} e^{-i\omega T}. \tag{19.14}$$

After a transient phase the motion is $X = X_0 e^{i\omega T}$ with $X_0 = \frac{eE_0/m}{\omega_0^2 - \omega^2 - i\omega\omega_0\tau_0}$. The 3-acceleration is $a = \omega^2 |X_0|$ and the radiated Larmor power given by (18.12) is, in lowest order, $dP/dT = 2e^2/3a^2$. The energy density of the wave is $W = E_0^2/(4\pi)$. Therefore, the cross section $\sigma \equiv (1/W)(dP/dT)$ is given by

$$\sigma = \frac{8\pi}{3} r_e^2 \frac{\omega^4}{(\omega_0^2 - \omega^2)^2 + \omega^2(\omega_0\tau_0)^2}, \tag{19.15}$$

where $r_e \equiv e^2/m$ is the classical electron radius (see Section 13.1). If the incident wave has low frequency $\omega \ll \omega_0$, then $\sigma \approx (8\pi/3)r_e^2(\omega/\omega_0)^4$ and the scattering is *Rayleigh scattering*. If the frequency is high, the electron is nearly free and we recover the Thomson cross section[2] $\sigma \approx (8\pi/3)r_e^2$; see Section 18.4.

19.2 The reaction force and the Larmor formulas

The Abraham–Lorentz–Dirac reaction force, the counter-effect of the radiation of an accelerated charge on its motion, is given by (19.10):

---

[2]For examples illustrating the use of the Thomson model in the study of atomic sources of radiation, see Raimond (2000).

$$g^\mu \equiv \frac{2q^2}{3}(\dot\gamma^\mu - \gamma^2 U^\mu). \tag{19.16}$$

Integrating this along the charge world line L, we find

$$\int_L g^\mu d\tau = \frac{2q^2}{3}\gamma^\mu\Big|_{\tau_1}^{\tau_2} - \frac{2q^2}{3}\int \gamma^2 U^\mu d\tau. \tag{19.17}$$

The first term is zero either because the accelerations vanish at spatial infinity if the motion is unbounded, or because its average is zero if the motion is bounded. The second term is just the *Larmor* formula giving the power radiated by the charge. Equation (19.17) can therefore be interpreted as an *energy balance*.

To understand this interpretation, we rewrite the zeroth component of the Abraham–Lorentz–Dirac equation $m\gamma^0 = F_{\text{ext}}^0 + g^0$ in the low-velocity limit (using the expression obtained in Section 6.2, footnote 2):

$$\frac{d}{dT}\left(\frac{1}{2}mV^2\right) \simeq F_{\text{ext}}^i . V_i + \frac{2q^2}{3}(\dot a^i V_i), \tag{19.18}$$

which states that the change of the (Newtonian) kinetic energy of the charge is equal to the work performed by the forces. Therefore, the change of the kinetic energy due to the work performed by the reaction force in this limit is

$$\Delta E_{\text{kin}} = \frac{2q^2}{3}\int dT(\dot a^i V_i) = \frac{2q^2}{3}(a^i V_i)\Big|_{T_1}^{T_2} - \frac{2q^2}{3}\int dT a^2. \tag{19.19}$$

The first term vanishes for the reasons given above, and the second is equal to the radiated power in the $V \to 0$ limit of the Larmor formula (18.12).

19.3 Caveats

The Abraham–Lorentz–Dirac equation (19.10) is a nonlinear differential equation of *third* order in $X^\mu(\tau)$. It therefore possesses solutions which do *not* approach those of the Lorentz equation when $q \to 0$, and these must be excluded. For example, in the absence of an external field (19.10) becomes (recalling that $U^\mu U_\mu = -1$)

$$\gamma^\mu = \frac{2q^2}{3m}(\dot\gamma^\mu - \gamma^2 U^\mu)$$

$$\implies \quad \gamma^2 = C^2 \exp\left(\frac{3m}{q^2}\tau\right) \tag{19.20}$$

$$\implies \quad U^i \propto \exp\left[\exp\left(\frac{3m}{q^2}\tau\right)\right] \quad \text{for large } \tau,$$

a result which is clearly unacceptable, because the theory is based on the fact that a free particle in an inertial frame undergoes uniform motion.[3]

---

[3]The time scale of the 'blowup' of the solution (19.20) is very short: $\tau_0 \equiv 2q^2/(3m) = 10^{-23}$ s for an electron.

These pathological solutions can be eliminated by using the Landau–Lifshitz trick of treating the reaction force as a perturbation of the Lorentz equation:

$$m\gamma^\mu = qF^{\mu\nu}U_\nu \qquad \Longrightarrow \qquad \begin{cases} m\dot\gamma^\mu = q(U_\nu U^\rho \partial_\rho F^{\mu\nu} + F^{\mu\nu}\gamma_\nu) \\ \qquad = q\left(U_\nu \partial_\rho F^{\mu\nu} + \frac{q}{m}F^{\mu\nu}F_{\nu\rho}\right)U^\rho \\ m^2\gamma^2 = q^2 F^{\mu\nu}F_{\mu\rho}U_\nu U^\rho \end{cases}$$

and replacing (19.10) by the nonlinear second-order differential equation

$$m\gamma^\mu = qF^{\mu\nu}U_\nu + \frac{2q^3}{3m}U^\rho U_\nu \partial_\rho F^{\mu\nu} + \frac{2q^4}{3m^2}\left(F^{\mu\nu}F_{\nu\rho}U^\rho + F^{\rho\sigma}F_{\nu\rho}U^\nu U_\sigma U^\mu\right), \qquad (19.21)$$

where $F_{\mu\nu}$ is *a priori* understood to be an *external* field. However, this modified Abraham–Lorentz–Dirac equation is not much better. Let us consider the example of a charge moving along a constant electric field where the Faraday tensor reduces to $F^{01} = E$. The second term of (19.21) then vanishes *as does the third term*, which implies that a charge uniformly accelerated by a constant electric field and which radiates energy according to (18.19) is not slowed down.[4]

It should also be noted that if the 'good' equation of motion is (19.10) or (19.21) rather than the 'original' Lorentz equation (11.14), then the energy conservation laws (12.27) must be modified because they *are based on* the Lorentz equation.

In conclusion, these *caveats* probably display the limits of the *external field approximation*, that is, they show that if we wish to describe the effect of a charge's radiation on its motion, we must take into account the other charges which create the field in which the charge of interest moves. Therefore, the Abraham–Lorentz–Dirac equation (19.10) or (19.21) can make sense only if we study a *system* of interacting charges.

---

[4]A deeper discussion can be found in Rohrlich (2007) and in Gralla, Harte, and Wald (2009).

20
Interacting charges I

We shall now address the problem of radiation by a *system* of point charges which, owing to the fact that the electromagnetic interaction propagates at finite speed, can only be solved iteratively, by assuming that all speeds are small compared to the speed of light. After deriving the *dipole* and *quadrupole* formulas giving the radiation field and the energy radiated by the system in the lowest orders, we find the equations of motion of the charges of the system to third order in the velocities, that is, including the effect of the reaction force at lowest order.

20.1 The dipole field and radiation

In Section 17.2 we obtained the Liénard–Wiechert electric and magnetic fields created by a moving charge in an inertial frame $\mathcal{S}$. Since the Maxwell equations are linear, the fields created by a *system* of charges will be the sum of the fields created by each charge. Using 3-vector notation, at the observation point (T, X^i) they can be written as [cf. (17.18)]

$$E = \sum_a \left[\frac{q(1-V^2)}{r^2(1-n.V)^3}(n-V) + \frac{q[(n.a)(n-V)-(1-n.V)a]}{r(1-n.V)^3} \right]_{\mathcal{R}},$$
$$B = \sum_a n_{\mathcal{R}} \wedge E_a, \tag{20.1}$$

where $E \equiv \sum_a E_a$ with the sum running over the charges q_a, and n, r, V, and the acceleration a carry the subscript $\mathcal{R}$. If $z_{\mathcal{R}}^i$ are the coordinates of the charge (a) at time $T_{\mathcal{R}}$, then $r_{\mathcal{R}}$ and the unit vector $n_{\mathcal{R}}$ are defined as $X^i - z_{\mathcal{R}}^i \equiv r_{\mathcal{R}} n_{\mathcal{R}}^i$. At the instant $T_{\mathcal{R}}$ we have $r_{\mathcal{R}} = T - T_{\mathcal{R}}$, and so $T - T_{\mathcal{R}}$ is the time taken by light to travel from the charge to the observation point. Finally, $V_{\mathcal{R}} \equiv \dot{z}_{\mathcal{R}}$ and $a_{\mathcal{R}} \equiv \dot{V}_{\mathcal{R}}$ are the 3-velocity and 3-acceleration of the charge at the retarded time $T_{\mathcal{R}}$; see Fig. 20.1.

Far from the charges, only the $1/r$ radiation part of the fields will contribute to the radiation of energy and momentum;[1] see (17.26). Moreover, if the velocities are small (which we have not assumed in the preceding chapters), in lowest order we have $r_{\mathcal{R}} n_{\mathcal{R}} = r n$, where $r n$ is the radius vector of the observation point (the origin of the frame is taken to be at the center of the charge distribution, which is assumed to be bounded). Equation (20.1) therefore reduces to

$$E = \frac{1}{r}\sum_a q\left[n(n.a_{\mathcal{R}_0}) - a_{\mathcal{R}_0}\right] + \cdots, \quad B = n \wedge E + \cdots, \tag{20.2}$$

where all the charge accelerations are evaluated at time $T_{\mathcal{R}_0}$, so that $T - T_{\mathcal{R}_0} = r$ is the time taken by light to travel from the origin of the frame to the observation point. Introducing the radius vectors $z_a(T)$ of the charges, the fields (20.2) can be rewritten as

---

[1] Here we do not deal with the radiation of angular momentum, the calculation of which involves also the Coulomb part of the field; see Section 18.1.

Relativity in Modern Physics. Nathalie Deruelle and Jean-Philippe Uzan.
© Oxford University Press 2018. Published in 2018 by Oxford University Press.
DOI: 10.1093/oso/9780198786399.001.0001

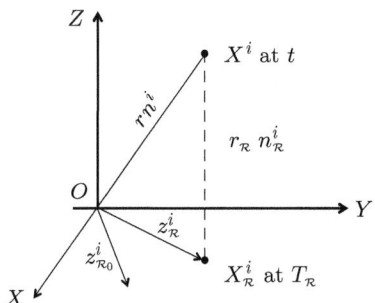

Fig. 20.1 Retarded quantities.

$$E = \frac{n(\ddot{d}_{\mathcal{R}_0}.n) - \ddot{d}_{\mathcal{R}_0}}{r} + \cdots, \quad B = \frac{\ddot{d}_{\mathcal{R}_0} \wedge n}{r} + \cdots, \tag{20.3}$$

where $d \equiv \sum_a qz$ is the dipole moment of the system and the dot denotes differentiation with respect to T. We see that E and B are perpendicular to the radius vector of the observation point $(n.E) = 0$: the field is that of a spherical wave which is nearly a plane wave at large r. Equations (20.3) are called the *first dipole formulas*.

To find the power lost by the system, we must return to the general expression (12.34), namely,

$$\frac{dP^0}{dT} = -\frac{1}{4\pi} \int_S (E \wedge B).n\, r^2\, do, \tag{20.4}$$

where do is the element of solid angle $[do = \sin\theta d\theta d\phi$ in spherical coordinates, where the components of n are $n = (\sin\theta\cos\phi, \sin\theta\sin\phi, \cos\theta)]$ and S is the sphere at infinity. Since $E \wedge B = E^2\, n$ we find that the power radiated by a system of slowly moving charges is given by the *second dipole formula*:

$$\frac{dP^0}{dT} = -\int \left[\ddot{d}_{\mathcal{R}_0}^2 - (n.\ddot{d}_{\mathcal{R}_0})^2 \right] do = -\frac{2\ddot{d}_{\mathcal{R}_0}^2}{3}, \tag{20.5}$$

where we have used (18.7) for the spatial averages.

We note that whereas the dipole moment depends in the general case $(\sum_a q \neq 0)$ on the choice of origin, its second derivative will not depend on the choice of inertial frame.

Finally, if the charge-to-mass ratio is the same for all the particles, the dipole moment is proportional to the radius vector ρ of the Newtonian center of mass of the system. In this case

$$d \equiv \sum_a qz = \frac{q}{m} \sum_a mz \equiv \frac{q}{m} \left(\sum_a m_a \right) \rho. \tag{20.6}$$

If the center of mass is in uniform translation, that is, if the system is not subject to an external force, then $\ddot{d} = 0$ and to obtain the radiated power it is necessary to keep the next-order term in the expansion of the field.

20.2   The quadrupole field and radiation

Finding the field and the radiation of a system of charges beyond the dipole approximation is rather more difficult but necessary in the absence of dipole radiation. It is also a useful exercise for studying the radiation of a *mass* system in theories of gravitation where the gravitational mass is equal to the inertial mass. Here we shall sketch the outlines of the approach.

We start from the three-dimensional expression for the Liénard–Wiechert potentials of a system of charges (17.18):

$$\Phi(T, X^i) = \sum_a \frac{q}{r_\mathcal{R}(1 - n_\mathcal{R} \cdot V_\mathcal{R})}, \quad A^j(T, X^i) = \sum_a \frac{q V_\mathcal{R}^j}{r_\mathcal{R}(1 - n_\mathcal{R} \cdot V_\mathcal{R})}. \tag{20.7}$$

A review of the definitions of the various quantities together with a figure were given in the preceding section.

We introduce the origin of the frame (taken to be at the center of the charge distribution, which is assumed bounded) and write $r_\mathcal{R} n_\mathcal{R} = r\, n - z_\mathcal{R}$, where $r\, n$ is the radius vector of the observation point with components X^i and length r, and $z_\mathcal{R}$ is the radius vector of the charge at $T_\mathcal{R}$. In addition, we assume that the charge velocities remain small. Then, since $r_\mathcal{R} n_\mathcal{R} = r\, n$ at lowest order in z/r, we have[2]

$$\begin{cases} \Phi = \sum_a \frac{q}{r} \left[1 + (n \cdot V_\mathcal{R}) + (n \cdot V_\mathcal{R})^2 + (n \cdot V_\mathcal{R})^3 + \mathcal{O}(V^4) + \mathcal{O}(z/r) \right], \\[2mm] A^j = \sum_a \frac{q V_\mathcal{R}^j}{r} \left[1 + (n \cdot V_\mathcal{R}) + (n \cdot V_\mathcal{R})^2 + \mathcal{O}(V^3) + \mathcal{O}(z/r) \right]. \end{cases} \tag{20.8}$$

Now let us take all the quantities at the time $T_{\mathcal{R}_0}$, so that $T - T_{\mathcal{R}_0} = r$ is the time for light to travel from the origin of the frame to the observation point. Then using the expansion of $V_\mathcal{R}$, see (20.10) below, we find

$$\begin{cases} \Phi = \sum_a \frac{q}{r} + \sum_a \frac{q}{r}(n \cdot V_{\mathcal{R}_0}) + \sum_a \frac{q}{r}\left[(n \cdot V)^2 + (n \cdot z)(n \cdot \dot{V}) \right]_{\mathcal{R}_0} \\[2mm] \quad + \sum_a \frac{q}{r}\left[(n \cdot V)^3 + 3(n \cdot z)(n \cdot V)(n \cdot \dot{V}) + \frac{1}{2}(n \cdot z)^2(n \cdot \ddot{V}) \right]_{\mathcal{R}_0} + \cdots, \\[3mm] A^j = \sum_a \frac{q V_{\mathcal{R}_0}^j}{r} + \sum_a \frac{q}{r}\left[V^j(n \cdot V) + \dot{V}^j(n \cdot z) \right]_{\mathcal{R}_0} \\[2mm] \quad + \sum_a \frac{q}{r}\left[\frac{\ddot{V}^j(n.z)^2}{2} + 2\dot{V}^j(n.z)(n.V) + V^j\left((n.V)^2 + (n.z)(n.\dot{V}) \right) \right]_{\mathcal{R}_0} + \cdots, \end{cases} \tag{20.9}$$

where $z_{\mathcal{R}_0}$ and $V_{\mathcal{R}_0}$ are the radius vector and speed of the charge at the instant $T_{\mathcal{R}_0}$, the dot denotes differentiation with respect to T, and the ellipsis stands for terms of order $(q/r)\, V^2(V^2 + z/r)$ and higher.

---

[2]Here, as in the preceding section, we are interested only in the radiation field.

Asymptotic expansion of retarded quantities

Here we outline the steps to arrive at (20.9).

First of all, we note that since $\dot{V} = \mathcal{O}(V/T)$ and $z = \mathcal{O}(VT)$, we have $\dot{V}z = \mathcal{O}(V^2)$, $\ddot{V}z^2 = \mathcal{O}(V^3)$, and so on.

Since $r_{\mathcal{R}} n_{\mathcal{R}} \equiv rn - z_{\mathcal{R}}$, it follows that $r_{\mathcal{R}} = r - (n.z_{\mathcal{R}}) + \mathcal{O}(z^2/r)$.

We introduce $T_{\mathcal{R}_0}$, so that $T - T_{\mathcal{R}_0} = r$.

We have $T - T_{\mathcal{R}} \equiv r_{\mathcal{R}}$, and therefore $T_{\mathcal{R}} = T_{\mathcal{R}_0} + r - r_{\mathcal{R}} = T_{\mathcal{R}_0} + (n.z_{\mathcal{R}}) + \mathcal{O}(z^2/r)$.

Consequently, $z_{\mathcal{R}} \equiv z(T_{\mathcal{R}}) = z_{\mathcal{R}_0} + (n.z_{\mathcal{R}})V_{\mathcal{R}_0} + \mathcal{O}(V^2 z(1 + z/r))$ [with $z_{\mathcal{R}_0} \equiv z(T_{\mathcal{R}_0})$ and $V_{\mathcal{R}_0} \equiv V(T_{\mathcal{R}_0})$], or, iterating to lowest order, $z_{\mathcal{R}} = z_{\mathcal{R}_0} + (n.z_{\mathcal{R}_0})V_{\mathcal{R}_0} + \mathcal{O}(V^2 z(1 + z/r))$.

We therefore find $V_{\mathcal{R}} \equiv V(T_{\mathcal{R}}) = V_{\mathcal{R}_0} + (n.z_{\mathcal{R}})\dot{V}_{\mathcal{R}_0} + \frac{1}{2}(n.z_{\mathcal{R}})^2 \ddot{V}_{\mathcal{R}_0} + \mathcal{O}(V^2(V^2 + z/r))$, or, substituting into this the expression for $z_{\mathcal{R}}$,

$$V_{\mathcal{R}} = V_{\mathcal{R}_0} + \dot{V}_{\mathcal{R}_0}(n.z_{\mathcal{R}_0}) + \left[\frac{1}{2}\ddot{V}(n.z)^2 + \dot{V}(n.V)(n.z)\right]_{\mathcal{R}_0} + \mathcal{O}(V^2(V^2 + z/r)). \quad (20.10)$$

The derivatives of the potential are easily found: Φ and A^j depend on the time T through the radius vectors $z_{\mathcal{R}_0}$ of the charges, evaluated at the time $T_{\mathcal{R}_0} = T - r$ for constant r. Moreover, if the observation point is shifted by an amount δX^i at constant r, this implies that the origin of the frame is shifted by the same amount, and so $\partial_i z_{\mathcal{R}_0}^j = -\delta_i^j$. First we verify that $\partial_\mu A^\mu = 0$ in the order under consideration. We then obtain

$$
\begin{cases}
F^{0i} = \dfrac{\sum_a q \left[n^i(n.\dot{V}) - \dot{V}^i\right]_{\mathcal{R}_0}}{r} \\[2ex]
\qquad + \sum_a \dfrac{q}{r}\Big[-2\dot{V}^i(n.V) - V^i(n.\dot{V}) + 3n^i(n.V)(n.\dot{V}) - \ddot{V}^i(n.z) \\[1.5ex]
\qquad\qquad + n^i(n.z)(n\ddot{V})\Big]_{\mathcal{R}_0} \cdots, \\[3ex]
F^{ij} = \dfrac{\sum_a q \left[n^j \dot{V}^i - n^i \dot{V}^j\right]_{\mathcal{R}_0}}{r} \\[2ex]
\qquad + \sum_a \dfrac{q}{r}\Big[(n.z)(n^j \ddot{V}^i - n^i \ddot{V}^j) + 2(n.V)(n^j \dot{V}^i - n^i \dot{V}^j) \\[1.5ex]
\qquad\qquad + (n.\dot{V})(n^j V^i - n^i V^j)\Big]_{\mathcal{R}_0} \cdots,
\end{cases}
\qquad (20.11)
$$

where the ellipsis now stands for terms of order $(q/r)(\dot{V}(V^2 + z/r))$ and higher. Of course, this expansion can also be derived from (17.18) for E and B.

Now let us assume that the charge-to-mass ratio is the same for all the particles, so that their dipole moment is proportional to the radius vector from the Newtonian center of mass of the system (see the preceding section). We also assume that the system is isolated: $\ddot{d} = 0$.

In the case of two charges q and q', a somewhat tedious but easy calculation gives the expression for the field in the inertial frame of the Newtonian center of mass ($mz + m'z' = 0$ or $qz + q'z' = 0$):

$$
\begin{cases}
F^{0i} = -\dfrac{qq'}{Qr}\Big[2\dot{\mathcal{V}}^i(n\,.\mathcal{V}) + \mathcal{V}^i(n\,.\dot{\mathcal{V}}) - 3n^i(n\,.\mathcal{V})(n\,.\dot{\mathcal{V}}) + \ddot{\mathcal{V}}^i(n\,.R) \\
\qquad\quad -n^i(n\,.R)(n\,.\ddot{\mathcal{V}})\Big]_{\mathcal{R}_0} + \cdots, \\[1.2em]
F^{ij} = \dfrac{qq'}{Qr}\Big[(n\,.R)(n^j\ddot{\mathcal{V}}^i - n^i\ddot{\mathcal{V}}^j) + 2(n\,\mathcal{V})(n^j\dot{\mathcal{V}}^i - n^i\mathcal{V}^j) \\
\qquad\quad +(n\,.\dot{\mathcal{V}})(n^j\,.\mathcal{V}^i - n^i\mathcal{V}^j)\Big]_{\mathcal{R}_0} + \cdots,
\end{cases}
\tag{20.12}
$$

where $Q = q + q'$, $R = z - z'$, and $\mathcal{V}$ is the relative velocity of the charges.

It is therefore a simple exercise [using (18.7) and (18.8) for the averages] to find the radiated power, which leads to the *quadrupole formula*:

$$
\begin{aligned}
\frac{dP^0}{dT} &= \int_S F^0{}_j F^j{}_i n^i r^2 do \\[0.8em]
&= \frac{2(qq')^2}{15Q^2}\Big[-7(R\,.\mathcal{V})(\dot{\mathcal{V}}\,.\ddot{\mathcal{V}}) - 8\mathcal{V}^2\dot{\mathcal{V}}^2 - (\mathcal{V}\,.\dot{\mathcal{V}})^2 - 2(R\,.\dot{\mathcal{V}})(\mathcal{V}\,.\ddot{\mathcal{V}}) \\[0.5em]
&\qquad\quad +3(R\,.\ddot{\mathcal{V}})(\mathcal{V}\,.\dot{\mathcal{V}}) - 2R^2\ddot{\mathcal{V}}^2 + (R\,.\ddot{\mathcal{V}})^2\Big]_{\mathcal{R}_0} \\[0.8em]
&= -\frac{1}{180}\left(\frac{d^3 D_{ij}}{dT^3}\right)_{\mathcal{R}_0}\left(\frac{d^3 D^{ij}}{dT^3}\right)_{\mathcal{R}_0},
\end{aligned}
\tag{20.13}
$$

where $D^{ij} = \sum_a q(3z^i z^j - z^2\delta^{ij}) = \frac{qq'}{Q}(3R^i R^j - R^2\delta^{ij})$ is the quadrupole moment of the system, and in obtaining the last line we have assumed that the motion is Coulombic[3]: $\dot{\mathcal{V}}^i = \frac{qq'}{\mu}\frac{R^i}{R^3}$, where $\mu \equiv mm'/(m+m')$.

20.3 The charge equations of motion

As we showed in the preceding chapter, any radiation of electromagnetic waves by accelerated charges must be compensated for by a reaction force. Here we shall find the equations of motion of a system of charges including the reaction force due to their dipole radiation.

---

[3] If the dipole moment is nonzero, the quadrupole correction to the dipole radiation is not given by (20.13). Instead, it is obtained by expanding the potential (20.9) to the next higher order in the velocities.

The method of this section can be used to find the power radiated by a system of two masses in the Nordström theory of gravitation. In that case it is necessary to expand the potential (10.30) and (10.31) to the next order in the velocities. For Newtonian motion we find

$$
\frac{dP^0}{dT} = \frac{1}{6}\left.\frac{dP^0}{dT}\right|_{RG} - \frac{4G^3(mm')^2}{9R^2}(R.\mathcal{V})^2(1 - 3a_2)^2,
$$

where a_2 characterizes the non-linearity of the interaction and where

$$
\left.\frac{dP^0}{dT}\right|_{RG} \equiv -\frac{G}{45}\left(\frac{d^3 Q_{ij}}{dT^3}\right)\left(\frac{d^3 Q^{ij}}{dT^3}\right) \quad \text{with} \quad Q^{ij} = \sum_a m(3z^i z^j - z^2\delta^{ij}),
$$

is the prediction, confirmed observationally, of general relativity.

In a given inertial frame the motion of a charge of mass m, charge q, and speed V interacting with other charges and in the absence of an external field is governed by the Lorentz equation (see Section 11.3)

$$m\frac{d}{dT}\frac{V}{\sqrt{1-V^2}} = q(E + V \wedge B) \quad \text{with} \quad E = -\nabla\Phi - \frac{\partial A}{\partial T}, \quad B = \nabla \wedge A, \quad (20.14)$$

where E and B are the electric and magnetic fields due to all the charges *including* the one we are studying. At time T and observation point X^i away from the charges, the Liénard–Wiechert potentials from which the fields are derived are given by

$$\Phi(T, X^i) = \sum_a \frac{q}{r_\mathcal{R}(1 - n_\mathcal{R}.V_\mathcal{R})}, \quad A^j(T, X^i) = \sum_a \frac{qV_\mathcal{R}^j}{r_\mathcal{R}(1 - n_\mathcal{R}.V_\mathcal{R})}. \quad (20.15)$$

These potentials depend on the positions $z_\mathcal{R} \equiv z(T_\mathcal{R})$ of the charges q at the retarded times $T_\mathcal{R}$ given by $T - T_\mathcal{R} = r_\mathcal{R}$, where $X^i - z_\mathcal{R}^i = r_\mathcal{R}n_\mathcal{R}^i$ with $n_\mathcal{R}$ a unit vector. For the Lorentz equation (20.14) to be an ordinary differential equation, it is necessary to express these retarded positions of the charges as a function of their position at time T. If the velocities V are small, we can find them using a Taylor series expansion, where the dot denotes the derivative with respect to T and $a \equiv \dot{V}$:

$$r_\mathcal{R}n_\mathcal{R}^i = rn^i + r_\mathcal{R}V - \frac{1}{2}r_\mathcal{R}^2 a + \frac{1}{6}r_\mathcal{R}^3 \dot{a} + \cdots, \quad \text{where} \quad rn^i \equiv X^i - z^i(T) \quad \text{and} \quad n.n = 1. \quad (20.16)$$

We can extract from this expression the expansions of $r_\mathcal{R}$ and $n_\mathcal{R}$, and then of $V_\mathcal{R} = V - r_\mathcal{R}a + \frac{1}{2}r_\mathcal{R}^2\dot{a} + \cdots$, so that for all r but $V \ll 1$ we have

$$\begin{cases} \Phi = \sum_a \frac{q}{r}\left\{1 + \frac{1}{2}\left[V^2 - (n.V)^2 - (n.a)r\right] + \frac{1}{3}(n.\dot{a})r^2 - r(V.a) + \cdots\right\}, \\ A^i = \sum_a \frac{q}{r}\left[V^i - ra^i + \cdots\right], \end{cases} \quad (20.17)$$

from which we derive the field expansions using $\partial_i(X^j - z^j) = \delta_i^j$ and $(X^j - z^j)\dot{} = -V^i$:

$$\begin{cases} E^i = \sum_a \frac{q}{r^2}\left\{n^i + \frac{1}{2}\left[V^2 n^i - 3n^i(n.V)^2 - ra^i - rn^i(n.a)\right] + \frac{2}{3}r^2\dot{a}^i + \cdots\right\}, \\ B^i = -e^{ijk}\sum_a \frac{q}{r^2}n_j V_k + \cdots \end{cases} \quad (20.18)$$

Now to obtain the Lorentz force we must evaluate these fields *at* the position z of the charge under study. The fields E and B then split into two parts: the part due to the other charges and the *proper* or *self* part due to the charge itself, which diverges as $1/r^2$. We decide by *fiat* to 'renormalize' the self part of the field by averaging over n and keeping only the term which is finite for $r \to 0$; see Chapter 19. This gives

$$E_{\text{self}}^i = \frac{2}{3}q\dot{a}^i + \cdots, \quad B_{\text{self}}^i = \mathcal{O}(V^3). \quad (20.19)$$

Once this point is settled the expansion of the charge equation of motion (20.14) follows. Here we shall limit ourselves to the case of two charges (m, q) and (m', q') to simplify the notation, and we set $RN^i \equiv z^i(T) - z'^i(T)$ with $(N.N) = 1$. All the calculations done, we find

$$ma^i = \mathcal{A}_0^i + \mathcal{A}_2^i + \mathcal{A}_3^i + \cdots \quad \text{with} \quad \mathcal{A}_0^i = \frac{qq'}{R^2} N^i \quad \text{and}$$

$$
\begin{cases}
\mathcal{A}_2^i = -m(V.a)V^i - \frac{1}{2}mV^2 a^i \\
\qquad + \dfrac{qq'}{2R^2}\left\{ N^i \left[V'^2 - 3(N.V')^2 - 2(V.V') - R(N.a')\right] + 2V'^i(N.V) - Ra'^i \right\}, \\
\mathcal{A}_3^i = \dfrac{2q}{3}(q\dot{a}^i + q'\dot{a}'^i) = \dfrac{2q}{3}\dfrac{d^3 d^i}{dT^3}.
\end{cases}
\tag{20.20}
$$

We see that the last term $\mathcal{A}_3^i$ is proportional to the third time derivative of the system dipole moment, and that the term involving $q\dot{a}$ is the lowest-order term of the expansion of the Abraham–Lorentz–Dirac force (19.10), and, finally, that this is a third-order differential equation because it involves $\dot{a}^i$. Its solutions therefore *a priori* depend on *three* initial conditions. Moreover, the term involving $\dot{a}^i$ is $\mathcal{O}(V^3)$ times smaller than the dominant term $(qq'/R^2)N^i$.

Therefore, to avoid the appearance of unwelcome unstable solutions, we proceed with an *order reduction*. This consists of transforming (20.20) (as well as the similar equation governing the motion of the charge q') by iterating it. More precisely, (20.20) reduces at lowest order to the Coulomb equation

$$ma^i = \frac{qq'}{R^2} N^i, \quad \text{from which we deduce that} \quad m\dot{a}^i = \frac{qq'}{R^3}\left[\mathcal{V}^i - 3N^i(N.\mathcal{V})\right], \tag{20.21}$$

where $\mathcal{V}^i \equiv V^i - V'^i$. Then, substituting these expressions along with the symmetric expressions for a'^i and $\dot{a}'^i$ into (20.20), we obtain a *reduced* equation of second order:

$$ma^i = A_0^i + A_2^i + A_3^i + \cdots \quad \text{with} \quad A_0^i = \frac{qq'}{R^2} N^i \quad \text{and}$$

$$
\begin{cases}
A_2^i = \dfrac{qq'}{2R^2}\left\{ N^i \left[V'^2 - V^2 - 2(V.V') - 3(N.V')^2\right] - 2\mathcal{V}^i(N.V) \right\} + \dfrac{(qq')^2}{m'R^3} N^i, \\
A_3^i = \dfrac{q^2 q'}{R^3}\left(\dfrac{q}{m} - \dfrac{q'}{m'}\right)\left[\dfrac{2}{3}\mathcal{V}^i - 2N^i(N.\mathcal{V})\right].
\end{cases}
\tag{20.22}
$$

Here A_3^i is the *dipole reaction force*. In contrast to A_2^i and A_0^i, it changes sign when the direction of the time arrow is reversed, *i.e.*, under the transformation $T \to -T$. This implies that it would have been absent if we had chosen the symmetric solution of the Maxwell equations (mentioned in Section 17.2) rather than the retarded solution (20.15).

Higher orders and Lagrange series

When the charge-to-mass ratio is the same for all the particles, in order to obtain the *quadrupole reaction force* it is necessary to continue the expansions two orders higher, which is a rather tedious calculation. A way of simplifying it and, above all, getting a clearer picture of the structure of the series at all orders is to use the *Lagrange method*.[4] For this, we start from the expression for the retarded potential created by a *continuous* charge distribution (17.14):

$$A^\mu(T, X^i) = \int d^3 X' \frac{j^\mu(T - r', X'^i)}{r'}, \quad \text{where} \quad r' = \sqrt{(X^i - X'^i)(X_i - X'_i)}, \quad (20.23)$$

and we make a Taylor–Lagrange series expansion of the current:

$$j^\mu(T - r', X'^i) = j^\mu(T) - \frac{\partial j^\mu}{\partial T}\bigg|_T r' + \frac{1}{2} \frac{\partial^2 j^\mu}{\partial T^2}\bigg|_T r'^2 + \cdots = \sum_{n=0}^{n=\infty} \frac{(-1)^n}{n!} \frac{\partial^n j^\mu}{\partial T^n}\bigg|_T r'^n, \quad (20.24)$$

so that

$$A^\mu(T, X^i) = \sum_{n=0}^{n=\infty} \frac{(-1)^n}{n!} \frac{d^n}{dT^n} \int d^3 X' r'^{n-1} j^\mu(T, X'^i). \quad (20.25)$$

At this stage we use the explicit expression for the current due to point charges, namely,

$$j^\mu(T, X^i) = \sum_a q \int d\tau\, \delta(T - z^0(\tau)) \delta_3(X^i - z^i(\tau)) \frac{dz^\mu}{d\tau}, \quad (20.26)$$

which when substituted into (20.25) gives, after integrating over all space, the expansion of the potential to all orders:

$$\Phi = \sum_a q \sum_{n=0}^{n=\infty} \frac{(-1)^n}{n!} \frac{d^n}{dT^n} r_a^{n-1}, \quad A^i = \sum_a q \sum_{n=0}^{n=\infty} \frac{(-1)^n}{n!} \frac{d^n}{dT^n} (V_a^i r_a^{n-1}), \quad (20.27)$$

with $r_a = \sqrt{(X^i - z_a^i(T))(X_i - z_i^a(T))}$. It can of course be checked that the expansion (20.27) does indeed reproduce (20.17) in the lowest orders. The most important thing to note is that the charge equation of motion (20.20), when the expansion is carried to all orders using (20.27) and after renormalization of the self term, becomes a differential equation of infinite order. This means that its solution a *priori* requires an infinite number of initial conditions at $T = 0$, or, in other words, its integration requires knowledge of the entire past history of the charge. And it is only after an order reduction involving an infinite number of steps that we can reduce it to a second-order differential equation. Here we see evidence of the operational limits of the Maxwell theory.

---

[4]Lagrange (1770).

21
Interacting charges II

The motion of a charge (m, q) in the field of another charge (m', q') and in its own field can of course be studied in the lowest orders of the velocities directly using the equation of motion obtained in the preceding chapter. However, the features of this motion are revealed more easily by deriving them from the *Darwin Lagrangian*. This allows us to rigorously establish a *balance* between the energy radiated by the system and the mechanical energy lost by the system.

We shall conclude our general study of the electromagnetic radiation of a system of charges by outlining the 'post-Minkowski' approach based on iteration in the 'coupling constant' qq' rather than in the velocities.

21.1   The Darwin Lagrangian and conserved quantities

In an inertial frame, the action of a system of interacting charges is the sum of the actions of the free charges (see Section 6.1), the action associated with the charge–field interaction (see Section 11.3), and the action of the field itself (see Section 12.3):

$$\mathbf{S} = -\sum_a m \int d\tau + \sum_a q \int A_\mu U^\mu \, d\tau - \frac{1}{16\pi} \int F_{\mu\nu} F^{\mu\nu} d\Omega \equiv \int dT \, L$$

$$\text{with} \quad L = -\sum_a m\sqrt{1 - V^2} + \sum_a q(-\Phi + V.A) + \frac{1}{8\pi} \int dV (E^2 - B^2).$$

(21.1)

In order to obtain a Lagrangian depending only on the charge world lines from which we can derive the equations of motion found in the preceding chapter, we must eliminate the fields[1] E and B in (21.1). We do this in two stages.

First we rearrange the last term of L in (21.1). Using the definition of the fields as a function of the potentials [cf. (11.7)] and applying the Leibniz rule and Gauss's theorem, we find

$$\int dV (E^2 - B^2) = \int dV \left[E. \left(-\nabla\Phi - \frac{\partial A}{\partial T} \right) - B.(\nabla \wedge A) \right]$$

$$= \int dV \left[\Phi \nabla.E - A.\nabla \wedge B + A.\frac{\partial E}{\partial T} \right] \qquad (21.2)$$

$$- \int_S (E\Phi + A \wedge B).n \, dS - \frac{d}{dT} \int dV \, E.A.$$

The last term is a total derivative with respect to the time. It can be ignored because it has no effect on the Euler–Lagrange equations. The next-to-last term is an integral over the

---

[1] Wheeler and Feynman (1949).

Relativity in Modern Physics. Nathalie Deruelle and Jean-Philippe Uzan.
© Oxford University Press 2018. Published in 2018 by Oxford University Press.
DOI: 10.1093/oso/9780198786399.001.0001

2-sphere at infinity; on the mass shell (that is, when the solution of the Maxwell equations is substituted into it) it will be given by the radiation parts of the field and the potential, both of which have $1/r$ behavior. It can therefore be ignored only if there is no radiation part in the solutions (because they have been integrated using a symmetric propagator), or if we only seek to extract from (21.1) a Lagrangian giving the equations of motion (20.22), *excluding* the reaction force A_3^i. Finally, we transform the first term using the field equations,[2] that is, the Maxwell equations (12.16) with point charges as the source, namely,

$$\nabla.E = 4\pi \sum_a q\delta_3(X^i - z^i(T)), \quad \nabla \wedge B = \frac{\partial E}{\partial T} + 4\pi \sum_a qV\delta_3(X^i - z^i(T)), \tag{21.3}$$

so that the Lagrangian (21.1) is replaced by

$$L = -\sum_a m\sqrt{1 - V^2} + \frac{1}{2}\sum_a q\left(-\Phi + V.A\right). \tag{21.4}$$

Once this first stage has been completed, we eliminate the fields by substituting into (21.4) the *solution* of the Maxwell equations, that is, the retarded potentials (20.17) truncated at order $\mathcal{O}(V^2)$. Since the self part of the potentials diverges as $1/r$ when it is evaluated on the charge world line, the final operation is to renormalize it by averaging over n and keeping only the finite terms.

The Lagrangian (21.1) of the two charges (m, q) and (m', q') located at $z^i(T)$ and $z'^i(T)$ and interacting only electromagnetically therefore in the end reduces to $L = L_{\text{Darwin}} + qq'[(N.v') - (N.v)]/4$, where L_{Darwin} is the *Darwin Lagrangian* (G. G. Darwin, 1920):

$$L_{\text{Darwin}} = \left(\frac{1}{2}mV^2 + \frac{1}{2}mV'^2 - \frac{qq'}{R}\right) + \frac{1}{8}mV^4 + \frac{1}{8}m'V'^4$$

$$+ \frac{qq'}{2R}\left[(N.V)(N.V') + (V.V')\right]. \tag{21.5}$$

Here we have set $z^i(T) - z'^i(T) \equiv RN^i$ with $(N.N) = 1$. Since this Lagrangian depends on the positions and velocities of the two charges, the conjugate momentum p_i of z^i and the force F^i are defined as

$$p^i \equiv \frac{\partial L_{\text{Darwin}}}{\partial V_i} = mV^i + \frac{1}{2}mV^2V^i + \frac{qq'}{2R}\left[N^i(N.V') + V'^i\right],$$

$$F^i \equiv \frac{dp^i}{dT} - \frac{\partial L_{\text{Darwin}}}{\partial z_i} = ma^i - A_0^i - A_2^i, \tag{21.6}$$

where A_0 and A_2 are given by (20.20). We thus verify that the Euler–Lagrange equations $F^i = 0$ are indeed identical to the equations of motion (20.20) through order $\mathcal{O}(V^2)$ inclusive. The use of the solution of the Maxwell equations to transform the Lagrangian (21.1) is therefore justified.[3]

---

[2]This operation must always be justified at the end.

[3]At higher orders it is necessary to substitute into (21.4) the Lagrange expansion (20.27) of the potentials (after regularizing the self part). The resulting Lagrangian then depends not only on the positions and

The action $\mathbf{S} = \int dT\, L_{\text{Darwin}}$ is a scalar, and so it must have the same numerical value in any inertial frame in which it is calculated, which makes it possible to derive conserved quantities from it (see Book 1, Section 8.3). Therefore, the invariance of $\mathbf{S}$ under spatial translations, spatial rotations, and time translations leads to the conservation of momentum P, angular momentum $\mathcal{M}$, and energy $\mathcal{E}$ of the system:

$$\frac{dP}{dT} = \sum_a F, \text{ where } P = \sum_a p, \quad \frac{d\mathcal{M}}{dT} = \sum_a z \wedge F, \text{ where } \mathcal{M} = \sum_a z \wedge p,$$

$$\frac{d\mathcal{E}}{dT} = \sum_a F.V, \text{ where } \mathcal{E} = \sum_a p.V - L_{\text{Darwin}}. \tag{21.7}$$

Let us give the explicit expression for $\mathcal{E}$:

$$\mathcal{E} = \frac{1}{2}mV^2 + \frac{1}{2}m'V'^2 + \frac{qq'}{R} + \frac{3}{8}mV^4 + \frac{3}{8}m'V'^4 + \frac{qq'}{2R}[(N.V)(N.V') + (V.V')]. \tag{21.8}$$

When the equations of motion are satisfied to order $\mathcal{O}(V^2)$, that is, when $F = 0$, then P, $\mathcal{M}$, and $\mathcal{E}$ are constants.

The center-of-mass system

It is easily shown that the system momentum $P = \sum_a p$ with p given in (21.6) can also be written as

$$P^i = \frac{dG^i}{dT} \quad \text{with} \quad G^i = \sum_a \left(m + \frac{1}{2}mV^2 + \frac{qq'}{2R}\right)z^i. \tag{21.9}$$

Since P is constant when the equations of motion are satisfied to order $\mathcal{O}(V^2)$, we have $G^i = P^i T + G^i_0$.

The *center-of-mass frame* (c.m. frame) is defined, as in Newtonian physics, as the inertial system in which $G^i = 0$. Iterating (21.9), we obtain the positions of the charges in this frame as a function of the relative position and velocity $RN^i = z^i - z'^i$, $\mathcal{V}^i = V^i - V'^i$:

$$z^i = \frac{m'}{M}RN^i + \frac{1}{2}RN^i\left(\frac{m-m'}{M^2}\right)\left(\mu\mathcal{V}^2 + \frac{qq'}{R}\right),$$

$$z'^i = -\frac{m}{M}RN^i - \frac{1}{2}RN^i\left(\frac{m'-m}{M^2}\right)\left(\mu\mathcal{V}^2 + \frac{qq'}{R}\right),$$

where we have set $M = m + m'$ and $\mu = mm'/M$. The vector RN^i is obtained by integrating the relative equation of motion written in the c.m. frame, $\dot{\mathcal{V}} = (A_0 + A_2) - (A'_0 + A'_2)$ with A_0 and A_2 from (20.22), and setting $\nu = \mu/M$:

$$\mu\dot{\mathcal{V}}^i = \frac{qq'}{R^2}N^i + \frac{qq'}{R^2}\left\{N^i\left[\mathcal{V}^2\left(3\nu - \frac{1}{2}\right) - \frac{3}{2}\nu(N.\mathcal{V})^2\right] - (1-2\nu)\mathcal{V}^i(N.\mathcal{V}) + 2\frac{qq'}{MR}N^i\right\}. \tag{21.10}$$

---

velocities of the charges, but also on their accelerations a [to order $\mathcal{O}(V^4)$], the derivative of the acceleration $\dot{a}$ [to order $\mathcal{O}(V^6)$], and so on. The Euler–Lagrange equations then are (see Book 1, Section 8.1) $\frac{\partial L}{\partial z} - \frac{d}{dT}\frac{\partial L}{\partial V} + \frac{d^2}{dT^2}\frac{\partial L}{\partial a} - \frac{d^3}{dT^3}\frac{\partial L}{\partial \dot{a}} \cdots = 0$. At this stage they can be iterated so as to reduce them to second-order differential equations, and it can be verified that they are identical, with the reaction force excluded, to the equations of motion obtained directly from the Lorentz equation using the method of Section 20.3.

Lorentz invariance of the equations of motion

Let us consider two different paths in an inertial frame $\mathcal{S}$, $z^i(T)$ and $\tilde{z}^i(T)$, which are related as (V_0 is a constant)

$$\tilde{z}^1(T) = \frac{z^1(T') + V_0 T'}{\sqrt{1 - V_0^2}}, \quad T = \frac{T' + V_0 z^1(T')}{\sqrt{1 - V_0^2}}, \quad \tilde{z}^2 = z^2(T'(T)), \quad \tilde{z}^3 = z^3(T'(T)).$$

This relation between $\tilde{z}^i$ and z^i can be interpreted as follows. If $X'^i = z^i(T')$ are the equations of a world line in the frame $\mathcal{S}'$, then $X^i = \tilde{z}^i(T)$ will be its equations in the frame $\mathcal{S}$ moving with velocity $-V_0$ along the X' axis. If the paths $\tilde{z}^i(T)$ and $z^i(T)$ are close to each other, we will have $T \sim T' + V_0 z^1$, $\tilde{z}^1 \sim z^1 - (V_0 z^1)V^1 + V_0 T$, and $\tilde{z}^2 \sim z^2 - (V_0 z^1)V^2$. Therefore, generalizing to a Lorentz transformation of velocity β^i in any direction and setting $\delta z^i \equiv \tilde{z}^i(T) - z^i(T)$, we have

$$\delta z^i = \beta^i T - (\beta.z)V^i \quad \Longrightarrow \quad \delta V^i = \beta^i - (\beta.V)V^i - (\beta.z)a^i.$$

The variation of the Darwin Lagrangian (21.5) under such a transformation follows immediately:

$$\delta L_{\text{Darwin}} = \sum_a (\beta.z)(F.V) + \left(\sum_a \frac{\partial L_{\text{Darwin}}}{\partial z}\right).\beta T - \frac{d}{dT}\left(\sum (p.V)(\beta.z)\right) + P.\beta.$$

The first term vanishes if $F = 0$, that is, if the equations of motion are satisfied to order $\mathcal{O}(V^2)$. The second also vanishes because $L_{\text{Darwin}}(z, z') = L_{\text{Darwin}}(z - z')$, and the third is already in the form of a total derivative with respect to time, which does not contribute to the equations of motion. Finally, we are left with the last term, where P is the total momentum. However, as we saw in (21.9), P is itself a total derivative with respect to time: $P^i = dG^i/dT$.

The equations of motion (20.20) and (20.22) are therefore indeed invariant under a Lorentz transformation.

21.2   Radiation reaction

Up to order $\mathcal{O}(V^2)$, the momentum, angular momentum, and energy of a system of two charges, defined in (21.7), are constant because the equations of motion (20.20) or (20.22) require that the force F vanish to this order. However, if now we include the reaction force $\mathcal{A}_3$ in the equations of motion, P, $\mathcal{M}$, and $\mathcal{E}$ will vary as

$$\frac{dP}{dT} = \sum_a \mathcal{A}_3, \quad \frac{d\mathcal{M}}{dT} = \sum_a z \wedge \mathcal{A}_3, \quad \frac{d\mathcal{E}}{dT} = \sum_a \mathcal{A}_3.V \text{ with } \mathcal{A}_3 = \frac{2q}{3}\frac{d^3 d}{dT^3}, \tag{21.11}$$

where $d^i = qz^i + q'z'^i$ is the dipole moment of the system. Then, in particular,

$$\frac{d\mathcal{E}}{dT} = \frac{2}{3}\frac{d}{dT}\left(\dot{d}\ddot{d}\right) - \frac{2}{3}\ddot{d}^2. \tag{21.12}$$

The first term is zero when averaged over time if the system remains confined (see Section 13.5, footnote 5). The second is equal to the dipole power (20.5) radiated to infinity. We have therefore established that the radiated energy and the mechanical energy lost by the system are equal.[4]

Radiation in the case of Coulomb motion

The motion of two charges q and q' of masses m and m' in lowest order in the velocities reduces to the motion of a charge q of mass $\mu = mm'/M$ (with $M = m + m'$) in the Coulomb field of a charge q' located at the origin of the c.m. frame. The equation of motion is [cf. (20.21)] $\dot{\mathcal{V}}^i = (qq'/\mu)(RN^i/R^2)$, where RN^i is the separation vector and $\mathcal{V}^i$ is its time derivative. In this frame we have $z^i = (m'/M)\,RN^i$ and $z'^i = -(m/M)\,RN^i$, and the dipole moment reduces to $d^i = \mu(q/m - q'/m')RN^i$. The average mechanical power lost by the system $d\mathcal{E}/dT = -2\ddot{d}^2/3$, which, as we just saw, is equal to the radiated dipole power dP^0/dT, then becomes

$$\frac{d\mathcal{E}}{dT} = -\frac{2(qq')^2}{3}\left(\frac{q}{m} - \frac{q'}{m'}\right)^2 \frac{1}{R^4}. \tag{21.13}$$

In the case of the hydrogen atom where $q = -q' = e$, $m' \gg m$, and $R = a_{\text{Bohr}} = e^2/(m\alpha^2)$, the expression reduces to that obtained in (18.27), namely, $d\mathcal{E}/dT = -\frac{2}{3}(m^2\alpha^8/e^2)$.

If $q/m = q'/m'$, the dipole term vanishes and the radiated power, which is quadrupolar, is given by (20.13) (see again footnote 4). An easy calculation using (20.21) as well as $qm'^2/M^2 + q'm^2/M^2 = qq'/Q$, where $Q = q + q'$, gives

$$\frac{dP^0}{dT} = -\frac{2(qq')^4}{5\mu^2 Q^2}\left[4\mathcal{V}^2 - \frac{11}{3}(N.\mathcal{V})^2\right]\frac{1}{R^4}.$$

The problem of the stability of the hydrogen atom

At lowest order in the velocities, the mechanical energy (21.8) of a system of two charges is the sum of the Newtonian kinetic energies and the electrostatic energy (13.5): $\mathcal{E} = \frac{1}{2}mV^2 + \frac{1}{2}m'V'^2 + qq'/R$. In the c.m. frame $\mathcal{E} = \frac{1}{2}\mu V^2 + qq'/R$, where μ is the reduced mass. If in addition we limit ourselves to circular motion, we have [cf. (13.16)] $\mathcal{V}^2 = -qq'/(\mu R)$. Therefore, $\mathcal{E} = \frac{1}{2}qq'/R$.

The loss of mechanical energy of the system due to the radiation was given in (21.13). We therefore have

$$-\frac{qq'}{2R^2}\dot{R} = -\frac{2(qq')^2}{3}\left(\frac{q}{m} - \frac{q'}{m'}\right)^2\frac{1}{R^4} \quad \Longrightarrow \quad T_{\text{fall}} = \frac{R_{\text{initial}}^3}{4(-qq')(q/m - q'/m')^2}. \tag{21.14}$$

For the hydrogen atom initially in its Bohr orbit, $R_{\text{initial}} = a_{\text{Bohr}}$, we have $T_{\text{fall}} \simeq e^2/(4m\alpha^6) \approx 1.5 \times 10^{-11}$ s. Therefore, the electron would fall onto the proton after about one hundred thousand orbits if classical electrodynamics still applied to the case where the ratio of the electron action to the action quantum $\hbar$ is less than 1; cf. (13.34).

---

[4]A similar balance can be established in quadrupole order; see (20.13). We find that on average,

$$\frac{d\mathcal{E}}{dT} = -\frac{1}{180}\left(\frac{d^3 D_{ij}}{dT^3}\right)\left(\frac{d^3 D^{ij}}{dT^3}\right),$$

but the calculations are rather lengthy.

The lifetime of the Rydberg states

As is well known, the radii R of the electron orbits in the Bohr model of the hydrogen atom are quantized: $R = n^2 a_{\text{Bohr}}$, where $a_{\text{Bohr}} = e^2/(m\alpha^2)$ is the Bohr radius and n is an integer. If we assume that R_{initial} is a *Rydberg state* for which $n \gg 1$, the time ΔT for an electron to fall from one orbit into another, that is, its lifetime in the state n, is given by (21.14):

$$\Delta T \approx 6n^5 T_{\text{fall}},$$

where we have used the fact that for large n, $\Delta R^3 \approx 3R^2 \Delta R = 6n^5 a_{\text{Bohr}}^3 \Delta n$. For example, for $n = 50$ we find $\Delta T = 2.9 \times 10^{-2}$ s, in good agreement with experiment, as well as with the criterion for classical electrodynamics to be valid, because the ratio of the electron action to the action quantum $\hbar$ given in (13.34) in this case is $\mathbf{S}/\hbar \sim n \gg 1$.

21.3 The exact equations of motion

The approach we have followed so far in this chapter is not suitable when studying a system of charges in rapid motion, even though the Lagrange series described in Section 20.3 can in principle be carried out to any order in the velocities.

To study this general case, we start as in Section 20.3 from the Lorentz equation determining the motion of a charge q of mass m in the field of the other charges and its own field, but we now use four-dimensional notation:

$$m\frac{dU^\mu}{d\tau} = qF^{\mu\nu}U_\nu, \quad \text{where} \quad F_{\mu\nu} = \partial_\mu A_\nu - \partial_\nu A_\mu \quad \text{and} \quad A^\mu = \sum_a \frac{qU_\mathcal{R}^\mu}{\varrho\mathcal{R}}. \tag{21.15}$$

Here $U^\mu = dz^\mu/d\tau$ is the 4-velocity of the charge and A^μ is the Liénard–Wiechert potential $\varrho_\mathcal{R} = -l_\mathcal{R}^\mu U_{\mu\mathcal{R}}$, where $l^\mu = X^\mu - X_\mathcal{R}^\mu$ is the null vector from the observation point X^μ to the event $X_\mathcal{R}^\mu = X^\mu(\tau_\mathcal{R})$, which is the intersection of the cone with apex X^μ and the charge world line. We again give the explicit expression (17.16) for the Faraday tensor:

$$F_{\mu\nu} = \sum_a \left\{ \frac{q}{\varrho^3}[U_\mu l_\nu - U_\nu l_\mu] + \frac{q}{\varrho^2}\left[\gamma_\mu l_\nu - \gamma_\nu l_\mu + (U_\mu l_\nu - U_\nu l_\mu)\frac{(l \cdot \gamma)}{\varrho}\right] \right\}_\mathcal{R}, \tag{21.16}$$

where $\gamma^\mu \equiv dU^\mu/d\tau$ is the charge 4-acceleration. As before, the Lorentz force $[qF^{\mu\nu}U_\nu$ evaluated at $X^\mu = z^\mu(\tau)]$ splits into a self part due to the charge itself and the part due to the other charges. The self part is the Abraham–Lorentz–Dirac force found in Section 19.1. To evaluate the other part, we introduce the following quantities (see Fig. 21.1):

$$\rho \equiv -(z^\mu - \hat{z}'^\mu)\hat{U}'_\mu, \quad \nu^\mu = -\hat{U}'^\mu + \frac{z^\mu - \hat{z}'^\mu}{\rho}, \quad \omega = U^\mu \hat{U}'_\mu, \quad \kappa = \nu_\mu U^\mu, \tag{21.17}$$

where $\hat{z}'^\mu$ is the intersection of the cone originating at z^μ and the world line of the charge q', and $\hat{U}'^\mu$ is the charge 4-velocity. Then the Lorentz equation is written as follows (for the case of two charges in order to simplify the notation):

$$m\gamma^\mu = \frac{2q^2}{3}\left(\dot{\gamma}^\mu - \gamma^2 U^\mu\right) + qq'\left(W^\mu + \hat{\gamma}'^\nu W_\nu^\mu\right)$$

$$\text{with} \quad W^\mu = \frac{\kappa\hat{U}'^\mu - \omega\nu^\mu}{\rho^2} \tag{21.18}$$

$$\text{and} \quad W_\nu^\mu = \frac{1}{\rho}\left[\delta_\nu^\mu(\omega + \kappa) + \hat{U}'^\mu(\kappa\nu_\nu - U_\nu) - \nu^\mu(U_\nu + \omega\nu_\nu)\right].$$

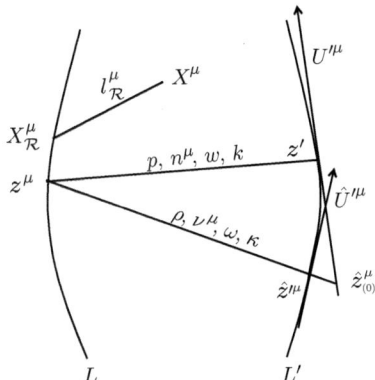

Fig. 21.1 Geometry of the exact 2-charge problem.

This equation as well as its homologue for the charge q' is exact and manifestly Lorentz invariant. It is of third order because it involves the derivative of the acceleration $\dot{\gamma}^\mu$. Moreover, it is *hereditary* because it depends on the position, velocity, and acceleration of the other charge at the retarded point $\hat{z}'^\mu$.

Once the 'hereditary' equation (21.18) has been reduced order by order following the steps outlined above to a differential equation which is always manifestly Lorentz invariant but is an ordinary equation, we can of course expand it in powers of $1/c$ and recover (20.22).

The 'predictivization' of hereditary equations

To transform a 'hereditary' equation into an ordinary differential equation, we proceed by iteration in the *coupling constant g* quadratic in the charges ($g = qq'$ or $g = q^2$).

At lowest order (21.18) reduces to $m\gamma^\mu = qq'W^\mu + \mathcal{O}(g^2)$, from which we find the derivative of the acceleration $\dot{\gamma}^\mu$ to the same order by differentiating W^μ as though the velocities were constant. This gives

$$\dot{W}^\mu = \frac{1}{\rho^3}\left\{\hat{U}'^\mu\left[3\kappa(\omega+1)-1\right] - \omega U^\mu - 3\nu^\mu\omega(\omega+1)\right\} + \mathcal{O}(g^2).$$

This *order reduction* allows us to replace (21.18) by an equation which is still hereditary, but is now of second order:

$$m\gamma^\mu = \frac{2q^2}{3}\dot{W}^\mu + qq'\left(W^\mu + \frac{qq'}{m'}W_\nu^\mu\,\hat{W}'^\nu\right) + \mathcal{O}(g^3). \tag{21.19}$$

Making (21.19) a *predictive* equation consists of getting rid of the retarded arguments in the second term order by order.

At lowest order (see Fig. 21.1), the retarded point $\hat{z}'$ is replaced by the point $\hat{z}'_0$, the intersection of the cone with apex z and the *straight line* parallel to U' with apex at some point z' of (L'). Equation (21.19) then reduces to

$$m\gamma^\mu = qq'W_1^\mu + \mathcal{O}(g^2), \tag{21.20}$$

where W_1^μ is obtained from W^μ given in (21.18) by

(a) replacing the retarded point $\hat{z}'$ by a point z' of (L'):

$$\rho \rightarrow p = -(z - z') \cdot U', \ \nu^\mu \rightarrow n^\mu = -U'^\mu + \frac{z^\mu - z'^\mu}{p}, \ \kappa \rightarrow k - (n \cdot U), \ \omega \rightarrow w = (U.U') ;$$

(b) translating this point z' to $\hat{z}'_0$: $z^\mu - z'^\mu \rightarrow z^\mu - z'^\mu - \lambda u'^\mu$, with λ such that $(z^\mu - z'^\mu - \lambda u'^\mu)^2 = 0$, that is, $\lambda = p(1 - n)$ with $n^2 = 1 + (z - z')^2/p^2$. Then

$$W_1^\mu = \frac{1}{p^2 n^3} \left[-w \frac{n^\mu + kU^\mu}{n} + \frac{k}{n} \left(U'^\mu + wU^\mu \right) \right]. \tag{21.21}$$

In the next order we perform the same operations (a) and (b) on $\hat{W}'^\nu$ and W_ν^μ. As for the quantity W^μ, it is calculated by replacing the retarded point $\hat{z}'$ by the point $\hat{z}'_1$, the intersection of the cone with apex z^μ and the curve with apex z' obtained assuming that the acceleration is equal to $W_1'^\mu$. This gives[5]

$$W^\mu = W_{(1)}'^\mu + p(1 - n) \int_0^1 d\lambda\, \mathcal{S} \left[W_{(1)}'^\nu \frac{\partial W_{(1)}^\mu}{\partial U'^\nu} \right] + \mathcal{O}(g^3), \tag{21.22}$$

where $\mathcal{S}$ is the operator which translates whatever follows it as[6] $z^\mu - z'^\mu \rightarrow z^\mu - z'^\mu - p(1-n)U'^\mu$.

*

The methods and results presented in the last part of this book are needed for predicting the motion and radiation of elementary charged particles in particle accelerators and in astrophysics. Moreover, they reveal the limits of the Maxwell theory, which predicts a catastrophic instability of matter (which is averted when the Maxwell theory is replaced by quantum electrodynamics). They also give an indication of how to deal with the problem of *gravitational* radiation in relativistic theories of gravity, like the Nordström theory whose features and limitations were discussed in Chapter 10, and also Einstein's general theory of relativity.

---

[5] A detailed discussion of *predictive mechanics* applied to electromagnetism can be found in, for example, Bel, Salas, and Sanchez (1973).

[6] If we then change to three-dimensional notation and expand in powers of the velocities, we recover the equations of motion (20.20) and (20.22), as we should.

22
Electromagnetism and differential geometry

22.1 p-forms and the exterior product

Let us consider[1] a $\binom{0}{p}$-type covariant tensor T constructed on the dual E_n^* of a vector space of dimension n and basis θ^i, $i = 1, \cdots, n$. T can be decomposed as $T = T_{i_1 i_2 \dots i_p}\, \theta^{i_1} \otimes \theta^{i_2} \dots \otimes \theta^{i_p}$. *Antisymmetrizing* T consists of associating with it the completely antisymmetric tensor T_a as

$$T_a = T_{[i_1 i_2 \dots i_p]}\, \theta^{i_1} \otimes \theta^{i_2} \dots \otimes \theta^{i_p}, \tag{22.1}$$

the completely antisymmetric components of which $T_{[i_1 i_2 \dots i_p]}$ can be defined by recursion:

$$T_{[ij]} \equiv \frac{1}{2!}(T_{ij} - T_{ji}), \quad T_{[ijk]} \equiv \frac{1}{3!}(T_{ijk} + T_{jki} + T_{kij} - T_{jik} - T_{ikj} - T_{kji}), \quad \text{and so on.} \tag{22.2}$$

Then, for example,

$$(\theta^i \otimes \theta^j)_a = \frac{1}{2}(\theta^i \otimes \theta^j - \theta^j \otimes \theta^i)$$

$$(\theta^i \otimes \theta^j \otimes \theta^k)_a = \frac{1}{3!}(\theta^i \otimes \theta^j \otimes \theta^k + \theta^j \otimes \theta^k \otimes \theta^i + \theta^k \otimes \theta^i \otimes \theta^j \tag{22.3}$$

$$-\theta^j \otimes \theta^i \otimes \theta^k - \theta^i \otimes \theta^k \otimes \theta^j - \theta^k \otimes \theta^j \otimes \theta^i).$$

Completely antisymmetric tensors of the type $\binom{0}{p}$ are called *p-forms* or *forms of degree p*. They form a subspace, denoted $\Lambda^p E_n^*$, of the ensemble of tensors of the type $\binom{0}{p}$ constructed on E_n^*. This subspace is invariant under a change of basis and has dimension $C_n^p = \frac{n!}{(n-p)!p!}$. The maximum dimension of a p-form is therefore $p = n$. If, for example, the dimension of E_n^* is 3, the dimensions of the spaces of $0, 1, 2, 3$-forms are respectively $1, 3, 3, 1$.

The *exterior product* (or *wedge product*) of a p-form α and a q-form β is a $(p+q)$-form defined as

$$\alpha \wedge \beta \equiv \frac{(p+q)!}{p!q!}(\alpha \otimes \beta)_a\,. \tag{22.4}$$

For example, the exterior product of two 1-forms θ^i and θ^j is the 2-form

---

[1] See Book 1, Chapters 2 and 3 for an introduction to the notions of vector and differential geometry used here. A colorful introduction to differential geometry can be found in Misner, Thorne, and Wheeler (1973), Chapter 4. A more comprehensive treatment is given in Choquet-Bruhat, DeWitt-Morette, and Dillard-Bleick (1977).

Relativity in Modern Physics. Nathalie Deruelle and Jean-Philippe Uzan.
© Oxford University Press 2018. Published in 2018 by Oxford University Press.
DOI: 10.1093/oso/9780198786399.001.0001

Book 2

$$\theta^i \wedge \theta^j = 2(\theta^i \otimes \theta^j)_a \, . \tag{22.5}$$

The exterior product possesses the following properties:

$$(c_1\,\alpha + c_2\,\beta) \wedge \gamma = c_1\,\alpha \wedge \gamma + c_2\,\beta \wedge \gamma, \quad (\alpha \wedge \beta) \wedge \gamma = \alpha \wedge (\beta \wedge \gamma),$$
$$\alpha \wedge \beta = (-1)^{pq} \beta \wedge \alpha, \tag{22.6}$$

where p and q are the degrees of the forms α and β, and c_1 and c_2 are constants. For example, we have that the exterior product of three 1-forms θ^i, θ^j, and θ^k is the 3-form

$$\theta^i \wedge \theta^j \wedge \theta^k = 3! \, (\theta^i \otimes \theta^j \otimes \theta^k)_a \, . \tag{22.7}$$

If θ^i is a basis of E_n^*, the natural basis of the space $\Lambda^p E_n^*$ of p-forms is $\theta^{i_1} \wedge \theta^{i_2} ... \wedge \theta^{i_p}$, with $i_1 < i_2 ... < i_p$. Any p-form α can therefore be decomposed as

$$\alpha = \frac{1}{p!} \alpha_{i_1 i_2 ... i_p} \, \theta^{i_1} \wedge \theta^{i_2} ... \wedge \theta^{i_p}, \tag{22.8}$$

where the θ^i are *not* ordered, so that its components in the ordered θ^i basis are the *ordered* antisymmetric part of[2] $\alpha_{i_1 i_2 ... i_p}$.

The ensemble of all the forms is the space $\Lambda^*(E_n) = \oplus_{p=0}^{n} \Lambda^p E_n^*$. The exterior product supplies this ensemble with a structure called a *graded algebra* or a *Grassmann algebra*.

22.2   The dual of a p-form

Let us consider a vector space E_n supplied with a Euclidean metric e or a Minkowski metric ℓ, and let us require that the basis θ^i of the dual E_n^* be (pseudo-)orthonormal, that is, that $e = \delta_{ij}\,\theta^i \otimes \theta^j$ or $\ell = \eta_{ij}\,\theta^i \otimes \theta^j$. This metric can serve as an index raiser and establishes a bijective map between the vectors of E_n and the forms of E_n^*.

Let us take a p-form $\alpha = \frac{1}{p!}\alpha_{i_1 ... i_p}\theta^{i_1} \wedge ... \theta^{i_p}$. Its *dual* is the $(n-p)$-form $*\alpha$ (where $*$ denotes the *Hodge operator*), defined as

$$*\alpha = \frac{1}{(n-p)!\,p!} e_{i_1 ... i_p i_{p+1} ... i_n} \alpha^{i_1 ... i_p} \theta^{i_{p+1}} \wedge ... \theta^{i_n}, \tag{22.9}$$

where $e_{i_1 ... i_n}$ is the Levi-Civita symbol of order n, and the indices have been raised using the metric, which has components η^{ij} or δ^{ij} (here the θ^i are not ordered). As an example we consider the Euclidean case and $n = 3$. The dual of the 0-form 1 is $\theta^1 \wedge \theta^2 \wedge \theta^3$, the dual of the basis 1-form θ^1 is $\theta^2 \wedge \theta^3$, the dual of the 2-form $\theta^1 \wedge \theta^2$ is θ^3, and finally the dual of $\theta^1 \wedge \theta^2 \wedge \theta^3$ is 1. In four dimensions and for Minkowski metric we have

$$*(\theta^0 \wedge \theta^1) = -\theta^2 \wedge \theta^3, \quad *(\theta^0 \wedge \theta^2) = +\theta^1 \wedge \theta^3, \quad *(\theta^0 \wedge \theta^3) = -\theta^1 \wedge \theta^2,$$
$$*(\theta^1 \wedge \theta^2) = +\theta^0 \wedge \theta^3, \quad *(\theta^1 \wedge \theta^3) = -\theta^0 \wedge \theta^2, \quad *(\theta^2 \wedge \theta^3) = +\theta^0 \wedge \theta^1, \tag{22.10}$$

or $*(\theta^i \wedge \theta^j) = \frac{1}{2} e^{ij}{}_{kl} (\theta^k \wedge \theta^l)$.

---

[2]Therefore, for example, a 2-form with $n = 3$ is written as $\alpha = \frac{1}{2}(\alpha_{12}\theta^1 \wedge \theta^2 + \alpha_{21}\theta^2 \wedge \theta^1 + ...) = \frac{1}{2}(\alpha_{12} - \alpha_{21})\theta^1 \wedge \theta^2 + \frac{1}{2}(\alpha_{13} - \alpha_{31})\theta^1 \wedge \theta^3 + \frac{1}{2}(\alpha_{23} - \alpha_{32})\theta^2 \wedge \theta^3 \equiv \alpha_{|[ij]|}\theta^i \wedge \theta^j$, where $\alpha_{|[ij]|}$ is the antisymmetrized *and* ordered ($i < j$) part of α_{ij}.

We can easily convince ourselves that if p is the degree of the form α, n is the dimension of the space, and sign g is the metric signature of this space ($+1$ for a Euclidean space, -1 for a Minkowski space), then

$$^{**}\alpha = (-1)^{p(n-p)}(\text{sign } g)\,\alpha\,. \tag{22.11}$$

The exterior product and the vector product

The Hodge operator can be used to relate the exterior product and the vector product. We recall that in Euclidean geometry in $n = 3$ dimensions, the vector product of two vectors v and w which are decomposed as $v = v^j h_j$ and $w = w^k h_k$ on a basis h_i of E_n is $v \wedge w \equiv e^i{}_{jk} v^j w^k h_i$. Using the Euclidean metric with components δ^{ij} to raise indices, we can associate with this vector $v \wedge w$ the form $e_{ijk} v^j w^k \theta^i$, where θ^i is the dual form of h_i: $\theta^i(h_j) = \delta^i_j$. This form is nothing but the dual of the exterior product of the forms $v_j\,\theta^j$ and $w_k\,\theta^k$ associated by raising the indices of the vectors v and w.

22.3 The exterior derivative

The *exterior derivative* is an operator, denoted d, which acts on a p-form to give a $(p+1)$-form and possesses the following defining properties: if f is a 0-form, $\mathrm{d}f(t) = t(f)$ (where t is a vector of E_n), which coincides with the definition of differential 1-forms[3] (see Book 1, Section 4.2). Moreover, $\mathrm{d}(\alpha + \beta) = \mathrm{d}\alpha + \mathrm{d}\beta$, where α and β are forms of the same degree. Finally,

$$\mathrm{d}(\alpha \wedge \beta) = \mathrm{d}\alpha \wedge \beta + (-1)^p \alpha \wedge \mathrm{d}\beta \qquad \text{and} \qquad \mathrm{d}^2 = 0, \tag{22.12}$$

where p is the degree of α.

This definition requires only the introduction of a vector space E_n (from which we construct its dual, and then its tensor products whose forms are the elements). However, if our building block is a (pseudo-)Euclidean space, we can introduce the natural bases $\frac{\partial}{\partial X^i}$ and $\mathrm{d}X^i$ associated with the Cartesian coordinates X^i, and then the exterior derivative is defined simply as follows: if $\alpha = \frac{1}{p!}\alpha_{i_1 i_2 \ldots i_p}\mathrm{d}X^{i_1} \wedge \ldots \mathrm{d}X^{i_p}$ is a p-form, its exterior derivative is

$$\mathrm{d}\alpha = \partial_l \alpha_{i_1 i_2 \ldots i_p}\mathrm{d}X^l \wedge \mathrm{d}X^{i_1} \wedge \ldots \mathrm{d}X^{i_p}, \tag{22.13}$$

which can be shown to possess all the properties in (22.12). The last property, called the *Poincaré lemma*, is therefore an immediate consequence of (22.13):

$$\mathrm{d}(\mathrm{d}\alpha) = \partial^2_{p\,l}\alpha_{i_1 i_2 \ldots i_p}\mathrm{d}X^p \wedge \mathrm{d}X^l \wedge \mathrm{d}X^{i_1} \wedge \ldots \mathrm{d}X^{i_p} = 0$$

owing to the symmetry of the second derivatives.

A form α whose exterior derivative is zero ($\mathrm{d}\alpha = 0$) is termed *closed*. A form α which is the exterior derivative of a form β ($\alpha = \mathrm{d}\beta$) is termed *exact*. An exact form is closed. Reciprocally, if a p-form is closed, then it is (at least locally) exact, that is, if α is such that $\mathrm{d}\alpha = 0$, then there exists a $(p-1)$-form β such that locally $\alpha = \mathrm{d}\beta$.

---

[3] In other words, the action of the 1-form $\mathrm{d}f$ on the vector t gives the same result as the action of the vector t on the function f.

When a metric is available, it is possible to define another derivative operator, the *codifferential*, denoted δ, which acts on a form α as

$$\delta\alpha = -(\text{sign } g)(-1)^{n(p+1)} \, {}^*\text{d}^* \, \alpha \,, \tag{22.14}$$

where sign g is the metric signature, n is the spatial dimension, p is the degree of the form α, and $*$ is the Hodge operator defined in (22.9). Since the operation $**$ is essentially the identity [see (22.11)] and owing to the Poincaré lemma ($d^2 = 0$), we have

$$\delta^2 = 0 \,. \tag{22.15}$$

22.4 A rewriting of the Maxwell equations

The exterior calculus can be used to obtain a compact and elegant formulation of Maxwell's equations. We have seen that in a system of Minkowski coordinates X^μ ($\mu = 0, 1, 2, 3$) they can be written as (see Sections 12.1 and 12.3)

$$F_{\mu\nu,\rho} + F_{\nu\rho,\mu} + F_{\rho\mu,\nu} = 0 \,, \qquad F^{\mu\nu}{}_{,\nu} = 4\pi j^\mu \,, \tag{22.16}$$

where $F_{\mu\nu}$ is the Faraday tensor $F_{\mu\nu} \equiv A_{\nu,\mu} - A_{\mu,\nu} \equiv \partial_\mu A_\nu - \partial_\nu A_\mu$, with A_μ the potential and j^μ the current vector, and the indices are raised using the coefficients of the inverse metric $\eta^{\mu\nu}$.

The potential is a 1-form, $A = A_\mu \text{d}X^\mu$. The Faraday tensor is a 2-form, the exterior derivative of A:

$$F = \text{d}A \,. \tag{22.17}$$

We indeed have[4] $\text{d}A \equiv \partial_\mu A_\nu \, \text{d}X^\mu \wedge \text{d}X^\nu = \frac{1}{2}(\partial_\mu A_\nu - \partial_\nu A_\mu)\text{d}X^\mu \wedge \text{d}X^\nu \equiv \frac{1}{2}F_{\mu\nu}\text{d}X^\mu \wedge \text{d}X^\nu$ $\equiv F$; *cf.* (22.13) and (22.8).

Since the Faraday tensor is an exact form, it is closed, that is,

$$\text{d}F = 0 \,, \tag{22.18}$$

which is just the first group of Maxwell's equations. Indeed,

$$0 = \text{d}F = \frac{1}{2}\partial_\rho F_{\mu\nu} \, \text{d}X^\rho \wedge \text{d}X^\mu \wedge \text{d}X^\nu = (\partial_0 F_{12} + \partial_1 F_{20} + \partial_2 F_{01})\text{d}X^0 \wedge \text{d}X^1 \wedge \text{d}X^2 + \ldots$$

The second group of Maxwell's equations is written as

$$\text{d}^*F = 4\pi({}^*j) \,. \tag{22.19}$$

Indeed,

$$F = F_{01} \, \text{d}T \wedge \text{d}X + \ldots + F_{12} \, \text{d}X \wedge \text{d}Y + \ldots, \quad {}^*F = -F_{01} \, \text{d}Y \wedge \text{d}Z + \ldots + F_{12} \, \text{d}T \wedge \text{d}Z + \ldots,$$

where we have used (22.10). From this we have

---

[4]This 'Cartan' formulation then leads to considering the *form* $A = A_\mu \text{d}X^\mu$ as the fundamental object of electromagnetism rather than the vector associated with it by 'raising an index' $A^\mu \frac{\partial}{\partial X^\mu}$.

$$\mathrm{d}^{*}F = \partial_{0}F^{01}\,\mathrm{d}T \wedge \mathrm{d}Y \wedge \mathrm{d}Z + \partial_{1}F^{01}\,\mathrm{d}X \wedge \mathrm{d}Y \wedge \mathrm{d}Z + ...$$

from applying the definition (22.13) or, regrouping the terms,

$$\mathrm{d}^{*}F = \partial_{\nu}F^{0\nu}\,\mathrm{d}X \wedge \mathrm{d}Y \wedge \mathrm{d}Z - \partial_{\nu}F^{1\nu}\,\mathrm{d}T \wedge \mathrm{d}Y \wedge \mathrm{d}Z + ...$$

As for the quantity $*j$, it is given by

$$*j = \frac{1}{6}e_{\mu\nu\rho\sigma}j^{\mu}\,\mathrm{d}X^{\nu} \wedge \mathrm{d}X^{\rho} \wedge \mathrm{d}X^{\sigma} = j^{0}\mathrm{d}X \wedge \mathrm{d}Y \wedge \mathrm{d}Z - j^{1}\mathrm{d}T \wedge \mathrm{d}Y \wedge \mathrm{d}Z + ...$$

Q.E.D.[5]

Taking the dual of (22.18) and (22.19), we obtain an equivalent formulation of the Maxwell equations:

$$\delta^{*}F = 0, \quad \delta F = 4\pi j. \tag{22.20}$$

Since $\delta^{2} = 0$ we have

$$\delta j = 0, \tag{22.21}$$

which corresponds to the law of current conservation $\partial_{\mu}j^{\mu} = 0$. Indeed,

$$\delta j = -{}^{*}\mathrm{d}^{*}j = -{}^{*}\mathrm{d}(j^{0}\mathrm{d}X \wedge \mathrm{d}Y \wedge \mathrm{d}Z - j^{1}\mathrm{d}T \wedge \mathrm{d}Y \wedge \mathrm{d}Z + ...)$$

$$= -{}^{*}(\partial_{0}j^{0}\mathrm{d}T \wedge \mathrm{d}X \wedge \mathrm{d}Y \wedge \mathrm{d}Z - \partial_{1}j^{i}\mathrm{d}X \wedge \mathrm{d}Y \wedge \mathrm{d}Z \wedge \mathrm{d}T + ...)$$

$$= -\partial_{\mu}j^{\mu}\,{}^{*}(\mathrm{d}T \wedge \mathrm{d}X \wedge \mathrm{d}Y \wedge \mathrm{d}Z) = \partial_{\mu}j^{\mu}.$$

22.5 Differential operators and the exterior derivative

The gradient of a function, ∇f, is the vector associated with the differential form $\mathrm{d}f$ by the metric (see Book 1, Section 4.6):

$$\nabla f = \partial^{i}f\frac{\partial}{\partial X^{i}}, \quad \mathrm{d}f = \partial_{i}f\,\mathrm{d}X^{i}. \tag{22.22}$$

Equation (22.21) indicates in addition the relation between the divergence of a vector and the codifferential of a 1-form:

$$\nabla \cdot \left(v^{i}\frac{\partial}{\partial X^{i}}\right) = \partial_{i}v^{i}, \quad \delta(v_{i}\mathrm{d}X^{i}) = -\partial_{i}v^{i}. \tag{22.23}$$

In Section 22.2 we also saw the relation between the vector product and the dual of the exterior product. In $n = 3$ dimensions and for a Euclidean metric, it can also be shown

---

[5]It is clear that here j is a 4-vector and not a 3-vector as in Section 12.3, eqn (12.16).

that the *curl* of a vector v, $\nabla \wedge v$, is the vector version of the dual of the differential of the associated 1-form. We have

$$\nabla \wedge \left(v^i \frac{\partial}{\partial X^i} \right) = (\partial_Y v^Z - \partial_Z v^Y) \frac{\partial}{\partial X} - (\partial_X v^Z - \partial_Z v^X) \frac{\partial}{\partial Y} + (\partial_X v^Y - \partial_Y v^X) \frac{\partial}{\partial Z} ,$$

$$^*\mathrm{d}(v_i \, \mathrm{d}X^i) = (\partial_Y v^Z - \partial_Z v^Y)\mathrm{d}X - (\partial_X v^Z - \partial_Z v^X)\mathrm{d}Y + (\partial_X v^Y - \partial_Y v^X)\mathrm{d}Z .$$
$$(22.24)$$

The Laplacian of a function f can be defined in two equivalent ways:

$$\triangle f = \nabla . \nabla f = \partial_i \partial^i f , \quad \delta \mathrm{d} f = -\partial_i \partial^i f . \tag{22.25}$$

Finally, the Laplacian (also called the *Laplace–Beltrami operator*) can be generalized as $\triangle \equiv \delta \mathrm{d} + \mathrm{d}\delta$, which is then referred to as the *Laplace–de Rham operator*.

22.6   Integration and the Stokes theorem

Let E be a vector space of dimension n, supplied with a Euclidean metric e or a Minkowski metric ℓ, and let (e_i) be an orthonormal basis of E (of dual basis ϵ^i), that is, $e = \delta_{ij} \epsilon^i \otimes \epsilon^j$ or $\ell = \eta_{ij} \epsilon^i \otimes \epsilon^j$. Let us consider the n-form $\epsilon^1 \wedge \epsilon^2 \wedge ... \wedge \epsilon^n \in \Lambda^n E^*$. If (e_i') [respectively (ϵ'^i)] is another orthonormal basis, we have $\epsilon'^1 \wedge ... \wedge \epsilon'^n = \det\Lambda \, \epsilon^1 \wedge ... \wedge \epsilon^n$, where Λ is the matrix taking us from the old basis to the new one. Now, $\det\Lambda = \pm 1$. We can eliminate this ambiguity in the sign by restricting ourselves to the ensemble of direct bases, and thus define the element of the oriented volume as

$$\omega = \epsilon^1 \wedge ... \wedge \epsilon^n \tag{22.26}$$

by taking the direct basis (e_i). Now if (h_i) is any direct basis with dual basis (θ^i), we have

$$\omega = \sqrt{\det(g_{ij})} \, \theta^1 \wedge ... \wedge \theta^n , \tag{22.27}$$

where $g = g_{ij}\theta^i \otimes \theta^j$ (with $g \equiv e$ or $g \equiv \ell$). Indeed, if P is the matrix taking us from (e_i) to (h_i), then $g_{ij} = g(h_i, h_j) = P_i^k P_j^l g(e_k, e_l) = \sum_k P_i^k P_j^k$, and therefore $\det(g_{ij}) = \det(P^t P) = (\det P)^2$.

This definition can be extended to metrics $g = g_{ij}(x^k)\mathrm{d}x^i \otimes \mathrm{d}x^j$, and we can thus define the integral of a function $f(x^k)$ over an open domain Ω as an elementary integral:

$$\int_\Omega f(x^k)\omega \equiv \int_\Omega f(x^k)\sqrt{\det(g_{ij})}\,\mathrm{d}x^1 \wedge ... \wedge \mathrm{d}x^n = \int_\Omega f(x^k)\sqrt{\det(g_{ij})}\,\mathrm{d}x^1...\mathrm{d}x^n . \tag{22.28}$$

Now let $\alpha = \frac{1}{p!}\alpha_{i_1...i_p}\mathrm{d}X^{i_1} \wedge ...\mathrm{d}X^{i_p}$ be a p-form decomposed on the natural basis of 1-forms $\mathrm{d}X^i$ associated with the (pseudo-)Euclidean coordinates X^i of a (pseudo-)Euclidean space of dimension n. Let us consider a hypersurface S of dimension p in this space. It can be defined by the parametric equations $X^i = X^i(\lambda^1, ..., \lambda^p)$. Substituting

$\mathrm{d}X^i = \frac{\partial X^i}{\partial \lambda^j}\mathrm{d}\lambda^j$ into the expression for α, we obtain its restriction to Σ_p, which is a differential form of order p in a space of dimension p, and is therefore of maximum degree[6]: $a(\lambda^i)\mathrm{d}\lambda^1 \wedge ... \wedge \mathrm{d}\lambda^p$. We then define the integral of α on S as the elementary integral

$$\int_S \alpha \equiv \int_S a(\lambda^i)d\lambda^1...d\lambda^p. \tag{22.29}$$

This allows us to assign an operational meaning to the *Stokes theorem*:

$$\int_V \mathrm{d}\alpha = \int_S \alpha, \tag{22.30}$$

where, if α is a p-form, S is the hypersurface of dimension p bounding the volume V of dimension $(p+1)$. The *Gauss theorem* is a special case of the Stokes theorem corresponding to $p = n - 1$, where n is the dimension of the space.

---

[6]For example, if $n = 3$ ($X^i = \{X, Y, Z\}$) and α is a 2-form, we have $a(\lambda^1, \lambda^2) = 2(\alpha_{[12]}\partial_{[1}X\partial_{2]}Y + \alpha_{[13]}\partial_{[1}X\partial_{2]}Z + \alpha_{[23]}\partial_{[1}Y\partial_{2]}Z)$, where $\partial_i = \partial/\partial\lambda^i$.

List of books and articles cited in the text

Bardeen, J. M. (1980). Gauge-invariant cosmological perturbations. *Phys. Rev. D*, **22**, 1882.

Basdevant, J.-L. (2007). *Lectures on quantum mechanics*. Springer, New York.

Bel, Ll., Salas, A., and Sànchez, J. M. (1973). Approximate solutions of predictive relativistic mechanics for the electromagnetic interaction. *Phys. Rev. D*, **7**, 1099–1106.

Berestetskii, V. B., Lifshitz, E. M. and Pitaevskii, L. P. (1971). *Relativistic quantum theory*, Vol. 4. Pergamon Press.

Choquet-Bruhat, Y., DeWitt-Morette, C., and Dillard-Bleick, M. (1977). *Analysis, manifolds and physics*. North-Holland, New York.

Compton, A. H. (1923). A quantum theory of the scattering of X-rays by light elements. *Phys. Rev.*, **21**, 483–502.

Damour, T. (1975). A new and consistent method for classical renormalization. *Nuovo Cim.*, **26B**, 157–164.

Darrigol, O. (2000). *Electrodynamics from Ampère to Einstein* (1st edn). Oxford University Press, Oxford.

Darrigol, O. (2005). *The Maxwell equations from MacCullagh to Lorentz* [in French]. Éditions Belin, Paris.

de Wit, B. (2008). *Lecture notes on quantum field theory*. http://www.staff.science.uu.nl /~wit00103/teaching.htm.

Dirac, P. A. M. (1964). *Lectures on quantum mechanics*. Yeshiva University Press, New York.

Einstein, A. (1905a). Zur Elektrodynamik bewegter Körper. *Annalen der Physik*, **17**, 891–921. [English translation by Jeffery, G. and Perrett, W. (1923). *The Principle of relativity*. London, Methuen.]

Einstein, A. (1905b). Ist die Trägheit eines Körpers von seinem Energieinhalt abhängig? *Annalen der Physik*, **18**, 639–641. [English translation by Jeffery, G. and Perrett, W. (1923). *The Principle of relativity*. London, Methuen.]

Ellis, G. F. R. and Uzan, J.-P. (2005). *c* is the speed of light, isn't it? *Am. J. Phys.*, **73**, 240–247; also, https://arxiv.org/abs/gr-qc/0305099.

Englert, F. and Brout, R. (1964). Broken symmetry and the mass of gauge vector mesons. *Phys. Rev. Lett.*, **13**, 321.

Englert, F. (2012). Symmetry breaking and the scalar boson—evolving perspectives. https://arxiv.org/abs/1204.5382.

Figueroa-O'Farrill, J. M. (1998). Electromagnetic duality for children. http://www.maths.ed.ac.uk/~jmf/Teaching/EDC.html.

Goldstone, J. (1961). Field theories with superconductor solutions. *Nuovo Cimento*, **19**, 154–164.

Gourgoulhon, É. (2013). *Special relativity in general frames: from particles to astrophysics.* Springer-Verlag, Berlin.

Gralla, S. E., Harte, Q. I., and Wald, R. M. (2009). A rigorous derivation of electromagnetic self-force. *Phys. Rev. D*, **80**, 024031; also, https://arxiv.org/abs/0905.2391.

Granovskii, Ya. I. (2004). Sommerfeld formula and Dirac theory. *Physics–Uspekhi*, **47**, 523.

Hawking, S. W. and Ellis, G. F. R. (1973). *The large scale structure of spacetime.* Cambridge University Press, Cambridge.

Henneaux, M. and Teitelboim, C. (1992). *Quantization of gauge systems.* Princeton University Press, Princeton.

Hernandez, O. F. (2005). Compton effect: interacting particles or interacting waves. https://arxiv.org/abs/physics/0508101.

Higgs, P. W. (1964). Broken symmetries and the masses of gauge bosons. *Phys. Rev. Lett.*, **13**, 508–509.

Itzykson, C. and Zuber, J. B. (2006). *Quantum field theory.* Dover, New York.

Jackson, J. D. (1975). *Classical electrodynamics* (2nd edn). Wiley, New York.

Jackson, J. D. (2002). From Lorenz to Coulomb and other explicit gauge transformations. *Am. J. Phys.*, **70**, 917–928; also, https://arxiv.org/abs/physics/0204034.

Lagrange, J.-L. (1770). Nouvelle méthode pour résoudre les équations littérales par le moyen de séries. *Mémoires de l'Académie Royale des Sciences et Belles-Lettres de Berlin* [in French], Vol. XXIV, 5–73.

Landau, L. and Lifshitz, E. (1980). *The classical theory of fields* (4th edn). Butterworth–Heinemann, Oxford.

Langlois, D. (2011). *Introduction to relativity* [in French]. Éditions Vuibert, Paris.

Lévy-Leblond, J.-M. (1976). One more derivation of the Lorentz transformation. *Am. J. Phys.*, **44**, 271–277.

Lévy-Leblond, J.-M. (1979). Additivity, rapidity, relativity. *Am. J. Phys.*, **47**, 1045.

Book 2

Martí, J. M. and Müller, E. (2003). *Numerical hydrodynamics in special relativity*; also, http://relativity.livingreviews.org/lrr-2003-7.

Maxwell, J. C. (1861). On physical lines of force. Available on the site https://en.wikisource.org/wiki/.

Minkowski, H. (1909). Raum und Zeit. *Jahresbericht der Deutschen Mathematiker-Vereinigung*. Teubner, Leipzig; *Phys. Z.*, **10**, 75–88 [English translation by Meghnad Saha: *Space and Time*. Wikisource translation: https://en.wikisource.org/wiki/].

Misner, C. W., Thorne, K. S., and Wheeler, J. A. (1973). *Gravitation*. W. H. Freeman, New York.

Ollitrault, J. Y. (2008). Relativistic hydrodynamics for heavy-ion collisions. *European Journal of Physics*, **29**, 275–302; also, https://arxiv.org/abs/0708.2433.

Pais, A. and Uhlenbeck, G. E. (1950). On field theories with non-localized action. *Phys. Rev.*, **79**, 145.

Pauli, W. (1981). *Theory of relativity* (1st edn, 1921). Dover, New York.

Raimond, J.-M. (2000). *Electromagnetism and relativity* [in French]. http://www.phys.ens.fr/cours/notes-de-cours/jmr/electromagnetisme.htm.

Rees, M. (1966). Appearance of relativistically expanding radio sources. *Nature*, **211**, 468–470.

Rindler, W. (1991). *Introduction to special relativity* (2nd edn). Clarendon Press, Oxford.

Rohrlich, F. (2007). *Classical charged particles* (3rd edn). World Scientific, Singapore.

Rougé, A. (2005). *Introduction to subatomic physics* [in French]. Éditions de l'École Polytechnique, Paris.

Ruegg, H. and Ruiz-Altaba, M. (2004). The Stueckelberg field. *Int. J. Mod. Phys. A*, **19**, 3265; also, https://arxiv.org/abs/hep-th/0304245.

Rybicki, G. B. and Lightman, A. P. (1985). *Radiative processes in astrophysics*. Wiley, New York.

Uzan, J.-P. *et al.* (2002). The twin paradox and space topology. *Eur. J. Phys.*, **23**, 277; also, https://arxiv.org/abs/physics/0006039.

Uzan, J.-P. and Lehoucq, R. (2005). *The fundamental constants* [in French]. Éditions Belin, Paris.

Wald, R. M. (1984). *General relativity*. University of Chicago Press, Chicago.

Weinberg, S. (2005). *The quantum theory of fields*, Vol. 1. Cambridge University Press, Cambridge.

Wheeler, J. A. and Feynman, R. P. (1949). Classical electrodynamics in terms of direct interparticle action. *Rev. Mod. Phys.*, **21**, 425–433.

Whitrow, G. J. (1935). On equivalent observers. *Quarterly Journal of Mathematics*, **6**, 249–260.

Whittaker, E. (1949). *From Euclid to Eddington*. Cambridge University Press, Cambridge.

Woodard, R. P. (2007). Avoiding dark energy with 1/R modifications of gravity. *Lect. Notes Phys.*, **720**, 403–433; also, https://arxiv.org/abs/astro-ph/0601672.

BOOK 3

General relativity and gravitation

Contents

Book 3

Book 3

Book 3

Part I

Curved spacetime and gravitation

There was difficulty reconciling the Newtonian theory of gravitation with its instantaneous propagation of forces with the requirements of special relativity; and Einstein working on this difficulty was led to a generalization of his relativity—which was probably the greatest scientific discovery that was ever made.

P. A. M. Dirac, cited by S. Chandrasekhar in *J. Astrophys. Astr.*, Vol. 5, 3 (1984)

1
The equivalence principle

In this introductory chapter we recall several relevant aspects of Newton's theory of gravity, as well as Maxwell's theory of electromagnetism, in order to describe the conceptual path that Einstein followed in going from the theory of special relativity to general relativity.

1.1   A 'general' relativity

Physicists of the 19th century sought to measure the speed of the Earth relative to the medium in which light propagates, the *aether*, which was at that time viewed as the incarnation of the absolute space of Newton. This is analogous to determining the velocity of an airplane relative to the ground by measuring the speed of sound from the plane (assuming that the amount of atmospheric drag due to the wake is known). This endeavor turned out to be fruitless: the speed of light was c relative to the reference frame where both the solar system and the aether were at rest (as indicated by stellar aberration), and it was *also* c relative to the Earth, as shown by the Michelson–Morley experiment.[1] As summarized by Henri Poincaré,[2] "*It seems at first sight that the aberration of light and the related optical and electrical phenomena will provide us a means of determining the absolute motion of the Earth, or rather its motion, not in relation to the other stars, but in relation to the ether. Fresnel had already tried it. . . , Michelson. . . failed as well.*

It seems that this impossibility of demonstrating an experimental evidence for absolute motion of the Earth is a general law of nature. . . ."

It was therefore necessary to make more and more hypotheses regarding the 'electro-dynamics of moving bodies' (such as the Lorentz–Poincaré 'fictitious' time, the FitzGerald–Lorentz contraction, and so on) in order to make the Lorentz and Maxwell equations, respect, within the framework of Newtonian physics, this impossibility of distinguishing the reference frame in which the aether is at rest (see, for example, Book 2, Sections 11.3 and 12.3). Einstein, in pondering the concept of time, showed that it was possible to reconcile the observed constancy of the speed of light in all inertial frames and the dynamical equivalence of these frames by means of a law relating velocities, under the condition that electromagnetism and mechanics be formulated within the framework of a new representation of space and time: special relativity. Thus, in 1905 the concepts of absolute space and the aether filling it retreated to limbo (see Book 2, Chapter 1).

Book 3

---

[1] See, for example, Book 1, Chapter 17 and Book 2, Chapter 1.

Everywhere in what follows we shall set $c = 1$ as in Book 2, at the beginning of which a Note on the Units can be found. Greek indices run from 0 to 3, and Latin ones from 1 to 3 (or n). The metric signature is chosen to be $(-1, +1, +1, +1)$.

[2] See Poincaré (1906).

Relativity in Modern Physics. Nathalie Deruelle and Jean-Philippe Uzan.
© Oxford University Press 2018. Published in 2018 by Oxford University Press.
DOI: 10.1093/oso/9780198786399.001.0001

However, the class of inertial reference frames remained privileged, as it is only in such frames that free particles can be at rest, thus allowing the reference frame to be determined, while also making it possible to see that such a particle is not subject to a force. In any other frame particles are accelerated, and can only be held at rest by the action of forces, that is, *inertial* forces, which indicate only that the frame with which their motion is associated is not an inertial frame. Such forces of inertia act as a sort of 'call to order' to the cohort of Galilean frames associated with the '*phantom*' corresponding to absolute space. One might therefore be dissatisfied with '*this actor who remains in the shadows, who acts on matter without in turn being acted upon by it*'.[3]

From 1907 on, the ambition of Einstein was to construct a theory in which *all* reference frames (and therefore none) were privileged, where there were no longer any inertial forces, no ordering of absolute space, and where, therefore, the laws of physics had the same form in all frames, inertial or not, so that no frame could be regarded as being privileged. In brief, he sought a theory of *general relativity*.[4]

This idea has several aspects. The first, called *general covariance*,[5] consists of writing the laws of dynamics in such a way that they preserve their form (that is, they are 'invariant') under a general change of coordinates. For example (as we have seen in Book 2, Section 6.2), in any frame, inertial or not, the equation of motion of a particle of mass m and velocity 4-vector u with components $u^\mu = dx^\mu/d\tau$ in the coordinates x^μ [$x^\mu = x^\mu(\tau)$ is the particle world line and τ is its the proper time] is written as

$$m\frac{\tilde{D}u^\mu}{d\tau} = f^\mu \quad \Longleftrightarrow \quad \frac{du^\mu}{d\tau} + \tilde{\Gamma}^\mu_{\nu\rho}u^\nu u^\rho = \frac{1}{m}f^\mu, \tag{1.1}$$

where the Christoffel symbols $\tilde{\Gamma}^\mu_{\nu\rho}$ characterizing the covariant derivative $\tilde{D}$ are determined by the coordinate system, that is, the chosen reference frame, and F is a contravariant vector describing the force (with components f^μ in the coordinates x^μ) acting on the particle.

Even though it is mathematically elementary within the framework of special relativity, where the time and space coordinates have the same status, this covariant formulation of the law of dynamics nevertheless has the interesting corollary (as we have stressed in Book 2, Section 5.3) of explicitly displaying the geometrical origin of the inertial forces: the noninertial nature of the reference frame is encoded in the Christoffel symbols $\tilde{\Gamma}^\mu_{\nu\rho}$, that is, in geometrical quantities.

A second facet of the principle of general relativity, of much richer content, is the attempt to assign a geometrical origin to *all* forces—inertial forces and actual forces, that is, to somehow absorb the vector F in an operator D, as mentioned in Book 2, Section 10.4. Then all forces would be manifestations of the geometry, and all motion would be free, but taking place in a spacetime which is geometrically more complex than the spacetimes considered up to now. We know that in 1915 Einstein succeeded in doing this for the gravitational force, but the other forces are still resisting. . . .

---

[3] Gilles Châtelet in *Les enjeux du mobile* [Châtelet (1993)]; translation by PFM.

[4] Ist es denkbar, dass das Prinzip der Relativität auch für Systeme gilt, welche relativ zueinander beschleunigt sind? (Is it conceivable that the principle of relativity also holds for systems that are accelerated relative to each other?). *Jahrb. Rad. Elektr.* **4**, 411 (1907), cited in Pais (1983).

[5] Or *indifference*, to use the term of T. Damour; see Damour (2005).

1.2 The equivalence principle

The tool that Einstein used to 'break into' absolute spacetime and extract gravitation[5] was the curious fact that all bodies fall in the same way in a gravitational field (see, for example, Book 1, Section 11.1).

Newtonian mechanics treats this fact as a fortuitous equality of gravitational and inertial mass. However, as Einstein revealed, this equality is not accidental. The acceleration of a free particle arising when its motion is attributed to use of a noninertial reference frame depends only on the motion of the frame, and not on the intrinsic properties of the particle such as its inertial mass. Therefore, all bodies 'fall' in the same way in an inertial field, just as in a gravitational field.

This similarity led Einstein to postulate the equality of inertial and gravitational mass, and to propose that inertial forces and gravitational forces are *identical*. This is the *equivalence principle*.[6]

This principle also has various aspects. First, since inertial forces are essentially geometrical, gravitation should be also (and therefore it should be possible to 'encode' it in the Christoffel symbols of a covariant derivative; see, for example, Book 2, Section 10.4). Moreover, it implies that an acceleration can simulate a gravitational field: for example, the observation of a frequency shift when a frequency is measured in an accelerated reference frame (for example, the Rindler frame; see Book 2, Section 5.2) can legitimately be attributed to a gravitational field rather than kinematical effects associated with the frame motion. Reciprocally, an actual gravitational field (for example, that of the Earth or the Sun) can be effaced by an acceleration: in a reference frame in 'free' fall and therefore accelerated relative to distant inertial frames, the gravitational field can legitimately be ignored because *co-moving* particles, that is, particles which are also in free fall, undergo uniform rectilinear motion, so that the reference frame is *de facto* inertial.[7]

An interesting corollary of this principle is that it clearly demonstrates that the concept of a free particle (at least with regard to gravitation) is not very meaningful. For example, particles 'attracted' by the Sun are 'free' in the frame which is moving along with them, which is a contradiction in terms. However, the definition of an inertial frame *is based on* the concept of a free particle. If no particle is free or if they are all free, the entire class of inertial frames then also retreats to limbo.

1.3 A mosaic of pieces of $\mathcal{M}_4$

The equivalence principle stipulates that any gravitational field can be effaced. Let us see what precisely this means.

The inertial forces to which free particles must be subjected in order for them to remain at rest in an accelerated reference frame become (by definition) unnecessary when we pass to

---

[6] *"Der glücklichste Gedanke meines Lebens"*, Einstein (*circa* 1920).

[7] *"I was sitting on a chair in the patent office in Bern when all of a sudden a thought occurred to me: 'If a person falls freely he will not feel his own weight.' I was startled. This simple thought made a deep impression on me. ..."*

"...for an observer falling freely from the roof of a house there exists—at least in his immediate surroundings—no gravitational field. Indeed, if the observer drops some bodies then these remain relative to him in a state of rest or of uniform motion .... The observer therefore has the right to interpret his state as 'at rest'." Albert Einstein, 1907, in *Fundamental Ideas and Methods of Relativity* (the Morgan Manuscript), cited by A. Pais (1982), pp. 178–179.

an inertial frame. Their disappearance is global in the sense that in the new, inertial, frame, *all* these particles undergo simple uniform rectilinear motion forever.

On the other hand, in a freely falling frame the gravitational field does not completely vanish: two particles initially at rest remain at rest and their separation remains fixed, but only for a certain length of time which is the shorter the more precise the measurements. The distance between the particles actually decreases little by little because Newton's force makes their trajectories converge toward the center of gravity of the body attracting them. Therefore, the 'effacement' property is only local: at any instant of time, in any place, there exist reference frames in which the gravitational field can be ignored, where the particles can be considered to be at rest, but these frames depend on the place and the time.

We are thus led to represent the spacetime framework in which the laws of gravity are formulated as a 'mosaic of small pieces of Minkowski spacetime'.[5] In each piece, that is, locally, in each 'tangent' inertial frame, spacetime is quasi-Minkowskian, gravity can be ignored, and it can be postulated that all the results of special relativity are valid there. This is the principle of *local relativity*.[8]

1.4   The reference 'mollusk'

Newtonian mechanics postulates the existence of material trihedra or *solids* for which the spatial separations of the constituents (measured using *rigid* rulers) are constant, and whose axes remain orthonormal throughout the motion. Such trihedra can (in principle) serve as global reference frames which impose a grid on all space, making it possible to test the laws of mechanics written in any accelerated frame obtained from the absolute frame by a rigid displacement. (And if the laws of mechanics are found not to be valid, it is *a priori* because the solid is distorted owing to inertial forces acting on it!)

On the other hand, in special relativity, solid bodies, global reference frames which impose a grid on all space, *de facto* materialize only inertial frames. The reason is simple. As we have seen in Book 2, Section 5.1, the passage to an accelerated frame is performed by a nonlinear change of coordinates involving the time, a transformation for which it would be unnecessarily restrictive to require that it preserve the orthogonality of the spatial axes (see, for example, Book 2, Section 5.4). Since, in addition, the equivalence principle robs inertial frames of their privileged status, the concept of a rigid reference body was ultimately abandoned by Einstein.[9]

In general relativity, spacetime thus becomes a simple *continuum*, an ensemble of events differentiated by their labeling in some reference frame. We can, for example, imagine this continuum and its reference *mollusk* as an ocean populated by bathyscaphes connected by ropes, whose relative motions are measured using the times of clocks undergoing arbitrary motion. In the vicinity of each point it is, however, possible to pass from this 'mollusk' to the

---

[8]It is also sometimes called the *Einstein equivalence principle*, while the identification of gravitational mass with inertial mass is, in contrast, called the *weak equivalence principle*.

[9] *"In gravitational fields there are no such things as rigid bodies with Euclidean properties; thus the fictitious rigid body of reference is of no avail in the general theory of relativity, . . .*

For this reason non-rigid reference-bodies are used which are as a whole not only moving in any way whatsoever, but which also suffer alterations in form ad lib. during their motion (. . .). This non-rigid reference-body, which might appropriately be termed a 'reference-mollusk,' is in the main equivalent to a Gaussian four-dimensional co-ordinate system chosen arbitrarily." Einstein (1954).

local 'grid' of Minkowskian coordinates of the bathyscaphe, that is, to pass to a momentarily inertial frame in free fall, the *tangent space*.

1.5 A curved spacetime

In Newtonian physics, physical space is represented by an 'absolute' Euclidean space. In general relativity, space and time are represented by a *manifold*, a continuum of points p labeled by four coordinates x^μ which locally reduce to Minkowski spacetime, as we have seen above.

A manifold[10] contains much less structure than a Euclidean space because, while geometrical quantities like vectors, tensors, and so on can be defined *locally* on a manifold (that is, they can be defined on the space tangent to a given point of the manifold), the concepts of parallelism and distance are not *a priori* defined on it.

To be able to compare quantities at different points and relate different pieces of $\mathcal{M}_4$ (that is, the tangent spaces) to each other, it is necessary to add an additional structure to a manifold. The role of a *connection* is to *define* the *parallel transport* of tensors. In particular, a connection makes it possible to construct *auto-parallel* curves (that is, the 'straightest' lines possible) of a manifold in the following way. If the world lines of such curves are $p = p(\lambda)$ or, in coordinates, $x^\mu = x^\mu(\lambda)$, and if $u = dp/d\lambda$ ($\Leftrightarrow u^\mu = dx^\mu/d\lambda$) are their tangent vectors, they satisfy the equation

$$\frac{du^\mu}{d\lambda} + \Gamma^\mu_{\nu\rho} u^\nu u^\rho = 0 \,, \tag{1.2}$$

where the 64 connection coefficients $\Gamma^\mu_{\nu\rho}$ (which are not necessarily symmetric in ν and ρ) are *a priori* arbitrary functions of the coordinates. As we have shown in Book 2, Section 5.3, if by a simple change of coordinates the connection coefficients can be made to vanish everywhere (which implies that they can be expressed in terms of only four functions of the coordinates), then the manifold is termed *flat*; otherwise, it is *curved*.

Now, in order to be able to compare at two significantly different times the distances between two particles and infer, for example, the existence of a central mass attracting them, a metric structure defining the concept of distance must also be introduced. This is the *metric tensor*, which defines the length element as

$$ds^2 = g_{\mu\nu} \, dx^\mu \, dx^\nu \,, \tag{1.3}$$

where the 10 functions $g_{\mu\nu}$ (symmetric in μ and ν) are *a priori* arbitrary. The specification of a metric makes it possible to define *geodesics*, that is, curves of extremal length. In the neighborhood of each point the coefficients $g_{\mu\nu}$ can be reduced by a change of coordinates to the coefficients $\eta_{\mu\nu}$ of the Minkowski metric, but in general it is not possible to do this globally (it only becomes possible if they can be expressed using only four functions; see, for example, Book 2, Section 5.3).

The concepts of parallelism and distance are conceptually distinct, but any metric can define a *Levi-Civita connection*. In this case, the 40 (symmetric) connection coefficients are

---

[10]It was in August 1912 that Einstein decided to represent spacetime by a space richer than Minkowski space. He called on his mathematician friend Marcel Grossmann for help, and thus was introduced to Riemannian geometry.

referred to as *Christoffel symbols* and are expressed as a function of the 10 metric components as (*cf.* Book 1, Section 3.4 or Book 2, Section 5.3)

$$\Gamma^{\mu}_{\nu\rho} = \frac{1}{2}g^{\mu\sigma}\left(\frac{\partial g_{\rho\sigma}}{\partial x^{\nu}} + \frac{\partial g_{\nu\sigma}}{\partial x^{\rho}} - \frac{\partial g_{\nu\rho}}{\partial x^{\sigma}}\right). \tag{1.4}$$

Auto-parallels and geodesics then become the same object. A manifold thus supplied with a metric and its associated Levi-Civita connection is a *(pseudo-)Riemannian manifold* ('pseudo' because the metric must reduce locally to $\eta_{\mu\nu}$ rather than to $\delta_{\mu\nu}$).

It is such pseudo-Riemannian manifolds which represent space and time in general relativity.[11] A space of this type is *curved* in the sense that the auto-parallels intersect each other (*cf.* the trajectories of 'free' particles in a co-moving reference frame, which come together at the gravitational center). In the special case where the spacetime is *flat*, there exists a global Minkowski reference frame (that is, all the 'little pieces'—the tangent spaces— can be 'glued' back together).

In general relativity, as in special relativity, the metric is a structure defined on spacetime. However, whereas in special relativity this structure, which may reduce to $\eta_{\mu\nu}$, is imposed *a priori*, in general relativity it is a physical quantity determined by the mass distribution. Therefore, in general relativity spacetime is not a passive receptacle of matter.

This is the form given by Einstein to a set of ideas promoted by Ernst Mach, the most relevant of which are (a) the inertia of a particle must be due to its interaction with all the masses of the universe, and (b) space has no existence in itself, independently of matter.

We see that while general relativity does incorporate in a certain manner the first idea (since forces of inertia and gravitation are no longer distinguishable), it does not include the second. Indeed, in the absence of matter, spacetime does not reduce to 'nothing', but rather to Minkowski spacetime. Therefore, Einstein did not completely realize his ambition.[12]

1.6 Gravitational redshift

In 1907, and again in 1911, Einstein illustrated the richness of the concepts of general relativity by a famous *gedankenexperiment* or thought experiment, which we shall discuss here using the calculations of Book 2, Section 5.2.

Let us consider, in an inertial reference frame (T, Z), a 'Rindler tower' of length h accelerated along the Z axis, that is, two hyperbolas with 'ground floor' given by $gT = \sinh g\tau$, $gZ = \cosh g\tau$ and 'top floor' by $g_h T = \sinh g_h\tau$, $g_h Z = \cosh g_h\tau$, with $h = 1/g_h - 1/g$. As we saw in Book 2, Section 5.2, light signals emitted at time intervals $\Delta\tau$ from the ground floor of the tower will be measured at the top floor at intervals of[13]

---

[11]A more detailed introduction to the spacetimes of general relativity along the lines we follow here can be found in Schrödinger (1950).

[12] *"I was hoping to show that space-time is not necessarily something to which one can ascribe a separate existence, independently of the actual objects of physical reality. Physical objects are not in space, but these objects are spatially extended. In this way the concept 'empty space' loses its meaning. (...) There is no such thing as an empty space, i.e., a space without field. Space-time does not claim existence on its own, but only as a structural quality of the field. Descartes was not so far from the truth when he believed he must exclude the existence of an empty space."* Einstein (1954).

[13]It is easily shown that when we limit ourselves to first order in gh, the result (1.5) for the 'rigid' Rindler tower will be the same as for a tower where the world lines from the ground floor and top floor are given by, for example, $(gT = \sinh g\tau, gZ = \cosh g\tau)$, $(gT = \sinh g\tau, gZ = \cosh g\tau + h)$, or $(Z = \frac{1}{2}gT^2, Z = \frac{1}{2}gT^2 + h)$.

$$\Delta\tau_h = (1 + gh)\Delta\tau. \tag{1.5}$$

It should be noted that $\Delta\tau_h$ is the period *as observed* from the top floor of the tower of the clock located at its ground floor, and also that if the clock were placed at the top floor, its signals would have period $\Delta\tau$, because we have postulated that devices measuring the time are not affected by their acceleration.

More generally, we have also seen in Book 2, Section 5.2, when calculating the Doppler effect in an accelerated frame where the tower is at rest, that

$$\Delta\tau_h = \sqrt{\frac{g_{00}(z = h)}{g_{00}(z = 0)}}\Delta\tau, \tag{1.6}$$

where $g_{\mu\nu}$ are the coefficients of the Minkowski metric in the accelerated frame and $z = 0$, $z = h$ are the world lines of the ground floor and the top floor of the tower in the coordinates in which the tower is at rest.

The tower is accelerated. However, the equivalence principle stipulates that an acceleration field is (locally) indistinguishable from a gravitational field. The same phenomenon of time dilation should then arise in a gravitational field, for example, that due to the Earth. In this case the ground floor of the tower is indeed the 'ground' floor and the top floor is its 'top', with h the height. The quantity g will be the acceleration due to the Earth, $g \equiv GM_\oplus/R_\oplus^2$. We then find

$$\Delta\tau_{\text{top}} = (1 + gh)\Delta\tau, \tag{1.7}$$

where we recall that $\Delta\tau_{\text{top}}$ is, for example, the period observed from the top of the tower of an atomic transition occurring at the bottom, and $\Delta\tau$ is the period of this transition, for the atom at either the bottom *or* the top of the tower.

Introducing the gravitational potential of the Earth $U(h) = -GM_\oplus/(R_\oplus + h)$, we find that $1 + gh = 1 + (U_0/U_h)(U_h - U_0)$, or in first order in $(U_h - U_0)/U_0$,

$$\Delta\tau_{\text{top}} \approx (1 + U_{\text{top}} - U_{\text{bottom}})\Delta\tau, \tag{1.8}$$

which can be generalized to any gravitational potential.

The equivalence principle therefore leads us to identify from (1.8) and (1.6)

$$g_{00} \approx -(1 + 2U), \tag{1.9}$$

where U is the Newtonian gravitational potential and the difference of the coefficient g_{00} from its Minkowski value is now interpreted as a deformation of spacetime due to gravitation.

The effect (1.7) was measured by Pound and Rebka in 1960. They placed a sample of radioactive iron at the bottom of a tower of height $h = 22.6$ m at Harvard University and observed the frequency of the emitted gamma rays from the top of the tower. The predicted redshift is $(\nu - \nu_{\text{top}})/\nu \approx (GM_\oplus/c^2R_\oplus)(h/R_\oplus) \approx 2.47 \times 10^{-15}$. They measured $(\nu - \nu_{\text{top}})/\nu = (2.57 \pm 0.26) \times 10^{-15}$.

Einstein had proposed measurement of the gravitational redshift of atomic transitions at the surface of the Sun, which in principle is much larger, of order 2×10^{-6}. However, the internal motion of the Sun gives rise to kinematical Doppler effects which are difficult to take into account, and it was only in 1991 that an accurate measurement could be made (by LoPresto *et al.*; the result agreed with the prediction to within 2 %).

Book 3

The currently most accurate measurement is that of Vessot and Levine (1976). They sent a maser clock to an altitude of 10,000 km, where they measured its frequency. They confirmed the prediction to within 2×10^{-4}. The atomic clocks of the network defining the International Atomic Time are not all located at sea level. For example, the clock in Boulder, Colorado is located at an altitude of 1600 m and gains about 5 microseconds per year compared to the clock in Greenwich. Since the accuracy of such clocks is currently of order 0.1 microseconds per year, they must be synchronized to correct for the difference.

The Hafele–Keating and Alley *et al.* experiments

Mention should also be made of the experiment of Hafele and Keating (1971), who compared the times of clocks which circled the globe from east to west and then from west to east with the time measured by a clock at rest on the Earth. A similar experiment was performed by Alley *et al.* (1975).

We note, however, that in these experiments it is necessary to include the time dilation due to the clock motion.

The inclusion of time dilation effects ('special and general') is now essential for achieving the proper functioning of engineering programs comparing the times of atomic clocks undergoing relative motion, such as clocks located on the Earth and on satellites of the GPS network; see Section 11.2.

The Minkowski spacetime of special relativity therefore exits the scene.

2
Riemannian manifolds

As we have argued in the preceding chapter, 'absolute, true, and mathematical' spacetimes representing 'relative, apparent, and common' space and time in Einstein's theory are Riemannian manifolds supplied with a metric and its associated Levi-Civita connection, where this metric simultaneously describes the coordinate system chosen to reference the events, that is, the positions of the matter elements at a given time, *and* the gravitational field acting on these elements.

Here we shall introduce the Riemann tensor characterizing curved spacetimes, and then the metric tensor, which allows lengths and durations to be defined. Finally, we shall discuss Riemannian manifolds.

2.1 The connection, parallel transport, and curvature

As we saw in our study of Euclidean space in curvilinear coordinates in Book 1, Chapter 3, the covariant derivative associates a tensor with another tensor defined at the *same* point p.[1] However, this operation, just like ordinary differentiation, allows, by means of a 'Taylor expansion', a tensor T to be transported *parallel* to itself from the point p to a nearby point q, thereby *connecting* the points p and q.

The *parallel transport* of a vector t^i (here $i = 1, ...n$) along a curve $x^j(\lambda)$ with tangent vector $u^j = dx^j/d\lambda$ is therefore defined by simple extension of the definition in Euclidean space:

$$\frac{Dt^i}{d\lambda} \equiv u^j D_j t^i = \frac{dt^i}{d\lambda} + \Gamma^i_{jk} u^j t^k = 0 \,, \tag{2.1}$$

where now the n^3 *connection coefficients* Γ^i_{jk} are *a priori* arbitrary functions of the n coordinates x^i. If the components t^i are known at $\lambda = \lambda_0$, the above differential equation determines them uniquely for all λ. Similarly, an *auto-parallel* is a curve such that $du^i/d\lambda + \Gamma^i_{jk} u^j u^k = 0$. In (pseudo-)Euclidean space this is the equation of a straight line, because in Cartesian coordinates all the Γ vanish.

Let us parallel-transport the vector of components t^i around a closed loop (see Fig. 2.1). In Euclidean space where there exist (Cartesian) coordinates for which all the Γ vanish, the integration of (2.1) is trivial and we obtain the well known result that the vector is the same at arrival as it was upon departure (see Book 1, Section 2.2). This is no longer true in the general case.

---

[1] In Book 1, Chapters 3 and 4 we presented an introduction to curvilinear coordinates and differential geometry in Euclidean space. Here we recall the transformation law for the components of a tensor under a change of coordinates $x^i \to x'^i$ (this was given in Book 1, Section 3.1):

$$T'^{ij\cdots}_{kl\cdots} = \frac{\partial x'^i}{\partial x^m} \frac{\partial x'^j}{\partial x^n} \cdots \frac{\partial x^p}{\partial x'^k} \frac{\partial x^q}{\partial x'^l} \cdots T^{mn\cdots}_{pq\cdots} \,.$$

For a more mathematical presentation of Riemannian geometry, see Part V of the present book (Book 3).

Relativity in Modern Physics. Nathalie Deruelle and Jean-Philippe Uzan.
© Oxford University Press 2018. Published in 2018 by Oxford University Press.
DOI: 10.1093/oso/9780198786399.001.0001

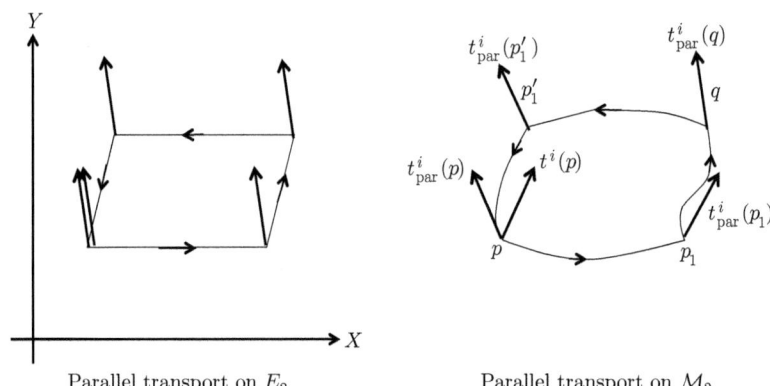

Parallel transport on E_2 $\qquad$ Parallel transport on $\mathcal{M}_2$

Fig. 2.1 Parallel transport.

Let us suppose that the loop consists of a first segment from p to p_1 along the coordinate line x^1, followed by a second segment from p_1 to q along the coordinate line x^2, a third from q to p_1' again along the coordinate line x^1, and, finally, to close the loop, a segment from p_1' to p along the coordinate line x^2. Equation (2.1) states that the components t_{par}^i of this parallel-transported vector are given implicitly at p_1, q, p_1', and p by the integro-differential equations

$$\begin{cases} t_{\text{par}}^i(p_1) = t^i(p) - \displaystyle\int_p^{p_1} \Gamma_{1k}^i t_{\text{par}}^k dx^1 \,, \quad t_{\text{par}}^i(q) = t_{\text{par}}^i(p_1) - \displaystyle\int_{p_1}^q \Gamma_{2k}^i t_{\text{par}}^k dx^2 \,, \\[3mm] t_{\text{par}}^i(p_1') = t_{\text{par}}^i(q) - \displaystyle\int_q^{p_1'} \Gamma_{1k}^i t_{\text{par}}^k dx^1 \,, \quad t_{\text{par}}^i(p) = t_{\text{par}}^i(p_1') - \displaystyle\int_{p_1'}^p \Gamma_{2k}^i t_{\text{par}}^k dx^2 \,, \end{cases} \tag{2.2}$$

so that, exchanging the limits on the last two integrals,

$$\Delta t^i \equiv t_{\text{par}}^i(p) - t^i(p) = \left(\int_{p_1'}^q - \int_p^{p_1} \right) \Gamma_{1k}^i t_{\text{par}}^k dx^1 - \left(\int_{p_1}^q - \int_p^{p_1'} \right) \Gamma_{2k}^i t_{\text{par}}^k dx^2 \,. \tag{2.3}$$

Since the coordinates of the various points are p: x^i, p_1: $x^i + \delta_1^i \Delta x^1$, q: $x^i + \delta_1^i \Delta x^1 + \delta_2^i \Delta x^2$, and p_1': $x^i + \delta_2^i \Delta x^2$, in lowest order we have

$$\Delta t^i = \Delta x^1 \Delta x^2 \left[\partial_2 (\Gamma_{1k}^i t_{\text{par}}^k) - \partial_1 (\Gamma_{2k}^i t_{\text{par}}^k) \right], \tag{2.4}$$

or, again using (2.1) (which gives $dt_{\text{par}}^k = -\Gamma_{ij}^k dx^i t^j$ or $\partial_i t_{\text{par}}^k = -\Gamma_{ij}^k t_{\text{par}}^j$),

$$\Delta t^i = R^i{}_{j21} \, t^j \Delta x^2 \Delta x^1 \,, \tag{2.5}$$

where

$$R^i{}_{jkl} \equiv \partial_k \Gamma_{lj}^i - \partial_l \Gamma_{kj}^i + \Gamma_{km}^i \Gamma_{lj}^m - \Gamma_{lm}^i \Gamma_{kj}^m \tag{2.6}$$

is the *curvature tensor* (or *Riemann–Christoffel tensor*) of the type $\binom{1}{3}$. It is a tensor because Δt^i, the difference of two vectors at the same point, is a vector, as are[2] Δx^1 and Δx^2. We

---

[2]It is therefore a useless (and tedious) exercise to prove this directly using the transformation law for the connection coefficients given in Book 1, Section 3.2 and recalled below in eqn (2.21).

obtain the same result if we compare at the point q the vectors parallel-transported *via* p_1 for one and *via* p_1' for the other. We therefore see that only if the Riemann tensor vanishes is the parallel transport of a vector (or any other tensor) independent of the path.[3]

Parallel transport on a 2-sphere

Let us consider a two-dimensional space with the coordinates (θ, ϕ), $\theta \in [0, \pi]$, $\phi \in [0, 2\pi]$ which is supplied with the connection

$$\Gamma^\theta_{\phi\phi} = -\sin\theta\cos\theta\,, \quad \Gamma^\phi_{\phi\theta} = \Gamma^\phi_{\theta\phi} = \frac{\cos\theta}{\sin\theta}\,,$$

with the other coefficients equal to zero. Let us take a vector with components $t^i(p) = (1, 0)$ at the point $p(\theta_0, \phi_0)$. Parallel-transporting it first from p to $p_1(\theta_0, \phi_1)$, and then to $q(\theta_1, \phi_1)$ and finally to $p_1'(\theta_1, \phi_0)$ to take it back to p, show that we have

$$\Delta t^\theta = \cos(\omega_1 - \omega_0) - 1\,, \quad \Delta t^\phi = \frac{\sin(\omega_1 - \omega_0)}{\sin\theta_0}\,,$$

with $\omega_1 = (\phi_1 - \phi_0)\cos\theta_1\,, \quad \omega_0 = (\phi_1 - \phi_0)\cos\theta_0\,.$

Show that the Riemann tensor reduces to $R^\theta_{\phi\theta\phi} = \sin^2\theta$, $R^\phi_{\theta\phi\theta} = 1$, so that, in agreement with the general expression (2.5), we have, in lowest order and setting $\Delta\theta = \theta_1 - \theta_0$ and $\Delta\phi = \phi_1 - \phi_0$: $\Delta t^\theta = 0$ and $\Delta t^\phi = -\Delta\theta\Delta\phi$.

2.2   Commutation of derivatives, torsion, and curvature

The covariant derivative of a function f with respect to the coordinate x^i is identical to the ordinary derivative $(D_j f = \partial_j f)$. The second derivative, $D_i D_j f = \partial_{ij} f - \Gamma^k_{ij}\partial_k f$, is therefore a 2-fold covariant tensor. Exchanging the indices i and j, we have

$$(D_i D_j - D_j D_i)f = -(\Gamma^k_{ij} - \Gamma^k_{ji})\partial_k f$$

$$\equiv -T^k_{ij}\partial_k f\,. \tag{2.7}$$

The quantities T^k_{ij} measure the antisymmetry of the connection. They are the components of a tensor T of the type $\binom{1}{2}$ called the *torsion*.

The connections of the spacetimes describing gravitation in general relativity are assumed to be torsion-free. We therefore henceforth require that

$$T^k_{ij} = 0\,, \tag{2.8}$$

in which case the number of connection coefficients is $n^2(n+1)/2$ (or 40 in four dimensions).

---

[3] In Part V of this book we present a more sophisticated discussion of the Riemann tensor. Among other things, we shall see that our choice of coordinate lines for defining a closed loop simplifies the discussion because their 'Lie brackets' are zero.

Let us calculate $D_i D_j v^k$. Here v^k is a vector, $D_j v^k$ is a tensor of the type $\binom{1}{1}$, and $D_i D_j v^k$ is a tensor of the type $\binom{1}{2}$. We therefore have $D_i D_j v^k = \partial_i (D_j v^k) - \Gamma^l_{ij} D_l v^k + \Gamma^k_{il} D_j v^l$, which, since $T^k_{ij} = 0$, leads to

$$(D_i D_j - D_j D_i) v^k = R^k{}_{mij} v^m . \tag{2.9}$$

Similarly, for a covariant vector λ_k,

$$(D_i D_j - D_j D_i) \lambda_k = -R^m{}_{kij} \lambda_m . \tag{2.10}$$

The Riemann tensor therefore measures the non-commutativity of the covariant derivative.

2.3 'Geodesic' deviation and curvature

Let us consider a family of auto-parallels $x^i_p(\lambda)$, where λ is the parameter along the curves and p is the index labeling them. If $u^i = dx^i/d\lambda$ is the vector tangent to the curves, we have $Du^i/d\lambda (\equiv u^j D_j u^i) = 0$. If in addition $n^i = \partial x^i/\partial p$ is the vector measuring the spacing of the curves, it is easy to show that, if the torsion vanishes,[4]

$$\frac{Du^i}{dp} (\equiv n^j D_j u^i) = \frac{Dn^i}{d\lambda} (\equiv u^j D_j n^i).$$

We can therefore calculate the relative 'acceleration' of two adjacent auto-parallels:

$$a^i \equiv \frac{D^2 n^i}{d\lambda^2} = \frac{D}{d\lambda} \frac{Dn^i}{d\lambda} = \frac{D}{d\lambda} \frac{Du^i}{dp} = u^j D_j (n^k D_k u^i)$$

$$= (u^j D_j n^k) D_k u^i + u^j n^k D_{jk} u^i \tag{2.11}$$

$$= (u^j D_j n^k) D_k u^i + u^j n^k (D_{kj} u^i + R^i{}_{mjk} u^m)$$

using (2.9). Now taking into account the fact that the curves are auto-parallels, we have

$$0 = \frac{D}{dp} \frac{Du^i}{d\lambda} = n^j D_j (u^k D_k u^i) = n^j u^k D_{jk} u^i + (n^j D_j u^k) D_k u^i \tag{2.12}$$

$$= (u^j D_j n^k) D_k u^i + u^j n^k D_{kj} u^i .$$

We thus obtain the expression often referred to as the *'geodesic' deviation* equation (even though it does not involve the metric):

$$a^i = R^i{}_{mjk} u^m u^j n^k , \tag{2.13}$$

which shows that if the Riemann tensor is nonzero, the parallels will eventually intersect.

---

[4]We have on the one hand

$$\frac{Du^i}{dp} = \frac{Du^i}{\partial x^j} \frac{\partial x^j}{dp} = n^j D_j u^i = n^j (\partial_j u^i + \Gamma^i_{jk} u^k) = \frac{du^i}{dp} + n^j u^k \Gamma^i_{jk} = \frac{\partial^2 x^i}{\partial p \partial \lambda} + n^j u^k \Gamma^i_{jk} ,$$

while on the other

$$\frac{Dn^i}{d\lambda} = \frac{Dn^i}{\partial x^j} \frac{\partial x^j}{d\lambda} = u^j D_j n^i = u^j (\partial_j n^i + \Gamma^i_{jk} n^k) = \frac{dn^i}{d\lambda} + u^j n^k \Gamma^i_{jk} = \frac{\partial^2 x^i}{\partial \lambda \partial p} + n^j u^k \Gamma^i_{kj} ,$$

and the two expressions are equal if $\Gamma^i_{jk} = \Gamma^i_{kj}$.

2.4 The metric tensor and the Levi-Civita connection

A connection allows the parallel transport of geometrical quantities from one point to another of a manifold to be defined. In order to also be able to define a *geodesic* or curve of extremal length between two points, we must now introduce the concept of distance.[5]

A *metric tensor* g is a 2-fold covariant [*i.e.*, of the type $\binom{0}{2}$], symmetric, and nondegenerate tensor field. In a system of coordinates x^i its $n(n+1)/2$ components $g_{ij}(x^k)$, which are symmetric ($g_{ij} = g_{ji}$) and invertible ($g^{ik}g_{kj} = \delta^i_j$), *define the length element as*[6]

$$ds^2 = g_{ij}\, dx^i\, dx^j \, . \tag{2.14}$$

As we have already seen in Book 1, Chapter 3, the metric tensor allows us to define a one-to-one correspondence between vectors and 1-forms. If v^i are the components of a vector v, then $v_i \equiv g_{ij}v^j$ is a 1-form isomorphic to v (and often also denoted as v). Reciprocally, if λ is a 1-form of components λ_i, then λ^i, where $\lambda^i \equiv g^{ij}\lambda_j$, is a vector. More generally, any tensor of the type $\binom{p}{q}$ can be associated with tensors of the type $\binom{p-1}{q+1}$ or $\binom{p+1}{q-1}$ in this way.

Finally, a metric defines the *scalar product* of two vectors v and v' as $(v \cdot v') = g_{ij}v^i v'^j$. In the case of a Lorentz metric, if the *norm* $(v \cdot v)$ of v is positive, v is termed space-like, while if it is negative v is time-like. If it is zero, v is termed null or light-like (it is also sometimes referred to as an isotropic vector).

The concept of a norm allows us to define the length of a time-like curve $\mathcal{C}$ of equation $x^i = x^i(\lambda)$ with tangent vector $u^i \equiv dx^i/d\lambda$, between the points p_1 and p_2, as

$$S[\mathcal{C}] = \int_{\lambda_1}^{\lambda_2} d\lambda \sqrt{-g_{ij}u^i u^j} = \int_{\lambda_1}^{\lambda_2} d\lambda \left(-g_{ij}\frac{dx^i}{d\lambda}\frac{dx^j}{d\lambda} \right)^{\frac{1}{2}} \tag{2.15}$$

(and similarly for a space-like curve). We note that S is reparametrization invariant, that is, it is invariant under $\lambda \mapsto \lambda(\tau)$.

Let us consider an ensemble of curves C_s labeled by s, with the equations $x^i = x^i_s(\lambda)$, all originating at p_1 and ending at p_2. We shall calculate the variation δS of the length of these curves when s is varied (this calculation generalizes that of Book 2, Section 2.2):

$$\delta S = \int_{\lambda_1}^{\lambda_2} \frac{1}{2\sqrt{-g_{ij}u_s^i u_s^j}} \left(-2g_{ij}\frac{d\delta x_s^i}{d\lambda}\frac{dx_s^j}{d\lambda} - \partial_k g_{ij}\delta x_s^k \frac{dx_s^i}{d\lambda}\frac{dx_s^j}{d\lambda} \right) d\lambda \, . \tag{2.16}$$

At this stage (but not before!) we can choose the parametrization, for example, $g_{ij}u_s^i u_s^j = -1$. Denoting the parameter by τ, dropping the index s and recalling that δx^i is zero at τ_1 and τ_2, we have

---

[5] In Books 1 and 2 all the geometrical properties of Newtonian and Minkowski spacetimes—in particular, the concept of parallel transport—were *deduced* from the existence of a metric (Euclidean or Minkowski). Here we consider the more general case where the metric and the connection are structures which are *a priori* independent.

[6] Indeed, if its components are known in the coordinates x^i, they are known in any other system of coordinates *via* the transformation law of 2-fold covariant tensors; *cf.* Book 1, Section 3.1 or footnote 1.

Book 3

$$\delta S = \int_{\tau_1}^{\tau_2} \left(-g_{ij} \frac{d\delta x^i}{d\tau} u^j - \frac{1}{2} \partial_k g_{ij} \delta x^k u^i u^j \right) d\tau$$

$$= \int_{\tau_1}^{\tau_2} \left(-\frac{d}{d\tau} (g_{ij} u^j \delta x^i) + \frac{d}{d\tau} (g_{ij} u^j) \delta x^i - \frac{1}{2} \partial_k g_{ij} \delta x^k u^i u^j \right) d\tau$$

$$= -g_{ij} u^j \delta x^i |_{\tau_1}^{\tau_2} + \int_{\tau_1}^{\tau_2} \left(\frac{d}{d\tau} (g_{ij} u^j) \delta x^i - \frac{1}{2} \partial_k g_{ij} \delta x^k u^i u^j \right) d\tau$$

$$= \int_{\tau_1}^{\tau_2} \delta x^k d\tau \left(g_{kj} \frac{du^j}{d\tau} + \partial_i g_{kj} u^i u^j - \frac{1}{2} \partial_k g_{ij} u^i u^j \right).$$

(2.17)

The curve is extremal if $\delta S = 0 \; \forall \; \delta x^k$. The integrand is then zero, which gives the *geodesic* equation determining it:

$$\frac{du^i}{d\tau} + \{^i_{jk}\} u^j u^k = 0 \quad \text{with} \quad g_{ij} u^i u^j = -1,$$

(2.18)

where

$$\{^i_{jk}\} = \frac{1}{2} g^{il} (\partial_j g_{kl} + \partial_k g_{lj} - \partial_l g_{jk})$$

are the *Christoffel symbols*.

If there exists a coordinate system in which all the metric coefficients are constant, the geodesics will be straight lines.

We see that for the concept of auto-parallel defined in Section 2.1 as the 'straightest possible' curve to coincide with the definition in (2.18) of a geodesic or the 'shortest (or longest) possible' curve, the connection coefficients of the covariant derivative must be the same as the Christoffel symbols. In this case the connection is referred to as the *Levi-Civita connection* or the *Riemannian connection*. The metric and connection are then compatible if

$$\Gamma^i_{jk} = \{^i_{jk}\} = \frac{1}{2} g^{il} (\partial_j g_{kl} + \partial_k g_{lj} - \partial_l g_{jk}).$$

(2.19)

Since the Christoffel symbols are symmetric, the Levi-Civita connection is also symmetric, that is, it is torsion-free. Therefore, the Riemann tensor defined in (2.6) is expressed as a function of the metric, and the manifold is said to be Riemannian.

There is another, equivalent and easily proved, way of requiring the compatibility of a metric and a connection, namely, that it be torsion-free and that[7]

$$D_i g_{jk} = 0.$$

(2.20)

Let us conclude by recalling the transformation law for the Christoffel symbols, which are not tensors, under a change of coordinates $x^i \to x'^i$ (cf. Book 1, Section 3.2):

$$\Gamma'^q_{pm} = \frac{\partial x^k}{\partial x'^m} \frac{\partial x^j}{\partial x'^p} \frac{\partial x'^q}{\partial x^i} \Gamma^i_{jk} + \frac{\partial x'^q}{\partial x^i} \frac{\partial^2 x^i}{\partial x'^p \partial x'^m}.$$

(2.21)

The difference of two Christoffel symbols $\delta \Gamma^i_{jk} = \Gamma^i_{jk} - \tilde{\Gamma}^i_{jk}$ corresponding to different metrics is a singly contravariant, 2-fold covariant tensor, because the last term in (2.21) cancels out.

---

[7]If we do not require that the torsion vanish, there will exist an infinite number of connections which are compatible with the metric. On the other hand, if the torsion is zero the metric connection is unique and given by (2.19), as was shown by Ricci.

2.5   Locally inertial frames

Let $g_{ij}(x^k)$ be the components of a metric tensor g in the coordinates x^k. In different coordinates $X^l = X^l(x^k)$ the components of g are $f_{ij}(X^k) = (\partial x^k/\partial X^i)(\partial x^l/\partial X^j)g_{kl}$. We make a Taylor series expansion about a point p_0 with coordinates X_0^k: $X^k = X_0^k + \epsilon^k$. Then

$$f_{ij}(X^k) = f_{ij}|_0 + \epsilon^m \frac{\partial f_{ij}}{\partial X^m}\Big|_0 + \cdots$$

$$\text{with} \qquad f_{ij}|_0 = \frac{\partial x^k}{\partial X^i}\frac{\partial x^l}{\partial X^j}g_{kl}|_0 \quad \text{and} \tag{2.22}$$

$$\frac{\partial f_{ij}}{\partial X^m}\Big|_0 = \left[g_{kl}\left(\frac{\partial^2 x^k}{\partial X^m \partial X^i}\frac{\partial x^l}{\partial X^j} + \frac{\partial^2 x^k}{\partial X^m \partial X^j}\frac{\partial x^l}{\partial X^i} \right) + \frac{\partial x^k}{\partial X^i}\frac{\partial x^l}{\partial X^j}\frac{\partial g_{kl}}{\partial X^m} \right]_0, \cdots$$

The question is, does there exist a change of coordinates such that $f_{ij} = \eta_{ij} + \mathcal{O}(\epsilon^p)$?

The answer is yes, but in general only at order $p = 2$.

The system $f_{ij}|_0 = \eta_{ij}$, with $n(n+1)/2 = 10$ equations for $n^2 = 16$ unknowns $(\partial x^k/\partial X^i)|_0$, has an infinite number of solutions involving six parameters, the six parameters of the Lorentz group.

Similarly, $(\partial f_{ij}/\partial X^m)|_0 = 0$ is a system of $n^2(n+1)/2 = 40$ equations for $n^2(n+1)/2 = 40$ unknowns $(\partial^2 x^k/\partial X^m \partial X^i)|_0$, and in general it has a unique solution.

We therefore have

$$f_{ij}(X^k) = \eta_{ij} + \frac{1}{2}\epsilon^m \epsilon^n \frac{\partial^2 f_{ij}}{\partial X^m \partial X^n}\Big|_0 + \cdots \tag{2.23}$$

However, it is not in general possible to cancel the second-order term because $(\partial^2 f_{ij}/\partial X^m \partial X^n)|_0 = 0$ is a system of $[n(n+1)/2]^2 = 100$ equations for only $n^2(n+1)(n+2)/3! = 80$ values of the third derivatives $(\partial^3 x^i/\partial X^m \partial X^j \partial X^k)|_0$. Therefore, the 20 second derivatives $(\partial^2 f_{ij}/\partial X^m \partial X^n)|_0$ remain undetermined. It is possible to relate them to the curvature tensor.

The components $\tilde{R}_{ijkl} \equiv f_{im}\tilde{R}^m{}_{jkl}$ of the Riemann tensor in the coordinates X^i in which the metric is expanded as in (2.23) are indeed given by

$$\tilde{R}_{ijkl} = \frac{1}{2}(\partial^2_{jk}f_{il} + \partial^2_{il}f_{jk} - \partial^2_{ik}f_{jl} - \partial^2_{jl}f_{ik})|_0 + \cdots \tag{2.24}$$

Owing to its symmetries, it has 20 independent components (see Section 2.6 for details). It is therefore possible to choose the change of coordinates $x^i \to X^i$ such that

$$\partial^2_{mn}f_{ij}|_0 = -\frac{1}{3}(\tilde{R}_{imjn} + \tilde{R}_{injm})|_0 \implies f_{ij}(X^k) = \eta_{ij} - \frac{1}{3}\epsilon^m \epsilon^n \tilde{R}_{imjn}|_0 + \cdots , \tag{2.25}$$

which respects the symmetries of $\partial^2_{mn}f_{ij}|_0$ and is compatible with (2.24).

In this system of coordinates X^i called *normal coordinates*, the Christoffel symbols and the geodesic equation are given by

$$\tilde{\Gamma}^i{}_{jk} = -\frac{1}{3}\epsilon^m (\tilde{R}^i{}_{jkm} + \tilde{R}^i{}_{kjm})|_0 + \cdots , \qquad \frac{d^2 X^i}{d\tau^2} = \frac{2}{3}\tilde{R}^i{}_{jkm}|_0 \frac{dX^j}{d\tau}\frac{dX^k}{d\tau}\epsilon^m + \cdots . \tag{2.26}$$

(The higher-order terms can be expressed as functions of the derivatives of the curvature tensor.) This is the *Riemann–Cartan theorem*.

We therefore see that, in agreement with the physical constraints described in Chapter 1, it is possible to construct at any point a system of quasi-Minkowski coordinates, that is, a quasi-inertial reference frame (defined up to Lorentz transformations). It is only when the Riemann tensor vanishes that this system can be extended to all of spacetime, in which case this is just the flat spacetime of Minkowski.

2.6   Properties of the Riemann tensor

In this section we shall present the very important properties of the curvature tensor. They are actually easy to prove when the manifold is Riemannian (that is, supplied with a metric connection). In that case it is possible to work in a frame which is locally Minkowskian, where the Christoffel symbols are zero, so that the curvature tensor reduces locally to (2.24). Then we easily see that

- $R^k_{\ lji} = -R^k_{\ lij}$ (this follows from the definition of the Riemann tensor and is true for any connection);
- $R^k_{\ lji} + R^k_{\ jil} + R^k_{\ ilj} = 0$ (the *first Bianchi identity*, which holds if the torsion vanishes);
- $D_m R^i_{\ jnp} + D_n R^i_{\ jpm} + D_p R^i_{\ jmn} = 0$ (the *second Bianchi identity*, which holds if the torsion vanishes).

When the connection, which is symmetric, is compatible with a metric tensor, the curvature tensor possesses the following additional properties:

$$R_{ijkl} = -R_{jikl} \quad \text{and} \quad R_{ijkl} = R_{klij}, \quad \text{where} \quad R_{ijkl} \equiv g_{ip} R^p_{\ jkl}. \tag{2.27}$$

Since the Riemann tensor is of the type $\binom{1}{3}$, one can derive from it tensors of the type $\binom{0}{2}$ by contraction. Owing to the above metric antisymmetry, the contraction $R^i_{\ ijk} \equiv 0$, and only the *symmetric Ricci tensor* R_{ij} remains:

$$R_{ij} \equiv R^l_{\ ilj} = -R^l_{\ ijl} = R_{ji}. \tag{2.28}$$

The *scalar curvature* is defined as

$$R \equiv g^{ij} R_{ij}. \tag{2.29}$$

Using the second Bianchi identity and the vanishing of the covariant derivative of the metric, it is easy to prove the following property, which plays an important role in general relativity:

$$D_i G^i_j = 0, \quad \text{where} \quad G^i_j \equiv R^i_j - \frac{1}{2}\delta^i_j R \tag{2.30}$$

is the *Einstein tensor*.

It is easily seen that the Riemann tensor in dimension $n = 2$ has a single independent component proportional to the scalar curvature: $R_{0101} = gR/2$, where g is the metric determinant, so that $R_{ij} = g_{ij}R/2$ (the Einstein tensor therefore vanishes identically in two dimensions).

In $n = 3$ dimensions the Riemann tensor possesses six independent components and is proportional to the Ricci tensor. In $n = 4$ dimensions it possesses 20 independent components; in general, the number of independent components is $n^2(n^2 - 1)/12$.

Finally, it is possible to isolate in the Riemann tensor the terms which depend on the Ricci tensor and the scalar curvature. The remaining term $C^{am}{}_{sq}$ is the *Weyl tensor*. More precisely, if n is the dimension of the manifold,

$$
R^{am}{}_{sq} = C^{am}{}_{sq} + \frac{1}{n-2} \left(\delta^a_s R^m_q + \delta^m_q R^a_s - \delta^m_s R^a_q - \delta^a_q R^m_s \right)
$$
$$
- \frac{1}{(n-1)(n-2)} \left(\delta^a_s \delta^m_q - \delta^a_q \delta^m_s \right) R \,.
$$

(2.31)

The Weyl tensor, which vanishes in dimensions less than four, possesses all the symmetries of the Riemann tensor and, in addition, is traceless[8]: $C^{am}{}_{aq} = 0$.

---

[8]It is worth noting that two Riemann spaces with metric tensors $\bar{g}_{ij}$ and g_{ij} which are 'conformally' related, that is, $\bar{g}_{ij} = F(x^k)\, g_{ij}$, where $F(x^k)$ is an arbitrary function of the coordinates, have the same Weyl tensor.

A more detailed discussion of the properties of the curvature tensor mentioned here can be found in, for example, Bourguignon (2005) or Straumann (2013).

3
Matter in curved spacetime

Here we shall study the laws of motion of matter—particles, fluids, or fields—in the presence of an external gravitational field. In accordance with the equivalence principle, this motion will be 'free', that is, constrained only by the geometry of the spacetime whose curvature represents the gravitation.

3.1 Geodesic motion of point masses

In special relativity, that is, in the absence of gravity, the world line of a free particle is a straight line in Minkowski spacetime which extremizes the proper time, that is, the length, of all the possible paths between two given events. According to the equivalence principle, the motion of a point mass in a gravitational field is 'free' motion, but in a spacetime curved by gravitation.[1] The *action* chosen to describe it will then be the length of all possible time-like paths between two events P_a and P_b, or

$$S_{\rm p} = -mc^2 \int_{P_a}^{P_b} \sqrt{-g_{\mu\nu}dx^\mu dx^\nu} \equiv \int_{\lambda_a}^{\lambda_b} L\, d\lambda \ \ \text{with}\ \ L = -mc^2\sqrt{-g_{\mu\nu}\frac{dx^\mu}{d\lambda}\frac{dx^\nu}{d\lambda}}\,, \qquad (3.1)$$

where m is the inertial mass of the particle, c is the speed of light (which we do not set equal to 1 in this section), λ is a parameter, and the metric coefficients $g_{\mu\nu}$ describe both the gravitational field and the coordinate system x^μ. As we have seen in Section 2.4, its extremization gives the geodesic equation:

$$\frac{Du^\mu}{d\tau} \equiv \frac{du^\mu}{d\tau} + \Gamma^\mu_{\nu\rho}u^\nu u^\rho = 0 \ \ \text{with}\ \ u^\mu u_\mu = -c^2\,, \qquad (3.2)$$

where $u^\mu = dx^\mu/d\tau$ is the 4-velocity of the mass point, $\lambda = \tau$ is its proper time, and $\Gamma^\mu_{\nu\rho}$ are the Christoffel symbols. This geodesic equation is the 'covariantized' version of the equation of motion in special relativity, obtained by replacing the ordinary derivative d by the covariant derivative D.

The Euler–Lagrange equations and the geodesic

The geodesic equation (3.2) can also be obtained from the action $S'_{\rm p} = \int L'd\lambda$ with $L' = g_{\mu\nu}u^\mu u^\nu$, which compared to (3.1) has the advantage of not being reparametrization-invariant.

---

[1]Here we assume that the 'point masses' under consideration have no proper rotation. The motion of 'spinning tops', that is, particles carrying a 'spin', will be studied in Section 11.7.

Relativity in Modern Physics. Nathalie Deruelle and Jean-Philippe Uzan.
© Oxford University Press 2018. Published in 2018 by Oxford University Press.
DOI: 10.1093/oso/9780198786399.001.0001

The extremization of S'_p when the path from P_a to P_b is varied gives

$$\delta S' = \frac{\partial L'}{\partial u^\mu}\delta x^\mu \Big|_{\tau_1}^{\tau_2} + \int_{\tau_1}^{\tau_2}\frac{\delta L'}{\delta x^\mu}\delta x^\mu d\tau \quad \text{with} \quad \frac{\delta L'}{\delta x^\mu} = \frac{\partial L'}{\partial x^\mu} - \left(\frac{\partial L'}{\partial u^\mu}\right)^{\cdot}.$$

The first term vanishes because the paths are fixed at the endpoints. The second term vanishes if the Euler–Lagrange equations $\delta L'/\delta x^\mu = 0$ are satisfied, that is, if

$$u^\nu u^\rho \partial_\mu g_{\nu\rho} = 2(g_{\mu\nu}u^\nu)^{\cdot} \quad \text{or} \quad \dot{u}_\mu = \frac{1}{2}u^\nu u^\rho \partial_\mu g_{\nu\rho}, \tag{3.3}$$

which is just the covariant version of the geodesic equation (3.2):

$$\frac{Du_\mu}{d\tau} = \dot{u}_\mu - \Gamma^\rho_{\mu\nu}u^\nu u_\rho = 0 \quad \text{with} \quad \Gamma^\rho_{\mu\nu} = \frac{1}{2}g^{\rho\lambda}(\partial_\mu g_{\nu\lambda} + \partial_\nu g_{\lambda\mu} - \partial_\lambda g_{\mu\nu}). \tag{3.4}$$

The advantage of (3.4) over (3.2) is that it also gives the light world lines whose tangent vectors are null lines ($u^\mu u_\mu = 0$).

In Newtonian mechanics and in an inertial frame, the action of a particle in a gravitational field is (*cf.* Book 1, Section 11.4)

$$S_{p(N)} = \int \left(-mc^2 + \frac{1}{2}mv^2 - mU\right)dt, \tag{3.5}$$

where t is the absolute time and U is the Newtonian potential (and we have added the constant $-mc^2$ so that S_p and $S_{p(N)}$ coincide in the absence of the field and when $c \to \infty$).

Since $S_p = -mc^2\int d\tau$ where τ is the proper time, we find that in the Newtonian limit

$$d\tau \sim dt\left(1 - \frac{v^2}{2c^2} + \frac{U}{c^2}\right) \quad \text{or} \quad ds^2 \equiv -c^2 d\tau^2 \sim -c^2 dt^2\left(1 + \frac{2U}{c^2}\right) + d\vec{r}\,^2. \tag{3.6}$$

We therefore recover in a general way the expression found in Section 1.6 for the metric component g_{00} as a function of the Newtonian potential in the limit of weak field and small velocities.

3.2   Equations of motion of fluids

As in special relativity (*cf.* Book 2, Section 8.3), the dynamics of a fluid assumed to be *perfect* is postulated to be encoded in its 'energy–momentum' tensor with components $T^{\mu\nu}$ given by

$$T^{\mu\nu} = (\epsilon + p)u^\mu u^\nu + p\, g^{\mu\nu}. \tag{3.7}$$

The velocity field of the fluid u^μ is normalized to unity, $g_{\mu\nu}u^\mu u^\nu = -1$, and the 'energy density' $\epsilon(x^\mu)$ and 'pressure' $p(x^\mu)$ are scalar functions. We see that this tensor is derived from its expression in special relativity by covariantization, that is replacement of the Minkowski metric by the metric of the Riemannian manifold describing the gravitational field in which the fluid is immersed.

In the phenomenological approach we are using here, the equation of motion of the fluid interacting with the gravitational field is obtained by *assuming* that its energy–momentum tensor is *conserved*, i.e., that

$$D_\nu T^{\mu\nu} = 0\,, \tag{3.8}$$

which is the convariantized version ($\partial_\mu \to D_\mu$) of the conservation law from which we can derive the conservation of energy, momentum, and angular momentum in special relativity (see, for example, Book 2, Section 8.2). Just as in special relativity, (3.8) can be decomposed as

$$u^\mu u^\nu \partial_\nu(\epsilon + p) + (\epsilon + p)(u^\nu D_\nu u^\mu + u^\mu D_\nu u^\nu) + \partial^\mu p = 0\,. \tag{3.9}$$

Contracting this with u^μ, we immediately obtain (since $u^\mu u_\mu = -1$)

$$D_\nu(\epsilon\, u^\nu) + p\, D_\nu u^\nu = 0\,, \tag{3.10}$$

so that (3.9) becomes equivalent to (3.10) plus

$$(\epsilon + p)u^\nu D_\nu u^\mu + u^\mu u^\nu \partial_\nu p + \partial^\mu p = 0\,. \tag{3.11}$$

Equation (3.11) is the *relativistic Euler equation* and (3.10) is the continuity equation.

Equations (3.9)–(3.11) are four equations in five unknowns, namely, ϵ, p, and three components of the vector u^μ (constrained by $u^\mu u_\mu = -1$). To completely determine the fluid motion in the gravitational field described by $g_{\mu\nu}$, we therefore need one more equation, the *equation of state*, which relates the pressure and the density (some examples are given in Book 2, Section 8.3).

The particle energy–momentum tensor

The 4-momentum of an ensemble of particles of mass m and 4-velocity u^μ can be defined as in special relativity as $P^\mu = \sum mu^\mu$ (with the sum running over the particles), but it cannot possibly be conserved before and after a 'collision' because gravitation is a long-range interaction. It is actually the energy–momentum tensor, introduced below, which describes the energy content of a system of particles in a gravitational field.

The energy–momentum tensor describing an ensemble of particles interacting only gravitationally is the covariantization of the energy–momentum tensor of special relativity (see, for example, Book 2, Section 8.4):

$$T^{\mu\nu} = \epsilon\, u^\mu u^\nu \quad\text{with}\quad \epsilon(x^\mu) \equiv \sum m \int \delta_4(x^\rho - x^\rho(\tau)) \frac{d\tau}{\sqrt{-g}}\,, \tag{3.12}$$

where $x^\mu(\tau)$ are the particle world lines and $u^\mu(x^\rho)$ is their velocity field [identified as their 4-velocities $u^\mu(\tau)$ owing to the presence of the Dirac delta functions $\delta_4(x^\mu) = \delta(t)\delta_3(x^i)$, for which $\int d^4x\, \delta_4(x^\mu) = 1$]. Since the 3-volume element is $\sqrt{-g}|_t d^3x$ (Book 1, Section 3.6), the factor $\sqrt{-g}$ guarantees that $\int_{t=\text{const}} T^{\mu 0}\sqrt{-g}\, d^3x = \sum mu^\mu$ is indeed the 4-momentum of the ensemble of particles. (We recall that to define it, it is necessary to specify the point on the world lines at which the 4-velocities are evaluated. It will therefore depend on the choice of time-like hypersurface; here $t = \text{const.}$)

We thus see from (3.7) that the pressure of a particle fluid is zero (which is to be expected, because the particles are 'free' in the gravitational field, and so they do not undergo collisions). The equations of motion (3.9)–(3.11) derived from the conservation of $T^{\mu\nu}$ reduce to

$$D_\nu(\epsilon\, u^\nu) = 0 \qquad \text{and} \qquad u^\nu D_\nu u^\mu = 0\,.$$

After integrating over all of spacetime and using the divergence theorem, the first, with the substitution of (3.12) for ϵ, expresses mass conservation ($\Sigma m = $ const), and the second again states that the particles move along geodesics.

3.3 The coupling of a field to gravity

Owing to the equivalence principle, the action describing a matter field in the presence of gravitation is derived from its expression in special relativity by covariantization, $\eta_{\mu\nu} \to g_{\mu\nu}$ et $\partial_\mu \to D_\mu$:

$$S_f[\Phi(x^\mu, s)] = \int \mathcal{L}(\Phi, D_\mu\Phi, g_{\mu\nu})\sqrt{-g}\, d^4x\,, \tag{3.13}$$

where g is the determinant of the metric coefficients $g_{\mu\nu}$ and $\sqrt{-g}\, d^4x \equiv \sqrt{-g}\, dx^0 dx^1 dx^2 dx^3$ is the volume element (*cf.* Book 1, Section 3.6), where the scalar $\mathcal{L}$ is the Lagrangian of the (possibly tensorial) field Φ, and s (which has three components) parametrizes at a given $x^0 \equiv t$ the various configurations $\Phi(x^i, s)|_t$. The *Lagrangian density* $\sqrt{-g}\, \mathcal{L}(\Phi, D_\mu\Phi, g_{\mu\nu})$ is a *functional* of the field Φ and depends on the coordinates x^μ through the metric $g_{\mu\nu}$, which is a given *function* of the coordinates. The choice (3.13), which respects the principle of local relativity, is called the *correspondence principle* or the *principle of minimal coupling*.[2]

We use $\delta\Phi \equiv (\partial\Phi/\partial s)|_0\, ds$ and $\delta\partial_\mu\Phi \equiv (\partial^2\Phi/\partial s\, \partial X^\mu)|_0\, ds = \partial_\mu\delta\Phi$ to denote the variations of Φ and its derivatives in going from the $s = 0$ configuration to a neighboring one. Since the metric remains fixed, the variation of the action $\delta_\Phi S_f \equiv (dS_f/ds)|_0 ds$ follows step-by-step that of special relativity (*cf.* Book 2, Section 8.1). Indeed, since $D_\mu\Phi^{\nu\cdots}_{\rho\cdots} = \partial_\mu\Phi^{\nu\cdots}_{\rho\cdots} + \Gamma^\nu_{\mu\alpha}\Phi^{\alpha\cdots}_{\rho\cdots} + \cdots - \Gamma^\alpha_{\mu\rho}\Phi^{\nu\cdots}_{\alpha\cdots} + \cdots$, it can be written as

$$\delta_\Phi S_f = \int \delta\mathcal{L}\sqrt{-g}\, d^4x = \int \left(\frac{\partial\mathcal{L}}{\partial\Phi}\delta\Phi + \frac{\partial\mathcal{L}}{\partial\partial_\mu\Phi}\delta\partial_\mu\Phi\right)\sqrt{-g}\, d^4x$$

$$= \int \frac{\delta\mathcal{L}}{\delta\Phi}\delta\Phi\sqrt{-g}\, d^4x + \int \partial_\mu\left(\frac{\sqrt{-g}\,\partial\mathcal{L}}{\partial\partial_\mu\Phi}\delta\Phi\right)d^4x\,, \tag{3.14}$$

$$\text{where} \qquad \frac{\delta\mathcal{L}}{\delta\Phi} \equiv \frac{\partial\mathcal{L}}{\partial\Phi} - \frac{1}{\sqrt{-g}}\partial_\mu\frac{\sqrt{-g}\,\partial\mathcal{L}}{\partial\partial_\mu\Phi}\,.$$

---

[2]It should, however, be noted that nothing except the 'simplicity principle' *a priori* forbids going beyond a simple covariantization and introducing terms proportional to the curvature into the action describing the matter. One could, for example, choose to add to (3.13) a term of the type $\int\Phi^2 R\sqrt{-g}\, d^4x$, where R is the scalar curvature. In that case the action (3.13) would depend on the metric and its first and second derivatives.

For a discussion of variational principles as well as the Lagrangian and Hamiltonian formalisms in the mechanics of point particles and in the classical theory of fields, see, for example, Book 1, Chapters 8 and 9, and Book 2, Chapters 8, 9, and 12.

The divergence theorem

Let us review the Gauss theorem (or divergence theorem) stated in Book 1, Sections 3.6 and 4.6 and also in Book 2, Section 8.5:

$$\int_{\mathcal{M}} \partial_\mu (\sqrt{-g}\, v^\mu) d^4 x = \int_{\partial\mathcal{M}} v^\mu dS_\mu$$

$$\text{with} \qquad dS_\mu = \sqrt{-g}\, e_{\mu\nu\rho\sigma} \frac{\partial x^\nu}{\partial y^1} \frac{\partial x^\rho}{\partial y^2} \frac{\partial x^\sigma}{\partial y^3} dy^1 dy^2 dy^3 = \sqrt{|h|} n_\mu d^3 y \,. \tag{3.15}$$

Here $\partial\mathcal{M}$ is the boundary of the 4-volume $\mathcal{M}$ (see Fig. 3.1), on which the equations are given in parametric form by $x^\mu = x^\mu(y^i)$; g is the determinant of the metric coefficients $g_{\mu\nu}$; $e_{\mu\nu\rho\sigma}$ is the Levi-Civita symbol, which is completely antisymmetric and for which $e_{0123} = 1$; h is the determinant of the metric induced on $\partial\mathcal{M}$ with components $h_{ij} = g_{\mu\nu}(\partial x^\mu/\partial y^i)(\partial x^\nu/\partial y^j)$; finally, n_μ is the 4-vector orthogonal to $\partial\mathcal{M}$, that is, $n_\mu(\partial x^\mu/\partial y^i) = 0$ and normalized such that $g^{\mu\nu} n_\mu n_\nu = \pm 1$ (depending on whether $\partial\mathcal{M}$ is time-like or space-like).

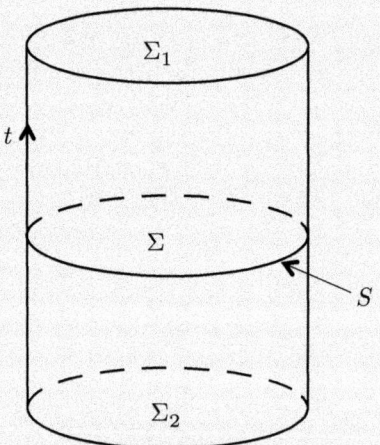

Fig. 3.1 The boundary $\partial\mathcal{M}$.

Now let us assume that coordinates adapted to $\partial\mathcal{M}$ have been chosen, that is, that its equation is $x^\mu = $ const for a given μ. If, for example, $\partial\mathcal{M} \equiv \Sigma$ is defined by $t = $ const, the metric will be written in these adapted coordinates as $ds^2 = g_{tt} dt^2 + h_{ij} dx^i dx^j$. We then will have $n_\mu = (\sqrt{-g_{tt}}, 0, 0, 0)$ and $\sqrt{-g_{tt}|h|} = \sqrt{-g}$, so that Gauss's theorem simplifies to

$$\int_{\partial\mathcal{M}} v^\mu dS_\mu = \int_\Sigma \sqrt{-g}\, v^t d^3 x \,. \tag{3.16}$$

Similarly, if $\partial\mathcal{M} \equiv \mathcal{B} = S \times L$, where S is a 2-sphere and $L = [t_1, t_2]$, and if the coordinates are chosen such that $ds^2 = g_{rr} dr^2 + h_{ij} dx^i dx^j$ where $x^i = \{t, \theta, \phi\}$, we will have

$$\int_{\partial\mathcal{M}} v^\mu dS_\mu = \int_{t_1}^{t_2} dt \int_S \sqrt{-g}\, v^r d\theta d\phi \,. \tag{3.17}$$

Applying the divergence theorem (3.15), the second term in (3.14) gives a boundary term which in adapted coordinates is written as [see (3.16) and (3.17)]

$$\int_{\partial\mathcal{M}} \frac{\partial\mathcal{L}}{\partial\partial_\mu\Phi} \delta\Phi\, dS_\mu = \int_\Sigma d^3x \left.\left(\frac{\sqrt{-g}\partial\mathcal{L}}{\partial\partial_0\Phi}\delta\Phi\right)\right|_{t_1}^{t_2} + \int_{t_1}^{t_2} dt \int_S \frac{\sqrt{-g}\partial\mathcal{L}}{\partial\partial_r\Phi}\delta\Phi\, d\theta d\phi, \qquad (3.18)$$

where Σ is a 3-volume at constant $x^0 \equiv t$ and S is its boundary, a 2-sphere at constant $x^1 \equiv r$.

The first term of (3.18) vanishes because the configurations are fixed at t_1 and t_2. The second also vanishes if we fix the configurations on S, or if we restrict ourselves to configurations which fall off sufficiently rapidly. Therefore, the boundary term in (3.14) is zero and the equations of motion of the field extremizing the action, $\delta_\Phi S_f = 0$, for any variation $\delta\Phi$ which vanishes on the boundary are the Euler–Lagrange equations. These are second-order differential equations which are covariant by construction[3]:

$$\frac{\delta\mathcal{L}}{\delta\Phi} \equiv \frac{\partial\mathcal{L}}{\partial\Phi} - \frac{1}{\sqrt{-g}}\partial_\mu \frac{\sqrt{-g}\partial\mathcal{L}}{\partial\partial_\mu\Phi} = 0. \qquad (3.19)$$

The Klein–Gordon and Maxwell equations

First of all, show that (see Book 1, Section 3.6)

$$\partial_\mu\sqrt{-g} = \frac{1}{2}\sqrt{-g}g^{\nu\rho}\partial_\mu g_{\nu\rho} = -\frac{1}{2}\sqrt{-g}g_{\nu\rho}\partial_\mu g^{\nu\rho}. \qquad (3.20)$$

Next, show that for any vector V^μ and any antisymmetric tensor $F^{\mu\nu}$ we have

$$\sqrt{-g}\, D_\mu V^\mu = \partial_\mu(\sqrt{-g}\, V^\mu), \quad \sqrt{-g}\, D_\mu F^{\mu\nu} = \partial_\mu(\sqrt{-g}\, F^{\mu\nu}). \qquad (3.21)$$

Since the Lagrangian of a scalar field is $\mathcal{L} = -\frac{1}{2}\partial_\mu\phi\,\partial^\mu\phi - V(\phi)$, where $\partial^\mu\phi \equiv g^{\mu\nu}\partial_\nu\phi$ and $V(\phi)$ is the self-interaction potential, show that the equation of motion (3.19) is the Klein–Gordon equation:

$$\Box\phi - \frac{dV}{d\phi} = 0 \quad \text{with} \quad \Box\phi \equiv D_\mu\,\partial^\mu\phi, \qquad (3.22)$$

where D_μ is the covariant derivative associated with the metric $g_{\mu\nu}$.

The Lagrangian of the electromagnetic field A_μ is $\mathcal{L} = -F_{\mu\nu}F^{\mu\nu}/16\pi$ with $F_{\mu\nu} = D_\mu A_\nu - D_\nu A_\mu = \partial_\mu A_\nu - \partial_\nu A_\mu$. Show in a similar way that the equations of motion (3.19) are the vacuum Maxwell equations:

$$D_\mu F^{\mu\nu} = 0. \qquad (3.23)$$

---

[3]Equations (3.19) are not changed when a divergence, $\partial_\mu\hat{V}^\mu$ or $\hat{V}^\mu \equiv \sqrt{-g}V^\mu(\Phi, \partial_\nu\Phi, \cdots)$, is added to $\mathcal{L}$, with the condition that the variations of the field configurations *and* their derivatives $\delta\Phi, \delta\partial_\nu\Phi\cdots$, vanish on $\partial\mathcal{M}$. Indeed,

$$\delta_\Phi \int d^4x\,\partial_\mu\hat{V}^\mu = \int d^4x\,\partial_\mu\delta_\Phi\hat{V}^\mu = \int d^4x\,\partial_\mu\left(\frac{\partial\hat{V}^\mu}{\partial\Phi}\delta\Phi + \frac{\partial\hat{V}^\mu}{\partial\partial_\nu\Phi}\delta\partial_\nu\Phi\cdots\right)$$

$$= \int_{\partial\mathcal{M}}\left(\frac{\partial V^\mu}{\partial\Phi}\delta\Phi + \frac{\partial V^\mu}{\partial\partial_\nu\Phi}\delta\partial_\nu\Phi\cdots\right)dS_\mu.$$

3.4   The energy–momentum tensors of a field

The concepts of energy, momentum, and angular momentum follow from the invariance of the solutions of the equations of motion under spatio-temporal translations or rotations (see, for example, Book 1, Chapters 7–9 and Book 2, Chapters 6–8). Here we shall see how the action is transformed, no longer under a modification of the field configuration, but instead under a displacement, or, in the 'passive' version, under a translation of the coordinate grid in the opposite direction.

In a displacement $x^\rho \to \tilde{x}^\rho = x^\rho + \xi^\rho$, where ξ^ρ is an infinitesimal vector, the field configuration changes by $\delta_\xi \Phi$ and the metric by $\delta_\xi g_{\mu\nu}$. We shall see more precisely below what these changes are. The action

$$S_{\mathrm{f}} = \int \mathcal{L}(\Phi, \partial_\mu \Phi, g_{\mu\nu}) \sqrt{-g}\, d^4 x\,, \tag{3.24}$$

where in order to simplify our discussion we assume that the Lagrangian $\mathcal{L}$ depends on Φ and its first derivatives as well as on the metric $g_{\mu\nu}$ but *not* its derivatives,[4] transforms as follows using (3.14) and (3.15):

$$\delta_\xi S_{\mathrm{f}} = \frac{1}{2} \int T^{\mu\nu} \delta_\xi g_{\mu\nu} \sqrt{-g}\, d^4 x + \int \frac{\delta \mathcal{L}}{\delta \Phi} \delta_\xi \Phi \sqrt{-g}\, d^4 x + \int_{\partial \mathcal{M}} \frac{\partial \mathcal{L}}{\partial \partial_\mu \Phi} \delta_\xi \Phi\, dS_\mu\,, \tag{3.25}$$

where $T^{\mu\nu}$ is the *energy–momentum* tensor of the field, which is 2-fold contravariant and symmetric by construction:

$$T^{\mu\nu} \equiv \frac{2}{\sqrt{-g}} \frac{\partial \sqrt{-g}\mathcal{L}}{\partial g_{\mu\nu}}\,. \tag{3.26}$$

The case of scalar and electromagnetic fields

First show that if $T^{\mu\nu}$ is defined by (3.26), then using (3.20) we will also have

$$T_{\mu\nu} = -\frac{2}{\sqrt{-g}} \frac{\partial(\sqrt{-g}\mathcal{L})}{\partial g^{\mu\nu}}\,.$$

The Lagrangian of a scalar field is $\mathcal{L} = -\frac{1}{2}\partial_\mu \phi \partial^\mu \phi - V(\phi)$ and that of an electromagnetic field is $\mathcal{L} = -F_{\mu\nu}F^{\mu\nu}/16\pi$. Show that their energy–momentum tensors are

$$T_{\mu\nu} = \partial_\mu \phi \partial_\nu \phi - g_{\mu\nu}\left(\frac{1}{2} g^{\rho\sigma}\partial_\rho \phi \partial_\sigma \phi + V(\phi)\right) \quad \text{and} \quad T_{\mu\nu} = \frac{1}{4\pi}\left(F_{\mu\rho}F_\nu{}^\rho - \frac{1}{4} g_{\mu\nu}F_{\rho\sigma}F^{\rho\sigma}\right).$$
$$\tag{3.27}$$

(Since $F_{\mu\nu} = D_\mu A_\nu - D_\nu A_\mu = \partial_\mu A_\nu - \partial_\nu A_\mu$, the Lagrangian depends only on the metric and not on its derivatives.)

Let us now specify how $g_{\mu\nu}$, the action itself, and Φ vary under a displacement.

---

[4]The generalization to the case where the Lagrangian also depends on derivatives of the metric is not difficult but quite tedious.

The metric components $g_{\mu\nu}$ in the coordinates $\{x^\rho\}$ are related to the metric components $\tilde{g}_{\mu\nu}$ in the system $\{\tilde{x}^\rho = x^\rho + \xi^\rho\}$ by the transformation law for 2-fold covariant tensors:

$$g_{\mu\nu}(x^\rho) = \frac{\partial \tilde{x}^\rho}{\partial x^\mu} \frac{\partial \tilde{x}^\sigma}{\partial x^\nu} \tilde{g}_{\rho\sigma}(\tilde{x}^\lambda) = (\delta^\rho_\mu + \partial_\mu \xi^\rho)(\delta^\sigma_\nu + \partial_\nu \xi^\sigma) \tilde{g}_{\rho\sigma}(\tilde{x}^\lambda), \qquad (3.28)$$

where x^ρ and $\tilde{x}^\rho$ are the coordinates of a point p in the first and the second coordinate system. However, the variation $\delta_\xi g_{\mu\nu}$ must be a function of only x^ρ (since it will be integrated over x^ρ). We can then make the decomposition $\tilde{g}_{\rho\sigma}(\tilde{x}^\lambda) = \tilde{g}_{\rho\sigma}(x^\lambda) + \xi^\lambda \partial_\lambda g_{\rho\sigma}$ to first order in ξ^μ, and we find

$$\delta_\xi g_{\mu\nu} \equiv g_{\mu\nu}(x^\rho) - \tilde{g}_{\mu\nu}(x^\rho) = \xi^\lambda \partial_\lambda g_{\mu\nu} + g_{\lambda\mu} \partial_\nu \xi^\lambda + g_{\mu\lambda} \partial_\mu \xi^\lambda = D_\nu \xi_\mu + D_\mu \xi_\nu, \qquad (3.29)$$

where the second equality follows from $D_\mu g_{\nu\rho} = 0$. (Here $\delta_\xi g_{\mu\nu}$ is the 'Lie derivative' of the metric with respect to ξ^μ.)

The first term on the right-hand side of (3.25) then becomes

$$\frac{1}{2} \int T^{\mu\nu} \delta_\xi g_{\mu\nu} \sqrt{-g}\, d^4x = \int T^{\mu\nu} D_\mu \xi_\nu \sqrt{-g}\, d^4x$$

$$= \int D_\mu (T^{\mu\nu} \xi_\nu) \sqrt{-g}\, d^4x - \int (D_\mu T^{\mu\nu}) \xi_\nu \sqrt{-g}\, d^4x \qquad (3.30)$$

$$= \int_{\partial\mathcal{M}} T^{\mu\nu} \xi_\nu\, dS_\mu - \int (D_\mu T^{\mu\nu}) \xi_\nu \sqrt{-g}\, d^4x,$$

because $D_\mu(T^{\mu\nu}\xi_\nu)\sqrt{-g} = \partial_\mu(T^{\mu\nu}\xi_\nu\sqrt{-g})$ [see (3.21)] and owing to the divergence theorem (3.15).

Moreover, since $\mathcal{L}$ is a scalar, we have $\delta_\xi \mathcal{L} = \xi^\mu \partial_\mu \mathcal{L}$, and from (3.29) and (3.20) we deduce that $\delta_\xi \sqrt{-g} = \sqrt{-g} D_\mu \xi^\mu = \partial_\mu(\sqrt{-g}\xi^\mu)$, so that the variation of the action [the left-hand side of (3.25)] is a surface term:

$$\delta_\xi S_{\mathrm{f}} = \int \delta_\xi(\mathcal{L}\sqrt{-g})\, d^4x = \int d^4x\, \partial_\mu(\sqrt{-g}\mathcal{L}\xi^\mu) = \int_{\partial\mathcal{M}} \mathcal{L}\,\xi^\mu dS_\mu. \qquad (3.31)$$

We note that the variation of the action vanishes if the vectors ξ^μ are taken to be zero on the boundary $\partial\mathcal{M}$ of the domain $\mathcal{M}$. We also see explicitly that the action (which is just a number for a given configuration Φ and metric $g_{\mu\nu}$) does not depend on the choice of coordinate system.

Finally, the configuration of the field Φ also changes. If $\Phi \equiv \phi$ is a scalar field, we have $\delta_\xi \phi = \xi^\mu \partial_\mu \phi$.

Therefore, (3.25) can be rewritten as follows in the case where the matter is a scalar field:

$$\int_{\partial\mathcal{M}} \left(\Theta_\mu^{\ \nu} - T_\mu^{\ \nu}\right) \xi^\mu dS_\nu = \int \left(\partial_\mu \phi \frac{\delta\mathcal{L}}{\delta\phi} - D_\nu T_\mu^{\ \nu}\right) \xi^\mu \sqrt{-g}\, d^4x, \qquad (3.32)$$

where $T^{\mu\nu}$ is defined in (3.26), $\delta\mathcal{L}/\delta\phi$ in (3.14), and

$$\Theta_\mu^{\ \nu} \equiv \delta_\mu^\nu \mathcal{L} - \frac{\partial\mathcal{L}}{\partial\partial_\nu\phi}\partial_\mu\phi \qquad (3.33)$$

is the *Noether canonical energy–momentum tensor*.

Book 3

At this stage we see that in fact the vector ξ^μ does not need to be infinitesimal. We could have introduced $\epsilon \xi^\mu$ with finite ξ^μ and $\epsilon \ll 1$. Then after dividing by ϵ we would have obtained (3.32) for a finite vector ξ^μ.

The vector ξ^μ is arbitrary. Its value in $\mathcal{M}$ is independent of its value on $\partial\mathcal{M}$, where it does not necessarily vanish. Therefore, for (3.32) to be satisfied, the integrands of the left- and right-hand sides must vanish separately. We then have

$$\Theta_\mu{}^\nu = T_\mu^\nu \qquad \text{and} \qquad \partial_\mu\phi\frac{\delta\mathcal{L}}{\delta\phi} = D_\nu T_\mu^\nu. \tag{3.34}$$

Thus, on the one hand, the energy–momentum tensors $\Theta_\mu{}^\nu$ and T_μ^ν of a scalar field must be equal, while on the other the Euler–Lagrange equations $\delta\mathcal{L}/\delta\phi = 0$ and the conservation equation for the energy–momentum tensor $D_\nu T_\mu^\nu = 0$ must be equivalent. (This demonstration can be generalized to fields of higher 'spin'; see the example of spin 1 that follows.)

Examples of Noether tensors

Show that for the Klein–Gordon Lagrangian $\mathcal{L} = -\frac{1}{2}\partial_\mu\phi\partial^\mu\phi - V(\phi)$ the associated Noether tensor (3.33) is identical to the tensor T_ν^μ defined in (3.26) and calculated in (3.27). Show that, in addition,

$$D_\nu T_\mu^\nu = \partial_\mu\phi\left(\Box\phi - \frac{dV}{d\phi}\right),$$

and that therefore the conservation of the energy–momentum tensor does indeed again give the Klein–Gordon equation of motion (if ϕ is not constant).

Now show that if the field Φ is a covariant vector field $\Phi \equiv A_\mu$, under an infinitesimal displacement $x^\rho \to \tilde{x}^\rho = x^\rho + \xi^\rho$ we will have[5] $\delta_\xi\Phi \equiv A_\mu(x^\rho) - \tilde{A}_\mu(x^\rho) = \xi^\nu\partial_\nu A_\mu + A_\nu\partial_\mu\xi^\nu = \xi^\nu D_\nu A_\mu + A_\nu D_\mu\xi^\nu$.

Next, show that in the case of massless 'spin-1' fields, where $\mathcal{L}$ is independent of A_ρ and $\partial\mathcal{L}/\partial\partial_\nu A_\rho$ is antisymmetric, eqn (3.32) can be written in the following form, setting $F_{\mu\nu} \equiv \partial_\mu A_\nu - \partial_\nu A_\mu$:

$$\int\left(\Theta_{\mu\text{sym}}^{\;\;\nu} - T_\mu^\nu\right)\xi^\mu dS_\nu = \int\left(F_{\mu\rho}\frac{\delta\mathcal{L}}{\delta A_\rho} - D_\nu T_\mu^\nu\right)\xi^\mu\sqrt{-g}\,d^4x, \tag{3.35}$$

where $\Theta_{\mu\text{sym}}^{\;\;\nu} \equiv \delta_\mu^\nu\mathcal{L} - \frac{\partial\mathcal{L}}{\partial\partial_\nu A_\rho}F_{\mu\rho}$ is the symmetrized version of the Noether tensor defined in (3.33) and $\sqrt{-g}\delta\mathcal{L}/\delta A_\rho$ here reduces to $\partial_\nu\sqrt{-g}\partial\mathcal{L}/\partial\partial_\nu A_\rho$.

Finally, show that in the specific case of the Maxwell Lagrangian $\mathcal{L} = -F_{\mu\nu}F^{\mu\nu}/16\pi$, the quantity $\Theta_{\mu\text{sym}}^{\;\;\nu}$ is indeed again equal to T_ν^μ calculated in (3.27), and moreover

$$D_\nu T_\mu^\nu = \frac{1}{4\pi}F_{\mu\rho}(D_\nu F^{\nu\rho}).$$

(Here we have exploited the antisymmetry of the Faraday tensor $F_{\mu\nu}$ and the first group of Maxwell equations which follow from it, namely, $D_\mu F_{\nu\rho} + D_\nu F_{\rho\mu} + D_\rho F_{\mu\nu} = 0$.) Therefore, again the conservation law for the energy–momentum tensor and the Euler–Lagrange equations of motion are equivalent.

---

[5]The variations $\delta_\xi A_\mu$, like the $\delta_\xi g_{\mu\nu}$ obtained in (3.30), are respectively the *Lie derivatives* of A_μ and the metric $g_{\mu\nu}$ with respect to the vector ξ^μ; see Chapter 23. We also have $\delta_\xi A^\mu = \xi^\nu D_\nu A^\mu - A^\nu D_\nu\xi^\mu$.

4

The Einstein equations

In the absence of matter there is no gravitational field, and the spacetime which represents this empty universe is Minkowski spacetime. (More precisely, if the gravitational field created by the matter can be neglected, the appropriate framework for describing the matter is that of special relativity.) The Einstein gravitational equations relate geometry and matter: specifically, they relate the Riemann tensor, or, more precisely, the Einstein tensor, to the geometrical object describing 'inertia', or energy content of the matter, that is, the energy–momentum tensor.

4.1   The Einstein equations

As we saw in the preceding chapter, the tensorial object which in general relativity describes matter in the presence of a gravitational field is the energy–momentum tensor $T_{\mu\nu}$.

A simple (that is, linear in the curvature) geometrical tensor of the same type as the energy–momentum tensor (that is, 2-fold covariant) is $R_{\mu\nu} - \frac{\alpha}{2} g_{\mu\nu} R$, where $R_{\mu\nu} = R^\rho{}_{\mu\rho\nu}$ is the Ricci tensor, $R = g^{\mu\nu} R_{\mu\nu}$ is the scalar curvature, and α is an *a priori* arbitrary constant. After various tentative versions (in particular, an $\alpha = 0$ version in 1914), the gravitational field equations proposed by Einstein in November 1915 were

$$G_{\mu\nu} = \kappa T_{\mu\nu}. \tag{4.1}$$

Here $G_{\mu\nu} = R_{\mu\nu} - \frac{1}{2} g_{\mu\nu} R$ is the Einstein tensor introduced in Section 2.6 and κ is the *Einstein constant*.

Ten equations for ten unknowns

The choice of the Einstein tensor on the left-hand side of (4.1) was made for the following reason.

These equations form a set of ten nonlinear partial differential equations. The coordinate system can be chosen arbitrarily. By means of a change of coordinates it is therefore possible to assign, at least locally, any value to four components of the metric tensor $g_{\mu\nu}$, so that the set is reduced to a set of six functions.

However, owing to the Bianchi identities (2.30), $D_\mu (R^{\mu\nu} - \alpha R g^{\mu\nu}/2) \equiv 0$ for $\alpha = 1$ only (Einstein and Hilbert did not know this in 1915...), and since $D_\mu g_{\nu\rho} \equiv 0$, it follows from (4.1) that the tensor $T^{\mu\nu}$ must be conserved, *i.e.*, the divergence must vanish: $D_\mu T^{\mu\nu} = 0$. The four equations (for example, the continuity and Euler equations; see Section 3.2) which determine the evolution of the four independent functions describing the matter (for example, its 4-velocity u^μ satisfying $u_\mu u^\mu = -1$ and its energy density or pressure following from an equation of state) are therefore included in the Einstein equations.

We thus have ten equations for ten unknowns and the system is not over-determined.

General relativity differs from, for example, electromagnetism, where the Lorentz equation giving the motion of the charges is not included in the Maxwell equations.

Relativity in Modern Physics. Nathalie Deruelle and Jean-Philippe Uzan.
© Oxford University Press 2018. Published in 2018 by Oxford University Press.
DOI: 10.1093/oso/9780198786399.001.0001

We still need to relate the constant κ to Newton's constant. In the Newtonian limit and in a quasi-inertial frame, $-g_{00} \sim 1 + 2U/c^2$, where U is the Newtonian potential (see Sections 1.6 and 3.1). The other components of the metric are at least of the same order. The Christoffel symbols $\Gamma^{\mu}_{\nu\rho}$ are therefore of order $1/c^2$, and their time derivatives are of order $1/c^3$. Consequently, at lowest order we have

$$R_{00} \sim \partial_i \Gamma^i_{00} \sim -\frac{1}{2}\partial_i\left(g^{ij}\partial_j g_{00}\right) \sim \frac{1}{c^2}\partial^i\left(\frac{\partial U}{\partial x^i}\right) \sim \frac{1}{c^2}\Delta U. \tag{4.2}$$

The energy–momentum tensor of a fluid then reduces to $T^{00} \sim \varrho c^2$, where ϱ is the proper mass density. These are the only terms of the Ricci and energy–momentum tensors that we need to calculate if we rewrite the Einstein equations in the form $R_{\mu\nu} = \kappa(T_{\mu\nu} - \frac{1}{2}g_{\mu\nu}T)$. Then in the limit of weak fields and small velocities they lead to the Poisson equation $\Delta U = 4\pi G\varrho$ (see Book 1, Section 11.4) if

$$\kappa = \frac{8\pi G}{c^4}, \tag{4.3}$$

where we recall that c is the speed of light and G is Newton's constant.[1]

The 'cosmological constant'

Since in general relativity the Riemannian spacetime is endowed with a Levi-Civita connection, the covariant derivative of the metric is zero: $D_\mu g_{\nu\rho} \equiv 0$; see Section 2.4. The Einstein equations (4.1) can therefore be generalized as

$$G_{\mu\nu} + \Lambda g_{\mu\nu} = \kappa T_{\mu\nu}, \tag{4.4}$$

which also includes the law of motion of the matter, $D_\mu T^\mu_\nu = 0$, if Λ is a constant of dimension L^{-2}.

Then in the Newtonian approximation we have [cf. (4.2) and (4.3)]

$$\Delta U = 4\pi G\varrho - 2\Lambda c^2,$$

the solution of which has the behavior $U/c^2 \approx -GM/c^2\ell - 2\Lambda\ell^2$, where ℓ is the characteristic length over which the potential varies. To recover the Poisson law we must have $\Lambda\ell^2 \ll GM/c^2\ell \ll 1$. Since Newtonian gravity has been well tested at stellar system and even galactic scales, the constant Λ can possibly play a role only at even larger scales, hence the name *cosmological constant*.

Having said this, we can rewrite (4.4) as $G_{\mu\nu} = \kappa(T_{\mu\nu} - \Lambda g_{\mu\nu}/\kappa)$, and interpret the term involving Λ as the energy–momentum tensor of the 'quantum vacuum', whose contribution to the gravitational field ('$\hbar\omega/2$ per virtual degree of freedom') is enormous. This is the *cosmological constant problem*.[2]

One rather cursory way of solving this problem is to postulate that the gravitation equations are $R_{\mu\nu} - \alpha R g_{\mu\nu}/2 = \kappa(T_{\mu\nu} - Tg_{\mu\nu}/4)$ with $\alpha = 1/2$ (Einstein, 1919), so that the traceless parts of the Ricci tensor and the energy–momentum tensor of the matter are related, and then to separately impose the matter equations of motion $D_\nu T^\nu_\mu = 0$. The vacuum energy, which

---

[1] The constant $G = 6.67 \times 10^{-11}$ m^3 kg^{-1} s^{-2} or, in the system with $c = 1$, $G = 2.477 \times 10^{-38}$ s kg^{-1}. We shall often use *geometrical units* in which $c = 1$, $G = 1$ and masses are expressed in meters or seconds. For example, $1M_\odot = 2 \times 10^{30}$ kg $= 1.5$ km $= 6$ μs and $M_\oplus = 6 \times 10^{24}$ kg $= 0.45$ cm.

[2] See, for example, Weinberg (1989).

is proportional to $g_{\mu\nu}$, then does not contribute to gravitation. Then, taking the divergence of the equations and using the Bianchi identities, we have $\partial_\nu R = -\kappa\,\partial_\nu T$, which implies that $R = -\kappa\,T + 4\Lambda$, where Λ is an integration constant. Substituting this into the original equation, we see that the latter is in fact strictly equivalent to (4.4), but now the cosmological constant is no longer related to the vacuum energy.[3]

4.2 The 1+3 decomposition

In Section 4.1 we mentioned that the Einstein equations, while leaving the choice of coordinate system free, are ten equations for ten unknowns. Now let us examine their structure more carefully.[4]

Let us distinguish one coordinate, which we call w, and denote the three others as x^i. It is an easy exercise to find the dependence of the Einstein tensor on the second derivatives of the metric with respect to w:

$$G^w_w = C^w_w\,, \quad G^w_i = C^w_i\,,$$
$$G^i_j = \frac{1}{2}\left[\delta^l_j(g^{iw}g^{kw} - g^{ww}g^{ik}) - \delta^i_j(g^{kw}g^{lw} - g^{kl}g^{ww})\right]\partial^2_{ww}g_{kl} + C^i_j\,,$$

(4.5)

where the C^μ_ν do not contain any second derivatives with respect to w.

We can therefore fix g_{ww} and g_{wi} (in agreement with the fact that the coordinate system can be freely chosen), but there exist four *constraint equations* on the g_{ij} and their first derivatives, namely, $G^w_\mu = \kappa T^w_\mu$. Here we rediscover the fact well known in gauge theory (for example, in the Maxwell theory; see Book 2, Section 14.1) that each *gauge invariance*, here *invariance under diffeomorphisms*, that is, under changes of coordinates, corresponds to a constraint. As for $G^i_j = \kappa T^i_j$, these are six *evolution equations* for the six second derivatives $\partial_{ww}g_{ij}$.

Let us push this '1+3 decomposition' a bit farther. To this end we shall at first use the special coordinates (w, x^i) called *Gauss coordinates* (or *synchronous coordinates* if w is the time coordinate). Then the metric takes the following form (we note that this coordinate system is not unique):

$$ds^2 = \epsilon\,dw^2 + h_{ij}(w, x^k)\,dx^i dx^j$$

(4.6)

with $\epsilon = -1$ if the coordinate w is the time and $\epsilon = +1$ if it is a spatial coordinate. Here h_{ij} are the components of the *induced metrics* on the surfaces $w = $ const. In addition,

$$K_{ij} \equiv \frac{1}{2}\partial_w h_{ij}$$

(4.7)

will be referred to as the *extrinsic curvature* of the surfaces $w = $ const. Then we can find the components of the Einstein tensor:

---

[3]See, for example, Ellis *et al.* (2011).

[4]For an introduction to the structure of the Einstein equations and, among other things, the solution of the 'Cauchy problem', that is, of the evolution equations, see Wald (1984), and also Hawking and Ellis (1973). For a deeper discussion see Choquet-Bruhat (2009) or J. Isenberg (2013).

$$\begin{cases} G_w^w = \dfrac{\epsilon}{2}(K^2 - K^{ij}K_{ij}) - \dfrac{1}{2}\bar{R}, \qquad G_j^w = \epsilon\bar{D}_i(K_j^i - \delta_j^i\,K), \\[2mm] G_j^i = -\epsilon\,\partial_w(K_j^i - \delta_j^i K) - \epsilon\left(KK_j^i - \dfrac{1}{2}\delta_j^i K^{kl}K_{kl} - \dfrac{1}{2}\delta_j^i K^2\right) + \bar{G}_j^i, \end{cases} \qquad (4.8)$$

where $\bar{D}$, $\bar{G}_j^i$, and $\bar{R}$ are the covariant derivative, the Einstein tensor, and the scalar curvature of the induced metric. We again find that G_w^w and G_j^w depend only on the first derivatives of the metric with respect to w, and so the equations $G_\mu^w = \kappa T_\mu^w$ are four constraint equations;[5] cf. (4.5).

Propagation of the constraints

Let us use the Bianchi identities to show that if the constraints are satisfied on a surface $w = w_0$, then they will be satisfied on *all* surfaces $w = \text{const}$.

In the system of Gaussian coordinates (4.6), (4.7) we have $\Gamma_{ij}^w = -\epsilon K_{ij}$, $\Gamma_{wj}^i = K_j^i$, and $\Gamma_{jk}^i = \bar{\Gamma}_{jk}^i$. The Bianchi identities $D_\mu G_\nu^\mu \equiv 0$ then can be written out as

$$\partial_w G_w^w + \bar{D}_i G_w^i + KG_w^w - K_j^i G_i^j \equiv 0 \quad \text{and} \quad \partial_w G_i^w + \bar{D}_j G_i^j + KG_i^w \equiv 0.$$

If now the evolution equations are satisfied on a given surface $w = w_0$, that is, if $G_j^i|_0 = 0$ and $\partial_j G_k^i|_0 = 0$ (for simplicity we work in the vacuum), the Bianchi identities give

$$(\partial_w G_w^w + +\bar{D}_i G_w^i + KG_w^w)_0 = 0 \quad \text{and} \quad (\partial_w G_i^w + KG_i^w)_0 = 0.$$

Therefore, if the constraints are satisfied at $w = w_0$, that is, if $G_w^w|_0 = G_w^w|_0 = G_i^w|_0 = \partial_i G_w^i|_0 = 0$, then they will be satisfied for all w (at least if they are analytic).

It will often prove useful to generalize this 1+3 decomposition by working in an arbitrary system of coordinates $x^\mu = (t, x^i)$ (not necessarily Gaussian) which, however, is still adapted to the foliation, that is, to the ensemble of hypersurfaces Σ defined by $t = \text{const}$, which we choose to be space-like.

Let $g_{\mu\nu}$ be the metric coefficients in these 'ADM coordinates'[6] x^μ. At each point of Σ it is possible to define three basis vectors V_i^μ of components δ_i^μ; see Fig. 4.1. The (covariant) vector n_μ normal to Σ and of unit length satisfies $g_{\mu\nu} n^\mu \delta_i^\nu = g_{\mu i} n^\mu = n_i = 0$ and $g^{\mu\nu} n_\mu n_\nu = -1$. Its components (after choosing the sign) are therefore $n_\mu = (-1/\sqrt{-g^{00}}, 0, 0, 0)$ and $n^\mu = (\sqrt{-g^{00}}, -g^{0i}/\sqrt{-g^{00}})$. Now let us decompose the vector tangent to the time lines, with components δ_0^μ, on the normal vector and the three basis vectors of Σ: $\delta_0^\mu = Nn^\mu + N^i\delta_i^\mu$, where $N = 1/\sqrt{-g^{00}}$ and $N^i = -g^{0i}/g^{00}$ are the *lapse* and the *shift*.

The lapse and the shift, along with the metrics $h_{ij} = g_{ij}|_t$ induced on the surfaces Σ defined by $t = \text{const}$, constitute the ADM variables. In terms of these variables the metric components $g_{\mu\nu}$ become

---

[5] We note that if the induced metric does not depend on w, then $K_{ij} = 0$ and the vacuum Einstein equations require that $\bar{G}_{ij} = 0$. Then the solution is just Minkowski spacetime, because the Riemann tensor is proportional to the Einstein tensor in three dimensions; see Section 2.6.

[6] Arnowitt, Deser, and Misner (1962). See also Misner, Thorne, and Wheeler (1973), as well as Wald (1984), Gourgoulhon (2012), and Poisson (2002). What are now called ADM coordinates were actually introduced by Yvonne Choquet-Bruhat in 1947.

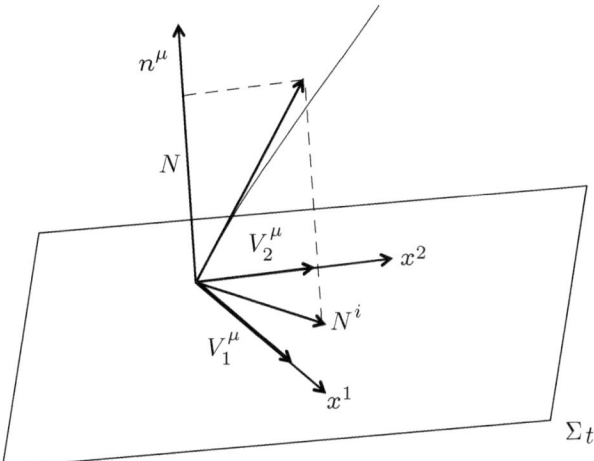

Fig. 4.1 The $1 + 3$ decomposition.

$$\begin{cases} g_{00} = -N^2 + N_i N^i \,, \quad g_{0i} = N_i \,, \quad g_{ij} = h_{ij} \,, \\[2mm] g^{00} = -\dfrac{1}{N^2} \,, \quad g^{0i} = \dfrac{N^i}{N^2} \,, \quad g^{ij} = h^{ij} - \dfrac{N^i N^j}{N^2} \end{cases} \tag{4.9}$$

(where all indices are moved using the induced metric h_{ij} and its inverse; we note that $\sqrt{-g} = N \sqrt{h}$). Finally, we introduce the extrinsic curvature

$$K_{ij} \equiv D_i n_j = \frac{1}{2N} (\dot{h}_{ij} - \bar{D}_i N_j - \bar{D}_j N_i) \,, \tag{4.10}$$

where the dot denotes the derivative with respect to the coordinate t and $\bar{D}$ is the covariant derivative associated with h_{ij}. [This expression does indeed reduce to (4.7) in Gaussian coordinates.]

A rather tedious but easy calculation allows us to write the components of the Riemann tensor in the ADM variables. Denoting $n^\mu T_{\mu\nu\ldots} \equiv T_{\mathbf{n}\nu\ldots}$, we find

$$\begin{cases} R_{ijkl} = \bar{R}_{ijkl} + K_{ik} K_{jl} - K_{il} K_{jk} \\[2mm] R_{ijk\mathbf{n}} = \bar{D}_i K_{jk} - \bar{D}_j K_{ki} \\[2mm] R_{i\mathbf{n}j\mathbf{n}} = -\dfrac{1}{N} \left(\dot{K}_{ij} - \mathcal{L}_{N^k} K_{ij} \right) + K_{ik} K^k_j + \dfrac{\bar{D}^2_{ij} N}{N} \,, \end{cases} \tag{4.11}$$

where $\mathcal{L}_{N^k} K_{ij} = N^k \partial_k K_{ij} + K_{ik} \partial_j N^k + K_{jk} \partial_i N^k$ is the 'Lie derivative' of the extrinsic curvature K_{ij} with respect to the shift N^k and $\bar{R}_{ijkl}$ is the Riemann tensor of the induced metric h_{ij}. The three equations in (4.11) are the *Gauss equations*, the *Codazzi equations*, and the *Ricci–York equations*, respectively. Another short calculation using the expressions $h^{ij} \dot{K}_{ij} = \dot{K} + 2N K^{ij} K_{ij} + 2K^{ij} \bar{D}_i N_j$ and $D_\mu n^\mu = K$ allows us to write the scalar curvature in the form

Book 3

$$R = \frac{2}{N}(\dot{K} - N^i \partial_i K) + K.K + K^2 + \bar{R} - \frac{2}{N}\triangle N$$
$$= K.K - K^2 + \bar{R} + \frac{2}{\sqrt{-g}}\,\partial_\mu(\sqrt{-g}\,n^\mu\,K) - \frac{2}{\sqrt{h}\,N}\partial_i(\sqrt{h}\,h^{ij}\,\partial_j N)\,,$$

(4.12)

where $K = h^{ij}K_{ij}$ and $K.K = K^{ij}K_{ij}$, and the quantity $\bar{R}$ is the scalar curvature of the metric h_{ij}. This expression for the scalar curvature is the starting point for the Hamiltonian formulation of general relativity; see footnote 6 and Section 4.5.

The induced metric and extrinsic curvature

Here we conclude by giving the definitions of the induced metric and extrinsic curvature valid in any coordinate system and for any surface Σ. They reduce to (4.6) and (4.7) in the Gaussian coordinates adapted to Σ.

We consider a length element $ds^2 = g_{\mu\nu}dx^\mu dx^\nu$ in the coordinates x^μ. We take a surface Σ in this space given by the equations $x^\mu = x^\mu(y^i)$, where y^i are three parameters. Three vectors tangent to this surface are $V_i^\mu = \partial x^\mu/\partial y^i$.

The *induced metric* on Σ is, in the *coordinates* y^i, $d\sigma^2 \equiv ds^2|_\Sigma = h_{ij}\,dy^i dy^j$ with $h_{ij} = g_{\mu\nu}V_i^\mu V_j^\nu$.

The vector normal to the surface, n^μ, is obtained by requiring that it is orthogonal to the tangent vectors, $g_{\mu\nu}n^\mu V_i^\nu = 0$, and that it is normalized, $g_{\mu\nu}n^\mu n^\nu = \epsilon$, where $\epsilon = -1$ if Σ is space-like and $\epsilon = 1$ if it is time-like (its orientation relative to Σ still needs to be fixed).

The *extrinsic curvature* of Σ measures the way Σ is 'curved' in the ambient space and is defined as

$$K_{ij} = V_i^\mu V_j^\nu D_\mu n_\nu\,,$$

(4.13)

where D is the covariant derivative associated with $g_{\mu\nu}$.

In a system of Gaussian coordinates (4.6) we have $V_i^\mu = \delta_i^\mu$ and, choosing $n^\mu = +\delta_w^\mu$, we find $K_{ij} = \frac{1}{2}\partial_w h_{ij}$. In ADM coordinates we recover (4.10).

4.3 Matching conditions

It is sometimes convenient to split spacetime into two distinct regions, an 'exterior' region V_e (the vacuum, for example) and an 'interior' region V_i (which may contain, for example, a perfect fluid), separated by a time-like 3-surface Σ representing, for example, the time evolution of the boundary of a star.

To find the matching conditions for these two regions, we use the Gaussian coordinates (4.6) with $\epsilon = +1$ (Hilbert, 1916; Darmois, 1927; Lichnerowicz, 1955). A possible discontinuity of $T_{\mu\nu}$ at Σ, located at $w = w_0$, would manifest itself at most, via the Einstein equations, as discontinuities in the second derivatives of h_{ij} with respect to w. Therefore, the first derivatives $\partial_w h_{ij}$ and h_{ij} themselves are continuous (respectively, C^0 and C^1 at least).

The matching conditions for the metric at Σ are therefore the continuity of the induced metric and the extrinsic curvature.

These matching conditions correspond to constraints on the energy–momentum tensor. Indeed, the components G_μ^w depend only on the metric and its first derivatives with respect to w; cf. (4.8). They are therefore continuous at Σ, and the same must hold for T_μ^w. For a perfect fluid $[T_\nu^\mu = (\epsilon + p)u^\mu u_\nu + p\,\delta_\nu^\mu$ with $g_{\mu\nu}u^\mu u^\nu = -1$; see Section 3.2], this implies that on Σ we have $T_w^w = (\epsilon + p)u^w u_w + p = 0$ and $T_i^w = (\epsilon + p)u^w u_i = 0$. The equation $u_i = 0$ is not a solution if the energy density $\epsilon \neq 0$, and so we must have

$$u^w = 0 \quad \text{and} \quad p = 0, \tag{4.14}$$

which states that the fluid is confined by Σ and that its pressure p is zero there.

Now let us assume that the two regions V_e and V_i are empty but that there exists on Σ a *thin layer* of matter, that is, that the components of the energy–momentum tensor are proportional to a Dirac delta distributions centered on a time-like Σ. Then, owing to the Einstein equations, the second derivatives of the h_{ij} contain delta distributions, and so the extrinsic curvature is discontinuous (proportional to a Heaviside distribution) and the induced metric is continuous (C^0). The constraint equations (4.8) then imply that only the components T_j^i contain delta distributions: $T_j^i = \tau_j^i \delta(w)$. The evolution equations (4.8) become (since $\epsilon = +1$) $-\partial_w (K_j^i - \delta_j^i K) + \cdots = \kappa \tau_j^i \delta(w)$ and, after integrating across Σ, relate the discontinuity of the extrinsic curvature to the energy–momentum tensor of the thin layer (Israel, 1966):

$$-[K_j^i - \delta_j^i K]_-^+ = \kappa \tau_j^i \tag{4.15}$$

(where any possible sign ambiguities in the definition of the jump $[K_j^i]_-^+$ are resolved case by case depending on the physics of the problem).

4.4 The Hilbert action and the Einstein equations

The Einstein equations (4.1) determining the gravitational field as a function of the matter which is present can also be derived from a principle of least action.

Since in general relativity the gravitational field is identified with a spacetime metric $g_{\mu\nu}$, the action of the gravitational field is a functional of $g_{\mu\nu}$, $S_H[g_{\mu\nu}] = \int \mathcal{L}_H \sqrt{-g} \, d^4x$, where the Lagrangian $\mathcal{L}_H$ is a scalar. Since the metric must be non-flat in order to describe a gravitational field rather than a simple inertia field, this Lagrangian $\mathcal{L}_H$ must be a function of the curvature tensor. Hilbert made the following choice (in November 1915):

$$S_H = \frac{1}{2\kappa} \int_{\mathcal{M}} (R - 2\Lambda) \sqrt{-g} \, d^4x. \tag{4.16}$$

The integral runs over a portion $\mathcal{M}$ of spacetime bounded by a 3-surface $\partial \mathcal{M}$; cf. Fig. 3.1. The constant $\kappa \equiv 8\pi G/c^4$ is the Einstein constant and Λ is the cosmological constant. Finally, R is the scalar curvature, the trace of the Ricci tensor $R_{\mu\nu}$, which itself is the contraction of the Riemann tensor:

$$R = g^{\mu\nu} R_{\mu\nu}, \quad R_{\mu\nu} = R^\rho{}_{\mu\rho\nu} = \partial_\rho \Gamma^\rho_{\mu\nu} - \partial_\nu \Gamma^\rho_{\mu\rho} + \Gamma^\rho_{\sigma\rho} \Gamma^\sigma_{\mu\nu} - \Gamma^\rho_{\sigma\nu} \Gamma^\sigma_{\mu\rho}, \tag{4.17}$$

where $\Gamma^\rho_{\mu\nu} = \frac{1}{2} g^{\rho\sigma} (\partial_\mu g_{\nu\sigma} + \partial_\nu g_{\rho\sigma} - \partial_\sigma g_{\mu\nu})$ are the Christoffel symbols of the metric $g_{\mu\nu}$. Therefore, $2\kappa \mathcal{L}_H = R - 2\Lambda$ depends on the metric and on both its first and its second derivatives.

Let us calculate the variation of S_H when the metric is varied. First we have (since $\delta \sqrt{-g} = -\sqrt{-g} g_{\mu\nu} \delta g^{\mu\nu}/2$)

$$\delta \int R \sqrt{-g} \, d^4x = \delta \int R_{\mu\nu} g^{\mu\nu} \sqrt{-g} \, d^4x$$
$$= \int \left(R_{\mu\nu} - \frac{1}{2} R g_{\mu\nu} \right) \delta g^{\mu\nu} \sqrt{-g} \, d^4x + \int g^{\mu\nu} \delta R_{\mu\nu} \sqrt{-g} \, d^4x. \tag{4.18}$$

To calculate the scalar $g^{\mu\nu}\delta R_{\mu\nu}$, it is useful to work in a locally inertial frame where the $\Gamma^{\mu}_{\nu\rho}$ vanish (see Section 2.5). Then from (4.17) we find

$$g^{\mu\nu}\delta R_{\mu\nu} = g^{\mu\nu}(\partial_{\rho}\delta\Gamma^{\rho}_{\mu\nu} - \partial_{\nu}\delta\Gamma^{\rho}_{\mu\rho}) = D_{\mu}V^{\mu} \quad \text{with} \quad V^{\mu} = g^{\nu\rho}\delta\Gamma^{\mu}_{\nu\rho} - g^{\mu\nu}\delta\Gamma^{\rho}_{\nu\rho}. \qquad (4.19)$$

Since this is a vector equation, it holds in any frame (V^{μ} is indeed a vector, because the difference $\delta\Gamma^{\rho}_{\mu\nu}$ of two Christoffel symbols is a tensor under coordinate transformations; see Section 2.4).

Then, recalling that for any vector V^{μ} we have $\sqrt{-g}D_{\mu}V^{\mu} = \partial(\sqrt{-g}V^{\mu})$, we find

$$2\kappa\,\delta S_{\mathrm{H}} = \int (G_{\mu\nu} + \Lambda g_{\mu\nu})\delta g^{\mu\nu}\sqrt{-g}\,d^4x + \int \partial_{\mu}(\sqrt{-g}V^{\mu})d^4x, \qquad (4.20)$$

where $G_{\mu\nu} = R_{\mu\nu} - \frac{1}{2}g_{\mu\nu}R$ is the Einstein tensor. The last term, a divergence, does not contribute to the equations of motion because it vanishes owing to Gauss's theorem if the variations of $g_{\mu\nu}$ and $\partial_{\rho}g_{\mu\nu}$ are taken to be zero on the boundary $\partial\mathcal{M}$ (in this the Hilbert action differs from the action of an ordinary field theory, where only the variations of the fields and not of their derivatives are taken to be zero on the boundary).[7]

The Gibbons–Hawking–York boundary term

The boundary term in (4.20) with V^{μ} given in (4.19) is written as $\int_{\mathcal{M}}\partial_{\mu}(\sqrt{-g}V^{\mu})d^4x = \int_{\partial\mathcal{M}}V^{\mu}dS_{\mu}$, where, expanding the $\delta\Gamma$ and then the $D_{\nu}\delta g_{\rho\sigma}$,

$$V^{\mu} = (g^{\mu\sigma}g^{\nu\rho} - g^{\mu\nu}g^{\rho\sigma})D_{\nu}\delta g_{\rho\sigma} = (g^{\mu\sigma}g^{\nu\rho} - g^{\mu\nu}g^{\rho\sigma})(\partial_{\nu}\delta g_{\rho\sigma} - \Gamma^{\lambda}_{\nu\rho}\delta g_{\lambda\sigma}).$$

Now let us work in a system of Gaussian coordinates (w, x^i), where the boundary equation is $w = w_0$ and the metric is $ds^2 = \epsilon\,dw^2 + h_{ij}dx^i dx^j$ with $\epsilon = \pm 1$ depending on whether the boundary is time-like or space-like. Introducing also the extrinsic curvature of $\partial\mathcal{M}$, $K_{ij} = \partial_w h_{ij}/2$ [cf. (4.7)], it is easy to see that the boundary term becomes

$$\int_{w=w_0}V^0\sqrt{-g}\,d^3x = -\epsilon\int_{w=w_0}\left[2\delta\left(\sqrt{|h|}\,K\right) - \sqrt{|h|}\left(K_{ij}-h_{ij}K\right)\delta h^{ij}\right]d^3x.$$

Therefore, if the Hilbert action (4.16) is replaced by the Gibbons–Hawking–York action,

$$2\kappa\,S_{\mathrm{GHY}} = \int_{\mathcal{M}}(R - 2\Lambda)\sqrt{-g}\,d^4x + 2\epsilon\int_{\partial\mathcal{M}}K\sqrt{|h|}\,d^3x, \qquad (4.21)$$

we will have

$$2\kappa\,\delta S_{\mathrm{GHY}} = \int_{\mathcal{M}}(G_{\mu\nu} + \Lambda g_{\mu\nu})\delta g^{\mu\nu}\sqrt{-g}\,d^4x + \epsilon\int_{\partial\mathcal{M}}(K_{ij} - h_{ij}K)\,\delta h^{ij}\sqrt{|h|}d^3x, \qquad (4.22)$$

and the last term will not contribute to the field equations if, in accord with the usual variational principles, only the variation of the metric δh^{ij} is required to vanish at the boundary.

---

[7]It should be noted that only the Hilbert Lagrangian leads, at least in four dimensions (as shown by Cartan), to second-order equations. The equations of motion derived from, for example, $\mathcal{L} = f(R)$ are of fourth order.

Let us now consider the source of the gravitational field, described by the action $S_{\rm m}$. The extremization of $S_{\rm m}$ with respect to the matter field gives the equations of motion; see Section 3.3. Its extremization with respect to the gravitational field $g_{\mu\nu}$,

$$\delta S_{\rm m} = -\frac{1}{2}\int T_{\mu\nu}\delta g^{\mu\nu}\sqrt{-g}\,d^4x$$
$$= \frac{1}{2}\int T^{\mu\nu}\delta g_{\mu\nu}\sqrt{-g}\,d^4x\,, \tag{4.23}$$

defines the energy–momentum tensor of the field, $T_{\mu\nu}(= g_{\mu\rho}g_{\nu\sigma}T^{\rho\sigma})$; see Section 3.4.

The extremization of the total action describing the system of gravitation plus matter, $S = S_{\rm m} + S_{\rm H}$, then leads naturally to the Einstein equations for the gravitational field: $\delta S_{\rm H}$ is given in (4.20) and $\delta S_{\rm m}$ in (4.23); the variation $\delta(S_{\rm H} + S_{\rm m})$ will vanish for any variation $\delta g^{\mu\nu}$ (which at the boundary is zero and has derivatives equal to zero) if the integrand is zero, that is, if

$$G_{\mu\nu} + \Lambda g_{\mu\nu} = \kappa\,T_{\mu\nu}\,. \tag{4.24}$$

The Palatini variation

Let us introduce the notation $\hat{g}^{\mu\nu} \equiv \sqrt{-g}g^{\mu\nu}$ and rewrite the Hilbert action (4.16) as (setting $\Lambda = 0$)

$$2\kappa\,S_{\rm H} = \int \hat{R}\,d^4x\,, \quad \text{where} \quad \hat{R} = \hat{g}^{\mu\nu}R_{\mu\nu} \quad \text{with} \quad R_{\mu\nu} = \partial_\rho\Gamma^\rho_{\mu\nu} - \partial_\nu\Gamma^\rho_{\mu\rho} + \Gamma^\rho_{\sigma\rho}\Gamma^\sigma_{\mu\nu} - \Gamma^\rho_{\sigma\nu}\Gamma^\sigma_{\mu\rho}\,, \tag{4.25}$$

which we view as a functional of the *independent* quantities $\hat{g}^{\mu\nu}$ and $\Gamma^\rho_{\mu\nu}$. Its variation will then be given by

$$2\kappa\,\delta S_{\rm H} = \int \left(R_{\mu\nu}\delta\hat{g}^{\mu\nu} + \frac{\delta\hat{R}}{\delta\Gamma^\rho_{\mu\nu}}\delta\Gamma^\rho_{\mu\nu}\right)d^4x + \int_{\partial\mathcal{M}} \frac{\partial R}{\partial\partial_\sigma\Gamma^\rho_{\mu\nu}}\delta\Gamma^\rho_{\mu\nu}dS_\sigma$$

$$\text{with} \quad \frac{\delta\hat{R}}{\delta\Gamma^\rho_{\mu\nu}} \equiv \frac{\partial\hat{R}}{\partial\Gamma^\rho_{\mu\nu}} - \partial_\sigma\frac{\partial\hat{R}}{\partial\partial_\sigma\Gamma^\rho_{\mu\nu}}\,.$$

If the action $S_{\rm m}$ describing the matter does not depend on $\Gamma^\rho_{\mu\nu}$, then the total action $S_{\rm H} + S_{\rm m}$ will be an extremum (for variations $\delta\Gamma^\rho_{\mu\nu}$ which cancel on the boundary of the domain) if

$$\frac{\delta\hat{R}}{\delta\Gamma^\alpha_{\beta\gamma}} = 0\,, \quad \text{that is, if} \quad \hat{g}^{\mu\nu}\Gamma^\beta_{\mu\nu}\delta^\gamma_\alpha + \hat{g}^{\beta\gamma}\Gamma^\epsilon_{\alpha\epsilon} - \hat{g}^{\gamma\epsilon}\Gamma^\beta_{\epsilon\alpha} - \hat{g}^{\epsilon\beta}\Gamma^\gamma_{\epsilon\alpha} - \partial_\alpha\hat{g}^{\beta\gamma} + \delta^\gamma_\alpha\partial_\epsilon\hat{g}^{\beta\epsilon} = 0\,. \tag{4.26}$$

Introducing the covariant derivative D with connection $\Gamma^\mu_{\nu\rho}$, it can easily be shown that

$$\partial_\mu\hat{g}^{\nu\rho} = D_\mu\hat{g}^{\nu\rho} + \hat{g}^{\nu\rho}\Gamma^\lambda_{\mu\lambda} - \hat{g}^{\lambda\rho}\Gamma^\nu_{\mu\lambda} - \hat{g}^{\nu\lambda}\Gamma^\rho_{\mu\lambda}\,. \tag{4.27}$$

Therefore, (4.26) reduces to $D_\alpha\hat{g}^{\beta\gamma} = 0$, which means (Palatini, 1919) that the connection must be the Levi-Civita connection (see Section 2.4):

$$\Gamma^\rho_{\mu\nu} = \frac{1}{2}g^{\rho\sigma}(\partial_\mu g_{\nu\sigma} + \partial_\nu g_{\rho\sigma} - \partial_\sigma g_{\mu\nu})\,.$$

In general relativity there is therefore no need to require *a priori* that the connection be the Levi-Civita connection; this will follow from the variation of the action, with the condition that the matter Lagrangian depends only on the metric and not on its derivatives.

Book 3

A 'special relativistic' interpretation of gravitation

The Palatini variation allows the Hilbert action to be interpreted as the action of a tensor field defined on a *flat* spacetime (Deser, 1970). Here is an alternative demonstration of this.

Let us work within special relativity and in an inertial frame where events are referenced by the Minkowski coordinates X^μ and the length element is written as $ds^2 = \eta_{\mu\nu} dX^\mu dX^\nu$. We consider the symmetric tensor fields $\hat{g}^{\mu\nu}(X^\sigma)$ and $\Gamma^\rho_{\mu\nu}(X^\sigma)$. The field $\hat{g}^{\mu\nu}$ is 2-fold contravariant, and so it transforms as $\hat{g}^{\mu\nu} \to \hat{g}'^{\mu\nu} = \Lambda_\rho{}^\mu \Lambda_\sigma{}^\nu \hat{g}^{\rho\sigma}$ under Lorentz transformations $X^\mu \to X'^\mu = \Lambda^\mu{}_\nu X^\nu$. The field $\Gamma^\rho_{\mu\nu}$ transforms in a similar manner; see Book 2, Section 1.3. Now the determinant $\hat{g}$ with field components $\hat{g}^{\mu\nu}$ (not to be confused with the trace $\eta_{\mu\nu}\hat{g}^{\mu\nu}$!) is easily seen to transform as a scalar: $\hat{g}' = \hat{g}$. Finally, the (ordinary) derivatives $\partial_\sigma \Gamma^\rho_{\mu\nu}$ are the components of a tensor which is singly contravariant and 3-fold covariant under Lorentz transformations.

The Hilbert action (4.25) can then be interpreted as the action describing the dynamics of the fields $\hat{g}^{\mu\nu}$ and $\Gamma^\rho_{\mu\rho}$ in an inertial frame where the volume element is d^4X. The action of the matter, a Klein–Gordon field, for example, can be written [in four dimensions; see (4.31) below] as $S_f = -\int \left(\frac{1}{2}\hat{g}^{\mu\nu}\partial_\mu\phi\partial_\nu\phi + \sqrt{-\hat{g}}\, V(\phi)\right) d^4X$.

The equations of motion are easily found. The equation for the field ϕ is

$$\partial_\nu(\hat{g}^{\mu\nu}\partial_\mu\phi) - \sqrt{-\hat{g}}\,\frac{dV}{d\phi} = 0 . \tag{4.28}$$

That for the field $\Gamma^\mu_{\nu\rho}$ is given by (4.26) and has the solution

$$\sqrt{-\hat{g}}\,\Gamma^\rho_{\mu\nu} = \frac{1}{2}\hat{g}^{\rho\sigma}\left(\partial_\mu\left(\sqrt{-\hat{g}}\hat{g}_{\nu\sigma}\right) + \partial_\nu\left(\sqrt{-\hat{g}}\hat{g}_{\mu\sigma}\right) - \partial_\sigma\left(\sqrt{-\hat{g}}\hat{g}_{\mu\nu}\right)\right), \tag{4.29}$$

where $\hat{g}_{\mu\nu}$ denotes the inverse of $\hat{g}^{\mu\nu}$, $\hat{g}_{\mu\rho}\hat{g}^{\nu\rho} = \delta^\nu_\mu$. [Notice that $\hat{g}_{\mu\nu} \neq \eta_{\mu\rho}\eta_{\nu\sigma}\hat{g}^{\rho\sigma}$.] The field $\hat{g}_{\mu\nu}$ transforms as a 2-fold covariant tensor under Lorentz transformations, as can be shown by explicit calculation. Equation (4.29) is not a dynamical equation because the action depends only linearly on the $\Gamma^\mu_{\nu\rho}$.

Finally, the equation of motion for the field $\hat{g}^{\mu\nu}$ is

$$R_{\mu\nu} = \kappa(\partial_\mu\phi\,\partial_\nu\phi + \sqrt{-\hat{g}}\hat{g}_{\mu\nu}V(\phi)) \quad \text{with} \quad R_{\mu\nu} \equiv \partial_\rho\Gamma^\rho_{\mu\nu} - \partial_\nu\Gamma^\rho_{\mu\rho} + \Gamma^\rho_{\sigma\rho}\Gamma^\sigma_{\mu\nu} - \Gamma^\rho_{\sigma\nu}\Gamma^\sigma_{\mu\rho}. \tag{4.30}$$

Now we introduce the auxiliary field $g^{\mu\nu}$ and its inverse $g_{\mu\nu}$, defined as

$$g^{\mu\nu} \equiv \hat{g}^{\mu\nu}/\sqrt{-\hat{g}}, \;\; g_{\mu\rho}g^{\nu\rho} = \delta^\nu_\mu, \;\; \text{so that } g \equiv \det g_{\mu\nu} = (\det g^{\mu\nu})^{-1} = \hat{g} \text{ (in four dimensions).} \tag{4.31}$$

The equations of motion (4.28)–(4.30) in terms of these auxiliary quantities become those of general relativity:

$$g^{\mu\nu}D_\mu D_\nu\phi - \frac{dV}{d\phi} = 0, \;\; \Gamma^\rho_{\mu\nu} = \frac{1}{2}g^{\rho\sigma}(\partial_\mu g_{\nu\sigma} + \partial_\nu g_{\rho\sigma} - \partial_\sigma g_{\mu\nu}), \text{ and } R_{\mu\nu} = \kappa(\partial_\mu\phi\,\partial_\nu\phi + g_{\mu\nu}V(\phi)),$$

where D_μ is the *covariant* derivative associated with the 'metric' $g_{\mu\nu}$, $R_{\mu\nu}$ is defined in (4.30) as a function of the associated 'Christoffel symbols' $\Gamma^\rho_{\mu\nu}$, and the free choice of the four field components $g_{\mu\nu}$ due to the Bianchi identities is interpreted as a gauge invariance, just as in electromagnetism.

4.5 The gravitational Hamiltonian

The action of the gravitational field, which here, as an example, we couple to a scalar field, is

$$S = \int \left(\frac{R}{2\kappa} - \frac{1}{2} g^{\mu\nu} \partial_\mu \phi \partial_\nu \phi - V(\phi) \right) \sqrt{-g}\, d^4x \,. \tag{4.32}$$

As we saw in Section 4.2, the length element, when written in terms of the ADM variables, that is, the lapse N, shift N_i, and 3-metric h_{ij}, reads

$$ds^2 = -(N^2 - N_i N^i)dt^2 + 2N_i dx^i + h_{ij} dx^i dx^j \,, \tag{4.33}$$

where the indices are raised using the inverse metric h^{ij}. The decomposition of the scalar curvature R was given in (4.12), and S, up to boundary terms which we ignore because they do not contribute to the equations of motion, becomes

$$S = \int \mathcal{L} \sqrt{-g}\, d^4x$$

$$\text{with} \quad \begin{cases} \mathcal{L} = \dfrac{1}{2\kappa}(K.K - K^2 + \bar{R}) + \dfrac{1}{2N^2}(\dot{\phi} - N^i \partial_i \phi)^2 - \dfrac{1}{2} h^{ij} \partial_i \phi \partial_j \phi - V(\phi) \\[2mm] K_{ij} = \dfrac{1}{2N}(\dot{h}_{ij} - \bar{D}_i N_j - \bar{D}_j N_i) \quad \text{and} \quad \sqrt{-g} = N\sqrt{h} \,, \end{cases} \tag{4.34}$$

where h_{ij} is the metric induced on the surfaces $t = \text{const}$, K_{ij} is their extrinsic curvature, and $\bar{R}$ is their scalar curvature. In addition, $K = h^{ij} K_{ij}$, $K.K = K^{ij} K_{ij}$, and the dot denotes differentiation with respect to t.

The Hamiltonian of the system is now constructed in the usual way (see, for example, Book 2, Sections 9.1 and 14.3 for an introduction to the Hamiltonian formalism in field theory).

The variables are $\{h_{ij}, N, N_i, \phi\}$, and N and N_i do not have conjugate momenta. The conjugate momenta of h_{ij} and ϕ are

$$p^{ij} \equiv \frac{\partial(\mathcal{L}\sqrt{-g})}{\partial \dot{h}_{ij}} = \frac{\sqrt{h}}{2\kappa}(K^{ij} - Kh^{ij})\,, \quad \pi \equiv \frac{\partial(\mathcal{L}\sqrt{-g})}{\partial \dot{\phi}} = \frac{\sqrt{h}}{N}(\dot{\phi} - N^i \partial_i \phi)\,. \tag{4.35}$$

The 'velocity extensions' $\dot{h}_{ij}$ and $\dot{\phi}$ are obtained by inversion:

$$\dot{h}_{ij} = \frac{4\kappa N}{\sqrt{h}}\left(p_{ij} - \frac{1}{2} p h_{ij} \right) + \bar{D}_i N_j + \bar{D}_j N_i \,, \qquad \dot{\phi} = \frac{N}{\sqrt{h}}\pi + N^i \partial_i \phi \,. \tag{4.36}$$

The Hamiltonian of the system is $H = \int (p^{ij} \dot{h}_{ij} + \pi \dot{\phi} - \sqrt{-g}\,\mathcal{L})d^4x \equiv \int \mathcal{H}\sqrt{-g}\, d^4x$, where the Hamiltonian density $\mathcal{H}\sqrt{-g}$ is, again up to a divergence, given by

$$\sqrt{-g}\,\mathcal{H} = \sqrt{h}(N\mathcal{C} + N^i \mathcal{C}_i)$$

$$\text{with} \quad \begin{cases} \mathcal{C} = \dfrac{2\kappa}{h}\left(p.p - \dfrac{1}{2} p^2 \right) - \dfrac{1}{2\kappa}\bar{R} + \dfrac{1}{2h}\pi^2 + \dfrac{1}{2} h^{ij} \partial_i \phi \partial_j \phi + V(\phi)\,, \\[3mm] \mathcal{C}_i = -2\bar{D}_j \left(\dfrac{p^j{}_i}{\sqrt{h}} \right) + \dfrac{\pi}{\sqrt{h}} \partial_i \phi \,. \end{cases} \tag{4.37}$$

The Hamilton equations then are

$$
\begin{cases}
\mathcal{C} = 0, \quad \mathcal{C}_i = 0, \\[2mm]
\dfrac{\delta(\sqrt{-g}\mathcal{H})}{\delta p^{ij}} = \dot{h}_{ij}, \quad
\dfrac{\delta(\sqrt{-g}\mathcal{H})}{\delta h_{ij}} = -\dot{p}^{ij}, \quad
\dfrac{\delta(\sqrt{-g}\mathcal{H})}{\delta \pi} = \dot{\phi}, \quad
\dfrac{\delta(\sqrt{-g}\mathcal{H})}{\delta \phi} = -\dot{\pi}.
\end{cases}
\tag{4.38}
$$

They are equivalent to the Einstein equations:

$$
\kappa^{-1} G_{\mu\nu} = \partial_\mu \phi \partial_\nu \phi - \left(\frac{1}{2}(\partial\phi)^2 + V(\phi) \right).
$$

The proof can be found in the references cited in footnote 6.

We see that on the mass shell, that is, when the equations of motion are satisfied, we have $\sqrt{-g}\mathcal{H} = 0$ up to divergences. The energy of the system of gravitational field plus matter, $H = \int \mathcal{H}\sqrt{-g}\, d^4x$, then reduces to a boundary term. This result is consistent with the equivalence principle, which states that a gravitational field can be effaced locally, and so its energy cannot be localized, as we shall see in detail in the following chapter.

5
Conservation laws

As already mentioned when we introduced matter energy–momentum tensors in Section 3.4, the concepts of energy, momentum, and angular momentum are related to the invariance properties of the solutions of the equations of motion under spacetime translations or rotations. Here we shall study in greater detail how these 'spacetime symmetries' can generate first integrals of the equations of motion which simplify their solution and also make it possible to define conserved quantities, or 'charges', characterizing the system.

5.1   Isometries and Killing vectors

A (pseudo-)Riemannian metric $g_{ij}(x^k)$ possesses an *isometry* if it is invariant under a displacement from point P to point $\tilde{P}$ along a certain path. Let us explain what this means. We consider the length element $ds^2 = g_{ij}dx^i dx^j$ between two neighboring points P and $P + dP$, and we transport P and $P + dP$ along the line integral of a vector field ξ^i at $\tilde{P}$ and $\tilde{P} + d\tilde{P}$. The length element between $\tilde{P}$ and $\tilde{P} + d\tilde{P}$ is $d\tilde{s}^2$. Since $\tilde{x}^k = x^k + \xi^k$, we have $\delta g_{ij} \equiv g_{ij}(\tilde{x}^k) - g_{ij}(x^k) = \xi^k \partial_k g_{ij}$ and $\delta(dx^k) \equiv d\tilde{x}^k - dx^k = (\partial_i \xi^k)dx^i$. The metric is invariant and therefore possesses an isometry defined by ξ^i if $\delta(ds^2) \equiv d\tilde{s}^2 - ds^2 = 0$, that is, if

$$\xi^m \partial_m g_{ij} + g_{mi}\partial_j \xi^m + g_{mj}\partial_i \xi^m = 0 \quad \Longleftrightarrow \quad D_j\xi_i + D_i\xi_j = 0 \,. \tag{5.1}$$

These equations, whose equivalence is easily shown, state that the *Lie derivative* $\mathcal{L}_\xi g_{ij}$ of the metric along the vector ξ^i is zero. The vector ξ^i along which the metric is invariant is a *Killing vector*, and (5.1) constrains both ξ^i and the metric.

Another (passive rather than active) way of expressing the conditions for the existence of symmetries of a space is to make an infinitesimal change of coordinate $x^k \to \tilde{x}^k = x^k + \xi^k$ (we have already seen this in Book 2, Section 8.5 and also Section 3.4). The components $g_{ij}(x^k)$ of the metric in the system $\{x^k\}$ at the point P with coordinates x^k are related to its components $\tilde{g}_{ij}(\tilde{x}^k)$ at P in the system $\{\tilde{x}^k\}$ by the transformation law for 2-fold covariant tensors:

$$g_{ij}(x^k) = \frac{\partial \tilde{x}^k}{\partial x^i}\frac{\partial \tilde{x}^l}{\partial x^j}\tilde{g}_{kl}(\tilde{x}^p) = (\delta_i^k + \partial_i\xi^k)(\delta_j^l + \partial_j\xi^l)\tilde{g}_{kl}(\tilde{x}^p)\,. \tag{5.2}$$

Now if we write $\tilde{g}_{kl}(\tilde{x}^p) = \tilde{g}_{kl}(x^p) + \xi^m \partial_m g_{kl}$, that is, if we introduce the metric at the point $\tilde{P}$, which in the system $\{\tilde{x}^k\}$ has the same coordinates as the point P in the system $\{x^k\}$, we find

$$g_{ij}(x^k) = \tilde{g}_{ij}(x^k) + D_j\xi_i + D_i\xi_j\,. \tag{5.3}$$

If $g_{ij}(x^k) = \tilde{g}_{ij}(x^k)$, the space possesses a symmetry, and we recover the Killing equation (5.1).

Book 3

Relativity in Modern Physics. Nathalie Deruelle and Jean-Philippe Uzan.
© Oxford University Press 2018. Published in 2018 by Oxford University Press.
DOI: 10.1093/oso/9780198786399.001.0001

Isometries of Minkowski spacetime

In Book 2, Section 8.5 we saw that Minkowski spacetime $\mathcal{M}_4$ possesses 10 Killing vectors: four translations, three spatial rotations, and three boosts.

The fact that the Minkowski metric possesses four translation Killing vectors is a consequence of the spatio-temporal *homogeneity* of flat space, and the fact that it possesses six rotational ones is a consequence of the *isotropy* of the space.

We can deduce from the Killing equations that $\tilde{D}_\alpha \tilde{D}^\alpha \xi^\beta = 0$ (because Minkowski covariant derivatives commute). To determine a Killing field everywhere, it is therefore sufficient to know the value of the field and its first derivatives at a point, which corresponds to 10 initial conditions (since the 16 first derivatives obey the 10 Killing equations). Therefore, the maximum number of Killing vectors that a metric can possess is 10 [or, more generally, $N(N+1)/2$ in N dimensions]. Minkowski spacetime is therefore *maximally symmetric*.

In curved space the Killing equation is very restrictive because we have[1]

$$D_i D_j \xi_k = R^l{}_{ijk} \xi_l. \tag{5.4}$$

Therefore, the values of ξ^i and its derivatives at a point determine all its higher derivatives at this point and thus its value at all points. Since in a manifold of N dimensions the vector ξ^i has N independent components and its derivative has $N(N-1)/2$ of them [owing to (5.1)], we see that an N-dimensional manifold possesses at most $N(N+1)/2$ independent Killing vectors.

A space may, of course, not possess the maximum number of possible Killing vectors. Indeed, we have in particular[2] that the ξ^i and their derivatives are constrained by

$$(D_i R^m{}_{jkl} - D_j R^m{}_{ikl})\xi_m + (\delta^n_i R^m{}_{jkl} - \delta^n_j R^m{}_{ikl} + \delta^n_l R^m{}_{kji} - \delta^n_k R^m{}_{lji})D_n \xi_m = 0. \tag{5.5}$$

5.2   First integrals of the geodesic equation

The motion of a test particle in a gravitational field is governed by the geodesic equation, which is written in covariant form as

$$u^\mu D_\mu u_\nu = 0 \quad \text{or} \quad \frac{du_\nu}{d\tau} = \frac{1}{2} u^\mu u^\rho \partial_\nu g_{\mu\rho}, \tag{5.6}$$

where $u^\mu = dx^\mu/d\tau$ is the 4-velocity and τ the proper time; see (3.4). We suppose that in the chosen coordinates x^μ, the 'adapted' coordinates, none of the metric coefficients depends on a given coordinate $x^{\underline{i}}$, which is termed *cyclic*. We then have the first integral

$$u_{\underline{i}} = \text{const along the trajectory because} \quad \partial_{\underline{i}} g_{\mu\rho} = 0, \quad \forall \mu, \rho. \tag{5.7}$$

For example, if the spacetime is *stationary*, that is, if in adapted coordinates none of the metric coefficients $g_{\mu\rho}$ depends on the time coordinate t, then the geodesic equation possesses

---

[1] In fact, for any form [*cf.* (2.10)] we have $(D_i D_j - D_j D_i)\xi_k = -R^m{}_{kij}\xi_m$, which, owing to the first Bianchi identity [$R^k{}_{(lji)} = 0$; see Section 2.6] and the fact that ξ^i is a Killing vector and therefore satisfies the condition (5.1), implies that $D_i D_j \xi_k + D_j D_k \xi_i + D_k D_i \xi_j = 0$ or $D_i D_j \xi_k - D_j D_i \xi_k = -D_k D_i \xi_j$, from which (5.4) follows.

[2] Using (5.1) and (5.4) and the fact that $D_i D_j D_k \xi_l - D_j D_i D_k \xi_l = -R^m{}_{kij} D_m \xi_l - R^m{}_{lij} D_k \xi_m$; see, for example, Stephani (1990).

A detailed discussion of maximally symmetric spaces will be given in Section 17.1.

a first integral $u_t = \text{const}$, which we identify as the energy per unit mass of the particle. Similarly, if none of the $g_{\mu\rho}$ depends on the azimuthal coordinate ϕ, then the angular momentum u_ϕ will be conserved.

Now, if none of the metric coefficients depends on the coordinate $x^{\underline{i}}$, the vector $\xi^\mu_{(i)} = \delta^\mu_{\underline{i}}$ is a Killing vector. Indeed, we then have $\xi_{\mu(i)} = g_{\mu\underline{i}}$, and so

$$D_\mu \xi_{\rho(i)} + D_\rho \xi_{\mu(i)} = \partial_\mu g_{\rho\underline{i}} + \partial_\rho g_{\mu\underline{i}} - 2\Gamma^\sigma_{\mu\rho}\xi_{\sigma(i)} = \partial_{\underline{i}} g_{\mu\rho} = 0\,, \tag{5.8}$$

because $2\Gamma^\sigma_{\mu\rho}\xi_{\sigma(i)} = \xi^\sigma_{(i)}(\partial_\mu g_{\rho\sigma} + \partial_\rho g_{\sigma\mu} - \partial_\sigma g_{\mu\rho})$ with $\xi^\sigma_{(i)} = \delta^\sigma_{\underline{i}}$.

In an arbitrary coordinate system, possible invariances of the spacetime will no longer be apparent because the metric coefficients can depend on all the coordinates. However, if there exists a Killing vector ξ^μ such that $D_\mu \xi_\nu + D_\nu \xi_\mu = 0$, then, taking its scalar product with the geodesic equation, we will have

$$\xi_\mu \frac{Du^\mu}{d\tau} = \frac{D(\xi_\mu u^\mu)}{d\tau} - u^\mu \frac{D\xi^\mu}{d\tau} = \frac{d(\xi_\mu u^\mu)}{d\tau} - u^\mu u^\nu D_\nu \xi_\mu = \frac{d(\xi^\mu u_\mu)}{d\tau} = 0\,, \tag{5.9}$$

because $u^\mu u^\nu D_\nu \xi_\mu = u^\mu u^\nu (D_\nu \xi_\mu + D_\mu \xi_\nu)/2 = 0$. In adapted coordinates where $\xi^\mu = \delta^\mu_{\underline{i}}$, we recover (5.7).

The case of a static, spherically symmetric field

As we shall see in detail in Section 6.1, the length element of a static, spherically symmetric spacetime can be written in adapted coordinates as

$$ds^2 = -e^{\nu(r)}dt^2 + e^{\lambda(r)}dr^2 + r^2(d\theta^2 + \sin^2\theta d\phi^2)\,. \tag{5.10}$$

The vectors $\xi^\mu_{(t)} = (1,0,0,0)$ and $\xi^\mu_{(\phi)} = (0,0,0,1)$ are two Killing vectors respectively representing the stationarity of the spacetime and the symmetry of the spacetime under rotation about the 'z axis'. From (5.7) and (5.9) we then have the two first integrals

$$u_t = -e^{\nu(r)}\frac{dt}{d\tau} \equiv -E \quad \text{and} \quad u_\phi = r^2 \sin^2\theta \frac{d\phi}{d\tau} \equiv L\,, \tag{5.11}$$

where the constants E and L are interpreted as the specific energy and the angular momentum of the particle in the field.

5.3 Isometries and energy–momentum

The equations of motion of a massive system in the presence of a gravitational field are derived from the conservation law of its energy–momentum tensor $D_\nu T^\nu_\mu = 0$, as we have seen in Section 3.2 in the case of fluids and in Section 3.3 in the case of matter fields.

If the gravitational field possesses symmetries, that is, if the spacetime possesses isometries and therefore Killing vectors ξ^μ, it is possible to extract first integrals from the law of motion $D_\nu T^\nu_\mu = 0$. Indeed, following the reasoning of Book 2, Section 8.5 for flat spacetime but an accelerated reference frame, we have

$$0 = \xi^\mu D_\nu T^\nu_\mu = D_\nu(\xi^\mu T^\nu_\mu) - T^{\mu\nu}D_\nu \xi_\mu = D_\nu(\xi^\mu T^\nu_\mu) \tag{5.12}$$

Book 3

because the Killing vector ξ^μ satisfies $D_\mu \xi_\nu + D_\nu \xi_\mu = 0$. Then, using (3.21) we can write

$$D_\nu(\xi^\mu T_\mu^\nu) = 0 \quad \Longleftrightarrow \quad \partial_\nu(\sqrt{-g}\,\xi^\mu T_\mu^\nu) = 0 \quad \Longleftrightarrow \quad \partial_t(\sqrt{-g}\,\xi^\mu T_\mu^0) = -\partial_i(\sqrt{-g}\,\xi^\mu T_\mu^i). \tag{5.13}$$

We integrate this over $dx^1 dx^2 dx^3 \equiv dr d\theta d\phi = d^3x$ and use the divergence theorem (3.15) to obtain

$$\frac{d}{dt}\int_\Sigma \xi^\mu T_\mu^0 \sqrt{-g}\, d^3x = -\int_S \xi^\mu T_\mu^r \sqrt{-g}\, d\theta d\phi. \tag{5.14}$$

The right-hand side, evaluated on the 2-sphere S bounding the 3-volume Σ, is zero if S is located outside the fluid or if the fields fall off sufficiently rapidly at spatial infinity. We then find that

$$Q_\xi \equiv -\int_\Sigma \xi^\mu T_\mu^0 \sqrt{-g}\, d^3x \tag{5.15}$$

is a constant of the motion, that is, $dQ_\xi/dt = 0$.

The case of a static, spherically symmetric field

As in (5.10), we consider a static, spherically symmetric spacetime in which the length element is written as

$$ds^2 = -e^{\nu(r)}dt^2 + e^{\lambda(r)}dr^2 + r^2(d\theta^2 + \sin^2\theta d\phi^2) \quad \Longrightarrow \quad \sqrt{-g} = e^{\frac{1}{2}(\nu+\lambda)}r^2 \sin\theta. \tag{5.16}$$

Let us consider a perfect fluid at rest in this field. Its energy–momentum tensor is [*cf.* Section 3.2] $T_{\mu\nu} = (\epsilon + p)u_\mu u_\nu + p g_{\mu\nu}$, where $\epsilon(r)$ and $p(r)$ are the energy density and pressure of the fluid, and, since the fluid is static, its 4-velocity is $u^\mu = (u^0, 0, 0, 0)$ with $u^0 u_0 = -1$. Therefore, $T_0^0 = -\epsilon(r)$.

The vector $\xi_{(t)}^\mu = (1, 0, 0, 0)$ is the Killing vector representing the static nature of the spacetime. The associated conserved quantity (5.15), which we can view as the energy of the fluid in the presence of the gravitational field, is then given by

$$Q_t = 4\pi \int_0^R \epsilon e^{\frac{1}{2}(\nu+\lambda)} r^2 dr. \tag{5.17}$$

It should be noted that this quantity is *not* what is called the *proper mass*, which is defined as

$$M_{\rm p} = 4\pi \int_0^R \epsilon e^{\frac{\lambda}{2}} r^2 dr, \tag{5.18}$$

where $e^{\frac{\lambda}{2}} r^2 \sin\theta\, dr d\theta d\phi$ is the volume element on the surface $t = \text{const}$.

5.4   Noether charges

In Section 3.4 we defined the (symmetric) energy–momentum tensor $T_{\mu\nu}$ of a matter field immersed in a gravitational field. We saw that it is conserved when the equations of motion are satisfied:

$$D_\nu T_\mu^\nu = 0 \quad \text{or} \quad \partial_\nu \hat{T}_\mu^\nu = \frac{1}{2}\hat{T}^{\nu\rho}\partial_\mu g_{\nu\rho}, \quad \text{where} \quad \hat{T}_\mu^\nu \equiv \sqrt{-g}\, T_\mu^\nu. \tag{5.19}$$

In general, that is, in the absence of a Killing vector, it is not possible to extract from (5.19) a quantity conserved in the motion (for example, the energy) by integrating over spacetime. Moreover, it is necessary to take into account the contribution of the gravitational field 'responsible' for the fact that $\partial_\mu g_{\nu\rho}$ cannot be cancelled out everywhere. In order to define conserved quantities of a gravitationally interacting system, we shall study how the *total* action $S = S_c + S_g$ of the matter *and* the gravitational field transforms under displacements (rather than considering just the matter action, as in Section 3.4).[3]

As we saw in Section 4.4, the Hilbert action of the gravitational field, $S_{\rm H} \propto \int \sqrt{-g} R \, d^4 x$, even though it contains second derivatives of the metric, does however lead to second-order equations, the Einstein equations. The reason is that the second derivatives of the metric present in $\sqrt{-g} R$ combine to form a divergence. Indeed, it is easily shown (using the relations $dg = -g g_{\mu\nu} dg^{\mu\nu}$ and $\partial_\rho g^{\mu\nu} = -\Gamma^\mu_{\lambda\rho} g^{\lambda\nu} - \Gamma^\nu_{\lambda\rho} g^{\mu\lambda}$) that

$$\hat{R} = \hat{G} + \partial_\mu \hat{v}^\mu \qquad \text{with} \qquad \begin{cases} G = g^{\mu\nu} \left(\Gamma^\lambda_{\mu\rho} \Gamma^\rho_{\nu\lambda} - \Gamma^\rho_{\mu\nu} \Gamma^\lambda_{\rho\lambda} \right), \\ v^\mu = g^{\nu\rho} \Gamma^\mu_{\nu\rho} - g^{\mu\nu} \Gamma^\rho_{\nu\rho}, \end{cases} \tag{5.20}$$

where from now on a hat will denote multiplication by $\sqrt{-g}$: $\hat{f} \equiv \sqrt{-g} \, f$. The first term $\hat{G}$ contains only first derivatives of the metric, and the second, $\partial_\mu \hat{v}^\mu$, is a divergence which does not contribute to the field equations. Therefore, the Euler–Lagrange equations obtained from extremizing the Hilbert action are indeed of second order.[4]

However, it is not a very good idea to choose (as Einstein did) $\int \hat{G} \, d^4 x$ as the gravitational action, because G is *not* a scalar, since the Christoffel symbols do *not* behave as tensors under general coordinate transformations. However, it is easy to covariantize G using the fact that *differences* of the Christoffel symbols are tensors; *cf.* (2.21). We therefore introduce another spacetime $\bar{\mathcal{M}}$, called the *reference spacetime*, with *given* metric $\bar{g}_{\mu\nu}$ in the selected coordinates x^μ.

We now consider the action [noting the resemblance between v^μ in (5.20) and the vector k^μ]

$$S_g = S_{\rm H} - \bar{S}_{\rm H} + \frac{1}{2\kappa} \int \partial_\mu \hat{k}^\mu \, d^4 x, \text{ where } S_{\rm H} = \frac{1}{2\kappa} \int \hat{R} \, d^4 x \text{ and } \bar{S}_{\rm H} = \frac{1}{2\kappa} \int \hat{\bar{R}} \, d^4 x,$$
$$\text{and where} \qquad k^\mu = - \left(g^{\nu\rho} \Delta^\mu_{\nu\rho} - g^{\mu\nu} \Delta^\rho_{\nu\rho} \right) \quad \text{with} \quad \Delta^\mu_{\nu\rho} \equiv \Gamma^\mu_{\nu\rho} - \bar{\Gamma}^\mu_{\nu\rho}. \tag{5.21}$$

The terms added to the Hilbert action do not contribute to the field equations, which thus remain the Einstein equations (the first, $\bar{S}_{\rm H}$, because S_g is extremized with respect to $g_{\mu\nu}$ and not $\bar{g}_{\mu\nu}$, and the second because the integrand is a divergence).

---

[3]We shall follow the line of attack of Katz *et al.* (1997), which is the covariantized version of the approach of Einstein (1918) and Freud (1939).

Other approaches can be found in, for example, Wald (1984) (who used the Komar integrals); Landau and Lifshitz (1972) [who used their 'pseudo-tensor' (1962) or the Einstein tensor (1918)]; and also Arnowitt, Deser, and Misner (1962) (the Hamiltonian formalism). See also the references cited in footnote 6 of Section 4.2.

For a comparison of the various approaches, see Poisson (2007), Jaramillo and Gourgoulhon (2010), or Blau (2016).

For a review, see, for example, Szabados (2004).

[4]Comparing (4.18) and (5.20), we necessarily have

$$\frac{\partial \hat{G}}{\partial g^{\mu\nu}} - \partial_\rho \frac{\partial \hat{G}}{\partial \partial_\rho g^{\mu\nu}} = \hat{G}_{\mu\nu} \qquad \text{and} \qquad \frac{\partial \hat{G}}{\partial g_{\mu\nu}} - \partial_\rho \frac{\partial \hat{G}}{\partial \partial_\rho g_{\mu\nu}} = -\hat{G}^{\mu\nu},$$

where $G_{\mu\nu}$ is the Einstein tensor.

The Einstein and Gibbons–Hawking–York actions

Setting $2\kappa\, S_g \equiv \int \hat{\mathcal{L}}_g \, d^4x$ with $\hat{\mathcal{L}}_g \equiv \hat{R} - \overline{\hat{R}} + \partial_\mu \hat{k}^\mu$, using (5.20) and (5.21) we can show that

$$\hat{\mathcal{L}}_g = \hat{G} + \partial_\mu(\hat{v}^\mu + \hat{k}^\mu) - \overline{\hat{R}} = \hat{G} + \partial_\mu\left(\hat{g}^{\nu\rho}\bar{\Gamma}^\mu_{\nu\rho} - \hat{g}^{\mu\nu}\bar{\Gamma}^\rho_{\nu\rho}\right) - \overline{\hat{R}}$$

$$= \hat{g}^{\mu\rho}\left(\Delta^\lambda_{\mu\sigma}\Delta^\sigma_{\rho\lambda} - \Delta^\sigma_{\mu\rho}\Delta^\lambda_{\sigma\lambda}\right) + \left(\hat{g}^{\mu\nu} - \overline{\hat{g}}^{\mu\nu}\right)\bar{R}_{\mu\nu}\,.$$

Therefore, $2\kappa\, S_g$ is indeed the covariantized version of the Einstein action $\int \hat{G}\, d^4x$. It has the advantage over the Hilbert action of leading to the Einstein equations when only the metric (and no longer its derivatives) is fixed at the boundary.

In Gaussian coordinates, where the length element is written as $ds^2 = \epsilon\, dw^2 + h_{ij}dx^i dx^j$ and the boundary equation is $w = w_0$, we have $\int \partial_\mu \hat{k}^\mu\, d^4x = \int_{w_0} k^w \sqrt{|h|}\, d^3x$ with, cf. (5.21), $k^w = 2\epsilon\left[K - \bar{K} - \frac{1}{2}(h^{ij} - \bar{h}^{ij})\bar{K}_{ij}\right]$, where $K_{ij} = \frac{1}{2}\partial_w h_{ij}$ and $\bar{K}_{ij} = \frac{1}{2}\partial_w \bar{h}_{ij}$ are the extrinsic curvatures of the boundary immersed respectively in $\mathcal{M}$ and $\overline{\mathcal{M}}$.

The boundary term appearing in the action (5.21) is therefore different from that of the Gibbons–Hawking–York action (4.21), $2\kappa\, S_{\mathrm{GHY}} = \int \hat{R}\, d^4x + 2\epsilon \int K \sqrt{|h|}\, d^3x$. The latter also leads to the Einstein equations when only the metric is fixed on the boundary; cf. (4.22).

In the transformation $x^\rho \to \tilde{x}^\rho = x^\rho + \xi^\rho$ the metrics of $\mathcal{M}$ and of $\overline{\mathcal{M}}$ change by $\delta_\xi g_{\mu\nu}$ and $\delta_\xi \bar{g}_{\mu\nu}$, respectively. Then using (4.18) and (4.19) we have

$$2\kappa\, \delta_\xi S_g = \int \left(\hat{G}_{\mu\nu}\delta_\xi g^{\mu\nu} - \overline{\hat{G}}_{\mu\nu}\delta_\xi \bar{g}^{\mu\nu} + \partial_\mu(\hat{V}^\mu - \overline{\hat{V}}^\mu + \delta_\xi \hat{k}^\mu)\right)d^4x\,. \tag{5.22}$$

In Section 3.4 we saw how a metric and a vector vary under a displacement [see (3.29) and footnote 5 of Section 3.4; these are their 'Lie derivatives']: $\delta_\xi g_{\mu\nu} = D_\mu \xi_\nu + D_\nu \xi_\mu$ (and so $\delta_\xi \bar{g}_{\mu\nu} = \bar{D}_\mu \xi_\nu + \bar{D}_\nu \xi_\mu$) and $\delta_\xi k^\mu = \xi^\nu D_\nu k^\mu - k^\nu D_\nu \xi^\mu$. Since moreover $\delta_\xi \hat{\mathcal{L}}_g = \partial_\mu(\hat{\mathcal{L}}_g \xi^\mu)$, an easy calculation discussed in detail below shows that (5.22) can be reduced to

$$\int \partial_\mu \left(\partial_\nu \hat{J}^{[\mu\nu]}\right)d^4x = 0 \quad \text{with} \quad \kappa \hat{J}^{[\mu\nu]} = D^{[\mu}\hat{\xi}^{\nu]} - \overline{D^{[\mu}\hat{\xi}^{\nu]}} + \hat{\xi}^{[\mu}k^{\nu]}\,, \tag{5.23}$$

where the brackets denote antisymmetrization: $J^{[\mu\nu]} \equiv \frac{1}{2}(J^{\mu\nu} - J^{\nu\mu})$. Here $J^{[\mu\nu]}$ is the *Katz superpotential*.

The Katz superpotential

Here we shall show how to transform (5.22) into (5.23) using the fact that $\delta_\xi g_{\mu\nu} = D_\mu \xi_\nu + D_\nu \xi_\mu$ and $\delta_\xi k^\mu = \xi^\nu D_\nu k^\mu - k^\nu D_\nu \xi^\mu$.

First of all we have $\hat{G}_{\mu\nu}\delta_\xi g^{\mu\nu} = -2G_{\mu\nu}D^\mu \hat{\xi}^\nu = -2D_\mu(G^\mu_\nu \hat{\xi}^\nu) + 2\hat{\xi}^\nu D_\mu G^\mu_\nu = -2\partial_\mu(G^\mu_\nu \hat{\xi}^\nu)$, because $\delta g^{\mu\nu} = -2D^{(\mu}\xi^{\nu)}$, $D_\mu G^\mu_\nu \equiv 0$ (the Bianchi identities), and $D_\mu \hat{V}^\mu = \partial_\mu \hat{V}^\mu$. (The fact that $D_\mu g_{\nu\rho} \equiv 0$ means that we can place the hat wherever we want.)

We can also show that V^μ defined in (4.19), $V^\mu = g^{\nu\rho}\delta\Gamma^\mu_{\nu\rho} - g^{\mu\nu}\delta\Gamma^\rho_{\nu\rho}$, can also be written as

$$V^\mu = (g^{\mu\nu}g^{\rho\sigma} - g^{\mu\rho}g^{\nu\sigma})D_\rho \delta_\xi g_{\nu\sigma}\,.$$

We then have $\hat{V}^\mu = 2\hat{g}^{\mu[\nu}g^{\rho]\sigma}D_\rho\delta g_{\nu\sigma} = 4\hat{g}^{\mu[\nu}g^{\rho]\sigma}D_\rho D_{(\nu}\xi_{\sigma)} = D_\nu D^\mu\hat{\xi}^\nu + \Box\hat{\xi}^\mu - 2D^\mu D_\nu\hat{\xi}^\nu = -2\partial_\nu(D^{[\mu}\hat{\xi}^{\nu]}) + 2R_\nu^\mu\hat{\xi}^\nu$ by the commutation rule for covariant derivatives (2.9), $2D_{[\mu}D_{\nu]}\xi^\rho = R^\rho{}_{\sigma\mu\nu}\xi^\sigma$, and because $D_\mu\hat{F}^{[\mu\nu]} = \partial_\mu\hat{F}^{[\mu\nu]}$.

Finally, we find $\delta_\xi\hat{k}^\mu = k^\mu D_\nu\hat{\xi}^\nu + \hat{\xi}^\nu D_\nu k^\mu - k^\nu D_\nu\hat{\xi}^\mu = -2\partial_\nu(\hat{\xi}^{[\mu}k^{\nu]}) + \hat{\xi}^\mu D_\nu k^\nu$.

Since moreover the Lie derivative of a scalar density $\hat{\mathcal{L}}_g$ is $\delta_\xi\hat{\mathcal{L}}_g = \partial_\mu(\mathcal{L}_g\hat{\xi}^\mu)$, using $\mathcal{L}_g$ given in (5.21) we can write (5.22) as

$$\int \partial_\mu\left(R\hat{\xi}^\mu - \overline{R\hat{\xi}^\mu} + \hat{\xi}^\mu D_\nu k^\nu\right)d^4x = 2\int\partial_\mu\left(-G_\nu^\mu\hat{\xi}^\nu + \overline{G_\nu^\mu\hat{\xi}^\nu}\right)d^4x$$

$$+2\int\partial_\mu\left(-\partial_\nu(D^{[\mu}\hat{\xi}^{\nu]}) + R_\nu^\mu\hat{\xi}^\nu\right)d^4x - 2\int\partial_\mu\left(-\partial_\nu(\overline{D^{[\mu}\hat{\xi}^{\nu]}}) + \overline{R_\nu^\mu\hat{\xi}^\nu}\right)d^4x$$

$$+\int\partial_\mu\left(-2\partial_\nu(\hat{\xi}^{[\mu}k^{\nu]}) + \hat{\xi}^\mu D_\nu k^\nu\right)d^4x.$$

Since $G_\nu^\mu = R_\nu^\mu - \frac{1}{2}\delta_\nu^\mu R$, putting everything on the left-hand side we see that (5.22) reduces to (5.23).

The identities (5.23) lead to conservation laws. Indeed, the first thing to note is that, as in Section 3.4, the vector ξ^μ does not need to be infinitesimal. Using the divergence theorem, (5.23) becomes, in adapted coordinates [see (3.15)–(3.18) and Fig. 3.1],

$$0 = \int\partial_\mu\left(\partial_\nu\hat{J}^{[\mu\nu]}\right)d^4x = \int_{\partial\mathcal{M}}\partial_\nu J^{[\mu\nu]}dS_\mu$$

$$= \int_\Sigma d^3x\,\partial_i\hat{J}^{[0i]}\Big|_{t_1}^{t_2} + \int_{t_1}^{t_2}dt\int_S\partial_\nu\hat{J}^{[r\nu]}\,d\theta\,d\phi. \tag{5.24}$$

If the second term in (5.24) vanishes, that is, if there is no flux through the 2-sphere S at infinity, then applying the divergence theorem again we find that

$$Q \equiv -\int_\Sigma d^3x\,\partial_i\hat{J}^{[0i]} = -\int_S\hat{J}^{[0r]}\,d\theta\,d\phi \tag{5.25}$$

is a constant [the expression for $\hat{J}^{[\mu\nu]}$ being given in (5.23)].[5]

The Freud superpotential and Einstein pseudo-tensor

It is easy to show (using $D_\mu\xi^\nu = \bar{D}_\mu\xi^\nu + \Delta_{\mu\rho}^\nu\xi^\rho$) that the Katz superpotential (5.23) can also be written as

$$\kappa\hat{J}^{[\mu\nu]} = \left(\hat{g}^{\rho[\mu} - \overline{\hat{g}^{\rho[\mu}}\right)\bar{D}_\rho\xi^{\nu]} + \hat{\xi}^{[\nu}g^{\nu]\rho}\Delta_{\rho\sigma}^\sigma + \hat{\xi}^{[\nu}g^{\rho\sigma}\Delta_{\rho\sigma}^{\mu]} + \hat{\xi}^\sigma g^{\rho[\mu}\Delta_{\rho\sigma}^{\nu]}.$$

---

[5]It may have been noticed that we have not added the matter contribution S_c to S_g. This is because S_c does *not* contribute to the charge Q since, as shown in Section 3.4 (at least in the cases of the Klein–Gordon and Maxwell fields), the integrands of both sides of (3.32) and (3.35) vanish identically (and the same occurs for perfect fluids).

The first term vanishes at spatial infinity (where $\hat{J}^{[\mu\nu]}$ is evaluated) if the reference spacetime $\bar{\mathcal{M}}$ is chosen to asymptotically approach $\mathcal{M}$, and the three following terms are the covariantized version of the *Freud superpotential* (1939).

Now the *current* $\hat{J}^\mu \equiv \partial_\nu J^{[\mu\nu]}$ derived from the superpotential $\hat{J}^{[\mu\nu]}$, which is conserved because $\hat{J}^{[\mu\nu]}$ is antisymmetric, reduces to the following when $\bar{g}_{\mu\nu} = \eta_{\mu\nu}$ and if ξ^μ is a constant vector:

$$\kappa \hat{J}^\mu = (G^\mu_\nu + t^\mu_\nu)\hat{\xi}^\nu \quad \text{with} \quad \partial_\mu \hat{J}^\mu = 0,$$

where the Einstein equations give $G^\mu_\nu = \kappa T^\mu_\nu$, and t^μ_ν is the *Einstein pseudo-tensor*, quadratic in the Christoffel symbols. Katz *et al.* (see footnote 3 above) showed that this pseudo-tensor is the (non-symmetrized) Noether canonical (pseudo)-tensor associated with the Einstein 'Lagrangian' $\hat{G}$:

$$2\kappa\, \hat{t}^\mu_\nu = \delta^\mu_\nu \hat{G} - \frac{\partial \hat{G}}{\partial \partial_\mu g_{\rho\sigma}} \partial_\nu g_{\rho\sigma}.$$

This Einstein pseudo-tensor is only useful for defining the energy (because ξ^μ must be constant in Minkowski coordinates) and, in contrast to the superpotential $\hat{J}^{\mu\nu}$, it does not permit the angular momentum of a gravitational system to be defined [see the discussion of this subject by Landau and Lifshitz (1972)].

In the absence of gravitational radiation, the *charge* Q, defined in (5.25) by an integral over the 2-sphere at infinity, is constant for *any* vector ξ^μ. However, let us suppose that spacetime is asymptotically flat. Then the vector ξ^μ representing time translations will be associated with a charge $Q \equiv Mc^2$, where M will naturally be identified as the *inertial mass* of the gravitational system in question. In addition, the vector representing rotations about the (Minkowski) 'axis' Oz will be associated with a charge J_z identified as the angular momentum of the system, and so on.

We again find that these conserved quantities are defined by integrals over the boundary of spacetime, as suggested by the analysis of the Hamilton equations made in Section 4.5.

The Komar integrals

The first term of the Katz superpotential (5.23) is also related to the *Komar integrals*; see footnote 3 above.

Let us consider the integrals over the 2-sphere at infinity:

$$\kappa I_\xi \equiv -\alpha_\xi \int_S D^{[0}\hat{\xi}^{r]}\, d\theta\, d\phi, \tag{5.26}$$

where α_ξ is a constant which may depend on the choice of vector ξ^μ. We can transform them into volume integrals using the 'inverse' Stokes theorem: $\kappa I_\xi = -\alpha_\xi \int_\Sigma D_\mu(D^{[0}\xi^{\mu]})\sqrt{-g}\, d^3x$.

If now spacetime is stationary *everywhere* and, for example, axially symmetric, we take the ξ^μ to be the Killing vectors corresponding to these isometries (see Section 5.1). Since now $D_{(\mu}\xi_{\nu)} = 0$, we have $D_\mu(D^{[\nu}\xi^{\mu]}) = -\Box\xi^\nu$. However, for any Killing vector we have $\Box\xi^\nu = -R^\nu_\lambda \xi^\lambda$, where R^μ_ν is the Ricci tensor; see Section 5.1. Therefore, $\kappa I_\xi = -\alpha_\xi \int_\Sigma R^0_\mu \xi^\mu \sqrt{-g}\, d^3x$.

The Einstein equations $R^\mu_\nu = \kappa\left(T^\mu_\nu - \frac{1}{2}\delta^\mu_\nu T\right)$ then give

$$I_\xi = -\alpha_\xi \int_\Sigma (T^0_\mu - \frac{1}{2}\delta^0_\mu T)\xi^\mu \sqrt{-g}\, d^3x. \tag{5.27}$$

Let us assume that the fluid is perfect $[T^{\mu\nu} = (\epsilon+p)u^\mu u^\nu + p\,g^{\mu\nu}]$ and static $(u^0 u_0 = -1)$. In the adapted coordinates where the metric coefficients are independent of time, the Killing vector ξ^μ expressing the static nature has the components $\xi^\mu_{(t)} = (1,0,0,0)$, and $I_{(t)}$ is then given by the *Tolman formula* (1930):

$$I_{(t)} = \int_\Sigma (\epsilon + 3p)\sqrt{-g}\, d^3x\,, \tag{5.28}$$

where we have chosen $\alpha_{(t)} = +2$ so that $I_{(t)}$ is equal to the mass of the source of the gravitational field in the Newtonian approximation, when $p \ll \epsilon$.

As we shall see in some examples below, the Komar integral (5.26) also gives the angular momentum of a rotating stationary body [the Killing vector $\xi^\mu_{(\phi)} = (0,0,0,1)$ is then the one which expresses the axial symmetry of the spacetime] if we choose $\alpha_{(\phi)} = -1$.

The strong equivalence principle

In general relativity the 'inertial' mass of matter, the sum of the masses of its elementary constituents and their non-gravitational interaction energy (electromagnetic, for example), which is what is measured in a locally inertial frame, is encoded in its energy–momentum tensor $T_{\mu\nu}$, which is a 'local' quantity since it is a function of the point.

This local energy density curves spacetime according to the Einstein equations, and this curvature is interpreted as the gravitational field created by the matter. The energy of this gravitational field cannot be described by a tensor. Indeed, it cannot be localized because the equivalence principle requires that the gravitation be locally effaced in freely falling reference frames.

Here we have revealed the remarkable fact that the *total* energy of the matter *and* the gravitational field that it creates, that is, the *inertial mass* $M_{\rm in}$ of the system, is in fact determined by the *asymptotic* behavior of the gravitational field.

On the other hand, the asymptotic behavior of the metric, in particular that of the component $g_{00} \approx -(1 + 2U/c^2)$, where U is the Newtonian potential, also determines the *gravitational mass* of the system, because in lowest order $U \approx -GM_{\rm grav}/r$.

In the following chapters we shall give some concrete examples which demonstrate that $M_{\rm in} = M_{\rm grav}$, indicating that general relativity satisfies the *strong equivalence principle*, namely, that the gravitational mass equals the inertial mass when all interactions, including gravity, are taken into account.

Part II

The Schwarzschild solution and black holes

I have read your paper with the utmost interest. I had not expected that one could formulate the exact solution of the problem in such a simple way. I liked very much your mathematical treatment of the subject. Next Thursday I shall present the work to the Academy with a few words of explanation.

A. Einstein to K. Schwarzschild, January 1916,
cited by J. Eisenstaedt, in *Einstein and the History of General Relativity:*
Einstein Studies, Vol. 1. Birkhauser, Boston (1989).

6

The Schwarzschild solution

To find the gravitational potential U produced by a spherically symmetric object in the Newtonian theory, one has to solve the Poisson equation $\triangle U = 4\pi G\varrho$, where the matter density ϱ and U depend only on the radial coordinate r and possibly on the time t. Outside the source the solution is $U = -GM/r$, where $M = 4\pi \int \varrho r^2 dr$ is the source mass. In general relativity the problem is to find the 'spherically symmetric' spacetime solutions of the Einstein equations, and the analog of the vacuum solution $U = -GM/r$ is the Schwarzschild metric.

6.1 The Schwarzschild metric

Static and spherically symmetric spaces

A spacetime is said to be stationary if there exist coordinate systems in which none of the metric coefficients depends on the time-like coordinate t. The length element is then written as (where $i = \{1, 2, 3\}$)

$$ds^2 = g_{00}(x^k)\, dt^2 + 2g_{0i}(x^k)\, dt\, dx^i + g_{ij}(x^k)dx^i dx^j \,.$$

A spacetime is said to be static if the length element is, in addition, invariant under time reversal, that is, if $g_{0i} = 0$:

$$ds^2 = g_{00}(x^k)\, dt^2 + g_{ij}(x^k)\, dx^i dx^j \,.$$

A spacetime is spherically symmetric if there exist spatial coordinates $x^i \equiv (R, \theta, \phi)$ (where the angles θ and ϕ respectively vary from 0 to π and from 0 to 2π) such that the spatial sections $t = \text{const}$ can be 'foliated' by 2-spheres labeled by R,

$$ds^2|_t = g_{RR}(R)dR^2 + f(R)(d\theta^2 + \sin^2\theta\, d\phi^2) \,,$$

where the coordinate R can be redefined such that, for example, $f(R) = r^2$ with $r \in [0, \infty[$.

Spacetimes whose length element can be cast into this form possess a Killing vector (see Section 5.1) corresponding to invariance under time translations and having components $\xi^\mu_{(t)} = (1, 0, 0, 0)$, and three other Killing vectors corresponding to the isometries of 2-spheres at constant t and r [if we limit ourselves to the $\theta = \pi/2$ 'plane', they reduce to $\xi^\mu_{(\phi)} = (0, 0, 0, 1)$; see (5.10)].

Let us consider a static, spherically symmetric spacetime. In *Schwarzschild–Droste* adapted coordinates, the length element is written as

$$ds^2 = -\mathrm{e}^\nu dt^2 + \mathrm{e}^\lambda dr^2 + r^2(d\theta^2 + \sin^2\theta d\phi^2) \,, \tag{6.1}$$

where ν and λ are functions only of r. The circumference of a sphere of radius r is $2\pi r$, but since spacetime is *a priori* curved, the ratio of the circumference of a sphere and its

Relativity in Modern Physics. Nathalie Deruelle and Jean-Philippe Uzan.
© Oxford University Press 2018. Published in 2018 by Oxford University Press.
DOI: 10.1093/oso/9780198786399.001.0001

proper radius, $\int dr \exp(\lambda/2)$, is not necessarily 2π. Moreover, the relation between the time coordinate t and the proper time τ of clocks at rest in the chosen system can vary from point to point as $\tau = t \exp(\nu/2)$. These functions ν and λ must be determined by the Einstein equations.

Outside the matter the Einstein equations are $G_{\mu\nu} = 0$. The Einstein tensor is calculated using its definition as a function of the Christoffel symbols (see the definitions of the Ricci and Einstein tensors in Section 2.6, the definition of the Riemann tensor in Section 2.2, and the definition of the Christoffel symbols in Section 2.4):

$$\Gamma^t_{tr} = \frac{\nu'}{2}; \quad \Gamma^r_{rr} = \frac{\lambda'}{2}; \quad \Gamma^\theta_{\theta r} = \Gamma^\phi_{\phi r} = \frac{1}{r}; \quad \Gamma^r_{tt} = \frac{1}{2}e^{\nu-\lambda}\nu';$$

$$\Gamma^r_{\theta\theta} = \frac{\Gamma^r_{\phi\phi}}{\sin^2\theta} = -re^{-\lambda}; \quad \Gamma^\varphi_{\theta\varphi} = -\frac{\Gamma^\theta_{\varphi\varphi}}{\sin^2\theta} = \frac{\cos\theta}{\sin\theta},$$

(6.2)

where the prime denotes the derivative with respect to r. From this we can extract the nonzero components of the Einstein tensor:

$$\begin{cases} G_{tt} = \frac{1}{r^2}e^\nu \frac{d}{dr}\left[r\left(1 - e^{-\lambda}\right)\right]; \quad G_{rr} = -\frac{1}{r^2}e^\lambda\left(1 - e^{-\lambda}\right) + \frac{1}{r}\nu'; \\[2mm] G_{\theta\theta} = \frac{1}{2}r^2 e^{-\lambda}\left[\nu'' + \frac{(\nu')^2}{2} + \frac{\nu'}{r} - \frac{\nu'\lambda'}{2} - \frac{\lambda'}{r}\right]; \quad G_{\phi\phi} = \sin^2\theta\, G_{\theta\theta}. \end{cases}$$

(6.3)

Integration of $G_{tt} = 0$ gives $e^{-\lambda} = 1 - r_g/r$, where r_g is an integration constant, and integration of $G_{rr} = 0$ gives $\nu = -\lambda + C$. We then see that the equation $G_{\theta\theta} = 0$ is satisfied identically (actually, owing to the Bianchi identities). Finally, the constant C can be absorbed in a redefinition of the coordinate t so that in the end

$$ds^2 = -\left(1 - r_g/r\right)c^2 dt^2 + \frac{dr^2}{1 - r_g/r} + r^2(d\theta^2 + \sin^2\theta d\phi^2). $$

(6.4)

This is the *Schwarzschild metric* (1916).[1]

When $r \to \infty$ this metric tends to the Minkowski metric in spherical coordinates. The coordinates (t, r, θ, ϕ) then measure the proper time of observers at infinity, as well as the proper distances separating them. In the Newtonian limit we must have $-g_{00} \sim 1 + 2U/c^2$, where $U = -GM/r$ is the Newtonian potential (see Sections 1.6 and 4.1). The *Schwarzschild*

---

[1]The derivation we have given here is that of Droste (December 1916). Schwarzschild himself had solved the Einstein equations using the coordinate system (x_0, x_1, x_2, x_3) in which the static, spherically symmetric metric is written as
$$ds^2 = -f_0 dx_0^2 + f_1 dx_1^2 + f_2[dx_2^2/(1 - x_2^2) + (1 - x_2^2)dx_3^2],$$
where the $f_i = f_i(x_1)$ satisfy $f_0 f_1 f_2^2 = 1$, so that the metric determinant is -1 (this condition had been imposed by Einstein and simplifies, among other things, the calculation of the Ricci tensor). Once the f_i were obtained, he made the change of coordinates $x_3 = \phi$, $x_2 = \cos\theta$, and $r = (3x_1 + r_g^3)^{1/3}$ so as to write the metric in the form (6.4). However, in his mind the 'center' of the field was at $x_1 = 0$ and *not* at $r = 0$.

See Schwarzschild (1916) and Schwarzschild (1916a) for the English translation of his two articles.

radius r_g must therefore be identified as $2GM/c^2$, where M is the (gravitational) mass of the central object giving rise to the field:

$$r_g = \frac{2GM}{c^2} \equiv 2m \,. \tag{6.5}$$

We note that the metric coefficients are singular at $r = r_g$, which is therefore called the *Schwarzschild singularity*. However, like Schwarzschild himself, we note that the radius of a star like the Sun ($R_\odot \approx 700{,}000$ km), within which the vacuum solution is no longer valid, is much larger than $r_g^\odot \approx 3$ km.

The Birkhoff–Jebsen theorem

Let us recall the hypotheses made by Schwarzschild in obtaining his solution. He assumed that the (four-dimensional) spacetime is static and spherically symmetric, and that it is 'Ricci-flat' ($R_{\mu\nu} = 0$). We note that it is not necessary to assume that spacetime is asymptotically flat.

Jebsen (in 1921) and then, independently, Birkhoff (in 1923) showed that the unique (local) solution of the vacuum Einstein equations which is only spherically symmetric (and not necessarily static) is in fact the Schwarzschild solution. This is the relativistic version of one of the aspects of Newton's theorem; see Book 1, Section 11.4.

Exercise 1: Show that any two-dimensional metric is diagonalizable and that therefore we can use the following ansatz for the metric of a spherically symmetric spacetime: $ds^2 = -g_{tt}(t,r)dt^2 + g_{rr}(t,r)dr^2 + r^2(d\theta^2 + \sin^2\theta d\phi^2)$.

Exercise 2: Now, by calculating the Einstein tensor, show that the vacuum equations require that λ and ν be independent of time, and so their solution is the Schwarzschild solution.

6.2 Static, spherically symmetric stars

In the study of stellar structure, that is, the stellar mass and radius as a function of the state of the matter composing a star, general relativity is needed only if the parameter r_g/R, where R is the stellar radius, is of order 1. (We note that this ratio is 1.4×10^{-9} for the Earth and 4×10^{-6} for the Sun, but about 0.2 for neutron stars.) We see from the expression for the Schwarzschild metric that this means that the spacetime must be sufficiently curved in order for relativistic effects to significantly modify the Newtonian results. Since it is these relativistic effects that we are interested in, here we shall limit ourselves to obtaining the equations determining the structure of a star composed of a perfect relativistic fluid, which we assume to be static, spherically symmetric, and non-rotating.

The metric of a static, spherically symmetric spacetime is written in Droste coordinates as [*cf.* (6.1)] $ds^2 = -e^\nu dt^2 + e^\lambda dr^2 + r^2(d\theta^2 + \sin^2\theta d\phi^2)$, where $\nu = \nu(r)$ and $\lambda = \lambda(r)$. The energy–momentum tensor of a perfect fluid is (see Section 3.2) $T_{\mu\nu} = (\epsilon + p)u_\mu u_\nu + pg_{\mu\nu}$, where $\epsilon = \epsilon(r)$ and $p = p(r)$ are the energy density and pressure of the star. Since the fluid is static, its velocity vector u^μ is $u^\mu = (u^0, 0, 0, 0)$, and since $u_\mu u^\mu = -1$ we have $u^0 = e^{-\nu/2}$. The nonzero components of $T_{\mu\nu}$ then are

$$T_{00} = \epsilon\, e^\nu \,; \quad T_{rr} = p\, e^\lambda \,; \quad T_{\theta\theta} = r^2 p \,; \quad T_{\phi\phi} = \sin^2\theta\, T_{\theta\theta} \,. \tag{6.6}$$

The stellar structure is therefore described by four functions ν, λ, ϵ, and p, which must be determined by the Einstein equations and the state of the matter making up the star.

Book 3

The first relation comes from the conservation of $T_{\mu\nu}$ imposed by the Bianchi identities $(D_\nu T^{\mu\nu} = 0)$, which in the case of spherical symmetry reduce to $D_\mu T_r^\mu = \partial_\mu T_r^\mu + \Gamma^\mu_{\mu\nu} T_r^\nu - \Gamma^\nu_{\mu r} T_\nu^\mu = 0$ with the Christoffel symbols given in (6.2). This relation is

$$(\epsilon + p)\frac{d\nu}{dr} = -2\frac{dp}{dr} . \tag{6.7}$$

The (00) and (rr) components of the Einstein equations $G_{\mu\nu} = \kappa T_{\mu\nu}$ give two more equations [see (6.3) for the nonzero components of the Einstein tensor in Droste coordinates]:

$$2\frac{d\mu}{dr} = r^2\epsilon , \quad \frac{d\nu}{dr} = \kappa\frac{2\mu + r^3 p}{r(r - 2\kappa\mu)} , \tag{6.8}$$

where $\quad \mu(r) \equiv \dfrac{r}{2\kappa}(1 - e^{-\lambda})\quad$ is the *mass function*.

Eliminating $d\nu/dr$ from (6.7) and (6.8), we can replace the second equation in (6.8) by the *Tolman–Oppenheimer–Volkov equation* (1939), and the system to be integrated becomes

$$2\frac{d\mu}{dr} = r^2\epsilon , \quad 2\frac{dp}{dr} = -\kappa(\epsilon + p)\frac{2\mu + r^3 p}{r(r - 2\kappa\mu)} , \tag{6.9}$$

for a given equation of state $p = p(\epsilon)$.

To integrate (6.9) numerically and find $\mu(r)$ and $\epsilon(r)$ and then $p(\epsilon(r))$, we actually need only a single initial condition, namely, the central density $\epsilon(0)$. Indeed, $\mu(0)$ must be zero because the spacetime, and $g_{rr} = e^\lambda = (1 - 2\kappa\mu(r)/r)^{-1}$ in particular, must be regular at $r = 0$. Therefore, to determine the structure of a relativistic star which is static and spherically symmetric, it is sufficient to specify an equation of state and a central density. The boundary of the star is the surface where the pressure and density vanish, and so the stellar radius R is defined by $p(R) = \epsilon(R) = 0$.

Outside the star the metric must be the spherically symmetric vacuum solution of the Einstein equations, that is, the Schwarzschild metric (6.4) for which $e^\nu = e^{-\lambda} = 1 - 2GM/r \equiv 1 - 2m/r$.

The matching conditions were determined in Section 4.3. In Gaussian coordinates the exterior and interior metrics are written as

$$d\sigma^2 = -e^\nu dt^2 + d\rho^2 + r^2(\rho)(d\theta^2 + \sin^2\theta d\phi^2),$$

with $e^{\lambda/2}dr = d\rho$. The metrics induced on the surface of the star must be continuous, and so ν must be continuous at $r = R$. Moreover, the extrinsic curvatures must also be continuous. Their components are

$$K_{tt} = -\frac{1}{2}\partial_\rho e^\nu = -\frac{1}{2}\nu' e^{(-\lambda/2 + \nu)} \quad \text{and} \quad K_{\theta\theta} = \frac{1}{2}\partial_\rho r^2 = r e^{-\lambda/2} . \tag{6.10}$$

The continuity of $K_{\theta\theta}$, that is, of λ, gives [using the definition of μ in (6.8) and the first equation in (6.9)]

$$m = \kappa\mu(R) = 4\pi G\int_0^R \epsilon\, r^2 dr . \tag{6.11}$$

The continuity of K_{tt} reduces to continuity of ν', which is guaranteed by the second equation in (6.8) evaluated at $r = R$, where $p = 0$.

6.3 A 'star' of constant density

The equation of state $\epsilon = $ const in the interior of a star (and $\epsilon = 0$ outside it), while of course not very realistic, does possess the features of many models. In this case, the mass function (6.9) is simply $\mu(r) = r^3 \epsilon/6$, which gives $e^\lambda = (1 - \kappa \epsilon r^2/3)^{-1}$. Equation (6.7) is written as $d\nu = -2dp/(\epsilon + p)$ and gives $\epsilon + p = Be^{-\nu/2}$, where B is a constant. Finally, the equation (6.8) for ν, after setting $z = e^{\nu/2} - 3B/2\epsilon$, gives $e^\nu = (3B/2\epsilon - D\sqrt{1 - \kappa \epsilon r^2/3})^2$, where D is another constant.

The conditions for matching to the exterior Schwarzschild metric are the same as before: λ, ν, and ν' must be continuous and p must vanish at $r = R$. This implies that $\kappa \epsilon/3 = 2m/R^3$, $B = \epsilon \sqrt{1 - 2m/R}$, and $D = 1/2$.

We thus obtain the metric in the interior of the star (Schwarzschild, 1916):

$$
\begin{cases}
ds^2 = -\left(\dfrac{3}{2}\sqrt{1 - 2m/R} - \dfrac{1}{2}\sqrt{1 - 2mr^2/R^3}\right)^2 dt^2 + \dfrac{dr^2}{1 - 2mr^2/R^3} \\
\qquad + r^2(d\theta^2 + \sin^2\theta d\phi^2)\,, \\[2ex]
p = \epsilon \dfrac{\sqrt{1 - 2mr^2/R^3} - \sqrt{1 - 2m/R}}{3\sqrt{1 - 2m/R} - \sqrt{1 - 2mr^2/R^3}}\,.
\end{cases}
\tag{6.12}
$$

This solution is valid if the stellar radius R is larger than the star's Schwarzschild radius, which is the case for objects such as the Earth, the Sun, white dwarfs, and neutron stars.[2]

The proper mass of a star

The Schwarzschild mass (6.11), namely, $M = 4\pi \int_0^R \epsilon(r) r^2 dr$, which is the *gravitational* mass of the star, is *not* equal to the star's *proper* mass, namely, the integral of the density over the proper volume: $M_{\mathrm{p}} = 4\pi \int_0^R \epsilon(r) r^2 \left[1 - 2\kappa\mu(r)/r\right]^{-\frac{1}{2}} dr$; cf. (5.18). The difference can be interpreted as a gravitational binding energy, because this is what it reduces to in the Newtonian limit. When the density is constant, we have $M = 4\pi R^3 \epsilon/3$ and

$$
M_{\mathrm{p}} = \frac{3M}{2(2GM/R)^{3/2}}\left(\mathrm{ArcSin}\sqrt{\frac{2GM}{R}} - \sqrt{\frac{2GM}{R}}\sqrt{1 - \frac{2GM}{R}}\right) = M\left(1 + \frac{3GM}{5R} + \cdots\right),
$$

so that $M - M_{\mathrm{p}} = -3GM^2/5R + \cdots$, which is indeed the Newtonian gravitational energy of the object; cf. Book 1, Section 15.2. This result is an indication that in general relativity the gravitational mass of a body, here its Schwarzschild mass M, is equal to its *inertial* mass, the sum of its proper mass M_{p} and its gravitational energy.

In addition, we note that the quantity $Q_t \equiv \int \epsilon \sqrt{-g}\, d^3x$, defined in (5.17) using $\xi^\nu_{(t)} D_\mu T^\mu_\nu = 0$, is $Q_t = M(1 - 3GM/5R + \cdots)$ for the Schwarzschild metric. The *Tolman mass*, equal to the *Komar mass* defined in (5.28), $I_t \equiv (\alpha/2)\int(\epsilon + 3p)\sqrt{-g}d^3x$, is $I_t = M$ for $\alpha = 2$.

---

[2] We note that the spatial sections are 3-spheres. As shown by Shepley and Taub (1967) as well as Buchdahl (1971) [see also Shankar and Whiting (2007)], this metric is also *conformally flat*, that is, it can be written as $ds^2 = \Omega^2[-dT^2 + dR^2 + R^2(d\theta^2 + \sin^2\theta d\phi^2)]$, where the *conformal factor* Ω is a known function of (T, R) under a suitable change of coordinates $(t, r) \to (T, R)$.

We also note that if $R = 2m$, then $p = -\epsilon$ and the metric becomes the de Sitter metric; see Section 17.4 (but it can never be matched to the Schwarzschild metric because p is never zero).

The inertial mass of a star

In Section 5.4 we saw how to associate a spacetime with Noether charges which, when the spacetime is asymptotically stationary, are identified as the 'inertial' mass. Let us calculate the mass $M_{\rm in}$ of a spacetime whose metric for $r \to \infty$ is written as

$$ds^2 \approx -(1 - 2GM/r + \cdots) + (1 + 2\gamma GM/r + \cdots)dr^2 + r^2(d\theta^2 + \sin^2\theta d\phi^2)$$

(which reduces to the asymptotic Schwarzschild metric for $\gamma = 1$) using the Katz superpotential; cf. (5.25):

$$M_{\rm in} = -\int_S \hat{J}^{[tr]}d\theta d\phi\,, \quad \text{where} \quad \kappa \hat{J}^{[\mu\nu]} = D^{[\mu}\hat{\xi}^{\nu]} - \overline{D^{[\mu}\hat{\xi}^{\nu]}} + \xi^{[\mu}\hat{k}^{\nu]}.$$

Here S is the 2-sphere at infinity; the hat denotes multiplication by $\sqrt{-g}$; a bar indicates that the quantity is evaluated in Minkowski spacetime in spherical coordinates; $\xi^\mu = (1,0,0,0)$ is the Killing vector associated with the stationarity of the spacetime and therefore satisfying $D_\mu \xi_\nu + D_\nu \xi_\mu = 0$; and, finally, k^μ is the Katz vector:

$$k^\mu = -(g^{\nu\rho}\Delta^\mu_{\nu\rho} - g^{\mu\nu}\Delta^\rho_{\nu\rho})\,, \quad \text{where} \quad \Delta^\mu_{\nu\rho} \equiv \Gamma^\mu_{\nu\rho} - \bar{\Gamma}^\mu_{\nu\rho}\,.$$

We have $D^{[0}\hat{\xi}^{r]} = -\sqrt{-g}\,g^{tt}g^{rr}\partial_r g_{tt}/2 = -GM\sin\theta + \mathcal{O}(1/r)$ and $\overline{D^{[0}\hat{\xi}^{r]}} = 0$.

Moreover, $k^r = -(g^{tt}\Delta^r_{tt} + 2g^{\theta\theta}\Delta^r_{\theta\theta}) + g^{rr}(\Delta^t_{rt} + 2\Delta^\theta_{r\theta})$ with $\Delta^r_{tt} = GM/r^2 + \cdots$, $\Delta^r_{\theta\theta} = 2\gamma GM + \cdots$, $\Delta^t_{rt} = GM/r^2 + \cdots$, and $\Delta^\theta_{r\theta} = 0$, so that $\hat{\xi}^{[t}k^{r]} = -GM(2\gamma - 1)\sin\theta + \cdots$.

Assembling these results, we find

$$M_{\rm in} = -\int_S \frac{d\theta d\phi}{\kappa}\, GM\sin\theta(-1 - (2\gamma - 1)) = \gamma M\,.$$

Therefore, since $\gamma = 1$ we find that the inertial mass of the star (that is, the energy it contains) is indeed equal to its gravitational mass, as the post-Newtonian calculation done above suggests (the first term is half the *Komar mass*; see Section 5.4). Therefore, at least in this particular case, general relativity satisfies the *strong equivalence principle*.

At a given time t and in the 'plane' $\theta = \pi/2$, the spatial geometry in Schwarzschild coordinates is determined outside and inside the star by

$$d\sigma^2_{\rm ext} = dr^2\left(1 - \frac{2m}{r}\right)^{-1} + r^2 d\phi^2\,, \quad d\sigma^2_{\rm int} = dr^2\left[1 - \frac{2m}{R}\left(\frac{r}{R}\right)^2\right]^{1} + r^2 d\phi^2\,. \tag{6.13}$$

These two-dimensional spaces can be embedded in a three-dimensional Euclidean space of metric $dS^2 = dr^2 + r^2 d\phi^2 + dz^2$. Indeed, the metric induced on a surface $z = z(r)$ is $d\sigma^2 = (1 + z'^2)dr^2 + r^2 d\phi^2$, where $z' \equiv dz/dr$.

On the exterior we have $1 + z'^2 = (1 - 2m/r)^{-1}$ and the surface is a paraboloid of equation $z^2 = 8m(r - 2m)$.

In the interior the geometry is that of a sphere, regular everywhere, including at $r = 0$. Indeed, setting $\sqrt{2m/R}\,(r/R) = \sin\theta$, we see that the metric (6.13) is written as $d\sigma^2_{\rm int} = (R^3/2m)(d\theta^2 + \sin^2\theta d\phi^2)$, which is the metric of a 2-sphere of radius $\sqrt{R^3/2m}$, as noted by Schwarzschild himself. Since $z'^2 = (1 - 2mr^2/R^3)^{-1}$ is continuous at $r = R$, the paraboloid and the sphere join together smoothly.

The global geometry can therefore be visualized as an elastic 'sheet' which is deformed at its center by the 'weight' of the star. This is called *Flamm's paraboloid* (1916) (see Fig. 6.1).

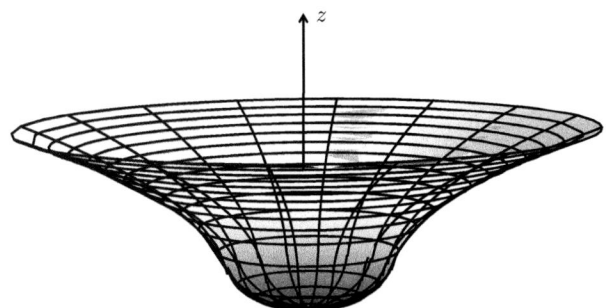

Fig. 6.1 A representation of the Flamm paraboloid.

The central pressure, *cf.* (6.12), is $p(0) = \epsilon(1 - \sqrt{1 - 2m/R})/(3\sqrt{1 - 2m/R} - 1)$. In order for it to remain finite we must have $m < 4R/9$, or, since $m = (4\pi/3)\epsilon R^3$, $m < m_{\rm crit}$ with $m_{\rm crit} = 4/(9\sqrt{3\pi\epsilon})$. If we take $\epsilon = 10^{15}$ g/cm^3, the density of a nucleus, we find $m_{\rm crit} \simeq 4M_\odot$. In the Newtonian limit the pressure $p = \epsilon m/2R$ remains finite.

The concept of critical mass

Within the framework of the Newtonian theory, a polytrope, that is, a body whose equation of state is $p = K\rho^\gamma$, has, if $\gamma = 4/3$, a mass which does not depend on the central density: $M = 2.02(4/\sqrt{\pi})(K/G)^{3/2}$ (see, for example, Book 1, Section 15.3). Frenkel in 1928, and also Anderson and Stoner in 1930, obtained the equation of state of a degenerate Fermi gas at zero temperature, which can describe white dwarfs. When the electron speed becomes close to the speed of light, the star turns out to be a polytrope with $\gamma = 4/3$. The coefficient K is known, $K = (3\pi^2)^{1/3}\hbar c Y_e^{4/3}/4m_{\rm B}^{4/3}$, where $Y_e \approx 0.5$ is the number of electrons relative to the number of baryons and $m_{\rm B}$ is the baryon (proton–neutron) mass. The mass is therefore also known: $M = 1.46M_\odot$. In 1930 Chandrasekhar understood that this is a critical mass: "*A star of large mass cannot pass into the white-dwarf stage, and one is left speculating on other possibilities.*"

The same, at least in order of magnitude, result can be obtained by an energy argument, which Landau made in 1931. We consider a degenerate Fermi gas of N electrons in a sphere of radius R which for simplicity we take to be homogeneous. The internal energy of these electrons is $E_{\rm int} = N\bar{E}$ with $\bar{E} = \int Ep^2 dp/\int p^2 dp$, where the integrals run up to $p_{\rm F}$, the Fermi momentum. In the ultrarelativistic limit $\bar{E} \propto p_{\rm F}$ with $p_{\rm F} \propto N^{1/3}/R$ at zero temperature. Therefore, $E_{\rm int} \propto Np_{\rm F} \propto N^{4/3}/R \propto M^{4/3}/R$. In addition, the gravitational energy of the star in the Newtonian limit is $E_{\rm grav} = -3GM^2/5R$; see Book 1, Section 15.2. The total energy of the star is therefore $E \propto (1 - (M/M_{\rm crit})^{2/3})(M^{4/3}/R)$. The precise calculation gives $M_{\rm crit} \approx 1.7M_\odot$. The equilibrium configuration minimizes E. If $M > M_{\rm crit}$, E is negative. It decreases as R decreases, and so the density increases and R continues to decrease. Thus, according to Landau, "*...the density of matter becomes so great that atomic nuclei come in close contact, forming one gigantic nucleus*", that is, a *neutron star*, a possibility already suggested by Baade and Zwicky in 1933 (we recall that the neutron was discovered by Chadwick in 1932).

To study the equilibrium of neutron stars, for which $2m/R \approx 0.3$, gravitation must be described using general relativity. If such a star is treated as a degenerate Fermi gas of neutrons at zero temperature, then the critical Chandrasekhar or Landau masses obtained when gravity is treated using Newtonian physics are four times larger than the critical mass of a white dwarf (because $M \propto Y_e^2$, where now $Y_e = 1$) or about $6M_\odot$, which does not agree with observation. If, on the other hand, the same gas is described using general relativity and the TOV equations (6.9) are integrated (numerically), we obtain $M_{\text{crit}} \approx 0.7M_\odot$ (Oppenheimer–Volkov, 1939), which is also in disagreement with observation .... The equations of state at these densities are in fact poorly known, but none of them admits a stable configuration beyond a neutron star, the mass of which is estimated to be of the order of a few solar masses.[3]

A star whose mass exceeds the critical mass, for example, owing to accretion processes, will collapse on itself. If the collapse is spherically symmetric, the gravitational field outside the star will still be represented by the Schwarzschild metric (6.4), owing to the Birkhoff–Jensen theorem. However, when the radius of the object becomes less than $2m$, this metric contains pathologies. It took about fifty years to understand the nature of these pathologies, and in the 1960s the idea of a 'black hole' was born.

Currently, about a dozen good candidates for black holes of the stellar type have been found. The best known is Cygnus X-1, discovered in the 1970s, and the currently most convincing candidate is GS 2023+338 (V 404 Cygni). Quasars are probably also black holes (of 10^6 to 10^9 solar masses), and many galaxies contain extremely massive black holes at their center, like our own Sagittarius A*.[4]

6.4   Gravitational collapse and black holes

Let us assume that a star has exceeded its critical mass and collapses on itself while preserving its spherical symmetry. The exterior gravitational field is then the Schwarzschild field and, since the pressure in the interior rapidly becomes negligible compared to the gravitational attraction, the surface can be described as an ensemble of particles in free fall. This corresponds to the Oppenheimer–Snyder model.[5]

The motion of a test particle in a gravitational field is governed, as we saw in Section 3.1, by the geodesic equation

$$\frac{Du^\mu}{d\tau} \equiv \frac{du^\mu}{d\tau} + \Gamma^\mu_{\nu\rho} u^\nu u^\rho = 0 \,, \tag{6.14}$$

where $u^\mu = dx^\mu/d\tau$ is the 4-velocity of the world line $x^\mu = x^\mu(\tau)$.

Let us consider radial motion in a static, spherically symmetric field. With the Christoffel symbols given in Section 6.1, eqn (6.14) for the component u^0 reduces to $du^0/d\tau + 2\Gamma^t_{tr} u^0 u^r = 0$ because $u^\theta = u^\phi = 0$, or $du^0/d\tau + \nu' u^0 u^r = 0$, or also $du^0/d\tau + u^0 d\nu/d\tau = 0$, since $u^r = dr/d\tau$. The solution is $u^0 = Ce^{-\nu}$, where C is a constant [the same result is obtained using the static nature of the solution; see (5.11)].

Moreover, if τ is the proper time along the world line, the 4-velocity u^μ is normalized, $g_{\mu\nu} u^\mu u^\nu = -1$, or $-e^\nu (u^0)^2 + e^\lambda (u^r)^2 = -1$.

If the metric is the Schwarzschild metric we have $e^\nu = e^{-\lambda} = 1 - 2m/r$, and then the radial geodesic equation reduces to

---

[3]To learn more about the structure of relativistic stars, see, for example, Collins (2003), Chapter 6, and also Grandclément (2008).

The quotations are from Chandrasekhar (1934) and Landau (1932).

[4]See, for example, Mueller (2007) for more details.

[5]Oppenheimer and Snyder (1939). Incidentally, the same issue (dated 1 September) contains the article by Bohr and Wheeler entitled *The Mechanism of Nuclear Fission*.

$$\left(1 - \frac{2m}{r}\right)\frac{dt}{d\tau} = \sqrt{1 - \frac{2m}{R}} \; ; \quad \left(\frac{dr}{d\tau}\right)^2 = -\frac{2m}{R} + \frac{2m}{r}, \tag{6.15}$$

where τ is the proper time of an 'observer' on the surface of the star and we have used the fact that the observer velocity is zero at R, the initial radius of the star. The second equation is integrated in parametric form:

$$r = \frac{R}{2}(1 + \cos\eta) \; ; \quad \tau = \frac{R}{2}\sqrt{\frac{R}{2m}}\,(\eta + \sin\eta), \tag{6.16}$$

where η varies from 0 to π. Therefore, the proper time effectively measured by an observer located on the surface of the star to fall from $r = R$ to $r = 0$ is finite and equal to

$$\tau_{\text{collapse}} = \tau(\pi) - \tau(0) = \frac{\pi}{2}R\sqrt{\frac{R}{2m}}. \tag{6.17}$$

Setting $R = 2mx$, we have $\tau_{\text{collapse}} = \pi m x^{3/2}$. Therefore, a star of the size and mass of the Sun ($m \approx 5 \; \mu s$, $x \approx 2.3 \times 10^5$), will collapse in about half an hour. If the initial radius of the star is of the order of the Schwarzschild radius $[x = \mathcal{O}(1)]$, then $\tau_{\text{collapse}} = \mathcal{O}(m)$, which is a few microseconds for m of the order of a solar mass. Finally, the collapse of the center of a supernova will take a few milliseconds.[6]

Let us now describe the collapse using the coordinate time t, the proper time of an observer at rest (constant r, θ, ϕ) at infinity. From (6.15) we obtain

$$t = -\int_R^r \frac{dr\sqrt{1 - 2m/R}}{(1 - 2m/r)\sqrt{2m/r - 2m/R}} \to \infty \quad \text{when} \quad r \to 2m. \tag{6.18}$$

Therefore, when described using the time t, the collapse takes infinitely long.

More precisely, let us imagine that an observer on the surface of the star and moving along with its collapse sends to the observer at rest a succession of light signals at regular infinitesimal intervals $\Delta\tau_{\text{em}}$ of his proper time. This time $\Delta\tau_{\text{em}}$ corresponds to a coordinate time given by (6.15): $\Delta t_{\text{em}} = \Delta\tau_{\text{em}}(\sqrt{1 - 2m/R})/(1 - 2m/r_{\text{em}})$, where $r_{\text{em}} = r\,(t_{\text{em}})$ is the stellar radius at the instant of emission. On the other hand, for the observer at rest at infinity $\Delta t_{\text{rec}} = \Delta\tau_{\text{rec}}$. The light signals follow the null radial geodesics of the Schwarzschild metric with equation $dr/dt = 1 - 2m/r$. We therefore have $\Delta t_{\text{rec}} = \Delta t_{\text{em}} - \Delta r_{\text{em}}/(1 - 2m/r_{\text{em}})$, where the second term is just the Doppler shift due to the fact that the emitter is moving. Here Δr_{em} and $\Delta\tau_{\text{em}}$ are related by the second expression in (6.15) (with $\Delta r_{\text{em}} < 0$) and so

$$\Delta\tau_{\text{rec}} = \frac{\Delta\tau_{\text{em}}}{1 - 2m/r_{\text{em}}}\left(\sqrt{1 - 2m/R} + \sqrt{2m/r_{\text{em}} - 2m/R}\right), \tag{6.19}$$

which tends to infinity when the star approaches its Schwarzschild radius $r_{\text{em}} = 2m$: the signals are shifted more and more toward the red.

---

[6]As Drumeaux noted in 1936, (6.15) governing the radial fall and its solution (6.16) are the same in Newtonian gravity (see for example, Book 1, Section 16.1), *with the condition that* the proper time τ is replaced by Newton's universal time t.

The description of the gravitational collapse of a dust cloud therefore depends critically on the observer. The observer who moves along with the collapse crosses the surface $r = 2m$ without any problem and reaches $r = 0$, where the density becomes infinite in a finite proper time. The observer who remains far away sees the star become black when it reaches the Schwarzschild radius $r = 2m$, which for this reason is called the *horizon* of the *black hole*.

We have seen that the equations of null radial geodesics are

$$\frac{dr}{dt} = \pm(1 - 2m/r).\tag{6.20}$$

The horizon is therefore a light-like surface where the photons are *at rest* as measured using the time t of an observer at rest at infinity.

6.5   The Lemaître–Tolman–Bondi solution

The spacetime outside a spherically symmetric dust cloud is the Schwarzschild spacetime. On the inside the metric is the solution of the Einstein equations in the presence of a perfect fluid at zero pressure.

Let us therefore consider a spherically symmetric cloud of incoherent matter ($p = 0$) which is not necessarily homogeneous. We seek the solution of the Einstein equations in a *synchronous* frame with coordinates $(\tau, \rho, \theta, \phi)$ such that the metric has the form

$$ds^2 = -d\tau^2 + e^{\lambda(\rho, \tau)}d\rho^2 + r^2(\rho, \tau)(d\theta^2 + \sin^2\theta\, d\phi^2),\tag{6.21}$$

where λ and r are functions of the radial coordinate ρ and the time τ. In this system of *co-moving* coordinates the trajectories $(\rho, \theta\, \phi) = $ const are geodesics and τ is their proper time.[7] The useful components of the Einstein tensor are (the others are either zero or are redundant owing to the Bianchi identities)

$$\begin{cases} G^\rho_\tau = e^{-\lambda}(-2\dot{r}' + \dot{\lambda}r')/r\,, \quad G^\rho_\rho = -(1 + 2r\ddot{r} + \dot{r}^2 - e^{-\lambda}r'^2)/r^2\,, \\ G^\tau_\tau = e^{-\lambda}(2rr'' + r'^2 - rr'\lambda')/r^2 - \dot{r}(r\dot{\lambda} + \dot{r})/r^2 - 1/r^2\,, \end{cases}\tag{6.22}$$

where the dot denotes the derivative with respect to τ and the prime denotes the derivative with respect to ρ. The components of the energy–momentum tensor are all zero except for $T^\tau_\tau \equiv -\epsilon(\rho, \tau)$. Integration of the first expression in (6.22), $G^\rho_\tau = 0$, then gives (unless $r' = 0$)

$$e^\lambda = \frac{r'^2}{1 + 2E}\,,\tag{6.23}$$

where E is an arbitrary function of ρ. The second expression, $G^\rho_\rho = 0$, can be written as $2r\ddot{r} + \dot{r}^2 - 2E = 0$, the first integral of which is

$$\dot{r}^2 = 2E + 2\mu/r\,,\tag{6.24}$$

where $\mu(\rho)$ is a new arbitrary function. Finally, it is usual to write the solutions of (6.24) in parametric form as

---

[7]Indeed, if the initial velocity is $u^\mu_{\text{in}} = (1, 0, 0, 0)$, then the acceleration, given by the geodesic equation, is $du^i/d\tau|_{\text{in}} = -\Gamma^i_{00} = 0$ in a synchronous frame.

$$
\begin{cases}
r = \dfrac{\mu}{-2E}(1 + \cos\eta)\,, & |\tau_0 - \tau| = \dfrac{\mu}{(-2E)^{3/2}}(\eta + \sin\eta)\,, \quad E < 0 \\[3mm]
r = \dfrac{\mu}{2E}(\cosh\eta - 1)\,, & |\tau_0 - \tau| = \dfrac{\mu}{(2E)^{3/2}}(\sinh\eta - \eta)\,, \quad E > 0 \\[3mm]
r^3 = (9\mu/2)(\tau_0 - \tau)^2\,, & E = 0\,,
\end{cases}
\tag{6.25}
$$

where $\tau_0(\rho)$ is a third arbitrary function. The desired metric then is

$$
ds^2 = -d\tau^2 + \frac{r'^2 d\rho^2}{1 + 2E} + r^2 d\Omega_2^2 \quad \text{with} \quad d\Omega_2^2 \equiv d\theta^2 + \sin^2\theta d\phi^2\,,
\tag{6.26}
$$

where $r(\rho, \tau)$ is given in (6.25). Since a redefinition of the coordinate ρ allows one of the functions, for example, $E(\rho)$, to be chosen arbitrarily, the solution will actually depend on two functions describing the physics of the problem, $\mu(\rho)$ and $\tau_0(\rho)$. The energy density follows from the last equation in (6.22), $G^\tau_\tau = 8\pi T^\tau_\tau$:

$$
\frac{4\pi\epsilon}{3} = \frac{\mu'}{(r^3)'}\,.
\tag{6.27}
$$

This solution was first obtained by Lemaître (1933), and then rediscovered by Tolman (1934) and later Bondi (1947).

6.6   The interior Friedmann solution

Just as Schwarzschild assumed a constant matter density in constructing his static stellar model (see Section 6.1), here we shall assume that a star undergoing collapse has zero pressure and is isotropic *but also* homogeneous, that is, its energy density ϵ depends only on the time: $\epsilon = \epsilon(\tau)$.

The Lemaître solution obtained in Section 6.5 then is simplified as follows. First of all, (6.27) implies that $r(\rho, \tau)$ has the form $r = a(\tau)(2\mu(\rho)/a_m)^{1/3}$, where $a(\tau) = (3a_m/8\pi\epsilon(\tau))^{1/3}$ and a_m is a constant. In view of (6.26), we next see that we can choose the coordinate ρ such that $2\mu(\rho) = a_m\rho^3$. The metric (6.26) then becomes $ds^2 = -d\tau^2 + a^2(\tau)[d\rho^2/(1+2E) + \rho^2 d\Omega_2^2]$.

Secondly, if we choose $-2E(\rho) = \rho^2$, then the spatial sections $\tau = $ const will be 3-spheres. Finally, if we choose $\tau_0(\rho) = 0$ and the case where a decreases with τ, the Lemaître solution will reduce to the Friedmann solution (1922):

$$
ds^2 = -d\tau^2 + a^2(\tau)\left(\frac{d\rho^2}{1 - \rho^2} + \rho^2 d\Omega_2^2\right), \quad \text{where} \quad
\begin{cases}
a = \dfrac{a_m}{2}(1 + \cos\eta) \\[3mm]
\tau = \dfrac{a_m}{2}(\eta + \sin\eta)
\end{cases}
\text{and} \quad \epsilon = \frac{3a_m}{8\pi a(\tau)^3}
\tag{6.28}
$$

(with $d\Omega_2^2 \equiv d\theta^2 + \sin^2\theta d\phi^2$).

To complete the description of this dust cloud undergoing collapse, we still need to match the exterior (Schwarzschild) and interior solutions. To do this, we calculate the circumference of the star in Droste coordinates: $C = 2\pi r$, where $r(\tau)$ is given by the geodesic equation (6.16). However, in the coordinates used in (6.28) it is written as $C = 2\pi\rho_0 a(\tau)$, where ρ_0, the coordinate radius of the star, is also a geodesic. Equating the two expressions for these coordinate choices, we obtain $R = \rho_0 a_m$ and $R\sqrt{R/2m} = a_m$, thus obtaining a_m and ρ_0 as functions of R and m, the initial radius and mass of the star. (It can be verified explicitly that the extrinsic curvature of the surface of the star, defined in Section 4.3, is indeed continuous.)

Therefore, the geometry and gravitation in the interior of this cloud are homogeneous and isotropic. There is no privileged direction (no more than on the surface of a 2-sphere). Nevertheless, it possesses a center because it has finite extent (like a spherical cap). As seen by a freely falling observer, its volume decreases from $2\pi^2 R^3$ to zero, while the density ϵ grows from $3m/4\pi R^3$ to infinity.

7

The Schwarzschild black hole

At the end of its thermonuclear evolution, a star collapses and, if it is sufficiently massive, does not become stabilized in a new equilibrium configuration. We have described an example of this collapse in the preceding chapter. The Schwarzschild geometry therefore represents the gravitational field of such an object *up to* $r = 0$. The Schwarzschild metric in its original form is however singular, not only at $r = 0$ where the curvature diverges,[1] but also at $r = 2m$, a surface which can nevertheless be crossed by geodesics. We shall see that by a judicious change of coordinates it is possible to eliminate this singularity and reveal an extended spacetime, that of a *black hole*.

7.1 The Lemaître coordinates

As we have seen in Section 6.5, the Lemaître–Tolman–Bondi metric (6.25)–(6.27) describes a spherically symmetric spacetime. It depends on three functions: $E(\rho)$, $\mu(\rho)$, and $\tau_0(\rho)$.

We shall require that it describe a space devoid of matter. Then the function μ reduces to a constant, $\mu = m$ [*cf.* (6.27)], and according to the Jebsen–Birkhoff theorem (see Section 6.1) the metric can be transformed into the Schwarzschild metric no matter what the functions $E(\rho)$ and $\tau_0(\rho)$ are.

Let us choose, as did Lemaître in 1933, $E = 0$ and $\tau_0 = \rho$. Then the metric reduces to [*cf.* (6.25)–(6.26)]

$$ds^2 = -d\tau^2 + \frac{d\rho^2}{(r/2m)} + r^2(d\theta^2 + \sin^2\theta d\phi^2) \text{ with } r(\rho, \tau) = 2m\left[\left(\frac{3}{4m}\right)(\rho - \tau)\right]^{\frac{2}{3}}. \quad (7.1)$$

The metric is now manifestly regular on the surface $r = 2m$ of equation $\rho - \tau = 4m/3$.

We still need to specify the change of coordinates $(\tau, \rho) \rightarrow (t, r)$.

First of all, from (7.1) we find $dr = \sqrt{2m/r}(d\rho - d\tau)$.

Next we need to find the functions $f(\tau, \rho)$ and $g(\tau, \rho)$ defining the differential $dt = f(\tau, \rho)d\tau + g(\tau, \rho)d\rho$.

For this we use the fact that in the co-moving system (τ, ρ), the coordinate lines $\rho = \rho_0$ are radial geodesics of proper time τ whose equations in Droste coordinates are known [see (6.15)]: $dr|_{\rho_0} = \pm\sqrt{2m/r - 2m/R} \, d\tau$ and $dt|_{\rho_0} = \sqrt{1 - 2m/R}(1 - 2m/r)^{-1}d\tau$.

The identification of $dr|_{\rho_0} = -\sqrt{2m/r} \, d\tau$ with $dr|_{\rho_0} = \pm\sqrt{2m/r - 2m/R} \, d\tau$ requires that we take the minus sign and choose $R = \infty$, so that we have $dt|_{\rho_0} = (1 - 2m/r)^{-1} d\tau$. This gives $f(\tau, \rho) = (1 - 2m/r)^{-1}$.

Finally, the integrability condition $\partial f/\partial\rho = \partial g/\partial\tau$ easily gives $g(\tau, \rho) = -2m/(r - 2m)$.

In the end, after inverting we obtain the desired result:

---

[1]The *Kretschmann invariant*, which will be calculated in Section 25.6, is $R_{\mu\nu\rho\sigma}R^{\mu\nu\rho\sigma} = 48m^2/r^6$.

Relativity in Modern Physics. Nathalie Deruelle and Jean-Philippe Uzan.
© Oxford University Press 2018. Published in 2018 by Oxford University Press.
DOI: 10.1093/oso/9780198786399.001.0001

$$\begin{cases} \tau = t + \displaystyle\int dr\, \frac{\sqrt{2m/r}}{1 - 2m/r} = t + 2m \left[2\sqrt{r/2m} + \ln\left(\frac{|\sqrt{r/2m} - 1|}{\sqrt{r/2m} + 1} \right) \right] , \\[3mm] \rho = t + \displaystyle\int \frac{dr}{(1 - 2m/r)\sqrt{2m/r}} \\[3mm] \quad = t + 2m \left[\sqrt{r/2m}\,(2 + r/3m) + \ln\left(\frac{|\sqrt{r/2m} - 1|}{\sqrt{r/2m} + 1} \right) \right] . \end{cases} \qquad (7.2)$$

The transformation is singular at $r = 2m$. The spacetime described by the metric (7.1) is therefore an 'extension' of that described by the Schwarzschild metric in its original form (just as the Euclidean plane in Cartesian coordinates is an extension of the plane in polar coordinates because it includes the origin).

The Schwarzschild 'singularity'

The singularity of the Schwarzschild metric (6.3) at $r = r_g = 2m$ has been studied since 1916. It is not eliminated by a change of coordinate $r \to r(\rho)$, as we can see from the following example:

$$r > 2m \;:\; ds^2 = -(1 - 2m/r(\rho))dt^2 + d\rho^2 + r^2(\rho)d\Omega_2^2$$
$$\text{with} \quad r = m(1 + \cosh u) \,, \;\; \rho = m(u + \sinh u)$$

$$r < 2m \;:\; ds^2 = -d\rho^2 + (2m/r(\rho) - 1)dt^2 + r^2(\rho)d\Omega_2^2$$
$$\text{with} \quad r = m(1 - \cos u) \,, \;\; \rho = m(u - \sin u) \,.$$

For $r < 2m$, $\rho \to \tau$ is a time coordinate and $t \to \rho$ is a radial coordinate. The metric, of the Kantowski–Sachs type, is a particular case of the Lemaître–Tolman–Bondi metric obtained by requiring that r depend only on $\rho \equiv \tau$; see Section 6.5.

In 1921 Painlevé and Gullstrand made the same change of time coordinate $t \to \tau(t, r)$ as Lemaître, *cf.* (7.2), without touching the radial coordinate and obtained $ds^2 = -(1 - 2m/r)d\tau^2 + 2\sqrt{2m/r}\,dr d\tau + d\vec{x}^2$. (In these coordinates the spatial sections are Euclidean.) They concluded that the Schwarzschild geometry is 'ambiguous'.

In 1924 Eddington suggested the form (rediscovered by Finkelstein in 1958)

$$ds^2 = -(1 - 2m/r)d\bar{t}^2 + (4m/r)dr d\bar{t} + (1 + 2m/r)dr^2 + r^2 d\Omega_2^2$$

with $\bar{t} = t + 2m \log(r/2m - 1)$, which by setting $\bar{t} = v - r$ can be simplified to

$$ds^2 = -(1 - 2m/r)dv^2 + 2dr dv + r^2 d\Omega_2^2 \,. \qquad (7.3)$$

In these systems the metric coefficients no longer diverge at $r = 2m$ (but some vanish). We note that when $r \to 2m$ and $t \to +\infty$, the value of the coordinate v can remain finite.

7.2   The Kruskal–Szekeres extension

Before describing the Kruskal–Szekeres extension,[2] let us do a preparatory exercise.

---

[2]Kruskal, 1960; Szekeres, 1960; among the precursors is Synge, 1949.

From Rindler spacetime to $\mathcal{M}_4$

We have stated that the spacetime of a black hole is much 'larger' than the spacetime covered by the Schwarzschild coordinates. As a preliminary exercise, we shall use the results of Book 2, Section 5.2 to show how the spacetime covered by the Rindler coordinates can be 'extended' to all of Minkowski spacetime.

We consider the two-dimensional metric

$$d\sigma^2 = -x^2 dt^2 + dx^2\,, \quad t \in [-\infty, +\infty]\,, \quad x \in]0, +\infty]\,. \tag{7.4}$$

Here we recognize the Rindler metric (see Book 2, Section 5.2; we have set $x = 1 + gz$ and $g = 1$). It is singular at $x = 0$. This singularity, which is the analog of the origin of the Euclidean plane in polar coordinates, is associated with the choice of coordinate system and not with the structure of spacetime (because the Riemann tensor is zero). The problem here is to find a new system of coordinates analogous to the Cartesian coordinates of the Euclidean plane which are regular everywhere (see Fig. 7.1).

To do this,[3] we consider the light-like $(d\sigma^2 = 0)$ geodesics of equation $t = \pm \ln x + \text{const}$, and introduce the so-called *null* coordinates

$$u = t - \ln x\,, \quad v = t + \ln x\,, \quad u, v \in [-\infty, +\infty]\,, \tag{7.5}$$

in which the metric (7.4) becomes

$$d\sigma^2 = -e^{v-u} du\, dv\,.$$

Next we make the new change of coordinates

$$U = -e^{-u}\,, \quad V = e^v\,, \quad U \in [-\infty, 0]\,, \quad V \in [0, +\infty]\,. \tag{7.6}$$

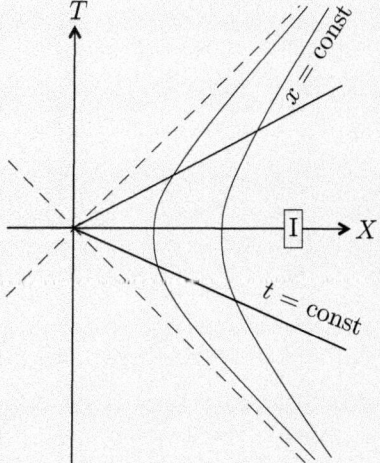

Fig. 7.1 Rindler and Minkowski spacetimes.

---

[3] Here we follow the presentation of Wald (1984), p. 149 *et seq.*

The metric then becomes $d\sigma^2 = -dU\,dV$. This metric is regular everywhere and we can *extend* the ranges of variation of U and V to *the entire* (U,V) *plane*. A final change of coordinates $T = (U+V)/2$, $X = (V-U)/2$ then leads to the familiar metric

$$d\sigma^2 = -dT^2 + dX^2, \quad T \in [-\infty, +\infty], \quad X \in [-\infty, +\infty]. \tag{7.7}$$

In passing from Rindler coordinates (t,x) to Minkowski coordinates (T,X), we have thereby not only eliminated the singularity at $x = 0$, but also extended the spacetime, because the Rindler coordinates cover only the region $X^2 - T^2 > 0$ of Minkowski spacetime. Inside this region we have

$$T = x \sinh t, \quad X = x \cosh t. \tag{7.8}$$

Since the singularity of the Schwarzschild metric at $r = 2m$ is not an intrinsic singularity, there exists a change of coordinates analogous to that taking Rindler coordinates to Minkowski coordinates which will eliminate it.

The sections $\theta = \text{const}$, $\phi = \text{const}$ of Schwarzschild spacetime have the metric

$$d\sigma^2 = -(1 - 2m/r)dt^2 + dr^2/(1 - 2m/r), \quad t \in [-\infty, +\infty], \quad r \in]2m, +\infty]. \tag{7.9}$$

Light-like radial geodesics of the Schwarzschild metric ($d\sigma^2 = 0$) have the equations $t = \pm r_* + \text{const}$, where r_* is the tortoise coordinate (the name is due to J. A. Wheeler) $r_* = r + 2m \ln(r/2m - 1)$. One then introduces the null coordinates (also called the Eddington–Finkelstein coordinates)

$$u = t - r_*, \quad v = t + r_*, \quad u, v \in [-\infty, +\infty], \tag{7.10}$$

in which the metric (7.9) becomes $d\sigma^2 = -(1 - 2m/r)du\,dv$, where r is given implicitly as a function of r^* and therefore of $(v - u)$.

Next we make the new change of coordinates

$$U = -e^{-u/4m}, \quad V = e^{v/4m}, \quad U \in [-\infty, 0], \quad V \in [0, +\infty], \tag{7.11}$$

which leads to a metric which is regular everywhere except at $r = 0$:

$$d\sigma^2 = -(32m^3/r)\,e^{-r/2m}dU\,dV.$$

The ranges of variation of U and V can thus be *extended* to the *entire* (U,V) plane.

The final coordinate change $T = (U+V)/2$, $X = (V-U)/2$ leads to the *Kruskal metric* (1960), which is regular everywhere except at $r = 0$:

$$ds^2 = \frac{32m^3}{r} e^{-\frac{r}{2m}} \left(-dT^2 + dX^2\right) + r^2 \left(d\theta^2 + \sin^2\theta d\phi^2\right),$$

$$T \in [-\infty, +\infty]; \quad X \in [-\infty, +\infty]. \tag{7.12}$$

In passing from Schwarzschild–Droste coordinates (t,r) to Kruskal coordinates (T,X) (see Fig. 7.2), we have not only eliminated the singularity at $r = 2m$, but also extended the spacetime because the coordinates (t,r) cover only the region $X^2 - T^2 > 0$. In this quadrant corresponding to $r > 2m$ the coordinate transformation is

$$\begin{cases} T = \sqrt{r/2m - 1}\,e^{r/4m}, \quad \sinh t/4m, \quad X = \sqrt{r/2m - 1}\,e^{r/4m} \cosh t/4m, \\[2mm] \left(\dfrac{r}{2m} - 1\right)\exp\dfrac{r}{2m} = X^2 - T^2, \quad \tanh\dfrac{t}{4m} = \dfrac{T}{X}. \end{cases} \tag{7.13}$$

In the other quadrants only the function $r(T,X)$ needs to be defined: we set $(r/2m - 1)\exp(r/2m) = X^2 - T^2$ for all r.

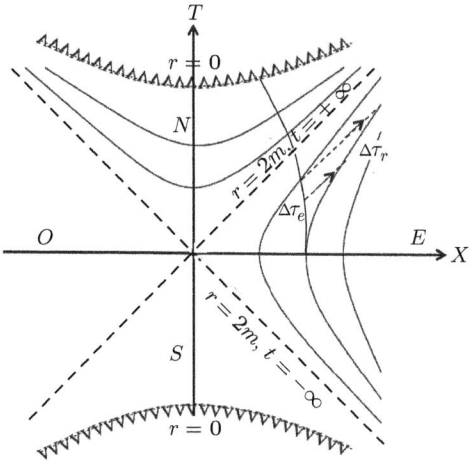

Fig. 7.2 The Kruskal diagram.

Regularization of a metric

Any two-dimensional metric with coefficients γ_{ab} in the coordinates x^a $(a = 1, 2)$ is, at least locally, *conformally flat*, that is, we can write

$$d\sigma^2 = \gamma_{ab}\, dx^a\, dx^b = e^{2U}\, \bar{\gamma}_{ab}\, dx^a\, dx^b\,,$$

where the scalar curvature of $\bar{\gamma}_{ab}$, which determines the curvature in two dimensions, is zero and $U(x^a)$ is the *conformal factor*. Indeed, a simple calculation gives[4] $\bar{R} = e^{2U}(R + 2\,\square\, U)$. Then if $\bar{\gamma}_{ab}$ is flat, we have $\bar{R} = 0$ and the conformal factor is a solution of $\square\, U = -R/2$, where $\square\, U \equiv \gamma^{ab} D_{ab} U = \sqrt{|\gamma|}^{-1} \partial_a(\sqrt{|\gamma|}\gamma^{ab}\partial_b U)$. (We write $D_a D_b \equiv D_{ab}$.)

If the metric $d\sigma^2$ is static and the coordinates $x^a = (t, x)$ are adapted coordinates, we can set $\gamma_{tt} \equiv -f(x)$ and $\gamma_{xx} \equiv 1/g(x)$. Then requiring that $U = U(x)$ (in this case $fg > 0$), we find

$$R = -\sqrt{\frac{g}{f}}\left(f'\sqrt{\frac{g}{f}}\right)'\,,\quad \square\, U = \sqrt{\frac{g}{f}}(\sqrt{fg}\,U')'\,;$$

$$\text{the solution of }\quad \square\, U = -\frac{R}{2}\quad\text{ then is }\quad e^{2U} = \frac{1}{(CC_1)^2}\, f\, e^{2C\int \frac{dx}{\sqrt{fg}}}\,, \tag{7.14}$$

where C and C_1 are constants of integration and the prime denotes differentiation with respect to x.

Now to find the transformation from the coordinates $x^a = (t, x)$ to the Minkowski coordinates $X^a = (T, X)$, where $d\sigma^2 = e^{2U}(-dT^2 + dX^2)$, we use the transformation laws of the Christoffel symbols and the metric coefficients:

---

[4]We have $\bar{\gamma}_{ab} = e^{-2U}\gamma_{ab}$ and $\bar{\gamma}^{ab} = e^{2U}\gamma^{ab}$. Then $\bar{\Gamma}^a_{bc} = \Gamma^a_{bc} - (\delta^a_b \partial_c U + \delta^a_c \partial_b U - \gamma_{bc}\partial^a U)$ and $\bar{R}^a{}_{bcd} = R^a{}_{bcd} + 2\delta^a_{[c} D_{bd]} U - 2\gamma_{b[c} D^a{}_{d]} U + 2\delta^a_{[c}\partial_b U \partial_{d]} U - 2\gamma_{b[c}(\partial^a U \partial_{d]} U - \delta^a_{d]}\partial_e U \partial^e U);\; \bar{R}_{bd} = R_{bd} + (D - 2)D_{bd}U + \gamma_{bd}\square U + (D - 2)(\partial_b U \partial_d U - \gamma_{bd}\partial_e U \partial^e U);\; \bar{R} = e^{2U}[R + 2(D - 1)\square U - (D - 1)(D - 2)\partial_e U \partial^e U]$, where D is the spatial dimension.

$$\frac{\partial^2 X^i}{\partial x^b \partial x^c} = \frac{\partial X^i}{\partial x^a} \bar{\Gamma}^a_{bc}, \quad \bar{\gamma}_{ab} = \frac{\partial X^c}{\partial x^a} \frac{\partial X^d}{\partial x^b} \eta_{cd}. \tag{7.15}$$

The first equation is written out as $\ddot{T} = \dot{T} \bar{\Gamma}^t_{tt} + T' \bar{\Gamma}^x_{tt}$, and so on, where a dot denotes differentiation with respect to t and a prime differentiation with respect to x. The $\bar{\Gamma}^a_{bc}$ are expressed as a function of the Γ^a_{bc} as (cf. footnote 4) $\bar{\Gamma}^t_{tt} = \Gamma^t_{tt} - (2\partial_t U - \gamma_{tt} \partial^t U)$, and so on. If the metric is static and the coordinates x^a are adapted, $\gamma_{tt} \equiv -f(x)$ and $\gamma_{xx} \equiv 1/g(x)$, the only nonzero Christoffel symbols will be, for the conformal factor found in (7.14),

$$\bar{\Gamma}^x_{xx} = -\frac{g'}{2g} - U' = -\frac{g'}{2g} - \frac{f'}{2f} + \frac{C}{\sqrt{fg}}, \quad \bar{\Gamma}^x_{tt} = \frac{f'g}{2} - fg\, U' = -C\sqrt{fg},$$

$$\bar{\Gamma}^t_{tx} = \frac{f'}{2f} - U' = -\frac{C}{\sqrt{fg}}.$$

The first equation in (7.15) is then easily integrated and gives $T(t,x)$ and $X(t,x)$, and the second equation allows us to fix the integration constants. In the end we obtain $d\sigma^2 = -f(x)dt^2 + dx^2/g(x) = e^{2U}(-dT^2 + dX^2)$, where U is given in (7.14) and

$$T = -C_1 e^{-C \int \frac{dx}{\sqrt{fg}}} \sinh Ct, \quad X = C_1 e^{-C \int \frac{dx}{\sqrt{fg}}} \cosh Ct \implies X^2 - T^2 = C_1^2 e^{-2C \int \frac{dx}{\sqrt{fg}}}. \tag{7.16}$$

Here C_1 is just a dimensionalization constant, but C can be used to regularize the metric. Indeed, if near $x = 1$ we have $f = (x-1)^p(1+\cdots)$ and $g = (x-1)^q(1+\cdots)$, with $p + q = 2$, then [cf. (7.14)] $e^{2U} \propto (x-1)^{p+2C}$, which does not vanish if we choose $C = -p/2$.

In the case where the metric is the Rindler metric, we have $f(x) = x^2$ and $g(x) = 1$, and so $p + q = 2$. Then [cf. (7.14)] $e^{2U} = x^{2+2C}/(CC_1)^2$. Let us choose $C = -1$ (and $C_1 = 1$). The conformal factor $e^{2U} = 1$ is then regular everywhere and (7.16) again gives the Rindler transformation: $T = x \sinh t$, $X = x \cosh t$.

If the metric is the Schwarzschild metric we have $f(x) = g(x) = 1 - 1/x$ (with $x \equiv r/2m$), and so again $p + q = 2$. Equation (7.14) then gives

$$e^{2U} = \frac{1}{(C_1 C)^2} \frac{e^{2Cr}}{r} (r - 2m)^{1+4mC},$$

which is regular at $r = 2m$ and equal to

$$e^{2U} = \frac{32m^3}{r} e^{-r/2m}$$

for $C = -1/4m$ (and $C_1 = 1/\sqrt{2m}$). Equations (7.16) then again give the Kruskal transformation (7.13).

Let us consider the surfaces $(\theta, \phi) = $ const. The curvature singularity $r = 0$ is represented in the Kruskal diagram (T, X) by the *two* space-like hyperbolas $T^2 - X^2 = 1$. In general, any curve in the (t, r) plane is represented two times in the (T, X) plane. Therefore, the straight lines $r = $ const $> 2m$ (trajectories of observers at rest in the Schwarzschild coordinates) are represented by hyperbolas in the east and west quadrants, and the straight lines $r = $ const $< 2m$ are represented by hyperbolas in the north and south quadrants. The coordinate singularity $r = 2m$ is represented by the light cone $T = \pm X$. It should also be noted that the 'origin' of the Kruskal diagram is actually a 2-sphere of radius $2m$ (this distinguishes the topology of the surface $T = 0$ from that of the Euclidean sections of the Rindler–Minkowski spacetime.) Finally, the lines $t = $ const correspond to the straight lines $T/X = $ const (we note that in the west quadrant t decreases when T grows), and the radial trajectories of photons are straight lines $T = \pm X + \text{const}$.

In this Kruskal diagram the graphical representation of light being redshifted as a star collapses, as described in Section 6.4, is particularly simple.

Let us expand on the analogy between the Rindler or Schwarzschild coordinates (t, r) and the Minkowski or Kruskal coordinates (T, X). An observer at $r = \text{const}$ in Schwarzschild coordinates can be compared to an accelerated observer in Rindler coordinates (neither one is undergoing free motion), and an observer freely falling toward the center is analogous to an inertial observer. Therefore, in the same way as an inertial observer *leaves the horizon* of a uniformly accelerated observer when he crosses the *Rindler future horizon* $T = X$ ($x = 0$, $t = \infty$), an observer freely falling toward the center of a black hole leaves the horizon of an exterior observer when he crosses the *Schwarzschild future horizon* $r = 2m$, $t = \infty$. In both cases the signals sent to the outside are shifted an infinite amount toward the red.

Similarly, the analog of an inertial observer who *enters the horizon* of an accelerated observer at $T = -X$ ($x = 0$, $t = -\infty$) is, in the spacetime of a black hole, an observer emerging from a *white hole* through the *past horizon* at $r = 2m$, $t = -\infty$. However, it is important to note that this last case is only possible if the black hole is *eternal* and not the product of the collapse of a star which was initially larger than its Schwarzschild radius.

Finally, these analogies between the Rindler and Schwarzschild horizons should not obscure the fact that whereas an inertial observer, after crossing the Rindler future horizon, remains in causal contact with all of Minkowski spacetime (only observers who are accelerated *ad aeternam* will be inaccessible), an observer who has crossed the Schwarzschild future horizon will no longer be able to communicate with the outside of the black hole, that is, with the region $r > 2m$. The light rays sent by this observer all converge toward the singularity at $r = 0$ which, since it is space-like, is not situated 'somewhere', but 'everywhere in the future'.

7.3 The Penrose–Carter diagram

As we have seen in Section 7.2, any two-dimensional metric is conformally flat. The *Penrose–Carter diagrams* exploit this fact and provide a powerful graphical tool for obtaining the 'maximal extension' of a manifold.[5]

Let us first consider the simple case of Minkowski spacetime in spherical coordinates: $ds^2 = -dt^2 + dr^2 + r^2 d\Omega_2^2$ with $d\Omega_2^2 = d\theta^2 + \sin^2\theta \, d\phi^2$, $t \in [-\infty, +\infty]$, $r \in]0, \infty]$. We introduce two new coordinates (Ψ, ξ) such that $t + r = \text{tg} \frac{1}{2}(\Psi + \xi)$ and $t - r = \text{tg}\frac{1}{2}(\Psi - \xi)$, with $(\Psi + \xi) \in [-\pi, +\pi]$ and $(\Psi - \xi) \in [-\pi, +\pi]$. The condition $r > 0$ restricts ξ to positive values. We then obtain

$$ds^2 = \Omega^2 d\tilde{s}^2 \quad \text{with} \quad d\tilde{s}^2 = -d\Psi^2 + d\xi^2 + \sin^2\xi \, d\Omega_2^2$$

$$\text{and} \quad \Omega^2 = \frac{1}{4\cos^2\frac{1}{2}(\Psi + \xi)\cos^2\frac{1}{2}(\Psi - \xi)}. \tag{7.17}$$

The metric $d\tilde{s}^2$ is that of the 'Einstein static universe' whose spatial sections are 3-spheres of constant radius $\xi = 1$. The sections $(\theta, \phi) = \text{const}$ are represented by the Penrose diagram of Fig. 7.3.

The Penrose–Carter diagram of Kruskal spacetime is obtained in a similar manner. First let us consider the east quadrant of the Kruskal diagram ($X > 0$, $T + X > 0$, and $T - X < 0$). We make the same coordinate change as before: $(T, X) \to (\Psi, \xi)$ such that $T + X = \text{tg}\frac{1}{2}(\Psi + \xi)$

---

[5]In this section we follow the now-classic presentations of Hawking and Ellis (1973), Chapter 5, and Misner, Thorne, and Wheeler (1973), Chapter 33.

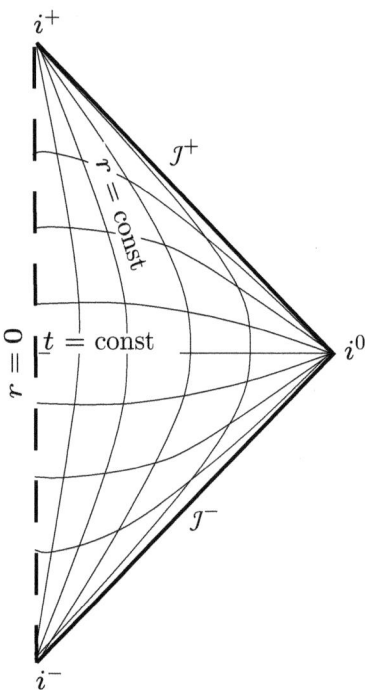

Fig. 7.3 The Penrose diagram of Minkowski spacetime.

and $T - X = \mathrm{tg}\frac{1}{2}(\Psi - \xi)$. The conditions on X and T imply that $(\Psi + \xi) \in [0, +\pi]$, $(\Psi - \xi) \in [-\pi, 0]$ and $\xi > 0$. The Kruskal metric in this quadrant is then written as

$$ds^2 = \Omega^2 d\tilde{s}^2 \quad \text{with} \quad d\tilde{s}^2 = -d\Psi^2 + d\xi^2 + r^2\Omega^{-2}d\Omega_2^2$$

$$\text{and} \quad \Omega^2 = \frac{32m^3}{r}\frac{e^{-r/2m}}{4\cos^2\frac{1}{2}(\Psi + \xi)\cos^2\frac{1}{2}(\Psi - \xi)}. \tag{7.18}$$

The sections $(\theta, \phi) = \text{const}$ of the metric $d\tilde{s}^2$ are therefore represented by compactification of the east quadrant of the Kruskal diagram; see Fig. 7.4. The west quadrant is obtained by changing the signs: $X \to -X$, $T \to -T$, $\Psi \to -\Psi$, $\xi \to -\xi$.

Now let us consider the north quadrant ($T > 0$, $T + X > 0$, $T - X > 0$, and $T^2 - X^2 < 1$), and introduce the coordinates (Ψ, ξ) by the same transformations as before. The sections $(\theta, \phi) = \text{const}$ of the metric $d\tilde{s}^2$ are represented by compactification of the north quadrant of the Kruskal diagram, where the curvature singularity $r = 0$ is space-like. The south quadrant is obtained by changing the signs.

Finally, to find the Penrose–Carter diagram of the entire Kruskal spacetime and thereby obtain the *maximal extension* of the Schwarzschild spacetime, we only need to 'glue' these two blocks together such that any time-like curve ends up either at infinity or at the curvature singularity; see Fig. 7.4.

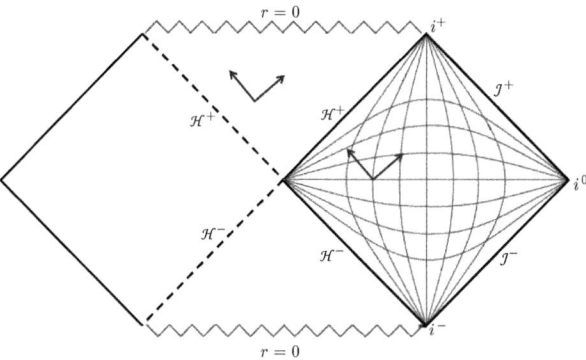

Fig. 7.4 The Penrose diagram of Kruskal spacetime.

The advantage of this diagram is that it clearly shows the causal structure of the spacetime of a Schwarzschild black hole. We see in particular that if the black hole is the result of the collapse of a star, only the east and north blocks are relevant; see Fig. 7.5. We also note that since the diagram is symmetric under time reversal, it can be imagined that a star could emerge from the past horizon: the black hole would then become a 'white hole' .... . The complete diagram applies only to an eternal black hole. Since the trajectories of light rays are straight lines $\Psi = \pm\xi + \text{const}$, we see that the north and south quadrants cannot communicate with each other.

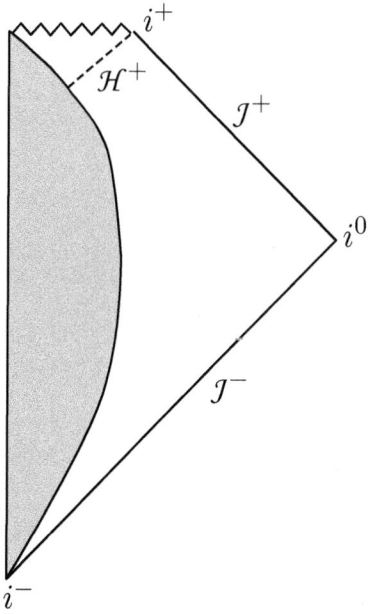

Fig. 7.5 The Penrose diagram of the spacetime of a collapsing star.

The Einstein–Rosen bridge

We have seen that the geometry of the sections ($t = $ const, $\theta = \pi/2$) for $r > 2m$ of the Schwarzschild metric in Schwarzschild coordinates [$d\sigma^2 = dr^2(1 - 2m/r)^{-1} + r^2 d\phi^2$] is that of a paraboloid $z^2 = 8m(r - 2m)$ embedded in a Euclidean space with metric $dS^2 = dz^2 + dr^2 + r^2 d\phi^2$ (see Section 6.3). If we consider a black hole, and therefore do not match this paraboloid to an interior metric (see Fig. 6.1), then we see that it connects *two* asymptotically flat regions, which is consistent with the above results. If we are interested only in the topology of this surface, that is, if we deform it, we can transform it into a *wormhole* (the terminology is that of Wheeler, 1962), also known as an *Einstein–Rosen bridge* (1935).

The following question then arises (see, for example, the discussion in Misner, Thorne, and Wheeler, p. 836 *et seq.*): if we take the two asymptotic regions to be *identical*, is it conceivable that a light signal might propagate from A to B, where these are two points located far from the black hole in a quasi-Minkowskian region of spacetime, by two different routes, one remaining in the asymptotic region and the other passing through the wormhole? Furthermore, is it possible that the second route could be faster than the first, in which case a signal could be sent from A to B at a speed greater than the speed of light?

Looking at the Penrose–Carter diagram of Fig. 7.4, we can see that the answer is 'no': a (radial) light signal emitted far from the black hole toward the black hole in the east quadrant will encounter the $r = 0$ singularity of the north quadrant and can never reach the west region. Therefore, a Schwarzschild black hole cannot serve as a 'time machine' to go back in time.

We can also answer this question by following a different argument. Namely, we can study how a wormhole (see Fig. 7.6), which is a 'photograph' of spacetime at a given time t, evolves as a photon sent by A at $t = 0$ approaches it. As long as the photon is outside the horizon, the coordinate t is regular and the wormhole is static: the route followed by the photon is a curve on the paraboloid. However, after crossing the horizon the geometry of the spatial section on which the photon propagates is no longer that of a paraboloid because the signature of the metric $d\sigma^2 = -dr^2/(2m/r - 1) + r^2 d\phi^2$ has changed: it is now that of the closed surface $z^2 = 8m(2m - r)$ embedded in the pseudo-Euclidean space of metric $dS^2 = -dz^2 + dr^2 + r^2 d\phi^2$. The route traveled by the photon inside the horizon is a curve on this surface, and we see that it ends up (r decreases) at the curvature singularity $r = 0$, in agreement with the preceding analysis.

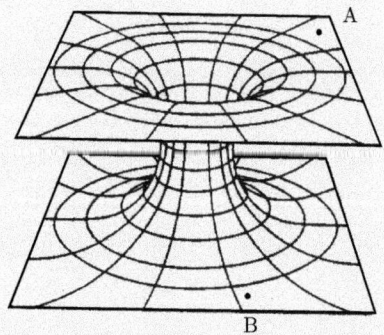

Fig. 7.6 The Einstein–Rosen wormhole.

The various notions of horizon

As we saw in Section 6.4, the surface $r = 2m$ is a surface of infinite redshift: light signals of a given proper frequency observed at $r > 2m$ are shifted more and more to the red the closer the source is. This is therefore a *horizon*.

As we saw from the Penrose–Carter diagrams in Figs. 7.4 and 7.5, this horizon is located on the past cone of the future infinity i_+. It bounds the universe visible to *all* external observers, and is therefore called the *event horizon*.

Since spacetime is stationary, the vector $\xi^\mu = (1, 0, 0, 0)$ is a Killing vector (such that $D_\mu \xi_\nu + D_\nu \xi_\mu = 0$, corresponding to invariance of the metric under time translations). Its norm $\xi_\mu \xi^\mu = (1 - 2m/r)$ vanishes at $r = 2m$. It therefore becomes light-like there and so $r = 2m$ is also a *Killing horizon*.

The Schwarzschild horizon is also a *light-like surface*, that is, a *null surface*: the covector n_μ normal to the 3-surfaces $r = \text{const}$ with components $(0, n_r, 0, 0)$ has norm $g^{rr}(n_r)^2 = (1 - 2m/r)(n_r)^2$ which vanishes at $r = 2m$.

We have also seen that photons there are at rest at the time t of an observer at infinity, [*cf.* (6.20)], and for this reason it is also called a *trapped surface*. Finally, it is a surface beyond which 'incoming' and 'outgoing' light signals are all directed toward the curvature singularity $r = 0$ (as illustrated by the Kruskal diagram of Fig. 7.2). Such surfaces are *apparent horizons*.[6]

7.4 The Reissner–Nordström black hole

The spherically symmetric solution of the Einstein equations in the presence of a radial electric field [$G_{\mu\nu} = \kappa T_{\mu\nu}$ with $T_{\mu\nu} = F_{\mu\rho} F_\nu{}^\rho - \frac{1}{4} g_{\mu\nu} F_{\rho\sigma} F^{\rho\sigma}$, where $F_{\mu\nu} = \partial_\mu A_\nu - \partial_\nu A_\mu$ and $A_\mu = (\Phi, 0, 0, 0)$] was obtained by Reissner (1916) and Nordström (1918). The metric and potential are written in Schwarzschild coordinates as

$$ds^2 = -(1 - 2m/r + q^2/r^2)dt^2 + \frac{dr^2}{1 - 2m/r + q^2/r^2} + r^2\, d\Omega_2^2, \quad \Phi = \frac{q}{\sqrt{G}\, r}, \qquad (7.19)$$

where $d\Omega_2^2 \equiv d\theta^2 + \sin^2\theta d\phi^2$, and $m \equiv GM$ and $q \equiv \sqrt{G}Q$ are the mass and electric charge of the black hole in geometrical units.

The origin $r = 0$ is a singularity where the curvature invariants diverge [the scalar curvature is zero because the Maxwell energy–momentum tensor is traceless, but $R_{\mu\nu} R^{\mu\nu} = 4q^4/r^8$ and $R_{\mu\nu\rho\sigma} R^{\mu\nu\rho\sigma} = 8(7q^4 - 12mqr + 6m^2 r^2)/r^8$].

If $q^2 > m^2$, the curvature singularity $r = 0$ is 'naked': particles visible from infinity can reach it and leave it. It is assumed that this never happens (a wish piously baptized as the *cosmic censorship hypothesis* by Penrose in 1969).[7]

If $q^2 < m^2$, the spacetime represents a black hole possessing two horizons $r_\pm$, the roots of $r^2 - 2mr + q^2 = 0$. Since in the region $0 < r < r_-$ the signature is $(-, +, +, +)$ (the same as for $r > r_+$), the curvature singularity $r = 0$ is time-like and not space-like as in the Schwarzschild case. We can introduce two sets of new coordinates (one for each horizon) according to the rules of Section 7.2, analogs of the Kruskal coordinates (T, X), in which the metric is regular at r_+ or r_-. We then obtain a Kruskal diagram covering the region $r_- < r < \infty$ and another covering the region $0 < r < r_+$. The introduction of new coordinates analogous to (Ψ, ξ) introduced in Section 7.3 in each of the regions $r_+ < r < \infty$, $r_- < r < r_+$, and $0 < r < r_-$ then allows us to construct the three blocks of the Penrose–Carter diagram in Fig. 7.7.

---

[6] For a review of these various concepts of horizon in the case where they do not coincide, see Gourgoulhon and Jaramillo (2008).

[7] See, for example, Penrose (1998). However, we note that if Q and M are the electron charge and mass, we have $q/m \approx 2 \times 10^{22} > 1$.

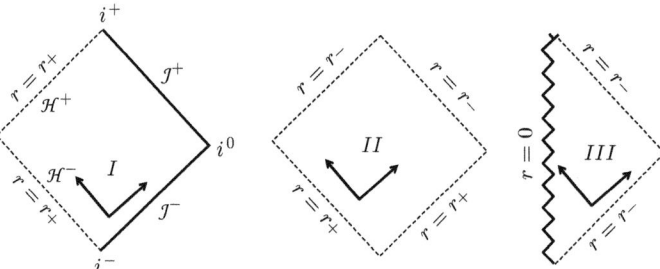

Fig. 7.7 The three blocks of the Penrose–Carter diagram.

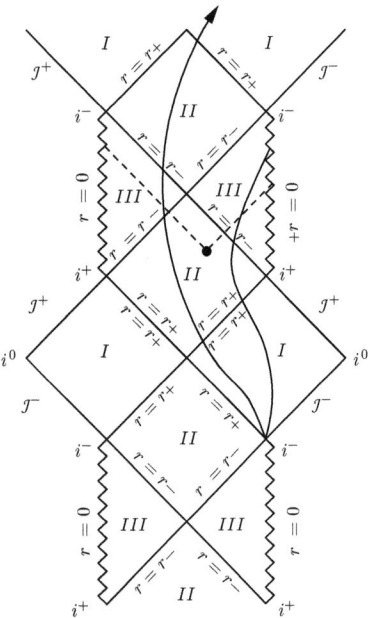

Fig. 7.8 The Penrose diagram of a Reissner–Nordström black hole.

Finally, to obtain a representation of the *maximal extension* of the Reissner–Nordström solution, we 'glue' these three blocks together such that any time-like curve will end up either at infinity or at the curvature singularity. We thus obtain the diagram of Fig. 7.8 (Graves and Brill, 1960 and Carter, 1966), where the region $0 < r < \infty$ is represented an infinite number of times, rather than 'only' twice as in the case of a Schwarzschild black hole. We see that a particle falling into the black hole can avoid the singularity and reemerge in another asymptotic region. The diagram also shows that signals from a light source received by an observer at infinity which cross the interior horizon are infinitely *blue*-shifted. Finally, the *Cauchy expansions* of space-like 3-surfaces Σ, that is, the ensemble of events located in the future cone of each of their points, do *not* cover the entire diagram, and so it is composed of causally disconnected regions.

Geodesics

The geodesics representing the world lines of objects in free fall (which are therefore electrically neutral) are obtained as in the Schwarzschild case (see Section 6.4), or by using the conservation laws (5.11) associated with the static, spherically symmetric nature of the solution: their 4-velocity is normed and the components u_0 and u_ϕ are conserved, $u_0 = -E$, $u_\phi = L$. Therefore [cf. (7.19)],

$$f\dot{t} = E \quad \text{and} \quad r^2\dot{\phi} = L \quad \text{with} \quad f \equiv 1 - 2m/r + q^2/r^2$$

$$\text{such that} \quad \dot{r}^2 = E^2 - U_e, \quad \text{where} \quad U_e = (1 - 2m/r + q^2/r^2)(1 + L^2/r^2). \tag{7.20}$$

As we can see from Fig. 7.9 showing the 'effective potential' U_e, any particle originating outside the black hole will penetrate the region $r < r_-$ but will never reach the singularity $r = 0$ where $U_e \to +\infty$ and will turn back; gravity appears to be *repulsive* in this region. This is another point on which the Reissner–Nordström solution differs from the Schwarzschild solution, where $U_e \to -\infty$ at $r = 0$ and particles, radial or with sufficient energy E, reach the singularity at $r = 0$ in a finite proper time; see, for example, (6.17) and Chapter 10.

As we shall see in the following chapter, the internal horizon $r = r_-$ of the Reissner–Nordström black hole is *unstable*, and so the geometry for $r < r_+$ is unknown. Therefore, the trajectories of particles trapped there are also unknown.

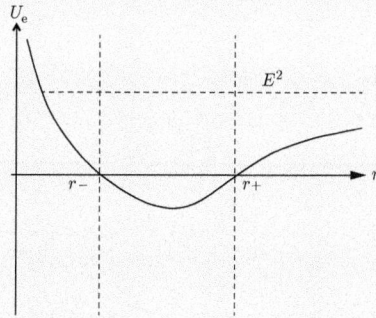

Fig. 7.9 The effective potential of radial geodesics.

The case $q^2 = m^2$

When $q^2 = m^2$, the Reissner–Nordström solution is

$$ds^2 = -\left(\frac{r-m}{r}\right)^2 dt^2 + \left(\frac{r}{r-m}\right)^2 dr^2 + r^2 d\Omega_2^2.$$

The surface $r = m$ is, as in the general case, a light-like surface of infinite redshift, and the Penrose diagram is constructed from only two blocks (Carter, 1966). However, it should be noted that the method described in Section 7.2 for regularizing a 2-metric does not apply in this case (because $q + p = 4$ rather than 2), and it is not possible to find coordinates which will allow us to write $d\sigma^2 \equiv -(1-m/r)^2 dt^2 + dr^2/(1-m/r)^2$ in the Kruskal form $d\sigma^2 = \Omega^2(-dT^2 + dX^2)$ with Ω nonzero[8] at $x = 1$.

Now, the metric near the horizon $r = m$ is

$$ds^2 \approx -(r/m - 1)^2 dt^2 + dr^2/(r/m - 1)^2 + m^2 d\Omega_2^2 \,.$$

The curvature of the 2-space (t, r) is $R = -2$. This is the anti-de Sitter hyperboloid AdS_2, which is perfectly regular everywhere. The spacetime near the horizon then is $AdS_2 \times S_2$. It turns out that this metric is an *exact* solution of the Einstein–Maxwell equations called the *Robinson–Bertotti solution* (1959). It is not spherically symmetric because the spatial sections are not foliated by 2-spheres.

Moreover, we note that the coordinate r is time-like between the two horizons $r_\pm$ of a generic Reissner–Nordström black hole, so that the proper time needed to go from one to the other at $t = \text{const}$ is

$$\Delta\tau = -\int_{r_+}^{r_-} \frac{r \, dr}{(r - r_+)(r - r_-)} = \pi m \,.$$

It does not depend on the ratio q/m, and so the limit $q \to m$ is singular.

---

[8]On this subject, see Carter (1972), as well as, for example, Carroll *et al.* (2009).

8

The Kerr solution

"It is a remarkable fact that there are roughly 10^{20} rotating black holes in the observable universe, and the spacetime near each one of them is given to a very good approximation by a simple explicit exact solution of the Einstein vacuum equations called, after its discoverer, the Kerr metric."

Gibbons, Lü, Page, and Pope, *The general Kerr–de Sitter metrics in all dimensions,*
hep-th/0404008.

8.1 Kerr–Schild metrics

The Einstein equations are nonlinear. However, there exists a class of "Kerr–Schild" metrics which linearize them. Using some appropriate coordinate system, let us consider metrics with the components

$$g_{\mu\nu} = \overline{g}_{\mu\nu} + l_{\mu\nu} \quad \text{with} \quad l_{\mu\nu} = f(x^\rho)l_\mu l_\nu , \tag{8.1}$$

where $\overline{g}_{\mu\nu}$ is a metric of a known 'background' spacetime, $f(x^\rho)$ is an *a priori* arbitrary function, and the vector field $l^\mu = \overline{g}^{\mu\nu}l_\nu$ is null and geodesic, that is, it satisfies

$$\overline{g}_{\mu\nu}l^\mu l^\nu = 0 , \quad l^\nu \overline{D}_\nu l^\mu = 0 , \tag{8.2}$$

where $\overline{D}$ is the covariant derivative associated with $\overline{g}_{\mu\nu}$. The calculation of the Ricci tensor of such a metric is a good exercise. We find

$$R^\mu_\nu = \overline{R}^\mu_\nu - l^{\mu\rho}\overline{R}_{\rho\nu} + \overline{D}_\rho(\overline{g}^{\mu\lambda}\Delta^\rho_{\nu\lambda}) , \quad \text{where} \quad \Delta^\mu_{\nu\rho} = \frac{1}{2}(\overline{D}_\nu l^\mu_\rho + \overline{D}_\rho l^\mu_\nu - \overline{D}^\mu l_{\nu\rho}) \tag{8.3}$$

(the indices can be raised or lowered with either metric $g_{\mu\nu}$ or $\overline{g}_{\mu\nu}$). This expression, which is exact, is *linear* in the 'perturbation' $l_{\mu\nu}$. The curvature scalar reduces to

$$R = \overline{R} - l^{\mu\nu}\overline{R}_{\mu\nu} + \overline{D}_\mu V^\mu , \quad \text{where} \quad V^\mu = l^\mu \overline{D}_\nu(fl^\nu) , \tag{8.4}$$

so that $\sqrt{-\overline{g}}\,\overline{D}_\mu V^\mu = \partial_\mu[l^\mu \partial_\nu(\sqrt{-\overline{g}}fl^\nu)]$.

Therefore, given a background metric $\overline{g}_{\mu\nu}$ and a null geodesic vector l^μ, the equation $R = 0$ will determine the function f. The metric $g_{\mu\nu}$ is then known, but it still must be checked to see whether or not it is a solution of the vacuum equations, that is, if it is Ricci-flat, $R^\mu_\nu = 0$. As we shall see, all the solutions of the Einstein–Maxwell equations (in four dimensions) describing black holes are of the Kerr–Schild type.

Relativity in Modern Physics. Nathalie Deruelle and Jean-Philippe Uzan.
© Oxford University Press 2018. Published in 2018 by Oxford University Press.
DOI: 10.1093/oso/9780198786399.001.0001

The case of spherical symmetry

Let us take as the background metric the Minkowski metric in spherical coordinates, $d\bar{s}^2 = -dT^2 + dr^2 + r^2(d\theta^2 + \sin^2\theta d\phi^2)$, and as the null geodesic vector $l^\mu = (1, -1, 0, 0)$ [of course, we could also choose the 'outgoing' vector $\bar{k}^\mu = (1, 1, 0, 0)$]. For a function f depending only on r, the equation (8.4) for $R = 0$ reduces to $(r^2 f)'' = 0$, the solution of which is $f(r) = (c + dr)/r^2$, where c and d are constants.

Now we see that $R_\nu^\mu = 0$ if $c = 0$. The desired length element then is (setting $d = 2m$)

$$ds^2 = -\left(1 - \frac{2m}{r}\right)dT^2 + \frac{4m}{r}dr\,dT + \left(1 + \frac{2m}{r}\right)dr^2 + r^2(d\theta^2 + \sin^2\theta d\phi^2)\,, \qquad (8.5)$$

which is just the Schwarzschild metric [in Eddington coordinates (1924); see (7.3)], as can be seen explicitly by making the change of coordinate $T \to t = T - T_0(r)$ with $T_0' = -2m/(r - 2m)$.

The trace of the Einstein–Maxwell equations is also $R = 0$ because the trace of the energy-momentum tensor of an electromagnetic field is zero in four dimensions, and the solution is again $f(r) = (c + dr)/r^2$. We then find that the traceless part of the equations $R_\nu^\mu = 8\pi T_\nu^\mu$ with $T_\nu^\mu = F^\mu_{\ \rho}F_\nu^{\ \rho} - \frac{1}{4}\delta_\nu^\mu F_{\rho\sigma}F^{\rho\sigma}$ and $F_{\mu\nu} = \partial_\mu A_\nu - \partial_\nu A_\mu$ is indeed satisfied for the vector potential $A_\mu = (q/r, 0, 0, 0)$ if we choose $c = q$. This is the Reissner–Nordström solution in Kerr–Schild coordinates.

The *tour de force* of R. Kerr in 1963 was to find a solution of the vacuum Einstein equations of the Kerr–Schild type which describes a rotating object.

8.2   The Kerr metric

Spheroidal coordinates

It has been known since the work of Maclaurin (1742) that in Newtonian physics a rotating self-gravitating body is not a sphere, but rather a spheroid, owing to the centrifugal force. The gravitational potential created by such a body possesses the same symmetry: the equipotentials are spheroids; see Book 1, Section 15.5. It is therefore useful to solve the Laplace equation determining the potential, $\triangle U = 0$, using *spheroidal coordinates* (r, θ, φ) related to Cartesian coordinates (X, Y, Z) as

$$X = \sqrt{r^2 + a^2}\sin\theta\cos\varphi\,, \quad Y = \sqrt{r^2 + a^2}\sin\theta\sin\varphi\,, \quad Z = r\cos\theta\,. \qquad (8.6)$$

Since $\frac{X^2 + Y^2}{r^2 + a^2} + \frac{Z^2}{r^2} = 1$, the surfaces $r = $ const are indeed spheroids. (The coordinate r is identified as the usual radial coordinate only when $a = 0$.)

The surface $r = 0$ is a disk of radius a in the $Z = 0$ plane. The origin $X = Y = Z = 0$ corresponds to $r = \theta = 0$. The locus of points $(r = 0, \theta = \pi/2)$ is the circle $X^2 + Y^2 = a^2$ in the $Z = 0$ plane.

In these coordinates the Euclidean length element is easily found to be

$$dl^2 = \frac{r^2 + a^2\cos^2\theta}{r^2 + a^2}dr^2 + (r^2 + a^2\cos^2\theta)d\theta^2 + (r^2 + a^2)\sin^2\theta\,d\varphi^2\,. \qquad (8.7)$$

The Laplacian then is $\triangle = (1/\sqrt{e})\partial_i\sqrt{e}e^{ij}\partial_j$, where e^{ij} is the inverse metric and $\sqrt{e} = (r^2 + a^2\cos^2\theta)\sin\theta$, so that the solution of the Laplace equation $\triangle U = 0$ which vanishes at infinity is, for $U = U(r)$,

$$U = \frac{GM}{a} \left(\text{arctg}\frac{r}{a} - \frac{\pi}{2} \right).$$

This is the gravitational potential of a *ring singularity* located at $r = 0$, where U is finite but has nonzero derivative.

To obtain the Kerr solution in Kerr–Schild coordinates, as the background metric we take the Minkowski metric in spheroidal coordinates:

$$d\bar{s}^2 = -dT^2 + \frac{\rho^2}{r^2 + a^2}dr^2 + \rho^2 d\theta^2 + (r^2 + a^2)\sin^2\theta\, d\varphi^2 \text{ with } \rho^2 \equiv r^2 + a^2\cos^2\theta, \quad (8.8)$$

where a is a constant, and as the null geodesic vector we take

$$l^\mu = \left(1, -1, 0, \frac{a}{r^2 + a^2} \right) \quad \Longrightarrow \quad l_\mu = \left(-1, -\frac{\rho^2}{r^2 + a^2}, 0, a\sin^2\theta \right). \quad (8.9)$$

[We could just as well have chosen the 'outgoing' vector $\bar{k}^\mu = (1, 1, 0, a/(r^2 + a^2))$.]

Solving the equation $R = 0$ [cf. (8.4)], we easily find $f = (c(\theta) + d(\theta)r)/\rho^2$. Next we must check to see if the metric we have found is Ricci-flat. The answer 'miraculously' turns out to be 'yes': we find $R^\mu_\nu = 0$ [with R^μ_ν given in (8.4)] if $c(\theta) = 0$ and $d(\theta) = 2m$, that is, if

$$f = \frac{2mr}{\rho^2}. \quad (8.10)$$

This is the Kerr solution (1963) in the form obtained by Kerr and Schild in 1964. Rearranging the intermediate results, it can be written as

$$ds^2 = -\left(1 - \frac{2mr}{\rho^2} \right) dT^2 + \frac{\rho^2}{r^2 + a^2}\left(1 + \frac{2mr}{r^2 + a^2} \right) dr^2 + \rho^2 d\theta^2$$

$$+ \left(r^2 + a^2 + \frac{2mra^2\sin^2\theta}{\rho^2} \right) \sin^2\theta d\varphi^2 \quad (8.11)$$

$$+ 4mr\left(\frac{dT\,dr}{r^2 + a^2} - \frac{a\sin^2\theta}{\rho^2}dT\,d\varphi - \frac{a\sin^2\theta}{r^2 + a^2}dr\,d\varphi \right)$$

$$\text{with} \quad \rho^2 \equiv r^2 + a^2\cos^2\theta.$$

For $m = 0$ it reduces, by construction, to the Minkowski metric (8.8) in spheroidal coordinates, and for $a = 0$ it reduces to the Schwarzschild metric (8.5) in Eddington coordinates.

We note that it is regular everywhere except at $\rho^2 = 0$, that is, at $r = 0$, $\theta = \pi/2$, which turns out to be a curvature singularity; see Section 8.4. We also see that light signals sent by an object at constant (r, θ, φ) are more shifted toward the red the closer the object is to the surface $\rho^2 = 2mr$.

The coefficients of the Kerr metric (8.11) do not depend on either T or φ. The spacetime is therefore stationary and axisymmetric, and the two Killing vectors associated with these isometries are $\xi^\mu_T = (1, 0, 0, 0)$ and $\xi^\mu_\varphi = (0, 0, 0, 1)$.

In addition, the Kerr spacetime possesses the remarkable property of being *circular*, that is, its Killing vectors satisfy the following equations[1]:

$$\mathrm{d}\xi^T \wedge \xi^T \wedge \xi^\varphi = 0\,, \quad \mathrm{d}\xi^\varphi \wedge \xi^T \wedge \xi^\varphi = 0\,. \tag{8.12}$$

This property guarantees that the cross terms involving $dT\,dr$ and $dT\,d\varphi$ present in (8.11) can be eliminated by a change of coordinates[2] $(T, \varphi) \to (t, \phi)$. Boyer and Lindquist (1967) found

$$T = t + 2m \int \frac{r\,dr}{\Delta}\,, \quad \varphi = \phi + 2ma \int \frac{r\,dr}{\Delta(r^2 + a^2)}\,, \quad \text{where} \quad \Delta = r^2 - 2mr + a^2\,. \tag{8.13}$$

This makes it possible to write the Kerr metric in a less forbidding form, for example (the algebra is elementary),

$$\begin{aligned}
ds^2 &= -\frac{\Delta\rho^2}{\Sigma}dt^2 + \frac{\Sigma \sin^2\theta}{\rho^2}(d\phi - \omega dt)^2 + \frac{\rho^2}{\Delta}dr^2 + \rho^2 d\theta^2 \\
&= -\left(1 - \frac{2mr}{\rho^2}\right)dt^2 + \frac{\rho^2}{\Delta}dr^2 + \rho^2 d\theta^2 + \frac{\Sigma}{\rho^2}\sin^2\theta\,d\phi^2 - \frac{4mra\sin^2\theta}{\rho^2}dt\,d\phi
\end{aligned} \tag{8.14}$$

with

$$\omega \equiv -\frac{g_{t\phi}}{g_{\phi\phi}} = \frac{2mra}{\Sigma} \quad \text{and} \quad \begin{cases} \Sigma \equiv \rho^2(r^2 + a^2) + 2mra^2 \sin^2\theta \\ = (r^2 + a^2)^2 - a^2\Delta\sin^2\theta\,. \end{cases} \tag{8.15}$$

We recall that $\Delta \equiv r^2 - 2mr + a^2$ and $\rho^2 \equiv r^2 + a^2\cos^2\theta$. In these coordinates we also have

$$g_{tt}g_{\phi\phi} - g_{t\phi}^2 = -\Delta\sin^2\theta\,, \quad \sqrt{-g} = \rho^2\sin\theta\,, \quad g^{t\phi} = -\frac{g_{t\phi}}{g_{tt}g_{\phi\phi} - g_{t\phi}^2} = -\frac{2mra}{\rho^2\Delta}\,,$$

$$g^{tt} = \frac{g_{\phi\phi}}{g_{tt}g_{\phi\phi} - g_{t\phi}^2} = -\frac{\Sigma}{\rho^2\Delta}\,, \quad g^{\phi\phi} = \frac{g_{tt}}{g_{tt}g_{\phi\phi} - g_{t\phi}^2} = \frac{\Delta - a^2\sin^2\theta}{\rho^2\Delta\sin^2\theta}\,. \tag{8.16}$$

The null vectors l_μ [defined in (8.9)] and k_μ have the following components in Boyer–Lindquist coordinates $x^\mu = (t, r, \theta, \phi)$:

$$\begin{cases} l^\mu\partial_\mu = \frac{r^2 + a^2}{\Delta}\partial_t - \partial_r + \frac{a}{\Delta}\partial_\phi \implies l_\mu dx^\mu = -dt - \frac{\rho^2}{\Delta}dr + a\sin^2\theta d\phi\,, \\ k^\mu\partial_\mu = \frac{r^2 + a^2}{\Delta}\partial_t + \partial_r + \frac{a}{\Delta}\partial_\phi \implies k_\mu dx^\mu = -dt + \frac{\rho^2}{\Delta}dr + a\sin^2\theta d\phi\,. \end{cases} \tag{8.17}$$

If $m^2 > a^2$, $\Delta = 0$ possesses two roots. Then the change of coordinates (8.13) is singular, which introduces coordinate singularities into the metric (8.14).

---

[1] These equations, written using concepts from differential calculus presented below in Part V, are easy to write out using the fact that the 'exterior' product $\wedge$ denotes antisymmetrization, so that, for example, $(\xi^T \wedge \xi^\varphi)_{\mu\nu} = \frac{1}{2}(\xi^T_\mu \xi^\varphi_\nu - \xi^T_\nu \xi^\varphi_\mu)$, while the 'exterior' derivative d satisfies $(\mathrm{d}\xi^T)_{\mu\nu} = \frac{1}{2}(\partial_\mu \xi^T_\nu - \partial_\nu \xi^T_\mu)$. It can therefore easily be shown (preferably with the aid of an algebra program!) that (8.12) is satisfied for any function $f = n(r)/\rho^2$ and therefore for the Kerr metric, where $n(r) = 2m$.

[2] Details can be found in, for example, Straumann (2013).

At infinity, $r \to \infty$, the asymptotic expansion of the Kerr metric (8.14) is

$$ds^2 \to - \left(1 - \frac{2m}{r} \right) dt^2 + dr^2 + r^2(d\theta^2 + \sin^2\theta d\varphi^2) - \frac{4ma\sin^2\theta}{r} dt\, d\phi\,. \tag{8.18}$$

Hence the solution turns out to be asymptotically flat, and the gravitational mass of the object whose gravitational field it represents is read off from the component g_{tt}: $M_{\mathrm{grav}} = m/G$. The cross term $g_{t\phi}$ is the same as that characterizing the rotation in the $\theta = \pi/2$ 'plane' of a star with Newtonian angular momentum $GJ_{\mathrm{N}} = ma$, as we shall see in Section 11.7, eqn (11.47) when we study relativistic celestial mechanics. The Kerr metric therefore describes the gravitational field of a rotating object.

It should, however, be noted that there is no metric solution of the Einstein equations in the presence of matter (with a realistic equation of state) which can be matched to the Kerr metric. For example, the metric outside a neutron star matches the Kerr metric only at lowest order (8.18).[3]

The Kerr–Newman solution

In 1965 Newman *et al.* generalized the Kerr solution and obtained an axially symmetric solution of the Einstein equations in the presence of an electromagnetic field. This solution is also of the Kerr–Schild type. In Boyer–Lindquist coordinates the metric is written as

$$ds^2 = -\frac{\Delta}{\rho^2} \left[dt - a\sin^2\theta\, d\phi \right]^2 + \frac{\sin^2\theta}{\rho^2} \left[(r^2 + a^2)d\phi - a\, dt \right]^2 + \frac{\rho^2}{\Delta} dr^2 + \rho^2 d\theta^2\,, \tag{8.19}$$

where $\Delta \equiv r^2 - 2mr + a^2 + q^2$ and $\rho^2 \equiv r^2 + a^2\cos^2\theta$. [It reduces to (8.14) for $q = 0$.] The Faraday tensor is given by[1]

$$F = \frac{q}{\rho^4} \left[(r^2 - a^2\cos^2\theta)\mathrm{d}r \wedge (dt - a\sin^2\theta\mathrm{d}\phi) + 2ar\cos\theta\sin\theta\, \mathrm{d}\theta \wedge [(r^2 + a^2)\mathrm{d}\phi - a\mathrm{d}t] \right]\,.$$

The magnetic field at infinity is $F_{\theta\phi}/r^2\sin\theta \approx 2aq\cos\theta/r^3$ and $F_{\phi r}/r\sin\theta \approx aq\sin\theta/r^3$. This is the field of a dipole of moment $\mu = qa$. Since the angular momentum of the object creating the field is $J = ma$, we have $\mu = qJ/m \equiv g(qJ/2m)$, where the *Landé g-factor* is 2 and therefore the same as that of an electron; see Book 2, Section 13.4. We note that for an electron of spin $\hbar/2$ we have $a/m \approx 3 \times 10^{46}$.

8.3   The geodesic equation

Since the Kerr solution is stationary and axially symmetric, the geodesic equation possesses two first integrals. In adapted coordinates, for example, Boyer–Lindquist coordinates (8.14), we have (see Section 5.2)

$$u_t \equiv g_{tt}\dot{t} + g_{t\phi}\dot{\phi} = -E\,, \quad u_\phi \equiv g_{\phi t}\dot{t} + g_{\phi\phi}\dot{\phi} = L\,, \tag{8.20}$$

where u_t and u_ϕ are the covariant components of the 4-velocity $u^\mu = (\dot{t}, \dot{r}, \dot{\theta}, \dot{\phi})$, the dot denotes differentiation with respect to the proper time τ, and where the constants E and

---

[3]See, for example, Hartle and Thorne (1969).

L are interpreted as the specific energy and angular momentum of the particle. Then by inverting and using (8.16) we find

$$
\begin{cases}
\dfrac{dt}{d\tau} = Lg^{t\phi} - Eg^{tt} \\[2mm]
\qquad = \dfrac{\Sigma E - 2mraL}{\rho^2 \Delta},
\end{cases}
\qquad
\begin{cases}
\dfrac{d\phi}{d\tau} = Lg^{\phi\phi} - Eg^{t\phi} \\[2mm]
\qquad = \dfrac{2mraE + (L/\sin^2\theta)(\rho^2 - 2mr)}{\rho^2 \Delta}.
\end{cases}
\tag{8.21}
$$

The normalization of u^μ, $g_{\mu\nu}u^\mu u^\nu = -\epsilon$ ($\epsilon = 1$ for time-like geodesics and $\epsilon = 0$ for light-like ones), then gives only one additional relation between $\dot{r}$ and $\dot{\theta}$, and so the variables θ and r are not obviously separable.

At this point it is useful to recall that, as we saw in Section 6.1, the Birkhoff–Jebsen theorem reveals an *a priori* unexpected property of spherically symmetric spacetime solutions of the vacuum Einstein equations, namely, that they are also static. They therefore possess an *additional* Killing vector associated with the invariance of the solution under time-like translations. A similar 'miracle' occurs for the Kerr solution.[4]

Indeed, the Kerr solution possesses, in addition to the two Killing vectors associated with its stationarity and axial symmetry, a *Killing tensor* $\xi_{\mu\nu}$ which is symmetric and satisfies

$$
D_{(\mu}\xi_{\nu\rho)} = 0,
\tag{8.22}
$$

namely (Penrose–Walker, 1969),

$$
\xi_{\mu\nu} = \Delta\, l_{(\mu}k_{\nu)} + r^2 g_{\mu\nu},
\tag{8.23}
$$

where l_μ and k_μ are the null vectors defined in Boyer–Lindquist coordinates by (8.17). [Verifying that (8.23) satisfies (8.22) is a good exercise.]

This allows us to reduce the integration of the geodesic equation to a quadrature. Indeed, if $\xi_{\mu\nu}$ is a Killing tensor, we have

$$
\begin{aligned}
\frac{d}{d\tau}(\xi_{\mu\nu}u^\mu u^\nu) &= u^\rho D_\rho(\xi_{\mu\nu}u^\mu u^\nu) \\
&= u^\rho u^\mu u^\nu D_\rho \xi_{\mu\nu} + \xi_{\mu\nu}\left[u^\nu(u^\rho D_\rho u^\mu) + u^\mu(u^\rho D_\rho u^\nu)\right] = 0,
\end{aligned}
\tag{8.24}
$$

where the first term is zero because the tensor is a Killing tensor, and the other terms are zero owing to the geodesic equation. Therefore, $\xi_{\mu\nu}u^\mu u^\nu = \mathcal{K}$, where $\mathcal{K}$ is the *Carter constant*. Combining this result with the normalization $g_{\mu\nu}u^\mu u^\nu = -\epsilon$, after some algebra using (8.23), (8.20), and (8.21) we find

$$
\begin{cases}
\rho^2 \dfrac{dr}{d\tau} = \pm\sqrt{[E(r^2 + a^2) - La]^2 - \Delta(\mathcal{K} + \epsilon r^2)}, \\[4mm]
\rho^2 \dfrac{d\theta}{d\tau} = \pm\sqrt{\mathcal{K} - \epsilon a^2 \cos^2\theta - (aE\sin\theta)^2 - (L/\sin\theta)^2 + 2ELa}.
\end{cases}
\tag{8.25}
$$

---

[4]Chandrasekhar wrote in *The Mathematical Theory of Black Holes* (1983): "*This discovery, by Carter, was the first of the many properties which have endowed the Kerr metric with the aura of the miraculous.*" [In 1966, B. Carter demonstrated the separability of the Hamilton–Jacobi equation rather than the existence of the Killing tensor, which is equivalent; see, for example, Frolov and Novikov (1998).]

Now we still need to integrate these equations and choose the initial conditions. Except in special cases, this is done numerically.[5]

Schwarzschild geodesics

The equations of motion of a free particle in the Schwarzschild field are obtained directly by using the fact that the spacetime is static and spherically symmetric (see Section 5.2). We can choose $\theta = \pi/2$, and then $u_t = -(1 - 2m/r)\dot{t} = -E$ and $u_\phi = r^2\dot{\phi} = L$. The equation for $r(\tau)$ is then obtained by normalizing the 4-velocity; see the special case of $L = 0$ in (6.15) and also (7.20). Of course, it can also be obtained by setting $a = 0$ (and $\mathcal{K} = L^2$) in eqns (8.21) and (8.25) above. The expression for $r(\tau)$ then is

$$\dot{r} = \pm\sqrt{E^2 - V_{\text{eff}}} \quad \text{with} \quad V_{\text{eff}} = \left(1 - \frac{2m}{r}\right)\left(\epsilon + \frac{L^2}{r^2}\right). \tag{8.26}$$

In Newtonian gravity, any particle of nonzero angular momentum can have a stable orbit about the central object. In general relativity it is necessary that $L^2 > 12m^2c^2$ (see Fig. 8.1). Moreover, we see that light does not travel in a straight line; it can even orbit a black hole at $r = 3m$.

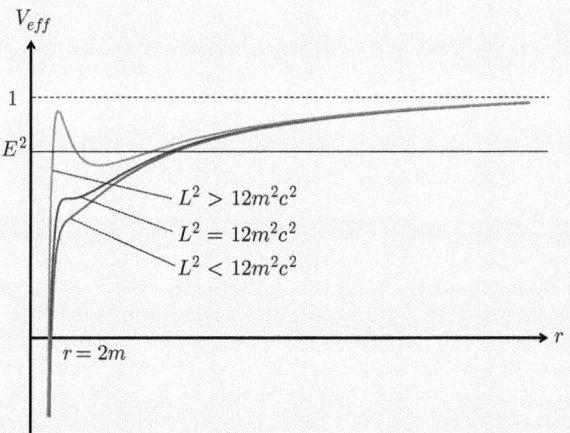

Fig. 8.1 The effective potential V_{eff}.

Here we shall just make a few remarks about the solutions (8.21) and (8.25) which will serve as an introduction to the following section.

• Geodesic motion can take place in the 'equatorial plane' $\theta = \pi/2$; it is sufficient to take $\mathcal{K} = (aE - L)^2$. If in addition we choose $L = aE$, the null geodesics ($\epsilon = 0$) are the line integrals of the null geodesic vectors k^μ and l^μ defined in (8.17) and we find

$$\frac{dr}{dt} = \pm\frac{\Delta}{r^2 + a^2}, \quad \frac{d\phi}{dt} = \frac{a}{r^2 + a^2}, \quad \text{where} \quad \Delta = r^2 - 2mr + a^2. \tag{8.27}$$

---

[5]See, for example, the website of Madore and also that of Riazuelo listed in the Bibliography. A detailed analytic study can be found in Chandrasekhar (1983).

We see that for $a^2 < m^2$ light travels in a circle in the 'plane' $\theta = \pi/2$ at $r_\pm = m \pm \sqrt{m^2 - a^2}$ with angular velocity $a/(r_\pm^2 + a^2)$ (measured with time t).

• The equations of motion of an object ($\epsilon = 1$) in 'radial' free fall ($L = 0$) from infinity ($E = 1$) in the 'plane' $\theta = \pi/2$ reduce to

$$\frac{dr}{d\tau} = -\sqrt{\frac{2m}{r}\left(1 + \frac{a^2}{r^2}\right)}, \quad \frac{dt}{d\tau} = \frac{\Sigma}{r^2\Delta}, \quad \frac{d\phi}{d\tau} = \frac{2ma}{r\Delta}, \tag{8.28}$$

where $\Sigma = (r^2 + a^2)^2 - a^2\Delta$. We thus see that the particle reaches $r = 0$. However, if $a^2 < m^2$, the coordinates t and ϕ diverge as $\ln(r - r_+)$ near $r_+ = m + \sqrt{m^2 - a^2}$. Now, we easily see from the definitions (8.13) that the equations (8.28) written in Kerr–Schild coordinates (T, φ) are regular at $r = r_+$. On the other hand, the equation for *outgoing* geodesics is singular in the Kerr–Schild coordinates (T, φ) constructed using the 'ingoing' null vector l^μ.

The innermost stable circular orbit (ISCO)

Certain astrophysical phenomena are modeled as rotating compact objects surrounded by an accretion disk lying in the equatorial plane, an ensemble of particles whose orbits become circular and which slowly drift toward the central object before falling onto it. Therefore, the concept of the innermost stable circular orbit (ISCO) is important.

An ISCO of the Kerr geometry is determined by the geodesic equation (8.25) with $\theta = \pi/2$ and $\mathcal{K} = (aE - L)^2$. The quartic radial equation becomes cubic and, setting $r = 1/u$ and $x = L - aE$, can be written as

$$u^{-4}\dot{u}^2 = F \quad \text{with} \quad F \equiv 2mu^3x^2 - u^2(2aEx + x^2 + a^2) + 2mu + E^2 - 1.$$

For the radial velocity to vanish we must have $F = 0$; for the orbit to be circular we must have $dF/du = 0$; and for the orbit to be the innermost one we must have $d^2F/du^2 = 0$. We therefore have three equations for the three unknowns E, L, and u. It is easily seen that the third one gives $E = (6mux^2 - x^2 - a^2)/(2ax)$. The second one then gives $x^2 = 1/(3u^2)$, and finally the first one reduces to a quartic equation for u: $9a^4u^4 - 28ma^2u^3 + 6(6m^2 - a^2)u^2 - 12mu + 1 = 0$.

When the metric is the Schwarzschild metric, $a = 0$ and we have $r = 6m$, $L^2 = 12m^2$, and $E = 2\sqrt{2}/3$. Therefore, the particle can radiate up to $E - 1 = 1 - 2\sqrt{2}/3 = 5.7\%$ of its mass as it falls toward the central object.

In the case $a = m$ there are two solutions, one retrograde with $r = 9m$, $E = 5/(3\sqrt{3})$, and $L = -22m/(3\sqrt{3})$ in which case only 3.8% of the mass can be radiated, and the other prograde with $r = m$, $E = 1/\sqrt{3}$, and $L = 2m/\sqrt{3}$ in which case up to 42% of the mass can be radiated.

These results show that the surroundings of a black hole can be the site of extremely energetic phenomena, as we shall see in detail in the following chapter.

8.4   The Kerr black hole

• *The curvature singularity*

As already mentioned in Section 8.2, there does *not* exist any object of finite size governed by a realistic equation of state whose external gravitational field can be given by the Kerr metric. This metric therefore can only describe a 'collapsed' object. Since the Kretschmann invariant $R_{\mu\nu\rho\sigma}R^{\mu\nu\rho\sigma} = 48m^2(r^2 - a^2\cos^2\theta)(\rho^4 - 16a^2r^2\cos^2\theta)/\rho^{12}$ (quadratic only in m because the metric is of the Kerr–Schild type) diverges at $r = 0$, $\theta = \pi/2$, this collapsed object is a curvature singularity. When r is the spheroidal coordinate of the Euclidean plane, it is a ring (see Section 8.1), hence the name *ring singularity*.[6]

The ratio a/m

The angular momentum of the Sun is estimated (by means of helioseismology) to be $J \approx 1.92 \times 10^{41}$ kg-m^2/s and the solar mass is $M \approx 1.99 \times 10^{30}$ kg, and so we have $a/m = Jc/GM^2 \approx 0.22$. If we view a neutron star as a homogeneous sphere, its Newtonian angular momentum will be given by $J \approx 2MR^2\omega/5$, and so $a/m = (4/5)(R\omega/c)/(2GM/c^2R)$. For $M = 1.4 M_\odot$, $R = 9$ km (and therefore $GM/c^2 R \approx 0.23$), and a period of the order of a millisecond, we find $a/m \approx 0.33$. The ratio a/m is difficult to estimate for astrophysical black holes, but it seems[7] that it may be about 1.

• *The horizons*

Let us consider the case $a^2 < m^2$, where $\Delta = (r - r_+)(r - r_-)$ with $r_\pm = m \pm \sqrt{m^2 - a^2}$. The geodesic motion is governed by (8.21) and (8.25). Whatever the values of E, L, and $\mathcal{K}$, they have the same asymptotic behavior near the zeros of Δ:

$$\frac{dr}{dt} \sim \pm \frac{\Delta}{r_\pm^2 + a^2}, \quad \frac{d\phi}{dt} \sim \frac{a}{r_\pm^2 + a^2} \quad \text{when} \quad r \to r_\pm \quad (8.29)$$

(and also $d\theta/dt \propto \Delta$). The radial velocity of light measured using the time of an observer at infinity is zero at $r = r_\pm$, which therefore are *trapped surfaces*. Any object, including light, rotates on these surfaces about the 'z axis' with uniform angular velocity

$$\Omega_\pm = \frac{a}{r_\pm^2 + a^2} = \frac{a}{2mr_\pm} \quad (8.30)$$

measured using the time t of a clock at rest at infinity. These are usually interpreted as the speeds of rotation of the surfaces $r = r_\pm$ themselves.

The determinant of the 3-metric induced on the surfaces $r = $ const is, cf. (8.16), $g|_\pm = -\rho^2 \Delta \sin^2 \theta$. It is zero at $r_\pm$. The equations $g_{ij}\zeta^j = 0$, where the g_{ij} with $i = \{t, \theta, \phi\}$ are given in (8.15), then have a non-trivial solution at $r = r_\pm$: $\zeta^j \propto (1, 0, -g_{t\phi}/g_{\phi\phi}|_\pm = \Omega_\pm)$. This defines the vector fields

$$\zeta_\pm^\mu = (1, 0, 0, \Omega_\pm) \quad (8.31)$$

tangent to the surfaces and of zero norm. The surfaces $r = r_\pm$ are therefore *light-like surfaces*.

The vectors $\zeta_\pm^\mu$ are Killing vectors because they can be written as $\zeta_\pm^\mu = \xi_{(t)}^\mu + \Omega_\pm \xi_{(\phi)}^\mu$, where $\xi_{(t,\phi)}^\mu$ are the Killing vectors associated with the stationarity and axisymmetry of the metric. Since their norms are zero at $r_\pm$, these surfaces are also *Killing horizons*.

We shall soon see that $r = r_+$ is also a surface of infinite redshift and an event horizon.

---

[6]Since the geodesic equations (8.21) and (8.25) are regular at $r = 0$ if $\theta \neq \pi/2$, geodesics can cross the singularity toward $r < 0$. ... We see from the metric (8.14) that $g_{\phi\phi}$ can then become negative, and the coordinate ϕ can become time-like. Since the latter is cyclic, there are *closed* time-like curves and regions where causality is violated. To learn more, see, for example, Carter (1972).

[7]See, for example, Thorne (1974) and McClintock, Narayan, and Steiner (2013).

The intrinsic geometry of the 2-surfaces $r = r_\pm$ at a given t is determined by the metric induced on them, namely

$$d\sigma_2^2 = \rho_\pm^2 d\theta^2 + \frac{4m^2 r_\pm^2}{\rho_\pm^2} \sin^2\theta \, d\phi^2 \qquad \text{with} \qquad \rho_\pm^2 \equiv r_\pm^2 + a^2 \cos^2\theta. \qquad (8.32)$$

Their area is

$$\mathcal{A}_\pm = 4\pi(r_\pm^2 + a^2) = 8\pi m r_\pm \qquad (8.33)$$

and their scalar curvature is $R_2 = 2(r_\pm^2 + a^2)(r_\pm^2 - 3a^2 \cos^2\theta)/\rho_\pm^6$. Even though they are often represented graphically as spheres or spheroids, they are neither of these.[8]

- *The static limit*

Let us consider an object at rest in the Boyer–Lindquist coordinate system (and therefore also in the Kerr–Schild system). Its 4-velocity $u^\mu = (u^0, 0, 0, 0)$ must be time-like, that is, it must have negative norm. Since this norm is $g_{tt}(u^0)^2 = -(1 - 2mr/\rho^2)(u^0)^2$, we must have

$$\rho^2 \geq 2mr \qquad \Longleftrightarrow \qquad r \geq r_e \quad \text{with} \quad r_e = m + \sqrt{m^2 - a^2 \cos^2\theta} > r_+. \qquad (8.34)$$

Therefore, it is not possible for any material object below the *static limit* $r = r_e$ to be at rest relative to infinity. For reasons which will be explained in the following chapter, this is also called the *ergosphere*.[8]

Now let us consider an object at $r = $ const, $\theta = $ const but possessing an angular velocity relative to infinity $d\phi/dt = \omega$. Its proper time τ is related to the time t of an observer at infinity through the length element $ds^2 = -d\tau^2$, that is,

$$d\tau^2 = -dt^2(g_{tt} + 2g_{t\phi}\omega + g_{\phi\phi}\omega^2) = -g_{\phi\phi}(\omega - \omega_+)(\omega - \omega_-)dt^2 \quad \text{with}$$

$$\omega_\pm = \frac{-g_{t\phi} \pm \sqrt{g_{t\phi}^2 - g_{tt}g_{\phi\phi}}}{g_{\phi\phi}} = \frac{2mra \pm (\rho^2/\sin\theta)\sqrt{\Delta}}{(r^2 + a^2)^2 - a^2\Delta\sin^2\theta}. \qquad (8.35)$$

Since $g_{\phi\phi}$ is positive for $r > r_+$, we must have $\omega \in [\omega_-, \omega_+]$ for the world line to be time-like. However, at $r < r_e$ where $g_{tt} > 0$, not only ω_+ but also ω_- is positive. Any object below the static limit therefore *must* rotate relative to infinity in the forward direction. In the limit $r = r_+$ where $\Delta = 0$ we recover the result (8.30), namely, that the object must rotate at the angular velocity $\Omega_+ = 2mr_+a/(r_+^2 + a^2)^2 = a/(r_+^2 + a^2)$.

We can see from (8.35) that surfaces ($r = $ const, $\theta = $ const) such that $\omega = \omega_\pm$ are surfaces of infinite redshift. This is true, in particular, for the horizon $r = r_+$.

- *Maximal extension*

The horizons $r = r_\pm$ are singularities of the coordinate system and not of the geometry, as we have already noted above. The extension of the $\theta = 0$ sections in which the metric for $\phi = $ const reduces to

$$ds^2 = -\frac{\Delta}{r^2 + a^2}dt^2 + \frac{r^2 + a^2}{\Delta}dr^2 \qquad (8.36)$$

is done using the Kruskal method presented in Section 7.1 and leads to a Carter–Penrose diagram very similar to that describing the spacetime of a Reissner–Nordström black hole;

---

[8]A more complete description of the geometry of the horizons (and of the ergosphere) of Kerr spacetime can be found, for example, in Visser (2009).

see Fig. 7.8. The past cone of the infinite future i_+ is bounded by $r = r_+$. Therefore, just like the horizon in the Schwarzschild case, the surfaces $r = r_+$ of the Reissner–Nordström and Kerr black holes are *event horizons*.[9]

Mass and angular momentum

In Section 8.2 we identified $M_{\rm g} \equiv m/G$ as the 'gravitational' mass of a black hole, and $J_{\rm N} \equiv a M_{\rm g}$ as the angular momentum, by identifying the asymptotic Kerr metric with the metric describing the gravitational field created by a rotating object with *Newtonian* angular momentum $J_{\rm N}$ in the post-Newtonian approximation.

The 'inertial' mass $M_{\rm in}$ and angular momentum $J_{\rm in}$ can be calculated using the conservation laws obtained in Section 5.4. The calculation of the mass is identical to that carried out in Section 6.3 to find the mass of Schwarzschild spacetime, and we find $M_{\rm in} = m/G$. Therefore, $M_{\rm in} = M_{\rm g}$, in agreement with the *strong equivalence principle*.

Let us now consider the Killing vector associated with axial rotations having components $\xi^\mu = (0, 0, 0, 1)$ in Boyer–Lindquist coordinates. The useful component of the Katz superpotential (5.23) then reduces to $\kappa \hat{J}^{[tr]} = D^t \hat{\xi}^r = -\frac{1}{2} \sqrt{-g}\, g^{rr} (g^{tt} g'_{t\phi} + g^{t\phi} g'_{\phi\phi})$. Using (8.19), we obtain its asymptotic form: $\kappa \hat{J}^{[tr]} \to 3\, ma \sin^3 \theta$, and so the 'inertial' angular momentum is $J_{\rm in} = -\int_S \hat{J}^{[tr]} d\theta d\phi = -a M_{\rm g} = -J_{\rm N}$. [The other definitions used in the literature, for example, the Komar integrals (5.26), give $J_{\rm in} = J_{\rm N}$, but after *ad hoc* changes of sign.]

To interpret this result we rewrite the Kerr metric (8.14) and (8.15) in the form $ds^2 = \eta_{ab} \vartheta^a \vartheta^b$, that is, in terms of the tetrad (see Book 1, Section 4.3 and Book 2, Section 5.2)

$$\vartheta^0 = \sqrt{\frac{\Delta \rho^2}{\Sigma}}\, dt\,, \quad \vartheta^1 = \sqrt{\frac{\Sigma \sin^2 \theta}{\rho^2}}\, (d\phi - \omega dt)\,, \quad \vartheta^2 = \sqrt{\frac{\rho^2}{\Delta}}\, dr\,, \quad \vartheta^3 = \sqrt{\rho^2}\, d\theta\,.$$

This tetrad defines a locally inertial frame: $\vartheta^a \sim dX^a$.

In this frame $dX^1 = 0$ defines the direction of a gyroscope pointing along the 'X^1 axis'. In terms of ϕ and t, this direction is given by $d\phi - \omega dt = 0$. Since ϕ gives the position of 'distant stars', we find that their angular velocity is $d\phi/dt = \omega > 0$ (measured using the time at infinity). A local inertial observer then sees himself rotating at angular velocity $-\omega$. On the horizon $\omega = \Omega_+$ with Ω_+ defined in (8.30). Therefore, the angular velocity of objects measured using the time t of an observer at rest at infinity is, when they reach the horizon, the *opposite* of that which a local inertial observer assigns to the black hole.

---

[9] For an in-depth study of the Kerr geometry see Carter (1972).

9

The physics of black holes I

In the preceding chapters we studied the Schwarzschild and Kerr solutions of the vacuum Einstein equations and introduced the concept of a *black hole*, a new object which is specific to this theory. Here we shall describe two related physical processes which can be induced by the gravitational field of a black hole: the Penrose process, which suggests that rotating black holes are large energy reservoirs, and superradiance, which is the first step in the study of black-hole stability.

9.1   The Penrose process and irreducible mass

The static limit of a Kerr black hole is the 2-surface inside which any object *must* rotate in the same direction as the black hole relative to an observer at rest at infinity. In Section 8.4 we saw that in adapted coordinates, for example, Boyer–Lindquist coordinates, it is given by $g_{tt} = 0$. In these coordinates the Killing vector associated with the stationarity of spacetime has the components $\xi^\mu_{(t)} = (1,0,0,0)$ and its norm is g_{tt}. It is therefore *space*-like below the static limit. The 4-velocity of any material object is time-like (or light-like) everywhere. Therefore, the scalar product $-\xi_{(t)}.u$ can be negative only inside the static limit.[1]

Let us consider an object A freely falling toward a Kerr black hole. It will travel along a geodesic, and so $-\xi_{(t)}.u_A = -g_{t\mu}u^\mu \equiv E_A > 0$ is a constant corresponding to its (specific) energy; see (8.20). We imagine that after passing the static limit it breaks up into two fragments, one of which, called B, has energy $-\xi_{(t)}.u_B \equiv E_B < 0$, which implies that it remains confined inside the static limit. The energy of the second fragment, C, is $E_C \equiv -\xi_{(t)}.u_C$, and owing to the law of energy conservation (the 'Einstein equivalence principle') $E_C = E_A - E_B > E_A > 0$. Therefore, this fragment can reach infinity, thereby *extracting* from the black hole an energy $E_C - E_A = |E_B|$. This is the *Penrose process*, and because of it the static limit is called the *ergosphere* (and the region located between the horizon and the static limit is called the *ergoregion*).[2]

To find the maximum energy that can be extracted from a Kerr black hole by the Penrose process, let us consider the vector introduced in (8.31) $\zeta = \xi_{(t)} + \Omega\,\xi_{(\phi)}$, where Ω is the constant angular velocity of rotation of the black hole and $\xi_{(t,\phi)}$ are the Killing vectors associated with the stationarity and axisymmetry. In Boyer–Lindquist coordinates its components are $\zeta^\mu = (1,0,0,\Omega)$ with $\Omega = -(g_{t\phi}/g_{\phi\phi})|_+ = a/(r_+^2 + a^2)$, where $r = r_+$, a solution of $\Delta = r^2 - 2mr + a^2$, is its exterior horizon; *cf.* (8.15). Its norm is given by (see Section 8.4)

---

[1]It is easy to see this in a local Minkowski frame. If $u^\mu = (1/\sqrt{1 - V^2}, V/\sqrt{1 - V^2})$ with $V < 1$ and if $\xi^\mu = (1, b)$, we have $-u.\xi = (1 - bV)\sqrt{1 - V^2}$, which cannot be negative unless $b > 1$, that is, if ξ is space-like.

[2]Penrose (1969); see also Ruffini and Wheeler (1971).

Relativity in Modern Physics. Nathalie Deruelle and Jean-Philippe Uzan.
© Oxford University Press 2018. Published in 2018 by Oxford University Press.
DOI: 10.1093/oso/9780198786399.001.0001

$$\zeta \cdot \zeta \equiv g_{\mu\nu}\zeta^\mu \zeta^\nu = g_{tt} + 2g_{t\phi}\Omega + g_{\phi\phi}\Omega^2 = g_{\phi\phi}(\Omega - \omega_+)(\Omega - \omega_-)$$

$$\text{with} \quad \omega_\pm = \frac{-g_{t\phi} \pm \sqrt{g_{t\phi}^2 - g_{tt}g_{\phi\phi}}}{g_{\phi\phi}} = \frac{2mra \pm (\rho^2/\sin\theta)\sqrt{\Delta}}{(r^2+a^2)^2 - a^2\Delta\sin^2\theta}. \tag{9.1}$$

Since $g_{\phi\phi}$ is positive outside the horizon, $\zeta\cdot\zeta$ is negative if $\Omega \in [\omega_-, \omega_+]$. This occurs near the horizon [and, as is easily shown, in the entire ergoregion if $m^2 > a^2(1 + 1/\sqrt{2})$]; see Fig. 9.1. On the horizon itself ζ has zero norm.

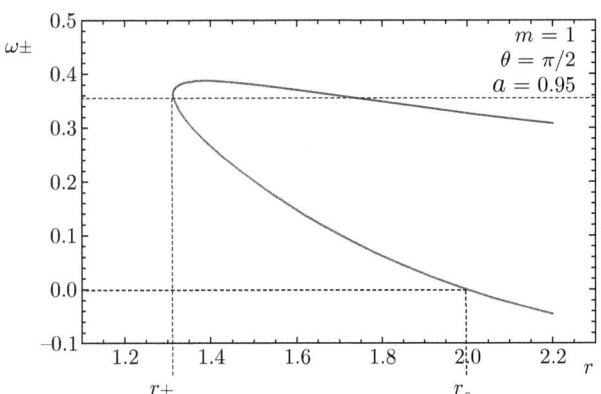

Fig. 9.1 $\omega_\pm$ (for $\theta = \pi/2$ and $a/m = 0.95$).

Since the scalar product of two time-like vectors is also time-like, we have

$$0 \geq \zeta \cdot u_B = (\xi_{(t)} \cdot u_B) + \Omega\left(\xi_{(\phi)} \cdot u_B\right) = -E_B + \Omega L_B, \tag{9.2}$$

where E_B and L_B are by definition the specific energy and angular momentum of the particle B of 4-velocity u_B and energy $E_B < 0$ created in the ergoregion by the Penrose process. Since particle B cannot leave the ergoregion it will decrease the effective mass of the black hole by $\delta M = E_B < 0$ and its angular momentum by $\delta J = L_B < E_B/\Omega < 0$. The mass and angular momentum of the black hole will therefore vary as

$$\delta M - \Omega\,\delta J \geq 0, \tag{9.3}$$

which translates the inequality (9.2) for the particle into an inequality for the black hole. The equality holds when the process occurs on the horizon.

In 1970, D. Christodoulou posed and then solved the following problem: is it possible to rewrite the inequality (9.3) in the form $\delta F \geq 0$, where $F(M, J)$ is a function of the intrinsic geometry of the black hole, for example, its area $F = F(\mathcal{A})$? The answer is 'yes'. Indeed, since the area of the black hole $\mathcal{A}(M, J)$ is $\mathcal{A} = 8\pi m r_+ = 8\pi GM(GM + \sqrt{G^2M^2 - J^2/M^2})$ because $m = GM$ and $a = J/M$ [*cf.* (8.33)], it can easily be shown that

$$\delta F = \frac{dF}{d\mathcal{A}}\frac{8\pi G}{\kappa}(\delta M - \Omega\,\delta J) \quad \Longrightarrow \quad \delta M - \Omega\,\delta J \geq 0 \quad \Longleftrightarrow \quad \delta F \geq 0, \tag{9.4}$$

where we have set $\kappa \equiv \sqrt{m^2 - a^2}/(2mr_+)$.

The surface gravity

The coefficient κ appearing in (9.4) can be given a geometrical meaning. As we saw in Section 8.4, the vector $\zeta^\mu = (1, 0, 0, \Omega)$, where $\Omega = a/(r_+^2 + a^2)$ is the velocity of the black hole, is a null vector, simultaneously tangent to and normal to the horizon r_+, and it is also a Killing vector. We define the *surface gravity* κ of the black hole as

$$\kappa^2 = -\frac{1}{2}(D_\mu \zeta_\nu \, D^\mu \zeta^\nu)|_{r_+} \quad \Longrightarrow \quad \kappa = \frac{\sqrt{m^2 - a^2}}{2mr_+} \quad \text{for the Kerr black hole.}^3$$

In the particular case of a Schwarzschild black hole, we can also give a 'physical' interpretation to $\kappa = 1/(4m)$ which explains its name. Since the 3-acceleration of a particle is $a^i = Du^i/d\tau$, if it is at rest we have $u^\mu = (1, 0, 0, 0)$ with $g_{tt}(u^t)^2 = -1$. Therefore, $a^i = -\Gamma^i_{tt}/g_{tt}$. Since $-g_{tt} = 1/g_{rr} = 1 - 2m/r$, we have $a^i = ((4m)^{-1}, 0, 0)$ on the horizon. Its modulus $|a| = \sqrt{g_{rr}}/(4m)$ diverges, but $|a|/u^t = 1/(4m) = \kappa$ is finite. Therefore, κ is the force per unit mass which must be applied to a particle for it to remain at rest on the horizon, after correction by a 'redshift' factor.

Now we need to choose the (increasing) function $F(\mathcal{A})$. With the goal of explaining quasars or gamma-ray bursts as processes of energy extraction from black holes, the Princeton school led by Wheeler[2] required that F have the dimensions of a mass and chose $F \equiv M_{\rm irr}$ with

$$GM_{\rm irr} = \sqrt{\frac{\mathcal{A}}{16\pi}} = \sqrt{\frac{r_+^2 + a^2}{4}} = \sqrt{\frac{mr_+}{2}} \quad \Longleftrightarrow \quad M^2 = M_{\rm irr}^2 + \frac{J^2}{4G^2 M_{\rm irr}^2}. \tag{9.5}$$

(The factor of 16π is chosen so that $M_{\rm irr} = M$ for $J = 0$, that is, for a Schwarzschild black hole.) The quantity $M_{\rm irr}$ is the *irreducible mass*. Owing to (9.4), it must generically increase when breakups, collisions, *etc.* occur near a black hole. If it remains constant, the transformation is said to be *reversible*.

Therefore, the maximum energy that a Kerr black hole initially of mass M and angular momentum J can lose is $M - M_{\rm irr}$. If the black hole is a Schwarzschild one, then $J = 0$ and $M = M_{\rm irr}$, and no energy can be extracted by a Penrose process. If the angular momentum is maximal, $J = aM$, the irreducible mass is minimal and equal to (since $r_+ = m$) $M_{\rm irr} = M/\sqrt{2}$, and so the energy that can in principle be extracted is optimal in this case:

$$\frac{M - M_{\rm irr}}{M} = 1 - 1/\sqrt{2} = 29\,\%. \tag{9.6}$$

We see that this energy can be huge when we recall that $1 \text{ g} = 5.6 \times 10^{32} \text{ eV} = 0.9 \times 10^{14} \text{ J}$.

---

[3]The concept of 'surface gravity' is due to Carter; see, for example, Carter (1972). The definition we have given here, see, for example, Wald (1984), allows κ to be calculated in Boyer–Lindquist coordinates. An alternative definition, $\zeta^\mu D_\mu \zeta^\nu = \kappa\zeta^\nu$, requires changing to coordinates which are regular on the horizon, for example, Eddington–Finkelstein coordinates.

Energy extraction from a black hole

To be able to extract from a Kerr black hole an energy equal to 29 % of its initial mass M by the Penrose process, the initial rotation of the black hole must be maximal: $J = M^2/G$. In addition, the breakup of the incident particle must take place on the horizon, and the ejected particle of energy $-E_B = -L_B\Omega > 0$ must be able to reach infinity. This is therefore more of a *gedankenexperiment* than a realistic model capable of explaining observed phenomena such as quasars, active galactic nuclei, gamma-ray bursts, and so on.

Collision processes occurring near the horizon are a more efficient means of extracting energy from a black hole. Indeed, the particle energy can be arbitrarily large, but the conditions for particles to be able to escape to infinity are very restrictive.[4]

Energy extraction from a charged black hole by the Penrose process is also possible. The irreducible mass is obtained as above, but now by studying the trajectories of *charged* particles in the Reissner–Nordström or Kerr–Newman metric. It turns out to also be given by $m_{\mathrm{irr}} = \frac{1}{2}(r_+^2 + a^2)^{\frac{1}{2}} = \frac{1}{2}[(m + (m^2 - q^2 - a^2)^{\frac{1}{2}}]$, so that the energy which is a priori available can reach 50 % of the initial mass (for $a = 0$ and $q = m$). However, the charge of a macroscopic body is rapidly neutralized by the environment unless there exists a mechanism preventing this.[5]

If the black hole is surrounded by an accretion disk, one can imagine that the ambient magnetic field could extract some of the rotational energy. This process, which was proposed by Blandford and Znajek in 1977, was studied within the framework of the *membrane paradigm*, in which locally inertial frames play an important role, with the goal of describing the horizon as a 'classical' object. The origin and efficiency of the process are nevertheless still under debate, and must be analyzed using relativistic magnetohydrodynamics.[6]

9.2 Superradiance

The study of the stability of black holes involves the linearization of the Einstein equations about the Schwarzschild or Kerr solution. As we shall see in what follows, the equations of motion for perturbations of the metric are wave equations. The problem then is to determine whether or not these solutions are bounded.

As a preliminary exercise, let us consider a massless scalar test field propagating in Schwarzschild spacetime (Price, 1971 and Thorne, 1972). Its equation of motion is the Klein–Gordon equation $\Box\Phi = 0$, or $\partial_\mu(\sqrt{-g}g^{\mu\nu}\partial_\nu\Phi) = 0$. In Droste coordinates the metric is $ds^2 = -(1 - 2M/r)dt^2 + dr^2/(1 - 2M/r) + r^2(d\theta^2 + \sin^2\theta d\phi^2)$ (here we use M to denote the geometric mass of the black hole in order to avoid confusion with the 'quantum number' m), and so Φ can be expanded in spherical harmonics:

$$\Phi(t, r, \theta, \phi) = \sum_{lm} Y_m^l(\theta, \phi)\frac{\Psi_{lm}(t, r)}{r}. \tag{9.7}$$

[4]Piran and Shaham (1977) and Bejger, Piran, *et al.* (2012).

[5]On this subject, see Ruffini *et al.* (2010).

[6]Blandford and Znajek (1977); see also Komissarov (2009) as well as Lasota, Gourgoulhon, *et al.* (2014). For an introduction to the 'membrane paradigm' see Damour (1979), and for recent developments see, for example, Gourgoulhon and Jaramillo (2005).

An introduction to relativistic magnetohydrodynamics can be found in, for example, Lichnerowicz (1967), or Gourgoulhon *et al.* (2011).

Now we easily find the equation for the modes Ψ_{lm}:

$$\frac{\partial^2 \Psi_{lm}}{\partial^2 r_*} - \frac{\partial^2 \Psi_{lm}}{\partial^2 t} - V_{\text{eff}} \Psi_{lm} = 0 \quad \text{with} \quad V_{\text{eff}} = \left(1 - \frac{2M}{r}\right)\left(\frac{2M}{r^3} + \frac{l(l+1)}{r^2}\right), \quad (9.8)$$

where $r_* \equiv r + 2M \ln(r/2M - 1)$ is the tortoise coordinate introduced in Section 7.2 which pushes away the horizon $r = 2M$ at $r_* \to -\infty$. The effective potential V_{eff} can be compared to that governing the motion of a massless particle; cf. (8.26). The essential feature here is that V_{eff} is positive everywhere outside the horizon; see Fig. 9.2.

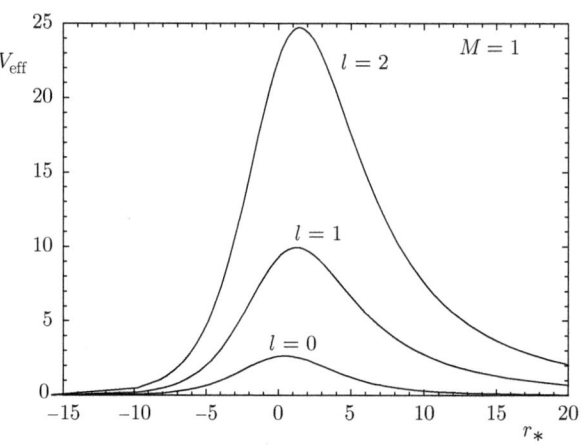

Fig. 9.2 The effective potential V_{eff}.

Since the metric is static, we can push the separation of variables farther and reduce (9.8) to an ordinary differential equation:

$$\frac{d^2 u}{dr_*^2} + \left(E^2 - V_{\text{eff}}\right) u = 0 \quad \text{with} \quad \Psi_{lm}(t, r) = \int dE \, e^{-iEt} \, u_{lm}(r, E) \quad (9.9)$$

with E real and positive. Here we have set $u_{lm}(r, E) = u$ in order to simplify the notation.

At spatial infinity and near the horizon, the effective potential vanishes and the solutions are waves:

$$r_* \to \infty, \quad u = Ae^{iEr} + Be^{-iEr}, \quad r_* \to -\infty, \quad u = Ce^{iEr_*} + De^{-iEr_*}, \quad (9.10)$$

where the constants A, B, C, and D can depend on l, m, and E.

Multiplying (9.9) by $\bar{u}$ and then subtracting the complex conjugate of the product, we find that the Wronskian $\bar{u}du/dr_* - u d\bar{u}/dr_*$ is constant, which for E real implies that

$$|A|^2 - |B|^2 = |C|^2 - |D|^2. \quad (9.11)$$

For example, for $C = 0$ we have $R + T = 1$ with $R \equiv |A|^2/|B^2|$ and $T \equiv |D|^2/|B^2|$. The 'transmission coefficient' T, which like R *a priori* depends on l, m, and E, is given by

$T \sim \exp[-2 \int \sqrt{V_{\text{eff}} - E^2} \, dr_*]$ in the WKB approximation, and is sometimes referred to as the 'greybody factor' (in the context of Hawking radiation; see Section 10.1).

If now we multiply (9.8) by $\partial \bar{\Psi}_{lm}^* / \partial t$, add to this product its complex conjugate, and integrate over r_*, we obtain (suppressing the indices l and m)

$$\frac{\partial}{\partial t} \int dr_* \left(\left| \frac{\partial \Psi}{\partial t} \right|^2 + \left| \frac{\partial \Psi}{\partial r_*} \right|^2 + V_{\text{eff}} |\Psi|^2 \right) = \left[\frac{\partial \bar{\Psi}}{\partial t} \frac{\partial \Psi}{\partial r_*} + \frac{\partial \Psi}{\partial t} \frac{\partial \bar{\Psi}}{\partial r_*} \right]_{-\infty}^{+\infty}. \tag{9.12}$$

As $r_* \to \pm\infty$, $V_{\text{eff}} \to 0$ and Ψ_{lm} is of the form $\Psi_{lm}(r,t) = e^{-iEt} u_{lm}(r, E)$, where $u \equiv u_{lm}(r, E)$ is given in (9.10) and satisfies the condition (9.11), so that the right-hand side of (9.12) is zero. Then for each mode (l, m) we find that $\int dr_* (|\dot{\Psi}|^2 + |\Psi'|^2 + V_{\text{eff}} |\Psi|^2))$, a sum of positive-definite terms, is a constant. Therefore, Ψ_{lm} cannot diverge in either space or time (at least as long as we restrict ourselves to the region outside the horizon).

Now let us carry out the same analysis using the Kerr metric. Although the spacetime is no longer spherically symmetric, the equation $\Box \Phi = 0$ is still separable. We decompose Φ as

$$\Phi(t, r, \theta, \phi) = \sum_{l,m} \int dE \, e^{im\phi} e^{-iEt} S_{lm}(\theta, E) \frac{u_{lm}(r, E)}{r}, \tag{9.13}$$

where (t, r, θ, ϕ) are the Boyer–Lindquist coordinates, E is real and positive, and the spheroidal harmonics $S_{lm}(\theta, E) \equiv S$ satisfy

$$\frac{1}{\sin \theta} \frac{d}{d\theta} \left(\sin \theta \frac{dS}{d\theta} \right) + S \left[\lambda - \left(aE \sin \theta - \frac{m}{\sin \theta} \right)^2 \right] = 0. \tag{9.14}$$

[When $aE = 0$, the eigenvalue λ reduces to $l/(l+1)$, $l \in \mathbb{N}$, and S becomes a spherical harmonic.] Then, using the explicit form of the Kerr metric given in (8.14)–(8.16), we easily find the equation for $u_{lm}(r, E) \equiv u$ [which generalizes (9.9)]:

$$\frac{d^2 u}{dr_*^2} + \frac{(r^2 + a^2)^2}{r^4} (E - E_0^+)(E - E_0^-) u = 0. \tag{9.15}$$

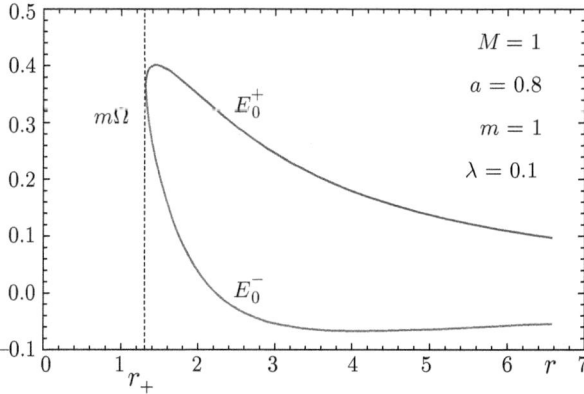

Fig. 9.3 The effective potentials $E_0^\pm$.

Here the 'effective potentials' $E_0^{\pm}$ are given by (see also Fig. 9.3)

$$E_0^{\pm} \equiv \frac{ma \pm \sqrt{\Delta}\sqrt{\lambda + 2(Mr - a^2)/r^2}}{r^2 + a^2}. \tag{9.16}$$

The tortoise coordinate r_* satisfies $dr_*/dr = r^2/\Delta$, and we recall that $\Delta = r^2 - 2Mr + a^2$. [The similarity of this equation to the geodesic equation (8.25) should be noted.]

At spatial infinity and near the horizon the solutions are waves:

$$\begin{cases} r^* \to \infty : \quad u = Ae^{iEr} + Be^{-iEr}, \\ r^* \to -\infty : \quad u = Ce^{ikr_*} + De^{-ikr_*} \quad \text{with} \quad k = \frac{r_+^2 + a^2}{r_+^2}(E - m\Omega), \end{cases} \tag{9.17}$$

where the constants A, B, C, and D may depend on l, m, and E, and $\Omega \equiv a/(r_+^2 + a^2)$ is the angular velocity attributed to the black hole; see (8.30). The fact that the Wronskian $\bar{u}du/dr_* - ud\bar{u}/dr_*$ is constant implies the generalization of (9.11):

$$E(|A|^2 - |B|^2) = k(|C|^2 - |D|^2). \tag{9.18}$$

For $C = 0$ and setting $R \equiv |A|^2/|B|^2$ and $T \equiv |k|\,|D|^2/(|B|^2 E)$ ($T > 0$), we then have

$$R = 1 + T > 1 \quad \text{if} \quad E - m\Omega < 0. \tag{9.19}$$

An incident wave can therefore be *amplified* by the gravitational field of the black hole (for the simple reason that the wave De^{-ikr_*} is actually traveling away from the black hole because $k < 0$). This phenomenon is called *superradiance*, and it is the 'wave' version of the Penrose process signaling an instability of the Kerr black hole.

On the stability of black holes

The linearization of the Einstein equations about the Schwarzschild solution leads to equations similar to (9.7)–(9.9) called the Regge–Wheeler (1957) and Zerilli equations (1970), and a proof of their stability more rigorous than the Detweiler–Ipser proof (1973) presented in (9.12) has been given by Kay and Wald (1987).

The linearization of the Einstein equations about the Kerr solution was done by Teukolsky (1972), and the study of superradiance (predicted by Zel'dovich in 1970) was begun by Starobinsky in 1973, as well as by Press and Teukolsky. The explicit calculation of the transmission coefficient T [see (9.19)] shows that the wave can be amplified by 0.3 %, and a similar analysis of the amplification of electromagnetic and gravitational waves indicates amplifications of 4.4 and 138 %, respectively. However, the fields of half-integer spins are not amplified. This is because the associated currents are positive-definite, in contrast to the Klein–Gordon current [that is, $-i(\Phi\partial_\mu\bar{\Phi} - c.c.)$; see Book 2, Section 9.3]. In 1989 Whiting showed that the superradiant modes cannot grow exponentially with time, but as yet there is no complete and rigorous proof of the stability of the solutions of $\Box\Phi = 0$ outside the Kerr horizon.

In 1970 Vishveswara began the study of *quasi-normal modes*, that is, solutions of the wave equations which are simultaneously purely incoming near the horizon and purely outgoing at infinity, that is, of the type (9.17) with $B = C = 0$. Their frequencies, which are necessarily complex, describe mode damping, in particular, damping of the gravitational waves emitted during stellar collapse to a black hole, or after the coalescence of a binary system.

The study of *massive* fields in Kerr spacetime leads to the prediction of a 'bomb' effect, the *black hole bomb* in the superradiant regime: the waves are amplified in each reflection on the wall of the potential which traps them near the black hole (Press and Teukolsky, 1972, and Damour *et al.* 1976).

Finally, as we saw in the preceding chapters, the inner horizon r_- of the Reissner–Nordström and Kerr black holes is a Cauchy horizon (see Fig. 7.8) which implies that it is unstable, as shown by Chandrasekhar and Hartle in 1982.[7]

9.3 Quantum superradiance

The phenomenon of superradiance can be understood as the spontaneous creation of particle–antiparticle pairs due to excitation of the quantum vacuum by the gravitational field, and the transmission coefficient T then is interpreted as the creation rate per unit time and unit volume.

This is the gravitational version of the Schwinger effect of vacuum instability due to a sufficiently strong electric field.

The Schwinger effect

Let us consider a particle of charge $-q < 0$ and mass μ (and spin 0 for simplicity) in a constant electric field $\mathcal{E}$ effective over a distance L along the z axis of an inertial frame (here we shall limit ourselves to two dimensions). Its classical motion is governed by the Lorentz equation (see Book 2, Section 11.4), which can be compared with the geodesic equation (8.25):

$$\mu^2 \left(\frac{dz}{d\tau} \right)^2 = (E - E_0^+)(E - E_0^-) \quad \text{with} \quad E_0^\pm = -q\mathcal{E}z \pm \mu \quad \text{for} \quad z \in [-L/2, L/2].$$

We consider the case where the field strength is above the critical value $2\mu/(|q|L)$. The effective potentials $E_0^\pm$ shown in Fig. 9.4 are the analogs of those of Fig. 9.3. For values of E such that $-q\mathcal{E}L/2 + \mu < E < q\mathcal{E}L/2 - \mu$, the particle will be reflected from the potential barrier. We note that on the right $E > E_0^+$ but on the left $E < E_0^-$. In addition, $-E_0^-(-E, +q) = E_0^+(E, -q)$.

In quantum field theory the particle is described by an operator $\hat{\Phi}$ satisfying the equation $(\partial_\mu - iqA_\mu)(\partial^\mu - iqA^\mu)\hat{\Phi} - \mu^2\hat{\Phi} = 0$ with $A^\mu = (-\mathcal{E}z, 0)$. We expand it in modes of the form $e^{-iEt}\Psi(z)$, where Ψ is the solution of an equation similar to (9.15):

$$\frac{d^2\Psi}{dz^2} = W\Psi \quad \text{with} \quad W = -(E - E_0^+)(E - E_0^-) = \mu^2 - (E + q\mathcal{E}z)^2.$$

The solutions at $z = \pm\infty$ are $\Psi \to Ae^{ik_+z} + Be^{-ik_+z}$ and $\Psi \to Ce^{ik_-z} + De^{-ik_-z}$, where $k_\pm = [(E \pm q\mathcal{E}L/2)^2 - \mu^2]^{1/2}$, with, according to the Wronskian theorem,

$$k_+(|A|^2 - |B|^2) = k_-(|C|^2 - |D|^2). \tag{9.20}$$

---

[7]For an introduction to the literature and recent work on the topics discussed here, see, for example, Dafermos and Rodnianski (2010); Berti, Cardoso, and Starinets (2009); Cardoso, Dias, Lemos, and Yoshida (2004); and Marolf and Ori (2012).

Book 3

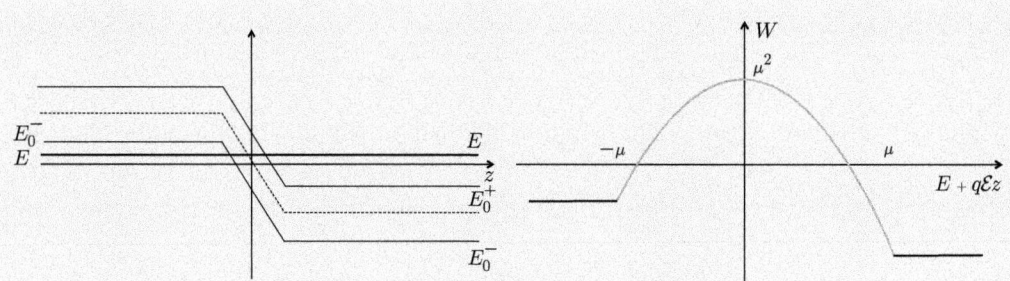

Fig. 9.4 The Schwinger critical field.

Since on the left $E < E_0^-$, the mode $Ce^{-iEt}e^{ik_-z}$ represents an *antiparticle* of energy $-E$ and charge $+q$ propagating to the left (and not to the right). Therefore, an antiparticle moving in the direction of decreasing z corresponds to $D = 0$ in (9.20). Setting $R = |A|^2/|B|^2$ and $T = k_-|C|^2/(k_+|B|^2)$, we then have

$$R = 1 + T > 1$$

as for the case of classical superradiance studied above. In the quantum treatment we are discussing here, this is interpreted as the spontaneous creation of particle–antiparticle pairs at a rate given by the transmission coefficient T. Using the WKB approximation, we easily find

$$T \approx \int \exp[-2\sqrt{W}dz] = e^{-\frac{\pi\mu^2}{|q|\mathcal{E}}} .$$

This phenomenon was predicted by Sauter in 1931, and also by Heisenberg and Euler in 1936, and interpreted in the context of quantum field theory by Schwinger in 1951.[8]

---

[8]As we saw in Book 2, Section 11.4, the effect is very weak for the electric field strengths presently accessible in the laboratory. However, it can be important in astrophysics if black holes are charged; see footnote 5. But the 'Schwinger' effect due to the rotation of a black hole can only play a role in the dynamics of microscopic black holes whose mass is of the order of the Planck mass, 10^{-5} g. See Starobinsky (1973) and Deruelle and Ruffini (1974).

10

The physics of black holes II

Here we give a brief description of Hawking radiation, which involves a combination of general relativity and quantum field theory and leads to a thermodynamical interpretation of the laws governing the evolution of black holes.

We conclude our study of black holes with the Israel theorem, which allows one to argue that if gravitation is described by general relativity, then not only do black holes exist, but all black holes are represented by the Kerr–Schwarzschild solution.

10.1 Hawking radiation

In the preceding chapter we limited ourselves to studying the various physical processes which occur *outside* the horizon of a black hole. However, we know that the spacetime of a black hole is larger than the east quadrant of the Kruskal diagram (see Fig. 7.2) to which external observers are confined. In quantum physics, the concepts of vacuum and particle are global and thus depend on the region of spacetime under consideration. Below we shall see that, as shown by S. W. Hawking in 1974, a consequence of this is that black holes represented by the east and north quadrants of the Kruskal diagram radiate like black bodies.[1]

The Unruh effect

Here, working within special relativity, we shall show that the concepts of vacuum and quantum particle depend on the portion of Minkowski spacetime accessible to the observer.

Let us use Rindler coordinates (t, z, x, y) in which the metric is written as

$$ds^2 = -(1 + az)dt^2 + \frac{dz^2}{4(1 + az)} + dx^2 + dy^2. \tag{10.1}$$

Making the transformation $aT = \sqrt{1 + az} \sinh at$, $aZ = \sqrt{1 + az} \cosh at$, we see that this is the Minkowski metric $ds^2 = -dT^2 + dZ^2 + dx^2 + dy^2$, that the coordinates $(t, z > -1/a)$ cover only quadrant I defined by $Z^2 - T^2 > 0$, and that the hyperbola $z = $ const represents the world line of a uniformly accelerated observer; see Book 2, Section 5.2.

Just as in Sections 9.2 and 9.3 where we studied fields propagating outside a black hole, here we shall consider a massless scalar field Φ defined in quadrant I. We decompose it in modes in Rindler coordinates (the analogs of Schwarzschild coordinates) as

$$\Phi = e^{-iEt} e^{i(p_x x + p_y y)} \psi_{E, p_x, p_y}(z).$$

Its Klein–Gordon equation of motion $\Box \Phi = 0$ then reduces to [we write $\psi_{E, p_x, p_y}(z) \equiv \psi$ to simplify the notation]

---

[1] Hawking (1974) and (1975).

Relativity in Modern Physics. Nathalie Deruelle and Jean-Philippe Uzan.
© Oxford University Press 2018. Published in 2018 by Oxford University Press.
DOI: 10.1093/oso/9780198786399.001.0001

$$\frac{d^2\psi}{dz_*} = -\psi \left[E^2 - (p_x^2 + p_y^2)e^{2az_*} \right],$$

where $z_* = \frac{1}{2a}\ln(1+az)$ is a tortoise coordinate which pushes away the horizon $z = -1/a$ at $z_* = -\infty$. The solution $\psi(z)$ is a linear combination of Bessel functions which near the horizon is a sum of incoming and outgoing modes [we choose $E > 0$]:

$$\Phi \propto \alpha e^{-iE(t-z_*)} + \beta e^{-iE(t+z_*)}. \tag{10.2}$$

The modes allow us to define the concepts of vacuum and particle in quadrant I in the standard manner of quantum field theory. To define now these concepts in quadrants I ($Z^2 - T^2 > 0$, $Z > 0$) and II ($Z^2 - T^2 < 0$, $Z > 0$) it is necessary to *continue* the modes (10.2) into quadrant II (see Fig. 10.1).

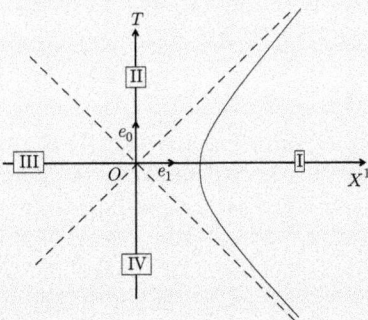

Fig. 10.1 Rindler coordinates and quadrants of Minkowski spacetime.

For this we introduce the advanced Eddington–Finkelstein coordinate, $v = t + z_*$, in which the metric (10.1) is written as

$$ds^2 = -(1+az)dv^2 + dv\, dz + dx^2 + dy^2.$$

It is *regular* on the future horizon, that is, for $z \to -1/a$, $t \to +\infty$ with v finite. As a function of v, eqn (10.2) becomes

$$\Phi \propto \alpha e^{-iEv}(1+az)^{iE/a} + \beta e^{-iDv}.$$

The second term is regular on the horizon. The first (the outgoing mode) can be analytically continued beyond $z = -1/a$ (writing $-1 = e^{\pm i\pi}$) as

$$\alpha e^{-iEv}(1+az)^{iE/a}$$

$$\to \quad \alpha e^{-iEv}\left[\Theta(1+az)\,(1+az)^{iE/a} + [\Theta(-1-az)\,(-1-az)^{iE/a}\exp(\pm\pi E/a) \right], \tag{10.3}$$

where Θ is the Heaviside distribution and the $+$ sign corresponds to Im $z < 0$. Since the vector $\partial/\partial v$ points toward the future, we choose this sign such that the mode describes an antiparticle state, just as in the 'standard' quantum field theory.

The mode (10.3) is then interpreted as follows. At $z > -1/a$ it represents a flux of $|\alpha|^2$ particles moving away from the horizon, and at $z < -1/a$ it represents a flux of $|\alpha|^2 \exp(2\pi E/a)$ particles moving toward increasing z. However, since the vector $\partial/\partial t$ is space-like at $z < -1/a$, these particles travel *backwards* in time. They are therefore *antiparticles* which follow the arrow of time and move toward decreasing z. Finally, since the mode describes an antiparticle state, it must be normalized to -1, which imposes that

$$|\alpha|^2 - |\alpha|^2 \exp(2\pi E/a) = -1 \qquad \Longrightarrow \qquad |\alpha|^2 = \frac{1}{\exp(2\pi E/a) - 1} . \qquad (10.4)$$

The coefficient $|\alpha|^2$ is interpreted as the rate of pair creation. The distribution is that of a black body at temperature

$$T = \frac{a}{2\pi} . \qquad (10.5)$$

If we now demand that the physical spacetime be represented by quadrants I and II of Minkowski spacetime, then an observer confined to region I (and therefore necessarily accelerated) must detect the particles, while an inertial observer does not detect any. This is the *Unruh effect*.[2]

Let us return to the study of a massless scalar field Φ propagating in Schwarzschild spacetime as in Section 9.2. Since the metric is $ds^2 = -(1 - 2M/r)dt^2 + dr^2/(1 - 2M/r) + r^2(d\theta^2 + \sin^2\theta d\phi^2)$, we can decompose it in the modes

$$\Phi(t, r, \theta, \phi) = Y_m^l(\theta, \phi) \frac{u_{lm}(r, E)}{r} e^{-iEt} ,$$

and the Klein–Gordon equation $\Box\Phi = 0$ for each mode reduces to [setting $u_{lm}(r, E) \equiv u$]

$$\frac{d^2u}{dr_*^2} + \left(E^2 - V_{\text{eff}}\right) u = 0 , \quad \text{where} \quad V_{\text{eff}} = \left(1 - \frac{2M}{r}\right)\left(\frac{2M}{r^3} + \frac{l(l+1)}{r^2}\right) \qquad (10.6)$$

and $r_* \equiv r + 2M \ln(r/2M - 1)$ is the tortoise coordinate which pushes away the horizon $r = 2M$ at $r_* \to -\infty$.

Near the horizon the effective potential vanishes and the solutions are a sum of outgoing and incoming waves:

$$\Phi \propto \alpha e^{-iE(t-r_*)} + \beta e^{-iE(t+r_*)} . \qquad (10.7)$$

Now if the physical spacetime is that of a star collapsing to form a black hole (see Fig. 10.2), it is necessary to continue these modes beyond the horizon from the west quadrant to the north quadrant of the Kruskal diagram; see Fig. 10.2. To do this we follow step by step the reasoning which led us to the Unruh effect above.

In terms of the Eddington–Finkelstein coordinate $v = t + r_*$, the Schwarzschild metric is written as

$$ds^2 = -\left(1 - \frac{2M}{r}\right)dv^2 + 2dv\,dr + r^2(d\theta^2 + \sin^2\theta\,d\phi^2) . \qquad (10.8)$$

It is regular on the future horizon, that is, for $r \to 2M$ and $t \to +\infty$ at finite v. The mode (10.7) becomes

---

[2] Unruh (1976); see also Davies (1975) and Fulling (1973). Here we follow the presentation of Damour and Ruffini (1976). See also Damour (2004).

We note that for $T = 1$ K, we must have $a \approx 1.5 \times 10^{19}$ g, where $g = 9.8$ m/s^2 is the acceleration of terrestrial gravity.

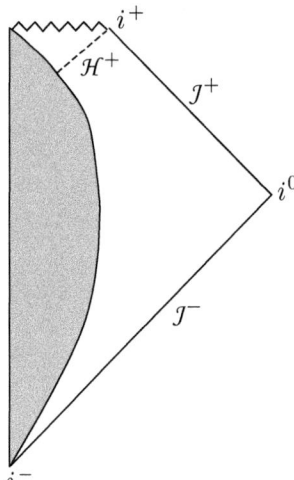

Fig. 10.2 The west and north quadrants of a star collapsing to form a black hole.

$$\Phi \propto \alpha e^{-iEv} e^{2iEr_*} + \beta e^{-iEv} = \alpha e^{-iEv} e^{2iEr} \left(\frac{r - 2M}{2M}\right)^{4iEM} + \beta e^{-iEv}. \tag{10.9}$$

The second term is regular on the horizon. The first term (the outgoing mode) can be analytically continued beyond the horizon as

$$\alpha e^{-iEv} e^{2iEr} \Big[\Theta\left(r/2M - 1\right) \left(r/2M - 1\right)^{4iEM}$$
$$+ \Theta\left(1 - 2M/r\right) \left(1 - 2M/r\right)^{4iEM} \exp(+4\pi EM)\Big], \tag{10.10}$$

where Θ is the Heaviside distribution and the $+$ sign was chosen so that the mode describes an antiparticle state. The normalization of the mode requires

$$|\alpha|^2 - \frac{1}{\exp(8\pi EM) - 1}. \tag{10.11}$$

The coefficient $|\alpha|^2$ is interpreted as the number of particle–antiparticle pairs created per mode and per unit time. The distribution is that of a black body at temperature

$$T = \frac{1}{8\pi M}. \tag{10.12}$$

The creation rate then is

$$\frac{dN}{dt} = \sum_l \int \frac{dE}{2\pi} \frac{(2l+1)T_l(E)}{\exp(8\pi EM) - 1}, \tag{10.13}$$

where $T_l(E)$ is the 'greybody factor' studied in Section 9.2, which describes the flux of created particles through the potential barrier V_{eff}. Therefore, an observer confined to the exterior of a 'black' hole will see it radiate.[3]

10.2 On the thermodynamics of black holes

The study of the Penrose process near a Kerr black hole leads to the conclusion that its irreducible mass can only increase; see Section 9.1. A similar but more general conclusion was reached by Hawking (1971), who, on the basis of a theorem of Penrose stating that light rays that generate a horizon cannot intersect, showed that the sum of the areas of the horizons of black holes interacting with matter can only increase, with the condition that the cosmic censorship hypothesis is valid and that (*via* the Einstein equations) the matter obeys the so-called weak energy condition (namely, that its energy–momentum tensor satisfies[4] $T_{\mu\nu}u^\mu u^\nu > 0$ with $u_\mu u^\mu < 0$).

In view of these results and also the fact that any information on the matter falling into a black hole other than its mass and angular momentum (here we ignore electric charge) is lost, in 1971 Bekenstein conjectured that black holes have an *entropy S* which, in general relativity, is proportional to their area:

$$S = \alpha \frac{\mathcal{A}}{l_{\text{P}}^2}, \tag{10.14}$$

where α is a dimensionless number and l_{P} (in units where the Boltzmann constant is taken equal to 1 so that S is dimensionless) has the dimensions of a length. Since it is constructed from fundamental quantities, it can only be the *Planck length*, $l_{\text{P}} = (\hbar G/c^3)^{1/2}$. This length involves, along with the constants $c (= 1)$ and G characterizing classical Einstein gravity, the Planck constant $\hbar$ characteristic of quantum phenomena. The fact that the area of a black hole (or its irreducible mass) can only increase is then interpreted as the *second law of thermodynamics*:

$$\delta S \geq 0. \tag{10.15}$$

The area of a black hole and therefore its entropy are functions of the quantities characterizing the black hole, namely its mass M and angular momentum J (we ignore electric charge). If the black hole is a Kerr black hole, but the law is valid for any black hole which is a solution of the Einstein equations, we will have [*cf.* (9.3)]

$$T\,\delta S = \delta M - \Omega\,\delta J \quad \text{with} \quad T = \frac{\kappa\,\hbar}{8\pi\alpha}, \tag{10.16}$$

where Ω and κ are the angular velocity and surface gravity of the black hole [$\Omega = a/(r_+^2 + a^2)$ and $\kappa = \sqrt{m^2 - a^2}/(2mr_+)$ for a Kerr black hole].

An important property of the surface gravity κ is that, as shown by Carter, it is *constant* on the horizon of the black hole. This is the *zeroth law*.[5] The quantity T (which has the

---

[3]There are many other ways of deriving Hawking radiation. A partial list can be found in, for example, Carlip (2009).

We note that the Hawking temperature of a black hole of mass comparable to the mass of the Sun is negligible: $T = 6 \times 10^{-8} (M/M_\odot)$ K.

[4]The proof can be found in Hawking (1972) and in, for example, Straumann (2013).

[5]It is proved in, for example, Straumann (2013).

dimensions of a mass in the units where the Boltzmann constant is equal to 1) then must be interpreted as the *temperature* of the black hole, and (10.16) will be the *first law* of thermodynamics for black holes.

The objection which has been raised to this *thermodynamical* interpretation of the laws of *dynamics* of black holes is that if a black hole is required to have a temperature, then it must be radiating (and this would also be true of the static Schwarzschild black hole). But no particle or field, at least a classical one, can emerge from the horizon. However, Hawking has shown (see Section 10.1) that since the quantum vacuum of the spacetime of a black hole differs from that of an observer at infinity, the hole does in fact radiate with a black-body spectrum at temperature $T = \kappa \hbar/(2\pi)$ [$T = \hbar/(8\pi GM)$ if it is a Schwarzschild black hole]. Therefore, the constant α must be

$$\alpha = 1/4 \,. \tag{10.17}$$

One suspects that the meaning and implications of these results are still the subject of intense discussions. . . .

10.3   The Israel uniqueness theorem

In Section 6.1 we saw that the only solution of the spherically symmetric vacuum Einstein equations is the Schwarzschild solution (this is the Birkhoff–Jebsen theorem). One can pose the question of whether or not black holes which are static but *not* spherically symmetric exist. The answer is 'no', as was shown by Israel, whose theorem we shall present in detail here. The Israel theorem thereby proves that the Schwarzschild black hole is unique.

- *Hypotheses and statement*

Spacetime is four-dimensional and static.

Its spatial sections can be foliated by closed 2-surfaces.

It satisfies the vacuum Einstein equations.

It is asymptotically flat.

It possesses a horizon, that is, a closed 2-surface of infinite redshift, where the curvature is finite.

The spacetime is therefore spherically symmetric. Therefore, it is the Schwarzschild spacetime.[6]

- *The equations of motion*

Spacetime is four-dimensional and static. In this case there exist adapted coordinates (t, x^i) in which the length element is written as

$$ds^2 = -V^2(x^i)dt^2 + h_{ij}(x^k)dx^i dx^j \tag{10.18}$$

and where the components of its Riemann and Ricci tensors are simply [*cf.* (4.11)]

$$R_{ijkl} = \bar{R}_{ijkl} \,, \quad R_{tijk} = 0 \,, \quad R_{titj} = V \bar{D}_{ij}^2 V \,,$$

$$R_{tt} = V \bar{\Delta} V \,, \quad R_{ti} = 0 \,, \quad R_{ij} = \bar{R}_{ij} - \frac{\bar{D}_{ij}^2 V}{V} \,, \tag{10.19}$$

where the bar indicates that the quantity is constructed using the spatial metric h_{ij}.

---

[6]See Israel (1967), and also (where some misprints are corrected) Frolov and Novikov (1998) or Straumann (2013).

Its spatial sections can be foliated by closed 2-surfaces, for example, the surfaces $V(x^i) =$ const. Then V can play the role of the radial coordinate and the length element becomes

$$ds^2 = -V^2 dt^2 + d\sigma^2 \quad \text{with} \quad d\sigma^2 = \rho^2(V, \theta^A)dV^2 + b_{AB}(V, \theta^C)d\theta^A d\theta^B, \tag{10.20}$$

where $\theta^A = (\theta, \phi)$. The $(2+1)$ decomposition of the spatial metric is done as in Section 4.2 [taking note of the changes of sign due to the fact that here the signature is $(+, +, +)$] and gives (we will not need the other components)

$$\bar{R}_{VV} = -\rho(\tilde{\Delta}\rho + \rho K.K + \partial_V K),$$

$$b^{AB} \bar{R}_{AB} = -\frac{1}{\rho}\partial_V K + 2K.K - K^2 - \frac{\tilde{\Delta}\rho}{\rho} + \tilde{R}, \tag{10.21}$$

where $K_{AB} \equiv \frac{1}{2\rho}\partial_V b_{AB}$ is the extrinsic curvature of the 2-surfaces $V = $ const, $K \equiv b^{AB} K_{AB}$ and $K.K \equiv K_{AB}K^{AB}$, and $\tilde{\Delta}$ and $\tilde{R}$ are the Laplacian and Gauss scalar curvature constructed with the 2-metric b_{AB}. In addition, since b is the determinant of the metric b_{AB}, we have

$$\bar{D}^2_{VV}V = -\frac{\partial_V \rho}{\rho}, \qquad \bar{D}^2_{VA}V = -\frac{\partial_A \rho}{\rho}, \qquad \bar{D}^2_{AB}V = \frac{K_{AB}}{\rho},$$

$$\bar{\Delta}V = \frac{1}{\rho\sqrt{b}}\partial_V\left(\frac{\sqrt{b}}{\rho}\right) = \frac{1}{\rho^3}(\rho^2 K - \partial_V \rho). \tag{10.22}$$

The spacetime satisfies the vacuum Einstein equations, namely, $R_{\mu\nu} = 0$. We then have, in particular, $R_{tt} = 0$, $R_{VV} = 0$, and $b^{AB}R_{AB} = 0$, or, using (10.19), (10.21), and (10.22),

$$\begin{cases} \partial_V\left(\frac{\sqrt{b}}{\rho}\right) = 0 \quad \Longleftrightarrow \quad \partial_V \rho = \rho^2 K, \\[2mm] \partial_V K = \frac{K}{V} - \tilde{\Delta}\rho - \rho K.K, \qquad \tilde{R} = \frac{2K}{\rho V} - K.K + K^2. \end{cases} \tag{10.23}$$

Using the first equation, the last two can be written in the following useful form in terms of $\Psi_{AB} \equiv \rho\left(K_{AB} - \frac{1}{2}b_{AB}K\right)$ [it can be checked that they are indeed equivalent to (10.23)]:

$$\begin{cases} \partial_V\left(\frac{\sqrt{b}K}{\sqrt{\rho}V}\right) + \frac{2\sqrt{b}}{V}\tilde{\Delta}(\sqrt{\rho}) = -\frac{\sqrt{b}}{2V\rho^{3/2}}\left(\partial_A\rho\,\partial^A\rho + 2\Psi.\Psi\right), \\[3mm] \partial_V\left[\frac{\sqrt{b}}{\rho}\left(KV + \frac{4}{\rho}\right)\right] + V\sqrt{b}\left(\tilde{\Delta}(\ln\rho) + \tilde{R}\right) = -\frac{V\sqrt{b}}{\rho^2}(\partial_A\rho\,\partial^A\rho + 2\Psi.\Psi). \end{cases} \tag{10.24}$$

We see from the form of the right-hand sides that the left-hand sides must be negative or zero. By integrating them over all space, we shall show that the conditions at infinity and on the horizon require that they vanish. The right-hand sides then will also vanish and we find $\partial_A \rho = 0$ and $\Psi_{AB} = 0$.

We note for what follows that

$$\int d\theta d\phi \sqrt{b}\, \tilde{\Delta} f(\rho) = 0\,, \qquad \int d\theta d\phi \sqrt{b}\tilde{R} = 8\pi\,, \qquad (10.25)$$

because the 2-surfaces $V = \text{const}$ are closed owing to the *Gauss–Bonnet theorem*.[7]

• *Spacetime is asymptotically flat*

In this case at spatial infinity we must have

$$ds^2 = -V^2 dt^2 + \rho^2 dV^2 + b_{AB} d\theta^A d\theta^B \rightarrow -(1 - 2m/r)dt^2 + dr^2 + r^2(d\theta^2 + \sin^2\theta d\phi^2)\,.$$

Therefore, $V^2 \approx 1 - 2m/r$, $\rho\, dV/dr \rightarrow 1$, and $\sqrt{b} \rightarrow r^2 \sin\theta$, or also [using $K = \partial_V \rho/\rho^2$; see (10.23)]

$$\rho \approx \frac{4m}{(1 - V^2)^2}\,, \qquad \sqrt{b} \approx \frac{4m^2 \sin^2\theta}{(1 - V^2)^2}\,, \qquad K \approx \frac{1 - V^2}{m} \quad \text{when} \quad V \rightarrow 1\,. \qquad (10.26)$$

• *Spacetime possesses a regular horizon*

This horizon, a closed 2-surface of infinite redshift where g_{tt} is zero, is defined by $V = 0$. It is regular if the Kretschmann invariant on it is finite. Then [see (10.19)] we have

$$R_{\mu\nu\rho\sigma}R^{\mu\nu\rho\sigma} = \bar{R}_{ijkl}R^{ijkl} + 4V^2 \bar{D}^2_{ij}V\, \bar{D}^{ij}V\,.$$

Since space is three-dimensional, the Riemann tensor $\bar{R}_{ijkl}$ is expressed as a function of the Ricci tensor and we have[8] $\bar{R}_{ijkl}R^{ijkl} = 4\bar{R}_{ij}\bar{R}^{ij} - \bar{R}^2$. Spacetime being Ricci-flat, we then find [cf. (10.19)] $\bar{R}_{ij} = \bar{D}^2_{ij}V/V$ and $\bar{R} = 0$, so that the Kretschmann invariant reduces to the following using (10.22):

$$\frac{1}{8}R_{\mu\nu\rho\sigma}R^{\mu\nu\rho\sigma} = \frac{\bar{D}_{ij}V\, \bar{D}^{ij}V}{V^2} = \frac{1}{V^2\rho^2}\left(K^2 + K.K + \frac{2\partial_A\rho\, \partial^A\rho}{\rho^2}\right)\,. \qquad (10.27)$$

In order for this sum of positive terms to be finite on the horizon $V = 0$, we must have

$$K_{AB} \rightarrow 0 \quad \text{and} \quad \rho \rightarrow \rho_{\mathrm{H}} \quad \text{when} \quad V \rightarrow 0 \qquad (10.28)$$

since ρ_{H} is a nonzero constant.

---

[7]Let $I = \int_S \sqrt{b}\, \tilde{\Delta} f\, d\theta\, d\phi = \int_S \partial_A \hat{v}^A\, d\theta\, d\phi$ with $\hat{v}^A = \sqrt{b}\, b^{AB}\partial_B f$. By the divergence theorem we then have $I = \int_{\partial S} V^A dS_A$. Since S is a closed surface, it has no boundary ∂_S and so $I = 0$.

Let $J = \int_S \sqrt{b}\, \tilde{R}\, d\theta d\phi$. It was first shown by Gauss for a special case, and then by Bonnet in 1848 and finally Chern in 1944 (using differential geometry, which is discussed in the last part of this book), that if S is a (regular) closed surface, then $J = 8\pi$. See Chern (1944).

[8]This is a consequence of the Gauss–Bonnet theorem generalized by Chern; see the last part of this book and footnote 7 above.

- *The theorem*

The first step is to integrate the field equations $R_{tt} = 0$, $R_{VV} = 0$, and $b^{AB}R_{AB} = 0$ over all space, the first equation in the form (10.23) and the other two in their form (10.24), using (10.25), (10.26), and (10.28). The first gives

$$\int_S \frac{\sqrt{b}}{\rho}\Big|_{V=1} d\theta d\phi = \int_S \frac{\sqrt{b}}{\rho}\Big|_{V=0} d\theta d\phi \quad \text{or} \quad \mathcal{A}_H = 4\pi m \rho_H\,, \tag{10.29}$$

where $\mathcal{A}_H \equiv \int \sqrt{b}|_0\, d\theta\, d\phi$ is the area of the horizon. The integrals of the left-hand sides of (10.24) are negative and, using (10.29), give

$$4m \leq \rho_H\,, \quad \mathcal{A}_H \geq \pi \rho_H^2 \quad \Longrightarrow \quad 4m \geq \rho_H\,. \tag{10.30}$$

The equality then must hold and we obtain the stated result, namely, that the left-hand sides of (10.24) vanish, and so the right-hand sides do, too. Therefore, the 2-surfaces $V = \text{const}$ must be such that

$$\partial_A \rho = 0\,; \qquad \Psi_{AB} = 0\,. \tag{10.31}$$

The second step of the proof is to show that the 2-surfaces $V = \text{const}$ must be spheres. We return to the field equations in the form (10.23). Using $\Psi_{AB} \equiv K_{AB} - b_{AB}K/2 = 0$, we find that the Gauss curvature of these 2-surfaces is given by $\tilde{R} = 2K/(V\rho) - K^2/2$ with $K = \partial_V \rho/\rho^2$. Since (10.31) requires that ρ depend only on the radial coordinate V, the same is true of $\tilde{R}$: the curvature of the (closed) 2-surfaces is constant, and so they are spheres.

The final step is to note that since our spacetime is a static solution of the Einstein vacuum equations and also is spherically symmetric, it can only be the Schwarzschild solution, as we saw in Section 6.1. *Q.E.D.* In the coordinates (t, V, θ, ϕ) the length element is written as

$$ds^2 = -V^2 dt^2 + \frac{16m^2}{(1-V^2)^4} dV^2 + \frac{4m^2}{(1-V^2)^2}(d\theta^2 + \sin^2\theta d\phi^2)\,. \tag{10.32}$$

The uniqueness of the Kerr black hole

The proof of the uniqueness of the Kerr black hole is a more difficult problem. The theorem, stated by Carter (1971) and Robinson (1975), is the following. Stationary and axisymmetric solutions of the vacuum Einstein equations, which possess a horizon which is regular and convex, which are asymptotically flat, and which have no curvature singularity outside the horizon, are Kerr solutions (specified by their mass M and angular momentum J). The theorem was extended to the charged Kerr–Newman case by Mazur (1982) and Bunting (1983).

The idea behind the proof is the following. The metric can be written as

$$ds^2 = -V dt^2 + 2W dt\, d\phi + X d\phi^2 + U[d\lambda^2/(\lambda^2 - c^2) + d\mu^2/(1 - \mu^2)]\,,$$

where the functions V, W, X, and U depend only on λ and μ, and $\lambda = c$ defines the horizon. The vacuum Einstein equations reduce to two second-order differential equations for the two 'Ernst potentials' $X(\lambda, \mu)$ and $Y(\lambda, \mu)$: $E(X, Y) = 0$ and $F(X, Y) = 0$. The behavior of Y for $\lambda \to \infty$ is then determined and depends on a single parameter J.

Now we assume that the solution possesses two *a priori* different solutions (X_1, Y_1) and (X_2, Y_2) parametrized by the same J.

In 1971, Carter showed that if these solutions are separated by an infinitesimal amount they must be identical. Then, in 1975 Robinson constructed an integral of the form $R = \int_c^\infty d\lambda \int_{-1}^{+1} d\mu \left[\frac{X_1}{X_2}(Y_2 - Y_1)F(X_1, Y_1) + \cdots \right]$ which is zero if the equations of motion, $E = 0$ and $F = 0$, are satisfied, and succeeded in writing the integrand as a sum of (eight!) positive-definite terms. These then must vanish separately, which implies that $X_1 = X_2$ and $Y_1 = Y_2$. Since the Kerr metric satisfies the equations of motion, it is therefore the only solution.[9]

The consequence of these theorems is that when a star collapses to form a black hole, it *must* lose *en route* everything that characterizes it (for example, its multipole moments, its baryon composition, and so on) in order to become in the end a Kerr black hole characterized only by its mass and angular momentum. As summarized in a famous statement by Wheeler, "*a black hole has no hair.*"

Of course, if the hypotheses on which these theorems are based are done away with, black holes can become 'hairy'. In fact, a number of solutions are now known in more than four dimensions, or in the presence of various gauge fields which generalize the Maxwell field or scalar fields ϕ of judiciously chosen potentials $V(\phi)$, and so on.[10]

---

[9]To learn more, see, for example, Carter (2004), and also Chandrasekhar (1983) and Chruściel *et al.* (2012), as well as Alexakis *et al.* (2013).

It should be noted that the proof assumes that the horizon is not degenerate, that is, $m^2 \neq a^2$, and that as yet there is no proof of the uniqueness of the Kerr–de Sitter solution (the solution of $G_{\mu\nu} = \Lambda g_{\mu\nu}$).

[10]See, for example, Emparan and Reall (2008).

Part III

General relativity and experiment

It was my good fortune to be present at the meeting of the Royal Society in London when the Astronomer Royal for England announced that the photographic plates of the famous eclipse, as measured by his colleagues at Greenwich Observatory, had verified the prediction of Einstein that rays of light are bent as they pass in the neighbourhood of the sun. The whole atmosphere of tense interest was exactly that of the Greek drama: we were the chorus commenting on the decree of destiny as disclosed in the development of a supreme incident. There was dramatic quality in the very staging:—the traditional ceremonial, and in the background the picture of Newton to remind us that the greatest of scientific generalisations was now, after more than two centuries, to receive its first modification. Nor was the personal interest wanting: a great adventure in thought had at length come safe to shore.

Alfred North Whitehead, description of the November 6, 1919 session
of the Royal Society, in *Science and the Modern World* (1925)

... we find that Einstein's theory passes this extraordinarily stringent test with a fractional accuracy better than 0.4% .... The clock-comparison experiment for PSR 1913 + 16 thus provides direct experimental proof that changes in gravity propagate at the speed of light, thereby creating a dissipative mechanism in an orbiting system. It necessarily follows that gravitational radiation exists and has a quadrupolar nature.

Joseph H. Taylor, Jr., in *Binary Pulsars and Relativistic Gravity*,
Nobel lecture, December 8, 1993

The Nobel Prize in Physics 2017 was awarded to Rainer Weiss, Barry C. Barish, and Kip S. Thorne 'for decisive contributions to the LIGO detector and the observation of gravitational waves'.

11
Tests in the solar system

In the solar system we can, as a first approximation, neglect the gravitational field of all celestial bodies except the Sun. In the Newtonian theory the planet trajectories are then Keplerian ellipses. Relativistic effects are weak because the dimensionless ratio characterizing them is everywhere less than $GM_\odot/c^2 R_\odot \simeq 2 \times 10^{-6}$, and so they can be added linearly to the Newtonian perturbations due to the other planets, the non-spherical shape of celestial bodies, and so on. The present chapter describes the observable relativistic effects in the solar system.

11.1 The solar system and the Schwarzschild metric

Here we shall describe the gravitational field of the Sun by a Schwarzschild spacetime whose metric in Droste coordinates is (see Section 6.1; we have changed the notation as $\nu \to 2\nu$, $\lambda \to 2\lambda$)

$$ds^2 = -e^{2\nu}dt^2 + e^{2\lambda}dr^2 + r^2(d\theta^2 + \sin^2\theta d\phi^2) \quad \begin{cases} \text{where} \quad e^{2\nu} = e^{-2\lambda} = 1 - 2m/r \\ \text{and} \quad m \equiv GM_\odot/c^2 \simeq 1.5 \text{ km}. \end{cases} \tag{11.1}$$

We can also use 'isotropic' radial coordinates $[r = \bar{r}(1 + m/2\bar{r})^2]$ or 'harmonic' coordinates $(\tilde{r} = r - m)$, in which the metric is written as

$$\begin{aligned} ds^2 &= -\left(\frac{1 - m/2\bar{r}}{1 + m/2\bar{r}}\right)^2 c^2 dt^2 + \left(1 + \frac{m}{2\bar{r}}\right)^4 (d\bar{r}^2 + \bar{r}^2 d\theta^2 + \bar{r}^2 \sin^2\theta d\phi^2) \\ &= -\left(\frac{\tilde{r} - m}{\tilde{r} + m}\right) c^2 dt^2 + \left[\left(1 + \frac{m}{\tilde{r}}\right)^2 \delta_{ij} + \left(\frac{\tilde{r} + m}{\tilde{r} - m}\right) \frac{m^2}{\tilde{r}^4} x_i x_i\right] dx^i dx^j. \end{aligned} \tag{11.2}$$

Book 3

Relativity in Modern Physics. Nathalie Deruelle and Jean-Philippe Uzan.
© Oxford University Press 2018. Published in 2018 by Oxford University Press.
DOI: 10.1093/oso/9780198786399.001.0001

where $r = r(\tilde{r})$, $r' = dr/d\tilde{r}$, and $\tilde{r}^2 = \delta_{ij}x^i x^j$. We also find $\sqrt{-\tilde{g}} = r^2 r'/\tilde{r}^2$ and

$$\sqrt{-\tilde{g}}\tilde{g}^{\mu 1} = (0, r' + x^1 x^1 f, x^1 x^2 f, x^1 x^3 f) \quad \text{with} \quad f = ((r/\tilde{r})^2(1 - 2m/r) - r'^2)/(\tilde{r}^2 r')$$

and analogous expressions for $\sqrt{-\tilde{g}}\tilde{g}^{\mu 2}$ and $\sqrt{-\tilde{g}}\tilde{g}^{\mu 3}$. The harmonic coordinate condition then reduces to $\tilde{\partial}_\mu \tilde{g}^{\mu 1} = 0$, or

$$4f + \tilde{r}f' + r''/\tilde{r} = 0,$$

a particular solution of which is $r'' = 0$, $4f + f' = 0$, that is, owing to the definition of f, $r = \tilde{r} + m$ and $f = -m/\tilde{r}^2$.

Substituting this solution into the expressions for $\tilde{g}_{tt}$ and $\tilde{g}_{ij}$, we see that the metric then is just (11.2).

The smallness of m relative to the typical distances in the solar system implies that only the first few lowest-order terms of the expansion of (11.1) or (11.2) will be needed. Indeed, if we are interested only in the post-Newtonian corrections to the Keplerian trajectories of the planets [which come from $g_{tt} = -(1 - 2m/r)$], it will prove useful to generalize the expansion of the Schwarzschild metric as

$$ds^2 = -\left(1 - 2\frac{m}{\tilde{r}} + 2\beta\frac{m^2}{\tilde{r}^2} + \cdots\right) dt^2$$

$$+ \left(1 + 2\gamma\frac{m}{\tilde{r}} + \cdots\right)(d\tilde{r}^2 + \tilde{r}^2 d\theta^2 + \tilde{r}^2 \sin^2\theta d\phi^2),$$

(11.4)

where $\tilde{r}$ is the isotropic coordinate (or harmonic coordinate $\tilde{r}$; they are equivalent at this order), and where the *Eddington parameters* β and γ are equal to 1 in general relativity. For arbitrary β and γ the metric (11.4) is a (simplified) form of the *parametrized post-Newtonian* (PPN) metric, and serves as a measure of the difference between general relativity and competing theories.[1]

11.2   The geodesic equation

The motion of a test particle in a gravitational field is governed by the geodesic equation $Du^\mu/d\lambda = 0$, where $u^\mu = dx^\mu/d\lambda$ and λ parametrizes the curve.

As we have seen in Section 5.2, the manifest static nature and spherical symmetry of the metric (11.1) give the first integrals of the motion. First we have

$$u_t = -e^{2\nu}\frac{dt}{d\lambda} = -\left(1 - \frac{2m}{r}\right)\frac{dt}{d\lambda} \equiv -E, \quad u_\phi = r^2 \sin^2\theta\frac{d\phi}{d\lambda} \equiv L,$$

(11.5)

where E and L are the energy and angular momentum per unit mass of the particle. In addition, the spherical symmetry implies that all sections $\theta = $ const are equivalent, and so we take $\theta = \pi/2$. We thus obtain the expression for the length element (11.1):

$$-\epsilon = -e^{-2\nu}E^2 + e^{2\lambda}\left(\frac{dr}{d\lambda}\right)^2 + \frac{L^2}{r^2},$$

(11.6)

---

[1] In Section 12.1 we shall see that this metric is also valid far away from a *system* of masses in the post-Newtonian approximation of general relativity (where $\beta = \gamma = 1$) when the system can be assumed stationary.

where $\epsilon = +1$ for time-like geodesics (in which case we take $\lambda = \tau$ so that $u_\mu u^\mu = -1$) and $\epsilon = 0$ for photons. As we have already seen in eqn (8.26) and Fig. 8.1, this equation can be rewritten as

$$\left(\frac{dr}{d\lambda}\right)^2 = E^2 - U_e \quad \text{with} \quad U_e = \left(1 - \frac{2m}{r}\right)\left(\epsilon + \frac{L^2}{r^2}\right). \tag{11.7}$$

An orbiting clock and GPS

Using the integrals (11.5) and (11.6) of geodesic motion in Schwarzschild spacetime, show that for a circular orbit we have, as in Newtonian gravity, $d\phi/dt = \sqrt{m/r^3}$.

[Answer: since $\dot{r} = 0$ from (11.2), we find $E^2 = V$ with $V = e^{2\nu}(1 + L^2/r^2)$. For the orbit to be circular we must in addition have $dV/dr = 0$, which gives L. We therefore obtain

$$u^\mu = (e^{-\nu}, 0, 0, \sqrt{\nu'/r})/\sqrt{1 - r\nu'},$$

or $d\phi/dt = e^\nu \sqrt{\nu'/r}$, and arrive at the result.]

Now show that the proper time $\Delta\tau_O$ of a clock in a circular orbit at r_O is related to the proper time $\Delta\tau_\oplus$ of a clock at rest at $r = r_\oplus$ as

$$\Delta\tau_O = \sqrt{\frac{1 - 3m/r_O}{1 - 2m/r_\oplus}}\,\Delta\tau_\oplus. \tag{11.8}$$

This is relevant to the GPS (global positioning system): using the fact that $2m_\oplus = 9$ mm, show that a clock on a GPS satellite (at an altitude of 20,000 km) runs about 40 µs per day fast relative to a clock located on the Earth ($r_\oplus = 6400$ km). Show that the positioning error accumulated over the period of a day will be about 12 km if these relativistic effects are not taken into account.

The standard method of finding the trajectory equations starting from (11.5)–(11.7) is to take the ratio of $(dr/d\lambda)^2$ and $(d\phi/d\lambda)^2$ and then set $u = 1/r$, which gives

$$\left(\frac{du}{d\phi}\right)^2 = \frac{1}{L^2}\left(E^2 - \epsilon\right) + \epsilon\frac{2m}{L^2}u - u^2 + 2m\,u^3. \tag{11.9}$$

Taking the derivative, we have that

$$\frac{d^2u}{d\phi^2} + u = \epsilon\frac{m}{L^2} + 3m\,u^2. \tag{11.10}$$

The term proportional to u^2 is the relativistic correction to the Binet equation of Newtonian gravity (see, for example, Book 1, Section 12.2). Equation (11.10) can be integrated in terms of Weierstrass elliptic functions.

11.3 The bending of light

The propagation of light in the gravitational field of the Sun is described by (11.10) with $\epsilon = 0$.

In Newtonian order the trajectories of light rays are straight lines $u = \sin\phi/b$, where b is the distance of closest approach (see Fig. 11.1). In first order, (11.10) then becomes $d^2u/d\phi^2 + u \simeq (3m/b^2)\sin^2\phi$, a particular solution to which at the required order is

$$u \simeq \frac{\sin\phi}{b} + \frac{3m}{2b^2}\left(1 + \frac{1}{3}\cos 2\phi\right). \tag{11.11}$$

For $r \to \infty$, that is, $u \to 0$, ϕ or $\pi - \phi$ are small and tend to $-2m/b$. A light ray coming in from infinity and passing the center at a distance b then is deflected by an angle

$$\Delta\phi \simeq \frac{4m}{b}. \tag{11.12}$$

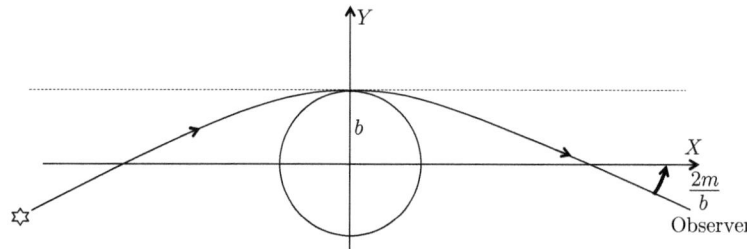

Fig. 11.1 The bending of light.

It is an instructive exercise to calculate $\Delta\phi$ using the system of isotropic coordinates and the PPN metric (11.4). In this case the trajectory of a light ray is determined by the vanishing of the interval and the geodesic equation. Let us assume that the ray is emitted in the plane $\theta = \pi/2$ along the $\bar{x}(= \bar{r}\cos\phi)$ axis with impact parameter $\bar{b}$. At first order in $m/\bar{r}$ we have [*cf.* (11.4)]

$$0 \simeq -\left(1 - \frac{2m}{\bar{r}}\right)dt^2 + \left(1 + \frac{2\gamma m}{\bar{r}}\right)(d\bar{x}^2 + d\bar{y}^2),$$

$$0 \sim \frac{d^2\bar{y}}{d\lambda^2} + \Gamma^y_{tt}\left(\frac{dt}{d\lambda}\right)^2 + \Gamma^y_{xx}\left(\frac{d\bar{x}}{d\lambda}\right)^2 \tag{11.13}$$

(for $\gamma=0$ the spatial sections are Euclidean). From the first equation we obtain in lowest order $dt/d\lambda \simeq d\bar{x}/d\lambda$; also, in the same order $d^2\bar{y}/d\lambda^2 \simeq (d^2\bar{y}/d\bar{x}^2)(d\bar{x}/d\lambda)^2$, and so the geodesic equation reduces to $d^2\bar{y}/d\bar{x}^2 + \Gamma^y_{tt} + \Gamma^y_{xx} = 0$. The Christoffel symbols can be calculated straightforwardly: $\Gamma^y_{tt} \simeq m\bar{y}/\bar{r}^3$ and $\Gamma^y_{xx} \simeq m\gamma\bar{y}/\bar{r}^3$. We then find

$$\Delta\phi \simeq \int_{-\infty}^{+\infty}\frac{d^2\bar{y}}{d\bar{x}^2}d\bar{x} \simeq 2m(1+\gamma)\int_0^{+\infty}\frac{\bar{b}\,d\bar{x}}{(\bar{x}^2 + \bar{b}^2)^{\frac{3}{2}}} \simeq \frac{2m(1+\gamma)}{\bar{b}}. \tag{11.14}$$

Since $\gamma = 1$ in general relativity, we recover (11.12) (at the order in which we are working $b \approx \bar{b}$), but we see that if we had ignored the curvature of space, as Einstein himself did in 1907, we would have obtained half the value, which is the predicted bending in the Newtonian theory (see Book 1, Section 17.1).

For a light ray grazing the surface of the Sun $m/b = 2 \times 10^{-6}$, and (11.12) gives (Einstein, 1915)

$$\Delta\phi = 1.75".$$ (11.15)

Measurements of light bending

The (rather crude) measurement of the effect (11.15) during the expeditions led by Arthur Eddington to the island of Príncipe in the Gulf of Guinea and the city of Sobral in Brazil in 1919, announced at a meeting of the Royal Society in London, sealed the triumph of the theory.[2]

The bending of light rays can also lead to 'gravitational mirage' effects. This phenomenon, predicted by Zwicky in the 1930s and studied by Einstein himself, was observed in 1979 by Walsh *et al.* for quasars. At present dozens of examples are known. The formation of multiple images and arcs can occur, depending on the geometry and the structure of the matter in the deflecting galaxy. The bending can also be interpreted as a 'lensing' effect due to the deflecting object, and it can be shown that the luminosity of the image is considerably enhanced if it is close to the line of sight of the deflecting object. This provides a means of detecting objects which are otherwise invisible.[3]

11.4 The Shapiro effect

Now let us calculate the time it takes for an electromagnetic signal to travel from a point 1 to a point 2 in the solar system. Using the first integrals (11.5), we rewrite (11.6) for the signal trajectory ($\epsilon = 0$) in the form

$$\frac{dr}{dt} = \left(1 - \frac{2m}{r}\right)\left[1 - \frac{1 - 2m/r}{1 - 2m/b}\left(\frac{b}{r}\right)^2\right]^{\frac{1}{2}},$$ (11.16)

where we have introduced b, the distance of closest approach to the Sun (such that $dr/dt|_b = 0$). In first order in m/r and m/b, (11.16) becomes

$$\frac{dt}{dr} = \frac{r}{\sqrt{r^2 - b^2}}\left\{1 + \frac{2m}{r}\left[1 + \frac{b}{2(r+b)}\right]\right\},$$ (11.17)

which can be integrated to obtain the coordinate time to travel from r to b:

$$t(r,b) \simeq \sqrt{r^2 - b^2} + 2m\ln\left(\frac{r + \sqrt{r^2 - b^2}}{b}\right) + m\sqrt{\frac{r-b}{r+b}}.$$ (11.18)

---

[2]See the quotation from Whitehead in the heading of the present part of this book. The newspaper *The Times* of November 7, 1919 carried the headline *Revolution in Science; New Theory of the Universe: Newtonian ideas overthrown.* And in the *New York Times* of November 9: *ECLIPSE SHOWED GRAVITY VARIATION; Diversion of Light Rays Accepted as Affecting Newton's Principles; HAILED AS EPOCHMAKING; British Scientist Calls the Discovery One of the Greatest of Human Achievements.* [Citations are from Isaacson (2008).]

For a complete discussion of tests of general relativity see Will (1993) and also Damour (2016).

[3]For an introduction, see Straumann (2013), and for a review see, for example, Bartelmann and Schneider (2001).

In isotropic coordinates $\bar{r}$ or harmonic coordinates $\tilde{r}$, which are equal to each other at first order in m and related to the Schwarzschild coordinate r as $\bar{r} \approx \tilde{r} = r - m$, the effect (11.18) becomes

$$t(\bar{r}, \bar{b}) \simeq \sqrt{\bar{r}^2 - \bar{b}^2} + 2m \ln \left(\frac{\bar{r} + \sqrt{\bar{r}^2 - \bar{b}^2}}{\bar{b}} \right) + 2m \sqrt{\frac{\bar{r} - \bar{b}}{\bar{r} + \bar{b}}}. \tag{11.19}$$

More generally, the PPN metric (11.4) in the required order and for $\theta = \pi/2$ can be written as

$$ds^2 \approx - \left(1 - \frac{2m}{\bar{r}} \right) dt^2 + \left(1 + \frac{2\gamma m}{\bar{r}} \right) (d\bar{r}^2 + \bar{r}^2 d\phi^2), \tag{11.20}$$

where the Eddington parameter γ is 1 in general relativity. The geodesic equation possesses two first integrals

$$\frac{dt}{d\tau} \approx E \left(1 + \frac{2m}{\bar{r}} \right), \qquad \frac{d\phi}{d\tau} = \frac{L}{\bar{r}^2} \left(1 - \frac{2m\gamma}{\bar{r}} \right), \tag{11.21}$$

and the vanishing of the interval (11.20) then gives

$$\frac{dt}{d\bar{r}} = \frac{\bar{r}}{\sqrt{\bar{r}^2 - \bar{b}^2}} \left[1 + \frac{m(1 + \gamma)}{\bar{r}} \left(1 + \frac{\bar{b}}{\bar{r} + \bar{b}} \right) \right], \tag{11.22}$$

where we have traded L/E for the distance of closest approach $\bar{b}$ as $L/(E\bar{b}) = 1 + m(1 + \gamma)/\bar{b}$. We then have

$$t(\bar{r}, \bar{b}) \simeq \sqrt{\bar{r}^2 - \bar{b}^2} + m(1 + \gamma) \ln \left(\frac{\bar{r} + \sqrt{\bar{r}^2 - \bar{b}^2}}{\bar{b}} \right) + m(1 + \gamma) \sqrt{\frac{\bar{r} - \bar{b}}{\bar{r} + \bar{b}}}, \tag{11.23}$$

and we indeed recover (11.19) for $\gamma = 1$.

The first term in (11.18), (11.19), or (11.23) is the equation for light propagating in a straight line at speed c. The terms involving the logarithm are the numerically dominant relativistic correction. The total time to go from point 1 to point 2 then is $t_{12} = t(r_1, b) + t(b, r_2)$ (for $|\phi_1 - \phi_2| > \pi/2$); see Fig. 11.2. If upon arriving at 2 the signal is reflected back to 1 and the motion of 1 is neglected, the duration of the round trip is $2t_{12}$. This effect was calculated by I. Shapiro in 1964.

Measurements of the Shapiro effect

The Shapiro effect was measured by Irwin Shapiro himself in 1968 and again in 1972 using radar signals emitted by the stations at Haystack, Massachusetts and Arecibo, Porto Rico in the direction of Mercury and Venus. The accuracy obtained was about 10 %. Since the delay is about 200 µs, its accurate measurement requires good knowledge of the topography of the planet. (For example, Venus presents altitude differences of order 1500 m or 5 µs.) Moreover, the solar corona, on which the signals undergo refraction, introduces additional delays which must be eliminated. Finally, since the time for a trajectory is about twenty minutes, an accuracy of 10 µs implies that the position of the planets must be known with an accuracy of at least 1.5 km. The effect was later measured more accurately using space probes such as Mariner 6 and 7 as targets, but owing, in particular, to the solar wind their motion is quite erratic, with leaps as high as 30 m or 0.1 µs. After the installation of a reflector on Mars by the Viking probe, the accuracy reached 10^{-3}. The best measurements today were obtained by following the Cassini probe sent to Saturn in 1997, which passed near Jupiter in 2003: the PPN parameter γ appearing in (11.23) is now constrained to the value 1 with an accuracy of better than 10^{-5} [see Will (1993) and Damour (2016)].

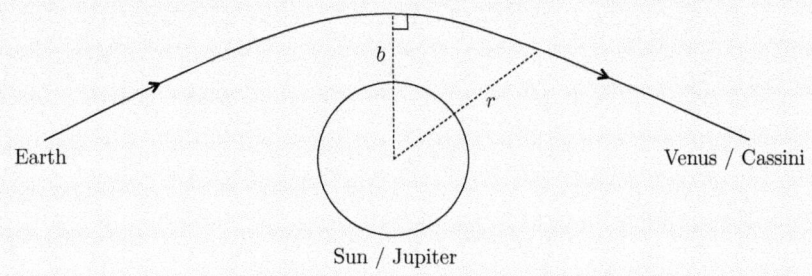

Fig. 11.2 The Shapiro effect.

It is important to note that the accuracy of the measurements is now good enough to distinguish between isotropic, harmonic, or Schwarzschild coordinates. We are thus led to global tests of general relativity, where the signal arrival times are compared with the expressions predicting them involving hundreds of parameters to take into account relativistic effects and Newtonian perturbations.

11.5 Advance of the perihelion

Here we shall carry out the calculations at post-Newtonian order using the PPN length element (11.4):

$$
\begin{aligned}
ds^2 = & -\left(1 - 2\frac{m}{\bar{r}} + 2\beta\frac{m^2}{\bar{r}^2} + \cdots\right) dt^2 \\
& + \left(1 + 2\gamma\frac{m}{\bar{r}} + \cdots\right)\left(d\bar{r}^2 + \bar{r}^2 d\theta^2 + \bar{r}^2 \sin^2\theta d\phi^2\right),
\end{aligned}
\tag{11.24}
$$

where we recall that the Eddington parameters β and γ are equal to 1 in general relativity.

The static nature and spherical symmetry imply that we can limit ourselves to geodesics in the 'plane' $\theta = \pi/2$, and the components u_t and u_ϕ of the 4-velocity are constant:

$$
\begin{aligned}
\left(1 - 2\frac{m}{\bar{r}} + 2\beta\frac{m^2}{\bar{r}^2} + \cdots\right)\frac{dt}{d\tau} &= 1 + E_n, \\
\left(1 + 2\gamma\frac{m}{\bar{r}} + \cdots\right)\bar{r}^2\frac{d\phi}{d\tau} &= L,
\end{aligned}
\tag{11.25}
$$

where τ, satisfying $d\tau^2 = -ds^2$, is the proper time and E_n and L are constants of the motion which can be identified as the energy and orbital angular momentum per unit mass of the planet.

We then can extract from the length element (11.24)

$$
\left(\frac{d\bar{r}}{d\tau}\right)^2 = \left(2E_n + \frac{2m}{\bar{r}} - \frac{L^2}{\bar{r}^2}\right)
$$
$$
+ \left(E_n^2 + \frac{4(1-\gamma)E_n m}{\bar{r}} + \frac{2(2-\beta-2\gamma)m^2}{\bar{r}^2} + \frac{4\gamma L^2 m}{\bar{r}^3}\right) + \cdots
$$

(11.26)

Setting $\bar{u} = 1/\bar{r}$, from (11.25) and (11.26) we find the equations for the trajectory $\bar{u}(\phi)$:

$$
\left(\frac{d\bar{u}}{d\phi}\right)^2 = \frac{2E_n}{L^2}\left(1 + \frac{E_n}{2}\right) + \frac{2m}{L^2}\left(1 + 2(1+\gamma)E_n\right)\bar{u}
$$
$$
- \left(1 - \frac{2(2-\beta+2\gamma)m^2}{L^2}\right)\bar{u}^2 + \cdots
$$

(11.27)

[If a radial coordinate different from $\bar{r} = \tilde{r} + \mathcal{O}(m^2)$ is used, a term proportional to u^3 arises.] Differentiating (11.27) then gives[4]

$$
\frac{d^2\bar{u}}{d\phi^2} + \bar{u}\left(1 - \frac{2(2-\beta+2\gamma)m^2}{L^2}\right) = \frac{m}{L^2}\left(1 + 2(1+\gamma)E_n\right).
$$

(11.28)

The solution of (11.28) satisfying (11.27) is an ellipse which precesses without deformation:

$$
\bar{r} = \frac{a(1-e^2)}{1 + e\cos\nu} \quad \text{with} \quad \nu = (\phi - \omega)\left(1 - \frac{(2-\beta+2\gamma)m^2}{L^2}\right),
$$

(11.29)

where the longitude of the periastron ω is an integration constant and the semi-major axis a and eccentricity e are related to E_n and L as

$$
\begin{cases}
a = -\frac{m}{2E_n}\left(1 + \frac{(3+4\gamma)E_n}{2}\right) \\[2mm]
e^2 = 1 + \frac{2E_n L^2}{m^2} - \frac{(7+8\gamma)L^2 E_n^2}{m^2} - 4(2-\beta+2\gamma)E_n.
\end{cases}
$$

(11.30)

The relativistic correction to the true anomaly ν in (11.29) is a secular term which makes the orbit non-periodic: the trajectory is an ellipse, which is not deformed when the isotropic coordinate $\bar{r}$ is used, the major axis of which turns slowly in the plane $\theta = \pi/2$ by an angle

$$
\Delta\omega = \frac{2\pi}{1 - (2-\beta+2\gamma)m^2/L^2} - 2\pi
$$
$$
\simeq \frac{6\pi m^2}{L^2}\frac{(2-\beta+2\gamma)}{3} \simeq \frac{6\pi m}{a(1-e^2)}\frac{(2-\beta+2\gamma)}{3}
$$

(11.31)

per period; see Fig. 11.3.

---

[4]The similarity of this equation to those governing the trajectory of a planet in the Nordström theory (see Book 2, Section 10.3), as well as to those for a charge in a Coulomb field in the Maxwell theory (see Book 2, Section 13.3), should be noted.

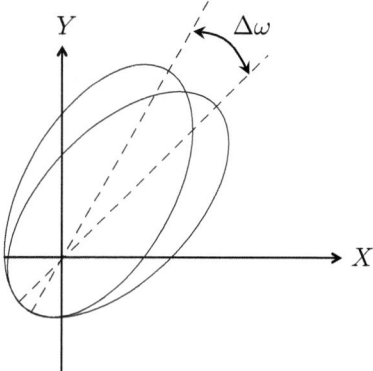

Fig. 11.3 Advance of the perihelion of Mercury.

The angle $\Delta\omega$ is a coordinate angle. It is effectively the angle measured on the Earth (which is far from the Sun) if the light coming from the planet travels in a straight line. This is not the case. Now, since the precession is a cumulative effect, after $N \gg 1$ revolutions $N\Delta\omega$ will be large enough that the distortion in the propagation of the light rays can be neglected, and $\Delta\omega$ can be compared directly to the observations. For Mercury, the orbital period is 88 days, $a = 5.8 \times 10^7$ km and $e = 0.2$. In addition, $m_\odot = 1.5 \times 10^3$ km and so

$$\Delta\omega = 42.98 \text{ arcsec/century} \tag{11.32}$$

in general relativity, where $\beta = \gamma = 1$.

The advance of the perihelion of Mercury

The result (11.32), obtained by Einstein in 1915, coincides with the precession, unexplained in Newtonian theory, discovered 70 years earlier by Le Verrier.[5] The astronomical observations of the last three centuries give the excess (11.32) with an accuracy of 10^{-2}. (We recall that the total observed precession is more than a hundred times larger; see, for example, Book 1, Section 13.3.) Radar measurements of the positions of the planets carried out since 1975 have led to large increases in the accuracy of the measurements, and the advance of the perihelion of Mercury is now known with an accuracy of 10^{-3}. Moreover, since the parameter γ is known to within 10^{-5} thanks to the Shapiro effect (see Section 11.4), the parameter β is also constrained to a value of 1 with an accuracy of 3×10^{-3}. In spite of this remarkable agreement between observation and the prediction of general relativity, the objection has been raised that part of the excess (11.32) may be due to the quadrupole moment of the Sun (Dicke, 1974; see Book 1, Section 14.1). However, the internal motion of the Sun is now well understood from study of the solar vibrational modes, and this objection has lost its plausibility; see Will (1993).

11.6 Post-Keplerian geodesics

To find the time dependence of the planet trajectories in the solar system when they are approximated by geodesics in the field of a central body, the simplest method is to start from

---

[5]This discovery seems to have caused Einstein's greatest emotion of his scientific life. See Pais (1982).

the parametrized post-Newtonian metric (11.24) and make a 'conchoidal' transformation of the radial coordinate $\bar{r} \to r_\delta$ with, to the required order, $\bar{r} = r_\delta + \delta\, m$, where for now δ is a free parameter. Then for $\theta = \pi/2$ we have

$$
ds^2 = -\left(1 - \frac{2m}{r_\delta} + \frac{2(\beta + \delta)m^2}{r_\delta^2} + \cdots\right) dt^2 + \left(1 + \frac{2m\gamma}{r_\delta} + \cdots\right) dr_\delta^2
$$
$$
+ r_\delta^2 \left(1 + \frac{2m(\gamma + \delta)}{r_\delta} + \cdots\right) d\phi^2,
\tag{11.33}
$$

where the Eddington parameters β and γ are 1 in general relativity.

Using the first integrals $-g_{tt}(dt/d\tau) = 1 + E_n$ and $g_{\phi\phi}(d\phi/d\tau) = L$ as well as $ds^2/d\tau^2 = -1$, we obtain

$$
\left(\frac{dr_\delta}{dt}\right)^2 \simeq A + \frac{2B}{r_\delta} + \frac{C_\delta}{r_\delta^2} + \frac{D_\delta}{c^2 r_\delta^3}, \qquad \frac{d\phi}{dt} \simeq \frac{H}{r_\delta^2} + \frac{I_\delta}{r_\delta^3},
\tag{11.34}
$$

where the coefficients $A, B, C_\delta...$ are given by

$$
\begin{cases}
A = 2E_n \left(1 - \dfrac{3E_n}{2}\right), \quad B = m\left(1 - 2(2+\gamma)E_n\right), \\[2mm]
C_\delta = -L^2 \left(1 - 2E_n + \dfrac{2(2+\beta+2\gamma+\delta)m^2}{L^2}\right), \quad D_\delta = 2(2+2\gamma+\delta)mL^2, \\[2mm]
H = L\left(1 - E_n\right), \quad I_\delta = -2(1+\gamma+\delta)mL.
\end{cases}
\tag{11.35}
$$

First choosing $\delta = -2(1+\gamma)$, we have $D_\delta = 0$ so that the equation for $r_\delta(t)$ takes a Newtonian form. Then if we choose $\delta = -(1+\gamma)$, we find $I_\delta = 0$ and it is the equation for $\phi(t)$ which takes a Newtonian form.

The integrations (see details below) then give, after reverting to the isotropic coordinate $\bar{r}$,

$$
n(t - T) = \eta - e_t \sin\eta, \qquad \bar{r} = a(1 - e\cos\eta),
$$
$$
\tan\left[\left(1 - \frac{(2-\beta+2\gamma)m^2}{L^2}\right)\frac{(\phi - \omega)}{2}\right] = \sqrt{\frac{1+e}{1-e}} \tan\frac{\eta}{2},
\tag{11.36}
$$

where T and ω are two constants of integration, the semi-major axis a and the eccentricity e are given in (11.30), and the *mean motion* n and eccentricity e_t are related to E_n and L (or a and e) as

$$
\begin{cases}
n = \dfrac{(-2E_n)^{3/2}}{m}\left(1 + \dfrac{(7+8\gamma)E_n}{4}\right) = \sqrt{\dfrac{m}{a^3}}\left[1 - \dfrac{(4+5\gamma)m}{2a}\right], \\[3mm]
e_t^2 = 1 + \dfrac{2E_n L^2}{m^2} + \dfrac{(9+8\gamma)L^2 E_n^2}{m^2} + 4\beta E_n \\[3mm]
\qquad = e^2\left(1 + 8(1+\gamma)E_n\right) = e^2\left[1 - \dfrac{4(1+\gamma)}{a}\right].
\end{cases}
\tag{11.37}
$$

The post-Newtonian trajectory of a planet can therefore be written in a form similar to its Newtonian limit where the two eccentricities e and e_t become equal to each other.[6]

---

[6]We shall see in Section 12.1 that the relative trajectory of two bodies with comparable masses can also be cast into a quasi-Keplerian form, bringing into play three rather than two eccentricities.

Integration of the post-Keplerian equations

Let us first consider (11.34) for $r_\delta(t)$ with $\delta \equiv \delta_1 = -2(1+\gamma)$ so that $dr_1/dt = A + 2B/r_1 + C_1/r_1^2$. This equation, of Newtonian form, can be integrated parametrically into $n(t-T) = \eta - e_t \sin\eta$, $r_1 = a_1(1 - e_t \cos\eta)$ with $n = (-A)^{3/2}/B$, $a_1 = -B/A$, and $e_t = \sqrt{1 - AC_1/B^2}$ [which are given explicitly in (11.37) with the use of (11.35)].

Since the coordinates $\bar{r}$ and r_δ are related as $\bar{r} = r_\delta + m\delta$, we find $\bar{r} = a_1(1 - e_t \cos\eta) + m\delta_1$, which can be written to the required order as $\bar{r} = a_r(1 - e_r \cos\eta)$ with $a_r = a_1 + m\delta_1$ and $e_r = e_t(1 - m\delta_1/a_1)$. Expanding these expressions using (11.35), we find that $a_r = a$ and $e_r = e$, where a and e are given in (11.30). We thus find the first equation in (11.36).

Now let us turn to (11.34) for $d\phi/dt$, this time setting $\delta \equiv \delta_2 = -(1+\gamma)$, so that it also reduces to Newtonian form: $d\phi_2/dt = H/r_2^2$. An efficient way to integrate this equation is to start from the quasi-Newtonian ansatz

$$\tan\left[\left(1 - \frac{m^2(2 - \beta + 2\gamma)}{L^2}\right)\frac{\phi}{2}\right] = \sqrt{\frac{1 + \bar{e}}{1 - \bar{e}}} \tan\frac{\eta}{2}$$

$$\text{with} \quad n(t - T) = \eta - e_t \sin\eta \quad \text{and} \quad r_1 = a_1(1 - e_t \cos\eta).$$

Differentiating the first equation and using the others to express $d\phi/d\eta$ as a function of r_1, which is then replaced by its value $r_1 = r_2 + (1+\gamma)m$, we see that to the required order in $1/c^2$ the term proportional to $1/r_2^3$ appearing in the expression for $d\phi/dt$ is eliminated if we choose $\bar{e} = e$. Finally, we check that the coefficient of the remaining term proportional to $1/r_2^2$ is indeed equal to H. Q.E.D. (The trajectory of a charge in a Coulomb field is obtained in the same way; see Book 2, Section 13.3.)

11.7 Spin in a gravitational field

We saw in Book 2, Chapter 7 that in the theory of relativity we are led to introduce the concept of a particle carrying, in addition to its inertial mass, an intrinsic angular momentum or (classical) *spin* j, a vector orthogonal to the particle 4-velocity u. In the absence of an external torque the equations of motion of the center of mass and the spin of an object in free fall in a gravitational field are

$$\frac{Du^\mu}{d\tau} = 0, \quad \frac{Dj^\mu}{d\tau} = 0 \quad \text{with} \quad j_\mu u^\mu = 0, \tag{11.38}$$

where the covariant derivative D describes the gravitational field in which the spin moves.[7]

- *Geodetic precession*

Let us consider a spin in the Schwarzschild field, in a circular orbit in the $\theta = \pi/2$ 'plane'. Then as we saw in Section 11.2, its 4-velocity has the components

---

[7]These are the Mathisson equations seen in Book 2, Section 7.3. Since an object carrying a (classical) spin cannot be considered as a point object, we are led to add to (11.38) a coupling to gravitational 'tidal forces', that is, to the Riemann tensor, which in addition causes the center of mass to deviate from the geodesic trajectory (Papapetrou, 1951 and Pirani, 1956). These corrections are negligible in the applications we consider in the present section, namely, the motion of a gyroscope in the field of the Earth.

A detailed description of the Gravity Probe B experiment can be found at http://einstein.stanford.edu/.

$$u^t \equiv \frac{dt}{d\tau} = \frac{e^{-\nu}}{\sqrt{1 - r\nu'}} = \frac{1}{\sqrt{1 - 3m/r}} \,,$$

$$u^\phi \equiv \frac{d\phi}{d\tau} = \frac{\sqrt{\nu'/r}}{\sqrt{1 - r\nu'}} = \sqrt{\frac{m}{r^3}} \sqrt{\frac{1}{1 - 3m/r}} \,,$$
(11.39)

and so $d\phi/dt = e^\nu \sqrt{\nu'/r} = \sqrt{m/r^3}$ (as in Newtonian gravity).

It is then straightforward to write the spin equation of motion (11.38) using the Christoffel symbols given in Section 7.2 (*modulo* the change $\nu \to 2\nu$ and $\lambda \to 2\lambda$). The component j^θ remains constant and

$$\frac{dj^\phi}{dt} + \frac{e^\nu}{r^2} \sqrt{r\nu'} j^r = 0 \,, \quad \frac{dj^r}{dt} - e^{-2\lambda + \nu} \sqrt{r\nu'} (1 - r\nu') j^\phi = 0$$
(11.40)

with $j^0 = re^{-\nu} \sqrt{r\nu'} j^\phi$, the solution of which, choosing $j^\phi(0) = 0$ and $j^r(0) = r^2$, is

$$j^\phi = -\frac{re^\lambda}{\sqrt{1 - r\nu'}} \sin \Omega t \,, \quad j^r = r^2 \cos \Omega t$$
(11.41)

$$\text{with} \quad \Omega = \frac{d\phi}{dt} e^{-\lambda} \sqrt{1 - r\nu'} = \sqrt{\frac{m}{r^3}} \sqrt{1 - \frac{3m}{r}} \,.$$

We then find that the angular velocity of the spin precession, called the *geodetic precession*, is (this was calculated by de Sitter in 1916)

$$\omega_{\text{geod}} = \sqrt{\frac{m}{r^3}} \left(1 - \sqrt{1 - \frac{3m}{r}} \right) \approx \frac{3}{2} \sqrt{\frac{m^3}{r^5}} \,.$$
(11.42)

(Therefore, $\omega_{\text{geod}} = -3\,\omega_{\text{Thomas}}$; see Book 2, Section 7.4.) For a gyroscope in a circular orbit about the Earth at an altitude of 642 km we have $\omega_{\text{geod}} \approx 6.6$ arcsec/yr, which is effectively the value measured in the Gravity Probe B experiment in 2011 with an accuracy of 1%, in excellent agreement with the prediction of general relativity.

The calculation of ω_{geod} in first order in m/r when the gravitational field is described by the parametrized post-Newtonian metric (11.24) is easily done by passing to the coordinate r and using eqns (11.39) and (11.40). We find

$$\Omega \approx \sqrt{\frac{m}{r^3}} \left[1 - \frac{m(1 + 2\beta)}{2r} \right] \,, \quad \frac{d\phi}{dt} \approx \sqrt{\frac{m}{r^3}} \left[1 + \frac{m(\gamma - \beta)}{r} \right]$$
(11.43)

$$\implies \quad \omega_{\text{geod}} = \left(\frac{1}{2} + \gamma \right) \sqrt{\frac{m^3}{r^5}} \,.$$

• *Lense–Thirring precession*

Now let us assume that the object creating the field also has a 'spin'. Its gravitational field now is no longer static, but only stationary, and the metric describing it must contain cross terms g_{0i} (the Kerr metric studied in Chapter 8 is an example). In lowest order, the correction to the precession of a gyroscope due to the presence of these terms (corresponding to 'spin–spin coupling') is added linearly to the geodetic precession and can therefore be calculated keeping *only* the corrections proportional to $g_{0i} \equiv 4g_i$ in the Minkowski metric:

$$ds^2 = -dt^2 + dx^2 + dy^2 + dz^2 + 8\,g_i\,dx^i\,dt\,. \tag{11.44}$$

Here, as we shall see below, g is a vector with components g_i given by $4g = (1+\gamma+\alpha_1/4)R \wedge J_{\mathrm{N}}/r^3$ (with $r^2 = R.R = \delta_{ij}x^ix^j$), J_{N} is the Newtonian angular momentum of the object creating the field, γ is the Eddington parameter, and α_1 is a new PPN parameter.

The calculation of the Christoffel symbols is straightforward: $\Gamma^x_{ty} = -\Gamma^y_{tx} = -2(\nabla \wedge g)^z$, $\Gamma^x_{tz} = -\Gamma^z_{tx} = 2(\nabla \wedge g)^y$, and $\Gamma^y_{tz} = -\Gamma^z_{tz} = -2(\nabla \wedge g)^x$.

In addition, in the lowest order in which we are concerned with here, we can assume that the gyroscope is at rest. Its equation of motion then reduces to $dj^\mu/dt + \Gamma^\mu_{ti}j^i = 0$ and is easily written as (see the discussion of vector calculus in Book 2, Section 11.1)

$$\frac{dj}{dt} = \omega_{\mathrm{LT}} \wedge j \quad \text{with} \quad \omega_{\mathrm{LT}} = -2\nabla \wedge g = \frac{1}{2}(1+\gamma+\alpha_1/4)\left(\frac{3R(R.J_{\mathrm{N}})}{r^5} - \frac{J_{\mathrm{N}}}{r^3}\right). \tag{11.45}$$

This effect was calculated by Lense and Thirring in 1918. It will decouple from the geodetic precession if the gyroscope has a polar orbit. Then on the average $\bar{\omega}_{\mathrm{LT}} = (1+\gamma+\alpha_1/4)J_{\mathrm{N}}/(4r^3)$.

If the vector J_{N} is the angular momentum of the Earth, we will have $J_{\mathrm{N}} = I\omega$, where $I = 0.33\,M_\oplus\,r_\oplus^2$ is the Earth's moment of inertia (see Book 1, Section 6.1) and ω is its diurnal speed. Then

$$\bar{\omega}_{\mathrm{LT}} \simeq 2(1+\gamma+\alpha_1/4) \times 10^{-2}\ \text{arcsec/yr}\,. \tag{11.46}$$

This effect was measured in the Gravity Probe B experiment in 2011, and is consistent with the value predicted by general relativity ($\gamma + \alpha_1/4 = 1$; see below) to within 20 %.

Therefore, the measurements of the geodetic and Lense–Thirring precessions test the equation of motion (11.38) of a spin in a gravitational field and constrain the PPN parameter γ (which, we recall, is known with an accuracy of 10^{-5} from measurement of the Shapiro effect; see Section 11.4) and α_1.

The gravitomagnetic field

It is easily seen that the linearized Einstein tensor of the (stationary) metric (11.44) reduces to $G_{0i} = -2\partial^2_{ij}g^j - 2\triangle g_i$.

It can also be seen that in the coordinate transformation $t = \tilde{t} + \xi$ we have $\tilde{g}_i = g_i - \partial_i\xi$. We can therefore have $\partial_i\tilde{g}^i = 0$ if ξ is chosen such that $\triangle\xi = \partial_ig^i$. In this 'Coulomb gauge' (see Book 2, Section 12.3) the Einstein equations are the same as those of magnetostatics, that is (suppressing the tildes),

$$\triangle g_i = -4\pi T_{0i}\,, \quad \text{the solution of which is} \quad g_i = \int \frac{T_{0i}(R')}{|R-R'|}dV'\,.$$

The role of the 'charge current' is played by the energy–momentum tensor $j_i = 4T_{0i}$, and so it is sufficient to 'copy and paste' the calculation of Book 2, Section 13.5 to obtain

$$g = \frac{1}{2r^3}R \wedge J_{\mathrm{N}}\,, \quad \text{where} \quad J_{\mathrm{N}i} \equiv e_{ijk}\int x^j(T^{k0})$$

is the angular momentum of the source; see Book 2, Section 7.2.

In lowest order we have $T^{0i} = \rho v^i$, where ρ is the mass density of the source and v^i is the velocity of its elements. If we imagine this as a symmetric top with angular velocity ω, we find $J_N = I\omega$, where I is the moment of inertia.

In spherical coordinates the metric (11.44) becomes (for $J_N \| OZ$)

$$ds^2 = -dt^2 + dr^2 + r^2(d\theta^2 + \sin^2 d\phi^2) - \frac{4J_N \sin^2 \theta}{r} dt \, d\phi. \tag{11.47}$$

(We have already used this in Section 8.2 [eqn (8.18)] to identify the parameter ma of the Kerr metric as the angular momentum of the black hole it describes.)

In the PPN formalism the coefficient of the vector g becomes arbitrary and so we can take it to be

$$4g = (1 + \gamma + \alpha_1/4)(R \wedge J_N)/r^3.$$

In several of the theories competing with general relativity, in particular, those based on a Lorentz-invariant PPN Lagrangian, $\alpha_1 = 0$ [see Will (1993) and references in footnote 2 above].

12

The post-Newtonian approximation

In the preceding chapter we studied the two-body problem in general relativity assuming that one of the bodies (the 'Sun') is much more massive than the other (a 'planet'). We could thus describe the gravitational field of the Sun by the Schwarzschild metric and the motion of the planet by the geodesic equation.

When the two bodies have comparable masses this 'Schwarzschild approximation' is obviously no longer valid. As we shall see in the present chapter as well as the four following ones, finding the gravitational field created by these bodies and their motion in this field is much more difficult than in Newtonian physics, where the two-body problem reduces to uniform rectilinear motion of the center of mass of the system combined with the Keplerian motion of an 'effective' particle (see Book 1, Chapter 12).

We shall begin our study of this two-body problem in general relativity by limiting ourselves to corrections proportional to $v^2 \sim m/R$, the so-called post-Newtonian or 1PN corrections to Newton's universal law of attraction.

First we will find the gravitational field, that is, the metric, created by the two bodies [eqns (12.1), (12.7), and (12.17) below, which generalize the Schwarzschild solution expanded through post-Newtonian order]. Then we will derive the equations of motion (12.25), and finally the actual motion, that is, the post-Keplerian trajectories (12.34), which generalize the post-Keplerian geodesics obtained in Section 11.6. Along the way we will have to make some simplifying hypotheses, which will be lifted in later chapters.

12.1 The metric at post-Newtonian order

Since our goal[1] is to describe the motion of two compact, self-gravitating bodies which are far-separated and moving slowly, we seek the metric in the region of spacetime where they are moving (this is called the *near zone*). The gravitational field is weak, and so the idea is to expand the metric about the Minkowski metric in powers of the small parameter v, the orbital velocity of the objects ($c = 1$). According to Newton's law, m/R is numerically comparable to v^2 (m is the mass of an object, R is the distance to its companion, and $G = 1$). The structure of the metric g through order $v^2 \sim m/R$ (called 1PN) therefore will be

$$g \approx 1 + (m/r) + (m/r)(v^2 + m/R) + \mathcal{O}(v^6) \,,$$

where r is the distance to the system. The first term m/r is the Newtonian potential. The second can be interpreted as arising from both the finiteness of the speed of propagation of the gravitational interaction and the nonlinearity of the Einstein equations.

---

[1]See also the now-classic discussion of Landau and Lifshitz (1972), as well as Weinberg (1972) or Straumann (2013). The method described here was initiated by Blanchet and Damour (1989), developed by Damour, Soffel, and Xu (1991), and then generalized to parametrized post-Newtonian metrics (using two Eddington parameters β and γ) by Damour and Esposito-Farese (1992).

Relativity in Modern Physics. Nathalie Deruelle and Jean-Philippe Uzan.
© Oxford University Press 2018. Published in 2018 by Oxford University Press.
DOI: 10.1093/oso/9780198786399.001.0001

It is convenient and sufficient to seek the metric describing the gravitational field of two compact objects in post-Newtonian order in the form

$$g_{tt} = -e^{2U} + \mathcal{O}(v^6)\,, \quad g_{ti} = 4g_i + \mathcal{O}(v^5)\,, \quad g_{ij} = \delta_{ij}e^{-2U} + \mathcal{O}(v^4)\,, \tag{12.1}$$

where it is understood that U is the sum of a term of order v^2 (the Newtonian potential) and a term of order v^4, and that g_i is a quantity of order v^3. The fact that g_{ij} is determined as a function of g_{tt} to order v^2 is a result of the Newtonian approximation; see below.

The metric at Newtonian order

We make the expansion (for the time being in an arbitrary coordinate system) $g_{tt} = -1 + g_{tt}^{(2)} + \cdots$, $g_{ij} = \delta_{ij} + g_{ij}^{(2)} + \cdots$ (g_{it} is of third order). The derivative with respect to the time t increases the order because $\partial/\partial x^i \approx 1/r \gg \partial/\partial t \approx v/r$. The Christoffel symbols then are (Latin indices are moved using δ^{ij})

$$\Gamma_{tt}^{(2)i} = \Gamma_{it}^{(2)t} = -\frac{1}{2}\partial_i g_{tt}^{(2)}\,, \quad \Gamma_{jk}^{(2)i} = \frac{1}{2}(\partial_j g_{ki}^{(2)} + \partial_k g_{ij}^{(2)} - \partial_i g_{jk}^{(2)})\,.$$

The components of the Ricci tensor are easily calculated and we find

$$R_{tt}^{(2)} = -\frac{1}{2}\triangle g_{tt}^{(2)}\,, \quad R_{ij}^{(2)} = \frac{1}{2}\left(\partial_i\xi_j + \partial_j\xi_i - \triangle g_{ij}^{(2)}\right) \quad \text{with} \quad \xi_i \equiv \frac{1}{2}\left(\partial_i g_{tt}^{(2)} - \partial_i g_{kk}^{(2)}\right) + \partial_k g_{ik}^{(2)}\,.$$

We see that a good choice of gauge is $\xi_i = 0$. Since at lowest order only the component T_{tt} of the energy–momentum tensor is nonzero, the Einstein equations $R_{\mu\nu} = 8\pi(T_{\mu\nu} - \frac{1}{2}g_{\mu\nu}T)$ reduce to

$$R_{tt}^{(2)} = 4\pi\,T_{tt}^{(0)}\,, \quad R_{ij}^{(2)} = 4\pi\,T_{tt}^{(0)}\delta_{ij}\,,$$

that is, $\triangle g_{tt}^{(2)} = -8\pi T_{tt}^{(0)}$ and $\triangle g_{ij}^{(2)} = -8\pi\delta_{ij}T_{tt}^{(0)}$, the solution of which is

$$g_{tt}^{(2)} = -2U\,, \quad g_{ij}^{(2)} = -2\delta_{ij}U \quad \text{with} \quad \triangle U = 4\pi T_{tt}^{(0)} \quad \text{or} \quad U(t,x^i) = -\int d^3x'\,\frac{T_{tt}^{(0)}(t,x'^i)}{r'} + \mathcal{O}(v^4)\,,$$

where $r' \equiv \sqrt{(x^i - x'^i)(x_i - x'_i)}$ is the distance from the field point x^i to a source point x'^i evaluated at time t, and U is just the Newtonian potential. Finally, we can check that the gauge condition $\xi_i = 0$ is satisfied identically by the solution. We see that $g_{ij}^{(2)}$ is determined as a function of $g_{tt}^{(2)}$, which justifies the ansatz (12.1) at post-Newtonian order.

The useful Christoffel symbols of the metric (12.1) are (the calculations of this section are elementary)

$$\Gamma_{tt}^t = \dot{U} + \mathcal{O}(v^4)\,, \quad \Gamma_{tt}^i = \partial_i U e^{4U} + 4\dot{g}_i + \mathcal{O}(v^6)\,, \quad \Gamma_{ti}^t = \partial_i U + \mathcal{O}(v^4)\,,$$
$$\Gamma_{tj}^i = -\delta_j^i \dot{U} + 2(\partial_j g^i - \partial^i g_j) + \mathcal{O}(v^5)\,, \quad \Gamma_{ik}^k = -3\partial_i U + \mathcal{O}(v^4)\,. \tag{12.2}$$

Here and below, $\dot{f} \equiv \partial_t f$.

The components of the Ricci tensor then are

$$R^{tt} = \triangle U + \partial_t(3\dot{U} + 4\partial_j g^j) + \mathcal{O}(v^6)\,, \quad R^{ti} = 2\triangle g^i - 2\partial^i(\dot{U} + \partial_j g^j) + \mathcal{O}(v^5)\,. \tag{12.3}$$

The components of the same order of the source $S^{\mu\nu} \equiv T^{\mu\nu} - \frac{1}{2}g^{\mu\nu}T$ are particularly simple:

$$S^{tt} = \frac{1}{2}\left(T^{tt} + T^{ii}\right) + \mathcal{O}(v^4), \quad S^{ti} = T^{ti} + \mathcal{O}(v^3). \qquad (12.4)$$

Therefore, the Einstein equations written in the form $R^{\mu\nu} = 8\pi\, S^{\mu\nu}$ reduce to two equations for the potentials U and g_i:

$$\begin{aligned}
\triangle U + \partial_t(3\dot{U} + 4\partial_j g^j) &= 4\pi\left(T^{tt} + T^{ii}\right) + \mathcal{O}(v^6),\\
\triangle g_i - \partial_i(\dot{U} + \partial_j g^j) &= 4\pi\, T^{ti} + \mathcal{O}(v^5).
\end{aligned} \qquad (12.5)$$

We still need to choose a gauge. The harmonic gauge, with $\partial_\nu(\sqrt{-g}\,g^{\mu\nu}) = 0$, requires at this order that $\dot{U} + \partial_j g^j = 0$, so that the equations (12.5) become (using the fact that $\triangle g^i = \square g^i$ in the order in which we are working)

$$\square U = 4\pi\left(T^{tt} + T^{ii}\right) + \mathcal{O}(v^6), \quad \square g^i = 4\pi\, T^{ti} + \mathcal{O}(v^5), \qquad (12.6)$$

the relevant solutions of which are, as in electromagnetism, the retarded Liénard–Wiechert potentials:

$$U(t, x^i) = -\int d^3x'\,\frac{T^{tt} + T^{ii}}{r'} + \mathcal{O}(v^6), \quad g^i = -\int d^3x'\,\frac{T^{ti}}{r'} + \mathcal{O}(v^5), \qquad (12.7)$$

where $r' \equiv \sqrt{(x^i - x'^i)(x_i - x'_i)}$ is the distance from the field point x^i to a source point x'^i evaluated at the retarded time $t - r'$. In the order in which we are working we can neglect all retardation effects in g^i, but not in U. We then again see that the metric (12.1) includes effects due to the finiteness of the speed of propagation of gravitation and effects due to the nonlinearity of the Einstein equations [because it is necessary to make the expansion $e^{2U} = 1 + 2U - 2U^2 + \mathcal{O}(v^6)$].

The asymptotic metric of a stationary system

Let us situate ourselves far from the system creating the gravitational field described by the metric (12.1) where the potentials are given in (12.7), and in addition assume that it can be considered stationary. Using r to denote the distance from the field point to an origin located inside the system, from (12.7) we can extract the asymptotic behavior of the potentials for $r \to \infty$:

$$g_i \approx -\frac{1}{r}\int d^3x\, T^{ti}, \quad U \approx -\frac{1}{r}\int d^3x\,(T^{tt} + T^{ii}) \equiv -\frac{M}{r},$$

where M is the *Tolman mass* defined in Section 5.4 (and, as we saw in Section 6.3, equal to the inertial mass of the system). The diagonal terms of the metric (12.1) then are

$$g_{tt} = -1 + 2M/r - 2(M/r)^2 + \mathcal{O}(v^6), \quad g_{ij} = \delta_{ij}(1 + 2M/r) + \mathcal{O}(v^4).$$

The cross term of the metric (12.1) is $g_{ti} = (4/r)\int d^3x\, T_{ti}$, and we thereby recover the gravitomagnetic term obtained in Section 11.7, which is also written as $g_{ti} = 2\epsilon_{ijk}x^j J^k/r^3$, where $J_{\mathrm{N}i} = \int d^3x\,\epsilon_{ijk}x^j T^{kt}$ is the angular momentum of the source.

If we choose the z axis to be parallel to the vector J_N, in the end the desired metric is found to be

$$ds^2 = -\left(1 - \frac{2M}{c^2 r} + \frac{2M^2}{c^4 r^2}\right) c^2 dt^2 + \left(1 + \frac{2M}{c^2 r}\right) d\vec{x}^2 - \frac{4J_N}{c^3 r^3}(x\,dy - y\,dx)dt\,. \tag{12.8}$$

We note that if $J_N = 0$, this expression is just the expansion of the Schwarzschild metric in harmonic (or isotropic) coordinates. However, whereas the latter was obtained assuming that the field is static and spherically symmetric, the metric (12.8) describes the field far from a system of arbitrary masses if the time scale associated with their motion is greater than r/v. For $J_N \neq 0$, (12.8) similarly generalizes the metric obtained in Section 11.7 which we used to calculate the Lense–Thirring effect.

12.2   The post-Newtonian field of compact bodies

To complete our study of the metric at post-Newtonian order, we must specify the nature of the sources, that is, their energy–momentum tensor.

We assume that the objects are compact enough that they can be treated as points and described by a distributional energy–momentum tensor (*cf.* Section 3.2)

$$T^{\mu\nu}(x^\rho) = \sum m \int \delta_4(x^\rho - z^\rho(\tau)) \frac{u^\mu u^\nu}{\sqrt{-g}} d\tau\,, \tag{12.9}$$

where m is their mass, $u^\mu = dz^\mu/d\tau$ is their 4-velocity normalized using the metric $g_{\mu\nu}$, and $\delta(x - z)$ is the Dirac distribution. Integrating on the world lines, (12.9) can also be written as

$$
\begin{aligned}
T^{\mu\nu}(t, x^i) &= \sum m \frac{u^\mu u^\nu}{\sqrt{-g}} \frac{d\tau}{dt} \delta_3(x^i - z^i(t)) \\
&= \sum m \frac{dz^\mu}{dt} \frac{dz^\nu}{dt} \frac{1}{\sqrt{g g_{\rho\sigma} \frac{dz^\rho}{dt} \frac{dz^\sigma}{dt}}} \delta_3(x^i - z^i(t))\,.
\end{aligned}
\tag{12.10}
$$

Since the metric is given by (12.1), we have $g g_{\rho\sigma}(dz^\rho/dt)(dz^\sigma/dt) = 1 - 2U - v^2 + \mathcal{O}(v^4)$, where $v^i \equiv dz^i/dt$ is the 3-velocity of the mass m. Therefore,

$$T^{ti} = -\sum m\,v_i\,\delta_3(x^j - z^j(t)) + \mathcal{O}(v^3)\,,$$

$$T^{tt} + T^{ii} = \sum m \left(1 + \frac{3}{2}v^2 + U\right) \delta_3(x^i - z^i(t)) + \mathcal{O}(v^4)\,. \tag{12.11}$$

Here to solve the equations (12.5) we use the 'Coulomb' gauge[2]

$$3\dot{U} + 4\partial_j g^j = 0\,, \tag{12.12}$$

which reduces the equations to

---

[2] We shall follow Straumann (2013) in showing how the choice of gauge affects the convergence properties of the solutions.

$$\triangle U = 4\pi (T^{tt} + T^{ii}),$$

and, if we set $g_i \equiv \zeta_i + \frac{1}{4}\partial_i \chi,$ then $\triangle \chi = U,$ $\triangle \zeta^i = 4\pi T^{ti}.$ (12.13)

Now we proceed by iteration. At lowest order we find $\triangle U = 4\pi \sum m \delta_3(x^i - z^i(t))$, the solution of which is the Newtonian potential $U(t, x^i) = -\sum m/|\vec{x} - \vec{z}(t)|$. The solution for ζ^i then follows immediately:

$$\zeta^i(t, x^i) = -\sum \frac{m v^i}{|\vec{x} - \vec{z}(t)|}, \tag{12.14}$$

and we verify that the gauge condition, which is now written as $\dot{U} + \partial_i \zeta^i = 0$, is satisfied, and is equivalent to the conservation of the energy–momentum tensor $\partial_t T^{(0)tt} + \partial_i T^{(1)ti} = 0$.
Now we find that χ satisfying $\triangle \chi = U$ is given by

$$\chi = -\frac{1}{2} \sum m |\vec{x} - \vec{z}(t)|. \tag{12.15}$$

We note that χ diverges at infinity, but only its derivatives appear in the metric. Then, using (12.13)–(12.15) and setting $n^i \equiv (x^i - z^i(t))/|\vec{x} - \vec{z}(t)|$, we have

$$g_i = -\frac{1}{8} \sum \frac{m}{|\vec{x} - \vec{z}(t)|} [7v_i + n_i(n.v)] + \mathcal{O}(v^5). \tag{12.16}$$

Lastly, we need to find U to the next order. To do this we substitute into the source $T^{tt} + T^{ii}$ given in (12.11) the solution for U at Newtonian order and evaluate it *on the* trajectories. Then divergent terms $\propto \sum m^2 \delta_3(x^i - z^i(t))/|\vec{x} - \vec{z}(t)|$ appear; these must be regularized to zero, as will be described in Chapter 15 in a more detailed analysis of the approximation schemes.[3] It is then possible to perform the integration and we find

$$U(t, x^i) = -\sum \frac{m}{|\vec{x} - \vec{z}(t)|} - \sum_a \frac{m}{|\vec{x} - \vec{z}_a(t)|} \left(\frac{3}{2} v_a^2 - \sum_{a' \neq a} \frac{m'}{|\vec{z}_a - \vec{z}_{a'}|} \right) + \mathcal{O}(v^6). \tag{12.17}$$

Therefore, in the post-Newtonian approximation of general relativity the gravitational field created by celestial bodies which are compact enough to be considered as point objects is described by the metric (12.1) with $g_{ti} \equiv 4g_i$ given by (12.16), $g_{tt} = -1 - 2U - 2U^2$, and $g_{ij} = \delta_{ij}(1 - 2U)$ with U given by (12.17).

12.3 The EIH equations of motion

We know that the motion of a test particle in a gravitational field is governed by its action $S = -m \int d\tau$, where τ is the curvilinear abscissa of its world line, that is, its proper time, and the extremization of this action leads to the geodesic equation; see Section 3.1.

---

[3] We note that this problem, which arises owing to the modeling of the bodies by a distributional energy–momentum tensor, is absent in the Maxwell theory, which is linear. However, we have already encountered this problem and solved it in the same way in our study of the gravitational field in the nonlinear Nordström theory; see Book 2, Section 10.4.

To obtain the equations of motion of bodies which are compact enough to be considered point objects, we *assume* that, just like test particles, they follow the geodesics of the metric describing the field that they create.[4] Therefore, the action of a body of mass m at $x^\mu = z^\mu(t)$ is postulated to be

$$S \equiv \int L_m \, dt = -m \int \sqrt{-g_{\mu\nu} \frac{dz^\mu}{dt} \frac{dz^\nu}{dt}} \, dt \,. \qquad (12.18)$$

Instead of expanding the geodesic equation found by extremization to post-Newtonian order, we shall first expand the integrand, that is, the Lagrangian L_m. Since the length element is $ds^2 = -e^{2U} dt^2 + 4g_i dt \, dx^i + e^{-2U} d\vec{x}^2$ [cf. (12.1)], we have

$$
\begin{aligned}
L_m &= -m e^U \left(1 - v^2 e^{-4U} - 8g_i v^i e^{-2U} \right)^{\frac{1}{2}} \\
&= -m \left(1 - \frac{v^2}{2} + U - \frac{1}{8}v^4 + \frac{3}{2}v^2 U + \frac{U^2}{2} - 4g_i v^i \right) + \mathcal{O}(v^6) \,,
\end{aligned}
\qquad (12.19)
$$

where the potentials U and g_i obtained in (12.16) and (12.17) must be evaluated *on the* trajectory, that is, on $\vec{x} = \vec{z}(t)$, where their self part $\propto m$ or m^2 diverges. In order to solve the Einstein equations and obtain the metric, we have ignored the self, divergent, parts; *cf.* Section 12.2. We shall do the same thing here, ignoring *all* the contributions to the field from the body m, a procedure which will be analyzed in Chapter 15. Then the potentials regularized at $\vec{x} = \vec{z}(t)$ are (in the case of two bodies, to which we restrict ourselves)

$$U(t, z^i(t)) = -\frac{m'}{R}\left(1 + \frac{3v'^2}{2}\right) + \mathcal{O}(v^6), \quad g_i = -\frac{m'}{8R}[7v_i' + N_i(N.v')] + \mathcal{O}(v^5), \quad (12.20)$$

where $z^i - z'^i \equiv RN^i$ and N^i is the unit vector, and so the regularized Lagrangian of the body m becomes

$$
\begin{aligned}
\frac{L_m}{m} &= -1 + \frac{1}{2}v^2 + \frac{m'}{R} + \frac{1}{8}v^4 + \frac{m'}{2R}[3(v^2 + v'^2) - 7(v.v') - (N.v)(N.v')] \\
&\quad - \frac{1}{2}\frac{m'^2}{R^2} + \mathcal{O}(v^6) \,.
\end{aligned}
\qquad (12.21)
$$

The Lagrangian of the two-body *system* is constructed so as to give the same equations of motion as (12.21) when $m \to 0$. This is the *Fichtenholz Lagrangian*[5] governing the motion of two compact, self-gravitating bodies:

---

[4] This hypothesis is 'natural' by 'continuity', but it must and can be proved. It is the *effacement property* which we will use below in Chapter 15.

[5] Here L_F is the gravitational analog of the *Darwin Lagrangian* of a system of two slowly moving charges in the Maxwell theory; see Book 2, Section 21.1. The most important difference between these two Lagrangians is that the Fichtenholz Lagrangian contains a term (the last term) which originates from the second iteration of the Einstein equations.

$$L_F = -(m + m') + \frac{1}{2}mv^2 + \frac{1}{2}m'v'^2 + \frac{mm'}{R}$$

$$+ \frac{1}{8}(mv^4 + m'v'^4) + \frac{mm'}{2R}[3(v^2 + v'^2) - 7(v.v') - (N.v)(N.v')] \qquad (12.22)$$

$$- \frac{mm'(m + m')}{2R^2} + \mathcal{O}(v^6).$$

The Euler–Lagrange equation of motion of the body m is obtained from (12.22) or (12.21): $\dot{p}_i \equiv dp_i/dt = \partial L_F/\partial z^i$, where $p_i \equiv \partial L_F/\partial v^i$. In lowest order this is just Newton's law:

$$\dot{v}^i = -\frac{m'}{R^2}N^i + \cdots. \qquad (12.23)$$

In the next order we have

$$p^i = mv^i + \frac{m}{2}v^2 v^i + \frac{mm'}{2R}\left[6v^i - 7v'^i - (N.v')N^i\right] + \cdots. \qquad (12.24)$$

When we calculate $\dot{p}$, terms proportional to $\dot{v}$ appear, and we replace them by their expression (12.23) in a consistent manner at the order under consideration. Therefore, the equations of motion of the body m in the gravitational field of its companion are finally given by

$$\dot{v}^i = -\frac{m'}{R^2}N^i + A_2^i + \mathcal{O}(v^6) \qquad \text{with}$$

$$A_2^i = \frac{m'}{R^2}\left\{ N^i \left[-v^2 - 2v'^2 + 4(v.v') + \frac{3}{2}(N.v')^2 + \frac{5m}{R} + \frac{4m'}{R} \right] \right. \qquad (12.25)$$

$$\left. + (v^i - v'^i)\left[4(N.v) - 3(N.v')\right] \right\},$$

which can be compared with the analogous equations governing the motion of two charges in the Maxwell theory; see Book 2, Section 20.3.

The equations (12.25) were obtained by Einstein, Infeld, and Hoffmann in 1938, hence the name EIH equations.[6]

---

[6] Einstein, Infeld, and Hoffmann (1938). See also Eddington and Clark (1938).

However, it turns out that the EIH equations (12.25) as well as the Lagrangian (12.22) attributed to Fichtenholz [Fichtenholz (1950)] were actually discovered by Lorentz and Droste in 1917; see Damour (1982).

The methods used in the EIH article involve very lengthy calculations. The authors themselves say, *"Unfortunately, as the work proceeds, the calculations become more and more extensive involving a great amount of technical detail which can have no intrinsic interest. To give all these calculations explicitly here would be quite impracticable and we are obliged to confine ourselves to stressing the general ideas of the work and merely announcing the actual results. For the convenience of anyone who may be interested in the details of the calculation, however, the entire computation of this part of our paper has been deposited with the Institute for Advanced Study so as to be available for reference."*

The fact that Sections 12.1–12.3 involve almost no calculation attests to the technical progress which has been made, perhaps to the detriment of rigor, because, as we have emphasized, here we have justified neither the method of regularizing the field equations and equations of motion, nor the validity of the geodesic equation; in this regard, see Chapter 15.

12.4 Conservation laws

As we have seen in Book 1, Section 8.3, study of the symmetries of a Lagrangian allows one to find, using the Noether theorem, the conservation laws for energy and angular momentum, as well as the motion of the center of mass. We recall that under an infinitesimal variation of the trajectories $\delta z(t) \equiv \tilde{z}(t) - z(t)$, $\delta z'(t) \equiv \tilde{z}'(t) - z'(t)$ the Lagrangian, a functional of the positions and velocities of the two bodies, varies as (this is the Noether identity)

$$\delta L_{\mathrm{F}} = \frac{d}{dt} \sum p_i \delta z^i - \sum F_i \delta z^i, \quad \text{where} \quad p_i \equiv \frac{\partial L_{\mathrm{F}}}{\partial v^i} \quad \text{and} \quad F_i \equiv \dot{p}_i - \frac{\partial L_{\mathrm{F}}}{\partial z^i}. \quad (12.26)$$

The Fichtenholz Lagrangian (12.22) is manifestly invariant under spatial translations ($\delta x^i = \xi^i$, which induces $\delta z^i = \xi^i$) because it depends only on the relative separation $z^i - z'^i \equiv RN^i$ of the bodies. It is also invariant under spatial rotations ($\delta x^i = \epsilon^{ijk}\xi_j x_k$) and changes of the time origin ($\delta t = \xi$), from which we find

$$\frac{dP_i}{dt} = \sum F_i, \quad \text{where} \quad P_i = \sum p_i; \quad \frac{dJ_i}{dt} = \sum \epsilon_{ijk} z^j F^k, \quad \text{where} \quad J_i = \epsilon_{ijk} \sum z^j p^k;$$

$$\frac{dE}{dt} = \sum F_i v^i, \quad \text{where} \quad E = \sum p_i v^i - L_{\mathrm{F}}.$$

$$(12.27)$$

When the EIH equations of motion (12.25) are satisfied, $F^i = 0$ and we see that the *momentum* P_i, the *angular momentum* J_i, and the *energy* E of the system are constants of the motion (as can also be verified by direct calculation).

Using the momenta p_i given in (12.24), we can check by identification (and using the equations of motion in Newtonian order) that

$$P_i = \frac{dG_i}{dt} \quad \text{with} \quad G^i = \sum m \left(1 + \frac{v^2}{2} - \frac{Gm'}{2R}\right) z^i. \quad (12.28)$$

Since P_i is constant, G_i represents the position of the *center of mass*. It is zero by definition in the *center-of-mass system*, and so by iteration we obtain

$$\begin{cases} z^i = \dfrac{m'}{m_{\mathrm{T}}} RN^i + \dfrac{1}{2} RN^i \left(\dfrac{m - m'}{m_{\mathrm{T}}^2}\right) \left(\mu V^2 - \dfrac{Gmm'}{R}\right), \\[3mm] z'^i = -\dfrac{m}{m_{\mathrm{T}}} RN^i - \dfrac{1}{2} RN^i \left(\dfrac{m' - m}{m_{\mathrm{T}}^2}\right) \left(\mu V^2 - \dfrac{Gmm'}{R}\right), \end{cases} \quad (12.29)$$

where $m_{\mathrm{T}} = m + m'$ and $\mu = mm'/m_{\mathrm{T}}$ are the total mass and the reduced mass, and $V^i = v^i - v'^i$. These expressions can be compared with those giving the motion of the center of mass of two charges in the Maxwell theory; see Book 2, Section 21.1.

Lorentz invariance of the Lagrangian

The Fichtenholz Lagrangian must also be invariant under Lorentz transformations $\delta_\beta x^i = \beta^i t$, $\delta_\beta t = (\beta.x)$. As we saw in Book 2, Section 21.1, such a transformation induces $\delta_\beta z^i = \beta^i t - (\beta.z)v^i$ and $\delta_\beta v^i = (\delta_\beta z^i)^{\cdot} = \beta^i - (\beta.v)v^i$. Explicit calculation then shows that

$$\delta_\beta L_{\rm F} = \frac{d}{dt}\left[(\beta.G) - \sum(p.v)(\beta.z)\right],$$

where G^i is given by (12.28). Therefore, $\delta_\beta L_{\rm F}$ is a total derivative which does not contribute to the equations of motion, and the Lagrangian is indeed Lorentz-invariant. Now by substituting $\delta_\beta L_{\rm F}$ into the general expression (12.26), we obtain

$$\frac{d}{dt}(G_i - tP_i) = \sum(F.v)z^i - t\sum F_i,$$

which again gives uniform rectilinear motion of the center of mass when the equations of motion are satisfied.

Therefore, as in Newtonian gravity, the problem of the motion of two bodies in the post-Newtonian approximation of general relativity can be studied in the center-of-mass system undergoing uniform rectilinear motion. It is only necessary now to find the equations of motion in this system using the EIH equations of motion (12.25). We shall see that, just as in Newtonian gravity, they reduce to the equation describing the relative motion.

12.5 Post-Keplerian trajectories

The relative acceleration of two compact bodies is obtained by calculating $\dot{v}^i - \dot{v}'^i$ using the EIH equations of motion (12.25). In the center-of-mass system defined in (12.29) we find (the post-Newtonian correction in fact does not play any role)

$$\dot{V}^i = -\frac{m_{\rm T}}{R^2}N^i$$

$$+\frac{m_{\rm T}}{R^2}\left\{N^i\left[\frac{m_{\rm T}}{R}(4+2\nu) - V^2(1+3\nu) + \frac{3}{2}\nu(N.V)^2\right] + V^i(N.V)(4-2\nu)\right\} + \mathcal{O}(v^6), \tag{12.30}$$

where $V^i = v^i - v'^i$ and $\nu = \mu/m_{\rm T} = mm'/m_{\rm T}^2$. This equation of motion can be compared to that of two charges in the Maxwell theory; see Book 2, Section 21.1, eqn (21.10).

Now to integrate (12.30) we can treat the second term of the acceleration as a perturbing force on the Newtonian motion and use the tools of perturbation theory developed by Gauss and others and discussed in, for example, Book 1, Section 13.3 to determine the motion of the osculating ellipse. However, this method amounts to a sort of 'epicycle multiplier', and it is more efficient to just perform the integration directly.

To do this we use the constants of the motion, namely, the energy and angular momentum defined in (12.27). In the center-of-mass system and setting $E - m_{\rm T} \equiv \mu E_n$ and $J_i = \mu L_i$, we find (again in the present section the calculations are elementary)

$$\begin{cases} E_n = \frac{1}{2}V^2 - \frac{m_{\rm T}}{R} + \frac{3}{8}(1-3\nu)V^4 + \frac{m_{\rm T}}{2R}\left[(3+\nu)V^2 + \nu(N.V)^2 + \frac{m_{\rm T}}{R}\right] + \mathcal{O}(v^6), \\ \\ L = R(N \wedge V)\left[1 + \frac{V^2}{2}(1 - 3\nu) + \frac{m_{\rm T}}{R}(3 + \nu)\right] + \mathcal{O}(v^5). \end{cases} \tag{12.31}$$

Given the dependence of $\dot{V}^i$ on N^i and V^i, we see that the motion is planar and we choose $\theta = \pi/2$. Then in polar coordinates (R, ϕ)

$$V^2 = \dot{R}^2 + R^2 \dot{\phi}^2, \quad |N \wedge V| = R\dot{\phi}, \quad \dot{R} = (N.V). \tag{12.32}$$

Substituting these results into (12.31) and introducing a new 'conchoidal' radial coordinate $r_\delta \equiv R - m_T \delta$, by iteration we find

$$\left(\frac{dr_\delta}{dt}\right)^2 \simeq A + \frac{2B}{r_\delta} + \frac{C_\delta}{r_\delta^2} + \frac{D_\delta}{r_\delta^3}, \quad \frac{d\phi}{dt} = \frac{H}{r_\delta^2} + \frac{I_\delta}{r_\delta^3}$$

with

$$\begin{cases} A = 2E_n \left(1 - \frac{3E_n}{2}(1 - 3\nu)\right), \quad B = m_T \left(1 - (6 - 7\nu)E_n\right), \\ C_\delta = -L^2 \left(1 - 2E_n(1 - 3\nu) + \frac{2m_T^2(5 - 5\nu/2 + \delta)}{L^2}\right), \quad D_\delta = m_T L^2(8 - 3\nu + 2\delta), \\ H = L\left(1 - E_n(1 - 3\nu)\right), \quad I_\delta = -2m_T(2 - \nu + \delta)L \end{cases} \tag{12.33}$$

(these expressions indeed reduce to those obtained in Section 11.6 for $\nu = 0$, $\gamma = \beta = 1$, and $m_T = m$). Next we proceed as in Section 11.6: first choosing $\delta = -4 + 3\nu/2$, we find $D_\delta = 0$ and the equation for $r_\delta(t)$ takes a Newtonian form. Then choosing $\delta = -2 + \nu$, we have $I_\delta = 0$ and the equation for $\phi(t)$ also takes Newtonian form. Returning to the coordinate R, the integrations then give[7]

$$n(t - T) = \eta - e_t \sin\eta, \quad R = a(1 - e\cos\eta),$$

$$\tan\frac{f}{2} = \sqrt{\frac{1 + e_\phi}{1 - e_\phi}} \tan\frac{\eta}{2} \quad \text{with} \quad f \equiv (\phi - \omega)\left(1 - \frac{3m_T^2}{L^2}\right). \tag{12.34}$$

The mean motion n, the semi-major axis a, and the three eccentricities e_t, e, and e_ϕ are expressed parametrically as a function of the constants of the motion E_n and L as (we recall that $m_T = m + m'$ and $\nu = mm'/m_T^2$)

$$\begin{cases} a = -\frac{m_T}{2E_n}\left(1 + \frac{(7 - \nu)E_n}{2}\right), \quad e^2 = 1 + \frac{2E_n L^2}{m_T^2} - \frac{5(3 - \nu)L^2 E_n^2}{m_T^2} - 2(6 - \nu)E_n, \\ n = \sqrt{\frac{m_T}{a^3}}\left[1 - \frac{m_T(9 - 2\nu)}{2a}\right], \quad e_t^2 = e^2\left[1 - \frac{(8 - 3\nu)m_T}{a}\right], \quad e_\phi^2 = e^2\left(1 + \frac{\nu m_T}{a}\right). \end{cases} \tag{12.35}$$

We recover the expressions obtained in Section 11.6 for the case $\nu = 0$ (where there are now only two eccentricities because $e_\phi = e$). The dominant relativistic effect, since it is secular, is the advance of the periastron [we recall that $L^2 = m_T a(1 - e^2)$]

$$\Delta\omega \simeq \frac{6\pi m_T}{a(1 - e^2)} \tag{12.36}$$

per period, where $m_T = m + m'$ is the *total* mass of the system (Robertson, 1933).

---

[7]See the details in Section 11.6 or in Damour and Deruelle (1985) and (1986). We have also used this method to obtain the motion of a charge in a Coulomb field; see Book 2, Section 13.3.

Finally, to obtain the relative orbit we eliminate η between the equations (12.34) for R and ϕ. A new conchoidal transformation in which R is written as $R = (ea/e_\phi)(1 - e_\phi \cos \eta) + a(1 - e/e_\phi)$ gives $R = (a - \mu/2)(1 - e_\phi^2)/(1 + e_\phi \cos f) + \mu/2$, which can also be put into quasi-Newtonian form[7]:

$$R = \frac{a(1-e)}{1 + e \cos f'} \quad \text{with} \quad f' = f - \nu e \frac{m_{\mathrm{T}}^2}{2L^2} \sin f, \tag{12.37}$$

where f is given in (12.34). We find $f = f'$ in the limit $\nu = 0$, in agreement with the results of Section 11.5.

For completeness, we still need to obtain the trajectories of the bodies themselves. These are derived from the relative orbit using (12.29). The polar angle of the body m is ϕ and that of m' is $\phi + \pi$. The radial motion can also be cast in quasi-Newtonian form (z is the modulus of the radius vector z^i of the body):

$$z = a_r(1 - e_r \cos \eta) \quad \text{with} \quad a_r = \frac{m'}{m_{\mathrm{T}}} a \quad \text{and} \quad e_r = e\left[1 - \frac{m(m - m')}{2m_{\mathrm{T}} a}\right], \tag{12.38}$$

and a similar expression describes the trajectory of the body[7] m'.

Let us summarize what we have learned. In Newtonian gravity the two-body problem can be solved exactly, by separating the center-of-mass motion and the relative motion, so that the problem is reduced to that of a test object in the field of a body effectively at rest. The relative trajectory is then given by Kepler's laws, from which the individual trajectories are derived. In general relativity the problem can only be dealt with using an iteration scheme. Here we have studied it in the post-Newtonian approximation, and have found that in the end the center-of-mass motion and the relative motion can again be separated, and that the laws of motion can be written in quasi-Keplerian form (12.34), (12.37), and (12.38).

12.6 The timing formula of a binary system

Let us consider a system of two compact stars, one of which emits electromagnetic signals at regular intervals. To predict the time of arrival of these signals at Earth, it is necessary to describe the motion of the emitter in the field of its companion using a theory of gravitation, and to calculate the trajectory of the signals using a theory of light. In this way we obtain a timing formula for predicting the time t_N of arrival of a signal N as a function of its time of emission T_N and the parameters describing the motion of the object, the motion of the Earth, the period of the signals, and so on: $t_N = F(T_N, \text{parameters})$. By fitting these predicted times to the observed times, we can in principle determine the parameters and test the theory of gravity underpinning the model. Here we shall present a brief derivation of the timing formula of a binary system based on general relativity.

To begin with, let us work within the framework of Newtonian theory, where the time is universal and geometry is Euclidean, where light propagates at speed c in a straight line, and where the trajectory of the emitter is a Keplerian ellipse. If $r_{\mathrm{b}} \vec{n}$ is the radius vector of a focal point of the ellipse to the center of mass of the solar system and $\vec{r}$ is the position vector of the emitter, the light travel time is $|r_{\mathrm{b}} \vec{n} - \vec{r}(t_{\mathrm{e}})|/c$, and so the arrival time t_{a} and emission time t_{e} of the signal N are related as (see the expressions and figures in Book 1, Section 12.2)

$$t_a \approx t_e + \Delta_R(t_e) + \text{const} \quad \text{with} \quad t_e = NP_e \quad \text{and} \quad \Delta_R = -\frac{\vec{n} \cdot \vec{r}(t_e)}{c} = \frac{r \sin\phi \sin i}{c}, \quad (12.39)$$

where P_e is the signal period, i is the inclination of the plane of the ellipse on the celestial sphere, and ϕ is the angle between $\vec{r}(t_e)$ and the line of nodes. Kepler's laws then give (starting from $\tan[(\phi - \omega_0)/2] = \sqrt{(1+e)/(1-e)}\tan\eta/2$ and then expanding in $\sin\phi = \sin[(\phi - \omega_0) + \omega_0]$; see the expressions and figures in Book 1, Section 12.2)

$$\Delta_R = x\left[\sin\omega_0(\cos\eta - e) + \sqrt{1 - e^2}\cos\omega_0\sin\eta\right] \quad \text{with} \quad \frac{2\pi}{P_b}(t_e - T) = \eta - e\sin\eta. \quad (12.40)$$

This is the Römer delay (*cf.* Book 1, Section 17.1), and it depends on the following five parameters which can be determined by a fit to the observed arrival times: $x \equiv a_e \sin i/c$ (where a_e is the semi-major axis of the ellipse traced by the emitter), the longitude of the periastron ω_0, the eccentricity of the orbit e, the orbital period P_b, and the moment of passage through the periastron T. In this Newtonian model the masses of the emitter and its companion therefore remain undetermined.

Now let us include (without going into the details of the calculations) the two dominant relativistic effects, namely, the precession of the periastron [given by the Robertson formula (12.36)] and the time dilation due to the fact that the proper time τ_e of the emitter differs from that of the solar system because the emitter moves in the gravitational field of its companion [$d\tau_e/dt_e = 1 + U(\vec{r}_e) - v_e^2/2$; see the GPS example discussed in Section 11.2]. Owing to (12.36), the inclusion of the first effect amounts to replacing ω_0 in the Römer delay by the following function of the time:

$$\omega \equiv \omega_0 + \frac{\Delta\omega}{2\pi}(\phi(\eta) - \omega_0) \quad \text{with} \quad \frac{\Delta\omega}{2\pi} = \left(\frac{P_b}{2\pi}\right)^{-\frac{2}{3}}\frac{3\,m_T^{2/3}}{(1 - e^2)}, \quad (12.41)$$

where the anomaly η is related to the time t by (12.34) and $m_T = m_e + m_c$, where m_e and m_c are the masses of the emitter and its companion. The inclusion of the second effect amounts to adding to the Römer delay an 'Einstein delay' given by (Blandford and Teukolsky, 1976)

$$t_e = \tau_e + \Delta_E \quad \text{with} \quad \Delta_E = \gamma\sin\eta, \quad \text{where} \quad \gamma = \left(\frac{P_b}{2\pi}\right)^{\frac{1}{3}}\frac{m_c(m_e + 2m_c)}{m_T^{4/3}}e. \quad (12.42)$$

The Römer delay is of order $1/c$ and Δ_E is of order $1/c^2$, and so we have $\Delta_E/\Delta_R \approx v$, where v is the orbital speed of the objects. Since the mass dependences of $\Delta\omega$ and Δ_E differ from each other, a careful fit to the observed arrival times thus permits m_e and m_c to be determined with an accuracy which increases with time (of course, if the underlying model is correct), because the advance of the periastron is a cumulative effect. Therefore, here general relativity is not a theory to be tested, but a tool for measuring masses.

In the following order $1/c^3$ it is necessary first to include the Shapiro effect [see Section 11.4], which introduces an additional delay given by

$$\begin{cases} \Delta_S = -2r\ln\left[1 - e\cos\eta + \sqrt{1 - e^2}\cos\omega\sin\eta - \sin i[\sin\omega(\cos\eta - e)]\right], \\[2mm] \text{where} \quad r = m_c \quad \text{and} \quad \sin i = \frac{x}{(P_b/2\pi)^{2/3}}\frac{m_T^{2/3}}{m_c}. \end{cases} \quad (12.43)$$

Secondly, we must include all the post-Keplerian corrections studied in Section 12.5 and rewrite the Römer delay (12.40) as

$$\Delta_{\mathrm{R}} = x \left[\sin\omega(\cos\eta - e_r) + \sqrt{1 - e_\phi^2}\,\cos\omega\,\sin\eta \right] \quad \text{with} \quad \frac{2\pi}{P_{\mathrm{b}}}(\tau_{\mathrm{e}} - T) = \eta - e_T \sin\eta. \qquad (12.44)$$

Here we recognize the eccentricities e_r and e_ϕ obtained in (12.38) and (12.35). The eccentricity e_T differs from e_t appearing in (12.34) because the time equation here is a function of the proper time τ_{e} of the emitter and not of t_{e}. An easy calculation gives $e_T = e_t(1+\delta) + e_\phi - e_r$, where $\delta = m_{\mathrm{c}}(m_{\mathrm{e}} + 2m_{\mathrm{c}})/(m_{\mathrm{T}}a)$, so that finally we have

$$\begin{cases} e_\phi = e_T(1+\delta_\phi) \quad \text{with} \quad \delta_\phi = \left(\frac{P_{\mathrm{b}}}{2\pi}\right)^{-\frac{2}{3}} \frac{1}{m_{\mathrm{T}}^{4/3}} \left(\frac{7}{2}m_{\mathrm{e}}^2 + 6m_{\mathrm{e}}m_{\mathrm{c}} + 2m_{\mathrm{c}}^2\right), \\[2ex] e_r = e_T(1+\delta_r) \quad \text{with} \quad \delta_r = \left(\frac{P_{\mathrm{b}}}{2\pi}\right)^{-\frac{2}{3}} \frac{1}{m_{\mathrm{T}}^{4/3}} \left(3m_{\mathrm{e}}^2 + 6m_{\mathrm{e}}m_{\mathrm{c}} + 2m_{\mathrm{c}}^2\right). \end{cases} \qquad (12.45)$$

We see that the *post-Keplerian parameters* r, $\sin i$, e_ϕ, and e_r given in (12.43) and (12.45) are completely fixed by the theory. Their measurement by very precise fitting to the observed arrival times, and the agreement of these results with the predictions, constitute further verification of the theory of general relativity. In Fig. 12.1 the intersection of the straight line $\Delta\omega = \Delta\omega(m_{\mathrm{c}}, m_{\mathrm{e}})$ [(12.41)] and the curve $\gamma = \gamma(m_{\mathrm{c}}, m_{\mathrm{e}})$ [(12.42)] giving the masses of the two bodies, and the intersections at the same point of the straight line $r = m_{\mathrm{c}}$ and the curve $\sin i = \sin i(m_{\mathrm{c}}, m_{\mathrm{e}})$ (dotted line), provide two tests of general relativity (the parameters are those of the double pulsar PSR J0937).[8]

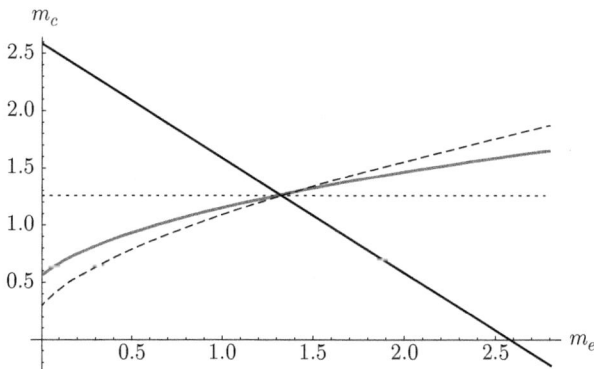

Fig. 12.1 Binary systems and post-Newtonian tests.

    Last but not least, in our analysis we have neglected an effect which is numerically as important as the post-Newtonian corrections: the radiation reaction on the motion of the emitter due to the *gravitational waves* emitted by the system. This will be discussed in the following chapters [see (14.33), which completes (12.44)].

13

Gravitational waves and the radiative field

Now we shall begin our study of the gravitational radiation produced by a system of massive objects. In this chapter we shall restrict ourselves to the *linear* approximation of general relativity, and compare it with the Maxwell theory of electromagnetism.

In Sections 13.1–13.3 we will study the properties of gravitational waves, which are the general solution of the linearized vacuum Einstein equations. In Section 13.4 we relate these waves to the energy–momentum tensor of the sources creating them. This will lead us to the 'first quadrupole formula', eqn (13.27), giving the gravitational radiation field of the sources when their motion is due to forces other than the gravitational force.

13.1 Linearization of the Einstein equations

We consider the metric

$$g_{\mu\nu}(x^\rho) = \eta_{\mu\nu} + h_{\mu\nu}(x^\rho)\,, \tag{13.1}$$

where $\eta_{\mu\nu}$ is the Minkowski metric and $h_{\mu\nu}$ are ten perturbations, $|h_{\mu\nu}| \ll 1$. These perturbations can simultaneously describe the presence of a weak gravitational field and an infinitesimal change of the coordinate system. Since the coordinates x^μ are quasi-Minkowskian, the reference frame is quasi-inertial.

It is easy to calculate the Einstein tensor in linear order. The Christoffel symbols, the Ricci tensor, and the scalar curvature are respectively given by (all indices are moved using $\eta^{\mu\nu}$, $h \equiv h^\mu_\mu$, and $\Box \equiv \eta^{\mu\nu}\partial^2_{\mu\nu}$)

$$\Gamma^\mu_{\nu\rho} = \frac{1}{2}(\partial_\nu h^\mu_\rho + \partial_\rho h^\mu_\nu - \partial^\mu h_{\nu\rho}) + \mathcal{O}(h^2_{\mu\nu})\,,$$

$$R_{\mu\nu} = \partial_\rho \Gamma^\rho_{\mu\nu} - \partial_\nu \Gamma^\rho_{\rho\mu} + \mathcal{O}(h^2_{\mu\nu}) = -\frac{1}{2}\left[\Box h_{\mu\nu} + \partial^2_{\mu\nu}h - \partial_\rho(\partial_\nu h^\rho_\mu + \partial_\mu h^\rho_\nu)\right] + \mathcal{O}(h^2_{\mu\nu})\,, \tag{13.2}$$

$$R = \partial^2_{\lambda\rho}h^{\lambda\rho} - \Box h + \mathcal{O}(h^2_{\mu\nu})\,.$$

The Einstein tensor $G_{\mu\nu} = R_{\mu\nu} - \frac{1}{2}g_{\mu\nu}R$ then is

$$G_{\mu\nu} = -\frac{1}{2}\left[\Box \gamma_{\mu\nu} + \partial_\rho(\eta_{\mu\nu}\partial_\lambda \gamma^{\rho\lambda} - \partial_\mu \gamma^\rho_\nu - \partial_\nu \gamma^\rho_\mu)\right] + \mathcal{O}(h^2_{\mu\nu})\,, \tag{13.3}$$

where we have set $h_{\mu\nu} = \gamma_{\mu\nu} - \frac{1}{2}\eta_{\mu\nu}\gamma$ in equation (13.3).

In Section 5.1 we saw how a metric transforms under an infinitesimal change of coordinates $x^\mu \to x^\mu = \tilde{x}^\mu - \xi^\mu$, where $\xi^\mu(x^\rho)$ is infinitesimally small: $\tilde{g}_{\mu\nu}(x^\rho) = g_{\mu\nu}(x^\rho) - (\partial_\mu \xi_\nu + \partial_\nu \xi_\mu)$

Relativity in Modern Physics. Nathalie Deruelle and Jean-Philippe Uzan.
© Oxford University Press 2018. Published in 2018 by Oxford University Press.
DOI: 10.1093/oso/9780198786399.001.0001

[here we have $\tilde{g}_{\mu\nu}(\tilde{x}^\rho) = \tilde{g}_{\mu\nu}(x^\rho)$, because at the order in which we are working $\tilde{g}_{\mu\nu} = \eta_{\mu\nu}$]. Therefore, $h_{\mu\nu} = \tilde{h}_{\mu\nu} + (\partial_\mu\xi_\nu + \partial_\nu\xi_\mu)$, $\gamma_{\mu\nu} = \tilde{\gamma}_{\mu\nu} + \partial_\mu\xi_\nu + \partial_\nu\xi_\mu - \eta_{\mu\nu}\partial_\rho\xi^\rho$, and

$$\partial_\rho\gamma_\mu^\rho = \partial_\rho\tilde{\gamma}_\mu^\rho + \Box\xi_\mu. \tag{13.4}$$

Let us choose a function ξ^μ which is a solution of $\Box\xi_\mu = \partial_\rho\gamma_\mu^\rho$ (it is defined up to a harmonic function $\xi^\mu = \xi_{\text{part}}^\mu + \zeta^\mu$ with $\Box\zeta^\mu = 0$). In such coordinate systems[1] the linearized Einstein equations $G_{\mu\nu} = \kappa T_{\mu\nu}$, where $T_{\mu\nu}$ is the matter energy–momentum tensor, are written as (suppressing the tildes)

$$\Box\gamma_{\mu\nu} = -2\kappa\,T_{\mu\nu} + \mathcal{O}(\gamma_{\mu\nu}^2), \quad \text{where} \quad g_{\mu\nu} = \eta_{\mu\nu} + \gamma_{\mu\nu} - \frac{1}{2}\eta_{\mu\nu}\gamma \quad \text{with} \quad \partial_\rho\gamma_\mu^\rho = \mathcal{O}(\gamma_{\mu\nu}^2). \tag{13.5}$$

They are the gravitational analog of the Maxwell equations for the electromagnetic potential in Lorenz gauges (see, for example, Book 2, Section 17.2), with the important difference that they are valid only at linear order.

13.2 Gravitational waves

The linearized vacuum Einstein equations are [*cf.* (13.3)]

$$\Box\gamma_{\mu\nu} + \partial_\rho(\eta_{\mu\nu}\partial_\lambda\gamma^{\rho\lambda} - \partial_\mu\gamma_\nu^\rho - \partial_\nu\gamma_\mu^\rho) = 0, \quad \text{where} \quad \gamma_{\mu\nu} = h_{\mu\nu} - \frac{1}{2}\eta_{\mu\nu}h_\rho^\rho.$$

Here $h_{\mu\nu}$ are the perturbations of a quasi-Minkowskian metric, and in this section it is understood that all equations are valid at order $\mathcal{O}(\gamma_{\mu\nu}^2)$. As we saw above, the invariance of the equations under coordinate transformations allows us to impose the harmonic condition on the metric: $\partial_\nu\gamma_\mu^\nu = 0$. Moreover, since $\Box\gamma_\mu^\mu = 0$, we can also require that[2] $\gamma_\mu^\mu = 0$. Therefore, the linearized vacuum Einstein equations can be reduced to

$$\Box h_{\mu\nu} = 0 \quad \text{with} \quad \partial_\nu h_\mu^\nu = 0 \quad \text{and} \quad h_\mu^\mu = 0. \tag{13.6}$$

Then, just like the Lorenz gauges in electromagnetism, the harmonic gauges, including (13.6), do not completely fix the coordinate system. For this it is necessary, just like in electromagnetism, to break the manifest Lorentz invariance of the field equations.

- *Six gauge-invariant perturbations*

We follow the procedure of Book 2, Chapter 14 and write the length element $ds^2 = (\eta_{\mu\nu} + h_{\mu\nu})dx^\mu dx^\nu$ in the form

---

[1] In linear order, coordinate systems for which $\partial_\rho\gamma_\mu^\rho = 0$ are referred to as Hilbert, de Donder, or Fock harmonic gauges.

If the metric is stationary, $\partial_t h_{\mu\nu} = 0$, in (13.5) we recover the gravitomagnetic field equation in the Lorentz–Coulomb gauge seen in Section 11.7.

[2] Under the change of coordinates $x^\mu \to x^\mu = \tilde{x}^\mu - \xi^\mu$ we have (see Section 13.1) $\gamma_{\mu\nu} = \tilde{\gamma}_{\mu\nu} + \partial_\mu\xi_\nu + \partial_\nu\xi_\mu - \eta_{\mu\nu}\partial_\rho\xi^\rho$. Let us choose $\xi^\mu = \partial^\mu\Lambda - \partial_\nu\sigma^{\mu\nu}$, where Λ is a particular solution of $\Box\Lambda = \gamma/2$ with $\gamma = \gamma_\mu^\mu$. The quantity $\sigma^{\mu\nu}$ is an antisymmetric tensor which is a particular solution of $\Box\sigma^{\mu\nu} = f^{\mu\nu}$, where $f^{\mu\nu}$ is also antisymmetric and $\partial_\nu f^{\mu\nu} = \partial^\mu\gamma/2$ ($f^{\mu\nu}$ exists because $\Box\gamma = 0$). We indeed have $\partial_\mu\xi^\mu = \Box\Lambda = \gamma/2$ and $\Box\xi^\nu = \partial^\mu\Box\Lambda - \partial_\nu f^{\mu\nu} = 0$; see Straumann (2013).

$$ds^2 = -(1 + 2A)dt^2 + 2B_i dx^i dt + (\delta_{ij} + h_{ij})\, dx^i dx^j \tag{13.7}$$

and then decompose B_i and h_{ij} as

$$B_i = \partial_i B + \bar{B}_i\,, \quad h_{ij} = 2C\delta_{ij} + 2\partial_{ij}^2 E + \partial_i \bar{E}_j + \partial_j \bar{E}_i + 2\bar{E}_{ij}\,, \tag{13.8}$$

where $\bar{B}_i$ and $\bar{E}_i$ are two divergence-free (Euclidean) vectors: $\partial_i \bar{B}^i = 0$ (the indices are moved using the Euclidean metric δ_{ij}). Here $\bar{E}_{ij}$ is a traceless transverse Euclidean tensor $\bar{E}_i^i = 0$ and $\partial_j \bar{E}^{ij} = 0$. In the Bardeen terminology,[3] we have thereby decomposed the ten components of the perturbation $h_{\mu\nu}$ into four scalar perturbations (A, B, C, E), four vector perturbations (the two independent components of the two divergence-free vectors $\bar{B}_i$ and $\bar{E}_i$), and two tensor perturbations (the two independent components of the traceless transverse tensor $\bar{E}_{ij}$).

Fourier decomposition of SVT perturbations

The above SVT decomposition of the metric is unique in the sense that, if B_i is a given function of the coordinates, B is the (regular, therefore unique) solution of $\triangle B = \partial_i B^i$, and so $\bar{B}_i$ follows: $\bar{B}_i = B_i - \partial_i B$. Likewise, we can obtain C and E from $\partial_{ij} h^{ij} = 2\triangle C + 2\triangle\triangle E$ and $h_i^i = 6C + 2\triangle E$. Then $\bar{E}_i$ is the solution of $\partial_j h_i^j = 2\partial_i C + 2\triangle\partial_i E + \triangle\bar{E}_i$, and $\bar{E}_{ij}$ follows.

We can arrive at the same result by decomposing any (square-integrable) perturbation $X(t, x^i)$ in Fourier modes as (see, for example, Book 2, Section 9.2)

$$X(t, x^i) = \int \frac{d^3k}{(2\pi)^{3/2}} X(t, k^j) e^{ik_i x^i} \quad \text{with} \quad X^*(t, -k^j) = X(t, k^j)\,.$$

(The traditional 'hats' on Fourier transforms are omitted.) Therefore, the Fourier components $X_i(t, k^j)$ of a 3-vector can be decomposed as $X_i = k_i X + \bar{X}_i$ with $k^i \bar{X}_i = 0$. Just as for electromagnetic waves, see Book 2, Section 15.1, we define (up to a rotation) a basis $\{e^1, e^2\}$ in the plane perpendicular to the vector k^i by $e_i^a k^i = 0$ and $e_i^a e_j^b \delta^{ij} = \delta^{ab}$ with $a, b \in \{1, 2\}$. We can then write the vector mode as $\bar{X}_i(t, k^j) = \sum_a X_a(t, \hat{k}^i) e_i^a(\hat{k}^j)$. This defines the two vector degrees of freedom X_a, which depend on $\hat{k}_i = k_i/k$ (because they are transverse). Next we define the projection operator $P_{ij} \equiv e_i^1 e_j^1 + e_i^2 e_j^2 = \delta_{ij} - \hat{k}_i \hat{k}_j$ (it can be checked that $P_j^i P_k^j = P_k^i$, $P_j^i k^j = 0$, and $P^{ij}\delta_{ij} = 2$), which allows us to extract the S and V components of a 3-vector as

$$X_i = (\hat{k}^j X_j)\hat{k}_i + P_i^j X_j\,.$$

Similarly, any symmetric tensor X_{ij} breaks up into SVT modes as $X_{ij} = T\delta_{ij} + \partial_{ij}^2 S + 2\partial_{(i}\bar{X}_{j)} + 2\bar{X}_{ij}$ with $\partial_i \bar{X}^i = 0$, $\bar{X}_i^i = 0$, and $\partial_i \bar{X}^{ij} = 0$. The tensor part $\bar{X}_{ij}$ has two independent components which we decompose as

$$\bar{X}_{ij}(t, k^i) = \sum_{\lambda=+,\times} X_\lambda(t, k^l)\, \varepsilon_{ij}^\lambda(\hat{k}^l) \quad \text{with} \quad \varepsilon_{ij}^\lambda = \frac{e_i^1 e_j^1 - e_i^2 e_j^2}{\sqrt{2}}\delta_+^\lambda + \frac{e_i^1 e_j^2 + e_i^2 e_j^1}{\sqrt{2}}\delta_\times^\lambda. \tag{13.9}$$

The *polarization tensor* ε_{ij}^λ is traceless ($\varepsilon_{ij}^\lambda \delta^{ij} = 0$) and transverse ($\varepsilon_{ij}^\lambda k^i = 0$), and the two polarizations are perpendicular ($\varepsilon_{ij}^\lambda \varepsilon_\mu^{ij} = \delta_\mu^\lambda$). Introducing the projector $P_{ij}^{ab} = P_i^a P_j^b - \frac{1}{2} P_{ij} P^{ab}$ and the trace extraction operator $\Theta_i^j = \hat{k}_i \hat{k}^j - \frac{1}{3}\delta_i^j$, we obtain the explicit SVT decomposition:

$$X_{ij} = \left(\frac{1}{3} X_{ab}\delta^{ab}\right)\delta_{ij} + \left(\frac{3}{2} X_{ab}\Theta^{ab}\right)\Theta_{ij} + 2\hat{k}_{(i}\left[P_{j)}^a \hat{k}^b X_{ab}\right] + P_{ij}^{ab} X_{ab}. \tag{13.10}$$

---

[3]Bardeen (1980), in the broader context of cosmology; see Chapters 19 and 20.

Under an infinitesimal change $x^\mu = \tilde{x}^\mu - \xi^\mu$ we have $h_{\mu\nu} = \tilde{h}_{\mu\nu} + (\partial_\mu \xi_\nu + \partial_\nu \xi_\mu)$. Decomposing ξ^μ into two scalars T, L and a divergence-free vector $\bar{L}^i$, $\xi^\mu = (T, \partial^i L + \bar{L}^i)$, we see that the perturbations transform as

$$A \to A - \dot{T}, \quad B \to B - \dot{L} + T, \quad C \to C, \quad E \to E - L,$$

$$\bar{B}_i \to \bar{B}_i - \dot{\bar{L}}_i, \quad \bar{E}_i \to \bar{E}_i - \bar{L}_i,$$

where the dot denotes the derivative with respect to t. Therefore, the linear combinations

$$\Phi = A + \dot{B} - \ddot{E}, \quad \Psi = -C, \quad \bar{\Phi}^i = -\bar{B}^i + \dot{\bar{E}}^i, \quad \bar{E}^{ij} \tag{13.11}$$

are six gauge-invariant quantities corresponding to two scalars (Φ and Ψ), the two components of the vector $\bar{\Phi}_i$, and the two components of the tensor $\bar{E}_{ij}$.

- *Two dynamical degrees of freedom*

First we note that the three groups of Bardeen perturbations, scalar (A, B, C, E), vector $(\bar{B}_i, \bar{E}_i)$, and tensor $(\bar{E}_{ij})$, can be considered separately, that is, the linearized Einstein equations for one of the three groups can be obtained by setting the two others to zero. This is because writing down the Einstein equations consists of taking derivatives with respect to x^μ, operations which do not change the type of perturbation (for example, the derivative of a vector $\bar{B}_i$ is $\partial_\mu \bar{B}_i = (\dot{\bar{B}}_i, \partial_j \bar{B}_i)$, all the components of which are of the vector type).

Next, we note that since the Einstein equations are invariant under coordinate changes, it must be possible to write their linearization about the Minkowski metric, just as for the Maxwell equations, as a function of only the six gauge-invariant quantities (Φ, Ψ), $\bar{\Phi}_i$, and $\bar{E}_{ij}$ defined above. (This fact is sometimes referred to as the *Stewart–Walker lemma*.)

Let us first consider scalar perturbations, where we notice that if, starting from a reference frame where they are (A, B, C, E) and making the coordinate change $x^\mu = \tilde{x}^\mu - \xi^\mu$ such that $L = E$ and $T = \dot{E} - B$, then we have $E \to 0$ and $B \to 0$, so that in this new so-called *longitudinal* or *Newtonian* gauge $\Phi = A$ and $\Psi = -C$. Therefore, an economical way of obtaining the equations of motion for Φ and Ψ is to work in the longitudinal gauge, where the length element reduces to

$$ds^2 = -(1 + 2\Phi)dt^2 + (1 - 2\Psi)\delta_{ij}dx^i dx^j. \tag{13.12}$$

To obtain the equations of motion for the two gauge-invariant vector perturbations $\bar{\Phi}_i$, we proceed in a similar manner. We work in a *vector* gauge where $\bar{E}_i = 0$, and the length element reduces to

$$ds^2 = -dt^2 - 2\bar{\Phi}_i dx^i dt + \delta_{ij} dx^i dx^j. \tag{13.13}$$

Finally, to find the equations of motion for the tensor perturbations $\bar{E}_{ij}$, there is no point in choosing a gauge, because these perturbations are identical to the traceless, transverse spatial perturbations of the metric:

$$ds^2 = -dt^2 + (\delta_{ij} + 2\bar{E}_{ij}) dx^i dx^j. \tag{13.14}$$

The Einstein tensor $G_{\mu\nu}$ in first order in the perturbations is then the sum of the three Einstein tensors calculated separately for the metrics (13.12)–(13.14). An easy calculation gives

$$G_{tt} = 2\triangle\Psi \,, \quad G_{ti} = 2\partial_i\dot\Psi + \frac{1}{2}\triangle\bar\Phi_i \,,$$

$$G_{ij} = \left[2\ddot\Psi + \triangle(\Phi-\Psi)\right]\delta_{ij} - \partial_{ij}^2(\Phi-\Psi) + \frac{1}{2}(\partial_i\dot{\bar\Phi}_j + \partial_j\dot{\bar\Phi}_i) + \ddot{\bar E}_{ij} - \triangle\bar E_{ij} \,. \tag{13.15}$$

[Of course, the result coincides with the SVT decomposition of (13.3). We also recover the general result of Section 4.2, namely, that the components G^0_μ do not contain second derivatives with respect to the time.]

The vacuum equations are $G_{\mu\nu} = 0$. The only solution of the scalar part which is regular everywhere and bounded then is $\Psi = \Phi = 0$. Similarly, the vector part gives $\bar\Phi_i = 0$. The equations then become

$$\Phi \equiv A + \dot B - \ddot E = 0 \,, \quad \Psi \equiv -C = 0 \,, \quad \bar\Phi^i \equiv -\bar B^i + \dot{\bar E}^i = 0 \,, \quad \Box\bar E^{ij} = 0 \,. \tag{13.16}$$

The first three represent, as in the general case, four constraint equations. Then, as in electromagnetism (see Book 2, Section 14.1), we find that the gauge 'hits twice': the first time in allowing the construction of $10 - 4 = 6$ perturbations which are invariant under changes of gauge (that is, changes of coordinates), and the second time *via* the field equations which contain four constraints (one per gauge invariance). We are thus left with the last equation, the propagation equation, for the two degrees of freedom of *gravitational waves*.

The solutions of (13.16) define an equivalence class. Fixing the gauge amounts to choosing an element of this class. The so-called TT (*transverse traceless*) gauge corresponds to the choice $B = E = \bar E_i = 0$, and we then see that $2\bar E_{ij} = h_{ij}^{\rm TT}$; it belongs to the class defined in (13.8) and is the analog of the Hamiltonian gauge in electromagnetism; see Book 2, Section 14.1. The metric coefficients are thereby completely determined and the length element is written as

$$ds^2 = -dt^2 + h_{ij}^{\rm TT}dx^i dx^j \,. \tag{13.17}$$

Any function $h(t - z)$ satisfies $\Box h = 0$ and describes a plane wave propagating at the speed of light along the z axis. In this particular case the length element (13.17) becomes, taking into account the two possible polarizations,

$$ds^2 = -dt^2 + (1 + h_+)dx^2 + (1 - h_+)dy^2 + 2h_\times dx dy + dz^2 \,. \tag{13.18}$$

If in addition this wave is monochromatic, the length element can be written as follows using the general expression (13.9):

$$ds^2 = -dt^2 + (1 + E_+\varepsilon_+)dx^2 + (1 - E_+\varepsilon_+)dy^2 + 2E_\times\varepsilon_\times dx dy + dz^2 \,,$$
$$\text{with} \quad E_{(+,\times)} = A_{(+,\times)}\cos[k(t - z) + \phi_{(+,\times)}] \,, \tag{13.19}$$

where $k^i = (0, 0, k)$ is the wave vector and we have set $\varepsilon_{xx}^+ = -\varepsilon_{yy}^+ \equiv \varepsilon_+$ and $\varepsilon_{xy}^\times = \varepsilon_{yx}^\times \equiv \varepsilon_\times^+$.

The Hamiltonian of gravitational waves

To obtain the equations of motion of the perturbations by extremizing the Einstein–Hilbert action, it is necessary to expand $\sqrt{-g}R$ through *second* order in the perturbations $h_{\mu\nu}$. The calculation is greatly simplified by using the SVT decomposition described above. It is done in three steps, one for each metric (13.12)–(13.14), and after integrating by parts and ignoring the total derivatives which do not contribute to the field equations we obtain $2\kappa S \equiv \int d^4x \sqrt{-g}R = 2\kappa(S_{\mathrm{T}} + S_{\mathrm{V}} + S_{\mathrm{S}})$ with

$$\kappa S_{\mathrm{S}} = \frac{1}{2}\int d^4x [2\partial_i \Psi (\partial^i \Psi - 2\partial^i \Phi) - 6\dot{\Psi}^2], \quad \kappa S_{\mathrm{V}} = \frac{1}{4}\int d^4x (\partial_i \bar{\Phi}_j \, \partial^i \bar{\Phi}^j),$$

$$\kappa S_{\mathrm{T}} = \frac{1}{2}\int d^4x (\dot{\bar{E}}_{ij} \, \dot{\bar{E}}^{ij} - \partial_k \bar{E}_{ij} \, \partial^k \bar{E}^{ij}).$$

Varying this action with respect to Φ, Ψ, $\bar{\Phi}_i$, and $\bar{E}_{ij}$ (and requiring that the solutions are bounded), we recover (13.16). Therefore, just as in electromagnetism (see Book 2, Section 14.2), the Einstein action is written in lowest order in terms of gauge-invariant quantities. We can then impose the constraints $\Phi = \Psi = \bar{\Phi}_i = 0$ and obtain the reduced Hamiltonian completely describing the two degrees of freedom of the field, which is the gravitational analog of the Hamiltonian of electromagnetism obtained in Book 2, Section 14.3. If we set $\bar{E}_{ij} = \sqrt{\kappa}\, q_{ij}$ so that the conjugate momentum is $\pi_{ij} = \dot{q}_{ij}$, it can be written in the canonical form

$$H_{\mathrm{reduced}} = \int d^3x \, \mathcal{H}_{\mathrm{reduced}} \quad \text{with} \quad \mathcal{H}_{\mathrm{reduced}} = \frac{1}{2}(\pi_{ij}\pi^{ij} + \partial_k q_{ij} \, \partial^k q^{ij}).$$

We know that in electromagnetism the on-shell Hamiltonian, that is, the Hamiltonian when a solution of the field equations is used, is equal to the field energy; see Book 2, Section 14.3. Let us take the case of a plane wave propagating along the z axis [*cf.* (13.18)]. The reduced Hamiltonian density then is (the dot denotes the derivative with respect to the argument $t - z$)

$$\mathcal{H}_{\mathrm{on\text{-}shell}} = \frac{1}{4\kappa}\dot{h}_{ij}^{\mathrm{TT}} \dot{h}_{\mathrm{TT}}^{ij} = \frac{1}{2\kappa}(\dot{h}_+^2 + \dot{h}_\times^2). \tag{13.20}$$

However, before claiming that this gives the energy density transported by the wave, it is necessary, as in electromagnetism, to include the total derivatives that we have so far ignored (see also Section 14.2 in the next chapter).

As the final step, in the reduced Hamiltonian we can make a Fourier expansion of q_{ij} using (13.10), which shows that gravitational waves can be viewed as two independent massless scalar fields. The quantization of these fields leads to the idea of the *graviton*. However, in contrast to the photon, the graviton will probably never be observed directly as a quantum object owing to the weakness of the coupling of gravitation to matter. But these fields may have played a crucial role at the very early universe; see Chapter 22.

13.3 The motion of a particle in a wave

Let us work in the TT gauge where the length element describing a gravitational wave is given by (13.17). A test particle follows a geodesic in this spacetime with the equation $d^2x^\mu/d\tau^2 + \Gamma^\mu_{\nu\rho}(dx^\nu/d\tau)(dx^\rho/d\tau) = 0$. Since at first order in h we have $2\Gamma^\mu_{tt} = \eta^{\mu\nu}(2\dot{h}_{t\nu}^{\mathrm{TT}} - \partial_\nu h_{tt}^{\mathrm{TT}}) = 0$ (because $h_{t\nu} = 0$), we see that a test particle initially at rest will remain at rest. However, this does not mean that a wave has no effect, but only that the grid of TT coordinates, said to be co-moving or synchronous, 'follows' the test particles.

The equivalence principle states that a freely falling observer whose trajectory in the TT gauge is $x^i = $ const (we shall take these constants to be zero) is locally inertial, and therefore can construct in his neighborhood a quasi-Minkowskian reference frame, the

Fermi coordinates (these were discussed in the context of special relativity in Book 2, Section 5.1). To find them we make an infinitesimal change of spatial coordinates $x^i \to \tilde{x}^i = x^i + \frac{1}{2}h^{\mathrm{TT}\,ij}|_0 x^j + \mathcal{O}(x^2)$. The metric then takes a Minkowskian form at first order in x^i [the calculation of the transformation of the metric tensor is particularly simple in the special case of a plane wave (13.18)]: $ds^2 \simeq -dt^2 + \delta_{ij}d\tilde{x}^i d\tilde{x}^j$. In this Fermi system the geodesic equation of a particle initially at rest becomes

$$\frac{d^2\tilde{x}^i}{d\tau^2} \simeq \frac{d^2\tilde{x}^i}{dt^2} \simeq -\tilde{\Gamma}^i_{tt}(\tilde{x}^i) \simeq -\tilde{\Gamma}^i_{tt}|_0 - \partial_j\tilde{\Gamma}^i_{tt}|_0\,\tilde{x}^j \simeq -\partial_j\tilde{\Gamma}^i_t|_0\tilde{x}^j\,. \tag{13.21}$$

(Here $\tilde{\Gamma}^i_{tt}|_0$ is zero because the Fermi system is locally inertial.) The Riemann tensor is of course nonzero and in first order is $\tilde{R}^i_{tjt} = \partial_j\tilde{\Gamma}^i_{tt}$; moreover, always in first order, $\tilde{R}^i_{tjt} = \dfrac{\partial\tilde{x}^i}{\partial x^\rho}\dfrac{\partial x^\mu}{\partial\tilde{t}}\dfrac{\partial x^\nu}{\partial\tilde{x}^j}\dfrac{\partial x^\sigma}{\partial\tilde{t}}R^\rho_{\mu\nu\sigma} \simeq R^i_{tjt}$. Finally, in the traceless transverse gauge $2R^i_{tjt} \simeq -\ddot{h}^{\mathrm{TT}i}_{\phantom{\mathrm{TT}i}j}|_0$, so that the geodesic equation (13.21) can also be written as

$$\frac{d^2\tilde{x}^i}{dt^2} \simeq -\tilde{R}^i_{tjt}|_0\,\tilde{x}^j \simeq \frac{1}{2}\ddot{h}^{\mathrm{TT}i}_{\phantom{\mathrm{TT}i}j}|_0\,\tilde{x}^j\,. \tag{13.22}$$

In this coordinate system a test particle initially at rest at $\tilde{x}^i = 0$ is accelerated relative to another test particle located at the coordinate origin. We recognize (13.22) as the geodesic deviation equation of Section 2.3: $D^2n^\mu/d\tau^2 = -R^\mu_{\nu\rho\sigma}u^\nu n^\rho u^\sigma$, where n^μ measures the spacing, for constant τ, of the geodesics. Indeed, with $\tilde{x}^i \equiv n^i$, at lowest order and for particles at rest [$u^\mu = (1,0,0,0)$], eqn (2.13) does reduce to (13.22).

Equation (13.22) can be integrated: $\tilde{x}^i = \frac{1}{2}h^{\mathrm{TT}i}_{\phantom{\mathrm{TT}i}j}|_0\tilde{x}^j$, and this allows, for example, the evolution of a circle of test particles in the field of the wave (13.19) to be described. (It is rather academic but instructive to compare the result to the motion of electric charges in the field of an electromagnetic wave as in Book 2, Section 15.4.)

Detection of a gravitational wave

The behavior of test particles in the field of a gravitational wave provides the basis for their detection using interferometry. Let us consider a Michelson interferometer in free fall (in practice this is achieved by hanging the mirrors as pendulums). The time taken by light signals to travel the length of the apparatus will be changed by the passage of a gravitational wave, resulting in a modification of the interference pattern, by which the wave can be detected.

More precisely, the trajectories of two mirrors in the TT gauge are $x = 0$ and $x = l_0$. Let us assume that a wave is incident on the system. The coordinate time taken by the light to make a round trip between the mirrors is such that $ds^2 = g_{00}dt^2 + 2g_{0x}dxdt + g^2_{xx}dx^2 = 0$ and we have $\delta t \simeq (2l_0/g_{00})\sqrt{(g_{0x})^2 - g_{00}g_{xx}}$, if l_0 is much smaller than the scale on which the metric varies. The proper time taken is $\delta\tau = \sqrt{g_{00}}\delta t$, and so it is modulated by the passage of the wave. If it is represented by (13.18) with $h_\times = 0$, we obtain, for example,

$$\delta\tau \simeq 2l_0\left(1 + \tfrac{1}{2}h_+\right).$$

The difference of the optical paths of the two arms and the maximum sensitivity (limited by photon noise) turn out to be $\propto h\sin(\pi l/\lambda)$ and $\propto \hbar\lambda_{\mathrm{em}}\nu^3/(c\pi\epsilon P)$, where h is the amplitude, λ is the wavelength, and ν is the frequency of the wave; $l = nl_0$ is the effective length of the arms of the interferometer, n being the number of reflections on the mirrors; λ_{em} and P are the wavelength and the power of the laser, and ϵ is the efficiency of the detector. Since the expected frequencies are $\nu \approx 1$ kHz, the path difference will be a maximum for $l \simeq 150$ km or $l_0 \approx 1$ km for $n = 150$. The maximum detectable amplitude (it is limited by the current level of laser technology) is $h \approx 10^{-21}$, and in principle this can be measured by the LIGO and VIRGO detectors.[4] In February 2016 the LIGO collaboration announced its first detection of a gravitational wave event, called GW150914, produced by the merging of two black holes.

The forced motion of a spring subject to the action of a gravitational wave provides the basis for a second type of detector. Adapting the geodesic deviation equation (13.22), we have

$$\ddot{x} + \frac{\omega}{Q}\dot{x} + \omega^2 x = \frac{1}{2} l\, \ddot{h}^{\mathrm{TT}}_{xx} \,,$$

where x is the extension of the spring, l is the spring length, Q is its quality factor, and ω is its frequency. The energy stored in the spring when it is in resonance with the wave is $E \propto ml^2\Omega^2 h^2 Q^2$, where Ω is the wave frequency and m is the spring mass.

In practice, the spring is an aluminum bar whose sensitivity is mainly limited by thermal noise. Using the numerical values $m = 1.5$ metric ton, $l = 1.5$ m, $\Omega = 1$ kHz, $Q = 10^5$, and $h \approx 10^{-20}$, we have $E \approx 10^{-20}$ J, or a lengthening of the bar of about 10^{-15} m, which is on the order of the size of an atomic nucleus.[5]

13.4 The first quadrupole formula

In the preceding sections we studied the propagation of gravitational waves and their interaction with test particles. Here we shall relate them, always in linear order, to the energy–momentum tensor of the source creating them.

The post-Minkowski approach developed here is different from the post-Newtonian approach of the previous chapter in that, while limiting ourselves to the *linear* approximation of general relativity, we do *not* assume a priori that the source velocities are small. Moreover, we will be interested in the gravitational field *far* from the sources, in the *wave zone*, where it is *radiative*.

In the linear approximation $g_{\mu\nu} = \eta_{\mu\nu} + h_{\mu\nu}$ the Einstein equations are written in harmonic coordinates as (see Section 13.1)

$$\Box \gamma_{\mu\nu} = -2\kappa\, T_{\mu\nu} + \mathcal{O}(\gamma^2_{\mu\nu}), \quad \text{where} \quad h_{\mu\nu} = \gamma_{\mu\nu} - \frac{1}{2}\eta_{\mu\nu}\gamma \quad \text{with} \quad \partial_\rho \gamma^\rho_\mu = \mathcal{O}(\gamma^2_{\mu\nu}). \tag{13.23}$$

As in electromagnetism, the relevant solutions are the retarded Liénard–Wiechert potentials

$$\gamma_{\mu\nu}(t, x^i) = 4G \int d^3x' \frac{T_{\mu\nu}(t - r', x'^i)}{r'} + \mathcal{O}(\gamma^2_{\mu\nu}), \tag{13.24}$$

where $r' \equiv \sqrt{(x^i - x'^i)(x_i - x'_i)}$ is the distance from the field point x^i to a source point x'^i evaluated at the retarded time $t - r'$ (recall that $\kappa = 8\pi G$).

---

[4]Justification for these estimates can be found in, for example, Kenyon (1990). For a detailed presentation of interferometry-based gravitational wave detectors, see the sites http://www.ligo.org/ and http://www.ego-gw.it/, as well as, for example, Pitkin *et al.* (2011).

[5]Schutz (2009), for example, gives more details about these *Weber bars*, named in homage to J. Weber, a pioneer in the construction of gravitational wave detectors in the 1960s.

Sufficiently far from the source the potential (13.24) can be approximated by a plane wave:

$$\gamma_{\mu\nu}(t, x^i) = \frac{4G}{r} \int d^3x'\, T_{\mu\nu}(t - r, x'^i) + \mathcal{O}(1/r^2) + \mathcal{O}(\gamma_{\mu\nu}^2),\tag{13.25}$$

where r is the distance from the field point to an origin located near the source. Using the conservation of the energy–momentum tensor $\partial_\nu T_\mu^\nu = 0$ (which follows from the harmonic condition), we can express all the γ_{ij} as functions of T_{tt} (see the following box). Moreover, as we saw, for example, in Section 4.1, in Newtonian order we have $T_{tt} \approx \varrho$, where ϱ is the mass density of the source. In addition, $T_{ti} \approx \varrho v_i$, where v^i is the 3-velocity of the fluid. In the end we can rewrite (13.25) at order $\mathcal{O}(\gamma_{\mu\nu}^2)$, $\mathcal{O}(1/r^2)$, and lowest order in the velocities as

$$\gamma_{tt}(t, x^i) = \frac{4G}{r} \int d^3x'\, \varrho, \quad \gamma_{ti}(t, x^j) = \frac{4G}{r} \int d^3x'\, \varrho\, v_i',$$

$$\gamma_{ij}(t, x^k) = \frac{2G}{r} \frac{d^2}{dt^2} \int d^3x'\, \varrho\, x_i' x_j',\tag{13.26}$$

where all quantities in the integrals are evaluated at the retarded time $t - r$.

The harmonic condition and conservation law

At the linear order in which we are working, the harmonic condition implies the conservation of the source energy–momentum tensor $\partial_\nu T_\mu^\nu = 0$, or

$$\partial_t T_{tt} = \partial_k T_t^k, \quad \partial_t T_{ti} = \partial_k T_i^k.$$

We multiply the second equation by x^j, integrate it over all space, and symmetrize it:

$$\frac{1}{2}\frac{d}{dt} \int d^3x\, (x_j T_{ti} + x_i T_{tj}) = \frac{1}{2} \int d^3x\, (x_j \partial_k T_i^k + x_i \partial_k T_j^k) = -\int d^3x\, T_{ij}.$$

[We have integrated by parts and used the divergence theorem and the fact that $(x_j T_i^k + x_j T_i^k)$ vanishes at infinity, because T_{ij} has compact support since the source is localized.] Next we multiply the first equation by $x_i x_j$ and again integrate over all space to obtain

$$\frac{d}{dt} \int d^3x\, T_{tt} x_i x_j = -\int d^3x (x_j T_{ti} + x_i T_{tj}).$$

Therefore,

$$\int d^3x\, T_{ij} = \frac{1}{2}\frac{d^2}{dt^2} \int d^3x\, T_{tt}\, x_i x_j.$$

Since $T_{tt} \approx \varrho$, we indeed see that (13.25) gives (13.26).

Now let us work in a given inertial frame, for example, a frame in which the source is at rest on the average. As we saw in Section 13.3, the gravitational waves that it emits are the transverse traceless part of the spatial part of the metric. In addition, they are gauge-invariant. We can therefore extract them directly from (13.26) using (13.10) and thereby obtain the *first quadrupole formula* (Einstein, 1918), that is, the asymptotic, radiative part of the linearized metric in the TT gauge at lowest order in the velocities:

$$ds^2_{\rm rad} = -dt^2 + h^{\rm TT}_{ij} dx^i dx^j$$

$$(13.27)$$

$$\text{with} \quad h^{\rm TT}_{ij} = \frac{2G}{3c^4 r} P^{kl}_{ij} \ddot{Q}_{kl} , \quad \text{where} \quad Q_{ij} = \int d^3x \, \varrho \left(3x_i x_j - \delta_{ij} r^2 \right)$$

is the quadrupole moment of the source (see Book 1, Section 14.1) evaluated at the retarded time $t - r$ (we have also reintroduced c).[6] The projection operator was given in (13.10):

$$P^{kl}_{ij} = P^k_i P^l_j - \frac{1}{2} P_{ij} P^{kl} , \quad \text{where} \quad P_{ij} = \delta_{ij} - n_i n_j \quad \text{with} \quad n^i = x^i / r .$$

$$(13.28)$$

The potentials $h^{\rm TT}_{ij}$ are the gravitational analog of the electromagnetic radiative potential in the Lorenz gauge, which, as we saw in Book 2, Sections 20.1 and 20.2, is expressed as a function of the second derivative of the dipole moment of the charges q creating the field, or of their quadrupole moment if the ratio q/m is the same for all the charges. It is therefore the equality of gravitational and inertial mass which explains why the radiative part of the metric is quadrupolar.

Let us conclude by specifying the domain of validity of the quadrupole formula (13.27).

For the perturbations (13.26) of the metric to indeed solve the linearized Einstein equations, they must satisfy the harmonic condition. Let us consider the case where the sources are point masses $m \equiv GM$. Then at orders $\mathcal{O}(m^2)$, $\mathcal{O}(1/r^2)$, and lowest order in the velocities the retarded potential (13.26) becomes

$$\gamma_{tt} = \sum \frac{4m}{r} , \quad \gamma_{ti} = \sum \frac{4m v_i}{r} , \quad \gamma_{ij} = \sum \frac{2m}{r} \left(2v_i v_j + \mathcal{O}(\dot{v}) \right) .$$

$$(13.29)$$

Written in this form, we clearly see that the harmonic condition, which reduces to $\dot{\gamma}_{ti} = \mathcal{O}(m^2)$, requires that $\dot{v}^i = \mathcal{O}(m)$. The radiation field is therefore given by (13.26) or (13.29) only in the 'adiabatic' approximation, that is, when it is possible to neglect the accelerations of the source. This is *not* possible for bodies interacting gravitationally: their accelerations are given in lowest order by Newton's law $\dot{v} \propto m/R^2$, and so $\dot{v} = \mathcal{O}(m)$ but also $v^2 \propto m/R$, so that the two terms in eqn (13.29) for γ_{ij} are of the same order, which invalidates the approximation. Therefore, the *first quadrupole formula* (13.27) derived here in the linear approximation of general relativity does not *a priori* give the gravitational waves emitted by a system of moving bodies, except when the bodies are moving under the action of forces *other* than gravity.

----

[6] Equation (13.27) can also be derived from (13.26) by a change of coordinates; see, for example, Stephani (1990).

Orders of magnitude

The order of magnitude of the metric components (13.27) is $h_{ij}^{\mathrm{TT}} = \mathcal{O}\left(mv^2/r\right)$, where m and v are the characteristic mass and velocity of the source and r is the distance from the source to the observation point. We see that the amplitude of the waves produced by any terrestrial object is *extremely* weak: indeed, 1 metric ton $= 7 \times 10^{-24}$ m. However, if the source is a binary star system, for example, the binary pulsar PSR 1913+16 (see footnote 7), we have $mv^2/r \approx 10^{-22}$, which is now in the domain of the measurable; see Section 13.3. However, as mentioned above, the formula then must be justified; we shall do this in Section 14.3.

---

[7]The characteristics of PSR 1913+16, composed of two neutron stars, are the following: $m_{\mathrm{total}} \approx 2 \times 1.4 M_\odot \approx 4.2 \times 10^3$ m; $v^2 \approx m/R$ (by the Newton–Kepler law) with a separation $R \approx R_\odot \approx 7 \times 10^8$ m, or $v \approx 0.25 \times 10^{-2}$; and $r \approx 5$ kpc $\approx 2 \times 10^{20}$ m. In Section 12.6 we saw how to determine these parameters by a timing analysis of signals emitted by the pulsar.

14
Gravitational radiation

Now that we have found the radiative potential of a source in the linear approximation of general relativity, we can calculate the energy radiated to infinity in lowest order. For this we will need to expand the Einstein equations to quadratic order in the metric perturbations, which is the subject of Section 14.1. We shall see that the radiated energy is then given by the (second) quadrupole formula (14.11), which is the gravitational analog of the dipole formula in the Maxwell theory. This formula is *a priori* valid only if the motion of the source is due to forces other than gravity.

In Section 14.3 we shall see that to prove this formula for the case of self-gravitating systems we need to *solve* the Einstein equations to quadratic order and obtain the radiative field in the post-linear approximation of general relativity.

14.1 The Einstein equations at post-linear order

When iterating the Einstein equations, it is often useful to write the Einstein tensor as (Fock, 1959)

$$2|g|G^{\mu\nu} = \mathcal{G}^{\rho\sigma}\partial^2_{\rho\sigma}\mathcal{G}^{\mu\nu} + \mathcal{D}^{\mu\nu}_\rho(\partial_\sigma\mathcal{G}^{\rho\sigma}) - \mathcal{Q}^{\mu\nu}, \quad \text{where} \quad \mathcal{G}^{\mu\nu} \equiv \sqrt{-g}g^{\mu\nu} \equiv \eta^{\mu\nu} - \gamma^{\mu\nu}, \quad (14.1)$$

g is the determinant of the metric $g_{\mu\nu}$, $\mathcal{D}^{\mu\nu}_\rho$ is a first-order differential operator acting on the divergence of $\mathcal{G}^{\rho\sigma}$ (so that the second term is zero in harmonic gauges with $\partial_\rho\mathcal{G}^{\rho\sigma} = 0$), and $\mathcal{Q}^{\mu\nu}$ is quadratic in the first derivatives of $g_{\mu\nu}$. [Its complete expression can be found in, for example, Bel *et al.* (1981).]

At post-linear order in $\gamma_{\mu\nu}$ the expansion of $g_{\mu\nu} = \eta_{\mu\nu} + h_{\mu\nu}$ is, indices being moved with $\eta_{\mu\nu}$

$$h_{\mu\nu} = \gamma_{\mu\nu} - \frac{1}{2}\eta_{\mu\nu}\gamma + \eta_{\mu\nu}\left(\frac{1}{8}\gamma^2 - \frac{1}{4}\gamma_{\alpha\beta}\gamma^{\alpha\beta}\right) - \frac{1}{2}\gamma\gamma_{\mu\nu} + \gamma_{\alpha\mu}\gamma^\alpha_\nu + \mathcal{O}(\gamma^3_{\mu\nu}). \quad (14.2)$$

The expansion of the Einstein tensor in harmonic coordinates for which $\partial_\sigma\mathcal{G}^{\rho\sigma} = 0$ then is

$$2|g|G^{\mu\nu} = -\Box\gamma^{\mu\nu} + (\gamma^{\alpha\beta}\partial^2_{\alpha\beta}\gamma^{\mu\nu} - \partial_\alpha\gamma^{\mu\beta}\partial_\beta\gamma^{\alpha\nu}) - 16\pi G|g|t^{\mu\nu}_{\text{LL}} + \mathcal{O}(\gamma^3_{\mu\nu}) \quad (14.3)$$

with $\partial_\nu\gamma^{\nu\mu} = 0$, where $\Box \equiv \eta^{\mu\nu}\partial^2_{\mu\nu}$ is the flat-space d'Alembertian and $t^{\mu\nu}_{\text{LL}}$ is the *Landau–Lifshitz pseudo-tensor*[1] (1941):

---

[1] Here we are keeping only its quadratic part; see Landau and Lifshitz (1971) and also, for example, Poisson (2007).

Note that Equations (14.2) and (14.3) can also be used to calculate the action $2\kappa S = \int d^4x\sqrt{-g}R$ of perturbations; we have seen its expression in terms of gauge-invariant quantities in Section 13.2.

Relativity in Modern Physics. Nathalie Deruelle and Jean-Philippe Uzan.
© Oxford University Press 2018. Published in 2018 by Oxford University Press.
DOI: 10.1093/oso/9780198786399.001.0001

$$2\kappa \, |g| t_{\text{LL}}^{\mu\nu} = \frac{1}{2} \partial^\mu \gamma^{\alpha\beta} \, \partial^\nu \gamma_{\alpha\beta} - \frac{1}{4} \partial^\mu \gamma \, \partial^\nu \gamma$$

$$+ \partial_\alpha \gamma^{\mu\beta} \, \partial^\alpha \gamma_\beta^\nu + \frac{1}{8} \eta^{\mu\nu} \, \partial_\alpha \gamma \, \partial^\alpha \gamma - \frac{1}{4} \eta^{\mu\nu} \, \partial_\alpha \gamma_{\beta\delta} \, \partial^\alpha \gamma^{\beta\delta} \tag{14.4}$$

$$+ \frac{1}{2} \eta^{\mu\nu} \, \partial_\alpha \gamma_{\beta\delta} \, \partial^\beta \gamma^{\alpha\delta} - \partial^\mu \gamma^{\alpha\beta} \, \partial_\alpha \gamma_\beta^\nu - \partial^\nu \gamma^{\alpha\beta} \, \partial_\alpha \gamma_\beta^\mu \,.$$

The factor 2κ has been introduced for dimensional reasons, and indices are lowered using $\eta_{\mu\nu}$.

Therefore, the Einstein equations in post-linear order are written as

$$\Box \gamma^{\mu\nu} = -2\kappa \, |g|(T^{\mu\nu} + \tau^{\mu\nu}) + \mathcal{O}(\gamma_{\mu\nu}^3) \qquad \text{with} \qquad \partial_\nu \gamma^{\nu\mu} = 0$$

$$\text{and} \qquad 2\kappa \, |g| \tau^{\mu\nu} = 2\kappa \, |g| t_{\text{LL}}^{\mu\nu} + \partial_\alpha \gamma^{\mu\beta} \, \partial_\beta \gamma^{\alpha\nu} - \gamma^{\alpha\beta} \, \partial_{\alpha\beta}^2 \gamma^{\mu\nu} \,, \tag{14.5}$$

where $T^{\mu\nu}$ is the source energy–momentum tensor and $|g| = 1 - \gamma + \mathcal{O}(\gamma_{\mu\nu}^2)$.

14.2   The second quadrupole formula

The harmonic condition $\partial_\nu \gamma_\mu^\nu = 0$ forces the second term of the Einstein equations (14.5) to have zero divergence.

Then, since $\partial_\nu \left(\partial_\alpha \gamma^{\mu\beta} \partial_\beta \gamma^{\alpha\nu} - \gamma^{\rho\sigma} \partial_{\rho\sigma}^2 \gamma^{\mu\nu} \right) = 0$, we have

$$\partial_\nu \left(|g|(T^{\mu\nu} + t_{\text{LL}}^{\mu\nu}) \right) = 0 \,. \tag{14.6}$$

After integrating over the 3-volume bounded by the 2-sphere at infinity and using the Gauss theorem and the fact that $T_{\mu\nu}$ is zero at infinity, we arrive at

$$\frac{d}{dt} E = - \lim_{r \to \infty} \int |g| t_{\text{LL}}^{ti} n_i r^2 d\Omega \qquad \text{with} \qquad E \equiv \int d^3 x \, |g|(T^{tt} + t_{\text{LL}}^{tt}) \,, \tag{14.7}$$

where $n^i = x^i/r$ and $d\Omega = \sin\theta \, d\theta d\phi$ is the element of solid angle. (In the order in which we are working we can set $|g| = 1$ in the surface integral.)

The calculation of $n_i t_{\text{LL}}^{ti}$ at infinity is easy when the radiative part, that is, the part proportional to $1/r$, of the metric is given by the first quadrupole formula given in Section 13.4. Then it is purely spatial, transverse, and traceless, and depends only on the retarded time $t - r$. Therefore, its derivatives $\partial_\rho \gamma_{\mu\nu}^{\text{TT}}$ are

$$\partial_\rho \gamma_{\mu\nu}^{\text{TT}} \equiv (\dot\gamma_{\mu\nu}, \partial_i \gamma_{\mu\nu})^{\text{TT}} = -n_\rho \dot\gamma_{\mu\nu}^{\text{TT}} \qquad \text{with} \quad n^\mu = (1, n^i)$$

$$\text{and} \quad \left(\gamma_{tt}^{\text{TT}} = 0 \,, \ \gamma_{it}^{\text{TT}} = 0 \,, \ \gamma_{ij}^{\text{TT}} = h_{ij}^{\text{TT}} \right) \,. \tag{14.8}$$

Since $n_\nu n^\nu = 0$, $\gamma_{t\mu}^{\text{TT}} = 0$, and $n^\nu \gamma_{\mu\nu}^{\text{TT}} = 0$, the radiative part of the Landau–Lifshitz pseudotensor given in (14.4) reduces to the first term:

$$2\kappa \, |g| \, t_{\text{LL}}^{\mu\nu} \simeq \frac{1}{2} n^\mu n^\nu \dot\gamma_{\alpha\beta} \dot\gamma^{\alpha\beta} |^{\text{TT}} \qquad \Longrightarrow \qquad 2\kappa \, |g| \, n_i t_{\text{LL}}^{ti} \simeq \frac{1}{2} \dot h_{\text{TT}}^{ij} \dot h_{ij}^{\text{TT}} \,. \tag{14.9}$$

Then, substituting the expression for h_{ij}^{TT} as a function of the second derivatives of the source quadrupole obtained in (13.27), eqn (14.7) becomes

$$\frac{dE}{dt} = -\frac{G}{72\pi} \int P^{ij}_{kl} \frac{d^3 Q^{kl}}{dt^3} P^{mn}_{ij} \frac{d^3 Q_{mn}}{dt^3} d\Omega, \tag{14.10}$$

where the projector P^{ij}_{kl} is given in (13.28). The integrand contains terms proportional to $n_i n_j$ and $n_i n_j n_k n_l$, and the integral over the angles is done using the standard expressions for spatial averages (see, for example, Book 2, Section 18.1). We thus obtain (with c restored)

$$\frac{dE}{dt} = -\frac{G}{45c^5} \frac{d^3 Q^{ij}}{dt^3} \frac{d^3 Q_{ij}}{dt^3}, \qquad \text{where} \qquad Q_{ij} = \int d^3x \, \varrho \left(3x_i x_j - \delta_{ij} r^2 \right). \tag{14.11}$$

This is the *second quadrupole formula* (Einstein, 1918). As we shall argue, E can be identified as the system energy. Therefore, (14.11) gives the energy lost by the source in the form of gravitational waves. We recall that to obtain this expression we had to write out the Einstein equations at post-linear order (Section 14.1) in order to obtain the conservation law (14.6), and then use the first quadrupole formula (13.27) giving the radiative field of a system of masses in motion due to non-gravitational forces at lowest order in the velocities.

The pseudo-tensor and energy of the system

In the expression for E in (14.7), the first term can represent the energy of the matter creating the field because $T_{\mu\nu}$ is its energy–momentum tensor and at lowest order $|g| \approx 1$. The radiative part of the integrand in the second term is given by [*cf.* (14.9)]

$$t^{tt}_{LL} \approx \frac{1}{4\kappa} \dot{h}^{ij}_{TT} \dot{h}^{TT}_{ij}. \tag{14.12}$$

It coincides with the energy density of the gravitational waves obtained in (13.20) from their Hamiltonian, which suggests that the second term in E can be identified as the field energy. However, the field energy is actually *not* the integral of (14.12) (where we can set $|g| = 1$) over all space, because close to and inside the sources the metric is not that of the radiative field, which is only asymptotic.

Now, according to the analysis of the conservation laws carried out in Chapter 5, E can in fact be written as a surface integral. Indeed, the pseudo-tensor (14.4) is derived from a superpotential because, as Landau and Lifshitz have shown,

$$\kappa \, \partial_\alpha H^{\mu\nu\alpha} = |g| G^{\mu\nu} + \kappa |g| t^{\mu\nu}_{LL}, \quad \text{where} \quad 2\kappa \, H^{\mu\nu\alpha} = \partial_\beta \left(|g| (\mathcal{G}^{\mu\nu} \mathcal{G}^{\alpha\beta} - \mathcal{G}^{\mu\alpha} \mathcal{G}^{\nu\beta}) \right).$$

[Using (14.3) and (14.4), this can be verified in post-linear order, to which we confine ourselves here.] Therefore, we have

$$E = \int H^{tti} n_i r^2 d\Omega.$$

Finally, to relate E to the gravitational mass of the source, we must calculate H^{tti} using the complete asymptotic metric and not only its radiative part.[2]

To specify the domain of validity of the quadrupole formula (14.11), let us consider the case where the source is an ensemble of point-like objects.

---

[2]The superpotential $H^{\mu\nu\alpha}$ gives the same result as the Freud–Katz potential defined in Section 5.4. To learn more, see Katz (1996).

Book 3

We then have $Q_{ij} = \sum m(3z_i z_j - \delta_{ij} z^2)$, where z_i is the location of the body m. The radiative field, proportional to $\dddot{Q}$, then is $\sim mvv + \mathcal{O}(m\dot{v}z)$, where, as we saw in Section 13.4, the accelerations of the source must be negligible in order to use a linear approximation. However, d^3Q/dt^3 is $\sim mv.\dot{v} + \mathcal{O}(m\ddot{v}z)$. The second quadrupole formula therefore is *a priori* applicable only if the terms proportional to $mv.\dot{v}$ numerically dominate those of order $m\ddot{v}z$. This will be the case for sources bound by non-gravitational forces, for which $v^2 \gg \dot{v}z$. However, it will *not* be true if the sources are gravitationally coupled, because then according to Newton's law we will have $\dot{v} \sim v^2/R \sim m/R^2$, where R is their separation, so that not only are the two terms of comparable magnitude, but in addition they are of the same order as the contributions $\sim m^2$ arising from the post-linear approximation to the radiative field.[3]

To describe in lowest order the radiation of gravitational waves by a system of gravitationally interacting bodies, it is therefore necessary to use, in the Landau–Lifshitz pseudo-tensor appearing in (14.7), an approximation of the radiative field which is better than the first quadrupole formula. As we shall see below, study of the radiative field at *post-linear* order will in fact allow us to extend the validity of the quadrupole formula (14.11) to self-gravitating systems, in particular, to binary systems of compact stars.

14.3   Radiation of a self-gravitating system

Here we shall show explicitly, following step-by-step the analysis of quadrupole radiation in electromagnetism carried out in Book 2, that the linear approximation only is not sufficient for finding the radiation of a self-gravitating system.

The steps in the analysis are the following:

- The linear expansion of the radiative field obtained in Section 13.4 is extended to the required order in the source velocities (rather than just the lowest order).

- The results of Chapter 15 are used to add the contribution from the second iteration of the Einstein equations.

- The asymptotic expression for the Landau–Lifshitz pseudo-tensor is derived, and it is noticed that (14.7) *again* gives the second quadrupole formula (14.11). We have thereby shown that it is valid *also* when the sources are self-gravitating.

- *Linear expansion of the radiative field*

Let us consider a gravitationally interacting system. As an example and to facilitate comparison with systems of slowly moving charges studied in Book 2, Section 6.1, we assume that the system consists of two point objects of masses m and m'. In the linear approximation, which we label by the subscript 1 for clarity,[4] their energy–momentum tensor is [see Section 4.1 and Book 2, Section 8.4]

---

[3]Of course, one could argue that by 'continuity' the quadrupole formula (14.11) must also hold for gravitationally coupled systems, which is the art of obtaining correct results using shaky arguments.

One could also formally integrate (14.5) as $\gamma_{\mu\nu} = 4G \int d^3x' \frac{(T_{\mu\nu} + \tau_{\mu\nu})}{r'}$, and then repeat the arguments of Section 13.4 to write $\int d^3x \, (T_{ij} + \tau_{ij}) = \frac{1}{2} \frac{d^2}{dt^2} \int d^3x \, (T_{tt} + \tau_{tt}) \, x_i x_j$. Then one could argue that at lowest order $T_{tt} + \tau_{tt} \simeq T_{tt} \simeq \varrho$, and conclude that the linear potential is valid beyond the adiabatic, linear approximation. However, the arguments of Section 13.4 are based on T_{ij} being bounded, which is not the case for τ_{ij}. Moreover, as we shall see in Chapter 15, the retarded integrals of $\tau_{\mu\nu}$ are *a priori* poorly defined.

[4]We also set $G = 1$ to simplify the notation.

$$T_1^{\mu\nu} = \sum m \int \delta_4(x^\rho - z^\rho(\tau))u^\mu u^\nu d\tau \,, \tag{14.13}$$

where δ_4 is the Dirac delta distribution.

The solution of the linearized Einstein equations in harmonic gauge $\partial_\nu \gamma^{\mu\nu} = 0$ is the Liénard–Wiechert potential obtained in (13.24) or

$$\gamma^1_{\mu\nu}(t, x^i) = \sum 4m \frac{u_\mu u_\nu}{\varrho}\Big|_{\mathcal{R}} \,, \quad \text{where} \quad \varrho_{\mathcal{R}} = -u_\mu l^\mu|_{\mathcal{R}} \quad \text{with} \quad l^\mu|_{\mathcal{R}} = x^\mu - x^\mu_{\mathcal{R}}$$

$$\text{and} \quad \eta_{\mu\nu} l^\mu_{\mathcal{R}} l^\nu_{\mathcal{R}} = 0 \,. \tag{14.14}$$

Here the subscript $\mathcal{R}$ indicates that the quantity is evaluated at the intersection of the past cone originating at the field point x^μ and the world line of the body in question; see Fig. 14.1. (This expression should be compared with the retarded potential created by charges in the Maxwell theory; see Book 2, Sections 17.1 and 17.2.)

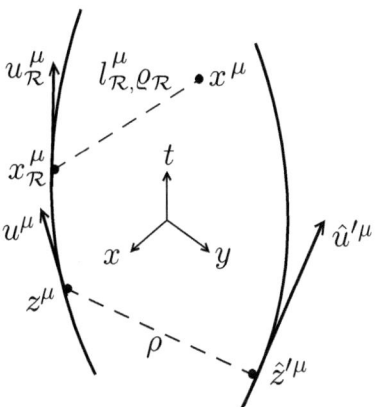

Fig. 14.1 Retarded quantities.

Retarded quantities and their derivatives

Here we recall some useful definitions and formulas obtained in Book 2 in the similar framework of the Maxwell theory.

We define $x^\mu_{\mathcal{R}}$ as the intersection of the retarded cone originating at the field point x^μ and the world line of the mass m, and we set $l^\mu_{\mathcal{R}} = x^\mu - x^\mu_{\mathcal{R}}$ (so that $\eta_{\mu\nu} l^\mu_{\mathcal{R}} l^\nu_{\mathcal{R}} = 0$). The proper time of the mass m at the retarded point $x^\mu_{\mathcal{R}}$ is denoted $\tau_{\mathcal{R}}$ and its 4-velocity is $u^\mu_{\mathcal{R}}$. Then, setting $\varrho_{\mathcal{R}} = -u_\mu l^\mu|_{\mathcal{R}}$ [see Book 2, Section 17.1], we have

$$\partial_\mu \tau_{\mathcal{R}} = -\frac{l^\mu}{\varrho}\Big|_{\mathcal{R}} \,, \quad \frac{\partial x^\nu_{\mathcal{R}}}{\partial x^\mu} = -\frac{u^\nu l_\mu}{\varrho}\Big|_{\mathcal{R}} \,, \tag{14.15}$$

$$\partial_\mu u^\nu_{\mathcal{R}} = -l_\mu \dot{u}^\nu|_{\mathcal{R}} \,, \quad \partial_\mu \varrho_{\mathcal{R}} = [n_\mu + (l_\mu/\varrho)(l \cdot \dot{u})]_{\mathcal{R}} \quad \text{with} \quad n^{\mathcal{R}}_\mu \equiv -u^{\mathcal{R}}_\mu + (l_\mu/\varrho)|_{\mathcal{R}} \,.$$

Since gravitation is a long-range interaction, the field and equations of motion of the bodies will be characterized by the 'Jacob's ladder' described in Book 2, Section 6.2 and will be given in terms of retarded quantities, doubly retarded quantities, and so on. In the post-linear order which we will discuss later on, we will find it necessary to introduce

$$\rho \equiv -(z^\mu - \hat{z}'^\mu)\hat{u}'_\mu \,, \quad \nu^\mu = -\hat{u}'^\mu + (z^\mu - \hat{z}'^\mu)/\rho \,, \quad w = (u.\hat{u}') \,,$$
$$\rho_\mathcal{R} \equiv -(x^\mu_\mathcal{R} - \hat{z}'^\mu_\mathcal{R})\hat{u}'^\mathcal{R}_\mu \,, \quad \nu^\mu_\mathcal{R} = -\hat{u}'^\mu_\mathcal{R} + (x^\mu_\mathcal{R} - \hat{z}'^\mu_\mathcal{R})/\rho_\mathcal{R} \,, \quad w_\mathcal{R} = (u.\hat{u}')_\mathcal{R} \,,$$

(14.16)

where $\hat{z}'^\mu$ is the intersection of the cone, originating at a point z^μ on the world line of the body m, and the world line of the body m', with $\hat{u}'^\mu$ the velocity of m'.

Using (14.15), from the expression for the metric (14.14) we find

$$\partial_\nu \gamma_1^{\mu\nu} = \sum 4m \left(\frac{\dot{u}^\mu}{\varrho}\right)_\mathcal{R} .$$

(14.17)

We therefore again reach the conclusion of Section 13.4: the harmonic condition is satisfied in linear order because $\dot{u}$ is of order m owing to Newton's equations of motion. We can also interpret (14.17) in a different way, namely, that the harmonic condition, which at this order is equivalent to the conservation of the energy–momentum tensor, *requires* that the equations of motion be of order m.

Expansion of retarded quantities

To change over to three-dimensional notation we set $l^0 = t - t_\mathcal{R} \equiv r_\mathcal{R}$ and $l^i = x^i - x^i_\mathcal{R} \equiv r_\mathcal{R} n^i_\mathcal{R}$, where $n^i_\mathcal{R}$ is a unit 3-vector. We then have (see Book 2, Section 17.1)

$$u^0_\mathcal{R} = \frac{1}{\sqrt{1 - v^2_\mathcal{R}}}, \quad u^i_\mathcal{R} = \frac{v^i_\mathcal{R}}{\sqrt{1 - v^2_\mathcal{R}}}, \quad \varrho_\mathcal{R} = r_\mathcal{R} \frac{1 - (n.v)_\mathcal{R}}{\sqrt{1 - v^2_\mathcal{R}}} .$$

(14.18)

Next, to find the asymptotic behavior of the field we introduce an origin located inside the system and write $r_\mathcal{R} n_\mathcal{R} = rn - z_\mathcal{R}$, where rn is the position vector of the field point x^i of length r and $z_\mathcal{R}$ is the position vector of the mass m at the retarded time $t_\mathcal{R}$. We then evaluate the retarded quantities at the time $t_{\mathcal{R}_0}$ such that $t - t_{\mathcal{R}_0} = r$ is the time taken for the gravitational interaction (as well as light) to travel from the origin of the frame to the field point. Then we have (see Book 2, Section 20.2)

$$v^i_\mathcal{R} = v^i_{\mathcal{R}_0} + \dot{v}^i_{\mathcal{R}_0}(n.z_{\mathcal{R}_0}) + \left[\tfrac{1}{2}\ddot{v}^i(n.z)^2 + \dot{v}^i(n.v)(n.z)\right]_{\mathcal{R}_0} + \cdots ,$$

(14.19)

where the dots denote terms of order v^4 and $v^2 z/r$.

Now the derivatives are easy to find. Quantities depend on the time through the positions $z^i_{\mathcal{R}_0}$ of the bodies evaluated at $t_{\mathcal{R}_0} = t - r$ with r a constant. Moreover, shifting the field point by an amount δx^i with r a constant implies shifting the origin by the same amount. Therefore,

$$\partial_t z^i_{\mathcal{R}_0} = v^i_{\mathcal{R}_0} \,, \quad \partial_i z^j_{\mathcal{R}_0} = -\delta^j_i \,.$$

(14.20)

Now, to find the linearized metric generated by an ensemble of slowly moving masses, it is sufficient to use the method described in Book 2, Sections 17.1 and 17.2 for the case of electromagnetism (and already used to study the gravitational field in the Nordström theory; see Book 2, Section 10.4). First we write (14.14) in three-dimensional notation using (14.18):

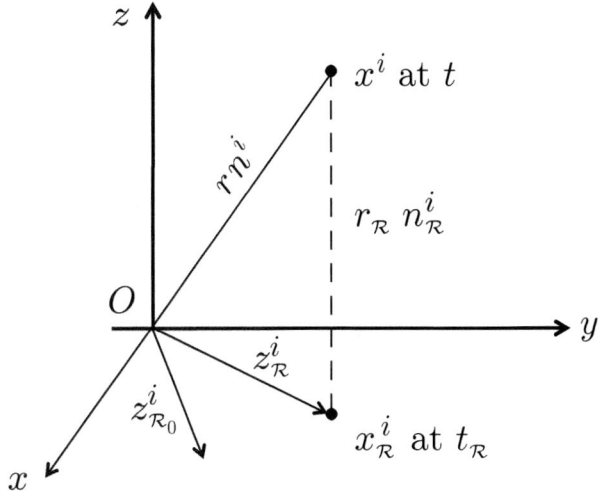

Fig. 14.2 Retarded quantities, three-dimensional notation.

$$\gamma^1_{\mu\nu} = \sum \frac{4m}{r_{\mathcal{R}}\sqrt{1-v_{\mathcal{R}}^2}\,(1-n_{\mathcal{R}}v_{\mathcal{R}})} \left(1\,,\, -v_i^{\mathcal{R}}\,,\, v_i^{\mathcal{R}}v_j^{\mathcal{R}}\right), \tag{14.21}$$

where the position $x_{\mathcal{R}}^i = x^i(t_{\mathcal{R}})$ and the 3-velocity $v_{\mathcal{R}}^i = v^i(t_{\mathcal{R}})$ of the mass m are evaluated at the retarded time $t_{\mathcal{R}}$ given implicitly by $t_{\mathcal{R}} = t - r_{\mathcal{R}}$ with $x^i - x^i(t_{\mathcal{R}}) \equiv r_{\mathcal{R}} n_{\mathcal{R}}^i$, where $n_{\mathcal{R}}$ is a unit 3-vector: $n_{\mathcal{R}}.n_{\mathcal{R}} = 1$; see Book 2, Section 17.2 and Fig. 14.2.

Next, we situate ourselves far from the system, introduce a fixed point inside the distribution, and write $r_{\mathcal{R}}n_{\mathcal{R}} = rn - z_{\mathcal{R}}$, where rn is the radius vector of the field point of components x^i and length r, and $z_{\mathcal{R}}$ is the radius vector of the mass m at $t_{\mathcal{R}}$. In addition, we assume that the speeds of the bodies remain small. Then, since $r_{\mathcal{R}}n_{\mathcal{R}} = rn$ to lowest order in z/r, we have (see Book 2, Section 20.2 for the analogous calculation in the Maxwell theory)[5]

$$\begin{cases} \gamma^1_{tt} = \sum \dfrac{4m}{r}\Big[1 + (n\,.v_{\mathcal{R}}) + \tfrac{1}{2}v_{\mathcal{R}}^2 + (n\,.v_{\mathcal{R}}^2) + (n\,.v_{\mathcal{R}})^3 \\[2mm] \qquad\qquad + \tfrac{1}{2}(n\,.v_{\mathcal{R}})v_{\mathcal{R}}^2 + \mathcal{O}(v^4) \mid \mathcal{O}(z/r)\Big], \\[3mm] \gamma^1_{ti} = -\sum \dfrac{4m\,v_i^{\mathcal{R}}}{r}\Big[1 + (n\,.v_{\mathcal{R}}) + \tfrac{1}{2}v_{\mathcal{R}}^2 + (n\,.v_{\mathcal{R}}^2) + \mathcal{O}(v^3) + \mathcal{O}(z/r)\Big], \\[3mm] \gamma^1_{ij} = \sum \dfrac{4m v_i^{\mathcal{R}} v_j^{\mathcal{R}}}{r}\Big[1 + (n\,.v_{\mathcal{R}}) + \mathcal{O}(v^2) + \mathcal{O}(z/r)\Big]. \end{cases} \tag{14.22}$$

Finally, we evaluate all quantities at the time $t_{\mathcal{R}_0}$ such that $t - t_{\mathcal{R}_0} = r$ is the time for the gravitational interaction to propagate from the origin of the frame to the field point. Then

---

[5]We shall not study the radiation of angular momentum, which requires expanding through order $1/r^2$.

using (14.19) we easily find (see the analogous calculation in Maxwell's theory in Book 2, Section 20.2)

$$
\begin{cases}
\gamma_{tt}^1 = \sum \dfrac{4m}{r} + \sum \dfrac{4m}{r}(n\,.v_{\mathcal{R}_0}) + \sum \dfrac{4m}{r}\left[(n\,.v)^2 + (n\,.z)(n\,.\dot{v}) + \tfrac{1}{2}v^2\right]_{\mathcal{R}_0} \\[2mm]
\quad + \sum \dfrac{4m}{r}\left[(n\,.v)^3 + 3(n\,.z)(n\,.v)(n\,.\dot{v}) + \tfrac{1}{2}(n\,.z)^2(n\,.\ddot{v}) + (v\,.\dot{v})(n\,.z) + \tfrac{1}{2}v^2(n\,.v)\right]_{\mathcal{R}_0} \\[2mm]
\quad + \cdots , \\[4mm]
\gamma_{tj}^1 = -\sum \dfrac{4mv_j|_{\mathcal{R}_0}}{r} - \sum \dfrac{4m}{r}\left[v_j(n\,.v) + \dot{v}_j(n\,.z)\right]_{\mathcal{R}_0} \\[2mm]
\quad -\sum \dfrac{4m}{r}\left[\tfrac{1}{2}\ddot{v}_j(n.z)^2 + 2\dot{v}_j(n.z)(n.v) + v_j\left((n.v)^2 + (n.z)(n.\dot{v})\right) + \tfrac{1}{2}v^2 v_j\right]_{\mathcal{R}_0} + \cdots , \\[2mm]
\gamma_{ij}^1 = \sum \dfrac{4mv_i v_j|_{\mathcal{R}_0}}{r} + \sum \dfrac{4m}{r}\left[(v_i\dot{v}_j + v_j\dot{v}_i)(n\,.z) + v_i v_j(n\,.v)\right]_{\mathcal{R}_0} + \cdots ,
\end{cases}
\tag{14.23}
$$

where $z_{\mathcal{R}_0}$ and $v_{\mathcal{R}_0}$ are the radius vector and velocity of the mass m at the time $t_{\mathcal{R}_0}$, a dot denotes the derivative with respect to t, and the dots stand for terms proportional to $(m/r)\,v^2(v^2 + z/r)$ and terms $\mathcal{O}(m^2)$, because we are working in the linear approximation. Since $\dot{v}$ appearing in (14.23) at $\mathcal{O}(v^3)$ is of order m owing to Newton's equations, we see explicitly that to obtain the radiative field of a self-gravitating system we *must* solve the Einstein equations at post-linear order, which we shall do in the following chapter.

The derivatives of the potential, which alone are involved in the calculation of the Landau–Lifshitz pseudo-tensor, are found using (14.20):

$$
\begin{cases}
\dot{\gamma}_{tt}^1 = \sum \dfrac{4m}{r}(n\,.\dot{v})|_{\mathcal{R}_0} + \sum \dfrac{4m}{r}\left[3(n\,.v)(n\,.\dot{v}) + (n\,.z)(n\,.\ddot{v}) + (v\,.\dot{v})\right]_{\mathcal{R}_0} + \cdots , \\[2mm]
\dot{\gamma}_{tj}^1 = -\sum \dfrac{4m}{r}\dot{v}_j|_{\mathcal{R}_0} - \sum \dfrac{4m}{r}\left[2(n\,.v)\dot{v}_j + v_j(n\,.\dot{v}) + \ddot{v}_j(n\,.z)\right]_{\mathcal{R}_0} + \cdots , \\[2mm]
\dot{\gamma}_{ij}^1 = \sum \dfrac{4m}{r}(\dot{v}_i v_j + v_i\dot{v}_j)|_{\mathcal{R}_0} + \cdots ,
\end{cases}
\tag{14.24}
$$

$$
\partial_i\gamma_{tt}^1 = -n_i\dot{\gamma}_{tt} , \qquad \partial_i\gamma_{tj}^1 = -n_i\dot{\gamma}_{tj} , \qquad \partial_k\gamma_{ij}^1 = -n_k\dot{\gamma}_{ij}
$$

[which coincides with the expansion of (14.17), as it must; in addition, we note that the calculation of the spatial derivatives requires knowledge of (14.23) at $\mathcal{O}(v^4)$]. Since $\dot{v} = \mathcal{O}(m)$, they are all $\mathcal{O}(m^2)$. More precisely, if we work in the center-of-mass system and use Newton's law of motion (at the required order), we have

$$\begin{cases} z^i = \dfrac{m'}{m_{\mathrm{T}}} RN^i \,, \quad v^i = \dfrac{m'}{m_{\mathrm{T}}} V^i \,, \quad \dot{v}^i = -m'\dfrac{N^i}{R^2} \,, \\[2mm] \qquad\qquad \ddot{v}^i = -m'\dfrac{V^i}{R^3} + 3m'\dfrac{(N.V)N^i}{R^3} \,, \\[3mm] z'^i = -\dfrac{m}{m_{\mathrm{T}}} RN^i \,, \quad v'^i = -\dfrac{m}{m_{\mathrm{T}}} V^i \,, \quad \dot{v}'^i = m\dfrac{N^i}{R^2} \,, \\[2mm] \qquad\qquad \ddot{v}'^i = m\dfrac{V^i}{R^3} - 3m\dfrac{(N.V)N^i}{R^3} \,, \end{cases} \qquad (14.25)$$

where $m_{\mathrm{T}} = m + m'$, $R^i \equiv z^i - z'^i$, $R^i \equiv RN^i$, $V^i \equiv v^i - v'^i = \dot{R}^i$, and $(N.V) \equiv N^i V_i$. Therefore, (14.24) becomes

$$\begin{cases} \dot{\gamma}^1_{tt} = -\dfrac{4mm'}{rR^2}\left[(N.V) + 4(n.V)(n.N) - 3(n.N)^2(N.V)\right]_{\mathcal{R}_0} + \cdots \,, \\[3mm] \dot{\gamma}^1_{ti} = \dfrac{4mm'}{rR^2}\left[2(n.V)N_i + 2(n.N)V_i - 3(n.N)(N.V)N_i\right]_{\mathcal{R}_0} + \cdots \,, \\[3mm] \dot{\gamma}^1_{ij} = -\dfrac{4mm'}{rR^2}\left[N_iV_j + N_jV_i\right]_{\mathcal{R}_0} + \cdots \,. \end{cases} \qquad (14.26)$$

• *Contribution of the post-linear order*

If, without further ado, we use (14.26) to calculate the flux at infinity $n_i t^{ti}_{\mathrm{LL}}$ with $t^{\mu\nu}_{\mathrm{LL}}$ given in (14.4), we will *not* obtain the quadrupole formula (14.11) because, as we have stressed repeatedly, the potentials (14.26) must be completed by the $\mathcal{O}(m^2)$ contributions from the second iteration of the Einstein equations. We shall calculate these contributions in Chapter 15 [see (15.11)], where we will find that we must add to (14.26) the following:

$$\begin{cases} \dot{\gamma}^{(2)}_{tt} = \dfrac{4mm'}{rR^2}(N.V)_{\mathcal{R}_0} \,, \quad \dot{\gamma}^{(2)}_{ti} = 0 \,, \\[3mm] \dot{\gamma}^{(2)}_{ij} = -\dfrac{4mm'}{rR^2}\left[N_iV_j + N_jV_i - 3N_iN_j(N.V)\right]_{\mathcal{R}_0} \,. \end{cases} \qquad (14.27)$$

Therefore, the derivatives of the asymptotic metric are not (14.26), but rather the sum of (14.26) and (14.27), that is

$$\begin{cases} \dot{\gamma}_{tt} = (n^i n^j \dot{\gamma}_{ij}) \,, \quad \dot{\gamma}_{ti} = -(n^j \dot{\gamma}_{ij}) \,, \\[3mm] \dot{\gamma}_{ij} = -\dfrac{4mm'}{rR^2}\left[2(N_iV_j + N_jV_i) - 3N_iN_j(N.V) + \mathcal{O}(1/r) + \mathcal{O}(v^2) + \mathcal{O}(m)\right]_{\mathcal{R}_0} \,. \end{cases} \qquad (14.28)$$

Using the spatial derivatives in (14.24), we then have $\partial_\alpha \gamma_{\mu\nu} \equiv (\dot{\gamma}_{\mu\nu}, \partial_i \gamma_{\mu\nu}) = -n_\alpha \dot{\gamma}_{\mu\nu}$ with $n^\mu = (1, n^i)$.

• *The second quadrupole formula revisited*

Although the metric we have obtained is not in the TT gauge, it is easy to calculate the flux at infinity $n_i t_{LL}^{ti}$ using (14.4):

$$
\frac{d}{dt}E = -\lim_{r\to\infty}\int |g| t_{LL}^{ti} n_i r^2 d\Omega = -\frac{r^2}{8\kappa}\int (2\dot{\gamma}_{\mu\nu}\dot{\gamma}^{\mu\nu} - \dot{\gamma}^2)d\Omega
$$

$$
= -\frac{r^2}{8\kappa}\int [n_i n_j n_k n_l \dot{\gamma}^{ij}\dot{\gamma}^{kl} - 2n_i n_j (2\dot{\gamma}^{ik}\dot{\gamma}_k^j - \dot{\gamma}^{ij}\dot{\gamma}_k^k) + 2\dot{\gamma}_{ij}\dot{\gamma}^{ij} - (\dot{\gamma}_i^i)^2]d\Omega
$$

$$
= -\frac{r^2}{60}[3\dot{\gamma}_{ij}\dot{\gamma}^{ij} - (\dot{\gamma}_i^i)^2] = -\frac{8m^2 m'^2}{15R^4}[12V^2 - 11(N.V)^2]_{\mathcal{R}_0}
$$

$$
= -\frac{G}{45c^5}\frac{d^3 Q^{ij}}{dt^3}\frac{d^3 Q_{ij}}{dt^3}.
$$

(14.29)

Here we have restored G and c, and Q_{ij} is the quadrupole moment (evaluated at the retarded time $t - r$) of a system of two point masses interacting gravitationally.[6]

Therefore, as announced, the quadrupole formula is valid even for bodies which interact gravitationally. [However, (14.27) still needs to be proved.]

Let us finish calculating $\dot{E}$. Since in the order considered the trajectories are Keplerian (*cf.* Book 1, Section 12.2), we have

$$
R = a(1 - \cos\eta) = \frac{a(1 - e^2)}{1 + e\cos\phi}, \quad nt = \eta - e\sin\eta \quad \text{with} \quad n^2 = \frac{m_T}{a^3},
$$

where a and e are the semi-major axis and the eccentricity of the relative orbit in the center-of-mass frame of the bodies of masses $m \equiv GM$ and $m' \equiv GM'$. We easily find

$$
\begin{cases}
(N.V) = \dfrac{na}{\sqrt{1-e^2}}e\sin\phi, \quad V^2 = \dfrac{n^2 a^2}{1-e^2}(1 + e^2 + 2e\cos\phi), \\[2mm]
\dot{\phi} = \dfrac{n}{(1-e^2)^{3/2}}(1 + e\cos\phi)^2.
\end{cases}
$$

(14.30)

Inserting (14.30) into (14.29), we obtain the average of $\dot{E}$ over a period $P = 2\pi/n$: $\langle \dot{E}\rangle \equiv P^{-1}\int_0^P \dot{E}\,dt = P^{-1}\int_0^{2\pi}\dot{E}\,d\phi/\dot{\phi}$, or

$$
\langle \dot{E}\rangle = -\frac{n}{2\pi}\int_0^{2\pi}\frac{8G^4 M^2 M'^2}{15a^2 c^5 (1-e^2)^{7/2}}(1 + e\cos\phi)^2[12(1 + e^2 + 2e\cos\phi) - 11e^2\sin^2\phi]\,d\phi
$$

$$
= -\frac{32G^4}{5c^5}\frac{M^2 M'^2(M + M')}{a^5}f(e)
$$

(14.31)

$$
\text{with}\quad f(e) = \left(1 + \frac{73}{24}e^2 + \frac{37}{96}e^4\right)(1 - e^2)^{-7/2}
$$

(Peters and Mathews, 1963).

---

[6]Indeed, we have $Q_{ij} = \sum m(3z_i z_j - \delta_{ij}z^2)$ or, in the center-of-mass frame, $Q_{ij} = (mm'/M)[3R_i R_j - \delta_{ij}(R.R)]$, which, using Newton's equations of motion (14.25), gives $d^3 Q_{ij}/dt^3 = (2mm'/R^2)[\delta_{ij}(N.V) + 9(N.V)N_i N_j - 6(N_i V_j + N_j V_i)]$. The square of this is $24(m^2 m'^2/R^4)[12V^2 - 11(N.V)^2]$ (we have already used this expression in Book 2, Section 21.2 to study the electromagnetic radiation of two charges).

Orders of magnitude

The energy (14.29) lost by a massive system in the form of gravitational waves is small.

Let us consider the example of a rod of mass m and length l, rotating about its center with frequency ω. We have $Q \sim Ml^2$, from which we find $d^3Q/dt^3 \sim Ml^2\omega^3$, and so $\dot{E} \sim -(G/c^5)M^2l^4\omega^6$. For $M = 1$ kg, $l = 1$ m, and $\omega = 1$ rad/s we obtain $\dot{E} \approx 10^{-54}$ W, which can be compared to, for example, thermal noise at 1 K: 10^{-23} W.

If the source is a stellar binary, we will have $\dot{E} \sim m^2R^4\omega^6$ with, according to Newton's law, $\omega^2 \sim m/R^3$ or $\dot{E} \sim (m\omega)^{10/3}$. Using the data on PSR 1913+16 (see footnote 7 of the preceding chapter), we obtain $\dot{E} \sim 10^{27}$ W. We can compare this, for example, to the (electromagnetic!) luminosity of the Sun: $L_\odot \sim 4 \times 10^{26}$ W (see, for example, Book 1, Section 15.2).

Back reaction of the radiation on the motion

The Newtonian energy of a binary system of two stars of equal masses in a circular orbit is $E_n = -m^2/(2R) \sim -(m^5\omega^2)^{1/3}$ (because $\omega^2 \sim m/R^3$ by Newton's law). We can therefore estimate the lifetime of the system by writing $\dot{E}_n = \dot{E}$, where $\dot{E}$ is given by the quadrupole formula (14.29). Then since $\dot{E} \sim -(m\omega)^{10/3}$ (see above), we have $\dot{R} \sim R^2\dot{E}/m^2 \sim -(m\omega)^2$, or also $\dot{\omega} \sim (m^5\omega^{11})^{1/3}$. For the data on PSR 1913+16 (see footnote 7 of Section 13.4) we find $\dot{\omega} \sim 10^{-20}$ s^{-2}. More precisely, we have $P_b = 2\pi/n$ and $a = GMM'/(-2E_n)$, and using (14.31) we obtain

$$\left\langle \frac{\dot{P}_b}{P_b} \right\rangle = -\frac{96G^{5/3}}{5c^5} \frac{MM'}{(M+M')^{1/3}} \frac{f(e)}{(P_b/2\pi)^{8/3}} \approx -\frac{2.4 \times 10^{-12}}{P_b} \text{ s}^{-1}. \tag{14.32}$$

As stated above in Section 12.6, this effect must be included in the timing formula of a binary system in which one of the masses emits electromagnetic signals at regular intervals. The relation (12.35) between the proper time of the emitter and the anomaly η then becomes

$$\eta - e_T \sin\eta = 2\pi \left(\frac{\tau_e - T}{P_b} - \frac{1}{2}\dot{P}_b \left(\frac{\tau_e - T}{P_b} \right)^2 \right). \tag{14.33}$$

However, it should be borne in mind that this expression still needs to be justified, because, *first of all*, it is based on the equations (14.28) for the radiative field including the contribution (14.27) of the second iteration of the Einstein equations which still needs to be found, and, *secondly*, it assumes that a balance has been established between the radiated power and the loss of mechanical energy of the system, the validity of which also needs confirmation; see the following chapter.

Book 3

15

The two-body problem and radiative losses

In the preceding chapter we showed that it is necessary to solve the Einstein equations in (at least) the post-linear approximation in order to be able to rigorously deal with the problem of the gravitational radiation of a system of self-gravitating bodies.

In this chapter we shall begin by finding the field created by compact objects in the post-linear approximation of general relativity. This will allow us, first of all, to obtain the radiative field of a self-gravitating system [eqn (15.11)], and thus to complete the analysis of the preceding chapter, where the contribution of the second iteration was left hanging. The second quadrupole formula (14.29) will then be completely proven.

Next, we shall find the equations of motion of the bodies in the field which they create to second order in the perturbations assuming that their velocities are small. We shall thereby see that to correctly describe the radiation reaction—the effect of the emitted gravitational radiation on their motion—at 2.5 PN order it will prove necessary to iterate the Einstein equations a third time. This will lead us to the equations of motion (15.17) and (15.18), which generalize to order $1/c^5$ the EIH equations of order $1/c^2$ already found in (12.25).

Finally, we shall study the effect of the radiation reaction force on the sources, and will show, within the framework of a Lagrangian formalism, that there is an energy balance at 2.5 PN order between the energy radiated to infinity and the mechanical energy lost by the system. This will justify the heuristic calculation of the radiation reaction effect on the source motion carried out at the end of the preceding chapter, eqns (14.32) and (14.33).

15.1 The gravitational field in the post-Minkowski approximation

To simplify our discussion,[1] we shall restrict ourselves to the case of two bodies which are sufficiently compact and far enough apart that their internal structure can be ignored. It is therefore heuristically convenient (as long as we limit ourselves to the post-linear approximation) to represent them by the distributional energy–momentum tensor

$$T^{\mu\nu}(x^\rho) = \sum m \int \delta_4(x^\rho - z^\rho(\tau)) \frac{\tilde{u}^\mu \tilde{u}^\nu}{\sqrt{-g}} d\tilde{\tau} = \sum m \int \delta_4(x^\rho - z^\rho(\tau)) \frac{u^\mu u^\nu}{\sqrt{g\, g_{\rho\sigma} u^\rho u^\sigma}} d\tau. \quad (15.1)$$

---

[1] Here we shall limit ourselves to only a discussion of the method. The method we choose is termed *post-Minkowskian* because it consists of iterating the Einstein equations in powers of G (or in the masses $m = GM/c^2$). It is only after this is done that a second expansion is made in the source velocities v, which are assumed to be small. And only at the end of the calculation do we use the equations of motion, which in the order considered here will be Newtonian: $\dot{v} \sim m/R^2$ and $v^2 \sim m/R$. This approximation scheme therefore differs from the *post-Newtonian* approximation schemes that we have described in Chapter 12 (at the lowest order), where we assumed *ab initio* that the velocities are small.

A complete discussion of the results can be found in Bel *et al.* (1981), as well as Damour (1982). See also Damour (1987).

Relativity in Modern Physics. Nathalie Deruelle and Jean-Philippe Uzan.
© Oxford University Press 2018. Published in 2018 by Oxford University Press.
DOI: 10.1093/oso/9780198786399.001.0001

In the first expression $\tilde{u}^\mu = dz^\mu/d\tilde\tau$ is their 4-velocity normalized using the metric $g_{\mu\nu}$ (see Section 3.2), and in the second, which is more useful for a post-Minkowski expansion, the 4-velocity $u^\mu = dz^\mu/d\tau$ is normalized using $\eta_{\mu\nu}$. The quantity δ_4 is the Dirac delta distribution. Finally, m and m' are their masses (we set $G = 1$).

We have solved this problem in linear order in Section 14.3, where the required definitions can be found. Let us assemble the results we have obtained. In first order in the perturbations where $g_{\mu\nu} = \eta_{\mu\nu} + \gamma_{\mu\nu} - \frac{1}{2}\gamma\eta_{\mu\nu}$, we have

$$T_1^{\mu\nu} = \sum m \int \delta_4(x^\rho - z^\rho(\tau))u^\mu u^\nu d\tau , \quad \gamma_1^{\mu\nu}(x^\rho) = \sum 4m\frac{u^\mu u^\nu}{\varrho}\bigg|_\mathcal{R} ,$$

$$\partial_\nu \gamma_1^{\mu\nu} = \sum 4m \left(\frac{\dot u^\mu}{\varrho}\right)_\mathcal{R} . \tag{15.2}$$

Now let us analyze the basic structure of the approximation scheme.

To linearize (in Section 13.1) and then iterate (in Section 14.1) the Einstein equations we have set $g = \eta + \gamma$ (here $=$ signs have only symbolic meaning). The Christoffel symbols therefore have the form $\Gamma = g\,\partial\gamma = (\eta + \gamma)\partial\gamma = \partial\gamma + \gamma\partial\gamma$, and the Einstein tensor is $G = \partial\Gamma + \Gamma\Gamma = \Box\gamma + (\partial\gamma)^2 + \gamma\partial^2\gamma + \mathcal{O}(\gamma^3)$. [Its exact expression is given in (14.2).] The energy–momentum tensor is written as $T = \sum m \int d\tau\, \delta_4 uu\left(1 + \gamma + \mathcal{O}(\gamma^2)\right)$.

In first order in γ the Einstein equations $G = T$ reduce to $\Box\gamma = -\sum m \int d\tau \delta_4 uu$, the solution of which is $\gamma_1 = (m/r)uu$, where uu is of order 1, v, or v^2 depending on the components [their exact expression is given in (15.2) or (14.21)].

At second order in γ the Einstein equations have the structure

$$\Box\gamma = (\partial\gamma_1)^2 + \gamma_1\Box\gamma_1 - 4\pi \sum m \int d\tau\, \delta_4(x - z)\,(1 + t_1) . \tag{15.3}$$

[Their exact expression is given in (14.3) and (14.4), and $t_1 = \frac{1}{2}\gamma_{\mu\nu}u^\mu u^\nu - \frac{1}{4}\gamma$.] As the example below shows, when the linear solution $\gamma_1 = (m/r)uu$ has been substituted into them, these equations are neither defined nor integrable.[2]

Regularization of the field equations

Let us study the paradigmatic example (15.3) more closely. At linear order $\gamma_1 = \sum m/\varrho_\mathcal{R}$, where $\varrho_\mathcal{R}$ was defined in (14.14) (see also Fig. 14.1). The quadratic part splits up as $\gamma_2 = \gamma_2^\times + \gamma_2^s$, where the cross terms and self terms must satisfy

---

[2] Taking into account the form of γ_1, we however see immediately (by integrating, for example, the regular term of order $m\gamma_1$ in the first term) that the metric at second order has the form $\gamma = \gamma_1 + \gamma_2$, where γ_2 contains terms of the type $(m/r)(m/R)$ with R the separation of the two masses. Therefore, the metric can contain terms of the type $(m/r)v^2$ originating from the spatial components of γ_1, as well as terms of the type $(m/r)(m/R)$ coming from γ_2. Since $v^2 \sim m/R$, we again find that we need to go to second order in the perturbations to describe the gravitational radiation of a self-gravitating body.

$$\begin{cases} \Box\gamma_2^\times = \sum mm' \left(\partial_\mu \frac{1}{\varrho_\mathcal{R}} \, \partial^\mu \frac{1}{\varrho'_\mathcal{R}} + \frac{1}{\varrho'_\mathcal{R}} \Box \frac{1}{\varrho_\mathcal{R}} - 4\pi \int d\tau \delta_4(x^\rho - z^\rho(\tau)) \frac{1}{\varrho'_\mathcal{R}} \right) , \\ \\ \Box\gamma_2^s = \sum m^2 \left(\partial_\mu \frac{1}{\varrho_\mathcal{R}} \, \partial^\mu \frac{1}{\varrho_\mathcal{R}} + \frac{1}{\varrho_\mathcal{R}} \Box \frac{1}{\varrho_\mathcal{R}} - 4\pi \int d\tau \delta_4(x^\rho - z^\rho(\tau)) \frac{1}{\varrho_\mathcal{R}} \right) . \end{cases} \tag{15.4}$$

These equations are neither defined nor integrable without the use of a regularization procedure.

The last source term for γ_2^s, which comes from $\sum m \int d\tau \delta_4 \gamma_1$, was dealt with in Book 2, Section 10.4 when we studied the Nordström theory: we regularized the quantity $\delta_4(x - z)/\varrho_\mathcal{R}$, which is undefined because $\varrho_\mathcal{R}$ vanishes at $x = z$, by setting it equal to zero. To be consistent, we must do the same with the penultimate term $(1/\varrho_\mathcal{R})\Box(1/\varrho_\mathcal{R})$ because $\Box(1/\varrho_\mathcal{R}) \propto \delta_4(x - z)$. The antepenultimate term, which can also be written as $1/\varrho_\mathcal{R}^4$ in the order in which we are working, is not integrable, but we see that if we require that the Leibniz rule hold, we will have $\Box(1/2\varrho_\mathcal{R}^2) = (1/\varrho_\mathcal{R})\Box(1/\varrho_\mathcal{R}) + 1/\varrho_\mathcal{R}^4 = 1/\varrho_\mathcal{R}^4$ by consistency with the previous regularizations.

The last two source terms for $\gamma_2^\times$ are equal and integrable. For the integration we use (see the example of the Nordström theory in Book 2, Section 10.4 and Fig. 14.1) $\Box(1/\rho_\mathcal{R}) = -4\pi \int d\tau \delta_4(x - z)/\varrho_\mathcal{R}$, where $\rho \equiv -(z^\mu - \hat{z}'^\mu)\hat{u}'_\mu$. Here $\hat{z}'^\mu$ is the intersection of the cone with apex at a point z^μ of the world line of the body m and the world line of m', and $\hat{u}'^\mu$ is the velocity (see the compendium of definitions of retarded quantities in Section 14.3). Finally, to integrate the first source term of $\gamma_2^\times$ we again impose the Leibniz rule, writing it as $2\partial_\mu(1/\varrho_\mathcal{R})\partial^\mu(1/\varrho'_\mathcal{R}) = \Box(1/(\varrho'_\mathcal{R}\varrho_\mathcal{R})) - (1/\varrho'_\mathcal{R})\Box(1/\varrho_\mathcal{R}) - (1/\varrho_\mathcal{R})\Box(1/\varrho'_\mathcal{R})$.

In the end we find

$$\gamma_2 = \sum m^2 \left(\frac{1}{\varrho^2} \right)_\mathcal{R} + \sum mm' \left(\frac{1}{2\varrho\varrho'} + \frac{1}{\varrho\rho} \right)_\mathcal{R} . \tag{15.5}$$

The lesson to be learned from this example is that an iteration scheme for the Einstein equations must be supplemented by coherent, unambiguous rules for the regularization and integration. At the order to which we limit ourselves here, the rather offhand *Hadamard regularization* illustrated by the above example (15.4) and (15.5) is sufficient.[3]

Having specified the regularization rules, after integration we find in the end that the metric describing the gravitational field of two gravitationally interacting compact bodies is, at second order in the post-Minkowski iteration scheme, $g_{\mu\nu} = \eta_{\mu\nu} + h_{\mu\nu}$, where the $h_{\mu\nu}$ are related to $\gamma_{\mu\nu}$ by the expansion (14.2) with $\gamma_{\mu\nu} = \gamma_{\mu\nu}^1 + \gamma_{\mu\nu}^2$. The linear term $\gamma_1^{\mu\nu}$ was given in (15.2), and the quadratic term is $\gamma_2^{\mu\nu} = \gamma_{2T}^{\mu\nu} + \gamma_{2s}^{\mu\nu} + \gamma_{2\times}^{\mu\nu}$ with [the main definitions are given in (14.15) and (14.16)]

$$\begin{cases} \gamma_{2T}^{\mu\nu} = 4mm' \sum (1 + 2w_\mathcal{R}^2) \left(\frac{u^\mu u^\nu}{\varrho\rho} \right)_\mathcal{R} , \\ \\ \gamma_{2s}^{\mu\nu} = \sum m^2 \left(\frac{7u^\mu u^\nu + n^\mu n^\nu}{\varrho^2} \right)_\mathcal{R} , \\ \\ \gamma_{2\times}^{\mu\nu} = 16mm' \sum \mathcal{D}^{\mu\nu} P . \end{cases} \tag{15.6}$$

---

[3]This is not the case at higher orders. To extract from the third iteration (of order m^3) the information needed to find the equations of motion at order $mv^5 \sim m^2v^3/R \sim m^3v/R^2$ (see below), an analytic continuation procedure based on the Riesz potentials must be used; see Damour (1982) and (1987). Beyond this, dimensional regularization procedures have been shown to be the most effective; see Damour (2006), and also Blanchet (2006).

The first term $\gamma_{2T}^{\mu\nu}$ is easily found: it is the retarded integral of the energy–momentum tensor ignoring the self term proportional to $\delta_4(x-z)/\varrho_{\mathcal{R}}$. At the order (m^2) to which we limit ourselves here, the velocities u^μ and u'^μ, as well as $w = (u\,.u')$, can be assumed constant.

The second term $\gamma_{2s}^{\mu\nu}$ is obtained just as easily when the velocities can be assumed constant, because, no matter which regularization procedure is used, near each body it must be identical to the expansion of the Schwarzschild metric (in harmonic coordinates, given in Section 11.1) owing to the *effacement principle*; see the following box.

In the final term $\gamma_{2\times}^{\mu\nu}$ the quantity $\mathcal{D}^{\mu\nu}$ is a second-order differential operator whose structure follows directly from that of $\tau^{\mu\nu}$ given in (14.4) and (14.5). It acts on the Fock function P satisfying $\Box P = 1/(\varrho_{\mathcal{R}}\varrho'_{\mathcal{R}})$ (P itself diverges, but only its derivatives are relevant here and they are finite); see the details in the following.

The effacement principle

General relativity shares a property with the Newton and Nordström theories: the *effacement principle*, according to which the gravitational field outside a spherical body as well as the motion of the body are independent of the internal structure of the body, and corrections due to the presence of a distant body arise only at higher orders.

In Newtonian gravity, where the correction to the $1/r^2$ law is $\delta F/F \sim (b/R)^5$, with b the radius of the bodies and R their separation (see, for example, Book 1, Section 14.3), the efface-ment follows from Newton's theorem stating that the field outside a spherical body is the same as if all the mass were concentrated at the center of the body (see Book 1, Section 11.4). As we have seen in Book 1, Sections 12.4 and 13.1, this theorem is valid only when the $1/r^2$ law holds and is related to the *strong equivalence principle*, that is, the equality of the gravitational and inertial masses for all bodies. This principle is itself related to the linearity of the theory; see the example of the Nordström theory in Book 2, Section 10.4.

To paraphrase an observation made by Brillouin in 1922, *in Newtonian gravitation we can efface the contribution of the moving body to determine the force which moves it. It does not appear evident that we can do the same in the Einstein theory when the body is an object or point particle of finite mass.*[4] The validity of the effacement principle in general relativity was proved in 1981 by Damour, who showed that corrections to the metric due to the internal structure of compact bodies arise only at order $(GM)^6$ (see the references cited in footnote 1). This result therefore justifies the description of sources by 'skeletonized', *i.e.*, distributional, energy–momentum tensors.

The relativistic Fock function

Let us specify the form of the cross term $\gamma_{2\times}^{\mu\nu}$ of the metric (15.6) in the post-linear, adiabatic approximation [that is, neglecting terms proportional to $mm'\dot{u}$ of order $\mathcal{O}(m^3)$].

Using $\gamma_1^{\mu\nu}$ from (15.2), the source of $\gamma_{2\times}^{\mu\nu}$ is a sum of various contractions of $mm'\partial_\rho(u'^\alpha u'^\beta/\varrho')\partial_\sigma(u^\gamma u^\delta/\varrho)|_{\mathcal{R}}$, the exact expression for which can be derived from (14.4) and (14.5). We introduce the *Fock function* P formally defined as $\Box P = 1/(\varrho_{\mathcal{R}}\varrho'_{\mathcal{R}})$. P diverges, but its derivatives $D_\alpha P$ and $D'_\alpha P$ (where $D_\alpha P$ is the variation of P when the worldline is displaced), defined as

$$\Box D_\alpha P = \left(\frac{1}{\varrho'_{\mathcal{R}}}\right)\partial_\alpha\left(\frac{1}{\varrho_{\mathcal{R}}}\right), \quad \Box D'_\alpha P = \left(\frac{1}{\varrho_{\mathcal{R}}}\right)\partial_\alpha\left(\frac{1}{\varrho'_{\mathcal{R}}}\right), \tag{15.7}$$

---

[4]Cited in Levi-Civita (1935).

are finite (we thus have $-\partial_\alpha + D_\alpha + D'_\alpha = 0$). Thus, the differential operator $\mathcal{D}^{\mu\nu}$, whose structure follows directly from the form of $\tau^{\mu\nu}$ given in (14.4) and (14.5), is

$$\mathcal{D}^{\mu\nu} \equiv \frac{w^2}{2} D'^\mu D^\nu - \frac{1}{4} D'^\mu D^\nu + w u'^\mu u^\nu D'_\alpha D^\alpha + \frac{1}{8} \eta^{\mu\nu} D'_\alpha D^\alpha - \frac{w^2}{4} \eta^{\mu\nu} D'_\alpha D^\alpha + \frac{w}{2} \eta^{\mu\nu} u'_\beta u^\alpha D'_\alpha D^\beta$$

$$- w u'^\alpha u^\nu D'^\mu D_\alpha - w u'^\alpha u^\mu D'^\nu D_\alpha + u'^\mu u^\nu u'^\beta u^\alpha D'_\alpha D_\beta - u'^\alpha u'^\beta u^\mu u^\nu D_\alpha D_\beta .$$

$$(15.8)$$

The explicit expression for the function $D_\alpha P$ is known;[5] see the references cited in footnote 1. It possesses the following properties [which can be derived from the example (15.4)]:

$$D_\alpha D^\alpha P = \left(\frac{1}{\varrho \rho} \right)_\mathcal{R} , \quad D'_\alpha D^\alpha P = \frac{1}{2} \left(\frac{1}{\varrho \varrho'} - \frac{1}{\varrho \rho} - \frac{1}{\varrho' \rho'} \right)_\mathcal{R} , \quad u^\alpha D_\alpha P = 0 .$$

We still must impose on the solution (15.6) the harmonic condition $\partial_\nu(\sqrt{-g}g^{\mu\nu}) = 0$ or $\partial_\nu \gamma^{\mu\nu} = 0$, which, we recall, allows the Einstein equations to be written in the form (15.3) and thus be integrated. We therefore must have $\partial_\nu(\gamma^1_{\mu\nu} + \gamma^2_{\mu\nu}) = \mathcal{O}(m^3)$. The derivatives of retarded quantities are calculated using (14.15) and (14.16) and the properties (15.7) and (15.8) of the Fock function. At the order we are working in, we can neglect the curvature of the world lines of the bodies, that is, the terms proportional to $\dot{u}$, and so on, in $\partial_\nu(\gamma^2_{\mu\nu})$. We finally obtain the following [where we recognize the first term as $\partial_\nu(\gamma^1_{\mu\nu})$ given above in (15.2)]:

$$\sum \frac{4m}{\varrho \mathcal{R}} (\dot{u}^\mu - \Gamma^\mu_1) \big|_\mathcal{R} = \mathcal{O}(m^3)$$

$$(15.9)$$

$$\text{with} \quad \Gamma^\mu_1 = \frac{m'}{\rho^2} \left[(1 - 2w^2)(\nu^\mu + (\nu.u)u^\mu) + 4w(\nu.u)(\hat{u}'^\mu + w\hat{u}^\mu) \right] .$$

We recall that $\dot{u}^\mu$ is the 4-acceleration of the body m at the point z^μ of its world line, and $\hat{u}'^\mu$ is the 4-velocity of the body m' (normalized to -1 using $\eta_{\mu\nu}$) at the retarded point $\hat{z}'^\mu$, the intersection of the cone with apex at z^μ and its world line L' (however, since at the order we are working in we can ignore terms proportional to $\dot{u}'^\mu$, $\hat{z}'^\mu$ can be located anywhere on L'). Finally, $w = (u.\hat{u}')$, $\rho = -(z^\mu - \hat{z}'^\mu).\hat{u}'_\mu$, and $\nu^\mu = -\hat{u}'^\mu + (z^\mu - \hat{z}'^\mu)/\rho$.

The harmonic condition therefore *imposes* the equation of motion of the bodies: $\dot{u}^\mu = \Gamma^\mu_1 + \mathcal{O}(m^2)$.

Now, because of the effacement principle, this equation of motion *must* also follow from the geodesic equation $\dot{u}^\mu = -u^\alpha u^\beta (\Gamma^\mu_{\alpha\beta} + u^\mu u_\rho \Gamma^\rho_{\alpha\beta})$, where the $\Gamma^\mu_{\nu\rho}$ are the Christoffel symbols or, at the order in which we are working,

---

[5] $D^\alpha P$ is a function of the field point x^μ and the associated retarded points on the world lines of the two bodies $x^\mu_\mathcal{R}$ and $x'^\mu_\mathcal{R}$ (see Fig. 14.1). It is an integral on the world line of m' taken from the doubly retarded point $\hat{z}'_\mathcal{R}$ associated with $x^\mu_\mathcal{R}$ to the retarded point $x'^\mu_\mathcal{R}$. At the adiabatic order we are considering, we can take the world line of m' to be a straight line, that is, set $u'^\mu_\mathcal{R} = \hat{u}'^\mu_\mathcal{R} \equiv u'^\mu$. The (somewhat experienced) reader will find that $D^\alpha P = \int_{\hat{z}'_\mathcal{R}}^{x'_\mathcal{R}} \frac{\beta^\alpha}{\varrho_\mathcal{R} \beta^2}$ with $\beta^\alpha \equiv R'^\alpha - \varrho_\mathcal{R} n^\alpha_\mathcal{R} - n^\alpha_\mathcal{R} \sqrt{(R' - n_\mathcal{R} \varrho_\mathcal{R})^2}$ and $R'^\alpha \equiv (x^\alpha - z'^\alpha(\tau')) + u^\alpha_\mathcal{R} \{ [x^\mu - z'^\mu(\tau')] .u^\mathcal{R}_\mu \}$.

$$\dot{u}^\mu = \Gamma_1^\mu + \mathcal{O}(m^2) \quad \text{with}$$

$$\Gamma_1^\mu = -u^\alpha u^\beta \left(\partial_\beta \gamma_\alpha^\mu - \frac{1}{2} \partial^\mu \gamma_{\alpha\beta} \right) + \frac{1}{4} \partial^\mu \gamma + u^\mu u^\alpha \left(\frac{1}{4} \partial_\alpha \gamma - \frac{1}{2} u^\beta u^\delta \partial_\alpha \gamma_{\beta\delta} \right),$$

(15.10)

where Γ_1^μ, which diverges when evaluated on the world line of m using the metric $\gamma_{\mu\nu}^1$, must be regularized in accordance with the previously established rules, which amounts to ignoring the entire contribution of the body m to the gravitational field it creates. Then the calculation is easy and shows that Γ_1^μ derived from (15.10) is indeed the same as that given in (15.9) originating from the harmonic condition.

Finally, it can be shown, always using the same regularization rules, that eqns (15.9) and (15.10) are also equivalent to the conservation of the energy–momentum tensor $D_\nu(|g|T^{\mu\nu}) = 0$, where $T^{\mu\nu}$ was given in (15.1).

In fact it is this consistency of the approximation scheme which justifies the regularization procedures we have used.

We shall conclude this section by giving the asymptotic expression for the metric (15.6) which we have used in Section 14.3. The self part $\gamma_{2\mathrm{s}}^{\mu\nu}$, proportional to $1/r^2$, does not contribute to the radiative field; this is not surprising since it represents the static, Schwarzschild part of the field. Finding the radiative part of $\gamma_{2\mathrm{T}}^{\mu\nu}$ at lowest order in the velocities does not pose any difficulty and is done using (14.18) and (14.19) giving the velocity expansions of the retarded quantities. Finding the radiative part of $\gamma_{2\times}^{\mu\nu}$ requires expansion of the function $D^\alpha P$, defined in (15.7) and footnote 5, in $1/r$ and in the velocities. At lowest order the calculation gives $D^i P = (N^i - n^i)/(2r)$, $D^t P = n_i D^i P$. Therefore, the contribution, at lowest order in the velocities, of the second iteration to the radiative field finally turns out to be

$$\gamma_{2\mathrm{T}}^{tt} = \frac{24 mm'}{rR}, \quad \gamma_{2\times}^{tt} = -\frac{28 mm'}{rR}, \quad \gamma_{2\times}^{ij} = -\frac{4 mm' N^i N^j}{rR},$$

(15.11)

with all the other components equal to zero. The time derivatives of these expressions were used in the preceding chapter [see (14.27)] to show that the second quadrupole formula is valid also for self-gravitating systems.

Finally, we note that if we assume that the velocities of the bodies are small, the expansion of Γ_1^μ in the equation of motion (15.10) (which is easily done because at the order in which we are working the four velocities can be assumed constant) again gives the EIH equations of motion obtained in Section 12.3 *except for* the terms of order (m^2).

15.2 Equations of motion at 2.5 PN order

Here we shall limit ourselves to summarizing the results and indicating the steps needed to find them.

The need for a third iteration

In Section 12.3 we obtained the equations of motion of two bodies in the 1 PN approximation, that is, in order $(v^2) \sim (m)$ beyond Newtonian order. We recall that this required taking into account the nonlinearity of the Einstein equations at lowest order in the velocities.

In Section 14.3 we also saw that the loss of gravitational energy of a self-gravitating system is given by the second quadrupole formula. Finally, in Sections 12.6 and 14.3 we saw that it is necessary to include the radiation reaction in the equations of motion of the bodies and their timing formulas.

As we learned from study of the motion of interacting charges using the Maxwell theory (see Book 2, Section 20.3), the fact that this radiation is quadrupolar rather than dipolar means that the radiation reaction will affect the motion at order (v^5) beyond Newtonian order, that is, it is a 2.5 PN-order effect:

$$\dot{v} = (m/R)(1 + v^2 + v^4 + v^5) + (m/R)^2(1 + v^2 + v^3) + (m/R)^3(1 + v) + \mathcal{O}(v^8).$$

Since terms of order m^3 appear, to obtain them it is necessary to iterate the Einstein equations to *post-post-linear* order, as Eddington realized in 1924.

- *Step 1: Expansion of the Einstein equations:* $G_{\mu\nu} = 8\pi T_{\mu\nu}$

We shall choose the small parameter to be $\gamma^{\mu\nu} = \eta^{\mu\nu} - \sqrt{-g}g^{\mu\nu}$, and we shall work in a system of harmonic coordinates such that $\partial_\nu \gamma^{\mu\nu} = 0$. The Einstein tensor reduced using the harmonic condition then is [see (14.1)]

$$2|g|G^{\mu\nu} = -\Box\gamma^{\mu\nu} + \gamma^{\alpha\beta}\partial^2_{\alpha\beta}h^{\mu\nu} - \sum_n Q^{\mu\nu}_{(n)} \quad \text{if} \quad \partial_\nu\gamma^{\mu\nu} = 0, \tag{15.12}$$

where $\Box$ is the d'Alembertian in flat spacetime and $Q_{(n)}$ is a product of order n of the $\gamma^{\mu\nu}$ and their first derivatives; $Q_{(2)}$ was given in Section 14.1. The calculation of $Q_{(3)}$ poses no difficulty.

The source, namely the two compact bodies whose motion we want to describe, is described by the distributional energy–momentum tensor (15.1) which we also expand in powers of $\gamma_{\mu\nu}$:

$$|g|T^{\mu\nu} = \sum m \int d\tau\, \delta_4(x - z)u^\mu u^\nu(1 + t_{(1)} + t_{(2)} + ...). \tag{15.13}$$

Here $\eta_{\mu\nu}u^\mu u^\nu = -1$, t_1 was given in (15.3), and t_2 is easy to calculate.

- *Step 2: Integration of the Einstein equations order by order*

We write $\gamma_{\mu\nu} = \gamma^1_{\mu\nu} + \gamma^2_{\mu\nu} + \gamma^3_{\mu\nu}$ and solve the Einstein equations order by order. The linear approximation was dealt with in Section 14.1, and the post-linear approximation in Section 15.1.

The third iteration is a very tedious calculation in the post-Minkowski scheme we are using here, and its explicit form has not been obtained; $\gamma^3_{\mu\nu}$ depends *a priori* on the motion of the sources at $x_\mathcal{R}$, $\hat{z}'_\mathcal{R}$, and $z_{\mathcal{R}\mathcal{R}}$, a triply retarded point.

However, since in the end only a velocity expansion will be necessary, it is possible to extract from it the required information,[6] namely, that the corresponding harmonic condition, of third order, equivalent to the conservation of $T^{\mu\nu}$ given in (15.2), leads to the equations of motion at order (m^2), which are just the geodesic equation in the metric at second order.

---

[6]Damour (1982), and the references cited in footnote 1 of the present chapter; see also the metric (115) at 2.5 PN order in Blanchet (2006).

This equation of motion at post-linear order is written as

$$\sum \frac{4m}{\varrho_{\mathcal{R}}} \left(\dot{u}^{\mu} - (\Gamma_1^{\mu} + \Gamma_2^{\mu}) \right) \big|_{\mathcal{R}} = \mathcal{O}(m^4) \,, \tag{15.14}$$

where Γ_1^{μ} was given in (15.9). The expression for Γ_2^{μ} is obtained from the metric (15.6). It contains terms proportional to $\dot{u}'^{\mu}$ and a term of the form $-\frac{11}{3} m \ddot{u}^i$ coming from the regularization of the metric on the world line, the analog of the Abraham radiation reaction force in electromagnetism (see Book 2, Section 19.1). Then, after these accelerations and derivatives of acceleration are reduced using the equation of motion at linear order (15.9), we obtain

$$\Gamma_2^{\mu} = \frac{m'}{\rho^3} \left[(m'\alpha_s + m\alpha_{\times})(\nu^{\mu} + (\nu.u)u^{\mu}) + (m'\beta_s + m\beta_{\times})(\hat{u}'^{\mu} + wu^{\mu}) \right] \tag{15.15}$$

$$\text{with} \quad \alpha_s = 2(2w^2 + (\nu.u)^2)\,, \quad \beta_s = -2w(\nu.u)\,, \quad \text{and}$$

$$\alpha_{\times} = -2A^{-5} - 5wA^{-4} + 5(1 - 2w^2)A^{-3} + 2w(3 - 4w^2)A^{-2} + 8(w^4 + 2w^2 - 1)A^{-1}$$
$$+ 28w(2w^2 - 1) + 16(2w^2 - 1)A - 12(\nu.u)(2w^2 - 1)\ln A\,,$$

$$\beta_{\times} = -2A^{-5} - 5wA^{-4} - 2(w^2 + 16/3)A^{-3} - 2w(13 + 4w^2)A^{-2} + (4w^4 + 40w^2 - 1)A^{-1}$$
$$+ 56w(2w^2 + 1) + 176w^2 A + 64wA^2 + 4w[3 - 2w^2 + (\nu.u)^2]\ln A\,,$$

where the various quantities introduced were defined in Section 14.3 and $A \equiv -(w + (\nu.u))$. This expression[7] gives an idea of the growing complexity of the calculations, and should be compared with the *exact* equation of motion of two charges in the Maxwell theory obtained in Book 2, Section 21.3. We note that at lowest order in the velocities (where $A = 1$ and $w = -1$), the terms involving β do not contribute and $\alpha_s = \alpha_{\times} = 4$.

The equation of motion (15.14) for the mass m at z (just like the equation of motion of a charge in the Maxwell theory; see Book 2, Section 21.3) is not an ordinary differential equation but rather a *hereditary* equation, because it depends on the location of the mass m' at the retarded time $\hat{z}'$, the intersection of the cone originating at z and the world line of m' (we can no longer assume that the velocities are constant in Γ_1^{μ}).

There exist known techniques for transforming such equations into ordinary equations which are functions only of z', where z' is any point on the trajectory of m' and is ordinarily chosen to be simultaneous with z $[z^{\mu} - z'^{\mu} = (0, RN)]$. These techniques—the Lagrange expansion of Book 2, Section 20.3 and the predictive mechanics of Book 2, Section 21.3— were developed for dealing with retardation effects due to the finiteness of the speed of light in electromagnetism. Bringing Γ_1^{μ} 'back' to the simultaneous point z' amounts, at lowest order, to replacing $\alpha_s = 4$ by $\alpha_s = 5$ [as is easily seen using (21.22) of Book 2], so that Γ_2^{μ} then reduces to

$$\Gamma_2^i = \frac{m'N^i}{R^3}(5m' + 4m + \mathcal{O}(v^2))\,, \tag{15.16}$$

which is just the order-(m^2) terms of the EIH equations of motion obtained in Section 12.3. (This result in itself is actually much ado about nothing, but it should be remembered that we will soon need expansions in higher orders in the velocities.)

---

[7]The equations (15.14) and (15.15) were obtained by Westpfahl and Göller (1979).

• *Step 3: Expansion in the velocities and inclusion of* (m^3) *order*

Only a bit of care is needed to change over to three-dimensional notation (see Section 14.3) and then expand the 'predictivized' (that is, transformed into ordinary differential equations) equations of motion (15.14) and (15.15) in powers of the 3-velocities v. However, including in them the contributions of the third iteration of the Einstein equations requires additional information which can be found in the references cited in the footnotes to this chapter. The final result is that the acceleration of the body m is given by

$$\frac{dv^i}{dt} = -\frac{Gm'}{R^2}N^i + \mathcal{A}_2^i + \mathcal{A}_4^i + \mathcal{A}_5^i + \mathcal{O}(v^8)\,. \tag{15.17}$$

The post-Newtonian term $\mathcal{A}_2^i$ was obtained by Einstein, Infeld, and Hoffmann, and also by Eddington and Clark in 1938 (also by Lorentz and Droste in 1917; see footnote 6 of Section 12.3). As we have seen in Section 12.3, it contains terms proportional to $m'v^2$ arising from the linear approximation, and terms proportional to mm' and m'^2 coming from the second iteration [given in (15.16)]. The terms $\mathcal{A}_4^i$ and $\mathcal{A}_5^i$ are the post-post-Newtonian corrections to the equations of motion and contain terms coming from the third iteration of the Einstein equations. The term $\mathcal{A}_4^i$ is derived from the Lagrangian given below, and $\mathcal{A}_5^i$ (to be compared with its Maxwellian homologue studied in Book 2, Section 20.3) is

$$\begin{aligned}
\mathcal{A}_5^i = &-\frac{4mm'}{5R^3}V^2[V^i - 3N^i(N.V)] \\
&+\frac{mm'}{R^4}\left[\left(\frac{208}{15}m' - \frac{24}{5}m\right)(N.V)N^i + \left(-\frac{32}{5}m' + \frac{8}{5}m\right)V^i\right],
\end{aligned} \tag{15.18}$$

where $V^i = v^i - v'^i$. The 2 PN and 2.5 PN corrections $\mathcal{A}_4^i$ and $\mathcal{A}_5^i$ were obtained[8] in 1982.

15.3 Conservation laws and energy balance

• *Definition of the mechanical energy of a system*

Just as the post-Coulomb equations of motion of two charges are derived from the Darwin Lagrangian in the Maxwell theory (see Book 2, Section 21.1), and just as the EIH equations are derived from the Fichtenholz Lagrangian (see Section 12.4), the equations (15.17) at 2 PN order, that is, up to and excluding the term $\mathcal{A}_5^i$, are also derivable from a Lagrangian, obtained either by expanding the action of a free particle (because the motion of the bodies is geodesic motion) or, more simply, by identification. It depends not only on the positions and velocities of the sources, but also on their accelerations a^i, a residue of the hereditary nature of the equations of motion. It is written as

$$L = \sum\left(\frac{1}{2}mv^2 + \frac{mm'}{2R}\right) + L_2 + L_4 + \mathcal{O}(v^8)\,, \tag{15.19}$$

where $R^i \equiv RN^i = z^i(t) - z'^i(t)$, L_2 is the Fichtenholz Lagrangian obtained in Section 12.3, and the explicit expression for L_4 is

---

[8]Damour and Deruelle (1981); Damour (1982a).

$$L_4 = \frac{1}{16}mv^6$$

$$+\frac{mm'}{R}\left[\frac{7}{8}v^4 + \frac{15}{16}v^2v'^2 - 2v^2(v.v') + \frac{1}{8}(v.v')^2 - \frac{7}{8}(N.v)^2v'^2\right.$$

$$\left. + \frac{3}{4}(N.v)(N.v')(v.v') + \frac{3}{16}(N.v)^2(Nv')^2\right]$$

$$+\frac{m^2m'}{R^2}\left[\frac{1}{4}v^2 + \frac{7}{4}v'^2 - \frac{7}{4}(v.v') + \frac{7}{2}(N.v)^2 + \frac{1}{2}(N.v')^2 - \frac{7}{2}(N.v)(N.v')\right]$$

$$+mm'\left[(N.a)\left\{\frac{7}{8}v'^2 - \frac{1}{8}(N.v')^2\right\} - \frac{7}{4}(v'.a)(N.v')\right]$$

$$+\frac{m^2m'}{R^3}\left[\frac{1}{2}m + \frac{19}{8}m'\right] \quad + m \to m'.$$

$$(15.20)$$

[It should be noted that only the last term $(19m^2m'^2/(8R^3))$ requires information about the third iteration.] This Lagrangian depends on the accelerations of the bodies, and the Euler–Lagrange equations are (see Book 1, Section 8.1)

$$F_i \equiv \dot{p}_i - \frac{\partial L}{\partial z^i} = \mathcal{O}(v^7) \quad \text{with} \quad p_i \equiv \frac{\partial L}{\partial v^i} - \frac{dq_i}{dt} \quad \text{and} \quad q_i = \frac{\partial L}{\partial a^i}, \quad (15.21)$$

and again give the equations of motion (15.17), including the explicit expression for $\mathcal{A}_4^i$ (excluding the term $\mathcal{A}_5^i$).

Owing to the Noether theorem, we then obtain the following as in Section 12.4, but now at 2.5 PN order:

$$\frac{dP_i}{dt} = \sum F_i, \quad \text{where} \quad P_i = \sum p_i,$$

$$\frac{dJ_i}{dt} = \sum \epsilon_{ijk} z^j F^k, \quad \text{where} \quad J_i = \epsilon_{ijk} \sum (z^j p^k + v^j q^k),$$

$$(15.22)$$

$$\frac{dE}{dt} = \sum F_i v^i, \quad \text{where} \quad E = \sum (p_i v^i + q_i a^i) - L.$$

When the equations of motion (15.21) are satisfied, that is $F^i = 0$, we find that the momentum P_i, the angular momentum J_i, and the energy E of the system are constants of the motion.

● *Radiative losses*

At 2.5 PN order the force F^i is no longer zero, $F^i = m\mathcal{A}_5^i$, and the system loses energy. If we work in the center-of-mass system (which is Newtonian at this order), using $V^i = v^i - v'^i$ and $m_T = m + m'$ we have

$$\frac{dE}{dt} = -\frac{8(mm')^2}{5m_T}\frac{d}{dt}\left[\frac{V^2(N.V)}{R^2}\right] - \frac{8(mm')^2}{15R^4}\left[12V^2 - 11(N.V)^2\right]. \quad (15.23)$$

Book 3

The first term, which comes from the second iteration of the Einstein equations, is a total derivative. Its integral over the Newtonian motion, which is periodic, does not contribute. The second term comes entirely from the third iteration. Therefore, the mechanical energy lost by the system is

$$E = -\int_0^T dt \, \frac{8(mm')^2}{15R^4} \left[12V^2 - 11(N.V)^2\right] = -\frac{G}{45c^5} \int_0^T dt \left(\frac{d^3Q_{ij}}{dt^3}\right)^2, \qquad (15.24)$$

where $Q_{ij} = \Sigma m \left(3z_i z_j - \delta_{ij} z^2\right)$ is the quadrupole moment of the system. In this expression we recognize the second quadrupole formula (14.29) giving the energy radiated to infinity in the form of gravitational waves. An energy balance is therefore established between the radiated energy and the mechanical energy which is lost.

Let us summarize what we have found. The study of the gravitational field carried out in the present chapter at post-linear order has, first of all, allowed us to completely justify the validity of the second quadrupole formula giving the energy radiated by a self-gravitating system. In addition, study of the motion of the bodies has allowed us, on the one hand, to justify the regularization procedures used to obtain the EIH equations and, on the other, to find the effect of the reaction force on the motion by including the relevant contributions of the third iteration. This force leads to a loss of mechanical energy which we have shown to be equal to the radiated energy. It results in a contraction of the relative orbit of the two bodies and a time delay in the return to the periastron, as we anticipated in Section 14.3.

The binary pulsar PSR 1913+16

In 1974 Hulse and Taylor discovered the pulsar PSR 1913+16 using the Arecibo radio telescope in Puerto Rico. This pulsar is a rotating neutron star which emits a radio beam with a period of 59 ms. The pulse arrival times display quasi-periodic fluctuations which suggest that it is orbiting an invisible companion. Analysis of the data immediately revealed that it is a very compact system consisting of two objects, each a few kilometers in diameter, taking about 8 hours to complete an orbit which would be entirely contained in the interior of the Sun. It is therefore an ideal laboratory for testing theories of gravity. As we saw in Section 12.6, measurement of the advance of the periastron (~ 4 degrees per year, to be compared with the advance of 43 arcoocconds per *century* of the perihelion of Mercury) combined with measurement of the 'Einstein effect' then allows the masses of the objects to be determined: they are of order $1.4 M_\odot$ each; see Fig. 15.1 below. After making an even more accurate measurement of the arrival times, Taylor announced in 1979 that he had measured a delay of the return to the periastron, signaling the loss of energy due to the emission of gravitational waves.

The fit to the measured arrival times is in perfect agreement with the results of the timing analysis based on general relativity,[9] described above in Sections 12.6 and 14.3. In Fig. 15.1 the intersection of the line $\Delta\omega = \Delta\omega(m_c, m_e)$ and the curve $\gamma = \gamma(m_c, m_e)$ gives the masses of the two bodies. The intersection, at the same point, of the curve (thick line) $\dot{P}_b = \dot{P}_b(m_c, m_e)$ [the explicit expression for which is given in (14.32)] is the first confirmation of general relativity in the strong-field limit, and the first proof of the existence of gravitational waves. This delay of the return to the periastron is currently measured with an accuracy of $\sim 10^{-2}$, which requires the orbital elements to be known with an accuracy of 10^{-6}.

---

[9]Damour and Taylor (1992).

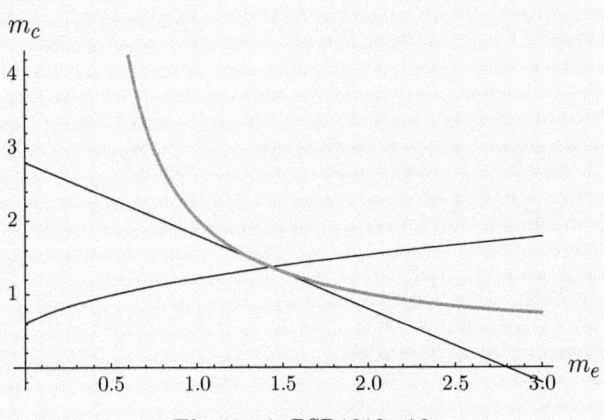

Fig. 15.1 PSR1913+16.

Hulse and Taylor were awarded the Nobel Prize in Physics in 1993 for their discovery of gravitational waves.[10]

---

[10]Since 1974, other systems of binary pulsars where general relativity can be tested have been discovered. The observational situation is reviewed in, for example, Kramer (2012). A review of the two-body problem pushed to the accuracy required for experiments to detect gravitational waves (order $1/c^7$ *beyond* the 2.5 PN approximation) can be found in, for example, Blanchet (2006).

See also the 'effective one-body approach' of Damour *et al.* in Buonanno and Damour (1999), and the discussion of current developments in Bini and Damour (2012), see also Chapter 16.

Finally, the approaches using numerical relativity are discussed in, for example, Grandclément and Novak (2009) and Baumgarte and Shapiro (2010).

All these developments were required to extract the gravitational wave signals GW150914 and those which followed from the LIGO/Virgo data and show that they were produced by the coalescence of black holes or neutron stars.

16

The two-body problem: an effective-one-body approach

Written in collaboration with Félix-Louis Julié

In this chapter we present the basics of the 'effective-one-body' approach to the two-body problem in general relativity and show that the 2PN equations of motion can be mapped, by means of an appropriate canonical transformation, to a geodesic motion in a static, spherically symmetric spacetime, thus considerably simplifying the dynamics. Then, including the 2.5PN radiation reaction force in the (resummed) equations of motion, we give the waveform (that is, the time dependence of the radiative field) during the inspiral, merger, and ringdown phases of the coalescence of two non-spinning black holes into a final Kerr black hole. We also comment on the current developments of this approach, which is instrumental in building the libraries of waveform templates that are needed to analyze the data collected by the current gravitational wave detectors.

16.1 The 2PN Hamiltonian

The two-body Lagrangian which describes the conservative part of the dynamics at 2PN order has been derived in harmonic coordinates in the previous chapter; see (15.20). Here we shall denote the positions, velocities, and accelerations by capital letters, $Z^i(t) - Z'^i(t) = RN^i$ with $(N.N) = 1$, $V^i = \frac{dZ^i}{dt}$, $A^i = \frac{dV^i}{dt}$, and we recall the expression for this Lagrangian:

$$L = -(m + m') + \frac{1}{2}mV^2 + \frac{1}{2}m'V'^2 + \frac{mm'}{R} + L_2 + L_4 + \mathcal{O}(V^8),$$

$$L_2 = \frac{1}{8}mV^4 + \frac{1}{8}m'V'^4 + \frac{mm'}{2R}\left[3(V^2 + V'^2) - 7(V.V') - (N.V)(N.V')\right] - \frac{mm'(m + m')}{2R^2},$$

$$L_4 = \frac{1}{16}mV^6 + \frac{mm'}{R}\left[\frac{7}{8}V^4 + \frac{15}{16}V^2V'^2 - 2V^2(V.V') + \frac{1}{8}(V.V')^2\right. \tag{16.1}$$

$$\left. -\frac{7}{8}(N.V)^2V'^2 + \frac{3}{4}(N.V)(N.V')(V.V') + \frac{3}{16}(N.V)^2(N.V')^2\right]$$

$$+\frac{m^2m'}{R^2}\left[\frac{1}{4}V^2 + \frac{7}{4}V'^2 - \frac{7}{4}(V.V') + \frac{7}{2}(N.V)^2 + \frac{1}{2}(N.V')^2 - \frac{7}{2}(N.V)(N.V')\right]$$

$$+mm'\left[(N.A)\left(\frac{7}{8}V^2 - \frac{1}{8}(N.V')^2\right) - \frac{7}{4}(V'.A)(N.V')\right]$$

$$+\frac{m^2m'}{R^3}\left[\frac{1}{2}m + \frac{19}{8}m'\right] + (m \leftrightarrow m') \,.$$

Relativity in Modern Physics. Nathalie Deruelle and Jean-Philippe Uzan.
© Oxford University Press 2018. Published in 2018 by Oxford University Press.
DOI: 10.1093/oso/9780198786399.001.0001

The Fokker Lagrangian

The two-body Lagrangian (16.1) can be derived in various ways. In Section 12.3 we built it from a symmetrized free-particle action (à la Fichtenholz), and in Section 15.3 we inferred it from the equations of motion obtained earlier. A third option, initiated by A. Fokker in 1929 and applied to general relativity by L. Infeld and J. Plebanski in 1960 (which we used in Book 2, Section 20.2 to obtain the Darwin Lagrangian for electromagnetism) consists in starting from the total action describing gravitational and matter fields:

$$S[g_{\mu\nu}, \Psi] = S_g[g_{\mu\nu}] + S_m[g_{\mu\nu}, \Psi].$$

Here the Einstein–Hilbert action for gravity $S_g[g_{\mu\nu}]$ is written 'à la Einstein', that is, up to a boundary term [see (5.20)], and includes a 'gauge fixing term' when working in harmonic coordinates:

$$S_g[g_{\mu\nu}] = \frac{1}{16\pi} \int \sqrt{-g} \left[g^{\mu\nu} \left(\Gamma^\lambda_{\mu\rho} \Gamma^\rho_{\nu\lambda} - \Gamma^\rho_{\mu\nu} \Gamma^\lambda_{\rho\lambda} \right) - \frac{1}{2} g_{\mu\nu} \Gamma^\mu \Gamma^\nu \right], \quad \text{where} \quad \Gamma^\mu \equiv g^{\nu\rho} \Gamma^\mu_{\nu\rho}.$$

As for the matter action, for point particles it reads as $S_m[g_{\mu\nu}, \Psi] = -\sum m \int \sqrt{-g_{\mu\nu} dx^\mu dx^\nu}$. The 'Fokker' action is

$$S_F[\Psi] \equiv S_g[\bar{g}_{\mu\nu}[\Psi]] + S_m[\bar{g}_{\mu\nu}[\Psi], \Psi],$$

where $\bar{g}_{\mu\nu}[\Psi]$ is a solution of Einstein's equations $\delta S/\delta g^{\mu\nu} = 0$ to the desired post-Newtonian order. In the case of point particles, the action $S_F = \int L_F[Z(t), Z'(t)] \, dt$, where L_F is the Fokker Lagrangian, depends only on the trajectories of the bodies. The extremization of $S_F[\Psi]$ with respect to the matter fields Ψ yields, at least formally, the matter equations of motion $\delta S/\delta \Psi = 0$, because

$$\frac{\delta S_F[\Psi]}{\delta \Psi} = \frac{\delta \bar{g}[\Psi]}{\delta \Psi} \left(\frac{\delta S}{\delta g} \right)_{g=\bar{g}} + \left(\frac{\delta S}{\delta \Psi} \right)_{g=\bar{g}},$$

where the first term vanishes when $\bar{g}_{\mu\nu}[\Psi]$ is a solution of the field equations $\delta S/\delta g = 0$. After proper regularization, this method gives back (16.1).[1]

The Lagrangian (16.1) depends on positions, velocities, and, linearly, on accelerations. Let us add to it a 2PN total time derivative,

$$L \to L_f = L + \frac{df}{dt}, \tag{16.2}$$

where f is the generic function

$$\begin{aligned}
\frac{f}{mm'} &= (f_1 V^2 + f_2 V.V' + f_3 V'^2)(N.V) - (f_4 V^2 + f_5 V.V' + f_6 V'^2)(N.V') \\
&\quad + f_7 (N.V)^3 + f_8 (N.V)^2 (N.V') - f_9 (N.V')^2 (N.V) \\
&\quad - f_{10}(N.V')^3 + f_{11}\left(\frac{m}{R}\right)(N.V) + f_{12}\left(\frac{m'}{R}\right)(N.V) \\
&\quad - f_{13}\left(\frac{m}{R}\right)(N.V') - f_{14}\left(\frac{m'}{R}\right)(N.V'),
\end{aligned} \tag{16.3}$$

---

[1] For a detailed presentation of the Fokker action through 4PN, see Bernard *et al.* (2016) together with Damour, Jaranowski, and Schäfer (2016).

depending on 14 parameters f_i. This total derivative generates a 2PN boundary term in the action that does not affect the equations of motion, which remain unchanged.

In order to reduce L_f to an 'ordinary' Lagrangian depending only on positions and velocities, one may 'naively' replace the accelerations by their leading order, that is, their Newtonian, on-shell expressions[2]:

$$L_f \to L_f^{\text{red}} = L_f \left(A^i \to -\frac{m'}{R^2} N^i \,, \; A'^i \to \frac{m}{R^2} N^i \right). \qquad (16.4)$$

The equations of motion derived from L_f^{red} do differ from the ones derived from L_f (i.e., from L) at the 2PN level. However, the *dynamics* is unchanged. The reason is that the reduction (16.4) amounts in fact to implicitly switching from the harmonic coordinates to new ones, defined, as can be checked, by the 2PN-level, f-dependent, contact transformation $Y^i = Z^i + \delta Z^i$, $Y'^i = Z'^i + \delta Z'^i$:

$$\delta Z^i = \frac{m'}{8} \left[14 V'^i (N.V') - N^i \left(7V'^2 - (N.V')^2 \right) \right]$$

$$- m' \left[2V^i \big(f_1(N.V) - f_4(N.V') \big) + V'^i \big(f_2(N.V) - f_5(N.V') \big) \right] - \frac{m' N^i}{R} \left(m f_{11} + m' f_{12} \right)$$

$$- m' N^i \left[f_1 V^2 + f_2 (V.V') + f_3 V'^2 + 3 f_7 (N.V)^2 + 2 f_8 (N.V)(N.V') - f_9 (N.V')^2 \right],$$

$$\delta Z'^i = \frac{m}{8} \left[-14 V^i (N.V) + N^i \left(7V^2 - (N.V)^2 \right) \right]$$

$$- m \left[V^i \big(f_2(N.V) - f_5(N.V') \big) + 2V'^i \big(f_3(N.V) - f_6(N.V') \big) \right] + \frac{m N^i}{R} \left(m f_{13} + m' f_{14} \right)$$

$$+ m N^i \left[f_4 V^2 + f_5 (V.V') + f_6 V'^2 - f_8 (N.V)^2 + 2 f_9 (N.V)(N.V') + 3 f_{10} (N.V')^2 \right].$$

We thus now have at our disposal an entire class of ordinary Lagrangians L_f^{red} (depending on the 14 parameters f_i), each one corresponding to a specific choice of coordinate system. The harmonic-coordinate, acceleration-dependent Lagrangians (16.2) do not belong to this class. The Arnowitt–Deser–Misner (ADM) coordinates (see Section 4.5) turn out to correspond to the choice (the other f_i being zero)

$$f_3 - f_4 - \frac{1}{4} \,, \quad f_{12} = f_{13} = \frac{1}{4} \,, \quad f_{11} = f_{14} = \frac{7}{4} \,. \qquad (16.5)$$

Now that the accelerations (A^i, A'^i) have been eliminated, it is a straightforward exercise to derive the associated class of Hamiltonians (see Book 1, Section 9.1):

$$P^i = \frac{\partial L_f^{\text{red}}}{\partial V^i} \,, \quad P'^i = \frac{\partial L_f^{\text{red}}}{\partial V'^i} \,, \quad H = P.V + P'.V' - L_f^{\text{red}} \,.$$

In the center-of-mass frame where $P^i + P'^i = 0$ (see Section 12.4), the conjugate variables are $R^i = Z^i - Z'^i$ and P^i. The relative motion lies in the equatorial section and is described using polar coordinates (R, Φ), with conjugate momenta $P_R = (N.P)$, $P_\Phi = R(N \wedge P)_z$,

---

[2]This reduction was done by Ohta *et al.* in 1973–74, and shown to be correct when Schäfer (1984) and Damour and Schäfer (1985) realized that it implies a coordinate change. For details see Damour and Schäfer (1991).

and from now on we use the notation $(Q, P) = (R, \Phi; P_R, P_\Phi)$. Introducing the dimensionless quantities

$$\hat{P}^2 = \hat{P}_R^2 + \frac{\hat{P}_\Phi^2}{\hat{R}^2} \ , \quad \hat{P}_R = \frac{P_R}{\mu} \ , \quad \hat{P}_\Phi = \frac{P_\Phi}{\mu M} \ , \quad \hat{R} = \frac{R}{M} \ ,$$

$$M = m + m' \ , \ \mu = \frac{mm'}{M} \ , \ \nu = \frac{\mu}{M} \ ,$$

we see that the two-body Hamiltonians H

$$\frac{H - M}{\mu} = \left(\frac{\hat{P}^2}{2} - \frac{h^{\mathrm{N}}}{\hat{R}} \right) + \hat{H}^{\mathrm{1PN}} + \hat{H}^{\mathrm{2PN}} + \cdots \tag{16.6}$$

depend, generically, on 17 coefficients:

$$\hat{H}^{\mathrm{1PN}} = \left(h_1^{\mathrm{1PN}} \hat{P}^4 + h_2^{\mathrm{1PN}} \hat{P}^2 \hat{P}_R^2 + h_3^{\mathrm{1PN}} \hat{P}_R^4 \right) + \frac{1}{\hat{R}} \left(h_4^{\mathrm{1PN}} \hat{P}^2 + h_5^{\mathrm{1PN}} \hat{P}_R^2 \right) + \frac{h_6^{\mathrm{1PN}}}{\hat{R}^2} \ ,$$

$$\hat{H}^{\mathrm{2PN}} = \left(h_1^{\mathrm{2PN}} \hat{P}^6 + h_2^{\mathrm{2PN}} \hat{P}^4 \hat{P}_R^2 + h_3^{\mathrm{2PN}} \hat{P}^2 \hat{P}_R^4 + h_4^{\mathrm{2PN}} \hat{P}_R^6 \right) \tag{16.7}$$

$$+ \frac{1}{\hat{R}} \left(h_5^{\mathrm{2PN}} \hat{P}^4 + h_6^{\mathrm{2PN}} \hat{P}_R^2 \hat{P}^2 + h_7^{\mathrm{2PN}} \hat{P}_R^4 \right) + \frac{1}{\hat{R}^2} \left(h_8^{\mathrm{2PN}} \hat{P}^2 + h_9^{\mathrm{2PN}} \hat{P}_R^2 \right) + \frac{h_{10}^{\mathrm{2PN}}}{\hat{R}^3} \ .$$

In general relativity, the seven coefficients h^{N} and h_i^{1PN} are found to be

$$h^{\mathrm{N}} = 1 \ ,$$

$$h_1^{\mathrm{1PN}} = -\frac{1}{8}(1 - 3\nu) \ , \quad h_2^{\mathrm{1PN}} = h_3^{\mathrm{1PN}} = 0 \ , \tag{16.8}$$

$$h_4^{\mathrm{1PN}} = -\frac{1}{2}(3 + \nu) \ , \quad h_5^{\mathrm{1PN}} = -\frac{1}{2}\nu \ , \quad h_6^{\mathrm{1PN}} = \frac{1}{2} \ .$$

At the 2PN level the coefficients h_i^{2PN} depend on the 14 parameters f_i, *i.e.*, on the coordinate systems introduced in (16.2) and (16.3). In general relativity, and in the ADM coordinates defined in (16.5), for example, they are

$$h_1^{\mathrm{2PN}} = \frac{1}{16} \left(1 - 5\nu + 5\nu^2 \right) \ , \quad h_2^{\mathrm{2PN}} = h_3^{\mathrm{2PN}} = h_4^{\mathrm{2PN}} = 0 \ ,$$

$$h_5^{\mathrm{2PN}} = \frac{1}{8}(5 - 20\nu - 3\nu^2) \ , \quad h_6^{\mathrm{2PN}} = -\frac{\nu^2}{4} \ , \quad h_7^{\mathrm{2PN}} = -\frac{3}{8}\nu^2 \ , \tag{16.9}$$

$$h_8^{\mathrm{2PN}} = \frac{5}{2} + 4\nu \ , \quad h_9^{\mathrm{2PN}} = \frac{3}{2}\nu \ , \quad h_{10}^{\mathrm{2PN}} = -\frac{1}{4}(1 + 3\nu) \ .$$

The equations of motion derived from the Hamiltonians (16.6) *et seq.* through Hamilton's equations $(dQ/dt = \partial H/\partial P, \ dP/dt = -\partial H/\partial Q)$ are strictly equivalent to those which can be obtained from the 2PN Lagrangian (16.1) after proper change of the coordinates given in (16.5).[3]

---

[3]This 2PN Hamiltonian was first obtained by Schäfer in 1985 and is currently known to 4PN order; see Damour, Jaranowski and Schäfer (2014).

16.2   The equations of motion of a test particle in a SSS metric

In view of the mapping to come of the two-body dynamics to that of a test particle in the field of an effective single body, let us here consider a static, spherically symmetric (SSS) metric in the equatorial section, written in Schwarzschild–Droste coordinates (t_e, r, ϕ) as

$$ds_e^2 = -A(r)\, dt_e^2 + B(r)\, dr^2 + r^2 d\phi^2 \,, \tag{16.10}$$

where the subscript 'e' stands for 'effective'.

The action of a test particle in the metric (16.10), of mass μ which we shall identify as the two-body reduced mass, that is, $\mu = mm'/(m + m')$, is (see Section 3.1)

$$S_e = \int L_e\, dt_e = -\mu \int \sqrt{-\frac{ds_e^2}{dt_e^2}}\, dt_e \,,$$

so that $L_e = -\mu\sqrt{A - B\,\dot{r}_e^2 - r^2\dot{\phi}_e^2}$ with $\dot{r}_e = \dfrac{dr}{dt_e}\,, \quad \dot{\phi}_e = \dfrac{d\phi}{dt_e}\,.$

The conjugate momenta (p_r, p_ϕ) of (r, ϕ) and the Hamiltonian H_e are defined as (see Book 1, Chapter 9)

$$p_r = \frac{\partial L_e}{\partial \dot{r}_e}\,, \quad p_\phi = \frac{\partial L_e}{\partial \dot{\phi}_e}\,, \quad H_e = p_r\dot{r}_e + p_\phi\dot{\phi}_e - L_e \,,$$

so that $$\frac{H_e}{\mu} = \sqrt{A(r)\left(1 + \frac{\hat{p}_r^2}{B(r)} + \frac{\hat{p}_\phi^2}{\hat{r}^2}\right)}, \tag{16.11}$$

where $$\hat{r} = \frac{r}{M}\,, \quad \hat{p}_r = \frac{p_r}{\mu}\,, \quad \hat{p}_\phi = \frac{p_\phi}{\mu M}\,, \quad \hat{p}^2 = \hat{p}_r^2 + \frac{\hat{p}_\phi^2}{\hat{r}^2}\,,$$

M being a mass which we will identify as the two-body total mass, that is, $M = m + m'$.

If we restrict our attention to the dynamics of the particle at 2PN order only, the metric potentials $A(r)$ and $B(r)$ can be expanded as

$$A(r) = 1 + \frac{a_1}{\hat{r}} + \frac{a_2}{\hat{r}^2} + \frac{a_3}{\hat{r}^3} + \cdots \,, \quad B(r) = 1 + \frac{b_1}{\hat{r}} + \frac{b_2}{\hat{r}^2} + \cdots \,, \tag{16.12}$$

where $a_1, a_2, a_3, b_1,$ and b_2 are five dimensionless coefficients characterizing the effective SSS spacetime at that order. The 2PN effective Hamiltonian then becomes

$$\frac{H_e}{\mu} - 1 = \hat{H}_e^{\mathrm{N}} + \hat{H}_e^{\mathrm{1PN}} + \hat{H}_e^{\mathrm{2PN}} + \cdots \tag{16.13}$$

where $\hat{H}_e^{\mathrm{N}} = \dfrac{\hat{p}^2}{2} + \dfrac{a_1}{2\hat{r}}\,, \quad \hat{H}_e^{\mathrm{1PN}} = -\dfrac{\hat{p}^4}{8} - b_1\dfrac{\hat{p}_r^2}{2\hat{r}} + a_1\dfrac{\hat{p}^2}{4\hat{r}} + \dfrac{a_2 - a_1^2/4}{2\hat{r}^2}\,,$

$$\hat{H}_e^{\mathrm{2PN}} = \frac{\hat{p}^6}{16} - \frac{\hat{p}^2(a_1\hat{p}^2 - 4b_1\hat{p}_r^2)}{16\hat{r}} + \frac{(4a_2 - a_1^2)\hat{p}^2 + 4(2b_1^2 - 2b_2 - a_1b_1)\hat{p}_r^2}{16\hat{r}^2} + \frac{a_1^3 - 4a_1a_2 + 8a_3}{16\hat{r}^3}.$$

The Hamilton equations of motion derived from (16.11), $dq/dt_e = \partial H_e/\partial p$ and $dp/dt_e = -\partial H_e/\partial q$ with $q = (r, \phi)$ and $p = (p_r, p_\phi)$, give back the geodesic equation and its first integrals obtained in Section 11.2:

$$H_e = \mu E \,, \quad \hat{p}_\phi = j \,, \quad \frac{d\phi}{dt_e} = \frac{j}{ME} A u^2 \,, \quad \left(\frac{dr}{dt_e} \right)^2 = \frac{A}{BE^2} F(u) \,,$$

$$\text{where} \quad F(u) = E^2 - A(u) \left(\epsilon + j^2 u^2 \right) \quad \text{with} \quad u = \frac{M}{r} \,,$$

and where E and j are the (dimensionless) energy and angular momentum of the particle (and $\epsilon = 1$ or $\epsilon = 0$ for time-like or null geodesics).

Circular time-like orbits are such that the radial velocity vanishes, $F = 0$, while circularity also requires $F' = 0$, where $F' = dF/du$. Hence j^2 and E are related to u by

$$j^2(u) = -\frac{A'}{(Au^2)'} \,, \quad E(u) = A\sqrt{\frac{2u}{(Au^2)'}} \,. \tag{16.14}$$

The innermost stable circular orbit (ISCO) requires the third (inflection point) condition $F'' = 0$, and its position, u_{ISCO}, is the root of the equation

$$\frac{A''}{A'} = \frac{(Au^2)''}{(Au^2)'} \,. \tag{16.15}$$

As for the position u_{LR} of the null circular orbit, or 'light ring', it is determined by the conditions $F_{\epsilon=0} = E^2 - j^2 A u^2 = 0$ and $dF/du|_{\epsilon=0} = 0$, and hence is given by the root of the equation

$$(Au^2)' = 0 \,. \tag{16.16}$$

This equation also gives the position of a circular time-like geodesic whose angular momentum formally goes to infinity. Note that in the Schwarzschild–Droste coordinates used here, circular orbits are described by the function $A(u)$ only, given in terms of three coefficients a_1, a_2, a_3 at 2PN order; see (16.12).

16.3 The EOB mapping

In Section 16.1 we obtained a class of two-body (center-of-mass) Hamiltonians $H(Q, P)$ depending on 17 coefficients h_i^{NPN} at 2PN order [see (16.6) *et seq.*]. We contrasted them with a much simpler effective Hamiltonian $H_e(q, p)$ depending on five coefficients a_i, b_i at 2PN order, which describes the geodesic motion of a test particle in an effective static and spherically symmetric metric; see (16.13). The 'effective-one-body' (EOB) mapping, proposed by Buonanno and Damour in 1998, consists in relating the two approaches by means of a canonical transformation together with a functional relation $H_e = f_{EOB}(H)$ between their Hamiltonians.[4]

● *The canonical transformation*

The first step of the mapping procedure is to relate the phase space coordinates (Q, P) of the two-body Hamiltonian $H(Q, P)$ to those, (q, p), of the effective Hamiltonian $H_e(q, p)$ by means of a canonical transformation.

---

[4]See Buonanno and Damour (1999).

Recalling that $(Q, P) = (R, \Phi; P_R, P_\Phi)$ and $(q, p) = (r, \phi; p_r, p_\phi)$, the canonical transformation is defined by a time-independent function (since the Hamiltonians are conservative) $F(q, Q)$ which shifts the two-body Lagrangian by the total derivative $L_f^{\mathrm{red}}(Q, \dot{Q}) = L'(q, \dot{q}) + dF/dt$ and $H'(q, p) = H(Q, P)$ (see Book 1, Section 9.2):

$$L' dt + dF = (p_r \, dr + p_\phi \, d\phi - H dt) + dF = P_R \, dR + P_\Phi \, d\Phi - H dt,$$

$$\text{and so}\quad dF = P_R \, dR + P_\Phi \, d\Phi - (p_r \, dr + p_\phi \, d\phi).$$

To avoid cumbersome algebra we shall instead consider the generating function $G(Q, p)$ such that

$$G = F + (p_r \, r + p_\phi \, \phi) - (p_r \, R + p_\phi \, \Phi),$$

$$\text{so that}\quad dG = dR \, (P_R - p_r) + d\Phi \, (P_\Phi - p_\phi) + dp_r \, (r - R) + dp_\phi \, (\phi - \Phi),$$

which yields the canonical transformation relating (q, P) to (Q, p):

$$r(Q, p) = R + \frac{\partial G}{\partial p_r}, \quad \phi(Q, p) = \Phi + \frac{\partial G}{\partial p_\phi}, \quad P_R(Q, p) = p_r + \frac{\partial G}{\partial R}, \quad P_\Phi(Q, p) = p_\phi + \frac{\partial G}{\partial \Phi}.$$

$$(16.17)$$

The generic ansatz for G, which generates 1PN and 2PN phase coordinate changes, depends on 9 parameters[5]:

$$\frac{G(Q, p)}{\mu M} = \hat{R}\hat{p}_r \left[\left(\alpha_1 \mathcal{P}^2 + \beta_1 \hat{p}_r^2 + \frac{\gamma_1}{\hat{R}} \right) + \left(\alpha_2 \mathcal{P}^4 + \beta_2 \mathcal{P}^2 \hat{p}_r^2 + \gamma_2 \hat{p}_r^4 + \delta_2 \frac{\mathcal{P}^2}{\hat{R}} + \epsilon_2 \frac{\hat{p}_r^2}{\hat{R}} + \frac{\eta_2}{\hat{R}^2} \right) + \cdots \right],$$

$$(16.18)$$

where we have (re)introduced the dimensionless quantities

$$\mathcal{P}^2 = \hat{p}_r^2 + \frac{\hat{p}_\phi^2}{\hat{R}^2}, \quad \hat{R} \equiv \frac{R}{M}, \quad \hat{p}_r \equiv \frac{p_r}{\mu}, \quad \hat{p}_\phi \equiv \frac{p_\phi}{\mu M}, \quad M = m + m', \quad \mu = \frac{mm'}{M}, \quad \nu = \frac{\mu}{M}.$$

This generating function does not depend on Φ (by isotropy), and so $P_\Phi = p_\phi$; see (16.17). Note also that for circular orbits for which $p_r = 0 \Leftrightarrow P_R = 0$, we have $\phi = \Phi$ and hence only the radial coordinates differ (but are both constant, because p_ϕ is then constant on shell).

The two-body Hamiltonian (16.6) *et seq.* is thus canonically transformed, $H(Q, p) \equiv H(Q, P(Q, p))$, using the last two relations in (16.17) (the use of G instead of F avoids having to perform inversions) and is of the form (its explicit expression is easily worked out)

$$\frac{H(Q, p) - M}{\mu} = \left(\frac{\mathcal{P}^2}{2} - \frac{h^{\mathrm{N}}}{\hat{R}} \right) + \hat{H}^{\mathrm{1PN}} + \hat{H}^{\mathrm{2PN}} + \cdots, \qquad (16.19)$$

where $h^{\mathrm{N}} = 1$ [see (16.8)], and where the 16 remaining coefficients appearing in $\hat{H}^{\mathrm{1PN}}$ and $\hat{H}^{\mathrm{2PN}}$ depend on the 14 parameters f_i introduced in (16.3) and on the 9 parameters entering the canonical transformation (16.18). Similarly, the effective Hamiltonian

---

[5]We already know from Newton's theory that when written in the center-of-mass frame, the Newtonian two-body and effective-one-body Hamiltonians are identical.

$H_e(Q, p) \equiv H_e(q(Q, p), p)$ (16.13) is transformed as follows using the first two relations in (16.17):

$$\frac{H_e(Q, p)}{\mu} - 1 = \left(\frac{\mathcal{P}^2}{2} + \frac{a_1}{2\hat{R}}\right) + \hat{H}_e^{1PN} + \hat{H}_e^{2PN} + \cdots, \tag{16.20}$$

where $\hat{H}_e^{1PN}$ and $\hat{H}_e^{2PN}$ depend on the 5 coefficients $(a_1, a_2, a_3, b_1, b_2)$ entering the effective metric coefficients at 2PN order [see (16.12)], as well as on the 9 parameters of the canonical transformation (16.18), and are also easily worked out.

- *The functional relation*

Now that we have expressed the two-body and effective Hamiltonians H and H_e in the same coordinate system (Q, p) [see (16.19)–(16.20)], the EOB mapping requires as a second step the imposition of a functional relation between them, $H_e = f_{EOB}(H)$, which at 2PN order will yield the desired relations between the five coefficients (a_i, b_i) entering H_e and the 17 coefficients h_i^{NPN} entering H.

This functional relation can *a priori* be expanded as follows, subtracting the rest-mass constants:

$$\frac{H_e(Q, p)}{\mu} - 1 = \left(\frac{H(Q, p) - M}{\mu}\right) \left[1 + \frac{\bar{\nu}_1}{2}\left(\frac{H(Q, p) - M}{\mu}\right) + \bar{\nu}_2\left(\frac{H(Q, p) - M}{\mu}\right)^2 + \cdots\right],$$

with the (mass-shifted) Hamiltonians being identical at Newtonian order [which already yields $a_1 = -2h^N = -2$, as can be seen from (16.19) and (16.20)]. Now, as has been proven up to 4PN in general relativity and even at all orders within a post-Minkowskian scheme, the relation must be quadratic *at all orders* with $\bar{\nu}_1 = \nu = \mu/M$ and $\bar{\nu}_2 = 0 \cdots$. That is, it must be[6]:

$$\frac{H_e(Q, p)}{\mu} - 1 = \left(\frac{H(Q, p) - M}{\mu}\right) \left[1 + \frac{\nu}{2}\left(\frac{H(Q, p) - M}{\mu}\right)\right]. \tag{16.21}$$

This functional relation is not guaranteed to hold a priori. Indeed, the generic 2PN Hamiltonians written in terms of the 17 coefficients h_i^{NPN} [see (16.6)] may now be considered as describing the two-body dynamics (in various coordinate systems parametrized by the 14 f_i) for an *arbitrary* theory (e.g., general relativity), which must be related to an effective Hamiltonian depending on 5 coefficients (a_i, b_i) [see (16.12)] by means of a canonical transformation depending on 9 parameters (16.18). We therefore expect $17 - 5 - 9 = 3$ conditions on the h_i^{NPN} coefficients.

No condition arises at Newtonian order. At 1PN order it turns out that an effective Hamiltonian H_e can be constructed provided that the coefficients h_i^{1PN} (which do not depend on the f_i) satisfy

$$2h_2^{1PN} + 3h_3^{1PN} = 0.$$

---

[6]See Damour (2016a). It should be noted that at 1PN order one can choose $\bar{\nu}_1 = 0$ (which then yields $a_2 = -\nu/4$ and $b_1 = (8 - \nu)/4$ and the effective Eddington parameters $\gamma = 1 - \nu/8$, $\beta = 1 - \nu/4$). This shows that the two-body post-Keplerian orbits, which depend on three eccentricities [see (12.34) and (12.35)], can be transformed into the post-Keplerian orbits of a test particle in a SSS metric, which depend only on two eccentricities [see (11.36) and (11.37)], by means of a (1PN) canonical transformation (16.17) [with $\alpha_1 = -3\nu/8$, $\beta_1 = 0$, and $\gamma_1 = (8 + \nu)/8$], where the (mass-shifted) Hamiltonians are identical at 1PN-Newtonian order inclusive.

Any theory (such as general relativity) for which the purely kinematical terms in the Lagrangian take the Lorentz-invariant form $m\sqrt{1-V^2} + m'\sqrt{1-V'^2}$ will have $h_2^{1\text{PN}} = 0$ and $h_3^{1\text{PN}} = 0$. This condition is therefore not restrictive. At 2PN order, the coefficients $h_i^{2\text{PN}}$ depend on the 14 parameters f_i, and the identification requires two further conditions. The first,

$$h_4^{2\text{PN}} = -\frac{2}{45}\left(12h_2^{2\text{PN}} + 18h_3^{2\text{PN}} + (h_2^{1\text{PN}})^2\right),$$

is no more restrictive than the 1PN condition, for the same reason. However, the second condition,

$$h_1^{2\text{PN}} + \frac{7}{3}h_2^{2\text{PN}} + h_3^{2\text{PN}} + h_5^{2\text{PN}} + h_6^{2\text{PN}} + h_7^{2\text{PN}} = \frac{h^{\text{N}}}{128}(5 + 2\nu + 5\nu^2)$$

$$-\frac{1}{8}(1+\nu)\left((3h_1^{1\text{PN}} + h_2^{1\text{PN}})h^{\text{N}} + h_4^{1\text{PN}} + h_5^{1\text{PN}}\right) + \frac{5}{2}h_1^{1\text{PN}}\left(7h_1^{1\text{PN}}h^{\text{N}} + 2(h_4^{1\text{PN}} + h_5^{1\text{PN}})\right)$$

$$+\frac{1}{6}h_2^{1\text{PN}}\left(13h_2^{1\text{PN}}h^{\text{N}} + 10(h_4^{1\text{PN}} + h_5^{1\text{PN}})\right) + \frac{35}{3}h_1^{1\text{PN}}h_2^{1\text{PN}}h^{\text{N}},$$

(16.22)

is restrictive, and the mapping of the two-body motion to an effective geodesic is possible for a subclass of theories only.

In general relativity, it can be checked that the coefficients $h_i^{2\text{PN}}$ [given, for example, in (16.8)–(16.9) when using ADM coordinates] *do* satisfy the condition (16.22) *whatever* the values of the 14 parameters f_i, that is, independently of the coordinate system in which the two-body Hamiltonian has been written, as required by the invariance of general relativity under diffeomorphisms.[7]

- *The effective metric*

Now inserting in the left-hand side of the functional relation (16.21) the explicit expressions for the coefficients h_i^{NPN} of the two-body Hamiltonians H (16.19) obtained in Section 16.1 [see (16.6) *et seq.*] and, in the right-hand side, the explicit expression for the effective Hamiltonian H_e (16.20) obtained in Section 16.2 [see (16.13)], the term-by-term identification yields a *unique* solution for the five coefficients a_i and b_i entering into H_e and therefore the effective SSS metric:

$$ds_\text{e}^2 = -A(r)\,dt_\text{e}^2 + B(r)\,dr^2 + r^2 d\phi^2,$$

$$A(r) = 1 + \frac{a_1}{\hat{r}} + \frac{a_2}{\hat{r}^2} + \frac{a_3}{\hat{r}^3}, \quad B(r) = 1 + \frac{b_1}{\hat{r}} + \frac{b_2}{\hat{r}^2} \quad \text{with} \quad \hat{r} = \frac{r}{M},$$

(16.23)

---

[7]The relation (16.22) also holds for scalar–tensor theories; see Julié and Deruelle (2017). In contrast, it is not satisfied by Maxwell's theory at second post-Coulombian order; see Buonanno (2000).

It is a straightforward exercise to extend the identification between generic two-body and effective Hamiltonians at 3PN or higher orders. One then sees that the identification at 3PN order [when imposing the quadratic relation (16.21) between the 2PN and effective Hamiltonians] requires a further condition which turns out *not* to be satisfied by general relativity. Therefore, the mapping of the two-body motion then has to be extended to a forced motion of a test particle in a SSS metric; see Damour, Jaranowski, and Schäfer (2000).

where $a_1 = -2$, $a_2 = 0$, $a_3 = 2\nu$, $b_1 = 2$, $b_2 = 2(2 - 3\nu)$ with $\nu = \dfrac{mm'}{M^2}$.

The simplicity of (16.23) is striking, since geodesic motion in this metric encompasses all the dynamics of the two-body problem at 2PN order. In particular, the 14 parameters f_i which parametrize the family of two-body ordinary, reduced Lagrangians at 2PN order [see (16.2)] are hidden in the canonical transformation (16.18), whose coefficients are found to be

$$\alpha_1 = -\frac{\nu}{2}, \quad \beta_1 = 0, \quad \gamma_1 = 1 + \frac{\nu}{2}, \quad \alpha_2 = \frac{1}{8}(1 - \nu)\nu, \quad \beta_2 = 0, \quad \gamma_2 = \frac{\nu^2}{2},$$

$$\delta_2 = f_6 \frac{m}{M} + f_1 \frac{m'}{M} - \nu \left(f_1 + f_6 + (-f_3 + f_5 + f_6)\frac{m}{M} + (f_1 + f_2 - f_4)\frac{m'}{M} - \frac{3}{2} + \frac{\nu}{8} \right),$$

$$\epsilon_2 = -\frac{\nu^2}{8} + f_{10}\frac{m}{M} + f_7 \frac{m'}{M} - \nu \left(f_7 + f_{10} + (f_9 + f_{10})\frac{m}{M} + (f_7 + f_8)\frac{m'}{M} \right),$$

$$\eta_2 = \frac{\nu}{4}(-19 + \nu) + f_{13}\frac{m}{M} + f_{12}\frac{m'}{M} + \nu(f_{11} - f_{12} - f_{13} + f_{14}).$$

16.4 The EOB Hamiltonian and the resummed dynamics

Now that H_e and the associated effective metric (16.23) have been constructed, we can invert the quadratic relation (16.21) to define an exact 'EOB Hamiltonian', which is a *resummed* version of the two-body Hamiltonian H and coincides, at 2PN, with its 2PN expansion given in (16.6) *et seq.* It reads [in the (q, p) phase-space coordinates of the effective problem and returning to dimensional coordinates] as

$$H_{\mathrm{EOB}} = M\sqrt{1 + 2\nu\left(\frac{H_e}{\mu} - 1\right)} \quad \text{with} \quad \frac{H_e}{\mu} = \sqrt{A\left(1 + \frac{p_r^2}{\mu^2 B} + \frac{p_\phi^2}{\mu^2 r^2}\right)}, \qquad (16.24)$$

where $A(r)$ and $B(r)$ are given in (16.23) and are now considered to be exact. (We again recall that $M = m + m'$, $\mu = mm'/M$, and $\nu = \mu/M$.)

As an illustration of the resummed dynamics defined by H_{EOB}, let us focus on the ISCO and light-ring orbital frequency $\Omega_{\mathrm{ISCO}} = (d\Phi/dt)|_{\mathrm{ISCO}}$ and $\Omega_{\mathrm{LR}} = (d\Phi/dt)|_{\mathrm{LR}}$ of the relative motion of the two bodies.[8]

From Hamilton's equations we have

$$\Omega = \frac{\partial H_{\mathrm{EOB}}}{\partial P_\Phi} \quad \text{with} \quad \Omega = \frac{d\Phi}{dt}, \quad \text{that is,} \quad \Omega = \frac{1}{\sqrt{1 + 2\nu(E - 1)}} \frac{\partial H_e}{\partial p_\phi},$$

where E is the energy per unit mass of the effective test particle and $P_\Phi = p_\phi$ because the generating function G does not depend on Φ; moreover, $\Phi = \phi$ for circular orbits [see the remark below (16.3)]. Therefore,

---

[8]In the standard post-Newtonian approach they are given by $\Omega_{\mathrm{PN}} = \partial H/\partial P_\Phi$, where H is the 2PN Hamiltonian (16.6) *et seq.* which depends on the 17 coefficients h_i^{NPN}. The analysis was performed in 1993 by Kidder, Will, and Wiseman (within the Lagrangian formalism and in harmonic coordinates) and by Wex and Schäfer (within the Hamiltonian formalism) and shown to lack robustness.

$$\Omega_{\mathrm{circ}} = \frac{\omega_{\mathrm{circ}}}{\sqrt{1 + 2\nu(E-1)}} \quad \text{with} \quad \omega_{\mathrm{circ}} = \frac{\partial H_{\mathrm{e}}}{\partial p_\phi}, \quad \text{that is,} \quad M\omega_{\mathrm{circ}} = \frac{j}{E} Au^2,$$

where $u = \frac{M}{r}$ and the dimensionless angular momentum j and energy E of the effective test particle are given by $E = A\sqrt{2u/(Au^2)'}$ and $j^2 = -A'/(Au^2)'$; see (16.14).[9] Now, at 2PN order the metric potential is $A = 1 - 2u + 2\nu u^3$ [see (16.23)], so that

$$E = \frac{1 - 2u + 2\nu u^3}{\sqrt{1 - 3u + 5\nu u^3}}, \quad j^2 = \frac{1 - 3\nu u^2}{u(1 - 3u + 5\nu u^3)}, \quad \text{and} \quad M\Omega_{\mathrm{circ}} = \frac{u^{3/2}\sqrt{1 - 3\nu u^2}}{\sqrt{1 + 2\nu(E-1)}}.$$
$$(16.25)$$

The ISCO and light-ring positions u_{ISCO} and u_{LR} satisfy $A''/A' = (Au^2)''/(Au^2)'$ and $(Au^2)' = 0$, respectively [see (16.15)–(16.16)], and therefore are the outermost roots of

$$1 - 6u + 3\nu u^2 + 20\nu u^3 - 30\nu^2 u^5 = 0 \ (\text{ISCO}) \quad \text{and} \quad 1 - 3u + 5\nu u^3 = 0 \ (\text{light ring}). \quad (16.26)$$

A remarkable feature of this resummed dynamics is that it is *exact* for $\nu = 0$ (since the ISCO and light ring are those of a geodesic in the Schwarzschild metric, $r_{\mathrm{ISCO}} = 6M$ and $r_{\mathrm{LR}} = 3M$; see Section 8.3). We also note that for all ν ($\in [0, 1/4]$), $r_{\mathrm{ISCO}} < 6M$ (which is at odds with the prediction based on the 2PN Hamiltonian H, but in agreement with the prediction based on an analysis at 3PN order). Finally, we see that $\Omega_{\mathrm{LR}} = 0$ because the light ring coincides with a circular time-like geodesic with infinite j (and hence p_ϕ), and that the canonical transformation (16.17)–(16.18) between (r, ϕ) and (R, Φ) then becomes singular.[10]

16.5   EOB dynamics including the radiation reaction force

In Sections 16.1–16.4 we developed a resummed EOB dynamics of the two-body problem, focusing on its conservative part. We now include radiation reaction effects which first circularize the orbit and then cause its radius to decrease. Our starting point will be the 2.5PN radiation reaction force obtained in harmonic coordinates in Section 15.2.

- *The 2.5PN radiation reaction force*

From the 2.5PN equations of motion (15.17)–(15.18) obtained in Section 15.2 we can derive the center-of-mass, 2.5PN radiation reaction force $\vec{\mathcal{F}}$ (with $\vec{N} = \vec{R}/R$, $\vec{V} = \dot{\vec{R}}$, where the dot denotes the derivative with respect to t):

$$\mu\dot{\vec{V}} = -\frac{M\mu}{R^2}\vec{N} + \cdots + \vec{\mathcal{F}} \quad \text{with} \quad M = m + m', \ \mu = \frac{mm'}{M},$$

---

[9]The relation between $\Omega = d\Phi/dt$ and ω means that ω is an angular velocity with respect to the rescaled 'effective' time: $\omega = d\phi/dt_{\mathrm{e}}$ with $t_{\mathrm{e}} = t/\sqrt{1 + 2\nu(E-1)}$.

[10]The analysis of circular orbits is just as straightforward at 3PN order; see Damour, Jaranowski, and Schäfer (2000). At that order the potentials A and B are given by $A = 1 - 2u + 2\nu u^3 + \nu a_4 u^4$ with $a_4 = 94/3 - 41\pi^2/32$, and $AB = 1 - 6\nu u^2 + b_3 u^3$ with $b_3 = 2(3\nu - 26)\nu$. The motion of the effective test particle is no longer geodesic but governed by $H_{\mathrm{e}} = \sqrt{\mu^2 A + A p_\phi^2/r^2 + (p_r^*)^2[1 + z_3(p_r^*)^2]}$ with $p_r^* = \sqrt{A/B}\, p_r$ and $z_3 = 2(4 - 3\nu)$. The value of z_3 does not affect the dynamics of circular orbits, for which $p_r = 0$.

and $\quad \vec{\mathcal{F}} = \dfrac{8(mm')^2}{5MR^3}\mathcal{V}^2\left[3\vec{N}(N.\mathcal{V}) - \vec{\mathcal{V}}\right] + \dfrac{8(mm')^2}{5R^4}\left[\dfrac{17}{3}\vec{N}(N.\mathcal{V}) - 3\vec{\mathcal{V}}\right].$

For circular motion $(N.\mathcal{V}) = 0$ and $(N \wedge \mathcal{V})_z = R\,\dot\Phi$, and so $\mathcal{F}_R|_{\text{circ}} = (N.\mathcal{F})$ vanishes and $\mathcal{F}_\Phi|_{\text{circ}} = R(N \wedge \mathcal{F})_z$ is given by

$$\mathcal{F}_\Phi|_{\text{circ}} = -\dfrac{8(mm')^2}{5MR}\dot\Phi\left(\mathcal{V}^2 + \dfrac{3M}{R}\right) \quad \text{with} \quad \mathcal{V}^2 = (R\dot\Phi)^2.$$

At the lowest, Newtonian, order to which we limit ourselves here, the contact and canonical transformations between the two-body and effective coordinates performed in (16.5) *et seq.* and (16.17) *et seq.* are the identity: $\Phi = \phi$, $R = r$. Therefore, for circular orbits for which $\mathcal{V}^2 = M/R$ so that $R/M = (M\dot\phi)^{-2/3}$ [see, for example, (14.30)], the leading-order, 'quadrupolar' radiation reaction force is[11]

$$\mathcal{F}_r|_{\text{circ}} = 0\,, \quad \mathcal{F}_\phi|_{\text{circ}}^{\text{quad}} = -\dfrac{32}{5}M\,\nu^2(M\dot\phi)^{7/3} \quad \text{with} \quad \nu = \dfrac{mm'}{M^2}\,. \tag{16.27}$$

- *The EOB equations of motion including the radiation reaction*

From the Lagrangian equations of motion including the radiation reaction effects [see, for example, (15.21)] we obtain the corresponding EOB Hamilton equations of motion (the dot denotes the derivative with respect to t):

$$\dot r = \dfrac{\partial H_{\text{EOB}}}{\partial p_r}\,, \quad \dot\phi = \dfrac{\partial H_{\text{EOB}}}{\partial p_\phi}\,, \quad \dot p_r = -\dfrac{\partial H_{\text{EOB}}}{\partial r} + \mathcal{F}_r\,, \quad \dot p_\phi = \mathcal{F}_\phi\,, \tag{16.28}$$

where $\quad H_{\text{EOB}} = M\sqrt{1 + 2\nu\left(\dfrac{H_e}{\mu} - 1\right)} \quad$ with $\quad \dfrac{H_e}{\mu} = \sqrt{A\left(1 + \dfrac{p_r^2}{\mu^2 B} + \dfrac{p_\phi^2}{\mu^2 r^2}\right)}$

and $\quad A(r) = 1 - 2u + 2\nu u^3\,, \quad B(r) = \dfrac{1}{A}\left(1 - 6\nu u^2\right) \quad$ with $\quad u = \dfrac{M}{r}\,.$

Note that B has been factored by $1/A$ in order to recover the exact Schwarzschild metric in the test mass limit $\nu = 0$. The functions $\mathcal{F}_i(q, \dot q) = \{\mathcal{F}_r(q, \dot q), \mathcal{F}_\phi(q, \dot q)\}$ are now to be considered as functions of $q = (r, \phi)$ and $p = (p_r, p_\phi)$ through Hamilton's equations:

$$\mathcal{F}_i(q, \dot q(q, p)) = \mathcal{F}_i(q, \partial H_{\text{EOB}}/\partial p)\,.$$

Let us consider a motion which initially is nearly circular (this is the case of physical interest because gravitational radiation tends to circularize orbits). The radial component of the reaction force is then nearly zero initially and will be assumed to remain negligible for the rest of the coalescence: $\mathcal{F}_r = \mathcal{F}_r|_{\text{circ}} = 0$. Similarly, $\mathcal{F}_\phi$ will be approximated by $\mathcal{F}_\phi = \mathcal{F}_\phi|_{\text{circ}}$

---

[11] Using the information presented in this book, it is not possible to derive more than the Newtonian expression (16.27) above for the radiation reaction force, which is a crude approximation. For an improved *resummed* expression of the radiative force including extra information from higher-order post-Newtonian expansion, see the seminal paper Buonanno and Damour (2000). For a review of the present state of the art, see Damour (2014).

[given in (16.27) in the Newtonian approximation] for all r. The equations of motion then become (here a prime denotes the derivative with respect to r)

$$\dot{r} = \frac{M}{H_e H_{\mathrm{EOB}}} \frac{A\, p_r}{B}, \quad \dot{p}_r = -\frac{M}{2 H_e H_{\mathrm{EOB}}} \left[A' + p_r^2 \left(\frac{A}{B} \right)' + p_\phi^2 \left(\frac{A}{r^2} \right)' \right], \quad \dot{p}_\phi = \mathcal{F}_\phi|_{\mathrm{circ}},$$

$$\dot{\phi} = \frac{M}{H_e H_{\mathrm{EOB}}} \frac{A\, p_\phi}{r^2}, \tag{16.29}$$

where H_e and H_{EOB} are given in (16.28), and $\mathcal{F}_\phi|_{\mathrm{circ}}$ is a function of $(M\dot{\phi})$, that is, of (r, p_r, p_ϕ) through the last (independent) Hamilton equation (16.29).

• *The initial conditions*

At $t = 0$, let us set, for example, $r_{\mathrm{in}} = 15M$ and $\phi_{\mathrm{in}} = 0$.

As for $p_r|_{\mathrm{in}}$ and $p_\phi|_{\mathrm{in}}$, they are chosen within the adiabatic, quasi-circular approximation. We recall that in the absence of radiative effects, $H_e = \mu E$ and $H_{\mathrm{EOB}} = M\sqrt{1 + 2\nu(E-1)}$, and for circular orbits [see (16.25)] we have (setting $u = M/r$)

$$p_\phi|_{\mathrm{circ}} = M^2 \nu \sqrt{\frac{1 - 3\nu u^2}{u(1 - 3u + 5\nu u^3)}}, \quad M\dot{\phi}|_{\mathrm{circ}} = \frac{u^{3/2}\sqrt{1 - 3\nu u^2}}{\sqrt{1 + 2\nu(E-1)}}, \quad E = \frac{1 - 2u + 2\nu u^3}{\sqrt{1 - 3u + 5\nu u^3}}. \tag{16.30}$$

We therefore choose $p_\phi|_{\mathrm{in}} = p_\phi|_{\mathrm{circ}}^{\mathrm{in}}$ with $u_{\mathrm{in}} = M/r_{\mathrm{in}}$.

The initial condition $p_r|_{\mathrm{in}}$ is obtained as follows. From Hamilton's equations (16.29) we have $\dot{p}_\phi = \mathcal{F}_\phi|_{\mathrm{circ}}$, that is, $(dp_\phi/dr)\dot{r} = \mathcal{F}_\phi|_{\mathrm{circ}}$, where $\mathcal{F}_\phi|_{\mathrm{circ}}$ can be approximated for large r by $\mathcal{F}_\phi|_{\mathrm{circ}}^{\mathrm{quad}}$ as in (16.27) and is a function of $\dot{\phi} = \dot{\phi}|_{\mathrm{circ}}$, that is, a function of $r = M/u$; see (16.30). Similarly, $p_\phi = p_\phi|_{\mathrm{circ}}$, and from eqn (16.30) we then know $(dp_\phi|_{\mathrm{circ}}/dr)$ in terms of r. Now $\dot{r}$ is given by Hamilton's equation $\dot{r} = \frac{MA}{BH_e H_{\mathrm{EOB}}} p_r$, where $H_e = \mu E$ and $H_{\mathrm{EOB}} = M\sqrt{1 + 2\nu(E-1)}$ with E in terms of r given by (16.30). Therefore, p_r is a known (but not very illuminating) function of r in the adiabatic, quasi-circular approximation (at lowest order in $u_{\mathrm{in}} = M/r_{\mathrm{in}}$ it reads $p_r|_{\mathrm{in}} = -\frac{64}{5} M\nu^2 u_{\mathrm{in}}^3$).

16.6   EOB waveform of two coalescing black holes

• *The EOB dynamics of two coalescing black holes*

The EOB dynamics derived above encapsulates the inspiraling of two compact bodies, for example, two black holes (or neutron stars if tidal effects are neglected), starting from an initially quasi-circular relative orbit.

The numerical integration of their equations of motion (16.29), with H_e, H_{EOB}, A, and B given in (16.28) and $\mathcal{F}_\phi|_{\mathrm{circ}} = \mathcal{F}_\phi|_{\mathrm{circ}}^{\mathrm{quad}}$ given in (16.27), where $\dot{\phi}$ is expressed in terms of (r, p_r, p_ϕ), is straightforward once a value of ν has been chosen and for the adiabatic, quasi-circular initial conditions given above. The result is given in Fig. 16.1, which shows that the inspiral of the two bodies remains quasi-circular even beyond their effective innermost stable circular orbit, and that the integration can be pushed all the way to the light ring.

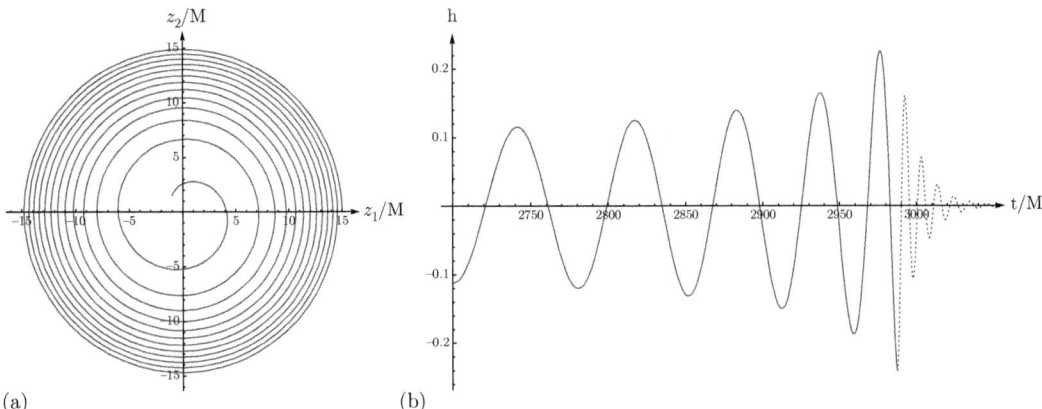

(a) (b)

Fig. 16.1 (a) The (effective) inspiral of two equal-mass ($\nu = 1/4$) black holes due to the (Newtonian) radiation reaction force (16.27), obtained by integration of the EOB equations of motion (16.29). The initial conditions at $\phi_{\rm in} = 0$, $r_{\rm in} = 15M$ are $p_\phi|_{\rm in} = 1.08\,M^2$ and $p_r|_{\rm in} = -3.61 \times 10^{-4}\,M$. The innermost stable circular orbit and light ring are located at $r_{\rm ISCO} = 5.72M$ and $r_{\rm LR} = 2.85M$. (b) The EOB waveform produced during the inspiral phase of the coalescence of two equal-mass black holes [solid line; see (16.31), setting $\mathcal{C} = 1$] up to merger, taken to occur at the effective light ring, where it is matched to the dominant quasi-normal mode of the final black hole [dotted line; see (16.32)]. The mass and angular momentum parameters of the final black hole are $M_{\rm BH} = 0.98M$ and $a_{\rm BH} = 0.79$. The quasi-normal mode pulsation and damping time are $M\omega_{\rm QNM} = 0.60$ and $M/\tau = 0.078$.

- *The gravitational waveform*

In Chapter 13 (and 14) we derived the 'first quadrupole formula', that is, the radiative gravitational field generated by a source at lowest (sometimes called 'Newtonian') order in terms of the second time derivative of its quadrupole moment; see (13.27). In the center-of-mass frame (at the lowest order, to which we restrict ourselves, the calculations are done using Newtonian physics) the quadrupole moment of two point-like bodies in circular orbits is $Q_{ij} = \mu(3z_i z_j - \delta_{ij} z^2)$ with $z_1 = r \cos\phi$ and $z_2 = r \sin\phi$, where the angular velocity $\dot\phi$ is taken to be constant. The time dependence of the amplitude of both modes of the gravitational wave is then given (up to a constant phase) by

$$h = \mathcal{C}(M\dot\phi)^{2/3} \cos(2\phi), \qquad (16.31)$$

where $\mathcal{C}$ is a dimensionless factor taking into account the position of the observer with respect to the plane of the orbit; it is proportional to μ/D, where D is the distance to the source. More precisely, up to a constant phase, the gravitational form (16.31) in a Cartesian coordinate system $\mathcal{R}' = (O', z_1', z_2', z_3')$ attached to the observer O' and in the TT gauge introduced in Section 13.2 is written as

$$h'^{TT}_{ij} = \begin{pmatrix} h_+ & h_\times & 0 \\ h_\times & -h_+ & 0 \\ 0 & 0 & 0 \end{pmatrix}, \quad \text{where} \quad h_+ = \frac{4\mu}{D}\left(\frac{1+\cos^2 i}{2}\right)\left(M\dot\phi\right)^{2/3}\cos(2\phi)$$

$$\text{and} \quad h_\times = \frac{4\mu}{D}(\cos i)\left(M\dot\phi\right)^{2/3}\sin(2\phi),$$

with i the inclination between the normal to the orbital plane and the line-of-sight OO'^i, along which we align the $(O'z'_3)$ axis for the sake of simplicity.

The inspiraling motion of the binary system due to the emission of gravitational waves is given by the EOB dynamics obtained above, which show that the amplitude and frequency of the waveform increase with time; see Fig. 16.1b.

The merging time is taken to be the moment the effective particle reaches the light ring.[12] The mass M_{BH} and angular momentum parameter a_{BH} of the final black hole formed after coalescence are thus estimated to be H_{EOB} (which is the energy of the binary system) as given in (16.28) and p_ϕ/M^2_{BH}, evaluated at the light ring.

From this point on the gravitational waveform is smoothly matched to

$$h = \mathcal{A}\, e^{-(t-t_{\mathrm{LR}})/\tau_{\mathrm{QNM}}}\cos\left(\omega_{\mathrm{QNM}}(t - t_{\mathrm{LR}}) + \mathcal{B}\right), \tag{16.32}$$

where $\mathcal{A}$ and $\mathcal{B}$ are two dimensionless constants determined by the matching conditions, and ω_{QNM} and τ_{QNM} characterize the dominant quasi-normal mode of the final black hole.[13]

The entire EOB waveform is plotted in Fig. 16.1b.

It should be recalled that obtaining this EOB waveform relies on a number of approximations and assumptions. First, the resummed EOB dynamics is built out of the 2PN Lagrangian, and the reaction force as well as the waveform are evaluated at leading, quadrupolar order. Second, the mass and angular momentum of the final black hole are identified with the energy and angular momentum of the two-body system, evaluated on the light ring. Third, the waveform at merger is matched to the dominant quasi-normal mode of the final black hole.

In order to produce useful waveform templates for analyzing the data collected by the current gravitational wave detectors, a far more elaborate EOB approach must be, and has been, developed. It is based on higher-order PN approximations, including the spins of the initial black holes, and defining merger by a fit to full numerical black hole coalescence simulations.[14]

---

[12]Identifying the moment of merger with the moment the effective particle reaches the light ring is supported by the fact that, in the test-particle limit, the light ring acts as a potential barrier below which the radial gravitational radiation emitted by the particle is strongly filtered, as first studied by Davis, Ruffini, Press, and Price (1971) and confirmed by present-day numerical relativity simulations.

[13]As noted in Section 9.2, the study of the quasi-normal (or 'ringing') modes which characterize a black hole was initiated by Vishveshwara in 1970. The values of ω_{QNM} and τ_{QNM} used here are those taken by Buonanno and Damour in their 2000 paper: $M_{\mathrm{BH}}\omega_{\mathrm{QNM}} = f_f[1 - 0.63(1 - a_{\mathrm{BH}})^{3/10}]$ and $\tau\,\omega_{\mathrm{QNM}} = 4f_Q(1 - a_{\mathrm{BH}})^{-9/20}$ with f_f and f_Q given in Echeverria (1989) (for $M_{\mathrm{BH}} = 0.98M$ and $a_{\mathrm{BH}} = 0.98$, $f_f = 0.9587$ and $f_Q = 0.9389$).

[14]The present state of the art is described in, for example, Nagar, Damour, Reisswig, and Pollney (2016) and Bohé et al. (2017).

Part IV

Friedmann–Lemaître solutions and cosmology

Cosmology: a space for thought on general relativity.

Jean Eisenstaedt, in *Foundations of Big Bang Cosmology*,
edited by F. W. Meyerstein, World Scientific, Singapore, 1989

Book 3

17
Cosmological spacetimes

Historically, in the first phase of its development the goal of relativistic cosmology was to construct spacetimes which could provide a good representation of the universe on a large scale, a first step in constructing a cosmological model. Here we shall present a few examples.

17.1 Maximally symmetric spaces

The Copernican and cosmological principles

In general relativity, geometry and matter determine each other, and *any* solution of the Einstein equations represents a possible 'universe'.

Therefore, a legitimate first approximation from the viewpoint of astronomical observations is to describe the distribution of matter surrounding us as being and remaining isotropic on a large scale. If, in addition, we stipulate that we do not occupy a privileged location in the universe, then the matter distribution must also be spatially homogeneous. However, nothing prevents it from evolving in time. This approximation/hypothesis about the matter distribution in the observable universe is called the *Copernican principle*.

The *cosmological principle* is a hypothesis about the geometry of the Riemannian spacetime representing the universe, which is assumed to be foliated by 3-spaces labeled by a cosmic time t which are homogeneous and isotropic, that is, 'maximally symmetric'.

The Einstein equations relating the Ricci tensor and the energy–momentum tensor then guarantee that the matter distribution is spatially homogeneous and isotropic.

In Section 5.1 we saw that isometries, the symmetries of a spacetime of dimension N, are defined by the existence of Killing vectors ξ^i, and that a maximally symmetric space,[1] described as being *homogeneous* and *isotropic*, possesses by definition the maximum number of Killing vectors, that is, $N(N+1)/2$ of them. Of these, N are *translations* and $N(N-1)/2$ are *rotations*. The geometry of such a space is obviously constrained, in particular, by (see Section 5.1)

$$(D_i R^m_{jkl} - D_j R^m_{ikl})\xi_m + (\delta^n_i R^m_{jkl} - \delta^n_j R^m_{ikl} + \delta^n_l R^m_{kji} - \delta^n_k R^m_{lji})D_n\xi_m = 0\,.$$

The coefficient of ξ_m and the antisymmetric part of the coefficient of $D_n\xi_m$ must therefore be zero. We then deduce[2] that the Riemann tensor of a maximally symmetric space must have the form

$$R_{ijkl} = \mathcal{K}(g_{ik}g_{jl} - g_{il}g_{jk})\,, \tag{17.1}$$

---

[1] An exhaustive presentation of anisotropic or inhomogeneous spaces of cosmological interest, which we do not discuss here, can be found in, for example, Ellis, Maartens, and MacCallum (2012).

[2] By (twice) contracting the antisymmetric part (equal to zero) of the coefficient of $D_n\xi_m$ and then using the fact that the coefficient of ξ_m is zero; see, for example, Stephani (1990).

Relativity in Modern Physics. Nathalie Deruelle and Jean-Philippe Uzan.
© Oxford University Press 2018. Published in 2018 by Oxford University Press.
DOI: 10.1093/oso/9780198786399.001.0001

where $\mathcal{K}$ is a constant with dimension L^{-2}. The space is said to have *constant curvature*, and its Ricci tensor and scalar curvature are

$$R_{ij} = \mathcal{K}(N-1)g_{ij}, \quad R = \mathcal{K}N(N-1). \tag{17.2}$$

It can be shown, as we assume, that maximally symmetric spaces which have constant curvature are uniquely determined by the dimension of the space N, the metric signature, and the value[3] of $\mathcal{K}$.

Since spaces of constant curvature are essentially unique, any method of obtaining their metric g_{ij} is good. The case $\mathcal{K} = 0$ is trivial because it corresponds to Euclidean space. To deal with the case $\mathcal{K} \neq 0$, we consider the (pseudo)-Euclidean space of dimension $N+1$ whose metric in Cartesian coordinates X^i and w (w is dimensionless) is

$$ds^2 = \mathcal{K}^{-1}dw^2 + f_{ij}dX^i dX^j, \tag{17.3}$$

where the metric f_{ij}, $(i,j = 1, \dots N)$, has the same signature as g_{ij} (for example, $f_{ij} = \delta_{ij}$ if the space is Riemannian, or $f_{ij} = \eta_{ij}$ if it is pseudo-Riemannian). Let us consider in this space the $(N-1)$-surface described by the equation

$$w^2 + \mathcal{K}f_{ij}X^i X^j = 1. \tag{17.4}$$

On this surface $dw^2 = \mathcal{K}^2(f_{ij}X^i dX^j)^2/(1 - \mathcal{K}f_{ij}X^i X^j)$, and the metric induced on it is

$$d\sigma^2 = g_{ij}dX^i dX^j \quad \text{with} \quad g_{ij} = f_{ij} + \mathcal{K}\frac{f_{ik}f_{jl}X^k X^l}{1 - \mathcal{K}f_{kl}X^k X^k}. \tag{17.5}$$

After an easy calculation (using the fact that $\Gamma^i_{jk} = \mathcal{K}X^i g_{jk}$), we see that the Riemann tensor has the form (17.1). Therefore, (17.5) is the desired metric.

In the case where $N = 2$, $f_{ij} = \delta_{ij}$, and $\mathcal{K} > 0$, we easily see, setting $X^1 = \mathcal{K}^{-1/2}\sin\theta\cos\phi$ and $X^2 = \mathcal{K}^{-1/2}\sin\theta\sin\phi$, that the metric (17.5) is then written as $d\sigma^2 = (d\theta^2 + \sin^2\theta\, d\phi^2)/\mathcal{K}$. The space is a 2-sphere of radius $1/\sqrt{\mathcal{K}}$.

When $N = 3$ and $f_{ij} = \delta_{ij}$, we set

$$(X^1, X^2, X^3) = \begin{cases} |\mathcal{K}|^{-1/2}\sin\chi(\sin\theta\cos\phi, \sin\theta\sin\phi, \cos\theta) \\ |\mathcal{K}|^{-1/2}\text{sh}\chi(\sin\theta\cos\phi, \sin\theta\sin\phi, \cos\theta) \end{cases} \text{if} \begin{cases} \mathcal{K} > 0 \\ \mathcal{K} < 0 \end{cases}, \tag{17.6}$$

and we find that if we set $\mathcal{K} \equiv K/R_c^2$, the metric (17.5) can be written as

$$d\sigma^2 = R_c^2\, d\Omega_3^2, \quad \text{where} \quad d\Omega_3^2 = d\chi^2 + f_K^2(\chi)(d\theta^2 + \sin^2\theta d\phi^2)$$

$$\text{with} \quad f_K(\chi) = \begin{cases} \sin\chi & \text{if } K = 1, \\ \sinh\chi & \text{if } K = -1, \\ \chi & \text{if } K = 0, \end{cases} \tag{17.7}$$

where it is understood that in the case $K = 0$, which we have added, the space is Euclidean, and the *radius of curvature* R_c becomes an arbitrary constant which gives the dimension of

---

[3] More precisely, if there exist two metrics with the same signature and dimension $g_{ij}(x^k)$ and $g'_{ij}(x^k)$ such that $R_{ijkl} = \mathcal{K}(g_{ik}g_{jl} - g_{il}g_{jk})$ and $R'_{ijkl} = \mathcal{K}(g'_{ik}g'_{jl} - g'_{il}g'_{jk})$, then there exists a change of coordinates which transforms g into g' (Eisenhart, 1949); see, for example, Weinberg (1972).

a length to the coordinate $R_c\chi$. The space is a 3-sphere $\mathcal{S}_3$ $(K = 1)$, a Euclidean 3-plane $\mathcal{E}_3$ $(K = 0)$, or a 3-hyperboloid $\mathcal{H}_3$ $(K = -1)$. Setting $r = f_K(\chi)$ (where the 'radial' coordinate r is dimensionless), (17.7) can also be written as

$$d\Omega_3^2 = \frac{dr^2}{1 - Kr^2} + r^2(d\theta^2 + \sin^2\theta d\phi^2). \tag{17.8}$$

Finally, with $\bar{r} = \dfrac{2r}{1 + \sqrt{1 - Kr^2}}$ we have

$$d\Omega_3^2 = \frac{1}{1 + K\bar{r}^2/4}[d\bar{r}^2 + \bar{r}^2(d\theta^2 + \sin^2\theta d\phi^2)]. \tag{17.9}$$

We note that the volume of $\mathcal{S}_3$ is finite:

$$\int d^3V = \int \sqrt{g}\, d\chi d\theta d\phi = R_c^3 \int \sin^2\chi \sin\theta\, d\chi d\theta d\phi = 2\pi^2 R_c^3.$$

The spaces $\mathcal{E}_3$ and $\mathcal{H}_3$ have infinite volume, unless they are supplied with a non-trivial topology (for example, a torus for a Euclidean space).

17.2 Spacetimes with homogeneous and isotropic sections

Let V_4 be a four-dimensional pseudo-Riemannian space. We assume that there exists a field of time-like vectors (which can be considered as tangents to the world lines) such that all the spatial sections of V_4 orthogonal to this field are maximally symmetric. A manifold with such a foliation is a *Friedmann (1922)–Lemaître (1927)–Robertson–Walker (1935)* space (called FLRW space), and its length element can be written as

$$ds^2 = -dt^2 + a^2(t)\, d\Omega_3^2 \quad \text{with} \quad d\Omega_3^2 = d\chi^2 + f_K^2(\chi)(d\theta^2 + \sin^2\theta d\phi^2), \tag{17.10}$$

where $f_K = \{\sin, I, \sinh\}$ corresponds to $K = \{-1, 0, 1\}$. Here $d\Omega_3^2$ is the metric of a 3-sphere, a 3-plane, or a 3-hyperboloid, and is given in other coordinate systems in eqns (17.7)–(17.9). (A term proportional to $dt\, dx^i$ will break the isotropy, and the coefficient of dt^2 can always be normalized to $c^2 = 1$.)

The time coordinate t is the *cosmic time*, and $a(t)$, which has the dimension of a length, is the *scale factor*. Finally,

$$H \equiv \frac{\dot{a}}{a}, \tag{17.11}$$

where the dot denotes the derivative with respect to t, is the *Hubble function*. It is also common to use the *conformal time* η, defined as $d\eta = dt/a(t)$.

By construction, the coordinate lines (r, θ, ϕ constant) are geodesics.

This can be seen directly. Since the equation of a geodesic with tangent vector u^μ is $du^\mu/d\tau + \Gamma^\mu_{\nu\rho}u^\nu u^\rho = 0$, if the initial 3-velocity is zero, $u^i|_{(\tau=0)} = 0$, it will remain zero because $(du^i/d\tau)|_{(\tau=0)} = 0$ since $\Gamma^i_{00} = 0$ for $i = (1, 2, 3)$.

These coordinate lines therefore represent the trajectories of test particles in free fall whose proper time is the cosmic time t ($dt \equiv u_\mu dx^\mu$), and the coordinate system (t, r, θ, ϕ) is said to be *comoving*.

The Riemann and Ricci tensors, the scalar curvature, and the Einstein tensor of the metric (17.10) are calculated using the definitions simplified by the fact that the spatial sections are maximally symmetric. We find [here $i = (1, 2, 3)$]

$$R_{ijkl} = \left(\frac{K}{a^2} + H^2\right)(g_{ik}g_{jl} - g_{il}g_{jk}), \quad R_{0i0j} = -\frac{\ddot{a}}{a}g_{ij}, \tag{17.12}$$

$$G^0_0 = -3\left(\frac{K}{a^2} + H^2\right), \quad G^i_j = -\left(\frac{2\ddot{a}}{a} + \frac{K}{a^2} + H^2\right)\delta^i_j, \quad R = 6\left(\frac{\ddot{a}}{a} + \frac{K}{a^2} + H^2\right), \tag{17.13}$$

where $K = (+1, 0, -1)$ for spherical, Euclidean, or hyperbolic spatial sections.

17.3 Milne spacetime

For an FLRW spacetime to have spatial sections which are not only homogeneous and isotropic but also to be maximally symmetric, that is, to possess four additional isometries, it is necessary and sufficient that its Riemann tensor can be written in the form (17.1), that is, owing to (17.12),

$$\left(\frac{\dot{a}}{a}\right)^2 + \frac{K}{a^2} = K_4. \tag{17.14}$$

(If $\dot{a} \neq 0$, this condition also implies that $\ddot{a}/a = K_4$.)

If $K_4 = 0$, the flat spacetime is just Minkowski spacetime $\mathcal{M}_4$.

If in addition we choose a foliation using Euclidean spatial sections $K = 0$, then the conditions (17.14) give the length element in a familiar form (after the dimensional constant a is absorbed in the coordinate r):

$$ds^2 = -dt^2 + dr^2 + r^2(d\theta^2 + \sin^2\theta d\phi^2). \tag{17.15}$$

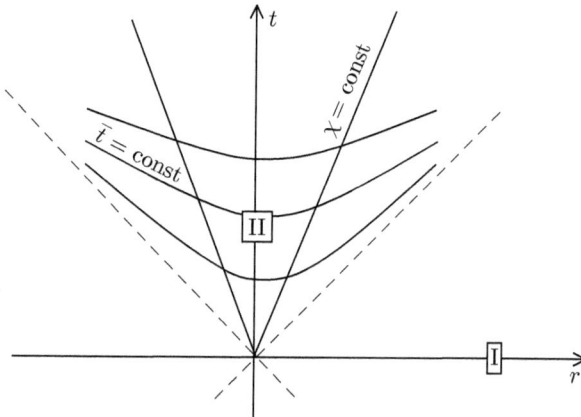

Fig. 17.1 Milne coordinates.

If we choose[4] $K = -1$, then $a = \bar{t}$ and the length element takes the *Milne* form (1935):

$$ds^2 = -d\bar{t}^2 + \bar{t}^2[d\chi^2 + \sinh^2 \chi(d\theta^2 + \sin^2 \theta d\phi^2)]. \tag{17.16}$$

Minkowski spacetime can therefore be viewed as an FLRW spacetime with hyperbolic spatial sections and scale factor proportional to $\bar{t}$ (Fig. 17.1). The change of coordinates $(t, r) = \bar{t}$ $(\cosh \chi, \sinh \chi)$ takes (17.16) to the previous form (17.15), which shows explicitly that $\bar{t} = 0$ is not a singularity of the spacetime (we mention in passing that the Milne coordinates do not cover all of $\mathcal{M}_4$). An observer comoving at $\chi = \chi_0$ is just an inertial observer undergoing displacement from the origin at speed $\tanh \chi_0$. It is useful to compare the Milne coordinates and the Rindler coordinates introduced in Section 6.2. The Milne coordinates cover quadrant II of the Minkowski spacetime, while the Rindler coordinates cover quadrant I.

17.4 de Sitter spacetime

A maximally symmetric FLRW spacetime, whose scale factor therefore satisfies (17.14) and for which the constant $K_4 \equiv H^2$ is positive, is called a *de Sitter* spacetime (1917). Several spatial foliations are possible [the first is due to Lemaître (1925), and the second to Klein (1918)]:

$$\begin{cases} K = 0 \implies ds^2 = -dt^2 + e^{2Ht}(dx^2 + dy^2 + dz^2), \\[2mm] K = +1 \implies ds^2 = -d\bar{t}^2 + \dfrac{\cosh^2 H\bar{t}}{H^2}[d\chi^2 + \sin^2 \chi(d\theta^2 + \sin^2 \theta d\phi^2)], \\[2mm] K = -1 \implies ds^2 = -d\tilde{t}^2 + \dfrac{\sinh^2 H\tilde{t}}{H^2}[d\tilde{\chi}^2 + \sinh^2 \tilde{\chi}(d\theta^2 + \sin^2 \theta d\phi^2)]. \end{cases} \tag{17.17}$$

In these coordinates the spatial sections of spacetime are manifestly maximally symmetric (they are 3-planes, 3-spheres, or 3-hyperboloids); however, the fact that de Sitter spacetime is maximally symmetric and therefore also static is not manifest.

Since de Sitter spacetime is maximally symmetric, it has the geometry of a 4-hyperboloid (it is non-flat and non-compact), and it can be embedded in a five-dimensional (pseudo-Euclidean) space. We can visualize this by suppressing two spatial dimensions. The intersections of the resulting 2-hyperboloid and the planes (parabolas, circles, or hyperbolas) represent the spatial sections of the three coordinate systems given above. It is therefore easy to see that only the foliation $K = +1$ covers the entire hyperboloid, that is, the entire de Sitter spacetime; see Fig. 17.2.

Finally, there exists another coordinate system $(\tau, \rho, \Theta, \Phi)$ (historically the first, introduced by de Sitter himself in 1917) in which the length element is written as

$$ds^2 = -(1 - \rho^2 H^2)d\tau^2 + \frac{d\rho^2}{1 - \rho^2 H^2} + \rho^2(d\Theta^2 + \sin^2 \Theta d\Phi^2). \tag{17.18}$$

---

[4]We note that $\mathcal{M}_4$ cannot be foliated by 3-spheres.

In this system the spatial sections are manifestly homogeneous and isotropic [they are 3-spheres; *cf.* (17.8)]. Since, in addition, the metric components are diagonal and independent of the time τ, the static nature of the spacetime is also manifest.

Foliations of de Sitter spacetime

We introduce the five-dimensional space with length element $ds^2 = dw^2 - dT^2 + dX^2 + dY^2 + dZ^2$. The de Sitter spacetime is the 4-hyperboloid of equation $w^2 - T^2 + X^2 + Y^2 + Z^2 = 1$. Then, using (17.4) and (17.5), we find that the induced metrics on this de Sitter 4-hyperboloid are written in the form (17.17) if the coordinates (t, x, y, z), $(\bar{t}, \chi, \theta, \phi)$, and $(\tilde{t}, \tilde{\chi}, \theta, \phi)$ are related to (T, X, Y, Z) as (setting $H = 1$)

$$
\begin{cases}
K = 0, \quad X = e^t x, \quad Y = e^t y, \quad Z = e^t z, \quad T = \sinh t + e^t (x^2 + y^2 + z^2)/2\,; \\
K = +1, \quad X = \cosh \bar{t} \sin \chi \sin \theta \cos \phi, \quad Y = \cosh \bar{t} \sin \chi \sin \theta \sin \phi, \\
\qquad\quad Z = \cosh \bar{t} \sin \chi \cos \theta, \quad T = \sinh \bar{t}\,; \\
K = -1, \quad X = \sinh \tilde{t} \sinh \chi \sin \tilde{\theta} \cos \tilde{\phi}, \quad Y = \sinh \tilde{t} \sinh \chi \sin \theta \sin \phi, \\
\qquad\quad Z = \sinh \tilde{t} \sinh \chi \cos \theta, \quad T = \sinh \tilde{t} \cosh \chi\,.
\end{cases}
$$

Next, we find that the 3-spaces of constant time are given by the following:

- $K = 0$: the surfaces $t = $ const are 3-paraboloids, intersections of the de Sitter 4-hyperboloid and the 3-planes $w + T = e^t$;

- $K = +1$: the surfaces $\bar{t} = $ const are 3-spheres, intersections of the de Sitter 4-hyperboloid and the 3-planes $T = \sinh \bar{t}$;

- $K = -1$: the surfaces $\tilde{t} = $ const are 3-hyperboloids, intersections of the 4-hyperboloid and the 3-planes $w = \cosh \tilde{t}$.

The coordinates $(\tau, \rho, \Theta, \Phi)$ are related to (T, X, Y, Z) as

$$
X = \rho \sin \Theta \cos \Phi, \quad Y = \rho \sin \Theta \sin \Phi, \quad Z = \rho \cos \Theta, \quad T = \sinh \tau \sqrt{1 - \rho^2}\,.
$$

They do not cover the entire de Sitter hyperboloid. The sections $\tau = $ const are 3-hemispheres; see Fig. 17.2.[5]

---

[5]See Moschella (2005), from which we have borrowed these figures.

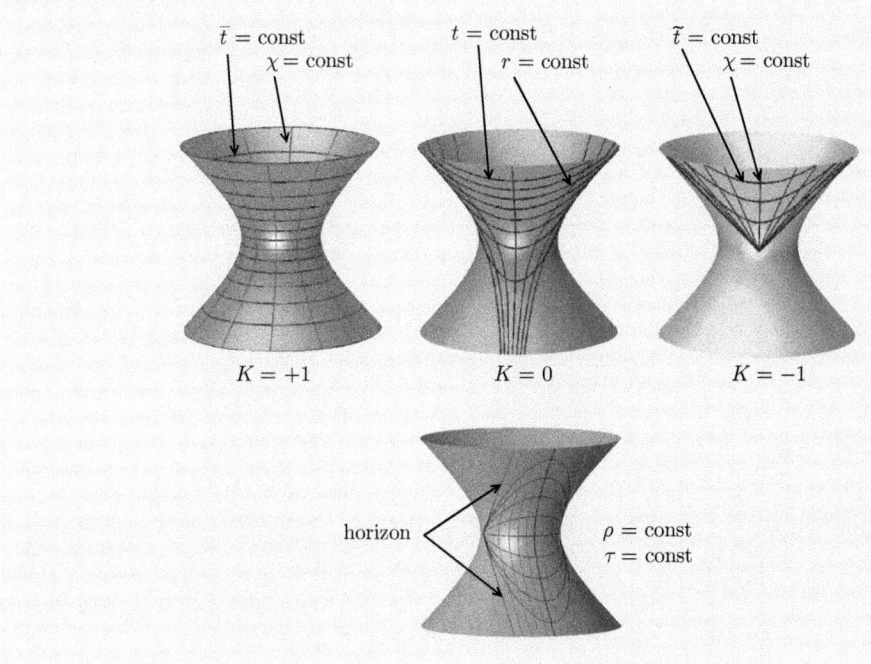

Fig. 17.2 Foliations of de Sitter spacetime.

De Sitter spacetime is often used as a simplified model of the universe, for example, in inflationary theories; see Chapter 20. It is also an ideal example for showing that complicated length elements may hide simple geometries, and that metric coefficients which are singular at certain points [for example, (17.18) at $\rho = 1/H$ or (17.17) at $\tilde{t} = 0$] sometimes signal a simple pathology of the coordinate system rather than a singularity of the curvature tensor and therefore of the spacetime itself.[6]

If $K_4 = -H^2$, the spacetime is called *anti-de Sitter* spacetime. Its geometrical structure is more complicated (it contains closed time-like curves).[7] It is useful for studying the AdS/CFT (anti-de Sitter/conformal field theory) correspondence.[8]

---

[6]The study of the geometry of de Sitter spacetime has historically played an important role in understanding the Schwarzschild singularity at $r = 2m$. See, for example, Eisenstaedt (1989).

[7]See, for example, Hawking and Ellis (1973).

[8]Maldacena (1998).

18
Friedmann–Lemaître spacetimes

The cosmological solutions on which the standard model is based satisfy the Copernican principle. They possess homogeneous and isotropic spatial sections, and the space is expanding. In this chapter we present the laws governing the evolution of the scale factor as well as Hubble's law, which is historically the first observational signature of cosmic expansion.

18.1 Redshift and luminosity distance

The Weyl postulate

We have seen that the decision to represent the universe by an FLRW spacetime is referred to as the 'cosmological principle': there is no privileged point in space or any privileged direction, and so the spatial sections must be maximally symmetric.

In the coordinate systems considered in the preceding sections where the metric of FLRW spacetimes is written, for example, in the form (17.10),

$$ds^2 = -dt^2 + a^2(t)\left[d\chi^2 + f_K^2(\chi)(d\theta^2 + \sin^2\theta d\phi^2)\right] \tag{18.1}$$

with $f_K = \{\sin, I, \sinh\}$ corresponding to $K = \{-1, 0, 1\}$, we have seen that the 'comoving' coordinate lines are geodesics.

The *Weyl postulate* (1923) stipulates that the 'cosmological fluid' consisting of galaxies, quasars, and so on, *visible or invisible*, follows such geodesics.

Let us consider a 'galaxy' which emits light. We shall represent it by a 'comoving' geodesic with $\chi = \chi_g$ (and, for example, $\theta = \pi/2$, $\phi = 0$) of the Friedmann–Lemaître metric given above (see Fig 18.1).

Let us consider the light arriving at an observer located at $\chi = 0$. Two successive peaks of the wave arrive at t_0 and $t_0 + \Delta t_0$, where Δt_0, the cosmic time interval, is also the proper time interval according to the observer's clock. The light was emitted by the galaxy located at χ_g at times t_e and $t_e + \Delta t_e$. It has travelled along a radial null geodesic: $ds^2 = 0 = -dt^2 + a^2 d\chi^2$. We therefore find that $dt/a(= -d\chi)$ is independent of the time, and so $\Delta t_e/a_e = \Delta t_0/a_0$, or also, introducing the light frequency $\nu = 1/\Delta t$,

$$1 + z \equiv \frac{\nu_e}{\nu_0} = \frac{a_0}{a_e}. \tag{18.2}$$

Let us make the meaning of eqn (18.2) more precise. A spectrum of rays of frequencies ν_0 is observed coming from a (comoving) galaxy. This spectrum is recognized as being that of an atomic transition of frequency ν on the (comoving) Earth which is systematically (and achromatically) red-shifted. If the Einstein equivalence principle is valid, that is, if the

Relativity in Modern Physics. Nathalie Deruelle and Jean-Philippe Uzan.
© Oxford University Press 2018. Published in 2018 by Oxford University Press.
DOI: 10.1093/oso/9780198786399.001.0001

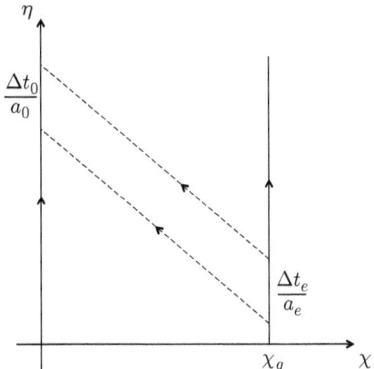

Fig. 18.1 Redshift.

gravitational field can be locally effaced,[1] so that the laws of non-gravitational physics apply in a freely falling frame, then this frequency ν will be the frequency of the transition which is measured locally *wherever the observer is*, and so $\nu_e = \nu$. Therefore, the *redshift* z of the galaxy,

$$1 + z = \frac{\nu_e}{\nu_0} = \frac{\nu}{\nu_0}, \tag{18.3}$$

becomes a measurable quantity which is related by (18.2) to the increase of the scale factor between $a(t_e) \equiv a_e$ and $a(t_0) \equiv a_0$.

The galaxy is referenced by its co-moving coordinates $\chi = \chi_g$ (and, for example, $\theta = \pi/2$, $\phi = 0$). This co-moving coordinate distance is an intermediary of the calculation (like the cosmic time t), well adapted to the integration of the geodesic equations, but not measurable.

We can define the *proper distance* D_p of the galaxy to the origin (the location of the observer) at the time t using the length element $ds|_t \equiv dl = a(t)d\chi$:

$$D_p = a(t)\chi. \tag{18.4}$$

This distance would be measurable if we had at our disposal a rigid ruler lying between the Earth and the galaxy, which is not the case in reality. Nevertheless, it can be stated that if $a(t)$ is an increasing function of the time, which is the case if the galactic spectra are systematically shifted toward the red, then the proper distance between two objects will increase with time. The universe (or, more correctly, its spatial sections) is therefore *expanding*.

However, a quantity which in principle is measurable is the *luminosity distance* D_L defined as

$$l = \frac{L}{4\pi D_L^2}, \tag{18.5}$$

where l is the measured apparent luminosity of the galaxy and L is its absolute luminosity (which is assumed to be known or measurable by relating it to the physical characteristics

---

[1] This does not occur in all theories of gravitation; for example, in the tensor–scalar theory the scalar field cannot be effaced. See, for example, Damour and Esposito-Farèse (1992).

of *standard candles*, for example, the period of the luminosity variation of Cepheid stars, or the light curve of type-Ia supernovae, and so on). This luminosity distance can be related to the coordinate distance χ_g using the following arguments. The absolute luminosity is given by $L = Nh\nu_e/\Delta t_e = Nh\nu_e^2$, where N is the number of photons emitted during a period Δt_e and $h\nu_e$ is their energy. The apparent luminosity is given by $l = N(h\nu_0/\Delta t_0)(1/S)$, where $S = 4\pi a_0 f_K(\chi_g)$ is the surface of the sphere reached by these photons at t_0 [the instant they are observed; see (18.1)]. Therefore, $D_L \equiv \sqrt{L/4\pi l} = a_0 f_K(\chi_g)\nu_e/\nu_0$, or, using (18.2),

$$D_L = a_0(1 + z_e)f_K(\chi_g). \tag{18.6}$$

18.2   Hubble's law

Hubble's law relates two measurable quantities, the redshift and the luminosity distance of a galaxy. Using (18.2) and (18.6) as well as the equation of the trajectory followed by the radiated photons, we obtain

$$\chi_g = \int_{t_e}^{t_0} \frac{dt}{a(t)}, \quad z_e = \frac{a_0}{a_e} - 1, \quad D_L = a_0(1 + z_e)f_K(\chi_g). \tag{18.7}$$

If the matter composition of the universe is known, then, as we shall see later on, the Einstein equations determine the function $a(t)$. Then the first equation above gives $a_0\chi_g$ as a function of t_0 and t_e, the second gives $a_0\chi_g$ as a function of z_e, and the third predicts the relation between D_L and z_e. Comparison of this prediction with observations then tells us whether or not the description of the matter content of the universe is correct.[2]

Scale factors and Hubble's law: examples

We assume that the spatial sections are Euclidean ($K = 0$) and that the scale factor obeys a power law: $a = \alpha t^q$, where α and q are two constants. Use (18.7) to show that

$$H_0 D_L = \frac{q}{1-q}(1+z)\left[1 - (1+z)^{\frac{q-1}{q}}\right] = z\left(1 + \frac{1-q_0}{2}z + \cdots\right), \tag{18.8}$$

where $H_0 = \dot{a}_0/a_0 = q/t_0$ and $q_0 \equiv -\ddot{a}_0 a_0/\dot{a}_0^2 = (1-q)/q$.
For $a = \alpha e^{H_0 t}$ we find

$$H_0 D_L = z(1+z). \tag{18.9}$$

Let us expand the scale factor $a(t)$ in a Taylor series:

$$a(t) = a_0\left[1 + H_0(t-t_0) - \frac{1}{2}q_0 H_0^2(t-t_0)^2 + \cdots\right] \quad \text{with} \quad H_0 \equiv \frac{\dot{a}_0}{a_0}, \quad q_0 \equiv -\frac{\ddot{a}_0 a_0}{\dot{a}_0^2}, \tag{18.10}$$

where H_0 is the *Hubble constant* and q_0 is the *deceleration parameter*. The first two equations in (18.7) then become

---

[2] We note that the relation between the luminosity distance D_L and the redshift z can be obtained directly by rewriting χ_g as $a_0\chi_g(z) = \int_0^z dz'/H(z')$. It is therefore sufficient to know $H(z)$, which is given by the Friedmann equations (Section 18.3), and it is useless in practice to calculate $a(t)$ explicitly, as we shall see in Section 19.3.

$$\begin{cases} a_0\chi_{\mathrm{g}} = (t_0 - t_{\mathrm{e}}) \left[1 + \frac{1}{2}H_0(t_0 - t_{\mathrm{e}}) + \cdots \right], \\[2ex] \dfrac{z_{\mathrm{e}}}{H_0} = (t_0 - t_{\mathrm{e}}) \left[1 + \left(1 + \frac{1}{2}q_0\right) H_0(t_0 - t_{\mathrm{e}}) + \cdots \right], \end{cases} \tag{18.11}$$

$$\implies \quad a_0\chi_{\mathrm{g}} = \frac{z_{\mathrm{e}}}{H_0}\left[1 - \left(\frac{1+q_0}{2}\right)z_{\mathrm{e}} + \cdots \right],$$

and so the third equation in (18.7) is written as

$$H_0 D_L = z_{\mathrm{e}}\left[1 - \frac{z_{\mathrm{e}}}{2}(q_0 - 1)\right] + \dots \tag{18.12}$$

At lowest order the redshift and the luminosity distance are proportional. This is *Hubble's law*, established observationally by Hubble in the 1930s. In principle, the deviations from linearity give q_0, as we shall see in Section 19.3.

18.3 The Friedmann–Lemaître equations

The Einstein equations are

$$G_{\mu\nu} + \Lambda g_{\mu\nu} = 8\pi G\, T_{\mu\nu}, \tag{18.13}$$

where we have included the cosmological constant Λ, and $T_{\mu\nu}$ is the energy–momentum tensor of the matter.

If the spacetime is maximally symmetric or possesses a family of maximally symmetric subspaces V_N of dimension N, coordinates x^a, and metric g_{ab}, the tensors describing the matter must possess the same symmetries. Just like those of the metric, their Lie derivatives with respect to the Killing vectors must then be zero. We thus see that the scalars must not depend on x^a, the vectors must have null components in V_N, and the components T_{ab} of symmetric 2-fold covariant tensors must be proportional to those of the metric tensor g_{ab}.

Therefore, owing to the symmetry imposed on FLRW models of the universe by the cosmological principle, the energy–momentum tensor representing the ensemble of matter constituents of the universe must have the form

$$T_{\mu\nu} = (\epsilon + p)u_\mu u_\nu + p g_{\mu\nu}, \tag{18.14}$$

where u^μ is a time-like vector field which can be normalized to unity ($g_{\mu\nu}u^\mu u^\nu = -1$) and which is perpendicular to the maximally symmetric 3-spaces, and ϵ and p are constant functions in these 3-spaces. If the coordinates are chosen such that the FLRW metric can be written in the standard form (17.10) or (18.1), then ϵ and p are functions of only the cosmic time t and $u^\mu = (1, 0, 0, 0)$.

We see that $T_{\mu\nu}$ is a perfect fluid (see Section 3.2). Then u^μ is interpreted as the 4-velocity of the 'cosmological fluid', ϵ is its energy density, and p its pressure. This fluid, just like a gas in thermodynamical equilibrium, is uniform throughout space and evolves only in cosmic time.

The components of the Einstein tensor for a Friedmann–Lemaître metric in the coordinates (17.10) have been given in (17.13). In addition [see (18.14)], $T_0^0 = -\epsilon$ and $T_j^i = p\,\delta_j^i$. The Einstein equations are then written as (Friedmann, 1922; Lemaître, 1925)

$$H^2 = \frac{8\pi G}{3}\epsilon - \frac{K}{a^2} + \frac{\Lambda}{3}\,, \quad \frac{\ddot{a}}{a} = -\frac{4\pi G}{3}(\epsilon + 3p) + \frac{\Lambda}{3}\,, \tag{18.15}$$

where we recall that $a(t)$ is the scale factor, t is the cosmic time, $H \equiv \dot{a}/a$, and $K = \{0, +1, -1\}$ for Euclidean, spherical, or hyperbolic spatial sections.

The Einstein equations contain the equations of motion of the matter by virtue of the Bianchi identities $D_\mu G^{\mu\nu} \equiv 0$. For a fluid, these equations, $D_\mu T^{\mu\nu} = 0$, are the conservation and Euler equations (3.10) and (3.11). In a Friedmann–Lemaître spacetime the Euler equation is satisfied identically and the conservation equation is $D_\mu T^{\mu 0} = \partial_\mu T^{\mu 0} + \Gamma^\mu_{\mu\nu} T^{\nu 0} + \Gamma^0_{\mu\nu} T^{\mu\nu} = 0$, or, writing out the Christoffel symbols explicitly, $\dot{T}^{00} + \Gamma^i_{i0} T^{00} + \Gamma^0_{ij} T^{ij} = 0$, or also

$$\dot{\epsilon} + 3H(\epsilon + p) = 0\,. \tag{18.16}$$

As we can see by taking the derivative of the first equation in (18.15) with respect to time, eqn (18.16) can replace the second equation in (18.15) as long as the scale factor is not constant.

It is sometimes practical to write the Friedmann–Lemaître equations (18.15) and (18.16) using the conformal time η related to the cosmic time as $dt = a\,d\eta$:

$$\mathcal{H}^2 = \frac{8\pi G}{3}\epsilon a^2 - K + \frac{\Lambda}{3}a^2\,, \quad \mathcal{H}' = -\frac{4\pi G}{3}(\epsilon + 3p)a^2 + \frac{\Lambda}{3}a^2\,, \quad \epsilon' + 3\mathcal{H}(\epsilon + p) = 0\,, \tag{18.17}$$

where a prime denotes the derivative with respect to η and $\mathcal{H} \equiv a'/a = aH$.

18.4   The first models of the universe (1917–1960)

• *The Einstein static model*

In 1917 the galaxies and their systematic recession were unknown (the Great Debate between Shapley and Curtis about the distance and nature of nebulae dates from the 1920s, and Hubble's law from the 1930s). Moreover, the advance judgment of the universe as being static was well established (the Newtonian cosmological models of expansion also date from the 1930s). Finally, the problems which arise in representing a homogeneous and isotropic universe of infinite extent in Newtonian physics were well known (see Book 2, Chapter 16).

It was probably for these reasons that Einstein sought a solution to his equations which was static, homogeneous, isotropic, and had spherical spatial sections (and therefore finite volume), where the matter is basically a pressure-less fluid (we recall that the cosmic microwave background radiation was not discovered until the 1960s).

The Einstein–Friedmann–Lemaître equations (18.15) then reduce to

$$8\pi G \epsilon_0^m + \Lambda - \frac{3}{a_0^2} = 0\,, \quad 4\pi G \epsilon_0^m - \Lambda = 0\,, \tag{18.18}$$

where ϵ_0^m is the constant energy density. We see that the presence of the cosmological constant is needed for the system to have a solution.

The introduction of this new fundamental constant was later viewed by Einstein as his 'biggest blunder', and he abandoned it definitively after accepting the validity of the work done by Friedmann and Lemaître, and after Hubble's discovery of the recession of the galaxies, an observation which indicated that the universe is expanding.[3]

• *The de Sitter and steady-state models*

An additional motivation that Einstein had for introducing the cosmological constant in his gravitational equations was the hope that the vacuum equations, $G_{\mu\nu} + \Lambda g_{\mu\nu} = 0$, would then not have *any* solution, neither the Schwarzschild solution (the only one known in 1917), nor even the Minkowski solution, so that the theory would not allow there to be any empty point in space (see footnote 12 of Section 1.6). However, in 1917 de Sitter found the solution bearing his name in the static form given in (17.18) with[4] $H^2 = \Lambda/3$. Therefore, the de Sitter universe is devoid of ordinary matter and curved only due to the cosmological constant, and the galaxies must be viewed as test particles following geodesics. The question is, which ones? Indeed, since the de Sitter spacetime is maximally symmetric, as we have seen in Section 17.4, it does not possess a privileged foliation, and various choices of cosmic time are possible.

In a system of adapted coordinates, the length element of de Sitter spacetime can be cast in the static form (17.18). First of all, we note that the light emitted by a particle (a galaxy) which is at rest at $\rho = \rho_g$ in this frame will be observed at $\rho = 0$ with a redshift given by $1 + z_e \equiv \nu_e/\nu_0 = (1 - (H\rho_g)^2)^{-1/2}$, from which we find $H\rho_g = \sqrt{z_e(2 + z_e)}/(1 + z_e)$. Since the luminosity distance is related to ρ_g as $HD_L = H\rho_g(1 + z_e)$ [cf. (18.5)], Hubble's law for such galaxies is $HD_L = \sqrt{z_e(2 + z_e)}$. It is not linear for short distances, and the 'galaxies' do not follow geodesics of the spacetime. However, if the galaxies are in radial free fall, we find that $HD_L = z_e$ from an easy calculation like that of Section 6.4 for studying the radial geodesics in the Schwarzschild metric. In this case Hubble's law is strictly linear. In both cases the galactic redshift is not interpreted in terms of expansion of the universe, but as gravitational redshift combined, in the second case, with a Doppler effect. It seems that this is also how Hubble himself understood his law, at least initially—proof, if needed, that the significance assigned to a cosmological observation depends on the theoretical framework in which it is interpreted.[5]

---

[3]The expression 'biggest blunder' attributed to Einstein by Gamow is probably apocryphal. However, Einstein did write in a letter to Lemaître on Septembre 26, 1947, "*The introduction of such a constant implies a considerable renunciation of the logical simplicity of the theory... Since I have introduced this term I had always a bad conscience... I cannot help to feel it strongly and I am unable to believe that such an ugly thing should be realized in nature.*"

This static model was defended until the 1970s by partisans of the 'tired light' theory, an attempt to explain the galactic redshift in terms of an interaction between photons and the intergalactic medium. But, on the one hand, such an interaction is not predicted by the standard laws of electromagnetism, and, on the other, it would imply a dispersion of light which is not observed. Finally, the existence of a diffuse cosmological background of radiation cannot be explained within such a scenario.

The cosmological constant itself has again become a component of the standard model; see the following chapter.

[4]In addition, the Schwarzschild length element can be generalized as

$$ds^2 = -(1 - 2m/r - \Lambda r^2/3)dt^2 + dr^2/(1 - 2m/r - \Lambda r^2/3) + r^2(d\theta^2 + \sin^2\theta d\phi^2),$$

which solves the equation $G_{\mu\nu} + \Lambda g_{\mu\nu} = 0$; Kottler (1918).

[5]See the conclusion of the article Hubble (1929).

In 1948, Bondi, Hoyle, Gold, and Narlikar postulated that the galaxies follow the geodesics $x^i = $ const of de Sitter spacetime described in the frame where the spatial sections are Euclidean and the length element takes the form $ds^2 = -dt^2 + e^{2Ht}(dx^2 + dy^2 + dz^2)$; see Section 17.4. The scale factor is $a(t) = e^{Ht}$ and the observed galactic redshift is interpreted as an expansion of the spatial sections. The Hubble parameter is then a constant, $H = H_0$, and as we saw in Section 18.2, the redshift/luminosity distance relation is $H_0 D_L = z_e(1 + z_e)$. The deceleration parameter is $q_0 = -1$.

In such a model the proper distance between the galaxies increases exponentially. To describe a universe in a steady state, that is, a universe which is globally identical to itself over the course of time t, Bondi *et al.* proposed the continuous creation of matter so as to compensate for the dispersion of the geodesic flow of the galaxies. In the eyes of its proponents, this idea is *a priori* no more shocking than the *ex nihilo* creation of all matter in a Big Bang (the expression is due to F. Hoyle, 1949, who used it in a scornful manner).

However, the steady-state model was abandoned after the 1965 discovery by Penzias and Wilson of the cosmic microwave background radiation. As we shall see below, the existence of this radiation strongly suggests that the present universe has emerged from a hot, dense phase in a state of thermodynamical equilibrium in the past, and so it has a history and is not in a steady state.[6]

---

[6]In 1978 Penzias and Wilson were awarded the Nobel Prize in Physics for this discovery.
For an introduction to the history of modern cosmology, see, for example, the American Institute of Physics site http://www.aip.org/history/cosmology/index.htm.

19
The Lambda-CDM model of the hot Big Bang

In 1948, under the impetus of George Gamow, Robert Hermann, Ralph Alpher, and Hans Bethe in particular, relativistic cosmology entered the second phase of its history. In this phase, physical processes, in particular, nuclear and atomic processes, are taken into account. This provides two observational tests of the model: primordial nucleosynthesis, which explains the origin of light nuclei, and the existence of the cosmic microwave background, and it establishes the fact that the universe has a thermal history. Study of the large-scale structure of the universe then indicates the existence of dark matter and a nonzero cosmological constant. This model, known as the ΛCDM model, is the standard model of contemporary cosmology.

19.1 The matter content of the universe and its evolution

We have seen that the cosmological principle requires the matter present in the universe as a whole to behave as a perfect fluid. But the cosmological fluid can be made up of several components which can interact in ways other than gravitationally. It is usual to describe these components also as perfect fluids, at least in a first approximation, and to characterize them by their equations of state, that is, their relations $p = p(\epsilon)$. The energy–momentum tensor of the cosmological fluid, which must be conserved according to the Einstein equations, is then viewed as the sum of the energy–momentum tensors of the fluid components, which separately might not be conserved.[1]

The ensemble of luminous sources—nebulae, galaxies, and so on—along with the surrounding non-luminous 'ordinary' matter (that is, baryonic matter) constitutes the most familiar component of the cosmological fluid and was historically the first to be studied. According to the Weyl postulate, each of these sources is assumed to be nearly at rest in the Friedmann–Lemaître frame (18.1). They therefore make up a pressure-less fluid. The energy–momentum tensor of the baryonic component then is $T_{\mu\nu}^{\rm b} = \epsilon_{\rm b}(t) u_\mu u_\nu$ with $T_0^{0\,\rm b} = -\epsilon_{\rm b}(t)$ and $T_{ij}^{\rm b} = T_{0i}^{\rm b} = 0$.

It appears necessary to add to this baryonic fluid a (large) amount of non-baryonic matter, that is, matter which interacts only gravitationally with baryonic matter and which is cold, that is, nonrelativistic, and also pressure-less. This is called *cold dark matter* (CDM).[2] We write this contribution as $T_{\mu\nu}^{\rm cdm} = \epsilon_{\rm cdm}(t) u_\mu u_\nu$ and set $\epsilon_{\rm m}(t) = \epsilon_{\rm b}(t) + \epsilon_{\rm cdm}(t)$. Then

---

[1]This is the case, for example, for ionized matter (protons and electrons) and photons which are coupled *via* Compton scattering.

[2]There are numerous indications of the existence of this matter component: the dynamics of galactic clusters (F. Zwicky, 1937), the rotation curves of spiral galaxies (1980), the failure of strictly baryonic models of structure formation (1980), gravitational lensing (1985), and temperature anisotropies in the cosmic microwave background. There also exist numerous candidates for this matter component in extensions of the Standard Model of particle physics. An alternative explanation based on modified Newtonian dynam-

Relativity in Modern Physics. Nathalie Deruelle and Jean-Philippe Uzan.
© Oxford University Press 2018. Published in 2018 by Oxford University Press.
DOI: 10.1093/oso/9780198786399.001.0001

$$T_0^{0\,\mathrm{m}} = -\epsilon_\mathrm{m}(t)\,, \quad T_j^{i\,\mathrm{m}} = 0\,. \tag{19.1}$$

Moreover, the universe contains a radiation fluid dominated by the 2.7 K cosmic microwave background radiation (CMB) discovered by Penzias and Wilson in 1965, along with hot dark matter consisting of three families of neutrinos. Its equation of state is[3] $p_\mathrm{r} = \epsilon_\mathrm{r}/3$ with

$$T_0^{0\,\mathrm{r}} = -\epsilon_\mathrm{r}(t)\,, \quad T_j^{i\,\mathrm{r}} = \frac{1}{3}\,\epsilon_\mathrm{r}(t)\delta_j^i\,. \tag{19.2}$$

Since these fluids making up the matter in the universe interact only gravitationally, their energy–momentum tensors are separately conserved: $D_\mu T^{\mu\nu} = 0$, which can be written as above, $\dot\epsilon + 3H(\epsilon + p) = 0$. For the cold matter (baryons plus CDM, $p = 0$) and the radiation ($p = \epsilon/3$), this equation can be integrated at sight to give

$$\epsilon_\mathrm{m} = \frac{\epsilon_0^\mathrm{m} a_0^3}{a^3} = \epsilon_0^\mathrm{m}(1+z)^3\,, \quad \epsilon_\mathrm{r} = \frac{\epsilon_0^\mathrm{r} a_0^4}{a^4} = \epsilon_0^\mathrm{r}(1+z)^4\,, \tag{19.3}$$

where $\epsilon_0^\mathrm{m} a_0^3$ and $\epsilon_0^\mathrm{r} a_0^4$ are integration constants and we have used the relation (18.2) between the scale factor and the redshift.

The dilution factor of $1/a^3$ for the cold matter density is simply due to the volume expansion. The $1/a^4$ decrease of the radiation density can be explained as a combination of the effects of expansion and the fact that the energy of a photon is proportional to its frequency, which decreases as $1/a$ owing to the cosmological redshift. Since the background radiation is black-body radiation, the Stefan–Boltzmann law implies that its energy density is proportional to the fourth power of its temperature, and so

$$T \propto \frac{1}{a} \quad \text{or} \quad T = T_0(1+z)\,, \tag{19.4}$$

where $T_0 \approx 2.7$ K is the temperature measured today. Therefore, as Gamow noticed in the 1940s, any expanding universe containing a radiation fluid was hotter and always dominated by radiation in the past.[4]

19.2   Evolution of the scale factor

If the matter is just cold matter and radiation interacting only gravitationally and therefore evolving according to (19.3), the Friedmann Lemaître equations (18.15) reduce to a differential equation for $a(t)$ (with the condition that $H_0 \neq 0$):

$$\frac{H^2}{H_0^2} = \Omega_0^\mathrm{m}\left(\frac{a_0}{a}\right)^3 + \Omega_0^\mathrm{r}\left(\frac{a_0}{a}\right)^4 + \Omega_0^\Lambda + \Omega_0^K\left(\frac{a_0}{a}\right)^2\,, \quad \text{where} \quad H \equiv \frac{1}{a}\frac{da}{dt} \tag{19.5}$$

---

ics (MOND) at extremely small accelerations (M. Milgrom, 1983) now appears to be quite forced; see, for example, Peter and Uzan (2009), Chapter 7.

[3] The *speed of sound* c_s of a fluid is given by $c_\mathrm{s}^2 = dp/d\epsilon$ (see, for example, Book 1, Section 16.4). If it is constant, the fluid equation of state is $p = c_\mathrm{s}^2\epsilon$. A fluid of light composed of photons propagating at speed c isotropically in three-dimensional space with a velocity dispersion $c_\mathrm{s} = \sqrt{\langle c^2 \rangle} = c/\sqrt{3}$ therefore has the equation of state $p = \epsilon/3$. This holds for any ultrarelativistic matter.

[4] Here we shall not discuss the thermal history of the universe (primordial nucleosynthesis, the cosmic microwave background radiation, the decoupling of radiation and matter, and so on), nor the tools used to study it (the Boltzmann equation, the freeze-out of interactions, and so on). These topics are discussed in, for example, Peter and Uzan (2009) or Mukhanov (2005).

and we have introduced the *density parameters*

$$\Omega_0^m \equiv \frac{8\pi G\epsilon_0^m}{3H_0^2}, \quad \Omega_0^r \equiv \frac{8\pi G\epsilon_0^r}{3H_0^2}, \quad \Omega_0^K \equiv -\frac{K}{H_0^2 a_0^2}, \quad \Omega_0^\Lambda \equiv \frac{\Lambda}{3H_0^2} \tag{19.6}$$

related to each other as

$$1 = \Omega_0^m + \Omega_0^r + \Omega_0^\Lambda + \Omega_0^K. \tag{19.7}$$

Therefore, in this model, which has been used in observational cosmology since the 1960s, the Friedmann–Lemaître equation (19.5) gives the time evolution of the scale factor as a function of the Hubble constant H_0 and three independent density parameters by quadrature.

We see from (19.5) that for sufficiently small $a(t)$ it is the radiation fluid which governs the evolution of the universe, and the scale factor grows as $t^{\frac{1}{2}}$. At $t = 0$ the energy densities and curvature tensor diverge. This is the Big Bang, a spacetime singularity at which the Einstein equations are no longer valid. Moreover, close to the Big Bang the matter is hot, because the temperature of the radiation, which is inversely proportional to the scale factor, becomes infinite there.[5]

Next, for $a > a_{eq}$, where $a_{eq}/a_0 \approx \Omega_0^r/\Omega_0^m$, the cold matter dominates and, as long as the curvature term and the cosmological constant can be ignored, the scale factor grows as $t^{\frac{2}{3}}$. This is the Einstein–de Sitter solution (1932).

The ultimate evolution of the universe depends on whether or not the cosmological constant and curvature of the spatial sections vanish. For $\Omega_0^\Lambda = 0$ and $\Omega_0^K < 0$ ($K = +1$), for example, the universe will collapse on itself after its expansion. However, for $\Omega_0^\Lambda > 0$ the universe will tend asymptotically to a de Sitter spacetime.

Scale factor of a ΛCDM model

In the case of a spatially Euclidean universe ($K = 0$) containing only pressure-less matter and a cosmological constant, so that $\Omega_0^m = 1 - \Omega_0^\Lambda$, it can be verified that the scale factor, which is dimensionless because $K = 0$, is explicitly given by

$$a(t) = \left(\frac{\Omega_0^m}{\Omega_0^\Lambda}\right)^{1/3} \sinh^{2/3}\left(\frac{3}{2}\sqrt{\Omega_0^\Lambda} H_0 t\right), \tag{19.8}$$

a solution which interpolates between the $t^{2/3}$ behavior in the matter-dominated era and the $\exp\left(3\sqrt{\Omega_0^\Lambda} H_0 t/2\right)$ behavior in the era dominated by the cosmological constant.

Einstein–de Sitter models

In the case where the universe is spatially flat and dominated by a fluid with equation of state $p = w\epsilon$ with w constant, eqn (18.16) implies that $\epsilon = \epsilon_0(a/a_0)^{-3(1+w)}$, and according to

---

[5]The geometry near a space-like singularity can be very different from its homogeneous and isotropic limit far from the singularity. See the work on this subject by Belinski, Khalatnikov, and Lifshitz, summarized in Landau and Lifshitz (1971). For recent developments within the larger framework of supergravity theories see, for example, Damour and Lecian (2011).

the Friedmann equation (19.5) $H = \Omega_0 H_0 (a/a_0)^{-3(1+w)/2}$. If we assume that $w \neq -1$ (which corresponds to a nonzero cosmological constant), this equation can be integrated to give

$$a \propto t^q \quad \text{with} \quad q = \frac{2}{3(1+w)} \,. \tag{19.9}$$

Using the definition of conformal time $dt = a\,d\eta$, we find that $a \propto \eta^{2/(1+3w)}$ as long as $w \neq -1/3$, and $a \propto e^\eta$ if $w = -1/3$.

Strictly speaking, the term *Einstein–de Sitter* model (1932) refers to the particular model with $p = 0$. It therefore assumes that the spatial sections are Euclidean ($K = 0$) and that the universe, without a cosmological constant ($\Lambda = 0$), contains only a pressure-less fluid. According to (19.9), $a = a_0(t/t_0)^{2/3}$, where the age of the universe today is given by $t_0 = 2/(3H_0)$.

19.3   Parameter values

The values of the Hubble constant H_0 and the density parameters Ω_0^m, Ω_0^r, and Ω_0^Λ qualitatively determining the evolution of the universe in the hot Big Bang model are derived from a variety of astronomical observations.

Hubble's law (18.12) relating the luminosity distance and redshift is linear for small redshifts $H_0 D_L \approx z_e$ and gives H_0. The value that astronomers agree upon at present is[6]

$$H_0 = 100h \text{ km/s/Mpc} \quad \text{with} \quad h \approx 0.7 \tag{19.10}$$

(where 1 pc = 3.26 light-years). The corresponding *Hubble time* $1/H_0$ is of order $9.78\ h^{-1} \times 10^9$ yrs ~ 14 billion yrs, and the *Hubble radius* c/H_0 is of order $3000\ h^{-1}$ Mpc ~ 4000 Mpc.

The density of electromagnetic radiation is dominated by the black-body radiation of the microwave background, of temperature

$$T_0 = 2.725 \text{ K} \,. \tag{19.11}$$

The Stefan–Boltzmann law ($\epsilon_0^{\mathrm{CMB}} = \sigma T_0^4$) combined with (19.10) then gives the corresponding density parameter ($\Omega_0^{\mathrm{CMB}} = 8\pi G \epsilon_0^{\mathrm{CMB}}/3H_0^2$). The neutrino temperature and density are directly related to the photon temperature and density [$T_\nu = T_\gamma (4/11)^{1/3}$, $\epsilon_\nu = 21(4/11)^{4/3}\epsilon_\gamma/8$ for three neutrino families], and so the total radiation density parameter is

$$\Omega_0^r \approx 4 \times 10^{-5} \,. \tag{19.12}$$

If pushed beyond linear order using distant supernovae as standard candles, Hubble's law can be used to find a relation between Ω_0^m and Ω_0^Λ. As we saw in (18.7), Hubble's law is derived from

$$\chi_g = \int_{t_e}^{t_0} \frac{dt}{a(t)} \,, \quad z_e = \frac{a_0}{a_e} - 1 \,, \quad D_L = a_0(1 + z_e)f_K(\chi_g) \,.$$

Using the second equation and the definition of the Hubble function $H = \dot{a}/a$, we can write $dt/a = da/a^2 H = -dz/(a_0 H)$. The first equation then becomes $\chi_g = \int dz/(a_0 H)$, and if H

---

[6]In practice, this determination involves a global fit of all the parameters to the observations. The result depends in particular on the number of free parameters considered, the data which is chosen, certain features of astrophysical models, and the statistical analysis. To learn more, see, for example, Peter and Uzan (2009), Chapters 4 and 7.

is replaced by (19.5) as a function of a and therefore of z, the third equation becomes, using (19.7) and neglecting the radiation contribution ($\Omega_0^r \ll 1$),

$$H_0 D_L = (1+z_e)\, f_K \left[\int_0^{z_e} \frac{dz}{\sqrt{1+(2+\Omega_0^m-2\Omega_0^\Lambda)z+(1+2\Omega_0^m-\Omega_0^\Lambda)z^2+\Omega_0^m z^3}} \right], \qquad (19.13)$$

where $f_K = \{\sin, I, \sinh\}$ for $\Omega_0^K = 1 - (\Omega_0^m + \Omega_0^\Lambda)$ negative, zero, or positive.

In the redshift range $z \approx [0, 1.5]$ the limited expansion of (19.13) obtained in (18.12) is insufficient, and a fit of the integral (19.13) to the observations gives

$$3\Omega_0^\Lambda - 4\Omega_0^m \approx 1, \quad \Omega_0^\Lambda - \Omega_0^m \approx 0.4, \qquad (19.14)$$

according to Perlmutter *et al.* (1998) or Riess *et al.* (1998), respectively. Of course, care must be taken to extract Ω_0^m and Ω_0^Λ separately, but it is clear that Ω_0^Λ, and therefore the cosmological constant Λ, cannot be zero.[7]

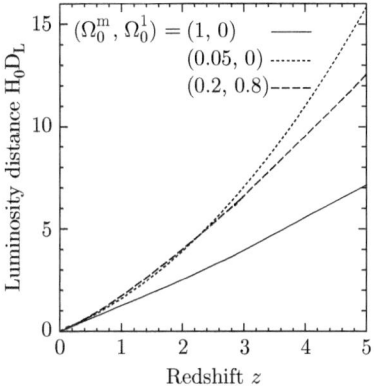

Fig. 19.1 Hubble's law in the ΛCDM model.

To go further, it is necessary to obtain additional observational data giving either a direct estimate of the actual cold matter density Ω_0^m, or a relation other than (19.14) between Ω_0^m and Ω_0^Λ. Such a relation is provided by the temperature anisotropy spectrum of the cosmic microwave background, which constrains Ω_0^K. All the current data are in agreement in giving[8]

$$|\Omega_0^K| < 10^{-2} \implies \Omega_0^m \approx 0.3, \quad \Omega_0^\Lambda \approx 0.7. \qquad (19.15)$$

Using these values and (19.12), we deduce that when matter and radiation are equally dominant we have (since $\Omega_0^m h^2 \approx 0.15$)

---

[7]The 2011 Nobel Prize in Physics was awarded to Perlmutter, Schmidt, and Riess for their 1998 discovery of the accelerated expansion of the universe by the observation of distant supernovae. See http://www.nobelprize.org/nobel-prizes/physics/laureates/2011.

[8]See, for example, Bahcall, Ostriker, Perlmutter, and Steinhardt (2008).

The discovery of anisotropies in the microwave background radiation and the confirmation of its black-body spectrum by the COBE satellite launched in 1989 resulted in the 2006 Nobel Prize in Physics being awarded to Mather and Smoot; see http://www.nobelprize.org/nobel-prizes/physics/laureates/2006/.

Currently, the best constraints on Ω_0^K come from analysis of the anisotropies of the CMB studied by the Planck satellite launched in 2009; see Planck Collaboration (2013).

$$z_{\text{eq}} \simeq 3600 \left(\frac{\Omega_0^{\text{m}} h^2}{0.15} \right), \quad T_{\text{eq}} \simeq 10^4 \left(\frac{\Omega_0^{\text{m}} h^2}{0.15} \right) \text{K} \simeq 0.85 \left(\frac{\Omega_0^{\text{m}} h^2}{0.15} \right) \text{eV} . \tag{19.16}$$

It is important to note that this is a temperature accessible in the laboratory, and the physics at this temperature is well understood.

Finally, we need to estimate the contribution of ordinary, baryonic, matter to Ω_0^{m}. The most reliable method is to measure the proportions (relative to hydrogen) of light elements (helium, deuterium, lithium, and so on) present in the universe. These elements (as shown by Gamow, Bethe, Hoyle, and Alpher) are products created in the first minutes of the history of the universe, during *primordial nucleosynthesis*. As long as the temperature is above several MeV, no nucleus can be formed because it would undergo instantaneous photodissociation. Therefore, the universe can contain only protons and neutrons in equilibrium owing to the weak interaction, and also electrons, neutrinos, and photons as well as their antiparticles. Starting from the time the temperature of the photon bath becomes of the order of the deuterium binding energy, $T \sim 0.066$ MeV, nuclear synthesis can begin. Since helium is more stable than deuterium, its production is favored by the nuclear reactions.[9] There is no significant production of heavier elements until much later during stellar nucleosynthesis.[10]

The proportions which are produced are the result of equilibrium between the rate of expansion of the universe and the density of protons present. It is found that

$$\Omega_0^{\text{b}} h^2 \approx 2 \times 10^{-2} \quad \text{or} \quad \Omega_0^{\text{b}} \approx 0.04 , \tag{19.17}$$

taking $h \approx 0.7$. Observations of CMB anisotropies (in particular, the position of the second acoustic peak) also provide a direct measurement of Ω_0^{b}. It should be noted that $\Omega_0^{\text{b}} < \Omega_0^{\text{m}}$, which is evidence in favor of the existence of dark matter.

The results (19.12), (19.15), and (19.17) of the ΛCDM cosmological model are, at the very least, surprising: the evolution of the universe is governed not, as believed up to the 1980s, by ordinary matter known in the laboratory, but by cold dark matter of an as-yet unknown nature, and a cosmological constant whose only 'natural' values are either strictly zero as Einstein favored, or, if it is a 'vacuum energy', $\epsilon_\Lambda \simeq \epsilon_{\text{Planck}}$, which is in disagreement by more than 120 orders of magnitude with the value predicted using the measured value of H_0.[11]

We stress the fact that all the cosmological observations used to determine these parameters concern epochs ranging from 0.1 s after the Big Bang to the present day, and involve only physical processes occurring at energies below 100 MeV, that is, physics at energies well understood from laboratory experiments and not at all speculative.

---

[9]We also note that this model predicts that the number of neutrino families is $N_\nu = 3$, which has been confirmed by accelerator experiments. However, the prediction for the lithium 7 abundance deduced from primordial nucleosynthesis combined with the observations of CMB anisotropies still disagrees, by a factor of about 3, with the astronomical observations. See Peter and Uzan (2009) and Weinberg (1972).

[10]As shown in the B^{2}FH paper of Burbidge, Burbidge, Fowler, and Hoyle (1957).

[11]We have $\epsilon_{\text{Planck}} \sim M_{\text{P}}^4$ and the cosmological value of the vacuum energy is $\epsilon_{\text{cosmo}} \sim 3\Omega_0^\Lambda H_0^2 M_{\text{P}}^2 \sim H_0^2 M_{\text{P}}^2$ using (19.15). Therefore, $\epsilon_{\text{Planck}} \sim (M_{\text{P}}/H_0)^2 \epsilon_{\text{cosmo}}$.

20
Inflationary models of the primordial universe

The hot Big Bang model described in the preceding chapter leaves certain problems unsolved, in particular, the problem of fine-tuning the initial conditions so that the universe has the observed properties, as well as the problem of the origin of large-scale structure.

We shall see that these problems are related to each other, and can be solved by assuming a period of accelerated expansion in the earliest history of the universe. Since the 1980s, the general acceptance of this idea of a primordial *inflationary* phase can be considered as the third phase in the history of the development of relativistic cosmology.

20.1 The hot Big Bang model: unanswered questions

The ΛCDM model of the hot Big Bang leaves a number of questions unanswered.

- *The flatness problem*

When the effect of the cosmological constant is not noticeable (as is the case for $a \ll a_0$, as seen from the observational data), the Friedmann–Lemaître equations (18.15) can also be written as

$$\Omega - 1 = \frac{K}{\dot{a}^2} = \frac{K}{(aH)^2}, \quad \frac{\ddot{a}}{a} = -\frac{4\pi G}{3}(\epsilon + 3p), \tag{20.1}$$

where $\Omega \equiv 1 - \Omega_K$ is the density parameter ($\Omega = 8\pi G\epsilon/3H^2$).

We see that in the standard model, where $\epsilon + 3p > 0$ during radiation- or cold matter-dominated periods, we have $\ddot{a} < 0$, and so $\dot{a} = aH$ decreases with time. Therefore, Ω must have been *very* close to 1 just after the Big Bang in order for the spatial sections of the universe not to be very curved at the present time. In other words, the solution $\Omega_K = 0$ is an unstable fixed point of the dynamics, and so $|\Omega_0^K| < 10^{-2}$ implies that $|\Omega_K| \ll 1$ in the primordial universe.

More precisely, we have $(\Omega - 1)/(\Omega_0 - 1) = (\dot{a}_0/\dot{a})^2 = (a_0 H_0/aH)^2$. Assuming that the actual universe is dominated by cold matter and neglecting the effect of the cosmological constant, we find (see Section 19.2) $\epsilon \sim \epsilon_{\rm m} \propto a^{-3}$ and $(H/H_0)^2 \approx \Omega_0^{\rm m}(a_0/a)^3$. Moreover, the scale factor at the radiation–matter transition epoch is given by $a_{\rm eq}/a_0 \approx \Omega_0^{\rm r}/\Omega_0^{\rm m}$. Therefore,

$$\Omega_{\rm eq} - 1 \approx -\frac{\Omega_0^K \Omega_0^{\rm r}}{(\Omega_0^{\rm m})^2}, \tag{20.2}$$

which is of order 10^{-5} using the parameter values given by recent observations; see Section 19.3. Before the transition period, when the universe is radiation-dominated, we have $a \propto \sqrt{t}$ (see Section 19.2) and so $\dot{a} \propto 1/a$, and in the end

Relativity in Modern Physics. Nathalie Deruelle and Jean-Philippe Uzan.
© Oxford University Press 2018. Published in 2018 by Oxford University Press.
DOI: 10.1093/oso/9780198786399.001.0001

Book 3

$$\Omega - 1 \approx (\Omega_{\text{eq}} - 1) \left(\frac{a}{a_{\text{eq}}}\right)^2 \approx -\frac{\Omega_0^K}{\Omega_0^{\text{r}}} \left(\frac{T_0}{T}\right)^2 , \tag{20.3}$$

where we have used (20.2) and replaced the scale factor by the microwave background temperature using $aT = a_0 T_0$. For any values of the parameters, Ω is arbitrarily close to 1 near the Big Bang. For example, during the epoch of primordial nucleosynthesis when $T \approx 0.05$ MeV $\approx 5 \times 10^8$ K, we have $\Omega_{\text{nucl}} - 1 = \mathcal{O}(10^{-14})$ for the parameter values given by current observations.

Therefore, the observed flatness of the universe demands a fine-tuning of the curvature parameter in the primordial universe, which seems rather unnatural.

- *The horizon and monopole problems*

The cosmological principle requires that the matter present in the universe as a whole be a perfect fluid of density ϵ and pressure p, and the Bianchi identities require that it be conserved (see Section 18.3). As long as the equation of state $w = p/\epsilon$ is constant, the conservation equation (18.16) implies that

$$\epsilon \propto a^{-3(1+w)} . \tag{20.4}$$

The density therefore decreases in an expanding universe if $1 + w > 0$, which is the case for all known matter.

In the Friedmann–Lemaître equation (18.15) the matter contribution ($\propto \epsilon$) dominates, for small a, the cosmological constant contribution ($\propto a^0$) and the contribution of the curvature term ($\propto a^{-2}$) if $3(1 + w) > 2$, or $\epsilon + 3p > 0$, which is also the case for all known matter. Therefore, the evolution of the scale factor as a function of the cosmic time t is given by (19.9):

$$a \propto t^q \quad \text{with} \quad q = \frac{2}{3(1 + w)} , \quad \text{where} \quad 0 < q < 1 . \tag{20.5}$$

The history of the universe then necessarily starts with a Big Bang, that is, with a singularity of the scale factor and also of the curvature. (To learn more about the geometry of spacetime near a singularity, see the references in footnote 5 of Section 19.2.)

This type of behavior of the scale factor near the Big Bang also leads to the *horizon problem*.[1] The proper distance traveled since the Big Bang by a photon and therefore by any causal phenomenon (also referred to as the *particle horizon*) is given by $D_{\text{p}} = a \int_0^t dt/a$; see (18.4) and (18.7). If we use (20.5) for the scale factor, the integral converges and

$$D_{\text{p}} = \frac{t}{1 - q} = \left(\frac{q}{1 - q}\right) \left(\frac{1}{H_0}\right) \left(\frac{T_0}{T}\right)^{\frac{1}{q}} , \tag{20.6}$$

where we have used the fact that $a = \alpha t^q$, $a_0 = \alpha(q/H_0)^q$, and $aT = a_0 T_0$, which relates the scale factor and the microwave background temperature.

Therefore, the spatial sections of the observable universe are at any instant composed of regions of volume $\mathcal{O}(D_{\text{p}}^3)$ which are causally independent, but have the same physics in each owing to the cosmological principle. This is unsatisfying: while the hypothesis of

---

[1]See Rindler (1953).

homogeneity and isotropy of our observable universe seems reasonable, it is clear that these conditions cannot have been established dynamically. This is therefore a strong but unrealistic hypothesis about the structure of space in its primordial phase.

To illustrate the possible observational consequences of the existence of a particle horizon, we assume, as predicted by grand unified theories,[2] that at the temperature T_{GUT}, 'monopoles' of mass $M_{\mathrm{GUT}} = T_{\mathrm{GUT}}$ are produced at the rate of about one per causal volume. These massive 'relics' are cold matter and therefore are diluted as $1/a^3$, that is, as T^3. Their current density is then given by the following, using (20.6) with $q = 1/2$:

$$\epsilon_{\mathrm{mGUT}}^0 \approx \left(\frac{M}{D_{\mathrm{p}}^3}\right)_{\mathrm{GUT}} \left(\frac{a_{\mathrm{GUT}}}{a_0}\right)^3 \approx \left(\frac{T_{\mathrm{GUT}}}{T_0}\right)^4 T_0 H_0^3, \tag{20.7}$$

corresponding to a density parameter of

$$\Omega_{\mathrm{mGUT}}^0 \equiv \frac{8\pi G \epsilon_{\mathrm{mGUT}}^0}{3H_0^2} = \frac{8\pi G}{3}\left(\frac{T_{\mathrm{GUT}}}{T_0}\right)^4 T_0 H_0 = \mathcal{O}(10^{13}) \tag{20.8}$$

for $T_0 = 3$ K, $H_0 = 70$ km/s/Mpc, and $T_{\mathrm{GUT}} = 10^{15}$ GeV. This is completely excluded by the observational data, which give $\Omega_{\mathrm{m\ total}}^0 \approx 0.3$. It is therefore necessary to choose one or the other: either the grand unified theories predicting the formation of such relics, or the description of the primordial universe given by the hot Big Bang model.

- *The problem of the origin of large-scale structure*

Another question that the hot Big Bang model leaves hanging is that of the origin of the large-scale structure in the universe (galaxies, galactic clusters, and also temperature anisotropies of the cosmic microwave background radiation). We shall put off the discussion of this topic to Chapters 21 and 22, because it requires study of the evolution of *perturbations* of the Friedmann–Lemaître metric.

However, the problem can be understood qualitatively as follows.

The 'proper' radius of the particle horizon is $D_{\mathrm{p}} = t/(1-q)$ if the scale factor grows as t^q, $q < 1$. However, the proper size of a fluctuation of a given co-moving size grows as the scale factor, that is, as t^q. We therefore see that sufficiently early in the history of the universe, any fluctuation has had a size greater than the size of the horizon. It therefore cannot have been produced by a causal process (see Fig. 20.1). This means that it is impossible within the standard model to explain the perturbation spectrum giving rise to large-scale structure—it must be postulated.

The horizon *vs.* the Hubble radius

The particle horizon defined in (20.6) depends on the history of the universe since its inception, which is poorly known. Instead, we characterize the 'size' of the universe by the *Hubble radius* $D_{\mathrm{H}} = 1/H$, where $H = \dot{a}/a$ is the Hubble parameter. Any physical length λ is associated with a co-moving length λ^{c} which is a constant given by $\lambda = a\lambda^{\mathrm{c}}$. Similarly, the *co-moving Hubble radius* is $1/(aH)$. A distance is termed sub-Hubble (respectively, super-Hubble) if it is smaller (larger) than D_{H}:

$$\text{super-Hubble:} \quad \lambda_{\mathrm{c}} > \frac{1}{aH}, \quad \text{sub-Hubble:} \quad \lambda_{\mathrm{c}} < \frac{1}{aH}. \tag{20.9}$$

---

[2]For a detailed discussion, see, for example, Linde (1990), and also Mukhanov (2005).

In terms of the co-moving wave number $k = 2\pi/\lambda_c$, this corresponds to

$$\text{super-Hubble:}\quad k < aH\,,\qquad \text{sub-Hubble:}\quad k > aH\,. \tag{20.10}$$

The problem of the origin of the large-scale structure of the universe can therefore be stated as follows. In the hot Big Bang model, the (co-moving) Hubble radius $1/(aH)$ always *grows*, while the co-moving distance between two 'structures' (galaxies or hot spots in the CMB anisotropy, for example) is constant (because they move on geodesics). Sufficiently far in the past, this distance was therefore larger than the Hubble radius, and it is impossible to see how objects separated by a distance larger than the 'size of the universe' characterized by the Hubble radius can have statistical properties similar to those observed; see Fig. 20.1.

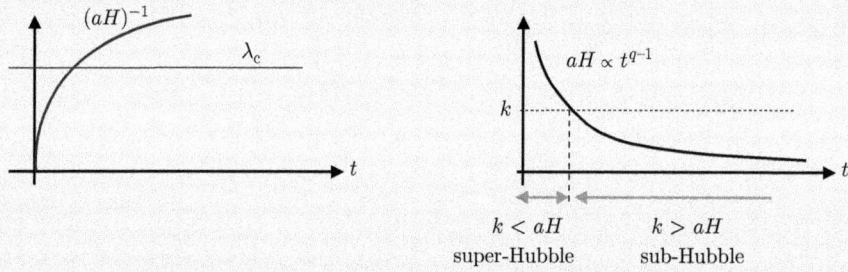

Fig. 20.1 The problem of the origin of large-scale structure when $a \propto t^q$, $0 < q < 1$.

We stress the fact that the Hubble radius is not associated with the notion of causality. It therefore differs from the concept of horizon. However, if $a \propto t^q$, with $0 < q < 1$ near the Big Bang, we have $D_p = qD_H/(1-q)$. The Hubble radius and the particle horizon then are of the same order of magnitude. We anticipate that this does not generalize to the inflationary phase.

20.2 Inflation

• *A brief period of accelerated expansion*

One way of solving the problems of the hot Big Bang model discussed above is to postulate the existence of an accelerated primordial expansion phase, or *inflation*.

If $\ddot{a} > 0$ between t_i and t_f, the co-moving Hubble radius $(aH)^{-1}$ *decreases* during this period; see Fig. 20.2. Moreover, if it decreases enough, the present-day observable universe, defined as $(aH)_0^{-1}$, could have been, in the distant past, *smaller* than the Hubble radius of the universe at that time. For this it is sufficient that $(aH)_0^{-1} < (aH)_i^{-1}$ or

$$\frac{(aH)_i^{-1}}{(aH)_f^{-1}} \frac{(aH)_f^{-1}}{(aH)_0^{-1}} > 1\,. \tag{20.11}$$

During the radiation era $a \propto t^{1/2}$, and so $(aH)^{-1} = 1/\dot{a} \propto a \propto T^{-1}$, where T is the temperature of the radiation. In addition, let us assume that the inflation is quasi-exponential, that is, that H is nearly constant between t_i and t_f. Then, if for simplicity we ignore the recent matter-dominated phase, the inequality (20.11) becomes

$$\frac{a_f}{a_i}\frac{T_0}{T_f} > 1 \quad\text{or}\quad N > \ln\frac{T_f}{T_0}\,, \quad \text{where we have set}\quad e^N \equiv \frac{a_f}{a_i}\,. \tag{20.12}$$

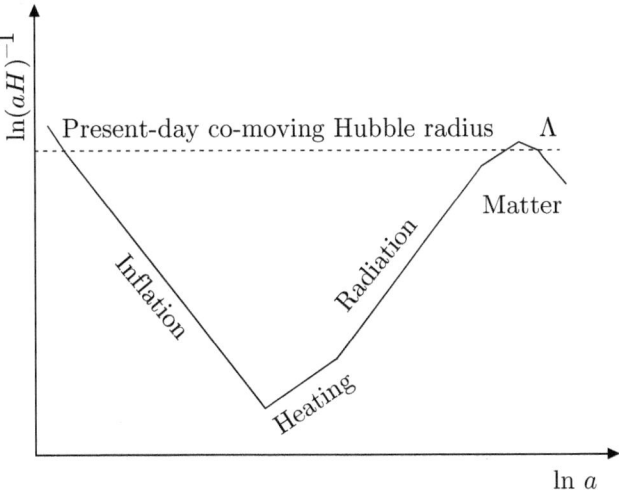

Fig. 20.2 Evolution of the Hubble radius.

Here N is the *number of nepers*, more commonly referred to as the number of *e-folds*, characterizing the duration of the inflationary phase. If we estimate that at the end of inflation the temperature has risen to grand-unified interaction scales, then $T_f \approx 10^{28}$ K. Since $T_0 \simeq 2.7$ K, we obtain roughly $N > 65$. Therefore, if the inflationary period lasts longer than 65 *e*-folds, it becomes possible to seek a physical origin of the formation of large-scale structure.[3]

Inflationary models also solve the horizon problem in the sense that between the beginning and the end of the inflationary period, the proper distance traveled by a light signal is given by (for $a \propto e^{Ht}$)

$$D_{\mathrm{p}} = a(t) \int_{t_i}^{t_f} \frac{dt}{a(t)} = \frac{1}{H} \left(e^N - 1 \right) \qquad (20.13)$$

and is exponentially large compared to the size of the Hubble radius predicted in the standard model, given in (20.6). For example, if at the end of the inflation monopoles are produced at the rate of one per causal volume D_{p}^3, their density will remain negligible. Any monopoles which might have been created *before* the start of the inflation will be completely dispersed by the exponential growth of the physical distance separating them. However, it is important to emphasize the fact that, strictly speaking, the horizon problem persists as long as the universe has a space-like initial singularity.

Finally, an inflationary phase also solves the flatness problem, because if $(aH)^{-1}$ decreases, then Ω_K will tend to 0; *cf.* (20.1). More precisely, if H is nearly constant between t_i and t_f so that $a \propto e^{Ht}$, we will have, *cf.* (20.1), $|\Omega^K(t_f)/\Omega^K(t_i)| \sim [a(t_f)/a(t_i)]^{-2} = \exp(-2N)$. For $N = 65$ we find $|\Omega^K(t_f)/\Omega^K(t_i)| \approx 10^{-55}$.

---

[3]A more accurate calculation including the matter-dominated era among other things gives $N \simeq 62$ for $T_f = 10^{16}$ GeV; see, for example, Liddle and Lyth (2000).

• *A scalar field as the inflation engine*

   The Friedmann equation (20.1) tells us that an inflationary phase (such that $\ddot{a} > 0$) can be generated only by the matter equation of state $\epsilon + 3p < 0$. A scalar field provides a simple and elegant device for accomplishing this.[4]

   Let us therefore add a homogeneous scalar field $\phi = \phi(t)$ to the matter components of the universe. Its energy–momentum tensor is, see Section 3.4,

$$-T^0_0{}^\phi \equiv \epsilon_\phi = \frac{1}{2}\dot{\phi}^2 + V(\phi), \quad T^i_j{}^\phi \equiv p_\phi \delta^i_j = \left(\frac{1}{2}\dot{\phi}^2 - V(\phi)\right)\delta^i_j,$$

$$\text{or} \quad p_\phi = w\epsilon_\phi \quad \text{with} \quad w = \frac{\frac{1}{2}\dot{\phi}^2 - V(\phi)}{\frac{1}{2}\dot{\phi}^2 + V(\phi)},$$

(20.14)

where $V(\phi)$ is the self-interaction potential of the field [$V(\phi) = m^2\phi^2/2$ for a field of mass m]. It is clear from this expression that the equation of state of the scalar field can vary between $w = +1$ if $\dot{\phi}^2 \gg V$ and $w = -1$ if $\dot{\phi}^2 \ll V$.

   If the field ϕ interacts only gravitationally with the other matter fluids, its energy–momentum tensor is conserved ($D_\mu T^{\mu\nu}_\phi = 0$). This gives the Klein–Gordon equation, which in the Friedmann–Lemaître metric is written as

$$\ddot{\phi} + 3H\dot{\phi} + V_\phi = 0, \quad \text{where we have set} \quad V_\phi \equiv \frac{dV}{d\phi}.$$

(20.15)

The Friedmann–Lemaître equations (18.15) then become

$$3H^2 = \kappa\left(\frac{1}{2}\dot{\phi}^2 + V(\phi) + \frac{\epsilon^r_0 a_0^4}{a^4}\right), \quad 3\frac{\ddot{a}}{a} = \kappa\left(V(\phi) - \dot{\phi}^2 - \epsilon^r_0\frac{a_0^4}{a^4}\right),$$

(20.16)

where, since we are interested in the primordial universe, we have neglected the contributions from the cold matter, the curvature term, and the cosmological constant Λ.

   Here we see the role that the scalar field can play. If during the evolution the field is in the *slow-roll regime*, $\dot{\phi}^2 \ll V$, it is nearly constant. If this regime lasts a sufficiently long time, the potential will also become constant and will rapidly dominate the radiation term. The scale factor then grows quasi-exponentially and the metric becomes $ds^2 = -dt^2 + e^{2Ht}dx^2$, which is just the de Sitter metric. During this *inflationary* period the scalar field can in a first approximation be treated as a 'primordial' cosmological constant $\Lambda_{\text{prim}} \equiv 3H^2 = \kappa V$.

   However, a cosmological constant can be viewed as a perfect fluid with equation of state $p = w\epsilon$ for $w = -1$ and $\epsilon = \Lambda_{\text{prim}}/\kappa$; see (20.4). Therefore, $\epsilon + 3p$ is negative and the flatness and horizon problems, and even the monopole problem, of the standard model can be solved or, at least, viewed from a new perspective, by relegating the period preceding dominance by the scalar field to the Planckian limbo of ignorance.[5]

---

[4]Among the precursors of this idea were Englert, Sato, and Starobinsky, and among the pioneers, Guth (who coined the term 'inflation'), Linde, Albrecht, and Steinhardt.

[5]We mention in passing that the cosmological constant present in the ΛCDM model of the hot Big Bang can also be replaced by a scalar field called 'quintessence'. For a review of these models, see, for example, Chongchitnan and Efstathiou (2007).

• *The slow-roll regime*

The question then comes down to how to construct a model in which the duration of the inflation is sufficiently long.

During this phase, the dynamical equations (20.15) and (20.16) reduce to $3H^2 = \kappa(\frac{1}{2}\dot{\phi}^2 + V)$ and $\ddot{\phi} + 3H\dot{\phi} + V_\phi = 0$ (from which we derive the useful relation $\dot{H} = -\frac{1}{2}\kappa\dot{\phi}^2$). They can be rewritten in terms of the parameters

$$\varepsilon \equiv -\frac{\dot{H}}{H^2} = \frac{\kappa}{2}\frac{\dot{\phi}^2}{H^2} \quad \text{and} \quad \delta \equiv -\frac{\ddot{\phi}}{H\dot{\phi}} \tag{20.17}$$

$$\text{as} \quad H^2(3 - \varepsilon) = \kappa V, \quad (3 - \delta)H\dot{\phi} = -V_\phi.$$

In the *slow-roll approximation* where $\varepsilon \ll 1$ and $\delta \ll 1$, we have $H \approx \sqrt{\kappa V/3}$, $\dot{\phi} \approx -V_\phi/\sqrt{3\kappa V}$, and $\dot{H} \approx -V_\phi^2/(6V)$, and so the parameters ε and δ can be written as functions of only the potential:

$$\varepsilon \approx \frac{1}{2\kappa}\frac{V_\phi^2}{V^2}, \quad \delta \approx \frac{1}{2\kappa}\left(2\frac{V_{\phi\phi}}{V} - \frac{V_\phi^2}{V^2}\right). \tag{20.18}$$

The number of e-folds, $N \equiv \ln(a_f/a_i)$ [see (20.12)] is calculated as a function of the initial and final values of the field ϕ_i and ϕ_f using $H \equiv \dot{a}/a = (d\ln a/d\phi)\dot{\phi}$, which gives $d\ln a/d\phi = H/\dot{\phi} = -\sqrt{\kappa/(2\varepsilon)}$, and so

$$N = \int_{\phi_f}^{\phi_i}\sqrt{\frac{\kappa}{2\varepsilon}}\,d\phi \approx \kappa\int_{\phi_f}^{\phi_i}\frac{V}{V_\phi}d\phi. \tag{20.19}$$

As we saw above, we need approximately $N > 60$ in order to solve the problems of the hot Big Bang model.

Therefore, at this stage we need to propose a plausible candidate for the scalar field. In 1981 Alan Guth suggested that it might be the scalar boson of grand unified theories, the same as that which, in the standard scenario, gives rise to the monopole problem. Unfortunately, this scenario of 'old inflation' is not viable, as it leads to a universe which is much too inhomogeneous.[6]

In the following section we shall present a simple model which for the present is consistent with the observations.

20.3 'Chaotic' inflation

In 1983 A. Linde[7] suggested *postulating* the existence of a scalar field called the *inflaton*, which is as simple as possible, that is, free and massive. The evolution equations of the primordial universe when governed by this field are then simply

$$3H^2 = \frac{1}{2}\dot{\phi}^2 + V(\phi), \quad \ddot{\phi} + 3H\dot{\phi} + \frac{dV}{d\phi} = 0, \quad V(\phi) = \frac{1}{2}m^2\phi^2 \tag{20.20}$$

(where the factor $\kappa = 8\pi G$ has been absorbed in a redefinition of the field and the potential). It is easy to see that the slow-roll parameters (20.18) are given by $\varepsilon \approx 2/\phi^2$ and $\delta \approx 0$.

---

[6] As Guth himself showed in Guth (1981).

[7] Linde (1983).

Therefore, if initially $\phi \gg 1$, the field will undergo a slow roll and the approximate solution of (20.20) will be

$$\ddot{\phi} \simeq 0, \quad \dot{\phi} \simeq -\sqrt{\frac{2}{3}}m, \quad H \simeq \frac{m\phi}{\sqrt{6}} \quad (\phi \gg 1), \tag{20.21}$$

which implies a quasi-exponential expansion of the scale factor. We can indeed integrate these equations to obtain

$$\phi(t) = \phi_i - \sqrt{\frac{2}{3}}mt, \quad a(t) = a_i \exp\left[\frac{1}{4}(\phi_i^2 - \phi^2(t))\right]. \tag{20.22}$$

The number of e-folds is given by (20.19) or

$$N(\phi_i) = \frac{1}{4}\phi_i^2 - \frac{1}{2}. \tag{20.23}$$

It is therefore sufficient that $\phi_i > 16$, that is, $\phi_i > 3M_P$ if we restore the units, in order to have $N > 70$. The larger ϕ_i is, the longer the duration of the inflationary phase.[8]

At the end of this slow-roll period the scalar field oscillates at the bottom of its potential well and the approximate solution of (20.20) becomes

$$\phi \simeq 2\sqrt{\frac{2}{3}}\frac{\sin mt}{mt}, \quad H \simeq \frac{2}{3t} \quad (\phi \simeq 0), \tag{20.24}$$

so that the scale factor grows on average as $t^{2/3}$, as though the universe were dominated by a pressure-less fluid.

The evolution of the scalar field and the scale factor is shown in Fig. 20.3 for a field which is initially in the slow-roll regime. If now we allow free initial conditions on $\dot{\phi}$, it can be shown that the slow-roll trajectory determined by (20.22) is an attractor in the dynamics. The trajectory in $(\phi, \dot{\phi})$ space converges rapidly to this solution, as shown in Fig. 20.3.

Other models

Of course, the basic scenario of chaotic inflation can be refined as much as one wants. The potentials can be made more complicated, several inflatons can be introduced, and so on.

---

[8]This is the origin of the idea (and the term) of *chaotic* inflation proposed by Linde: the universe can be imagined as a juxtaposition of independent regions, each governed by (20.20), but with arbitrary initial values of the field ϕ (a rather bold hypothesis!). Given this situation, certain regions, like our own, will have already left the inflationary period, while others are still in it. It can also be argued that the regions where ϕ_i is the largest will 'inflate' the most and will exponentially dominate the physical volume of the universe at the end of the inflation, and so it is probable that most observers will inhabit such a zone. Finally, quantum fluctuations of the field can be invoked to predict that in certain regions ϕ will *grow*, which leads to the idea of *eternal* chaotic inflation. However, it must be borne in mind that global cosmic time will disappear when the universe is represented by an inhomogeneous geometry, and that the concept of probability in this context is debatable.

A scenario of this type combined with the fact that string theory defined in 10- or 11-dimensional spacetime predicts a large number of compactifications ($10^{500} \ldots$), that is, possible fundamental geometries, has led to the idea of a *landscape*, in which all these geometries are realized in the universe and selected by inflation, and the *anthropic principle*. See Susskind (2009).

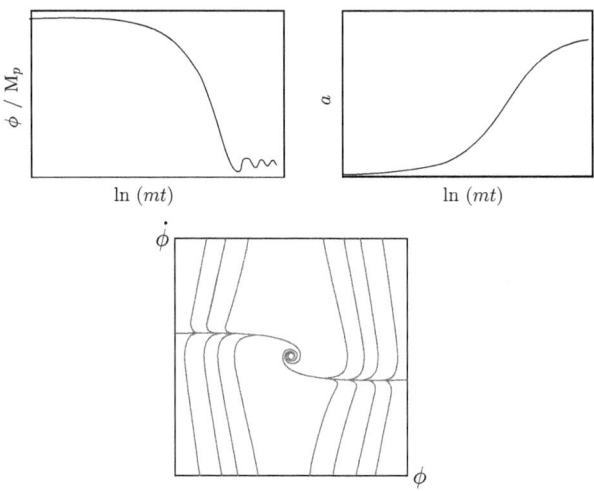

Fig. 20.3 Dynamics of the field and the scale factor for a $m^2\phi^2/2$ chaotic inflationary model.

At present there exist more than one hundred inflationary models which differ in their potential, their coupling—minimal or non-minimal—to gravity, and the number of scalar fields involved.

The simplest models with one field generalize the potential $V = m^2\phi^2/2$ to power-law potentials $V = \lambda\phi^n$. In such models the slow-roll regime is reached at large values of the field because $\varepsilon = n^2/16\pi G\phi^2$. Models of the type $V \propto \exp\sqrt{16\pi G/p}\,\phi$ lead to a power-law expansion $a \propto t^p$, but have the flaw that they never stop inflating, because $\varepsilon = \delta = 1/p$. The class of models with small field value develop inflation for $\phi \sim 0$. This is the case of $V \propto 1 - (\phi/m)^p$, for which it can be shown that $\delta < -\varepsilon$.

Among the models with more than one field, we should mention hybrid inflation constructed using the potential $V = \lambda(\sigma^2 - v^2)/4 + g^2\phi^2\sigma^2/2 + V(\phi)$. As long as $\phi > \phi_c = \sqrt{\lambda v^2/g^2}$, the potential has a valley along $\sigma = 0$. The field ϕ then undergoes a slow roll along this valley until it reaches ϕ_c, where the mass of the field σ, $m^2 = g^2(\phi^2 - \phi_c^2)$, becomes negative, giving rise to an instability which halts the inflation.[9]

Exiting inflation

The problem with all inflationary scenarios is how to 'make a graceful exit', that is, how to find a mechanism which converts the field ϕ into a radiation fluid in order to match this primordial inflationary evolution to the standard scenario governed by the mixture of radiation, cold matter, and the cosmological constant. Since the nature of the inflation is unknown, its couplings to matter are modeled on the ones commonly used in field theory, for example, $\phi^2\psi^2$, where ψ stands for ordinary matter. The Klein–Gordon equation of motion for ψ then turns out to belong to the class of Matthieu differential equations, whose solutions are interpreted as the production of bursts of particles ψ. This 'pre-heating' phase must be followed by a thermalization period whose mechanisms are still not well understood.[10]

---

[9]See, for example, Lyth and Riotto (1999) for an exhaustive description of inflationary models inspired by high energy physics.

[10]This mechanism was introduced in Shtanov, Traschen, and Brandenberger (1995); Kofman, Linde, and Starobinsky (1997); and Greene *et al.* (1997).

21
Cosmological perturbations

Structures are observed in the universe at all scales—galaxies, groups of galaxies, galactic clusters. Here we describe the first steps toward an understanding of their formation by studying the evolution of perturbations at linear order in Friedmann–Lemaître spacetimes.

To simplify our discussion we limit ourselves to the textbook case where the spatial sections of the background space are Euclidean ($K = 0$), and anisotropic perturbations and entropy perturbations are absent, which basically means that the matter reduces to a single fluid [see eqns (21.11) and (21.14) for definitions].

The relativistic and Newtonian theories of cosmological perturbations differ. In the final section of this chapter we shall discuss the limit in which they converge.

21.1 Perturbations of the geometry

Following the approach adopted in Section 13.2 in our study of gravitational waves, we write the length element of a perturbed Friedmann–Lemaître spacetime as

$$ds^2 = a^2(\eta) \left[-(1 + 2A)d\eta^2 + 2B_i dx^i d\eta + (\delta_{ij} + h_{ij})dx^i dx^j \right], \qquad (21.1)$$

where the perturbations A, B_i, and h_{ij} are 10 functions of η and x^i which must be determined using the Einstein equations. We shall consider only the case $K = 0$, and so the background co-moving spatial metric is δ_{ij}.

As in Section 13.2, we make an SVT (scalar–vector–tensor) decomposition of the perturbations. We then have

$$B_i = \partial_i B + \overline{B}_i \quad \text{with} \quad \partial_i \overline{B}^i = 0. \qquad (21.2)$$

Similarly, h_{ij} is decomposed as

$$h_{ij} = 2C\delta_{ij} + 2\partial_i\partial_j E + 2\partial_{(i}\overline{E}_{j)} + 2\overline{E}_{ij} \quad \text{with} \quad \partial_i \overline{E}^{ij} = 0, \ \overline{E}^i_i = 0. \qquad (21.3)$$

The ten original perturbations are thus replaced by four scalars (A, B, C, and E), two vectors ($\overline{B}^i$ and $\overline{E}^i$) corresponding to $2 \times 2 = 4$ perturbations, and one tensor ($\overline{E}^{ij}$) corresponding to $6 - 1 - 3 = 2$ perturbations. Since we are limiting ourselves to linear order, the indices on these quantities are raised and lowered using δ_{ij}: $\overline{B}^i = \delta^{ij}\overline{B}_j$. We note that these SVT quantities (for example, B, the solution of $\triangle B = \partial_i B^i$) are unique only when the boundary conditions are specified.

To extract the perturbations which remain invariant under a change of coordinates, we consider the active transformation $x^\mu \to \tilde{x}^\mu = x^\mu + \xi^\mu$, with the four components of the displacement field ξ^μ decomposed as

$$\xi^0 = T, \quad \xi^i = L^i = \partial^i L + \overline{L}^i, \qquad (21.4)$$

Relativity in Modern Physics. Nathalie Deruelle and Jean-Philippe Uzan.
© Oxford University Press 2018. Published in 2018 by Oxford University Press.
DOI: 10.1093/oso/9780198786399.001.0001

where T and L are two scalars and $\overline{L}^i$ with $\partial_i \overline{L}^i = 0$ is a vector representing two perturbations.

The metric then transforms as

$$g_{\mu\nu} \to \tilde{g}_{\mu\nu} = g_{\mu\nu} - \mathcal{L}_\xi g_{\mu\nu} = g_{\mu\nu} - D_\mu \xi_\nu - D_\nu \xi_\mu = g_{\mu\nu} - \partial_\mu \xi_\nu - \partial_\nu \xi_\mu + 2\Gamma^\rho_{\mu\nu} \xi_\rho,$$

from which we easily extract $A \to A - T' - \mathcal{H}T$, $B_i \to B_i + \partial_i T - L'_i$, and $h_{ij} \to h_{ij} - \partial_i L_j - \partial_j L_i - 2\mathcal{H}T\delta_{ij}$ (a prime corresponds to differentiation with respect to η, and $\mathcal{H} = a'/a$). The scalar variables therefore transform as

$$A \to A - T' - \mathcal{H}T, \quad B \to B + T - L', \quad C \to C - \mathcal{H}T, \quad E \to E - L, \tag{21.5}$$

and the vector and tensor variables as

$$\overline{B}^i \to \overline{B}^i - \overline{L}^{i\prime}, \quad \overline{E}^i \to \overline{E}^i - \overline{L}^i, \quad \overline{E}_{ij} \to \overline{E}_{ij}. \tag{21.6}$$

Therefore, the *Bardeen potentials*[1]

$$\Psi \equiv -C - \mathcal{H}(B - E'), \quad \Phi \equiv A + \mathcal{H}(B - E') + (B - E')', \quad \overline{\Phi}^i \equiv \overline{E}^{i\prime} - \overline{B}^i, \quad \overline{E}^{ij} \tag{21.7}$$

are gauge-invariant, that is, they are invariant under a change of coordinates. They are made up of $2 = 4 - 2$ scalars Ψ and Φ, $1 = 2 - 1$ vector $\overline{\Phi}^i$ corresponding to two perturbations, and one tensor $\overline{E}^{ij}$ corresponding to two perturbations. These six quantities cannot be made to vanish by any change of coordinates if they are not zero initially.

The potentials (21.7) become those of Section 13.2 if the scale factor is constant, that is, for $\mathcal{H} = 0$, because the background space then reduces to Minkowski space.

It is important to stress the fact that these gauge-invariant perturbations are useful, as they allow the calculations to be simplified and make it easy to pass from one coordinate system to another. However, they are not quantities which *a priori* are directly observable.

Linearization of the Einstein tensor

The components of the Einstein tensor of the metric (21.1) are given at linear order by $G^\mu_\nu = G^\mu_\nu|_{\text{FL}} + \delta G^\mu_\nu$ with (when $K = 0$)

$$a^2 \delta G^0_0 = 2 \left[3\mathcal{H}^2 A - 3\mathcal{H}C' + \triangle(C + \mathcal{H}B - \mathcal{H}E') \right], \tag{21.8}$$

$$a^2 \delta G^0_i = -2\partial_i \left(\mathcal{H}A - C' \right) - \frac{1}{2}\triangle(\overline{E}'_i - \overline{B}_i), \tag{21.9}$$

$$a^2 \delta G^i_j = \partial^i \partial_j \left[(E' - B)' + 2\mathcal{H}(E' - B) - (C + A) \right] + \delta^i_j \left[-\triangle(E' - B)' - 2\mathcal{H}\triangle(E' - B) \right.$$

$$-2C'' - 4\mathcal{H}C' + \triangle C + 2\mathcal{H}A' + \triangle A + 2(2\mathcal{H}' + \mathcal{H}^2)A \Big]$$

$$+ \delta^{ik}\partial_{(k} \left\{ \left[\overline{E}'_{j)} - \overline{B}_{j)} \right]' + 2\mathcal{H} \left[\overline{E}'_{j)} - \overline{B}_{j)} \right] \right\} + \overline{E}''^i_j + 2\mathcal{H}\overline{E}'^i_j - \triangle\overline{E}^i_j, \tag{21.10}$$

where $\triangle$ is the Euclidean Laplacian.

---

[1] These variables were introduced in Bardeen (1980); see also Stewart (1990).

21.2 Matter perturbations

The perturbations of the energy–momentum tensor (18.14) describing the matter present in the universe can be written in the general form

$$\delta T_{\mu\nu} = (\delta\epsilon + \delta p)\bar{u}_\mu\bar{u}_\nu + \delta p\,\bar{g}_{\mu\nu} + 2(\epsilon + p)\bar{u}_{(\mu}\delta u_{\nu)} + p\,\delta g_{\mu\nu} + a^2 p\,\pi_{\mu\nu}\,, \qquad (21.11)$$

where $u^\mu = \bar{u}^\mu + \delta u^\mu$ is the vector tangent to the geodesics of co-moving observers. The normalization condition on $\bar{u}^\mu$ ($\bar{g}_{\mu\nu}\bar{u}^\nu\bar{u}^\mu = -1$) implies that $\bar{u}^\mu = a^{-1}\delta_0^\mu$ and $\bar{u}_\mu = -a\delta_\mu^0$. Since u^μ is normalized as $g_{\mu\nu}u^\nu u^\mu = -1$, we deduce that $2\bar{u}_\mu\delta u^\mu + \delta g_{\mu\nu}\bar{u}^\mu\bar{u}^\nu = 0$ and so $\delta u^0 = -A/a$. Defining $v^i \equiv a\delta u^i$ and $u_\mu = \bar{u}_\mu + \delta u_\mu$, we then have

$$\delta u^\mu = a^{-1}(-A, v^i)\,, \quad \delta u_\mu = a(-A, v_k + B_k)\,, \qquad (21.12)$$

and so (always for $K = 0$)

$$\delta T_{00} = \epsilon a^2 (\delta + 2A)\,, \quad \delta T_{0i} = -\epsilon a^2 [(1 + w)v_i + B_i]\,,$$

$$\delta T_{ij} = pa^2 \left(h_{ij} + \frac{\delta p}{p}\delta_{ij} + \pi_{ij} \right)\,, \qquad (21.13)$$

$$\delta T_0^0 = -\epsilon\delta\,, \quad \delta T_i^0 = \epsilon(1 + w)[\partial_i(v + B) + \bar{v}_i + \bar{B}_i]\,, \quad \delta T_j^i = \delta p\,\delta_j^i + p\,\bar{\pi}_j^i\,,$$

after introducing the density contrast $\delta \equiv \delta\epsilon/\epsilon$ and setting $w = p/\epsilon$. It is also usual to decompose δp as

$$\delta p = c_{\rm s}^2\delta\epsilon + p\Gamma\,, \qquad (21.14)$$

where $c_{\rm s}$ is the *adiabatic speed of sound* and Γ is the *entropy perturbation*.

We can decompose v_i into a scalar part and a vector part as

$$v_i = \partial_i v + \bar{v}_i \quad \text{with} \quad \partial_i\bar{v}^i = 0\,.$$

The quantity $\pi_{\mu\nu}$ is the anisotropic stress tensor. It is chosen to be traceless ($g^{\mu\nu}\pi_{\mu\nu} = 0$) because it can be absorbed in the isotropic pressure δp. It is a symmetric tensor orthogonal to u^μ, that is, $u^\mu\pi_{\mu\nu} = 0$, which implies that $\pi_{00} = \pi_{0i} = 0$. Its spatial part can then be decomposed into SVT perturbations as

$$\pi_{ij} = \triangle_{ij}\bar{\pi} + \partial_{(i}\bar{\pi}_{j)} + \bar{\pi}_{ij} \quad \text{with} \quad \triangle_{ij} \equiv \partial_i\partial_j - \frac{1}{3}\delta_{ij}\triangle\,.$$

The ten components of $\delta T_{\mu\nu}$ are therefore regrouped into four scalars (δ, δp, v, π), two vectors ($\bar{v}^i$, $\bar{\pi}^i$) representing four perturbations, and one tensor ($\bar{\pi}_{ij}$) representing two perturbations.

Now we can construct gauge-invariant perturbations of the matter by following the same procedure as in Section 21.1. Under a change of coordinates (21.4) we have $\delta Q \to \delta Q - \mathcal{L}_\xi\overline{Q}$ with $\mathcal{L}_\xi\overline{Q} = \xi^\alpha\partial_\alpha\overline{Q} = T\overline{Q}'$ and $\delta u^\mu \to \delta u^\mu - \mathcal{L}_\xi\bar{u}^\mu$ with $\mathcal{L}_\xi\bar{u}^\mu = \xi^\alpha\partial_\alpha\bar{u}^\mu - \bar{u}^\alpha\partial_\alpha\xi^\mu$. From the SVT decomposition we then deduce that

$$\delta\epsilon \to \delta\epsilon - \epsilon'T\,, \quad \delta p \to \delta p - p'T\,, \quad v \to v + L'\,, \quad \bar{v}_i \to \bar{v}_i + \overline{L}_i'\,. \qquad (21.15)$$

We observe that the quantities π, $\bar{\pi}_i$, $\bar{\pi}_{ij}$, and Γ are gauge-invariant. One can define various gauge-invariant perturbations of the *density contrast*, for example,

$$\delta^N = \delta + \frac{\epsilon'}{\epsilon}(B - E'), \quad \delta^C = \delta + \frac{\epsilon'}{\epsilon}(v + B).\tag{21.16}$$

The gauge-invariant pressure perturbations δp^N and δp^C are defined in the same way.[2] Finally, we define the gauge-invariant velocity perturbations as

$$V = v + E', \quad \bar{V}_i = \bar{v}_i + \overline{B}_i.\tag{21.17}$$

In what follows we shall consider only the case of an anisotropic, pressure-less fluid $\pi_{\mu\nu} = 0$, and we shall also assume that the entropy perturbation Γ is zero. In this case the speed of sound is $c_s^2 = \delta p/\delta\epsilon$ with, we recall, $p/\epsilon = w$. If, in addition, w is constant, then[3] $c_s^2 = w$.

The Friedmann equations

We shall often need to use the Friedmann–Lemaître equations governing the unperturbed geometry. Here we recall these equations, and give some useful expressions derived from them. We set $w = \epsilon/p$ and $c_s^2 = p'/\epsilon'$. In our case where $K = \Lambda = 0$ we have

$$\kappa\epsilon a^2 = 3\mathcal{H}^2, \quad 2\mathcal{H}' = -\mathcal{H}^2(1+3w), \quad \epsilon' = -3\mathcal{H}(1+w)\epsilon, \quad w' = -3\mathcal{H}(1+w)(c_s^2-w).\tag{21.18}$$

For $w = \text{const}$, $a \propto \eta^n$ with $n = \dfrac{2}{1+3w}$, and $\epsilon \propto a^{-3(1+w)}$.

21.3 Evolution of vector and tensor perturbations

The Stewart–Walker lemma

Under a change of coordinates, any scalar Q transforms as $\delta Q \to \delta Q - \mathcal{L}_\xi \overline{Q}$ in first order in the perturbations. Since each vector field ξ engenders a gauge transformation, we conclude that the only gauge-invariant quantities are those satisfying $\mathcal{L}_\xi \overline{Q} = 0 \, \forall \, \xi$. This is the Stewart–Walker lemma. Since the relativistic equations are covariant, they can always be cast in the form $Q = 0$, where Q is a tensor field. We thus conclude that it is always possible to write them in linear order as a function of only gauge-invariant variables.

As we saw in Section 13.2, since scalar, vector, and tensor perturbations decouple at linear order, each sector can be treated independently. The Stewart–Walker lemma then allows us to study them in any particular gauge, for example, the gauge in which $B = E = \bar{E}_i = 0$, where scalar and vector perturbations reduce to $A = \Phi$, $C = -\Psi$, and $\bar{B}_i = -\bar{\Phi}_i$; $\delta = \delta^N$, $v = V$, and $\bar{v}_i = \bar{V}_i - \bar{\Phi}_i$; see (21.7), (21.16), and (21.17).

---

[2]Of course, these variables are not independent: $\delta^C = \delta^N + \frac{\epsilon'}{\epsilon}V$.

[3]These approximations are not always good. For example, the anisotropic pressure of the photons and neutrinos should be taken into account, which requires adopting a kinetic approach and writing down a perturbed Boltzmann equation [see Peter and Uzan (2009), Chapter 7]. Moreover, topological defects and magnetic fields produce a nonzero anisotropic pressure. The entropy perturbation Γ is zero for a perfect fluid, but not for a mixture (for example, a matter–radiation fluid). See Peter and Uzan (2009), Chapter 5 for a more general discussion ($K \neq 0$, nonzero π^{ij} and Γ, a mixture of fluids).

The vector perturbations are the easiest to deal with. (We recall that we are assuming that there are no anisotropic matter perturbations, $\pi^{\mu\nu} = 0$, and no entropy perturbation, $\Gamma = 0$, that the spatial sections are Euclidean, $K = 0$, and that $\Lambda = 0$.) The conservation equation $\delta[D_\nu T^{\mu\nu}] = 0$ contains only a single vector component (the vector part of the Euler equation), which is written as

$$\bar{V}_i' + \mathcal{H}\left(1 - 3c_{\mathrm{s}}^2\right)\bar{V}_i = 0. \tag{21.19}$$

Two other equations come from the $0i$ and ij parts of the Einstein equations which, using (21.9), (21.10), and (21.18), reduce to

$$\triangle\overline{\Phi}_i = -6\mathcal{H}^2(1+w)\bar{V}_i, \qquad \overline{\Phi}_i' + 2\mathcal{H}\overline{\Phi}_i = 0. \tag{21.20}$$

Equation (21.20) gives $\overline{\Phi}_i = f(x^k)a^{-2}$ and $\bar{V}_i = -\triangle f[6\mathcal{H}^2 a^2(1+w)]^{-1}$, and using the Friedmann equations (21.18) we can verify that (21.19) is satisfied identically. If, in addition, $c_{\mathrm{s}}^2 < 1/3$, the vector perturbations will be damped out during the expansion and will play only a negligible role in the formation of large-scale structure in the universe (as long as $\bar{\pi}_i$ is zero).

The case of tensor perturbations is more interesting. These appear only in the transverse, traceless part of the spatial components of the Einstein equations, $\delta G_{ij} = \kappa \delta T_{ij}$ [see (21.8)–(21.10) and (21.13)], which are written as (for $K = \bar{\pi}_j^i = 0$)

$$\overline{E}_{kl}'' + 2\mathcal{H}\overline{E}_{kl}' - \triangle\overline{E}_{kl} = 0. \tag{21.21}$$

This is a propagation equation similar to that for gravitational waves derived in Section 13.2, but with a damping term due to the cosmological expansion.

Let us consider a Fourier mode $\bar{E}_{ij}(\eta, x^k) = \bar{E}_{ij}(\eta)\,e^{ik_k x^k}$ and assume that the scale factor behaves as $a \propto \eta^n$, where n is related to the equation of state of the cosmic fluid as $n = 2/(1 + 3w)$; see (21.18). Equation (21.21) then takes the form

$$\frac{d^2\overline{E}_{ij}}{dx^2} + \frac{2n}{x}\frac{d\overline{E}_{ij}}{dx} + \overline{E}_{ij} = 0, \tag{21.22}$$

with $x \equiv k\eta$, the solution of which is

$$\overline{E}_{ij} = x^{1/2-n}\left[A_{ij}\,J_{n-\frac{1}{2}}(x) + B_{ij}\,N_{n-\frac{1}{2}}(x)\right], \tag{21.23}$$

where A_{ij} and B_{ij} are two constant, transverse, traceless tensors and J_ν and N_ν are Bessel functions (see the following box). The solution which is regular for $x \to 0$ is the growing mode: $\overline{E}_{ij} = x^{1/2-n}J_{n-1/2}(x)\,A_{ij}$. Therefore, as can also be seen directly from (21.22), the mode $\overline{E}_{ij}(\eta)$ is constant as long as it is a super-Hubble mode ($k\eta < 1$), then it undergoes damped oscillations as soon as it becomes sub-Hubble (see Fig. 21.1).

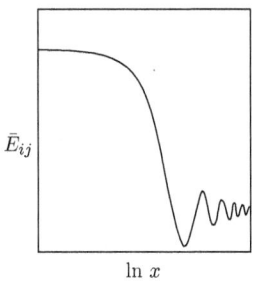 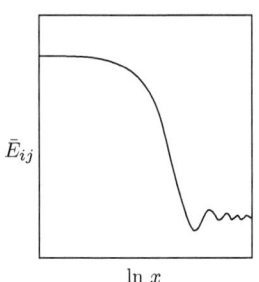

Fig. 21.1 Evolution of gravitational waves in a radiation-dominated universe $n = 1$ (left) and a matter-dominated universe $n = 2$ (right).

Bessel functions

The Bessel functions $Z_\nu(z)$ are solutions of the differential equation

$$\frac{d^2 Z_\nu}{dz^2} + \frac{1}{z}\frac{dZ_\nu}{dz} + \left(1 - \frac{\nu^2}{z^2}\right) Z_\nu = 0 \,. \tag{21.24}$$

The Bessel functions of the first (J_ν) and second (N_ν) kinds are two independent solutions of this equation. We also define the Hankel functions

$$H_\nu^{(1)}(z) = J_\nu(z) + iN_\nu(z) \,, \qquad H_\nu^{(2)}(z) = J_\nu(z) - iN_\nu(z) \,.$$

They satisfy the recursion relations

$$zZ_{\nu-1} + zZ_{\nu+1} = 2\nu Z_\nu \,, \quad Z_{\nu-1} - Z_{\nu+1} = 2Z_\nu' \,.$$

Near $z = 0$ they behave as

$$J_\nu(z) \sim \frac{1}{\Gamma(\nu+1)} \left(\frac{z}{2}\right)^\nu \,, \quad N_\nu(z) \sim \frac{1}{\Gamma(-\nu+1)\sin\nu\pi} \left(\frac{z}{2}\right)^{-\nu} \,,$$

while at infinity

$$J_{\pm\nu}(z) \sim -\sqrt{\frac{2}{\pi z}}\frac{\Gamma\left(\nu+\frac{3}{2}\right)}{\Gamma\left(\nu-\frac{1}{2}\right)} \sin\left(z \pm \frac{\pi}{2}\nu - \frac{\pi}{4}\right) \,,$$

$$N_{\pm\nu}(z) \sim -\sqrt{\frac{2}{\pi z}}\frac{\Gamma\left(\nu+\frac{3}{2}\right)}{\Gamma\left(\nu-\frac{1}{2}\right)} \cos\left(z \pm \frac{\pi}{2}\nu - \frac{\pi}{4}\right) \,,$$

$$H_\nu^{(1)}(z) \sim \sqrt{\frac{2}{\pi z}}e^{i\left(z-\frac{\pi}{2}\nu-\frac{\pi}{4}\right)} \,, \quad H_\nu^{(2)}(z) \sim \sqrt{\frac{2}{\pi z}}e^{-i\left(z-\frac{\pi}{2}\nu-\frac{\pi}{4}\right)} \,.$$

21.4 Evolution of scalar perturbations

The equations of motion of the scalar perturbations follow from the Einstein equations and the conservation of the energy–momentum tensor. Owing to the Bianchi identities, these equations are not independent, but from the technical point of view they are all useful.

The Stewart–Walker lemma assures us that they can be written in terms of gauge-invariant quantities. We can therefore either start from the general metric (21.1) and show that the equations involve only the variables (21.7), which is an excellent exercise, or we can from the beginning work in a particular gauge. We shall adopt the second approach and work in the *Newtonian gauge* defined by the condition $B = E = 0$. Then from (21.7) we have $A = \Phi$ and $C = -\Psi$, and so the metric for the scalar perturbations is

$$ds^2 = a^2(\eta) \left[-(1 + 2\Phi)d\eta^2 + (1 - 2\Psi)\delta_{ij}dx^i dx^j \right]. \tag{21.25}$$

The scalar part of the Einstein equations gives four equations. The combinations $\delta G_0^0 + 3\mathcal{H}\partial_i^{-1}\delta G_i^0$, the traceless part of δG_j^i, δG_i^0, and $\delta G_i^i + 3c_s^2\delta G_0^0$ prove the most useful. First, using (21.8)–(21.10), (21.13), and (21.18), we obtain two constraint equations which, in our case where $K = \Lambda = 0$, $\bar{\pi}_j^i = \Gamma = 0$, can be written as

$$\triangle\Psi = \frac{3\mathcal{H}^2}{2} \delta^C \quad \text{and} \quad \Psi - \Phi = 0, \tag{21.26}$$

where the density contrast δ^C was defined in (21.16). Next we have

$$\Psi' + \mathcal{H}\Phi = -\frac{3\mathcal{H}^2}{2}(1 + w)V, \tag{21.27}$$

where the velocity perturbation V was defined in (21.17). Then, using $\Psi = \Phi$,

$$\Phi'' + 3\mathcal{H}\left(1 + c_s^2\right)\Phi' + 3\mathcal{H}^2(c_s^2 - w)\Phi - c_s^2\triangle\Phi = 0. \tag{21.28}$$

The conservation equations give a continuity equation obtained starting from $u_\mu D_\nu T^{\mu\nu} = 0$:

$$\left(\frac{\delta^N}{1 + w}\right)' = -\triangle V + 3\Psi', \tag{21.29}$$

where the density contrast $\delta^N = \delta^C - (\epsilon'/\epsilon)V = \delta^C + 3\mathcal{H}(1 + w)V$ [from (21.18)] was defined in (21.16), and an Euler equation obtained starting from $(g_{\mu\alpha} + u_\mu u_\alpha)D_\nu T^{\mu\nu} = 0$:

$$V' + \mathcal{H}V = -\Phi - \frac{c_s^2}{1 + w}\delta^C. \tag{21.30}$$

After decomposing the perturbations in Fourier modes, these equations lead to second-order linear ordinary differential equations. The unknowns are Φ, Ψ, V, and δ^C. Once Φ is known from the solution of (21.28), we trivially have $\Psi = \Phi$. Then we extract δ^C algebraically from the Poisson equation (21.26). Equation (21.27) then gives V. The conservation equations (21.29) and (21.30) are redundant: they follow from (21.26)–(21.28), as can be shown explicitly using the Friedmann equations (21.18).

To solve (21.28) we need to have an equation of state $w \equiv p/\epsilon$. If we assume that w is constant, which according to (21.18) implies that $c_s^2 = w$, and $a \propto \eta^n$ with $n = 2/(1 + 3w)$, then (21.28) reduces to (for $w \neq 0$)

$$\frac{d^2 \Phi}{dy^2} + \frac{6(1+w)}{1+3w} \frac{1}{y} \frac{d\Phi}{dy} + \Phi = 0 \quad \text{with} \quad y \equiv \sqrt{w}\, k\eta, \tag{21.31}$$

where k is the comoving wave number of the Fourier mode. The solutions of this equation are the Bessel functions (see Section 21.3)

$$\Phi = x^{-\nu}[C_1 J_\nu(y) + C_2 N_\nu(y)] \quad \text{with} \quad \nu = \frac{5 + 3w}{2(1 + 3w)}. \tag{21.32}$$

The Poisson equation (21.26) and then eqn (21.27) [or (21.30)] give the density contrast and the velocity perturbation:

$$\delta^C = -\frac{(1+3w)^2}{6w} y^2 \Phi, \quad \sqrt{w}\, kV = -\frac{(1+3w)^2}{6(1+w)} y^2 \left[\frac{d\Phi}{dy} + \frac{2}{1+3w} \frac{\Phi}{y}\right]. \tag{21.33}$$

In the radiation era when $w = 1/3$, the solutions (21.32) and (21.33) become simply

$$\Phi = \frac{1}{y^2}\left[\Phi_+ \left(\frac{\sin y}{y} - \cos y\right) + \Phi_- \left(\frac{\cos y}{y} + \sin y\right)\right],$$
$$\sqrt{3}kV = -\frac{y}{2}\left(y\frac{d\Phi}{dy} + \Phi\right), \quad \delta^C = -2y^2\Phi, \quad \delta^N = -2\left[y\frac{d\Phi}{dy} + (1+y^2)\Phi\right], \tag{21.34}$$

where we have also given the expression for $\delta^N = \delta^C + 3\mathcal{H}(1 + w)V$; cf. (21.16). The super-Hubble modes ($y \ll 1$) which dominate in the end correspond to $\Phi_- = 0$. Their long-term behavior is shown in Figs. 21.2 and 21.3.

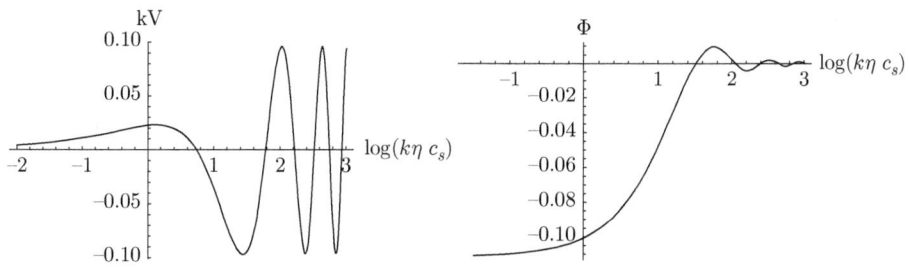

Fig. 21.2 Evolution of the Bardeen potential and the velocity (the case of a radiation fluid).

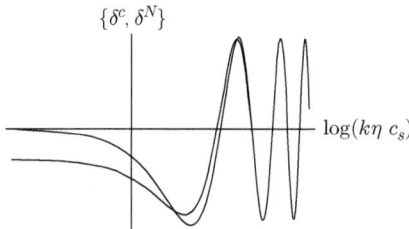

Fig. 21.3 Evolution of the density contrasts.

Let us conclude with the even simpler case of a pressure-less matter fluid $w = c_s^2 = 0$. In this case $a \propto \eta^2$, $\mathcal{H} = 2/\eta$, and the solution of (21.28), which reduces to $\Phi'' + 6\Phi'/\eta = 0$, is found immediately:

$$\Phi = \Phi_+ + \frac{\Phi_-}{x^5} \quad \text{with} \quad x \equiv k\eta. \tag{21.35}$$

Equations (21.26)–(21.30) then give

$$\delta^C = -\frac{1}{6}x^2 \Phi = Ca + \frac{D}{a^{3/2}} \quad \text{and} \quad -kV = \frac{1}{3}\Phi_+ x - \frac{1}{2}\frac{\Phi_-}{x^4}. \tag{21.36}$$

Therefore, for the growing mode we have $\delta_{\text{mat}}^C \propto a$, independently of the wavelength, which corresponds to the Newtonian result;[4] see Book 1, Section 16.4.

Matching of the perturbations in different eras

Using (21.28) as well as the Friedmann background equations (21.18) (after passing to Fourier space), it can be verified that the quantity

$$\zeta \equiv \Phi + \frac{2}{3\mathcal{H}}\frac{\Phi' + \mathcal{H}\Phi}{1 + w} \quad \text{is such that} \quad \zeta' = -\frac{2}{3}\frac{\mathcal{H}}{1 + w}c_s^2\left(\frac{k}{\mathcal{H}}\right)^2\Phi. \tag{21.37}$$

The evolution of the super-Hubble modes $(k/\mathcal{H} \ll 1)$ therefore has ζ as the first integral, that is, $\zeta = \text{const}$. This conserved quantity is very useful for following the evolution of the perturbations.

As an example, let us consider the evolution of the gravitational potential as the equation of state of the cosmic fluid changes. We assume that w passes from w_1 for $\eta < \eta_*$ (the first era) to w_2 for $\eta > \eta_*$ (the second era). During each phase, $a \propto t^{q_i} \propto \eta^{n_i}$ with $n_i = q_i/(1 - q_i)$ and $1 + w_i = 2/(3q_i)$. The general solution of (21.28) in each era then is $\Phi = \frac{\mathcal{H}}{a^2}\left[A_- + A_+ \int a^2(1 + w)d\eta\right]$, where A_+ and A_- are two constants characterizing the growing and decaying modes. If we assume that the decaying mode in the first era is negligible at η_*, then $\Phi(\eta < \eta_*) \sim A_+ (1 + w_1) q_1/(q_1 + 1) = 2A_+/[3(1 + q_1)]$. For $\eta \gg \eta_*$, it is the growing mode in the second era which dominates: $\Phi(\eta \gg \eta_*) \sim 2A_+/[3(1 + q_2)]$. The conservation of ζ then implies that

$$\frac{\Phi(\eta \gg \eta_*)}{\Phi(\eta < \eta_*)} = \frac{1 + q_1}{1 + q_2}. \tag{21.38}$$

For the matter–radiation transition, q changes from $q_1 = 1/2$ to $q_2 = 2/3$, and so $\frac{\Phi(\eta \gg \eta_{\text{eq}})}{\Phi(\eta < \eta_{\text{eq}})} = \frac{9}{10}$ for the super-Hubble modes.

---

[4]We recall that here we are limiting ourselves to the particularly simple case where the spatial sections are Euclidean $(K = 0)$, and the cosmological fluid has zero anisotropic pressure $(\pi^{\mu\nu} = 0)$ and zero entropy $(\Gamma = 0)$. This has in fact restricted us to the case where a single fluid dominates, and, moreover, in the end we have assumed that its equation of state is $p/\epsilon = w = \text{const}$.

However, it is just as easy to analytically study the more realistic case of a mixture of photon, neutrino, baryon, and cold dark matter fluids, with the baryons and photons initially coupled. See, for example, Mukhanov (2005) or Langlois (2003).

To get an idea of how the anisotropies of the cosmic microwave background created by these perturbations are calculated, see, for example, Uzan, Deruelle, and Riazuelo (2000), where the anisotropic pressure of possible topological defects is also included, and a simplified numerical code for calculating the microwave background anisotropies is given.

In order to be useful for analyzing the observational data, the evolution of the perturbations and anisotropies of the cosmic microwave background needs to be calculated within the framework of the kinetic theory using efficient numerical codes. On this subject see, for example, Peter and Uzan (2009), Chapter 6.

21.5 The sub-Hubble limit

Let us focus on the regime of small wavelengths in a universe dominated by a pressure-less fluid ($w = 0$) and a cosmological constant. This regime is relevant to understanding the growth of large-scale structure in the universe. As we have seen in Fig. 21.3, in this regime all gauges become identical, and the evolution equation of the density perturbations is written as (the dot denotes the derivative with respect to the cosmic time t with $dt = a\, d\eta$)

$$\ddot\delta + 2H\dot\delta = 4\pi G\epsilon\delta\,, \tag{21.39}$$

where we recall that $\delta = \delta\epsilon/\epsilon$. This expression is obtained from the conservation equations (21.29) and (21.30) (which do not depend on the presence of a cosmological constant, and where Ψ' can be neglected), namely, $\dot\delta = k^2 V/a$ and $\dot V + HV = -\Phi/a$, together with the constraint equation (21.26), which in the general case is written as $-k^2\Phi = \kappa a^2\epsilon\delta/2$. This is the same equation as that governing the evolution of the density contrast in the Newtonian theory of cosmological perturbations; see Book 1, Section 16.4.

The general solution of (21.39) is

$$\delta(\mathbf{x}, t) = D_+(t)\varepsilon_+(\mathbf{x}) + D_-(t)\varepsilon_-(\mathbf{x})\,,$$

where the two functions $\varepsilon_+(\mathbf{x})$ and $\varepsilon_-(\mathbf{x})$ are determined by the initial conditions. The functions $D_\pm$ are two independent solutions of

$$\ddot D + 2H\dot D - \frac{3}{2}H^2(t)\Omega^{\mathrm{m}}(t)D = 0\,, \tag{21.40}$$

where we have used the fact that $4\pi G\epsilon_{\mathrm{m}}(t) = \frac{3}{2}H^2\Omega_{\mathrm{m}}(t)$. In the particular case of the Einstein–de Sitter universe [$\Lambda = K = 0$, $\Omega^{\mathrm{m}}(t) = \Omega_0^{\mathrm{m}} = 1$ so that $a \propto t^{2/3}$], we find that $D_+(t) \propto t^{2/3} \propto a$ and $D_-(t) \propto t^{-1} \propto a^{-3/2}$, as in the Newtonian theory.

It is convenient to use the scale factor $x = a/a_0$ as the time variable. Since $\Omega^{\mathrm{m}}(t) = \Omega_0^{\mathrm{m}} x^{-3}(H_0/H)^2$, we deduce that

$$\frac{d^2 D}{dx^2} + \left(\frac{1}{H}\frac{dH}{dx} + \frac{3}{x}\right)\frac{dD}{dx} - \frac{3}{2}\frac{\Omega_0^{\mathrm{m}}}{x^5}\left(\frac{H_0}{H}\right)^2 D = 0\,. \tag{21.41}$$

We can then check that $D_- = H$ is indeed a solution as long as the growing mode is given by

$$D_+(x) = \frac{5}{2}\frac{H}{H_0}\Omega_0^{\mathrm{m}}\int_0^x \frac{dx'}{[x'H(x')/H_0]^3}\,. \tag{21.42}$$

The arbitrary normalization coefficient has been fixed so as to recover $D_+ = a/a_0$ for Einstein–de Sitter space. In general, this integral must be calculated numerically. However, in the case of a universe with Euclidean spatial sections containing only pressure-less matter and a cosmological constant ($K = w = 0$, $\Lambda \neq 0$), the scale factor $a(t)$ is given by (19.8) and we obtain

$$D_+(t) = a(t)\,{}_2F_1\left[1, \frac{1}{3}; \frac{11}{6}; -\sinh^2\left(\frac{3\alpha t}{2}\right)\right]$$

$$= {}_2F_1\left[1, \frac{1}{3}; \frac{11}{6}; -\frac{\Omega_0^\Lambda}{\Omega_0^{\mathrm{m}}(1+z)^3}\right]\frac{1}{1+z} \tag{21.43}$$

with $\alpha = H_0\sqrt{\Omega_0^\Lambda} = \sqrt{\Lambda/3}$, where ${}_2F_1$ is a hypergeometric function (see Fig. 21.4).

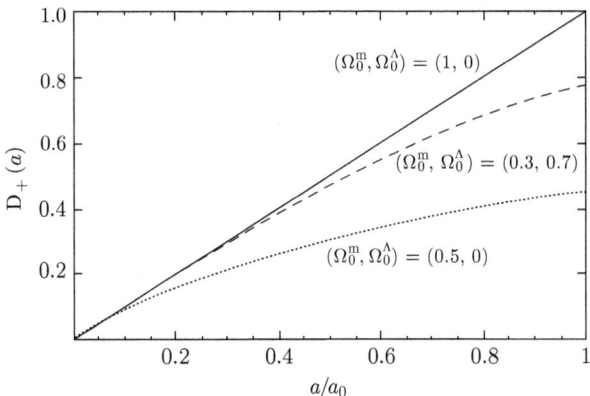

Fig. 21.4 Growth rate of the density perturbation.

The Newtonian continuity equation, see Book 1, Section 16.4, implies that the divergence of the velocity field is given by $\nabla \cdot \mathbf{u} = -a\dot{\delta}$. Since $\delta \propto D(t)$, this implies that

$$\theta(\mathbf{x}, t) \equiv (aH)^{-1}\nabla \cdot \mathbf{u} = -f(\Omega_0^{\mathrm{m}}, \Omega_0^{\Lambda})\delta \quad \text{with} \quad f(\Omega_0^{\mathrm{m}}, \Omega_0^{\Lambda}) \equiv \frac{d \ln D(a)}{d \ln a}. \tag{21.44}$$

The quantity θ represents the local fluctuation of the Hubble expansion rate. Since $\triangle\Phi = \kappa a^2 \epsilon \delta/2$, we have $\nabla \cdot \mathbf{u} = -faH\delta = -(faH/4\pi G\epsilon_{\mathrm{m}}a^2)\nabla \cdot \nabla\Phi$, from which, after integration, we find that the velocity field takes the general form

$$\mathbf{u} = -\frac{2}{3}\frac{f_+(\Omega_0^{\mathrm{m}}, \Omega_0^{\Lambda})}{aH\Omega_0^{\mathrm{m}}}\nabla\Phi + \nabla \wedge \mathbf{U}, \tag{21.45}$$

where $\mathbf{U}$ is an arbitrary function representing the vorticity of the fluid. The Euler equation can be used to check that the vorticity does not have a source and that it is attenuated by the expansion. For an Einstein–de Sitter universe, we can also verify that

$$f_+ = 1, \quad \mathbf{u}_+ = -t^{1/3}\nabla\Phi, \quad f_- = -\frac{3}{2}, \quad \mathbf{u}_- = t^{-\frac{1}{3}}\nabla\Phi. \tag{21.46}$$

Therefore, the growing mode corresponds to a velocity field describing particles falling into the potential wells, which enhances the density contrast, while the decaying mode corresponds to a configuration where the fluid escapes from the potential wells, thereby erasing the density contrasts.

Finally, the gravitational potential is given by the Poisson equation $\triangle\Phi = 4\pi G\epsilon_{\mathrm{m}}a^2\delta$ and evolves as $\Phi \propto D_+(t)/a(t)$. It is therefore constant in an Einstein–de Sitter universe, as we have already seen in (21.35). In the general case the gravitational potential decreases with time. The typical amplitude of the density fluctuations is usually characterized by the quantity σ_8 giving the variance of δ in a sphere of radius $8h^{-1}$ Mpc. The Poisson equation can be used to find the typical amplitude of the gravitational potential: $\sigma_\Phi = \frac{3}{2}\Omega_0^{\mathrm{m}}\left(\frac{8 \text{ Mpc}}{H_0^{-1}}\right)^2\sigma_8$. Therefore, $\sigma_\Phi \sim 10^{-5}$ if $\sigma_8 \sim 1$. This tells us that even if the density contrast becomes of order unity, which corresponds to a nonlinear regime for the matter, the gravitational potential will remain weak.

These results can be used to relate the relativistic perturbation theory to the Newtonian theory. They make it possible to include a spatial curvature and a cosmological constant, and put the perturbation theory on a transparent footing.[5]

Nonlinear structures form over the course of time, and to understand them we need to resort to other techniques (higher-order perturbation theory or numerical simulations).

Book 3

---

[5]The foundations were first laid by Lifshitz (1946).

22

Primordial quantum perturbations

Perturbation theory can be used to understand how large-scale structures evolve, but it is not predictive if the initial conditions are not determined. The goal of this chapter is to show that fluctuations of quantum origin are generated during inflation and that this process supplies initial conditions compatible with the observations. These fluctuations are therefore an important prediction of inflationary models.

22.1   Perturbations during inflation

We recall that during inflation the equations describing the dynamics as a function of conformal time are (see Section 20.2)

$$\kappa \phi'^2 = 2(\mathcal{H}^2 - \mathcal{H}'), \quad \kappa a^2 V = 2\mathcal{H}^2 + \mathcal{H}' \implies \phi'' + 2\mathcal{H}\phi' + a^2 V_\phi = 0,$$

with $V_\phi \equiv dV/d\phi$ and $\kappa = 8\pi G \equiv 8\pi/M_{\rm P}^2$, where $M_{\rm P}$ is the Planck mass (in units where $c = \hbar = 1$).

The study of perturbations during inflation follows the same lines as in Chapter 21. We must first find the perturbation of the energy–momentum tensor of the scalar field. For this we decompose the field ϕ into a homogeneous part and a perturbation, $\phi = \phi(t) + \delta\phi(x^i, t)$. Inserting this decomposition into eqn (3.27) for the energy–momentum tensor, we find

$$\delta T_{\mu\nu} = 2\partial_{(\nu}\phi\partial_{\mu)}\delta\phi - \left(\frac{1}{2}g^{\alpha\beta}\partial_\alpha\phi\partial_\beta\phi + V\right)\delta g_{\mu\nu}$$

$$-g_{\mu\nu}\left(\frac{1}{2}\delta g^{\alpha\beta}\partial_\alpha\phi\partial_\beta\phi + g^{\alpha\beta}\partial_\alpha\delta\phi\partial_\beta\phi + V_\phi\delta\phi\right).$$

Since the perturbed metric is decomposed as in (21.1), we have

$$a^2\delta T_0^0 = -\phi'\delta\phi' - a^2 V_\phi\delta\phi + A\phi'^2, \quad a^2\delta T_j^i = \left(\phi'\delta\phi' - \phi'^2 A - a^2 V_\phi\delta\phi\right)\delta_j^i, \quad (22.1)$$

and

$$a^2\delta T_i^0 = -\partial_i\left(\phi'\delta\phi\right), \quad a^2\delta T_0^i = \partial^i\left(\phi'\delta\phi + B\phi'^2\right) + \overline{B}^i\phi'^2. \quad (22.2)$$

The perturbation of the scalar field $\delta\phi$ is not invariant under the coordinate change (21.4), $\delta\phi \to \delta\phi - \phi'T$, and so, owing to (21.5), it is possible to construct various gauge-invariant perturbations of the field ϕ, for example,

$$\chi = \delta\phi + \phi'(B - E'), \quad Q = \delta\phi - \phi'\frac{C}{\mathcal{H}}, \quad (22.3)$$

which are related as $Q = \chi + \phi'\Psi/\mathcal{H}$. Here χ represents the fluctuation of the scalar field in the Newtonian gauge (in which $E = B = 0$).

Relativity in Modern Physics. Nathalie Deruelle and Jean-Philippe Uzan.
© Oxford University Press 2018. Published in 2018 by Oxford University Press.
DOI: 10.1093/oso/9780198786399.001.0001

Let us quickly take a look at the vector perturbations. Using (21.9) and (21.10), the vector parts of the Einstein equations $\delta G^i_j = 0$ and $\delta G^0_i = 0$ give

$$\overline{\Phi}'_i + 2\mathcal{H}\overline{\Phi}_i = 0\,, \quad \triangle\overline{\Phi}_i = 0\,. \tag{22.4}$$

The second of these means that $\overline{\Phi}_i$ can only depend on the time, and the first implies that it must decrease as a^{-2}. There is therefore no vector perturbation. The difference from the fluid case should be noted; *cf.* (21.20).

The case of gravitational waves is more interesting. As for a fluid, there is only a single equation coming from the transverse, traceless part of the spatial component of the Einstein equations. It is therefore identical to (21.21) because the spatial part of the energy–momentum tensor (22.1) is a pure trace. Introducing the auxiliary variable [see (13.10)]

$$\sqrt{8\pi G}\mu_\lambda \equiv aE_\lambda\,, \tag{22.5}$$

where $\lambda = +, \times$ indicates the polarization, (21.21) can be rewritten as

$$\mu''_\lambda - \triangle\mu_\lambda - \frac{a''}{a}\mu_\lambda = 0\,. \tag{22.6}$$

Therefore, the two polarizations are decoupled and satisfy the same propagation equation.

The equations of motion for the three scalar perturbations Φ, Ψ, and χ are obtained by writing the linearized Einstein equations in Newtonian gauge (where $B = E = 0$). Using eqns (21.8)–(21.10), (22.1), and (22.2), they can be written as

$$-3\mathcal{H}\Psi' + \triangle\Psi = \frac{\kappa}{2}(\phi'\chi' + 2a^2V\Phi + a^2V_\phi\chi)\,,$$

$$\Psi' + \mathcal{H}\Phi = \frac{\kappa}{2}\phi'\chi\,, \tag{22.7}$$

$$\Phi = \Psi\,.$$

The first is the 00 component, the second is the $0i$ component, and the third is the traceless part of the ij component (and we see that the equation supplied by the trace is redundant). The last two equations are constraints. When they are substituted into the first, we arrive at the single equation governing the only remaining dynamical perturbation, that is, Φ. Using the background equations recalled at the beginning of this section, this equation can be written in various equivalent forms, for example,[1]

$$\Phi'' + 2\left(\mathcal{H} - \frac{\phi''}{\phi'}\right)\Phi' + 2\left(\mathcal{H}' - \mathcal{H}\frac{\phi''}{\phi'}\right)\Phi - \triangle\Phi = 0 \tag{22.8}$$

or

$$u'' - \frac{\theta''}{\theta}u - \triangle u = 0 \quad \text{with} \quad \theta \equiv \frac{\mathcal{H}}{a\phi'} \quad \text{and} \quad u \equiv \frac{a}{\phi'}\Phi \tag{22.9}$$

or also (after some algebra using the constraints)

---

[1] See Mukhanov, Feldman, and Brandenberger (1992); Sasaki (1983); and Kodama and Sasaki (1984). These equations should be compared with eqns (21.26)–(21.28) describing the evolution of the perturbations when the matter is a fluid.

$$v'' - \frac{z''}{z}v - \triangle v = 0 \quad \text{with} \quad z \equiv \frac{a\phi'}{\mathcal{H}} \quad \text{and} \quad \begin{cases} v \equiv a\left(\frac{\phi'}{\mathcal{H}}\Psi + \chi\right) = \frac{2}{\kappa z}\left(\frac{a^2\Phi}{\mathcal{H}}\right)' \\ \\ \Longleftrightarrow \quad \triangle\Phi = \frac{\kappa}{2}\frac{\phi'^2}{\mathcal{H}}\left(\frac{v}{z}\right)'. \end{cases} \tag{22.10}$$

The qualitative behavior of the solutions for Φ can easily be seen from (22.9) and (22.10). When the spatial variations of Φ (or u or v) dominate the time variations (the sub-Hubble regime), their Fourier modes behave as harmonic oscillators. In the super-Hubble regime, on the other hand, the perturbations are zero modes which no longer depend on the time and are the sum of a growing and a decaying mode. (See Section 22.4 below on the quantitative study of the solution for v.)

Of course, perturbations other than Φ may be interesting to study, for example, the *curvature perturbation*

$$\mathcal{R}_c = -\Psi - \frac{\mathcal{H}\chi}{\phi'} = -\frac{v}{z} \quad \text{with} \quad \mathcal{R}_c' = -\frac{2\mathcal{H}}{\kappa\phi'^2}\triangle\Phi. \tag{22.11}$$

Just like the perturbation ζ introduced in (21.37) (in fact, $\mathcal{R}_c = -\zeta$), it is useful for studying the evolution of the perturbations in the super-Hubble regime because then it is constant. Moreover, since $\Psi = -C - \mathcal{H}(B - E')$ and $\chi = \delta\phi + \phi'(B - E')$ [see (21.7) and (22.3)], $\mathcal{R}_c$ is related to the metric and scalar field perturbations as

$$\mathcal{R}_c = C - \frac{\mathcal{H}}{\phi'}\delta\phi. \tag{22.12}$$

Similarly, we can introduce

$$A_c \equiv \Phi - \frac{1}{a}\left(\frac{a\chi}{\phi'}\right)' = A - \frac{1}{a}\left(\frac{a\delta\phi}{\phi'}\right)', \tag{22.13}$$

and so $\mathcal{R}_c$ and A_c are identified as the perturbations C and A of the metric in the co-moving gauge where $\delta\phi = 0$.

We end by noting that when $\phi' = 0$, the universe is described by a de Sitter space. The Einstein equations then imply that $\Phi = \Psi = 0$, and the fluctuations of the field ϕ are not coupled to the geometry. The equations above are therefore only valid for $\phi' \neq 0$.

22.2 The action of the perturbations

Another method of deriving the equations of motion of the perturbations is to start from the action of general relativity coupled to a scalar field, corresponding to the Lagrangian $\sqrt{-g}(R/2\kappa - \partial_\mu\phi\partial^\mu\phi/2 - V)$, and expand to second order in the metric and scalar field perturbations.[2] The calculation is greatly simplified by treating the scalar, vector, and tensor perturbations separately and using the Newtonian gauge.

---

[2]When $a = 1$, we of course recover the action of the perturbations in Minkowski spacetime obtained in Section 13.2. The action of the perturbations of FLRW spacetime is derived in detail by Mukhanov, Feldman, and Brandenberger (1992).

For the vector perturbations the action at second order is

$$\delta^{(2)} S^{(\mathrm{V})} = \frac{1}{4\kappa} \int d^4 x \, a^2 \partial_i \bar{\Phi}_j \partial^i \bar{\Phi}^j \,, \tag{22.14}$$

whose extremization indeed gives (22.4).

As far as gravitational waves are concerned, we recall that μ is defined in (22.5) and we obtain

$$\delta^{(2)} S^{(\mathrm{T})} = \frac{1}{2\kappa} \int d^4 x \, a^2 \left(\bar{E}'_{ij} \bar{E}'^{ij} - \partial_k \bar{E}_{ij} \partial^k \bar{E}^{ij} \right)$$

$$= \frac{1}{2} \sum_\lambda \int d^4 x \left[(\mu'_\lambda)^2 - \partial_i \mu_\lambda \partial^i \mu_\lambda + \frac{a''}{a} \mu_\lambda^2 \right]. \tag{22.15}$$

It is easily seen that extremization with respect to E_{ij} or μ_λ again gives (21.21) or (22.6), respectively.

For the scalar modes we find

$$\delta^{(2)} S^{(\mathrm{S})} = \frac{1}{2\kappa} \int d^4 x \, a^2 [-6\Psi'^2 - 12\mathcal{H}\Phi\Psi' - 2\partial_i \Psi (2\partial^i \Phi - \partial^i \Psi) - 2(\mathcal{H}' + 2\mathcal{H}^2)\Phi^2$$

$$+ \kappa(\chi'^2 - \partial_i \chi \partial^i \chi - a^2 V_{,\phi\phi} \chi^2 + 6\phi' \Psi' \chi - 2\phi' \chi' \Phi - 2a^2 V_\phi \Phi \chi)]. \tag{22.16}$$

The action and evolution equations of the scalar modes

It is a rather tedious exercise to show that the extremization of the action (22.16) relative to the three variables Φ, Ψ, and χ again gives the equations of motion found in Section 22.1. It can be checked that the variation with respect to Φ gives

$$2\triangle\Psi - 6\mathcal{H}\Psi' - (\phi'\chi' + 2a^2 V\Phi + a^2 V_\phi \chi) = 0\,, \tag{22.17}$$

which is just the first equation in (22.7). The variation with respect to χ gives

$$\chi'' + 2\mathcal{H}\chi' - \triangle\chi + a^2 V_{\phi\phi}\chi - \phi'(\Phi' + 3\Psi') + 2a^2 V_\phi \Phi = 0\,, \tag{22.18}$$

and the variation with respect to Ψ gives

$$\triangle(\Phi - \Psi) = 3(a^2 F)'/a^2 \quad \text{with} \quad F = \kappa\phi'\chi/2 - (\Psi' + \mathcal{H}\Phi)\,. \tag{22.19}$$

Then some further manipulations are needed (a computer algebra program is useful):

(1) We write χ in terms of F, Φ, and Ψ: $\chi = (2/\kappa\phi')(F + \Psi' + \mathcal{H}\Phi)$.

(2) We extract Φ as a function of F and Ψ from (22.19) and then find χ in terms of F and Ψ.

(3) Now it is possible to express (22.17) in terms of only F and Ψ, to isolate Ψ'', and thus to calculate Ψ''' as a function of F and Ψ.

(4) Replacing χ and Φ in (22.18) by their expressions obtained in steps (1) and (2), we obtain an equation involving F and $\Psi \ldots \Psi'''$.

(5) Finally, we replace Ψ'' and Ψ''' by the expression obtained in step (3), and find $\triangle F = 0$, or $F = 0$. We thereby recover the second equation in (22.7), and (22.19) reduces to the third equation in (22.7): $\Phi = \Psi$.

This shows that the gauge can be fixed before varying the action.

If we now substitute into (22.16) the constraints $\Phi = \Psi$ and $\Psi' + \mathcal{H}\Phi = \kappa\phi'\chi/2$, it can be shown[3] (after a lengthy calculation) that this action can be rewritten as a function of only $v \equiv a\left(\phi'\Psi/\mathcal{H} + \chi\right)$ (up to total derivatives):

$$\delta^{(2)}S^{(\mathrm{S})} = \frac{1}{2}\int d^4x \left[(v')^2 - \delta^{ij}\partial_i v\partial_j v + \frac{z''}{z}v^2\right] \equiv \int \mathcal{L}_\mathrm{S}d^4x. \tag{22.20}$$

The extremization of this action does indeed again give (22.10). This displays the special status of the perturbation v, whose action is identical to that of a scalar field of variable mass propagating in Minkowski spacetime.

It is clear that the actions (22.15) and (22.20) for gravitational waves and the scalar mode are analogs of each other (we can pass from one to the other simply by exchanging z''/z and a''/a). We can therefore focus our attention on the evolution of the scalar modes.

22.3   Determination of the initial conditions

The evolution equation of the scalar modes (22.10) is a second-order differential equation whose solutions depend on two initial conditions. In Fourier space it is written as

$$v_k'' + \left(k^2 - \frac{z''}{z}\right)v_k = 0 \tag{22.21}$$

and its general solution is

$$v(k^i, \eta) = A(k^i)v_\uparrow(k^i, \eta) + B(k^i)v_\downarrow(k^i, \eta),$$

where $v_\uparrow$ and $v_\downarrow$ are two linearly independent solutions. Inflationary theory is said to be predictive because it suggests a way of determining the constants $A(k^i)$ and $B(k^i)$.

We recall that during inflation, the wavelength of each Fourier mode grows more quickly than the comoving Hubble radius, so that each mode is sub-Hubble at the start of the inflation (see Fig. 22.1). It is in this regime where $k^2 \gg z''/z$, which does not exist in the standard model of the hot Big Bang, that the initial conditions will be fixed by quantizing the perturbation v.

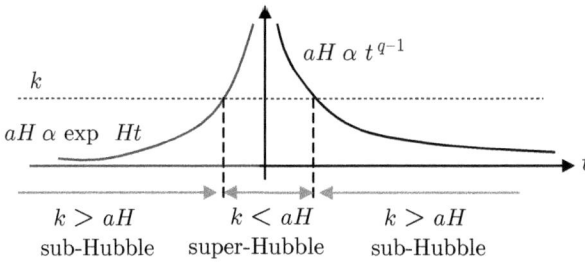

Fig. 22.1 Evolution of a co-moving mode k and the co-moving Hubble radius during the inflationary and hot Big Bang phases.

---

[3]Mukhanov and Chibisov (1981).

The action (22.20) for v is the canonical action of a scalar field of variable mass $m^2 = -z''/z$ in a Minkowski spacetime with the standard quantization.[4]

In the Heisenberg representation, the field $v(x^i, \eta)$ is promoted to the status of a quantum operator which, upon second quantization, can be decomposed as

$$\hat{v}(x^i, \eta) = \int \frac{d^3k}{(2\pi)^{3/2}} \left[v_k(\eta) e^{ik_j x^j} \hat{a}_{\mathbf{k}} + v_k^*(\eta) e^{-ik_j x^j} \hat{a}_{\mathbf{k}}^\dagger \right], \qquad (22.22)$$

where $\hat{a}_{\mathbf{k}}^\dagger$ and $\hat{a}_{\mathbf{k}}$ are the creation and annihilation operators, and the modes v_k are solutions of (22.21).

The first step in the canonical quantization is to introduce the conjugate momentum $\pi = \partial \mathcal{L}_{\mathrm{S}} / \partial v' = v'$, where $\mathcal{L}_{\mathrm{S}}$ was defined in (22.20). Then π is also promoted to operator status, and the Hamiltonian takes the form (omitting the hats)

$$H = \int (v'\pi - \mathcal{L}_{\mathrm{S}}) \, d^3x = \frac{1}{2} \int \left(\pi^2 + \delta^{ij} \partial_i v \partial_j v - \frac{z''}{z} v^2 \right) d^3x. \qquad (22.23)$$

The operators $\hat{v}$ and $\hat{\pi}$ must satisfy the commutation relations

$$[\hat{v}(x^i, \eta), \hat{v}(y^i, \eta)] = [\hat{\pi}(x^i, \eta), \hat{\pi}(y^i, \eta)] = 0, \quad [\hat{v}(x^i, \eta), \hat{\pi}(y^i, \eta)] = i\delta(x^i - y^i) \qquad (22.24)$$

on any hypersurface of constant time.

The equation of motion for $\hat{v}$ is given by the Heisenberg equations $\hat{v}' = i\left[\hat{H}, \hat{v}\right]$ and $\hat{\pi}' = i\left[\hat{H}, \hat{\pi}\right]$ and is just (22.21).

Just as in quantization in Minkowski spacetime, the creation and annihilation operators satisfy the commutation relations

$$[\hat{a}_{\mathbf{k}}, \hat{a}_{\mathbf{k}'}] = [\hat{a}_{\mathbf{k}}^\dagger, \hat{a}_{\mathbf{k}'}^\dagger] = 0, \qquad [\hat{a}_{\mathbf{k}}, \hat{a}_{\mathbf{k}'}^\dagger] = \delta(\mathbf{k} - \mathbf{k}'). \qquad (22.25)$$

These relations are compatible with (22.24) if v_k is normalized as

$$W(k) \equiv v_k {v'}_k^* - v_k^* v_k' = i, \qquad (22.26)$$

because $[\hat{v}(x^i, \eta), \hat{\pi}(y^i, \eta)] = \int \frac{d^3k}{(2\pi)^3} e^{ik_j(x^i - y^i)} W(k)$. This condition does not determine v_k completely, because $u_k = \alpha_k v_k + \beta_k v_k^*$ is also normalized if $|\alpha_k|^2 - |\beta_k|^2 = 1$. We therefore require that the sub-Hubble modes, that is, the modes of frequency high compared to the Hubble expansion rate, for which (22.21) reduces to $v'' + k^2 v_k = 0$, be the positive-frequency modes:

$$v_k(\eta) \to \frac{1}{\sqrt{2k}} e^{-ik\eta}, \quad k\eta \to -\infty, \qquad (22.27)$$

just as in quantization in flat spacetime. These initial conditions completely determine the solution, that is, the two arbitrary functions $A(k^i)$ and $B(k^i)$ appearing in (22.21).

---

[4]For the original articles, see Mukhanov and Chibisov (1981), and also Hawking (1982); Starobinsky (1982); Guth and Pi (1982); and Bardeen, Steinhardt, and Turner (1983).

Next we construct a Fock representation of the Hilbert space in which $\hat{v}$ and $\hat{\pi}$ operate. The vacuum state $|0\rangle$ is defined in the standard manner by the condition

$$\forall \mathbf{k}, \quad \hat{a}_{\mathbf{k}} |0\rangle = 0. \tag{22.28}$$

In the cosmological context, this state $|0\rangle$ is called the *Bunch–Davies vacuum*.[5] The other states can then be constructed by repeated action using the creation operators $\hat{a}_{\mathbf{k}}^{\dagger}$.

The *correlation function* of the operator $\hat{v}$, $\xi_v \equiv \langle 0|\hat{v}(x^i, \eta)\hat{v}(y^i, \eta)|0\rangle$, is then given by

$$\xi_v = \int \frac{d^3k}{(2\pi)^3} |v_k|^2 e^{ik_j(x^j - y^j)} = \int \frac{dk}{k} \frac{k^3}{2\pi^2} |v_k|^2 \frac{\sin kr}{kr}, \tag{22.29}$$

where the isotropy of the background space has been used to integrate over the angles. Now we can read off the expression for the power spectrum of v:

$$\mathcal{P}_v(k) \equiv \frac{k^3}{2\pi^2} |v_k|^2, \tag{22.30}$$

and deduce the spectrum of the curvature perturbation $\mathcal{R}_c$ defined in (22.11):

$$\mathcal{P}_\mathcal{R}(k) = \frac{k^3}{2\pi^2} \left| \frac{v_k}{z} \right|^2. \tag{22.31}$$

On super-Hubble scales the field v must be classical. However, it is derived from a quantum field, which explains why it is described by stochastic variables, but the quantum to classical transition is a mechanism far from being understood.[6] Therefore, quantum mechanics and the choice of the Bunch–Davies vacuum are essentially a method of determining the free functions $A(k)$ and $B(k)$ of the general solution (22.21). This choice completely fixes the amplitude and k dependence of the spectrum.

The case of gravitational waves is identical, and each polarization corresponds to an independent scalar field μ which is quantized in the same way.

22.4   Slow-roll inflation

Perturbation spectrum in the adiabatic approximation

Let us consider a massless scalar test field φ propagating in a cosmological space of metric $ds^2 = a^2(\eta)(-d\eta^2 + d\bar{x}^2)$. Its action is $S = -\frac{1}{2}\int d^4x \sqrt{-g}g^{\mu\nu}\partial_\mu\varphi\partial_\nu\varphi$. Introducing $\psi = a\varphi$, this action becomes, up to a total derivative, the action of a scalar field propagating in flat spacetime: $S = \int d^4x \left(\eta^{\mu\nu}\partial_\mu\psi\partial_\nu\psi + (a''/a)\psi^2\right)$. In de Sitter spacetime where $a = -(H\eta)^{-1}$ with H constant, the effective mass of ψ is $m_{\text{eff}}^2 = -a''/a = -2/\eta^2$.

---

[5]For an introduction to field theory in curved spacetime, see Birrel and Davies (1984) and Mukhanov and Winitzki (2007).

[6]See, for example, Polarski and Starobinsky (1996).

The field ψ is quantized like v in the preceding section, with the function z replaced by the scale factor[7] a. The equation of motion of the modes $\psi_k(\eta)$ is $\psi_k'' + \left(k^2 - 2/\eta^2\right)\psi_k = 0$, and we keep the positive-frequency solution

$$\psi_k = \sqrt{\frac{\hbar}{2k}}e^{-\mathrm{i}k\eta}\left(1 - \frac{\mathrm{i}}{k\eta}\right).$$

The correlation function of φ, namely, $\langle 0|\hat\varphi(\eta,x^i)\hat\varphi(\eta,y^i)|0\rangle$, is easily calculated by expanding φ as in (22.22) using the commutation properties (22.25) of the operators $\hat a_k$ and $\hat a_k^\dagger$ and their action on the vacuum state: $\hat a_k|0\rangle = 0$. As in (22.29), we obtain

$$\langle 0|\hat\varphi(\eta,x^i)\hat\varphi(\eta,y^i)|0\rangle \equiv \int d^3k\, e^{\mathrm{i}k_j(x^j-y^j)}\frac{\mathcal{P}_\varphi(k)}{4\pi k^3} \quad \text{with} \quad \mathcal{P}_\varphi(k,\eta) = \frac{k^3|\psi_k|^2}{2\pi^2 a^2},$$

$$\text{or} \quad \mathcal{P}_\varphi(k,\eta) = \hbar\left(\frac{H}{2\pi}\right)^2\frac{1+x^2}{x^2}, \quad \text{where} \quad x \equiv \frac{Ha}{k} = -(k\eta)^{-1}.$$

We therefore see that the spectrum of φ initially decreases as $1/x^2$ in the sub-Hubble limit where $x \ll 1$, then it tends to a constant in the super-Hubble limit $x \gg 1$.

The case of gravitational waves is treated in the same way, recalling that they are equivalent to two scalar fields normalized as $h = 2\sqrt{\kappa}\varphi$. Their spectrum then is

$$\mathcal{P}_{\mathrm{T}}(k,\eta) = 8\hbar\kappa\left(\frac{H}{2\pi}\right)^2\frac{1+x^2}{x^2} \rightarrow 8\hbar\kappa\left(\frac{H}{2\pi}\right)^2 \quad \text{in the super-Hubble limit}.$$

In de Sitter space the spectra are therefore scale-invariant in the super-Hubble regime: they are independent of k. However, an inflationary period is not strictly a de Sitter phase because $\dot H = -\dot\phi^2/(2\kappa)$. The spectra obtained in fact depend on the time.

The *spectral index* n_{T} of gravitational waves is then defined as

$$n_{\mathrm{T}} - 1 \equiv \left.\frac{d\ln\mathcal{P}_{\mathrm{T}}}{d\ln k}\right|_{k=aH}.$$

Since, on the one hand, $d\ln k = d\ln(aH) \sim d\ln a = da/a = H dt$ while, on the other, $\mathcal{P}_{\mathrm{T}} \propto H^2$ so that $d\ln\mathcal{P}_{\mathrm{T}} = 2dH/H = 2\dot H dt/H$, we find

$$n_{\mathrm{T}} - 1 \sim \frac{2\dot H}{H^2} \sim -2\varepsilon,$$

where we have introduced the slow-roll parameter $\varepsilon \equiv -\dot H/H$. A similar calculation,[7] which will be discussed below, for the spectral index of the curvature perturbation $\mathcal{R}$ whose spectrum is given in (22.31) leads to

---

[7]See, for example, D. Langlois (2010).

In Section 10.1 we saw that the quantizations of a field in inertial and accelerated frames are not equivalent. The quantization of a field in de Sitter spacetime depends also on the choice of the spatial sections. See, for example, Mukhanov and Winitzki (2007).

$$n_{\rm s} - 1 \equiv \left.\frac{d\ln\mathcal{P}_\mathcal{R}}{d\ln k}\right|_{k=aH} = 2\delta - 4\varepsilon, \quad \text{where} \quad \delta \equiv -\frac{\ddot\phi}{H\dot\phi}.$$

In Section 20.3 we saw that in the slow-roll regime $\varepsilon \approx (1/2\kappa)(V_\phi/V)^2$, $\eta = V_{\phi\phi}/(\kappa V)$ (where $\eta = \delta + \varepsilon$), and $N = \kappa \int (V/V_\phi)d\phi$. Let us take the example of the potential $V \propto \left(1 - \exp[-\sqrt{2\kappa/3}\,\phi]\right)^2$. We then find

$$n_{\rm s} - 1 = \frac{2}{N}, \quad n_{\rm T} - 1 = -\frac{3}{2N^2}, \quad \tilde r \equiv \frac{\mathcal{P}_{\rm T}}{\mathcal{P}_\mathcal{R}} = \frac{12}{N^2} \implies \tilde r = 3(n_{\rm s} - 1)^2,$$

relations which can be tested directly by observations of the cosmic microwave background anisotropy.[8]

In order to put the formalism described in the preceding sections to work, let us consider the case of inflation in the slow-roll regime (see Section 20.2).

The first step is to determine the dynamics of the background FLRW spacetime as a function of the slow-roll parameters ε and δ defined in (20.17), which can also be written as

$$\varepsilon = 1 - \frac{\mathcal{H}'}{\mathcal{H}^2}, \quad \delta = 1 - \frac{\phi''}{\mathcal{H}\phi'}. \tag{22.32}$$

In the first-order slow-roll approximation, we see that ε' and δ' are of second order, and so the parameters ε and δ can be treated as constants. Integration of the first equation (22.32) assuming that ε is constant and small gives

$$\mathcal{H} = -\frac{1+\varepsilon}{\eta} + \mathcal{O}(2) \quad \left(\implies a = -\frac{1+\varepsilon}{H\eta} + \mathcal{O}(2)\right), \tag{22.33}$$

in which $H \equiv \mathcal{H}/a$ can be assumed constant. [The term $\mathcal{O}(2)$ corresponds to second-order contributions in the slow-roll parameters, and we recall that $\eta < 0$ during the inflationary phase.] This equation can be used first of all to calculate $a''/a \equiv \mathcal{H}' + \mathcal{H}^2$:

$$\frac{a''}{a} = \frac{2+3\varepsilon}{\eta^2} + \mathcal{O}(2). \tag{22.34}$$

The term z''/z, where $z \equiv a\phi'/\mathcal{H}$ [see (22.10)], is obtained from $z'/z = \mathcal{H} + \phi''/\phi - \mathcal{H}'/\mathcal{H} = \mathcal{H}(1 + \varepsilon - \delta)$, which is found using the second equation in (22.32) and (22.34). We deduce that $z''/z = \mathcal{H}^2[2 + 2\varepsilon - 3\delta]$, which, using (22.33), implies that

$$\frac{z''}{z} = \frac{2+6\varepsilon - 3\delta}{\eta^2} + \mathcal{O}(2). \tag{22.35}$$

The evolution equations for the scalar and tensor modes are therefore, at first order in the slow-roll approximation,

$$v'' + \left(k^2 - \frac{2+6\varepsilon - 3\delta}{\eta^2}\right)v = 0, \quad \mu_\lambda'' + \left(k^2 - \frac{2+3\varepsilon}{\eta^2}\right)\mu_\lambda = 0. \tag{22.36}$$

---

[8]See, for example, Planck Collaboration (2013), figure on p. 10.

The solutions of these two equations are similar. They are expressed as linear combinations of Hankel functions (see Section 21.3):

$$v(\mathbf{k}, \eta) = A_1(\mathbf{k})\sqrt{-\eta}H_\nu^{(1)}(-k\eta) + A_2(\mathbf{k})\sqrt{-\eta}H_\nu^{(2)}(-k\eta) \tag{22.37}$$

with

$$\nu = \frac{3}{2} + 2\varepsilon - \delta \ \ (\text{scalar modes}), \quad \nu = \frac{3}{2} + \varepsilon \ \ (\text{tensor modes}). \tag{22.38}$$

Now we can fix the initial conditions by the procedure described in Section 22.3. Using the asymptotic behavior of the Hankel functions (given in Section 21.3), the general solution (22.37) satisfying the initial condition (22.27) is

$$v(\mathbf{k}, \eta) \equiv v_k(\eta) = \frac{\sqrt{\pi}}{2}e^{i\left(\nu - \frac{1}{2}\right)\frac{\pi}{2}}\sqrt{-\eta}H_\nu^{(1)}(-k\eta). \tag{22.39}$$

The super-Hubble ($|k\eta| \ll 1$) behavior of this solution is (see Section 21.3)

$$v_k(\eta) \to 2^{\nu - \frac{3}{2}}\frac{\Gamma(\nu)}{\Gamma\left(\frac{3}{2}\right)}e^{i\left(\nu - \frac{1}{2}\right)\frac{\pi}{2}}\frac{1}{\sqrt{2k}}(-k\eta)^{-\nu + \frac{1}{2}} \quad \text{for} \quad |k\eta| \ll 1, \tag{22.40}$$

so that in this regime the curvature perturbation $\mathcal{R}_k = v_k/z$ is given by

$$\mathcal{R}_k(\eta) \to \frac{H}{M_P\sqrt{\varepsilon}}\frac{\sqrt{4\pi}}{\sqrt{2k^3}}2^{\nu - \frac{3}{2}}\frac{\Gamma(\nu)}{\Gamma\left(\frac{3}{2}\right)}e^{i\left(\nu - \frac{1}{2}\right)\frac{\pi}{2}}(1+\varepsilon)^{-\nu + \frac{1}{2}}\left(\frac{k}{aH}\right)^{-\nu + \frac{3}{2}} \quad \text{for} \quad |k\eta| \ll 1, \tag{22.41}$$

where η has been eliminated using (22.33). The Planck mass $M_P = G^{-1/2}$ is used to replace G because, as we recall, we are working in units where $\hbar = c = 1$. The fact that the universe is not strictly a de Sitter universe is translated into a scale dependence given by $-\nu + 3/2 = \delta - 2\varepsilon$, as is given by the adiabatic approximation discussed above.

The calculation for gravitational waves is identical: μ_λ has the form (22.39) with $\nu = 3/2 + \varepsilon$, and the scale dependence of the spectrum becomes $-\nu + 3/2 = -\varepsilon$, in agreement with the adiabatic approximation presented above.

Perturbations after inflation

The quantity ζ defined in (21.37) remains constant for the super-Hubble modes. This makes it possible to relate the curvature perturbations during inflation (22.41) to those during the radiation era, that is, to the initial conditions for large-scale structure formation. Indeed, for the super-Hubble modes $\zeta = -\mathcal{R}_c$ [see (21.37)], and its conservation implies that

$$\Phi_{\text{rad}}(k\eta \ll 1) = \frac{2}{3}\frac{1+\varepsilon}{\varepsilon}\Phi_{\text{inf}}(k\eta \ll 1) \tag{22.42}$$

if the inflationary phase is well described by the slow-roll regime. Therefore, it is the change of the equation of state from $-1 + \frac{2}{3}\varepsilon$ to $\frac{1}{3}$ and the fact that the curvature perturbations of the super-Hubble modes are constant that lead to the amplification of the gravitational potential between the two eras.

22.5   Predictions of slow-roll inflation

The results of the preceding section allow us to give generic predictions about slow-roll inflation, namely, the properties of the perturbations at the end of the inflation for the super-Hubble modes.

- *Vector perturbations are vanishingly small*

- *The spectrum of the scalar perturbations can be obtained*

Equation (22.41) can be used to find the power spectrum of the curvature perturbations. To do this we use (22.38) for ν and the fact that $\Gamma(1+h)/\Gamma(1) \sim 1 + h\gamma_{\rm E}$, where $\gamma_{\rm E} \sim 0.58$ is the Euler constant and $2^h \sim 1 + h\ln 2$. We find

$$\mathcal{P}_{\mathcal{R}} = \frac{1}{\pi}\frac{H^2}{M_{\rm P}^2 \varepsilon}\left[1 - 2(2C+1)\varepsilon + 2C\delta\right]\left(\frac{k}{aH}\right)^{2\delta - 4\varepsilon} \tag{22.43}$$

for the super-Hubble modes, where $C = \gamma_{\rm E} + \ln 2 - 2$. The scalar modes therefore have a nearly scale-invariant spectrum with spectral index (easily obtained above in the adiabatic approximation)

$$n_{\rm s} - 1 \equiv \frac{d\ln\mathcal{P}_{\mathcal{R}}}{d\ln k} = 2\delta - 4\varepsilon\,. \tag{22.44}$$

The value $n_{\rm s} = 1$ characterizes a scale-invariant spectrum reached in the de Sitter limit (in which $\varepsilon = \delta = 1$). The super-Hubble fluctuations have a typical amplitude $H/M_{\rm P}\sqrt{\varepsilon}$ (we recall that $H = H_{\rm inflation}$ is nearly constant). If we push the expansion to the next highest order in the slow-roll parameters, the spectral index will acquire a weak dependence on k, because the slow-roll parameters then will evolve during the inflationary phase.[9]

- *The scalar perturbations have Gaussian statistics*

The curvature perturbation (22.41) is written as $\mathcal{R}_k(\eta) \sim \frac{H}{M_{\rm P}\sqrt{\varepsilon}}\frac{A_\nu}{\sqrt{2k^3}}\left(\frac{k}{aH}\right)^{-\nu+\frac{3}{2}}$ in the super-Hubble regime, and so the operator $\hat{\mathcal{R}}_{\rm c}$ behaves as

$$\hat{\mathcal{R}}_{\rm c} \approx \int \frac{d^3k}{(2\pi)^{3/2}}\hat{\mathcal{R}}_{\mathbf{k}}\, e^{i\mathbf{k}\cdot\mathbf{x}} \sim \frac{H}{M_{\rm P}\sqrt{\varepsilon}}A_\nu \int \frac{d^3k}{(2\pi)^{3/2}}\frac{1}{\sqrt{2k^3}}\left(\frac{k}{aH}\right)^{-\nu+\frac{3}{2}}\left(\hat{a}_{\mathbf{k}} + \hat{a}_{-\mathbf{k}}^\dagger\right)e^{i\mathbf{k}\cdot\mathbf{x}}\,. \tag{22.45}$$

Therefore, each mode is proportional to the operator $(\hat{a}_{\mathbf{k}} + \hat{a}_{-\mathbf{k}}^\dagger)$. The commutation relations of this operator are the same as those of a Gaussian field.[10] Therefore, at super-Hubble scales the operator $\hat{\mathcal{R}}$ can be replaced by a Gaussian stochastic classical field. Introducing the unit

---

[9]See Lyth and Stewart (1993) and Stewart and Gong (2001).

[10]Indeed, all the correlation functions can be expressed in terms of the two-point correlation function $\langle 0|\hat{\mathcal{R}}_{\mathbf{k}}\hat{\mathcal{R}}_{\mathbf{k}'}|0\rangle = P_{\mathcal{R}}(k)\delta(\mathbf{k}+\mathbf{k}')$ as

$$\langle 0|\hat{\mathcal{R}}_{\mathbf{k_1}}\dots\hat{\mathcal{R}}_{\mathbf{k_{2p}}}|0\rangle = \sum_{\rm perm\ (i,j)}\prod \langle 0|\hat{\mathcal{R}}_{\mathbf{k_i}}\hat{\mathcal{R}}_{\mathbf{k_j}}|0\rangle,\quad \langle 0|\hat{\mathcal{R}}_{\mathbf{k_1}}\dots\hat{\mathcal{R}}_{\mathbf{k_{2p+1}}}|0\rangle = 0\,.$$

Gaussian random variable $e_v(\mathbf{k})$ satisfying $\langle e_v(\mathbf{k}) \rangle = 0$ and $\langle e_v(\mathbf{k}) e_v^*(\mathbf{k}') \rangle = \delta(\mathbf{k} - \mathbf{k}')$, we can formally make the replacement $\hat{v}_{\mathbf{k}} \to v_{\mathbf{k}} = v_k(\eta) e_v(\mathbf{k})$. The quantum expectation value $\langle 0|...|0 \rangle$ is then replaced by an expectation value of a classical ensemble $\langle ... \rangle$. This is an effective way of taking into account the quantum to classical transition.

- *The scalar perturbations are 'adiabatic'*

In the case of a single scalar field there is only one scalar degree of freedom, and so the metric perturbations are directly related to, for example, $\mathcal{R}_c$, which therefore completely determines the initial perturbations of the various fluids after inflation.

- *Gravitational waves*

The same procedure can be used for gravitational waves. We recall, *cf.* (22.5), that

$$\mu_\lambda(\mathbf{k}, \eta) = \sqrt{\frac{M_P^2}{8\pi}} a(\eta) \overline{E}_\lambda(\mathbf{k}, \eta)$$

for each polarization. The solution becomes formally the same as for v if we set $\delta = 0$ and $\varepsilon \to \varepsilon/2$. Defining the power spectrum as $\langle \overline{E}_\lambda(\mathbf{k}, \eta_f) \overline{E}_{\lambda'}^*(\mathbf{k}', \eta_f) \rangle = \frac{2\pi^2}{k^3} \mathcal{P}_T(k) \delta(\mathbf{k} - \mathbf{k}') \delta_{\lambda\lambda'}$, at super-Hubble scales we obtain

$$\mathcal{P}_T = \frac{k^3}{2\pi^2} \frac{64\pi}{M_P^2} \left| \frac{\mu_k}{a} \right|^2 = \frac{16}{\pi} \frac{H^2}{M_P^2} [1 - 2(C+1)\varepsilon] \left(\frac{k}{aH} \right)^{-2\varepsilon}. \tag{22.46}$$

Just like the scalar modes, the gravitational waves develop from super-Hubble correlations with a nearly scale-invariant spectrum whose spectral index is

$$n_T - 1 \equiv \frac{d \ln \mathcal{P}_T}{d \ln k} = -2\varepsilon \tag{22.47}$$

and which has typical amplitude H/M_P, in agreement with the adiabatic approximation presented in Section 22.4. Their statistics is also Gaussian.

- *The consistency relation*

The results (22.43), (22.44), (22.46), and (22.47) give rise to a relation between the two types of perturbation. Renormalizing the power spectra as

$$A_S^2 \equiv \frac{4}{25} \mathcal{P}_\mathcal{R}(k), \quad A_T^2 \equiv \frac{1}{100} \mathcal{P}_T(k), \tag{22.48}$$

it can be checked that the ratio of the scalar and the tensor powers is $r = A_T^2/A_S^2 = \varepsilon$. Comparing this with (22.47), we find the consistency relation between the amplitudes of the spectra and the spectral index for gravitational waves:

$$r = \frac{1 - n_T}{2}. \tag{22.49}$$

If we manage to detect primordial gravitational waves, this relation will in principle allow us to reject the scenario of single field, slow-roll, inflation if it is found not to hold, or, if it does hold, it will reinforce the credibility of this scenario.

The relations (22.43) and (22.46) completely specify the properties of scalar and tensor perturbations at the end of the inflation. It is important to stress the fact that these initial conditions are statistical and allow only a general description of the large-scale structure distribution and the temperature anisotropies of the cosmic microwave background.

The standard model of cosmology: current status

Figure 22.2 summarizes the present theoretical representation of our universe. It has a hot Big Bang phase, where the physics is not speculative starting from 0.1 s after the Big Bang. The description of the universe after this time requires the use of only well understood microphysics and gravitation at linear order. As time advances, nonlinear structures form, and their understanding requires resorting to other techniques (higher-order perturbation theory, numerical simulations). However, it should be noted that the existence of dark matter indicates that the Standard Model of particle physics must be extended.

Processes occurring in the more primordial universe such as baryogenesis or matter production during the reheating phase following inflation are based on more speculative physics, and at present no model is really satisfactory enough to be considered as the standard model.

The inflationary phase also is based on speculative physics (even though after the detection of the Higgs boson at the LHC we know that scalar fields do exist), but it is compatible with observations of the cosmic microwave background.

Finally, the description of the state of the universe preceding this inflationary phase is extremely speculative and open to numerous conjectures. Primordial cosmology is the playground of the phenomenology of quantum theories of gravity and, to paraphrase the words of Jean Eisenstaedt cited above, remains a space for thought not only on general relativity, but also on theoretical physics.

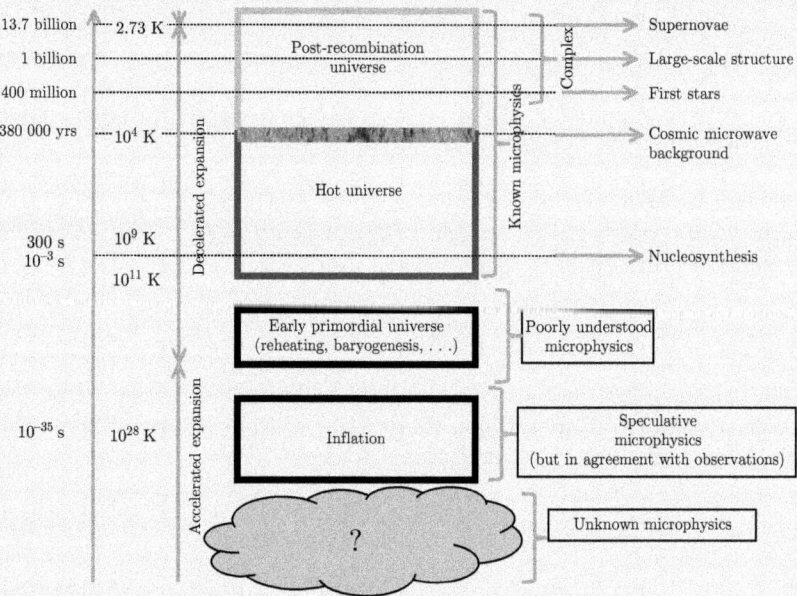

Fig. 22.2 The standard model of cosmology.

Part V

Elements of Riemannian geometry

. . . as time goes on it becomes increasingly evident that the rules which the mathematician finds interesting are the same as those which Nature has chosen.

P. A. M. Dirac, Lecture delivered on presentation of the James Scott prize, February 6, 1939; Proc. R. Soc. (Edinburgh), Vol. 59, 122 (1939)

Book 3

23

The covariant derivative and the curvature

23.1 Tangent spaces of a non-connected manifold

Let us consider[1] a non-connected manifold $\mathcal{M}$, that is, a set of points p labeled by their coordinates x^i, for the time being without a connection or a metric structure.[2] The *tangent space* at p, denoted $E(p)$, is a vector space (of dimension n) whose 'natural' basis associated with the coordinates x^i is the set of n vectors denoted $\partial/\partial x^i$; see Fig. 23.1. A *tangent* (or vector) t at p is thus an element of $E(p)$ and can be decomposed as $t = t^i \partial/\partial x^i$, where the t^i are the 'natural components' (summation on i is understood). It acts on differentiable functions $f(x^i)$ as

$$t(f) = t^i \frac{\partial f}{\partial x^i} \,, \tag{23.1}$$

where the derivatives $\partial f/\partial x^i$ are evaluated at p. If as the function f we take the kth coordinate of the point p, $f = x^k$, then $t(x^k) = t^k$: the action of a tangent on the coordinates of the point where the tangent space is 'located' gives its components in these coordinates. N.B. We also will use ∂_t to denote a tangent, hence the notations:

$$\partial_t f \equiv \partial_{t^i \frac{\partial}{\partial x^i}} f \equiv \partial_{t^i \partial_i} f \equiv t^i \partial_i f \equiv t(f) \,,$$

$$\text{where} \quad \frac{\partial}{\partial x^i} \equiv \partial_i \,. \tag{23.2}$$

One way of visualizing the tangent spaces, or 'planes', of a manifold is to imagine a 2-sphere embedded in $\mathcal{E}_3$ along with the 2-planes tangent to its surface. No matter which coordinate system is used to label the points of this 2-sphere, its tangent planes cannot be made to lie 'flat', one on top of the other, because the sphere is a curved space. However, it is important to note that the machinery we are describing here does not require that the manifold be embedded in a higher-dimensional space. It should also always be borne in mind that it can be generalized to the case where the 'points' p are not labeled by the coordinates x^i, in which case we will use h_i to denote a basis of $E(p)$.

---

[1] For an elementary introduction to vector and differential geometry, see, for example, Book 1, Chapters 2 and 4. A discussion with vivid imagery can be found in Misner, Thorne, and Wheeler (1973).

[2] The precise definitions of the terms *manifold* (a continuum of points distinguished by n coordinates x^i), *chart* (coordinate system), and *atlas* (set of charts needed to cover the entire manifold) can be found in, for example, Straumann (2013). For more details see also Bishop and Goldberg (1980) or Choquet–Bruhat, DeWitt–Morette, and Dillard–Bleick (1978).

Relativity in Modern Physics. Nathalie Deruelle and Jean-Philippe Uzan.
© Oxford University Press 2018. Published in 2018 by Oxford University Press.
DOI: 10.1093/oso/9780198786399.001.0001

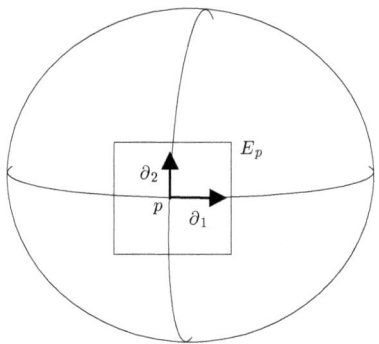

Fig. 23.1 Tangent space of a manifold.

We associate with a tangent t acting at p on a function f a *form* denoted $\mathrm{d}f$, an element of the *cotangent* space $E^*(p)$ dual to $E(p)$, which acts on a vector t to give the result of the action of t on f:

$$\mathrm{d}f(t) = t(f). \tag{23.3}$$

If we take as the function f the kth coordinate of the point p, $f = x^k$, then $t(x^k) = t^k$ according to (23.1) and, from the definition of t and linearity, (23.3) gives $\mathrm{d}x^k(t) \equiv \mathrm{d}x^k \left(t^i \partial/\partial x^i\right) \equiv t^i \mathrm{d}x^k \left(\partial/\partial x^i\right) = t^k$. We therefore have $\mathrm{d}x^k \left(\partial/\partial x^i\right) = \delta_i^k$. The n forms $\mathrm{d}x^k$ thus form a basis, the *natural basis*, of the cotangent space $E^*(p)$.

Any form λ of E_p^*, called a *differential form*, then can be decomposed as $\lambda = \lambda_i \, \mathrm{d}x^i$. If there exists a function $f(x^i)$ such that $\lambda_i = \partial f/\partial x^i$, we denote $\lambda \equiv \mathrm{d}f = \left(\partial f/\partial x^i\right) \mathrm{d}x^i$, and the form λ, of which $\mathrm{d}x^i$ are examples, is termed *exact*.

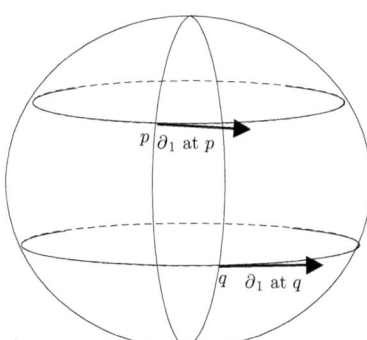

Fig. 23.2 Example of vector fields.

Let us now consider not a tangent space $E(p)$ associated with a particular point p, but an ensemble of such spaces $\{E(p)\}$. The natural basis vector $\partial/\partial x^i$ (i fixed) associated with p then becomes a *function* of the point p. The application which associates with any point p the object $\partial/\partial x^i$ is the *vector field* $\partial/\partial x^i$.

Let us consider p vectors $\partial/\partial x^i$ and q forms $\mathrm{d}x^j$ from the natural bases $E(p)$ and $E^*(p)$ associated with the coordinates x^k. Their tensor products (where i and j vary from 1 to n)

define a basis of the space of multilinear forms, or *tensors*, of the type $\binom{p}{q}$, which are p-fold contravariant and q-fold covariant. Any tensor $\binom{p}{q}$ can then be written as

$$T = T^{i_1 \dots i_p}_{j_1 \dots j_q} \partial_{i_1} \otimes \dots \otimes \partial_{i_p} \otimes \mathrm{d}x^{j_1} \otimes \dots \otimes \mathrm{d}x^{j_q} , \quad \text{where} \quad \partial_i \equiv \frac{\partial}{\partial x^i} . \tag{23.4}$$

The $T^{i_1 \dots i_p}_{j_1 \dots j_q}(x^i)$ are the components of the tensor field $T(p)$ in the natural basis associated with the coordinates x^i. It is such objects which in general relativity describe the elements of physical reality.

Let us recall the transformation law for the components of a tensor under a change of coordinates $x^i \mapsto x^i = x^i(x'^j)$:

$$T'^{i_1 \dots i_p}_{j_1 \dots j_q} = T^{k_1 \dots k_p}_{l_1 \dots l_q} \frac{\partial x'^{i_1}}{\partial x^{k_1}} \dots \frac{\partial x'^{i_p}}{\partial x^{k_p}} \frac{\partial x^{l_1}}{\partial x'^{j_1}} \dots \frac{\partial x^{l_q}}{\partial x'^{j_q}} , \tag{23.5}$$

where all quantities are expressed as functions of $x^m(x'^n)$. In fact, the law (23.5) serves as the *definition* of a tensor.

23.2 The exterior derivative

In this section we summarize the elementary introduction to the exterior calculus of Book 2, Chapter 5. To learn more, see the works cited in footnote 2.

A tensor, which we denote as α, of the type $\binom{0}{p}$ and completely antisymmetric, that is, whose components $\alpha_{i_1 \dots i_p}$ are antisymmetric in all their indices in a basis θ^i of the cotangent space $E^*(p)$, and therefore in any basis, is called a *p-form* or a *form of degree p*. Such objects form a subspace of the ensemble of tensors of the type $\binom{0}{p}$ constructed on $E^*(p)$, invariant under a change of basis, of dimension $C^p_n = n!/(n-p)!p!$, where n is the dimension of $E^*(p)$. The maximum dimension of a p-form then is $p = n$. If, for example, $n = 4$, the dimensions of the spaces of $0, 1, 2, 3, 4$-forms are $1, 4, 6, 4, 1$, respectively.

The *exterior product* of a p-form α and a q-form β is a $(p+q)$-form defined by

$$\alpha \wedge \beta \equiv \frac{(p+q)!}{p!q!} (\alpha \otimes \beta)_a , \tag{23.6}$$

where the index a denotes antisymmetrization.[3] If θ^i is a basis of $E^*(p)$, the natural basis of the space of p-forms is $\theta^{i_1} \wedge \theta^{i_2} \dots \wedge \theta^{i_p}$, with $i_1 < i_2 \dots < i_p$. Therefore, any p-form α can be decomposed as

$$\alpha = \frac{1}{p!} \alpha_{i_1 i_2 \dots i_p} \theta^{i_1} \wedge \theta^{i_2} \dots \wedge \theta^{i_p} , \tag{23.7}$$

---

[3] The antisymmetrization can be defined by recursion in a basis as $T_{[ij]} \equiv (T_{ij} - T_{ji})/2$, $T_{[ijk]} \equiv (T_{ijk} + T_{jki} + T_{kij} - T_{jik} - T_{ikj} - T_{kji})/3!$, and so on. For example, the antisymmetrized product of two 1-forms θ^i and θ^j is $(\theta^i \otimes \theta^j)_a = (\theta^i \otimes \theta^j - \theta^j \otimes \theta^i)/2$, and their exterior product is the 2-form $\theta^i \wedge \theta^j = 2(\theta^i \otimes \theta^j)_a = \theta^i \otimes \theta^j - \theta^j \otimes \theta^i$.

where the θ^i are *not* ordered, and so its components in the basis of the ordered θ^i are the *ordered* antisymmetric part[4] of $\alpha_{i_1 i_2 \ldots i_p}$.

The *exterior derivative* is an operator, denoted d, acting on a p-form to give a $(p+1)$-form. It possesses the following properties, which define it: if f is a 0-form (that is, an ordinary function) and t is a vector of $E(p)$, then $\mathrm{d}f(t) = t(f)$, which coincides with the definition of differential 1-forms given in (23.3). Moreover, $\mathrm{d}(\alpha + \beta) = \mathrm{d}\alpha + \mathrm{d}\beta$, where α and β are forms of the same degree. Finally,

$$\mathrm{d}(\alpha \wedge \beta) = \mathrm{d}\alpha \wedge \beta + (-1)^p \alpha \wedge \mathrm{d}\beta \quad \text{and} \quad \mathrm{d}^2 = 0, \tag{23.8}$$

where p is the degree of α. In the natural bases $\partial/\partial x^i$ and $\mathrm{d}x^i$ associated with the coordinates x^i, the exterior derivative is defined simply as

$$\mathrm{d}\alpha = \partial_l \alpha_{i_1 i_2 \ldots i_p} \mathrm{d}x^l \wedge \mathrm{d}x^{i_1} \wedge \ldots \mathrm{d}x^{i_p}. \tag{23.9}$$

A form α whose exterior derivative vanishes ($\mathrm{d}\alpha = 0$) is said to be *closed*. A form α which is the exterior derivative of a form β ($\alpha = \mathrm{d}\beta$) is said to be *exact*.

23.3 The Lie bracket and Lie derivative

A vector is a directional derivative operator which can be written as

$$v \equiv \partial_v = v^i \partial_i \tag{23.10}$$

(in the coordinates x^i); it acts on a function f to give the function $v^i(\partial f/\partial x^i)$. The commutator or *Lie bracket* $[v, w]$ of two vector fields v and $w = w^j \partial_j$ is the antisymmetrized composition of v and w, viewed not as elements of a vector space or as singly-contravariant tensors, but as derivative operators:

$$[v, w] \equiv v \circ w - w \circ v \equiv \partial_v \circ \partial_w - \partial_w \circ \partial_v$$
$$= v^i \partial_i \circ w^j \partial_j - w^i \partial_i \circ v^j \partial_j = (v^i \partial_i w^j - w^i \partial_i v^j) \partial_j. \tag{23.11}$$

We note that the Lie bracket of two basis vectors is zero ($[\partial_i, \partial_j] = 0$), and we can easily prove the *Jacobi identity*

$$[u, [v, w]] + [v, [w, u]] + [w, [u, v]] = 0. \tag{23.12}$$

The Lie bracket and closed paths

Let us consider a point p of coordinates x^i and the integral curve of a vector field u originating at p (see Fig. 23.3). We move forward by an amount $\mathrm{d}\lambda_1$ along this curve to arrive at the point

---

[4]For example, a 2-form with $n = 3$ is written as

$$\alpha = \frac{1}{2}(\alpha_{12}\theta^1 \wedge \theta^2 + \alpha_{21}\theta^2 \wedge \theta^1 + \ldots)$$
$$= \frac{1}{2}(\alpha_{12} - \alpha_{21})\theta^1 \wedge \theta^2 + \frac{1}{2}(\alpha_{13} - \alpha_{31})\theta^1 \wedge \theta^3 + \frac{1}{2}(\alpha_{23} - \alpha_{32})\theta^2 \wedge \theta^3 \equiv \alpha_{|[ij]|}\theta^i \wedge \theta^j,$$

where $\alpha_{|[ij]|}$ is the antisymmetrized *and* ordered ($i < j$) part of α_{ij}.

$p+d_1p$ with coordinates $x^i + u^i d\lambda_1 + \dot{u}^i (d\lambda_1)^2/2 + \cdots$. At this point the components of a vector field v are, to first order in $d\lambda_1$, $v^i(x^j) + (\partial_j v^i)u^j d\lambda_1$. Then moving forward again by $d\lambda_2$ along the integral curve of the field v originating at $p+d_1p$, we reach the point $p+d_{12}p$ with coordinates $x^i + u^i d\lambda_1 + \dot{u}^i (d\lambda_1)^2/2 + [v^i(x^j) + (\partial_j v^i)u^j d\lambda_1]d\lambda_2 + \dot{v}^i (d\lambda_2)^2/2 + \cdots$. If now we reverse the operations, passing from p to $p+d_2p$ along the field v, then to $p+d_{21}p$ along u, we arrive at the point with coordinates $x^i + v^i d\lambda_2 + \dot{v}^i (d\lambda_2)^2/2 + [u^i(x^j) + (\partial_j u^i)v^j d\lambda_2]d\lambda_1 + \dot{u}^i (d\lambda_1)^2/2 + \cdots$. The two points $p+d_{12}p$ and $p+d_{21}p$ are the same if $u^j \partial_j v^i - v^j \partial_j u^i = 0$, that is, if the Lie bracket of the vectors u and v is zero. There then exist coordinates x^1 and x^2 tangent to the vector fields[5] u and v: $u = \partial/\partial x^1$ and $v = \partial/\partial x^2$.

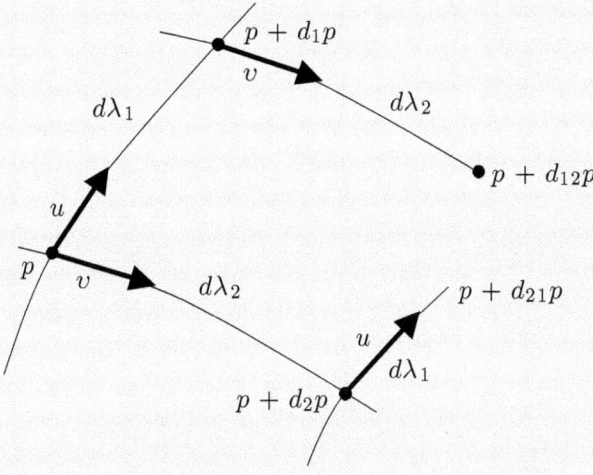

Fig. 23.3 The Lie bracket.

The Lie bracket is a vector, which can be verified by looking at how it transforms under a change of coordinates. It therefore defines a derivative operator, the *Lie derivative*, which can be introduced independently in the following manner.

We consider a point p of coordinates x^i and the integral curve of a vector field u originating at p. We move an amount $d\lambda$ along this line to arrive at the point $\tilde{p} = p + dp$ with coordinates $\tilde{x}^i = x^i + u^i(x^j)d\lambda$ at first order in $d\lambda$. At this point, another vector field v can be decomposed as $v(\tilde{p}) = v^i(x^j + u^j d\lambda)(\partial/\partial \tilde{x}^i) = (v^i(x^j) + (\partial_j v^i)u^j d\lambda)(\partial/\partial \tilde{x}^i)$, at first order in $d\lambda$.

The *pullback* of the field v is the vector field v such that $\tilde{v}(p) = v(\tilde{p})$. We then have $\tilde{v}^i(\partial/\partial x^i) = (v^i(x^j) + (\partial_j v^i)u^j d\lambda)(\partial/\partial \tilde{x}^i)$.

Since $(\partial/\partial \tilde{x}^i) = (\partial x^j/\partial \tilde{x}^i)(\partial/\partial x^j)$ with $(\partial x^j/\partial \tilde{x}^i) = \delta_i^j - \partial_i u^j d\lambda$ because $\tilde{x}^j = x^j + u^j(x^i)d\lambda$, we find

$$\tilde{v}^i(x^j)(\partial/\partial x^i) = \left(v^i(x^j) + (\partial_j v^i)u^j d\lambda\right)\left(\delta_i^j - \partial_i u^j d\lambda\right)(\partial/\partial x^j)$$
$$= \left(v^i(x^j) + d\lambda(u^j \partial_j v^i - v^j \partial_j u^i)\right)(\partial/\partial x^i).$$

---

[5] Instead of moving in $\mathcal{M}$ along integral curves of u or v, it is possible to move in the tangent planes, which are then affine, along the *straight lines* of the tangent vectors u or v and then project the points $p+d_1p$, and so on, onto the manifold. The conditions for, on the one hand, $p+d_{12}p$ and $p+d_{21}p$ to be identical and, on the other, to return to p after arriving at $p+d_{12}p$ and making a complete loop, are, at second order, that the Lie bracket of u and v vanish.

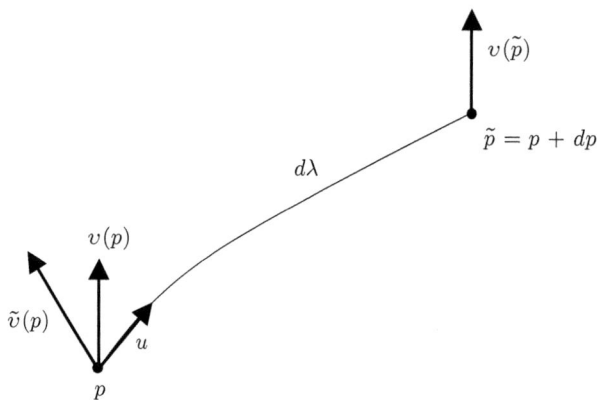

Fig. 23.4 The Lie derivative.

The Lie derivative of the vector field v with respect to the vector field u is then defined as

$$\mathcal{L}_u v = \frac{\tilde{v}(p) - v(p)}{d\lambda} = (u^j \partial_j v^i - v^j \partial_j u^i)\partial_i . \qquad (23.13)$$

Therefore, the Lie bracket and the Lie derivative of a vector field are equivalent:

$$\mathcal{L}_u v = [u, v] , \qquad (23.14)$$

where the operator $\mathcal{L}_u$ gives the bracket $[u, v]$ by acting on v. The Jacobi identity is then written equivalently as either

$$[u, [v, w]] + [v, [w, u]] + [w, [u, v]] = 0 \quad \text{or} \quad \mathcal{L}_{[u,v]} = [\mathcal{L}_u, \mathcal{L}_v] . \qquad (23.15)$$

We note that the Lie derivative is *not* a directional derivative which would allow the vector v to be transported along the integral curve of u.

If we define the Lie derivative with respect to the vector u of a scalar field f as $\mathcal{L}_u f = u^i \partial_i f$, we easily obtain the components of the Lie derivative of a form ω with respect to the vector u:

$$\mathcal{L}_u \omega = (u^i \partial_i \omega_j + \omega_i \partial_j u^i)dx^j , \qquad (23.16)$$

as well as those of any tensor. For example, the Lie derivative with respect to the vector u of a 2-fold covariant tensor, $g = g_{ij}dx^i \otimes dx^j$, can be decomposed as

$$\mathcal{L}_u g = (u^i \partial_i g_{jk} + g_{ij}\partial_k u^i + g_{ik}\partial_j u^i)dx^j \otimes dx^k . \qquad (23.17)$$

'Passive' definition of the Lie derivative

Under an infinitesimal change of coordinates $x^i \to \tilde{x}^i = x^i + u^i d\lambda$, the components of a vector field at a point p are $v^i(x^j)$ in the system x^i and $\tilde{v}^i(\tilde{x}^j)$ in the system $\tilde{x}^i$. From the vector transformation law we have $v^i = (\partial x^i/\partial \tilde{x}^j)\tilde{v}^j$ or, at first order in $d\lambda$, $v^i(x^j) = \tilde{v}^i(\tilde{x}^j) - v^j \partial_j u^i = \tilde{v}^i(x^j) + d\lambda(u^j \partial_j v^i - v^j \partial_j u^i)$.

The quantities $\tilde{v}^i(x^j)$ are the components of the vector field v in the coordinate system $\tilde{x}^i$ at the point $\tilde{p}$ which in the system $\tilde{x}^i$ has the same coordinates x^i as the point p in the system without the tilde.

In this so-called *passive* approach, the Lie derivative of v with respect to u is defined as having, in the system x^i, the components

$$(\mathcal{L}_u v)^i = \frac{v^i(x^j) - \tilde{v}^i(x^j)}{d\lambda} = u^j \partial_j v^i - v^j \partial_j u^i. \tag{23.18}$$

23.4   The covariant derivative and connected manifold

The derivative operators introduced in the two preceding sections have limited effectiveness: the exterior derivative acts only on p-forms, that is, completely antisymmetric tensors; the vector fields and Lie brackets act only on functions; finally, the Lie derivative with respect to a vector field involves not only this field, but also its derivative, and therefore is not the generalization of a directional derivative which would allow tensors to be transported from one point to another. For this we must introduce an additional object, the *covariant derivative*.

Let us recall its operational definition.[6]

As an example, let us take a singly contravariant and singly covariant tensor field, that is, a $\binom{1}{1}$ field: $T = T^i_j \, \partial_i \otimes dx^j$, where T^i_j are its components in the coordinates x^i. Its covariant derivative with respect to a vector $v = v^i \partial_i$ is a tensor of the same type, denoted $D_v T$:

$$D_v T = D_{v^i \partial_i} T = v^i D_{\partial_i} T \equiv v^i D_i T = v^i D_i (T^j_k \, \partial_j \otimes dx^k)$$

$$= v^i \left[(D_i T^j_k) \partial_j \otimes dx^k + T^j_k (D_i \partial_j) \otimes dx^k + T^j_k \partial_j \otimes (D_i dx^k) \right]$$

$$= v^i \left[(\partial_i T^j_k) \partial_j \otimes dx^k + T^j_k (\Gamma^l_{ij} \partial_l) \otimes dx^k + T^j_k \partial_j \otimes (-\Gamma^k_{il} dx^l) \right], \tag{23.19}$$

$$\text{because} \quad D_i \partial_j \equiv \Gamma^l_{ij} \partial_l, \quad D_i dx^k \equiv -\Gamma^k_{il} dx^l,$$

$$\equiv v^i (D_i T^j_k) \, \partial_j \otimes dx^k, \quad \text{where} \quad D_i T^j_k = \partial_i T^j_k + \Gamma^j_{il} T^l_k - \Gamma^l_{ik} T^j_l.$$

The *connection coefficients* Γ^i_{jk} are n^3 functions of the coordinates which *define* the covariant derivative D, and the $D_i T^j_k$ are the *natural components* of $D_{\partial/\partial x^i} T \equiv D_{\partial_i} T \equiv D_i T$. We note that in (23.19) we have used the usual properties of derivative operators, in particular, the Leibniz rule for differentiating composite functions.

We also recall that an *auto-parallel* is a curve $x^i = x^i(\lambda)$ with tangent vector $u^i = dx^i/d\lambda$ such that

$$D_u u = 0, \quad \text{that is,} \quad \frac{du^i}{d\lambda} + \Gamma^i_{jk} u^k u^k = 0. \tag{23.20}$$

Finally, we recall the transformation law for the connection coefficients under a change of coordinates[6]:

---

[6] For more detail, see Book 1, Chapters 3 and 4, or the studies cited at the beginning of the present chapter.

$$\Gamma'^{j}_{ki} = \frac{\partial x^r}{\partial x'^i} \frac{\partial x^l}{\partial x'^k} \frac{\partial x'^j}{\partial x^p} \Gamma^p_{lr} + \frac{\partial^2 x^l}{\partial x'^k \partial x'^i} \frac{\partial x'^j}{\partial x^l} . \tag{23.21}$$

23.5   Torsion of a covariant derivative

Let us consider the action of the covariant derivative with respect to the coordinate x^i [that is, with respect to the basis vector ∂_i of the tangent space $E(p)$] on the basis vector ∂_j. The result is a vector:

$$D_j \partial_i = \Gamma^k_{ji} \partial_k \quad \Longrightarrow \quad D_j \partial_i - D_i \partial_j = (\Gamma^k_{ji} - \Gamma^k_{ij}) \partial_k \equiv -T^k_{ij} \partial_k . \tag{23.22}$$

The quantities T^k_{ij} measure the antisymmetry of the connection. They are the components of a tensor T of the type $\binom{1}{2}$, called the *torsion*. The fact that T is a tensor can be verified at sight by considering how its components transform under a change of coordinates, because the second term in (23.21) is symmetric in k and i, and so T transforms as (23.5).

In a more intrinsic, that is, coordinate-independent manner, the *torsion* T^D of a covariant derivative D is a $\binom{1}{2}$-type tensor which acts on pairs of vectors (v, w) to give another vector z [$T_D : (v, w) \to z$] according to

$$T^D(v, w) \equiv D_v w - D_w v - [v, w] , \tag{23.23}$$

where $[v, w]$ is the Lie bracket defined in Section 23.3. To make the connection with the definition (23.22), it is sufficient to decompose v and w on a basis, for example, $D_v w = v^i D_i(w^j \partial_j)$, and use the fact that the Lie bracket of two basis vectors is zero. Then from the definition of the connection coefficients $D_i \partial_j = \Gamma^k_{ij} \partial_k$ we find

$$T^D(v, w) \equiv T^k_{ij} v^i w^j \partial_k \quad \text{with} \quad T^k_{ij} = \Gamma^k_{ij} - \Gamma^k_{ji} . \tag{23.24}$$

The fact that $T^D(v, w)$ is linear in v^i and w^j is sufficient for showing that it is indeed a type-$\binom{1}{2}$ tensor.

Therefore, the torsion measures the difference between the antisymmetrized covariant derivative and the Lie bracket, or also the Lie derivative.

The spacetime connections describing gravitation in general relativity are assumed to be torsion-free.

23.6   Curvature of a covariant derivative

Similarly, the *Riemann–Christoffel curvature* R^D of a covariant derivative D is a type-$\binom{1}{3}$ tensor which acts on a vector triplet (u, v, w) to give another vector z [$R_D : (u, v, w) \to z$] as

$$R^D_{u,v} w \equiv [D_u, D_v] w - D_{[u,v]} w , \quad \text{where} \quad [D_u, D_v] w \equiv D_u(D_v w) - D_v(D_u w) . \tag{23.25}$$

It measures the non-commutativity of the covariant derivatives.

If now for u, v, and w we use the basis vectors ∂_i, ∂_j, and ∂_k of the tangent space $E(p)$, since $D_i D_j \partial_k = D_i(\Gamma^l_{jk}\partial_l) = (\partial_i\Gamma^m_{jk} + \Gamma^l_{jk}\Gamma^m_{il})\partial_m$ and the Lie bracket of two basis vectors is zero, we immediately find

$$R^D_{\partial_i,\partial_j}\partial_k = (D_iD_j - D_jD_i)\partial_k \equiv R^m{}_{kij}\partial_m$$

$$\text{with}\quad R^m{}_{kij} = \partial_i\Gamma^m_{jk} - \partial_j\Gamma^m_{ik} + \Gamma^m_{il}\Gamma^l_{jk} - \Gamma^m_{jl}\Gamma^l_{ik}\,. \tag{23.26}$$

The curvature is a type-$\binom{1}{3}$ tensor, as can be verified rather laboriously by studying the transformation law of its components using (23.5), and more easily by showing (a useful exercise) that

$$R^D_{u,v}w = u^i v^j w^k R^m{}_{kij}\partial_m\,. \tag{23.27}$$

In a similar manner, see (23.19), we obtain the action of the operator $(D_iD_j - D_jD_i)$ on the 1-form of basis $\mathrm{d}x^k$:

$$(D_iD_j - D_jD_i)\mathrm{d}x^k = -R^k{}_{lij}\mathrm{d}x^l\,, \tag{23.28}$$

and, by the Leibniz rule, its action on any tensor. For example,

$$(D_iD_j - D_jD_i)(\mathrm{d}x^k \otimes \partial_l) = -R^k{}_{mij}\mathrm{d}x^m \otimes \partial_l + R^m{}_{lij}\mathrm{d}x^k \otimes \partial_m\,. \tag{23.29}$$

Now to make the connection with the 'component' approach, we consider $R^D_{\partial_i,\partial_j}v = D_iD_jv - D_jD_iv$, cf. (23.25). This is a vector given by, cf. (23.26), $R^D_{\partial_i,\partial_j}v = v^k R^m{}_{kij}\partial_m$. Therefore, $v^k R^m{}_{kij}$ is the mth component of the vector $(D_iD_j - D_jD_i)v$. We can then write

$$(D_iD_j - D_jD_i)v^m = R^m{}_{kij}v^k \tag{23.30}$$

and thus recover the formula obtained in Section 2.2.

Now let us study the relative 'acceleration' of two closely spaced auto-parallels, defined as $a = D_uD_un$, with components $a^i = D^2n^i/d\lambda^2$. We have

$$D_uD_un = -R^D_{n,u}u\,. \tag{23.31}$$

This result follows, first of all, from the fact that $D_un = D_nu$ because $[n, u] = 0$; secondly, it follows from the definition of the curvature tensor:

$$R^D_{n,u}u = D_n(D_uu) - D_u(D_nu) - D_{[n,u]}u = -D_u(D_nu)\,, \quad \text{because} \quad D_uu = 0\,.$$

If we make (23.31) explicit on a basis, we recover the 'geodesic' deviation equation derived in Section 2.3: $a^i = -R^i{}_{mkj}u^mu^jn^k$.

The vector fields n satisfying (23.31) are called *Jacobi fields*.

24

Riemannian manifolds

24.1   The metric manifold and Levi-Civita connection

A *metric tensor* g is a tensor field on $\mathcal{M}$ which is 2-fold covariant [*i.e.*, of the type $\binom{0}{2}$], symmetric, and nondegenerate. In a coordinate system x^i it is written as

$$g = g_{ij}\mathrm{d}x^i \otimes \mathrm{d}x^j \, . \tag{24.1}$$

The $n(n+1)/2$ functions $g_{ij}(x^k)$, which are symmetric ($g_{ij} = g_{ji}$) and invertible ($g^{ik}g_{kj} = \delta^i_j$), define it.

Given g, it is always possible to find, at a point, a basis of *orthonormal* vectors h_i of the tangent space such that

$$g(h_i, h_j) = 0 \quad \text{if} \quad i \neq j, \quad g(h_i, h_i) = \pm 1. \tag{24.2}$$

If $n = 4$, the set of h_i is called a *tetrad* or a *vierbein*, or, more generally, a *frame field* or *moving frame*. If all the signs are positive, the metric is *Riemannian*, and if one sign is negative it is *Lorentzian*.[1]

We know that the metric tensor allows us to define a one-to-one correspondence between vectors and 1-forms. Indeed, if $v = v^i h_i$ is a vector, then $v_i \theta^i$, where $v_i \equiv g_{ij}v^j$ [here $g_{ij} \equiv g(h_i, h_j)$], is a 1-form isomorphic to v (and often also denoted by v). Reciprocally, if $\lambda = \lambda_i \theta^i$ is a 1-form, then $\lambda^i h_i$, where $\lambda^i \equiv g^{ij}\lambda_j$, is a vector ($g^{ij}$ is the inverse of the matrix g_{ij}). More generally, with any tensor of the type $\binom{p}{q}$ we thus associate tensors of the type $\binom{p-1}{q+1}$ or $\binom{p+1}{q-1}$.

Finally, a metric defines the *scalar product* of two vectors v and v' as $\langle v, v' \rangle \equiv g(v, v') = g_{ij}v^i v'^j$. In the case of a Lorentzian metric, if the *norm* $\langle v, v \rangle$ of v is positive, v is termed space-like; if it is negative, v is time-like; if it is zero, v is isotropic or null.

A connection D is a *metric connection* if for any vector u

$$D_u g = 0 \, . \tag{24.3}$$

Decomposing g on the natural basis x^i, this condition becomes

$$0 = D_u g = u^i D_i(g_{jk}\mathrm{d}x^j \otimes \mathrm{d}x^k) = u^i(D_i g_{jk})\mathrm{d}x^j \otimes \mathrm{d}x^k \, ,$$

or

---

[1]N.B. It is sufficient that the manifold be 'paracompact' to allow for a Riemannian metric. On the other hand, a Lorentzian metric can only be defined if the Euler characteristic of the manifold is zero; therefore, a 2-sphere whose Euler characteristic is two does not admit a Lorentzian metric. Definitions and examples can be found in any textbook on differential geometry.

Relativity in Modern Physics. Nathalie Deruelle and Jean-Philippe Uzan.
© Oxford University Press 2018. Published in 2018 by Oxford University Press.
DOI: 10.1093/oso/9780198786399.001.0001

$$D_i g_{jk} \equiv \partial_i g_{jk} - \Gamma^l_{ij} g_{lk} - \Gamma^l_{ik} g_{jl} = 0 \,. \tag{24.4}$$

The condition (24.3) can also be written as

$$\partial_u g(v, w) = g(D_u v, w) + g(v, D_u w) \,, \tag{24.5}$$

[where $\partial_u g(v, w) = u(g(v, w))$], a very useful relation which can be proved by expanding the various terms on a basis and then using (24.4).[2]

A theorem due to Ricci states that given a metric g, there exists one and only one metric connection of zero torsion, which is called the Levi-Civita connection (or covariant derivative). Equation (24.4) is sufficient for proving this and finding the explicit expression for the coefficients Γ^i_{jk} of this connection as a function of the metric components g_{ij}. Then if $\Gamma^i_{jk} = \Gamma^i_{kj}$, we find

$$\Gamma^i_{jk} = \frac{1}{2} g^{il} (\partial_j g_{kl} + \partial_k g_{lj} - \partial_l g_{jk}) \,. \tag{24.6}$$

The geodesic and geodesic deviation equations

In Section 2.4 we arrived at the 'auto-parallel' equation (23.20) by finding the equation for the 'geodesics', curves of extremal length, in terms of the Christoffel symbols, which we have identified as the connection coefficients by requiring that geodesics and auto-parallels be identical.

Here to obtain the geodesic equation in a more geometrical manner, we shall extremize not the length of the curves $\mathcal{C}_s$ [with the equations $x^i = x^i_s(\lambda)$], which has the inconvenience of being reparametrization-invariant, but instead their 'energy'

$$E(\mathcal{C}_s) = \frac{1}{2} \int_{\lambda_1}^{\lambda_2} g(u, u) \, d\lambda \,, \tag{24.7}$$

where u is the vector tangent to the curve $\mathcal{C}_s$ (that is, $u^i = dx^i_s/d\lambda$). Using n to denote the vector with components $n^i = \partial x^i_s/\partial s$, we obtain

$$\begin{aligned}
\left. \frac{dE}{ds} \right|_{s=0} &= \frac{1}{2} \int_{\lambda_1}^{\lambda_2} \frac{\partial g(u, u)}{\partial s} \, d\lambda \equiv \frac{1}{2} \int_{\lambda_1}^{\lambda_2} \partial_n g(u, u) \, d\lambda = \int_{\lambda_1}^{\lambda_2} g(D_n u, u) \, d\lambda \quad \text{using (24.5)} \\
&= \int_{\lambda_1}^{\lambda_2} g(D_u n, u) \, d\lambda \quad \text{from } D_u n = D_n u, \text{ the torsion and Lie bracket } [u, n] \text{ being zero} \\
&= \int_{\lambda_1}^{\lambda_2} [\partial_u g(n, u) - g(n, D_u u)] \, d\lambda \quad \text{again using (24.5)} \\
&= \int_{\lambda_1}^{\lambda_2} \frac{dg(n, u)}{d\lambda} \, d\lambda - \int_{\lambda_1}^{\lambda_2} g(n, D_u u) \, d\lambda \\
&= -\int_{\lambda_1}^{\lambda_2} g(n, D_u u) \, d\lambda \quad \text{if the vector } n \text{ vanishes at } \lambda_1 \text{ and } \lambda_2.
\end{aligned} \tag{24.8}$$

---

[2]Indeed, if $u = \partial_i$, $v = \partial_j$, and $w = \partial_k$, we have $\partial_u g(v, w) = \partial_i g_{jk}$ because $\partial_u g(v, w) = u(g(v, w))$, since $\partial_u f = u^i \partial_i f$ and $g(v, w) = v^j w^k g_{jk}$; $g(D_u v, w) = g(D_i \partial_j, \partial_k) = g(\Gamma^l_{ij} \partial_l, \partial_k) = \Gamma^l_{ij} g_{lk}$ and, similarly, $g(v, D_u w) = g(\partial_j, D_i \partial_k) = g(\partial_j, \Gamma^l_{ik} \partial_l) = \Gamma^l_{ik} g_{jl}$, from which (24.5) follows via (24.4).

Since the vector n is arbitrary between the extremities of the curves, we have $(dE/ds)|_{s=0} = 0$ if the curve C_0 is a geodesic, that is, if

$$D_u u = 0 \iff u^i D_i(u^j \partial_j) = 0 \iff u^i (D_i u^j) \partial_j = 0 \iff \frac{du^j}{d\lambda} + \Gamma^j_{ik} u^i u^k = 0. \quad (24.9)$$

The vector u tangent to the geodesic has constant norm since

$$\frac{dg(u,u)}{d\lambda} = D_u g(u,u) = 2g(D_u u, u) = 0 \quad \text{owing to (24.5) and because } D_u u = 0. \quad (24.10)$$

Thus, for time-like curves we can choose $g(u,u) = -1$ (that is, $g_{ij} u^i u^j = -1$), in which case the vector u is called the 4-velocity.

24.2 Properties of the curvature tensor

Here we state without proof the intrinsic versions of the properties of the Riemann–Christoffel tensor of a covariant derivative already given in Section 2.6.[3]

For any vectors u, v, w, and t,

- $R^D_{u,v} w = -R^D_{v,u} w$, or in component form $R^i{}_{jkl} = -R^i{}_{jlk}$, which follows from the definition;

- $D_u R^D_{v,w} + D_v R^D_{w,u} + D_w R^D_{u,v} = 0$, or in component form $D_m R^i{}_{jnp} + D_n R^i{}_{jpm} + D_p R^i{}_{jmn} = 0$, which is proved using the fact that $D_u R^D_{v,w} t = D_u(R^D_{v,w} t) - R^D_{D_u v, w} t - R^D_{v, D_u w} u - R^D_{v,w} D_u t$;

- if the torsion is zero, $R^D_{u,v} w + R^D_{v,w} u + R^D_{w,u} v = 0$, or in component form $R^l{}_{ijk} + R^l{}_{jki} + R^l{}_{kij} = 0$ (the *first Bianchi identity*, which follows from the Jacobi identity satisfied by the Lie bracket).

When the connection, which is symmetric, is compatible with a metric tensor, the curvature tensor possesses the additional property

$$g(R_{X,Y} Z, U) = -g(R_{X,Y} U, Z), \quad (24.11)$$

or, in component form $R_{ijkl} = -R_{jikl}$, where $R_{ijkl} \equiv g_{ip} R^p{}_{jkl}$ owing to the fact that $D_X g = 0$. Moreover, it can easily be shown that[3]

$$g(R_{X,Y} Z, U) = g(R_{Z,U} X, Y) \quad \text{or, in components,} \quad R_{lkij} = R_{jikl}. \quad (24.12)$$

Let us recall the definitions of the Ricci tensor R_{ij} and the scalar curvature R:

$$R_{ij} \equiv R^l{}_{ilj} = -R^l{}_{ijl}, \quad \text{which is symmetric:} \quad R_{ij} = R_{ji} \quad \text{and} \quad R \equiv g^{ij} R_{ij}. \quad (24.13)$$

[Writing out the intrinsic form of the definitions (24.13) is quite tedious.[3]]

---

[3]For the proofs, see, for example, Bourguignon (2005) or Straumann (2013).

The Einstein tensor

$$G_{ij} \equiv R_{ij} - \frac{1}{2} g_{ij} R \quad \text{satisfies the Bianchi identities} \quad D_i G^{ij} \equiv 0. \qquad (24.14)$$

We conclude by giving the dimension $\mathcal{D}R$ of the Riemann tensor, that is, the number of its independent components[3] in dimension n:

$$\mathcal{D}R = \frac{1}{12} n^2 (n^2 - 1). \qquad (24.15)$$

For $n = 2$, $\mathcal{D}R = 1$; for $n = 3$, $\mathcal{D}R = 6$; and for $n = 4$, $\mathcal{D}R = 20$.

24.3 Variation of the Hilbert action

In Section 4.4 we obtained the Einstein equations of general relativity by varying the Hilbert action. We shall do the same derivation here,[4] but in the intrinsic manner, using the tools developed in the present and the preceding chapters.

Let us consider the integral over a domain Ω

$$S_{g_t} = \int_\Omega R_t \, \omega_t, \qquad (24.16)$$

where R_t is the scalar curvature of the metric g_t and $\omega_t = \sqrt{-\det g_t} \, d\Omega$ with $d\Omega = \mathrm{d}x^1 \wedge \mathrm{d}x^2 \wedge \mathrm{d}x^3 \wedge \mathrm{d}x^4$ being the volume element. Since $R = c(g^{-1}.Ric)$, where Ric is the Ricci tensor, where g^{-1} is the 2-fold contravariant tensor which is the inverse of the metric g, and where c is the contraction operator, (24.16) is written in component form as $S_{g_t} = \int_\Omega g^{ij} R_{ij} \sqrt{-g} \, d\Omega$ (where in this notation g is the metric determinant).

The metric g_t depends on a parameter t, and for small t we can write

$$g_t = g + th. \qquad (24.17)$$

First of all we have

$$\left. \frac{d\omega_t}{dt} \right|_{t=0} = \frac{1}{2} c(g^{-1}.h) \, \omega. \qquad (24.18)$$

In tensor notation $\omega = \sqrt{-g} \, d\Omega$, h is a tensor with components δg_{ij}, $c(g^{-1}.h) = g^{ij} \delta g_{ij}$, and so (24.18) then reads $\delta \sqrt{-g} = \frac{1}{2} \sqrt{-g} g^{ij} \delta g_{ij}$. We note that $\delta g^{ij} = -g^{ik} g^{jl} \delta g_{kl}$.

Next we have

$$\left. \frac{dR_t}{dt} \right|_{t=0} = -(h, Ric) + g^{-1} \left. \frac{dRic_t}{dt} \right|_{t=0}. \qquad (24.19)$$

In tensor notation $\delta R = R_{ij} \delta g^{ij} + g^{ij} \delta R_{ij} = -R^{ij} \delta g_{ij} + g^{ij} \delta R_{ij}$.

Now let us introduce some quantities which will prove useful in dealing with the second term in (24.19).

---

[4]Following Bourguignon (2005).

• The codifferential[5] of the 2-fold covariant tensor h, δh, is the 1-form defined as $\delta h = -c(g^{-1}Dh)$, where D is the covariant derivative associated with the metric g. In component form, $(\delta h)_i = -g^{kl}D_k\delta g_{li}$. We also have (since $Dg = 0$)

$$\delta\delta h = g^{ij}g^{kl}D_jD_k\delta g_{li}\,. \tag{24.20}$$

• The Laplacian is such that $\Delta\mathrm{Trace}_g h = \Delta[c(g^{-1}h)]$, where $\Delta = \delta\mathrm{d}$. In component form, $c(g^{-1}h) = g^{ij}\delta g_{ij}$, $\mathrm{d}[c(g^{-1}h)] = \partial_k(g^{ij}\delta g_{ij})\mathrm{d}x^k$, and $\delta\{\mathrm{d}[c(g^{-1}h)]\} = -D^k[\partial_k(g^{ij}\delta g_{ij})]$, so that

$$\Delta\mathrm{Trace}_g h = -g^{kl}D_kD_l(g^{ij}\delta g_{ij})\,. \tag{24.21}$$

Next we proceed by adopting the Landau and Lifshitz method of working in a system of locally normal coordinates, that is, coordinates in which the components of the metric g_{ij} are constant through first order inclusive. Then the components of the Ricci tensor reduce to

$$R_{ij} = \frac{1}{2}g^{kl}(\partial_{ik}g_{jl} + \partial_{jl}g_{ik} - \partial_{ij}g_{kl} - \partial_{kl}g_{ij})\,, \tag{24.22}$$

and an easy calculation shows that the second term in (24.19) can be written as

$$g^{-1}\frac{dRic_t}{dt}\bigg|_{t=0} \equiv g^{ij}\delta R_{ij} = g^{ij}g^{kl}(\partial_{jk}\delta g_{il} - \partial_{ij}\delta g_{kl}) + \delta g^{ij}g^{kl}\partial_{jl}g_{ik}$$
$$-\frac{1}{2}\delta g^{ij}g^{kl}(\partial_{ij}g_{kl} + \partial_{kl}g_{ij})\,. \tag{24.23}$$

Moreover, always in a locally inertial frame (expanding D_k in terms of Γ^i_{jk} and then $g\partial g$) we have

$$\begin{cases} \delta\delta h = g^{ij}g^{kl}\partial_{jk}\delta g_{il} + \delta g^{ij}g^{kl}\partial_{jl}g_{ki} - \dfrac{1}{2}\delta g^{ij}g^{kl}(\partial_{ij}g_{kl} - \partial_{kl}g_{ij})\,, \\[2mm] \Delta\mathrm{Trace}_g h = -g^{ij}g^{kl}\partial_{ij}\delta g_{kl} + \delta g^{ij}g^{kl}\partial_{kl}g_{ij}\,. \end{cases} \tag{24.24}$$

Therefore,

$$g^{-1}\frac{dRic_t}{dt}\bigg|_{t=0} = \delta\delta h + \Delta\mathrm{Trace}_g h = \delta(\delta h + \mathrm{d}\mathrm{Trace}_g h)\,. \tag{24.25}$$

Since the codifferential of a 1-form and the divergence of a vector are related as $\delta(v_i\mathrm{d}x^i) = -D_iv^i$, we have thereby shown that (24.25) is a divergence. In component form,

$$g^{ij}\delta R_{ij} = g^{ij}D_jq_i = \frac{1}{\sqrt{-g}}\partial_i(\sqrt{-g}q^i) \quad \text{with} \quad q_i = g^{kl}D_k\delta g_{li} - g^{kj}D_i\delta g_{kj}\,. \tag{24.26}$$

Now that we have finished these preliminaries, we return to (24.16) and find that

---

[5]N.B. Do not confuse the codifferential δh with the increment δg_{ij} ...

$$\frac{dS_t}{dt}\Big|_{t=0} = \int_\Omega \left(\omega \frac{dR}{dt}\Big|_{t=0} + R \frac{d\omega}{dt}\Big|_{t=0} \right)$$

$$= \int_\Omega \left(\omega \left[-(h, Ric) + g^{-1} \frac{dRic_t}{dt}\Big|_{t=0} \right] + \frac{1}{2} c(g^{-1}.h) \omega R \right)$$

$$= -\int_\Omega \left((h, Ric) - \frac{1}{2} Rg \right) \omega + \int_\Omega \omega g^{-1} \frac{dRic_t}{dt}\Big|_{t=0} \qquad (24.27)$$

$$= -\int_\Omega \left((h, Ric) - \frac{1}{2} Rg \right) \omega + \int_\Omega \omega \delta(\delta h + d\mathrm{Trace}_g h),$$

where we have seen that the integrand of the last term is a divergence, and so the integral is zero if the h and their derivatives vanish on the boundary, that is, if the variations δg_{ij} and $\partial_k \delta g_{ij}$ are zero on the boundary of the domain Ω. Therefore, in the end,

$$\frac{dS_t}{dt}\Big|_{t=0} = -\int_\Omega \left((h, Ric) - \frac{1}{2} Rg \right) \omega = -\int_\Omega \delta g_{ij} G^{ij} \sqrt{-g} d\Omega,$$

$$\text{where} \quad G_{ij} \equiv R_{ij} - \frac{1}{2} g_{ij} R \qquad (24.28)$$

is the Einstein tensor. The Hilbert action then is an extremum if the vacuum Einstein equations, $G_{ij} = 0$, are satisfied.

25

The Cartan structure equations

25.1 A useful relation

Let[1] ω be a 1-form ($\omega = \omega_i \mathrm{d}x^i$ in the coordinates x^i), and let $\mathrm{d}\omega$ be its exterior derivative (that is, $\mathrm{d}\omega = \partial_j \omega_i\, \mathrm{d}x^j \wedge \mathrm{d}x^i$, where $\mathrm{d}x^j \wedge \mathrm{d}x^i \equiv \mathrm{d}x^j \otimes \mathrm{d}x^i - \mathrm{d}x^i \otimes \mathrm{d}x^j$ is the exterior product of the forms $\mathrm{d}x^j$ and $\mathrm{d}x^i$. We recall that owing to the antisymmetry of the exterior product, $\mathrm{d}^2\omega = 0$, which is the Poincaré lemma; see Section 23.2). Since $\mathrm{d}\omega$ is a 2-form, it acts on pairs of vectors (u, v) to give a function. We can write

$$\mathrm{d}\omega(u, v) = u\,\omega(v) - v\,\omega(u) - \omega([u, v])\,. \tag{25.1}$$

Indeed, decomposing $u = u^i \partial_i$, $v = v^j \partial_j$, and $\omega = \omega_k \mathrm{d}x^k$, we have $\mathrm{d}\omega = \partial_l \omega_k \mathrm{d}x^l \wedge \mathrm{d}x^k$ and

(1) $\mathrm{d}\omega(u, v) = \partial_l \omega_k (u^l v^k - u^k v^l) = (\partial_l \omega_k - \partial_k \omega_l) u^l v^k$;

(2) $u\,\omega(v) = u[\omega_k \mathrm{d}x^k(v^j \partial_j)] = u(\omega_k v^k) = u^i \partial_i(\omega_k v^k) = u^i v^k \partial_i \omega_k + u^i \omega_k \partial_i v^k$;

(3) similarly, $v\,\omega(u) = v^i u^k \partial_i \omega_k + v^i \omega_k \partial_i u^k$;

(4) $\omega([u, v]) = \omega_k(u^j \partial_j v^k - v^j \partial_j u^k)$, using the definition of the Lie bracket.

The terms involving derivatives of the vector components cancel out and we arrive at (25.1).

If we take the vectors v and w to be vectors of a basis which is not necessarily holonomic (that is, $v = h_i$ and $w = h_j$, where $h_i = \partial_i$ only if the basis is natural), and if we choose ω to be a vector of the associated dual basis $\omega = \theta^k$ [with $\theta^k(h_i) = \delta_i^k$], then (25.1) becomes

$$\{\mathrm{d}\theta^k(h_i, h_j)\}\, h_k = -[h_i, h_j] \tag{25.2}$$

(the right-hand side is a vector, and the coefficient of h_k on the left-hand side is a scalar).[2]

25.2 The connection and torsion forms

Let h_i be a basis of the tangent space and θ_j the associated dual basis $[\theta^k(h_i) = \delta_i^k]$. We supply the manifold with an affine connection D defined as

$$D_{h_i} h_j = \gamma_{ij}^k\, h_k\,, \tag{25.3}$$

where the functions γ_{ij}^k are the *Ricci rotation coefficients* [if the basis h_i is a natural basis ($h_i = \partial_i$), the rotation coefficients become the connection coefficients Γ_{ij}^k].

---

[1] A more complete discussion of the topics covered in this chapter can be found in Straumann (2013).

[2] Indeed, (25.1) immediately gives $\mathrm{d}\theta^k(h_i, h_j) = h_i[\theta^k(h_i)] - h^j[\theta^k(h_i)] - \theta^k([h_i, h_j])$. Now, since $\theta^k(h_j) = \delta_j^k$ and h_i is a derivative operator, the first two terms vanish and the expression reduces to $\mathrm{d}\theta^k(h_i, h_j) = -\theta^k([h_i, h_j])$. The right-hand side is the kth component of the vector $[h_i, h_j]$ (since $[h_i, h_j] \equiv ([h_i, h_j])^l h_l$, and so $\theta^k([h_i, h_j]) = ([h_i, h_j])^l \theta^k(h_l) = ([h_i, h_j])^k)$. Equation (25.2) then follows.

Relativity in Modern Physics. Nathalie Deruelle and Jean-Philippe Uzan.
© Oxford University Press 2018. Published in 2018 by Oxford University Press.
DOI: 10.1093/oso/9780198786399.001.0001

Following Cartan, we treat the γ_{ij}^k as the result of the action of 1-forms ω_j^k, called *connection forms*, on the vectors h_i:

$$\gamma_{ij}^k = \omega_j^k(h_i) \quad \Longleftrightarrow \quad \omega_j^k = \gamma_{ij}^k\, \theta^i \,. \tag{25.4}$$

Armed with these definitions, we can express the torsion as a function of the connection forms. Indeed, since h_i and h_j are two basis vectors, from the definitions of the torsion (23.23) and the connection (25.3) and (25.4) and using (25.2) we have

$$T^D(h_i, h_j) \equiv D_{h_i} h_j - D_{h_j} h_i - [h_i, h_j] = \{\omega_j^k(h_i) - \omega_i^k(h_j) + \mathrm{d}\theta^k(h_i, h_j)\}\, h_k \,. \tag{25.5}$$

The scalar $\omega_j^k(h_i)$ can be viewed as a 2-form $(\omega_l^k \otimes \theta^l)$ acting on the pair of vectors (h_i, h_j): $\omega_j^k(h_i) = (\omega_l^k \otimes \theta^l)(h_i, h_j)$. Similarly, $\omega_i^k(h_j) = (\theta^l \otimes \omega_l^k)(h_i, h_j)$. Therefore, $T(h_i, h_j)$ can be rewritten as

$$T^D(h_i, h_j) = \{(\omega_l^k \wedge \theta^l + \mathrm{d}\theta^k)(h_i, h_j)\}\, h_k \,, \tag{25.6}$$

or

$$T^D = \Omega^k \otimes h_k \quad \text{with} \quad \Omega^k \equiv \mathrm{d}\theta^k + \omega_j^k \wedge \theta^j \,, \tag{25.7}$$

where the Ω^k are the *torsion* differential 2-forms (that is, antisymmetric covariant tensors).

Equation (25.7) is Cartan's *first structure equation*.

In a natural basis where $\theta^k = \mathrm{d}x^k$, we have $\omega_j^k = \Gamma_{ij}^k \mathrm{d}x^i$ and $\Omega^k = \Gamma_{ij}^k \mathrm{d}x^i \wedge \mathrm{d}x^j = (\Gamma_{ij}^k - \Gamma_{ji}^k)\mathrm{d}x^i \otimes \mathrm{d}x^j = T_{ij}^k \mathrm{d}x^i \otimes \mathrm{d}x^j$. Finally, $T^D = T_{ij}^k \mathrm{d}x^i \otimes \mathrm{d}x^j \otimes \partial_k$.

25.3 The curvature forms

If now we take ω in (25.1) to be the connection 1-forms ω_j^i, then by a calculation analogous to that leading to (25.6) we find that the curvature defined in (23.25) can be rewritten as

$$R_{h_i, h_j}^D h_k \equiv D_{h_i} D_{h_j} h_k - D_{h_j} D_{h_i} h_k - D_{[h_i, h_j]} h_k = \{(\mathrm{d}\omega_k^m + \omega_l^m \wedge \omega_k^l)(h_i, h_j)\} h_m \tag{25.8}$$

or, more compactly,

$$R^D = \Omega_k^m \otimes h_m \otimes \theta^k \quad \text{with} \quad \Omega_k^m \equiv \mathrm{d}\omega_k^m + \omega_l^m \wedge \omega_k^l \,. \tag{25.9}$$

This is Cartan's *second structure equation*, and the quantities Ω_j^i are the *curvature* 2-forms.

In a natural basis where $h_i = \partial_i$, $\theta^i = \mathrm{d}x^i$, and $\omega_j^i = \Gamma_{ij}^k \mathrm{d}x^i$, we have $\Omega_k^m = R^m{}_{kij} \mathrm{d}x^i \otimes \mathrm{d}x^j$, where $R^m{}_{kij} = \partial_i \Gamma_{jk}^m - \partial_j \Gamma_{ik}^m + \Gamma_{il}^m \Gamma_{jk}^l - \Gamma_{jl}^m \Gamma_{ik}^l$. In any basis we can write

$$\Omega_k^m = \frac{1}{2} r^m{}_{kij} \theta^i \wedge \theta^j \,, \tag{25.10}$$

where $r^m{}_{kij} = -r^m{}_{kji}$ are the components of the curvature tensor in the basis h_i.

25.4 The Levi-Civita connection

Now let us assume that the manifold is supplied with a metric g. If the connection D is a metric connection (that is, $D_u g = 0 \; \forall\, u$), then, as we saw in (24.5),

$$\partial_u g(v, w) = g(D_u v, w) + g(v, D_u w) \,. \tag{25.11}$$

Let us take for the vectors u, v, w three basis vectors of the tangent space h_i, h_j, h_k and set $g_{ij} \equiv g(h_i, h_j)$ as well as [see (23.2)] $\partial_{h_i} g_{jk} = h_i(g_{jk}) \equiv g_{jk,i}$. Then, using the definition

(23.19) of the covariant derivative D as a function of the Ricci rotation coefficients and their relation (25.4) to the connection forms, we have

$$g_{jk,i} = g(D_{h_i}h_j, h_k) + g(h_j, D_{h_i}h_k)$$

$$= g(\gamma_{ij}^l h_l, h_k) + g(h_j, \gamma_{ik}^l h_l)$$

$$= \gamma_{ij}^l g_{lk} + \gamma_{ik}^l g_{jl} \qquad (25.12)$$

$$= \omega_j^l(h_i)g_{lk} + \omega_k^l(h_i)g_{jl}$$

$$= (\omega_{kj} + \omega_{jk})(h_i), \quad \text{where} \quad \omega_{kj} \equiv \omega_j^l g_{lk} \, .$$

The metric condition (25.11) then is rewritten as

$$dg_{jk} = \omega_{kj} + \omega_{jk} \, , \qquad (25.13)$$

where it is understood that $dg_{jk} \equiv \partial_{h_l} g_{jk} \theta^l$, and so we indeed have $dg_{jk}(h_i) = \partial_{h_i} g_{jk} \equiv g_{jk,i}$.

In a natural basis where $h_i = \partial_i$, (25.13) is written in tensor form as $g_{jk,i} = g_{lk}\Gamma_{ij}^l + g_{jl}\Gamma_{ik}^l$, where now $g_{jk,i} = \partial g_{jk}/\partial x^i$.

If the basis h_i is such that the components g_{ij} are constants, then $\partial_{h_i} g_{jk} = 0$ because h_i is a derivative operator, and so the metric condition on the connection then requires that the connection forms ω_{kj} be antisymmetric.

25.5 Components of the Riemann tensor

Let us consider a Riemannian manifold, that is, a metric manifold with a torsion-less connection. Let h_i be a basis of the tangent space and θ^i the dual basis.

The Cartan equations determining its curvature tensor then are

$$\begin{cases} d\theta^i + \omega_j^i \wedge \theta^j = 0 \, , \\ \omega_{ij} + \omega_{ji} = dg_{ij} \, , \\ d\omega_j^i + \omega_k^i \wedge \omega_j^k = \frac{1}{2}r_{jkl}^i \theta^k \wedge \theta^l \, . \end{cases} \qquad (25.14)$$

The first equation states that the connection is torsion-free [*cf.* (25.7)]. Using $\omega_j^i = \gamma_{kj}^i \theta^k$ and decomposing the 2-forms $d\theta^i$ as

$$d\theta^i = -\frac{1}{2}C_{jk}^i \theta^j \wedge \theta^k, \qquad (25.15)$$

it gives

$$C_{jk}^i = \gamma_{jk}^i - \gamma_{kj}^i \, . \qquad (25.16)$$

If the basis is natural, that is, if $\theta^i = dx^i$, then all the C_{jk}^i are zero by the Poincaré lemma and we again find that the Ricci rotation coefficients γ_{jk}^i identified as the connection coefficients Γ_{jk}^i are symmetric.

The second equation in (25.14) states that the connection is a metric connection [*cf.* (25.13)]. Writing $dg_{ij} = g_{ij,k}\theta^k$ [where we recall that $g_{ij,k} \equiv \partial_{h_k} g_{ij} = h_k(g_{ij})$; *cf.* (23.2)] and

$\omega_{ij} = g_{ik}\omega^k_j = g_{ik}\gamma^k_{lj}\theta^l$, by a suitable combination of cyclic permutations and using (25.16) we extract the expression

$$\gamma^i_{jk} = \frac{1}{2}\left(C^i_{jk} - g_{kl}g^{im}C^l_{jm} - g_{jl}g^{im}C^l_{km}\right) + \frac{1}{2}g^{il}\left(g_{kl,j} + g_{jl,k} - g_{jk,l}\right). \tag{25.17}$$

In a natural basis the C^i_{jk} vanish, $g_{kl,j} = \partial_j g_{kl}$, and we recover the familiar expression for the Christoffel symbols Γ^i_{jk} as a function of the derivatives of the metric components. Conversely, if the basis h_i is such that the metric components $g_{ij} = g(h_i, h_j)$ are constant, then only the first term of (25.17) survives.

Finally, writing

$$d\omega^i_j = d(\gamma^i_{kj}\theta^k) = \gamma^i_{kj,l}\theta^l \wedge \theta^k - \frac{1}{2}\gamma^i_{kj}C^k_{lm}\theta^l \wedge \theta^m, \tag{25.18}$$

where $\gamma^i_{kj,l} \equiv h_l(\gamma^i_{kj})$, from the third equation in (25.14) we extract the components r^i_{jkl} of the curvature tensor:

$$r^i_{jkl} = \gamma^i_{lj,k} - \gamma^i_{kj,l} + \gamma^i_{km}\gamma^m_{lj} - \gamma^i_{lm}\gamma^m_{kj} - \gamma^i_{mj}C^m_{kl}. \tag{25.19}$$

In a natural basis where $C^m_{kl} = 0$ we recover the familiar expression for the Riemann tensor as a function of the Christoffel symbols Γ^i_{jk} (which are symmetric) and their derivatives $\Gamma^i_{jk,l} = \partial_l \Gamma^i_{jk}$.

25.6 The Riemann tensor of the Schwarzschild metric

We consider a static, spherically symmetric spacetime. In Droste coordinates (t, r, θ, ϕ) the metric tensor has the form

$$g = -e^{2\nu}dt^2 + e^{2\lambda}dr^2 + r^2(d\theta^2 + \sin^2\theta d\phi^2), \tag{25.20}$$

where ν and λ are two arbitrary functions of r, $dx^\mu = (dt, dr, d\theta, d\phi)$ is the basis of the cotangent space canonically associated with the natural basis $\partial_\mu = (\partial_t, \partial_r, \partial_\theta, \partial_\phi)$ of the tangent space at (t, r, θ, ϕ), and $dt^2 \equiv dt \otimes dt$, and so on.

To calculate the Riemann tensor of this metric by the Cartan method, we choose the basis of the cotangent space to be the four 1-forms

$$\theta^0 = e^\nu dt, \quad \theta^1 = e^\lambda dr, \quad \theta^2 = rd\theta, \quad \theta^3 = r\sin\theta d\phi, \tag{25.21}$$

and so in this (nonholonomic) basis the metric (25.20) is written as

$$g = \eta_{\mu\nu}\theta^\mu \otimes \theta^\nu \tag{25.22}$$

with $\eta_{\mu\nu} = (-1, +1, +1, +1)$.

The exterior derivatives of these forms are easily calculated[3] and we find

$$d\theta^\mu = -\frac{1}{2}C^\mu_{\nu\rho}\theta^\nu \wedge \theta^\rho \quad \text{with} \quad C^0_{01} = \nu'e^{-\lambda}, \quad C^2_{21} = C^3_{31} = \frac{e^{-\lambda}}{r}, \quad C^2_{32} = \frac{\cos\theta}{r\sin\theta}, \tag{25.23}$$

where a prime denotes the derivative with respect to r, $C^\mu_{\nu\rho} = -C^\mu_{\rho\nu}$, and the other coefficients $C^\mu_{\nu\rho}$ are either zero or are found by antisymmetry.

---

[3]For example, $d\theta^0 = d(e^\nu dt) = \partial_\mu e^\nu dx^\mu \wedge dt = \nu'e^\nu dr \wedge dt = \nu'e^\nu(e^{-\lambda}\theta^1) \wedge (e^{-\nu}\theta^0) = \nu'e^{-\lambda}\theta^1 \wedge \theta^0$, from the definition of the exterior derivative and the Poincaré lemma. On the other hand, $d\theta^0 = -\frac{1}{2}C^0_{\mu\nu}\theta^\mu \wedge \theta^\nu = -\frac{1}{2}(C^0_{01}\theta^0 \wedge \theta^1 + C^0_{10}\theta^1 \wedge \theta^0) = -C^0_{10}\theta^1 \wedge \theta^0$, because C^0_{01} is antisymmetric. Therefore, $C^0_{10} = -\nu'e^{-\lambda}$.

The Ricci rotation coefficients can then be obtained by direct calculation using (25.17) and the fact that $g_{\mu\nu,\rho} = 0$ since the basis is orthonormal.[4]

However, it is faster to find the connection 1-forms directly starting from the structure equations expressing the absence of torsion and the metric nature of the connection, namely [*cf.* (25.14)], $\omega_{\mu\nu} + \omega_{\nu\mu} = 0$ and $d\theta^\mu + \omega^\mu_\nu \wedge \theta^\nu = 0$. The first equation gives

$$\omega^0_i = \omega^i_0 \quad \text{and} \quad \omega^i_j = -\omega^j_i . \tag{25.24}$$

The second gives

$$\nu' e^{-\lambda}\theta^1 \wedge \theta^0 + \omega^0_1 \wedge \theta^1 + \omega^0_2 \wedge \theta^2 + \omega^0_3 \wedge \theta^3 = 0 ,$$

$$\omega^0_1 \wedge \theta^0 + \omega^1_2 \wedge \theta^2 + \omega^1_3 \wedge \theta^3 = 0 ,$$

$$\frac{e^{-\lambda}}{r}\theta^1 \wedge \theta^2 + \omega^0_2 \wedge \theta^0 - \omega^1_2 \wedge \theta^1 + \omega^2_3 \wedge \theta^3 = 0 , \tag{25.25}$$

$$\frac{e^{-\lambda}}{r}\theta^1 \wedge \theta^3 + \frac{\cos\theta}{r\sin\theta}\theta^2 \wedge \theta^3 + \omega^0_3 \wedge \theta^0 - \omega^1_3 \wedge \theta^1 - \omega^2_3 \wedge \theta^2 = 0 .$$

These are four equations for six unknowns (ω^0_i, ω^1_2, ω^1_3, ω^2_3), but we know that there is a single solution. With a bit of insight we find

$$\omega^0_1 = \nu' e^{-\lambda}\theta^0 , \quad \omega^0_2 = \omega^0_3 = 0 , \quad \omega^1_2 = -\frac{e^{-\lambda}}{r}\theta^2 , \quad \omega^1_3 = -\frac{e^{-\lambda}}{r}\theta^3 ,$$

$$\omega^2_3 = -\frac{\cos\theta}{r\sin\theta}\theta^3 . \tag{25.26}$$

The curvature 2-forms, $\Omega^\mu_\nu = d\omega^\mu_\nu + \omega^\mu_\rho \wedge \omega^\rho_\nu$, then follow directly[5]:

$$\Omega^0_1 = -e^{-2\lambda}(\nu'\lambda' - \nu'' - \nu'^2)\theta^0 \wedge \theta^1 , \quad \Omega^0_2 = -\frac{\nu' e^{-2\lambda}}{r}\theta^0 \wedge \theta^2 ,$$

$$\Omega^0_3 = -\frac{\nu' e^{-2\lambda}}{r}\theta^0 \wedge \theta^3 , \tag{25.27}$$

$$\Omega^1_2 = \frac{\lambda' e^{-2\lambda}}{r}\theta^1 \wedge \theta^2 , \quad \Omega^1_3 = \frac{\lambda' e^{-2\lambda}}{r}\theta^1 \wedge \theta^3 , \quad \Omega^2_3 = \frac{1 - e^{-2\lambda}}{r^2}\theta^2 \wedge \theta^3 .$$

Since $\Omega^\mu_\nu = \frac{1}{2}r^\mu_{\nu\rho\sigma}\theta^\rho \wedge \theta^\sigma$, we can read off the components of the Riemann tensor from (25.27):

$$r^0_{101} = e^{-2\lambda}(\nu'\lambda' - \nu'' - \lambda'^2) , \quad r^0_{202} = r^0_{303} = -\frac{\nu' e^{-2\lambda}}{r} ,$$

$$r^1_{212} = r^1_{313} = \frac{\lambda' e^{-2\lambda}}{r} , \quad r^2_{323} = \frac{1 - e^{-2\lambda}}{r^2} . \tag{25.28}$$

If now the metric is the Schwarzschild metric, that is, if $e^{-2\lambda} = e^{2\nu} = 1 - 2m/r$, we have

---

[4]For example, $\gamma^0_{01} = \frac{1}{2}(C^0_{01} + C^1_{00} - C^0_{10}) = C^0_{01} = \nu' e^{-\lambda}$. Similarly, we find $\gamma^0_{11} = \gamma^0_{21} = \gamma^0_{31} = 0$, and so $\omega^0_1 = \gamma^0_{k1}\theta^k = \nu' e^{-\lambda}\theta^0$.

[5]For example, $\Omega^0_2 = d\omega^0_2 + \omega^0_\rho \wedge \omega^\rho_2 = \omega^0_\rho \wedge \omega^\rho_2 = \omega^0_1 \wedge \omega^1_2 = -(\nu' e^{-2\lambda}/r)\theta^0 \wedge \theta^2$.

$$r^0_{101} = r^2_{323} = -2r^0_{202} = -2r^1_{212} = \frac{2m}{r^3}\,, \tag{25.29}$$

and so the Kretschmann curvature invariant $K \equiv r^\mu_{\ \nu\rho\sigma} r_\mu^{\ \nu\rho\sigma}$, where the indices are raised and lowered using the metric $\eta_{\mu\nu}$, is

$$K = 4(r^0_{101})^2 + 8(r^0_{202})^2 + 8(r^1_{212})^2 + 4(r^2_{323})^2 = \frac{48m^2}{r^6}\,. \tag{25.30}$$

List of books and articles cited in the text

Alexakis, S., Ionescu, A. D., and Klainerman, S. (2013). *Rigidity of stationary black holes with small angular momentum on the horizon.* https://arxiv.org/abs/1304.0487.

Arnowitt, R., Deser, S., and Misner, C. (1962). *Gravitation: an introduction to current research*, edited by L. Witten. Wiley, New York; also, https://arxiv.org/abs/gr-qc/0405109.

Bahcall, N., Ostriker, J., Perlmutter, S., and Steinhardt, P. (2008). *The cosmic triangle: assessing the state of the universe.* https://arxiv.org/pdf/astro-ph/9906463.pdf.

Bardeen, J. M. (1980). Gauge-invariant cosmological perturbations. *Phys. Rev. D*, **22**, 1882.

Bardeen, J., Steinhardt, P., and Turner, M. (1983). Spontaneous creation of almost scale-free density perturbations in an inflationary universe. *Phys. Rev. D*, **28**, 679–693.

Bartelmann, M. and Schneider, P. (2001). Weak gravitational lensing. *Phys. Rep.*, **340**, 291–472.

Baumgarte, T. W. and Shapiro, S. L. (2010). *Numerical relativity: solving Einstein's equations on the computer.* Cambridge University Press, Cambridge.

Bejger, M., Piran, T. *et al.* (2012). Collisional Penrose process near the horizon of extreme Kerr black holes. *Phys. Rev. Lett.*, **109**, 121101; also, https://arxiv.org/abs/1205.4350.

Bel, Ll. *et al.* (1981). Poincaré-invariant gravitational field and equations of motion of two pointlike objects: the postlinear approximation of general relativity. *Gen. Rel. Grav.*, **13**, 963.

Bernard, L., Blanchet, L. *et al.* (2016). Fokker action of non-spinning compact binaries at the fourth post-Newtonian approximation. *Phys. Rev. D*, **93**, 084037; also, https://arxiv.org/abs/1512.02876; see also (2017) *Phys. Rev. D*, **96**, 104043.

Berti, E., Cardoso, V., and Starinets, A. O. (2009). *Quasinormal modes of black holes and black branes.* https://arxiv.org/abs/0905.2975.

Bini, D. and Damour, T. (2012). Gravitational radiation reaction along general orbits in the effective one-body formalism. *Phys. Rev. D*, **86**, 124012; also, https://arxiv.org/abs/1210.2834.

Birrell, N. D. and Davies, P. C. W. (1982). *Quantum fields in curved space.* Cambridge University Press, Cambridge.

Bishop, R. and Goldberg, S. (1980). *Tensor analysis on manifolds.* Dover, New York.

Blanchet, L. and Damour, T. (1989). Post-Newtonian generation of gravitational waves. *Ann. Inst. H. Poincaré*, **50** (4), 377–408.

Blanchet, L. (2006). Gravitational radiation from post-Newtonian sources and inspiralling compact binaries. *Living Rev. Relativity*, **9**, 4; also, http://www.livingreviews.org/lrr-2006-4.

Blandford, R. D. and Znajek, R. L. (1977). Electromagnetic extraction of energy from Kerr black holes. *Mon. Not. R. Astr. Soc.*, **179**, 433–456.

Blau, M. (2016). *Lecture notes on general relativity.* http://www.blau.itp.unibe.ch/Lecturenotes.html.

Bohé, A. *et al.* (2017). An improved effective-one-body model of spinning, nonprecessing binary black holes for the era of gravitational-wave astrophysics with advanced detectors. *Phys. Rev. D*, **95**, 044028; also, https://arxiv.org/abs/1611.03703.

Bourguignon, J.-P. (2005). *General relativity* [in French]. Éditions de l'École Polytechnique, Paris.

Buonanno, A. and Damour, T. (1999). Effective one-body approach to general relativistic two-body dynamics. *Phys. Rev. D*, **59**, 084006; also, https://arxiv.org/pdf/gr-qc/9811091.pdf.

Buonanno, A. and Damour, T. (2000). Transition from inspiral to plunge in binary black hole coalescence. *Phys. Rev. D*, **62**, 064015; also, https://arxiv.org/abs/gr-qc/0001013.

Buonanno, A. (2000). Reduction of the two-body dynamics to a one-body description in classical electrodynamics. *Phys. Rev. D*, **62**, 104022; also, https://arxiv.org/abs/hep-th/0004042.

Burbidge, E. M., Burbidge, G. R., Fowler, W. A., and Hoyle, F. (1957). Synthesis of the elements in stars. *Rev. Mod. Phys.*, **29**, 547.

Cardoso, V., Dias, O., Lemos, J., and Yoshida, S. (2004). *The black hole bomb and super-radiant instabilities.* https://arxiv.org/abs/hep-th/0404096.

Carlip, S. (2009). Black hole thermodynamics and statistical mechanics. *Lect. Notes Phys.*, **769**, 89–123; also, https://arxiv.org/abs/0807.4520.

Carroll, S. M., Johnson, M. C., and Randall, L. (2009). Extremal limits and black hole entropy. *JHEP*, 0911, 109; also, https://arxiv.org/abs/0901.0931.

Carter, B. (1972). In *Black Holes–Les Astres Occlus*, edited by B. deWitt and C. deWitt, Les Houches 1972 Summer School. Gordon and Breach, New York.

Carter, B. (2004). *Mechanics and equilibrium geometry of black holes, membranes, and strings.* https://arxiv.org/abs/hep-th/0411259.

Chandrasekhar, S. (1931). The maximum mass of ideal white dwarfs. *Astrophys. J.*, **74**, 81.

Chandrasekhar, S. (1934). Stellar configurations with degenerate cores. *The Observatory*, **57**: 373–377.

Chandrasekhar, S. (1983). *The mathematical theory of black holes*. Oxford University Press, Oxford.

Châtelet, G. (1993). English translation: *Figuring space: philosophy, mathematics, and physics*. Springer, Dordrecht (1999).

Chern, S. S. (1944). A simple intrinsic proof of the Gauss–Bonnet formula for closed Riemannian manifolds. *Ann. Math.*, **45**, 747–752.

Chongchitnan, S. and Efstathiou, G. (2007). Can we ever distinguish between quintessence and a cosmological constant? *Phys. Rev. D*, **76**, 043508; also, https://arxiv.org/abs/0705.1955.

Choquet-Bruhat, Y., DeWitt-Morette, C., and Dillard-Bleick, M. (1978). *Analysis, manifolds and physics*. North-Holland, Amsterdam.

Choquet-Bruhat, Y. (2009). *General relativity and the Einstein equations*. Oxford University Press, Oxford.

Chruściel, P., Lopes Costa, J., and Heusler, M. (2012). Stationary black holes: uniqueness and beyond. *Living Rev. Relativity*, **15**, 7; also, https://arxiv.org/abs/1205.6112.

Collins II, G. W. (2003). *The fundamentals of stellar astrophysics*. http://bifrost.cwru.edu/personal/collins/astrobook/.

Dafermos, M. and Rodnianski, I. (2010). *The black hole stability problem for linear scalar perturbations*. https://arxiv.org/abs/1010.5137.

Damour, T. and Ruffini, R. (1976). Black-hole evaporation in the Klein–Sauter–Heisenberg–Euler formalism. *Phys. Rev. D*, **14**, 332. Damour, T. (2004). *The entropy of black holes: a primer*. https://arxiv.org/abs/hep-th/0401160.

Damour, T. (1979). *Some mechanical, electromagnetic, thermodynamical, and quantum properties of black holes* [in French]. Doctoral thesis, available at http://www.ihes.fr/~damour/.

Damour, T. and Deruelle, N. (1981). Radiation reaction and angular momentum loss in small angle gravitational scattering. *Phys. Lett.*, **87A**, 81–84.

Damour, T. (1982). Gravitational radiation and the motion of compact bodies. In *Les Houches Summer School on Gravitational Radiation*, edited by N. Deruelle and T. Piran. North-Holland, Amsterdam.

Damour, T. (1982a). The two-body problem and *bremsstrahlung* in general relativity [in French]. *C. R. Acad. Sci. Paris Sér. II*, **294**, 1355.

Damour, T. and Deruelle, N. (1985). General relativistic celestial mechanics of binary systems. I. The post-Newtonian motion. *Ann. de l'I. H. P.*, section A, **43**, 107–132 (1985); II.

The post-Newtonian timing formula. *Ann. de l'I. H. P.*, section A, **44**, 263 (1986); also, http://www.ihes.fr/~damour/Conferences/DamourDeruelleAIHP85.pdf.

Damour, T. (1987). The problem of motion in Newtonian and Einsteinian gravity. In *Three hundred years of gravitation*, edited by Hawking, S. and Israel, W. Cambridge University Press, Cambridge.

Damour, T. and Schäfer, G. (1991). Redefinition of position variables and the reduction of higher-order Lagrangians. *J. Math. Phys.*, **32**, 127.

Damour, T., Soffel, M., and Xu, C. (1991). *Phys. Rev. D*, **43**, 3271; *Phys. Rev. D*, **45**, 1017.

Damour, T. and Esposito-Farese, G. (1992). Tensor-multi-scalar theories of gravitation. *Class. Quant. Grav.*, **9**, 2093–2176.

Damour, T. and Taylor, J. (1992). Strong-field tests of relativistic gravity and binary pulsars. *Phys. Rev. D*, **45**, 1840.

Damour, T., Jaranowski, P., and Schäfer, G. (2000). On the determination of the last stable orbit for circular general relativistic binaries at the third post-Newtonian approximation. *Phys. Rev. D*, **62**, 084011; also, https://arxiv.org/abs/gr-qc/0005034.

Damour, T. (2005). *Einstein aujourd'hui* [in French]. CNRS Éditions, Paris; *Once upon Einstein*. CRC Press, Boca Raton, Florida.

Damour, T. (2006). Gravitational waves, relativistic astrophysics and cosmology. Notes for the course given at the Institut Henri Poincaré, available at http://www.luth.obspm.fr/IHP06/.

Damour, T. and Lecian, O. M. (2011). Statistical properties of cosmological billiards. *Phys. Rev. D*, **83**, 044038; also, https://arxiv.org/abs/1011.5797.

Damour, T. (2014). The general relativistic two body problem, in *Brumberg Festschrift*, edited by S. M. Kopeikin. de Gruyter, Berlin; also, https://arxiv.org/abs/1312.3505.

Damour, T., Jaranowski, P., and Schäfer, G. (2014). Nonlocal-in-time action for the fourth post-Newtonian conservative dynamics of two-body systems. *Phys. Rev. D*, **89**, 064058; alsohttps://arxiv.org/abs/1401.4548.

Damour, T. (2016). *Experimental tests of gravitational theory*. http://pdg.lbl.gov/ 2016/reviews/rpp2016-rev-gravity-tests.pdf.

Damour, T., Jaranowski, P., and Schäfer, G. (2016). Conservative dynamics of two-body systems at the fourth post-Newtonian approximation of general relativity. *Phys. Rev. D*, **93**, 084014; also, https://arxiv.org/pdf/1601.01283.pdf.

Damour, T. (2016a). Gravitational scattering, post-Minkowskian approximation and Effective One-Body theory. *Phys. Rev. D*, **94**, 104015; also, https://arxiv.org/abs/1609.00354.

Davies, P. C. W. (1975). Scalar particle production in Schwarzschild and Rindler metrics. *J. Phys. A*, **8**, 609–616.

Davis, M., Ruffini, R., Press, W. H., and Price, R. H. (1971). Gravitational radiation from a particle falling radially into a Schwarzschild black hole. *Phys. Rev. Lett.*, **27**, 1466.

Deruelle, N. and Ruffini, R. (1974). Quantum and classical relativistic energy states in stationary geometries. *Phys. Lett.*, **52B**, 437–441.

Echeverria, F. (1989). Gravitational-wave measurements of the mass and angular momentum of a black hole. *Phys. Rev. D*, **40**, 3194.

Eddington, A. and Clark, G. (1938). The problem of N bodies in general relativity theory. *Proc. Roy. Soc. A*, **166**, 465–475.

Einstein, A. (*c.* 1920). Fundamental Ideas and Methods of Relativity. *The Morgan manuscript.*

Einstein, A. (1954). *Relativity, the special and general theory.* Methuen, London.

Einstein, A., Infeld, L., and Hoffmann, B. (1938). The gravitational equations and the problem of motion. *Ann. Math.*, **39**, 65–100.

Eisenstaedt, J. (2005). *Before Einstein: relativity, light, and gravitation* [in French]. Éditions du Seuil, Paris.

Eisenstaedt, J. (1989). Cosmology: a space for thought on general relativity, in *Foundations of Big-Bang cosmology*, edited by F. W. Meyerstein. World Scientific, Singapore. [French version in *Cosmologie et philosophie, hommage à Jacques Merleau-Ponty. Epistémologiques*, **1**, 197–218.]

Ellis, G. F. R. *et al.* (2011). On the trace-free Einstein equations as a viable alternative to general relativity. *Class. Quant. Grav.*, **28**, 225007; also, https://arxiv.org/abs/1008.1196.

Ellis, G. F. R., Maartens, R., and MacCallum, M. (2012). *Relativistic cosmology.* Cambridge University Press, Cambridge.

Emparan, R. and Reall, H. S. (2008). Black holes in higher dimensions. *Living Rev. Relativity*, **11**, 6; also, http://relativity.livingreviews.org/Articles/lrr-2008-6/.

Fichtenholz, I. G. (1950). Lagrangian form of the equations of motion at the second order approximation of Einstein's gravitation theory [in Russian]. *Zh. Eksp. Teor. Fiz.*, **20**, 233–242.

Frolov, V. and Novikov, I. (1998). *Black hole physics.* Kluwer, Dordrecht.

Fulling, S. A. (1973). Nonuniqueness of canonical field quantization in Riemannian space-time. *Phys. Rev. D*, **7**, 2850.

Gourgoulhon, E. and Jaramillo, J. L. (2005). A 3+1 perspective on null hypersurfaces and isolated horizons. *Phys. Rep.*, **423**, 159–294; also, https://arxiv.org/abs/gr-qc/0503113.

Gourgoulhon, E. (2012). *3+1 formalism and bases of numerical relativity.* Springer-Verlag, Berlin; also, https://arxiv.org/abs/gr-qc/0703035.

Gourgoulhon, E. and Jaramillo, J. L. (2008). New theoretical approaches to black holes. *New Astron. Rev.*, **51**, 791; also, https://arxiv.org/abs/0803.2944.

Gourgoulhon, E. *et al.* (2011). Magnetohydrodynamics in stationary and axisymmetric spacetimes: a fully covariant approach. *Phys. Rev. D*, **83**, 104007; also, https://arxiv.org/abs/1101.3497.

Grandclément, P. (2008). *Compact objects* [in French]. http://luth.obspm.fr/fichiers/enseignement/objets_compacts_F1.pdf.

Grandclément, P. and Novak, J. (2009). Spectral methods for numerical relativity. *Living Rev. Relativity*, **12**, 1; also, https://arxiv.org/abs/0706.2286.

Greene, P. *et al.* (1997). Structure of resonance in preheating after inflation. *Phys. Rev. D*, **56**, 6175–6192.

Guth, A. (1981). *The inflationary universe: a possible solution to the horizon and flatness problems. Phys. Rev. D*, **23**, 347.

Guth, A. and Pi, S.-Y. (1982). Fluctuations in the new inflationary universe. *Phys. Rev. Lett.*, **49**, 1110–1113.

Hartle, J. B. and Thorne, K. S. (1969). Slowly rotating relativistic stars. III. Static criterion for stability. *Astrophys. J.*, **158**, 719–726.

Hawking, S. W. (1972). *Black Holes–Les Astres Occlus*, edited by B. deWitt and C. deWitt, Les Houches 1972 Summer School. Gordon and Breach, New York.

Hawking, S. W. and Ellis, G. F. R. (1973). *The large scale structure of spacetime*. Cambridge University Press, Cambridge.

Hawking, S. W. (1975). Black hole explosions. *Nature*, **248**, 30–31 (1974); Particle creation by black holes. *Comm. Math. Phys.*, **43**, 199–220 (1975).

Hawking, S. W. (1982). The development of irregularities in a single bubble inflationary universe. *Phys. Lett.*, **B115**, 295–297.

Hubble, E. (1929). A relation between distance and radial velocity among extra-galactic nebulae. *Proc. Nat. Acad. Sci.*, **15**, 168–173.

Isaacson, W. (2008). *Einstein, his life and universe*. Simon & Schuster, New York.

Isenberg, J. (2013). The initial value problem in general relativity, in *The Springer handbook of spacetime*, edited by A. Ashtekar and V. Petkov. Springer-Verlag, Berlin; also, https://arxiv.org/abs/1304.1960.

Israel, W. (1967). Event horizons in static vacuum spacetimes. *Phys. Rev.*, **164**, 1776.

Jaramillo, J. L. and Gourgoulhon, E. (2010). Mass and angular momentum in general relativity. https://arxiv.org/abs/1001.5429.

Julié, F. L. and Deruelle, N. (2017). Two-body problem in scalar–tensor theories as a deformation of general relativity: an effective-one-body approach. https://arxiv.org/abs/1703.05360.

Katz, J., Bičák, J., and Lynden-Bell, D. (1997). Relativistic conservation laws and integral constraints for large cosmological perturbations. *Phys. Rev. D*, **55**, 5957–5969; also, https://arxiv.org/abs/gr-qc/0504041.

Katz, J. (1996). Energy in general relativity. *In Gravitational dynamics*, edited by Lahav, O. *et al.* Cambridge University Press, Cambridge.

Kenyon, I. R. (1990). *General relativity*. Oxford University Press, Oxford.

Kodama, H. and Sasaki, M. (1984). Cosmological perturbation theory. *Prog. Theor. Phys. Suppl.*, **78**, 1–166.

Kofman, L., Linde, A., and Starobinsky, A. (1997). Towards the theory of reheating after inflation. *Phys. Rev. D*, **56**, 3258–3295.

Komissarov, S. (2009). Blandford-Znajek mechanism versus Penrose process. *J. Korean Phys. Soc.*, **54**, 2503–2512; also, https://arxiv.org/abs/0804.1912.

Kottler, F. (1918). Über die physikalischen Grundlagen der Einsteinschen Gravitations theorie [On the physical foundations of Einstein's gravitational theory (in German)]. *Ann. Phys. (Berlin)*, **56**, 401.

Kramer, M. (2012). Probing gravitation with pulsars. https://arxiv.org/abs/1211.2457.

Landau, L. D. (1932). On the theory of stars. *Phys. Z. Sowjet.*, **1**, 285.

Landau, L. D. and Lifshitz, E. (1971). *The classical theory of fields* (3rd edn). Pergamon Press, Oxford.

Langlois, D. (2003). *Inflation, quantum fluctuations and cosmological perturbations*. https://arxiv.org/abs/hep-th/0405053.

Langlois, D. (2010). *Lectures on inflation and cosmological perturbations*. https://arxiv.org/abs/1001.5259.

Lasota, J. P., Gourgoulhon, E. *et al.* (2014). Extracting black hole rotational energy: The generalized Penrose process. *Phys. Rev. D*, **89**, 024041; also, https://arxiv.org/abs/1310.7499.

Levi-Civita, T. (1935). Matter points and celestial bodies in general relativity [in French], in *Jubilé de M. Marcel Brillouin*. Éditions Gauthier-Villars.

Lichnerowicz, A. (1967). *Relativistic hydrodynamics and magnetohydrodynamics: Lectures on the existence of solutions*. W. A. Benjamin, New York.

Liddle, A. R. and Lyth, D. H. (2000). *Cosmological inflation and large-scale structure*. Cambridge University Press, Cambridge.

Lifshitz, E. (1946). On the gravitational stability of the expanding universe. *J. Phys. (USSR)*, **10**, 116.

Linde, A. (1990). *Particle physics and inflationary cosmology*. Harwood, Chur; reproduced in https://arxiv.org/abs/hep-th/0503203.

Linde, A. (1983). Chaotic inflation. *Phys. Lett.*, **B129**, 177.

Lyth, D. and Riotto, A. (1999). Particle physics models of inflation and the cosmological density perturbation. *Phys. Rep.*, **314**, 1–146.

Lyth, D. and Stewart, E. D. (1993). A more accurate analytic calculation of the spectrum of cosmological perturbations produced during inflation. *Phys. Lett.*, **B302**, 171–175.

McClintock, J. E., Narayan, R., and Steiner, J. F. (2013). *Black hole spin via continuum fitting and the role of spin in powering transient jets*. https://arxiv.org/abs/1303.1583.

Madore, D. *Kerr black holes images and videos*. http://www.madore.org/david/math/kerr.html.

Maldacena, J. (1998). The large N limit of superconformal field theories and supergravity. *Adv. Theor. Math. Phys.*, **2**, 231–252; also, https://arxiv.org/abs/hep-th/9711200.

Marolf, D. and Ori, A. (2012). *Outgoing gravitational shock-wave at the inner horizon: The late-time limit of black hole interiors*. https://arxiv.org/abs/1109.5139.

Misner, C. W., Thorne, K. S., and Wheeler, J. A. (1973). *Gravitation*. W. H. Freeman, New York.

Moschella, U. (2005). The de Sitter and anti-de Sitter sightseeing tour. *Séminaire Poincaré*, **1**, 1–12.

Mueller, A. (2007). *Experimental evidence of black holes*. http://arxiv.org/abs/astro-ph/0701228.

Mukhanov, V. (2005). *Physical foundations of cosmology*. Cambridge University Press, Cambridge.

Mukhanov, V., Feldman, H., and Brandenberger, R. (1992). Theory of cosmological perturbations. *Phys. Rep.*, **215**, 203–233.

Mukhanov, V. and Chibisov, G. (1981). Quantum fluctuations and a nonsingular universe. *JETP Lett.*, **33**, 532–535.

Mukhanov, V. and Winitzki, S. (2007). *Introduction to quantum effects in gravity*. Cambridge University Press, Cambridge.

Nagar, A., Damour, T., Reisswig, C., and Pollney, D. (2016). Energetics and phasing of nonprecessing spinning coalescing black hole binaries. *Phys. Rev. D*, **93**, 044046; also, arXiv:1506.08457.

Oppenheimer, J. R. and Snyder, H. (1939). On continued gravitational contraction. *Phys. Rev.*, **56**, 455.

Pais, A. (1983). *Subtle is the Lord.* Oxford University Press, Oxford.

Penrose, R. (1969). Gravitational collapse: the role of general relativity. *Riv. Nuovo Cimento*, **1**, 252; reproduced in *Gen. Rel. Gravit.*, **34**, 1141 (2002).

Penrose, R. (1998). The question of cosmic censorship, in *Black holes and relativistic stars*, edited by R. M. Wald. University of Chicago Press, Chicago.

Peter, P. and Uzan, J.-P. (2009). *Primordial cosmology.* Oxford University Press, Oxford.

Piran, T. and Shaham, J. (1977). Upper bounds on collisional Penrose processes near rotating black hole horizons. *Phys. Rev. D*, **16**, 1615.

Pitkin, M. *et al.* (2011). Gravitational wave detection by interferometry (ground and space). *Living Rev. Relativity*, **14**, 5; also, http://relativity.livingreviews.org/Articles/lrr-2011-5/.

Planck Collaboration (2013). *Planck 2013 results. XVI. Cosmological parameters.* https://arxiv.org/abs/1303.5076; also, http://www.esa.int/SPECIALS/Planck/ index.html.

Poincaré, H. (1906). *Sur la dynamique de l'électron*, Palermo Rendiconti **21**, 129 [English translation: *On the dynamics of the electron*, https://en.wikisource.org/wiki/Translation:On_the_Dynamics_of_the_Electron_(July)].

Poisson, E. (2002). *An advanced course in general relativity.* http://www.physics.uoguelph.ca/poisson/research/notes.html.

Poisson, E. (2007). *Post-Newtonian theory.* Available at http://www.physics.uoguelph.ca/poisson/research/notes.html.

Polarski, D. and Starobinsky, A. (1996). Semiclassicality and decoherence of cosmological perturbations. *Class. Quant. Grav.*, **13**, 377–392.

Riazuelo, A. *Some details about the APOD of 2010 December 7.* http://www2.iap.fr/users/riazuelo/bh/APOD.php.

Rindler, W. (1953). Visual horizons in world-models. *Month. Not. R. Astron. Soc.*, **116**, 662.

Ruffini, R. and Wheeler, J. A. (1971). Introducing the black hole. *Physics Today*, **24**, 30–41.

Ruffini, R., Vereshchagin, G., and Xue, S. (2010). Electron–positron pairs in physics and astrophysics: from heavy nuclei to black holes. *Phys. Rep.*, **487**, 1–140.

Sasaki, M. (1983). Gauge-invariant scalar perturbations in the new inflationary universe. *Prog. Theor. Phys.*, **70**, 394–411.

Schrödinger, E. (1950). *Space-time structure.* Cambridge University Press, Cambridge.

Schutz, B. (2009). *A first course in general relativity* (2nd edn). Cambridge University Press, Cambridge.

Schwarzschild, K. (1916). *On the gravitational field of a mass point according to Einstein's theory* [English translation by S. Antoci and A. Loinger. https://arxiv.org/abs/physics/ 9905030].

Schwarzschild, K. (1916a). *On the gravitational field of a sphere of incompressible fluid according to Einstein's theory* [English translation by S. Antoci. https://arxiv.org/abs/ physics/ 9912033].

Shankar K. and Whiting, B. F. (2007). *Conformal coordinates of a constant density star.* https://arxiv.org/abs/0706.4324.

Shtanov, Y., Traschen, J., and Brandenberger, R. (1995). Universe reheating after inflation. *Phys. Rev. D*, **51**, 5438–5455.

Starobinsky, A. A. (1973). Amplification of waves during reflection from a rotating black hole. *Sov. Phys.–JETP*, **37**, 28–32.

Starobinsky, A. (1982). Dynamics of phase transition in the new inflationary scenario and generation of perturbations. *Phys. Lett.*, **B117**, 175–178.

Stephani, H. (1990). *General relativity* (2nd edn). Cambridge University Press, Cambridge.

Stewart, J. (1990). Perturbations of Friedmann–Robertson–Walker cosmological models. *Class. Quant. Grav.*, **7**, 1169.

Stewart, E. D. and Gong, J.-O. (2001). The density perturbation power spectra to second order corrections in the slow-roll expansion. *Phys. Lett.*, **B510**, 1–9.

Straumann, N. (2013). *General relativity* (2nd edition). Springer, Dordrecht.

Susskind, L. (2009). The anthropic landscape of string theory, in *Universe or multiverse?*, edited by B. Carr. Cambridge University Press, Cambridge; also, https://arxiv.org/abs/ hep-th/0302219.

Szabados, L. B. (2004). Quasi-local energy-momentum and angular momentum in general relativity. *Living Rev. Relativity*, **7**, 4; also, http://www.livingreviews.org/lrr-2004-4.

Thorne, K. S. (1974). Disk-accretion onto a black hole. II. Evolution of the hole. *Astrophys. J.*, **191**, 507–519.

Unruh, W. (1976). Notes on black hole evaporation. *Phys. Rev. D*, **14**, 870.

Uzan, J.-P., Deruelle, N., and Riazuelo, A. (2000). CMB anisotropies seeded by coherent defects. *Annalen Phys.*, **9**, 288–298; also, https://arxiv.org/abs/astro-ph/ 9810313.

Visser, M. (2009). The Kerr spacetime: a brief introduction, in *The Kerr spacetime: rotating black holes in general relativity*, edited by D. L. Wiltshire *et al.* Cambridge University Press, Cambridge; also, https://arxiv.org/abs/0706.0622.

Book 3

Wald, R. M. (1984). *General relativity*. University of Chicago Press, Chicago.

Weinberg, S. (1972). *Gravitation and cosmology*. John Wiley, New York.

Weinberg, S. (1989). The cosmological constant problem. *Rev. Mod. Phys.*, **61**, 1.

Westpfahl, K. and Göller, M. (1979). *Lett. Nuovo Cimento*, **26**, 573.

Will, C. M. (1993). *Theory and experiment in gravitational physics*. Cambridge University Press, Cambridge; The confrontation between general relativity and experiment. *Living Rev. Relativity*, **9**, 3 (2006); also, http://relativity.livingreviews.org/Articles/lrr-2006-3/.

Index